Handbook of
Neurochemistry

SECOND EDITION

Volume 10
PATHOLOGICAL NEUROCHEMISTRY

Handbook of
Neurochemistry
SECOND EDITION

Edited by Abel Lajtha
Center for Neurochemistry, Wards Island, New York

Handbook of
Neurochemistry

SECOND EDITION

Volume 10
PATHOLOGICAL NEUROCHEMISTRY

Edited by
Abel Lajtha
Center for Neurochemistry
Wards Island, New York

PLENUM PRESS • NEW YORK AND LONDON

Library of Congress Cataloging in Publication Data

Main entry under title:

Handbook of neurochemistry.

Includes bibliographies and indexes.
Contents: v. 1. Chemical and cellular architecture—[etc.]—v. 7. Structural elements of the nervous system— —v. 10. Pathological neurochemistry.
1. Neurochemistry—Handbooks, manuals, etc. I. Lajtha, Abel. [DNLM: 1. Neurochemistry. WL 104 H434]
QP356.3.H36 1982 612'.814 82-493

ISBN 978-1-4757-0799-1 ISBN 978-1-4757-0797-7 (eBook)
DOI 10.1007/978-1-4757-0797-7

Softcover reprint of the hardcover 1st edition 1985

A Division of Plenum Publishing Corporation
233 Spring Street, New York, N.Y. 10013

Contributors

Naren L. Banik, Department of Neurology, Medical University of South Carolina, Charleston, South Carolina 29425

Norman H. Bass, Department of Neurology, Lexington Veterans Administration Medical Center and University of Kentucky College of Medicine, and Sanders Brown Research Center on Aging, Lexington, Kentucky 40536

L. Battistin, Clinica delle Malattie Nervose e Mentali, Universita di Padova, 35100 Padova, Italy

Charles J. Billington, Division of Endocrinology, Department of Medicine, University of Minnesota, Minneapolis, Minnesota 55455

Samuel Bogoch, Department of Biochemistry, Foundation for Research on the Nervous System and Boston University School of Medicine, Boston, Massachusetts 02215

S. Winston Bogoch, College of Physicians and Surgeons, Columbia University, New York, New York 10032

Roscoe O. Brady, Developmental and Metabolic Neurology Branch, National Institute of Neurological and Communicative Disorders and Stroke, National Institutes of Health, Bethesda, Maryland 20205

Pak Hoo Chan, Department of Neurology, University of California School of Medicine, San Francisco, California 94143

Jørgen Clausen, The Neurochemical Institute, DK-2200 Copenhagen, Denmark, and the Institute for Biology and Chemistry, Roskilde University, DK-4000 Roskilde, Denmark

M. Dam, Clinica delle Malattie Nervose e Mentali, Università di Padova, 35100 Padova, Italy

Steven T. DeKosky, Department of Neurology, Lexington Veterans Administration Medical Center and University of Kentucky College of Medicine, and Sanders Brown Research Center on Aging, Lexington, Kentucky 40536

Lynn Eleanor DeLisi, Neurogenetics Branch, Intramural Research Program, National Institute of Mental Health, Bethesda, Maryland 20205

Darryl C. De Vivo, Division of Pediatric Neurology, Columbia University College of Physicians and Surgeons, New York, New York 10032

Salvatore DiMauro, Department of Neurology, Columbia University College of Physicians and Surgeons, New York, New York 10032

Pierre M. Dreyfus, Department of Neurology, University of California School of Medicine, Davis, California 95616

Barney E. Dwyer, Epilepsy and Neurology Research Laboratories, Veterans Administration Medical Center, Sepulveda, California 91343; and Department of Neurology and The Brain Research Institute, University of California School of Medicine, Los Angeles, California 90024

K. A. C. Elliot, Department of Biochemistry, McGill University, Montreal, Quebec H3G 1Y6 Canada

Robert A. Fishman, Department of Neurology, University of California School of Medicine, San Francisco, California 94143

E. Geller, Neurobiochemistry Laboratory, Veterans Administration Medical Center, Los Angeles, California 90073; and Department of Psychiatry, University of California, Los Angeles, California 90024

Mark S. Gold, Fair Oaks Hospital and Psychiatric Diagnostic Laboratory of America, Summit, New Jersey 07901

Ashley T. Haase, Departments of Medicine and Microbiology and Immunology, University of California School of Medicine, San Francisco, California 94143; and Veterans Administration Hospital, San Francisco, California 94121

Theodore A. Hare, Department of Pharmacology, Jefferson Medical College, Thomas Jefferson University, Philadelphia, Pennsylvania 19107

Henrik A. Hartmann, Department of Pathology, University of Wisconsin Medical School, Madison, Wisconsin 53706

George A. Hashim, The Jane Forbes Clark Surgical Research Laboratories, Departments of Surgery and Microbiology, St. Luke's–Roosevelt Hospital Center and Columbia University, New York, New York 10032

Edith Heilbronn, Unit of Neurochemistry and Neurotoxicology, University of Stockholm, S-172 46 Sundbyberg, Sweden

Edward L. Hogan, Department of Neurology, Medical University of South Carolina, Charleston, South Carolina 29425

F. A. Hommes, Department of Cell and Molecular Biology, Medical College of Georgia, Augusta, Georgia 30912

D. Karcher, Laboratory of Neurochemistry, Born-Bunge Foundation, Universitaire Instelling Antwerpen, B-2610 Antwerp, Belgium

Yasuo Kishimoto, John F. Kennedy Institute and Department of Neurology, Johns Hopkins University School of Medicine, Baltimore, Maryland 21205

A. Lowenthal, Laboratory of Neurochemistry, Born-Bunge Foundation, Universitaire Instelling Antwerpen, B-2610 Antwerp, Belgium

Erik Lycke, Department of Virology, Institute of Medical Microbiology, University of Göteborg, 413 46 Göteborg, Sweden

Anne M. Morin, Epilepsy and Neurology Research Laboratories, Veterans Administration Medical Center, Sepulveda, California 91343; and Department of Neurology and The Brain Research Institute, University of California School of Medicine, Los Angeles, California 90024

John E. Morley, Neuroendocrine Research Laboratory, Veterans Administration Medical Center, Minneapolis, Minnesota 55417

Hugo W. Moser, John F. Kennedy Institute, Baltimore, Maryland 21205

F. Petty, Veterans Administration Medical Center and Department of Psychiatry, University of Texas, Dallas, Texas 75216

Stanley B. Prusiner, Departments of Neurology and Biochemistry and Biophysics, University of California, San Francisco, California 94143

E. Ritvo, Department of Psychiatry, University of California, Los Angeles, California 90024

Susan Y. Schmidt, Berman-Gund Laboratory for the Study of Retinal Degenerations, Harvard Medical School, Massachusetts Eye and Ear Infirmary, Boston, Massachusetts 02114

A. D. Sherman, Department of Psychiatry, University of Iowa College of Medicine, Iowa City, Iowa 52242

Kunihiko Suzuki, Rose F. Kennedy Center and Department of Neurology, Albert Einstein College of Medicine, Bronx, New York 10461

Ara Tourian, Neurogenetics and Cell Biology Laboratory, Department of Medicine, Division of Neurology, Duke University Medical Center, Durham, North Carolina 27710

Karl Verebey, New York State Division of Substance Abuse Services Testing and Research Laboratory; and Department of Psychiatry, Downstate Medical School, Brooklyn, New York 11217

Claude G. Wasterlain, Epilepsy and Neurology Research Laboratories, Veterans Administration Medical Center, Sepulveda, California 91343; and Department of Neurology and The Brain Research Institute, University of California School of Medicine, Los Angeles, California 90024

G. Winokur, Department of Psychiatry, University of Iowa College of Medicine, Iowa City, Iowa 52242

James H. Wood, Cerebral Blood Flow Laboratories, Division of Neurosurgery, Emory University Clinic, Atlanta, Georgia 30322

Richard Jed Wyatt, Neuropsychiatry Branch, Intramural Research Program, National Institute of Mental Health, Saint Elizabeths Hospital, Washington, D.C. 20032

M. B. H. Youdim, Rappaport Family Institute for Research in the Medical Sciences, Department of Pharmacology, Technion, Haifa, Israel

A. Yuwiler, Neurobiochemistry Laboratory, Veterans Administration Medical Center, Los Angeles, California 90073; and Department of Psychiatry, University of California, Los Angeles, California 90024

Preface

More than for any other volume of the *Handbook of Neurochemistry,* the chapters in this volume on *Pathological Neurochemistry* deal with the interface of the laboratory bench with the patient's bedside. Most of the chapters reflect the confluence of basic scientists, clinical investigators, and physicians. Considered here are many of the more important disorders that afflict the nerves, muscles, spinal cord, and/or brain of mankind throughout the world. There are well over 500 such disorders. And our understanding of their nature and of measures for effective prevention or treatment depends significantly on application of the biochemical disciplines that characterize neurochemistry.

Before World War II, any attempt to compile a volume on pathological neurochemistry would have been largely descriptive and very rudimentary, as such "handbooks" by Hans Winterstein (1929), Irvine Page (1937), and others demonstrate. But thanks to the many major advances in research and technology in the postwar decades, we now stand at the threshold of understanding how to manage many of the major neurological disorders, and we may expect more such delineations in the immediate decades ahead. Neurochemistry, defined broadly, has played a central role in this extraordinary turn of events, progressing from what J. L. W. Thudichum in 1884 called objects of anxious empiricism to his anticipation of the proud exercise of chemical precision. Indeed, over the next decades we may reasonably expect future editions of this *Handbook* to devote progressively greater space to the pathological and less to the physiological aspects of neurochemistry.

In the present volume, there are notable examples of successful or promising interfaces between the laboratory bench and the clinical bedside. One major advance has been among the genetically determined disorders of the nervous system—in the groups of lipid and mucopolysaccharide storage diseases. The recognition that these disorders arise as consequences of deletions or attenuations of specific lysosomal hydrolases has made possible the development of relatively simple diagnostic and screening procedures and effective prenatal counseling, so that what were once baffling and hopelessly fatal conditions are now susceptible to clinical management and many may soon prove to be treatable by enzyme replacement or analogous therapies.

Major advances have also taken place in the field of movement disorders. The discovery of the neurochemical lesion in Parkinson's disease—the failure of dopamine transmission in nigral neurons—led to the introduction of therapy with L-DOPA and congeners and the rescue of a sizable adult population group from progressive disability and isolation. In a different way, the even larger problem of epilepsy has been significantly diminished: first, by the devising of practical clinical methods for measuring the circulating levels of anticonvulsant drugs to enhance greatly the clinical management of seizure patients; and second, by the application of the positron-emission tomographic (PET) technique with ^{18}F-labeled 2-deoxyglucose to evaluation of seizure patients *in vivo*. The latter studies have uncovered significant differences between ictal and interictal periods and between seizure foci and adjacent normal brain areas that could not have been anticipated or indeed studied in any other way. The observations of a relatively depressed glucose uptake in areas of seizure foci suggest new research directions for one of the most complex of neurological disorders, epilepsy.

The isolation and purification of receptors for acetylcholine and the exploitation of immunochemical techniques have now provided us with a solution to the problem of myasthenia gravis and with a more rational therapy. Myasthenia gravis may now be clearly classified as an autoimmune disorder. Other potential candidates such as multiple sclerosis seem likely to follow. Surely, the resort to immunochemical techniques carries much promise for many basic and clinical problems of the nervous system. Yet there are challenges as well in such examples as the neurotropic and slow and latent viruses. Not only do viruses or viral particles conceal themselves in neural cells (e.g., the herpesviruses), but others, like the viruses of poliomyelitis and rabies, may elude general bodily defenses by entering axonal terminals and utilizing axonal transport systems to invade and destroy nerve cells. These are special but important facets of the relevance of newer knowledge about axonal transport to clinical problems. And the analogies provided by such neurotropic viruses seem to point to one of the major trophic mechanisms responsible for the maintenance of connective integrity between pre- and postsynaptic elements.

Another most fruitful area of research concerns the neuropeptides. Here, almost from the beginning, the interface between basic and clinical was clearly perceived. On the one hand, the opiate receptors, enkephalins, drug addiction, and pain mechanisms provided an obvious example that continues to intrigue us with its expanding complexities. We ought not to overlook the fact that pain (its mechanisms and their management) represents one of the most prevalent and important clinical and public health problems facing the world today. On the other hand, there are such examples as the hypothalamic–pituitary systems in which a neurotransmitter such as dopamine regulates the elaboration of certain hypothalamic peptide releasing factors, which in turn control the release of specific pituitary hormones. Moreover such hormones are now known to exhibit significant feedback and other modulatory effects on the central nervous system itself. Thus, the triad of neurotransmitter, neuromodulator, and neurohormone in tandem or even in combination (in one molecule) begins to provide the neurochemical perspective with which to comprehend neuroendocrine, behavioral, and other complex phenomena of central processing.

In fact, the whole problem of central processing still poses many fundamental challenges. Sensory input from specialized receptors for touch, position sense, pain, light, sound, and the like must be transduced by mechano-, chemo-, photo-, and other specialized receptors into encoded nerve signals that can be centrally analyzed, stored in and recalled from memory, and responded to appropriately. Much more research must be accomplished before we can adequately contend with such disorders of central processing as autism, learning and language disabilities, deafness, blindness, and perhaps even the dementias of the Alzheimer type.

One final area deserves particular attention because of its socioeconomic importance and our as yet rudimentary understanding of the nature of such disorders. Included are cerebrovascular disorders or strokes (the third leading cause of death, after heart disease and cancer, in the United States) and trauma to the central nervous system, primarily head and spinal cord injuries (which afflict the young adult male in particular and are major factors in nearly three-fourths of the fatal accidents in the United States). In fact, these are universal, worldwide problems of a magnitude comparable to that for the United States. For the clinician, and for the neurochemist as well, much of the problem concerns the consequences of these insults to the central nervous system: edema and coma in particular are persisting challenges in terms of mechanisms and of clinical management.

With respect to stroke, recent promising research approaches include the role of prostaglandins in aggregation of platelets, in the control of the caliber of blood vessels, and in the effects of acetylsalicylic acid to inhibit the initial cyclooxygenase step of prostaglandin synthesis from arachidonic acid—all focused on the puzzling problem of pathological mechanisms and prophylaxis in cerebral ischemia and thrombosis. Additionally, more knowledge is needed concerning unique neurochemical aspects of cerebral vessels, which are embryologically and pharmacologically distinct from blood vessels elsewhere in the body.

With respect to central nervous system trauma, a major unsolved problem is why the brain and spinal cord fail to repair damaged areas to achieve spontaneous restoration of function. The tragedy of the wheelchair-bound paraplegic paralyzed below the level of spinal cord injury and unable to regenerate functional connections across the injury gap is the dramatic example. But there are comparable situations for patients with head injury, stroke, multiple sclerosis, and the like. In all cases, the original developmental potential of central axonal outgrowth and establishment of appropriate functional connections among central neurons does not operate when mature brain or spinal cord tissue is damaged. Yet peripheral nerves do regenerate under like circumstances, central connections in nonmammalian species regrow, and in mammals the potential in the central machinery remains. Thus, the neurochemical events and factors involved in developmental and regenerative processes in the mammalian and especially the human central nervous system pose an unusually urgent challenge. In such types of research investigators at the laboratory bench may be involved without realizing the clinical implications and needs.

It seems appropriate to conclude this Preface with a brief review of the dimensions of such clinical problems in the United States today (data for other countries would be quite comparable). Developmental disorders of the nervous system affect one out of every ten children; there are some one-half million new strokes each year, with 2.5 million Americans continuing to be disabled by stroke; 2 million are legally deaf (and another 12 million have significant hearing impairment); 2 or more million have epilepsy; more than 1.5 million elderly are demented (and account for nearly one-half of all nursing home patients in this country); about 0.5 million have Parkinson's disease; there are 400,000 new cases of head trauma and 10,000 new cases of spinal cord injury each year (with some 100,000 continuing paraplegics); and much more. For example, we have yet to comprehend or to begin to contend with the immense problem of neurotoxicology, especially in the industrial sector. In all, one out of every five persons, or 50 million Americans, are afflicted with neurological or communicative (hearing, speech, language) disorders. The cost to the United States in terms of care, lost income, and the like exceeds 65 billion dollars each year. And the costs to patients in terms of quality of life and to their families in terms of disruptions of home life as well as financial hardship or outright disaster are incalculable. In the case of stroke alone, only 10% of all survivors can return to full activity, whereas fully one-half are so severely disabled that full-time home or institutional care is required.

Yet the expenditures for research by the public and private sectors together amount to considerably less than 400 million dollars annually, or only about 0.5% of the annual cost to society of these disorders. Clearly, a modest increase in the support of research, particularly in the Federal and industrial sectors, would allow us to capitalize on the current upsurge in the neurosciences, as new knowledge and new technologies continue to impinge on us. Two aspects in particular warrant special support and encouragement. One is the training of new investigators (with a special attention to those with clinical orientation) to fill the gaps created by attrition, new fields and technologies, and new academic teaching institutions. The other is the provision of research funding at the interface between the laboratory bench and the patient's bedside. Human research is extraordinarily difficult, demanding, and expensive, and it does not compete well with nonclinical or basic research. Both types are needed, but research on man deserves special attention.

In all of these considerations there is a continuing and vital central role for neurochemistry. Physiological phenomena—electrophysiology, if you will—depend on membrane structure, on ionic gradients and conductance, and on energy metabolism in their support; and on specific transmitters, receptors, and intracellular interactions to generate and display the observed functional phenomena. The same may be said by analogy for pharmacological, immunological, toxicological, endocrinological, and especially behavioral disciplines. In this and preceding volumes in this *Handbook* series, many chapters underscore the central importance of neurochemistry for our understanding of the nervous system, for the management of its disorders, and, indeed, for man himself. For it is his nervous system that makes man human; and it is in its composition, circuitry, awareness of the surrounding world from sensory inputs, storage and recall of such experiences, and responses appropriate to its central processing that the human nervous system operates primarily as a chemical entity of marvelous design and superb performance. May careful perusal of this volume bring such understanding and appreciation.

Donald B. Tower

Director-Emeritus, NINCDS, NIH
Asst. Surgeon General (ret), USPHS

Contents

Chapter 3

Peptides and Brain Pathology

John E. Morley and Charles J. Billington

Chapter 4

Neurochemistry of Brain Tumors

Samuel Bogoch and S. Winston Bogoch

Chapter 5

Lipidoses

Roscoe O. Brady

Chapter 8

Adrenoleukodystrophy

Yasuo Kishimoto, Hugo W. Moser, and Kunihiko Suzuki

Chapter 9

Brain Edema

Pak Hoo Chan and Robert A. Fishman

Chapter 10

Multiple Sclerosis

Jørgen Clausen

Chapter 11

Multiple Sclerosis and Allergic Encephalomyelitis

George A. Hashim

Chapter 12

The Biology of Huntington's Chorea

Ara Tourian

Chapter 13

Diseases and Induced Lesions of the Neuromuscular Endplate

Edith Heilbronn

Chapter 14

Biochemistry of the Spinal Cord

Edward L. Hogan and Naren L. Banik

Chapter 15

The Epilepsies

Claude G. Wasterlain, Anne M. Morin, and Barney E. Dwyer

Chapter 21

Neurochemical Aspects of Schizophrenia

Lynn Eleanor DeLisi and Richard Jed Wyatt

Chapter 22

Endorphins and Mental Disease

Karl Verebey and Mark S. Gold

Chapter 24

Depression

F. Petty, A. D. Sherman, and G. Winokur

Chapter 25

Biochemical Studies on Autism

A. Yuwiler, E. Geller, and E. Ritvo

Chapter 26

Metabolic Encephalopathies

L. Battistin and M. Dam

Chapter 27

Brain Iron Metabolism: Biochemical and Behavioral Aspects in Relation to Dopaminergic Neurotransmission

M. B. H. Youdim

Chapter 28

The Neurochemistry of Vitamin Deficiencies

Pierre M. Dreyfus

1

Disorders of Glycogen Metabolism

Salvatore DiMauro and Darryl C. De Vivo

1. INTRODUCTION

The concentration of glycogen in the central nervous system (CNS) is approximately 0.1 g/100 g fresh tissue as compared to 1.0 g/100 g in muscle and 6–10 g/100 g in liver. Because of this low content and its rapid disappearance after death, the very presence of glycogen in the brain remained controversial until the late 1930s, when Kerr documented the presence of a polysaccharide indistinguishable from liver glycogen.[1] Major fluctuations of tissue glycogen content are accompanied by even greater changes of water content (and volume).[2] The advantages afforded the brain by the protection of the bony skull and the fibrous dura mater are attenuated by the metabolic vulnerability of the brain. The low brain tissue glycogen content renders this organ vulnerable to insult within several minutes of onset of hypoglycemia or hypoxia.

The functional significance of glycogen in the brain is not completely understood, but it is generally assumed that it represents a reserve of readily available energy to be tapped in conditions of glucose depletion.[2–7] Glycogen stores in rat brain would sustain metabolic activity only for 3 min in the total absence of blood glucose, but if glucose influx were to decrease instead of cease completely, the energy reserve would sustain activity for much longer periods.[4]

Several lines of evidence support the role of glycogen as a reserve of cerebral energy: glycogen concentration is increased during hibernation, slow-wave sleep, hypnosis, diabetes mellitus, chronic ketosis, and after administration of hypnotic drugs and tranquilizers; conversely, it is decreased in hyperthermia, in animals deprived of slow-wave sleep, and in conditions of anoxia or hypoglycemia.[3–10] The effects of denervation, radiation, trauma, and ischemia are more complex, but they usually result in glycogen accumulation, possibly related to decreased oxidative metabolism of injured tissue.[4] The main biosynthetic and degradative pathways for glycogen are similar in brain and in other tissues and are only briefly outlined here (Fig. 1).

Salvatore DiMauro and Darryl C. De Vivo • Department of Neurology, and Division of Pediatric Neurology, Columbia University College of Physicians and Surgeons, New York, New York 10032.

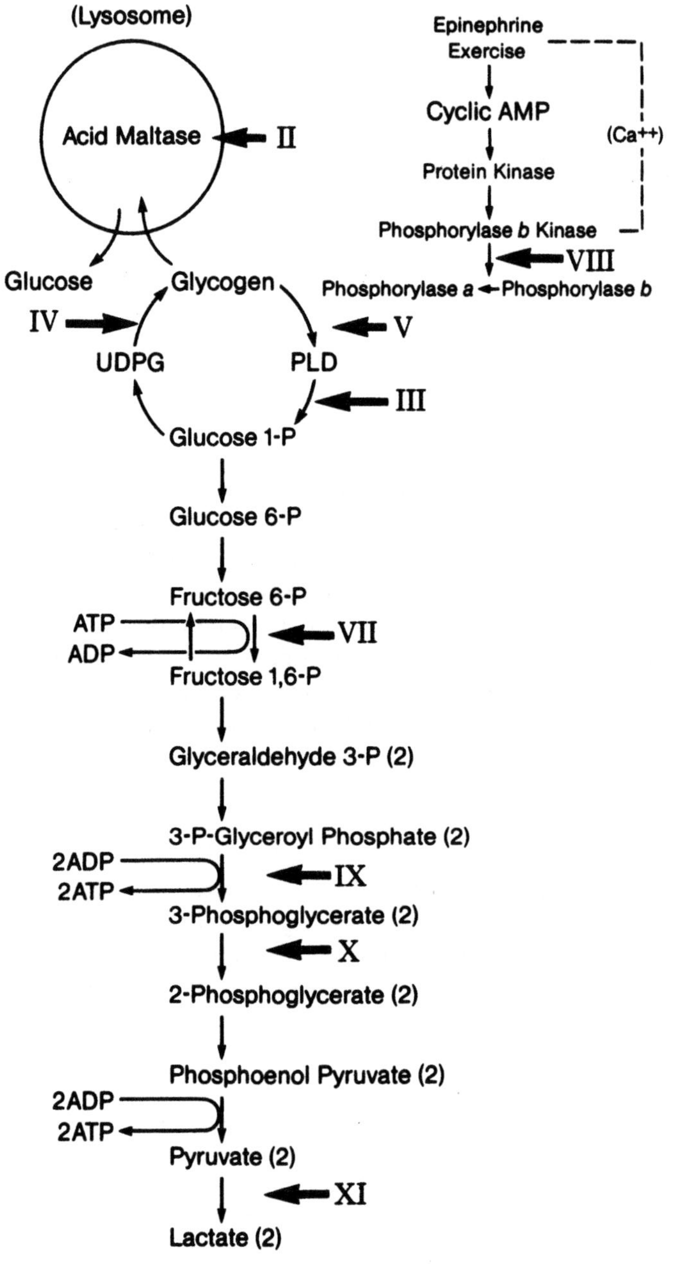

Fig. 1. Scheme of glycogen metabolism and glycolysis. The arrows point to known enzyme defects, and the Roman numerals refer to the following glycogenoses: I, glucose-6-phosphatase (von Gierke); II, acid maltase (Pompe and later onset variants); III, debrancher enzyme (Cori–Forbes); IV, brancher enzyme (Andersen); V, myophosphorylase (McArdle); VI, liver phosphorylase (Hers); VII, phosphofructokinase (PFK: Tarui); VIII, phosphorylase kinase (Hug); IX, phosphoglycerate kinase (PGK); X, phosphoglycerate mutase (PGAM, DiMauro); XI, lactate dehydrogenase (LDH, Kanno).

Glycogen synthesis is catalyzed by two main enzymes, glycogen synthetase and branching enzyme. Glycogen synthetase (UDPG-glycogen transglucosylase) catalyzes the transfer of a glucosyl unit from uridine diphosphate-glucose (UDPG) onto the peripheral chain of a preexisting glycogen molecule in an α-1,4-glucosidic link. This stepwise elongation of peripheral chains proceeds until approximately 15 or 18 glucosyl units have been linked in a linear fashion. At this point, the branching enzyme [1,4-α-glucan: 1,4-α-glucan 6-α - D-(1,4-α-D-glucano)transferase] transfers a short glucosyl chain (seven or eight units long) from one peripheral chain onto another in α-1,6-glucosidic link, thus starting a new peripheral branch.

Glycogenolysis is controlled by two pathways, one cytosolic, the other lysosomal. In the cytosol, glycogen phosphorylase removes glucosyl units stepwise from the peripheral chains of glycogen in a phosphorolytic reaction with liberation of glucose-1-phosphate. After the peripheral chains have been shortened to about four glucosyl units (this resulting polysaccharide on which phosphorylase has acted maximally is phosphorylase-limit dextrin or PLD), the debranching enzyme, in a two-step reaction, removes the residual short chains, thus "debranching" the glycogen molecule. The first step of this reaction involves the transfer of a maltotriosyl unit from one site to another of the PLD molecule. The single glucosyl unit left behind (linked by α-1,6-glucosidic bond) is then hydrolyzed by an α-1,6-glucosidase. Both transferase and glucosidase appear to be properties of the same protein. Lysosomal glycogenolysis is catalyzed by acid maltase, an acid hydrolase with both α-1,4 and α-1,6-glucosidase activities, capable of digesting glycogen completely to glucose. The functional role of this enzyme is not completely understood. Clearly, one function is the degradation of glycogen trapped within autophagic vacuoles, salvaging glucose in the process.

The content of glycogen in the brain is small and varies within a narrow range. However, the turnover of the polysaccharide is quite active, suggesting that glycogen synthesis and breakdown are delicately controlled.

Glycogen synthetase and phosphorylase exist in liver and muscle in two interconvertible forms, one phosphorylated, the other dephosphorylated. Phosphorylation, which activates phosphorylase but deactivates synthetase, is mediated by a cyclic-AMP-dependent sequence initiated by a hormone-dependent adenylate cyclase. Brain phophorylase is also activated by cyclic AMP, and adenylate cyclase is present in the brain, suggesting that the same regulation mechanism exists in nervous tissue. Hormonal effects on brain glycogen metabolism, however, have not been documented conclusively.[5] The "inactive" forms of the two enzymes are not, in fact, devoid of activity but, rather, are dependent for activity on the concentration of various metabolites. A third form of control, documented in skeletal muscle, may be exerted by glycogen itself through a feedback mechanism on the activation–deactivation cycles of phosphorylase and synthetase.

2. GLYCOGEN STORAGE DISEASES

Of the 11 distinct enzyme defects known to affect muscle or liver (Fig. 1), only one has been documented in the brain, acid maltase deficiency (glycogenosis type II). However, there is indirect evidence that other enzyme defects may be expressed in nervous tissue, including debranching enzyme deficiency (glycogenosis type III), branching enzyme deficiency (glycogenosis type IV), and phosphoglycerate kinase (PGK) deficiency (glycogenosis type IX). Finally, the brain may be indirectly affected in disorders of glycogen metabolism that are biochemically expressed in other tissues, particularly the liver.

2.1. Glycogenoses Affecting the Brain Directly

2.1.1. Acid Maltase Deficiency (AMD; Glycogenosis Type II; Pompe Disease)

Acid maltase deficiency causes two major clinical pictures: a generalized, rapidly progressive and fatal disease of infancy and a more benign neuromuscular disorder manifesting in childhood or adult life.[11]

In infantile AMD, the clinical picture is dominated by profound weakness and hypotonia, massive cardiomegaly, moderate hepatomegaly, and macroglossia. Death, as a result of cardiac and respiratory failure, occurs almost invariable before 1 year of age. Although it is difficult to assess the mental status of these profoundly weak infants, they are described in most reports as alert and interested in their environment.

Numerous histopathological studies of the nervous system have been reported, but only few ultrastructural studies.[11] By light microscopy, large amounts of glycogen are seen in the perykaryon of glial cells in both gray and white matter, whereas cortical neurons show much smaller quantities of glycogen. Moderate diffuse gliosis was reported in some cases. In the spinal cord, the neurons of the anterior horns appear ballooned and contain abundant PAS-positive material that is digested by diastase. Schwann cells of both anterior and posterior spinal roots and of peripheral nerves also contain excessive glycogen. By electron microscopy, the most striking feature is the presence of glycogen granules within membrane-bound vacuoles.[12,13] These glycogen-laden vacuoles are particularly abundant in neurons of anterior horn cells and of motor nuclei of the brainstem and in Schwann cells. They are scarce in cortical neurons but abundant in glial cells. Abundant free glycogen has been found in the same cells in one ultrastructural study[12] but not in another[13]: this discrepancy may have arisen from the different intervals between death and fixation of the nervous tissue in the two studies (3 hr and 12 hr). Both intralysosomal and free glycogen appeared to be composed of normal β particles, which stained intensely with silver methenamine, a preferential stain for polysaccharides.

Glycogen concentration was measured in brains obtained at autopsy from several patients. Although it was increased in all cases, the values reported varied over a wide range (0.2 to 2.3 g/100 g wet tissue): these differences can be attributed to the different methods used to isolate glycogen, to different intervals between death and autopsy, and, possibly, to different stages of the disease. In one study,[12] the highest concentration of glycogen was found in the gray matter of the spinal cord (0.82 g/100 g wet tissue), but glycogen content also was markedly increased in cerebral gray and white matter (0.37 and 0.57 g/100 g, approximately ten times greater than normal).

Acid maltase activity was virtually undetectable in several studies, whereas other enzymes, including phosphorylase and acid phosphatase, had normal activities.

The lack of acid maltase and the predominance of intralysosomal glycogen in both central and peripheral nervous system is in agreement with Hers' concept of "inborn lysosomal disease."[14] However, the presence of much free glycogen in the brain as in other tissues, is difficult to explain.

The severe involvement of spinal and brainstem motor neurons may contribute, together with the massive accumulation of glycogen in muscle, to the profound hypotonia, weakness, and hyporeflexia of Pompe disease. On the other hand, the much milder involvement of cortical neurons may explain why affected babies do not appear grossly retarded throughout their brief life-span.

In the childhood and adult forms of AMD, the clinical picture is usually limited to the musculature, with progressive weakness. Truncal and proximal muscles are more involved than distal muscles.[11] In young boys, the disorder closely simulates Duchenne muscular dystrophy, and in adults it is often incorrectly diagnosed as limb-girdle dystrophy or polymyositis. Early and severe involvement of respiratory muscles is often present and should alert the clinician to the possible diagnosis of AMD. Lack of cardiac involvement is an important feature distinguishing the late-onset forms from infantile AMD. Mental retardation was described in two patients with childhood AMD[15,16] but never in patients with adult AMD.

There are four postmortem studies of patients with childhood AMD.[15–18] In the two patients with mental retardation, excessive glycogen was found in the central nervous system.[15,16] In one of them, glycogen storage predominated in spinal motoneurons and brainstem nuclei,[16] whereas in the other, glycogen appeared to be more abundant in the brain than in the spinal cord.[15] Glycogen concentration and acid maltase activity were measured in the first of these two cases[16]: glycogen was 0.2 g/100 g in cerebrum and cerebellum, and acid maltase was undetectable.

In the only patient with adult ADM studied at autopsy, all visceral organs, including the nervous system, were normal by light and electron microscopy.[19] Accordingly, biochemical studies showed no accumulation of glycogen in nervous tissue. However, acid maltase activity was as markedly decreased in the brain as in all other organs. The small residual activity (about 7% of normal) may be enough to prevent accumulation of glycogen in the brain and in other viscera, but it is not easy to explain why the same relative residual activity should not protect skeletal muscle from excessive glycogen storage. Another limitation is that acid maltase activity was measured with an artificial substrate (methylumbelliferyl-α-D-glucoside) rather than with the natural substrate (glycogen).

2.1.2. Debrancher Deficiency

The typical presentation of debrancher deficiency (glycogenosis type III; Cori–Forbes disease) consists of hepatomegaly, growth retardation, fasting hypoglycemia, and ketonuria in childhood, spontaneously remitting around puberty.[11] In a few patients, slowly progressive weakness becomes apparent in adult life, and muscle biopsy shows severe vacuolar myopathy with glycogen storage. Although weakness affects proximal more than distal muscles, atrophy of distal muscles is not uncommon, and the electromyograph (EMG) usually shows a mixture of myopathic features and signs of increased muscle irritability with fibrillations, positive sharp waves, and myotonic potentials.[11] The presence of distal wasting and a "mixed" EMG in these patients often suggests the diagnosis of motor neuron disease or Charcot–Marie–Tooth disease. Motor nerve conduction velocities were decreased in three patients with debrancher deficiency "myopathy."

There is no evidence of mental retardation or any other sign of brain dysfunction in patients with either clinical form of debrancher enzyme deficiency. However, there is no evidence for tissue-specific isoenzymes of debrancher. Therefore, the defect should be generalized. In agreement with this concept, decreased debrancher activity has been documented in muscle, liver, heart, erythrocytes, leukocytes, fibroblast, and muscle cultures.[11] Although neither pathology nor debrancher activity have been reported, increased glycogen concentration (0.3 g/100 g wet weight) has been found in the brain of a patient,[20] suggesting that the enzyme defect is expressed in the nervous tissue.

Thus, in debrancher enzyme deficiency, the nervous system appears to be involved biochemically, although clinical signs of brain dysfunction are limited to hypoglycemic seizures in early childhood. The slow motor nerve conduction velocities seen in some patients may be secondary to glycogen storage in Schwann cells; nerve biopsies have not been examined. Similarly, the distal wasting and the presence of abundant fibrillations in the EMG of patients with late-onset neuromuscular disorder may reflect glycogen accumulation in anterior horn cells, but this speculation remains to be documented.

2.1.3. Branching Enzyme Deficiency

Branching enzyme deficiency (glycogenosis type IV; Andersen disease) is a rare, rapidly progressive disease of infancy characterized by hepatosplenomegaly, progressive cirrhosis, and chronic hepatic failure.[11] Death, as a result of liver or heart insufficiency, occurs in the first years of life. Although the clinical picture is usually dominated by liver dysfunction, severe hypotonia, decreased or absent reflexes, contractures, and marked wasting have been described in three patients, suggesting the diagnosis of spinal muscular atrophy.[11]

It is uncertain whether there is a single molecular form or multiple isoenzymes of branching enzyme. In morphological studies, a basophilic and intensely PAS-positive material partially resistant to α-amylase digestion was seen in all tissues, but the relative amount of this abnormal polysaccharide in different tissues varied from case to case. In the electron microscope, the storage material consisted of filamentous and finely granular material, often associated with glycogen particles of normal appearance.[21]

The central nervous system was studied in five cases and appeared normal in two. In the others, PAS-positive spheroids were seen in subpial and perivascular zones of the brainstem and spinal cord but never within neurons. Electron micoscopy showed that the spheroids were composed of branched osmiophilic filaments, 600 nm in diameter and were located within distended astrocytic processes.[21] No such storage material was seen in neuronal perikarya or processes.

The enzyme defect has never been documented directly in nervous tissue, and the structure of the polysaccharide stored in the brain has not been chemically analyzed, but the staining characteristics and ultrastructural features of the stored material are similar to those of the amylopectinlike polysaccharide present in liver. This glycogen has abnormally long outer chains and relatively few branching points.

2.1.4. Myophosphorylase Deficiency

The clinical manifestations of myophosphorylase deficiency (glycogenosis type V; McArdle disease) are confined to skeletal muscle: intolerance to intense exercise, cramps, and myoglobinuria. Between episodes of myoglobinuria, most patients are normal, but about one-fourth of them have some degree of fixed weakness, and almost all have increased serum creatine kinase at rest.[11] The incidence of seizures seems to be higher in patients with McArdle disease than in the general population (5 of 112 patients), and two patients had mental retardation. Seizures have been attributed to transient hypoglycemia induced by a disproportionate demand of glucose by exercising muscles that cannot utilize glycogen. However, seizures may occur without preceeding exercise, and blood glucose does not consistently decrease during exercise in patients with McArdle disease.

The isoenzyme composition of phosphorylase in human brain has not been studied in detail, but preliminary studies showed 20 to 30% inhibition of brain phosphorylase activity by antibodies against purified muscle phosphorylase, suggesting that the muscle isoenzyme is a minor component of brain phosphorylase. A partial defect of brain phosphorylase activity would then be expected to occur in McArdle disease, but it is unlikely that this would have functional consequences. A combination of exercise-induced hypoglycemia and hyperventilation could explain seizures occurring in patients with myophosphorylase deficiency.

2.1.5. Phosphoglycerate Kinase Deficiency

The most common clinical presentation of this X-linked recessive disease includes nonspherocytic hemolytic anemia and central nervous system dysfunction. Neurological problems have varied in severity: all patients showed some degree of mental retardation with delayed language acquisition and behavioral abnormalities, and some had seizures.[22,23]

Human PGK is believed to be a single polypeptide controlled by a gene locus on the X chromosome for all tissues except sperm.[22] Because of the apparent lack of tissue-specific isoenzymes (except for PGK 2, which is limited to spermatogenic cells), PGK deficiency should be expressed in all tissues including the brain. Although the enzyme defect has not been directly documented in the brain, the severe brain involvement in most patients can be explained by impairment of the glycolytic pathway.

Secondary involvement of the brain in PGK deficiency is unlikely because the enzyme defect does not cause hypoglycemia.

The lack of symptoms of brain dysfunction in some patients may be attributable to the presence of sufficient residual enzyme activity to prevent severe energy shortage.

2.1.6. Other Glycogenoses

In some patients with direct or indirect evidence of glycogen storage in the brain, the enzyme defect remains obscure. Four boys, two of them brothers, manifested in their early teens a clinical syndrome characterized by progressive weakness, cardiomyopathy, and mental retardation.[24,25] Two of them died at age 16 of cardiac failure.[24] The intellectual impairment varied in severity in these four patients.

Muscle biopsies in all cases showed histological and ultrastructural changes identical to those of acid maltase deficiency, but acid maltase activity was normal when measured with both artificial and with natural substrates. Kinetic studies of the enzyme in crude muscle extracts failed to show any abnormality. Glycogen concentration in muscle was mildly increased, and the glycogen had normal structure.

Genetic transmission appeared to be autosomal recessive in the family of the two affected brothers, whose parents were asymptomatic.[25] In another family, the mother of the propositus was weak and died of cardiac failure at age 37, and several siblings of both sexes were similarly affected, suggesting autosomal dominant transmission.[24] The brain was not studied in the two patients who died.

Severe progressive encephalopathy dominated the clinical picture in a girl with hepatomegaly since infancy and stunted growth but no hypoglycemia.[26] She could never walk unassisted, and at 5 years of age she appeared grossly retarded, had severe spasticity and bilateral Babinski signs. Vision seemed to be preserved. Numerous liver biopsies showed consistently increased glycogen and variable but consistently decreased phosphorylase activity. Activation of liver phosphorylase *in vivo* by administration of glucagon occurred normally. Because the child had chronically elevated urinary excretion of epinephrine and norepinephrine, it was postulated that high levels of catecholamines may have caused chronic deactivation of liver phosphorylase.[27] A brain biopsy obtained when the patient was 7 years old showed marked accumulation of free glycogen in three forms: individual β particles (as in normal brain), small aggregates of β particles similar to the rosettes (or α particles) normally seen in liver, and larger aggregates of α particles.[20] Glycogen concentration was 0.12 g/100 g tissue, compared to normal values below 0.05 g/100 g. Glycogen content was normal in muscle.

Although this patient clearly had a glycogen storage disease affecting both liver and brain, the precise biochemical defect remains to be defined, and the pathogenesis of the brain disease is obscure.

The pathology of the brain was very similar in a girl with a rapidly progressive neurological disease presenting soon after birth with difficulty sucking and swallowing, dyspnea, hypotonia followed by spastic tetraplegia, mental retardation, and amaurosis.[28] The child died at the age of 11 months because of respiratory infection. In contrast to the previous case, this patient did not have hepatomegaly, and liver biopsy was normal. A biopsy of the frontal cortex showed accumulation of glycogen β particles and of small and "giant" α-particles. The β particles predominated in glial cells, whereas α particles were seen in neuronal processes.

In one case of cerebrohepatorenal syndrome (Zellweger disease), excessive glycogen in the form of normal-looking β particles was found in neuronal perikarya and processes and in glial cells, predominantly in the cerebral cortex.[29] The glycogen was free in the cytoplasm and within nuclei. This autosomal recessive multisystem disorder is characterized by severe hypotonia, craniofacial dysmorphia, cerebral, renal, and skeletal abnormalities, cirrhosis, increased tissue stores of iron, and death in infancy. The metabolic defect is not known, but a primary mitochondrial dysfunction has been suggested by biochemical studies of brain and liver mitochondria showing impaired electron transport.[30] Two possible explanations have been offered for the increased brain glycogen in cerebrohepatorenal sysndrome: (1) a nonspecific response to prolonged barbiturate administration similar to that demonstrated in experimental animals or (2) the consequence of a block in mitochondrial metabolism; both glycogen and triglyceride tend to accumulate in skeletal muscle of patients with "mitochondrial myopathies".[11]

2.2. *Glycogenoses Affecting the Brain Indirectly*

Hypoglycemia may produce lethargy, coma, seizures, and brain damage in glucose-6-phosphatase and glycogen synthetase deficiencies.

Glucose-6-phosphatase deficiency (glycogenosis type 1; Von Gierke disease) causes hepatomegaly, growth retardation, hypoglycemia, ketosis, hyperlipemia, hyperuricemia, and hyperlactacidemia in childhood. The enzyme defect is expressed in liver, kidney, and intestinal mucosa. There is evidence of genetic heterogeneity, but the disease is usually transmitted as an autosomal recessive trait with predominance in males. Those patients who survive beyond childhood tend to have milder clinical manifestations later in life. Mental development is normal in these patients. The relatively mild involvement of the brain despite chronic severe hypoglycemia reflects the ability of the brain to use lactate, pyruvate, and ketone bodies as alternative energy sources. The enzyme is present in low activity in brain but presumably has no functional role.[10]

Glycogen synthetase deficiency has been described in three families.[31–33] It caused stunted growth and severe fasting hypoglycemia with ketonuria. Mental retardation was reported in the three affected children who survived past infancy. The liver was virtually devoid of glycogen and showed fatty degeneration in all cases. In two patients, the brain showed diffuse, nonspecific changes of the white matter (presence of reactive astrocytes and increased microglia), which were considered secondary to prolonged hypoglycemia or anoxia. Biochemical studies showed that glycogen synthetase activity was markedly decreased in liver but normal in muscle, erythrocytes, and leukocytes,[31–34] suggesting the existence of multiple tissue-specific isoenzymes under separate genetic control. It is not known whether brain glycogen synthetase is different from the liver enzyme.

In liver phosphorylase deficiency (glycogenosis type VI; Hers' disease) and in the two genetic forms of phosphorylase kinase deficiency (one X-linked recessive, the other autosomal recessive), hypoglycemia is either absent or mild, and symptoms of brain dysfunction do not usually occur.

3. LAFORA DISEASE AND OTHER POLYGLUCOSAN STORAGE DISEASES

3.1. Myoclonus Epilepsy with Lafora Bodies (Lafora Disease)

This is a hereditary disease transmitted as an autosomal recessive trait and affecting both sexes equally. In the classical (Unverricht) type, the clinical picture is characterized by the triad of epilepsy, myoclonus, and dementia. Other neurological manifestations include ataxia, dysarthria, spasticity, and rigidity. Onset is in adolescence, and the course is rapidly progressive; death occurs in most patients between 17 and 24 years of age. The most common initial manifestation is epilepsy, usually in the form of generalized seizures. Myoclonus and progressive intellectual regression appear 2 or 3 years after the onset of epilepsy. In terminal stages there is severe dementia, and status epilepticus is common. Laboratory findings are normal except for electroencephalographic changes; bilaterally synchronous discharges of spike and wave formations are usually seen in association with myoclonic jerks.

The pathological hallmark of the disease is the presence in the CNS of the bodies first described by Lafora in 1911[35]: round, basophilic, PAS-positive intracellular inclusions that vary in size from small "dustlike" bodies less than 3 nm in diameter to large bodies up to 30 nm in diameter.[36] The larger bodies often show a dense core and a less dense periphery. Lafora bodies are typically seen in neuronal perikarya and processes, not in glial cells, and are more abundant in cerebral cortex, substantia nigra, thalamus, globus pallidus, and dentate nucleus. Ultrastructural studies have shown that Lafora bodies are not limited by a membrane and consist of two components in various proportions: amorphous, electron-dense granules and irregular, often branched filaments.[36] The filaments, about 60 nm in diameter, are often continuous with the granular material. Accumulations of a material similar to that of the Lafora bodies are seen in liver, heart, skeletal muscle, skin, and retina, suggesting that Lafora disease is a generalized storage disorder.

Histochemical and biochemical studies of crude tissues and of isolated Lafora bodies have shown that the storage material is a branched polysaccharide composed of glucose (polyglucosan) similar to the amylopectinlike polysaccharide that accumulates in branching enzyme deficiency.[37,38] Despite these similarities, branching enzyme activity was normal in brain and muscle from one patient,[36,39] and the biochemical defect in Lafora disease remains obscure.

The Lundborg type of myoclonus epilepsy has a slower and more protracted course, with onset in the late teens and death usually in the fifth or sixth decade. Polyglucosan bodies are rare in neuronal perikarya and more abundant within axons.

3.2. Other Polyglucosan Body Disease

A characteristic neurological syndrome has been described in seven patients, consisting of progressive upper and lower motor neuron involvement, sensory loss, neurogenic bladder, and, in five patients, dementia without myoclonus or epilepsy.[40–43] Onset is in the fifth or sixth decade, and the course varies between 3 and 20 years. Throughout the central and peripheral nervous system there are numerous polyglucosan bodies in processes of neurons and astrocytes but not in perikarya. Other tissues are also affected, including liver, heart, and skeletal and smooth muscle. The metabolic error is not known.

In six patients with double athetosis associated with other fixed or progressive neurological disorders such as torsion dystonia or cerebral palsy, polyglucosan bodies were confined to the lateral segment of the pallidum.[44] These bodies were present in neuronal perikarya and even more abundantly in neuronal processes. These deposits, also called Bielschowsky bodies after the author who first described them,[45] have the same staining and ultrastructural characteristics as Lafora or other polyglucosan bodies but have a more variable, often elongated, shape.[44] Besides their exclusive localization in the pallidum and variable shape, another important difference of the Bielschowsky bodies from other polyglucosan bodies is that they are not found outside the central nervous system. The pathogenic significance of these bodies and their relationship to status marmoratus of the putamen, which was seen in three of the six patients, are unknown.

Another form of polyglucosan body is represented by corpora amylacea, which accumulate progressively and nonspecifically with age. These deposits are more commonly found within astrocytic processes in the hippocampus, the subependymal regions of the lateral and third ventricles and the floor of the fourth ventricle, and also in intramuscular nerves. The deposits accumulate gradually in individuals above 40 years of age and seem to be part of the normal aging process. Corpora amylacea have been isolated from human brains and characterized chemically.[46] As in the case of Lafora bodies, only glucose was identified after acid hydrolysis; protein and a relatively high amount of phosphate were also recovered. The presence of phosphate and sulfate groups may explain the metachromasia of corpora amylacea. Also like Lafora bodies, the polyglucosan of corpora amylacea is relatively resistant to α-amylase digestion but sensitive to prolonged exposure to enzymes that hydrolyze both α-1,4- and α-1,6-glucosidic bonds.

Recently, starchlike granules, called noncongophilic granules (NCG), have been isolated from human cortical gray matter[47]: in unstained preparations, they appeared as cuboidal crystals varying in diameter between 3 and 30 μm. They reacted positively with stains for polysaccharides and were brightly birefringent in the polarizing microscope. In the electron microscope, the granules had well-defined edges and a granular appearance. Chemical analysis showed the presence of glucose and some uronic acid but no hexosamine or protein. The granules were isolated from brains of patients and normal individuals of all ages. The yield increased with age. The granules were visualized best *in situ* in thick (20 μm) sections stained with PAS or examined in the polarizing microscope. It has been suggested that visualization of these granules in thin (5 μm) sections of the cerebral cortex may be difficult because of their hardness and the resulting displacement by the microtome knife. The precise localization of the noncongophilic granules (whether intracellular or extracellular) and their relationship to normal glycogen and to corpora amylacea remain to be defined.

REFERENCES

1. Kerr, S. E., 1983, *J. Biol Chem.* **123**:443–449.
2. Cahill, G. F., and Aoki, T. T., 1980, *Cerebral Metabolism and Neural Function* (J. V. Passonneau, R. A. Hawkins, W. D. Lust, and F. A. Welsh, eds.), Williams and Wilkins, Baltimore, pp. 234–242.
3. Karnovsky, M. L., Burrows, B. L., and Zoccoli, M. A., 1980, *Cerebral Metabolism and Neural Function* (J. V. Passonneau, R. A. Hawkins, W. D. Lust, and F. A. Welsh, eds.), Williams and Wilkins, Baltimore, pp. 359–366.
4. Siesjo, B. K., 1978, *Brain Energy Metabolism*, John Wiley and Sons, New York.
5. Coxon, R. V., 1970, *Handbook of Neurochemistry*, Volume 3 (A. Lajhta, ed.), Plenum Press, New York.
6. Miguel, J., and Haymaker, W., 1982, *Histology and Histopathology of the Nervous System* (W. Haymaker and R. O. Adams, eds.) Charles C Thomas, Springfield, Illinois, pp. 920–972.
7. Lowry, O. H., Passonneau, J. V., Hasselberger, F. X., and Schulz, D. W., 1964, *J. Biol. Chem.* **239**:18–30.
8. Halowach-Thurston, J., Hauhart, R. E., Jones, E. M., and Ater, J. L., 1975, *J. Biol. Chem.* **250**:1751–1758.
9. De Vivo, D. C., Leckie, M. P., Ferrendelli, J. S., and McDougal, D. B., 1978, *Ann. Neurol.* **3**:331–337.
10. Hawkins, R. A., 1980, *Cerebral Metabolism and Neural Function* (J. V. Passonneau, R. A. Hawkins, W. D. Lust, and F. A. Welsh, eds.), Williams and Wilkins, Baltimore, pp. 367–381.
11. DiMauro, S., 1979, *Handbook of Clinical Neurology*, Volume 41, Part II (P.J. Vinken, G. W. Bruyn, and S. P. Ringel, eds.), North-Holland, Amsterdam, pp.175–233.
12. Gambetti, P., DiMauro, S., and Baker, L., 1971, *J. Neuropathol. Exp. Neurol.* **30**:412–430.
13. Martin, J. J., deBarsy, T., Van Hoff, F., and Palladini, G., 1973, *Acta Neuropathol.* **23**:229–344.
14. Hers, H. G., 1965, *Gastroenterology* **48**:625–633.
15. Smith, H. L., Amick, L. D., and Sidbury, J. B., 1966, *Am. J. Dis. Child.* **111**:475–481.
16. Smith, J., Zellweger, H., and Afifi, A. K., 1967, *Neurology (Minneap.)* **17**:537–549.
17. Martin, J. J., deBarsy, T., De Schrijver, F., Leroy, J. G., and Palladini, G., 1976, *J. Neurol. Sci.* **30**:155–166.
18. Martin, J. J., deBarsy, T., and den Tandt, W. R., 1976, *J. Neurol.* **213**:105–118.

19. DiMauro, S., Stern, L. Z., Mehler, M., Nagle, R. B., and Payne, C., 1978, *Muscle Nerve* **1**:27–36.
20. Hug, G., and Schubert, W. K., 1966, *Clin. Res.* **14** (abstr. 441).
21. Schochet, S. S., McCormick, W. F., and Zellweger, H., 1970, *Arch. Pathol.* **90**:354–363.
22. Valentine, W. N., Hsieh, H. S., Paglia, D. E., Anderson, H. M., Baughan, M. A., Jaffe, E. R., and Garson, O. M., 1960, *N. Engl. J. Med.* **280**:528–534.
23. Konrad, P. N., McCarthy, D. J., Mauer, A. M., Valentine, W. N., and Paglia, D. E., 1973, *J. Pediatr.* **82**:456–460.
24. Danon, M. J., Oh, S. J., DiMauro, S., Manaligold, J. R., Eastwood, A., Naidu, S., and Schliselfeld, L. H., 1981, *Neurology (N.Y.)* **31**:51–57.
25. Riggs, J. E., Schochet, S. S., Gutmann, L., Shanske, S., Neal, W. A., and DiMauro, S., 1983 *Neurology (N.Y.)* **33**:873.
26. Hug. G., Schubert, W. K., and Schwachman, H., 1965, *J. Pediatr.* **67**:741–751.
27. Hug. G., Schubert, W. K., Chuck, G., and Garancis, J. G., 1967, *Am. J., Med.* **42**:139–145.
34. Resibois-Gregoire, A., and Dourov, N., 1966, *Acta Neuropathol.* **6**:70–79.
29. Agamanolis, D., and Patre, S., 1979, *J. Neurol. Sci.* **41**:325–342.
30. Goldfischer, S., Moore, C. L., Johnson, A. B., Spiro, A. J., Valsamis, M. P., Wisniewski, H. K., Ritch, R. H., Norton, W. T., Rapin, I., and Gartner, L. M., 1973, *Science* **182**:62–64.
31. Lewis, G. M., Spencer-Peet, J., and Stewart, K. M., 1963, *Arch. Dis. Child.* **3**:40–47.
32. Parr, J., Teree, T. M., and Larner, J., 1976, *Pediatrics* **35**:770–777.
33. Aynsley-Green, A., Williamson, D. H., and Gitzelmann, R., 1977, *Arch. Dis. Child.* **52**:573–579.
34. Gitzelmann, R., Aynsley-Green, A., and Williamson, D. H., 1977, *Clin. Chim. Acta* **79**:219–221.
35. Lafora, G. R., 1911, *Vichows Arch. [Pathol. Anat.]* **205**:295–303.
36. Gambetti P. L., DiMauro, S., Hirt, L., and Blume, R. P., 1971, *Arch. Neurol.* **25**:483–493.
37. Yokoi, S., Austin, J., Witmer, F., and Sakai, M., 1968, *Arch. Neurol.* **19**:15–33.
38. Sakai, M., Austin, J., Witmer, F., and Trueb, L., 1970, *Neurology (Minneap.)* **20**:160–176.
39. Ponzetto-Zimmerman, C., and Gold, A. M., 1982, *Biochem. Med.* **28**:83–93.
40. Suzuki, K., David, E., and Kutschman, B., 1971, *Arch. Neurol.* **25**:69–80.
41. Peress, N. S., DiMauro, S., and Roxburgh, V. A., 1979, *Arch. Neurol.* **36**:840–845.
42. Robitaille, Y., Carpenter, S., Karpati, G., and DiMauro, S., 1980, *Brain* **103**:315–336.
43. Okamoto, K., Llena, J. F., and Kirano, A., 1982, *Acta Neuropathol.* **58**:73–77.
44. Probst, A., Sandoz, P., Vanoni, C., and Baumann, J. U., 1980 *Acta Neuropathol.* **51**:119–126.
45. Bielschowsky, M., 1912, *J. Psychol. Neurol.* **18**:513–521.
46. Sakai, M., Austin, J., Witmer, F., and Trueb, L., 1969, *Arch. Neurol.* **21**:526–544.
47. Bobin, S. A., Wisniewski, H. M., Kieras, F. J., and Iqbal, K., 1981, *Acta Neuropathol.* **55**:47–51.

2

Aminoacidemias and Brain Dysfunction

F. A. Hommes

1. INTRODUCTION

Many of the inborn errors of amino acid metabolism are associated with mental retardation or other central nervous system involvement. The chemical imbalance imposed on the brain by the accumulating amino acid or its metabolites may not only interfere with the normal development of the central nervous system but may also interfere with the normal development of the brain, which is then held responsible for the mental retardation.

The interaction between amino acids and brain metabolism, especially in relation to development, has been the subject of several reviews.[1–3] Since phenylketonuria is perhaps the most studied of the inborn errors of amino acid metabolism, it is not surprising that most of these reviews deal specifically with the effects of phenylalanine and its metabolites on the growing brain.

Dobbing[4,5] has introduced the concept of the vulnerable period in brain development, a period of active growth and differentiation of the brain. During this period, the brain is especially vulnerable to chemical imbalance, which can lead to delayed maturation or no maturation at all. The mechanism for this delay is poorly understood, to say the least. Many compounds, including some of the amino acids and their metabolites, are known to interfere with brain development, but the complexity of the central nervous system and its functions has not yet permitted a detailed theory on the mechanism for this interference. To be sure, the study of brain development and its pathology may not be the only specialty that suffers from such a lack of basic theories. But full understanding of the interaction between amino acids and central nervous system metabolism may have to wait until a clear understanding of brain differentiation and maturation becomes available. A theory for this differentiation has been proposed by Changeux and co-workers.[6,7] Any such theory should take into account the fact that the enormous complexity of the central nervous system

F. A. Hommes • Department of Cell and Molecular Biology, Medical College of Georgia, Augusta, Georgia 30912.

all but forbids a completely genetically determined specification of the brain and its many functions. The number of genes is probably too small to allow for a complete specification. Epigenetic factors must therefore play an important role as well.

It has been postulated[6,7] that a genetic basis (the genotype) programs the interaction between main categories of neurons and that these first synaptic contacts exist under three states: labile, stable, and regressed. The labile state may mature to a stable state or may, irreversibly, regress. It is furthermore postulated that this maturation and regression are controlled by epigenetic factors leading to the specific phenotype. These epigenetic factors include the activity of the postsynaptic cell. The stabilization process of the early synaptic network is the basis for the diversity and specificity. The period of this stabilization process is the vulnerable period in the development of the nerve terminal.

There is a considerable body of evidence (for review see ref. 7) that the genetic basis provides an excessive production of labile and functional contacts at early stages of development that are not observed in the mature brain. Regression has apparently taken place during maturation, leaving only the functional contacts.

It is conceivable that amino acids and their metabolites interfere with the stabilization process, causing more of the labile contacts to regress, although experimental data to support this notion are lacking. It is nevertheless an attractive hypothesis because it could provide a unifying basis for understanding the mental retardation that is observed in many of the inborn errors of amino acid metabolism.

2. *Aminoacidemias*

In the following sections, some of the inborn errors of amino acid metabolism associated with mental retardation or neurological dysfunction are reviewed.

2.1. *Argininemia*

Arginase (E.C. 3.5.3.1) deficiency, resulting in argininemia, is an autosomal recessive disorder.[8] The clinical picture is rather homogeneous. Early infancy is usually unremarkable, but motor difficulties arise at 2 to 4 months of age. Spasticity of the lower extremities develops, progressing to spastic dysplegia. Seizures, psychosis, and severe mental retardation are frequently observed.

It is not clear why brain dysfunction is associated with this disease. Despite a low or virtually absent activity of arginase in the liver, blood ammonia levels are not consistently elevated. It cannot be excluded, however, that transient periods of hyperammonemia, especially during early phases of brain development, will leave the patients with permanent brain damage. An intriguing feature of this condition is the orotic aciduria even in the absence of hyperammonemia.[9–11] However, there is no evidence for orotic acid toxicity in humans.[12] Dietary treatment of this condition is only successful when the arginine levels are sufficiently reduced.[8,13] This would suggest that arginine *per se* does have some effect on the function of the central nervous system, but since no autopsied cases have been reported, no information is available as to what the substrate of this interference with brain function would be.

2.2. Argininosuccinic Aciduria

Clinically, three types of argininosuccinic aciduria can be distinguished, the distinction being made on the basis of time of onset: neonatal, subacute, and delayed-onset type. The mental retardation associated with this autosomal recessive trait may be related to the hyperammonemia, especially in the neonatal type. It remains to be established whether argininosuccinate itself is neurotoxic. The report of one case of maternal argininosuccinic aciduria with a normal offspring would speak against such a hypothesis.[14] Pathological investigations on the brain of diseased patients have been reported [15–20]: the abnormalities observed do not seem to be specific and resemble very much those seen in ornithine transcarbamylase deficiency, [21] the appearance of Alzheimer type II astrocytes being the most prominent feature. Rats with experimental hyperammonemia show the same phenomenon, and it has been shown that this correlates with the blood ammonia level.[22,23] These data thus do not allow a conclusion to be drawn as to whether argininosuccinate itself is neurotoxic.

2.3. Citrullinemia

As with the other inherited disorders of the urea cycle, it is difficult to distinguish between a neurotoxic effect of the compounds accumulating as a result of the enzyme deficiency and neurotoxic effects of the hyperammonemia frequently associated with these disorders. Many, but not all, of the citrullinemia patients reported have central nervous system involvement.[24] Pathological findings have been reported in only a few such patients. The brain was found to be edematous,[25–27] which is also observed in hyperammonemia resulting from ornithine transcarbamylase deficiency.[24] Nerve cell degeneration, demyelination, and/or delayed myelin formation and enlarged glial cells have been observed as well,[27] abnormalities that are also observed in hyperammonemia. This does not necessarily point to a specific neurotoxic effect of citrulline.

It has been observed that *in situ* perfusion of the rat brain resulted in a 25% inhibition of glucose uptake and lactate production when 5 mM citrulline was added to the perfusion medium. Brain pyruvate kinase (E.C. 2.7.1.40) proved to be inhibited by citrulline with an apparent K_i of 3.2 mM.[28] Blood citrulline levels may approach 5 mM in untreated citrullinemia patients, especially in patients with the neonatal type of the disease. Citrulline levels in the cerebrospinal fluid are, however, usually lower than those in blood. Citrulline levels in the brain of autopsied cases have not been reported. Inhibition of a basic metabolic pathway in the brain, such as glycolysis at the level of pyruvate kinase, may account for the high rate of death among patients with the neonatal form of the disease. Those with the subacute type or the late-onset type have, in general, lower blood citrulline levels, and the survival rate of early infancy is significantly higher.

2.4. Histidinemia

Histidinemia has long been considered to be an inborn error of histidine catabolism associated with mental retardation. It is caused by the absence, or very low levels, of histidase activity (histidine ammonia lyase, E.C. 4.3.1.3.).[29] A mouse model has been identified that is completely identical to the human condition.[30–33] The availability of this truly genetic animal model has greatly facilitated the study of this condition. As far as viability, size, weight, mating ability, skeletal growth, and all other gross phenotypes are concerned, affected animals are indistinguishable from normal littermates.[34] This would seem to indicate that the histidinemic condition does not interfere with brain development (and brain function) after birth. That may not be entirely true for the period before birth, at least in mice.

Offspring born to histidinemic mice, showed abnormalities that developed during the second trimester of pregnancy.[35] Abnormal structures were observed in the otoliths, utriclus, sacculus, semicircular canals, ductus reuniens, endolymphatic sacs, and cochlea of the inner ear. However, this was not observed in all offspring of the histidinemic mothers, and the incidence of abnormalities decreased over the years this particular strain of mice was maintained.[35] Back crosses with the original strain of histidinemic mice, maintained and selected for a high frequency of inner ear abnormalities of the offspring, increased the frequency of mid-ear abnormalities. Intercrosses showed a segregation of the susceptibility to inner ear abnormalities. There is apparently another genetic system, distinct from the gene coding for histidase, that induces susceptibility to inner ear abnormalities when the development of the inner ear takes place in an environment with a high histidine level. The implication is that it is not the high histidine level *per se* that induces the developmental abnormalities but that a specific, additional genetic make-up is required.

No reports have appeared on deafness of children of histidinemic mothers. The initial reports of a high incidence of brain dysfunction in histidinemic children may have been biased, because a careful retrospective, prospective, and interpretive study showed that histidinemic individuals with IQ scores below 70 comprised 1% of the total histidinemic population.[36] It was therefore concluded that histidinemia did not significantly interfere with brain development. This would be in agreement with the results obtained with the mouse model.

The animal model nevertheless illustrates what may be an important feature of the pathogenesis of brain dysfunction. It was demonstrated that a second genetic system is required for the abnormalities to manifest themselves. The offspring of the histidinemic female mice must have a genetically determined predisposition for the increased histidine levels to become harmful. Such genetically determined predispositions at loci distinct from the ones responsible for the primary defects may contribute significantly to the varying degrees of mental retardation or brain dysfunction observed with many of the inborn errors of metabolism.

2.5. *Homocystinuria*

Homocystinuria is an autosomal recessive disorder of the conversion of homocysteine to cystathionine. The affected enzyme, cystathionine-β-synthase (E.C. 4.2.1.13),[37] requires pyridoxal-5-phosphate as a cofactor. About 40% of homocystinuria patients respond favorably to a treatment with very high doses of pyridoxine.[38] These patients and the nonresponders can be treated with a diet low in methionine. Although it has not as yet been demonstrated unequivocally that this dietary treatment prevents all of the sequelae observed in the untreated cases, it is evident that the treated patients have less severe complications. These complications include, but are not limited to, vascular abnormalities and central nervous system abnormalities. Many of the untreated patients are mentally retarded; EEG abnormalities, spasticity, and psychiatric disturbances are less frequently observed.

The question of whether this central nervous system involvement is caused by a chemical imbalance in the brain or is secondary to frequent microinfarctions in the vascular system of the brain is an unresolved one.[39] Cerebral vascular thrombosis, especially during early childhood, may lead to mental retardation. Such infarcts have repeatedly been reported in autopsied brains of homocystinuria patients,[39–43] but such an observation does not necessarily point to a cause-and-effect relationship. Other changes in the brain, such as spongy degeneration and demyelination in the white matter of cerebrum, cerebellum, and spinal cord, without any evidence of vascular changes[44] or neuronal loss in the cerebral cortex and hippocampus[45] have been reported as well.

It has been suggested that cystathionine, the product of the cystathionine-β-synthase-catalyzed reaction, is a neurotransmitter.[46] Cystathionine is indeed present in human brain in high concentration.[47] The role of cystathionine as an inhibitory neurotransmitter has been questioned because the concentrations of cystathionine in gray and white matter of the spinal cord are virtually the same.[48] The intracellular localization of cystathionine-β-synthase in brain is suggestive of a synaptic function. It is localized in the cytosol as well as in the mitochondria.[49] This dual localization permits a separation of a biosynthetic role for cystathionine-β-synthase (cytosolic) from a biosynthetic role specifically for transmitter function (particulate). Further experimentation failed, however, to provide evidence for a high-affinity uptake of cystathionine by synaptosomes isolated from various regions of the brain.[50] Cystathionine, therefore, does not fulfill one of the criteria[51] for acceptance as a neurotransmitter. Cystathionine was notably deficient in three autopsied brains of homocystinuria patients.[52,53] On the other hand, there are a number of known homocystinuria patients without mental retardation.[54,55] It is therefore questionable whether cystathionine is neurotoxic.

Homocystinuria is frequently associated with hypermethioninemia. Although methionine may be toxic to the liver,[56,57] patients with hypermethioninemia as a result of methionine–adenosyl transferase (E.C. 2.5.1.6) deficiency[58,59] are not mentally retarded. It is therefore unlikely that accumulated methionine contributes to the mental retardation observed in some homocystinuria patients.

2.6. Hyperammonemia

Hyperammonemia is not the result of a single enzyme deficiency but is observed in a number of inborn errors of amino acid metabolism, particularly of the enzymes of the urea cycle. Theories on ammonia neurotoxicity center around a decrease in available energy to the brain in the form of ATP. Two ammonia-consuming reactions known to take place in the brain affect the operation of the Krebs cycle. The first is the conversion of α-ketoglutarate, NADH, and NH_3 to glutamate, catalyzed by glutamate dehydrogenase (E.C. 1.4.1.2), and the second is the formation of glutamine from glutamate, ATP, and NH_3, catalyzed by glutamine synthetase (E.C. 6.3.1.2). Depletion of cerebral α-ketoglutarate would result in an impairment of the Krebs cycle. It has been shown in dogs and in mice that the concentration of α-ketoglutarate in cerebral cortex is decreased after injection of ammonia.[60,61] Interpretations of data on cerebral α-ketoglutarate are not easy in view of the different pools of glutamate and α-ketoglutarate in the brain: a small one with a high turnover and a larger one with a less active turnover,[62] the smaller one being about 20% of the larger one. Changes in the level of α-ketoglutarate restricted to the small pool may evade detection because such changes may be lost in the total pool that is analyzed. This may be the reason for conflicting reports on changes in the α-ketoglutarate level of brain in ammonia intoxication.[65,66]

The second reaction involving ammonia, the synthesis of glutamine, consumes ATP. Although brain glutamine synthesis increases considerably after ammonia administration,[66] it is questionable whether such a drain on ATP is sufficient to lower the ATP levels enough to affect the function of the central nervous system.[60] In fact, no change in total brain ATP could be detected after administration of ammonia to rats, although a small decrease in creatine phosphate was observed.[66] There is apparently some drain on the total energy stores of the brain but only of such a small magnitude that the creatine phosphate system can easily buffer the changes.

A more plausible explanation for the ammonia neurotoxicity may be found in the ion shifts that take place when the ammonia concentration increases. It was observed that the plasma K^+ increased while the plasma Na^+ remained constant after administration of ammonia. The extrusion of K^+ from the brain may reflect exchange of K^+ for NH_4^+.[67,68] It has been calculated that these ion shifts can result in a decrease of about 15 mV in the resting membrane potential,[7] thereby bringing the potential closer to the threshold for firing and resulting in an increased nerve cell excitability and activity and, consequently, convulsions.

2.7. Hyperlysinemia

Two types of hyperlysinemia have been described. The first type shows periodic hyperlysinemia associated with hyperammonemia; the second type shows a persistent hyperlysinemia without hyperammonemia. The neurological symptoms associated with the first type can be largely if not completely explained by the hyperammonemia (*vide infra*).[69–71] The question remains why this type of hyperlysinemia is associated with hyperammonemia. Competitive inhibition of arginase by lysine has been offered as an explanation.[69–71] One would expect the ornithine level in plasma to be low, which was not the case. It has recently been shown that lysine inhibits the uptake of ornithine by liver mitochondria,[72] thereby limiting the availability of intramitrochondrial ornithine for reaction with carbamylphosphate. Although a plausible explanation, this falls short of explaining the absence of hyperammonemia in the second type of hyperlysinemia. The plasma lysine levels are about the same (three to four times the normal value) in both conditions.[73] Five of the seven patients with persistent hyperlysinemia are mentally retarded. It is inlikely that hyperammonemia did contribute to the mental retardation, because lysine loading tests in these patients failed to induce a hyperammonemia. Increased blood ammonia levels early in the lives of these patients that may have escaped detection are therefore unlikely.

A salient observation is that a sibling and a cousin of one of the patients had a normal development but nevertheless showed all the biochemical characteristics of the persistent hyperlysinemia syndrome.[74,75] It is therefore questionable whether lysine is neurotoxic.

2.8. Maple Syrup Urine Disease

Maple syrup urine disease (MSUD) is an autosomal recessive disorder resulting from a deficiency of the branched chain α-keto acid (BCKA) dehydrogenase.[76] Clinically, five different phenotypes can be distinguished,[77] with varying neurological involvement. The classic type, with residual enzyme activity of 0–2%, has an early onset with ketoacidosis, seizures, coma, and mental retardation if the patient survives the first weeks of life. The intermittent type, with residual enzyme activity of 2–40%, has a later onset, which is usually precipitated by intercurrent infections. Such patients generally have a normal psychomotor development. In the intermediate type, the residual enzyme activity is 5–25% that of normals. Mental retardation is common. Patients with the thiamine-responsive form of the disease have about 40% residual enzyme activity and have a normal psychomotor development. A fifth type, caused by dihydrolipoyl dehydrogenase, one of the components of the BCKADH, with residual enzyme activity of about 10%, is a variant form of MSUD.The dihydrolipoyl dehydrogenase of the BCKADH complex is the same enzyme as that of other α-keto acid dehydrogenases.

A deficiency of the BCKADH leads to accumulation of the branched-chain α-keto acids as well as to accumulation of the corresponding amino acids, leucine, isoleucinc, and valine. Few autopsied cases have been analyzed, but the information available suggests that the branched-chain amino acids and their α-keto metabolic products accumulate in the brain at high levels. Values of ten-, four-, and two-fold above control values have been reported for leucine, isoleucine, and valine, respectively.[78] The total BCKA in the gray matter was found to be 0.8 mM in one case.[79] Such concentrations are well within the range for inhibition of enzymes, leading to acute manifestations of the disease. Leucine or a metabolite of leucine seems to be responsbile for these acute manifestations, since a loading test in a MSUD patient with this amino acid resulted in neurological symptoms, whereas no such symptoms were observed during a loading test with either isoleucine or valine.[80] 2-Ketoisocaproate (KIC), the transamination product of leucine, has been shown to inhibit, competitively with respect to acetyl-CoA, the fatty acid synthetase of rat brain.[6] The K_i for KIC was 0.9 mM, which is within the range of KIC levels in the brain of MSUD patients. The transamination product of isoleucine, 2-keto-3-methylvalerate (KMV), showed the same inhibition (K_i 3.5 mM), whereas the transamination product of valine, 2-ketoisovalerate (KIV), was not inhibitory. The K_i for KMV seems to be too high to be of physiological significance.

Although it is difficult to visualize how inhibition of fatty acid synthetase could give rise to acute neurological symptoms, it may be involved in the long-term effects of this condition on central nervous system development. Decreased amounts of lipids have been reported in the brain of autopsied cases of MSUD[78,82] if the patients lived for some months. No such changes are observed in patients with the classical type of the disease or in patients in whom dietary therapy had been instituted in early infancy.[82,83]

Degradation products of myelin have never been observed in the autopsied cases, which could suggest a defect in myelin synthesis rather than an increased myelin breakdown. It is, however, doubtful whether the myelin deficiency can fully explain the chronic mental and neurological sequelae of the disease. At least one patient is known who died at the age of 3 years and who had normal brain lipids.[82] Dietary treatment was instituted when the patient was 35 days old, and the child was mentally retarded. It can certainly not explain the acute manifestations of the disease.

Several sites of inhibition of major cerebral metabolic pathways by KIC have been identified. Earlier reports on the inhibition of pyruvate oxidation are conflicting and may have been resolved by the demonstration of inhibition of the uptake of pyruvate by brain mitochondria.[84,85] The inner mitochondrial membrane has a translocator specific for pyruvate and β-hydroxybutyrate.[86–88] In brain, the specificity of this translocator for β-hydroxybutyrate is higher than for pyruvate,[85] which has some significance in view of the fact that the immature brain uses ketone bodies for both synthetic and general energy purposes.[89] Inhibition of the uptake of ketone bodies by the mitochondria will result in a decreased production of ATP, a decreased availability of citrate in the cytosol for fatty acid synthesis, and a decreased availability of precursors for neurotransmitter synthesis, especially glutamate for GABA. A decreased content of glutamate and GABA has indeed been observed in brain of two patients with MSUD.[78]

2.9. *Nonketotic Hyperglycinemia*

Nonketotic hyperglycinemia is an autosomal recessive disorder of the interconversion of glycine and serine. This interconversion takes place in the liver and brain and consists of a glycine cleavage system, yielding carbon dioxide, ammonia, and a tetrahydrofolate derivative from glycine. The tetrahydrofolate derivative condenses with a second glycine molecule to serine.

The glycine cleavage system consists of four enzymes, which have been named P, H, T, and L protein.[90–92] The P protein is a pyridoxyl-phosphate-dependent glycine decarboxylase, which in cooperation with

$$E_1\text{—CHO} + H_2N\text{—}CH_2\text{—COOH} \rightleftharpoons E_1\text{—CH=N—}CH_2\text{—COOH}$$

the lipoic-acid-containing H protein carries out the decarboxylation:

$$E_1\text{—CH=N—}CH_2\text{—COOH} + E_2\begin{matrix}\diagup S \\ | \\ \diagdown S\end{matrix} \rightleftharpoons E_1\text{—C=N—}CH_2\text{—}\underset{\underset{H}{\overset{|}{S}}}{S}\text{—}E_2 + CO_2$$

This aminomethyl-carrying protein complex reacts with the T protein in a tetrahydrofolate-requiring reaction to form ammonia and 5,10-methylenetetrahydrofolate:

$$E_1\text{—CH=N—}CH_2\text{—}\underset{\underset{SH}{|}}{S}\text{—}E_2 + FH_4 + H_2O \rightarrow 5{,}10\text{—}CH_2FH_4$$
$$+ NH_3 + E_1\text{—CHO} + NH_3 + E_2\begin{matrix}\diagup SH \\ \diagdown SH\end{matrix}$$

The L protein, which is a lipoate dehydrogenase, reoxidizes the SH groups:

$$E_2\begin{matrix}\diagup SH \\ \diagdown SH\end{matrix} + NAD^+ \rightleftharpoons E_2\begin{matrix}\diagup S \\ | \\ \diagdown S\end{matrix} + NADH + H^+$$

The enzyme serine hydroxymethyltransferase (E.C. 2.1.2.1) catalyzes the condensation of the second glycine molecule with the 5,10-methylenetetrahydrofolate to serine:

$$5{,}10\text{—}CH_2FH_4 + H_2N\text{—}CH_2\text{—COOH} + H_2O \rightarrow HO\text{—}CH_2\text{—}CHNH_2COOH + FH_4$$

Experiments *in vivo*[93] and *in vitro*[94] have shown that there is virtually no conversion of carbon atom 2 of glycine into carbon atom 3 of serine in nonketotic hyperglycinemia patients. Subsequently, Hiraya and co-workers[95] showed that both the P protein and the H protein were reduced in activity. Since the P protein was immunologically and kinetically indistinguishable from that of normal controls, it was suspected that the H protein was mutated in nonketotic hyperglycinemia. Indeed, it was found that the H protein of the patient had only two instead of four SH goups as in the H protein of normal controls. It was suggested that it was devoid of lipoic acid.

Such a block in the glycine cleavage system leads to a considerable accumulation of glycine in blood, cerebrospinal fluid, and brain.[96] Several investigations have shown that glycine acts as a neurotransmitter agent in certain inhibitory neurons of the spinal cord.[97-99] A neurotoxic action of glycine is therefore likely. Clinical evidence has been presented that newborns with nonketotic hyperglycinemia are normal during the first few hours of life, until the glycine level starts to build up.[100,101] Then, hypothermia, apathy, and respiratory depression begin to appear. If the newborn survives this period, a partial recovery can be observed, but the patients gradually deteriorate to a severe mental retardation and have continuous low-grade convulsions with hypsarrythmia.

This biphasic response to increased cerebral levels of glycine may point to two different mechanisms of glycine neurotoxicity. Animal experiments support this hypothesis.[101,102] When newborn rats were injected with high doses of glycine (3 mg/g body weight, a relatively high dose, which was necessary to obtain forebrain glycine levels comparable to those observed in nonketotic hyperglycinemia patients), a characteristic sequence of abnormal behavior was observed, starting with rotatory behavior, followed by prolonged periods of standstill, convulsions, respiratory depression, and cyanosis. Thirty minutes after the glycine injection, 21 out of 23 newborn rats had died. Control newborn rats treated with an equimolar amount of serine did not show this sequence of behavioral abnormalities, although the blood ammonia levels were elevated to the same extent. The increased blood ammonia is therefore not responsbile for the abnormal behavior. Rats of 8 days of age subjected to the same treatment showed these abnormalities only to a limited degree, whereas rats 15 days of age and older did not show this at all, although the glycine content of the forebrain of these older rats was the same.[101] These data demonstrate that the glycine neurotoxicity is age related. A more developed brain is less sensitive to glycine toxicity. It should be emphasized that these were short-term experiments. Curtis and co-workers[99] have shown that strychnine is a rather specific antagonist for glycine, competing with glycine for the receptor site on the postsynaptic membrane. Strychnine at a dose of 0.2 μg/g body weight proved to be quite effective in increasing the survival rate of the newborn rats treated with glycine. Without the simultaneous administration of strychnine, the survival rate was 13%, which increased to 90% when the newborn rats were treated with a high dose of strychnine.

The development of the glycine synaptic receptor has been studied by Benanides and co-workers.[103] Strychnine was used for the binding studies. The intrinsic hypothesis in this study is, therefore, that the glycine and the strychnine binding sites are completely identical. The specific binding of strychnine is low at birth in rat brain and in rat spinal cord and rapidly increases to adult levels in the brain by day 5 in both tissues when expressed as strychnine bound per milligram protein. The total amount of strychnine bound per brain or spinal cord reached adult levels at 15 and 30 days after birth, respectively. This divergence was interpreted as a faster synthesis of receptors than of protein in both tissues. Experimental hyperglycinemia induced in newborn rats increased the number of strychnine binding sites, but the same treatment in 10-day-old rats did not result in an increase in strychnine binding. Although these results are difficult to correlate with the neurological symptoms observed in the newborn nonketotic hyperglycinemia patient or with the data obtained with experimental hyperglycinemia, it is clear that developmental changes in glycine receptors in brain and spinal cord do take place and that the development of these receptors is influenced by glycine itself. Changeux and Danchin[6,7] have introduced the concept of a labile state of the synapse and its transition to a stable state during development. This transition is epigenetically controlled by the postsynaptic cell, including the activity of the synapse itself. If glycine can be considered to be a part of the activity of the synapse, then such increased glycine levels may indeed lead to a deranged development and mental retardation.

The early phase of glycine neurotoxicity seems therefore to be associated with the inhibitory neurotransmitter action of glycine. Strychnine therapy has been tried in nonketotic hyperglycinemia patients with modest success[104,105] or with no success at all.[106] It can be expected that strychnine ameliorates the early phase of glycine neurotoxicity in view of the data obtained with the animal experiments referred to above.

No studies have as yet been reported on the second and apparently more permanent phase of glycine neurotoxicity. It remains to be established whether this second phase follows the first in time or whether the second phase is superimposed the first, escaping clear recognition because of the prevailing symptoms associated with the first phase. A deficiency of one-carbon groups has been postulated,[94] but attempts to resupply the tissue with one-carbon groups via methionine or via N^5-formyltetrahydrofolate did not result in neurological improvement, although the plasma glycine levels were temporarily decreased.[94,107]

An intriguing phenomenon is the valine sensitivity of at least some of the nonketotic hyperglycinemia patients. Valine induces a comatose state with EEG changes in these patients.[91,100,109] These effects are specific for valine. A possible explanation may be derived from the work reported by O'Brien[110] on the inhibition of glycine oxidation by branched-chain α-keto acids. It was found that α-ketoisovaleric acid, α-ketoisocaproic acid, and α-keto-β-methylvaleric acid (the transamination products of valine, leucine, and isoleucine, respectively) inhibited glycine oxidation by 40–60%. This inhibition was specific for the branched-chain α-keto acids and was not observed with either the amino acids or with pyruvate or α-ketoglutarate. It was suggested that the lipoate dehydrogenase is common to the branched-chain α-keto acid dehydrogenase complex and the glycine oxidase system (L protein). However, in that case it would be expected that glycine would inhibit the oxidation of branched-chain α-keto acids, which could not be demonstrated.[110] Moreover, it has been demonstrated that the lipoate dehydrogenase of the branched-chain α-keto acid dehydrogenase and of the pyruvate dehydrogenase are identical[111,112] and that inhibition of the glycine cleavage system was not observed with pyruvate.[110] The inhibition of the oxidation of glycine by the branched-chain α-keto acids is, however, real, whatever the mechanism is. A low residual activity of the glycine cleavage system may then be further reduced by inhibition by branched-chain α-keto acids, resulting in higher intracerebral glycine levels.

2.10. Phenylketonuria

Phenylketonuria (PKU) is perhaps the most studied inborn error of amino acid metabolism. Despite the fact that the enzyme defects of PKU and its variant forms are known and a therapy that prevents the mental retardation and techniques for mass newborn screening are available and widely applied, very little is known about the causes of the mental retardation in patients afflicted with this condition if untreated. The reason for this discrepancy between the vast amount of clinical knowledge and the scanty evidence on the biochemical level may be that no true genetic animal model of this disease is available. The initial report on the dilute lethal mouse as a possible model for PKU[113] could not be substantiated in further investigations.[114] Chemically induced models have therefore been used. Currently, two such models are available, both based on the administration in the diet of an inhibitor of phenylalanine hydroxylase (E.C. 1.14.3.1) and excess phenylalanine. The first model, developed by Berry and co-workers,[115] followed the observation of Koe and Weismann[116] on the inhibition of phenylalanine hydroxylase by *p*-chlorophenylalanine (PCPA); the second model, developed by Greengard,[117] uses α-methylphenylalanine (αMP) as an inhibitor of phenylalanine hydroxylase.

With both models, a sustained elevated level of phenylalanine (Phe) in blood can be obtained, but on a molar basis about three to four times more αMP is needed to obtain the same blood Phe levels as with PCPA. This may in part be because αMP does not inhibit the kidney phenylalanine hydroxylase.[118] Otherwise, both methods are fairly equivalent, although several reports have indicated that PCPA does inhibit other enzyme systems as well, such as tyrosine hydroxylase (E.C. 1.14.16.2[a]),[116] tryptophan hydroxylase (E.C. 1.14.16.4),[119] and pyruvate kinase.[120] Other side effects observed with the PCPA model, such as cataracts, growth retardation, skin lesions, and loss of body hair, are less frequently observed with the αMP model or are present to a smaller degree. Both models decrease the brain serotonin[121,122] and catecholamine levels,[122] a characteristic shared with the human PKU condition.[123] Growth retardation has also been observed with the αMP model[121,122] (H.K. Berry, personal communication). An exhaustive review on the merits and drawbacks of these models has been presented by Voorhees *et al.*[3]

An important aspect of working with these models is the fact that rats are usually used as the experimental animal. The rat is born much more immature than the human. The newborn rat is, as far as brain maturation is concerned, comparable to the human fetus of about 30 weeks, gestation.[124] A hyperphenylalaninemic condition in the newborn rat is therefore not comparable to the PKU condition of the human neonate. The intrauterine environment protects the human PKU fetus sufficiently against a buildup of phenylalanine. The hyperphenylalaninemic condition should be induced in rats at least 5 days of age in order to mimic the human PKU condition. Studies on younger rats may reveal effects of phenylalanine on brain development, but extrapolation to the human condition has to be made cautiously in view of the difference in brain maturation between the two species at the moment of birth.

On the basis of studies on these models and/or biochemical studies on PKU patients, several theories on the biochemical abnormalities leading to the mental retardation observed in untreated PKU patients have been presented: (1) disturbance of vitaimin B_6 metabolism, (2) disturbance of glutamate metabolism, (3) interference with cerebral energy metabolism, (4) disturbance of cerebral serotonin metabolism, (5) disturbance of cerebral catecholamine metabolism, (6) interference with cerebral protein synthesis, and (7) disturbance of myelin metabolism.

It has been shown that rats fed a diet high in phenylalanine and pyridoxine performed poorly in the water maze learning test.[125] Since it had earlier been shown that pyridoxilene-β-phenylethylamine occurred in urine and brain of hyperphenylalaninemic rats and in the urine of patients with PKU,[126,127] it was suggested that the abnormal water maze learning test was caused by an accumulation of this compound rather than a deficiency of the B_6 vitamins. Indeed, no deficiency of the B_6 vitamins could be demonstrated in hyperphenylalaninemic rat brain, but a lower turnover of pyridoxal phosphate was evident in the hyperphenylalaninemic rats.[128] In order to maintain the same level of pyridoxal phosphate with a slower turnover, the rate of synthesis has to be decreased as well. Pyridoxyl kinase (E.C. 2.7.1.35) is inhibited by pyridoxilene-β-phenethylamine under *in vitro* conditions,[129] but evidence for inhibition of this enzyme under *in vivo* conditions could not be obtained.[128] The data are therefore incompatible.

Subsequent studies revealed another role for pyridoxamine phosphate and/or pyridoxamine, because it was found that pretreatment of hyperphenylalaninemic rats with pyridoxamine prevented the accumulation of, particularly, phenylacetate in the brain.[130,131] It was suggested that pyridoxamine and/or its phosphate is actively associated with the removal of excess keto acids and aldehydes from the brain. Implicit in this suggestion is that the mental retardation associated with untreated PKU is caused by increased cerebral phenylacetate levels. A model to simulate the biochemical and behavioral abnormalities of human PKU based on this idea has been developed[132,133] by injection of phenylacetate in rats starting at the age of 4 days. That age may indeed be sufficiently young for the subcutaneously injected phenylacetate to reach the brain, because the blood–brain barrier may not have developed sufficiently to exclude foreign compounds. Using intraperitoneal injection of phenylalanine (1 g/kg) every 2 hr for 14 hr, Goldstein[134] was able to show the presence of phenyllactic acid in brains of 18-day-old and younger rats. However, this metabolic product of phenylalanine was not detectable in brains of 25-, 40-, and 75-day-old rats subjected to the same treatment. This emphasizes the importance of the development of the blood–brain barrier. Phenylacetate is slightly cytotoxic to rat cerebellar cells in culture at 2.2 mM and 4.4 mM and lethal at 6.6 mM. However, no effect was seen at 1.1 mM and lower levels.[135] The levels of phenylacetate in the brains of rats subcutaneously injected with phenylacetate are well below the concentrations at which cytotoxicity can be observed.[136] It should, however, be mentioned that an *in vitro* system does not necessarily directly equate with the *in vivo* system.

Phenylacetate does have an effect on behavior: hypoactivity in an open field and poor performance in the water maze and shuttle box tests were observed.[136] It also delays the maturation of synapses as measured by the high-affinity uptake of choline and GABA, the former being more reduced than the latter.[137]

The ganglioside content of the cerebral cortex, as a marker of synaptic membranes, was significantly reduced. The treatment of the rats used in this study started at day 2 after birth and lasted until the 21st day of life, after which the rats received the normal laboratory food. Behavioral testing was done at the age of 8–9 weeks, whereas the biochemical tests were carried out when the animals were 4–5 months of age. Phenylacetate treatment can therefore give rise to long-lasting effects. That treatment was, however, started at the second day of life, which seems to be a little too young to be quite comparable to the human newborn PKU. Similar deficits in behavior and in the biochemical parameters studied were also observed in rats with experimental hyperphenylalaninemia induced with PCPA and phenylalanine.[136,137] Although both regimens resulted in the same long-lasting effects, this does not necessarily prove that phenylacetate is the main toxic agent of human PKU.

Abnormal glutamate metabolism has been implicated in the neuropathogenesis of PKU. Glutamic acid decarboxylase (E.C. 4.1.1.15) was found to be inhibited *in vitro* by some of the phenylalanine metabolites.[138] This could result in a reduced rate of formation of GABA, whose role as putative inhibitory neurotransmitter has been well established.[139] However, none of the experimental PKU models showed a reduced level of GABA in brain, whether the hyperphenyalaninemic condition was induced with PCPA (R. Berger and F. A. Hommes, unpublished data) or with αMP[140,141] as inhibitor of phenylalanine hydroxylase. Plasma levels of glutamic acid are indeed reduced in human PKU[142] as well as in experimental hyperphenylalaninemia[140,141] but that does not result in decreased brain levels.[140,141,143] It is therefore unlikely that a disturbed metabolism of glutamic acid contributes to the neuropathogenesis of PKU.

There is considerable evidence that phenylalanine and phenylpyruvate inhibit glycolysis in brain at the level of pyruvate kinase.[144–146] The K_i values are of the order of 10 mM,[144] which is considerably higher than the levels of these compounds in the brain in human PKU[147,148] or in the brain of rats with experimental hyperphenylalaninemia.[140,141,143] The brain pyruvate kinase is the M-type isoenzyme.[145,149] Inhibition of this isoenzyme by phenylalanine is reversed by alanine,[150] which would counteract an inhibition by phenylalanine. Although the inhibition can be demonstrated *in vitro, in vivo* measurements did not show a significant effect on the glycolytic flux.[151,152] No other enzymes of carbohydrate metabolism are known to be inhibited by phenylalanine and/or its metabolites, at least not at concentrations comparable to those found in brain. Inhibition of such basic pathways of cerebral energy metabolism would be expected to result in severe, acute symptoms, which are not characteristic of human PKU.

It is a well-established fact that serotonin metabolism is affected in PKU patients. Such patients excrete less 5-hydroxyindole acetate[153–162] and 5-hydroxytryptamine,[154,155,157,158,160,162] whereas the 5-hydroxytryptamine level in serum [154,155,159,161] and cerebrospinal fluid[163,164] is decreased. These abnormalities are corrected on institution of the low-phenylalanine diet.

There are probably two sites of inhibition of 5-hydroxytryptamine synthesis in the brain: decreased transport of tryptophan into the brain and inhibition of tryptophan hydroxylase.

Tryptophan is actively taken up by synaptosomes by a high-affinity system with a K_m of 0.01 mM and a low-affinity system with a K_m of 1 mM[165] The overall system is inhibited by phenylalanine as well as by the other neutral amino acids, leucine, isoleucine, valine, tyrosine, and methionine. The transfer of tryptophan from blood into synaptosomes involves not only uptake by these synaptosomes but also transport across the blood–brain barrier. The apparent K_m for tryptophan for the latter transport has been determined at 0.7 mM.[166] Like the uptake by synaptosomes, the transport of tryptophan across the blood–brain barrier is inhibited by the same neutral amino acids. The apparent K_i for phenylalanine was found to be 0.09 mM.[166] It can therefore be expected that the tryptophan level in brain in hyperphenylalaninemia is decreased. This has indeed been observed in human PKU[58] as well as in experimental hyperphenylalaninemia induced by injection of phenylalanine.[168] Hyperphenylalaninemia induced by adding αMP and phenylalanine to the diet failed to produce changes in the brain tryptophan content.[122,141]

A decreased content of 5-hydroxytryptamine in the central nervous system has been observed in human PKU[167] as well as in experimental hyperphenylalaninemia.[121,122,141,169–171] Hyperphenylalaninemia induced with PCPA models are of little value in this respect because PCPA has been shown to inhibit tryptophan hydroxylase. The observation that the tryptophan content of the brain of rats with hyerphenylalaninemia induced with the αMP system is not different from that of controls[122,141] suggests that the limited supply of substrate as a result of inhibition of tryptophan transport may not be the only factor that controls 5-hydroxytryptamine levels in the brain. It has been pointed out that the reduction in brain tryptophan content may be too small to account for the decreased 5-hydroxytryptamine levels in view of the K_m of tryptophan for tryptophan hydroxylase.[2,172] Phenylalanine can act as a competing substrate for tryptophan hydroxylase with a K_m of 290 μM versus a K_m of 32 μM for tryptophan.[65] That value for the K_m for phenylalanine is sufficiently low compared to the brain phenylalanine levels of around 1000 μM to result in inhibition of tryptophan hydroxylation.

Evidence equating 5-hydroxytryptamine depletion with mental deficiency is inconsistent.[61,66–70] Moreover, it is difficult if not impossible to sort out the effects of 5-hydroxytryptamine depletion *per se* from other biochemical imbalances induced by the hyperphenylalaninemia regimen in experimental animals.

Catecholamines were likewise found to be reduced in brains of autopsied cases of PKU.[167] Evidence for a decreased flux through tyrosine hydroxylase *in vivo* in PKU patients has been obtained by administering deuterated L-tyrosine and measuring the excreted metabolites of dopamine and norepinephrine.[184] A lower flux through tyrosine hydroxylase was found when the plasma phenylalanine level was high than when it was low. Inhibition by phenylalanine of tyrosine hydroxylase, which is the rate-limiting step in catecholamine biosynthesis, has been demonstrated at sufficiently low levels of phenylalanine that such inhibition is of physiological importance.[185]

Such studies on experimental animal models are less conclusive since PCPA and αMP, used as the inhibitors of phenylalanine hydroxylase in these models, also inhibit tyrosine hydroxylase.[116,186,187]

That biogenic amines including 5-hydroxytryptamine do have an effect on the mental status is evident from patients with variant forms of PKU caused by dihydropteridine reductase deficiency[179] or a defect in the biosynthesis[180] of this cofactor for phenylalanine hydroxylase, tryptophan hydroxylase, and tyrosine hydroxylase. A small number of such patients have been treated with L-DOPA, 5-hydroxytryptophan, and a peripheral aromatic amino acid decarboxylase inhibitor in addition to treatment with a diet low in phenylalanine. In one study,[181] substitution therapy was started at the age of 9 months, the patient having received a phenylalanine-restricted diet from the age of 2 weeks.[182] Considerable clinical improvement was observed during a 9-month period of treatment. A second reported case was treated for 1.5 years, starting at the age of 14 months, after having been on a low-phenylalanine diet since the age of 7 months, with considerable clinical improvement. [83] Two other patients, siblings, have been reported.[164] In the older patient, substitution therapy was started at age of 24 months, but no clinical improvement could be observed. Her younger brother was treated with a phenylalanine-restricted diet from the age of 3 weeks and placed on substitution therapy at the age of 6.5 months. Considerable clinical improvement was observed after a treatment period of 8 months.

Although the number of patients treated with this substitution therapy in addition to a phenylalanine-restricted diet is small and the length of treatment too short to allow conclusions as to the ultimate efficacy of the treatment, it is clear that an early institution of the substitution therapy is beneficial for the patients.

This could be taken as evidence that a deficiency of biogenic amines is indeed involved in the pathogenesis of the mental defect of hyperphenylalaninemia. The reports on these patients do not allow a conclusion as to which of the biogenic amines, 5-hydroxytryptamine or DOPA or both, are responsible for the corrective action.

It can be expected that such defects in the biosynthesis of the biogenic amines lead to a more severe deficiency of these compounds in the brain than is the case in classical PKU, although no data are available to substantiate this hypothesis. The level of metabolites of the biogenic amines in cerebrospinal fluid would suggest this, since the concentrations of homovanillic acid and 5-hydroxyindoleacetic acid in cerebrospinal fluid of untreated classical PKU patients were considerably higher than those of the untreated variant form of PKU.[164] The question of whether the mental deficiency as observed in these variant forms of PKU has the same pathogenesis as in classical PKU is an open one, and, consequently, whether biogenic amine depletion in classical PKU contributes to the mental deficiency is likewise unsettled.

The effect of hyperphenylalaninemia on brain protein synthesis continues to be an area of conflicting reports. It has been reported that hyperphenylalaninemia causes polysome disaggregation in the brain,[118–190] thereby inhibiting protein synthesis, but the exact mechanism remains obscure. Older animals are more resistant[118–191] but the nature of this maturation process has not been elucidated.

It has been suggested that the polysome disaggregation is in fact an artifact of the release of ribonucleases during the preparation of the polyribosome fraction from brain homogenates.[192] Analysis of the resulting monoribosomes has, however, shown that they were not associated with mRNA or peptidyl-tRNA,[191] which makes the disaggregation unlikely to be a result of ribonuclease action. Earlier experiments using the PCPA-based model to induce hyperphenylalaninemia have to be judged with caution, since it has been shown that PCPA itself can give rise to polyribosome disaggregation, in contrast to αMP.[193] The transport of phenylalanine and methionine across the blood–brain barrier is mediated by the same carrier.[166] Mutual competitive inhibition of uptake has been observed not only in hyperphenylalaninemic rats[166] but in PKU children as well.[195] A possible explanation for polysome disaggregation in brain could therefore be an inhibition of the uptake of methionine, resulting in the relative deficiency of this amino acid. This could lead to inhibition of protein synthesis, since methionine plays a specific role in the initiation of translation.[196] Phenylalanine did indeed inhibit the uptake of methionine into rat brain and the incorporation of methionine into rat brain protein, but no difference could be observed between rats 13 days of age and 25 days of age,[197] whereas at the age of 25 days, there is considerable less sensitivity of polyribosome disaggregation for phenylalanine.[191] The rat did show an age-dependent rate of methionine incorporation into cerebral protein with a maximum at 5 days after birth, coinciding with the highest fractional rate of protein synthesis.[196] Such an age dependence of methionine uptake has also been observed in humans, the highest rate being observed at birth.[195]

The transport of amino acids across the blood–brain barrier consists of exchange of amino acids as well as a net influx during the period of growth of the brain, especially at younger ages. Inhibition of the exchange of amino acids by phenylalanine in the mature rat may not affect the rate of protein synthesis, since no depletion of amino acid pools may occur. In the rapidly growing rat, the major part of the amino acids taken up will be used for protein synthesis. With such a drain on amino acid pools for protein synthesis, inhibition of the uptake of amino acids may lead to inhibition of protein synthesis as well. Indeed, the incorporation of lysine into brain protein was found to be smaller in the hyperphenylalaninemic rat 3 days of age than in the saline-treated control of the same age, but such a difference was not observed in rats of 24 days of age.[197]

Studies on myelin from autopsied PKU cases have shown that the amount of myelin is less than normal.[198] The myelin isolated from such brains had essentially the same composition as the myelin isolated from control brains.[198] These observations have been confirmed in rats with experimental hyperphenylalaninemia irrespective of whether this condition was induced with phenylalanine alone,[199–201] with the PCPA model,[202,203] or with the αMP model.[121,141,204–206] A decreased amount of myelin with an apparently normal gross composition has also been observed with the phenylacetate model.[207] Small variations in the chemical composition of myelin of hyperphenylalaninemic rats may not be demonstrated when the gross composition is analyzed. Such variations have indeed been found. Hyperphenylalaninemia induced early in the neonatal period caused myelin to be abnormal with respect to sulfatides[203,208,209] and galactolipids.[199] Inhibition of brain sulfatide formation has been attributed to diminished availability of adenosine-3-phosphate-5-phosphosulfate, a substrate necessary for the formation of sulfatide from cerebroside.[210] No difference was seen in the PCPA rat model in cholesterol and cerebrosides at the 15th postnatal day, just prior to the burst in myelination, but a difference was seen at the 23rd day in rats treated from day 7 with a high-phenylalanine-containing diet.[208] The *in vivo* rate of incorporation of [^{14}C] glucose into brain lipids in a PCPA model was found to be reduced,[211] as was the incorporation of mevalonate into sterols.[212] A defect in the elongation[213] as well as in the desaturation of fatty acids has been suggested.[202,205] A decreased amount of unsaturated fatty acids has also been observed in autopsied human PKU brain.[214–216]

A common denominator in these studies is a decreased amount of myelin per brain in both human PKU and experimental hyperphenylalaninemia. The question of whether this is caused by a decreased synthesis or an increased rate of breakdown or both may be a futile one at this stage of the investigations, because small changes in one of the many components of the myelin sheath may lead to a decreased stability of the myelin membrane and, consequently, to a higher turnover of myelin.

An increased rate of myelin breakdown has indeed been observed in experimental hyperphenylalaninemia. Initial studies using the PCPA model showed a dramatically reduced half-life of the fast component of myelin as compared to normal controls.[202,217] Studies on myelin of the normal rat and mouse have demonstrated that a fraction of myelin (the fast component) has a turnover rate considerably higher than the rest of the myelin.[218–221] Whether this remaining myelin is a homogeneous pool as far as turnover is concerned is an open question.[219] Subsequent studies showed that the increased turnover of the fast component of myelin in experimental hyperphenylalaninemia was indeed a result of the hyperphenylalaninemic condition *per se*, because it was not observed in weight-matched controls or in rats that received PCPA only.[115,223] It furthermore proved to be independent of the inhibitor of phenylalanine hydroxylase used to induced the hyperphenylalaninemic condition, because identical decreases in the half-life of the fast component of myelin were observed with the αMP model.[141] The decreases in half-life were observed in the light, medium, and heavy fractions (cf. ref. 223) of myelin[141] as well as in the individual protein components of myelin.[222,223] A summary of these results is presented in Table I. The hyperphenylalaninemic condition was induced

Table I
Summary of Turnover of Myelin, Myelin Subfractions, and Myelin Proteins in Experimental Hyperphenylalaninemia[a]

Diet	C	WMC	PCPA	αMP	PCPA-HyPhe	αMP-HyPhe
Whole myelin	37/n.d.	30–36/n.d.	33/n.d.	26/n.d.	2/n.d.	3/n.d.
Small basic protein	—	—	—	—	9/n.d.	—
Large basic protein	—	—	—	—	3/16	—
DM-20	57/n.d.	—	—	—	9/66	—
Proteolipid protein	33/n.d.	—	—	—	2/16	—
Wolfgram proteins	43/n.d.	—	—	—	5/76	—
Light myelin	—	67/n.d.	—	—	—	4/n.d.
Medium myelin	—	43/n.d.	—	—	—	3/n.d.
Heavy myelin	—	33/n.d.	—	—	—	3/n.d.

[a] Half-lives are given in days of fast/slow components of normal (C), hyperphenylalaninemic rats (PCPA-HyPhe for the hyperphenylalaninemic condition induced by PCPA plus Phe, αMP-HyPhe induced by αMP plus Phe), weight-matched controls (WMC), and rats receiving the inhibitor only (PCPA and αMP, respectively) in their diet. The specific diets were given to rats 24 days of age injected with [^{3}H]lysine. n.d., not determined. All half-lives for the fast components are uncorrected for the contribution by the slow components.

in these rats when they were 25 days of age, that is, well beyond the peak in myelination.[225] This stage of brain development in the rat model is similar to that of a PKU child of 5–8 years old, at which time the low-phenylalanine dietary therapy is relaxed or terminated. If it is possible to extrapolate the results obtained with these rat models to the human condition, they would strongly suggest a continuation of dietary therapy, although an age at which excess phenylalanine no longer interferes with myelin metabolism has yet to be established, if such an age does exist.

The effect on the decreased half-life is specific for myelin, since no effect of the hyperphenylalaninemic condition on total brain protein turnover could be observed either with the PCPA model[202,217] or with the αMP model.[141] The effect on myelin synthesis is small and somewhat dependent on the type of inhibitor used. With the PCPA model, a slight inhibition was observed,[197,202,217,222,223] whereas the αMP model showed the same rate of incorporation of label into myelin as normal controls, weight-matched controls, or rats receiving αMP alone. The increased turnover of the fast component of myelin is therefore not compensated by an increased rate of synthesis. The net result is then a loss of myelin, which has also been observed in the human PKU brain. No data are presently available that could shed light on a mechanism for the increased turnover of at least part of the myelin. It has, however, been shown that a hyperphenylalaninemic condition reduces the rate of sulfate incorporation into cerebrosides.[209,210] This would mean that less cerebroside sulfatide is available to interact with myelin basic protein. Myelin basic protein with less cerebroside sulfatide attached to it is more susceptible to proteolytic attack.[226,227] This could be the basis for an explanation of the increased rate of turnover of the fast component of myelin in experimental hyperphenylalaninemia.

2.11. Tyrosinemia

Tyrosinemia I or tyrosinosis is an autosomal recessive disorder caused by a deficiency of fumarylacetoacetate hydratase (E.C. 3.7.1.2)[228] an enzyme of tyrosine catabolism. Such a deficiency leads to accumulation of and excretion of succinylacetoacetate and its decarboxylated product succinylacetate. The latter compound is a powerful inhibitor of δ-aminolevulinic acid dehydratase (E.C. 4.2.1.24)[228,229] the enzyme deficient in acute intermittent porphyria. The polyneuropathy associated with tyrosinemia I is therefore related to the biochemical abnormalities of acute intermittent porphyria.[232]

Tyrosinemia II or Richner–Honhart syndrome is likewise an autosomal recessive disorder and results from a deficiency of cytosolic tyrosine aminotransferase (E.C. 2.6.1.5).[233] Mental retardation is frequently observed. Noteworthy are three patients whose diagnoses of tyrosinemia II were made at the ages of 26, 29, and 51 years, respectively, who exhibited normal mental state.[234,235] It is therefore questionable whether tyrosine is neurotoxic. Transient tyrosinemia occurs frequently in the neonate, especially in prematures. Although the enzymic basis of this form of tyrosinemia is unclear, many affected neonates react favorably to high doses of ascorbic acid, which would suggest an overloading of *p*-hydroxyphenylpyruvate oxidase. Ascorbic acid protects this enzyme from substrate inhibition.[236] Mild mental retardation and impaired psycholinguistic abilities have been associated with this condition.[237] Until the enzyme defect has been fully classified, it is difficult to assess whether these neurological symptoms are really caused by the accumulated tyrosine.

3. CONCLUDING COMMENTS

The association of brain dysfunction with many of the inborn errors of amino acid metabolism is a striking one. Some caution about a causative relationship between amino acid (or its metabolites) accumlation and brain dysfunction is, however, warranted. Many of the patients have come to the attention of clinicians because of neurological symptoms, and in subsequent work-up, abnormalities in amino acid metabolism were then identified. Perhaps it has been too easily accepted that the disturbance of amino acid metabolism is responsible for the brain dysfunction. The case of histidinemia is a good example. The high degree of variability of clinical expression of the neurological symptoms with the same enzyme defect, seemingly unrelated to the level of residual activity of the affected enzyme, points to other factors responsible for the degree of central nervous system involvment. Here again, the case of histidinemia is a good example, since it was observed that a second genetic locus is involved in the expression of midear abnormalities in addition to the histidase locus leading to high histidine levels. This area of additional genetic make-up that may amplify or, to a degree, neutralize the damaging effects of accumulating metabolites is virtually unexplored. It is nevertheless an important area, because it may provide better insight into the mechanism of interference of accumulated metabolites with central nervous system function and development, which should ultimately lead to a more rational decision as to which patients are in need of treatment.

Inclusion of the additional genetic make-up as one of the determinants for the expression of brain dysfunction implies that the primary enzyme defect is not solely responsible for the phenotypic expression of the mutation, in other words, that the phenotypic expression is in fact multifactorial.

The identification of loci in addition to the primary enzyme defects of amino acid metabolism contributing to brain dysfunction will be a difficult task and may not be possible unless a better understanding becomes available of the mechanisms of central nervous system differentiation, development, and maturation.

REFERENCES

1. Gaull, G. E., Tallan, H. H., Lajtha, A., and Rassin, D. K., 1975, *Biology of Brain Dysfunction*, Vol. III (G. E. Gaull, ed.), Plenum Press, New York, pp. 47–143.
2. Kaufman, S., 1977, *Adv. Neurochem.* **2**:1–132.
3. Voorhees, C. V., Butcher, R. E., and Berry, K. K., 1980, *Neurosci. Behav. Res.* **5**:177–190.
4. Dobbing, J., 1968, *Applied Neurochemistry* (A. N. Davison and J. Dobbing, eds.), Blackwell, Oxford, pp. 287–316.
5. Dobbing, J., 1974, *Pediatrics* **53**:2–6.
6. Changeux, J. P., Courrege, P., and Danchin, A., 1973, *Proc. Natl. Acad. Sci. U.S.A.* **70**:2974–2978.
7. Changeux, J. P., and Danchin A., 1976 *Nature* **264**:705–712.
8. Synderman, S. E., Sansoricg, C., Chen, W. Y., Norton, P. M., and Phonsalkar, S. V., 1977, *J. Pediatr.* **90**:563–568.
9. Qureski, I. A., Letorte, J., Ouellet, R., Lelievre, M., and Laberge, C., 1981 *Diabet. Metab.* **7**:5–10.
10. Naglas, E. W., and Cederbaum, S. D., 1981, *J. Inherit. Metab. Dis.* **4**:207–210.
11. Cederbaum, S. D., Meodjono, S. J., Shaw, K. N. F., Carter, M., Naylar, E., and Walser, M., 1982, *J. Inherit. Metab. Dis.* **5**:95–99.
12. Kelley, W. N., 1982, *The Metabolic Basis of Inherited Disease* (J. B. Stanbury, J. B. Wyngaarden, D. S. Frederickson, J. L. Goldstein, and M. S. Brown, eds.), McGraw-Hill, New York, pp. 1202–1226.
13. Cederbaum, S. D., Shaw, K. N. F., and Valenti, M., 1977, *J. Pediatr.* **90**:569–573.
14. Shih, V. E., 1978, *The Metabolic Basis of Inherited Disease* (J. B. Stanbury, J. B. Wyngaarden, and D. S. Frederickson, eds.), McGraw-Hill, New York, pp. 362–386.
15. Baumgartner, R., Scheidegger, S., Stalder, G., and Hattinger, A., 1968, *Helv. Pediatr. Acta* **23**:77–106.
16. Carton, S., DeSchryver, F., Kint, J., Van Deurne, J., and Hooft, C., 1969, *Acta Paediatr. Scand.* **58**:528–534.
17. Levin, B., and Dobbs, R. H., 1968, *Proc. Soc. Med.* **61**:773–776.
18. Glick, N. R., Snodgrass, P. J., and Schafer, I. A., 1976, *Am. J. Hum. Genet.* **28**:22–30.
19. Solitare, G. B., Shih, V. E., Nelligan, D. J., and Dolan, T. F., 1969, *J. Ment. Defic. Res.* **13**:153–170.
20. Lewis, P. D., and Miller, A. L., 1970, *Brain* **93**:413–422.
21. Bruton, C. J., Corsellis, J. A. N., and Russell, A., 1970, *Brain* **93**:423–434.
22. Cavanagh, J. B., and Kyu, M. H., 1969, *Lancet* **2**:620–622.
23. Cavanagh, J. B., and Kyu, M. H., 1971, *J. Neurol. Sci.* **12**:63–72.
24. Walser, M., 1982, *The Metabolic Basis of Inherited Disease* (J. B. Stanbury, J. B., Wyngaarden, D. S. Frederickson, J. L. Goldstein, and M. S. Brown, eds.), McGraw-Hill, New York, pp. 402–438.

25. VanderZee, S. P. M., Trybels, J. M. F., Monnens, L. A. H., Hommes, F. A., and Schretlen, E. D. A. M., 1971, *Arch. Dis. Child.* **46**:847–851.
26. Ghisalfi, J., Augier, D., Martinez, J., Barthe, P., Andrieu, P., Besse, P., and Regnier, A., 1972, *Pediatrie* **28**:55–65.
27. Wick, H., Bachmann, C., Baumgartner, R., Brechbuhler, T., Colombo, J. P., Weismann, U., Mihatsch, M. J., and Ohnacker, H., 1973, *Arch. Dis. Child.* **48**:636–641.
28. Okken, A., Vander Bly, J. F., and Hommes, F. A., 1973, *Pediatr. Res.* **7**:52.
29. LaDu, B. N., Howell, R. R., Jacoby, G. A., Seegmiller, J. E., and Zannani, V. G., 1962, *Biochem. Biophys. Res. Commun.* **7**:398–402.
30. Kacser, H., Bulfield, G., and Wallace, M. E., 1973, *Nature* **244**:77–79.
31. Bulfield, G., and Kacser, H., 1974, *Arch. Dis. Child.* **49**:545–552.
32. Bulfield, G., and Kacser, H., 1975, *J. Neurochem.* **24**:403–405.
33. Kacser, H., Khin Mya Mya, Duncker, M., Wright, A. F., Bulfield, G., Lyon, M., and McLaser, A., 1977, *Nature* **264**:262–263.
34. Kacser, H., Bulfield, G., and Wright, A., 1979, *Models for the Study of Inborn Errors of Metabolism* (F. A. Hommes, ed.), Elsevier/North Holland Biomedical Press, Amsterdam, pp. 33–41.
35. Kacser, H., Khin Mya Mya, and Bulfield, G., 1979, *Methods for the Study of Inborn Errors of Metabolism* (F. A. Hommes, ed.), Elsevier/North Holland Biomedical Press, Amsterdam, pp. 43–53.
36. Scriver, C. R., and Levy, H. L., 1982 *Am. J. Hum. Genet.* **34**:62A.
37. Mudd, S. H., Finkelstein, J. D., Irreverre, F., and Laster, L., 1964, *Science* **143**:1443–1445.
38. Mudd, S. H., and Levy, H. L., 1982, *The Metabolic Basis of Inherited Disease* (J. B. Stanbury, J. B. Wyngarrden, D. S. Frederickson, J. L. Goldstein, and M. S. Brown, eds.), McGraw-Hill, New York, pp. 522–559.
39. Dunn, H. G., Perry, T. L., and Dolman, C. L., 1966, *Neurology* (*Minneap.*) **16**:407–420.
40. White, H. H., Rowland, L. P., Araki, S., Thompson, H. L., and Cowen, D., 1965, *Arch. Neurol.* **13**:455–470.
41. Hopkins, I., Taunley, R. R. Q., and Shipman, R. T., 1969, *J. Pediatr.* **75**:1082–1083.
42. Kaeser, A. C., Rodnight, R., and Ellis, B. H., 1969, *J. Neurol. Neurosurg. Psychiatry* **32**:88–93.
43. Gibson, J. B., Carson, N. A. J., and Neill, D. W., 1964, *J. Clin. Pathol.* **17**:427–431.
44. Chou, S. M., and Waisman, H. A., 1965, *Arch. Pathol.* **79**:357–363.
45. Gaull, G. E., Carson, N. A. J., Dent, C. E., and Field, C. M. B., 1964, *Proceedings of the International Copenhagen Congress for the Scientific Study of Mental Retardation*, Volume I (J. Oster, ed.), Copenhagen, p. 91.
46. Weisman, R., Davidoff, R. A., and Aprison, M. H., 1966, *Life Sci.* **5**:1431–1440.
47. Tallan, H. H., Moore, S., and Stein, W. H., 1958, *J. Biol. Chem.* **230**:707–716.
48. Johnston, G. A. R., 1968, *J. Neurochem.* **15**:1013–1017.
49. Rassin, D. K., and Gaull, G. E., 1975, *J. Neurochem.* **24**:969–978.
50. Tudball, N., and Beaumont, A., 1979, *Biochim. Biophys. Acta* **588**:285–293.
51. Barchas, J. D., Akil, H., Elliott, G. R., Holman, R. B., and Watson, S. J., 1978, *Science* **200**:964–973.
52. Brenton, D. P., Cushworth, D. C., and Gaull, G. E., 1965, *Pediatrics* **35**:50–56.
53. Gerritsen, T., and Waisman, H. A., 1964, *Science* **145**:588.
54. Brenton, D. P., and Cushworth, D. C., 1971, *Inherited Disorders of Sulphur Metabolism* (N. A. J. Carson and D. N. Raine, eds.), Churchill Livingstone, Edinburgh, London, pp. 264–272.
55. Pullan, D. H. H., 1980 *Neonatal Screening for Inborn Errors of Metabolism* (H. Bichel, R. Guthrie, and G. Hammersen, eds.), Springer-Verlag, Berlin, pp. 29–44.
56. Hardwick, D. F., Frederick, E. Q., Clayton, J. E., and Smith, L. H., 1970, *Metabolism* **19**:381–391.
57. Bachmann, C., Mihatsch, J. J., Baumgartner, R. E., Brechbuhler, T., Buhler, U.K., Olafson, A., Ohnacker, H., and Wick, H., 1971, *Helv. Paediatr. Acta* **26**:228–243.
58. Finkelstein, J. D., Kyle, W. E., and Martin, J. J., 1975, *Biochem. Biophys. Res. Commun.* **66**:1491–1497.

59. Gaull, G. E., Tallan, H. H., Lonsdale, D., Przyrembel, H., Schaffner, F., and Von Bossewitz, D. B., 1981, *J. Pediatr.* **98**:734–741.
60. Bessman, S. P., 1958, *Chemical Pathology of the Nervous System* (J. Folch-Pi, ed.), Pergamon Press, New York, pp. 370–376.
61. Clark, G. M., and Eiseman, B., 1958, *N. Eng. J. Med.* **159**:178–180.
62. Berl, S., Tayaki, S., Clarke, D. D., and Waelsch, H., 1962, *J. Biol. Chem.* **237**:2562–2569.
63. Shorey, J., McCandless, D. W., and Schenker, S., 1967, *Gastroenterology* **53**:706–711.
64. Van den Berg, C. J., 1973, *Compartmentation in Brain Metabolism* (R. Balász and J. E. Cremer, eds.), Macmillan, London, pp. 137–166.
65. Hindfelt, B., and Siesjo, B. K., 1971, *Scand. J. Clin. Lab. Invest.* **28**:365–374.
66. Hawkins, R. A., Miller, A. L., Nielsen, R. C., and Veech, R. L., 1973, *Biochem. J.* **134**:1001–1008.
67. Urba, R., Falberger, J., and Kanturek, V., 1958, *J. Neurochem.* **2**:186–187.
68. Rybova, J., 1959, *J. Neurochem.* **4**:304–310.
69. Colombo, J. P., Richterich, R., Donath, A., Spohr, A., and Rossi, E., 1964, *Lancet* **1**:1014–1017.
70. Colombo, J. P., Bengi, W., Richterich, R., and Rossi, E., 1967, *Metabolism* **16**:910–925.
71. Colombo, J. P., 1971, *Congenital Disorders of the Urea Cycle and Ammonia Detoxification*, S. Karger, Basel, pp. 100–124.
72. Metoki, K., Kitchings, L., and Hommes, F. A., 1982, *Am. J. Hum. Genet.* **34**:58A.
73. Ghadimi, H., 1978, *The Metabolic Basis of Inherited Disease* J. B. Stanbury, J. B. Wyngaarden, and D. S. Frederickson, eds.), McGraw-Hill, New York, pp. 387–396.
74. Woody, N. C., 1964, *Am. J. Dis. Child.* **108**:543–553.
75. Woody, N. C., Hutzler, J., and Dancis, J., 1966, *Am. J. Dis. Child.* **112**:577–580.
76. Dancis, J., Hutzler, J., and Levitz, M., 1963, *Pediatrics* **32**:234–238.
77. Tanaka, K., and Rosenberg, L. E., 1982, *The Metabolic Basis of Inherited Disease* (J. B. Stanbury, J. B. Wyngaarden, D. S. Frederickson, J. L. Goldstein, and M. S. Brown, eds.), McGraw-Hill, New York, pp. 440–473.
78. Prensky, A. L., and Mason, H. W., 1966, *J. Neurochem.* **13**:863–874.
79. Dreyfus, P. M., and Prensky, A. L., 1964, *Nature* **214**:276.
80. Snyderman, S. E., Norton, P. M., Raitman, E., and Holt, J. E., 1964, *Pediatrics* **34**:454–472.
81. Clark, J. B., and Land, J. M., 1974, *Biochem. J.* **140**:25–29.
82. Prensky, A. L., Carr, S., and Moser, H. W., 1968, *Arch. Neurol.* **19**:552–558.
83. Menkes, J. H., and Solcher, H., 1967, *Arch. Neurol.* **16**:486–491.
84. Land, J. M., and Clark, J. B., 1974, *FEBS Lett.* **44**:348–351.
85. Clark, J. B., and Land, J. M., 1975 *Normal and Pathological Development of Energy Metabolism* (F. A. Hommes, ed.), Academic Press, London, pp. 177–188.
86. Halestrap, A. P., and Denton, R. M., 1974, *Biochem J.* **138**:313–316.
87. Mowbray, J., 1974, *FEBS Lett.* **44**:344–347.
88. Brouwer, A., Smits, G. G., Tas, J., Meyer, A. J., and Tager, J. M., 1973, *Biochemie* **55**:717–725.
89. Cremer, J. E., Teal, H. M., and Heath, D. F., 1975, *Normal and Pathological Development of Energy Metabolism* (F. A. Hommes, ed.), Academic Press, London, pp. 133–141.
90. Yoshida, T., and Kikuchi, G., 1972, *J. Biochem (Tokyo)* **72**:1503–1516.
91. Kikuchim, G., 1973, *Mol. Cell Biochem.* **1**:169–187.
92. Matokawa, Y., and Kikuchi, G., 1974, *Arch. Biochem. Biophys.* **164**:624–633.
93. Ando, T., Nyhan, W. L., Gerritsen, T., Gong, L., Heiner, D. A., and Bray, P. F., 1968, *Pediatr. Res.* **2**:254–291.
94. DeGroot, C. J., Troelstra, H. A., and Hommes, F. A., 1979, *Pediatr. Res.* **4**:238–243.
95. Hiraya, H., Kochiki, H., Howasaka, K., Kikuchi, G., and Nyhan, W. L., 1981, *J. Clin. Invest.* **68**:525–534.
96. Bachman, C., Mihatsch, J. J., Baumgartner, R. E., Brechbühler, T., Bühler, U. K., Olafsson, A., Ohnacker, A., and Wick, H., 1971, *Helv. Paediatr. Acta* **26**:228–243.
97. Werman, R., Didoff, R., and Aprison, M. H., 1968, *J. Neurophysiol.* **31**:81–95.
98. Aprison, M. H., and Werman, R., 1965, *Life Sci.* **4**:2075–2083.
99. Curtis, D. R., Hösli, L., and Johnston, G. A. R., 1967, *Nature* **215**:1502–1503.

100. DeGroot, C. J., Touwen, B. C. L., Huisjes, H. J., and Hommes, F. A., 1978, *Prog. Brain Res.* **48**:199–205.
101. DeGroot, C. J., Everts, R., Touwen, B. C. L., and Hommes, F. A., 1978, *Prog. Brain Res.* **48**:199–205.
102. DeGroot, C. J., Everts, R. S., Gramsbergen, A., and Hommes, F. A., 1979, *Models for the Study of Inborn Errors of Metabolism* (F. A. Hommes, ed.), Elsevier/North-Holland Biomedical Press, Amsterdam, pp. 183-189.
103. Benanides, J., Lopez-Lahaya, J., Valdivieso, F., and Ugarte, M., 1981, *J. Neurochem.* **37**:315–320.
104. Gitzelmann, R., Steinmann, B., Otten, A., Dumeruth, G., Herdan, M., Reubi, C. J., and Cuenod, M., 1977, *Helv. Paediatr. Acta* **32**:517–525.
105. Arneson, D., Ch'ien, L. T., Chance, P., and Wilroy, R. S., 1979, *Pediatrics* **63**:369–373.
106. MacDermolt, K. D., Nelson, W., Reichert, C. M., and Schulman, J. D.,1980, *Pediatrics* **65**:61–64.
107. Trybels, J. M. F., Monnens, L. A. H., VanderZee, S. P. M., Vrencken, J. A. T., Sengers, R. C. A., and Schretlen, E. D. A. M., 1974, *Pediatr. Res.* **8**:598–605.
108. Levy, H. L., Nishimura, R. N., Erickson, A. M., and Janowska, S. E., 1972, *Pediatr. Res.* **6**:395.
109. Krieger, I., and Hart, Z. N., 1974, *J. Pediatr.* **85**:43–48.
110. O'Brien, W. E., 1978, *Arch. Biochem. Biophys.* **189**:291–297.
111. Robinson, B. H., Taylor, J., and Sherwood, W. G., 1977, *Pediatr. Res.* **11**:1198–1202.
112. Taylor, J., Robinson, B. H., and Sherwood, G., 1978, *Pediatr. Res.* **12**:60–62.
113. Searle, A. F., 1952, *Heredity* **6**:395–401.
114. Treiman, D. M., and Tourian, M., 1973, *Biochim. Biophys. Acta* **313**:163–169.
115. Berry, H. K., Butcher, T. E., Kasmaier, K. J., and Poncet, J. B., 1975, *Biol. Neonate* **26**:88–101.
116. Koe, B. K., and Weismann, A., 1966, *J. Pharmacol. Exp. Ther.* **154**:499–516.
117. Greengard, O., Yoss, M. S., and Delvalle, J. A., 1976, *Science* **192**:1007–1008.
118. Delvalle, J. A., Diemel, G., and Greengard, O., 1978, *Biochem. J.* **170**:449–459.
119. Jequier, E., Lowenberg, W., and Sjoerdsma, A., 1967, *Mol. Pharmacol.* **3**:274–278.
120. Schwark, W. S., Singal, R. L., and Ling, G., 1970, *Life Sci.* **9**:939–945.
121. Lane, J. B., Schone, B., Langenbeck, U., and Neuhoff, V., 1980, *Biochim. Biophys. Acta.* **627**:144–156.
122. Taylor, E. H., Hommes, F. A., and Stewart, D. E., 1983, *Biochem. Med.* **29**:307–317.
123. McKean, C. M., 1972, *Brain Res.* **47**:469–476.
124. Agrawal, H. C., Bone, A. H., and Davison, A. N., 1971, *Phenylketonuria, Biochemistry, Genetics, Diagnosis, Therapy* (H. Bichel, F. P. Hudson, and L. I. Woolf, eds.), Georg Thieme Verlag, Stuttgart, pp. 121–125.
125. Loo, Y. H., and Ritman, P., 1967, *Nature* **213**:914–916.
126. Loo, Y. H., 1967, *J. Neurochem.* **14**:813–821.
127. Loo, Y. H., and Ritman, P., 1964, *Nature* **203**:1237–1239.
128. Loo, Y. H., and Mack, K., 1972, *J. Neurochem.* **19**:2377–2383.
129. Loo, Y. H., and Whittaker, V. P., 1967, *J. Neurochem.* **14**:997–1011.
130. Loo, Y. H., and Mack, K., 1972, *J. Neurochem.* **19**:2385–2394.
131. Loo, Y. H., Scotto, L., and Horning, M. G., 1977, *J. Neurochem.* **29**:411–415.
132. Loo, Y. H., Scotto, J., and Wisniewski, H. M., 1978, *Myelination and Demyelination* (J. Polo, ed.), Plenum Press, New York, pp. 453–469.
133. Wen, G. Y., Wisniewski, H. M., Shek, J. W., Loo, Y. H., and Fulton, T. F., 1980, *Ann. Neurol.* **7**:557–566.
134. Goldstein, F. B., 1961, *J. Biol. Chem.* **236**:2656–2661.
135. Silberberg, D. H., 1967, *Arch. Neurol.* **17**:524–529.
136. Fulton, T. R., Triano, T., Robe, A., and Loo, Y. H., 1980, *Life Sci.* **27**:1271–1281.
137. Loo, Y. H., Fulton, T. R., Miller, K., and Wisniewski, H. M., 1980, *Life Sci.* **27**:1283–1290.
138. Tashian, R. E., 1961, *Metabolism* **10**:393–402.
139. Curtis, D. R., 1973, *Proc. Aust. Assoc. Neurol.* **9**:145–153.
140. Isaacs, C. E., and Greengard, O., 1980, *Biochem J.*, **192**:441–448.

141. Taylor, E. H., 1982, *Effect of Experimental Hyperphenylalaninemia on Myelin Metabolism and Neurotransmitter Synthesis at Later Stages of Brain Development*, Ph.D. Thesis, Medical College of Georgia, Augusta.
142. Perry, T. L., Hansen, S., Tischler, B., Bunting, R., and Diamond, S., 1970, *N. Eng. J. Med.* **282**:761–766.
143. Dienel, G. A., 1981, *J. Neurochem.* **36**:34–43.
144. Weber, G., 1969, *Proc. Natl. Acad. Sci. U.S.A.* **63**:1365–1369.
145. Weber, G., Glaser, R. I., and Ross, R. A., 1970, *Adv. Enzyme Regul.* **8**:13–36.
146. Glaser, R. I., and Weber, G., 1971, *Brian Res.* **33**:439–450.
147. Adriaenssens, K., Allen, R. J., Lowenthal, A., Mardens, Y., and Tourtelotte, W. W., 1969, *J. Genet. Hum.* **17**:223–230.
148. McKean, C. M., and Pelisson, N. A., 1970 *N. Engl. J. Med.* **283**:1364–1367.
149. Tanaka, T., Harano, Y., Sue, F., and Marimura, H., 1967, *J. Biochem.* **62**:71–91.
150. Vijayvargiya, R., Schwark, W. S., and Singhal, R. L., 1969, *Can. J. Biochem.* **47**:895–898.
151. Miller, A. L., Hawkins, R. A., and Veech, R. L., 1973, *Science* **179**:904–906.
152. Blass, J. P., Lewis, C. A., and Frost, S. R., 1973, *Clin. Res.* **21**:261A.
153. Armstrong, M. D., and Robinson, K. S., 1954, *Arch. Biochem. Biophys.* **52**:287–288.
154. Pare, C. M. B., Sandler, M., and Stacey, R. S., 1957, *Lancet* **1**:551–553.
155. Pare, C M. B., Sandler, M., and Stacey, R. S., 1958, *Lancet* **2**:1099–1101.
156. Nadler, H. L., Berman, J. L., and Hsia, D. Y. Y, 1965, *J. Pediatr.* **67**:710–711.
157. Caredda, P., Apollino, T., Giavanini, M., and Tenconi, L., 1974, *Helv. Paediatr. Acta* **19**:267–278.
158. Berendes, H., Anderson, J. A., Ziegler, M. R., and Ruttenberg, D. , 1958, *Am. J. Dis. Child.* **96**:430.
159. Pare, C M. B., Sandler, M., and Stacey, R. S., 1959, *Arch. Dis. Child.* **34**:422–425.
160. Baldridge, R. C., Barofsky, L., Baird, H., Reichle, F., and Bullock, D., 1959, *Proc. Soc. Exp. Biol. Med.* **100**:529–531.
161. Pare, C. M. B., Sandler, M., and Stacey, R. S., 1960, *J. Neurol. Neurosurg. Psychiatry* **23**:341–346.
162. Perry, T. L., Hansen, S., Tischler, B., and Hestrin, M., 1964, *Proc. Soc. Exp. Biol. Med.* **115**:118–123.
163. Tu, J., and Partington, M. W., 1972, *Dev. Med. Child. Neurol.* **14**:457–466.
164. Butler, I. J., O'Flynn, M. E., Seifert, W. E., and Howell, R. R., 1981, *J. Pediatr.* **98**:729–733.
165. Mandell, A. J., Knapp, S., and Hsu, L. L., 1974, *Life Sci.* **14**:1–17.
166. Pardridge, W. M., 1977, *J. Neurochem.* **28**:103–108.
167. McKean, C. M., 1972, *Brain Res.* **47**:469–476.
168. McKean, C. M., Baggs, D. E., and Peterson, N. A., 1968, *J. Neurochem.* **15**:235–241.
169. Green, H. S., Greenberg, S. M., Erickson, R. W., Sawyer, J. L., and Ellison, T., 1962, *J. Pharmacol. Exp. Ther.* **136**:174–178.
170. Yuwiler, A., and Louttit, R. T., 1961, *Science* **134**:831–832.
171. McKean, C. M., Schonberg, S. M., and Giarman, N. J., 1967, *Science* **157**:213–241.
172. Yuwiler, A., and Geller, E., 1969, *J. Neurochem.* **16**:999–1005.
173. Fong, J. H., and Kaufman, S., 1975, *J. Biol. Chem.* **250**:4152–4158.
174. Woolley, D. W., and Vander Hoeven, T., 1964, *Science* **144**:883–884.
175. Woolley, D. W., and Vander Hoeven, T., 1964, *Science* **144**:1593–1594.
176. Butcher, R. E., Voorhees, C. V., and Berry, H., 1970, *Life Sci.* **9**:1261–1268.
177. Butcher, R. E., Voorhees, C. V., Kindt, C. W., Kazmaier-Novak, K.J., and Berry, H. K., 1977, *Pharmol. Biochem. Behav.* **7**:129–133.
178. Schaefer, G. J., Barrett, R. J., Sanders-Bush, E., and Voorhees, C. V., 1974, *Pharmacol. Biochem. Behav.* **2**:783–789.
179. Kaufman, S., Holtzman, N. A., Milstein, S., Butler, I. J., and Krumholz, A., 1975, *N. Engl. J. Med.* **293**:785–790.
180. Kaufman, S., Berlow, S., Summer, G. K., Milstein, S., Schulman, J. D., Orloff, S., Spielberg, S., and Pueschel, S., 1978, *N. Engl. J. Med.* **299**:673–679.
181. Bartholomé, K., and Byrd, D. J., 1975, *Lancet* **2**:1042–1043.

182. Barthalomé, K., 1974, *Lancet* **2:**1580.
183. Tanaka, T., Aihara, K., Iwai, K., Kohashi, M., Tomita, K., Narisawa, K., Arai, N., Yoshida, H., and Usui, T., 1981, *Eur. J. Pediatr.* **136:**275–280.
184. Curtius, H. C., Baerlocher, K., and Völlmin, J. A., 1972, *Clin. Chim. Acta* **42:**235–239.
185. Nagatsu, T., Levitt, M., and Udenfriend, S., 1974, *J. Biol. Chem.* **239:**2910–2917.
186. Udenfriend, S., Zaltzman-Nizenberg, P., and Nagatsu, T., 1975, *Biochem. Pharmacol.* **14:**837–845.
187. Tardiana, M. L., Porter, C. C., Stone, C. A., and Hanson, H., 1970, *Biochem. Pharmacol.* **19:**1601–1614.
188. Aoki, K., and Siegel, F. L., 1977, *Science* **168:**129–130.
189. Siegel, F. L., Aoki, K., and Calwell, R. E., 1971, *J. Neurochem.* **18:**537–547.
190. MacInnes, J. W., and Schlesinger, K., 1971, *Brain Res.* **29:**101–1101.
191. Taub, F., and Johnson, T. C., 1975, *Biochem. J.*, **151:**173–180.
192. Roberts, S., 1974, *Aromatic Amino Acids in the Brain, Ciba Foundation Symposium*, Elsevier, Amsterdam, pp. 299–318.
193. Binek, P. A., Johnson, T. C., and Kelly, C. J., 1981, *J. Neurochem.* **36:**1476–1484.
194. Kelly, C. J., and Johnson, T. C., 1978, *Biochem. J.* **174:**931–938.
195. Comar, D., Saudubray, J. M., Duthilleal, A., Delfarge, J., Maziere, M., Berger, G., Charpentier, C., and Todd-Pakropek, A., 1981, *Eur. J., Pediatr.* **136:**13–19.
196. Hughes, J. V., and Johnson, T. C., 1977, *Biochem. J.* **163:**527–537.
197. Berger, R., Dias, T., and Hommes, F. A., 1979, *Models for the Study of Inborn Errors of Metabolism* (F. A. Hommes, ed.), Elsevier/North-Holland Biomedical Press, Amsterdam, pp. 11–122.
198. Shah, S. N., Peterson, N. A., and McKean, C. M., 1972, *J. Neurochem.* **19:**2369–2376.
199. Shah, S. N., Peterson, N. A., and McKean, C. M., 1972, *J. Neurochem.* **19:**479–485.
200. Prensky, A. L., Fishman, M. A., and Daftari, B., 1971, *Brain Res.* **33:**181–191.
201. Koshaka, S., and Tsukada, Y., 1979, *Keio J. Med.* **28:**97–108.
202. Berger, R., Springer, J., and Hommes, F. A., 1980, *Mol. Cell Biol.* **26:**31–36.
203. Shah, S. N., and Johnson, R. C., 1978, *Exp. Neurol.* **61:**370–379.
204. Lane, J. D., Schone, B., Langenbeck, U., and Neuhoff, V., 1979, *Models for the Study of Inborn Errors of Metabolism* (F.A. Hommes, ed.), Elsevier/North-Holland Biomedical Press, Amsterdam, pp. 141–148.
205. Johnson, R. C., and Shah, S. N., 1980, *Neurochem. Res.* **5:**709–718.
206. Figlewicz, D. A., and Druze, M. J., 1980, *Exp. Neurol.* **67:**315–329.
207. Loo, Y. H., Scotto, J., and Wisniewski, H. M., 1978, *Adv. Exp. Med. Biol.* **160:**453–469.
208. Grundt, I. K., and Hale, K., 1974, *Brain Res.* **74:**269–277.
209. Sprinkle, T. J., and Rennert, O. M., 1976, *J. Neurochem.* **26:**499–504.
210. Chase, H. P., and O'Brien, D., 1980, *Pediatr. Res.* **4:**96–102.
211. Shah, S. N., Peterson, N. A., and McKean, C. M., 1970, *J. Neurochem.* **17:**279–284.
212. Shah, S. N., Peterson, N. A., and McKean, C. M., 1969, *Biochim. Biophys. Acta* **187:**236–242.
213. Johnson, R. C., and Shah, S. N., 1973, *J. Neurochem.* **21:**1225–1240.
214. Foute, J. L., Allen, R. J., and Agranoff, B. W., 1965, *J. Lipid Res.* **6:**518–524.
215. Cumings, J. N., Grundt, I. K., and Yanagihara, T., 1978, *J. Neurol. Neurosurg. Psychiatry* **31:**334–337.
216. Gerstl, B., Malamud, N., Eng, L. F., and Hagman, R. B., 1967, *Neurology (Minneap.)* **17:**51–58.
217. Berger, R., Springer, J., and Hommes, F. A., 1977, *Hum. Hered.* **27:**167.
218. Soleri, M. J., Bone, A. H., and Davison, A. N., 1974, *Biochem. J.* **142:**499–507.
219. Singh, H., and Jungalwala, F. B., 1979, *Int. J. Neurosci.* **9:**123–131.
220. Lajtha, A., Lutskovitz, L., and Toth, J., 1976, *Biochim. Biophys. Acta* **425:**511–520.
221. Lajtha, A., Toth, J., Fujimoto, K., and Agrawal, H. C., 1977, *Biochem. J.* **164:**323–329.
222. Hommes, F. A., Eller, A. G., and Taylor, E. H., 1982, *J. Inherit. Metab. Dis.* **5:**21–27.
223. Hommes, F. A., Eller, A. G., and Taylor, E. H., 1982, *Inborn Errors of Metabolism in Humans* (F. Cockburn and R. Gitzelmann, eds.), MTP Press, Lancaster, pp. 193–199.
224. Mattieu, J. M., Quarles, R. H., Brady, R. O., and Webster, H., 1973, *Biochim. Biophys. Acta* **329:**305–317.

225. Norton, W. T., and Poduslo, S. E., 1973, *J. Neurochem.* **21**:759–773.
226. London, Y., and Vossenberg, F. G. H., 1973, *Biochim. Biophys. Acta* **307**:478–490.
227. London, Y., Demel, R. A., Geurts van Vessel, W. S. M., Vossenberg, F. G. A., and Van Deenen, L. L. M., 1973, *Biochim. Biophys. Acta* **311**:520–530.
228. Lindblad, B., Lindstet, S., and Steen, G., 1977, *Proc. Natl. Acad. Sci. U.S.A.* **74**:4641–4645.
229. Ebert, P. S., Hess, R. A., Frykholm, B. C., and Tschudy, D. P., 1979, *Biochem. Biophys. Res. Commun.* **88**:1382–1390.
230. Strife, C. F., Zuroweste, E. L., Emmet, E. H., Finelli, V. N., Petering, H. G., and Berry, H. K., 1977, *J. Pediatr.* **90**:400–404.
231. Kennaway, N. G., and Buist, N. R. M., 1971, *Pediatr. Res.* **5**:287-297.
232. Bardelli, A. M., Borgogni, P., Farnetani, M. A., Fois, A., Frezzotti, R., Matei, R., Molinelli, M., and Sargentini, I., 1977, *Ophthalmologica* **175**:5–9.
233. Goldsmith, L. A., Thorpe, J. M., and Roe, C. R., 1979, *J. Invest. Dermatol.* **73**:530.
234. Fellman, J. H., Fujita, T. S., and Roth, E. S., 1972, *Biochim. Biophys. Acta* **284**:90–100.
235. Danks, D. M., Tippett, P., and Rogers, J., 1975, *Acta Paediatr. Scand.* **64**:209–214.

3

Peptides and Brain Pathology

John E. Morley and Charles J. Billington

1. METHODS AND PROBLEMS OF THE INVESTIGATION OF THE ROLE OF PEPTIDES IN BRAIN PATHOLOGY

Over 30 different peptides have been identified in the central nervous system.[1] Thus, any investigation of the role of neuropeptides in pathology is certain to produce variety and diversity. The explosive rate at which new neuropeptides have been discovered over the last decade has meant that our understanding of their possible functions in the central nervous system has lagged far behind. It seems clear that the neuropeptides along with amino acids and monoamines play an integral role as neurotransmitters in the communication network of the brain. Thus, the discovery of abnormalities in the synthesis, release, or receptor binding of the neuropeptides appears to offer the keys to unlocking many of the secrets of neurological and psychiatric disease. At present, however, our knowledge of the pathophysiological processes in which neuropeptides are involved is rudimentary. In this chapter we attempt to place in perspective some of the exciting glimpses of the role of peptides in brain pathology that have been obtained over the past few years. Also, making use of the known effects of neuropeptides in animals, we attempt to pinpoint the areas in which it appears most likely that neuropeptides will be found to play an important role in pathological processes.

Because of the difficulty of access to the central nervous system during life, the majority of methods to approach the diagnosis of neuropeptide disorders are indirect. Measurement of neuropeptides in the peripheral circulation or the urine provides little useful information concerning brain pathology, as it is now clear that the majority of neuropeptides are synthesized in a variety of peripheral tissues as well as in the brain. The method most widely used to date to pinpoint central neurotransmitter dysfunction has been to examine the release of pituitary hormones under a variety of stimulatory and inhibitory conditions. As pituitary hormone release is under direct control of hypothalamic

John E. Morley • Neuroendocrine Research Laboratory, Veterans Administration Medical Center, Minneapolis, Minnesota, 55417. *Charles J. Billington* • Division of Endocrinology, Department of Medicine, University of Minnesota, Minneapolis, Minnesota, 55455.

neurotransmitters, it has been argued that these manipulations allow the pituitary to act as a window to the brain.

This approach has been most aggressively pursued in the attempts to unravel the mysteries of psychiatric disease, spawning a huge literature in which extravagant claims have been made for the utility of dexamethasone suppression and TRH tests in subtyping mental disease. Unfortunately, as this testing is indirect, the findings are highly susceptible to minor alterations in the peripheral endocrine system (a common concomitant of almost any systemic disease). Thus, this form of indirect testing, although a useful research tool, has too many potential false positives and negatives to ever become a highly useful diagnostic approach. In addition, it seems that most neurological and psychiatric disease is liable to involve highly localized alterations in neuropeptides in areas remote from the hypothalamus, and these alterations are unlikely to be reflected in alterations in the hypothalamic–portal–pituitary system.

During life the most direct access to the central nervous system appears to be through the measurement of neuropeptides in the cerebrospinal fluid. Many of the identified neuropeptides have been measured in the CSF of humans.[2] The physiological role of these neuropeptides in the CSF is unknown, but knowledge concerning their levels in health and disease may prove useful. Many of these neuropeptides are secreted actively into the CSF, and their levels appear to be regulated independently of their peripheral levels. Unfortunately, the diffuse distribution of neuropeptides in the brain makes it unlikely that the CSF will be useful in detecting circumscribed defects in neuropeptides. Another problem is that studies on biogenic amines have shown a marked concentration gradient between ventricular and lumbar CSF.[3] Such gradients have not been found in the CSF for immunoreactive substance P[4] and arginine vasopressin.[5] This problem can also be partially circumvented by taking fairly large amounts of CSF (15–30 ml) and only using the later aliquots for peptide analysis. A major problem in examining neuropeptides in the CSF appears to relate to the possibility that a variety of neurological syndromes that produce direct damage to CNS tissue may lead to fairly non-specific increases in SCF neuropeptides.[6]

At present, one of the areas that seems to hold the most promise in defining the role of neuropeptides in brain pathology appears to be their measurement in postmortem tissues. There is remarkably little alteration in neuropeptide concentrations when brain tissue is stored at 4°C,[3,7] suggesting that the variability in time taken to obtain nervous tissue after death is relatively unimportant. The clinical status of the patient before death can make profound differences in the measurements obtained post-mortem, and both the severity and the length of the disease process need to be taken into account before clean comparisons between groups can be made.[8]

In addition, the effects of drugs administered premortem on neuropeptide levels need to be kept in mind. As neuropeptides are found predominately in neurons, it is not surprising that their concentrations should be found to be decreased in conditions in which there is neuronal destruction. At present there is no appropriate method available to express data corrected for neuronal cell loss with associated proportional increase in glial tissue. Finally, it needs to be stressed that the measurement of neuropeptides in postmortem tissue represents a single static measurement of substances with extremely short-lives and rapid turnovers. Determination of peptide content alone is a crude method by which to seek pathophysiological correlations. Measurements of turnover rate or secretion would be preferable, but techniques for such measurements are not feasible. Thus, the alterations found in peptide concentrations may reflect alterations in secretion, synthesis, or degradation rates. Balanced alterations in these factors may well lead to potentially profound pathophysiological change without net change in tissue concentrations.

Neuropeptide concentrations are much lower than the concentrations typically found of monoamine and amino acid neurotransmitters. This creates problems in the measurements of these neuropeptides; in particular, the specificity of the substance being measured is often in doubt. This is further complicated by the fact that most neuropeptides appear to be present in multiple forms. Thus, without characterization of the immunoreactive substance being measured by chromatography, there is always uncertainty as to the exact nature of the substance being measured. Immunocytochemistry may prove to be an extremely useful tool in pinpointing localized areas of peptide deficiency, but it needs to be stressed that this technique fails to give quantitive information and has marginal specificity. The use of autoradiographic techniques for the demonstration of receptors in postmortem tissue[9] and the direct measurement of ligand binding coupled with Scatchard analysis[10] both appear to have potential use in unraveling the pathophysiological significance of brain peptides.

2. *ENDOGENOUS OPIOID PEPTIDES AND PATHOLOGY*

The demonstration of the existence of opiate receptors and the subsequent isolation of their endogenous opioid ligands represent one of the most exciting discoveries of the last decade. These discoveries caught the imagination of both the public and the scientific communities, resulting in a deluge of scientific and quasiscientific literature on the subject. This has led to numerous extravagant claims and a barrage of misinformation on the pivotal role of endogenous opioid peptides and their receptors in the regulation of the brain in health and disease. The problems associated with the initial enthusiasm surrounding the discovery of the opioid peptides were neatly summarized by Beaumont and Hughes[11]: "The opioid peptides appear to create euphoria not only in experimental animals but also indirectly in the investigators themselves."

2.1. *Opioid Peptides and Analgesia*

It seems clear that the opioid peptides are involved in the regulation of analgesia, although it needs to be recognized that they represent only one part of a complex modulatory system(s) responsible for filtering of information concerning tissue damage. Present evidence suggests that opiates produce their analgesic effects by different mechanisms at three or more levels: (1) inhibiting or modulating somatosensory afferents at supraspinal levels, (2) inhibiting primary somatosensory afferents in the dorsal horn of the spinal cord, and (3) activating descending inhibitory pathways.[12] A variety of studies in humans and animals have demonstrated that electrical stimulus of the periaqueductal gray can lead to analgesia[13] that can be at least partially antagonized by the opiate antagonist naloxone.[14] At present, much evidence now supports the concept that stimulus-produced analgesia results from the release of an endogenous opiate-like factor.[15] Similarly, multiple studies have suggested a role for endogenous opioids in acupuncture analgesia.[16–20]

Patients with chronic pain represent a heterogeneous group in whom psychological factors are bound to be very marked. For this reason, it is not surprising that studies on the role of the endogenous opiates in chronic pain have led to a variety of conflicting results. Terenius and his colleagues[21,22] have clearly demonstrated elevated endogenous opiate levels as measured by radioreceptor assay in patients with neurogenic pain compared to those levels found in healthy volunteers or in patients with psychogenic chronic pain. In patients with organic chronic pain, there is a significant correlation between CSF opiate levels and the serotonin metabolite 5-hydroxyindoleacetic acid[21] Other studies have found elevated enkephalinlike material in the CSF of chronic pain patients[23] and have confirmed the elevated CSF endogenous opiates in patients with chronic pain.[24]

Studies in patients with cancer and intractable pain have shown that infusion of intrathecal β-endorphin produces profound and long-lasting analgesia, with an average of 33.4 hr of pain relief from a single 3-mg injection.[25] There were no undesirable side effects as are normally associated with opiate analgesia, such as respiratory depression, hypotension, hypothermia, or catatonia.

The endogenous opiate system appears to exist predominantly to modulate pain, and in patients with chronic pain, it appears that tolerance to endogenous opiates develops fairly rapidly. The possibility that psychogenic chronic pain may develop in some people who have an excess opioid response to life's stressors and thus become tolerant to the body's own analgesic system is a fascinating but as yet untested possibility.

Recently, Dunger *et al.*[26] described a 13-year-old boy with insensitivity to pain associated with disordered hypothalamic function in whom a disturbance of the opioid peptide system was postulated. Naloxone decreased his ability to withstand pain evoked by a radiant heat stimulus. He had normal levels of β-endorphinlike and methionine-enkephalinlike immunoreactivity in the CSF and serum. The authors felt that this finding of normal opioid peptide levels did not invalidate their hypothesis, because local concentrations of these peptides in the brain may not be reflected by blood or CSF levels. In a second subject with congenital indifference to pain reported by Fraioli *et al.*,[27] high levels of total opioid activity as measured by radioreceptor assay were found with minimal changes in CSF β-endorphin as measured by radioimmunoassay.

2.2. *Hypertension, Diabetes, and Pain*

A number of studies have implicated the endogenous opioid systems in the altered sensitivity to pain demonstrated by patients with hypertension and diabetes mellitus. Zamir and Shuber[28] showed that hypertensive humans have a higher threshold for sensation of pain in the tooth pulp than normotensive controls. Animal experiments in genetically hypertensive rats have shown that they were also less responsive to noxious stimuli and that this decreased sensitivity to pain was naloxone reversible.[29,30] Hypertensive animals have also been shown to have an increased spinal cord opioid activity.[31] As sodium is well recognized to alter binding to the opiate receptor *in vitro*,[12] it is possible that the salt and water retention that is an etiologic factor in the development of hypertension in many subjects may be playing a role.

Clinically, it is well recognized that patients with diabetes mellitus appear to have increased perception of pain, particularly when they develop peripheral neuropathies. There is now considerable *in vivo* evidence in animals that alterations in serum glucose modulate the responsiveness of animals to opiates.[31–35] Recently, we have found that the administration of glucose lowers the pain threshold in normal humans (J. E. Morley, G. A. Morley, and A. S. Levine, unpublished observations). These findings strongly supoort a role for the endogenous opioid peptides or their receptors in the altered pain threshold seen in diabetics and raise questions concerning a possible pathophysiological role for opiate receptors in reactive hypoglycemia.

2.3. Endogenous Opioid Peptides and Circulatory Pathology

A variety of studies in animals have clearly established a role for endogenous opioid peptides in the pathogenesis of circulatory collapse (shock) in animals.[36] Peters *et al.*[37] reported an increase in blood pressure lasting for at least 45 min in patients with septic shock who received 0.4 to 1.2 mg naloxone. At present, there is no evidence that naloxone will improve survival in humans. Animal studies have also suggested that naloxone can reduce the incidence of stress ulceration by enhancing gastric mucosal blood flow.[38]

Naloxone has been reported to temporarily improve the neurological deficits in two patients experiencing transcient ischemic attacks.[39] At present, whether or not naloxone can reverse experimentally induced neurological deficits in animals is controversial.[40–43] Faden, Jacobs, and Holaday[44] have demonstrated that administration of opiate antagonists can prevent neurological deficits associated with injury-induced ischemia following spinal trauma. Young[45] has shown that this is associated with increased blood flow to the ischemic spinal cord following experimental spinal contusion.

One claimed effect of β-endorphin has been that it is released during jogging[46] and thus is responsible for producing "jogger's high" by a direct action on the brain. In view of the well-recognized effects of endogenous opioids on blood flow, it seems more likely that if β-endorphin is involved in the production of the "jogger's high," it does this by producing mild hypotension and/or shunting of blood from the brain. This would lead to mild anoxia, a well-known method for stimulating a high.

Although at present the majority of studies concerning cerebral circulation have centered on the endogenous opioids, it will not be surprising to find that a number of other neuropeptides play a role in the autoregulation of the cerebral circulation. A number of peptides present in the central nervous system, including vasopressin, thyrotropin-releasing hormone (TRH), somatostatin, substance P, vasoactive intestinal peptide, neurotensin, and bradykinin are well recognized as being potently vasoactive, and axonal processes that terminate on or near blood vessels contain some of these peptides. Thus, these and other peptides may be found to play a role in the vasospasm that occurs following subarachnoid hemorrhage or to be involved in the evolution of cerebrovascular accidents.

2.4. Subacute Necrotizing Encephalomyelopathy and Endogenous Opiates

Subacute necrotizing encephalomyelopathy (Leigh's syndrome) is an autosomal recessive degenerative disorder that occurs in infancy and early childhood and is characterized by progressive deterioration of brainstem functions, ataxia, convulsions, hypotonia, and failure to thrive. The histopathological changes are present in the brainstem, basal ganglia, cerebellar cortex, and dentate nucleus and consist of bilateral areas of rarefaction, loss of myelin, and cell proliferation in the capillary and glial elements. Brandt *et al.*[47] reported an infant who had repeated attacks characterized by lethargy, depression, and unconsciousness. He was generally resistant to pain but showed hypersensitivity to pain during the prodrome. Some of the symptoms were reversed by naloxone. At postmortem, the histological picture was characteristic of Leigh's syndrome. He had elevated levels of endogenous opioids as measured by radioreceptor assay in his spinal fluid and increased levels of methionine- and leucine-enkephalin but not β-endorphin in the cortex. In addition, norepinephrine levels were also greatly elevated in the CSF.

3. HUNTINGTON'S DISEASE

Huntington's disease is an autosomal dominantly inherited degenerative disorder in which abnormal movements occur with onset in the majority of patients during the fourth decade. Behavioral changes usually precede the onset of the movement disorder. In about 10% of cases, the onset occurs under 15 years of age, and in these cases the disease runs a more rapid course. Although there is generalized atrophy of neurons in this disorder, with whole brain weight being decreased by 20%, the brunt of the disease falls on the basal ganglia, whose weight may be reduced by as much as 50%.[48] Major neuronal damage is seen in the caudate, putamen, globus pallidus, and substantia nigra with relative sparing of the anterior striatal complex, which contains the nucleus accumbens. Within the substantia nigra there is usually a marked degree of atrophy in the zona reticulata with only moderate atrophy in the zona compacta. There is an increase in total glial cells when compared to the loss of neuronal cells.[33]

The abnormalities in neuropeptides found in the brains of patients with Huntington's disease have recently been extensively reviewed by Bird.[8] Briefly, there are marked decreases in substance P,[49] and CCK[50] concentrations are markedly decreased in the globus pallidus and substantia nigra with normal levels in the caudate and frontal cortex. The maintenance of levels of these neuropeptides in the substantia nigra may be related to the fact that both of these peptides are related to dopaminergic neurons,[51,52] which are spared in the striatum of patients with Huntington's disease.[53] There is a significant increase in GnRH in median eminence of female patients, [54] which may be related to the increased dopamine levels found in the choreic brain.[53] This increase in GnRH has been related to the increased fertility[55] and libido[56] reported in these patients.

As patients with Huntington's disease have been shown to have hyperresponsiveness to the neuroendocrine system that controls GH secretion,[57] it is somewhat surprising to find that somatostatin concentrations are increased in the choreic hypothalamus compared to normal.[8] Concentrations of VIP were reported to be normal in the cerebral cortex and caudate of control and choreic brains.[58] There is a decrease in angiotensin-converting enzyme activity in the corpus striatum[59] and substantia nigra[60] of postmortem tissue from Huntington's patients, suggesting possible involvement of a striatonigral angiotensin pathway. There is also a decrease in the GABA biosynthetic enzyme glutamic acid decarboxylase and in GABA itself in patients with Huntington's disease.[61,62]

Studies by Hays and Paul[63,64] have demonstrated reduced CCK receptor binding in the frontal cortex and putamen of patients with Huntington's disease. The 40–60% loss of CCK receptor in the cortex is greater than the 10–20% neuronal degeneration in the cortex of Huntington's patients,[65] suggesting a possible specific involvement of cortical CCK-receptor-containing neurons in Huntington's disease.

In summary, not surprisingly, neuropeptide levels were found to be decreased in the areas that bear the brunt of the neuroanatomic damage in Huntington's disease (substantia nigra and globus pallidus), and thus it is difficult to assign a specific pathophysiological role to these peptides in this disease. In view of the multiple vegetative symptoms such as hyperphagia, progressive cachexia, hyperhidrosis, and increased libido[57] reported in this disease, the preliminary evidence of increased values of somatostatin and GnRH perhaps has greater significance.

4. *NEUROPEPTIDES, MEMORY, AND AGING*

Animal studies have clearly shown that ACTH, β-endorphin, and related peptides enhanced motivational and attentional processes,[66–69] whereas vasopressin facilitates consolidation and retrieval memory processes.[70–74] Recent studies have provided evidence suggesting that these neuropeptides are involved in the regulation of mental performance in humans as well.

Patients lacking vasopressin, i.e., those with central diabetes insipidus, have deficits in attention, concentration, and memory.[75,76] Administration of vasopressin by nasal spray has been reported to improve memory in Korsakow's syndrome[77] elderly subjects,[78] patients with senile dementia,[79] patients with posttraumatic amnesia,[80] and patients with cognitive impairment[81,82] as well as in some normal subjects.[83] It should be pointed out, however, that not all studies have produced positive results, particularly those carried out in patients with severe impairment.[84] Animal studies have suggested that some of the benefit of peripherally administered arginine vasopressin may arise from the increase in blood pressure it produces.[85] High levels of vasopressin have been reported in the CSF of patients with presenile and senile dementia,[86] and low concentrations of immunoreactive vasopressin have been found in the globus pallidus, but not in other areas of the brain, in subjects with senile dementia compared to levels in the brains of mentally unimpaired people of a similar age.[87] Ballenger *et al.*[88] found that the digit span on the WAIS was significantly correlated with CSF vasopressin.

In a recent review, deWied and van Ree[89] suggested that although it seemed unlikely that vasopressin would be useful in patients with severe neurological damage such as Alzheimer's disease or chronic alcoholism, it may well be beneficial in patients with diabetes insipidus, in amnesia following trauma or ECT, and in elderly people with mild cognitive disturbances.

Studies with fragments of the ACTH peptides have shown improved attention and short-term, but not long-term, memory in humans.[90–93] The highly potent ACTH 4–9 analogue (Org 2766) has been shown to improve mood and increase motivation and ward behavior after chronic oral administration in aged subjects.[94]

A number of studies in animals have suggested that both natural and synthetic opioid peptides facilitate either storage or retrieval of memory.[95] Stein and Belluzzi[96] have suggested that immediate postlearning activation of the opiate receptors facilitates memory consolidation. At present there are no studies in humans on the role of the opiate receptors in memory.

Substance P produces amnesia when injected into the substantia nigra and amygdala and facilitates learning when injected into the medial nucleus of the septum and the lateral hypothalamus,[97] and substance P has been demonstrated to be decreased in some hypothalamic areas of 2-year-old rats.[98] At present there are no human studies on substance P levels in the brain or CSF of subjects with senile dementia. However, the animal studies suggest that the relationship of substance P to memory should be explored in humans.

Somatostatin-like immunoreactivity has been reported to be reduced in the cortex and hippocampus of patients suffering from senile and presenile dementia of the Alzheimer type.[88,99–101] The levels of somatostatin are decreased in parallel with those of choline acetyltransferase. Wood *et al.*[217] have reported a 50% reduction in CSF somatostatin in patients with Alzheimer's disease and mixed dementia. Cholecystokinin receptor binding has been reported to be normal in the temporal and cingulate cortex of patients with Alzheimer's disease.[99]

These studies suggest that the decline in cognitive function in the elderly may be related to a decreased bioavailability of these neuropeptides and suggest that as methods for improved delivery of neuropeptides across the blood–brain barrier become available, dramatic improvements in the mental ability of the aged could be obtained by treatment with neuropeptides.

5. *PARKINSON'S DISEASE, MULTIPLE SCLEROSIS, AND SPINOCEREBELLAR DYSFUNCTION*

Parkinson's disease is characterized by a central dopamine deficiency.[102] The close anatomic[103] and biochemical[104] relationship between dopaminergic and enkephalinergic systems in the central nervous system prompted Taquet *et al.*[105] to study the methionine-enkephalin content in brains obtained from patients with Parkinson's disease. There were marked decreased (80–95%) in methionine-enkephalin levels in the substantia nigra and the ventral tegmental area. There were no differences in methionine-enkephalin content in the caudate nucleus and putamen, which is in constrast to the marked reduction in dopamine levels in these areas. These findings are suggestive of a specific deficit in methionine-enkephalin-containing neurons in Parkinson's disease.

Low levels of CSF somatostatin have been reported in patients with Parkinson's disease and multiple sclerosis.[106,107] This is in sharp contrast to the report by Patel *et al.*[6] that a variety of neurological syndromes known to damage neuronal tissue result in elevated somatostatin levels in the CSF. Levels of SCF somatostatin in patients with multiple sclerosis appear to correlate with the degree of activity of the disease.[107] The significance of these findings is unknown at present. It is possible that the decreased levels of somatostatin in the CSF of patients with multiple sclerosis is related to the abnormal antibodies[108] and/or proteolytic enzymes[108] found in the CSF of these patients.

Thyrotropin-releasing hormone has been shown to produce improved coordination in an ataxic mouse strain.[109] For this reason, Sobrie *et al.*[110] investigated the effect of TRH on the ataxia and eye movement abnormalities of patients with spinocerebellar ataxia. In their original report, they found a significant transient effect of TRH on both the ataxia and abnormal eye movements. They extended these studies to several hundred patients affected with cerebellar disorders in a large collaborative study and concluded that doses as low as 500 μg of TRH intravenously could produce transient improvement in cerebellar signs and symptoms.[111] The responses were, however, idiosyncratic and were more often seen in the mildly affected patients. A smaller study in America (only 13 patients) by LeWitt and Ehrenkranz[112] failed to find any effect of TRH (1–6.5 mg) on cerebellar ataxia. There is a need for long-term trials of TRH, perhaps using more potent analogues, to assess the potential role of this neuropeptide in cerebellar disorders.

6. *SUBSTANCE P AND SHY–DRAGER SYNDROME*

Shy–Drager syndrome (multiple-system atrophy) is characterized clinically by extrapyramidal, cerebellar, and corticospinal tract signs associated with profound autonomic dysfunction.[113,114] The autonomic dysfunction appears to be primarily related to central dysfunction rather than to failure of the postganglionic noradrenergic neurons.[114–117] Pathologically, there is neuronal degeneration in brainstem autonomic nuclei (especially the dorsal nucleus and locus coeruleus), the intermediolateral columns of the spinal cord, and occasionally in the peripheral autonomic ganglia.[113,114,117]

Nutt *et al.*[118] reported a significant decrease in substance P concentrations in the CSF of patients with Shy–Drager syndrome compared to normals. Patients with Huntington's disease and Parkinson's disease were not significantly different from normal. Patients with peripheral neuropathy from a variety of causes also had significantly decreased substance P levels. None of these patients had evidence of autonomic dysfunction. No evidence of decreased substance P levels was seen in patients with myopathy. It should be pointed out that there was substantial overlap between the different groups, so CSF levels of substance P represent a poor diagnostic discriminator.

Anatomically, substance-P-containing neurons are intimately linked to the brainstem catecholamine cell groups, neurons of the intermediolateral columns of the cord, and noradrenergic neurons of the peripheral autonomic ganglia.[119–121] In addition, in animals there is evidence for a functional interrelationship between the autonomic nervous system and substance P. Central administration of brain norepinephrine and dopamine[122] and substance P excites neurons of the locus coeruleus[123] and peripheral autonomic ganglia.[124] Thus, it seems reasonable to postulate a role for the degeneration of substance-P-containing neuronal systems in the pathogenesis of the Shy–Drager syndrome.

Multiple studies have suggested that substance P is a neurotransmitter in primary afferent neurons and in the transmission of painful impulses.[125] It is thus not surprising that substance P was found to be reduced in neuropathic processes.

7. *HYPOTHALAMIC–PITUITARY SYNDROMES*

The regulation of pituitary secretion by releasing and inhibitory neuropeptides from the hypothalamus is now well recognized. Numerous hormonal syndromes caused by isolated deficiency or excess of these hypothalamic factors have now been described. Examples of hypersecretory syndromes include (1) Cushing's disease, in which primary excess of CRF is postulated to cause pituitary hyperplasia and excess ACTH and cortisol production,[126] (2) precocious puberty, which appears in many cases to be cause by disinhibition of LRH secretion[127] and in some cases by hamartomas or other tumors excreting an excess of LRH,[128] and (3) hypothalamic hyperthyroidism, which may be caused by an excess of TRH secretion or a lack of one or more of the hypothalamic factors responsible for inhibiting TRH secretion.[129] Isolated deficiency of LRH release appears to be the problem in Kallman's syndrome, a condition characterized by anosmia and hypothalamic hypogonadism.[130] Deficiency of TRH secretion results in the syndrome of "hypothalamic hypothyroidism," also known as tertiary hypothyroidism.[129] A deficiency of GHRF appears to be the primary defect in the majority of patients with idiopathic dwarfism associated with growth hormone deficiency.[127] Children under severe emotional deprivation have been shown to have growth failure secondary to impairment of growth hormone secretion.[127] Whether this is a result of suppression of GHRF release or an excess of somatostatin production is not established.[127] In rats, the inhibition of GH release in response to stress appears to be caused by excess somatostatin secretion and can be prevented by treatment with somatostatin antisera.[130]

8. DEPRESSION AND BIPOLAR AFFECTIVE DISORDERS

A central neuropharmacological basis for depression and other affective disorders has long been suspected. For many years, the theory of the psychobiology of these disorders has been largely related to two major hypotheses: the catecholamine or monoaminergic hypothesis and the cortisol theory. More recently, measurements have been made of the many other neuropeptides known to exist in the brain to determine if quantitative alterations can help describe the pathophysiology of affective disorders.

With regard to neuropeptide alterations, however, the cortisol abnormalities in depression may have relevance. It is well known now that many patients suffering from endogenous depression do not suppress cortisol secretion normally when challenged with the exogenous steroid dexamethasone in a standard overnight test.[131] Elevations in aggregiate daily cortisol secretion[132] and elimination of the normal diurnal rhythm of cortisol release[133] have also been found. The relationship of these phenomena to depression is strengthened by the observation that successful treatment of the psychiatric disorder by standard methods results in normalization of the corisol excretion pattern. The explanation for the cortisol excretion abnormalities is thought to lie with the hypothalamic–pituitary unit, where it is presumed that excess CRF leads to heightened ACTH release and thus to hypercortisolism.[134] However, rigorous proof of CRF and ACTH elevations has not yet been established. Nonetheless, alterations of the hypothalamic and pituitary peptides have been implicated, and the central abnormality has been referred to as disinhibition[135] of the hypothalamic–pituitary–adrenal (HPA) axis. This latter description takes into account the normal response of the HPA axis to stimulation testing with metyrapone. Elucidation of this observed hypercortisolism's relationship to a similar elevation in stress awaits further study.

More recently, interest in the role endogenous opioid peptides may be playing in the affective disorders has been increasing. Measurements of β-endorphin levels in cerebrospinal fluid have been made and have shown no difference between depressives and controls.[136] However, clinical trials of β-endorphin administration have been carried out and appear to demonstrate significant, albeit transient, improvement in some depressive symptoms.[137] In bipolar affective illness, elevations of certain CSF endorphin fractions (not β-endorphin) have been found by one grouup of investigators.[138] Naloxone, the opiate antagonist, has not reliably affected symptoms of mania in bipolar disorders.[139]

The impaired thyrotropin response seen in some patients with endogenous depression[1410] and the ability of TRH to enhance dopamine turnover[141] led to studies of effect of TRH in the treatment of depression. Preliminary anecdotal reports suggested that TRH may have some benefit,[142] but this could not be confirmed in more rigorous double-blind trials.[143] Kirkegaard *et al.*[144] have reported elevated levels of TRH in the CSF of depressed patients both before and after therapy. At present, the role of TRH in depression is controversial and definitely not proven.

Many other neuropeptides have been measured in the cerebrospinal fluid of depressed patients. When compared to controls, CSF levels of somatostatin[145] and arginine vasopressin[146] are decreased in depressed patients. Elevations of CSF angiotensin I have been suggested[146] in depression. In bipolar affective disorders or mania, elevations of arginine vasopressin,[146] lower levels of oxytocin,[146] and lower levels of calcitonin[146] have been observed in the CSF. Peptides found to be unaffected by these disorders include bombesin and cholecystokinin.[145] Limited postmortem studies have been carried out so far on the brains of depressed patients. Cortical CCK and VIP levels[147] and TRH and neurotensin concentrations in the amygdala[3] have been reported to be normal. The functional significance of these observations awaits explanation based on future research.

9. SCHIZOPHRENIA AND NEUROPEPTIDES

Observations on the effectiveness of the phenothiazine class of antipsychotic drugs in clinically improving schizophrenic patients and in blocking the neurotransmitter dopamine led to promulgation of the dopamine hypothesis of the cause of schizophrenia.[148] The postulate states that excess activity of dopaminergic neural tracts, principally in the limbic system, leads to schizophrenia. Successive formulations of the hypothesis have implicated either excess of dopamine itself or hypersensitivity or -reactivity of the dopamine receptor. Much of the neuropharmacological research performed in recent years has been directed at testing this hypothesis.

Support for the hypothesis has been found in the similarity of amphetamine-induced psychosis, a condition in which increased brain dopamine has been found, to paranoid schizophrenia.[149] Further support stems from the observation that the clinical effectiveness of antipsychotic drugs may be correlated with a rise of serum prolactin, a neurohormone under tonic dopaminergic inhibition.[150]

Pathological studies have not been in accord with the dopamine hypothesis thus far. Examination of the cerebrospinal fluid for homovanillic acid, a dopamine metabolite, has shown no difference between schizophrenics and controls.[151] Postmortem studies have not shown an increase in dopamine metabolites in either the caudate, putamen, or nucleus accumbens, a portion of the mesolimbic system.[152] More recent studies do suggest an increase in the number of [^{3}H] haloperidol binding sites in the caudate and nucleus accumbens of schizophrenics, an observation that may represent evidence for altered sensitivity to dopamine in these individuals.[153] Certainly, the work on dopamine and its receptors has not ruled out an important role for other neurotransmitters and modulators.

Endogenous opioid peptides have received much attention as putative etiologic substances in schizophrenia. Theories have been advanced suggesting that an excess and/or a deficiency of endogenous opioids play a role in the pathogenesis of schizophrenia.[154,155] Following the initial euphoria and rapid publication of many single-blind or nonblind trials, a clearer role for the endogenous opioid peptides in schizophrenia is emerging. Since the original single-blind trial by Gunne *et al.*[156] showing a decrease in auditory hallucinations after naloxone in four of six neuroleptic-treated schizophrenics, four double-blind trials that studies ten or more patients have shown an improvement in auditory hallucinations following naloxone administration.[157–160] The effects of naloxone appear to be more prominent in schizophrenic patients concurrently treated with neuroleptic medication.[160] Administration of β-endorphin and synthetic enkephalin analogues have produced mixed results that, in general, suggest that parental administration of these agents produces no effect in schizophrenics.[161–165]

Des-Try[1]-β-endorphin is the destyrosine derivative of β-endorphin (β-LPH_{61-77}) and occurs naturally in the human CSF.[166] It has no opiatelike activity, and animal studies have suggested that it shares a number of properties in common with the neuroleptic haloperidol.[167] It, however, lacks the locomotor and sedative effects seen with haloperidol. For these reasons, its role as an antipsychotic has been investigated. The initial studies showed a reduction or total disappearance of symptoms in schizophrenics partly or completely resistant to convential neuroleptics and suffering from long-lasting psychoses.[168] Other studies failed to confirm these effects.[169,170] There is a need for more elaborate animal research with this compound before further clinical trials are undertaken.

Elevated opioid peptides were first reported to be present in the CSF of schizophrenics, as analyzed with a radioreceptor in 1976 by Terenius *et al.*[138]. These increases were partly reversible with therapy. Naber *et al.*[171] found lower levels of endogenous opioid activity using a radioreceptor assay in schizophrenic men than in normal controls. Value in schizophrenic women did not differ from those in normals. Domschke and colleagues[172] reported significantly elevated levels of β-endorphin in acute schizophrenics on neuroleptics and lower levels in chronic aged patients. Four subsequent studies failed to deminstrate any alterations in CSF levels of β-endorphin in schizophrenic patients.[173,174] At present, no conclusion can be made on the levels of endogenous opioid peptides in the CSF.

Gerner and Yamada[175] examined the levels of somatostatin, bombesin, and cholecystokinin in the CSF of schizophrenics. They found a decrease in bombesinlike activity with no change in the other peptides. Normal levels and a normal ratio of angiotensin I to angiotensin II in schizophrenics compared to controls have been reported.[176] Levels of angiotensin I appeared to increase with the longer duration of illness. Carman and associates[177,178] reported lower CSF calcitonin levels in agitated schizophrenics and manic patients compared to controls. Gjerris *et al.*[179,180] found no differences in VIP, gastrin, and CCK levels in schizophrenic CSF compared to normals.

Postmortem studies have shown elevated methionine-enkephalin and substance P levels in the caudate nuclei of chronic undifferentiated schizophrenics.[181] Neurotensin levels have been reported to be unaltered in a number of areas[181,182]; TRH levels are normal in the amygdala,[3] as are VIP and CCK levels in the entorhinal cortex.[147] A survey of six peptides (VIP, CCK, substance P, neurotensin, somatostatin, and bombesin) in ten brain areas from 14 schizophrenic patients showed reduced levels of CCK and somatostatin in the hippocampus and a decreased CCK content in the amygdala.[182] Other peptides were unaltered in the total schizophrenic group.

10. NEUROPEPTIDES AND OBESITY

The initiation and termination of feeding comprise a complex process involving a variety of factors including the hedonic qualities and physicochemical properties of food as well as the organism's response to food ingestion. In addition to these short-term regulators of appetite, there exists a second system that monitors body weight and attempts to maintain adipose tissue mass at a closely contolled set-point. Over the last decade, a variety of neuropeptides have been demonstrated to play a role at every level of appetite regulation.[183]

The major region for the integration of all the impulses responsible for the regulation of appetite is the hypothalamus. The hypothalamus acts as a transducer responsible for integrating the multiple sensory inputs describing the *milieu interier* and maintaining nutritional homeostasis of the organism by activating or deactivating the food-seeking behaviors of the animal. The hypothalamus performs this task through a complex interrelationship of monoamines and neuropeptides on the backdrop of hypothalamic neurons.[183] The same monamines and neuropeptides that regulate appetite also play a role in other hypothalamic vegetative functions.[183] This overlap in regulatory substances allows a close degree of coordination of the hypothalamus over related life-sustaining processes.

Since the original description of cholecystokinin inhibition of feeding,[184] a variety of neuropeptides have been shown to decrease food intake. It has become clear that neuropeptides decrease food intake both through central and peripheral mechanism. Thus, in the case of cholecystokinin, its major effect appears to be by activating ascending vagal fibers in the periphery.[185,186] whereas bombesin, another endogenous peptide incriminated in appetite regulation, appears to produce its effects by directly activating neurons in the hypothalamus.[186] Recent studies have shown that peripherally administered cholecystokinin reduces food intake in humans.[187] Calcitonin, which is released from the thyroid gland after a meal, is a potent anorectic agent whose effect appears to be mediated by inhibiting calcium ion transport into hypothalamic cells.[188] Recently, some neuropeptides (such as substance P, which occurs in high concentrations in taste fibers) have been shown to produce modulatory effects on appetite regulation by altering the hedonic properties of foods.[189] Within the hypothalamus itself, there appears to be a cascade system of neuropeptides and monoamines involved in the fine tuning of appetite regulation.[183] This system is similar to the well-recognized cascade systems involved in blood clotting and complement activation.

We have previously postulated that there is a tonic signal driving the animal to eat and that appetite regulation is predominantly involved with attempting to inhibit this constant food drive.[183] Recent evidence has suggested that one of the major groups of substances responsible for producing this food drive is the endogenous opiates.[190] A specific endogenous opiate, dynorphin, which has its own opiate receptor (the κ receptor), appears to be the opiate involved in the initiation of feeding.[191] Studies in humans support the concept, developed in lower animals, that endogenous opiates play an integral role in feeding behavior. In particular, naloxone has been demonstrated to decrease feeding in obese humans[192] and 2-deoxyglucose-induced feeding in humans.[193] β-Endorphin levels were elevated in human obesity,[194] although the significance of this finding is uncertain. It has been postulated that the major effect of endogenous opiates is to induce the feeding drive, with their analgesic properties representing an epiphenomenon. Much evidence exists linking the nutritional status of the animal to changes in pain perception.[195,196] Animal models have clearly suggested that stress-induced overeating is related to the endogenous opiate system.[197,198]

In conclusion, neuropeptides have been demonstrated to play multiple roles in the regulation of appetite. The explosion of substances known to be involved in appetite regulation clearly suggest that no single substance will be found to be an entirely satisfactory anorectic agent and that the most sensible approach to the treatment of obesity will represent the use of multiple therapeutic modalities.

11. ANOREXIA NERVOSA

Anorexia nervosa is a syndrome characterized by profound weight loss, amenorrhea, and refusal to eat associated with an abnormal perception of body image. The cause of this syndrome is unknown. Multiple endocrine abnormalities have been described in this syndrome, including abnormalities of the hypothalamic–pituitary–ovarian axis,[199,200] thyroid function,[201] and the hypothalamic–pituitary–adrenal axis.[202] Whether these changes are related to the anorectic syndrome itself or to the weight loss and malnutrition that result from the syndrome is unclear.

Direct pathological study of the neurochemistry of this disorder has been limited. Kay *et al.*[203] have reported elevations in total CSF opioid activity measured by radioreceptor assay in patients with anorexia nervosa at their minimum weight compared with levels at their restored weight, with levels from weight-recovered anorectic subjects, and with normal controls. Gerner and Sharp[136] found no significant differences in CSF β-endorphin immunoreactivity in anorexia subjects compared to normal subjects and to those with other mental disorders. Anorectic subjects had depressed somatostatinlike immunoreactivity and normal CCK-like and bombesinlike immunoreactivity compared to normals.[175]

The recent isolation of corticotropin-releasing factor (CRF), a 41-amino-acid peptide,[204] which appears to be Selye's mythical "stress-coordinating substance," has led to potential insight into stress-related anorexias. Administration of CRF to animals produces a syndrome similar to that seen in patients with anorexia nervosa.[205] The hypercortisolism[206] and poor suppressibility of the hypothalamic–pituitary–adrenal axis[207] seen in patients with anorexia nervosa is suggestive of an increase in CRF, although part of the cortisol elevation is explained by decreased metabolism.[208] The possibility that CRF excess represents the primary defect in anorexia nervosa needs to be explored.

12. PEPTIDES AND THE PERIPHERAL NERVOUS SYSTEM

Although the bulk of work on peptidergic neurohumoral systems has concentrated on the brain, there has been important work beginning to detail the role peptides play in the autonomic and peripheral nervous systems.

In particular, the peptides seem to provide a significant portion of the regulation of gastrointestinal motility. The presence of peptides such as VIP, substance P, glucagon, Met-enkephalin, and somatostatin has been known for some time. Deficiencies of some of these have recently been described in pathological states. Analysis by immunohistochemistry has revealed in Hirschsprung's diseased colons a marked reduction of VIP and substance P content as well as a diminution in cells containing enteroglucagon and somatostatin.[209] A similar deficiency of all four of these peptides was also found in Chagas' disease, a condition caused by infection with the flagellate protozoan *Trypanosoma cruzi*.[210] Partly on the basis of this work, an inhibitory role on motility has been postulated for VIP in the gut, compatible with the apropulsive megacolon seen in these diseases. Similar studies of VIP content in Crohn's inflammatory bowel disease found an elevation of this peptide,[211] possibly consistent with the diarrhea seen in this condition. The VIP nerves are grossly abnormal in Crohn's disease, being both more numerous and distorted.[212] These changes are also present at areas distant from the areas of active involvement.

Recently, VIPergic nerves have also been found in the penis,[213] suggesting possible involvement in impotence syndromes.

Finally, the production of peptide hormones by tumors has been recognized for years. The classic case of such production occurs in the Cushing's syndrome caused by ectopic ACTH production.[214] Ectopic production of growth hormone-releasing factor (GRF) by a pancreatic tumor has led to the isolation and characterization of this substance using cDNA techniques and the demonstration that it appears to be highly similar or identical to hypothalamic GRF.[215,216] Expanding knowledge of peptide has led to the discovery that almost all neuropeptides are produced ectopically by at least one tumor. Thus, somatostatin, neurotensin, and bombesin have all been demonstrated to be produced by carcinoma of the lung, particularly the small-cell variety (oat cell).[217] This peptide production continues in tumor or cell cultures even after numerous subcultures.[218] The potential significance of this ectopic production of neuropeptides is unknown, but it is tempting to speculate a role for them in the paraneoplastic syndromes associated with malignancy.

13. CONCLUSION

There appear to be numerous alteration in neuropeptides in a variety of central nervous system disorders. At present, the most accessible route to diagnosis appear to be through the cerebrospinal fluid. Further studies are ncessary to establish the potential diagnostic usefulness of this approach. Over the next decade, we can expect to see increasing use of CSF neuropeptide measurement in the diagnosis of neurological and psychiatric diseases. In addition, it seems likely that a number of neuropeptide analogues will become available for the therapy of these conditions. A start in this direction has been made with the use of the LRH analogues in the treatment of precocious puberty.[219] We can expect to see a variety of methods used in an attempt to improve delivery of neuropeptides across the blood–brain barrier such as the packaging in lysosomes. We can expect our understanding of the pathology of the brain to become more confused before the light of understanding dawns, because, in the words of Emerson Pugh, "If the human brain were so simple that we could understand it, we would be so simple that we couldn't."

REFERENCES

1. Iverson, L. L., 1982, *Lancet* **2**:914–918.
2. Post, R. M., and Gold, P., Rubinow, D. R., Ballenger, J. C., Bunney, W. E., Jr., and Goodwin, F. K., 1982, *Life Sci.* **31**:1–15.
3. Edwardson, J. A., and McDermott, J. R., 1982 *Br. Med. Bull.* **38**:259–264.
4. Nutt, J. G., Mroz, E. A., Leeman, S. E., WIlliams, A. C., Engel, W. K., and Chase, T. N., 1980, *Neurology (N.Y.)* **30**:1280–1285.
5. Jenkins, J. S., Mather, H. M., and Arg, V., 1980, *J. Clin. Endocrinol. Metab.* **50**:364–367.
6. Patel, Y. C., Rao, K., and Reichlin, S., 1977, *N. Engl. J. Med.* **296**:259–533.
7. Cooper, P. E., Fernstrom, M. H., Ronstad, O. P., Leeman, S. E., and Martin, J. B., 1981, *Brain Res.* **218**:219–232.
8. Bird, E. D., 1981, *Neurosecretion and Brain Peptides* (J. B. Martin, S. Reichlin, and K. L. Bick, eds.), Raven Press, New York, pp. 657–672.
9. Pert, C. B., 1981, *Neurosecretion and Brain Peptides* (J. B. Martin, S. Reichlin, and K. L. Bick, eds.), Raven Press, New York, pp. 117–131.
10. Hay, S. E., Goodwin, F. K., and Paul S. M., 1981, *Peptides* **2**:21–26.
11. Beaumont, A., and Hughes, J., 1979, *Annu. Rev. Pharmacol.* **19**:245–267.
12. Bunney, W. E., Jr., Pert, C. B., Klee, W., Costa, E., Pert, A., and Davis, G. C., 1979, *Ann. Intern. Med.* **91**:239–250.
13. Yeung, J. C., Yaksh, T. L., and Rudy, T. A., 1977, *Pain* **4**:23.
14. Akil, H., Richardson, D. E., Hughes, J., and Barchas, J. D., 1978, *Science* **201**:463–465.
15. Watkins, L. R., and Mayer, D. J., 1982 *Science* **216**:1185–1192.
16. Mayer, D. J., Price, D. D., and Rafii, A., 1975, *First World Congress on Pain* (J. M. R. Besson, ed.), Raven Press, New York, p. 276.
17. Jiang, Z. Y., Ye, Q., Shen, Y. T., Zhu, F. X., Tang, S. Q., Liang, N. J., and Zeng, X. C., 1978, *Acta Zool. Sin.* **24**:1–8.
18. Pomeranz, B., Cheng, R., and Law, P., 1977, *Exp. Neurol.* **54**:172–178.
19. Huang, Y., Zhen, J. Z., Li, D. R., and Xie, G. Y., 1978, *Natl. Med. J. China* **4**:193–195.
20. Sjolund, B., Terenius, L., and Eriksson, M., 1977, *Acta Physiol. Scand.* **100**:382–384.
21. Terenius, L., 1979, *Advances in Pain Research and Therapy*, Volume 3 (J. J. Bonica, ed.), Raven Press, New York, pp. 459–471.
22. Sjolund, B., Terenius, L., and Eriksson, M., 1977 *Acta Physiol. Scand.* **100**:382–384.

23. Akil, H., Richardson, D. E., Hughes, J., and Barchas, J. D., 1978, *Science* **201**:463–465.
24. Sicuteri, F., Anselmi, B., Curradi, C., Michelacci, S., and Sassi, A., 1978, *Adv. Biochem. Psychopharmacol.* **18**:363–366.
25. Oyama, T., Jun, T., Yamaya, R., Ling, N., and Guillemin, R., 1980, *Lancet* **1**:122–124.
26. Dunger, D. B., Leonard, J. V., Wolff, O. H., and Preece, M. A., 1980, *Lancet* **1**:1277–1281.
27. Fraioli, F., Fabbri, A., Moretti, C., Santoro, C., and Isidori, A., 1981, *Endocrinology* **108**:238A.
28. Zamir, N., and Shuber, E., 1980, *Brain Res.* **201**:471–474.
29. Zamir, N., and Segal, M., 1979, *Brain Res.* **160**:170–173.
30. Zamir, N., Simanto, R., and Segal, M., 1980, *Brain Res.* **184**:299–310.
31. Davis, W. M., Miya, T. S., and Edwards, L. D., 1956, *J. Am. Pharm. Assoc.* **45**:60–62.
32. Levine, A. S., Morley, J. E., Brown, D. M., and Handwerger, B. S., 1981, *Clin. Res.* **29**:261A.
33. Levine, A. S., and Morley, J. E., 1981, *Peptides* **2**:261–264.
34. Lowy, M. T., Maickel, R. P., and Yim, G. K. W., 1980, *Life Sci.* **26**:2113–2118.
35. Levine, A. S., Morley, J. E., McClain, C. J., Handwerger, B. S., and Brown, D. M., 1981, *Clin. Res.* **29**:703A.
36. Holaday, J. W., and Faden, A. I., 1981, *Neurosecretion and Brain Peptides* (J. B. Martin, S. Reichlin, and K. L. Bick, eds.), Raven Press, New York, pp. 421–434.
37. Peters, W. P., Johnson, M. W., Friedman, P. A., and Mitch, W. E., 1981, *Lancet* **1**:529–532.
38. Morley, J. E., Levine, A. S., and Silvis, S. E., 1982, *Life Sci.* **31**:693–699.
39. Baskin, D. S., and Hosobuchi, Y., 1981, *Lancet* **2**:272–275.
40. Levy, R., Feustel, P., Severinghaus, J., and Hosobuchi, Y., 1982, *Life Sci.* **31**:2205–2206.
41. Hosobuchi, Y., Baskin, D. S., and Woo, S. K., 1982, *Science* **215**:69–71.
42. Baskin, D. S., Kieck, C. F., and Hosobuchi, Y., 1982, *Life Sci.* **31**:2201–2204.
43. Holaday, J. W., and D'Amato, R. J., 1982, *Science* **218**:593–594.
44. Faden, A. I., Jacobs, T. P., and Holaday, J. W., 1981, *Science* **211**:493–494.
45. Young, W., 1981, *J. Neurosurg.* **55**:209–219.
46. Carr, D. B., Ballen, B. A., Skrinar, G. S., Arnold, M. A., Rosenblatt, M., Beitnis, I. Z., Martin, J. B., and McArthur, J. N., 1981, *N. Engl. J. Med.* **305**:560–562.
47. Brandt, N. J., Terenius, L., Jacobson, B. B., Klinken, L., Nordius, A., Brandt, S., Blegvad, K., and Yssing, M., 1980, *N. Engl. J. Med.* **303**:914–916.
48. Campbell, A. M. G., Corner, B., Norman, R. M., and Urich, H., 1961, *J. Neurol. Neurosurg. Psychiatry* **24**:71–77.
49. Gale, J. S., Bird, E. D., Spokes, E. G. S., Iversen, L. L., and Jessell, T., 1978, *J. Neurochem.* **30**:633–634.
50. Emson, P. C., 1980, *Transmitter Biochemistry of Human Brain Tissue Symposium,* (P. C. Emson, ed.), Pergamon Press, Oxford, p. 80.
51. Ljungdahl, A., Hokfelt, T., Nilsson, G., and Goldstein, M., 1978, *Neuroscience* **3**:945–976.
52. Schultzberg, M., Hokfelt, T., and Lundberg, J. M., 1982, *Br. Med. Bull.* **38**:309–313.
53. Bird, E. D., 1976, *Biochemistry and Neurology* (H. F. Bradford and C. D. Marsden, eds.), Academic Press, London, pp. 83–91.
54. Bird, E. D., Chiappa, S. A., and Fink, G., 1976, *Nature* **260**:536–538.
55. Reed, S. C., and Palm, J. D., 1951, *Science* **113**:294–296.
56. Huntington, G. S., 1872, *Med. Surg. Rep.* **26**:317–321.
57. Martinez-Campos, A., Giovannini, P., Cocchi, D., Zanardi, P., Parati, E. A., Caraeeni, T., and Muller, E. E., 1981, *Neurosecretion and Brain Peptides* (J. B. Martin, S. Reichlin, and K L. Bick, eds.), Raven Press, New York, pp. 521–540.
58. Emson, P. C., Fahrenkrug, J., and Spokes, E. G. S., 1979, *Brain Res.* **173**:174–178.
59. Arregui, A., Bennett, J. P., Jr., Bird, E. D., Yamamura, H. T., Iversen, L. L., and Snyder, S. H., 1977, *Ann. Neurol.* **2**:294–298.
60. Arregui, A., Emson, P., Iverson, L. L., and Spokes, E. G. S., 1979, *Adv. Neurol.* **23**:517–525.
61. Bird, E. D., and Iversen, L. L., 1974, *Brain* **97**:457–472.
62. Perry, T. L., Hansen, S., and Kloster, M., 1973, *N. Engl. J. Med.* **288**:337–342.
63. Hays, S. E., and Paul, S. M., 1982. *Life Sci.* **31**:319–322.

64. Hays, S. E., Goodwin, F. K., and Paul, S. M., 1981, *Brain Res.* **225**:452–456.
65. Emson, P. C., Rehfeld, J. F., Langern, H., and Rossor, M., 1980, *Brain Res.* **198**:497–500.
66. de Wied, D., 1969, *Frontiers in Neuroendocrinology* (W. F. Ganong and L. Martini, eds.), Oxford University Press, London, New York, pp. 97–140.
67. Stratton, L. O., and Kastin, A. J., 1974, *Horm. Behav.* **5**:149–155.
68. Rigter, H., and Van Riezen, H., 1975, *Physiol Behav.* **14**:563–566.
69. Isaacson, R. L., Dunn, A. J., Rees, H. D., and Waldock, B., 1976, *Physiol. Psychol.* **4**:159–162.
70. Bohus, B., Hendrickx, H. H. L., Van Kolfschoten, A. A., and Krediet, T. G., 1975, *Sexual Behavior: Pharmacology and Biochemistry* (M. Sandler and G. L. Gessa, eds.), Raven Press, New York, pp. 269–275.
71. de Wied, D., 1971, *Nature* **232**:58–60.
72. Schulz, H. Kovacs, G. L., and Telegdy, G., 1976, *Cellular and Molecular Basis of Neuroendocrine Processes* (E. Endroczi, ed.), Akademai Kiado, Budapest, pp. 555–564.
73. de Wied, D., and Bohus, B., 1978, *Prog. Brain Res.* **48**:327–334.
74. Lande, S., Flexner, J. B., and Flexner, L. B., 1972, *Proc. Natl. Acad. Sci. USA* **69**:558–560.
75. Gilot, P., Crabbe, J., and Legros, J., 1980, *Acta Psychiatr. Belg.* **80**:775–761.
76. Wagner, A., Jardanhazy, T., Laczi, F., Szilard, J., Telegdy, G., and Laszlo, F. A., 1979, *Acta Med. Hung.* **36**:81.
77. Leboeuf, A., Lodge, J., and Eames, P. G., 1978, *Lancet* **2**:1370.
78. Legros, J. J., Gilot, P., Seron, X., Claessens, J., Adam, A., Moeglen, J. M., Audibert, A., and Berchier, P., 1978, *Lancet* **1**:41–42.
79. Delwaide, P. J., Devoitille, J. M., and Ylieff, M., 1980 *Acta Psychiatr. Belg.* **80**:748–754.
80. Oliveros, J. C., Jandali, M. K., Timsit-Berthier, M., Remy, R., Benghezal, A., Audibert, T., and Moeglen, J. M., 1978, *Lancet* **1**:42.
81. Weingartner, H., Gold, P., Ballenger, J. C., Smallberg, S. A., Summers, R., Rubinow, D. R., Post, R. M., and Goodwin, F. K., 1981, *Science* **211**:601–603.
82. Weingartner, H., Kaye, W., Gold, P., Smallberg, S., Peterson, R., Gillin, J. C., and Ebert, M., 1981, *Life Sci.* **29**:2721–2726.
83. Medvedev, V. I., Bakharev, V. D., Grechko, A. T., and Nezovikab'ko, V. N., 1980, *Hum. Physiol.* **6**:307–310.
84. Koch-Hendriksen, N., and Nielsen, H., 1981, *Lancet* **1**:38–39.
85. LeMoal, M., Koob, G. F., Koda, L. Y., Bloom, F. E., Manning, M., Sawyer, W. H., and Rivier, J., 1981, *Nature* **291**:491–493.
86. Tsuji, M., Takahasi, S., and Akazawa, S., 1981, *Psychoneuroendocrinology* **6**:171–176.
87. Rossor, M. N., Iversen, L. L., Mountjoy, C. W., Roth, M., Hawthorn, J., Ang, V. J., and Jenkins, J. S., 1980, *Lancet* **2**:1367–1368.
88. Ballenger, J. C., Post, R. M., Gold, P. N., Robertson, G. O., Bunney, W. E., Jr., and Goodwin, F. K., 1980, *Sci. Proc. Am. Psychiatr. Assoc.* **733**:144.
89. deWied, D., and van Ree, J. M., 1982, *Life Sci.* **31**:709–719.
90. Kastin, A. J., Sandman, C. A., Stratton, L. O., Schally, A. V., and Miller, L. H., 1975, *Prog. Brain Res.* **42**:143–150.
91. van Reizen, H., Rigter, H., and de Wied, D., 1977, *Behav. Biol.* **20**:311–324.
92. van Praag, H. M., and Verhoeven, W. M. A., 1980, *Adv. Biol. Psychiatry* **5**:20–45.
93. Branconnier, R. J., Cole, J. O., and Gardos, G., 1979, *Psychopharmacology* **61**:161–165.
94. Pigache, R. M., and Rigter, H., 1981, *Front. Horm. Res.* **8**:193–207.
95. Olson, G. A., Olson, R. D., Kastin, A. J., and Coy, D. H., 1981, *Neurosci. Biobehav. Rev.* **3**:285–299.
96. Stein, L., and Belluzzi, J. D., 1979, *Fed. Proc.* **38**:2468–2472.
97. Huston, J. P., and Staubli, U., 1981, *Endogenous Peptides and Learning and Memory Processes* (J. L. Martinez, R. A. Jensen, R. B. Messing, H. Rigter, and J. L. McGaugh, eds.), Academic Press, New York, pp. 521–540.
98. Dupont, A., Savard, P., Merand, Y., Labrie, F., and Boissier, J. R., 1981, *Life Sci.* **29**:2317–2322.
99. Wood, P. L., Etienne, P., Lul, S., Gauthier, S., Cahal, S., and Nair, N. P. V., 1982 *Life Sci.* **31**:2073–2079.

100. Rossor, N. M., Emson, P. C. E., Mountjoy, C. Q., Roth, S. M., and Iversen, L. L., 1980, *Neurosci. Lett.* **20**:373–377.
101. Davies, P., and Terry R. D., 1981, *Neurobiol. Aging* **2**:9–14.
102. Hornykiewicz, O., 1979, *The Neurobiology of Dopamine* (A. S. Horn, J. Korf, and B. H. C. Westerink, eds.), Academic Press, New York, pp. 633.
103. Johnson, R. P., Sar, M., and Stumpf, W. E., 1980, WBrain Res. **194**:566–568.
104. Schwartz, J. C., Pollard, H., Llorens, C., Malfroy, B., Gros, C., Pradelles, P., and Dray, F., 1978, *The Endorphins* (E. Costa and M. Trabuckhi, eds.), Raven Press, New York, pp. 245–254.
105. Taquet, H., Javoy-Agid, F., Cesselin, F., and Agid, Y., 1981, *Lancet* **1**:1367–1368.
106. Sorenson, K. V., Christensen, S. E., Dupont, E., Hansen, A. P., Pedersen, E., and Orskov, H., 1980, *Acta Neurol. Scand.* **61**:186–191.
107. Orskov, H., Hansen, A. P., de Fine Olivarius, B., Pedersen, E., Sorensen, K., Dupont, E., and Ingerslev, J., 1982, *Proceedings 3rd World Congress of Biological Psychiatry* (B. Jansson, C. Perris, and G. Struew, eds.), Elsevier, Amsterdam.
108. McFarlen, D. E., and McFarland, H. F., 1982, *N. Engl. J. Med.* **307**:1246–1251.
109. Adachi, K., Konagaya, M., Muroga, T., and Toshima, H., 1977, *Igaku no Agumi* **101**:74–75.
110. Sobrie, I., Yamamoto, M., Kongaya, M., Iida, M., and Takayonagi, T., 1980, *Lancet* **1**:418–419.
111. Sobue, I., Muroga, T., Konagaya, M., Iida, M., and Takayanagi, T., 1980, *Spinocerebellar Degenerations* (I. Sobue, ed.), University of Tokyo Press, Tokyo, pp. 83–94.
112. LeWitt, P. A., and Ehrenkranz, J. R. L., 1982, *Lancet* **2**:981.
113. Shy, G. M., and Drager, G. A., 1960, *Arch. Neurol.* **2**:511–527.
114. Bannister, R., and Oppenheimer, D. R., 1972, *Brain* **95**:457–474.
115. Ziegler, M. G., Lake, C. R., and Kopin, I. J., 1977, *N. Engl. J. Med.* **296**:293–297.
116. Diamond, M. A., Murray, R. H., and Schmid, P. G., 1970, *J. Clin. Invest.* **49**:1341–1348.
117. Petito, C. K., and Black, I. B., 1978, *Ann. Neurol.* **4**:6–17.
118. Nutt, J. G., Mroz, E. A., Leeman, S. E., Williams, A. C., Engel, W. K., and Chase, T. N., 1980, *Neurology (Minneap.)* **30**:1280–1285.
119. Hokfelt, T., Elfvin, L. G., Schultzberg, M., and Wiesel, F. A., 1977, *Brain Res.* **132**:29–41.
120. Ljungdahl, A., Hokfelt, T., Nilsson, G., and Goldstein, M., 1978, *Neuroscience* **3**:945–976.
121. Pickel, V. M., Joh, T. H., Reis, D. J., Leeman, S. E., and Miller, R. J., 1979, *Brain Res.* **160**:387–400.
122. Magnusson, T., Carlsson, A., Fisher, G. H., Chang, D., and Folkers, K., 1976, *J. Neural Transm.* **38**:9–93.
123. Guyenet, P. G., and Aghajanian, G. K., 1977, *Brain Res.* **136**:178–184.
124. Dun, N. J., and Karczmar, A. G., 1979, *Neuropharmacology* **18**:215–218.
125. Otsuka, M., and Konishi, S., 1976, *Cold Spring Harbor Symp. Quant. Biol.* **40**:135–143.
126. Gold, E. M., 1979, *Ann. Intern. Med.* **90**:829–844.
127. Reichlin, S. R., 1981, *Textbook of Endocrinology* (R. H. Williams, eds.), W. B. Saunders, Philadelphia, pp. 589–645.
128. Bierich, J. R., 1975, *J. Clin. Endocrinol. Metab.* **4**:107–112.
129. Morley, J. E., 1981, *Endocrine Rev.* **2**:396–436.
130. Lieblich, J. M., Rogol, A. D., White, B. J., and Rosen, S. W., 1982, *Am. J. Med.* **73**:506–519.
131. Carroll, B. J., Feinberg, M., Gredon, J. F., Tarika, J., Albala, A. A., Haskett, R. F., James N. M., Kronfol, Z., Lohr, N., Steiner, M., deVigne, J. P., and Young, E., 1981, *Arch. Gen. Psychiatry* **38**:15–22.
132. Sachar, E. J., Hellman, L., Fukushima, D. K., and Gallagher, T. F., 1970, *Arch. Gen. Psychiatry* **23**:289–298.
133. Sachar, E. J., Hellman, L., Roffwarg, H. P., Halpern, F. S., Fukushima, D. K., and Gallagher, T. F., 1973, *Arch. Gen. Psychiatry* **28**:19–24.
134. Walsh, B. T., 1982, *Psychosom. Med.* **44**:85–91.
135. Brown, W. A., and Qualls, C. B., 1981, *Psychol. Res.* **4**:115–128.
136. Gerner, R. H., and Sharp, B., 1982, *Brain Res.* **237**:244–247.

137. Gerner, R. H., Catlin, D. H., Gorelik, D. A., Hui, K. K., and Li, C. H., 1980, *Arch. Gen. Psychiatry* **37**:642–647.
138. Terenius, L., Wahlstrom, A., Lindstrom, L., and Widerlor, E., 1976, *Neurosci. Lett.* **3**:157–162.
139. Berger, P. A., Akil, H., Watson, S. J., and Barchas, J. D., 1982, *Annu. Rev. Med.* **33**:397–415.
140. Loosen, P. T., and Prange, A. J., 1982, *Am. J. Psychiatry* **139**:405–416.
141. Sternbach, H., Gerner, R. H., and Gwirtsman, H. E., 1982, *J. Clin. Psychiatry* **43**:4–6.
142. Kastin, A. J., Ehrenring, R. H., Schalch, D. S., and Anderson, M. S., 1972, *Lancet* **2**:740–742.
143. Morley, J. E., 1979, *Life Sci.* **25**:1539–1550.
144. Kirkegaard, C., Faber, J., Hummer, L., and Rogowski, P., 1979, *Psychoneuroendocrinology* **4**:227–235.
145. Gerner, R. H., and Yamada, T., 1982, *Brain Res.* **238**:298–302.
146. Post, R. M., Gold, P., Rubinow, D. R., Ballenger, J. C., Bunney, W. E., Jr., and Goodwin, F. K., 1982, *Life Sci.* **31**:1–15.
147. Perry, R. H., Dockray, G. J., Dimaline, R., Perry, E. K., Blessed, G., and Tomlinson, B. E., 1981, *J. Neurol. Sci.* **51**:465–472.
148. van Kammen, D. P., 1979, *Psychoneuroendocrinology* **4**:37–46.
149. Horrobin, D. F., 1979, *Lancet* **1**:529–530.
150. Meltzer, H. Y., Busch, D., and Fang, V. S., 1981, *Psychoneuroendocrinology* **6**:17–36.
151. Post, R. M., Fink, E., Carpenter, W. T., and Goodwin, F. K., 1975, *Arch. Gen. Psychiatry* **32**:1063–1069.
152. Crow, T. J., Baker, H. F., Cross, A. J., Joseph, M. H., Lofthouse, R., Longden, A., Owen, F., Riley, G. V., Glorer, V., and Killpack, W. S., 1979, *Br. J. Psychiatry* **134**:249–256.
153. Owen, F., Cross, A. Y., Crow, T. J., Longden, A., Poulter, M., and Riley, G. V., 1978, *Lancet* **2**:223–226.
154. Watson, S. J., Akil, H., and Berger, P. A., 1979, *Arch. Gen. Psychiatry* **36**:220–223.
155. Verbey, K., Volavka, J., and Clouet, D., 1978, *Arch. Gen. Psychiatry* **35**:871–888.
156. Gunne, L. M., Lindstom, L., and Terenius, L., 1977, *J. Neural Transm.* **40**:13–19.
157. Davis, G. C., Bunney, W. E., Jr., DeFraites, E. G., Kleinman, J. E., van Kammen, D. P., Post, R. M., Wyatt, R. J., 1977, *Science* **197**:74–77.
158. Emrich, H. M., Cording, C., Pirce, S., Krolling, A., von Zemen, D., and Herz, A., 1977, *Pharmakopsychiatrie* **10**:265–270.
159. Watson, S. J., Berger, P. A., Akil, H., Mills, M. S., and Barchas, J. D., 1978, *Science* **201**:73–76.
160. Pickar, D., Vartanian, F., Bunney, W. E., Jr., Maier, H. P., Gastpar, M. T., Prakash, R., Sethi, B. B., Lideman, R., Belyaev, B. S., Tsutsulkovskaya, V. A., Jungkunz, G., Nedopil, N., Verhoeven, W., van Fraag, H., 1982, *Arch. Gen. Psychiatry* **39**:313–319.
161. Berger, P. A., Watson, S. J., Akil, H., Elliot, C. R., Rubin, R. T., Pfefferbaum, A., Davis, K. L., Barchas, J. D., and Li, C. H., 1980, *Arch. Gen. Psychiatry* **37**:635–639.
162. Gerner, R. H., Catlin, D. H., Goerlick, D. A., Hui, K. K., and Li, C. H., 1980, *Arch. Gen. Psychiatry* **37**:642–647.
163. Pickar, D., Dabis, G. C., Schulz, S. C., and Gold, P. W., 1980, *Am. J. Psychiatry* **37**:635–639.
164. Jorgensen, A., Fog, R., and Veillis, B., 1979, *Lancet* **1**:935.
165. Ruther, E., Jungkunz, G., and Nedopil, N., 1980, *Prog. Neuropsychopharmacol.* **4**:304–305.
166. Loeber, J., Verhaef, J., Burbach, J. P. H., and van Ree, J. M., 1979, *Acta. Endocrinol. (Kbh.)* **91**(Suppl. 255):74.
167. de Wied, D., Kovacs, G. L., Bohus, B., van Ree, J. M., and Greyen, H. M., 1978, *Eur. J. Pharmacol.* **49**:427–430.
168. Verhoeven, W. M. A., van Praag, H. M., van Ree, J. M., and de Wied, D., 1979, *Arch. Gen. Psychiatry* **36**:294–302.
169. Tamminga, C. A., Tighe, P. J., Chase, T. N., DeFraites, G., and Schaffer, M. H., 1981, *Arch. Gen. Psychiatry* **38**:167–168.
170. Emrich, H. M., Zandig, M., Kissling, W., Dirlich, G., Zerssen, D. V., and Herz, A., 1980, *Pharmakopsychiatrie* **13**:290–298.

171. Naber, D., Pickar, D., Post, R. M., van Kammen, D. P., Waters, R. N., Ballenger, J. C., Goodwin, F. K., and Bunney, W. E., 1981, *Am. J. Psychiatry* **138:**1457–1461.
172. Domschke, W., Dickschas, A., and Mitznegg, P., 1979, *Lancet* **1:**1024.
173. Emrich, H. M., Hollt, V., Kissling, W., Fischler, M., and Herz, A., 1979, *Pharmakopsychiatrie* **12:**269–276.
174. Ross, M., Berger, P. A., and Goldstein, A., 1979, *Science* **205:**1163–1164.
175. Gerner, R. H., and Yamada, T., 1982, *Brain Res.* **238:**298–302.
176. van Kammen, D. P., Waters, R. N., Gold, P. W., Sternberg, D. E., Pickar, D., Naber, D., Robertson, C., Ganten, D., Reichlin, S., Ballenger, J. C., Post, R. M., and Bunney, W. E., Jr., 1982, *Proceedings 3rd World Congress of Biological Psychiatry* (B. Jannson, C., Perris, and G. Struew, eds.), Elsevier, Amsterdam, p. 48.
177. Carman, J. S., Post, R. M., Ballenger, J. C., Becker, K. L., and Silva, O., 1980, *Soc. Biol. Psychiatry* **10:**105.
178. Carman, J. S., Post, R. M., Ballenger, J. C., and Goodwin, F. K., 1982, *Proceedings 3rd World Congress of Biological Psychiatry* (B. Jansson, C. Perris, and G. Struew, eds.), Elsevier, Amsterdam, p. 48.
179. Gjerris, A., Fahrenkrug, J., Hammer, M., Schaffalitzky de Muchadell, O. P., Bojholm, S., Vendsborg, P., and Rafaelsen, O. J., 1980, *Abstracts of the 12th Collegium Internationale Neuro-Psychopharmacolgicum,* (P. Riederer and E. Usdin, eds.), Macmillan, London, p.115.
180. Gjerris, A., and Fahrenkrug, J., *Proceedings 3rd World Congress of Biological Psychiatry* (B. Jansson, C. Perris, and G. Struew, eds.), Elsevier, Amsterdam, p. 84.
181. Kleinmann, J. E., Karoum, F., Rosenblatt, J., and Freed, W. J., 1981, *Biological Psychiatry* (C. Perris, G. Struew, and B. Jansson, eds.), Elsevier, Amsterdam, pp. 711–714.
182. Roberts, G. W., Ferrier, I. N., Lee, Y. C., Crow, I. N., and Bloom, S. R., 1982, *Regul. Peptides* **3:**81.
183. Morley, J. E., 1980, *Life Sci.* **27:**355–368.
184. Gibbs, J., Young, R. C., and Smith, G. P., 1973, *Nature* **245:**323–325.
185. Smith, G. P., Jermone, C., Cushin, B. J., Eterro, R., and Simansky, K. J., 1981, *Science* **213:**1036–1037.
186. Morley, J. E., Levine, A. S., Kneip, J., and Grace, M., 1982, *Life Sci.* **30:**1943–1947.
187. Kisseleff, H. R., Pi-Sunyer, F. X., Thornton, J., and Smith, G. P., 1981, *Am. J. Clin. Nutr.* **34:**154–160.
188. Levine, A. S., and Morley, J. E., 1981, *Brain Res.* **222:**187–191.
189. Morley, J. E., Levine, A. S., and Murray, S. S., 1981, *Brain Res.* **226:**334–338.
190. Morley, J. E., Levine, A. S., 1982, *Am. J. Clin. Nutr.* **35:**757–761.
191. Morley, J. E., and Levine, A. S., 1981, *Life Sci.* **29:**1901–1903.
192. Atkinson, R. L., 1982 *J. Clin. Endocrinol. Metab.* **55:**196–198.
193. Thompson, D. A., Welle, S. L., Lilavivat, U., Penicaud, L., and Campbell, R. G., 1982, *Life Sci.* **31:**847–852.
194. Given, J. R., Wiedmann, E., Anderson, R. N., and Kitabachi, A. E., 1980, *J. Clin. Endocrinol. Metab.* **50:**975–976.
195. McGivern, R., Berka, C., Berston, G. B., Walker, J. M., and Sandman, C. A., 1979, *Life Sci.* **25:**885–888.
196. Levine, A. S., Morley, J. E., Brown, D. M., and Handwerger, B. S., 1982, *Physiol. Behav.* **28:**987–989.
197. Morley, J. E., and Levine, A. S., 1980, *Science* **209:**1259–1261.
198. Lowy, M. T., Maickel, R. P., and Yim, G. K. W., 1980, *Life Sci.* **26:**2113–2118.
199. Boyar, R. M., Katz, J., Finkelstein, J. W., Kapen, S., Weiner, H., Weitzman, E. D., and Hellman, L., 1974, *N. Engl. J. Med.* **291:**861–865.
200. Beaumont, P. V. V., Carr, P. J., and Gelder, M. G., 1973, *Psychol. Med.* **3:**495–501.
201. Leslie, R. D. G., Isaacs, A. J., Gomez, J., Raggatt, P. R., and Baylers, R., 1978, *Br. Med. J.* **5:**526–528.
202. Walsh, B. T., 1982, *Psychosom. Med.* **44:**85–91.
203. Kaye, W. H., Pickar, D., Naber, D., and Ebert, M. H., 1982, *Am. J. Psychiatry* **139:**643–645.
204. Spiess, J., Rivier, J., Rivier, C., and Vale, W., 1981, *Proc. Natl. Acad. Sci. U.S.A.* **78:**6517–6521.

205. Morley, J. E., and Levine, A. S., 1982, *Life Sci.* **31**:1459–1464.
206. Walsh, B. T., Katz, J. L., Levin, J., Kream, J., Fukushima, D. K., Hellman, C. D., Weiner, H., and Zumoff, B., 1978, *Psychosom. Med.* **40**:499–506.
207. Rose, R. M., and Sacher, E., 1981, *Textbook of Endocrinology* (R. H. Williams, ed.), W. B. Saunders, Philadelphia, p. 645.
208. Boyar, R .M., Hallman, L. D., Roffwarg, H., Katz, J., Zumoff, B., O'Connor, J., Bradlow, L., and Fukushima, D. K., 1977, *N. Engl. J. Med.* **296**:190–193.
209. Bishop, A. E., Polak, J. M., Lake, B. D., Bryant, M. G., and Bloom, S. R., 1981, *Histopathology* **5**:679–688.
210. Long, R. G., Barnes, A. V., O'Shaughnessy, Bannister, R., Bishop, A. E., Albuquerque, R. H., McGregor, G. P., Polak, J. M., and Bloom, S. R., 1980, *Lancet* **1**:559–562.
211. Polak, J. M., Bishop, A. E., Long, R. G., Bryant, M. G., McGregor, G. P., Albuquerque, R. H., and Bloom, S. R., 1979, *Gut* **20**:A942.
212. Bloom, S. R., and Polak, J. M., 1982, *Br. Med. Bull.* **38**:233–238.
213. Polak, J. M., Mina, S., Gu, V., and Bloom, S. R., 1981, *Lancet* **2**:217–219.
214. Liddle, G. W., Nicholson, W. E., Island, D. P., Orth, D. N., Abe, K., and Lowder, S. C., 1969, *Recent Prog. Horm. Res.* **25**:283–316.
215. Guillemin, R., Brazeau, P., Bohlen, P., Esch, F., Ling, N., and Wehrenberg, W. B., 1982, *Science* **218**:585–586.
216. Rivier, J., Spiess, J., Thorner, M., and Vale, W., 1982, *Nature* **300**:276–278.
217. Wood, S. M., Wood, J. R., Ghater, J. R., Lee, Y. C., O'Shaughnessy, D., and Bloom, S. R., 1981, *J. CLin. Endocrinol. Metab.* **53**:1310–1312.
218. Sorenson, G. D., Bloom, S. R., Ghater, M. A., Del Prete, S. A., Cate, C. C., and Pettergil, O. S., 1984, *Regul. Peptides* (in press).
219. Comite, F., Cutler, G. B., Jr., Rivier, J., Vale, W. W., Loriaux, D. L., and Crowley, W. F., Jr. 1981, *N. Engl. J. Med.* **305**:1546–1550.

4

Neurochemistry of Brain Tumors

Samuel Bogoch and S. Winston Bogoch

1. INTRODUCTION

The attempt to define a unique chemistry of malignant growth has been stimulated in recent years by progress in the chemical methodology available to examine both cell products and the genetic codes for their production. Immunochemistry has provided a powerful set of methods and a conceptual framework with which to begin to understand cancer. However, as is usual for the early stages of understanding in any area, dogma has impeded progress. Thus, the failure over six decades to define a chemical pathology of cancer in terms of an immune theory, which Ehrlich formulated at the beginning of the century, or any other theory led to the dictum that since no unique cancer chemistry had been found, it did not exist.

Some hope was kindled when it was demonstrated that certain substances commonly observed in fetal tissues but not in appreciable concentrations in normal adult tissues could be found in abundance in cancer. Then, with monoclonal antibodies, certain as yet undefined cell constituents were then found to be characteristic of certain cancer cell types. In addition to these cell-specific constituents, the first family of general transformation antigens was described. Finally, recent work on the oncogenes has provided evidence for both cell-specific and more general transformation genes.

The findings to be reviewed already have uses in the diagnosis of brain tumors and important implications as well for their therapy.

2. MEMBRANES AND MEMBRANE RECEPTORS

For over a century, pathologists have been successfully distinguishing under the microscope normal from neoplastic cells on the basis of uniform and reliable morphological criteria. These criteria include the size, shape, and po-

Samuel Bogoch • Department of Biochemistry, Foundation for Research on the Nervous System, and Boston University School of Medicine, Boston, Massachusetts 02215. *S. Winston Bogoch* • College of Physicians and Surgeons, Columbia University, New York, New York 10032.

sition of the cell and the appearance and staining of its intra- and extracytoplasmic membranes. The obvious nature of these changes, visible to the eye, has encouraged the search for biochemical changes that underlie these changes to the malignant state.

Neuroectodermal tumors may arise following the transplacental administration of ethylnitrosourea[1] and N-methyl-N-nitrosourea; the latter has been shown to methylate bases in nuclear DNA that brain cells are unable to excise.[2] The induction of brain tumors by chemical agents including the nitrosoureas and polycyclic hydrocarbons has been reviewed by Lantos,[3] and their induction by viral means has been reviewed by Bullard and Bigner.[4] Recent advances in the preparation of brain tumor cell cultures for chemical study are illustrated in the study of Bressler *et al.*[5] of transformed newborn rat oligodendrocytes, which were followed by studying light and scanning electron microscopic morphological features as well as chemical markers: the induction of lactate dehydrogenase by dibutyryl cyclic adenosine monophosphate, the presence of cyclic nucleotide phosphohydrolase, and the hydrocortisone induction of glycerol phosphate dehydrogenase.

Rankin, Shapiro, and Posner[6] detailed stability and chromosomal evolution of early passages of freshly resected human gliomas and concluded that if the complexity of karyotypic evolution is to be studied and correlated to specific phenotypic characteristics of clinical significance, it must be examined in early-passage cells rather than in established cell lines. The relevance of this evidence and of that to be discussed later under monoclonal antibodies on the heterogeneity of antigenic composition and its variability over time as different clones gain dominance in a given tumor cell population makes it imperative that the neurochemist who cares to have a stable and reproducible characteristic under examination understand the nature of the disappearing and reappearing target being studied.

Lectin-induced mitogenesis in glioma but not in meningioma, schwannoma, or adenocarcinoma cell lines has been described.[8] The effect of a viral glycoprotein in initiating cell fusion in mouse neuroblastoma N-18 cells but not in BHK-21 cells has been studied.[9]

Will *et al.* in Mandel's group[10] have described two methods to isolate cell surface membranes from cultured glioblastoma cells. Their specific ATPase activity was seven- to eightfold that of the homogenate. Both fractions, however, were still contaminated by smooth endoplasmic reticulum as judged by the activity of NADPH-dependent cytochrome *c* reductase. Butyrate-induced morphological changes, which to some extent represent reversal of the transformation process, have been studied in rat glioma cells *in vitro*,[11] and Black *et al.* in Kornblith's group[12] have added electrophysiological determinations to the immunologic, biochemical, and ultrastructural procedures that they employ to characterize their human glioblastoma-derived cell culture line.

The function of cell membranes of cultured brain tumor lines as well as neuroblastoma–glioma hybrids has been explored for a number of metabolites and receptors. For example, the membrane as a possible control point for glucose utilization has been studied in C1300 neuroblastoma and C6 glioma cells,[13] regulation of muscarinic acetylcholine receptor concentration has been studied in cloned N1E-115 neuroblastoma cells,[14] and intracellular activity and release of acetylcholinesterase have been studied in mouse neuroblastoma cells.[15] Hamprecht's group have studied lithium ion flux through the action potential sodium channel,[16] sodium-dependent uptake and release of taurine,[17] synthesis of enkephalinlike opioid peptides,[18] opioid receptors,[19] and the relationship among the action of calcium ions, opioids, and prostaglandin E_1 on the level of cyclic AMP[20] in neuroblastoma–glioma hybrids. Similarly, Haga and Haga have studied α-adrenergic receptors in membranes of the Sabol and Nirenberg neuroblastoma–glioma hybrid NG108,[21] and Green and Clark have studied specific muscarinic-cholinergic desensitization.[22] That such binding studies in addition to providing basic information on neuronal receptors may provide insights into the nature of the malignant cell is illustrated in the studies of Spinelli and Ishii on the fact that neurite outgrowth is specifically and reversibly inhibited by phorbol and other tumor promotors in cultured embryonic sensory and sympathetic ganglia but is enhanced in cultured human SH-SY5Y neuroblastoma cells.[23] The finding by Feder and Gilbert[24] of clonal evolution and karyotype heterogeneity in neuroblastoma cells derived from one patient at two points in his illness, in agreement with the findings in glioma already cited,[6] must be kept in mind, however.

3. *NUCLEIC ACIDS AND POLYAMINES*

DNA distributions of human brain tumor cells have been analyzed by flow cytometry on fixed cells and isolated nuclei by Knebel and Hoshino.[7] Vital sorts were done to determine clonogenicity. The relative DNA content of 56 human intracranial tumors was also investigated by Mork and Laerum.[25] Biosynthetic pathways of nucleotides were studied in mouse neuroblastoma clonal line M1 and rat C6 glioma cells grown in tissue culture,[26] and the relative rates of synthesis and turnover of poly(A)-containing mRNAs were determined in the S-20 (cholinergic) and NIE-115 (adrenergic) neuroblastoma clones after treatment with dibutyryl cyclic AMP or the phosphodiesterase inhibitor Ro20-1724.[27] In the latter study, the results showed that neither increased intracellular cyclic AMP levels nor growth inhibition necessarily results in a relative increase in the synthesis of the poly(A)-containing mRNAs.

The first examinations of oncogene structure have been in other than brain tumors, but evidence on malignin from brain glioma cells, the cancer recognins from breast and lymphoma cell lines, and sharing of the specificities for monoclonal antibodies to unknown antigens suggest that shared specificities will be the case for the oncogene structure as well (see Sections 6, 7).

The polyamines putrescine, spermidine, and spermine were determined in CSF of patients with medulloblastoma[28] and in glioblastoma and anaplastic astrocytoma[29] by Fulton *et al.* and Marton *et al.*, respectively, of Wilson's group. Although the polyamines correlated well with the clinical data in medulloblastomas, they were found to be of little use for monitoring tumor progression in glioblastomas and astrocytomas. The authors suggest that this difference may reflect the frequent location of medulloblastomas within or adjacent to the ventricular system or subarachnoid space. Harik and Sutton[30] found the putrescine content of astrocytomas obtained at surgery to be elevated when compared to normal brain but that spermidine and spermine levels varied widely in the tumors assayed and did not correlate with criteria of malignancy.

4. GLYCOLIPIDS AND LIPIDS

Bogoch[31–34] demonstrated the inverse relationship between cell replication and cell membrane macromolecular carbohydrate concentration in a series of studies between 1965 and 1972 on the concentration of carbohydrates in gangliosides and glycoproteins in human glioma brain tumor tissue compared to normal human brain and in C6 glioma grown in mice. Thus, malignant cells had both lower concentrations of carbohydrate and simpler chains in terms of the number and type of constituent hexoses, hexosamines, and sialic acid. The complex carbohydrates were related to greater complexity of cell–cell interaction.[31]

Liepkalns *et al.*[35] referred to the work of Fishman and Brady, Manuelides, and others who found human gliomas to have increased proportions of gangliosides structurally simpler than G_{M1}, most pronouncedly in the least differentiated tumors, and then showed that even in glioma cells in tissue culture, higher cell densities were associated with greater proportions of complex gangliosides species. These authors suggested that these changes might result from cell contact and relate to cell differentiation. Related findings were reported by Srinivas and Colburn,[36] who showed in the transformation of mouse epidermal cells that tumor-promoting phorbol decreased precursor incorporation into trisialoganglioside and increased incorporation into disialoganglioside. In addition, these ganglioside responses were antagonized by the antipromoter retinoic acid. A related study by McLawhon *et al.*[37] on mouse neuroblastoma cell line N4TG1 showed that inhibition of cyclic AMP synthesis prevented the formation of G_{M2} from G_{M3} and G_{M1} from G_{M2}, whereas the elevation of cyclic AMP had the opposite effect. Calcium and manganese concentrations appear to regulate some of these changes in NB41A neuroblastoma cells.[38] D-Glucosamine and papaverine inhibit human neuroblastoma growth in culture,[39] and G_{M2} and interferon may each retard growth of glioma cells in culture[40] (see also Section 6).

The molar percent of cerebrosides in various neural tumors was significantly reduced,[41] and 25-hydrocholesterol, a potent inhibitor of sterol synthesis, was shown to kill proliferating C6 glioma cells in culture.[42]

5. GLYCOPROTEINS AND PROTEINS

In this and the next section, those proteins and glycoproteins that have been isolated from brain tumor tissue or cells and to some degree characterized are discussed. In Section 6, those substances (which may or may not be proteins) are discussed whose presence in tumors can only be inferred from the fact that polyclonal or monoclonal antibodies have been generated to them.

Bogoch *et al.*[33,43] first reported in 1968 the isolation, separation by quantitative Cellex D column chromatography, and characterization by quantitative determination of hexose, hexosamine, and sialic acid as well as gel electrophoresis, of unique glycoproteins from human astrocytomas and gliomas removed at surgery. Tumor glycoprotein 10B was different from its normal counterpart in that its carbohydrate constituents were much reduced in total concentration and in variety, whereas the concentration of the protein moiety was increased seven- to tenfold over that in normal brain. It was the further study of glycoprotein 10B that led to the isolation of the cleaved 10,000-dalton peptide product astrocytin and its counterpart in tissue culture of tumor glial cells, malignin (see Section 7).

Lubitz *et al.*[44] iodinated cell surface glycoproteins of normal and malignant glial cell lines, isolated them by lectin affinity chromatography, and compared them by gel electrophoresis. Seven glycoproteins in the molecular weight range between 70K and 220K as well as several minor components of lower molecular weight were distinguished.

McKeever *et al.*[45] and Shitara *et al.*[46] of Kornblith's group have begun to analyze the proteins and glycoproteins of cultured human astrocytoma cells and C6 glioma cells in terms of the differences between proteins retained and those released into the culture medium. After radiolabeling the cells with focuse, they detected a major extracellular 85K glycoprotein by gel electrophoresis.

Heikinheimo *et al.*[47] described the synthesis of the pregnancy-specific β-1-glycoprotein(SP1)-like material in normal brain-derived and malignant glial-cell cultures. The active biosynthesis of SP1 was confirmed by metabolic labeling and subsequent immune precipitation of the culture medium.

Using lectin affinity chromatography on the [^{35}S]-radiolabeled P24 glycoprotein released into the culture media, Peiper *et al.*[48] have provided evidence that the carbohydrate moiety of this glycoprotein is different on lymphoblasts and neuroblastoma cells.

Other studies of interest are those on the stimulation by dibutyryl cyclic adenosine monophosphate of ectopic production of the free β subunit of chorionic gonadotropin by human brain tumor cell line CBT,[49] phosphorylation of endogenous proteins by cyclic-AMP-dependent protein kinase in mouse neuroblastoma cells,[50] the increase in acid-insoluble protein in C6 glial cells during the growth-inhibitory response elicited by dexamethasone,[51] the appearance of carcinoembryonic antigen in CSF of adult brain tumor patients,[52] and the presence of progesterone receptors in meningiomas.[53]

Neither S-100 protein nor GFA protein nor myelin basic protein are tumor-associated antigens since they are abundant constituents of normal nervous system, but studies on these proteins in relation to brain tumors up to about 1979 have been reviewed by Eng.[54] More recent studies on GFA include *in vitro* synthesis studies of GFA by Bigbee *et al.* of Eng's group[55,56] and studies of the expression of GFA in human gliomas[57] and in childhood brain tumors.[58,59] S-100 distribution has also been studied in pediatric brain tumors.[60] Actin and

Table I
Polyclonal Antibodies to Brain Tumors

Name	Source	Specific	Nonspecific	References
Antimalignin antibody (immune adsorpn., cytotoxicity, *in vivo* labeling)	Patients	+		34,82–98
None (cytotoxic)	Patients	+		99
None (cytotoxic)	Patients	+		100
M5-4 (cytotoxic)	Monkey		+	101
M5-30, M5-2 (cytotoxic)	Monkey	+		101
None (cytotoxic)	Patients	+		102
None (cytotoxic)	Patients		+	103
None (cytotoxic)	Rats	+		104
None (hemadsorption)	Normals	7/200		105
None (cytotoxicity, etc.)	Patients	+		106
None (migration inhib.)	Rats	+		107
None (electron micr.)	Patients	+		108

myosin synthesis during differentiation of neuroblastomas also has been examined.[61]

Interest in brain tumor isoenzymes includes lactic dehydrogenase,[62] dopamine β-hydroxylase,[63,64] alkaline phosphatase,[65] pyruvate kinase,[66] and protein kinases.[67] The high acid proteinase activity observed by Lajtha and Marks[68] has also been observed by Qavi *et al.* of Allen's group for lysosomal brain tumor proteins.[69]

6. ANTIBODIES TO BRAIN TUMORS

There is no longer any doubt that there are antibodies generated to brain tumors. Tables I and II list some of the monoclonal (II) and polyclonal (I) antibodies that have been identified and studied. It is also clear that there are both specific and nonspecific activities represented. The nonspecific activities may be shared with other tumors or with normal adult or fetal tissues. These last are of less interest, although those shared with fetal tissues are of interest not only because of the possible light that may be cast on the relationship of the malignant process to cell "regression" (oncofetal antigens) but also because both diagnostic and therapeutic efforts in the adult may not be impeded.

It is important to note that specificity is of two types: (1) cell type and (2) specific to the process of malignancy (as distinct from the normal, hyperplastic, or other nonmalignant state). There are many examples in Tables I and II of cell-type specificity, and in several the epitope (antigenic structure) is present in more than one type of cell. The only example to date of specificity for the malignant state rather than the cell type involved is the antimalignin antibody, although it is expected that more of this type will be found. Obviously, both

Table II
Monoclonal Antibodies to Brain Tumors

Name	Specificity[a] N	M	G	E	C	S	L	A	F	References
None	+	+	+	−	−	−	−	−	−	71
Ab 165		+	+	−	−	+		−,	−	72
Ab 376	+	+	+	−	+		−	Lu	−	72
HSAN 1.2	+	−		−	−	−	−	+	+	73
Anti-Thy-1		+								74
	All clones of glioma									
Nine monoclonals	Heterogeneity									74
MA 21H12	+	+	+		−	+	−	+	+	75
MA 27A10	+	+	+		−	−	−	+	+	75
MA 22B8	+	−	+		−	−	−	−	+	75
MA 1H8cl 3	+	+	+	−	−	−	−	−	+	76
MA 1H8cl 2	+	+	+	−	−	−	−	−	+	76
MA 4D2cl 6	+	+	+	−	−	−	−	−	+	77
MA 7H10cl 4	+	+	+	−	−	−	−	−	+	77
UA1,U13A,D1H7,A2B5	+						−			78
A2B5	+						−	−		79
4.2		+						+		80[b]
UJ13A	+	+	+	+	−	−	−	+	+	81
Antimalignin antibody	+	+	+	+	+	+	+	−	−	70[c]
	General transformation antigen malignin and related other cancer recognins regardless of cell type									

[a] N, neuroblastoma; M, melanoma; G, glioma; E, neuroectodermal; C, carcinoma; S, sarcoma; L, lymphoid malignancy; A, normal adult; F, normal fetus; Lu, lung positive; +, positive for some but not necessarily all cell lines of the designated type.

[b] Preliminary characterization of antigen as "associated with the ganglioside fraction."

[c] Only monoclonal antibody for which antigen has been characterized.

types of specificity should prove useful both diagnostically and, potentially, for therapy.

This field is only just beginning, as evidenced by the fact that identification of, let alone structural information on, the antigens is scant. Methods such as immunoprecipitation with specific antibody combined with synthesis in *in vitro* systems should keep many neurochemists busy for the next several years.

The companion field of the general immunology of brain tumors with both cellular and indirect evidence of humoral mechanisms of defense has led to several dozen recent papers of interest, which are beyond the scope of this chapter. A useful recent orientation is the review by Brooks and Roszman.[109]

7. *MALIGNIN AND RELATED CANCER RECOGNINS*

The isolation and characterization of brain glycoproteins from cell surface membranes with the thought that they might be involved in the cell–cell interaction of normal "high-information" states in brain led to their examination in the "low-information" state of brain tumors in which the normal constraints

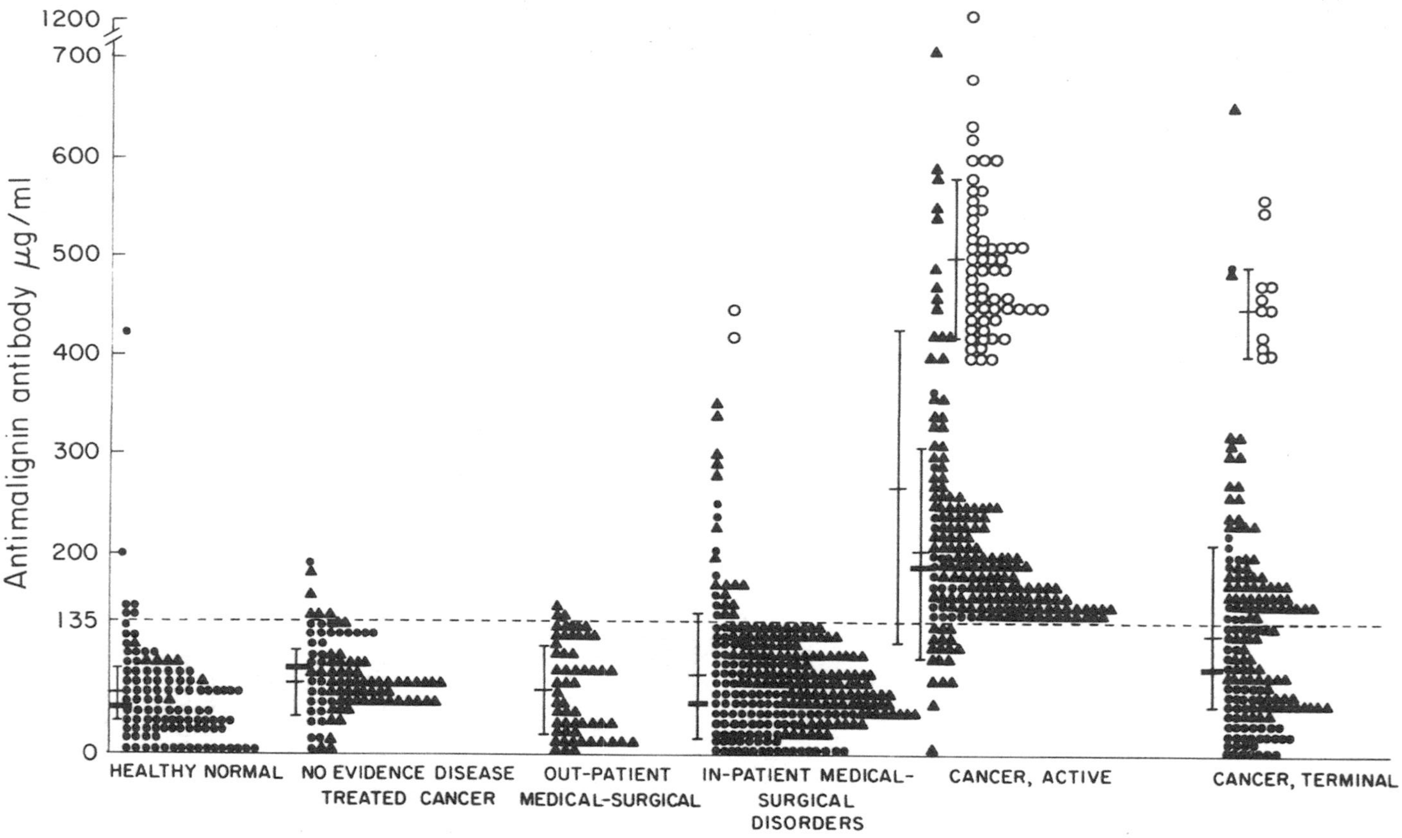

Fig. 1. Superimposition of the results of the independent blind SmithKline Clinical Laboratories (SK) study[96] on those of the earlier Bogoch et al study.[13] The concentrations of Anti-Malignin Antibody in the SK study[96] on those of the earlier Bogoch et al study.[93] The concentrations of Anti-Malignin Antibody in the SK study (●) are superimposed on those of the earlier study[93] (▲ STAG only elevated; ○ FTAG also elevated). Mean values are indicated by horizontal bars, (▬) for the present study, and (—) for the earlier study.

Table III[a]
Amino Acid Composition of Polypeptides[b] *Produced from Human Malignant Glial Tumor and from Three Different Types of Malignant Cell Grown in Tissue Culture: Residues per Molecule of Protein*

	Astrocytin (*in vivo* brain glioma)	Malignin (*in vitro*[c] brain glioma)	Recognin *M* (*in vitro*[c] mammary MCF-7 carcinoma)	Recognin L (*in vitro*[c] lymphocytic P_3J lymphoma)
Threo	5	5	5	5
Ser	6	5	5	5
½Cys	2	1	1	1
Meth	1	2	1	1
Val	4	6	6	6
Ileu	2	4	4	4
Phe	3	3	3	3
Lys	8	6	6	6
His	2	2	2	2
Arg	4	5	5	5
Asp	9	9	9	8
Glu	13	13	11	10
Leu	8	8	8	7
Tyr	2	3	2	1
Pro	4	4	4	5
Gly	6	6	9	13
Ala	9	7	9	10
Total number of residues	88	89	90	92
Molecular weight calculated	9,690	10.067	9,870	9,606

[a] From Bogoch and Bogoch.[90]
[b] Specimens were hydrolyzed *in vacuo* with 6 N HCl at 108°C for 12 hr. The nearest integer for the mole number of each amino acid is the average of two separate determinations. All above determinations were performed "blind" by the Boston University Medical Center central facility for amino acid analysis. Results obtained with 24-hr and 72-hr hydrolysis of malignin at two additional laboratories, respectively, were not significantly different when serine, threonine, tyrosine, and cysteine were corrected for additional destruction by acid.
[c] Cells grown in tissue culture.

limiting cell division and interaction did not pertain. As pointed out in Section 5, this led to the isolation of tumor glycoprotein 10B, as its constituent cleaved 10,000-dalton polypeptide astrocytin, and, subsequently, from tissue culture of malignant glial cells, the closely related product malignin.[32–34,43,70,82–96]

When malignin was bound to bromacetylcellulose as an inert carrier, affinity binding made it possible to isolate and quantitate its elevated antibody in immunized animals and brain tumor patients' serum.[84] It soon became apparent, however, that antimalignin antibody was elevated not only in brain tumors but in every malignancy tested.[88,91–96]

This finding made it necessary to examine other cancer cells to see if malignin could be demonstrated by immunofluorescent means[84,97,98] and by extraction.[90] The peptides that were extracted from other than brain glial tumor cells were neither identical to malignin nor totally different from it but, quite

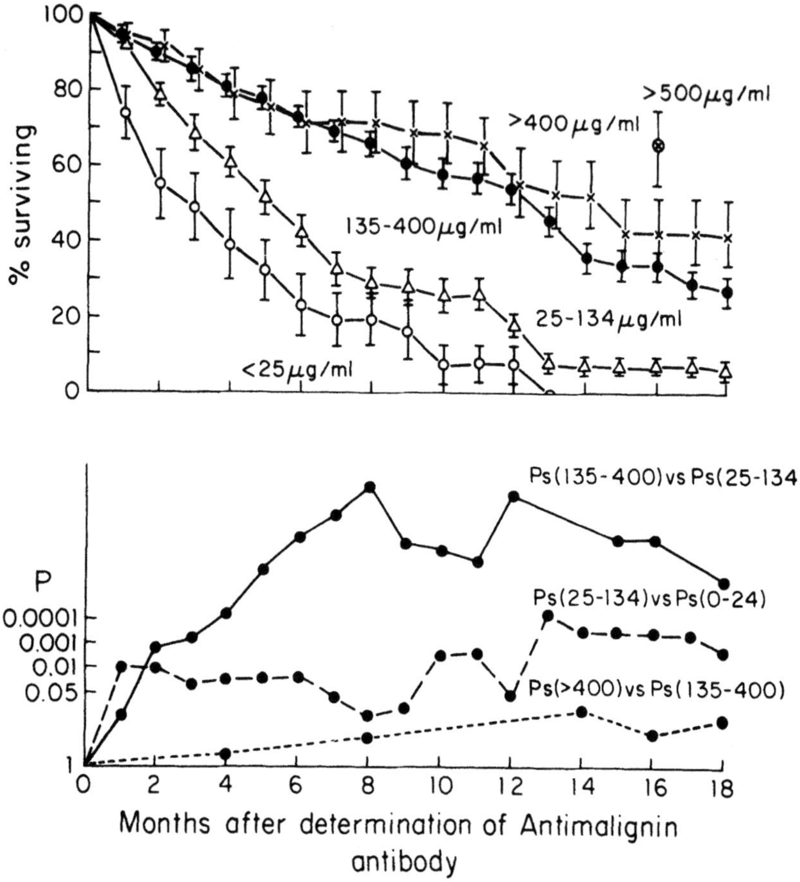

Fig. 2. Actuarial survival of 511 cancer patients grouped according to their concentration of antimalignin antibody: 0–25 μg/ml ○, 25–134 μg/ml △, 135–400 μg/ml ●, >400 μg/ml ×, and the subgroup of those above 400 whose antibody concentration was above 500 μg/ml ⊗ (combined series I and II). *P*, the probability, indicates the level of significance of the difference between the probabilities of survival, *Ps*, by month for each of the pairs of groups being compared, e.g., Ps(135–400) *vs*. Ps(25–134). See text for further details. (From Bogoch *et al.*[96])

unexpectedly, formed a family of closely related peptides (Table III).[90] The sequence of these substances, now being studied, should give information on the shared epitopes.

Antimalignin antibody has been independently shown in four blind studies involving 1,878 cancer patients and controls to be elevated in 93% of active cancer patients, with a 5.2% overall incidence of asymptomatic or false-positive results (Fig. 1). The quantitative determination of the antibody can therefore be of value as a diagnostic aid.[88,93,96]

The level of the antibody is quantitatively related to survival (Fig. 2).[96] Also, in successfully treated cancer patients with no further evidence of disease, antimalignin antibody returns to normal levels in 97% of cases.[88,93,96] The antibody can therefore be useful in monitoring the effect of treatment and patient progress.

REFERENCES

1. Jones, E. L., Searle, C. E., and Smith, W. T., 1973, *J. Pathol.* **109:**123.
2. Margison, G. P., and Kleihues, P., 1975, *Biochem. J.* **148:**521.
3. Lantos, P. L, 1980, *Brain Tumours* (D. G. T. Thomas and D. I. Graham, eds.), Butterworths, London, pp. 85–108.
4. Bullard, D. E., and Bigner, D. D., 1980, *Brain Tumors* (D. G. T. Thomas and D. I. Graham, eds.), Butterworths, London, pp. 51–84.
5. Bressler, J. P., Cole, R., and de Vellis, 1983, *Cancer Res.* **43:**709–715.
6. Shapiro, J. R., Yung, W. K. A., and Shapiro, W. R., 1981, *Cancer Res.* **41:**2349–2359.
7. Knebel, K., and Hoshino, T., 1980, *Proc. Am. Assoc. Cancer Res.* **21:**56.
8. Liwnicz, B. H., 1981, *J. Neuropathol. Exp. Neurol.* **40:**337.
9. Dille, B. J., McGee, J. E., and Johnson, T. C., 1982, *J. Neurochem.* **38:**422–426.
10. Will, H., Benenson, A., Devilliers, G., and Mandel, P., 1982, *J. Neurochem.* **39:**924–932.
11. Ko, L., and Koestner, A., 1980, *J. Natl. Cancer Inst.* **65:**1017–1027.
12. Black, P. M., Kornblith, P. L., Davison, P. F., Liszczak, T. M., Merk, L. P., Smith, B. H., McKeever, P. E., and Quindlen, E. A., 1982, *J. Neurosurg.* **56:**62–72.
13. Keller, K., Lange, K., and Noske, W., 1981, *J. Neurochem.* **36:**1012–1017.
14. Shifrin, G. S., and Klein, W. L., 1980, *J. Neurochem.* **34:**933–999.
15. Kimhi, Y., Mahler, A., and Saya, D., 1980, *J. Neurochem.* **34:**554–559.
16. Reiser, G., Scholz, F., and Hamprecht, B., 1982, *J. Neurochem.* **39:**228–234.
17. Kurzinger, K., and Hamprecht, B., 1981, *J. Neurochem.* **37:**956–967.
18. Glaser, T., Hubner, K., and Hamprecht, B., 1982, *J. Neurochem.* **39:**59–69.
19. Glaser, T., Hubner, K., Castiglione, R. de, and Hamprecht, B., 1981, *J. Neurochem.* **37:**1613–1617.
20. Brandt, M., Buchen, C., and Hamprecht, B., 1980, *J. Neurochem.* **34:**643–651.
21. Haga, T., and Haga, K., 1981, *J. Neurochem.* **36:**1152–1159.
22. Green, D. A., and Clark, R. B., 1982, *J. Neurochem.* **39:**1125–1131.
23. Spinelli, W., and Ishii, D. N., 1983, *Proc. Am. Assoc. Cancer Res.* **24:**10.
24. Feder, M. K., and Gilbert, F., 1983, *J. Natl. Cancer Inst.* **70:**1051–1056.
25. Mork, S. J., and Laerum, O. D., 1980, *J. Neuarsurg.* **53:**198–204.
26. Dierich, A., Wintzerith, M., Sarlieve, L., and Mandel, P., 1980, *J. Neurochem.* **34:**(5):1126–1129.
27. Morrison, M. M., Hall, C. L., Pardue, S., Brodueur, R., Baskin, F., and Rosenberg, R. N., 1980, *J. Neurochem.* **34:**50–58.
28. Marton, L., Edwards, M. S., Levin, V. A., Lublich, W. P., and Wilson, C. B., 1979, *Cancer Res.* **39:**993–997.
29. Fulton, D. S., Levin, V. A., Lublich, W. P., Wilson, C. B., and Marton, L. J., 1980, *Cancer Res.* **40:**3293–3296.
30. Harik, S. I., and Sutton, C. H., 1979, *Cancer Res.* **39:**5010–5015.
31. Bogoch, S., 1969, *The Biochemistry of Memory with an Inquiry into the Function of the Brain Mucoids*, Oxford University Press, New York.
32. Bogoch, S., 1971, *Abstracts 3rd International Meeting International Society of Neurochemistry*, Pergamon Press, Budapest. p. 101.
33. Bogoch, S., 1972, *Functional and Structural Proteins of the Nervous System* (A. N. Davison, P. Mandel, and I. G. Morgan, eds.), Plenum Press, New York, pp. 39–52.
34. Bogoch, S., 1975, *Current Trends in Sphingolipidoses and Allied Disorders* (B. W. Volk and L. Schneck, eds.), Plenum Press, New York, pp. 555–566.
35. Liepkalns, V. A., Icard, C., Yates, A. J., Thompson, D. K., and Hart, R. W., 1981, *J. Neurochem.* **36:**1959–1965.
36. Srinivas, S., and Colburn, N. H., 1982, *J. National Cancer Inst.* **68:**469–473.
37. McLawhon, R. W., Schoon, G. S., and Dawson, G., 1981, *J. Neurochem.* **37:**132–139.
38. Bremer, E. G., Sapirstein, V. S., Savage, T., and McCluer, R. H., 1982, *J. Neurochem.* **38:**333–341.
39. Helson, L., Helson, P., and Sordillo, P., 1979, *Proc. Am. Assoc. Cancer Res.* **20:**248.

40. Yates, A. J., Stephens, R. E., Elder, P. E., and Rice, J. R., 1981, *J. Neuropathol. Exp. Neurol.* **40**:302.
41. Chou, K.-H., and Jungalwala, F. B., 1981, *J. Neurochem.* **36**:394–401.
42. Maltese, W. A., Reitz, B. A., and Volpe, J. J., 1981, *Cancer Res.* **41**:3448–3452.
43. Bogoch, S., Belval, P. C., Sweet, W. H., Sacks, W., and Korsch, G., 1968, *Protides of Biological Fluids* (H. Peeter, ed.), Elsevier, Amsterdam, p. 129.
44. Lubitz, W., Westermark, B., and Peterson, P. A., 1980, *Int. J. Cancer* **25**:53–58.
45. McKeever, P. E., Quindlen, E., Banks, M. A., Williams, U., Kornblith, P. L., Laverson, S., Greenwood, M. A., and Smith, B. A., 1981, *Neurology (N.Y.)***31**:1445–1452.
46. Shitara, N., McKeever, P. E., Smith, B. H., Pleasants, R. E., Banks, M. A., and Kornblith, P. L., 1982, *J. Neurochem.* **39**:948–953.
47. Heikinheimo, M., Paasivuo, R., and Whlstrom, T., 1981, *Br. J. Cancer* **43**:654.
48. Peiper, S., Komada, Y., Melvin, S., and Tarnowski, S. S., 1983, *Proc. Am. Assoc. Cancer Res.* **24**:221.
49. Rosen, S. W., Calvert, I., Weintraub, B. D., Tseng, J. S., and Rabson, A. S., 1980, *Cancer Res.* **40**:4325–4328.
50. Prashad, N., Evetts, C., and Wischmeyer, B., 1980, *J. Neurochem.* **35**(1):38–46.
51. Grasso, R. J., Holbrook, N. J., and Woodzinski, S. F., 1981, *J. Neurochem.* **37**:515–517.
52. Hill, S., Martin, E., Ellison, E. C., and Hunt, W. E., 1980, *J. Neurosurg.* **53**:627–632.
53. Vaquero, J., Marcos, M. L., Martinez, R., and Bravo, G., 1983, *Surg. Neurol.* **19**:11–13.
54. Eng., L. F., 1980, *Brain Tumors* (D. G. T. Thomas and D. I. Graham, eds.), Butterworths, London, pp. 109–120.
55. Bigbee, J. W., and Eng, L. F., 1982, *J. Neurochem.* **38**:130–134.
56. Bigbee, J. W., Bigner, D. B., and Eng, L. F., 1983, *J. Neuropathol. Exp. Neurol.* **42**:80–86.
57. Jones, T. R., Ruoslahti, E., Schold, S. C., and Bigner, D. D., 1983, *Cancer Res.* **42**:168–177.
58. Marsden, H. B., Kumar, S., Kahn, J., and Anderton, B. J., 1983, *Int. J. Cancer* **31**:439–445.
59. Roessmann, U., Velasco, M. E., Gambetti, M. D., and Autilio-Gambetti, L., 1983, *J. Neuropathol. Exp. Neurol.* **42**:113–121.
60. Nakamura, Y., Becker, L. E., and Marks, A., 1983, *J. Neuropathol. Exp. Neurol.* **42**:136–145.
61. Rein, D., Gruenstein, E., and Lessard, J., 1980, *J. Neurochem.* **34**:1459–1469.
62. Vivekanandan, S., Rao, A. P. K., Selvam, R., and Kanaka, T. S., 1982, *Acta. Neurol. Scand.* **66**:347–354.
63. Fraeyman, N. H., Van de Velde, E. J., Smet, F. H. D., and De Schaepdryver, A. F., 1982, *J. Neurochem.* **39**:1179–1184.
64. Eldeeb, B. B., Burns, S., Robinson, R., Hammond, E. M., and Mann, J. R., 1983, *Br. J. Cancer* **47**:115–121.
65. Takahara, N., Herz, F., Singer, R. M., Hirano, A., and Koss, L. G., 1982, *Cancer Res.* **42**:563–568.
66. Veelen, C. W. M. van, Herbiest, H., Zulch, K. J., Ketel, B. A. van, Vlist, M. J. M. van der, Vlug, A. M. C., Rijksen, G., and Staal, G. E. J., 1979, *Cancer Res.* **39**:4263–4269.
67. Imashuku, S., Fossett, M., Green, A., and Hayes, A., 1979, *Proc. Am. Assoc. Cancer Res.* **20**:195.
68. Lajtha, A., and Marks, N., 1969, *Dis. Nerv. Syst.* [*Suppl.*] **30**:36.
69. Qavi, H. B., Allen, N., and Clendenon, N. R., 1982, *Eur. J. Cancer Clin. Oncol.* **18**:463–469.
70. Bogoch, S., Bogoch, E., and Tsung, Y.-K., 1981, *Lancet* **1**:141–142.
71. Liao, S. K., Clarke, B. J., Gallie, B. L., and Dem, P. B., 1981, *Proc. Am. Assoc. Cancer Res.* **22**:282.
72. Seeger, R. C., Rosenblatt, H. M., Imai, K., and Ferrone, S., 1981, *Cancer Res.* **41**:2714–2717.
73. Reynolds, C. P., and Smith, R. G., 1981, *Proc. Am. Assoc. Cancer Res.* **22**:402.
74. Wilkstrand, C. J., Bigner, S. H., and Bigner, D. D., 1983, *Cancer Res.* **43**:3327–3334.
75. Bourdon, C. S., Wilkstrand, C. J., Coleman, R. E., and Bigner, D. D., 1982, *J. Neuropathol. Exp. Neurol.* **41**:367.

76. Wilkstrand, C. J., Bourdon, M. A., Pegram, C. N., and Bigner, D. D., 1982, *J. Neuroimmunol.* **3**:43–62.
77. Wilkstrand, C. J., and Bigner, D. D., 1982, *Cancer Res.* **42**:267–275.
78. Kemshead, J. T., Greaves, M. F., Walsh, F., Chayen, A., and Parkhouse, M., 1981, *Proc. Am. Assoc. Cancer Res.* **22**:399.
79. Kemshead, J. T., Walsh, F., Pritchard, J., and Greaves, M., 1981, *Br. J. Cancer* **27**:447–452.
80. Yeh, M. Y., Hellstrom, I., Abe, K., Hakomori, S., and Hellstrom, K. E., 1982, *Br. J. Cancer* **29**:269–275.
81. Allan, P. M., Garson, J. A., Harper, E. I., Asser, U., Coakham, H. B., Brownell, B., and Kemshead, J. T., 1983, *Int. J. Cancer* **31**:591–598.
82. Bogoch, S., 1974, *Biological Diagnosis of Brain Disorders* (S. Bogoch, ed.), Spectrum-Halstead-Wiley, New York, pp. 358–361.
83. Bogosh, S., 1976, *Trans. Am. Soc. Neurochem.* **7**:239.
84. Bogoch, S., 1977, *Natl. Cancer Inst. Monogr.* **46**:133–137.
85. Bogoch, S., 1977, *Mechanisms, Regulation and Special Functions of Protein Synthesis in the Brain* (S. Roberts, A. Lajtha, and W. H. Gispen, eds.), Elsevier, Amsterdam, New York, pp. 433–440.
86. Bogoch, S., 1977, *Behavioral Neurochemistry* (J. M. R. Delgado and F. V. DeFeudis, eds.), Spectrum–Wiley, New York, pp. 197–222.
87. Bogoch, S., and Bogoch, E., 1979, *Trans. Am. Soc. Neurochem.* **10**:78.
88. Bogoch, S., Bogoch, E., Fager, C. A., Goldensohn, E. S., Harris, J. H., Hickock, D. F., Lowden, J. A., Lux, W. E., Ransohoff, J., and Walker, M. D., 1979, *Neurology (Minneap.)* **29**:584.
89. Bogoch, S., and Bogoch, E. S., 1979, *Lancet* **1**:987.
90. Bogoch, S., and Bogoch, E. S., 1979, *Neurochem. Res.* **4**:467–473.
91. Bogoch, S., and Bogoch, E. S., 1980, *Neurochemistry and Clinical Neurology* (L. Battistin, G. Hashim, and A. Lajtha, eds.), Alan R. Liss, New York, pp. 407–424.
92. Bogoch, S., and Bogoch, E. S., 1980, *Perspectives of Immunology* (S. A. Rosenberg, ed.), Academic Press, New York, pp. 693–696.
93. Bogoch, S., Bogoch, E., Fager, C. A., Harris, J. H., Ambrus, J. L., Lux, W. E., and Ransohoff, J. A., 1982, *J. Med.* **13**:49–69.
94. Bogoch, S., and Bogoch, E. S., 1983, *Protides of the Biological Fluids*, Vol. 30 (H. Peeters, ed.), Pergamon Press, New York, pp. 337–352.
95. Bogoch, S., Bogoch, E. S., Fager, C. A., Harris, J. H., Ambrus, J. L., Lux, W. E., and Ransohoff, J. A., 1982, *Abstract 13th International Cancer Congress*, Intl. Union Against Cancer, Seattle, p. 5.
96. Bogoch, S., Bogoch, E. S., Antich, P., Dungan, S. M., Harris, J. H., and Ambrus, J. L., 1984, *Protides of the Biological Fluids*, Vol. 31 (H. Peeters, ed.), Pergamon Press, New York, p. 739–747.
97. Harris, J. H., Gohara, A., Redmond, F., Bogoch, S., and Bogoch, E. S., 1980, *Perspectives of Immunology*: Serologic Analysis of Human Cancer Antigens, (S. A. Rosenberg, ed.), Academic Press, New York, pp. 571–582.
98. Redmond, F. A., Harris, J. H., Loeb, T. L., Bogoch, S., Bogoch, E. S., and Gohara, A., 1980, *Fed. Proc.* **39**:462.
99. Kornblith, P. L., Pollock, L. A., Coakham, H. B., Quindlen, E. A., and Wood, W. C., 1979, *J. Neurosurg.* **51**:47–52.
100. Coakham, H. B., Kornblith, P. L., Quindlen, E. A., Pollick, L. A., Wood, W. C., and Hartnett, L. C., 1980, *J. Natl. Cancer Inst.* **64**:223–233.
101. Wilkstrand, C. J., and Bigner, D. D., 1979, *Cancer Res.* **39**:3235–3243.
102. Mitchell, M. S., Lerner, A. B., Hebert, G. Z., and Tonsgard, J. H., McIntosh, S., and Slater, J., 1979, *Proc. Am. Assoc. Cancer Res.* **20**:331.
103. Martin-Achard, A., Diserens, A.-C., Tribolet, N. De, and Carrel, S., 1980, *Int. J. Cancer* **25**:219–224.
104. Stavrou, D., Hulten, M., Anzil, A. P., and Bilzer, T., 1980, *Int. J. Cancer* **26**:629–637.
105. Pfreundschuh, M., Rohrich, M., Piotrowski, W., Berlit, P., and Penzholz, H., 1982, *Int. J. Cancer* **29**:517–521.

106. Sevadei, F., Parente, R., Spagnolli, F., Gaist, G., Radovni, R., Bucci, M.,and Steiner, L., 1982, *Acta Neurochir.* **60**:71–80.
107. Oda, Y., and Tokuriki, Y., Handa, H., and Kieler, J., 1980, *Neurol. Med. Chir.* (*Tokyo*) **20**:651–658.
108. Lauro, G. M., Solheid, C., Medolago-Albani, L., Lorenzo, N. di, and Guidetti, B., 1983, *Acta Neuropathol.* (*Berl.*) **59**:127–132.
109. Brooks, W. H., and Roszman, T. L., 1980, *Brain Tumors* (D. G. T. Thomas and D. I. Graham, eds.), Butterworths, London, pp. 121–132.

5

Lipidoses

Roscoe O. Brady

1. INTRODUCTION

Following the elucidation of the enzymatic defects in the sphingolipid storage disorders in the late 1960s and early 1970s, the predictions concerning the development of rapid diagnostic and carrier detection tests and the ability to monitor pregnancies at risk for any of these disorders[1] have now been fulfilled. Accurate genetic counseling is in world-wide use, and its application has led to the control of the incidence of many of these life-threatening conditions.[2] Current research in sphingolipids is evolving in three primary directions. The first is the elucidation of pertinent aspects of the pathological biochemistry that underlies these conditions. The second is devoted to discovering the genetic alterations in diseases. The third is the development of effective therapy for these disorders. Because the medical aspects of the sphingolipidoses have been reviewed so frequently and these summaries are readily available,[1,3–5] specific clinical manifestations of the various disorders are indicated only where required to indicate an approach to the abnormal biochemistry, genetic mechanism, or strategy for treating these conditions.

2. PATHOLOGICAL BIOCHEMISTRY

2.1. Gaucher's Disease

This disorder is the most frequently encountered metabolic storage disorder. It is caused by insufficient activity of glucosylceramide-β-glucosidase (glucocerebrosidase). Patients with this disorder have been divided into three categories. Most patients, classified as type 1 Gaucher's disease, exhibit hepatosplenomegaly and bone involvement. Although this category was previously called the adult form, clinical manifestations may be evident in the very early years. A principal characteristic is the absence of CNS involvement. Patients

Roscoe O. Brady • Developmental and Metabolic Neurology Branch, National Institute of Neurological and Communicative Disorders and Stroke, National Institutes of Health, Bethesda, Maryland 20205.

with organomegaly and rapidly progressing brain damage are classified as type 2 (infantile). Blindness, deafness, opisthotonus, progressive dementia, and death between 12 and 18 months are characteristic of this form of Gaucher's disease. Patients with later onset of brain involvement comprise type 3 (juvenile). These patients generally do not have indications of brain damage until their mid- to late teens, although exceptions have been reported. Characteristically, myoclonic seizures become progressively frequent and increasingly difficult to control with antiepileptic medication, and there is gradually but inexorably worsening mentation.

Considerable difficulty has been encountered in attempting to obtain a satisfactory understanding of the neurological damage in type 2 and 3 patients. Depletion of neurons and neuronophagia by phagocytic cells has frequently been described in type 2 patients.[6,7] Accumulation of lipid deposits in neurons is not a characteristic of this condition. With time, collections of rather typical "Gaucher cells" in the Virchow–Robin spaces become apparent. In type 3, the latter manifestation has been described as the dominant histopathological finding. These considerations have led to the speculation that the glucocerebroside that causes neuronal destruction is derived from turnover of gangliosides, whereas the material stored in the Virchow–Robin spaces is glucocerebroside that arises primarily from leuko- and erythocytorrhexis.[8] Support for this contention is derived from analysis of the fatty acid patterns in glucocerebroside in the two forms. It has been reported that the dominant fatty acid in glucocerebroside from type 2 patients is 18 carbon atoms in length whereas in type 3 there are more C_{22} and C_{24} fatty acids.[8] The fact that fatty acids in gangliosides are characteristically stearic acid and those in neutral sphingolipids in red cells center around C_{24} seems consistent with this deduction.

Nilsson and Svennerholm have also reported that patients with types 2 and 3 Gaucher's disease accumulate striking quantities of glucosylsphingosine (psychosine) in the brain.[8] This material is said to be cytotoxic, and it is currently assumed to play a significant role in the pathogenesis of the CNS damage. It is presumed that glucosylsphingosine is synthesized from UDP-glucose and sphingosine analogous to the reaction for the formation of galactosylsphingosine.[9] Its accumulation is caused by the deficiency of glucocerebrosidase, which also catalyzes the hydrolysis of glucosylsphingosine but at about 1/200 the rate of glucocerebroside.[10]

Although this hypothesis is attractive, one must also take into account the fact that glucocerebroside itself is a toxic substance that has been shown to stimulate the secretion of lymphocyte-activating factor by microphages and release of lysosomal enzymes from these cells.[11] Thus, one cannot determine the relative contribution of it or glucosylsphingosine to neurocytotoxicity at this time. This statement seems particularly relevant since we have found a close correlation of the neurological signs with the distribution of accumulated glucocerebroside in various regions of the brain (J. A. Barranger, M. D. Ullman, E. Wilson, and R. O. Brady, unpublished data).

2.2. *Krabbe's Disease*

This disorder is caused by insufficient activity of galactocerebroside-β-galactosidase. Hyperirritability, gradual onset of blindness, deafness, and progressive mental retardation are characteristic of the disorder. Nests of "globoid bodies" appear in the white matter of the brain. Although it has not been possible to demonstrate a net accumulation of galactocerebroside, Vanier and Svennerholm have reported increased quantities of galactosylsphingosine (psychosine) in brains from patients with Krabbe's disease.[12,13] The reported cytotoxicity of galactosylsphingosine[14,15] seems consistent with some of the neuropathology in this disorder. However, once again, it is not possible at this time to discriminate completely between the damage caused by psychosine and galactocerebroside, since the latter compound has been shown to induce the formation of "globoid bodies" characteristic of this disorder.[16–18] It is conceivable that in time the disparate reactions will be identified. Thus, psychosine may be intimately involved in the loss of neurons whereas galactocerebroside induces reactive proliferation of cells that become globoid bodies. However, this delineation does not seem to be applicable to the pathogenesis of Gaucher's disease.

2.3. Neimann–Pick Disease

One would imagine that similar investigations might be undertaken in Niemann–Pick disease. Here the accumulation of sphingomyelin is caused by deficiency of sphingomyelinase activity.[19] Patients with this disorder have also been subdivided into various clinical classes depending on the presence or absence of CNS involvement and on the time of onset of the manifestations of the disorder.[20] Again, organomegaly and, in most patients, brain damage are the dominant signs of this disease. It has been difficult to demonstrate specific toxic reactions to sphingomyelin, probably because exogenous administration of this lipid to Niemann–Pick cells in culture does not result in its accumulation because of its hydrolysis by nonlysosomal isozymes of sphingomyelinase.[21] One wonders whether the deacylated compound sphingosylphosphorylcholine also accumulates in Niemann–Pick disease and contributes to cellular damage.

2.4. Generalized (G_{MI}) Gangliosidosis

This sphingolipid storage disorder is caused by deficiency of the β-galactosidase that initiates the breakdown of monosialotetrahexosyl gangliosides. Patients have been divided into infantile, juvenile, and adult categories called types 1–3, respectively. Mental and motor retardation are constant features, and 50% of type 1 patients have a cherry-red spot in the macula. Hepatosplenomegaly is also found in type 1. Ganglioside G_{M1} accumulates in neurons in the form of multilamellated cytoplasmic bodies.[22]

An impressive correlation between ganglioside accumulation and pathological manifestations has been reported by Purpura and his associates.[23] In particular, they have identified a particular trio of responses to ganglioside that have been most thoroughly studied in a feline analogue of human G_{M1} gangliosidosis. These reactions include meganeurite formation, secondary outgrowth of neurites, and aberrant differentiation of dendrites. These investigators emphasize that only certain neurons appear to respond in this fashion. Even so, the stimulation of membrane synthesis in this disorder seems in sharp contrast with the conditions discussed previously, where the accumulating sphingolipids were strongly cytotoxic. Ganglioside G_{M1} has long been implicated as an important receptor in plasma cell membranes for external effector molecules.[24] The most widely studied effects have been cellular responses to cholera toxin. When G_{M1} is added to ganglioside-deficient cells, the biochemical changes are proportional to the quantity of supplemental G_{M1}.[25] It therefore seems likely that increasing the quantity of this receptor would enhance the responsiveness of such "plastic" neurons to a nerve growth factor involved in neurite formation.

3. *GENETIC ALTERATIONS*

3.1. *Proteins That Cross React with Antibodies Raised against Sphingolipid Hydrolases*

3.1.1. *Gaucher's Disease*

3.1.1a. Detection of Cross-Reacting Material (CRM) with Monospecific Polyvalent Antibody. Although mutations of the primary structure of enzymes in metabolic storage disorders have long been suspected to be the major causes of the decreased catalytic activity, to date no such specific change has been demonstrated in any of the sphingolipid storage disorders. The most instructive information concerning fundamental changes in a sphingolipid hydrolase has just become available in Gaucher's disease. Investigations with monospecific polyvalent antibody raised against homogeneous human placental glucocerebrosidase have revealed specific mutations of glucocerebrosidase in the various clinical types of this disorder.[26] In normal mature cells such as confluent cultures of skin fibroblasts, glucocerebrosidase exists in three isozyme forms. The apparent molecular weights are 63,000, 61,000, and 56,000, with the last being the dominant species. These cross-reacting materials are readily detected by immunoblotting after electrophoresis of proteins in tissue extracts (Fig. 1).

Patients with type 1 Gaucher's disease show CRM to the three isozymes, but the strikingly diminished catalytic activity of glucocerebrosidase in these cells is conclusive identification of the disorder. Cells from patients with types 2 and 3 Gaucher's disease with early and late onset of CNS damage, respectively, show only the larger isozyme forms. Other experiments indicate that the 56,000 form is the result of processing of the 63,000 precursor molecule.[27] Thus, in patients with neurological forms of Gaucher's disease, not only is there mutation of the catalytic site causing diminished glucocerebrosidase activity, but the altered enzyme cannot be properly converted to the mature 56,000 isozyme through proteolysis or modification of the oligosaccharide moieties. The importance of this impediment appears to be twofold. First, the 56,000-dalton isozyme is the major isozyme in normal mammalian brain tissue.[26] Second, this species appears to be the form that is present in lysosomes and therefore is appropriately situated to carry out its intracellular catalytic function.[28]

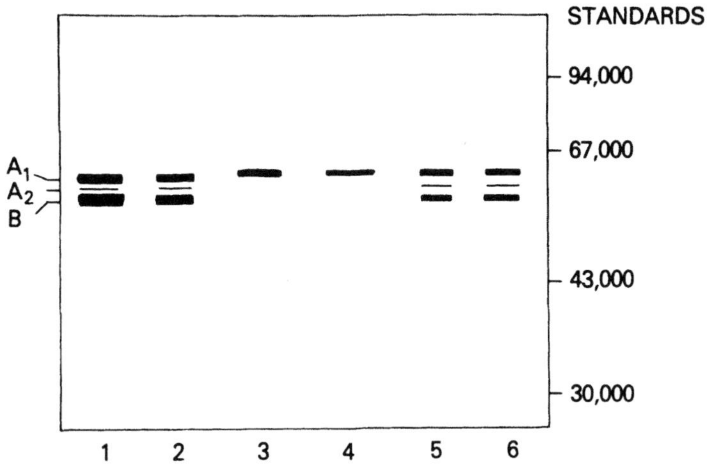

Fig. 1. Radioimmunoblotting patterns of glucocerebrosidase isozymes in extracts of cultured skin fibroblasts using monospecific antibody against homogeneous placental glucocerebrosidase. Lane 1, normal control; lanes 2–4 from types 1, 2, and 3 Gaucher's disease, respectively; lanes 5 and 6 from types 2 and 3 Gaucher heterozygotes.

It therefore appears that (1) the mutated glucocerebrosidase in the neuronopathic forms of Gaucher's disease is prevented from exhibiting a necessary lysosome-targeting signal such as mannose-6-phosphate[29] or other molecule(s)[30,31] because of impaired processing of the precursor protein, or (2) it may be incapable of acquiring them because of a mutation in the genetic code at a critical asparagine residue to which the necessary oligosaccharide is linked. These abnormalities appear to have some resemblance to, but are distinct from, the defect in I-cell disease, in which the mannose-6-phosphate signal is not generated because of deficient activity of UDP-N-acetylglucosaminylphosphate transferase.[32,33]

3.1.1b. Investigations with Monoclonal Antibodies. An extraordinarily important development concerning the molecular basis of these hereditary diseases has resulted from studies with monoclonal antibodies raised against placental glucocerebrosidase.[34] Using monoclonal antibody 8E4 in an immunoblotting procedure, we observed that this antibody reacted with a specific epitope that is present on the mutated glucocerebrosidase in patients with type 3 Gaucher's disease but not with the isozymes from type 2 Gaucher cells[35] (Fig. 2). This discovery permits identification of the precise phenotype in all three forms of Gaucher's disease. The test can be performed on uncultured skin biopsy specimens.[27] Furthermore, it is highly likely that this discrimination will also be possible using extracts of amniotic cells that have been grown to confluency. This accurate identification of Gaucher phenotypes provides a high degree of sophistication and added enlightenment concerning genetic counseling in Gaucher's disease. It is assumed that similar techniques using polyvalent and monoclonal antibodies directed against homogeneous human sphingolipid hydrolases will be applicable to many other hereditary metabolic

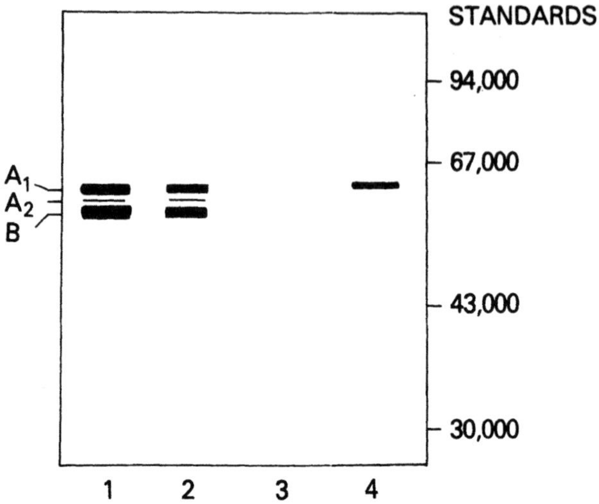

Fig. 2. Radioimmunoblotting patterns of glucocerebrosidase isozymes in extracts of cultured skin fibroblasts using monoclonoal antibody 8E4. Lane 1, normal control; lane 2 Gaucher's disease type 1; lane 3 Gaucher's disease type 2; lane 4 Gaucher's disease type 3.

disorders for which there is wide variability in the clinical presentation. This approach appears particularly likely to provide the ability to discriminate among the various clinical forms of Niemann–Pick disease and metachromatic leukodystrophy. A whole new era of diagnostic precision appears imminent on the basis of these demonstrations.

3.1.2. Tay–Sachs Disease (B-Variant)

This lipid storage disorder whose primary manifestations are a cherry-red spot in the macula and progressively severe brain damage is caused by a deficiency of one of the two hexosaminidase isozymes known as Hex A.[36] These isozymes are characteristically assayed using 4-methylumbelliferyl or *p*-nitrophenyl conjugates of N-acetylglucosamine or N-acetylgalactosamine, since the enzymes exhibit relaxed specificity with regard to the anomeric configuration on carbon 4 of these aminosugars. Although Hex A activity is drastically reduced in tissues from these patients, Hex B is increased manyfold over normal. Hexosamimidase A is the isozyme that is required for the hydrolytic cleaveage of N-acetylgalactosamine from the accumulating ganglioside G_{M2}.[37]

Hexosaminidase A is comprised of protein subunits called α and β that are coded by genes on chromosomes 15 and 5, respectively.[38,39] Considerable controversy exists at this time concerning the presence or absence of CRM in tissues from Tay–Sachs patients. Several groups of investigators reported that they were unable to demonstrate CRM to hexosaminidase A in extracts of tissues from patients with Tay–Sachs disease.[40–42] However, Srivastava and co-workers subsequently reported that hexosaminidase A CRM was detectable in liver extracts from Tay–Sachs patients if they cross linked the hexosaminidase A that was used as antigen with glutaraldehyde.[43] This report requires confirmation.

Considerable effort has been devoted to investigations of the synthesis and processing of the α and β hexosaminidase polypeptides. The α chain is initially synthesized as a protein with an M_r of 67,000 and is then trimmed to an M_r of 54,000. The nascent β chain has an M_r of 63,000 and is cleaved to an M_r of 29,000 and smaller fragments.[44–46] The β chain polypeptides have been reported to be slightly dissimilar in size,[47] and the current proposed structure of Hex A is $\alpha\beta_a\beta_b$. The structure of hexosaminidase B may be $2(\beta_a\beta_b)$. Cultured skin fibroblasts from five typical Tay–Sachs patients of Ashkenazic Jewish ancestry did not contain immunoprecipitable α chain, although this chain was discovered in normal quantity in fibroblasts from a non-Jewish Tay–Sachs infant that were extracted with sodium dodecylsulfate.[48] It is presumed that the mutation in the latter family is distinct from that in the usual Tay–Sachs patients, where there may be an absence of translatable messenger RNA or α polypeptide is formed that is not immunoprecipitable.

3.1.3. Sandhoff's Disease (O-Variant Form of Tay–Sachs Disease)

Both hexosaminidase A and B isozymes are deficient in this form of Tay–Sachs disease because of the absence of β chains. To date, no CRM to β chain has been demonstrated in cells from Sandhoff patients. The progression of the clinical signs and mental deterioration is more rapid in patients with Sandhoff's disease than in classic Tay–Sachs disease. Considerably more asialo-G_{M2} (N-acetylgalactosaminylgalactosylglucosylceramide) accumulates in the brain of these infants than in conventional Tay–Sachs disease. This observation is consistent with the demonstration that there are two routes for ganglioside G_{M2} catabolism in normal brain.[49,50] One pathway is initiated by the cleavage of the molecule of N-acetylgalactosamine by hexosaminidase A, forming ganglioside G_{M3}. The second route involves the initial hydrolysis of sialic acid, forming asialo-G_{M2}. Both hexosaminidase A and hexosaminidase B catalyze the cleavage of N-acetylgalactosamine from asialo-G_{M2}.

3.1.4. Generalized (G_{M1}) Gangliosidosis

In contrast with the situation in Tay–Sachs disease, all of the tissues from patients with generalized (G_{M1}) gangliosidosis that have been tested have shown to have normal or supranormal quantities of CRM to anti-β-galactosidase, indicating the synthesis of a structurally altered enzyme.[22] The gene coding for this protein is on chromosome 3. The mutated enzyme in some of these patients may have diminished affinity for ganglioside G_{M1}.[51] Although K_m alterations have been reported in other situations,[52,53] it is difficult to predict how common such a change will be in the sphingolipidoses, since this was not found to be the case when mutated glucocerebrosidase was isolated from the spleen of a patient with Gaucher's disease and comprehensive kinetic studies were performed.[54]

3.2. Activator-Deficiency Diseases

3.2.1. AB-Variant of Tay–Sachs Disease

Patients with this form of Tay–Sachs disease exhibit most of the classic signs of the B variant. Here both Hex A and Hex B isozymes exhibit normal activity when artificial substrates are used for these assays, and no reduction of G_{M2} catabolism could be shown *in vitro* in the presence of detergent. It has been known for many years that detergents are required to assay the activity of sphingolipid hydrolases *in vitro*. Furthermore, it was clearly realized that this situation represented a somewhat artificial condition, and considerable speculation has been directed to the possibility that natural counterparts of detergents were involved in the solubilization of these substrates *in vivo*.[55] These considerations received substantial impetus from the discovery by Li and his co-workers that liver contained a heat-stable nondialyzable factor that promoted the hydrolysis of ganglioside G_{M2} by purified hexosaminidase A.[56] These experiments took on added attractiveness in view of the frequent observation that the activity of this enzyme with ganglioside G_{M2} was extraordinarily low *in vitro* compared with its ability to catalyze the hydrolysis of chromogenic and fluorogenic derivatives of N-acetylgalactosamine and N-acetylglucosamine.[57] A most intriguing finding was the demonstration by Conzelmann and Sandhoff that absence of the activating factor appeared to be the basis of the pathogenesis of a patient with the AB variant form of Tay–Sachs disease.[58] Another case of absent activator has recently been reported by Hechtman and his co-workers.[59] The factor appears to combine with the ganglioside substrate, and this combination can now interact with the active site of hexosaminidase A.[60]

3.2.2. Other Disorders

Evidence has been obtained by Stevens and co-workers that a deficiency of a similar heat-stable activator protein is involved in a rare form of metachromatic leukodystrophy.[61] Another activator seems to promote the enzymatic hydrolysis of several lipids including glucocerebroside, galactocerebroside, and sphingomyelin.[62] As far as I am aware, there has been no report of a deficiency of this activator, although such a situation might be anticipated.

Considerable controversy has been raised by the report that a heat-stable factor is missing in patients with type C Niemann–Pick disease.[63] My own experience in assaying sphingomyelinase activity in cultured skin fibroblasts from these patients is inconsistent with this deduction.[64] A significant decrease of sphingomyelinase activity was noted when [^{14}C]sphingomyelin was used in the presence of an optimal concentration of detergent. Furthermore, diminished sphingomyelinase activity was also evident when the water-soluble chromogenic substrate 2-hexadecanoylamino-4-nitrophenylphosphorycholine was used to assay the enzyme. This situation is in contrast with the AB variant of Tay–Sachs disease, where completely normal hexosaminidase activity is observed with chromogenic and fluorogenic derivatives of N-acetylglucosamine.

Besley *et al.* have reported that genetic complementation occurs in cells formed by the fusion of cultured skin fibroblasts from type A and C Niemann–Pick patients, suggesting that two different genes are involved.[65] This finding contrasts with the situation in Gaucher's disease, where complementation did not occur when cells derived from patients with various phenotypes were fused (J. A. Barranger, C. Chang, and R. O. Brady, unpublished observations). The genetic mutations in that disorder have been shown to be allelic.[26] The pathogenetic basis of type C Niemann–Pick disease requires further resolution.

3.3. Deficiency of a Stabilizer Glycoprotein in Patients with Combined β-Galactosidase and Neuraminidase Deficiencies

Some patients whose symptomatology resembles the skeletal abnormalities, retinal cherry-red spot, myoclonus, and dementia seen in generalized (G_{M1}) gangliosidosis have a combined deficiency of β-galactosidase and neuroaminidase.[66] Complementation studies indicated that the genetic mutation was different from that in G_{M1}-gangliosidosis[67] and that involved in isolated neuraminidase deficiency in patients classified as mucolipidosis type I.[68] It was also observed that β-galactosidase activity in cells from heterozygotes was normal, whereas neuraminidase activity was intermediate between normal and affected.[66,68] It was subsequently found that the turnover time of β-galactosidase in fibroblasts from patients with the combined deficiencies was dramatically shortened from normal.[69] On the basis of experiments with inhibitors of lysosomal protease activity, it was postulated that the β-galactosidase and neuraminidase were synthesized normally but were especially sensitive to degradation by lysosomes.[70,71] It was subsequently reported that cells from patients with combined β-galactosidase–neuraminidase deficiency lack a glycoprotein whose M_r is 32,000 that is believed to be required to unite and bind monomeric forms of these enzymes to lysosomal membranes and to prevent their intralysosomal degradation.[72]

3.4. Significance of Recognizing Types of Genetic Mutations

In the preceding sections, we have seen that genetic mutations in various patients with sphingolipid storage disorders can vary in dramatic ways. Although the majority of patients by far appear to have specific mutations in the genetic code for the enzyme polypeptide, the pathogenetic basis in other cases is clearly distinctive. Thus, the demonstration of deficient aminosugar transferase in I-cell disease, missing activator proteins in variant forms of Tay–Sachs disease and metachromatic leukodystrophy, and protective glycoprotein in β-galactosidase–neuraminidase deficiency are obviously extraordinarily important for devising correct therapy for these disorders. Other forms of heritable genetic storage disorders are certainly conceivable. In particular, it is worth mentioning that a lysosomal storage disorder has been demonstrated in Balb/c mice that exhibit combined sphingomyelinase and glucocerebrosidase deficiencies with profound neurological involvement.[73] The disorder is transmitted as an autosomal recessive trait. Here again, one suspects that the enzymes are synthesized properly but do not reach or become properly inserted into lysosomes because of a missing signal. On the other hand, it is conceivable that there has been a mutation of a structural or transport protein in the lysosomal membrane in these animals in analogy with the recent demonstration in cystinosis.[74] Other unconventional etiologies in heritable metabolic disorders should be anticipated.

4. *ENZYME REPLACEMENT THERAPY*

4.1. *Replacement of Enzymes in Metabolic Disorders without CNS Involvement*

4.1.1. *Fabry's Disease*

In view of the finding in the pioneering investigation of enzyme replacement in a patient with Sandhoff's disease in whom none of the intravenously administered hexosaminidase A reached the CNS,[75] investigators turned their attention to the possibility of replacing deficient enzymes in disorders such as type 1 Gaucher's disease and Fabry's disease, where the brain is not involved.

The first enzyme that became available in suffient purity and quantity for a replacement trial was the ceramidetrihexosidase that is lacking in patients with Fabry's disease. The enzyme was isolated from human placental tissue, a source that is in wide use at this time.[76] When this enzyme was injected into two patients with the disorder, the threefold elevation of ceramidetrihexoside in the blood returned to the normal level.[77] However, the quantity of this lipid in the circulation returned to the preinfusion value by 72 hr after administration of enzyme. No data on tissue lipid levels were obtained in this investigation. Several years later, a similar response was reported when ceramidetrihexosidase isolated from plasma and human spleen was injected.[78] It was obvious from the rapid return of ceramidetrihexoside in the circulation in these patients that a much larger quantity of enzyme would be required in order to achieve a clinical response. This investigation was delayed for a considerable time because of a pyrogenic contaminant in placental ceramidetrihexosidase preparations. However, this difficulty has been surmounted, and it is expected that enzyme replacement trials will soon begin again in this disorder.

4.1.2. *Gaucher's Disease*

Since most patients with a heritable metabolic storage disorder belong to this group, it is not surprising that the most extensive investigations of enzyme replacement centered on this condition. Glucocerebrosidase is an extraordinarily hydrophobic enzyme, and its purification was delayed for a considerable time until biochemists developed satisfactory purification schemes for its isolation. Eventually, a procedure was devised that led to a sufficiently pure preparation of glucocerebrosidase from human placenta to examine its effectiveness.[10]

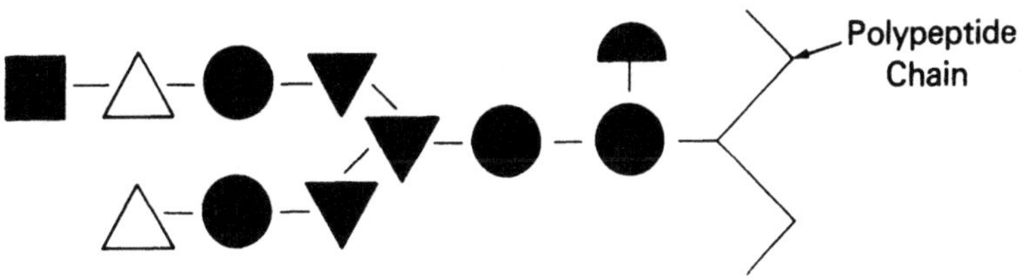

Fig. 3. Structure of glycoproteins with complex oligosaccharide residues. ■ = N-acetylneuraminic acid; △ = galactose; ● = N-acetylglucosamine; ▼ = mannose; and ⏶ = fucose.

When this enzyme was infused intravenously into two patients with Gaucher's disease, it was cleared from the circulation with a half-time of 20 min. Examination of glucocerebroside in liver biopsy specimens obtained prior to infusion and 24 hr after administration of the enzyme revealed that there was a 26% decrease in the quantity of stored glucocerebroside in the livers of both recipients.[79] Furthermore, the elevated level of glucocerebroside in the circulation gradually returned to normal in both patients, suggesting redistribution of glucocerebroside from the blood to the vacated tissue stores. This reduction in circulating lipid lasted a comparatively long time,[80,81] suggesting that enzyme replacement might be required on an infrequent basis if satisfactory clearance of accumulated sphingolipid could be achieved.

It was apparent from infusion of glucocerebrosidase in a subsequent patient that comparatively large quantities of this enzyme would be required for sizable reductions of tissue glucocerebroside.[82] Because the procedure originally developed to isolate glucocerebrosidase could not be scaled up sufficiently, another enzyme purification scheme was developed that incorporated hydrophobic column chromatography to take advantage of the strong hydrophobicity of glucocerebrosidase.[83] Enzyme prepared in this fashion has been administered to 12 young Gaucher patients on a prospective replacement trial. We have the impression that the recipients are improved with regard to their general health and growth patterns; progression of the hepatosplenomegaly has been arrested, and platelet levels have been stabilized in these recipients.[84–86] This investigation is now in the control phase, where injection of enzyme or only vehicle (4% human serum albumin) is randomized.

In spite of these encouraging findings, the effects of this enzyme in adult patients with type 1 Gaucher's disease have been less consistent. Half of the recipients have had a significant reduction of hepatic glucocerebroside; in others, there was no detectable change. We believe a major reason for this inconsistency is the fact that most of the injection glucocerebrosidase that becomes localized in the liver is taken up by hepatocytes rather than by Kupffer cells, where glucocerebroside accumulates. This cellular predilection has been shown to be caused by the presence of terminal molecules of galactose on the oligosaccharide chains of glucocerebrosidase that bind with great avidity to the galactose lectin on hepatocyte plasma membranes[88] (Fig. 3).

We have therefore embarked on an investigation to target the enzyme to Kupffer cells as well as to other macrophages by modifying the terminal sugars of the oligosaccharide moieties. Three approaches were undertaken. The first was to covalently link mannose oligosaccharides to the native enzyme, since macrophages have a lectin that is specific for this sugar. The results of this technique have not been encouraging, probably because a triantennary oligosaccharide is required for maximal binding.[89] We therefore attached triantennary trimannosyldilysyl residues to glucocerebrosidase. Here we achieved improved delivery to macrophages,[90] but the increase was not judged to be enough to warrant clinical trials with this form of the enzyme. We have therefore concentrated on a third tactic that involves sequential enzymatic cleavage of N-acetylneuraminyl, galactosyl, and N-acetylglucosaminyl residues, leaving mannose-terminated glucocerebrosidase.[91] This procedure has resulted in greatly increased delivery of glucocerebrosidase to Kupffer cells[91] (F. S. Furbish, R. O. Brady, and J. A. Barranger, unpublished observations). We plan to carry out a clinical trial of enzyme replacement as soon as sufficient quantities of the appropriately modified enzyme become available.

4.2. Replacement of Enzymes in Metabolic Disorders with CNS Involvement: Tay–Sachs Disease

In addition to the previously cited enzyme replacement trial in Sandhoff's disease, there has been a major investigation in patients with the B-variant form of Tay–Sachs disease. Human placental hexosaminidase A was injected intravenously, intrathecally, and intracisternally into two[92] or possibly three[93] Tay–Sachs patients. The enzyme was administered as a mixture of free and polyvinylpyrrolidone-bound hexosaminidase A. Infusions were accompanied by pyrogenic reactions on several occasions. There was no evidence of clinical improvement. The investigators noted that no humoral immune response was detected, an important determination if, indeed, there is no cross-reacting material to hexosaminidase A in these individuals. Furthermore, there appeared to be a noticeable decrease in the quantity of ganglioside G_{M2} in the serum of one of the recipients shortly after infusion of the enzyme.

4.3. Replacement of Enzyme When Both Central and Peripheral Nervous Systems Are Involved: Metachromatic Leukodystrophy

The earliest examinations of the effect of enzyme replacement in a human lipid storage disorder were carried out in patients with metachromatic leukodystrophy. Partially purified arylsulfatase A obtained from human urine was injected intrathecally in one patient[94]; enzyme from beef brain was administered to another.[95] Both patients experienced severe pyrogenic reactions, and no clinical benefit was noted. Both groups made the noteworthy observation that intrathecally infused enzyme did not penetrate into the substance of the brain. Neither group detected antibody formation, a remarkable finding in view of the use of animal tissue in the second experiment.

In contrast to these negative findings, two groups have reported that the addition of arylsulfatase A to cultured skin fibroblasts derived from patients with metachromatic leukodystrophy increased the degradation of labeled sulfatide in these cells.[96,97] This observation has provided much incentive for us to continue to develop effective means to deliver the required catalysts to neural tissues. A particularly important consideration in this situation is that both the CNS and PNS are pathologically involved in metachromatic leukodystrophy. Here, in addition to the requirement to deliver enzyme to the brain, one must devise means to correct the PNS lesions. Although no conclusive experiments are yet available, it is conceivable that exogenous enzyme might be taken up by PNS synapses or nerve terminals and transported retrogradely to the neuronal cell body. Conceivably, one could link an appropriate ligand that has high affinity for presynaptic membranes to arylsulfatase A to enhance uptake of the enzyme. However, one still has to consider that the deposition of sulfatide is in the myelin sheath rather than in neural or glial cell bodies, and delivery of enzyme to these stores may require the development of additional techniques.

5. *STRATEGIES FOR ENZYME REPLACEMENT IN METABOLIC DISORDERS OF THE CNS*

Although it seems likely that the solution to the problem of effective enzyme delivery to sphingolipid-storing macrophages in peripheral tissue is near at hand, neuroscientists are still laboriously involved in developing effective procedures to deliver the necessary enzymes to the brain. In view of the inability to reverse the clinical process in Sandhoff's disease, Tay–Sachs disease, and metachromatic leukodystrophy largely because the exogenous enzymes did not reach the involved cells, careful consideration must be given to devising procedures to provide sufficient enzyme to the CNS. The method that appears most likely to be successful at this time is the temporary alteration of the blood–brain barrier by intraarterial infusion of hyperosmolar solutions of solutes such as mannitol or arabinose.[98] Conditions must be carefully adjusted with regard to the concentration and velocity of infusion so that permanent neurological deficits are not created. It has been shown that a physiologically normal amount of a lysosomal enzyme can be delivered to the brain of rodents using this technique. This investigation was extended to primates, and no evidence of brain damage was detected in these animals.[99]

Several additional significant observations have been made in the course of these studies. It has been found that mannose-terminated glycoprotein enzymes such as horseradish peroxidase are specifically endocytosed by neurons in the CNS after intravenous injection and modification of the blood–brain barrier.[100] The importance of this is emphasized by the observation that hexosaminidase A is a mannose-terminated enzyme.[101] Further, it has been demonstrated that radioiodinated hexosaminidase A is specifically bound by a high-affinity receptor on rat brain synaptosomes[102] and synaptic plasma membranes.[103] Thus, if the present limitation of the comparatively small quantity of hexosaminidase A that can be delivered to the brain even after barrier modification can be overcome,[104] one would have a rational basis for enzyme replacement in disorders of this type.

Two final comments deserve mention. First, the blood–brain barrier has been modified in humans without causing neurological damage.[105] Second, recent work indicates that hexosaminidase A that reaches the neuronal cells in the brain following alteration of the blood–brain barrier is endocytosed by these cells and becomes packaged in lysosomes.[106] This finding is extraordinarily significant, since the pathological membranous cytoplasmic bodies in the neurons of Tay–Sachs patients are formed from these subcellular particles.[107] Thus, in spite of the difficulty in predicting when and by what mode the delivery of enzymes to the brain will eventually become clinically useful, these studies should be vigorously pursued since there appears to be a reasonable chance for the eventual success of these endeavors.

REFERENCES

1. Brady, R. O., 1972, *Handbook of Neurochemistry*, Volume 7 (A. Lajtha, ed.), Plenum Press, New York, pp. 34–46.
2. Kaback, M. M., 1981, *Lysosomes and Lysosomal Storage Diseases* (J. A. Lowden and J. W. Callahan, eds.), Raven Press, New York, pp. 331–342.
3. Brady, R. O., 1980, *Metabolic Control and Disease* (P. K. Bondy and L. E. Rosenberg, eds.), W. B. Saunders, Philadelphia, pp. 523–543.
4. Brady, R. O., 1981, *Basic Neurochemistry*, 3rd ed. (G. Siegel, R. W. Albers, R. Katzman, and B. W. Agranoff, eds.), Little, Brown, Boston, pp. 615–626.
5. Stanbury, J. B., Wyngaarden, J. B., Fredrickson, D. S., Goldstein, J. L., and Brown, M. S., 1983, *The Metabolic Basis of Inherited Disease*, 5th ed., McGraw-Hill, New York.
6. Norman, R. M., Ulrich, J., and Lloyd, O. C., 1956, *J. Pathol. Bacteriol.* **72:**121–131.
7. Banker, B. Q., Miller, J. C., and Crocker, A. C., 1962, *Cerebral Sphingolipidoses* (S. M. Aronson and B. W. Volk, eds.), Academic Press, New York, pp. 73–99.
8. Nilsson, O., and Svennerholm, L., 1982, *J. Neurochem.* **39:**709–718.
9. Cleland, W. W., and Kennedy, E. P., 1960, *J. Biol. Chem.* **235:**45–51.
10. Pentchev, P. G., Brady, R. O., Hibbert, S. R., Gal, A. E., and Shapiro, D., 1973, *J. Biol. Chem.* **248:**5256–5261.
11. Gery, I., Zigler, J. S., Brady, R. O., and Barranger, J. A., 1981, *J. Clin. Invest.* **68:**1182–1189.
12. Vanier, M.-T., and Svennerholm, L., 1976, *Adv. Exp. Med. Biol.*, **68:**115–126.
13. Svennerholm, L., Vanier, M.-T., and Mansson, J. E., 1980, *J. Lipid Res.* **21:**53–64.
14. Taketomi, T., and Nishimura, K., 1964, *Jpn. J. Exp. Med.* **34:**255–265.
15. Suzuki, K., Tanaka, H., and Suzuki, K., 1976, *Current Trends in Sphingolipidoses and Allied Disorders* (B. W. Volk and L. Schneck, eds.), Plenum Press, New York, pp. 99–114.
16. Austin, J., Lehfeldt, C., and Maxwell, W., 1961, *J. Neuropathol. Exp. Neurol.* **20:**284–285.
17. Austin, J., and Lehfeldt, D., 1965, *J. Neuropathol. Exp. Neurol.* **24:**265–289.
18. Olsson, R., Sourander, P., and Svennerholm, L., 1966, *Acta Neuropathol. (Berl.)* **6:**153–163.
19. Brady, R. O., Kanfer, J. N., Mock, M. B., and Fredrickson, D. S., 1966, *Proc. Natl. Acad. Sci. U.S.A.* **55:**366–369.
20. Brady, R. O., 1983, *The Metabolic Basis of Inherited Disease*, 5th ed. (J. B. Stanbury, J. B. Wyngaarden, D. S. Fredrickson, J. L. Goldstein, and M. S. Brown, eds.), McGraw-Hill, New York, pp. 831–841.
21. Spence, M. W., Clarke, J. T. R., and Cook, H. W., 1984, *J. Biol. Chem.* **258:**8595–8600.
22. O'Brien, J. S., 1983, *The Metabolic Basis of Inherited Disease*, 5th ed. (J. B. Stanbury, J. B. Wyngaarden, D. S. Fredrickson, J. L. Goldstein, and M. S. Brown, eds.), McGraw-Hill, New York, pp. 945–969.
23. Purpura, D. P., and Walkey, S. U., 1981, *Gangliosides in Neurological and Neuromuscular Function, Development and Repair* (M. M. Rapport and A. Gorio, eds.), Raven Press, New York, pp. 1–16.

24. Fishman, P. H., and Brady, R. O., 1976, *Science* **194**:906–915.
25. Moss, J., Fishman, P. H., Manganiello, V. C., Vaughan, M., and Brady, R. O., 1976, *Proc. Natl. Acad. Sci. U.S.A.* **73**:1034–1037.
26. Ginns, E. I., Brady, R. O., Pirruccello, S., Moore, C., Sorrell, S., Furbish, F. S., Murray, G. J., Tager, J., and Barranger, J. A., 1982, *Proc. Natl. Acad. Sci. U.S.A.* **79**:5607–5610.
27. Ginns, E. I., Erickson, A., Barnevald, R., Brady, R. O., Tager, J. M., and Barranger, J. A., 1984, *Isozymes, Current Topics in Biological and Medical Research* (M. C. Rattazzi, J. G. Scandalios, and G. S. Whitt, eds.), Alan R. Liss, New York, pp. 83–94.
28. Furbish, F. S., Oliver, K. L., Zirzow, G. C., Murray, G. J., Ginns, E. I., Brady, R. O., and Barranger, J. A., 1983, *Fed. Proc.* **42**:1825.
29. Kaplan, A., Achord, D. T., and Sly, W. S., 1977, *Proc. Natl. Acad. Sci. U.S.A.* **74**:2026–2030.
30. Owada, M., and Neufeld, E. F., 1982, *Biochem. Biophys. Res. Commun.* **105**:814–820.
31. Waheed, A., Pohlman, R., Hasilik, A., von Figura, K., van Elsen, A., and Leroy, J. G., 1982, *Biochem. Biophys. Res. Commun.* **105**:1052–1057.
32. Hasilik, A., Waheed, A., and von Figura, K., 1981, *Biochem. Biophys. Res. Commun.* **98**:761–767.
33. Reitman, M. L., and Kornfeld, S., 1981, *J. Biol. Chem.* **256**:4275–4281.
34. Barneveld, R. A., Tegelaers, F. P. W., Ginns, E. I., Visser, P., Laanen, E. A., Brady, R. O., Galjaard, H., Barranger, J. A., Reuser, A. J. J., and Tager, J. M., 1983, *Eur. J. Biochem.* **134**:585–589.
35. Ginns, E. I., Tegelaers, F. P. W., Barneveld, R., Galjaard, H., Reuser, A. J. J., Tager, J. M., Brady, R. O., and Barranger, J. A., 1983, *Clin. Chim. Acta* **131**:283–287.
36. Okada, S., and O'Brien, J. S., 1969, *Science* **165**:698–700.
37. Kolodny, E. H., Brady, R. O., and Volk, B. W., 1969, *Biochem. Biophys. Res. Commun.* **37**:526–531.
38. Lalley, P. A., Rattazzi, M. C., and Shows, T. B., 1974, *Proc. Natl. Acad. Sci. U.S.A.* **71**:1569–1573.
39. Gilbert, F., Kucherlapati, R., Creagan, R. P., Murnane, M. J., Darlington, G. I., and Ruddle, F. H., 1975, *Proc. Natl. Acad. Sci. U.S.A.* **72**:263–267.
40. Carrol, M., and Robinson, D., 1973, *Biochem. J.* **131**:91–96.
41. Srivastava, S. K., and Beutler, E., 1974, *J. Biol. Chem.* **249**:2054–2057.
42. Bartholomew, W. R., and Rattazzi, M. C., 1974, *Int. Arch. Allergy Appl. Immunol.* **46**:512–524.
43. Srivastava, S. K., Ansari, N. H., Hawkins, L. A., and Wiktorowicz, J. E., 1979, *Biochem. J.* **179**:657–664.
44. Hasilik, A., and Neufeld, E. F., 1980, *J. Biol. Chem.* **255**:4937–4945.
45. Hasilik, A., and Neufeld, E. F., 1980, *J. Biol. Chem.* **255**:4946–4950.
46. Frisch, A., and Neufeld, E. F., 1981, *J. Biol. Chem.* **256**:8242–8246.
47. Mahuran, D. J., Tsui, F., Gravel, R. A., and Lowden, J. A., 1982, *Proc. Natl. Acad. Sci. U.S.A.* **79**:1602–1605.
48. Proia, R. O., and Neufeld, E. F., 1982, *Proc. Natl. Acad. Sci. U.S.A.* **79**:6360–6364.
49. Brady, R. O., and Kolodny, E. H., 1972, *Progress in Medical Genetics*, Volume VIII (A. G. Steinberg and A. G. Bearn, eds.), Grune and Stratton, New York, pp. 225–241.
50. Brady, R. O., and Barranger, J. A., 1981, *The Molecular Basis of Neuropathology* (A. N. Davison and R. H. S. Thompson, eds.), Edward Arnold, London, pp. 188–220.
51. Norden, A. G. W., and O'Brien, J. S., 1975, *Proc. Natl. Acad. Sci. U.S.A.* **72**:240–244.
52. O'Brien, J. S., and Warner, T. G., 1980, *Clin. Genet.* **17**:35–38.
53. Ben-Yoseph, Y., Momoi, T., Baylerian, M. S., and Nadler, H. L., 1982, *Clin. Chim. Acta* **123**:233–240.
54. Pentchev, P. G., Brady, R. O., Blair, H. E., Britton, D. E., and Sorrell, S. H., 1978, *Proc. Natl. Acad. Sci. U.S.A.* **75**:3970–3973.
55. Brady, R. O., 1978, *Annu. Rev. Biochem.* **47**:687–713.
56. Li, Y.-T., Mazzotta, M. Y., Wan, C.-C., Orth, R., and Li, S.-C., 1973, *J. Biol. Chem.* **248**:7512–7515.
57. Tallman, J. F., Brady, R. O., Quirk, J. M., Villalba, M., and Gal, A. E., 1974, *J. Biol. Chem.* **249**:3489–3499.

58. Conzelmann, E., and Sandhoff, K., 1978, *Proc. Natl. Acad. Sci. U.S.A.* **75**:3979–3983.
59. Hechtman, P., Gordon, B. A., and Ng Ying Kin, N. M. K., 1982, *Pediatr. Res.* **16**:217–222.
60. Conzelmann, E., and Sandhoff, K., 1979, *Hoppe Seylers Z. Physiol. Chem.* **360**:1837–1849.
61. Stevens, R. L., Fluharty, A. L., Kihara, H., Kaback, M. M., Shapiro, L. J., March, B., Sandhoff, K., and Fischer, G., 1981, *Am. J. Hum. Genet.* **33**:900–906.
62. Wenger, D. A., Sattler, M., and Roth, S., 1982, *Biochim. Biophys. Acta* **712**:639–649.
63. Christomanou, H., 1980, *Hoppe Seylers Z. Physiol. Chem.* **361**:1489–1502.
64. Gal, A. E., Brady, R. O., Hibbert, S. R., and Pentchev, P. G., 1975, *N. Engl. J. Med.* **293**:632–636.
65. Besley, G. T. N., Hoogeboom, A. J. M., Hoogeveen, A., Kleijer, W. J., and Galjaard, H., 1980, *Hum. Genet.* **54**:409–412.
66. Wenger, D. A., Tarby, T. J., and Wharton, C., 1978, *Biochem. Biophys. Res. Commun.* **82**:589–595.
67. Galjaard, H., Hoogeveen, A., Kleijer, W., de Wit-Verbeek, H. A., Reuser, A. J. J., Ho, M. W., and Robinson, D., *Nature* **257**:60–62.
68. Hoogeveen, A. T., Verheijen, F. W., D'Azzo, A., and Galjaard, H., 1980, *Nature* **285**:500–502.
69. Van Diggelen, O. P., Galjaard, H., Sinnott, M. L., and Smith, P. J., 1980, *Biochem. J.* **188**:337–343.
70. Suzuki, Y., Sakuraba, H., Hayashi, K., Suzuki, K., and Imahori, K., 1981, *J. Biochem. (Tokyo)* **90**:271–273.
71. Galjaard, H., Hoogeveen, A., Verheijen, F., Van Diffelen, O. P., Konings, A., D'Azzo, A., and Reuser, A. J. J., 1981, *Perspectives in Inherited Metabolic Diseases,* Volume 4 (B. Berraand S. DiDonato, eds.), Milan, pp. 317–333.
72. D'Azzo, A., Hoogeveen, A., Reuser, A. J. J., Robinson, D., and Galjaard, H., 1982, *Proc. Natl. Acad. Sci. U.S.A.* **79**:4535–4539.
73. Pentchev, P. G., Gal, A. E., Booth, A. D., Omodeo-Sale, F., Fouks, J., Neumeyer, B. A., Quirk, J. M., Dawson, G., and Brady, R. O., 1980, *Biochim. Biophys. Acta* **619**:669–679.
74. Gahl, W. A., Bashan, N., Tietze, F., Bernardini, I., and Schulman, J. D., 1982, *Science* **217**:1263–1265.
75. Johnson, W. G., Desnick, R. J., Long, D. M., Sharp, H. L., Krivit, W., Brady, R., and Brady, R. O., 1973, *Enzyme Therapy in Genetic Diseases* (R. J. Desnick, R. W. Bernlohr, and W. Krivit, eds.), Williams and Wilkins, Baltimore, pp. 120–124.
76. Johnson, W. G., and Brady, R. O., 1972, *Methods Enzymol.* **28**:849–856.
77. Brady, R. O., Tallman, J. F., Johnson, W. G., Gal, A. E., Leahy, W. E., Quirk, J. M., and Dekaban, A. S., 1973, *N. Engl. J. Med.* **289**:9–14.
78. Desnick, R. J., Dean, K. J., Grabowski, G. A., Bishop, D. F., and Seeley, C. C., 1979, *Proc. Natl. Acad. Sci. U.S.A.* **76**:5326–5330.
79. Brady, R. O., Pentchev, P. G., Gal, A. E., Hibbert, S. R., and Dekaban, A. S., 1974, *N. Engl. J. Med.* **291**:989–993.
80. Pentchev, P. G., Brady, R. O., Gal, A. E., and Hibbert, S. R., 1975, *J. Mol. Med.* **1**:73–78.
81. Brady, R. O., 1977, *Metabolism* **26**:329–345.
82. Brady, R. O., Pentchev, P. G., Gal, A. E., Hibbert, S. R., Quirk, J. M., Mook, G. E., Kusiak, J. W., Tallman, J. F., and Dekaban, A. S., 1976, *Current Trends in the Sphingolipidoses and Allied Disorders* (B. W. Volk and L. Schneck, eds.), Plenum Press, New York, pp. 523–532.
83. Furbish, F. S., Blair, H. E., Shiloach, J., Pentchev, P. G., and Brady, R. O., 1977, *Proc. Natl. Acad. Sci. U.S.A.* **74**:3560–3563.
84. Brady, R. O., Barranger, J. A., Gal, A. E., Pentchev, P. G., and Furbish, F. S., 1980, *Enzyme Therapy in Genetic Diseases, 2* (R. J. Desnick, ed.), Alan R. Liss, New York, pp. 361–368.
85. Brady, R. O., Barranger, J. A., Furbish, F. S., Murray, G. J., Stowens, D. W., and Ginns, E. I., 1982, *Advances in the Treatment of Inborn Errors of Metabolism* (M. D'A Crawford, D. A. Gibbs, and R. W. E. Watts, eds.), John Wiley and Sons, London, New York, pp. 53–63.
86. Brady, R. O., Barranger, J. A., Furbish, F. S., Stowens, D. W., and Ginns, E. I., 1982, *Gaucher Disease: A Century of Delineation and Research* (R. J. Desnick, S. Gatt, and G. A. Grabowski, eds.), Alan R. Liss, New York, pp. 669–680.

87. Furbish, F. S., Steer, C. J., Barranger, J. A., Jones, E. A., and Brady, R. O., 1980, *Biochem. Biophys. Res. Commun.* **81**:1047–1053.
88. Ashwell, G., and Morell, A., 1974, *Adv. Enzymol.* **41**:99–128.
89. Brady, R. O., Barranger, J. A., Pentchev, P. G., Furbish, F. S., Gal, A. E., 1983, *Inborn Errors of Metabolism in Humans* (F. Cockburn and R. Gitzolman, eds.), MTP Press, London pp. 139–154.
90. Doebber, T. W., Wu, M. S., Bugianesi, R. L., Ponpipom, M. M., Furbish, F. S., Barranger, J. A., Brady, R. O., and Shen, T. Y., 1982, *J. Biol. Chem.* **257**:2193–2199.
91. Furbish, F. S., Steer, C. J., Krett, N. L., and Barranger, J. A., 1981, *Biochim. Biophys. Acta* **673**:425–434.
92. von Specht, B. U., Geiger, B., Arnon, R., Passwell, J., Keren, G., Goldman, B., and Padeh, B., 1979, *Neurology (Minneap.)* **29**:848–854.
93. Godel, V., Blumenthal, M., Goldman, B., Keren, G., and Padeh, G., 1978, *Metab. Ophthalmol.* **2**:27–31.
94. Austin, J. H., 1967, *Inborn Disorders of Sphingolipid Metabolism* (S. M. Aronson and B. W. Volk, eds.), Academic Press, New York, pp. 359–387.
95. Greene, H. L., Hug, G., and Schubert, W. K., 1969, *Arch. Neurol.* **20**:147–153.
96. Porter, M. T., Fluharty, A. L., and Kihara, H., 1971, *Science* **172**:1263–1265.
97. Wiesmann, U. N., Rossi, E. E., and Herschkowitz, N. N., 1972, *Acta Paediatr. Scand.* **61**:196–302.
98. Barranger, J. A., Rapoport, S. I., Fredericks, W. R., Pentchev, P. G., MacDermot, K. D., Steusing, J. K., and Brady, R. O., 1979, *Proc. Natl. Acad. Sci. U.S.A.* **76**:481–485.
99. Smith, M. T., Girton, M., Rapoport, S. I., Brady, R. O., and Barranger, J. A., 1980, *J. Neuropathol. Exp. Neurol.* **39**:389.
100. Brady, R. O., and Barranger, J. A., 1981, *Trends Neurosci.* **4**:265–267.
101. Steer, C. J., Kusiak, J. W., Brady, R. O., and Jones, E. A., 1979, *Proc. Natl. Acad. Sci. U.S.A.* **76**:2774–2778.
102. Kusiak, J. W., Toney, J. H., Quirk, J. M., and Brady, R. O., 1979, *Proc. Natl. Acad. Sci. U.S.A.* **76**:982–985.
103. Kusiak, J. W., Quirk, J. M., and Brady, R. O., 1980, *Enzyme Therapy in Genetic Diseases, 2* (R. J. Desnick, ed.), Alan R. Liss, New York, pp. 93–102.
104. Neuwelt, E. A., Barranger, J. A., Brady, R. O., Pagel, M., Furbish, F. S., Quirk, J. M., Mook, G. E., and Frenkel, E., 1981, *Proc. Natl. Acad. Sci. U.S.A.* **78**:5838–5841.
105. Neuwelt, E. A., Frenkel, E. P., Diehl, J., Vu, L. H., Rapoport, S. I., and Hill, S., 1980, *Neurosurgery* **7**:44–52.
106. Neuwelt, E. A., Barranger, J. A., Pagel, M. A., Quick, J. A., Brady, R. O., and Frenkel, E. P., 1984, *Neurology* **34**:1012–1019.
107. Tallman, J. F., Brady, R. O., and Suzuki, K., 1971, *J. Neurochem.* **18**:1775–1777.

6

Neuropathies with Deranged Metabolism

Hugo W. Moser

1. CLASSIFICATION AND EPIDEMIOLOGY OF NEUROPATHIES

The WHO Task Force on Neuropathology has suggested the classification of neuropathies shown in Table 1.[1] World-wide, the most common causes of neuropathy are leprosy, trauma, nutritional deficiency and alcoholism, and diabetes. Several comprehensive reviews are available.[2–4] The first three categories are beyond the scope of this chapter. We focus here mainly on the inherited neuropathies and on the diabetic neuropathies. Although the inherited neuropathies do not approach in frequency those resulting from the main environmental causes (infection, trauma, nutritional deficiency), they are highly significant causes of disability. In a recent survey of 205 patients with neuropathy referred to the Mayo Clinic, it was found that 42% suffered from inherited neuropathies.[5] In addition to their etiologic classification, neuropathies are subdivided in accordance with basic pathological mechanisms. These mechanisms are discussed in detail in standard reference works[2,3] and are summarized in Fig. 1. *Wallerian degeneration* results from transection of the axon through trauma. In *segmental demyelination*, there is primary destruction of the myelin sheath, although the axon is left intact. In *axonal degeneration*, there is a metabolic derangement of the whole neuron, and the myelin sheath breaks down concomitantly with the axon. Most commonly, the distal portion of the axon is most severely affected ("dying back phenomenon").

2. SPECIAL TESTS AVAILABLE FOR THE STUDY OF HUMAN PERIPHERAL NERVES

Even though peripheral nerves are obviously more accessible for study than is the central nervous system, the etiologic diagnosis of peripheral nerve

Hugo W. Moser • John F. Kennedy Institute, Baltimore, Maryland 21025.

Table I
Classification of Peripheral Neuropathies[a]

I. Axonopathies
  A. Genetically determined
  B. Acquired
    1. Exogenous toxins and drugs
    2. Metabolic disorders
    3. Deficiency states
    4. Miscellaneous
II. Myelinopathies
  A. Genetically determined
  B. Acquired
    1. Idiopathic, infectious, or postinfectious
    2. Toxic
    3. Metabolic
III. Other types
  A. Infectious
  B. Ischemic
  C. Mechanical
  D. Miscellaneous

[a] From Schoenberg.[1]

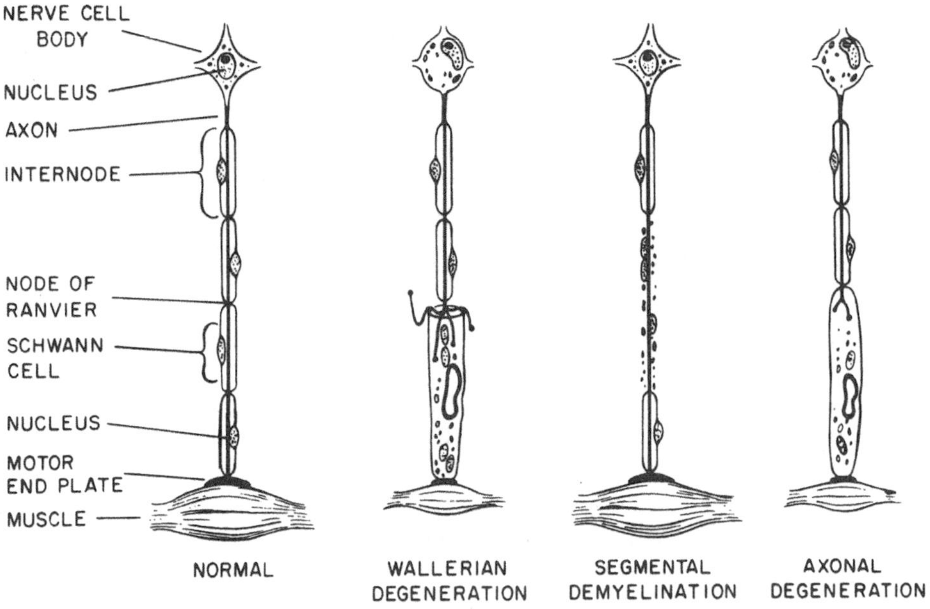

Fig. 1. Three forms of neuropathy (see text). (From Asbury and Johnson.[3])

disease still is a source of great difficulty. It is stated that with usual approaches exact diagnosis is achieved in only one-half of the cases but that with intensive evaluation the percentage of classifiable cases can be increased to 76%.[5] This improvement in diagnostic precision is often based on the use of relatively newly developed special tests, some of which are described here. Several of these tests are of very recent origin, and others are at a border zone between clinical practice and research. It is anticipated that further refinement and application will not only increase knowledge about diagnostic classifications but also have the potential of increasing basic knowledge about nervous system function and structure, and it is for this reason that they are summarized here.

2.1. *Percutaneous Neurophysiological Techniques*

Determination of motor conduction velocities often allows distinction between nerve disorders that damage the axon mainly (axonopathies) and those in which the damage is to the myelin sheaths (myelinopathies),[6] the motor conduction velocity being slowed to a greater extent in the latter. Techniques for sensory conduction velocity measurement are also being refined, as are a variety of computer-based techniques to quantify temperature and vibration sense.[6]

2.2. *Sural Nerve Biopsy*

The sural nerve is the nerve that is selected most frequently for biopsy. It is located on the back of the leg above the lateral malleolus. It has a fairly superficial position, so that the biopsy is not technically difficult or dangerous, and it is a sensory nerve only. Interruption causes a sensory deficit that involves an area approximately 3 cm in diameter just below the lateral malleolus and does not cause a significant handicap. Being a distal nerve segment, the sural nerve is affected in many forms of neuropathy so that it has a high chance of providing significant information about the disease process. Some authors favor removal of a 3 to 4 cm segment of the entire nerve,[3] whereas others remove only a fascicle.[7] Significant complications from the procedure are rare. In a series involving 103 patients, there was one wound infection, and one patient developed a painful neuroma that required a second procedure for its removal.[8] About one-third of the patients had mild or, rarely, moderate discomfort or numbness lasting up to 2 months. Thus, although the procedure clearly should not be undertaken in the absence of a significant clinical problem that cannot be resolved in other ways, the risk and morbidity are sufficiently low that it is carried out frequently in centers that specialize in the study of peripheral nerve disease. It is emphasized that it is wisest for the biopsy to be performed by a surgeon with experience in this procedure and that great care must be taken to coordinate the biopsy and the tests to be performed.

2.2.1. *Morphological Techniques*

Apart from usual light and electron microscopic techniques, the sural nerve is studied by quantitative morphometric techniques.[3,9,10] Another valuable

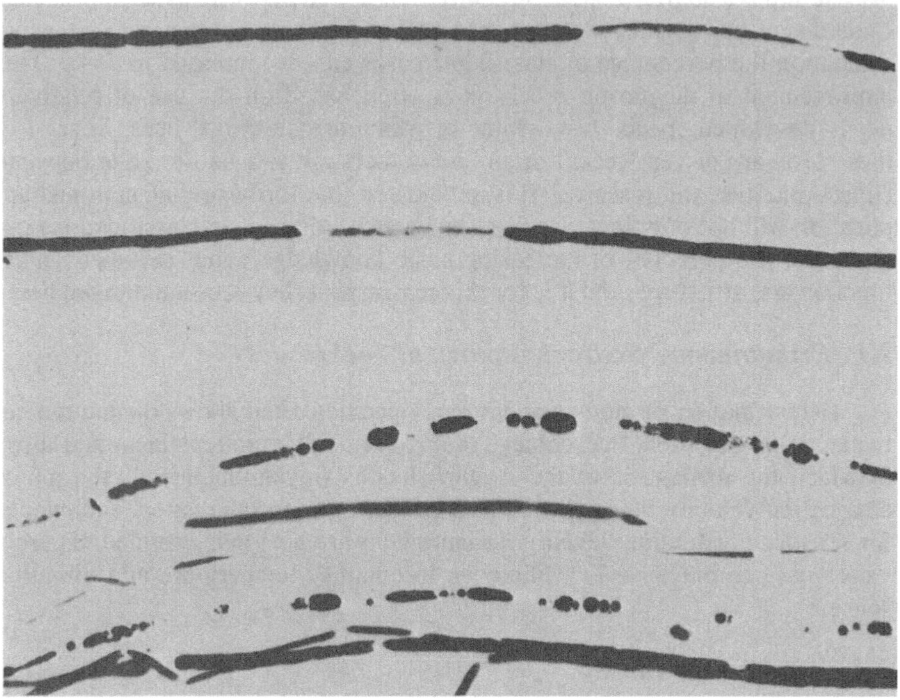

Fig. 2. Teased fiber preparation stained with osmic acid and magnified 150×. The upper two fibers demonstrated segmental demyelination, and the lower two fibers Wallerian degeneration. (From Asbury and Johnson.[3])

technique is the "teased" single fiber. With this process, single fibers as long as 1 cm can be separated from fascicular bundles. Figure 2 illustrates such a preparation and demonstrates the differences between segmental demyelination and Wallerian degeneration.

2.2.2. Biochemical Techniques

In the past, the interpretation of biochemical assays of peripheral nerve has been complicated by the fact that the relatively uninformative epineurium and perineurium make up about two-thirds of most peripheral nerves. A simple microdissection technique is now available that permits separation of endoneurial contents[11] (Fig. 3). The endoneurial fraction contained 90% of the cerebroside, sulfatide, and sterol of the total nerve sample (Fig. 4). The amount of endoneural material that can be dissected in this way from a biopsy sural nerve specimen is sufficient to allow quantitative analysis of lipids and other constituents.[12,13] By different techniques, preparations freed of epineurium suitable for metabolic studies have also been obtained from tibial nerve fascicles of experimental animals.[14]

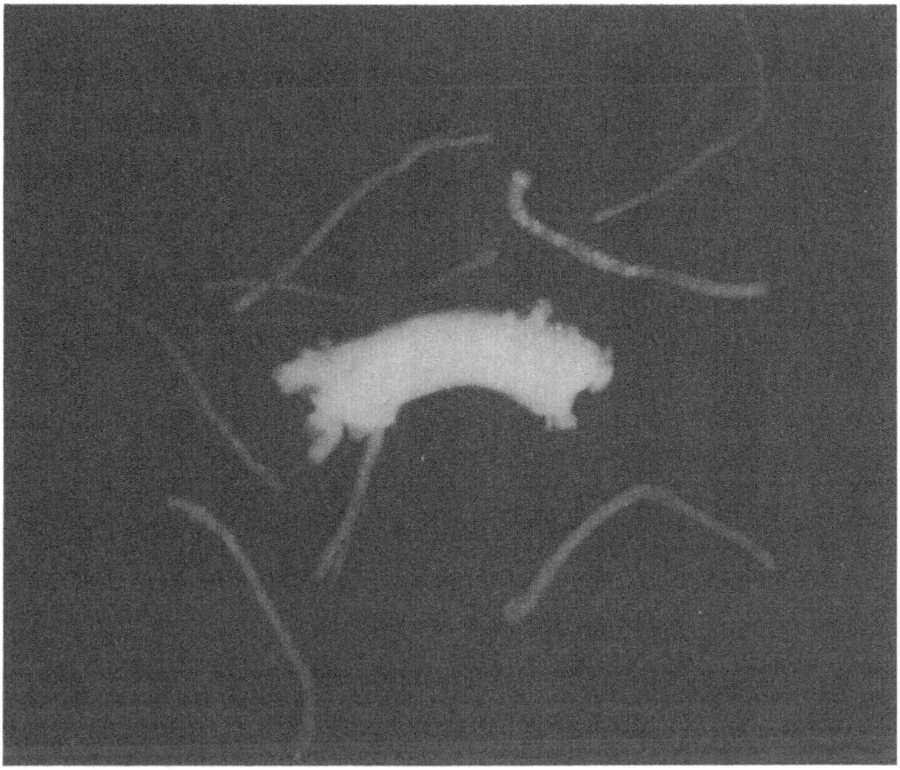

Fig. 3. Preparation of endoneurial fascicles from a human sural nerve biopsy specimen. A 1-cm segment of nerve was immersed in ice-cold saline and dissected with watchmakers' forceps under a dissecting microscope. Endoneurial fascicles can be grasped and pulled away. The figure shows seven fascicles obtained in this way. These fascicles contain most of the myelin lipids and only small amounts of collagen (Fig. 4). (From Asbury and Johnson.[3])

2.2.3. *Physiological Studies*

Sural nerve biopsy samples can be used to study conduction velocity[2] and axonal transport.[15]

2.2.4. *Xenografts*

Aguayo and associates have developed an ingenious xenograft technique for the study of human sural nerve biopsy specimens. They obtained a 5-mm sural nerve fascicle from a patient with metachromatic leukodystrophy and laid it between severed and retracted ends of the sciatic nerves of immunosuppressed mice.[16] Three weeks after the graft and later, there were numerous myelinated and unmyelinated nerve fibers within the grafted segment and the distal stump, indicating that the mouse nerve had regenerated through the region of the graft and beyond. When immunosuppression was discontinued, there was rejection of Schwann cells and perineurial cells in the region of the

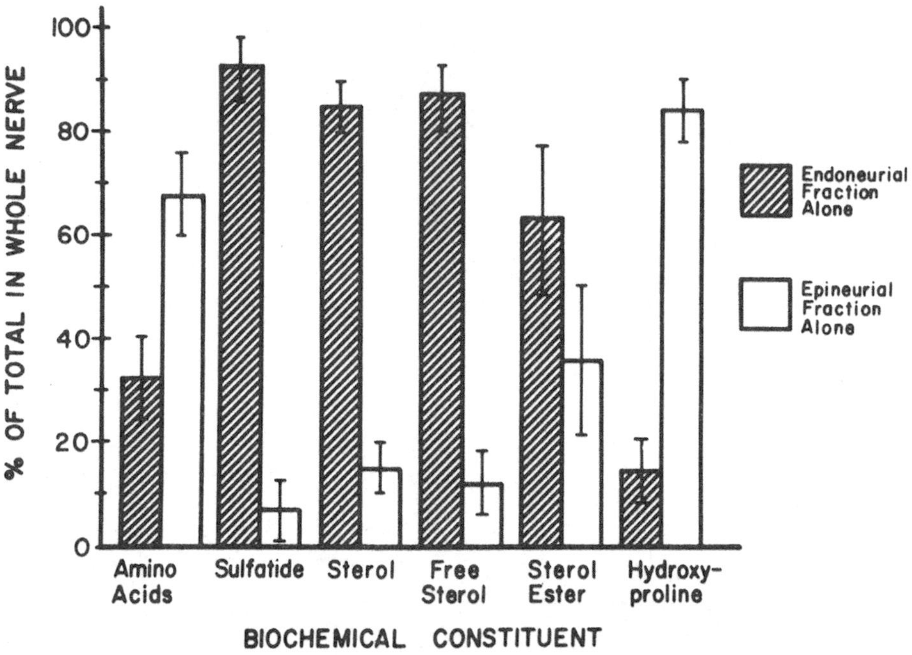

Fig. 4. Comparison of the chemical composition of endoneurial fraction (obtained as shown in Fig. 3) and the epineurial fraction, which remains behind after the endoneurial fascicles have been plucked away. Note that the endoneurial fascicles contain most of the myelin lipids (sulfatides and free sterol). (From Asbury and Johnson.[3])

graft only, indicating that these were the only cells foreign to the host. Furthermore, metachromatic lipids characteristic of metachromatic leukodystrophy were present in the Schwann cells of the grafted segment only, indicating that the genetic defect characteristic of this disorder was still expressed in the xenograft.

This system provides a powerful new method to study the pathogenesis of human peripheral nerve disease. It has been applied to a variety of disorders. Appenzeller performed such a xenograft with a sural nerve biopsy specimen from a patient with an unknown type of neuropathy associated with loss of myelin.[17] In contrast to what Aguayo *et al.* had observed with the metachromatic leukodystrophy nerve, here the grafted segment showed a normal number of fully myelinated mouse axons. This result indicates that the engrafted human Schwann cells had a normal capacity to form myelin and that the loss of myelin observed in the patient most likely was attributable to an as yet undefined pathological process in the axon.

2.2.5. Human Schwann Cell Cultures

Askanas *et al.* have developed a procedure that permits the cultivation of Schwann cells from human sural nerve biopsy specimens.[18] A critical feature is an explant-reexplantation technique that reduces the number of non-Schwann cells without the use of cytotoxic agents. The cultures could be maintained up to 14 weeks. The potential of this system for the study of human disease was demonstrated in studies with Schwann cells obtained from adrenomyeloneuropathy patients.[19] When hexacosanoic acid was added to the medium in which these cells were growing, they developed the cytoplasmic inclusions characteristic of this disease, whereas Schwann cell cultures from normal individuals did not. These studies show that the cultured Schwann cells had retained the metabolic defect and can be used to study the pathogenesis of this disorder.

3. *HUMAN NEUROPATHIES ASSOCIATED WITH DISORDERED METABOLISM*

This discussion focuses on the genetically determined disorders of peripheral nerve. In addition, we discuss the diabetic neuropathies and the neuropathies associated with paraproteinemias and dysproteinemias. We have chosen to include these last two topics because of recent advances that provide interesting new information about disease mechanisms.

3.1. Genetic Neuropathies

Even though the genetic neuropathies have been recognized for over 100 years and are of frequent occurrence, their classification is still a matter of controversy. Because they follow Mendelian patterns of inheritance, it is presumed that all or almost all are associated with a discrete enzymatic defect, but the nature of the biochemical defect has been defined in only a few instances.

We follow here the classification schemes of Dyck *et al.*[2] and Asbury and Johnson.[3] In keeping with the purpose of the chapter, emphasis is placed on those disorders in which at least some aspects of the metabolic derangement have been defined.

3.1.1. Hereditary Sensory Neuropathy Including Dysautonomia

This disorder is subdivided into four subgroups[3]: HSN I, autosomal dominant mode of inheritance; HSN II, autosomal recessive mode of inheritance; HSN III, familial dysautonomia; HSN IV, congenital insensitivity to pain. Information about biochemical mechanisms exists only in respect to HSN III, familial dysautonomia, a disorder with an autosomal recessive mode of inheritance that affects mainly autonomic and sensory function.[20] Symptoms include autonomic instability, impaired perception of pain, temperature, and taste, absent overflow tears, lack of fungiform papillae on the tongue, and loss of histamine flare response. The sural nerve lacks large-caliber myelinated axons and is depleted of nonmyelinated ones.[21]

A relationship to an abnormality of nerve growth factor has been suggested because of the resemblance to the physiological and pathological abnormalities observed in experimental animals following the administration of antibodies to nerve growth factor.[22] Further support for this relationship was provided by the demonstration that patients with familial dysautonomia had a threefold increase in serum levels of materials that cross reacted wtih the β chain of nerve growth factor in a radioimmunoassay.[23] Schwartz and Breakefield[24] tested the production of nerve growth factors by cultured skin fibroblasts of patients with dysautonomia. They found that there was no difference in the levels of nerve growth factor as measured by radioimmunoassay but found that the biological activity of β nerve growth factor was reduced to 10% of control. Although both of these studies implicate β nerve factor, the specific results are at variance: Siggers *et al.*[23] showed normal serum activity of nerve growth factor by bioassay and increased levels by radioimmunoassay, whereas the converse was found in the fibroblast assay.[24] Additional studies are required to resolve this. It has also been reported that patients with dysautonomia have a depletion of substance-P-containing axons in the substantia gelatinosa[25] and increased levels of tyrosine hydroxylase per sympathetic neuron even though the total number of these neurons was diminished.[26]

3.1.2. Hereditary Motor and Sensory Neuropathies

According to the classification of Dyck *et al*,[2] this category includes six types: HMSN type 1 includes dominantly inherited hypertrophic neuropathies including the Charcot–Marie–Tooth type; HMSN type 2 differs from type 1 in that onset of symptoms is later and nerves are not hypertrophied; type 3 is the recessively inherited disorder first described by Dejerine and Sottas and associated with marked enlargement of nerves; type 4 is Refsum's disease, to be described below; in type 5 HMNS, the neuropathy is associated with spastic paraplegia, and in type 6, with optic atrophy. Although the clinical and morphological aspects of these disorders are beyond the scope of this chapter, it is important to be aware of them because they are common, much more so than the relatively small number of genetic neuropathies about the biochemical basis of which we now have some knowledge. These are discussed below. A recent study provides important information about the relative frequency of these disorders[27]: HMSN I comprised more than half of the 287 cases studied, and HMSN II (60 cases) is the second most common.

3.1.2a. Possible Role of Abnormalities of Axonal Transport. Although the mechanisms of the most common forms of neuropathies remain obscure, there is speculation that at least some of them may be related to abnormalities of axonal transport. Insights about these mechanisms are being gained with studies of the mechanisms of neurotoxins. For example, it has been shown that the toxin B,B′-iminodiproprionitrile (IDPN) interferes with the slow axonal transport of neurofilament proteins.[29] Very recently, the same transport defect has been shown in a hereditary form of a canine spinal muscular atrophy, a probable animal analogue of human amyotrophic lateral sclerosis (ALS). This finding is of particular interest because of the resemblance between the neuropathology of human ALS and experimental IDPN intoxication.[31] A variety of neurotoxins such as acrylamide cause distal axonal neuropathies and are

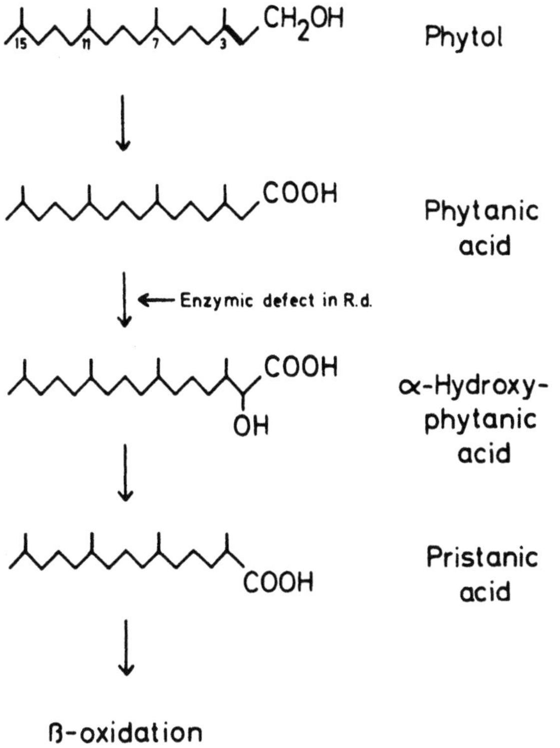

Fig. 5. The metabolic defect in Refsum's disease. (From Dyck *et al.*[2])

important experimental models of the common "dying-back" neuropathies.[28] In respect to the genetic neuropathies, it is of great interest that there is a strong resemblance between the morphological changes induced by acrylamide and a genetically determined human disorder, namely, giant axonal neuropathy.[32]

3.1.2b. Refsum's Disease. Refsum's disease is listed as HMSN IV in the classification of Dyck and Lampert.[5] It is of singular importance here, since it is the one form of neuropathy whose biochemical mechanism is beginning to be understood and for which specific therapy is now available.

It is characterized by pigmentary retinal degeneration, chronic polyneuropathy, ataxia, an increased cerebrospinal fluid protein level and, in most cases, anosmia, hearing impairment, and, at times, lens opacities and ichthyosis.[33,34] The disease is associated with the tissue accumulation of phytanic acid (3,5,7,11-tetramethylhexadecanoic acid).[35]

The metabolism of phytanic acid in mammalian tissues involves an initial conversion by α-oxidation to a fatty acid with one fewer carbon atom, pristanic acid. The pristanic acid is then oxidized by β-oxidation (Fig. 5). Formation of α-hydroxyphytanic acid is the first step. This reaction is carried out in liver mitochondria. It is stimulated by NADPH and required molecular oxygen. It is stimulated by ferric iron and inhibited by ferrous iron.[36] This liver α-oxidation system differs from that in brain, which is in the microsomal fraction[37] and is stimulated by ferrous iron.[38] The liver α-oxidation system is thus entirely distinct from the brain α-oxidation system. In Refsum's disease, the metabolic defect is the inability to convert phytanic acid to pristanic acid. This was demonstrated in cultured skin fibroblasts, and presumably this system resembles the liver α-oxidation system and differs from that in brain. Fibroblasts of Refsum's disease patients were unable to oxidize phytanic acid but oxidized pristanic acid normally (Fig. 3). α-Hydroxyphytanic was also oxidized normally, thus making it likely that the basic defect involves the α-hydroxylation of phytanic acid.[35]

It is of great therapeutic significance that in man phytanic acid is of dietary origin only. This forms the basis for the successful dietary therapy for Refsum's disease[34] and provides an analogy to phenylketonuria. The mode of inheritance is autosomal recessive. Diagnosis depends on demonstration of increased plasma levels of phytanic acid. In Refsum's disease, this makes up 5–30% of total fatty acids compared to less than 0.3 mg/dl in normal plasma.[35]

The peripheral nerves in Refsum's disease show hypertrophic neuropathy with the "onion bulb" as its pathological hallmark. This is thought to be the consequence of repeated episodes of segmental demyelination and subsequent repair and remyelination. The number of myelinated fibers is greatly reduced, and the myelin sheaths are thinner than expected for the caliber of the axon with which they are associated.[3] Motor conduction velocity is reduced greatly. Deep tendon reflexes are diminished or absent, and atrophy and sensory loss are most intense in the distal parts of the extremities. The neuropathy and other aspects of the disease are subject to exacerbations, which may be life threatening.[39]

It is of both practical and theoretical importance that dietary therapy improves the peripheral neuropathy, muscular strength, and cardiac function in Refsum's disease. The convincing evidence for this is summarized in a brief recent article by Professor Refsum.[34] Dietary restriction of phytanic acid and phytol intake lowers plasma phytanic acid levels (Fig. 6). The degree of improvement clearly correlates with the degree to which plasma phytanic acid levels have been reduced.[34,39] Most impressive are the results with one of Professor Refsum's original patients, who, 34 years after the onset of his illness, is in good general health and employed full time. Ulnar nerve conduction velocity is 30 m/sec compared to 7 before therapy, and the cerebrospinal fluid protein is now normal (Table II).

Refsum's Norwegian patients may be the only ones in whom dietary therapy has reduced phytanic acid levels to normal for long periods of time. Most likely, this reflects the great care and commitment to dietary supervision. Our own more limited experience with dietary therapy alone has not been as successful in that we were unable to reduce the phytanic acid level to less than 20 mg/dl. We found that the combination of diet therapy and plasma exchange at 2 to 3 week intervals has allowed us to maintain plasma phytanic acid levels at 10 mg/dl,[40] and a similar experience was reported by Gibberd *et al.*[41] The observation that the reduction of phytanic acid levels alone leads to improvement in Refsum's disease indicates that this substance or a metabolite derived from it is the toxic agent. The cross-sectional area of phytanic acid is 50%

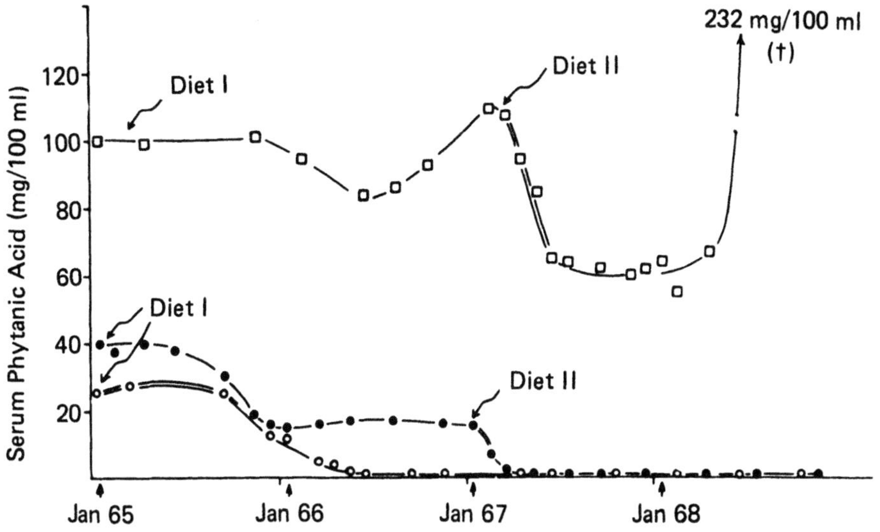

Fig. 6. Three patients were started on a diet that restricts phytanic acid and phytol intake. The patients whose blood values are shown with the open and closed circles achieved improvement in peripheral nerve and cardiac function and stabilization of hearing and vision. The patient shown with the open squares was unable to adhere to the diet and died as a result of cardiac and peripheral nerve disease complications. (From Dyck *et al.*[2])

greater than that of hexadecanoic acid devoid of methyl groups, and O'Brien has proposed that the accumulation of the phytanic acid in the myelin sheath of Refsum's disease patients disrupts myelin structure because of this steric factor.[42]

3.1.3. Adrenoleukodystrophy

Adrenoleukodystrophy (ALD), which is discussed in Chapter 8, is described here because it probably represents another disorder of fatty acid metabolism. There is tissue accumulation of saturated very-long chain fatty acids as a result of a genetically determined inability to degrade them.[43] The most convenient diagnostic method is the demonstration of abnormally high levels of saturated very-long-chain fatty acids, particularly C26:0 and C24:0, in plasma.[44]

Table II
Effects of Dietary Treatment in a Patient with Refsum's Disease[a]

	1965	1966	1967	1968	1972
Phytanic acid (mg/100 ml)	37	17	2	2	<0.5
Nerve conduction velocity (m/sec)	7	19	20	25	30
Cerebrospinal fluid protein (mg/100 ml)	135	120	82	70	70

[a] Values are for serum phytanic acid, conduction velocity in the ulnar nerve of the left forearm, and the protein content of the cerebrospinal fluid of patient K.M. during dietary treatment. From Dyck *et al.*[2]

Adrenoleukodystrophy presents most commonly in childhood, with behavioral disturbances, dementia, visual disturbances, and motor disability as a result of demyelination in the cerebral hemispheres. Peripheral neuropathy is an infrequent occurrence in childhood ALD. Less commonly, ALD presents in adulthood as a more slowly progressive disorder that involves mainly the spinal cord. This disorder, which is probably not fundamentally distinct from ALD, is referred to as adrenomyeloneuropathy (AMN).[45] Mild to moderate peripheral neuropathy may be present and manifests by diminished ankle jerks and distal sensory impairment. Motor conduction velocities are mildly slowed. Sural nerve biopsy shows slight to moderate loss of myelinated fibers. It also shows clusters of small myelinated fibers surrounded by loose Schwann cell basal lamina and, occasionally, small Schwann cell "onion-bulbs," both of which are evidence of axonal and myelin regeneration. The characteristic ultrastructural inclusions that are seen regularly in adrenal cortex and in brain macrophages are present only infrequently in the peripheral nervous system. Unfortunately, in contrast to Refsum's disease, in which dietary phytanic restriction can reverse the neuropathy, dietary restriction of very-long-chain fatty acids does not appear to benefit patients with adrenoleukodystrophy.[46] This is probably because very-long-chain fatty acids are both of endogenous and exogenous origin.

3.1.4. A New Peripheral Nerve Disorder Associated with Adrenal Insufficiency

Recently, Dyck *et al.* described two brothers with what appears to be a new entity in which adrenal insufficiency is associated with hepatosplenomegaly, dementia, and peripheral neuropathy. Changes in plasma and nerve fatty acids were also demonstrated but involved the polyunsaturated fatty acids rather than the saturated very-long-chain fatty acids, which are abnormal in adrenoleukodystrophy.[47] The disorder thus appears to be entirely distinct from ALD.

3.1.5. Fabry's Disease

Fabry's disease is an X-linked disorder associated with the accumulation of ceramide trihexoside in lysosomes of endothelial and perithelial cells. The biochemical basis of the disorder is discussed in Chapter 5.

Neuropathy is a prominent part of the clinical picture, characterized by autonomic dysfunction and crises of pain.[48] Metric studies of sural nerve biopsy specimens have shown loss of larger unmyelinated and small myelinated fibers. The large myelinated fibers are relatively preserved.[49] Abnormal lipid accumulations are demonstrated in the cytoplasm of perineural and endothelial cells of the vasa nervorum. The axons are swollen, and their internal organelles are lost.[50] In contrast to what is observed in metachromatic leukodystrophy, the Schwann cells are not involved, and this may account for the sparing of the large myelinated fibers. It has been postulated that Schwann cells and a large myelin sheath act as a net barrier to the deposition of the trihexoside.[50] The distinctive distribution of peripheral nerve fiber damage may result both from involvement of autonomic ganglia[49] and from the higher susceptibility of large myelinated fibers to the deposition of lipid.[50] As a practical point, it has been found that phenytoin is of value for the therapy of neuropathy-related pain in patients with Fabry's disease.[51]

3.1.6. Metachromatic Leukodystrophy

Peripheral neuropathy is an important clinical manifestation and diagnostic sign in metachromatic leukodystrophy. Metachromatic leukodystrophy is an autosomal recessive disorder in which there is a deficiency of the enzyme arylsulfatase A or cerebroside sulfatase. This defect causes the accumulation of sulfatide, an acidic lipid that has metachromatic staining properties. Sulfatide accumulates in central and peripheral myelin.

Peripheral nerve involvement is a constant feature in metachromatic leukodystrophy, and the reliability of sural nerve biopsy for the diagnosis of metachromatic leukodystrophy is well established.[52] There is segmental demyelination and the characteristic accumulation of metachromatic lipids in macrophages and in areas of myelin disintegration with preservation of the axon. The electron microscope reveals several types of inclusions in macrophages and Schwann cells.[53] The most specific of these are "tuffstone" bodies, which contain irregularly arranged material with a periodicity of 5.8 nm, and which probably represent sulfatide. It is of interest that they are also present in Schwann cells surrounding unmyelinated axons.[53]

The pattern of nerve involvement in metachromatic leukodystrophy stands in sharp contrast to that seen in Fabry's disease. It is the large myelinated fibers that are involved, and study of teased fibers has shown that the pattern is characteristic of segmental demyelination,[54] that is, the myelin disturbance relates to a Schwann cell rather than to the axon. The prime involvement of the Schwann cell is demonstrated by the xenograft studies of Aguayo,[16] which we have already referred to. When a sural nerve fascicle from an MLD patient was engrafted into mouse sciatic nerve, the characteristic nerve pathology of MLD was demonstrated by the human MLD Schwann cells that had ensheathed a normal mouse axon. Xenograft studies with Fabry's disease sural nerve fascicles again highlight the difference between the two diseases: here the abnormal material was found in the perineurial cells.[55]

Apart from its importance for the morphological confirmation of the diagnosis of MLD, the demonstration of the neuropathy is of diagnostic value in two other settings. First, the combination of neuropathy and dementia may be the first lead to the diagnosis of the adult form of MLD.[56] Second, the presence or absence of peripheral neuropathy helps to distinguish between adult MLD and the entity of "pseudoarylsulfatase deficiency." The latter term refers to persons who have low arylsulfatase A (in the range of the ALD homozygote) but who show no other sign of disease and who do not accumulate excess sulfatide.[57] The mechanism and nature of pseudoarylsulfatase A deficiency are not yet clear. It probably has little or no clinical significance except that it adds immensely to the problems of genetic counseling. One of the challenges is to distinguish presymptomatic adult MLD (an entity of great clinical importance) from the clinically insignificant pseudoarylsulfatase deficiency. It is likely that all patients with adult MLD do have peripheral neuropathy demonstrable by electrophysiological studies or sural nerve biopsy,[57] and this would not be true in pseudoarylsulfatase deficiency.

3.1.7. Globoid Leukodystrophy

The basic defect in globoid leukodystrophy is the deficiency of the enzyme galactocerebroside-β-galactosidase (see Chapter 5). The central nervous system shows accumulation of characteristic multinucleate globoid cells, deficiency and destruction of myelin, and loss of oligodendrocytes. This is associated with progressive spastic paralysis and blindness and usually leads to death before the age of 2 years. The central nervous system manifestations of globoid leukodystrophy are so severe that they overshadow peripheral nerve involvement. Nevertheless, involvement of peripheral nerve is a constant feature in the most common infantile form of the disease[58] but probably not in the late-onset globoid cell leukodystrophy. There is reduction of large myelinated fibers with preservation of unmyelinated fibers. Studies of teased preparations have revealed intermittent shortening of myelinic internodes with resulting disparity of internodal lengths along individual fibers, changes that are characteristic of segmental demyelination and remyelination.[59] Schwann cells and macrophages in peripheral nerve contain characteristic ultrastructural inclusions with the appearance of hollow twisted tubules, which probably represent deposits of galactocerebroside. Peripheral nerves do not contain the multinucleate globoid cells so characteristic of the central nervous lesions. Motor conduction velocity is decreased in infantile globoid leukodystrophy. This has been noted as early as 7 weeks of age.[60] The "twitcher" mouse represents an animal analogue of human globoid leukodystrophy. Transplants have shown that the neuropathy of this globoid leukodystrophy is caused by Schwann cell dysfunction.[61]

3.1.8. Disorders of Lipoprotein Metabolism: Tangier Disease (Familial HDL Deficiency)

Tangier disease is a rare disorder characterized by absence or severe deficiency of plasma high-density lipoproteins (HDL), hepatosplenomegaly, yellowish discoloration of tonsils, and abnormal accumulation of cholesteryl esters in a large number of tissues. There is an autosomal recessive mode of inheritance. A distinctive type of peripheral neuropathy was present in 13 of the 26 reported cases.[62] Diagnosis is made by demonstrating a low blood cholesterol level and the absence of HDL (or α-lipoprotein) by electrophoresis.

The neuropathy is quite distinctive. There is diminution or loss of pain and temperature sensation over the face, trunk, and proximal extremities, weakness of facial muscles, and weakness and atrophy of intrinsic hand muscles. Deep tendon reflexes and proprioception are relatively preserved.[63,64] This constellation of signs may lead to the erroneous diagnosis of syringomyelia. However, in Tangier disease there is no evidence of spinal cord involvement, and, once suspected, the diagnosis is readily established by measurement of blood cholesterol and HDL levels.

Nerve biopsy also yields striking and characteristic results. In the involved nerves, there is a striking reduction in the number of small-diameter myelinated nerves. The elegant serial studies by Dyck *et al.*[64] have shown a progressive reduction of these fibers in the relatively mildly involved sural nerve. A characteristic finding was the accumulation of sudanophilic droplets in degenerating fibers and in the cytoplasm of Schwann cells. The cholesteryl ester content of a sural nerve biopsy specimen was increased. It is likely that the lipid droplets in degenerating fibers and in Schwann cells represent cholesteryl esters or other lipids and that their accumulation relates in some way to the deficiency of plasma HDL. High-density lipoproteins are thought to play a role in the removal of lipids from cells, but the exact mechanisms are far from clear.

3.1.9. Abetalipoproteinemia (Bassen–Kornweig Syndrome)

Abetalipoproteinemia is associated with diminished deep tendon reflexes, impaired vibration and position sense, and ataxia. Most likely this is caused by degeneration of the posterior columns in the spinal cord. Peripheral nerves have not been studied in detail, and the extent of peripheral nerve involvement is uncertain.[62,64]

3.1.10. Disorders of Sterol Metabolism: Cerebrotendinous Xanthomatosis

Cerebrotendinous xanthomatosis (CTX) is a rare progressive disorder characterized by xanthomas of Achilles or other tendons, juvenile cataract, dementia, progressive cerebellar involvement, and peripheral neuropathy.[65] The mode of inheritance is autosomal recessive. Fifty cases have been described.[66] There is pathognomonic accumulation of cholestanol in nervous tissues and in the tendon xanthoma and a subtle but diagnostic increase of the same substance in plasma. The basic defect appears to be an impairment of the production of certain bile acids, which, possibly through defective feedback inhibition, leads to overproduction of both cholesterol and cholestanol.[67] It is encouraging to note that oral administration of chenodeoxycholic acid reduces plasma cholestanol levels and appears to stabilize (at least) the clinical abnormalities.[66] Cerebrotendinous xanthomatosis, along with Refsum's syndrome, may represent a treatable lipidosis.

Mild peripheral nerve involvement was demonstrable in three out of four patients and was evidenced mainly by slowed motor sensory conduction.[69] Sural nerve biopsy showed a slight decrease in the density of large myelinated fibers. Segmental demyelination and remylination and "onion bulbs" were demonstrated (a result of repeated cycles of demyelination and remyelination).[70]

3.1.11. Genetic Disorders of Carbohydrate Metabolism

Deposition of glycogen and of lipofuscin has been reported in Pompe's disease (glycogenosis II). However, there is little or no disturbance of peripheral nerve myelin.[71,72] Neurogenic atrophy of muscle was observed in a 15-year-old patient with Hunter syndrome (mucopolysaccharidosis II) and was attributed to a combination of neuronal ganglioside storage and a distal neuropathy.[73] Compression neuropathies such as the carpal tunnel syndrome occur with increased frequency in the mucopolysaccharidoses.

3.1.12. Porphyric Neuropathy

Attacks of neuropathy are a complication of the hepatic porphyrias and may be severe and fatal. It is generally thought to be an axonal neuropathy along with some degree of segmental demyelination.[3] A recent morphometric study showed a severe reduction in myelinated fibers in the sural and tibial nerves with a normal number of fibers in ventral and dorsal nerve roots. Unmyelinated fibers were also abnormal, and study of a sural nerve biopsy obtained during the early stage of the attack indicated that the initial damage was to the axon.[74] The mechanism of porphyric neuropathy is poorly understood. The toxic action is thought to be related in some way to the accumulation of porphyrin precursors such as δ-aminolevulinic acid or porphobilinogen, since neuropathy occurs only in those porphyrias in which the levels of these substances are elevated.[75]

3.2. Nongenetic Neuropathies

3.2.1. Diabetic Neuropathies

Neuropathy is a frequent complication of diabetes and thus represents an important clinical problem. The neuropathy takes a variety of forms. Asbury and Johnson[3] subdivide them into four types: (I) distal symmetrical primarily sensory neuropathy, (II) autonomic neuropathy, (III) proximal asymmetric painful primarily motor neuropathy, and (IV) cranial mononeuropathy. The last two types are caused by vascular lesions and are not discussed here.

In type I (distal symmetrical primarily sensory neuropathy), the major pathological finding is loss of axons, most pronounced in distal nerves,[3] although in chronic insidious cases, there is concomitant segmental demyelination as evidenced by slowing of nerve conduction velocity and teased nerve fiber studies.

Recent laboratory and clinical reports provide some intriguing new insights into the biochemical mechanisms of diabetic neuropathy. These reports focus on the role of *myo*-inositol, phosphoinositides, and sorbitol and are based on *in vivo* studies in animals made experimentally diabetic with alloxan, on metabolic studies with nerve fascicle preparations derived from these animals,[76] and on a multicenter clinical trial.[77] The results and rationale are summarized in a recent editorial.[78] It had been proposed for some time that *myo*-inositol may have a role in diabetic neuropathy. *myo*-inositol levels in axons and Schwann cells are normally 100 times higher than in plasma, and they are decreased in animals with experimental diabetes. The slowed nerve conduction velocity observed in these animals can be prevented if normal *myo*-inositol levels can be maintained. Recent studies have led to the hypothesis that the lowered nerve *myo*-inositol level in diabetic animals may act by interfering with the rapid turnover of phosphoinositide, which involves the release and reincorporation of arachidonic acid.[79] When albumen-bound arachidonic acid is added to the medium of a normal nerve fascicle preparation, oxygen consumption is increased by 22 to 31%. This increase in oxygen consumption fails to occur when the tissue is depleted of *myo*-inositol and is restored when *myo*-inositol is added back. The depletion of *myo*-inositol also decreased energy use from Na^+, K^+-ATPase activity.

Other studies on the biochemical mechanisms of diabetic neuropathy have pointed to a role of the sugar alcohol sorbitol, which is present in increased concentration in nerve and lens of some diabetic patients and in animals with experimental diabetes. The sorbitol is derived from glucose by the action of aldose reductase and then converted to fructose. Long-term administration of an aldose reductase inhibitor to animals with experimental diabetes lowers nerve sorbitol and restores normal nerve conduction velocity,[80] and very recently, a multicenter double-blind crossover clinical trial of an aldose reductase inhibitor reported small but significant improvements in motor conduction velocity in diabetic patients.[77] It has been postulated that the aldose reductase inhibitor may act by preventing the reduction in nerve *myo*-inositol levels.[78]

3.2.2. *Neuropathies Associated with Monoclonal Gammopathies*

Neuropathies associated with paraproteinemias and dysproteinemias represent troublesome complications of these disorders. Dyck *et al.*[5] have emphasized the importance of screening patients with unclassified neuropathies for these disorders. Two recent reports have contributed to our understanding of these disorders by demonstrating that the myelin-associated glycoprotein of peripheral nerve was the antigen for the monoclonal IgM in the serum of two patients with this type of neuropathy.[81,82]

4. *CONCLUDING REMARKS*

The human neuropathies represent an extremely diverse group of disorders. In spite of intensive clinical interest and the relative accessibility of the peripheral nervous system, the classification and understanding of these disorders are still very limited. This also applies to large groups of genetically determined neuropathies. For a small fraction of these disorders, the basic biochemical defect is beginning to be understood, and these disorders have been reviewed in this chapter. A summary was also presented about new morphological, physiological, biochemical, and tissue culture techniques that provide the hope that our understanding of the mechanism and therapy of these disorders will improve.

REFERENCES

1. Schoenberg, B. S., 1982, *International Conference on Peripheral Neuropathies* (S. Refsum, C. L. Bolis, and A. Portera Sanchez, eds.), Elsevier, Amsterdam, pp. 49–61.

2. Dyck, P. J., Thomas, P. K., and Lampert, E. H., 1975, *Peripheral Neuropathy*, W. B. Saunders, Philadelphia.
3. Asbury, A. K., and Johnson, P. C., 1978, *Pathology of Peripheral Nerve*, W. B. Saunders, Philadelphia.
4. Dyck, P. J., 1982, *N. Engl. J. Med.* **307**:283–286.
5. Dyck, P. J., Oviatt, K. F., and Lampert, E. H., 1981, *Ann. Neurol.* **10**:222–226.
6. Blom, S., 1981, *Neuropediatrics* **12**:3–8.
7. Dyck, P. J., and Lofgren, E. P., 1968, *Med. Clin. North. Am.* **52**:885–893.
8. Asbury, A. J., and Connolly, E. H., 1973, *J. Neurosurg.* **38**:391–392.
9. O'Sullivan, D. J., and Swallow, M., 1968, *J. Neurol. Neurosurg. Psychiatry* **31**:464–470.
10. Behse, F., Buchthal, F., Carlsen, F., and Knappeis, G. G., 1974, *Brain* **97**:773–784.
11. Brown, M. J., Pleasure, D. E., and Asbury, A. K., 1976, *J. Neurol. Sci.* **29**:361–369.
12. Brown, M. J., Iwamori, M., Kishimoto, Y., Rapoport, B., Moser, H. W., and Asbury, A. K., 1978, *Ann. Neurol.* **5**:245–252.
13. Yao, J. K., Dyck, P. J., VanLoon, J. A., and Moyer, T. P., 1981, *J. Neurochem.* **36**:1211–1218.
14. Greene, D. A., Winegrad, A. I., Carpentier, J. L., Brown, M. J., Fukuma, M., and Orci, L., 1979, *J. Neurochem.* **33**:1007–1018.
15. Brimijoin, S., Capek, P., and Dyck, P. J., 1973, *Science* **180**:1295–1297.
16. Aguayo, A. J., Kasarjian, J., Skamene, E., Kongshavn, P., and Bray, G. M., 1977, *Nature* **268**:753–755.
17. Appenzeller, O., Kornfeld, M., and Atkinson, R., 1979, *Ann. Neurol.* **7**:251–261.
18. Arkanas, V., Engel, W. K., Dalakas, M. C., Lawrence, J. V., and Carter, L. S., 1980, *Arch. Neurol.* **37**:329–337.
19. Askanas, V., Moser, H. W., Engel, W. K., Moser, A. B., Siddique, T., and Bosch, P., 1981, *Neurology (N.Y.)* **31**:128.
20. Riley, C. M., Day, R. L., Greeley, D. M., and Langford, W. S., 1949, *Pediatrics* **3**:468–478.
21. Aguayo, A. J., Nair, C. P. V., and Bray, G. M., 1971, *Arch. Neurol.* **24**:106–116.
22. Pearson, J., Axelrod, F., and Dancis, J., 1974, *Ann. N.Y. Acad. Sci.* **228**:288–300.
23. Siggers, D. C., Rogers, J. E., Boyer, S. H., Margolet, L., Dorkin, H., Banerjee, S. P., and Shooter, E. C., 1976, *N. Engl. J. Med.* **259**:629–634.
24. Schwartz, J. P., and Breakefield, X. O., 1980, *Proc. Natl. Acad. Sci. U.S.A.* **77**:1154–1158.
25. Pearson, J., Brandeis, L., and Cuello, C., 1979, *Nature* **295**:61–63.
26. Pearson, J., Brandeis, L., and Goldstein, M., 1979, *Science* **206**:71–72.
27. Hagberg, B., and Lyon, G., 1981, *Neuropediatrics* **12**:9–17.
28. Spencer, P. S., and Schaumburg, H. H., (eds), 1980, *Experimental and Clinical Neurotoxicology*, Williams and Wilkins, Baltimore.
29. Griffin, J. W., Hoffman, P. N., Clark, A. W., Carroll, P. T., and Price, D. L., 1978, *Science* **202**:633–635.
30. Griffin, L. W., Cork, L. C., Adams, R. J., and Price, D. L., 1982, *J. Neuropathol. Exp. Neurol.* **41**:370.
31. Griffin, J. W., and Price, D. L., 1976, *Amyototrophic Lateral Sclerosis* (J. M. Andrews, R. T. Johnson, M. A., Brazier, eds.), Academic Press, New York, pp. 33–60.
32. Asbury, A. K., Gale, M. K., Cox, S. C., Baringer, J. R., and Berg, B. O., 1972, *Acta Neuropathol.* **20**:237–247.
33. Refsum, S., 1946, *Acta Psychiatr. Neurol. Scand.* [*Suppl.*] **38**:1–303.
34. Refsum, S., 1981, *Arch. Neurol.* **38**:605–606.
35. Steinberg, D., 1982, *Metabolic Basis of Inherited Disease*, 5th ed. (J. B. Stanbury, J. B. Wyngaarden, D. S. Fredrickson, J. L. Goldstein, and M. S. Brown, eds.), McGraw-Hill, New York, pp. 731–747.
36. Tsai, S. C., Arigan, J., and Steinberg, D., 1969, *J. Biol. Chem.* **244**:2682–2692.
37. Hoshi, M., and Kishimoto, Y., 1973, *J. Biol. Chem.* **248**:4123.
38. Davies, W. E., Hajra, A. K., Parmar, S. S., Radin, N. S., and Mead, J. F., 1966, *J. Lipid Res.* **7**:270–276.
39. Lundberg, A., Lilja, L. G., Lundberg, P. O., and Try, K., 1972, *Eur. Neurol.* **8**:309–324.

40. Moser, H. W., Braine, H., Pyeritz, R. E., Ullman, D., Murray, C., and Asbury, A. K., 1980, *Enzyme Therapy in Genetic Diseases*, 2 (R. J. Desnick, ed.), Alan R. Liss, New York, pp. 491–497.
41. Gibberd, F. B., Billimoria, J. D., Page, N. G. R., and Retsas, S., 1979, *Lancet* **1**:575–578.
42. O'Brien, J. D., 1967, *J. Theor. Biol.* **15**:207–327.
43. Singh, I., Moser, H. W., Moser, A. B., and Kishimoto, Y., 1981, *Biochem. Biophys. Res. Commun.* **102**:1223–1229.
44. Moser, H. W., Moser, A. B., Frayer, K. K., Chen, W. C., Schulman, J. D., O'Neill, B. P., and Kishimoto, Y., 1981, *Neurology (N.Y.)* **31**:1241–1249.
45. Griffin, J. W., Goren, E., Schaumburg, H., Engel, K., and Loriaux, L., 1977, *Neurology (Minneap.)* **27**:1107–1113.
46. Brown, F. R., III, Van Duyn, M.A.S., Moser, A. B., Schulman, J. D., Rizzo, W. B., Snyder, R. D., Murphy, J. V., Kamoshita, S., Migeon, C. J., and Moser, H. W., 1982, *Johns Hopkins Med. J.* **151**:164–170.
47. Dyck, P. J., Yao, J. K., Knickerbocker, D. E., Holman, R. T., Gomes, M. R., Hayles, A. B., and Lambert, E. R., 1981, *Neurology (N.Y.)* **31**:925–934.
48. Cable, W. J. L., Kolodny, E. H., and Adams, R. D., 1982, *Neurology (N.Y.)* **32**:498–502.
49. Sima, A. A. F., and Robertson, D. M., 1978, *Arch. Neurol.* **35**:291–299.
50. Cable, W. J. L., Dvorak, A. M., Osage, J. E., and Kolodny, E. H., 1982, *Neurology (N.Y.)* **32**:347–353.
51. Lockman, L. A., Hunninghake, D. B., Krivit, W., and Desnick, R. J., 1973, *Neurology (Minneap.)* **23**:871–875.
52. Olsson, Y., and Sourander, P., 1969, *Acta Pediatr. Scand.* **58**:15–24.
53. Thomas, P. K., King, R. H. M., Kocen, R. S., and Brett, E. M., 1977, *Acta Neuropathol.* **39**:237–245.
54. Dayan, A. D., 1967, *J. Neurol. Neurosurg. Psychiatry* **30**:311–318.
55. Ohnishi, A., Tateishi, J., Matsumoto, T., Shida, K., and Kuroiwa, Y., 1979, *Neurology (Minneap.)* **29**:899–901.
56. Bosch, E. P., and Hart, N. H., 1978, *Arch. Neurol.* **35**:475–477.
57. Lott, I., Dulaney, J. T., Milunsky, A., Hoefnagel, D., and Moser, H. W., 1976, *J. Pediatr.* **89**:438–440.
58. Dunn, H. G., Lake, B. D., Dolman, G. C., and Wilson, J., 1969, *Brain* **92**:329–344.
59. Schlaepper, W. W., and Prensky, A. L., 1972, *Acta Neuropathol.* **20**:55–66.
60. Lieberman, J. L., Oshtory, M., Taylor, R. G., and Dreyfus, P. M., 1980, *Arch. Neurol.* **37**:446–447.
61. Scaravilli, F., and Jacobs, J. M., 1981, *Nature* **290**:56–58.
62. Herbert, P. N., Assman, G., Gorro, A. M. Jr., and Fredrickson, D. S., 1982, *Metabolic Basis of Inherited Disease* (J. B. Stanbury, J. B. Wyngaarden, D. S. Fredrickson, J. L. Goldstein, and M. S. Brown, ed.), McGraw-Hill, New York, pp. 589–621.
63. Kocen, R. S., Lloyd, J. K., Lascelles, P. T., Fosbrooke, A. S., and William, D., 1967, *Lancet* **1**:1341–1345.
64. Dyck, P. J., Ellefson, R. D., Yao, J. K., and Herbert, P. N., 1978, *J. Neuropathol. Exp. Neurol.* **37**:119–137.
65. VanBogaert, L., Scherer, H. J., and Epstein, E., 1937, *Une Forme Cerebrale de la Cholesterinose Generalisee*, Masson, Paris.
66. Berginer, V. M., and Abeliovich, D., 1981, *Am. J. Med. Genet.* **10**:151–157.
67. Salen, G., Shefer, S., Cheng, F. W., Dayal, B., Batta, A. K., and Tint, G. S., 1979, *J. Clin. Invest.* **63**:38–44.
68. Salen, G., Miewether, T. W., and Nicolou, G., 1975, *Biochem. Med.* **14**:57–74.
69. Kuritzky, A., Berginer, V. M., and Korczyn, A. D., 1979, *Neurology (Minneap.)* **29**:880–881.
70. Ohnishi, A., Yamashita, Y., Goro, I., Kuroiwa, Y., Murakami, S., and Ikeda, M., 1979, *Acta Neuropathol.* **45**:43–45.
71. Araoz, C., Sun, C. N., Shenefelt, R., and White, H. J., 1974, *Neurology (Minneap.)* **24**:739–742.
72. Goebel, H. H., and Lenard, H. G., Kohlschutter, A., and Pilz, H., 1977, *Ann. Neurol.* **2**:111–115.

73. Schmitt, H. P., 1981, *Neuropediatrics* **12:**83–91.
74. Thorner, P. S., Bilbao, J. M., Sima, A. A. F., and Briggs, S., 1981, *Can. J. Sci. Neurol.* **8:**281–287.
75. Sima, A. A. F., Kennedy, J. C., Blakeslee, D., and Robertson, D. M., 1981, *Can. J. Neurol. Sci.* **8:**105–114.
76. Greene, D. A., and Winegrad, A. I., 1981, *Diabetes* **30:**967–974.
77. Judzewitsch, R. G., Jaspan, J. B., Polonsky, K. S., Weinberg, C. R., Halter, J. B., Halar, E., Pfeifer, M. A., Vukadinovic, C., Bernstein, L., Schneider, M., Liang, K. Y., Gabbay, K. H., Rubenstein, A. H., and Porte, D., Jr., 1982, *N. Engl. J. Med.* **308:**119–125.
78. Winegrad, A. I., Simmons, D. A., and Martin, D. B., 1982, *N. Engl. J. Med.* **308:**152–154.
79. Simmons, D. A., Winegrad, A. I., and Martin, D. B., 1982, *Science* **217:**848–851.
80. Yue, D. K., Hanwell, M. A., Satchell, P. M., and Turtle, J. R., 1982, *Diabetes* **31:**789–794.
81. Braun, P. E., Frail, D. E., and Latov, N., 1982, *J. Neurochem.* **39:**1261–1265.
82. Steck, A. J., Murray, N., Meier, C., Page, N., and Perruisseau, G., 1983, *Neurology (N.Y.)* **33:**19–23.

7

RNA in Degenerative Diseases, Aging, and Intoxication

Henrik A. Hartmann

1. INTRODUCTION AND METHODS

In bringing together information about RNA in degeneration, aging, and intoxication of the nervous system, the predominant thought has been that changes in RNA must be of crucial importance to protein synthesis and other vital functions of the neuron and that, therefore, a lack of the proper type of RNA may be the cause of neurological diseases. Many investigators have established that changes in RNA do take place. One of the difficulties in comparing the results from different laboratories is that they have been obtained with different methods so that the data are not immediately commensurable.

An extremely accurate method was developed by Edstrom.[1] It uses microdissection of single nerve cells by micromanipulation followed by extraction of RNA by ribonuclease and measurement of absorption at 260 nm. It measures RNA in picograms per neuron, and the base ratios can be determined after hydrolysis and microelectrophoresis on a nitrocellulose fiber.

Another type of method measures RNA absorption directly in a tissue section. Shea[2] developed a method measuring absolute amounts of RNA bound stoichiometrically to a dye, azure B. There is no doubt that this method is quantitatively accurate; however, when it is used on tissue sections such as done by Mann *et al.*[3] at 16 μm thickness, only part of a cell will be measured. This explains why by their method a nerve cell from the hypoglossal nerve only contains 32.9 pg. Using the Edstrom method, Uemura and Hartmann[4] found 222.7 pg in the same cell. At any rate, both methods have been used to measure differences in RNA in various diseases.

Today most investigators use polyacrylamide gel electrophoresis[5] to separate different types of RNA. This is usually done after extraction of homogenates with phenol to obtain large enough amounts of RNA, although bulk isolation of neurons[6] can be achieved.

Henrik A. Hartmann • Department of Pathology, University of Wisconsin Medical School, Madison, Wisconsin 53706.

To what extent incorporation studies with isotopes such as [^{3}H]uridine can be used in postmortem human diseased tissue remains an open question.

2. RNA IN MOTOR NEURON DISEASES OF MAN

Apparently, Koenig was the first[7] to suggest that reduced RNA might be of etiologic significance in amyotrophic lateral sclerosis (ALS). He produced a myelopathy in cats by intrathecal injection of fluoropyrimidines (5-fluoroorotic acid and 5-fluorouridine) or actinomycin D.[8]

Later, Mann and Yates found that nerve cells from patients with motor neuron disease showed a marked reduction of RNA measured as absorption at 260 nm directly on the tissue section.[9]

Applying the single-cell method of Edstrom, Davidson *et al.*[10,11] found that in ALS there was a 42% reduction of RNA from 513 pg to 300 pg in anterior motor horn neurons from the cervical enlargement and a 31% reduction of RNA from 557 pg to 386 pg in the lumbar enlargement. The neurons of nucleus dorsalis, a sensory nucleus, showed no such change, averaging 280 pg versus 259 in controls. When base ratios of RNA were determined by microelectrophoresis, adenine was significantly reduced to 15.52% from 17.47% in anterior motor neurons from the cervical and lumbar spinal cord.[12]

To determine what type of RNA might be primarily affected, it is necessary to obtain larger amounts of RNA from human spinal cords. Hartmann and Davidson[13] used the gray anterior horn substance from the cervical and lumbar enlargement and isolated RNA by the technique of Dingman and Sporn[14] utilizing phenolic extraction. The RNA was separated by polyacrylamide electrophoresis according to the method of Loening.[5] In several patients, Hartmann *et al.* found that there is a reduction of rRNA, namely, both 28 S and 18 S, whereas tRNA, 4 S, appears increased.[15] This is in keeping with electronmicroscopic studies by Hirano,[16] which reveal a marked loss of ribosomes. The reason for the loss of rRNA is unclear. It may be speculated that the enzyme that is responsible for the synthesis of preribosomal RNA, RNA polymerase I, is not functioning properly.[13]

Bradley has offered a so-called unifying hypothesis claiming that ALS is caused by a deficiency of DNA repair mechanisms.[17] At the present time, there is a great need for more work to test this hypothesis.

3. RNA IN AGING OF THE CENTRAL NERVOUS SYSTEM

The contention that nerve cells are lost with age is common, although reliable data showing that aging is always associated with cell loss are difficult to find. Colon,[18] in a study of human brain, found a 44% loss of neurons at the age of 80 years, which in itself would lead to a loss of RNA.

Hydén analyzed the anterior motor horn cell from human spinal cord[19] with the single-cell technique of Edstrom and found an average of 670 pg in 40- to 50-year-old men and only 540 pg in 60- to 70-year-old men.

Mann *et al.* measured RNA absorption at 260 nm in the brains of 82 persons of age range 2–91 years.[3] They found a decrease of only about 15% in the neurons of the hippocampus and the dentate nucleus at the age of 90 years. The olivary neurons, however, showed a loss of RNA of about 60%. The nucleolar volume showed a reduction of similar magnitude. Lipofuscin pigment increased in a reverse proportion, a finding for which they had no explanation.

Uemura and Hartmann,[4] in a study using the Edstrom single-cell technique, found 222.7 pg RNA per hypoglossal neuron at the age of 49 years and only 117 pg RNA at 92 years, a reduction of 48%. There was a slowly sloping curve for the ages in between. They found somewhat different values in the prefrontal cortex, where each neuron remained at 27 pg RNA until the age of 66 years and then dropped to 18 pg after the age of 80 years.[20] In the subiculum, adjacent to the hippocampus, they found an average of 44 pg up to the age of 52 years, after which there was an increase. They suggested that the increased amount of RNA preceded the formation of neurofibrils.[21]

In conclusion, it might be stated that although RNA content decreases with age in most areas, this decrease is by no means synchronous in all regions of the brain.

4. *RNA IN SENILE DEMENTIA*

Although morphological studies from the earliest times have described changes in many areas of the brain, the hippocampal area with its important role in recent memory has always received most attention.[22,23] Electron microscopic studies by Kidd[24] showed paired helical filaments replacing the ribosomes within the nerve cells. In the adjacent subiculum, Uemura and Hartmann[25] found that the nerve cell bodies that contained neurofibrillary tangles in the phase microscope contained only 28 pg RNA compared to 41 pg in controls. This extensive reduction of RNA in senile dementia compares well with the data of Mann *et al.*,[26] who found a 30–50% reduction of RNA in neurons from the hippocampus in senile dementia. They also found losses of RNA in numerous other locations such as Betz cells of the motor cortex, Purkinje cells of the cerebellum, and facial nerve cells of the brainstem and concluded that the widespread reductions in cytoplasmic RNA indicate a general decline in metabolic levels, which may have followed, or indeed produced, a reduction in dendritic arborization in affected nerve cells.[27]

5. *RNA IN MERCURY INTOXICATION*

When one studies the effect of toxic substances on the nervous system, it may be useful to have information on the symptoms that are the first to appear. As an example, it was known that mercury first gives sensory symptoms, so the attention could be directed to the most sensitive cell type.

Studying the rat, Chang *et al.*,[28] using the Edstrom method, found a gradual reduction of RNA in spinal ganglion neurons from 1248 pg to only 544 pg in 6 weeks when CH_2HgCl was given to rats at a daily dose of 1 mg/kg body weight. This is a loss of 56%. A smaller loss, to 895 pg, i.e., 28%, was found in the same cells when $HgCl_2$ was administered instead. On the other hand, when the anterior motor horn neurons from the spinal cord were analyzed, there was a slight increase in RNA both after CH_3HgCl, from 650 pg to 807 pg, 24%, and after $HgCl_2$, to 800 pg, 23%. All of these changes were statistically significant.

When the base ratios of RNA were analyzed,[29] guanine was increased from 32 to 37%, and cytosine decreased from 28 to 21% after 11 weeks of HgCl intoxication. This corresponded to a shift of the G/C ratio from 1.16 to 1.73 and the AG/CU ratio from 1.12 to 1.33.

When the rats were given [^{3}H]uridine and sacrificed 1 or 24 hr afterwards, there was a marked reduction of incorporation into the nuclear as well as the cytoplasmic RNA in animals intoxicated with CH_3HgCl as compared to controls.[30] They interpreted their findings as a decrease in RNA synthesis.

Carmichael and Cavanagh[31] essentially confirmed these findings and also found that the nucleolus showed a severe reduction in incorporation of [^{3}H]uridine. Since preribosomal RNA is synthesized in the nucleolus, one would think that their data are in favor of decreased rRNA synthesis; however, they claim that the depression of labeling follows the structural changes in the ribosomes by several days.

Syversen,[32] working with bulk-isolated neurons from rats that had been injected with [^{14}C]leucine and [^{3}H]uridine 1/2 hr before sacrifice, favored a primary effect by methyl mercury on RNA synthesis. This is in keeping with the data of Gruenwedel and Cruikshank,[33] who tested the effect of methyl mercury on HeLa S3 cells and found that DNA and RNA syntheses were considerably more inhibited than protein synthesis.

Simpson[34] determined the sites of binding of methylmercuric ions with nucleosides and found that the first sites of reaction were the N_3-H bond of uridine and the N_1-H bond of guanosine. These are the bonds that are involved in Watson–Crick hydrogen bonding for DNA.

6. *DISCUSSION OF HYPOTHESES*

In a paper on motor neuron disease, Mann and Yates[9] speculated that the pathogen in this disease acted primarily on the nucleus, where it caused an inhibition of mRNA synthesis. They claimed that there is initially a slowly advancing condensation of the DNA, leading to a progressive restriction in the rate of transcription of m-, r-, and tRNA.

Their ideas were in part influenced by Koening,[7,8] who had produced a disease in rats that allegedly resembled motor neuron disease by intrathecal injection of actinomycin, which preferentially inhibits the synthesis of rRNA precursors in the nucleolus. More recently, Bradley[17] has proposed a hypothesis that claims that a whole group of disease result from deficiency of DNA repair enzymes. He includes ALS and Alzheimer's and Parkinson's diseases and suggests that etiologic factors such as aging, toxins, virus, and heredity operate via the same mechanism.

Such hypotheses are interesting and may be useful guidelines for future research. Objective evidence in support of these views is presently very limited. In ALS, RNA is decreased,[9,10] and rRNA is largely responsible.[13] The decrease is restricted to motor neurons and does not involve sensory neurons.[11] In methylmercuric chloride intoxication, RNA is decreased in sensory neurons but not in motor neurons, where, in fact, it is increased.[28] In aging, most investigators report reduction of RNA, but the rate is more extensive in some regions than in others.[4,20,21,26] These differences are strongly reminiscent of the well-known neuropathological enigma of "selective vulnerability of neurons." Could it be that transcription of RNA is turned off more in some neurons than in others? Perhaps the strongest argument for these hypotheses is the work that shows that methylmercuric hydroxide can alter the secondary structure in superhelical DNA, a region that serves a role in transcription.[35]

Note added in proof:

Since this chapter was submitted, Sulkowska and Marotta (1984, *Science* **225**:947–949) have reported substantial decreases in RNA and increases in ribonuclease activity in brains from patients with senile dementia (Alzheimer's disease). They believe that the result is a decrease in protein synthesis which influences neuronal functioning including level of transmitters.

REFERENCES

1. Edstrom, J. E., 1964, *Methods in Cell Physiology*, Volume I (D. M. Prescott, ed.), Academic Press, New York, pp. 417–447.
2. Shea, J. R., 1970, *J. Histochem. Cytochem.* **18**:143–152.
3. Mann, D. M. A., and Yates, P. O., 1977, *J. Neurol. Neurosurg. Psychiatry* **40**:299–302.
4. Uemura, E., and Hartmann, H. A., 1978, *Brain Res. Bull.* **3**:207–211.
5. Loening, V. E., 1967, *Biochem. J.* **102**:251–257.
6. Okazaki, H., Abe, S., and Satake, M., 1978, *J. Neurochem.* **31**:1149–1155.
7. Koenig, H., 1969, *Motor Neuron Diseases: Research on Amyotrophic Lateral Sclerosis and Related Disorders* (F. H. Norris, Jr., and L. T. Kurland, eds.), Grune and Stratton, New York, pp. 245–249.
8. Koenig, H., 1969, *Motor Neuron Diseases: Research on Amyotrophic Lateral Sclerosis and Related Disorders* (F. H. Norris, Jr., and L. T. Kurland, eds.), Grune and Stratton, New York, pp. 347–368.
9. Mann, D. M. A., and Yates, P. O., 1974, *J. Neurol. Neurosurg. Psychiatry* **37**:1047–1052.
10. Davidson, T. J., Hartmann, H. A., and Johnson, P. C., 1981, *J. Neuropathol. Exp. Neurol.* **40**:32–36.
11. Davidson, T. J., and Hartmann, H. A., 1981, *J. Neuropathol. Exp. Neurol.* **40**:187–192.
12. Davidson, T. J., and Hartmann, H. A., 1981, *J. Neuropathol. Exp. Neurol.* **40**:193–198.
13. Hartmann, H. A., and Davidson T. J., 1982, *Adv. Neurol.* **36**:89–103.
14. Dingman, W., and Sporn, M. B., 1962, *Biochim. Biophys. Acta.* **61**:164–177.
15. Hartmann, H. A., Smathers, P. Z., and White, S. K., 1983, *Molecular Aspect of Neurological Disorders* (L. Austin and P. L. Jeffrey, eds.), Academic Press, Sydney, pp. 183–192.

16. Hirano, A., 1982, *Adv. Neurol.* **36**:75–88.
17. Bradley, W., 1982, *Adv. Neurol.* **36**:493–502.
18. Colon, E. J., 1972, *Psychiatr. Neurol. Neurochir.* **75**:261–270.
19. Hydén, H., 1964, *RNA and Brain Function, Memory and Learning* (M. Brazier, ed.), University of California Press, Berkeley, pp. 29–68.
20. Uemura, E., and Hartmann, H. A., 1978, *J. Neuropathol. Exp. Neurol.* **37**:487–496.
21. Uemura, E., and Hartmann, H. A., 1979, *Exp. Neurol.* **65**:107–117.
22. Alzheimer, A., 1907, *Allg. Z. Psychiatr.* **64**:146–148.
23. Fisher, O., 1910, *Z. Ges. Neurol. Psychiatrie* **3**:371–471.
24. Kidd, M., 1963, *Nature* **197**:192–193.
25. Uemura, E., and Hartmann, H. A., 1979, *Brain Res. Bull.* **4**:301–305.
26. Mann, D. M. A., Yates, P. O., Barton, 1977, *J. Neurol. Neurosurg. Psychiatry* **40**:299–302.
27. Scheibel, M. E., Lindsay, R. D., Tomiyasu, V., and Scheibel, A. B., 1975, *Exp. Neurol.* **47**:392–403.
28. Chang, L. W., Desnoyers, P. A., and Hartmann, H. A., 1972, *J. Neuropathol. Exp. Neurol.* **32**:489–501.
29. Chang, L. W., Desnoyers, p. A., and Hartmann, H. A., 1973, *Acta Neuropathol.* **32**:77–83.
30. Chang, L. W., Martin, A. H., and Hartmann, H. A., 1972, *Exp. Neurol.* **37**:62–67.
31. Carmichael, N., and Cavanagh, J. B., 1976, *Acta. Neuropathol.* (*Berl.*) **34**:137–148.
32. Syversen, T. L. M., 1982, *Toxicol. Lett.***10**:31–34.
33. Gruenwedel, D. W., and Cruikshank, M. K., 1979, *Biochem. Pharmacol.* **28**:551–655.
34. Simpson, R. B., 1964, *J. Am. Chem. Soc.* **86**:2059–2065.
35. Beerman, T. A., and Lebowitz, J., 1973, *J. Mol. Biol.* **79**:451–470.

8

Adrenoleukodystrophy

Yasuo Kishimoto, Hugo W. Moser, and Kunihiko Suzuki

1. INTRODUCTION

Adrenoleukodystrophy (ALD) is a fatal X-linked genetic disease associated with progressive demyelination and adrenal insufficiency. Since discovery of the accumulation of very-long-chain fatty acids in brain white matter and adrenal cortex[1] in patients, this disease has been of significant interest for neurochemists. As a result of the development of a specific diagnostic method based on a subtle but significant increase in very-long-chain fatty acids in cultured skin fibroblasts[2,3] and plasma,[4] the number of known ALD patients has increased considerably, and this disease is now considered to be one of the more frequently occurring inherited diseases involving the lipid metabolic system. In this chapter, we review the recent progress that has been made on the study of the biochemical deficiency in this disease.

2. DEFINITION OF THE DISEASE

The association between diffuse cerebral demyelination and adrenal deficiency was first described by Siemerling and Creutzfeldt in 1923.[5] Cerebral demyelination and adrenal disease were called melanodermic leukodystrophy of bronze Schilder's disease. Blaw[6] first introduced the term ALD to describe this condition. The important discovery that there is a specific inclusion material in the brain, adrenal gland, peripheral nerve, and testis[7–9] of ALD patients and the subsequent finding that the inclusion material is cholesterol esters of very-long-chain fatty acids[1] contributed to the identification of this disease as a lipid storage disease.

Yasuo Kishimoto • John F. Kennedy Institute and Department of Neurology, Johns Hopkins University School of Medicine, Baltimore, Maryland 21205. *Hugo W. Moser* • John F. Kennedy Institute, Baltimore, Maryland 21205. *Kunihiko Suzuki* • Rose F. Kennedy Center and Department of Neurology, Albert Einstein College of Medicine, Bronx, New York 10461.

As has been true for many other inborn errors of metabolism, there are four types of ALD categorized according to the time of the onset of the disease, as shown in Table I[10]. The inclusion material and presence of abnormal very-long-chain fatty acids have been confirmed in all four types of ALD. Childhood ALD is by far the most common type. It and adrenomyeloneuropathy, which is the adult type of ALD,[11,12] occur in the same family and are considered to be caused by the same genetic defect. Neonatal ALD,[10,13–15] on the other hand, does not have the same traits as common ALD, and its occurrence in the female and certain other features suggest that it has an autosomal recessive mode of inheritance.

3. IDENTIFICATION OF ABNORMAL VERY-LONG-CHAIN FATTY ACIDS IN ALD TISSUES

3.1. Brain

3.1.1. Cholesterol Esters

The early studies of the brain lipids of ALD patients resulted in identification of only grossly reduced myelin-characteristic lipids, thus indicating a nonspecific lipid change secondary to demyelination.[16–19] Cholesterol esters were also found to be high in ALD brain, although it is not unusual to find higher concentrations of cholesterol esters in brain with active demyelination. The amounts of cholesterol esters were generally found to be 6–16% of the free cholesterol in gray matter and 13–80% of the free cholesterol in white matter of ALD brain. In control brains, the value was generally reported as 1–3%.

Igarashi *et al.*[1] were the first to report the striking abnormality in the fatty acid composition of cholesterol esters from ALD brain. Whereas cholesterol esters in control brain contained C_{14}–C_{22} but mostly C_{16}–C_{20} fatty acids, cholesterol esters in ALD brain contained large amounts of very-long-chain fatty acids (C_{24}–C_{30} or more), as illustrated in Fig. 1. These very-long-chain fatty acids were present in cholesterol esters in all five ALD brains examined but not in the four control brains. Most of the fatty acids were saturated and unbranched. Small amounts of unsaturated acids with corresponding carbon chains were also reported in most cases. The specific occurrence of abnormal cholesterol esters was later confirmed by many other groups for all types of ALD.[20–23] Table II summarizes the abnormalities in fatty acids found by different investigators. The proportion of very-long-chain fatty acids in total cholesterol fatty acids varies from patient to patient, but on the average one-quarter are abnormal very-long-chain fatty acids. It should be emphasized that 25:0 and 26:0 are the major fatty acids. This phenomenon is in contrast with the normal pattern of very-long-chain fatty acids in mammalian tissues. Usually very-long-chain fatty acids occur as components of sphingolipids and consist of C_{22} to C_{27} with 22:0, 24:0, and 24:1 as the major fatty acids.[24,25]

Table I
The Four Clinical Forms of Adrenoleukodystrophy

	Neonatal ALD	Childhood ALD	Adrenomyeloneuropathy	Symptomatic heterozygote
Sex	Male (3), female (1)	Male	Male	Female
Most common age of onset	First few days of life	4–8 years	Third decade	Third decade
Most common duration	1–6 years	1–9 years	Two decades or more	More than two decades
Most common neurological symptoms	Severe psychomotor retardation, opisthotonus, seizures	Dementia, visual, hearing, motor disturbances	Progressive spastic paraparesis, peripheral neuropathy bowel, bladder disturbance	Progressive spastic paraparesis
Endocrine abnormality	Adrenal insufficiency	Adrenal insufficiency—mild to severe	Adrenal insufficiency—severe, hypogonadism	Not demonstrated so far

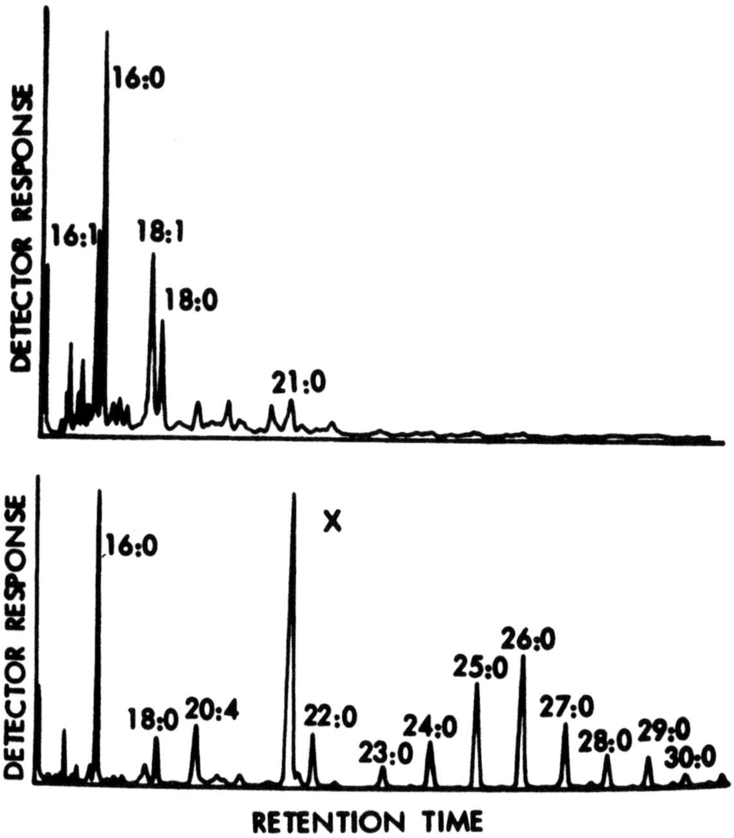

Fig. 1. Gas chromatographic tracings of fatty acid methyl esters of brain cholesterol esters. Technical details are described in ref. 1. The top is from a normal brain, and the bottom from ALD case 1. Note the large amounts and the bell-shaped distribution of very-long-chain fatty acids in the ALD specimen. The major peak marked X in the ALD sample is unidentified. (From Igarashi *et al.*[1] with permission from the International Society of Neurochemistry.)

The abnormal cholesterol esters containing very-long-chain fatty acids occur mostly in the white matter, where gross demyelination is apparent. Electron microscopy showed that the inclusion, which apparently contains abnormal cholesterol esters, is found exclusively in invading macrophages.[7] When brain was homogenized and fractionated into subcellular fractions by the standard procedure, the yield of myelin was drastically reduced as compared to the control. However, a large amount of less dense material, called a floating fraction, which apparently included myelin fragments, was obtained. This fraction contained relatively large amounts of cholesterol esters.[19] The fatty acid composition of the cholesterol esters in this fraction was reported in recent studies showing that they contained large amounts of very-long-chain fatty acids.[25a] Whether the abnormal cholesterol esters are intrinsic to the myelin or represent contamination by the inclusion material found in the macrophages must be determined.

Table II
The Abundance of Very-Long-Chain Fatty Acids in Cholesterol Esters in ALD White Matter

	Igarashi *et al.*[1]		Rumsey *et al.*[19]	Menkes and Corbo[21]		Manz *et al.*[14]	Molzer *et al.*[22,23]			
Fatty acids[a]	ALD (5)[b]	Control (4)	ALD (2)	ALD (2)	Control (2)	neonatal ALD (1)	ALD (1)	atypical ALD (1)	AMN (1)	Control (9)[c]
24:0	4.1 ± 2.2	0.0	3.2	5.8	—	1.8	3.9	4.6	10.0	0.6 ± 0.2
25:0	6.6 ± 3.2	0.0	7.1	8.8	3.9	2.4	8.9	13.8	13.9	0.1 ± 0.2
26:0	6.3 ± 3.1	0.0	5.7	7.9	—	4.1	10.6	26.0	14.5	0.03 ± 0.10
27:0	2.5 ± 1.6	0.0	1.5	—	—	4.7	3.5	9.3	3.5	—
28:0	1.1 ± 0.9	0.0	0.9	—	—	2.2	2.6	6.5	2.6	—
29:0	0.9 ± 0.6	0.0	1.9	—	—	0.6	2.9	2.1	0.9	—
30:0	0.7 ± 0.4	0.0	—	—	—	0.5	2.6	2.2	0.3	—
>24[d]	25.8 ± 11.1	0.0	20.3	22.5	3.9	24.0	55.4	67.0	50.8	4.8 ± 2.4

[a] Percent of total fatty acids in cholesterol esters.
[b] The number of parentheses indicates the number of patients studied.
[c] Controls with myelin breakdown.
[d] These include unsaturated fatty acids.

Table III
Abundance of Very-Long-Chain Fatty Acids in Gangliosides in ALD Brain White Matter[1,26]

	Total gangliosides		Individual ganglioside in an ALD brain[1]					
Fatty acids[a]	ALD (4)	Control (2)	G_{M1}	G_{D3}	G_{D1a}	G_{D2}	G_{D1b}	G_{T1}
22:0	1.9 ± 1.0	0	3.6	3.5	3.7	4.0	3.8	3.4
23:0	2.4 ± 2.2	0	6.2	6.9	7.2	9.4	8.6	9.1
24:1	7.5 ± 9.4	0	10.0	10.0	10.6	18.7	15.9	16.6
24:0	3.4 ± 4.5	0	5.4	5.3	6.6	9.7	8.6	11.2
25:0	1.4 ± 1.8	0	1.5	1.4	2.0	2.4	4.5	4.0
>22[b]	19.0 ± 20.5	2.5	27.8	28.2	31.4	47.4	42.8	50.4

[a] Percent of total ganglioside fatty acids.
[b] Includes all other minor fatty acids.

3.1.2. Gangliosides

The abnormal very-long-chain fatty acids occur in ALD brain not only as cholesterol esters but also in many other lipids found in smaller quantities. Among the other lipids, the amount of very-long-chain fatty acids found in gangliosides is large and close to the amount found in cholesterol esters. The abnormality in gangliosides was identified by the TLC pattern.[1,26] Examination of the fatty acids revealed the gross abnormality shown in Table III. All of these fatty acids were attached to gangliosides by amide linkages. The major very-long-chain fatty acid was 24:0, with only a very small quantity of 25:0 and 26:0. In normal brain, gangliosides consist of only regular-chain fatty acids, mainly 18:0 with lesser amounts of 16:0 and 20:0.[27,28] The abnormality in gangliosides, especially those highly sialilated, indicates that the fatty acid abnormality is not confined to white matter but is present in gray matter as well. The occurrence of abnormal cholesterol esters in gray matter has also been reported.[20]

3.1.3. Cerebrosides and Sulfatides

In normal brain, most very-long-chain fatty acids occur as components of cerebrosides and sulfatides, which are the major lipids of the myelin sheath.[24,29] As mentioned previously, the major fatty acids of these lipids are 22:0, 24:0, and 24:1. This is quite different from the abnormal fatty acids in the cholesterol esters but does resemble the abnormal fatty acids in the gangliosides in ALD brain. In addition, cerebrosides and sulfatides in brain contain large amounts of very-long-chain α-hydroxy fatty acids.[24]

Examining the fatty acids in cerebrosides and sulfatides in ALD brain is complicated because of the gross decrease of these lipids when demyelination occurs. Nevertheless, there is a subtle increase in nonhydroxy very-long-chain fatty acids ($\geq C_{25}$), as shown in Table IV. The concentration of 24:1, a major fatty acid of cerebroside, significantly decreased, whereas that of 24:0, another

Table IV
Abundance of Very-Long-Chain Fatty Acids in Cerebrosides in ALD Brain White Matter

	Igarashi *et al.*[1]		Miyatake *et al.*[31]		Ramsey *et al.*[32]	
Fatty acids[a]	ALD (4)	Control (2)	ALD (1)	Control (1)	ALD (2)	Control (1)
24:1	24.4 ± 5.2	42.2	17.6	31.7	20.4	35.2
24:0	18.2 ± 3.7	13.5	22.5	18.7	13.9	18.2
25:1	10.5 ± 0.9	10.5	8.5	6.6	10.3	6.2
25:0	7.4 ± 3.0	4.9	14.2	5.4	9.4	4.0
26:1	13.7 ± 1.6	10.1	8.9	6.2	12.8	12.1
26:0	4.4 ± 0.9	1.5	5.5	2.1	4.0	1.4
$\geqq C_{25}$	36.0	27.0	37.1	20.3	36.5	23.7

[a] Percent of total cerebroside nonhydroxy fatty acids.

major acid, decreased in one case but increased slightly in all others. On the other hand, the total amount of fatty acids with a carbon number of 25 or more increased 30–80% in ALD brain. This is especially significant because cerebrosides from most demyelinating brain generally contain a smaller ratio of these fatty acids.[30] A similar decrease in 24:0 and 24:1 and an increase in the longer-chain α-hydroxy fatty acids was observed (Table V). Essentially similar changes in sulfatides were also reported.[31,32]

3.1.4. Other Lipids

In two ALD brains, very-long-chain fatty acids, C_{25}–C_{34}, were present as free acids as well as in triglycerides, although they are in much smaller quantities compared to cholesterol esters.[32] Both white and gray matter contained these. Unfortunately, the fatty acid compositions were not compared with a control. No very-long-chain fatty acids were reported in glycerophospholipids.

Table V
Abundance of α-Hydroxylated Very-Long-Chain Fatty Acids in Cerebrosides in ALD Brain White Matter

	Igarashi *et al.*[1]		Miyatake *et al.*[31]		Ramsey *et al.*[32]	
Fatty acids[a]	ALD (3)	Control (2)	ALD (1)	Control (1)	ALD (2)	Control (1)
24h:1	11.9 ± 1.4	17.9	7.7	16.7	12.7	19.6
24h:0	30.7 ± 7.3	31.0	43.7	43.7	32.5	47.0
25h:1	5.1 ± 1.5	6.5	16.9	8.7	5.2	1.1
25h:0	12.1 ± 1.8	10.7	3.6	3.8	13.1	2.9
26h:1	9.2 ± 1.3	9.3	6.5	1.4	9.8	4.6
26h:0	4.8 ± 1.3	2.9	3.2	4.7	4.9	1.0
$\geqq C_{25}$	31.2	29.4	30.2	18.6	23.0	9.6

[a] Percent of total cerebroside α-hydroxy fatty acids.

Table VI
Lipid Content of Adrenal Glands[a]

	ALD Patient (mg/g tissue)	Control (mg/g tissue)
Total lipids	239	216
Nonpolar lipids	214	180
Cholesterol esters	50.8	13.0
Triglyceride	121	160
Cholesterol	4.2	2.4
Glycolipids	2.1	8.0
Phospholipids	22.8	31.7

[a] Y. Kishimoto, unpublished results.

3.2. Adrenal Cortex

Despite the severity of the pathological changes in the adrenal cortex of ALD patients, sometimes even greater than those in ALD brain, the biochemical abnormality has been shown only in a few cases.[1,33] This paucity results mainly from the small amount of tissue available because of atrophy resulting from steroid therapy. Nevertheless, as in the case of brain, the concentration of cholesterol esters in the adrenal glands of ALD patients was nearly fourfold higher than the control, as shown in Table VI. Examination of the fatty acid composition of the cholesterol esters clearly disclosed a profound abnormality similar to that in brain cholesterol esters (Table VII). The gangliosides from ALD also contained two- to sixfold higher concentrations of very-long-chain fatty acids as compared to the control adrenal gangliosides. A similar abnormality was also found in alkaline-stable lipids, mostly sphingomyelin. On the other hand, triglycerides, free fatty acids, and glycerophospholipids were normal.

Table VII
Abundance of Very-Long-Chain Fatty Acids in Cholesterol Esters in ALD Adrenal Gland[1]

Fatty acids[a]	ALD (4)	Control (4)
24:1	3.0 ± 1.9	0
24:0	7.9 ± 3.2	0
25:0	2.6 ± 1.2	0
26:1	1.2 ± 0.5	0
26:0	5.2 ± 5.1	0
27:0	1.2 ± 1.3	0
28:0	0.6 ± 0.3	0
29:0	0.6 ± 0.3	0
≧24	22.3	0

[a] Percent of total cholesterol ester fatty acids.

Table VIII
Abundance of Very-Long-Chain Fatty Acids in Cerebrosides from ALD Peripheral Nerves[34]

Fatty acids[a]	ALD (3)	Control (2)
24:1	19.1 ± 2.6	22.2
24:0	36.9 ± 3.6	36.5
25:0	10.5 ± 0.3	6.8
26:0	3.6 ± 1.1	2.1
24h:1	2.8 ± 0.6	2.8
24h:0	38.5 ± 2.0	38.9
25h:0	7.9 ± 0.5	5.8
26h:0	1.9 ± 0.3	1.4

[a] Percent of total cerebroside nonhydroxy and α-hydroxy fatty acids, respectively.

3.3. Peripheral Nerves

As described above, Powers and Schaumburg[8] also found the characteristic inclusion in peripheral nerves. However, examination of biopsy specimens of two sural nerves and one peroneal nerve from ALD patients showed that the cholesterol esters from these nerves did not contain any detectable abnormal very-long-chain fatty acids.[34] These nerves contained similar amounts of cerebrosides and sulfatides as the two controls examined. Examination of the homologue compositions of these cerebrosides by reverse-phase HPLC did not show any significant abnormality in the fatty acid composition in ALD nerves except for a small increase in the proportion of C_{25} and C_{26} fatty acids (Table VIII).

3.4. Cultured Skin Fibroblasts

An enzyme defect in inborn errors of metabolism can often be detected in the skin fibroblasts of patients. Therefore, once abnormal cholesterol esters were discovered in the brain and adrenal glands of ALD patients, these were sought in cultured skin fibroblasts from ALD patients. However, as in the case of the peripheral nerves, the very-long-chain fatty acids were not found in the cholesterol esters from fibroblasts. On the other hand, careful examination of the total lipid fatty acids indicated that there was a subtle but consistent increase in 25:0 and 26:0 acids,[2] as shown in Fig. 2. The increase is best expressed by the ratio of 26:0/22:0, as shown in Table IX. The concentration of 22:0 was found to be fairly constant in fibroblasts from both ALD and control. Sphingomyelin from ALD fibroblasts contained a smaller proportion of 26:0, as shown in the upper part of Table IX. This indicates that there is an as yet uncharacterized lipid that contains a higher proportion of 26:0/22:0 in ALD fibroblasts.

This study was extended to many more skin fibroblast specimens (Table IX lower part).[3] This not only confirmed the initial findings but also showed that a diagnosis can be made with certainty by this technique. This method of

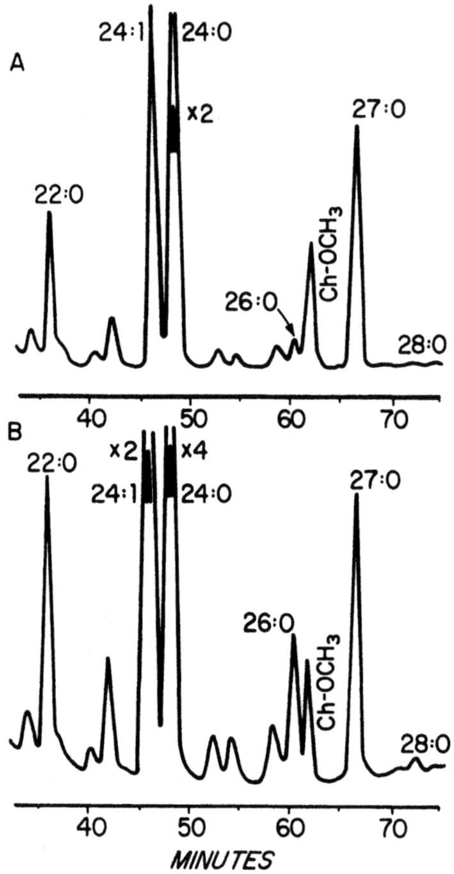

Fig. 2. A portion of the GLC tracing of the total lipid fatty acid esters from skin fibroblasts. A is the control, and B is ALD. See ref. 2 for preparation of samples and GLC conditions. Ch-OCH_3 is cholesterol methyl ether. (From Kawamura *et al.*[2] with permission from Academic Press.)

detecting a fatty-acid abnormality has thus become a reliable and relatively simple method for diagnosing ALD. With this procedure and the determination of the ratio in plasma (see below), the diagnosis of ALD and AMN can be made with extreme accuracy, and a total of 173 patients has now been identified world wide. In addition, carriers of ALD have intermediate values for the 26:0/22:0 ratio, as shown in Table IX; thus, it is possible to identify carriers of the disease.

3.5. Blood

Careful examination of total lipid fatty acids from plasma indicated that detection of the fatty-acid abnormality is also possible by using capillary gas chromatography (Fig. 3).[4] The 26:0/22:0 ratio was definitely higher in ALD than in controls (Table X). The values for obligated heterozygotes were again between ALD and controls and clearly distinguishable from both values. This assay, which requires only 0.5 ml of plasma, provides a rapid and reliable assay for ALD diagnosis and carrier detection.

The concentration of 26:0 and, to a lesser extent, 24:0 increased over the control in all lipids examined: glycerolipids, sphingomyelin, and glycolipid.

Table IX
Relative Abundance of Hexacosanoic Acid in Cultured Skin Fibroblasts from ALD Patients

Reference	Preparation	Ratio 26:0/22:0
Kawamura *et al.*[2]	Total lipids	
	ALD (6)	0.88 ± 0.06
	Control (18)	0.05 ± 0.02
	Sphingomyelin	
	ALD (4)	0.48 ± 0.16
	Control (3)	0.03 ± 0.01
Moser *et al.*[3]	Total lipids	
	ALD (autopsy proven) (5)	0.78 ± 0.14
	ALD (clinically typical) (6)	0.89
	AMN (autopsy proven) (2)	0.76 ± 0.09
	AMN (clinically typical) (2)	0.56
	ALD carrier (5)	0.26 ± 0.14
	Control[a] (29)	0.06 ± 0.03

[a] The control included 12 fibroblasts from homozygotes or heterozygotes of other metabolic diseases and five from unknown progressive neurological diseases.

Curiously, the only exception was in cholesterol esters, where the 26:0 concentration was lower than the control value. This finding is in agreement with the initial report of Igarashi *et al.*[1]

A fatty-acid abnormality was also found in the erythrocytes.[35] The ratio of 26:0/22:0 was 2.6-fold higher than the control in sphingomyelin isolated from erythrocyte plasma membrane (Table XI). Gas chromatography–mass spectrometry was used to make the determination. Similarly, the concentration of hexacosanoic acid in the total lipids of white blood cells was found to be fivefold higher in ALD than in controls.[36]

3.6. Other Tissues

When total lipid fatty acids were examined by capillary gas chromatography, almost all of the ALD tissues examined showed a distinctly higher ratio of 26:0/22:0 than the control (H. W. Moser and A. B. Moser, personal communication). When the cultured skeletal muscle from an AMN patient was cultured in medium containing a large amount of hexacosanoic acid, these muscle cells accumulated the hexacosanoic acid 5–10 times more than the control cells.[37] The increase was observed in fatty acids from nearly all lipids examined, namely, triglycerides, phospholipids, and glycolipids. Strangely, the increases in cholesterol esters and free fatty acids were minimal.

3.7. Summary

The very-long-chain fatty acids (C_{24}–C_{30}) that accumulate in ALD tissues were first found in cholesterol esters in white matter and adrenal cortex, which

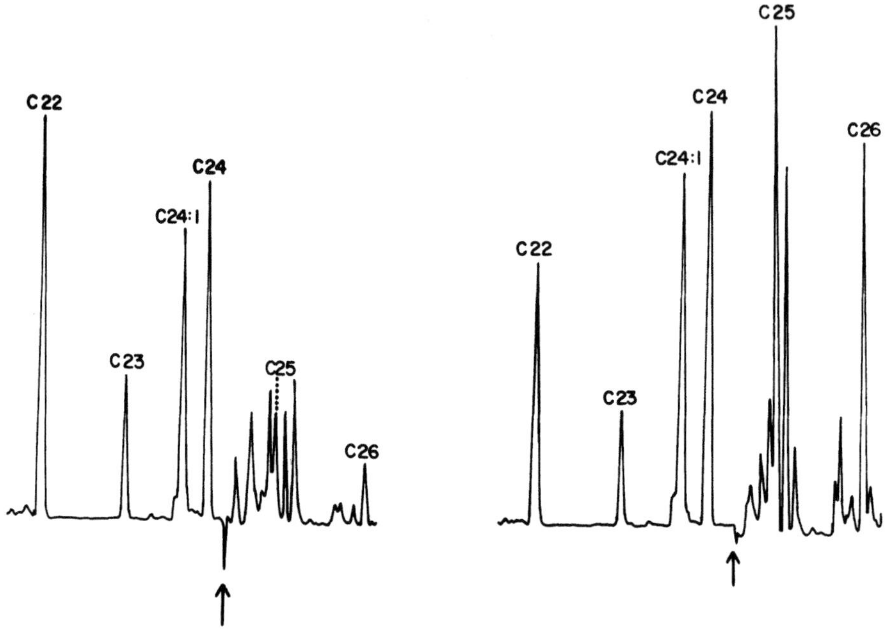

Fig. 3. Saturated very-long-chain fatty acids in the plasma of (left) a 16-year-old normal male and (right) of a 12-year-old patient with adrenoleukodystrophy. At the time indicated by the arrow, the sensitivity of the recording system was increased 16-fold, thus increasing the size of the peaks for the C_{25} and C_{26} fatty acids by that factor. Note that the C_{25} and C_{26} peaks are much higher in the ALD sample and that in the ALD sample the C_{24} peak is larger than C_{22}, whereas the reverse is true of the control sample. Although the C24:1 peak in the ALD sample is higher than in the control in this tracing, this was not true for the other ALD patients (From Moser *et al.*[4] with permission from Harcourt Brace Jovanovich.)

are the most pathologically affected tissues. However, on further examination of the various lipids in different tissue samples, it became obvious that the abnormal very-long-chain fatty acids were distributed more or less throughout many lipid classes and many other tissues. Nevertheless, brain white matter and adrenal cortex contain the largest concentrations of abnormal very-long-chain fatty acids, mostly as cholesterol esters. A rough estimate of the distribution of the abnormal very-long-chain fatty acids in ALD brain is shown in Table XII. The total amount of very-long-chain fatty acids in ALD brain is

Table X
Relative Abundance of Hexacosanoic Acid in Plasma of ALD Patients[4]

	Ratio 26:0/22:0
ALD hemizygotes (18)[a]	0.064 ± 0.005
ALD heterozygotes (17)	0.035 ± 0.002
Control without neurological or metabolic diseases (22)	0.014 ± 0.002
Control with disorders other than ALD (28)	0.012 ± 0.002

[a] Included two cases of AMN, whose values were essentially similar to those in ALD.

Table XI
Relative Abundance of Hexacosanoic Acid in Erythrocyte Sphingomyelin of ALD Patients[35]

	Ratio 26:0/22:0
ALD (8)	0.26 ± 0.04
Control (16)	0.10 ± 0.02

about 3 mg/g dry weight. It is interesting to note that the cholesterol esters containing very-long-chain fatty acids were found only in brain and adrenal gland, both of which are pathologically affected areas. All other tissues, including the peripheral nerves, contained small amounts of very-long-chain fatty acids in lipids other than cholesterol esters.

Despite the lack of an apparent morphological abnormality, a subtle increase in very-long-chain fatty acids in cultured skin fibroblasts, blood plasma, erythrocytes, white blood cells, and aminocytes provides an easy diagnostic test to identify the disease with high accuracy.

4. IDENTIFICATION OF THE ENZYME DEFECT

4.1. Introduction

Because of the genetic nature of the disease, ALD was thought to be caused by a metabolic defect. Before the discovery of the accumulation of abnormal very-long-chain fatty acids, cholesterol and other steroids were the primary focus, mainly because of the involvement of the adrenal cortex. In fact, an excess accumulation of cholesterol in cultured skin fibroblasts from ALD patients was reported.[38] However, a later study did not confirm this and indicated normal fibroblast sterol metabolism.[39]

Table XII
Distribution of Abnormal Very-Long-Chain Fatty Acids (VLCFA) in ALD Brain[a]

Lipids	Estimated mol. wt.	Concentration of lipids/g dry wt.		Proportion of VLCFA in lipids (%)	Concentration of VLCFA (μmol/g dry wt.)
		mg	μmol		
Cholesterol esters	730	9.72	13.3	31	4.1
Cerebrosides[b]	830	5.90	7.1	5[c]	0.35[c]
Gangliosides	2000	2.50	12.7	10	1.3

[a] Estimated from the data from many investigators.[1,15,20–23,26]
[b] Includes both hydroxy and nonhydroxy fatty acids.
[c] In excess over control value.

4.2. Metabolism of Cholesterol Esters in ALD

The most striking biochemical abnormality is the accumulation of abnormal very-long-chain fatty acids in cholesterol ester fractions of brain white matter and adrenal cortex. This finding initially led to the hypothesis that there was a deficiency in a cholesterol ester hydrase.

Brain normally contains three different enzymes for the hydrolysis of cholesterol esters, one in the mitochondrial fraction with an optimal pH of 4.2, one in the microsomal fraction with an optimal pH of 6.0, and the last in the myelin fraction with an optimal pH of 6.8.[40,41] The enzyme in the mitochondrial fraction appears to be a lysosomal enzyme and the most active of the three. When the postnuclear fractions from the gray and white matter of ALD patients were assayed for all three cholesterol esterase activities with cholesterol ester of oleic acid (18:1), lignoceric acid (24:0), and cerotic acid (26:0), their activities were essentially similar to that of the control (Table XIII).[42] Generally, the hydrolysis of cholesterol lignocerate and cholesterol cerotate was less than 2% and 0.2%, respectively, of the hydrolyses of cholesterol oleate.

Brain contains a cholesterol-esterifying enzyme in microsomes.[40] This enzyme has a pH optimum of 5.2 and is different from esterifying enzymes in other tissues in that the activity does not require CoA and ATP, and free fatty acid seems to be the substrate. When this enzyme was assayed in whole homogenates from gray and white matter, there were no significant differences between ALD and control with either oleic acid or cerotic acid as substrate, as also shown in Table XIII.[43] An interesting observation is that the ratio of enzyme activity for cerotic acid to oleic acid is more than tenfold higher than the ratio for the cholesterol ester hydrolysis mentioned above. Studies with rat brain also demonstrate similar differences in the fatty acid specificity for the synthesis and hydrolysis of cholesterol esters. Therefore, it has been proposed that the very-long-chain fatty acids produced in ALD brain are converted to cholesterol esters faster than they are degraded. This would explain why the abnormal very-long-chain fatty acids in ALD brain were found predominantly in the cholesterol fraction. Two other enzymes in brain, which are apparently different from the microsomal enzyme, were reported recently. The cholesterol ester synthetase present in myelin has a pH optimum identical to that of the microsomal enzyme and also acts on free fatty acids.[44] On the other hand, the enzyme reported more recently[45] has a pH optimum of 7.4, is activated by CoA and ATP, and appears to be localized in microsomes. The activity of this enzyme has not been determined in ALD brain.

In addition to brain, cholesterol esterase activity was also examined in cultured skin fibroblasts using cholesterol lignocerate as a substrate. The hydrolytic activity had an optimum pH of 3.0 and was deficient in fibroblasts from patients with cholesterol ester storage disease and Wolman's disease. However, the activity in ALD fibroblasts was found to be normal.[46] Normal synthesis and hydrolysis of cholesterol cerotate as well as cholesterol oleate were also reported in the adrenal gland obtained from the biopsy of an ALD patient.[47]

In summary, the synthesis and hydrolysis of very-long-chain fatty acid esters of cholesterol were not deficient in ALD tissues. However, the relatively

Table XIII
Cholesterol-Ester-Metabolizing Enzyme Activities in ALD Brain[43]

			Substrate[a]			Ratio × 100	
Enzyme activity	Subcellular fraction		C18:1	C24:0	C26:0	C24:0/C18:1	C26:0/C18:1
Cholesterol esterase	Mitochondria[b] (pH 4.2)	ALD (3)	3920 ± 783	72.6 ± 12.0	4.83 ± 3.24	1.85	0.12
		Control (2)	3010	50.6	4.21	1.68	0.14
	Microsomes[b] (pH 6.0)	ALD (3)	176 ± 53	6.23 ± 1.40	—[c]	3.54	—
		Control (2)	177	3.76	—	2.12	—
	Myelin[d] (pH 6.8)	ALD (3)	100	—	—	—	—
		Control (2)	40	—	—	—	—
Cholesterol ester synthetase	Gray matter[b]	ALD (3)	410 ± 108		77 ± 7		18.8
		Control (2)	254		91		35.8
	White matter[d]	ALD (3)	594 ± 235		132 ± 49		22.2
		Control (2)	400		99		24.6

[a] Cholesterol ester containing the fatty acids for cholesterol esterase assay and free fatty acid for cholesterol ester synthetase assay.
[b] Postnuclear fraction from gray matter was used without further fractionation using optimal conditions for enzyme present in each subcellular fraction.
[c] Undetectable.
[d] Postnuclear fraction from white matter was used without further fractionation using optimal conditions for each enzyme.

fast conversion of cerotic acid to cholesterol ester and the relatively slow hydrolysis of cholesterol cerotate in normal human and rat brain explains the striking accumulation of very-long-chain fatty acids in ALD brain. It is interesting to point out that the enzyme for cholesterol ester synthesis in adrenal cortex is similar to the brain enzyme in that free fatty acid appears to be the preferred substrate.[48]

4.3. In Vivo Metabolism of Very-Long-Chain Fatty Acids in Cultured Skin Fibroblasts

Since cultured ALD skin fibroblasts contain a small but significantly greater amount of very-long-chain fatty acids than normal, it would be expected that the abnormality lies in the metabolic function of the tissue. In fact, the uptake of [1-^{14}C]lignoceric acid and [1-^{14}C]cerotic acid by ALD fibroblasts was, on average, slightly higher than for control cells, whereas the uptake of [1-^{14}C]stearic acid was essentially similar in both tissues, as shown in Fig. 4.[42] On the other hand, the level taken up of [1-^{14}C]cerotic acid did not change at all for several days, whereas that of [1-^{14}C]stearic acid decreased considerably in both fibroblasts.

A similar experiment indicated that there was a considerable decrease in the uptake of radioactivity from [1-^{14}C]lignoceric acid (Y. Kishimoto and A. B. Moser, unpublished results). The radioactivity in five cultures of ALD cells decreased to 70 $\pm$ 10% in 3 days, whereas that in three cultures of control cells decreased to 59 $\pm$ 14% of the original level (Fig. 5). More recently radioactivity from [1-^{14}C]cerotic acid complexed with bovine serum albumin was incorporated into complex lipids at a significantly higher rate in ALD fibroblasts than in controls.[49] These subtle differences would explain the difference in the level of very-long-chain fatty acids in cultured skin fibroblasts.[2]

The study of the distribution of radioactivity taken up from [1-^{14}C]-lignoceric acid indicated that the radioactivity is similarly distributed among various lipids classes in both cells (Y. Kishimoto and A. B. Moser, unpublished results). More than one-half of the radioactivity was recovered in sphingolipids, and the rest in the phospholipid fractions. Free fatty acids contained only a small amount of radioactivity, and cholesterol esters nearly none. The radioactivity in each individual sphingolipid was consistently higher in ALD fibroblasts than in control (Fig. 6), as was observed in the total lipid extract.

In summary, the uptake of very-long-chain fatty acids by ALD fibroblasts was slightly higher than that by any control cells, and the turnover of lignoceric acid was slightly slower in ALD cells. These differences, though small, may account for the increase in the levels of very-long-chain fatty acids in ALD fibroblasts. The difference in radioactivity was observed in all lipid classes examined and not in one particular lipid. This observation confirms the idea that the defect in ALD is in the metabolism of very-long-chain fatty acids and not in the metabolism of a specific lipid.

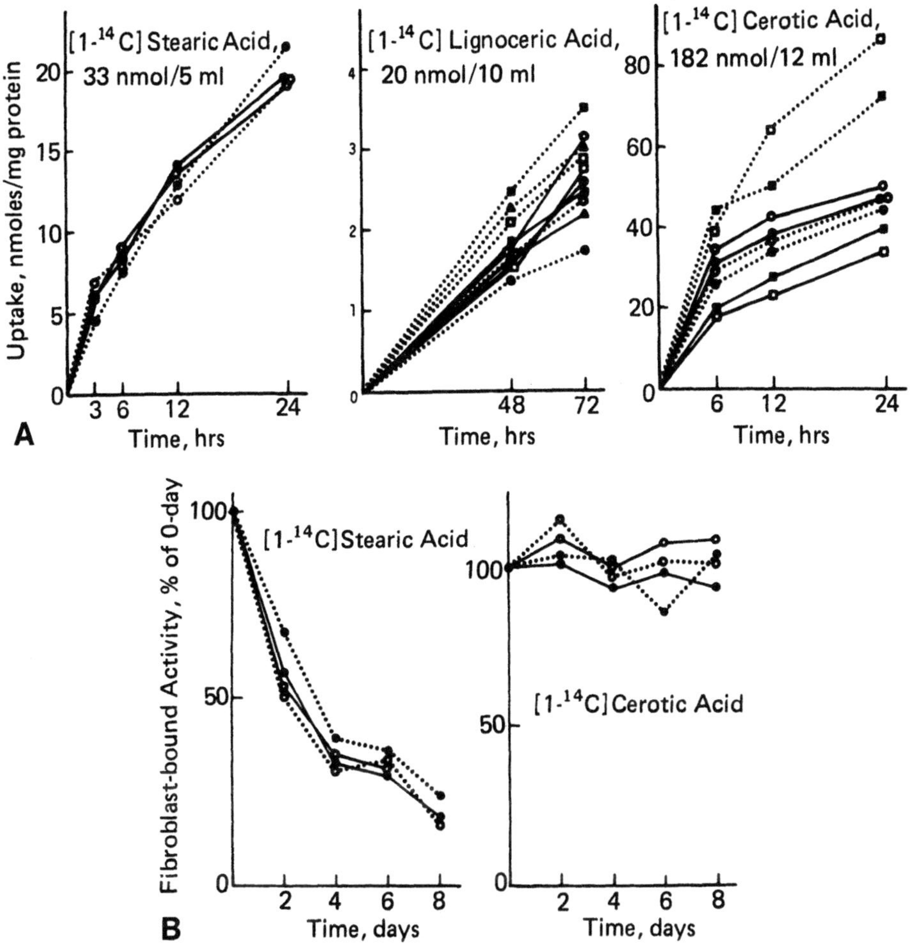

Fig. 4. (A) Uptake of fatty acids by cultured fibroblasts. Uptake of stearic, lignoceric, and cerotic acids was studied according to the experimental designs described in detail in the text. Since the amounts of added fatty acids were different, the results are not comparable among different fatty acids. Solid lines represent the control cell lines, and dotted lines the ALD cell lines. Some ALD cell lines consistently took up the very-long-chain fatty acids at greater rates than controls. Note the different time scale for the experiment with lignoceric acid. (B) Exclusion of fatty acids by cultured fibroblasts. Exclusion of the radioactivity from the cells was determined for 8 days after 24-hr uptake. Refer to the text for experimental details. Stearic acid was rapidly excluded from cultured fibroblasts, but cerotic acid appeared to remain cell bound once taken up. Solid lines represent the control cell lines, and dotted lines the ALD cell lines. (From Ogino *et al.*[42])

4.4. Defect in Very-Long-Chain Fatty Acid Oxidation in Cultured Skin Fibroblasts from ALD Patients

Many of the aforementioned observations point to a possible enzymic defect in the metabolism of very-long-chain fatty acids. In fact, a decrease in the oxidation of fatty acids, especially of very-long-chain fatty acids, in cultured skin fibroblasts[47] and white blood cells[50] from ALD patients has been demonstrated.

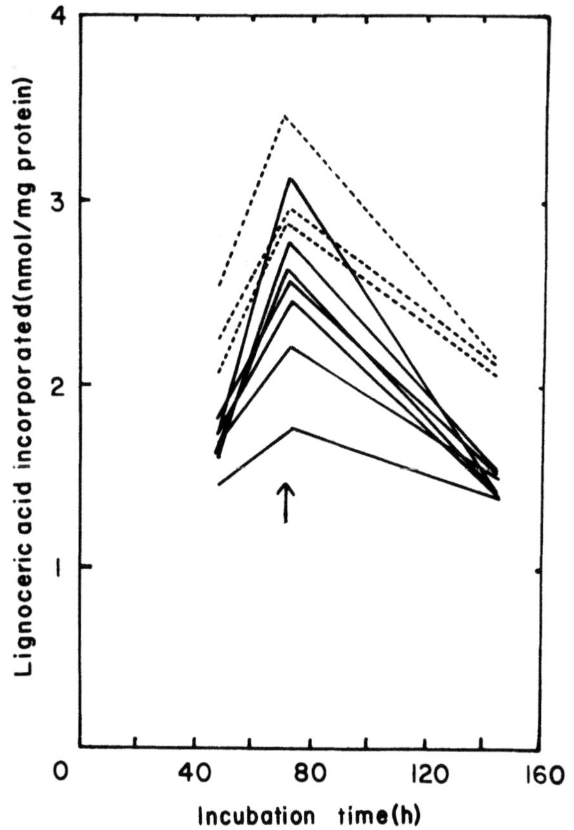

Fig. 5. Uptake and turnover of lignoceric acid in ALD skin fibroblasts. Twenty nanomoles of [1-^{14}C]lignoceric acid (5 mCI/mmol) was fed to cultured skin fibroblasts from three ALD patients (data shown by broken line) and seven controls (solid line), and radioactivity uptake was measured after 48 and 72 hr. The radioactivity was chased with nonradioactive lignoceric acid, and the radioactivity remaining after 72 hr was measured. (From unpublished data by A. B. Moser and Y. Kishimoto.)

In cultured skin fibroblast homogenates, CO_2 production from palmitic acid, which indicates the overall process of fatty acid oxidation, was the same for ALD and controls (Table XIV). Palmitic acid is the principal substrate of the β-oxidation system in mammalian tissues. However, in ALD fibroblasts, the oxidation of stearic acid, which has two more carbon atoms than palmitic acid, was reduced to 59%, and the oxidation of lignoceric and cerotic acids, C_{24} and C_{26} saturated fatty acids, was reduced to only 12 and 23% of the control values,[47] respectively. When the ratios of the oxidation of lignoceric acid to palmitic and stearic acid were compared, ALD tissues had only 8.6% and 21% of the control ratios, respectively. Similarly, the oxidation ratios of cerotic to palmitic and stearic acids were 16% and 40% of the controls, respectively. These ratios are useful for the diagnosis of ALD because protein determination and simultaneous assay of control cells can be eliminated.

A similar difference was also demonstrated with white blood cells (Table XV). The oxidation of palmitic acid was not different in ALD and control cells,

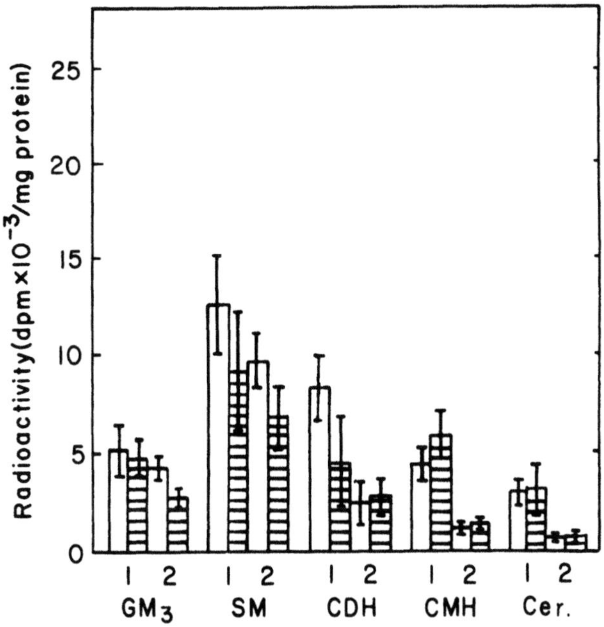

Fig. 6. Distribution of radioactivity taken up by ALD skin fibroblasts. The lipids extracted from fibroblasts shown in Fig. 5 were fractionated, and radioactivity in each sphinogolipid was determined. Open bars represent the activity in ALD fibroblasts, and shaded bars represent the activity in control cells. The numbers under each column indicate the cells obtained after 72 hr exposure to radioactive lignoceric acid (1) and after 72 hr chase period (2). GM_3, SM, CDH, CMH, and Cer. indicate GM_3 ganglioside, sphingomyelin, ceramide dihexoside, ceramide monohexoside, and ceramide, respectively. (From unpublished data by A. B. Moser and Y. Kishimoto.)

but the oxidation of lignoceric acid in ALD was only 28% of that in the control cells. The ratio of the oxidation of lignoceric to palmitic acid in ALD was 28% of the ratio in control cells. This can also be used for the diagnosis of ALD.

A similar defect in fatty acid oxidation was also indicated in the biopsy sample of adrenal gland from an ALD patient.[47] The oxidation of [1-^{14}C]oleic acid was only 0.29 pmol/mg protein per hr, whereas the six control tissues showed activities ranging from 2.6 to 37 pmol/mg protein per hr. The control tissues were obtained from autopsies, and the time lapse between death and the assay was 4–15 times longer for the controls than for the ALD patient. This indicates that the low activity in ALD was not caused by postmortem change.

The abnormalities in the metabolism of very-long-chain fatty acids in ALD tissues may not only be in the oxidation and incorporation into a complex lipid, exemplified by cholesterol ester synthesis as mentioned above, but may also arise from an increased synthesis. In fact, a significant increase in the incorporation of radioactivity in hexacosanoic acid from [1-^{14}C]stearic acid by ALD skin fibroblasts as compared to control cells was demonstrated.[51] Although an increase in the synthesis of hexacosanoic acid by ALD cells was indicated for the increase in radioactivity, another possibility is that the increase in radioactivity is caused by a defect in the oxidation of hexacosanoic acid, since

Table XIV

$^{14}CO_2$ Produced from [1-^{14}C] Fatty Acids by Cultured Skin Fibroblast Homogenate from ALD Patients

	[1-^{14}C]Fatty acids (pmol/mg protein per hr)[a]				Ratio (%)			
	Palmitic acid (16:0)	Stearic acid (18:0)	Lignoceric acid (24:0)	Cerotic acid (26:0)	24:0/16:0	24:0/18:0	26:0/16:0	26:0/18:0
ALD	35 ± 9 (4)	24 ± 8 (18)	0.9 ± 0.1 (26)	0.7 ± 0.2 (4)	2.6	3.8	2.0	2.9
Control	32 ± 7 (5)	41 ± 11 (16)	7.3 ± 1.0 (20)	3.0 ± 0.5 (5)	30.4	17.8	12.5	7.3
ALD/control	1.09	0.59	0.12	0.23	8.6	21	16	40

[a] Mean ± S.D. The number in parentheses indicates the number of patients examined.

Table XV
$^{14}CO_2$ *Produced from [1-^{14}C]Fatty Acids by White Blood Cell Homogenate from ALD Patients*

	[1-^{14}C]Fatty acids (pmol/mg protein per hr)[a]		Ratio
	Palmitic acid (16:0)	Lignoceric acid (24:0)	24:0/16:0 (%)
ALD	91 ± 19 (8)	2.1 ± 0.2 (17)	2.3
Control	90 ± 22 (7)	7.5 ± 1.3 (12)	8.3
ALD/control	1.01	0.28	0.28

[a] Mean ± S.D. The number in parentheses indicates the number of patients examined.

impaired oxidation would lead to an increase in the radioactivity in hexacosanoic acid. In fact, the *de novo* synthesis of palmitic acid was much less in ALD cells. This observation indicates that the degradation of [1-^{14}C]stearic acid in ALD cells was slower than in the controls.

In summary, the evidence obtained thus far strongly suggests that the oxidation of fatty acids is impaired in ALD and AMN. The defect in oxidation appears to be proportionally more severe for very-long-chain fatty acids than for other fatty acids. This agrees with the observation that the pathological damage is most apparent in ALD brain and adrenal cortex, where the fatty acid accumulation has occurred. Brain contains unusually large amounts of very-long-chain fatty acids (C_{22}–C_{26}) as components of cerebrosides and sulfatides, which in turn are the chief components of the myelin sheath.[24]

Two pathways for the degradation of very-long-chain fatty acids have been demonstrated. One involves α-oxidation, in which one carbon at a time is removed.[24] α-Hydroxylation is the initial step and appears to be rate determining.[52] α-Hydroxy fatty acids, the products of α-hydroxylation, and odd-numbered-carbon-chain fatty acids, the products of α-oxidation,[24] are largely incorporated into cerebrosides and sulfatides to form a unique fatty acid composition. The other pathway is β-oxidation, which is more common in other mammalian tissues. However, the brain β-oxidation system responsible for the oxidation of very-long-chain fatty acids appears to be different from the one found in other tissues.[53] It appears not to need CoA but requires the heat-stable and heat-labile factors that are obtained from brain cytosol and are also essential for α-hydroxylation.[52] The CoA system oxidizes palmitic and other regular-chain fatty acids actively but does not seem to oxidize very-long-chain fatty acids to any significant extent.[54] On the other hand, relatively high activity of the CoA-dependent oxidation of very-long-chain fatty acids was found in rat liver.[55]

Interestingly, the subcellular distribution and the effects of inhibitors suggest that the oxidation of very-long-chain fatty acids may be related to peroxisomes. The β-oxidation system must be investigated, and the specific step that is defective in ALD determined. However, the α-oxidation system appears to be functioning normally, because the fatty acid composition of cerebrosides and sulfatides is largely unaffected in ALD brain.

4.5. Localization of the ALD Gene

Migeon *et al.*[56] cloned skin fibroblasts from heterozygotes in three families segregating the ALD mutation. They found that the clones were of two types: clones with a normal ratio of 26:0/22:0 and clones with an excess of 26:0 similar to that found in fibroblasts of affected males. This study also demonstrated a close relationship of the ALD gene with that of the phenotypic expression of glucose-6-phosphate dehydrogenase (G6PD). Within one kindred, the ALD abnormality was always with the same G6PD isozyme. These observations indicated not only that the ALD locus is X-linked but also that there is a close linkage between ALD and G6PD loci. The ALD locus thus can be mapped on the human X-chromosomes near the G6PD locus at Xg28.

In a study that took advantage of the proximity of the G6PD and ALD loci, childhood ALD cell lines that also exhibited the G6PD A phenotype were cocultured with normal cells or with neonatal ALD cells, both of which exhibited the G6PD B variant (B. R. Migeon, A. B. Moser, and H. W. Moser, unpublished data). Comparison of the proportion of G6PD A and B with the 26:0/24:0 ratio on each cultured cell suggests that there was no cross correction by coculturing. This study suggests that unlike mucopolysaccharidoses, the deficient enzyme in ALD is not communicable between cells.

5. THERAPEUTIC ATTEMPTS

5.1. Nutritional Approach

The abnormal very-long-chain fatty acids that accumulate in brain and adrenal cortex of ALD patients have four possible origins. First, very-long-chain fatty acids may be synthesized in brain by a chain elongation system.[57,58] Second, they may be synthesized in other tissues such as liver and transported to brain and adrenal tissue. Liver is the major site of fatty acid synthesis. Third, they may come from nutritional sources such as plant and yeast products. These materials are known to contain small amounts of very-long-chain fatty acids as components of waxes and sphingolipids.[59] And finally, the very-long-chain fatty acids may be synthesized by bacteria present in the digestive system, absorbed, and delivered to the tissue.

As mentioned previously, the composition of very-long-chain fatty acids in ALD is different from that normally found in mammalian tissues. Whereas the latter has a pyramidal composition made up mostly of C_{22}–C_{26} fatty acids with C_{24} as the major component, the abnormal fatty acids in ALD consist of C_{24}–C_{30} with C_{25} and C_{26} as major components. This composition is similar to that found in plants and microbes.[59] Therefore, it was thought that the source of the abnormal fatty acids was dietary. In order to determine whether these acids are actually exogenous, deuterium-labeled [3,3,5,5-$^{2}H_4$)hexacosanoic acid was synthesized and administered orally to a patient for 100 days, 10 mg per day.[60] Examination of postmortem brain tissue by gas chromatography–mass spectrometry demonstrated that a considerable amount of the deuterated hexacosanoic acid was deposited as cholesterol esters. From the concentration of deuterated hexacosanoic acid found in the patient's brain, it could be roughly calculated that the total amount deposited in the brain was 10–20 mg, which represented 1–2% of the total hexacosanoate administered.

Although other possible sources of very-long-chain fatty acids cannot be excluded, this observation clearly indicates that at least part of the hexacosanoic acid deposited in the ALD brain was exogenous. On the basis of this finding, a therapeutic trial was conducted in which very-long-chain fatty acids were restricted in the patient's diet.[61] The content of very-long-chain fatty acids was determined, and only foods containing a minimum amount of very-long-chain fatty acids were given to several ALD patients at relatively early stages of the disease. Unfortunately, the hexacosanoic acid level in plasma, which was used to indicate the presence of very-long-chain fatty acids in tissue, was not lowered. On the contrary, the progress of the disease was not slowed in any of the patients in therapy.

Our failure to reduce the hexacosanoate level by dietary means indicated that the abnormal very-long-chain fatty acids deposited in ALD brain also come from one of the other three sources. In order to discover whether very-long-chain fatty acids synthesized in the patient's body also contributed to the abnormal acids, daily doses of 50 ml of deuterated water were given to several ALD patients for the last 196 days of their lives.[62] The deuterated water, after being diluted with body water, is incorporated into fatty acids during their synthesis.[63] Examination of the brain and other tissues by gas chromatography–mass spectrometry revealed that very-long-chain fatty acids in cholesterol esters and other lipids contained small amounts of deuterium. Although the site of synthesis could not be determined, this observation clearly shows that very-long-chain fatty acids are being synthesized in the patient's body and explains the unreduced levels of these acids when dietary restrictions were imposed.

5.2. *Attempt to Stimulate Oxidation*

The enzyme defect causing ALD has been shown to be in fatty acid oxidation.[47] In addition, there is evidence that the oxidation of very-long-chain fatty acids in liver may be taking place in peroxisomes rather than in mitochondria.[55] These two observations suggest that ALD may be caused by a defect in the fatty acid oxidation in peroxisomes. The activity of fatty acid oxidation in peroxisomes increases significantly in liver with administration of hypolipidemic drugs.[64] On the basis of these observations, an ALD patient was treated with clofibrate. Although the level of very-long-chain fatty acids in the patient's plasma decreased significantly, the decrease was only similar to the decrease in other fatty acids. The lack of any specific effect on the level of very-long-chain fatty acids and severe side effects forced the investigators to abandon the experiment. A similar investigation was done substituting carnitine for the hypolipidemic, since carnitine is required for the oxidation of fatty acids in mitochondria[58] and possibly in peroxisomes.[64] This also proved ineffective.

5.3. Attempt to Remove Very-Long-Chain Fatty Acids

As described previously, the level of very-long-chain fatty acids in plasma also increases in ALD patients. Because of the high level in plasma, it is clear that the fatty acids accumulated in brain and adrenals are transported by plasma. In order to cut this supply off, plasmapheresis was performed on an ALD patient. Unfortunately, the level of hexacosanoic acid in the plasma returned to the abnormal level only a few days after every plasmapheresis, and the treatment had to be discontinued. Another possible method for removing abnormal fatty acids from plasma and possibly from brain and adrenal gland is by a bone marrow transplant. It has already been shown that the enzyme that oxidizes very-long-chain fatty acids is present in human white blood cells. Since this enzyme activity appears to be defective, it would be worthwhile to examine the effect of a bone marrow transplant whereby normal white blood cells would be supplied continuously.

6. PATHOGENESIS OF ADRENOLEUKODYSTROPHY

Despite the identification of the pathogenic compound as very-long-chain fatty acids (C_{24}–C_{30}) and the demonstration that the enzymic defect is in the oxidation of these fatty acids, the cause of the drastic pathological changes observed in brain white matter and adrenal cortex of ALD patients is not known. Brain normally contains large amounts of very-long-chain fatty acids as components of cerebrosides and sulfatides, but the abnormality in the metabolism of these sphingolipid fatty acids appears to be minimal at most. Instead, the very-long-chain fatty acids are accumulated as components of cholesterol esters in brain as well as in the adrenal cortex. It is noteworthy that the brain and adrenal gland are the only tissues severely affected by this disease and also the only tissues that contain the abnormal very-long-chain fatty acids as components of cholesterol esters. It is possible, therefore, that the cholesterol esters with these very-long-chain fatty acids may cause a distortion in the membrane structure because of the extra-long carbon chains. In fact, the presence of up to 5% of these cholesterol esters in myelin isolated from ALD patients was recently reported.[65] This myelin showed an abnormal X-ray diffraction pattern. In normal myelin, the content of cholesterol esters is negligible, and there are several contrary reports that indicate the absence of cholesterol esters in ALD myelin.

Another possible cause of the damage in ALD tissues is a circulating toxic factor. A similar cause has been proposed for multiple sclerosis.[66] The inflammation and demyelination observed in ALD brain resemble multiple sclerosis. It is hypothesized that the same toxic factor that causes demyelination and adrenal insufficiency also damages the oxidation of very-long-chain fatty acids in the liver. The defect in oxidation could result in the rise in the plasma level of very-long-chain fatty acid and subsequent accumulation in brain and adrenal gland for unknown reasons. In this connection, it is interesting to note that the nature of the cholesterol-synthesizing enzyme is similar in both brain and adrenal gland and is unique among all other tissues examined. It should be reemphasized here that the metabolism of the very-long-chain fatty acids normally present in brain appears to be unaffected. This observation strongly suggests that the enzyme involved in the oxidation of fatty acids that is defective in ALD is different from that involved in sphingolipid fatty acid metabolism.

7. *SUMMARY*

A decade ago adrenoleukodystrophy was an ambiguous, ill-defined disease, and it is now well-characterized as a genetic disorder. The discoveries of the characteristic inclusion in brain white matter and adrenal cortex of ALD patients and the abnormal very-long-chain fatty acids in the cholesterol esters from ALD patients were the springboard for all subsequent progress toward the discovery of the cause of this disease. The development of a simple and reliable diagnostic tool helped identify many new patients and also enabled carrier and prenatal diagnosis. Furthermore, the enzymic defect in very-long-chain fatty acid oxidation was demonstrated in cultured skin fibroblasts and white blood cells obtained from ALD patients. Thus, it is now clear that adrenoleukodystrophy is caused by a defective X-linked gene that affects the oxidation of very-long-chain fatty acids. Further elucidation of the enzyme step deficient in ALD patients should provide further insight into the disease mechanism.

ACKNOWLEDGMENT. The authors are grateful for skillful editorial and typing assistance from Ms. Geneva R. Davis and Janice White.

REFERENCES

1. Igarashi, M., Schaumburg, H. H., Powers, J., Kishimoto, Y., Kolodny, E., and Suzuki, K., 1976, *J. Neurochem.* **26:**851–860.
2. Kawamura, N., Moser, A. B., Moser, H. W., Ogino, T., Suzuki, K., Schaumburg, H., Milunsky, A., Murphy, J., and Kishimoto, Y., 1978, *Biochem. Biophys. Res. Commun.* **82:**114–120.
3. Moser, H. W., Moser, A. B., Kawamura, N., Murphy, M., Suzuki, K., Schaumburg, H., and Kishimoto, Y., 1980, *Ann. Neurol.* **7:**542–549.
4. Moser, H. W., Moser, A. B., Frayer, K. K., Chen, W., Schulman, J. D., O'Neill, B. P., and Kishimoto, Y., 1981, *Neurology (N.Y.)* **31:**1241–1249.
5. Siemerling, E., and Creuzfeldt, H. G., 1923, *Arch. Psychiatr. Nervenkr.* **68:**217–244.
6. Blaw, M. E., 1970, *Handbook of Clinical Neurology*, Volume 10 (P. G. Vinken and G. W. Bruyn, eds.), American Elsevier, New York, pp. 128–133.
7. Schaumburg, H. H., Richardson, E. P., Suzuki, K., and Raine, C. S., 1974, *Arch. Neurol.* **31:**210–213.
8. Powers, J. M., and Schaumburg, H. H., 1974, *Am. J. Pathol.* **76:**481–491.
9. Powell, H., Tindall, R., Schultz, P., Paa, D., O'Brien, J., and Lampert, P., 1975, *Arch. Neurol.* **32:**250–260.
10. Moser, H. W., Moser, A. B., Kawamura, N., Migeon, B., O'Neill, B. P., Fenselau, C., and Kishimoto, Y., 1980, *Johns Hopkins Med. J.*, **147:**217–224.
11. Griffin, J. W., Goren, E., Schaumburg, M. D., Engel, W. K., and Loriaux, L., 1977, *Neurology (Minneap.)* **27:**1107–1113.

12. Schaumburg, H. H., Powers, J. M., Raine, C. S., Spencer, P. S., Griffin, J. W., Prineas, J. W., and Boehme, D. M., 1977, *Neurology (Minneap.)* **27**:1114–1119.
13. Ulrich, J., Herschkowitz, N., Heitz, P., Sigrist, T., and Baelocher, P., 1978, *Acta Neuropathol.* **43**:77–83.
14. Manz, H. J., Schuelein, M., McCullough, D. C., Kishimoto, Y., and Eiben, R. M., 1980, *J. Neurol. Sci.* **45**:245–260.
15. Jaffe, R., Crumbrine, P., Hashida, Y., and Moser, H. W., 1982, *Am. J. Pathol.* **108**:100–111.
16. Suzuki, Y., Tucker, S. H., Rorke, L. B., and Suzuki, K., 1970, *J. Neuropathol. Exp. Neurol.* **29**:405–419.
17. Eto, Y., and Suzuki, K., 1971, *J. Neurochem.* **19**:1007–1016.
18. Eviatar, L., Harris, D. R., and Menkes, J. H., 1973, *Biochem. Med.* **8**:268–279.
19. Ramsey, R. B., Banik, N. L., Scott, T., and Davison, A. N., 1976, *J. Neurol. Sci.* **29**:277–294.
20. Ramsey, R. B., Banik, N. L., and Davison, A. N., 1979, *J. Neurol Sci.* **40**:189–196.
21. Menkes, J. H., and Corbo, L. M., 1977, *Neurology (Minneap.)* **27**:928–932.
22. Molzer, B., Bernheimer, H., and Toifl, K., 1981, *Acta Neuropathol. (Berl.) [Suppl.]* **7**:211–214.
23. Molzer, B., Bernheimer, H., Budka, H., Pilz, P., and Toifl, K., 1981, *J. Neurol. Sci.* **51**:301–310.
24. Kishimoto, Y., and Radin, N. S., 1966, *Lipids* **1**:47–61.
25. Eichberg, J., Hauser, G., and Karnovsky, M. L., 1969, *The Structure and Function of Nervous Tissue*, Volume 3 (G. H. Bourne, ed.), Academic Press, New York, pp. 185–287.
25a. Brown, F. R., Chen, W. W., Kirschner, D. A., Frayer, K. L., Powers, J. M., Moser, A. B., and Moser, H. W., 1983, *J. Neurochem.* **41**:341–348.
26. Igarashi, M., Belchis, D., and Suzuki, K., 1976, *J. Neurochem.* **27**:327–328.
27. Suzuki, K., 1965, *J. Neurochem.* **12**:969–979.
28. Kishimoto, Y., and Radin, N. S., 1965, *J. Lipid Res.* **7**:141–145.
29. Nonaka, G., and Kishimoto, Y., 1979, *Biochim. Biophys. Acta* **572**:432–441.
30. Kishimoto, Y., and Kawamura, N. 1979, *Mol. Cell Biochem.* **23**:17–25.
31. Miyatake, T., Ariga, T., Atsumi, T., and Komiya, Y., 1977, *Myelination and Demyelination* (J. Palo, ed.), Plenum Press, New York, pp. 585–599.
32. Ramsey, R. B., Banik, N. L., and Davison, A. N., 1977, *J. Neurol. Sci.* **32**:69–77.
33. Moser, H. W., Moser, A. B., Powers, J. M., Nitowsky, H. M., Schaumburg, H. H., Norum, R. A., and Migeon, B. R., 1982, *Pediatr. Res.* **16**:172–175.
34. Yahara, S., Moser, H. W., Kolodny, E. H., and Kishimoto, Y., 1980, *J. Neurochem.* **34**:694–699.
35. Tsuji, S., Suzuki, M., Ariga, T., Sekine, M., Kuriyama, M., and Miyatake, T., 1981, *J. Neurochem.* **36**:1046–1049.
36. Singh, I., Moser, H. W., Moser. A. B., and Kishimoto, Y., 1984, *Pediatr. Res.* **18**:286–290.
37. McLaughlin, J., Askanas, V., and Engel, W. K., 1980, *Biochem. Biophys. Res. Commun.* **92**:1202–1207.
38. Burton, B. K., and Nadler, H. L., 1974, *Pediatr. Res.* **8**:170–175.
39. Yavin, E., Milunsky, A., DeLong, G. R., Nash, A. H., and Kolodny, E. H., 1976, *Pediatr. Res.* **10**:540–543.
40. Eto, Y., and Suzuki, K., 1971, *Biochim. Biophys. Acta* **239**:293–311.
41. Eto, Y., and Suzuki, K., 1973, *J. Biol. Chem.* **248**:1986–1991.
42. Ogino, T., Schaumburg, H. H., Suzuki, K., Kishimoto, Y., and Moser, A., 1978, *Myelination and Demyelination* (J. Palo, ed.), Plenum Press, New York, pp. 601–619.
43. Ogino, T., and Suzuki, K., 1981, *J. Neurochem.* **36**:776–779.
44. Choi, M. U., and Suzuki, K., 1978, *J. Neurochem.* **31**:879–885.
45. Jagannatha, H. M., and Sastry, P. S., 1981, *J. Neurochem.* **36**:1352–1360.
46. Michels, V. V., and Beaudet, A. L., 1980, *Pediatr. Res.* **14**:21–23.
47. Singh, I., Moser, H. W., Moser, A. B., and Kishimoto, Y., 1981, *Biochem. Biophys. Res. Commun.* **102**:1223–1229.
48. Shyamala, G., Lossow, W. J., and Chaikoff, I. L., 1965, *Proc. Soc. Exp. Biol. Med.* **118**:138–142.

49. Chen. W. W., Lue, N., Muralidharan, V. B., Kishimoto, Y., and Moser, H. W., 1982, *J. Cell Biol.* **95**:453A.
50. Singh, I., Moer, H. W., Moser, A. B., and Kishimoto, Y., 1982, *Trans. Am. Soc. Neurochem.* **13**:187.
51. Tsuji, S., Sano, T., Ariga, T., and Miyatake, T., 1981, *J. Biochem. (Tokyo)* **90**:1233–1236.
52. Kishimoto, Y., Akanuma, H., and Singh, I., 1979, *Mol. Cell Biochem.* **28**:93–105.
53. Uda, M., Singh, I., and Kishimoto, Y., 1981, *Biochemistry* **20**:1295–1300.
54. Kawamura, N., and Kishimoto, Y., 1981, *J. Neurochem.* **36**:1786–1791.
55. Kawamura, N., Moser, H. W., and Kishimoto, Y., 1981, *Biochem. Biophys. Res. Commun.* **99**:1216–1225.
56. Migeon, B. R., Moser, H. W., Moser, A. B., Axelman, J., Sillence, D., and Norum, R. A., 1981, *Proc. Natl. Acad. Sci. U.S.A.* **78**:5066–5070.
57. Bourre, J.-M., Paturneau-Jouas, M. Y., Daudu, D. L., and Baumann, N. A., 1977, *Eur. J. Biochem.* **72**:41–47.
58. Murad, S., and Kishimoto, Y., 1978, *Arch. Biochem. Biophys.* **185**:300–306.
59. Markley, K. S., 1960, *Fatty Acids: Their Chemistry, Properties, Production and Uses* (K. S. Markley, ed.), Interscience, New York, pp. 23–250.
60. Kishimoto, Y., Moser, H. W., Kawamura, N., Platt, M., Pallante, S. L., and Fenselau, C., 1980, *Biochem. Biophys. Res. Commun.* **96**:69–76.
61. Brown, F. R., III., Van Duyn, M. A. S., Moser, A. B., Schulman, J. D., Rizzo, W. B., Snyder, R. D., Murphy, J. V., Kamoshita, S., Migeon, C. J., and Moser, H. W., 1982, *Johns Hopkins Med. J.* **151**:164–172.
62. Moser, H. W., Pallante, S. L., Moser, A. B., Rizzo, W. B., Schulman, J. D., and Fenselau, C., 1983, *Ped. Res.* **17**:293A.
63. Wakil, S. J., 1970, *Lipid Metabolism* (S. J. Wakil, ed.), Academic Press, New York, pp. 1–48.
64. Lazarow, P. B., and deDuve, C., 1976, *Proc. Natl. Acad. Sci. U.S.A.* **73**:2043–2046.
65. Brown, F. R., Chen, W. W., Kirschner, D. A., Frayer, K. L., Powers, J. M., Moser, A. B., and Moser, H. W., 1983, *J. Neurochem.* **41**:341–348.
66. Wolfgram, F., 1972, *Multiple Sclerosis: Immunology, Virology, and Ultrastructure* (F. Wolfgram, G. W. Ellison, J. G. Stevens, and J. M. Andrews, eds.), Academic Press, New York, pp. 173–182.

9

Brain Edema

Pak Hoo Chan and Robert A. Fishman

1. INTRODUCTION

Brain edema accompanies a wide variety of pathological processes. It contributes to the morbidity and mortality of many neurological diseases, including head injury, stroke, brain tumor, cerebral infections (e.g., brain abscess, encephalitis, and meningitis), lead encephalopathy, hypoxia, hypoosmolality, disequilibrium syndromes associated with dialysis and diabetic ketoacidosis, and some forms of obstructive hydrocephalus. Although many aspects of the pathophysiology of brain edema have been clarified,[1-7] the molecular mechanisms and biochemical events that underlie the formation of edema are poorly understood. This chapter reviews current hypotheses concerning the development of brain edema and its potential drug therapy.

2. DEFINITIONS AND CLASSIFICATIONS

Brain edema is defined as an increase in brain volume caused by an increase in water and sodium content, generally associated with a fall in brain potassium content. When brain edema is localized or mild in degree, it is associated with little or no clinical evidence of brain dysfunction; when severe, it causes major focal or generalized signs of dysfunction (or both), including various types of cerebral herniation, which eventually result in failure of medullary centers for respiration and circulation.

Brain edema has been divided into the following three categories based on neuropathological and experimental observations: (1) vasogenic, (2) cellular (cytotoxic), and (3) interstitial (hydrocephalic).[8-11] A recent monograph[10] summarizes knowledge concerning pathogenesis, the location and composition of the edema fluid, and changes in capillary permeability in the various forms of edema.

Pak Hoo Chan and Robert A. Fishman • Department of Neurology, University of California School of Medicine, San Francisco, California 94143.

2.1. Vasogenic Edema

Vasogenic edema is the most common form of brain edema observed in clinical practice. Characteristics of this edema are (1) an increased brain water and sodium content, (2) an expanded extracellular space (the extracellular fluid volume is increased by the edema fluid, which has the features of a plasma filtrate containing plasma protein), and (3) an increased permeability of brain capillary endothelial cells to macromolecules (e.g., albumin, whose entry into brain is normally severely limited by the capillary endothelial cells).

In the vasogenic edemas, the functional integrity of the endothelial cells is altered. Increased endothelial permeability has been established in various experimental models such as freezing lesions, stab wounds, brain compression, anoxia, experimental brain tumors, and allergic encephalomyelitis.[2,12] This increased permeability varies inversely with the molecular weight of the various markers, with a greater increase in the entry of inulin (mol. wt. 5,000) than of albumin (mol. wt. 69,000). There is evidence for both ultrastructural defects in the tight endothelial cell junctions (characteristic of the blood–brain barrier) and an increase in the number of pinocytotic vesicles in the endothelial cells that appear responsible for the transport of macromolecules. Although morphological evidence suggests that increased vesicular transport is characteristic of the vasogenic edemas,[13–16] the quantitative importance of each of these changes is not known. The fact that entry of tracers into brain varies inversely with molecular weight suggests diffusion through channels rather than bulk transport in vesicles. The cerebral white matter is far more vulnerable to vasogenic edema than is the gray matter, but the reason for this is unknown. It has been suggested that this vulnerability is caused by the low capillary density and blood flow is normal white matter as opposed to the cortical and subcortical gray matter.

2.2. Cellular Edema

Cellular (cytotoxic) edema is characterized by swelling of all the cellular elements of the brain (neurons, glia, and endothelial cells) with a concomitant reduction in the volume of the brain's extracellular fluid space.[8–10,17,18] The occurrence of cellular edema and a decrease in brain ECF space has been demonstrated with an electrical impedance method in experimental animals with acute brain ischemia.[19,20]

One cause of cellular edema is cerebral energy depletion as a result of hypoxia, as seen after cardiac arrest or asphyxia. The cellular swelling is osmotically determined by the appearance of increased intracellular osmoles (largely sodium, lactate, and hydrogen ions), which induce the rapid entry of water into cells.[18,21] Another cause—acute hypoosmolality of the plasma and extracellular fluid—can arise from acute dilutional hyponatremia, inappropriate secretion of antidiuretic hormone, or acute sodium depletion.[22–24] A third cause is osmotic disequilibrium, i.e., the syndromes observed with hemodialysis or diabetic ketoacidosis, wherein excessive intracellular solutes result in excessive hydration when the plasma osmolality is rapidly reduced with therapy.[25]

2.3. *Interstitial (Hydrocephalic) Edema*

Interstitial edema occurs in obstructive hydrocephalus when the increased intraventricular hydrostatic pressure forces CSF across the ventricular walls into the periventricular white matter.[26,27] Obstruction of the circulation of the CSF within the ventricular system or of CSF absorption in the subarachnoidal spaces can result in this transependymal movement of CSF and, thereby, an absolute increase in the volume of the brain's extracellular fluid.

3. *MEASUREMENT OF EDEMA*

3.1. *In Vivo Edema Measurement*

3.1.1. *Wet–Dry Weight Method*

The usual procedure for quantitating edema *in vivo* is to sample the edematous area of brain and a control area (preferably the homologous contralateral area) and to determine the respective dry-weight percentages. This is most conveniently done by weighing each fresh tissue sample (wet weight) in a tared container, drying the sample at approximately 100°C to constant weight (usually within 12–24 hr), and reweighing to obtain the weight of the residue (dry weight). The percentage dry weight (P) is simply calculated as:

$$P = (\text{dry weight/wet weight}) \times 100$$

The percentage of swelling (edema) or of shrinkage can then be calculated by the formula of Elliot and Jasper.[6]

$$[P_{\text{cont}} - P_{\text{exp}}] \times 100 = \text{percent swelling (or shrinkage)}$$

3.1.2. *Specific Gravity*

By using a liquid gradient column consisting of a mixture of the organic solvents bromobenzene and kerosene of known specific gravity, the tissue water content of small brain samples (5–15 mg) can be calculated according to Nelson *et al.*[28]

$$\text{g H}_2\text{O/g tissue} = 1 - [(\text{Spgr}_t - 1)/(1 - 1/\text{Spgr}_s)\ \text{Spgr}_t]$$

where Spgr_s is the specific gravity of the tissue solid.

$$\text{Spgr}_s = 1/1 - [(\text{Spgr}_t - 1)g_t/(g_s \times \text{Spgr}_t)]$$

in which g_t is fresh tissue weight, g_s is dry tissue weight, and Spgr_t is specific gravity of tissue.

The specific gravity method has been refined by Marmarou *et al.*,[29,30] who developed a new formula to compensate for the protein content of edema fluid and to reduce the experimental error. Recently, density gradients of nontoxic Percoll, a form of colloidal silica, soaked with polyvinylpyrrolidone have been used to replace toxic bromobenzene–kerosene.[31]

3.1.3. Computerized Tomography, Positron Emission Tomography, and Nuclear Magnetic Resonance

Computerized tomography (CT), position emission tomography (PET), and nuclear magnetic resonance (NMR) are noninvasive *in vivo* techniques that can be used to measure brain edema in experimental animals and human subjects. Computerized tomographic scanning is a technique in which X-ray data recorded from many different directions are reconstructed mathematically to yield cross-sectional views of brain. This technique has been proven to be an extremely useful diagnostic tool for brain edema.[32–34]

Positron emmission tomography (PET) reconstructs an image of the radioactivity in the brain following an injection of a radionuclide in the living animal. This technique gives results similar to quantitative autoradiography without damaging the tissue. It has been used extensively to measure regional cerebral blood volume (RCBV) and, subsequently, the concentration of a particular radiopharmaceutical in the vascular compartment compared to be extravascular compartment of tissue.[35,36] In addition, PET can be used to measure the regional brain water as well as the brain water permeability.[37]

Nuclear magnetic resonance (NMR) is a new technique for obtaining cross-sectional images of the brain without exposing the subject studied to ionizing radiation.[38–41] When certain atomic nuclei, for example, protons, are placed in a magnetic field and stimulated by radio waves of a particular frequency, they reemit some of the absorbed energy in the form of radio signals. After the radio-frequency pulse, the water protons return to their equilibrium state parallel to the static magnetic field, and the emitted signal diminishes. These phenomena are characterized by two time constants—the longitudinal (the direction of the static magnetic field) relaxation time (T_1) and the transverse (perpendicular to the static magnetic field) relaxation time (T_2). The relaxation times of protons in brain tissue depend on their local environment. Since the protons in a liquid environment tend to relax more slowly than those in another environment (e.g., greater T_1 and T_2), relaxation times are prolonged in edematous tissues. With this powerful technique, preliminary studies of brain edema both in humans and in experimental animals have been made.[42–49]

3.1.4. Other Methods

Other techniques, including studies of electrical impedance, radioactive markers, intracranial pressure, and edema fluid analysis, may be useful for the study of cerebral edema *in vivo*.[50]

3.2. *In Vitro Edema Measurement*

3.2.1 *Brain Slices as a Tool for Study of Cellular (Cytotoxic) Swelling*

Brain slices have been used in numerous laboratories to study cellular edema. Single first cortical slices are composed chiefly of gray matter and are preferable to mixed gray and white matter samples because one obtains reproducible tissue water and electrolyte contents. The blood–brain barrier does not exist in brain slices, and the incubation media is contiguous with the extracellular fluid space of the tissue. Our laboratory has chosen the single first cortical slice with pia intact to minimize the physical damage from cutting. These must be less than 0.35 mm in thickness to allow sufficient oxygen to diffuse throughout the slice. The wet–dry weight method used to quantitate edema *in vivo* models is applicable to *in vitro* studies as well.[6,18,51,52] The intracellular space can be determined indirectly by employing various extracellular markers.[53] An increase in intracellular swelling is characteristically accompanied by increased intracellular sodium and decreased intracellular potassium; therefore, the measurement of the changes of intracellular cations has been used as an indicator of cellular swelling.[3,18,54–56] It is noteworthy that optimally prepared cortical slices require 20–30 min of reconstitution in oxygenated buffer to restore the tissue stores of ATP and energy charge. However, even with optimal handling and oxygenation, normal cortical slices are somewhat edematous compared to normal cortex *in vivo*.

3.2.2. *Cell Cultures*

Brain cell cultures provide another useful tool for the study of cellular edema *in vitro*. The extracellular space of brain slices no longer exists in cultured brain cells; it is replaced by an extracellular "trapped space" whose volume is dependent on the experimental conditions. This "trapped space" has been defined by subjecting the cell cultures to direct filtration[57] or to rapid density gradient filtration.[58] By using rapidly diffusable markers such as ^3H_2O or [^{3}H]methyl-D-glucose and extracellular markers such as [carboxyl-^{14}C]inulin (mol. wt. 5000) or [1,2-^{14}C]polyethylene (mol. wt. 4000), the intracellular volume of pellets of cultured neoplastic brain cells (including rat C-6 glioma cells, mouse N18 TG-2 neuroblastoma cells, and NG108-15 neuroblastoma–glioma hybrid cells) has been measured.[59] The intracellular volume was calculated according to the method of Holian *et al.*[58]

$$\text{Intracellular volume } (\mu\text{l}) = \frac{\text{Pellet cpm } (^3\text{H}) - [EV \times \text{supernatant cpm}/\mu\text{l } (^3\text{H})]}{\text{cpm}/\mu\text{l of supernatant } (^3\text{H})}$$

$$EV\ (\mu\text{l}) = \frac{\text{cpm in pellet } (^{14}\text{C})}{\text{cpm}/\mu\text{l in supernatant } (^{14}\text{C})}$$

The volume of cultured brain cells (μm^3) has also been determined electrically by the Coulter method. The cell volume of C-6 glioma, N-115 neuroblastoma, and primary glial cells has been studied with media of various osmolalities and with edema-inducing factors.[60,61]

4. BIOCHEMISTRY AND MOLECULAR MECHANISMS OF BRAIN EDEMA

4.1. Ion Flux and Na^+,K^+-ATPase

The mechanisms of cellular swelling have been studied in brain slices and in glial cell cultures.[18,50–52,54–56,62] Early studies showed that Na^+,K^+-ATPase is involved in regulating Na^+ and K^+ cation flux in astroglial and other mammalian cells.[63,64] Metabolic inhibitors, hypoxia, and cardioactive glycosides, which affect the normal function of the sodium pump, induce cellular swelling. In each case, there is a failure of energy-dependent active transport of sodium and water from the cell with a fall in brain ATP levels.[21,54] The involvement of the Na^+ pump in cellular edema has been demonstrated in all cellular elements in nervous tissue including neurons, glia, and endothelial cells.

The use of media containing high concentration of K^+ or HCO_3^-/CO_2 buffer also causes cellular swelling in brain slices and in glial cell cultures.[55,56,62,65–67] In fact, high external K^+ concentrations (up to 30 to 70 mM) occur in brain tissue during ischemia or hypoxia.[68] It is well known that high external concentrations of K^+ cause depolarization of presynaptic membranes and the release of various neurotransmitters into synaptic clefts. The accumulation of extracellular excitatory neurotransmitters, glutamic acid and aspartic acid in particular, has been shown to inhibit the Na^+ pump and to cause cellular swelling.[52] The mechanisms of cation transport and Na^+,K^+-ATPase activity have been described in an earlier volume of this *Handbook*.[69]

Besides Na^+,K^+-ATPase, carbonic anhydrase and SITS-sensitive anion-exchange systems are also involved in HCO_3^-/CO_2-stimulated swelling in primary glial cell cultures.[62] In this system (Fig. 1), CO_2 rapidly equilibrates across the cell membrane and is hydrated, forming $HCO_3^- + H^+$; this process is greatly accelerated by intracellular carbonic anhydrase. Then H^+ leaves the cells in exchange for Na^+, and HCO_3^- leaves in exchange for Cl^-; this increase in Na^+ and Cl^- intracellularly causes cellular swelling. Evidence for the Na^+–H^+ exchange system is provided by the observation that amiloride, which inhibits $^{22}Na^+$ uptake, also reduces intracellular pH. The mechanisms of anion transport in brain tissues have been described in Volume 1 of this *Handbook*.[71]

4.2. Arachidonic Acid and Polyunsaturated Fatty Acids

By the use of cortical brain slices as an *in vitro* bioassay system, it has been shown that the membranes of polymorphonuclear leukocytes (WBC) can induce cellular swelling.[54] In these experiments, the edematous brain slices were characterized by an increase in sodium and a decrease in inulin space and potassium content. The factor in WBC membranes responsible for edema formation was further defined as the lipid-soluble fraction, especially the free

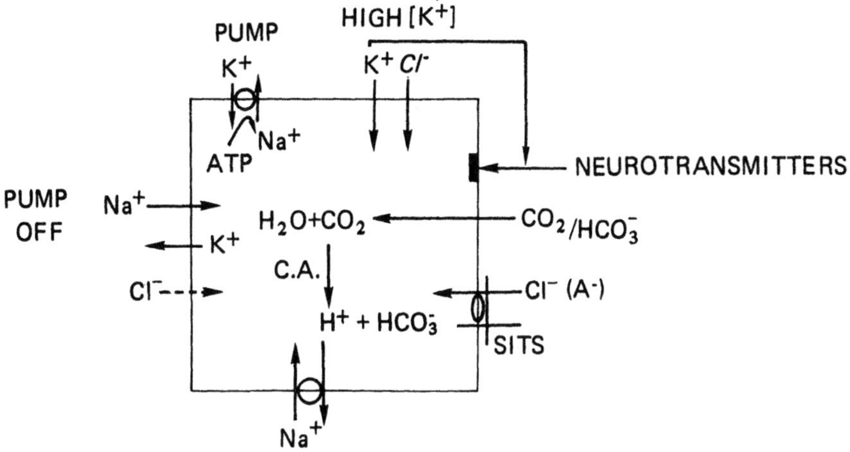

Fig. 1. Ions flux and cellular edema. (This figure was reproduced from Kimelberg[62] with permission of Raven Press.)

fatty acids.[51] Further analysis has indicated that only the polyunsaturated fatty acids (PUFA), especially arachidonic acid (20:4), an intrinsic constituent of the phospholipids of cellular membranes, are potent inducers of cellular edema. The saturated fatty acids [e.g., nonanoic acid (9:0), lauric acid (12:0), and palmitic acid (16:0)] and the monounsaturated oleic acid (18:1) did not induce edema (Table I). Cortical slices incubated in 0.5 mM arachidonic acid had swollen 12.6% at 15 min and continued to swell during the 90 min of incubation to reach a maximum of 35.2%.[72] Studies of various molar ratios of 20:4 to bovine serum albumin (BSA) suggested that free or unbound 20:4 activates the process of brain swelling, whereas BSA-bound 20:4 was ineffective. The swelling in this system was irreversible; i.e., it was not modulated by BSA,

Table I
Effects of Fatty Acid on Cellular Edema in Cortical Slices[a]

Fatty acid	Swelling (%)	Inulin space (%)	Lactate (mmol/kg dry wt.)	Na^+ (mEq/kg dry wt.)	K^+ (mEq/kg dry wt.)
Control	11.3	46.11	236	599	393
Nonanoic acid (9:0)	12.6	49.42	269*	519	312
Lauric acid (12:0)	11.6	49.22	320*	651	315
Palmitic acid (16:0)	9.4	46.12	258	838*	387
Oleic acid (18:1)	11.0	41.07**	278**	694**	419
Linoleic acid (18:2)	32.5*	33.47*	519*	1340*	92*
Linolenic acid (18:3)	26.0*	42.88*	391*	949*	188*
Arachidonic acid (20:4)	35.2*	32.6*	527*	1239*	140*
Docosahexaenoic acid (22:6)	33.5*	41.47*	543*	1215*	77*

[a] Data obtained from ref. 51. * $P < 0.01$, ** $P < 0.02$, Student's *t*-test.

Table II
In Vivo Effects of Free Fatty Acids on Cerebral Water, Sodium, Potassium, and Blood–Brain Barrier Permeability[a]

Fatty acid	H_2O content (%)	Na^+ (mEq/kg dry wt.)	K^+ (mEq/kg dry wt.)	[^{125}I]Bovine serum albumin space (%)
Krebs–Ringer	79.5 ± 0.2(36)	241.1 ± 1.3(36)	555.6 ± 2.0(36)	2.04 ± 0.1(9)
Nonanoate (9:0)	79.1 ± 0.07(9)	248.4 ± 0.7(9)	553.8 ± 1.0(9)	2.17 ± 0.02(6)
Palmitate (16:0)	79.3 ± 0.07(17)	254.8 ± 0.5(17)	545.7 ± 0.7(17)	2.30 ± 0.13(6)
Oleate (18:1)	79.1 ± 0.08(6)	264.5 ± 0.8(6)	536.3 ± 1.6(6)	2.40 ± 0.01(6)
Linolenate (18:3)	79.8 ± 0.04(6)	278.2 ± 1.2(6)*	319.6 ± 0.8(6)*	4.25 ± 0.02(6)*
Arachidonate (20:4)	80.6 ± 0.2(29)*	283.8 ± 1.1(29)*	258.0 ± 1.7(29)*	6.68 ± 0.29(12)*

[a] Data obtained from ref. 74. (*n*), number of animals, * $P < 0.01$ compared to control right hemisphere, Student's *t*-test.

although BSA could extract 46% of the arachidonic acid from brain slices.[72,73] These data suggested that the arachidonic-acid-induced brain swelling could not be reversed by BSA.

Cellular edema induced by PUFA has been confirmed *in vivo*.[74,75] Intracerebral infusion of PUFAs, including arachidonic acid (20:4), linolenic acid (18:3), and docosahexaeonic acid (22:6), caused a significant increase in cerebral water and sodium content concomitant with a decrease in potassium content. Saturated fatty acids including nonanoic acid (9:0) and palmitic acid (16:0) and a monounsaturated fatty acid, oleic acid (18:1), were not effective (Table II).[74] The induction of brain edema by arachidonic acid was dose dependent and maximal between 24 and 48 hr after infusion. This time course of edema development is similar to the time course of the edema induced by cold injury.[76] Both Evans blue permeability and the [^{125}I]bovine serum albumin (BSA) space of the thalamus were significantly increased at 24 hr after the infusion of arachidonic acid. Morphological studies confirm that arachidonic acid, when infused into brain in nanomolar quantities, induces edema. This edema has the pathological characteristics of vasogenic brain edema, namely, breakdown of the blood–brain barrier to protein tracers and a tendency for edema fluid to spread along white matter.[4]

The basic mechanisms underlying the PUFA-induced cellular edema *in vitro* and the cellular and vasogenic edemas *in vivo* are largely unknown. Polyunsaturated fatty acids readily undergo autooxidation by forming lipid peroxides[77–79] or by conversion to radical intermediates by cyclooxygenase and lipoxygenase.[80,81] Moreover, the metabolism of arachidonic acid by cytochrome P-450 could lead to the formation of superoxide anions and other arachidonic acid hydroperoxides.[82] These processes suggest that free radicals are capable of mediating the cellular swelling. When rat brain cortical slices were incubated with PUFA, transient formation of superoxide radicals and lipid peroxide was observed.[83] The malondialdehyde content of glioma cells was also markedly increased by incubation with exogenous arachidonic acid.[84]

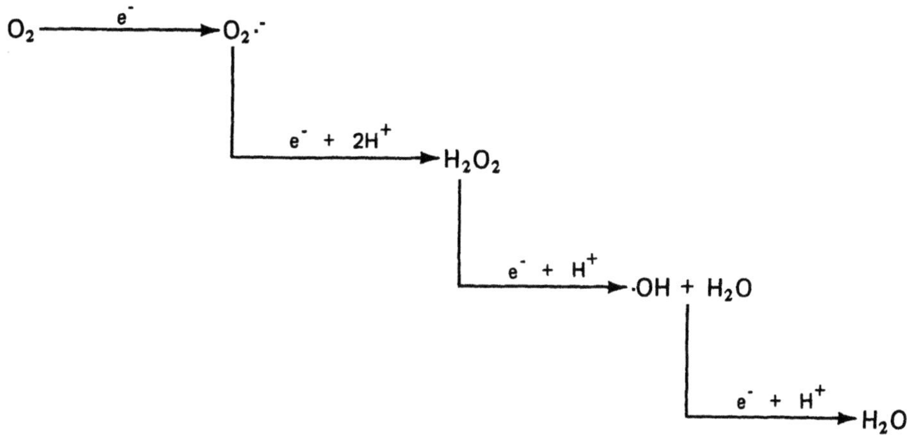

Fig. 2. Production of free radicals by single-electron reduction of oxygen in biological systems. $O_2\cdot^-$, superoxide; H_2O_2, hydrogen peroxide; $\cdot OH$, hydroxyl radical.

These studies suggest that free PUFA-mediated radical intermediates may play an important role in the formation of edema.

4.3. *Oxygen-Derived Free Radicals*

Superoxide anion ($O_2\cdot^-$), hydrogen peroxide (H_2O_2), and hydroxyl radicals ($\cdot OH$) are produced in biological systems by the single-electron reduction of oxygen[85] (Fig. 2). It has been suggested that these highly reactive oxygen-derived free radicals may play an important role in CNS ischemia and trauma.[78,86–88] Also, it is very likely that the brain's abundant membrane phospholipids undergo lipid peroxidation and degradation induced by free radicals because of their high content of polyunsaturated fatty acids.[77] In addition to the indicated involvement of free radicals in CNS injury, it has been demonstrated that changes in vascular permeability in spinal cord and hamster cheek pouch are also induced by free radicals.[89,90] Although the mechanisms for the permeability changes are not clear, it is most likely that vascular membranes are highly susceptible to free-radical-induced peroxidative damage.

Using brain slices as an *in vitro* bioassay system, we have demonstrated that cellular injury and edema were induced by xanthine oxidase/hypoxanthine/ADP-Fe^{3+}[91] (a widely used *in vitro* free-radical-generating system[92,93]). The cellular injury and edema were characterized by increased intracellular sodium and lactic acid levels concomitant with a decreased potassium level. Furthermore, membrane phospholipids were degraded, and there was an increase in lipid peroxidation and a release of free fatty acids, i.e., oleic acid (18:1), arachidonic acid (20:4), and docosahexaenoic acid (22:6).[91] These data indicate that oxygen-derived free radicals cause membrane perturbation *in vitro*.

In vivo studies have confirmed the *in vitro* findings.[94] Xanthine oxidase/hypoxanthine/ADP-Fe^{3+} solutions (0.1 ml) was infused into caudate–putamen, and the brain was frozen rapidly *in situ* at various times thereafter; brain water and sodium content were significantly increased concomitant with a decreased potassium content at 24 and 48 hr after the infusion. The degree of brain edema and injury was dependent on the dose of xanthine oxidase. Histological studies showed the development of spongy neuropil and neuronal cytoplasmic vacuoles at 2 hr and infiltration of polymorphonuclear leukocytes at 24 hr, following by lipid-laden macrophages and reactive astrocytes. Time-course studies showed profound leakage of fluorescent Evans blue into neuropil at 2 hr (an indication of increased blood–brain barrier permeability). There was little or no extravasation of the dye at later times. These data suggest that oxygen-derived free radicals cause damage to the endothelial cells of the blood–brain barrier and also directly injure brain, causing edema and structural changes in neurons and glia.[94]

The types of free radicals produced by the enzymatic system that are associated with cellular injury and phospholipid degradation are not yet clear. Whenever superoxide anions are formed from the xanthine oxidase system, they will react with Fenton's reagent (iron salt) to form hydroxyl radicals ($\cdot OH$). The mechanism is a superoxide-driven, iron-promoted Haber–Weiss reaction.[95]

$$O_2\cdot^- + ADP\text{-}Fe^{3+} \rightarrow ADP\text{-}Fe^{2+} + O_2$$

$$2O_2\cdot^- + 2H^+ \rightarrow H_2O_2 + O_2$$

$$ADP\text{-}Fe^{2+} + H_2O_2 \rightarrow ADP\text{-}Fe^{3+} + OH^- + \cdot OH$$

Furthermore, infusion of uric acid, the end product of the reaction of xanthine oxidase and hypoxanthine, did not promote tissue injury and phospholipid degradation. Therefore, these data suggest that $O_2\cdot^-$ anions and $\cdot OH$ radicals are the primary radical species involved in the membrane degradative processes.

4.4. Other Edema-Inducing Factors

It has been proposed that, in addition to PUFA, PUFA intermediates, and oxygen-derived free radicals, lysosomal enzymes, the kallikrein–kininogen–kinin system, biogenic amines, and excitatory amino acids can induce edema.[52,96]

4.4.1. Lysosomes

It has been demonstrated that lysosomes lose their membranes and release their proteolytic enzymes into the surrounding cytoplasm in damaged spinal cord. In the time-course studies of focal cerebral ischemia, however, the release of lysosomal enzymes is a late event in ischemic nerve cell damage.[97] Although lysosomal enzymes are released in cold-injured brain,[98] the significance of lysosomal enzymes in edema formation is not clear.

4.4.2. The Kallikrein–Kininogen–Kinin System

Kallidin or bradykinin, the nona- or decapeptide, is the active principle of the kallikrein–kininogen–kinin (KKK) system. The formation of these kinins is inhibited in plasma and in brain under normal conditions. However, when the brain is injured, the KKK system is activated, and kinins are released. Kallidin or bradykinin will enter the cerebral parenchyma through the blood–brain barrier and cause further damage to brain cells.[96] Since the components of the KKK system are found chiefly in plasma, the physiological or pathological functions exclusively controlled by the KKK system (without the presence of other plasma factors) have not yet been clearly established.

4.4.3. Glutamate and Biogenic Amines

By use of the single rat brain cortical slice as an *in vitro* bioassay system, it has been demonstrated that the dicarboxylic amino acid neurotransmitters L-glutamic acid and L-aspartic acid greatly increase intracellular brain swelling while increasing intracellular Na^+, water content, and lactate production and decreasing inulin space and intracellular K^+.[52,99,100] Equimolar GABA, taurine, glycine, the putative inhibitory neurotransmitter amino acids, and equimolar α-aminoisobutyric acid had no effect.[52] It has been demonstrated by Baethmann *et al.* that the perfusion of 10–20 mM glutamate into the ventricular system of cats (to bypass the blood–brain barrier) leads to cerebral edema after 24 hr.[101,102] The increase in brain water content was paralleled by an increase in tissue Na^+ concomitant with loss of K^+. These studies indicate that the accumulation of extracellular glutamate has cytotoxic effects on brain cells. It has been demonstrated recently that a brief incubation (5 min) in 0.5 mM arachidonic acid caused a dramatic decrease in glutamate uptake in brain slices and in synaptosomes.[103] Thus, the inhibition of the reuptake of glutamate and other neurotransmitters may be an important factor in the development of cellular edema. Besides glutamate and excitatory neurotransmitter amino acids, biogenic amines such as serotonin and norepinephrine are involved in vasogenic edema.[103–105] Although some evidence has been found for serotonin-induced alteration of blood–brain barrier permeability,[106] the exact role of serotonin and other biogenic amines in brain edema requires further elucidation.

5. POSSIBLE MECHANISMS FOR THE DEVELOPMENT OF BRAIN EDEMA

There are extensive experimental data that suggest that the alteration of membrane integrity of brain cells and endothelial cells initiates the development of brain edema.[75,107,108] Following ischemia and cold injury, endogenous arachidonic acid and other PUFA are rapidly released (in seconds or minutes) from membrane phospholipids.[109,110] Normally, brain concentrations of free arachidonic acid are very low. However, the level of arachidonic acid in the brain can increase 10- to 100-fold following insults such as ischemia,[88,109–114]

Table III
Arachidonic Acid Release Induced by Pathological Insults Associated with Development of Edema

Pathological conditions	Animal	Arachidonic acid release[a]	Edema[a]	References
Ischemia	Rat	+	+	109,110,112
Postdecapitation ischemia	Rat	+	N.D.	111,113,149,176
Carotid occusion ischemia	Gerbil, rat	+	+	88,114,175
Brain compression	Rat	+	+	112,169
Hypoxia	Rat	+	N.D.	115,182
Hypoglycemic injury seizure	Rat	+	N.D.	183,184
Cold injury	Rat	+	+	76,185
Oxygen-derived free radicals	Rat	+	+	94

[a] +, increased significantly; N.D., not determined.

cold injury,[76] and hypoxia.[115] Furthermore, prostaglandins, the cyclooxygenase products of arachidonic acid, were significantly increased along with oxygen free radicals in cerebral cortex after experimental brain injury.[116–119] The correlation of the release of arachidonic acid and the development of brain edema is summarized in Table III. On the basis of *in vitro* and *in vivo* studies of arachidonic acid and brain edema, it has been proposed that the following steps may lead to the development of vasogenic and cellular edema (Fig. 3).[75]

1. Brain edema that is induced by various pathological insults begins to develop with the activation of phospholipase A_2 and C.[110,120,121] The activation of phospholipase A_2 requires the influx of Ca^{2+}.[122] These enzymes hydrolyze membrane phospholipids, forming arachidonic acid and other lipid compounds. Other hydrolyzing enzymes may also be involved.
2. Once arachidonic acid is released from phospholipids, it is readily converted to prostaglandins, thromboxanes, and oxygen radicals by cyclooxygenase. Free radicals and arachidonic acid induce structural disturbances of cellular membranes of various target cells. The role of lipoxygenase metabolites on the structural integrity of membranes in the ensuing functional disturbance is not known.[81]
3. The membrane perturbation of neurons and glia induced by arachidonic acid causes a reduction in uptake of the neurotransmitters GABA and glutamate as well as a reduction of Na^+,K^+-ATPase activity.[123,124] These events lead to a shift in cations and water and the development of cellular edema.
4. The membrane perturbation induced by arachidonic acid may activate the vesicular transport of macromolecules across brain endothelial cells (pinocytosis). On the other hand, arachidonic acid may injure the endothelial cells directly.[94] The increased permeability of solutes and water from blood to brain will lead to the development of vasogenic edema.

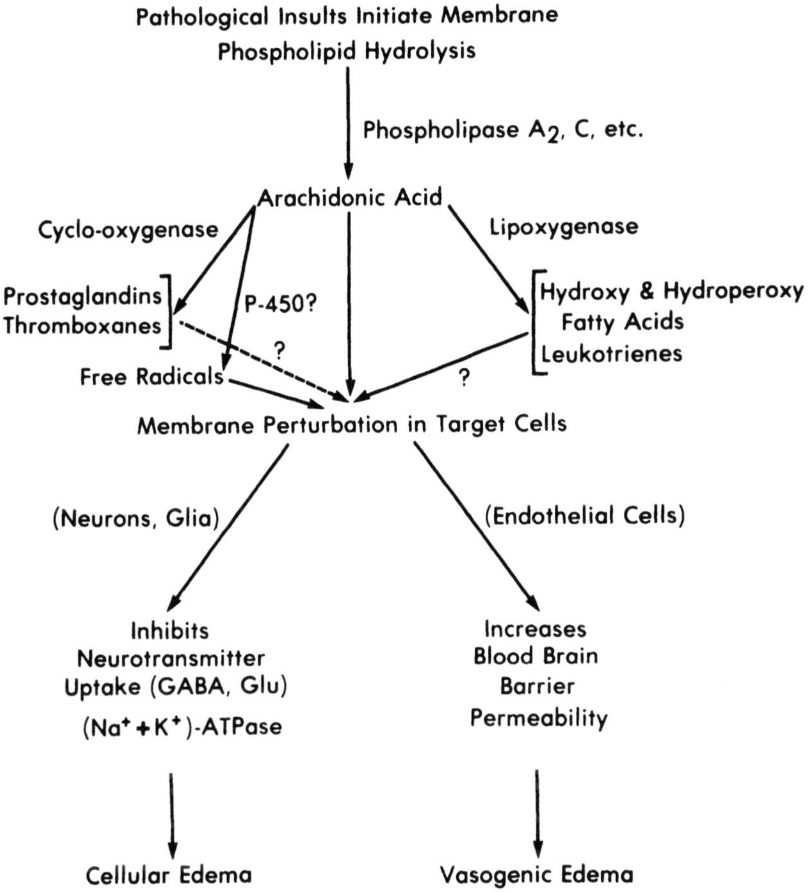

Fig. 3. Hypothetical mechanisms for the development of brain edema.

A clear understanding of these biochemical events is imperative in order to understand the mechanisms of brain edema and to allow the development of more rational therapeutic measures.

6. *THERAPEUTIC APPROACHES TO TREATMENT OF BRAIN EDEMA*

The treatment of brain edema in clinical practice includes glucocorticoids, osmotherapy, acetazolamide, furosemide, hypothermia, and barbiturates.[10,125] Most of these therapeutic approaches are empirical, and the basic molecular mechanisms of the beneficial effects of these drugs on brain edema are still poorly understood. As we have stated earlier, the fundamental mechanisms in the development of both cellular and vasogenic edema are (1) alteration of membrane integrity, (2) release and accumulation of arachidonic acid (PUFAs), and (3) radical metabolites. The following review of the action of potentially therapeutic drugs is based on their ameliorating effects on arachidonic acid cascades.

6.1. Glucocorticoids

Since 1961, various high-potency glucocorticoids, chiefly dexamethasone, have been used widely in the management of intracranial hypertension and brain edema.[4,5,126] The antiedema activity of glucocorticoids may be mediated by their inhibitory effects on membrane phospholipase A_2 activity.[127,128] Numerous studies also suggest the antiinflammatory action of phospholipase A_2 inhibitors in various other biological systems.[129–133]

Dexamethasone inhibits the release of arachidonic acid from cellular membranes.[127] In view of the ability of arachidonic acid to induce both vasogenic and cellular edema, it is possible that one beneficial effect of dexamethasone may be the inhibition of release of edema-producing PUFAs from cellular membranes. Thus, the action of glucocorticoids on inflammatory and immune mechanisms is a fundamental issue in modern biology.[126,134,135] Glucocorticoid bound to cytoplasmic receptor forms a glucocorticoid–receptor complex, which then becomes bound to the nucleus.[136,137] This binding between glucocorticoid and cell nuclei induces the transcriptive and translative processes of various enzymes and a phospholipase A_2 inhibitory protein, lipomodulin.[128,138–140] Lipomodulin inhibits the release of arachidonic acid and the subsequent inflammatory processes.[141] Lipomodulin has not yet been identified in brain, but if it can be proven that these inhibitory proteins are produced there, they may have therapeutic potential for the treatment of brain edema. Furthermore, nonglucocorticoid steroids such as testosterone and estrogen do not inhibit inflammation and have little or no effect on arachidonic acid release in cultured cells.[127] Therefore, the inhibition of arachidonic acid release and the subsequent cascade of biologically potent molecules may be an important molecular mechanism for the antiedema activity of glucocorticoids.

On the other hand, glucocorticoids have been proposed as membrane-stabilizing agents,[142,143] and so the basic mechanisms of the direct or indirect membrane effect or the indirect effect on the induction of phospholipase A_2 inhibitory protein(s) require further elucidation. Moreover, it has been reported that dexamethasone reverses the changes of local cerebral glucose utilization of injured brain: this suggests that functional disturbances in traumatized brain are also modified by steroid treatment.[144,145] It is noteworthy that glucocorticoids have been shown to have a therapeutic role only in the treatment of vasogenic edema and that they are ineffective in the cellular edemas. The reasons for this disparity are not known.

6.2. Amphiphilic and Membrane-Active Agents

Besides glucocorticoids, other kinds of phospholipase inhibitors may also have some therapeutic value on brain edema. Both chloroquine and mepacrine, the amphiphilic and membrane active agents, are effective inhibitors of both lysosomal phospholipase A_2 and C activities.[146,147] Preliminary studies have shown that both chloroquine and mepacrine reverse the release of arachidonic acid from brain microvessels induced by oxygen-derived free radicals.[148] Further biochemical and pharmacological studies of the effects of these agents on blood–brain barrier permeability and vasogenic edema are necessary.

6.3. CDP-Choline and CDP-Ethanolamine

It has been demonstrated by Dorman *et al.*[149] that after rat brain lipids were labeled with [^{3}H]acetate, 5 min of cerebral ischemia caused a 2.2-fold increase in radioactivity in the free fatty acids concomitant with a loss of more than 20% of the radioactivity from phosphatidylcholine and phosphatidylethanolamine. However, an intracerebral injection of 0.6 mol each of cytidine diphosphocholine (CDP-choline) and cytidine diphosphoethanolamine (CDP-ethanolamine) prevented the phospholipid degradation and the release of free fatty acids.[144] Furthermore, intravenous injections of CDP-choline reversed brain edema and restored Na^+,K^+-ATPase activity in rabbits induced by cold injury.[150] Therefore, CDP-choline and CDP-ethanolamine may have important therapeutic values in the treatment of brain edema.

6.4. Calcium Antagonists

It is well known that the activation of phospholipase A_2 requires the presence of calcium.[122,151–153] The Ca^{2+}-free fatty acid cascade has been involved in hypoxia/ischemia-induced brain damage and edema.[115,154] Furthermore, it has been suggested that the influx of Ca^{2+} into cells may be the triggering event that leads to brain cell damage following ischemia and hypoglycemia.[108] The action of calcium antagonists, including nitrendipine, nifedipine, nisoldipine, verapamil, and D-600, as potent inhibitors of calcium transport in various biological systems[155,156] has been reported. The antiedema effects of these Ca^{2+} antagonists are not known and require further study.

6.5. Antioxidants and Free Radical Scavengers

Since free radicals may be involved in brain injury and edema, the application of exogenous antioxidants or free radical scavengers may provide therapeutic values in brain edema. Normally, brain cells and other mammalian cells have their own defense mechanisms against free-radical-induced lipid peroxidation.[157] Both enzymatic and nonenzymatic free radical scavenging systems exist in cellular membranes as well as in cytoplasmic compartments (Fig. 4). Superoxide dismutase (SOD) is a specific scavenger for superoxide radicals,[158,159] whereas catalase and glutathionine peroxidase are specific for the hydrolysis of hydrogen peroxides.[159–162] The effects of enzymes, especially SOD, on inflammatory processes have been studied in various biological systems.[159,163–165] However, the therapeutic use of SOD in brain edema has not been reported.

It has been demonstrated that oxygen free radicals and arachidonic acid metabolites are formed during the recirculation of blood to ischemic brain.[166]

1. Enzymatic systems

$$O_2\cdot^- \xrightarrow{\text{Superoxide dismutase}} O_2 + H_2O_2$$

$$H_2O_2 \xrightarrow[\text{Glutathione peroxidase (other peroxidases)}]{\text{Catalase}} H_2O + O_2$$

2. Nonenzymatic systems

Vitamin E (tocopherol)
Vitamin C (ascorbate)
Glutathione
Selenium
β-Carotene

Fig. 4. Enzymatic and nonenzymatic free-radical-scavenging systems in brain. $O_2\cdot^-$, superoxide radical; H_2O_2, hydrogen peroxide.

These oxygen free radicals initiate the peroxidative processes that may lead to the subsequent cellular damage and edema.[157,167] By using cold injury to produce vasogenic edema in cat brain, it was demonstrated that the antioxidant dipheny-*p*-phenylenediamine ameliorates edema production.[168] Earlier, Demopoulos *et al.*[143] demonstrated that pretreatment with dexamethasone significantly decreased the level of edema and of lipoperoxide products.[142,143] Pretreatment with the antioxidant vitamin E also eliminates compression-induced brain edema in experimental animals.[169] Chronic deficiency of vitamin E caused progressive neuropathological lesions and accumulation of lipopigment in neuronal perikarya and CNS endothelial cells in rhesus monkeys.[170] It has been shown that vitamin E modulates the lipoxygenation of arachidonic acid in leukocytes.[171] Further study is needed to evaluate the possible therapeutic value of vitamin E on various forms of brain edema.

Barbiturates may have beneficial effect on experimental cerebral ischemia.[172–174] High doses have been required, and in most instances treatment is effective only if given prior to the ischemia. The beneficial effects of barbiturates on brain injury may be mediated by their abilities to scavenge free radicals and attenuate the release of PUFAs.[78,175–177] The mode of the antiperoxidative action of barbiturates and their possible beneficial effects on brain edema require further elucidation.

6.6. Other Possible Therapeutic Agents

Hyperosmolar solutions were introduced to the treatment of brain edema because they cause an acute transient drop in intracranial pressure[178] resulting from an acute decrease in brain volume.[179–180] The drop is greatest in normal brain and least in areas of vasogenic brain edema. Hypertonic mannitol and glycerol are used most commonly in clinical practice. Osmotic gradients obtained with hypertonic solutes are short-lived because each of the solutes in use reaches an equilibrium concentration in the brain after a delay of only a few hours.[9] This may result in an undesirable rebound increase in intracranial pressure.[9–11]

Acetazolamide, an inhibitor of carbonic anhydrase, causes about 50% reduction in the rate of CSF formation within the ventricles, presumably by reducing the availability of H^+ ions to exchange with Na^+ ions within the epithelial cells of choroid plexus and perhaps at brain capillary endothelial cells, where CSF is also formed.[178] Furosemide inhibits the final steady-state level of intracellular Cl^-, and it partially inhibits K^+ uptake.[181] It also inhibits the formation of CSF by 25% by a mechanism independent of carbonic anhydrase. However, the beneficial effects of acetazolamide and furosemide on various forms of brain edema are probably not significant apart from the periventricular (interstitial) brain edema that accompanies hydrocephalus. Elucidation of the biochemical mechanisms of the various forms of brain edema should provide more rational and effective therapies in the future.

ACKNOWLEDGMENT. This work was supported by N.I.H. grant NS-14543. We thank Ms. Nancy Petrisko for preparing the manuscript and Dr. J. W. Schmidley for reviewing the manuscript.

REFERENCES

1. Beks, J. W. F., Bosch, D. A., and Brock, M. (eds.), 1976, *Intracranial Pressure*, Volume III, Springer-Verlag, New York.
2. Katzman, R., Clasen, R., Klatzo, I., Meyer, J. S., Pappius, H. M., and Waltz, A. G., 1977 *Stroke* **8**:509–540.
3. Katzman, R., and Pappius, H. M., 1973, *Brain Electrolytes and Fluid Metabolism*, Williams and Wilkins, Baltimore.
4. Pappius, H. M., and Feindel, W. (eds.), 1976, *Dynamics of Brain Edema*, Springer-Verlag, New York.
5. Reulen, H. J., and Schurmann, K. (eds.), 1972, *Steroids and Brain Edema*, Springer-Verlag, Berlin.
6. Elliott, K. A. C., and Jasper, H., 1949, *Am. J. Physiol.* **157**:122–129.
7. Stewart-Wallace, A. M., 1939, *Brain* **62**:426–438.
8. Klatzo, I., 1967, *J. Neuropathol. Exp. Neurol.* **26**:1–14.
9. Fishman, R. A., 1980, *Cerebrospinal Fluid in Diseases of the Nervous System*, W. B. Saunders, Philadelphia.
10. Fishman, R. A., 1982, *Basic Neurochemistry*, (G. J. Siegel, R. W. Albers, B. W. Agranoff, and R. Katzman, eds.), Little, Brown, Boston.
11. Fishman, R. A., 1975, *N. Engl. J. Med.* **293**:706–711.
12. Rapoport, S. I., 1976, *Blood–Brain Barrier in Physiology and Medicine*, Raven Press, New York.
13. Petito, C. K., 1979, *J. Neuropathol. Exp. Neurol.* **38**:222–234.
14. Westergaard, E., Hertz, M. M., and Bolwig, T. G., 1978, *Acta Neuropathol.* **41**:73–80.
15. Westergaard, E., 1980, *Adv. Neurol.* **28**:55–74.
16. Houthoff, H. J., and Go, K. G., 1980, *Adv. Neurol.* **28**:75–81.
17. Fishman, R. A., 1974, *Res. Publ. Assoc. Res. Nerv. Ment. Dis.* **53**:159:171.
18. Fishman, R. A., Reiner, M., and Chan, P. H., 1977, *J. Neurochem.* **28**:1061–1067.
19. Van Harreveld, A., 1966, *Brain Tissue Electrolytes*, Butterworths, London, Washington.
20. Hossmann, K. A., 1976, *Dynamics of Brain Edema* (H. M. Pappius, and W. Feindel, eds.), Springer-Verlag, New York, pp. 219–227.
21. Siesjö, B. K., 1978, *Brain Energy Metabolism*, John Wiley and Sons, New York.

22. Andreoli, T. E., Grantham, J. J., and Rector, F. C., Jr. (eds.), 1977, *Disturbances in Body Fluid Osmolality*, American Physiological Society, Bethesda.
23. Dila, C. F., and Pappius, H. M., 1972 *Arch. Neurol.* **26**:85–90.
24. Rymer, M. M., and Fishman, R. A., 1973, *Arch. Neurol.* **28**:49–54.
25. Arieff, A. I., and Kleeman, C. R., 1973, *J. Clin. Invest.* **52**:571–583.
26. Fishman, R. A., and Greer, M., 1963, *Arch. Neurol.* **8**:156–161.
27. Mihorat, T. H., 1972, *Hydrocephalus and the Cerebrospinal Fluid*, Williams and Wilkins, Baltimore.
28. Nelson, S. R., Mantz, M. L., and Maxwell, J. A., 1971, *J. Appl. Physiol.* **30**:268–271.
29. Marmarou, A., Poll, W., Shulman, K., and Bha-avan, H., 1978, *J. Neurosurg.* **49**:530–537.
30. Marmarou, A., Tanaka, K., and Shulman, K., 1982, *J. Neurosurg.* **56**:246–253.
31. Tengvar, C. H., Forssen, M., Hultstrom, D., Olsson, Y., Pertoft, H., and Pettersson, Å, 1982, *Acta Neuropathol.* (*Berl.*) **57**:143–150.
32. Penn, R. D., 1980, *Adv. Neurol.* **28**:383–394.
33. Clasen, R. A., Huckman, M. S., VonRoenn, K. A., Pandolfi, S., Laing, I., and Clasen, J. R., 1980, *Adv. Neurol.* **28**:395–412.
34. Miller, J. D., Gudeman, S. K., Kishore, P. S., and Becker, D. P., 1980, *Adv. Neurol.* **28**:413–422.
35. Raichle, M. E., 1979, *Brain Res. Rev.* **1**:47–68.
36. Raichle, M. E., 1980, *Adv. Neurol.* **28**:423–427.
37. Rapoport, S. I., Olino, K., Fredericks, W. R., and Pettigrew, K. D., 1978, *Brain Res.* **150**:653–657.
38. Kaufman, L., Crooks, L. E., and Margulis, A. R., (eds.), 1982, *Nuclear Magnetic Resonance Imaging in Medicine*, Igaku-Shoin, Tokyo.
39. Lauterbur, P. C., 1973, *Nature* **242**:140–191.
40. Pykett, I. L., Newhouse, J. H., Buonanno, F. S., Brady, T. J., Goldman, M. R., Kistler, J. P., and Pohost, G. M., 1982, *Radiology* **143**:157–168.
41. Burt, C. T., 1982, *Life Sci.* **31**:2793–2808.
42. Go, K. G., and Edges, H. T., 1975, *Arch. Neurol.* **32**:462–465.
43. Thulborn, K. R., DuBoulay, G. H., Duchen, L. W., and Radda, G., 1982, *J. Cereb. Blood Flow Meta.* **2**:299–306.
44. Naruse, S., Horikawa, Y., Tanaka, C., Hirakawa, K., Nishikawa, H., and Toshizaki, K., 1982, *J. Neurosurg.* **56**:747–752.
45. Schmidley, J. W., Brito, A., Chan, P. H., Fishman, R. A., and Crooks, L., (*N.Y.*) **33**(Suppl. 2):144.
46. Pykett, I. L., Buonanno, F. S., Brady, T. J., and Kistler, J. P., 1983, *Stroke* **14**:173–177.
47. Buononno, F. S., Pykett, I. L., Brady, T. J., Vielma, J., Burt, C. T., Goldman, M. R., Hinshaw, W. S., Pohost, G. M., and Kislter, J. P., 1983 *Stroke* **14**:178–184.
48. Spetzler, R. F., Zabramski, J. M., Kaufman, B., and Yeung, H. N., 1983, *Stroke* **14**:185–190.
49. Asato, R., Handa, H., Hashi, T., Hatta, J., Komoike, M., and Yazaki, T., 1983, *Stroke* **14**:191–197.
50. Go, K. G., 1980, *Adv. Neurol.* **28**:1–8.
51. Chan, P. H., and Fishman, R. A., 1978, *Science* **201**:358–360.
52. Chan, P. H., Fishman, R. A., Lee, J. L., and Candelise, L., 1979, *J. Neurochem.* **33**:1309–1315.
53. Pappius, H. M., 1982, *Handbook of Neurochemistry*, Volume 1, 2nd ed. (A. Lajtha, ed.), Plenum Press, New York, pp. 139–149.
54. Fishman, R. A., Sligar, K., and Hake, R. B., 1977, *Ann. Neurol.* **2**:89–94.
55. Bourke, R. S., and Tower, D. B., 1966, *J. Neurochem.* **13**:1071–1097.
56. Bourke, R. S., and Tower, D. B., 1966, *J. Neurochem.* **13**:1099–1117.
57. Gliemann, J., Osterlind, K., Vinten, J., and Gammeltoft, S., 1972, *Biochim. Biophys. Acta* **286**:1–9.
58. Holian. A., Deutsch, C. J., Holian, S. K., Daniele, R. P., and Wilson, D. F., 1979, *J. Cell Physiol.* **98**:137–144.
59. Chan, P. H., Kerlan, R., and Fishman, R. A., 1982, *J. Neurosci. Res.* **8**:67–72.

60. Olson, J., and Holtzman, D., 1982, *Brain Res.* **246**:273–279.
61. Kempski, O., Chaussy, L., Gross, U., Zimmer, M., and Baethmann, A., 1983, *Brain Res.* **279**:217–228.
62. Kimelberg, H. K., 1979, *Neural Trauma* (J. Popp, R. S. Bourke, L. R. Nelson, and H. K. Kimelberg, eds.), Raven Press, New York, pp. 137–153.
63. Hertz, L., and Schousboe, A., 1975, *Int. Rev. Neurobiol.* **18**:141–211.
64. MacKnight, A. D. C., and Leaf, A., 1977, *Physiol. Rev.* **57**:510–562.
65. Bourke, R. S., Kimelberg, H. K., and Daze, M. A., 1978, *Brain Res.* **154**:196–202.
66. Bourke, R. S., Kimelberg, H. K., West, C. R., and Bremer, A. M., 1975, *J. Neurochem.* **25**:323–328.
67. Lund-Anderson, H., and Hertz, L., 1970, *Exp. Brain Res.* **11**:199–212.
68. Blank, W. F., Jr., and Kirshner, H. S., 1977, *Brain Res.* **123**:113–124.
69. Latzkovits, L., and Fajszi, C., 1982, *Handbook of Neurochemistry*, Volume 1, 2nd ed. (A. Lajtha, ed.), Plenum Press, New York, pp. 1-30.
70. Kimelberg, H. K., and Semenoff, D., 1982, *Trans. Am. Soc. Neurochem.* **13**:112.
71. Kimelberg, H. K., and Bourke, R. S., 1982, *Handbook of Neurochemistry*, Volume 1, 2nd ed. (A. Lajtha, ed.), Plenum Press, New York, pp. 31–67.
72. Chan, P. H., Fishman, R. A., Lee, J. L., and Quan, S. C., 1980, *Neurochem. Res.* **5**:629–640.
73. Chan, P. H., Fishman, R. A., Longar, S., Chen, S., and Chew, S., 1983, *Neural Membranes*, (G. Sun, N. Bazan, J-Y. Wu, G. Porcellati, and A. Y. Sun, eds.), Humana Press, Clifton, New Jersey, pp. 141–152.
74. Chan, P. H., Fishman, R. A., Caronna, J., Schmidley, J. W., Prioleau, G., and Lee, J., 1983, *Ann. Neurol.* **13**:625–632.
75. Chan, P. H., and Fishman, R. A., 1984, *Fed. Proc.* **43**:210–213.
76. Chan, P. H., Longar, S., and Fishman, R. A., 1983, *Brain Res.* **277**:329–337.
77. Mead, J. F., 1976, *Free Radicals in Biology*, Volume 1 (W. A. Pryor, ed.), Academic Press, New York, pp. 51–68.
78. Demopoulos, H. B., Flamm, E. S., Seligman, M. L., *Neural Trauma*, (J. Popp, R. S. Bourke, L. R. Nelson, and H. K. Kimelberg, eds.), Raven Press, New York, pp. 63–78.
79. Fridovich, S. E., and Porter, N. A., 1981, *J. Biol. Chem.* **256**:260–265.
80. Wolfe, L. S., 1982, *J. Neurochem.* **38**:1–14.
81. Samuelsson, B., 1983, *Science* **220**:568–575.
82. Capdevila, J., Chacos, N., Werringloer, J., Prouch, R. A., and Estabrook, R. W., 1981, *Proc. Natl. Acad. Sci. U.S.A.* **78**:5362–5366.
83. Chan, P. H., and Fishman, R. A., 1980, *J. Neurochem.* **35**:1004–1007.
84. Chan, P. H., and Fishman, R. A., 1982, *Brain Res.* **248**:151–157.
85. Fridovich, I., 1982, *Pathology of Oxygen* (A. P. Autor, ed.), Academic Press, London, pp. 1–17.
86. Flamm, E. S., Demopoulos, H. B., Seligman, M. L., Poser, R. G., and Ransohoff J., 1978, *Stroke* **9**:445–447.
87. Demopoulos, H. B., Flamm, E., Seligman, M., and Pietronigro, D., 1982, *Apthology of Oxygen* (A. P. Autor, ed.), Academic Press, London, pp. 127–155.
88. Yoshida, A., Abe, K., Busto, R., Watson, B. D., Kogure, K., and Ginsberg, M. D., 1982, *Brain Res.* **245**:307–316.
89. Del Maestro, R. F., Bjork, J., and Arfors, K. E., 1982, *Microvac. Res.* **22**:239–254.
90. Del Maestro, R. F., Bjork, J., and Arfors, K. E., 1982, *Microvac. Res.* **22**:255–270.
91. Chan, P. H., Yurko, M., and Fishman, R. A., 1982, *J. Neurochem.* **38**:525–531.
92. Pederson, T. E., and Aust, S. D., 1973, *Biochem. Biophys. Res. Commun.* **52**:1071–1078.
93. Kellogg, E. W., and Fridovich, I., 1975, *J. Biol. Chem.* **250**:8812–8817.
94. Chan, P. H., Schmidley, J. W., Fishman, R. A., and Longar, S. M., 1984, *Neurology (N.Y.)* **34**:315–320.
95. Aust, S. D., and Sringer, B. A., 1982, *Free Radicals in Biology*, Volume 5 (W. A. Pryor, ed.) Academic Press, New York, pp. 1–28.
96. Baethmann, A., Oettinger, W., Rothenfußer, Kempski, O., Unterberg, A., and Geiger, R., 1980, *Adv. Neurol*, **28**:171–195.

97. Little, J. R., Kerr, F. W. L., and Sundt, T. M., 1974, *Arch. Neurol.* **30**:448–455.
98. Bingham, W. G., Paul, S. E., and Sastry, K. S. S., 1971, *Neurology (Minneap.)* **21**:111–121.
99. Banay-Schwartze, M., Gergeley, A., and Lajtha, A., 1974, *Brain Res.* **65**:265–276.
100. Moller, M., Mollgard, K., Lund-Anderson, H., and Hertz, L., 1974, *Exp. Brain Res.* **65**:265–276.
101. Baethmann, A., Oettinger, W., Rothenfußer, W., Kempski, O., and Unterberg, A., 1980, *Intracranial Pressure*, Volume IV (K. Shulman, A. Marmaroü, J. D. Miller, D. P. Becker, G. M. Hochwald, and M. Brock, eds.), Springer-Verlag, New York, pp. 291–297.
102. Oettinger, W., Baethmann, A., Rothenfußer, W., Geiger, R., and Mann, K., 1976, *Dynamics of Brain Edema*, (H. M. Pappius, and W. Frindel, eds.), Springer-Verlag, New York, pp. 161–163.
103. Bulle, P. H., 1957, *Proc. Soc. Exp. Biol. Med.* **94**:553–556.
104. Osterholm, J. L., Bell, J., Meyer, R., and Pyenson, J., 1969, *J. Neurosurg.* **31**:408–421.
105. Costa, J. L., Ito, L. L., Spatz, M., Klatzo, I., and Demirjian, C., 1974, *Anture* **248**:135–136.
106. Westergaard, E., 1975, *Acta Neuropathol. (Berl.)* **32**:27–42.
107. Fishman, R. A., and Chan, P. H., 1981, *Trans. Am. Neurol. Assoc.* **106**:1–4.
108. Siesjö, B. K., 1981, *J. Cereb. Blood Flow Metab.* **1**:155–185.
109. Bazan, N., 1970, *Biochim. Biophys. Acta.* **218**:1–10.
110. Bazan, N., and Tureo, R., 1980, *Adv. Neurol.* **28**:197–215.
111. Marion, J., and Wolfe, L. S., 1979, *Biochim. Biophys. Acta* **574**:25–32.
112. Rehncrona, S., Westerberg, E., Akesson, B., and Siesjö, B. K., 1982, *J. Neurochem.* **38**:84–93.
113. Tang, W., and Sun, G. Y., 1982, *Neurochem. Int.* **4**:269–273.
114. Kuwashima, J., Fujitani, B., Nakamura, K., Kadokawa, T., Toshida, K., and Shimizu, M., 1976, *Brain Res.* **110**:547–557.
115. Gardiner, M., Nilsson, B., Rehncrona, S., and Siesjö, B. K., 1981, *J. Neurochem.* **36**:1500–1505.
116. Gaudet, R. J., and Levine, J., 1980, *Stroke* **11**:648–652.
117. Gaudet, R. J., Alam, I., and Levine, L., 1980, *J. Neurochem.* **35**:653–658.
118. Kontos, H. A., Wei, E. P., Povlishock, J. T., Dietrich, W. D., Magieri, C. J., and Ellis, E. F., 1980, *Science* **209**:1242–1245.
119. Ellis, E. F., Wright, K. F., Wei, E. P., and Kontos, M. A., 1981, *J. Neurochem* **37**:892–896.
120. Chan, P. H., Chen, S., Schmidley, J. W., and Fishman, R. A., 1983, *J. Neurochem.* **41**(Suppl.):S150A.
121. Edgar, A. D., Strosnajder, J., and Horrocks, L. A., 1982, *J. Neurochem.* **39**:111–116.
122. Van Den Bosch, H., 1980, *Biochim. Biophys. Acta* **604**:191–246.
123. Chan, P. H., Kerlan, R., and Fishman, R. A., 1983, *J. Neurochem.* **40**:309–316.
124. Chan, P. H., Fishman, R. A., Chen, S., and Chew, S., 1983, *J. Neurochem.* **41**:1550–1557.
125. Marshall, L. F., 1980, *Adv. Neurol.* **28**:459–469.
126. Fishman, R. A., 1982, *N. Engl. J. Med.* **306**:359–360.
127. Hong, S. L., and Levine, L., 1976, *Proc. Natl. Acad. Sci. U.S.A.* **73**:1730–1734.
128. Flower, R. J., and Blackwell, G. J., 1979, *Nature* **278**:456–459.
129. Hammarstrom, S., Hamberg, M., Duell, E., Stawiski, M. A., Anderson, T. F., and Voorhees, J. J., 1977, *Science* **197**:994–996.
130. Vadas, P., Wasi, S., Movat, H. Z., and Hay, J. B., 1981, *Nature* **293**:583–585.
131. Dey, S. K., Hoversland, R. C., and Johnson, D. C., 1982, *Prostaglandins* **23**:619–630.
132. Beyer-Mears, A., and Barnett, A., 1980, *Exp. Neurol.* **68**:240–248.
133. Kaplan-Harris, L., and Elsbach, P., 1980, *Biochim. Biophys. Acta* **618**:318–326.
134. Baxter, J. B., and Roussear, G. G. (eds.), 1979, *Glucocorticoid Hormone Action*, Springer-Verlag, New York.
135. McEwen, B. S., Davis, P. G., Parson, B., and Pfaff, D. W., 1979, *Annu. Rev. Neurosci.* **2**:65–112.
136. Baxter, J. B., and Tomkins, G. M., 1971, *Proc. Natl. Acad. Sci. U.S.A.* **68**:932–937.
137. Higgins, S. J., Baxter, J. D., and Rousseau, G. G., 1979, *Glucocorticoid Hormone Action* (J. B. Baxter, and G. G. Rousseau, eds). Springer-Verlag, New York, pp. 135–160.
138. Hirata, F., Schiffmann, E., Venkatasubramanian, K., Salomon, D., and Axelrod, J., 1980, *Proc. Natl. Acad. Sci. U.S.A.* **77**:2533–2536.

139. Hirata, F., 1981, *J. Biol. Chem.* **256**:7730–7733.
140. Blackwell, G. J., Carnuccio, R., DiRosa, M., Flower, R. J. Parente, L., and Persico, P., 1980, *Nature* **287**:147–149.
141. Kuehl, F. A., and Egan, R. W., 1980, *Science* **210**:978–984.
142. Seligman, M. L., Mitamura, J., Shera, N., and Demopoulos, H. B., 1979, *Photochem. Photobiol.* **29**:549–558.
143. Demopoulos, H. B., Milvy, P., Kakari, S., and Ransohoff, J., 1972, *Steroids and Brain Edema* (H. J. Reulen, and K. Schurmann, eds.), Springer-Verlag, New York, pp. 29–39.
144. Pappius, H. M., 1982, *Ann. Neurol.* **12**:157–162.
145. Pappius, H. M., 1981, *Ann. Neurol.* **9**:484–491.
146. Matsuzawa, Y., and Hostetler, K. Y., 1980, *J. Biol. Chem.* **255**:5190–5194.
147. Dise, C. A., Burch, J. W., and Goodman, D. B., 1982, *J. Biol. Chem.* **257**:4701–4704.
148. Au, A. M., Chan, P. H., Chen, S. F., and Fishman, R. A., 1983, *Fed. Proc.* **42**:2007.
149. Dorman, R. V., Dabrowiecki, Z., and Horrocks, L. A., 1983, *J. Neurochem.* **40**:276–279.
150. Rigoulet, M., Guerin, B., Cohadon, F., and Vandendreissche, M., 1979, *J. Neurochem.* **32**:535–541.
151. Hofmann, S., and Majerus, P. W., 1982, *J. Biol. Chem.* **257**:14359–14364.
152. Billah, M. M., Lapetina, E. G., and Cuatrecasas, P., 1980, *J. Biol. Chem.* **255**:10227–10231.
153. Billah, M. M., Lapetina, E. G., and Cuatrecasas, P., 1981, *J. Biol. Chem.* **156**:5399–5403.
154. Harris, R. J., Symon, L., Branston, N. M., and Bayham, M., 1981, *J. Cereb. Blood Flow Metab.* **1**:203–209.
155. Hagiwara, S., and Byerly, L., 1981, *Annu. Rev. Neurosci.* **4**:69–125.
156. Toll, L., 1982, *J. Biol. Chem.* **257**:13189–13192.
157. Freeman, B. A., and Crapo, J. D., 1982, *Lab. Invest.* **47**:412–426.
158. Fridovich, I., 1975, *Annu. Rev. Biochem.* **44**:147.
159. Fridovich, I., 1982, *Pathology of Oxygen* (A. P. Autor, ed.), Academic Press, New York, pp. 1–13.
160. Tappel, A. L., 1980, *Free Radicals in Biology*, Volume 4 (W. A. Pryor, ed.), Academic Press, New York, pp. 1–47.
161. Flohe, L., 1982, *Free Radical in Biology*, Volume 5 (W. A. Pryor, ed.), Academic Press, New York, pp. 1–47.
162. Cohen, G., 1982, *Pathology of Oxygen* (A. P. Autor, ed.), Academic Press, New York, pp. 115–126.
163. McCord, J. M., Wong, K., Stokes, S. H., Petrone, W. F., and English, D., *Pathology of Oxygen* (A. P. Autor, ed.), Academic Press, New York, pp. 75–83.
164. McLennan, G., and Autor, A. P., 1982, *Pathology of Oxygen* (A. P. Autor, ed.), Academic Press, New York, pp. 84–97.
165. Lund-Olesen, K., 1982, *Pathology of Oxygen* (A. P. Autor, ed.), Academic Press, New York, pp. 339–353.
166. Yoshida, S., Inoh, S., Asano, T., Sano, K., Kubota, M., Shimazaki, H., and Ueta, N., 1980, *J. Neurosurg.* **53**:323–331.
167. Willmore, L. J., and Rubin, J. J., 1982, *Brain Res.* **246**:113–119.
168. Ortega, B. D., Demopoulos, H. B., and Ransohoff, J., 1972, *Steroids and Brain Edema* (H. J. Reulen, and K. Schurmann, eds.), Springer-Verlag, New York, pp. 167–175.
169. Yoshida, S., Busto, R., Myron, D., Ginsberg, M. D., Abe, K., Martinez, E., Watson, B. D., and Scheinberg, P., 1983, *Neurology (N.Y.)* **33**:166–172.
170. Nelson, J. S., Fiteh, C. D., Fischer, V. W., Brown, G. O., and Chou, A. C., 1981, *J. Neuropathol. Exp. Neurol.* **40**:166–186.
171. Goetzl, E. J., 1980, *Nature* **188**:183–185.
172. Hoff, J. T., Smith, A. L., Hankinson, H. L., and Nielson, S. L., 1975, *Stroke* **6**:28–33.
173. Michenfelder, J. D., Milde, J. H., and Sundt, T. M., 1976, *Arch. Neurol.* **33**:345–350.
174. Hossmann, K. A., 1978, *International Conference on Atherosclereosis*, L. A. Carlson, R. Pauletti, C. R. Sirtori, and G. Weber, eds.), Raven Press, New York, 251–256.
175. Smith, D., Rehncrona, S., and Siesjö, B. K., 1980, *Anesthesiology* **53**:186–194.
176. Shiu, G. K., Nemmer, J. P., and Nemoto, E. M., 1983, *J. Neurochem.* **40**:880–884.
177. Yoshida, S., Inoh, S., Asano, T., Sano, K., Shimasaki, H., and Ueta, N., 1983, *J. Neurochem.* **40**:1278–1286.

178. Schmidley, J. W., 1984, *Neurological Pathophysiology—An Introduction to Neuroscience of Neurological Disease* (R. Collins and A. Pearlman, eds.), Oxford University Press, Oxford (in press).
179. Chan, P. H., and Fishman, R. A., 1979, *Brain Res.* **161**:293–301.
180. Chan, P. H., Pollack, E., and Fishman, R. A., 1981, *Brain Res.* **225**:143–153.
181. Kimelberg, H. K., and Hirata, H., 1981, *Soc. Neurosci. Abstr.* **7**:698.
182. Sun, G. Y., Manning, K., and Strosznajder, J., 1980, *Neurochem. Res.* **5**:1211–1219.
183. Agardh, C. D., and Siesjö, B. K., 1981, *J. Cereb. Blood Flow Metab.* **1**:267–275.
184. Siesjö, B. K., Ingvar, M., and Westerberg, E., 1982, *J. Neurochem.* **39**:796–802.
185. Chan, P. H., and Fishman, R. A., 1984, *Recent Progress in the Study and Therapy of Brain Edema* (K. G. Go and A. Baethmann, eds.), Plenum Press, New York, pp. 193–202.

10

Multiple Sclerosis

Jørgen Clausen

1. INTRODUCTION

The term "demyelinating disease" is conventionally given to a group of diseases of the central nervous system (CNS) characterized by focal or diffuse areas of demyelination and neuroglial sclerosis, usually unrelated to tracts and fiber systems. Demyelination is of two types, the so-called primary demyelination and secondary demyelination. The primary demyelinating diseases are multiple sclerosis (MS), Schilder's diffuse cerebral sclerosis, post- and parainfectious encephalomyelitis, postvaccinal and vaccine encephalomyelitis, acute necrotizing hemorrhagic leukoencephalitis, subacute combined degeneration, central pontine myelinoclasis, Marchiafava–Bignani syndrome, and progressive multifocal leukoencephalopathy. A characteristic finding in primary demyelination is that the myelin sheets disintegrate initially, leaving the axon intact.

In secondary demyelination (Wallerian degeneration)[1] e.g., caused by cutting a nerve, this lesion will be followed by a fragmentation of myelin on the first day followed on the third or fourth day by swellings in the sheet. Between the seventh and the tenth day numerous spherules and fatty droplets appear, and at the end of the second week, a large part of the myelin has been removed by phagocytosis. In Wallerian degeneration, the peripheral part of the neuron degenerates.

The most widespread demyelinating disease is multiple sclerosis (i.e., disseminated sclerosis). This disease was originally described by Cruveillier[2] and Carswell.[3] Supplementary descriptions were later made by Valentiner[4] and Rindfleisch.[5] However, Charcot[6] was the first to recognize that the pathological findings were related to demyelination without destruction of axons. The areas of demyelination were given the term "*sclérose en plaques.*" Many plaque areas are visible by the naked eye in sections of the brain and medulla. They appear as gray, slightly depressed areas, asymmetrically and sharply outlined. In an advanced stage of the disease, the brain may show shrinkage of the cerebral hemispheres with widening of the sulci and slight or moderate en-

---

Jørgen Clausen • The Neurochemical Institute, DK-2200 Copenhagen, Denmark; and the Institute for Biology and Chemistry, Roskilde University, DK-4000 Roskilde, Denmark.

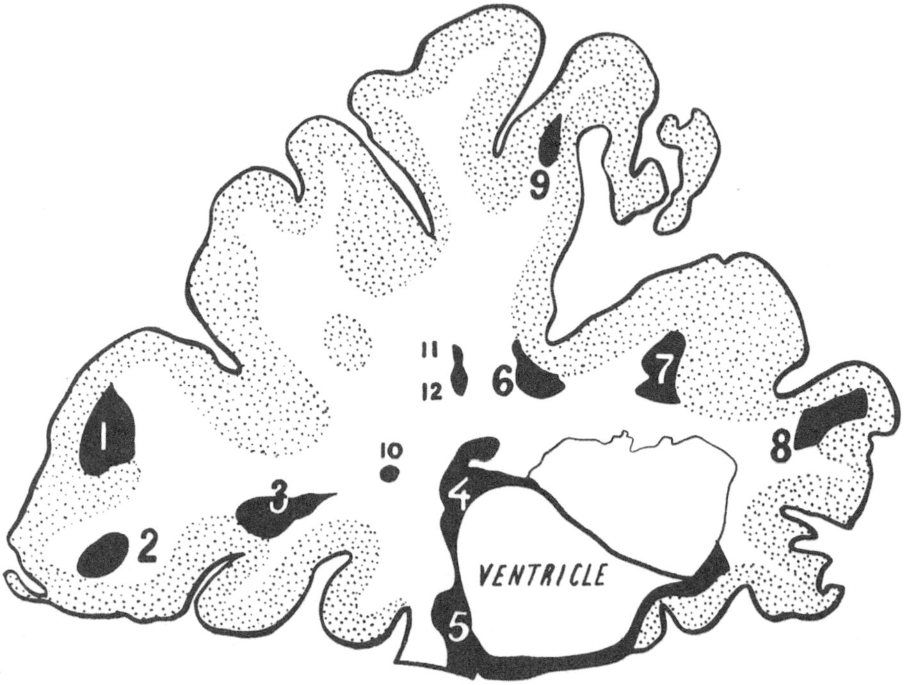

Fig. 1. Typical sites of lesions in multiple sclerosis. The plaque areas are numbered from 1 to 12. Plaques occur especially around the ventricles. (From Fog.[21])

largement of the ventricles. On touching the tissue in simple palpation, it is obvious that the MS brain is more solid than the normal autopsy brain. The plaques may occur all over the brain, but in advanced stages of the disease especially, the wall of the lateral ventricles may contain numerous plaques. In the brainstem and spinal cord, favored sites are around iter Sylvius, under the floor of the fourth ventricle, and especially in the spinal cord. Characteristic lesions are below the pial surface and extend inwards as hemispherical or conical areas (Fig. 1).

Clinically, the disease is characterized by symptoms varying in time and place.[7,8] Thus, retrobulbar neuritis, i.e., blindness of one eye lasting a few weeks, may be one of the initial symptoms.[9,10] Of patients with retrobulbar neuritis, 30% develop MS, and 30% probably develops retrobulbar neuritis following MS. Other symptoms may be paresthesias of days or weeks in duration, vertigo, or paresis of muscular groups, thus producing speech and gait disturbances, neuralgia, spasticity, disturbed function of urinary bladder and defecation, and, finally, lack of coordination of muscular groups. The clinical course may be unpredictable and can be of different types.[11] Electrophysiologically, the demyelination may hamper nerve conduction in specific CNS pathways including those examined by visual evoked potentials, somatosensory responses, and auditory evoked potentials.[12–14] Assay of functions of these pathways may contribute to tracing the topographic localization of the plaques.

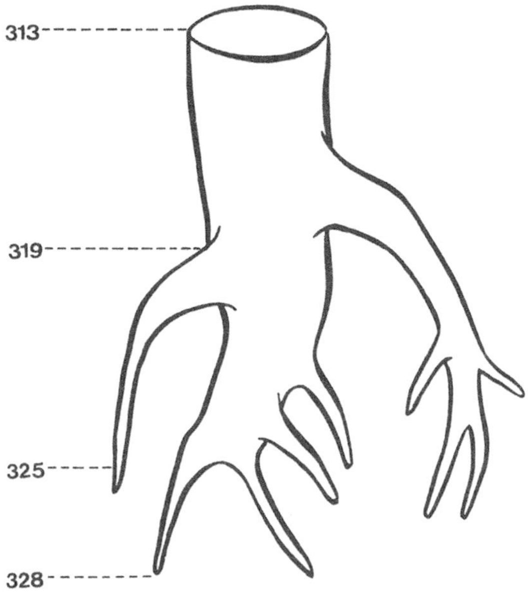

Fig. 2. A central vessel with ramifications where plaque has occurred (confirmed by histological studies).

2. *HISTOLOGICAL FINDINGS*

Under low magnification, the outlines of most plaques appear sharply defined from the white matter. Ludwin[15] indicates that there are two types of plaques, i.e., the active and the chronic plaque. The chronic plaque is sharply demarcated from the white matter. Here, the density of neurons is slightly diminished, and macrophages and oligodendrocytes are nearly lacking.[16] The active plaque represents the early lesion in MS; here, numerous lipid-laden

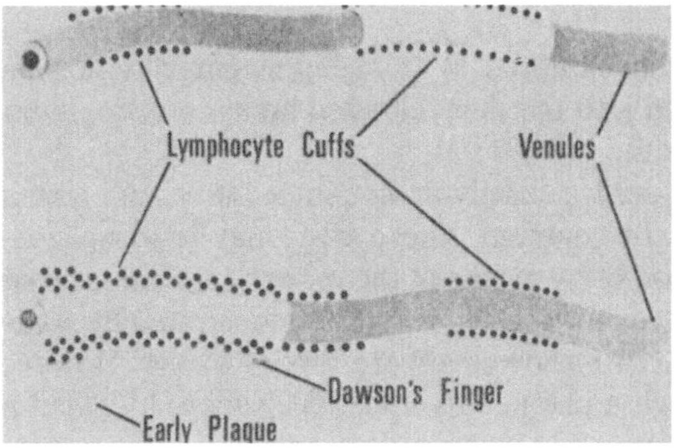

Fig. 3. A drawing showing the early plaque formation growing along the venules. In these areas, lymphocytes are accumulating. (From Adams.[21a])

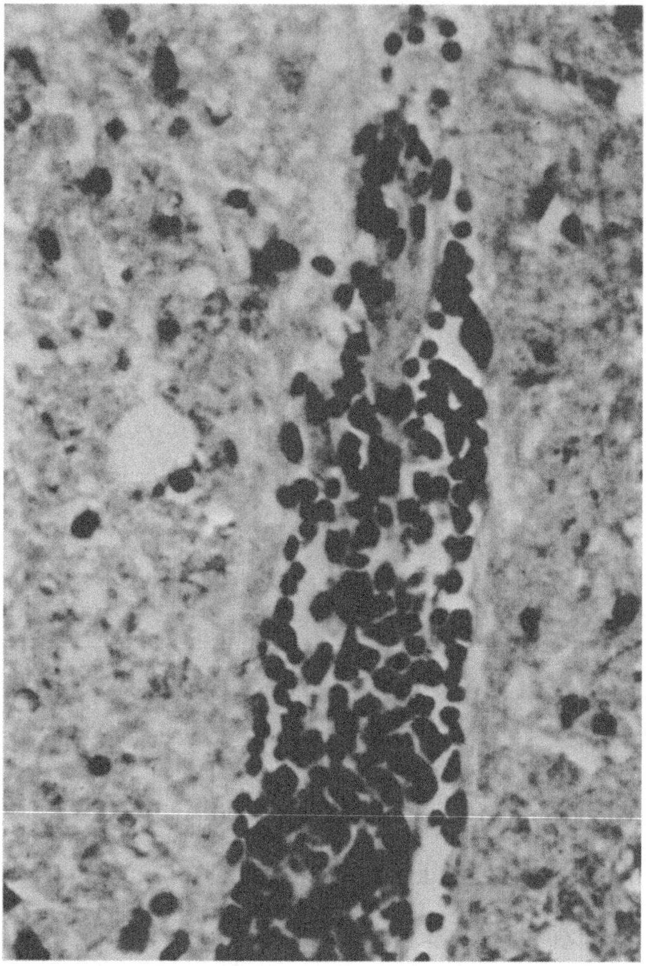

Fig. 4. An infiltrate of lymphocytes extending out of a plaque along the course of a vein (Dawson's finger) (stain, hematoxylin and eosin; magnification ×400). (From Adams.[21a])

macrophages are encountered. With higher magnification, some myelin sheaths are seen to extend into the demyelinated areas, creating a border zone about 50–100 μm in width.

In the spinal cord, partially demyelinated areas are common. The demyelinated areas may be common. These areas may be visualized by means of the aniline blue method or by means of the reduced ammoniacal silver technique.[17] Often the axons are intact; thus, King[18] demonstrated by studying 125 cerebral plaques that only 10% showed severe loss of axons. However, axons do not pass intact through a plaque. By 1903, Bielschowsky[17] had noticed evidence of regeneration of axons in plaques. Itoyama *et al.*[19] have even shown evidence of remyelination of axons in the spinal cord by Schwann cells. Silver impregnation[20] has traced the accumulation of reticulin fibers in neuroglia.

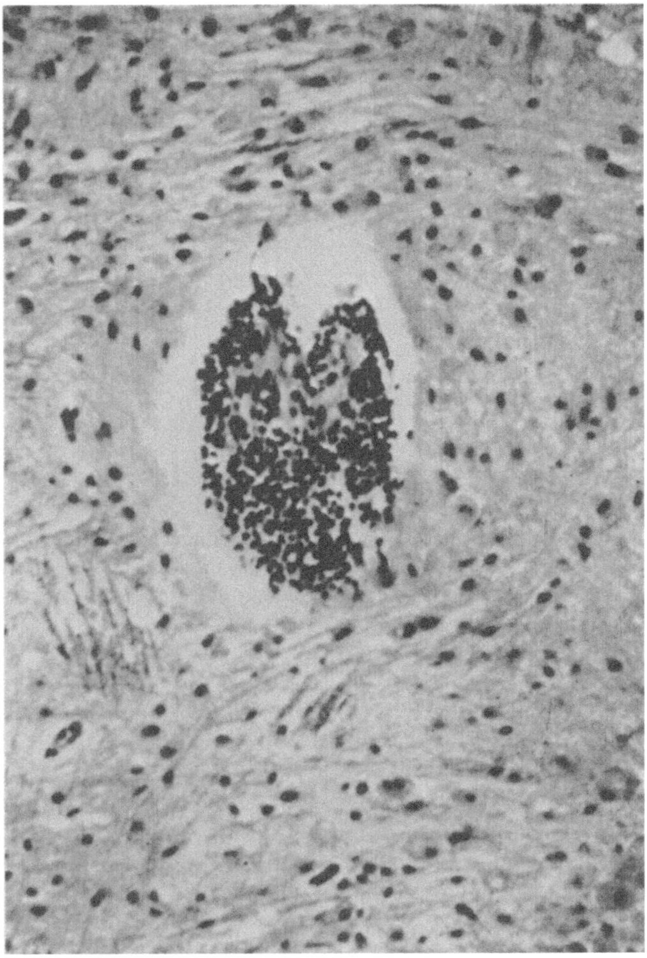

Fig. 5. Perivenous cuff formed almost exclusively of lymphocytes in an area of normal myelin in active MS (stain, hematoxylin and eosin; magnification ×100). (From Adams.[21a])

From an anatomic point of view, it is interesting that most plaques are localized around venules and seem to grow and spread along the vascular system[21] (Figs. 2,3). In early plaques, lymphocytes accumulate perivenously as "perivenous cuffs" (Figs. 4–6). These perivenous lymphocytes most probably accumulate prior to plaque formation (Fig. 5).

Attempts to draw conclusions about the etiology of MS on the basis of the histological findings have been hampered by the failure to demonstrate patterns of demyelination in MS resembling those found in experimental animal models, whether toxic, viral, or autoimmune in nature.[22–25] Thus the so-called "vesicular" demyelination found in experimental allergic encephalomyelitis (EAE)[26] has only been demonstrated in one autopsy brain 4 hr after death.[27]

Furthermore, virus inclusion bodies (e.g., measles virus, paramyxovirus, virus 6/94, multiple-sclerosis-associated agent, simian SV-5-virus, and coron-

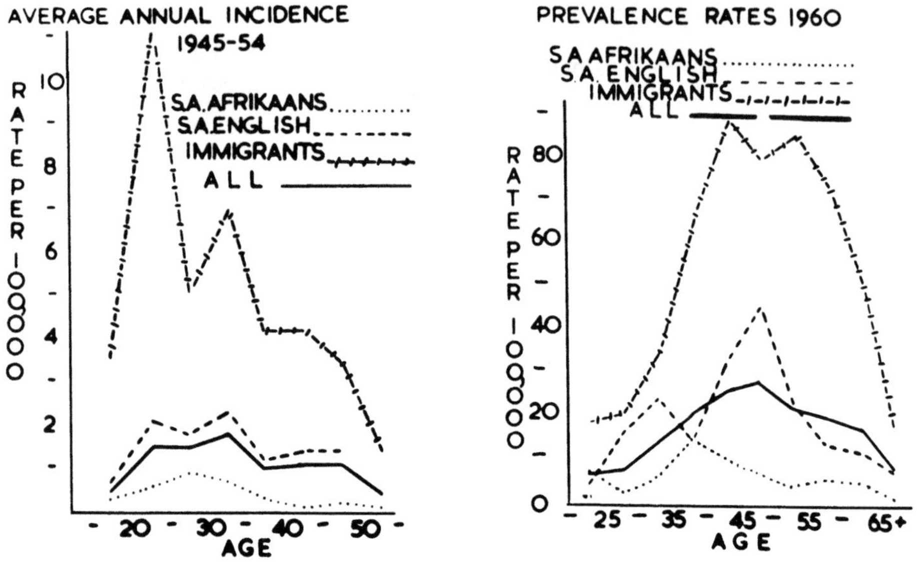

Fig. 6. The incidence and prevalence of multiple sclerosis in white South African ethnic groups (including immigrants from England). It is obvious that in all age groups above 15 years, the incidence and prevalence are much higher among immigrants than among native-born South Africans. (From Dean.[48])

avirus) have rarely been traced, and such reports are contradicted by other laboratories.[28–39] The viruslike bodies found have never been confirmed by reinvestigations in other laboratories, but it may very well be that in MS as in the Theiler virus infection (associated with demyelination) of the rat brain, the virus is only present in the brain tissue in minor amounts.[40]

In the border zones of plaques, there seems to be an accumulation of microglia cells, and the astrocytes have swollen cell bodies and are binucleated. Here, the demyelination is associated with liberation of lysosomal enzymes, especially proteinases.[41] The proteolytic enzymes may well be liberated within the myelin sheath in early stages of the degeneration, resulting in disruption of lipids bound to trypsin-digestible proteins. It is probable that the increased proteolytic activity in the degenerating nerves is not mainly a result of increased cellular secretion but of liberation or activation of enzymes preexisting in or around the normal nerve, and only at later stages of the demyelination may the increased proteolytic activity be related to liberation from macrophages and microglial cells.[42] *In vitro* studies have furthermore revealed that soluble enzymes of leukocytes are able to degradate basic protein and Wolfgram protein, both of which are structural proteins of the myelin sheath.[43] The degeneration of the myelin sheath proteins is associated with a change in the lipid composition in the plaque areas. Thus already in 1945, Rossiter[44] demonstrated increased esterified cholesterol.

As is mentioned below, in EAE and MS, macrophages and subpopulations of lymphocytes accumulate in the perivascular space. Here they form lymphoid accumulations and lymphaticlike capillaries, probably channeling specific antibodies into the brain.[45]

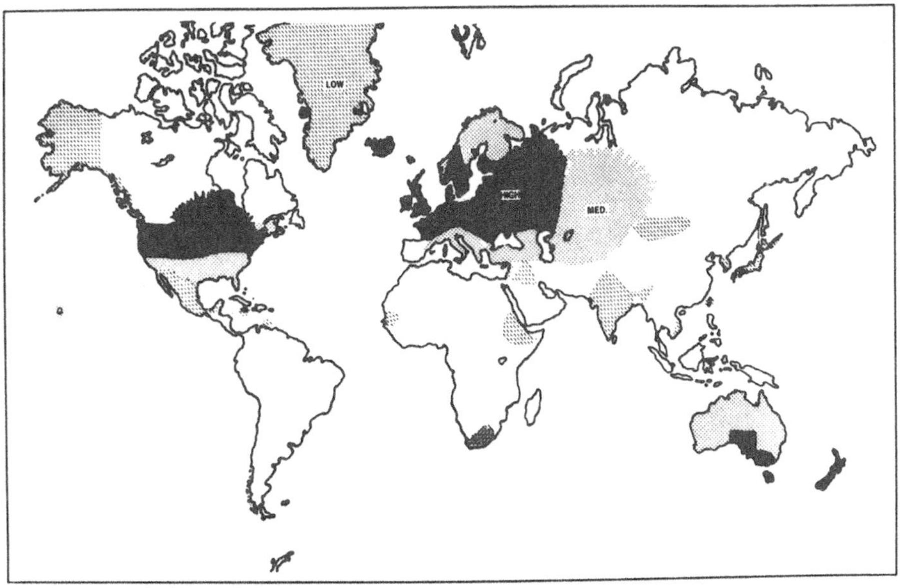

Fig. 7. Worldwide distribution of MS according to high- (solid), medium- (dotted), and low- (diagonal dashed) risk areas as of 1974. The high-risk area of Europe may reach the head of the Adriatic Sea in the Balkans. (From Kurtzke.[53])

3. *EPIDEMIOLOGIC STUDIES*

3.1. Topographic Studies

Several authors[46–52] have demonstrated that the prevalence of the disease among immigrants to an area is dependent on the age at which they migrate. Thus, Englishmen migrating to South Africa have nearly the same prevalence and incidence of multiple sclerosis in South Africa as in England if the individuals are above 15 years of age at the time of migration (Fig. 6). Similar studies have been performed by Leibowitz,[47] who studied immigrants in Israel originating from Arabic countries compared to those coming from central Europe. However, if the age at immigration is below 15 years, the incidence decreases to a lower level. These differences are explained by the theory that MS patients are infected by an unknown agent (virus?) in childhood and because of a very long incubation period, the disease does not become apparent until after adulthood. Clearly, the disease has the highest incidence and prevalence in industrial countries in the northern latitudes corresponding to central Europe and the northern parts of the United States[53] (Fig. 7). These are areas of high hygienic standards. In this respect, there are many similarities between multiple sclerosis and poliomyelitis.

Recently, two separate studies performed by Kurtzke and Hyllested[54] and by Poskanzer *et al.*[55] gave further support to the infectious theory. Thus, Kurtzke and Hyllested[54] demonstrated that the incidence of MS on the Faroe

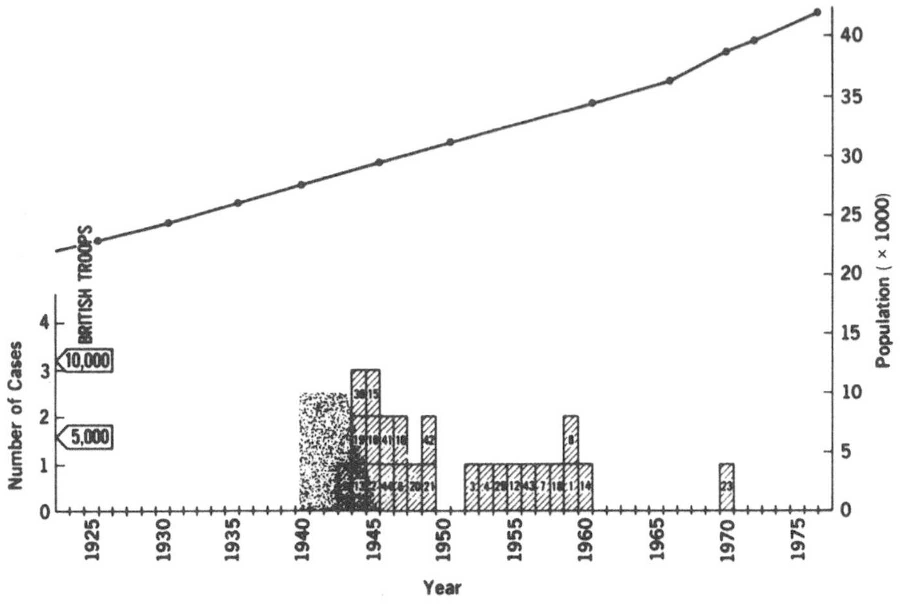

Fig. 8. Distribution of all cases of MS among native-born resident Faroese according to calendar year of clinical onset, 1920 to 1977. Resident population for the Faroes is also noted. (From Kurtzke.[53])

Islands sharply rose to a peak value 4–5 years after these islands were occupied by English troops during World War II, and the epidemic curve declined 3–4 years after the English troops had left the Faroes after second World War (Figs. 8,9). Furthermore, a retrospective study[55] on the social connections between MS patients living on the Orkney and Shetland Islands demonstrated significantly high incidences of either previous personal contact or common contact persons. These data are supported by Gudmundsson,[56] who, on Iceland, found accumulation of MS in three local areas. Finally, however, the data by Isager *et al.*[57] seem to indicate that, opposite to what may be the case for Hodgkin's disease, MS does not seem to be related to increased school contact.

3.2. Genetic Implications in Multiple Sclerosis

In MS patients of the western European populations, certain tissue types seem to occur with a significantly higher rate than in the normal population, especially the tissue types HL-B7 and DW2 and, to a smaller extent, HL-A3.[58,59] However, in other parts of the world, other HL types may be related to MS. Thus, in Canada, Pathy *et al.*[60] found MS related to HL-A7 and W-18. Compared to other HL types related to diseases such as ankylosing spondylitis, where the HL types at risk have 140 times more risk for disease than other HL types,[61] in MS, the risk is only about two times (Table I). The hypothesis has been put forward that the association between MS and HL types may result from a gene linked to the HLA complex.[62] Haile *et al.*[63] have conducted a

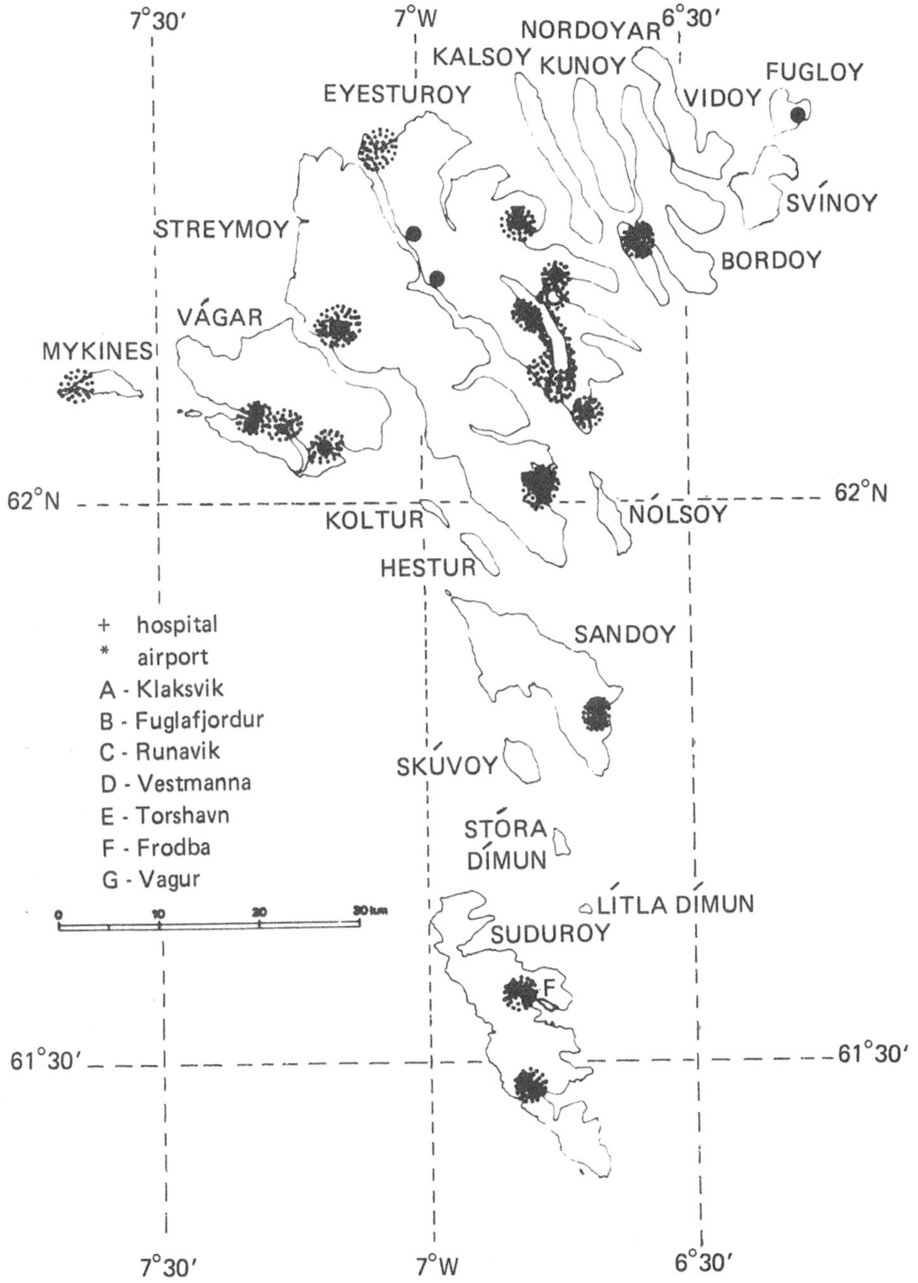

Fig. 9. Residences of MS patients during World War II (large solid circles) superimposed on locations of British troops (dotted areas) during the 1940–1945 occupation of the Faroe Islands. (From Kurtzke.[53])

Table I
Linkages between HL Antigens and Various Syndromes[a]

Disease/syndrome	HL-A antigen	Occurrence of HL types in patients (%)	Occurrence of HL types in controls (%)	Relative risk
Spondylitis ankylopoetica	HL-A27	90	7	141.0
Reiters syndrome	HL-A27	76	6	46.6
Acute uveitis anterior	HL-A27	55	8	16.7
Myasthenia gravis	HL-A8	52	24	4.6
Multiple sclerosis	HL-A3	36	25	1.7
	HL-A7	36	25	1.5
Acute lymphocytic leukemia	HL-A2	63	37	1.7
	4c(W5)	25	16	1.6
Hodgkin's disease	HL-A1	39	32	1.3
	HL-A8	26	22	1.3
Ragweed hayfever	HL-A7	50	19	4.0

[a] The table demonstrates a comparison of the occurrence of certain HL antigens in patients with different diseases and in a control population (expressed as percent). From these data the relative risk for development of the diseases can be calculated. From Lewin.[61a]

formal linkage analysis in 40 families. If one assumes an autosomal dominant model of inheritance, the results are supported by a linkage when the presumed penetrance of the MS gene is low. These data are in part supported by Drachman *et al.*,[64] who studied the HLA types in families with more than two MS cases among first-degree relatives. No HL-A chromosome or type had any direct causal relationship to MS susceptibility, but indirect association could not be ruled out.

3.3. Epidemiologic Studies on the Virus Antibody Titers in MS

Sigurdsson[65] introduced the term "slow virus disease" to refer to a disease characterized by

1. A protracted period of latency (for months or years) after the virus has entered the body.
2. The latency period being followed by clinical symptoms. The disease often develops to a progressive fatal outcome.
3. The infection may often affect a single species and even often one specific organ.

Since several animal diseases such as scrapie and visna in sheep, mouse hepatitis caused by a corona virus (JHM), and canine distemper are caused by slow virus infections of the CNS, clinically active MS has been suggested to be a slow virus disease. Several authors[66] have tried by epidemiologic means to trace a probable virus as the etiologic agent. In 1962, Adams and Imagawa[67] demonstrated that groups of MS patients showed increased serum antibody titers against measles virus. Several studies have confirmed this finding[68–71] (Figs. 10,11).

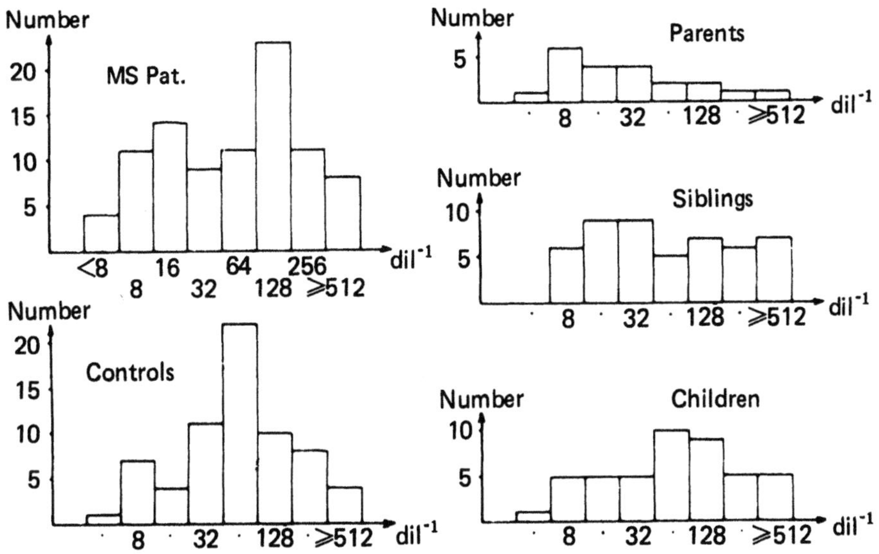

Fig. 10. The measles antibody titers in multiple sclerosis patients, their relatives, and the control population. Abscissa, reciprocal value of serum dilution; ordinate, number of patients in the respective groups. It is obvious that the MS group includes a group with increased measles antibody titers. (From Ammitzbøll and Clausen.[69])

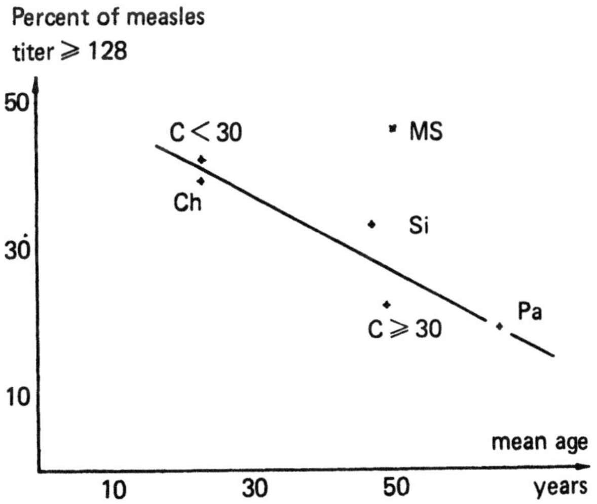

Fig. 11. Plot of the frequency of measles titer ≥128 versus mean age. The regression line for the groups without MS has the slope -0.52 ± 0.09. MS, MS patients; Ch, children of MS patients; Si, siblings of MS patients; Pa, parents of MS patients; C < 30, control <30 years old; C ≥ 30, control ≥30 years old. (From Ammitzbøll and Clausen.[69])

At the moment it is not clear if the high measles antibody titers in MS are an epiphenomenon or directly related to the disease. Thus, on one hand, annealing experiments[72] have given evidence that in MS brains restricted areas of RNA express a measles virus genome, whereas on the other hand, Ammitzbøll *et al.*[73] were not able by lysolecithin fusion of cells from MS patients with Vero cells to demonstrate virus particles, e.g., measles virus. Furthermore Jersild *et al.*[58,59] demonstrated that among MS patients those with HL-A types prone for MS (HL-A3, A7, and/or W18) exhibited higher measles antibody titers than patients with other HL-A types. These studies have in part been confirmed by Whitaker *et al.*[74] who, furthermore, excluded any other relationship of HL-1 types and serum antibody titers for rubella, herpes simplex, cytomegalovirus, and parainfluenza virus. Since canine distemper virus (CDV) has antigens in common with measles virus, and since it may well be that the increased incidence of MS on the Faroe Islands during World War II was related to canine distemper, epidemiologists have compared incidences of CD and MS and the CDV titers in MS and non-MS patients; however, no connection has been established.[75,76]

Other viruses have also been implicated in MS, i.e., vaccinia, herpes, parainfluenza,[77–79] (cf. Section 4) and even bacterias and rickettsia.[80] In regard to herpes virus, Martin[81] has put forward the theory that MS could be caused by a herpes type 2 infection in an adult who did not have any childhood infection with herpes type 1. The overlapping antigenic specificities of the two viruses lead to antibodies against type 1 in part protecting against type 2 infection. The material on which the above-mentioned studies were made are too small for a safe conclusion to be drawn as to a relationship.[82]

4. *BIOCHEMICAL AND PATHOLOGICAL FINDINGS IN TISSUE AND BIOLOGICAL FLUIDS*

4.1. Changes in Nervous Tissue

Since the pathological changes in MS include both demyelination and changes in the cellular composition in the plaques and border zone areas, biochemical abnormalities have been found in lipid, protein, and carbohydrate composition.

Changes in lipids have been reported both in plaque areas and so-called macroscopically normal white matter of MS patients. Thus, a characteristic finding is the decrease in myelin lipids and increase in the cholesterol ester concentration (cf. survey in ref. 82). This accumulation may be a mechanism for complexing the fatty acids liberated from the polar lipids degraded during the demyelination, thus preventing the free fatty acids from precipitating in the plaque area as calcium soaps. The cholesterol esters formed[84] are digested and transported by microglia and macrophages. Similarly, the digestion of the myelin sheath is followed by a decrease in the amount of proteins. Generally, in MS there is an increased fragility of cerebral lysosomes, giving rise to increased autolysis.[85] Recently, Yahara *et al.*,[86] by a careful dissection and separation

Table II
Comparison of Fatty Acid Composition of Normal and MS Myelin[a]

	Total myelin	PC	PE	PS	Sph	Cerebroside II
C 14:0			I > II			
C 16:0					I > II**	
C 16:1	I < II		I > II			
C 18:0				I < II	I > II	I > II†
C 18:1			I > II			
C 20:4		I > II*	I < II			
C 22:6			I < II			
C 24:0	I < II*					
C 25:1					I < II	
C 24:1						I < II
C 26:1						I < II

[a] Where nothing is mentioned, the groups are significantly different ($t > t_{0.995}$). * $P < 0.01$; ** $P < 0.025$; † $P < 0.05$. I, fatty acid percentage of normal myelin; II, fatty acid percentage of MS myelin. PC, phospatidylcholin; PE, phosphatidyl ethanolamine; PS, phosphatidylserin; Sph: Sphingomyelin. From Clausen and Hansen.[93]

of plaque areas from periplaque and apparently normal white matter, compared the lipid composition in these zones. It was possible to demonstrate an increase in the cerebroside and sulfatide content from the central plaque area to the normal-appearing white matter.

In 1953, Swank[87] suggested a close epidemiologic correlation between MS and the dietary intake of saturated fat and/or dairy products (cf. also ref. 86). This suggestion led workers to focus on changes in the fatty acid content of CNS and blood, especially with regard to the content of essential fatty acids (EFA).[89] Thus, Cumings *et al.*,[90] Gerstl *et al.*,[91] and Thompson[92] demonstrated in "normal white matter" of MS patients that the fatty acid composition was characterized by increased saturation.

Clausen and Hansen[93] found unrelated changes in saturation in different phospho- and glycolipid fractions of MS brains, thus arguing that in MS tissue there are no signs of general EFA deficiency (Table II). On the other hand, Clausen and Møller[94] found an increased suceptibility of rats to development of experimental allergic encephalomyelitis in EFA-deficient animals compared to normal control animals. These facts may be related to the disputed findings of a decreased content of linoleic acid in MS serum lipids,[95] which have not been confirmed by other workers[96–100] or have been reported to be unspecific because similar changes are found in other neurological diseases and serious illnesses.[101,102]

The controversy about the significance of changes in the linoleate level in MS has been resolved by the demonstration of a low level of glutathione peroxidase activity in MS. Glutathione peroxidase (GSH-Px) eliminates the peroxides formed by peroxidation of polyunsaturated fatty acids. Therefore, low levels of GSH-Px may give rise to peroxidation of linoleate. Glutathione peroxidase is a selenium-containing enzyme. Therefore, in areas of the world with low dietary intake of selenium causing low GSH-Px activity, low linoleate levels may be seen. Clausen (unpublished data) has also noticed that the GSH-Px activity changes with the disease activity of MS (Section 4.3.2).

Myelin can be isolated from human brains by gradient ultracentrifugation of homogenate, usually in 0.32 M sucrose.[103] Discontinuous centrifugation in gradients increasing from 0.32 M sucrose has revealed the myelin to be distributed into different subfractions from 0.5 to 0.9 M sucrose.[103,104] Because of this, and because of lysosomal digestion and different cellular distribution in MS tissue, even in macroscopically normal appearing matter, where microplaques may be situated, the myelin and its subfractions isolated from MS brains may be partially dismanteled and may even contain nonmyelin particulate fractions. It is unclear to what extent myelin fractions isolated from MS brains thus represent an anatomically pure fraction. This may explain the conflicting results obtained in earlier studies. Thus, Cumings and Goodwin[102] found a decrease in the cerebroside/sulfatide ratio. On the other hand, Clausen and Hansen[93] and Woelk and Borri[105] found a decreased sulfatide content (expressed per dry weight) and an increase in the cerebroside/sulfatide ratio of MS myelin. Phosphotidylserine and -inositol have also been found to be decreased.[93,105] Since the myelin of MS brains originates from nonplaque areas, these findings argue for biochemical abnormalities even in nonplaque areas.[86,106] The myelin, furthermore, shows an abnormal distribution with regard to its buoyant densities in sucrose gradients.[107,108] Thus, Yu *et al.*[108] state that the "pathological" myelin in MS is less dense than normal myelin. They even found evidence from their findings on the P_0-protein in the upper layer from MS spinal cord that "remyelinated" myelin is present in MS. Thus, it must be concluded that MS myelin shows an abnormal structural density.[109]

The finding by Riekkinen *et al.*[110] of a lack of basic protein in MS myelin has not been corroborated by Wolfgram and Tourtelotte.[111] Eng *et al.*,[112] or Wolfgram.[106] Thus, apart from changes in the buoyant densities of MS myelin, most authors conclude that the MS myelin from normal-appearing white matter of MS brain may be quite normal, at least in its protein pattern. The lipid changes found mainly concern fatty acid chain length and probably changes in the proportion of cerebrosides to sulfatides.

With regard to changes in CNS proteins in MS, Riekkinen *et al.*[113] found increased proteinase activities in border zones to plaques, probably explaining the decreased content here of myelin basic protein.[114] No changes were found in other myelin proteins.

Although, until now, no conclusive answers have been drawn by histological studies as to whether a microbiological agent is involved in the pathogenesis of MS, early immunochemical studies revealed circulating complement-fixing antibodies in MS serum against an antigen extractable with ethanol.[115] Although conflicting results have later been accumulated,[116] this has stimulated further work on the question of whether "MS-specific antigens" are present in MS autopsy brains. Thus, it has been possible by means of salting out procedures, column ion exchange, and sepharose affinity chromatography to isolate from MS brains a glycoprotein fraction in which two glycoprotein antigens could be traced by crossed immunelectrophoresis. Neither of the two

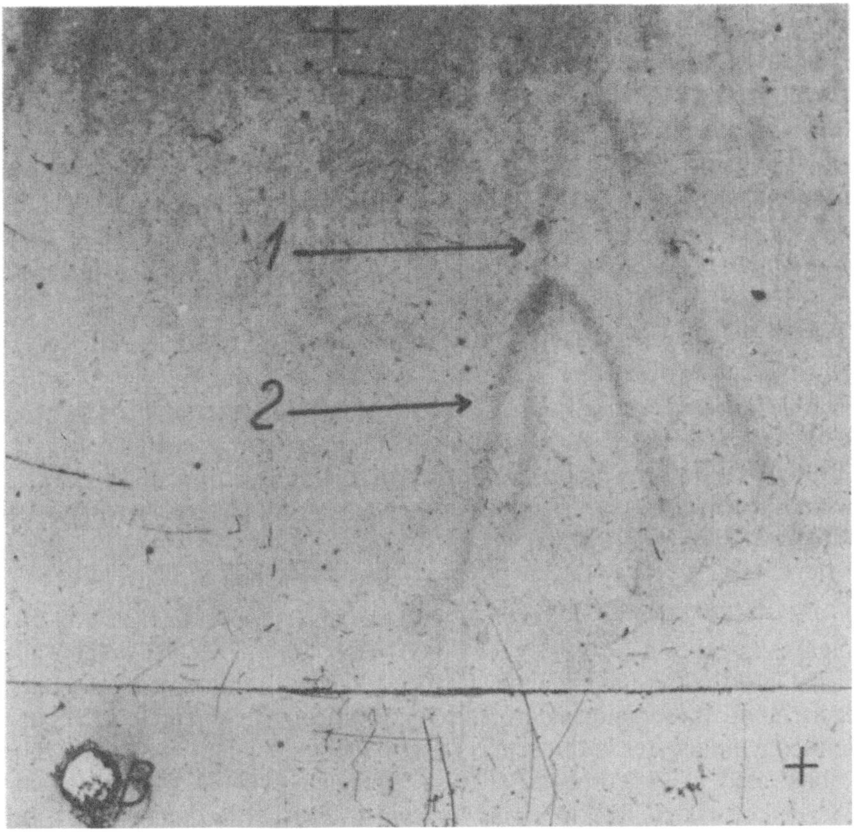

Fig. 12. Crossed immunoelectrophoresis of a glycoprotein fraction isolated from a MS brain. The fraction was obtained from the cytosol (100,000 g_{max} 1 hr supernatant) by stepwise molecular filtration, DEAE column chromatography, followed by con A sepharose column affinity chromatography (for details see refs. 101–103). The glycoprotein fraction was then subjected to crossed immunoelectrophoresis [40 μg of the glycoprotein fraction was applied in hole B in 1% buffered agarose (0.05 M sodium veronal buffer, pH 8.6)] and electrophoresed from left to right. Then, the electrophoretic fraction was electrophoresed perpendicularly to the first dimension into an antibody-containing gel. The antibody was manufactured by immunizing rabbits with the cytosol fraction of MS brain tissue. The antibody was absorbed with normal brain cytosol fraction (for details see refs. 101–104). Two precipitates appear corresponding to two antigens that are specific for MS brain tissue.

glycoprotein antigens was present in control brains.[117–120] A third antigen was further traced in the MS cytosol fraction (Fig. 12, Table III).

A blinded preparative study on 42 brains[120] revealed these antigens only to be present in MS and not in non-MS brains. In MS autopsy brains, the antigens were traced in 80% of all samples studied and in 100% of all spinal cord MS samples studied (Table IV).

Immunoglobulin G isolated from MS serum furthermore seems to contain antibody specificity against the "MS-specific antigens"[120] and to the cytosol proteins of MS brains but not to other particulate fractions of MS brains.[121,122]

Table III
Quantitation of the Yields of Crude Glycoprotein Fraction (G) and Glycoprotein Fraction 2 (G2) Isolated from Multiple Sclerosis (MS) and Control (K) Brain Cytosol (60 mg Protein, Approximately 3 g Brain Wet Weight)[a]

Brain type	No. of brains used	Total G isolated (mg protein) (Mean ± S.D.)	Total G2 isolated (μg protein) (Mean ± S.D.)
MS	7	2.50 ± 0.32	262.5 ± 53.8
K	9	2.04 ± 0.22	141.3 ± 30.7
		$P < 0.025$	$P < 0.001$
Percent recovery from			
MS cytosol		4.2	0.44
K cytosol		3.4	0.24
Percent of G2 isolated from			
MSG			10.5
KG			6.9

[a] *P* values are calculated for Student's *t* test. G was isolated by means of Con A-Sepharose column and G2 by DEAE cellulose column chromatography of G. From Rastogi and Clausen.[119]

Chemically, two of these three antigens may be characterized as glycoproteins since they have affinity for Con A. They have molecular weights above 100,000, and the third of them may be formed in measles-infected Vero cells.[118] Furthermore, these antigens are able to specifically to stimulate MS (but not non-MS) lymphocytes in the active Rosette method similarly to what has been demonstrated in experimental allergic encephalomyelitis, where Hashim *et al.*[123] demonstrated that the basic protein increases the number of rosette-forming T cells. It may well be the MS-specific antigens that cause crude brain extracts of MS brains to increase the number of active rosetting T cells from the peripheral blood of MS patients.[124–127] This reaction seems specific for MS lymphocytes, although lymphocytes from patients with rheumatoid arthritis and systemic lupus erythematosus are also significantly stimulated.

The MS-specific antigens are able to induce a significantly higher loss of neutral proteinase from MS leukocytes than from control lymphocytes, which

Table IV
Presence of MS-Specific CNS Antigens (MGS 2) in CNS Sections[a]

Diagnosis	Number of patients		Sections containing MSG2/total number of sections		
	Total	With MSG2	Cerebral hemispheres	Brainstem	Spinal cord
Multiple Sclerosis	23	18	13/23	3/6	6/6
Other Neurologic Diseases (OND)	12	0	0/10	0/2	0/2
Controls	10	0	0/10	0/1	0/0

[a] From Rastogi *et al.*[120]

are able to degrade myelin proteins[128] (Fig. 13). The supernatant of stimulated leukocytes hampers the uptake of thymidine of MS lymphocytes significantly more than non-MS lymphocytes.[129] The identity and origin of the MS specific proteins are, however, still unknown.

4.2. Changes in the Cerebrospinal Fluid

In 1942, Kabat *et al.*[130] demonstrated by means of free electrophoresis that γ-globulins were elevated in the CSF of MS patients. Later, by means of agar,–agarose and polyacrylamide (PAGE) electrophoresis,[131–137] it has been shown that the increase affects discrete intervals of mobilities in the γ areas, i.e., that the γ-globulins appear as oligoclonal bands representing products of few lymphocyte clones (B cells) (Figs. 14,15). In contrast to the findings in subacute sclerosing encephalitis (SSPE), these bands do not represent antibodies against one type of virus (measles virus in SSPE) but contain antibodies against measles, rubella, vaccinia virus, and/or bacterias that differ from MS patient to MS patient[135,138,140] corresponding to the increased viral antibody titers in CSF of MS patients.[70] The oligoclonal bands have also been traced in plaques, but the bands of different plaques from the same patient do not correspond.[141] The oligoclonal immunoglobulins are synthesized by lymphocytes (B cells) that have migrated to CNS.[142] Recent data collected on the immunologic mechanisms involved in experimental allergic encephalomyelitis (EAE) may explain at least some of the findings mentioned above. Thus, EAE is an autoimmune disease (considered to be an animal model of MS) provoked by challenging experimental animals with myelin basic protein (BP) or by peptide remnants of BP treated with proteinases.[123,124]

During the development of EAE, lymphocytes of peripheral blood and lymph nodes migrate to CNS. They are thus localized in the perivascular and perineural space as well as in CNS tissue. Recent studies by Traugott *et al.*[145] have documented that B and T lymphocytes in EAE have different localizations. Thus, the B cells were found in the perivascular space, whereas the T cells were traced to the CNS tissue. This different distribution may mean that the B cells do not communicate with subsets of T cells, and, locally, they are challenged by different viruses to which the patient has previously been infected and which probably are activated for unknown reasons. Because the B cells are not communicating with T cells, the B cells may uncontrollably synthesize individual clones of immunoglobulins of different immunologic specificity. This is in agreement with the findings of Ammitzbøll, Clausen, and Fog[146] that oligoclonal bands occur preferentially in the CSF of MS patients with high measles virus titers. Kam-Hansen[147] demonstrated that among the cells of CSF in MS, the T cells are proportionally higher and the B cells lower than in blood. Furthermore, she found a lower number of active T cells in CSF compared to blood in MS. Since, as described above, an uneven distribution and lack of communication among different subsets of lymphocytes may well occur in MS, the biological meaning of the abnormal lymphocyte distribution in CSF of MS patients may be difficult to evaluate. Also, other facts may influence the proportions among subsets of lymphocytes. Thus, Coyle *et al.*[148] found immune

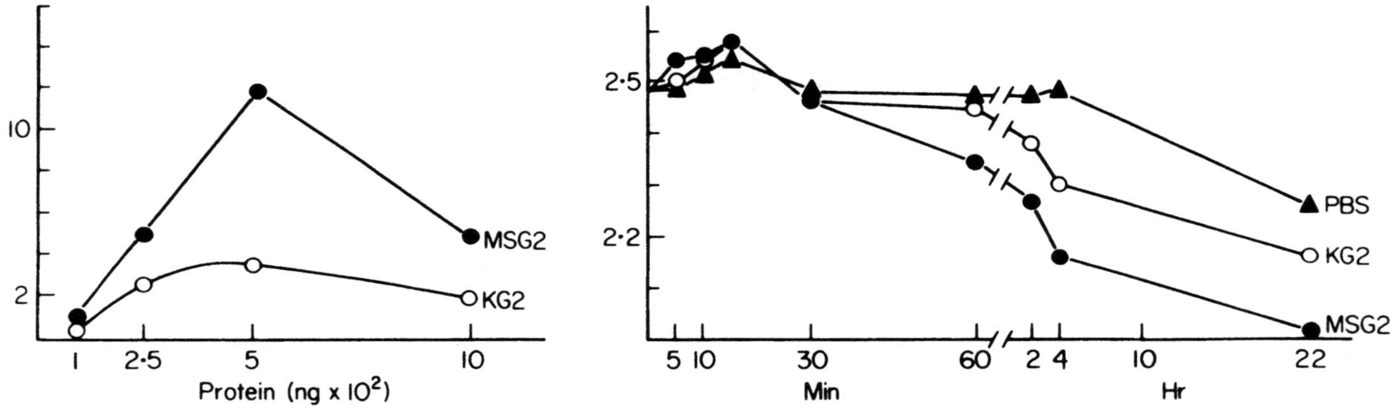

Fig. 13. The glycoprotein fractions of cytosol from MS and control brains (MSG2 and KG2, respectively) were isolated as indicated in Fig. 12. Left: Percent loss of intracellular neutral proteinase from leukocytes of an MS patient incubated with various amounts of MSG2 and KG2 for 4 hr *in vitro*. Abscissa, nanograms protein MSG2 or KG2; ordinate, percent loss of neutral proteinase activity. Right: Lysosomal neutral proteinase activity of leukocytes of a normal subject after incubation with KG2 and MSG2 in PBS for various periods of time. Abscissa, incubation period; ordinate, neutral proteinase activity.

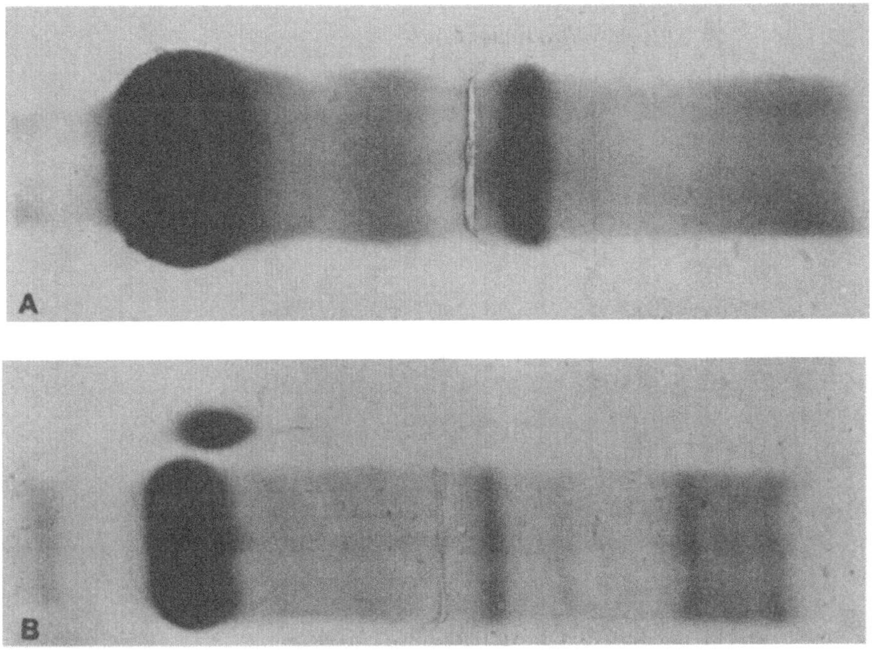

Fig. 14. Oligoclonal bands. Photo of agar-gel microelectrophoresis of 100-fold concentrated CSFs from a patient with lymphocytic meningitis and from a patient with multiple sclerosis. A: Case H.F. no. 633, lymphocytic meningitis. B: Case A.A. no. 3439, multiple sclerosis. A volume of 10 μl was applied in the trough.

complexes in CSF from 50% of all MS patients associated with a relative decrease in the number of T-γ cells (suppressor cells) (*vide infra*). The lack of communication among different subsets of lymphocytes in CNS may also explain the finding of oligoclonal bands in cerebrovascular diseases.[149]

The spinal fluid is formed by the plexus choroideus and as an effusion from the surface of the brain. Therefore, the demyelination process in MS gives rise to liberation of lipids into CSF. In normal CSF, the concentrations of lipids and lipoproteins are very low. Thus, the level of serum β-lipoprotein is in the range of 0.010 to 0.062 mg/100 ml.[150] In the demyelination process, the concentration of β-lipoprotein is raised, probably because of a decreased blood–brain barrier. Compared to the normal CSF, the MS CSF has a raised glycolipid/lecithin ratio, probably as a result of liberation into CSF of cerebrosides from the CNS.[151]

As an alternative to the theory that MS may be caused by a microbiological agent, several authors have put forward the theory that MS, like EAE, is an autoimmune disease caused by serum or spinal fluid autoantibodies against myelin basic protein, myelin lipids, e.g., gangliosides, or muscular or oligodendroglial antigens.[152–157] Since membrane antigens, basic protein, and several other antigens such as the staphylococcal antigen A bind to the Fc moiety and not to the antibody site of immunoglobulins, such studies cannot demonstrate any specificity before work with the $(Fab)_2$ piece has proven the spec-

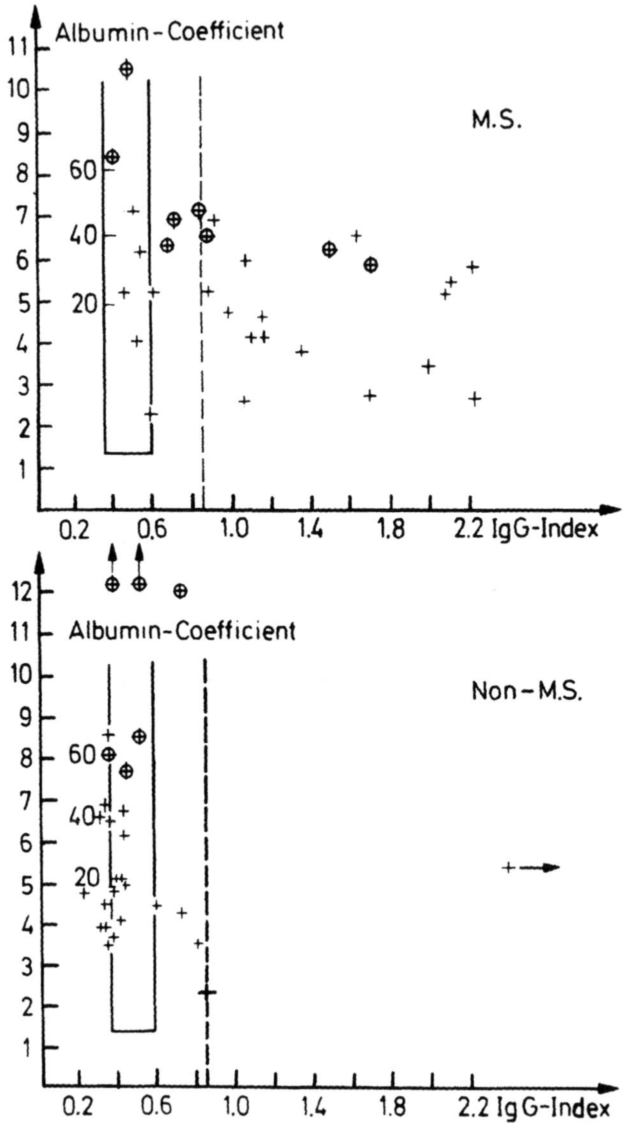

Fig. 15. This figure demonstrates the specific intrathecal synthesis of IgG in multiple sclerosis by relating the IgG index, i.e., the ratio of concentration of IgG in spinal fluid and serum to the corresponding ratio for albumin:

$$\frac{(\text{CSF IgG})/(\text{serum IgG})}{(\text{CSF albumin})/(\text{serum albumin})} = \text{IgG Index}$$

to the albumin coefficient (concentration of albumin of spinal fluid to that of serum). The albumin coefficient reflects the increased permeability of the blood–brain barrier, which may cause a passive diffusion of IgG from serum to spinal fluid compartment. This factor is included and compensated for in the IgG index. Obviously, the MS group demonstrates a significantly raised IgG index, arguing for specific intrathecal IgG globulin synthesis.

ificity of the reaction.[159] Using the $(Fab)_2$ piece, Traugott and Raine[153] demonstrated that the autoantibodies to the basic protein in the spinal fluid in MS are an unspecific phenomenon, since other patients also have these autoantibodies. On the other hand, by means of the complement fixation test, Ryberg[158] was able to trace antibrain antibodies of IgG type in both MS serum and CSF (92% of all MS patients studied). The titers of these antibodies correlated with the malignancy and disability of the disease.

Recently, Gerhard *et al.*[158] and Nagelkerken *et al.*[159] elucidated the oligoclonal bands in CSF from MS patients by means of antiidiotypic antibodies. The levels of these antibodies have been demonstrated to be 8–15 times higher in CSF than in serum. Gerhard *et al.*[158] further demonstrated that the antiidiotypic antibodies have specificity towards Theiler murine encephalomyelitis virus. Only more extensive studies can evaluate the implications of these findings.

4.3. Changes in Blood

Although MS is a disease of the central nervous system, several biochemical and immunologic parameters are changed outside the CNS. These changes involve the formed elements of blood, lipids, proteins (enzymes), and trace elements.

Broman[161] demonstrated by means of supravital staining that there is an open communication between the blood compartment and the plaque areas. This may explain some of the changes traced in blood of MS patients.

The extracerebral changes in MS have been shown to be either changes in immunoregulatory mechanisms or changes in membrane function. At the present state of the research, it is not possible to evaluate if these changes are epiphenomena or directly related to the etiology.

4.3.1. Changes in Immunoregulatory Mechanisms

Studies on the proportion of total number of T cells to that of B cells in MS compared to normal controls and related to phase and type of multiple sclerosis have not given consistent results.[147,148,162] As already mentioned in Section 4.2, the immunoregulatory control of B cell secretion of antibodies may be lacking in MS as a result of different topographic localization of B cells and T cells in MS (Tμ, Tγ, and Tk cells). Thus, in phases of acute exacerbations of MS, Arnason and Antel[163] demonstrated reduced concanavalin-A-induced suppressor cell activity. The specificity of these changes is difficult to evaluate because the immunologic abnormalities related to the demyelinating process occur in CNS as in EAE.[164] Thus, the lymphocytes from the blood in MS may migrate to the CNS, and the remaining circulating cells of the blood may thus only reflect an epiphenomenon. In acute attacks of MS, most authors[163–168] have traced a deficient suppressor function (increase in Tγ cells).

These data are, however, difficult to explain from data demonstrating a decreased reactivity of lymphocytes towards measles virus antigens[169,170] and other virus antigens.[171] Neighbor and Bloom[172] even demonstrated a reduced virus (measles)-dependent lymphocyte suppressor and interferon production in MS. The Tγ cells are, however, labile and, when exposed to immunocomplexes, redistribute their receptors by forming receptors for the μ chain of IgM, whereby the Tγ cells are transformed to Tμ cells.[173] Therefore, since Goust *et al.*[174] traced immunocomplexes in 14 out of 19 MS patients, the abnormal proportion of Tγ to Tμ cells in MS may simply reflect circulating immunocomplexes, and this may explain why Huddlestone and Oldstone[175] only traced in MS a transient decrease in suppressor cells during disease exacerbations.

In contrast to these findings, Dore-Duffy *et al.*[176] demonstrated an increased adherence of MS blood lymphocytes to epithelial cells persistently infected with measles virus. Frey *et al.*[177] and Kinnman and Link[178] could, however, not confirm this phenomenon as specific for MS. Furthermore, the reactivity of blood lymphocytes towards a crude brain extract and towards MS-specific brain antigens (*vide supra*) is increased when lymphocyte reactivity is assayed by means of the active rosette method of Fudenberg *et al.*[124–127] The lymphocytes rosetting with sheep red blood cells within a few minutes of reaction are those T cells involved in the delayed hypersensitivity—the so-called active T cells. This increased activity was, however, lower in the group of patients with relapses.[127] At the moment, it is not clear how to explain the discrepancy between the deficient suppressor function activity and the increased number of active T cells. Gangliosides liberated from CNS may influence the T-cell reactions.[179]

Studies on a possible autosensitization of lymphocytes in MS patients to myelin basic protein or its fragments as well as studies on the presence of cytotoxic antibodies against the basic protein or glia cells or on the possibility that this protein induces secretion of interferon or leukocyte or macrophage migration inhibition or enhancement factors have given conflicting results.[179–184]

The communications among different types of leukocytes are mediated by secretion products. Among these are proteinases. Since, as stated above, the plaques are in open communication with the blood compartment, the proteinases probably involved in the demyelination are liberated, e.g., from macrophages to the blood, where they are partially complexed to α_2-macroglobulin (causing a decreased electrophoretic mobility of serum α_2M in MS[184]). The α_2-M-bound proteinases in MS contain a fraction with an acidic isoelectric point of 4.2.[186] At the moment, it is not clear which role that factor plays in either the demyelination or the immunoregulation. Since proteinases may be secreted like lymphokines from lymphocytes and macrophages and may regulate the function of subsets of lymphocytes (and leukocytes), the liberation of proteinases in MS brains may well contribute to the above-mentioned abnormal distribution of lymphocyte subpopulation.

4.3.2. Changes in Membranes

Early in MS research (*vide supra*), it was suggested that the disease was related to a high intake of saturated fat and/or a deficiency in essential fatty acids (EFA).[87,187,188] These changes in fatty acid composition, although not specific for MS,[96,101] may lead to changes in membrane fluidity, "sludging" of the red cells with increased fragility of these cells,[189] and increased agglutinability of thrombocytes.[89] Indeed, Prosiegel *et al.*,[190] in a careful study of thrombocyte agglutinability in MS and control patients that took into account dietary and pharmacological factors, have recently demonstrated that in acute exacerbations of MS the thrombocytes reveal an increased agglutinability induced by ATP and serotonin. At the moment, it is not possible to decide if this finding is an epiphenomenon or directly related to the etiology. However, the increased agglutinability may be related to an increased peroxidation activity in MS as a result of decreased activity of glutathione peroxidase.[191–193] The increased peroxidation rate in MS may also explain the decreased levels of EFA. Glutathione peroxidase is a selenium-containing enzyme. In agreement with the fact that Jensen *et al.*[194] found changes in serum selenium levels in MS that were proportional to the lowered GSH-Px activity of MS lymphocytes, the low GSH-Px levels found in MS lymphocytes were also traced in granulocytes and erythrocytes of MS patients.[194] In MS patients with low GSH-Px activities of hematogenous cells and low serum selenium levels as well as low linoleate levels in the plasma membranes of peripheral lymphocytes, antioxidative medication with 3 mg selenium (as sodium selenite) combined with 2 g vitamin C and 500 mg vitamin E per day completely normalized the above mentioned biochemical abnormalities. At the moment, it is unclear if this normalization process is associated with clinical improvement (J. Clausen, preliminary results). The low level of glutathione peroxidase may cause a low rate of prostaglandin formation,[195] since the 15-hydroperoxyprostaglandin R_2 is decomposed by GSH peroxidase; thus, the low level of decomposition of peroxyprostaglandin intermediates may induce platelet aggregation.[196]

5. *TREATMENT OF MULTIPLE SCLEROSIS*

Neither preventive measures nor curative treatments of MS are available. The evaluation of clinical effects of treatment of a disease such as MS with a more or less progressive development and with spontaneous remissions is difficult and should always follow a strictly defined line.[197,198] From what has been mentioned above about theories about the pathogenesis of MS, clinical trails to stop or retard the disease have concerned:

1. Immunologic intervention.
2. Intervention with the proteolytic degradation of myelin.
3. Dietary treatments.
4. Neuropharmacological treatment.

5.1. *Immunologic Intervention*

Immunologic intervention has implied use of ACTH, corticosteroid hormones, IgG, cyclophosphamide, transfer factor treatment, and plasmapheresis.[199,202]

Steroid hormones and ACTH seem to restore clinical performance during acute exacerbations but do not influence the major clinical course, even if intrathecal application is used.[203] Neither the transfer factor[204] nor cyclophosphamide seems to be able to arrest the disease, although a recent study seems to indicate that the transfer factor can retard cases with moderate activity.[205] However, studies on a limited number of MS patients of the effect of plasmapheresis seem to indicate a beneficial effect on a selected group of patients refractory to prednisone (cases with acute severe exacerbations). At the moment, the clinical data are too small to give any safe conclusions.

Initial studies of the beneficial effect of human immunoglobulins[199] on a limited number of MS patients seem to indicate improvement of clinical scores that measure physical symptoms.

5.2. Intervention with the Proteolytic Degradation of Myelin

As mentioned above (cf. also ref. 124), the demyelination in EAE and MS may be caused by the activation and liberation *in situ* of proteinases. Experimentally, Boehme *et al.*[206,207] and Sibley *et al.*[208] were able to suppress EAE by means of proteinase inhibitors. Therefore, it may be relevant to use such inhibiters in MS treatment.

Only few studies have been performed in this field. Amaducci *et al.*[209] elucidated the effect of ϵ-aminocaproic acid (EACA), a proteinase inhibitor, i.e., inhibitor of the conversion of plasminogen to plasmin in a short-term open clinical trial. Small probable improvements in clinical scores were noticed.

5.3. Dietary Treatment

Originally, Swank[187,188] treated MS patients with a low-fat diet in long-term experiments. Clinical improvement but no arrest of the disease has been quoted. Following demonstration of low EFA levels in MS serum, three double-blind studies of the effect of dietary linoleic acids on the clinical course of MS have been performed. In two double-blind studies,[210,211] a significant beneficial effect was found on severity and duration of relapses, whereas a third study[212] was unable to demonstrate any effect. Dworkin,[213] however, has reinvestigated all three studies and drawn attention to the fact that in all three studies the beneficial effect of linoleic acid is found mainly in patients with illness of short duration and/or low disability.

The question about the eventual beneficial effect of dietary treatment of MS patients with essential fatty acids should, however, be reinvestigated from the point of view of whether the low linoleate level is related to increased autoperoxidation of essential fatty acids as a result of low glutathione peroxidase and selenium levels in MS.[191–194] An antiperoxidative treatment such as that indicated above with selenium (as sodium selenite), vitamin C, and vitamin E may more easily normalize low essential fatty acid levels (J. Clausen, unpublished data).

5.4. Neuropharmacological Treatment

This treatment is only symptomatic and concerns treatment of spasticity, urinary bladder function, and defecation. These treatments are not the topic of the present communication, but the reader is referred to classical pharmacological textbooks.

6. CONCLUSION

Multiple sclerosis is a demyelination disease characterized by a perivenous demyelination in areas of the brain where lymphocytes and macrophages also initially accumulate. Most authors believe it to be an inflammatory disease, probably of viral origin. Most abnormalities demonstrated are nonspecific. Thus, the finding of oligoclonal bands in the spinal fluid is nonspecific since this is also found in cerebrovascular diseases, and the variation in pattern of oligoclonal bands in MS tissue (CNS) also argues that this finding is nonspecific. Further, the demonstration that CSF oligoclonal bands in MS contain polyspecific antibodies against different viruses and even bacteria argue for either the nonspecificity of this finding or lack of immunologic communication between subsets of lymphocytes in CNS. Changes in subsets of lymphocytes in serum and CSF in MS may either be explained by a redistribution phenomenon with migration of specific subsets to CNS or by redistribution of lymphocyte receptors as a result of their exposure to immunocomplexes identified in both MS serum and CSF.

The finding of autoantibodies in MS to myelin proteins and lipids is also nonspecific, since these antibodies are found in other neurological diseases as well.

The pathogenic role and specificity of the so-called MS-specific antigens found in CNS of MS patients need more studies; the same applies to both the increased level of measles virus antibodies in biological fluids of MS patients and the annealing experiments that identified a measles virus genome in MS CNS.

The changes in fatty acid composition of cellular membranes and the low levels of essential fatty acids of serum lipids in MS as well as the increased peroxidation rate may well be epiphenomena, but, on the other hand, they argue for systemic cellular changes in MS.

REFERENCES

1. Waller, A. V., 1850, *Phil. Trans. R. Soc.* **1**:423.
2. Cruveillier, J., 1835–42, *Atlas d'Anatomie Pathologique du Corps Humain* 2, Baillière, Paris.
3. Carswell, R., 1838, *Pathological Anatomy, Illustrations of the Elementary Forms of Disease*, London.
4. Valentiner, W., 1856, *Dtsch. Klin.* **8**:147.
5. Rindfleisch, 1863, *Virchows Arch.* **26**:474.
6. Charcot, J. M., 1868, *Gaz. Hop. (Paris)* **41**:554,557,566.

7. Kurtzke, J. F., 1970, *Acta Neurol. Scand.* **46**:484–492.
8. McAlpine, D., 1973, *Br. Med. J.* **2**:292–295.
9. Parinaud, H., 1884, *Prog. Med. (Paris)* **12**:641.
10. Hyllested, K., 1950, *Ugeskrift Laeger* **112**:853–866.
11. Fog, T., and Linnemann, F., 1970, *Acta Neurol. Scand.* [*Suppl.*] **47**:1–175.
12. Regan, D., Milner, B. A., and Heron, J. R., 1976, *Brain* **99**:43–66.
13. Maurer, K., Schäfer, E., Hopf, H. C., and Leitner, H., 1980, *J. Neurol.* **223**:43–58.
14. Eisen, A., Stewart, J., Nuddleman, K., and Cosgrove, J. B. R., 1979, *Neurology (Minneap.)* **29**:827–834.
15. Ludwin, S. K., 1981, *Adv. Neurol.* **31**:123–168.
16. Bartlett, P. F., and Mackay, I. R., 1983, *J. Clin. Lab. Immunol.* **11**:1–7.
17. Bielschowsky, M., 1903, *Neurol. Zentralbl.* **22**:770.
18. King, L. S., 1937, *Arch. Pathol.* **23**:338.
19. Itoyama, Y., Webster, H., Richardson, E. P., and Trapp, B. D., 1983, *Ann. Neurol.* **14**:339–346.
20. Peters, G., 1936, *Z. Ges. Neurol. Psychiatrie* **155**:178.
21. Fog, T., 1965, *Acta Neurol. Scand.* (*Suppl.*] **15**:41.
21a. Adams, C. V., 1977, *Br. Med. Bull.* **33**:15–20.
22. Weiner, L. P., 1973, *Arch. Neurol.* **28**:298–303.
23. Prineas, J. W., and Raine, C. S., 1976, *Neurology (Minneap.)* **26**(2):29–32.
24. Prineas, J. W., and Connell, F., 1978, *Neurology (Minneap.)* **28**(2):68–75.
25. Lampert, P. W., 1978, *Am. J. Pathol.* **91**:175–208.
26. Dal Canto, M. C., Wisniewski, H. M., and Johnson, A. B., 1975, *J. Neurol. Sci.* **24**:313–319.
27. Kirk, J., 1979, *Neuropathol. Appl. Neurobiol.* **5**:289–294.
28. Prineas, J., 1972, *Science* **178**:760–763.
29. Ter Meulen V., Koprowski, H., Iwasaki, Y., Käckell, Y. M., and Müller, D., 1972, *Lancet* **2**:1–5.
30. Field, E. J., Cowshall, S., Narang, H. K., and Bell, T. M., 1972, *Lancet* **2**:280–281.
31. Barbosa, L. H., and Hamilton, R., 1973, *Lancet* **1**:1415–1417.
32. Lewandowski, L. J., Lief, F. S., Verini, M. A., Pienkowski, M. M., Ter Meulen, V., and Koprowski, H., 1974, *J. Virol.* **13**:1037–1045.
33. Carp, R. I., Licursi, P. C., Merz, P. A., and Merz, G. S., 1972, *J. Exp. Med.* **136**:618.
34. Pathak, S., and Webb, H. E., 1976, *Lancet* **2**:311.
35. Woyciechowska, J. L., Madden, D. L., and Sever, J. L., 1977, *Lancet* **2**:1046–1049.
36. Mitchell, D. N., Porterfield, J. S., Micheletti, R., Lange, L. S., Goswami, K. K. A., Taylor, P., Jacobs, J. P. Hockley, D. J., and Salsbury, A. J., 1978, *Lancet* **2**:387–391.
37. Cobill, J. M., Hugehes, D., Keith, A. B., Cameron, K. R., Mason, J., and Caspary, E. A., 1979, *J. Neurol. Sci.* **41**:11–16.
38. Burks, J. S., DeVald, B. L., Jankovsky, L. D., and Gerdes, J. C., 1980, *Science* **209**:933–934.
39. Raine, C. S., Powers, J. M., and Suzuki, 1974, *Arch. Neurol.* **30**:39–46.
40. Lipton, H. L., and Dal Canto, M. C., 1976, *J. Neurol. Sci.* **30**:201–207.
41. Adams, C. W. M., 1965, *Neurohistochemistry*, Elsevier, Amsterdam.
42. Abercrombie, M., and Johnson, M. L., 1946, *J. Neurol. Neurosurg. Psychiatry* **9**:113.
43. Rastogi, S. C., and Clausen, J., 1980, *Clin. Exp. Immunol.* **42**:50–56.
44. Rossiter, R. J., 1965, *Neurochemistry* (K. A. C. Elliott, I. H. Page, and J. H. Qastel, eds.), Charles C Thomas, Springfield, Illinois, p. 696.
45. Prineas, J. W., 1979, *Science* **203**:1123–1125.
46. Alter, M., Liebowitz, U., and Speer, J., 1966, *Arch. Neurol.* **15**:234–237.
47. Leibowitz, U., Halpern, L., and Alter, M., 1964, *Arch. Neurol.* **10**:502–512.
48. Dean, G., 1972, *Multiple Sclerosis, Progress in Research* (E. J. Field, T. M. Bell, and P. R. Carnegie, eds.), North-Holland, Amsterdam, London, pp. 196–203.
49. Acheson, E. D., 1972, *Multiple Sclerosis: A Reappraisal*, 2nd ed. (D. McAlpine, C. E., Lumsden, and E. D. Acheson, eds.), Churchill Livingstone, Edinburgh, pp. 22.
50. Kurtzke, J. F., 1972, *Multiple Sclerosis, Progress in Research* (E. J. Field, T. M. Bell, and P. R. Carnegie, eds.), North-Holland, Amsterdam, London, pp. 208–228.

51. Acheson, E. D., 1977, *Br. Med. Bull.* **33**:9–14.
52. Kurtzke, J. F., and Kurland, L. T., 1973, *Clinical Neurology*, Volume 3 (A. B. Baker and L. H. Baker, eds.), Harper and Row, New York. pp. 1–80.
53. Kurtzke, J. F., 1980, *Neurology* (*N.Y.*) **30**:61–79.
54. Kurtzke, J. F., and Hyllested, K., 1979, *Ann. Neurol.* **5**:6–21.
55. Poskanzer, C. D., Walker, A. M., Prenney, L. B., and Sheridan, J. L., 1981, *Neurology* (*N.Y.*) **31**:708–713.
56. Gudmundsson, K. R., 1971, *Acta Neurol. Scand.* (Suppl.]**48**:47.
57. Isager, H., Larsen, S., and Hyllested, K., 1980, *Int. J. Epidemiol.* **9**:145–147.
58. Jersild, C., Ammitzbøll, T., Clausen, J., and Fog, T., 1973, *Lancet* **1**:151–152.
59. Jersild, C., 1978, *The HLA System and Multiple Sclerosis*, The National Foundation, Alan R. Liss, New York.
60. Pathy, D. W., Mervart, H., Campling, B., Rand, C. G., and Stiller, C. R., 1974, *Can. J. Neurol. Sci.* **1**:211–213.
61. Svejgaard, A., Hauge, M., Jersild, C., Platz, P., Ryder, L. P., Nielsen, L. S., and Thomsen, M., 1975, *Monographs in Human Genetics, The HLA System*, Volume 7 (L. Beckman and M. Hauge, eds.) S. Karger, Basel.
61a. Lewin, R., 1975, *Laboratoriumsblatter* **1**:1.
62. Terasaki, P. I., and Mickey, M. R., 1976, *Neurology* (*Minneap.*) **26**:56–58.
63. Haile, R. W., Hodge, S. E., Visscher, B. R., Spence, M. A., Detels, R., McAuliffe, T. L., Park, M. S., and Dudley, J. P., 1980, *Clin. Genet.* **18**:160–167.
64. Drachman, D. A., Davison, W. C., and Mittal, K. K., 1976, *Arch. Neurol.* **33**:406–413.
65. Sigurdsson, B., 1954, *Br. Vet. J.* **110**:341–354.
66. Cook, S. D., and Dowling, P. C., 1980, *Neurology* (*N.Y.*] **30**(2):80–91.
67. Adams, J. M., and Imagawa, D. T., 1962, *Proc. Soc. Exp. Biol. Med.* **111**:562–566.
68. Panelius, M., 1969, *Acta Neurol. Scand.* [Suppl.] **39**:45.
69. Ammitzbøll, T., and Calusen, J., 1972, *Acta Neurol. Scand.* **48**:47–56.
70. Salmi, A., Panelius, M., and Vainiopää, R., 1974, *Acta Neurol. Scand.* **50**:183–193.
71. Beck, H. W., and Clausen, J., 1977, *Zentralbl. Bakteriol. Hyg.* [*Orig. A*] **338**:431–443.
72. Haase, A. T., Ventura, P., and Tourtelotte, W. W., 1981, *Science* **212**:672–674.
73. Ammitzbøll, T., Offner, H., Clausen, J., Kobayasi, T., Asboe-Hansen, G., Hyllested, K., and Fog, T., 1976, *Acta Neurol. Scand.* **53**:137–151.
74. Whitaker, J. N., Herrmann, K. L., Rogentine, G. N., Stein, S. F., and Kollins, L. L., 1976, *Arch. Neurol.* **33**:399–403.
75. Kurtzke, J. F., and Priester, W. A., 1979, *Acta Neurol. Scand.* **60**:312–319.
76. Krakowka, S., and Koestner, A., 1978, *Lancet* **1**:1127–1128.
77. Johnson, K. P., Likosky, W. H., Nelson, B. J., and Fein, G., 1980, *Arch. Neurol.* **37**:537–541.
78. Cremer, N. E., Johnson, K. P., Fein, G., and Likosky, W. H., 1980, *Arch. Neurol.* **37**:610–615.
79. Forghani, B., Cremer, N. E., Johnson, K. P., Fein, G., and Likosky, W. H., 1980, *Arch. Neurol.* **37**:616–619.
80. Szekeres, J., Palffy, Gy., and Paradi, J., 1980, *Lancet* **2**:1089–1090.
81. Martin, J. R., 1981, *Lancet* **2**:777–781.
82. Fraser, K. B., 1977, *Br. Med. Bull.* **33**:34–39.
83. McAlpine, D., Lumsden, C. E., and Acheson, E. D., 1965, *Multiple Sclerosis, A Reappraisal*, E. and S. Livingstone, Edinburgh.
84. Friede, R. L., 1966, *Topographic Brain Chemistry*, Academic Press, New York, London.
85. McKeown, S. R., and Allen, I. V., 1979, *Neuropathology* **5**:405–415.
86. Yahara, S., Kawamura, N., Kishimoro, Y., Saida, T., and Tourtelotte, W. W., 1982, *J. Neurol. Sci.* **54**:303–315.
87. Swank, R. L., 1953, *Arch. Neurol. Psychiatry.* **69**:91–103.
88. Agranoff, B. W., and Goldberg, D., 1974, *Lancet* **2**:1061.
89. Mertin, J., and Meade, C. J., 1977, *Br. Med. Bull.* **33**:67–71.
90. Cumings, J. N., Shortman, R. C., and Skibic, T., 1965, *J. Clin. Pathol.* **18**:641–644.
91. Gerstl, B., Eng, L. F., Tavaststjerna, M. G., Smith, J. K., and Kruse, S. L., 1970, *J. Neurochem.* **17**:677–689.

92. Thompson, R. H. S., 1966, *Proc. R. Soc. Med.* **59**:269–276.
93. Clausen, J., and Hansen, I. B., 1970, *Acta Neurol. Scand.* **46**:1–17.
94. Clausen, J., and Møller, J., 1967, *Acta Neurol. Scand.* **43**:375–388.
95. Baker, R. W. R., Thompson, R. H. S., and Zilka, K. J., 1966, *J. Neurol. Neurosurg. Psychiatry* **29**:95–98.
96. Wolfram, G., Eckhart, J., and Zöllner, N., 1977, *Nutr. Metab.* **21(Suppl. 1)**:127–129.
97. Yoshida, M., Takase, S., Itahara, K., and Nakanishi, T., 1983, *Acta Neurol. Scand.* **68**:362–364.
98. Wolfgram, F., Myers, L., Ellison, G., and Knipprath, W., 1975, *Neurology (Minneap.)* **25**:766–768.
99. Heipertz, R., Klauke, W., Pilz, H., and Ritter, G., 1977, *J. Neurol.* **214**:153–157.
100. Callaghan, N. C., Kearney, B., and Love, W. C., 1973, *J. Neurol. Neurosurg. Psychiatry* **36**:668–673.
101. Love, W. C., Cashell, A., Reynolds, M., and Callaghan, N., 1974, *Br. Med. J.* **462**:18–21.
102. Cummings, J. N., and Goodwin, H., 1968, *Lancet* **2**:664.
103. Norton, W. T., 1981, *Demyelinating Disease: Basic and Clinical Electrophysiology* (S. G. Waxman and J. M. Richie, eds.), Raven Press, New York, pp. 93–121.
104. Rhein, L. D., and Sampugna, J., 1981, *Lipids* **7**:502–507.
105. Woelk, H., and Borri, P., 1973, *Eur. Neurol.* **10**:250.
106. Wolfgram, F., 1972, *Chemical Theories of the Demyelination in Multiple Sclerosis, Immunology, Virology and Ultrastructure* (F. Wolfgram, G. W. Ellison, J. G. Stevens, and J. M. Andrews, eds.), Academic Press, New York, pp. 172–182.
107. Suzuki, K., Kamoshita, S. Eto, Y., Tourtelotte, W. W., and Gonatas, J. O., 1973, *Arch. Neurol.* **28**:293–297.
108. Yu, R. K., Ueno, K. Glaser, G. H., and Tourtelotte, W., 1982, *J. Neurochem.* **39**:464–477.
109. McKhann, M., 1982, *Annu. Rev. Neurosci.* **5**:219–239.
110. Riekkinen, P. R., Palo, J., Arstila, A. U., Savolainen, H. J., Rinne, U. K., Kivalo, E. K., and Frey, H., 1971, *Arch. Neurol.* **24**:545.
111. Wolfgram. F., and Tourtelotte, W. W., 1972, *Neurology (Minneap.)* **22**:1044.
112. Eng, L. F., Chao, F.-C., Gerstl, B., Pratt, D., and Travastsjerna, M. G., 1968, *Biochemistry* **7**:4455.
113. Riekkinen, P. J., Clausen, J., Frey, H. J., Fog, T., and Rinne, U. K., 1970, *Acta Neurol. Scand.* **3**:349–353.
114. Cuzner, M. L., Bernard, R. O., McGregor, B. J. L., Borshell, N. J., and Davison, A. N., 1976, *J. Neurol. Sci.* **29**:323–334.
115. Sachs, H., and Steiner, G., 1934, *Klin. Sochenschr.* **13**:1714.
116. Roemer, G. B., and Mai, K., 1967, *Z. Immun. Allg. Klin. Immunol.* **134**:370–390.
117. Clausen, J., Fog, T., Offner, H., and Rastogi, S. C., 1978, *Acta Neurol. Scand.* **57**(Suppl. 67):247–248.
118. Rastogi, S. C., Clausen, J., Offner, H., Konat, G., and Fog, T., 1979, *Acta Neurol. Scand.* **59**:281–296.
119. Rastogi, S. C., and Clausen, J., 1980, *Clin. Chim. Acta* **101**:85–92.
120. Rastogi, S. C., Clausen, J., and Tourtelotte, W. W., 1983, *Eur. Neurol.* **22**:17–21.
121. Clausen, J., 1983, *J. Neurol. Sci.* **60**:205–216.
122. Hukkanen, V., Salmi, A., and Frey, H., 1982, *J. Neuroimmunol.* **3**:295–305.
123. Hashim, G. A., Lee, D. M., Pierce, J. C., Braun, C. W., and Fitzpatrick, H. F., 1978, *Neurochem. Res.* **3**:37–48.
124. Offner, H., Konat, G., Legg, N. J., Raun, N. E., Winterberg, H., and Clausen, J., 1980, *Immunopathology* **16**:367–373.
125. Offner, H., Rastogi, S. C., Konat, G., and Clausen, J., 1979, *J. Neurol.* Sci. **42**:349–355.
126. Offner, H., Fog, T., Rastogi, S. C., Konat, G., and Clausen, J., 1979, *Acta Neurol. Scand.* **59**:49–54.
127. Turner, A., Cuzner, M. L., Davison, A. N., and Rudge, P., 1980, *J. Neurol. Neurosurg. Psychiatry* **43**:305–309.
128. Rastogi, S. C., and Clausen, J., 1980, *Clin. Exp. Immunol.* **42**:50–56.
129. Rastogi, S. C., and Clausen, J., 1982, *Acta Neurol. Scand.* **67**:218–221.

130. Kabat, E. A., Moore, D. H., and Landow, H., 1942, *J. Clin. Invest.* **21**:571.
131. Löwenthal, A., v. Sande, M., and Karcher, D., 1960, *J. Neurochem.* **6**:51–56.
132. Clausen, J., Matzke, J., and Gerhardt, W., 1964, *Acta Neurol. Scand.* **40**(Suppl. 10]:49–56.
133. Link, H., 1967, *Acta Neurol. Scand* **43**(Suppl. 28]:1–136.
134. Christensen, O., Clausen, J., and Fog, T., 1978, *J. Neurol.* **218**:237–244.
135. Laurenzi, M. A., 1981, *Immunochemical Characterization of Immunoglobulins and Viral Antibodies Synthesized within the Central Nervous System in Patients with Multiple Sclerosis and Controls*, Linköping University Medical Dissertation, Linköping.
136. Hosein, Z. Z., and Johnson, K. P., 1981, *Neurology (N.Y.)* **31**:70–76.
137. Schuller, E., 1981, *Les Proteines du Liquide Céphalo-Rachidien et les Maladies Immunitaires du Systeme Nerveux*, Hoechst-Behring, Paris.
138. Kempe, C. H., Takabayashi, K., Miyamoto, H., McIntosh, K., Tourtellotte, W. W., and Adams, J. M., 1973, *Arch. Neurol.* **28**:278–279.
139. Roström, B., 1981, *Acta Neurol. Scand.* **86**[Suppl.]:63.
140. Nordal, H. J., Vandvik, B., and Norrby, E., 1978, *Scand. J. Immunol.* **7**:473–479.
141. Mattson, D. H., Ross, R. P., and Arnason, B. G. W., 1980, *Nature* **287**:335–337.
142. Sandberg-Wollheim, M., 1975, *Studies on Immunoglobulin Synthesis and Lymphocyte Subpopulations in the Cerebrospinal Fluid*, Grafo Tryk, Malmö.
143. Nakao, A., Davis, W. J., and Einstein, E. R., 1966, *Biochim. Biophys. Acta.* **130**:163–171.
144. Govindarajan, K. R., Rauch, H. C., Clausen, J., and Einstein, E. R., 1974, *J. Neurol. Sci.* **23**:295–306.
145. Traugott, U., Schevach, E., Chiba, J., Stone, J. H., and Raine, C. S., 1981, *Science* **214**:1251–1252.
146. Ammitzbøll, T., Clausen, J., and Fog, T., 1977, *Acta Neurol. Scand.* **56**:153–158.
147. Kam-Hansen, S., 1980, *Distribution and Function of Lymphocytes from the Cerebrospinal Fluid and Blood in Patients with Multiple Sclerosis*, Linköping University, Linköping.
148. Coyle, P. K., Brooks, B. R., Hirsch, R. L., Cohen, S. R., O'Donnell, P., Johnson, R. T., and Wolinsky, J. S., 1980, *Lancet* **2**:229–232.
149. Roström, B., and Link, H., 1981, *Neurology (N.Y.)* **31**:590–596.
150. Clausen, J., 1966, *Acta Neurol. Scand.* **42**:153–160.
151. Clausen, J., and Fog, T., 1969, *Pathogenesis and Etiology of Demyelinating Diseases*, S. Karger, Basel, pp. 648–656.
152. Nagai, Y., Sakakibara, K., and Uchida, T., 1979, *Immunomodulatory Roles of Gangliosides in EAE and EAN*, Hertie Foundation, Frankfurt/Main.
153. Traugott, U., and Raine, C. S., 1981, *Neurology (N.Y.)* **31**:695–700.
154. Sheremata, W., Eylar, E. H., and Cosgrove, J. B. R., 1977, *J. Neurol. Sci.* **32**:255–263.
155. Arnon, R., Crisp, E., Kelley, R., Ellison, G. W., Myers, L. W., and Tourtellotte, W. W., 1980, *J. Neurol. Sci.* **46**:179–186.
156. Nordal, H. J., and Vandvik, B., 1977, *Scand. J. Immunol.* **6**:327–334.
157. Bernard, C. C. A., Randell, V. B., Horvath, L. B., Carnegie, P. R., and Mackay, I. R., 1981, *Immunology* **43**:447–457.
158. Gerhard, W. A., Taylor, Z., Wroblewska, M., Sandberg-Wollheim, M., and Koprowski, H., 1981, *Proc. Natl. Acad. Sci. U.S.A.* **78**:3225–3229.
159. Nagelkerken, L. M., Aalberse, B. C., Walbeek, H. K., and Out, T. A., 1980, *J. Immunol.* **125**:384–389.
160. Ryberg, B., 1981, *Studies on Complement-Fixing Antibrain Antibodies in Multiple Sclerosis*, Infotryk, Malmö.
161. Broman, T., 1947, *Acta Psychiatr. Neurol.* **46**:58–71.
162. Lisak, R. P., Levinson, A. I., Zweiman, B., and Abdou, N. I., 1975, *Clin. Exp. Immunol.* **22**:30–34.
163. Arnason, B. G. W., and Antel, J. P., 1978, *Ann. Immunol.* **129C**:159–170.
164. Traugott, U., Scheinberg, L. C., and Raine, C. S., 1981, *Ann. Neurol.* **11**:182–186.
165. Dropcho, E. J., Richman, D. P., Antel, J. P., and Arnason, B. G. W., 1978, *Fed. Proc.* **37**:1764.
166. Booss, J., and Dwyer, J. M., 1979, *Neurology (Minneap.)* **29**:537.
167. Antel, J. P., and Weinrich, M., and Arnason, B. G. W., 1978, *Neurology (Minneap.)* **28**:993–1003.

168. Gonzalez, R. L., Dau, P. C., and Spitler, L. E., 1979, *Clin. Exp. Immunol.* **36**:78–84.
169. Offner, H., 1976, *Studies on Pathogenic Factors in Multiple Sclerosis*, Thesis, Fossum, Birkerød, Denmark.
170. McFarland, H. F., and McFarlin, D. E., 1979, *Ann. Neurol.* **6**:101–106.
171. Schauf, V., Schauf, C. L., Mizen, M., and Davis, F. A., 1981, *Acta Neurol. Scand.* **64**:383–393.
172. Neighbor, P. A., and Bloom, B. R., 1979, *Proc. Natl. Acad. Sci. U.S.A.* **76**:476–480.
173. Pichler, W. J., and Broder, S., 1981, *Immunol. Rev.* **56**:163–197.
174. Goust, J. M., Chenais, F., Carnes, J. E., Hames, C. G., Fudenberg, H. H., and Hogan, E. L., 1978, *Neurology (Minneap.)* **28**:421–425.
175. Huddlestone, J. R., and Oldstone, M. B. A., 1979, *J. Immunol.* **123**:1615–1618.
176. Dore-Duffy, P. Zurier, R. B., Donaldson, J., Nystrom, S. S., Viola, M. V., Rothman, B., and Thompson, H. G., 1979, *Neurology* (Minneap.) **29**:232–235.
177. Frey, H., Salmi, A., and Svedström, E., 1979, *J. Neurol.* **220**:99–104.
178. Kinnman, J., and Link, H., 1979, *Acta Neurol. Scand.* **59**:241–247.
179. Davison, A. N., and Ilyas, A. A., 1982, *Int. Arch. Allergy Appl. Immunol.* **69**:393–396.
180. Colby-Germinario, S. P., Sheremata, W., Bain, B., and Eylar, E. H., 1977, *J. Neurol. Sci.* **33**:111–129.
181. Mar, P., Gradl, T., and Dörner, C., 1979, *J. Neurol. Sci.* **41**:369–377.
182. Frick, E., and Stickl, H., 1980, *J. Neurol. Sci.* **46**:187–197.
183. Santoli, D., Hall, W., Kastrukoff, L., Lisak, R. P., Rerussia, B., Trinchieri, G., and Koprowski, H., 1981, *J. Immunol.* **126**:1274–1278.
184. Mar, P., 1980, *J. Neurol. Sci.* **47**:285–303.
185. Rastogi, S. C., Clausen, J., and Fog, T., 1981, *Eur. Neurol.* **20**:33–39.
186. Rastogi, S. C., and Clausen, J., 1980, *Clin. Chim. Acta* **107**:141–144.
187. Swank, R. L., 1955, *Arch. Neurol. Psychiatry* **73**:631–644.
188. Swank, R. L., 1956, *Ann. Intern. Med.* **45**:812–824.
189. Schauf, C. L., Frischer, H., and Davis F. A., 1980, *Neurology (N.Y.)* **30**:323–325.
190. Neu, I. S., Prosiegel, M., and Pfaffenrath, V., 1982, *Acta Neurol. Scand.* **66**:497–504.
191. Shukla, V. K. S., Jensen, G. E., and Clausen, J., 1977, *Acta Neurol. Scand.* **56**:542–550.
192. Jensen, G. E., Gissel-Nielsen, G., and Clausen, J., 1980, *J. Neurol. Sci.* **48**:61–67.
193. Szeinberg, A., Golan, R., Ben-Ezzer, J., Sarova-Pinhas, I., and Kindler, D., 1981, *Acta Neurol. Scand.* **63**:67–75.
194. Jensen, G. E., and Clausen, J., 1984, *J. Neurol. Sci.* **63**:45–53.
195. Srivastava, K. C., Fog, T., and Clausen, J., 1975, *Acta Neurol. Scand.* **51**:193–199.
196. Flohé, L., Günzler, W. A., and Ladenstein, R., 1976, *Kroc Foundation Glutathione: Metabolism and Function* (I. M. Arias, and W. B. Jakoby, eds.), *Series*, No. 6, Raven Press, New York, pp. 115–135.
197. Brown, J. R., Beebe, G. W., Kurtzke, J. F., Loewenson, R. B., Silberberg, D. H., and Tourtellotte, W. W., 1979, *Neurology (Minneap.)* **29**:3–23.
198. Bauer, H. J., 1978, *Neurology (Minneap.)* **28**:8–19.
199. Rothfelder, U., Neu, I., and Pelka, R., 1982, *Munch. Med. Wochenschr.* **124**:46–54.
200. Ammitzbøll, T., Clausen, J., and Fog, T., 1975, *Acta Neurol. Scand.* [*Suppl.*]**63**:99–111.
201. Theys, P., Gosseye-Lissoir, F., Ketelaer, P., and Carton, H., 1981, *J. Neurol.* **225**:119–133.
202. Dau, P. C., Petajan, J. H., Johnson, K. P., Panitch, H. S., and Bornstein, M. B., 1980, *Neurology (N.Y.)* **30**:1023–1028.
203. Neu, I., 1982, *Munch. Med. Wochenschr.* **124**:77.
204. Fog, T., Pedersen, L., Raun, N. E., Kam-Hansen, S., Mellerup, E., Platz, P., Ryder, L. P., Jakobsen, B. K., and Grob, P., 1978, *Lancet* **1**:851–853.
205. Basten, A., Pollard, J. D., Steward, G. J., Frith, J. A., McLeod, J. G., Walsh, J. C., Garrick, R., and v. der Brink, C. M., 1980, *Lancet* **2**:931–934.
206. Boehme, D. H., Fordice, M. W., and Marks, N., 1974, *Brain Res.* **75**:153–162.
207. Boehme, D. H., Umezawa, H., Hashim, G., and Marks, N., 1978, *Neurochem. Res.* **3**:185–194.
208. Sibley, W. A., Kiernat, J., and Laguna, J. F., 1978, *Neurology (Minneap.)* **28**:102–105.
209. Amaducci, L., Inzitari, D., Sita, D., Antuono, P., Capparelli, R., and Arfaioli, C., 1979, *Acta Neurol.* **34**:214–258.

210. Millar, J. H. D., Zilkha, K. J., Langman, M. J. S., Payling-Wright, H., Smith, A. D., Belin, J., and Thompson, R. H. S., 1973, *Br. Med. J.* **1973:**765–768.
211. Bates, D., Fawcett, P. R. W., Shaw, D. A., and Weightman, D., 1978, *Br. Med. J.* **1978:**1390–1391.
212. Paty, D. W., Cousin, H. K., Read, S., and Adlakha, K., 1978, *Acta Neurol. Scand.* **58:**53–58.
213. Dworkin, R. H., 1981, *Lancet* **1:**1153–1154.

11

Multiple Sclerosis and Allergic Encephalomyelitis

George A. Hashim

1. INTRODUCTION

During the past 40 years, research in multiple sclerosis gained worldwide interest since this progressively crippling disease of the central nervous system is not localized in any particular area of the world. Multiple sclerosis is a human disorder with varying clinical manifestations resulting from interference in neuronal conduction. The multiplicity of clinical symptoms of the disease and the unpredictable prognosis of individual cases suggested multiple forms caused by a number of agents rather than a single clue to a single cause. Pathological studies of tissues from MS patients revealed destruction of nerve fiber myelin and axis cylinders leading to the formation of "sclerotic plaque," generally distributed around vessels. Although many exceptions to the general distribution of these lesions have been reported,[1–3] it is clear that myelin degeneration resulting in impaired conductivity remains a general finding in multiple sclerosis. In the acute stage of the disease, the vessels become dilated with lymphocytes and plasma cells infiltrating into the perivascular spaces, suggesting cellular involvement and cell-mediated immunity in this disorder. Although a number of theories regarding the etiology and pathogenesis of multiple sclerosis have been proposed, the negative evidence against environmental factors,[4] inherited factors,[5,6] malnutrition,[2,7] infections,[8] trauma,[9] vaccination against smallpox,[10] and viruses[11,12] outweigh the positive.

2. MULTIPLE SCLEROSIS AND EXPERIMENTAL ALLERGIC ENCEPHALOMYELITIS

A vast amount of literature has been published concerning the possible relationship between experimental allergic encephalomyelitis (EAE) and mul-

George A. Hashim • The Jane Forbes Clark Surgical Research Laboratories, Departments of Surgery and Microbiology, St. Luke's–Roosevelt Hospital Center and Columbia University, New York, New York 10032.

tiple sclerosis.[13–16] That EAE is a cell-mediated autoimmune disease of myelin has been documented well,[17,18] and one of the agents responsible for initiating cellular events leading to demyelination is known to be the encephalitogenic basic protein (BP) found in myelin of the central nervous system.[19–21]

2.1. Clinical and Pathological Features

Characteristic clinical and pathological signs of EAE follow the development of a cell-mediated immune response to myelin BP[22–24] and closely resemble acute multiple sclerosis in man.[25] Causative agents responsible for multiple sclerosis have not been clearly defined; the myelin BP that plays a role in the development of rabies postvaccinal encephalomyelitis[26] suggests that the basic protein is capable of inducing demyelination in man. This suggestion was confirmed when EAE was induced in man following a laboratory accident.[27] The encephalitogenic BP is a normal componant of CNS myelin, and the purified protein induces cell-mediated immune rejection of normal myelin in homologous and heterologous species.[28–33] The BP from human CNS myelin is capable of inducing disease in monkeys, rabbits, guinea pigs, and rats. In view of these relationships, the assumption is made that the human myelin BP, like BP from other species, is capable of inducing demyelinating disease in man.

2.2. The Acute Form

The acute form of the animal disease, EAE, resembles acute multiple sclerosis,[25] and the chronic relapsing EAE[34] parallels the progressive form of multiple sclerosis. From these relationships among the myelin BP, EAE, and the human demyelinating disease, a worldwide interest was generated in the possible use of BP in the treatment of multiple sclerosis in man. In this chapter, I review some of the evidence generated from EAE research in animals and from studies in multiple sclerosis patients using the myelin BP.

The myelin BP from brain or spinal cord tissues of a number of species induces EAE in laboratory animals.[19,30,35–40] The disease is caused by a delayed hypersensitivity reaction to myelin BP,[17,41] which precedes the onset of clinical signs of EAE. Depending on the amount of BP, the nature of the adjuvant, the amino acid sequence of the antigenic determinant, and the animal species used, several forms of EAE have been reported.

2.3. The Hyperacute Form

The acute or ordinary form of EAE is monophasic, severe, and usually fatal.[16] Hyperacute EAE takes an earlier onset with a rapid and severe course characterized by seizures, coma, and spasticity along with high incidence of death.[42,43] Both acute and hyperacute EAE present similar immunologic and pathological expressions of disease[43] and together with their respective clinical signs provide a parallel for the human disorder ranging from chronic multiple sclerosis to acute hemorrhagic necrotizing leukoencephalopathy.[42–44]

2.4. Chronic and Chronic Relapsing EAE

These are two similar forms of this animal disease.[34,45] They are elicited in "juvenile" strain 13 guinea pigs[34,46–49] with suspensions of spinal cord tissues emulsified with large amounts of (2.5 mg *Mycobacterium*) of complete Freund's adjuvant. The disease is characterized by a latent period lasting 8 to 12 weeks, and the absence of clinical signs of disease during this period is not matched by negative pathology.[48] Although this chronic, progressive, and relapsing disease may not resemble all aspects of multiple sclerosis in man, the finding of a wide spectrum of CNS lesions and the presence of recent, chronically demyelinating lesions with active inflammation and Schwann cells parallel some of the features seen in multiple sclerosis.[50]

2.5. Localized Form

Three localized forms of EAE were produced following passive transfer of disease in Lewis rats.[51] These forms, which occur around a focal injury,[52–54] are characterized by perivascular cuffing around capillaries and vessels, infiltration by inflammatory cells of adjoining parenchyma, enlargement of extracellular spaces, edema, and fibrin deposits. The unique feature of localized EAE is the presence of basophils within lesions and adjacent parenchyma.[51]

3. THE RESPONSIBLE ANTIGENS

The five forms of EAE provide a spectrum of animal models for the study of multiple sclerosis in man. The clinical and pathological manifestations in multiple sclerosis appear to be similar to those seen in EAE.[55] This being the case, the myelin basic protein occupies a central role in the development of events leading to demyelination. What do we know about this protein? What are the immunologic events that the protein induces or inhibits? And what justifies its use for the treatment of demyelinating diseases in man? Finally, is it the only myelin or CNS antigen that induces demyelination?

3.1. Lipophilin

To undertake this discussion, we chose to look at two specific myelin components, both of which are capable of inducing demyelination, and perhaps both are capable of preventing as well as arresting the demyelinating process. The two specific components in question are myelin basic protein (MBP) and myelin lipophilin. The latter is known as a proteolipid protein of myelin.[56] The administration of purified lipophilin, a hydrophobic proteolipid protein, induces delayed-type hypersensitivity and demyelination in experimental animals.[57] Both clinical and pathological courses of the disease were similar, if not identical, to EAE induced by MBP. Although detailed structural analysis of the disease-inducing regions of lipophilin has not been completed, studies have shown that the disease-inducing region for guinea pigs is similar to that described for MBP.[57] The disease-inducing capacity of lipophilin is matched by its ability to potentiate MBP-induced EAE in experimental animals.[58]

The physical chemistry of the two myelin components underwent intensive investigation in years past. Rightly so, for the chemical properties of the proteins underlie their biological activity. Myelin lipophilin undergoes conformational flexibility in different environments. The water-soluble lipophilin[59] showed a helical configuration of between 14 and 40%. In chloroform–menthanol, lipophilin assumed an α-helical structure covering almost the entire molecule. By using dialysis from phenol–acetic acid–urea to water solution,[60] an α-helical or β conformation may be obtained. Further, the α and β conformational structures of lipophilin gave molecular weights of 86,000 and 500,000, respectively,[60] suggesting a high degree of aggregation with a monomeric molecular weight value of about 28,000.[61] Whether these conformational changes contribute to myelin stability or instability is not clear; however, it is clear that the transition from the α to the β configuration does not alter the ability of lipophilin to induce EAE in guinea pigs.[57]

The myelin basic protein, unlike lipophilin, undergoes very few known conformational changes in aqueous media. The absence of cysteine and the presence of over 25% basic amino acid residues randomly distributed over the entire molecule provided some basis for the conclusion that the MBP was devoid of secondary structure. The MBP has an open conformation with considerable and unique structure, though it is devoid of α-helical or β structures.[62,63] Further, X-ray scattering data[64] are consistent with a rodlike structure (15×150 Å) and are in agreement with a prolate elipsoid model for MBP.

3.2. Chemical Properties

The chemical properties of myelin lipophilin and MBP provide information bearing on possible interaction and organization of the myelin membrane. Lipophilin immobilizes fatty acid chains, especially boundary or neighboring lipids.[64] Because of its basic charge and the random distribution of 38 basic amino acid residues along the length of the polypeptide chain, MBP binds acidic lipids and thereby forms clusters conducive to myelin organization and stability. Indeed, binding to membrane lipids might result in exposure of determinants that are normally sequestered within the parent myelin basic protein.[65]

The relationship between the chemical properties of the two myelin components and their respective biological activity cannot be overemphasized. The EAE-inducing property of lipophilin has been suspected for some years[66]; however, documentation of this activity is a recent finding reported from our laboratory.[57] Details of the biological activity of this myelin surface-bound lipoprotein have not been matched by studies bearing on its disease-regulatory capacity. Although this work is at an intensive stage in its development, it is possible that this particular antigen, like MBP, is endowed with the ability not only to induce disease but also to regulate disease development. Thus, the nature of the "fail-safe" determinants to lipophilin remains undefined.

3.3. Myelin Basic Protein

Myelin basic protein is bound to the cytoplasmic side of the myelin membrane and is therefore normally inaccessible either as a receptor or as a focus for immune attacks directed against myelin. Given the conditions for its exposure and release during normal myelin turnover, myelin injury, or myelin pathology, the MBP not only becomes available for interaction with the immune system but also undergoes several molecular changes, outstanding among which is enzyme degradation from within and from without the brain (cathepsin D), giving rise to MBP fragments[67] capable of expressing one or more immunologic properties ascribed to the parent protein molecule.[68]

What is the fate of these fragments, their sequence boundaries, their respective biological activities, and how do they influence the immune response against normal myelin? Answers to some of these questions remain incomplete at the present time; however, a detailed study of the primary structure of the MBP and fragments derived therefrom has provided important information relevant to the biological activity of this molecule.

3.4. Chemical Properties

The human and the bovine BPs have a calculated molecular weight near 18,200[62,69]; the high content of arginine, lysine, and histidine renders a net basic charge on this molecule,[69] and the absence of sulfhydryl groups distinguishes the BP from globular proteins, suggesting that tertiary structure mediated by sulfur bridges is less important than the primary structure in defining the determinants responsible for its biological activities. These findings necessitated the elucidation of the complete amino acid sequence of this protein.[31,69]

3.5. Fragments from MBP

The intact BP induces the formation of humoral as well as cell-mediated immunity in experimental animals. The presence of a number of antigenic determinants located in various regions of the BP molecule (for review see ref. 68) suggests that the development of a particular immunologic response specific for a particular region of the MBP might follow the release of the intact BP or BP fragments from CNS myelin. The myelin BP is released into the CSF during episodes of acute exacerbation,[70] from where it might gain access into the circulation.

The presence of intact BP or fragments of BP[71] in either the blood or the CSF is expected to generate immune responses, especially when components of the CNS have the capacity to mount immune responses.[72,73] Comparative studies of CSF from multiple sclerosis patients and sheep challenged with EAE-inducing CNS tissue revealed the presence of BP in amounts ranging from 12 to 75 ng/ml CSF. The BP was found in 42 patients with active MS, whereas none was found in 30 patients with inactive MS and in only three of 600 patients with nondemyelinating neurological diseases.[74] Further, these investigators failed to demonstrate anti-basic protein antibodies in CSF of patients with active or inactive MS; however, free and bound antibodies were found in the CSF of all challenged sheep regardless of the severity of EAE.[74]

3.6. Myelin Basic Protein and Cerebrospinal Fluid

The absence of antibodies and the documented presence of the antigen in CSF fluids from MS patients may be explained by the episodic release of the antigen compared to the continued flux of antigen from endogenous as well as exogenous sources in EAE, thus permitting the development of antibodies to a rather weak immunogen.

Approximately 90% of MS patients studied displayed oligoclonal IgG bands when concentrated CSF was analyzed by agarose electrophoresis.[75–78] Between two and ten separate cathodic bands unique for each patient were detected.[79] These abnormal IgGs, which have been detected in CSF but not in sera from MS patients, were reported present in CSF from patients with other disease conditions.[80–83] The antigenic specificity of CSF oligoclonal IgG bands to myelin BP and their relationship to multiple sclerosis have not been documented with clarity. In some MS patients, the majority of these bands can be absorbed with measles antigens, whereas in others, the measles antigens were either partially or totally ineffective.[84,85] Indeed, in none of these studies was there a demonstration of oligoclonal IgG specificity to myelin BP.[86] It may be concluded that the humoral phase of the immune response to BP in MS patients has not been very informative, and the available data generated from MS research did not permit the drawing of definitive conclusions bearing on humoral immunity to BP and disease processes. These findings are not at variance with those found in EAE research. Although the development of humoral immunity can be documented in this acute form of disease (EAE), a relationship between antibody to BP and disease induction could not be established. In contrast to cell-transfer experiments in EAE, serum or antibody transfer from diseased animals failed to induce disease in normal recipients.

3.7. Myelin Basic Protein and Immune Complexes

Failure to demonstrate immune complexes in serum or CSF of MS patients is not a surprising finding. Indeed, the kinetics of BP release from myelin during the "active" phase and its absence in the "inactive" phase[77] suggest that the occasional release of minute amounts of antigen for possible complex formation might escape the sensitivity of the assay systems used. Similarly, immune complexes, especially precipitative complexes, are continually withdrawn from the circulation, reducing their titer below the sensitivity of current detection methods. Alternatively, the basically charged BP is known to complex with serum components[87] and with CNS tissue-bound IgG regardless of IgG specificity[88,89] and thus to be unavailable for detection. The presence of free BP in the CSF during the acute phase of MS is matched by the absence of anti-BP antibodies. In contrast, the virtual absence of free BP from CSF in EAE is matched by the presence of BP complexes as well as free antibodies. Thus, the kinetics of

BP release from endogenous sources, which appear to vary with the phase of the MS disease, influence the level of both free BP and BP complexes.

3.8. Myelin Basic Protein and Cell-Mediated Immunity

Cell-mediated immunity to myelin basic protein has been documented in both EAE and multiple sclerosis. The inherent pitfall of this observation is the assumption that the development of delayed-type hypersensitivity (DTH) invariably leads to disease. Indeed, this is not the case. As we delve into the molecular basis for antigens and antigenic determinants responsible for DTH and disease, we are inclined to conclude that the DTH may or may not lead to demyelination.[90] In fact, a particular type of DTH response caused by antigen-specific suppressor cells is known to immunoregulate disease development.[91,92]

Studies have shown that the encephalitogenic BP induces DTH responses in experimental animals[41] that correlate with the eventual development of disease. This generalization is true only if the intact BP[41] or encephalitogenic determinants derived from BP[90] are used. In contrast, the subsequent development of disease did not materialize following the development of DTH induced by nonencephalitogenic fragments derived from BP.[93] These studies[41,93] and others[94,95] demonstrate that modification of the disease-inducing sequence by enzymatic degradation of the intact BP,[95] by substitution of specific amino acids in the encephalitogenic sequence,[91,92,94] or by deletion of certain amino acids[90] renders the sequence nonencephalitogenic but maintains the necessary structure for recognition by the lymphoid system and leads to the development of delayed-type hypersensitivity in animals that remained free of clinical and pathological signs of disease.

The development of cell-mediated immunity to intact BP and to fragments from the BP has been documented by a number of assay procedures, although lymphoblastic transformation studies in MS gave conflicting results.[96–99] Modifications[100,101] of the macrophage migration-inhibition assay[102] did not prove to be specific for MS patients and fell short of demonstrating hypersensitivity in all patients studied.[103–105] By this assay six of 15 MS patients showed positive response to myelin BP,[106] whereas other studies demonstrated a considerable overlap when the magnitude of migration-inhibition responses was compared to those obtained for control patients.[107]

Recently, we described an assay for antigen-stimulated rosette-forming T cells (the AgARFC assay) in peripheral blood of guinea pigs challenged with myelin BP.[108] The kinetics of the AgARFC in sensitized animals correlated with delayed-type hypersensitivity reactions to myelin BP when the latter was measured by the skin test assay.[109] Subsequent studies with the AgARFC assay demonstrated[110] the presence of lymphocytes sensitized to myelin BP in the peripheral blood of MS patients. The AgARFC response was specific to myelin BP and was not demonstrated by other basically charged proteins such as histones. However, peripheral blood T-cell sensitization to myelin BP was demonstrated in patients with central nervous system tumors or cerebrovascular accidents, suggesting that other disorders of the CNS that destroy myelin might result in BP release leading to the development of cell-mediated immunity. Whether such BP-release episodes lead to the eventual development of CNS pathology even in MS patients remains undocumented.

4. PREVENTION, SUPPRESSION, AND TREATMENT

The relationship between EAE and MS has been documented, and interest in prevention, suppression, and treatment of EAE became the concern of both basic scientists and clinicians. The ease with which the disease may be produced in experimental animals together with the reported amino acid sequence of the BP[69] as well as the species-specific encephalitogenic determinants found in the parent protein (for review see refs. 68,111) provided the background information for *in vivo* studies of cellular mechanisms underlying disease induction. Certainly, knowledge of the amino acid sequences of the disease-inducing determinants permitted their respective synthesis to render them nonencephalitogenic.[94]

The results of these studies permitted *in vivo* experiments aimed at suppressing the development of cellular sensitivity in challenged animals in the hope of preventing development of BP-sensitized T lymphocyte subsets without the compromise of immune competence generally observed in other treatment modalities such as steroids, cytotoxic drugs, or antilymphocyte therapies. What are these antigens that have been used to prevent and suppress EAE development? How effective are they? And how can they be used for the treatment of multiple sclerosis in man?

4.1. The Use of MBP for Treatment

Several modalities have been described for prevention, suppression, and treatment of EAE. There is general agreement about the encephalitogenic property of the myelin BP, which can be demonstrated only when the MBP is administered in CFA but not in IFA. On the basis of these observations, several investigators concluded that the BP might be useful for prevention and treatment of both EAE and MS. Pretreatment of animals with large amounts of CNS tissue prevented subsequent development of disease when the pretreated animals were challenged with encephalitogenic emulsions.[112–114] Large amounts of the BP, injected before challenge, were effective in preventing onset of clinical signs of EAE.[115–117]

We have argued against these conclusions and wondered how an encephalitogenic protein can be used to prevent and to suppress a disease produced by the same encephalitogen. Obviously, the BP preparations used for treatment lacked the immunologic primer such as *M. butyricum* required for inducing disease. Also, larger amounts of BP were used and were administered at more frequent intervals. We challenged these studies as well as the conclusions drawn by demonstrating the development of EAE even when rabbits were treated with immunizing doses of myelin BP provided a small percent of the total BP dose was administered as an emulsion in CFA.[118] This was not a surprising finding, since it has been experienced in many laboratories[119,120]; however, the results suggested that the BP potentiated the disease when it was administered in IFA to immunologically primed animals. In suppressing the development of clinical signs of EAE, the BP was not totally effective, and the results varied with experimental designs. In addition, the BP did not prevent the development of CNS lesions characteristic of EAE when it was given before[116,117] or after challenge.[121]

We reexamined the effectiveness of the intact BP in altering the course of clinical and histological signs of EAE when various dosages of the homologous BP were administered before challenge with encephalitogenic emulsions. A BP treatment course of 15 daily injections failed to prevent the development of disease when the pretreated animals were challenged with encephalitogenic emulsions shortly or 30 days after the end of the treatment.[122] Similarly, BP treatments (15 daily doses) initiated on day 1, 4, 8, or 11 after the challenge did not prevent development of disease irrespective of the dose. Treatment with BP delayed disease onset, reduced the severity of both clinical and histological signs of disease, and perhaps was responsible for eventual reversal of clinical signs of disease in some of the treated animals.[122] The results of this study further showed that the BP was responsible for prolonged periods of paralysis, suggesting a milder form of disease. It is possible that the prolonged periods of paralysis were caused by a BP-induced reduction in the number of EAE cells, leading to eventual recovery of some of the treated animals. This conclusion is supported by the finding of reduced prevalence of histological lesions in the CNS of treated and recovered animals.

4.2. Myelin Basic Protein and Clinical Trials

Failure of the intact myelin BP to arrest multiple sclerosis in man[123,124] and to prevent and suppress development of EAE in experimental animals[122] is an expected finding in view of the well-documented ability of the BP to induce demyelinating disease regardless of the species from which it was derived. Also, careful screening of published work did not reveal suppressor cell function that may be attributed to BP regardless of the method by which it was administered. The administration of BP is known to induce both humoral and cell-mediated immune responses. The anti-basic-protein antibodies were associated neither with blocking the development of cell-mediated immunity nor with the onset of clinical or histological signs of EAE.[125] The administration of BP emulsified in adjuvants induced the formation of cell-mediated immunity, and the incorporation of *Mycobacterium* into the BP–adjuvant emulsion potentiated the immune response and led to the development of CNS pathology characteristic of EAE. Lymphocytes from animals sensitized to BP (more than 10^9 lymphocytes) are capable of inducing EAE when transferred to syngeneic normal and unmodified recipients.[17,126] It is interesting to note that recipient animals develop clinical and histological signs of EAE regardless of whether these animals were challenged with encephalitogenic emulsions of the BP.

4.3. Myelin Basic Protein Determinants

Failure to prevent, suppress, or arrest demyelination was ascribed to the intact MBP molecule but not to submolecular regions of this protein. This conclusion is supported by a massive amount of experimental evidence. The ability of the BP to induce disease in a particular species is the property of specific regions made up of eight to ten amino acids.[111] The rest of the molecule does not appear to be contributory, since the small and defined portions induce a full-blown disease similar if not identical to that induced by the entire molecule.

Similarly, the MBP does not contain readily recognizable sequences that activate suppressor cell function capable of regulating disease development.[91] Lymphocytes from donors immunized with MBP plus CFA transferred disease, and those from donors immunized with MBP plus IFA did not transfer protection to normal recipients. These results do not suggest that EAE-regulating sequences do not exist in the parent BP molecule. These determinants are sequestered within the BP molecule and are capable of inducing suppressor cell function only after the parent protein has been fragmented in such a way as to delete one or more amino acids from the region responsible for disease.[91,92] To illustrate this phenomenon, let us consider the Lewis rat model and the antigens responsible for both disease and disease regulation.

4.4. The Lewis Rat EAE Model

The sequence that induces disease in the Lewis rat has been studied in detail. Casual inspection of the sequence does not reveal any special features except the noted absence of both tyrosine and tryptophan. However, the sequence of this determinant undergoes a number of phylogenetic variations, none of which is known to interfere with the ability of this sequence to induce disease:

	75 76 77 78 79 80 81 82 83 84	
Bovine:	H–Ala–Gln–Gly–His–Arg–Pro–Gln–Asp–Glu–Asn–OH	......+EAE
Rat:	H–Ser–Gln–(–)–(–)–Arg–Thr–Gln–Asp–Glu–Asn–OH	......+EAE
G. pig:	H–Ser–Gln–(–)–(–)–Arg–Thr–Gln–Asp–Glu–Asn–OH	......+EAE
Human:	H–Ser–His–Gly–His–Arg–Thr–Gln–Asp–Glu–Asn–OH	......pending

The presence or absence of Gly-His as in the sequences from bovine and guinea pig MBP (or rat), respectively, does not influence the encephalitogenic potency, and neither does a serine substituted for Ala or Pro at either or both positions 75 and 80.

Bovine:	H–Ala–Gln–Gly–His–Arg–Pro–Gln–Asp–Glu–Asn–OH	......+EAE
Bovine:	H–Ala–Gln–(–)–(–)–Arg–Pro–Gln–Asp–Glu–Asn–OH	......+EAE
G. pig:	H–Ser–Gln–(–)–(–)–Arg–Ser–Gln–Asp–Glu–Asn–OH	......+EAE
G. pig:	H–Ser–Gln–Gly–His–Arg–Ser–Gln–Asp–Glu–Asn–OH	......+EAE

Further studies of this determinant for disease in the Lewis rat revealed that the Gly-His sequence, which is a rare occurrence in the BP molecule, influences recognition of this determinant by the lymphoid system. Its presence in the intact BP renders the bovine BP virtually nonencephalitogenic, whereas its absence from the guinea pig sequence renders the guinea pig MBP a potent encephalitogen. In an effort to determine the causal relationship for the differences in the potency of the two proteins, peptide sequences covering this particular region of the MBP were synthesized in our laboratory[127,128]:

69 70 71 72 73 74 75 76 77 78 79 80 81 82 83 84
H–Gly–Ser–Leu–Pro–Gln–Lys–Ala–Gln–Gly–His–Arg–Pro–Gln–Asp–Glu–Asn–OH
H–Gly–Ser–Leu–Pro–Gln–Lys–Ala–Gln–(–)–(–)–Arg–Pro–Gln–Asp–Glu–Asn–OH
H–Pro–Gln–Lys–Ala–Gln–Gly–His–Arg–Pro–Gln–Asp–Glu–Asn–OH
H–Ala–Gln–Gly–His–Arg–Pro–Gln–Asp–Glu–Asn–OH
H–Ala–Gln–(–)–(–)–Arg–Pro–Gln–Asp–Glu–Asn–OH
H–Ala–Gln–Gly–His–Arg–Pro–Gln–Asp–Glu–Gly–OH

The sequence of the N-terminal six-amino-acid residues, H-Gly-Ser-Leu-Pro-Gln-Lys, is native to both the bovine and guinea pig BP. The presence of these six amino acids at the N-terminal end of the disease-inducing sequence prevented cellular recognition of this determinant and rendered the sequence nonencephalitogenic; however, deletion of Gly-His (residues 77 and 78) restored the disease-inducing potency of this determinant. Clearly, a high degree of conformational change occurs in this region of the BP molecule, which may be detected when this region is reproduced in the form of synthetic peptide. Thus, the absence of Gly-His releases the conformational restrictions imposed on the disease-inducing determinant and renders it recognizable by the lymphoid system and capable of inducing disease. Alternatively, sequestration of this determinant may be abolished by deletion of up to six amino acid residues from the N-terminal end of the 69–84 sequence. The sequence of the C-terminal ten amino acids (residues 75–84) is a potent encephalitogen in the presence or in the absence of Gly-His. In contrast, deletion or substitution of the C-terminal residue (Asn) destroyed the disease-inducing property of this determinant.[129]

4.5. Induction and Regulation

We have shown that the encephalitogenic activity and potency of the disease-inducing determinants are related to amino acid sequences of specific regions of the parent BP molecule. Recognition of these determinants and their ability to activate a specific T lymphocyte subset responsible for disease is contingent on the presence of the complete amino acid sequence of the disease-inducing region for a particular species. Thus, the most interesting aspect of BP conformation is its role in making available or unavailable biologically active regions of the MBP molecule. We have shown that the determinant for Lewis rats is sequestered in the MBP, and its ability to induce disease is dependent on fragmentation of the parent molecule. The BP fragmentation must proceed in such a manner that the end result is the presentation and recognition of the complete amino acid sequence responsible for disease. Should fragmentation be limited to specific sites along the polypeptide chain, the encephalitogenic sequence might still be intact but sequestered within the fragment and unavailable to activate the effector phase of the immune response. A clear example of this process is the fragmentation of BP, and, depending on the enzyme used, the resulting fragments are either disease inducing[130] or disease preventing[131] when they are tested in respondent animals. Fragmentation of BP might occur at a particular linkage in the region responsible for disease; the resulting fragment, albeit incapable of inducing disease, was capable of preventing disease when it was administered to animals before challenge.[131]

Just how the nonencephalitogenic sequence regulates the immune response to encephalitogenic antigens was demonstrated in the Lewis rat model. The native disease-inducing determinant for Lewis rats was modified by a Gly substituted for Asn at the C-terminal end[92]:

S6:	H–Ala–Gln–Gly–His–Arg–Pro–Gln–Asp–Glu–Asn–OH....	+EAE
S79:	H–Ala–Gln–Gly–His–Arg–Pro–Glu–Asp–Glu–Gly–OH....	–EAE

Modification of a single amino acid residue at the C-terminal end of this sequence destroyed the ability of this determinant to induce disease even when peptide S79 was administered at high doses. In contrast, Lewis rats pretreated with a single dose of 12.5 μg S79 were protected from disease. Increasing the pretreatment dose to 100 μg S79 suppressed the development of both clinical and histological signs of EAE. Peptide S79-induced protection from encephalitogenic challenge was transferred to naive recipients with donor lymph node or splenic lymphocytes. Again, successful transfer of unresponsiveness to encephalitogenic challenge was aborted by pretreatment with a single dose of cyclophosphamide.[92]

The results of these studies show that the disease-inducing determinant has a number of immunologic functions that may be related not only to the complete amino acid sequence but also to portions of the sequence responsible for disease. Should the parent protein be presented to the immune system, the complete sequence of the disease-inducing region is recognized by the effector phase of the immune response. Indeed, recognition of a portion of the disease-inducing sequence does not occur when either the intact basic protein or the isolated encephalitogenic region is administered to test animals. This conclusion is supported by the fact that the BP is incapable of preventing or suppressing disease development[122] or activating suppressor cells that can protect animals from EAE. Thus, the disease-inducing sequence is a powerful and dominant antigen and is preferentially recognized at the cellular level.

Our studies show that a second antigenic activity may be expressed by the disease-inducing sequence. Such activity may be demonstrated only by a partial but not the complete sequence of the disease-inducing determinant. The partial or modified sequence activates suppressor cell function capable of regulating the immune response to an encephalitogenic challenge. It is clear that the determinant for suppressor cell function is built into the complete sequence required for disease; EAE is not expected to develop should the generated fragments carry a portion of the disease-inducing sequence. The same fragment was capable of conferring protection on the pretreated animals.

5. SUPPRESSOR LYMPHOCYTES IN EAE AND MS

Regulation of the immune response to encephalitogenic antigen by suppressor T lymphocytes in EAE[91,92] parallels the reported fluctuating levels of suppressor cell activity in multiple sclerosis patients.[132] Changes in suppressor cell activity in multiple sclerosis correlated with disease status. The activity was lower in patients with clinically active disease than in controls and patients with clinically stable disease.[132] The fluctuating levels of suppressor cell activity in multiple sclerosis patients were informative in terms of the presence or absence of disease-inducing lymphocytes and in terms of antigenic specificity of the suppressor cell subsets; the latter was documented by concanavalin A (Con A) activation *in vitro*[132] in the absence of neural antigens. Although longitudinal studies of the kinetics of the Con A-activated suppressor cells in multiple sclerosis patients are not available, the patients invariably go into relapses. These results suggest the development of an imbalance between suppressor cell and effector cell activity favoring disease development.

This suggestion gains further support from the results of a longitudinal study of MS patients and from animals that have recovered from EAE. In acute EAE, challenged rats that have survived the first acute EAE episode recovered completely from clinical signs of EAE. The recovery phase correlated with the development of high levels of suppressor cells as shown by their ability to adaptively transfer unresponsiveness to an encephalitogenic challenge[133]; however, those rats that were found to be unresponsive to a second challenge[133,134] harbored EAE-inducing or T effector lymphocytes that respond to BP *in vitro*[134–136] and induce EAE when they are transferred to naive syngeneic recipients.[137,138] The increased level of suppressor cells in rats that have recovered from EAE parallels similar increases of suppressor cell activity in multiple sclerosis patients who have recovered from an exacerbating episode.[139] Thus, in EAE as in multiple sclerosis, the development of suppressor cells, even in the recovery (EAE) or in the remitting phases (MS), does not suggest clonal deletion of T effector lymphocytes; rather, the suppressor cell activity delayed disease onset by controlling the effector, the proliferative, or both phases of the immune response to neural antigens.

6. CONCLUDING REMARKS

With respect to the antigens responsible for modulating the immune response, the overwhelming evidence demonstrates that EAE induction, like EAE regulation, is the product of the immune response to encephalitogenic and nonencephalitogenic regions of the myelin basic protein. Studies of suppressor cell activity in multiple sclerosis patients fell short of demonstrating antigenic specificity; however, the BP that activates the effector phase of the immune response in EAE was shown to detect MBP-specific rosette-forming T cells after *in vitro* exposure of peripheral blood lymphocytes from multiple sclerosis patients to human myelin basic protein.[140] The presence of varying numbers of circulating BP-sensitive lymphocytes in multiple sclerosis was documented in patients whose disease condition was stationary for months and years; however, the level of BP-sensitive lymphocytes varied with the disease status of individual patients.[141] In this longitudinal study of seven patients,[141] the rise of BP-sensitive T lymphocytes correlated with the development of neurological signs of disease. These results suggest that cellular sensitivity to BP, which induces neurological disease in animals and in man, may be responsible for demyelination in multiple sclerosis.

ACKNOWLEDGMENTS. This work was supported by grants from the National Institutes of Health, The National Multiple Sclerosis Society, The M. T. Biddle Foundation, and The Clark Foundation.

REFERENCES

1. Dawson, J. W., 1916, *Trans. R. Soc. Edinburgh.* **1**:517–740.
2. Putnam, T. J., 1937, *Arch. Neurol. Psychiatry* **37**:1298–1321.
3. Dow, R. S., and Berglund, G., 1942, *Arch. Neurol. Psychiatry* **47**:1–18.
4. Allison, R. S., 1961, *Proc. R. Soc. Med.* **54**:1–4.
5. Pratt, R. T. C., Compston, N. D., and McAlpine, D., 1951, *Brain* **74**:191–232.
6. McAlpine, D., Lumsden, C. E., and Acheson, E. D., 1959, *Multiple Sclerosis, A Reappraisal*, E. and S. Livingstone, London.
7. Swank, R. L., 1950, *Am. J. Med. Sci.* **220**:420–435.
8. Margulis, M. S., Soloviev, V. D., and Shubladze, A. K., 1946, *J. Neurol. Neurosurg. Psychiatry* **9**:63–74.
9. McAlpine, D., and Compston, N., 1952, *Q. J. Med.* **21**:135–167.
10. Miller, H., Cendrowski, W., and Schapira, K., 1967, *Br. Med. J.* **2**:210–213.
11. Poskanzer, D. C., Schapira, K., and Miller, H., 1963, *Lancet* **2**:917–921.
12. Field, E. J., 1966, *Br. Med. J.* **2**:564–565.
13. Miller, H., and Schapira, K., 1959, *Br. Med. J.* **1**:737–740,811–815.
14. Lumsden, C. E., 1961, *Proc. R. Soc. Med.* **54**:11–15.
15. Alvord, E. C., Jr., 1966, *J. Neuropathol. Exp. Neurol.* **25**:1–17.
16. Paterson, P. Y., 1966, *Adv. Immunol.* **5**:131–208.
17. Paterson, P. Y., 1960, *J. Exp. Med.* **111**:119–136.
18. Levine, S., Hoenig, E. M., and Wenk, E. J., 1967, *Proc. Soc. Exp. Biol. Med.* **126**:454–458.
19. Roboz-Einstein, E., and Henderson, N., 1959, *Allergic Encephalomyelitis* (M. W. Kies and E. C. Alvord, Jr., eds.), Charles C Thomas, Springfield, Illinois, pp. 281–292.
20. Lumsden, C. E., 1949, *Brain* **72**:517–537.
21. Eylar, E. H., Salk, J., Beveridge, G. C., and Brown, L. V., 1969, *Arch. Biochem. Biophys.* **132**:34–48.
22. Alvord, E. C., Jr., 1968, *The Central Nervous System* (O. T. Baily, and D. E. Smith, eds.), William and Wilkins, Baltimore, pp. 52–70.
23. Shaw, C. M., Alvord, E. C. Fahlberg, W. J., and Kies, M. W., 1962, *J. Immunol.* **89**:54–61.
24. Paterson, P. Y., 1976, *Textbook of Immunopathology* (P. A. Miescher, and H. J. Miller-Eberhard, eds.), Grune and Stratton, New York, p. 179.
25. Adams, R. D., 1959, *Allergic Encephalomyelitis* (M. W. Kies and E. C. Alvord, eds.), Charles C Thomas, Springfield, Illinois, pp. 183–209.
26. Shiraki, H., and Otani, S., 1959, *Allergic Encephalomyelitis* (M. W. Kies and E. C. Alvord, Jr., eds.), Charles C Thomas, Springfield, Illinois, pp. 281–292.
27. Drachman, D. A., Paterson, P. Y., and Bornstein, M. B., 1974, *Neurology (Minneap.)* **24**:364.
28. Kies, M. W., and Alvord, E. C., Jr. (eds), 1959, *Allergic Encephalomyelitis*, Charles C Thomas, Springfield, Illinois, pp. 293–299.
29. Waksman, B. H., 1959, *Int. Arch. Allergy Suppl.* **14**:1–87.

30. Lumsden, C. E., 1966, *Excerpta Medica Int. Cong. Ser.* **100:**231–239.
31. Eylar, E. H., and Hashim, G. A., 1969, *Transactions of the Second International Meeting of the International Society for Neurochemistry*, (R. Paoletti, R. Fumagalli and C. Galli, eds.), Tamburini, Milan, Italy, Sept. 1–5.
32. Levine, S., Prineas, J., and Scheinberg, L. C., 1969, *Proc. Soc. Exp. Biol. Med.* **131:**986–990.
33. Alvord, E. C., Jr., 1970, *Handbook of Clinical Neurology* Volume 9, (P. J. Vinken and G. W. Bruyn, eds.), North Holland, Amsterdam, pp. 500–571.
34. Wisniewski, H. M., and Keith, A. B., 1977, *Ann. Neurol.* **1:**144–148.
35. Kibler, R. F., Fox, R. H., and Shapira, R., 1964, *Nature* **204:**1273–1275.
36. Caspary, E. A., and Field, E. J., 1963, *Nature* **197:**1218.
37. Wolfgram, F., 1965, *Ann. N.Y. Acad. Sci.* **122:**104–115.
38. Martenson, R. E., and Lebaron, F. N., 1966, *J. Neurochem.* **13:**1469–1479.
39. Nakao, A., Davis, W. J., and Roboz-Einstein, E., 1966, *Biochim. Biophys. Acta.* **130:**163–170.
40. Carnegie, P. R., Bencina, B., and Lamoureux, G., 1967, *Biochem. J.* **105:**559–568.
41. Shaw, C. M., Alvord, E. C., Jr., Kaku, J., and Kies, M. W., 1965, *Ann. N.Y. Acad. Sci.* **122:**318–331.
42. Levine, S., and Wenk, E. J., 1964, *Science* **146:**1681–1682.
43. Levine, S., 1974, *Acta Neuropathol.* **28:**179–189.
44. Alvord, E. C., Jr., 1970, *Handbook of Clinical Neurology*, Volume 9 (P. J. Vinken and G. W. Bruyn, eds.), North-Holland, Amsterdam, pp. 500–571.
45. Raine, C. S., and Stone, S. H., 1977, *N.Y. State J. Med.* 77:1693–1696.
46. Raine, C. S., 1976, *Progress in Neuropathology* Volume 3, (H. M. Zimmerman, ed.), Grune and Stratton, New York, p. 225.
47. Stone, S. H., and Lerner, E. M., II, 1965, *Ann. N.Y. Acad. Sci.* **122:**227–241.
48. Raine, C. S., Snyder, D. H., Valsamis, M. P., and Stone, S. H., 1974, *Lab Invest.* **31:**369–380.
49. Snyder, D. H., Valsamis, M. P., Stone, S. H., and Raine, C. S., 1975, *J. Neuropathol. Exp. Neurol.* **34:**209–221.
50. Ghatak, H., Hirano, A., Doron, Y., and Zimmerman, H. M., 1973, *Arch. Neurol.* **29:**262–267.
51. Hoenig, E. M., and Levine, S., 1974, *Neuropathol. Exp. Neurol.* **23:**251–259.
52. Clark, G., and Bogdanove, L. H., 1955, *J. Neuropathol. Exp. Neurol.* **14:**433–437.
53. Levine, S., 1960, *J. Neuropathol. Exp. Neurol.* **19:**238–247.
54. Levine, S., and Hoenig, E. M., 1968, *J. Immunol.* **100:**1310–1318.
55. Adams, C. W. M., 1975, *J. Neurol. Sci.* **25:**165–182.
56. Folch, J., and Lees, M., 1951, *J. Biol. Chem.* **191:**807–817.
57. Hashim, G. A., Wood, D. D., and Moscarello, M. A., 1980, *Neurochem. Res.* **5:**1137–1145.
58. Madrid, R. E., Wisniewski, H. M., Hashim, G. A., Fredane, L., Wood, D. D., and Moscarello, M., 1982, *Trans. Am. Soc. Neurochem.* **13:**211.
59. Stoffyn, P., and Folch-Pi, J., 1971, *Biochem. Biophys. Res. Commun.* **44:**157–161.
60. Moscarello, M. A., Anthony, J., and Epand, R. M., 1973, *Biochemistry* **12:**3402–3406.
61. Boggs, J. M., and Moscarello, M. A., 1978, *Biochim. Biophys. Acta* **515:**1–21.
62. Eylar, E. H., and Thompson, M., 1969, *Arch. Biochem. Biophys.* **129:**468–479.
63. Chao, L. P., and Einstein, E. R., 1970, *J. Neurochem.* **17:**1121–1132.
64. Boggs, J. M., Vail, W. J., and Moscarello, M. A., 1976, *Biochim. Biophys. Acta.* **448:**517–530.
65. Boggs, J. M., Samji, N., Moscarello, M. A., Hashim, G. A., and Day, E. D., 1983, *J. Immunol.* **130:**1687–1694.
66. Waksman, B. H., Porter, H., Lees, M. B., Adams, R. D., and Folchi-Pi, J., 1954, *J. Exp. Med.* **100:**451–471.
67. Benuck, M., Marks, N., and Hashim, G. A., 1975, *Eur. J. Biochem.* **52:**615–621.
68. Hashim, G. A., 1978, *Immunol. Rev.* **39:**61–107.
69. Eylar, E. H., Borstoff, S. W., Hashim, G. A., Caccam, J., and Burnett, P., 1971, *J. Biol. Chem.* **246:**5770–5784.

70. Cohen, S. R., Herndon, R. M., and Mekhann, G. M., 1976, *N. Engl. J. Med.* **295**:1455–1457.
71. Whitaker, J. N., 1977, *Neurology (Minneap.)* **27**:911–920.
72. Tourtellot, W. W., 1972, *Immunology, Virology and Ultrastructure in Multiple Sclerosis*, (F. Wolfgram, Ellison, G., J. G. Stevens and J. M. Andrews, eds.), Academic Press, New York, pp. 285–333.
73. Cutler, R. W. P., Merler, E., and Hammerstad, J. P., 1968, *Neurology* **18**(2):129–132.
74. Gutstein, H. S., and Cohen, S. R., 1978, *Science* **199**:301–302.
75. Lowenthal, A., Van Sande, M., and Karcher, D., 1960, *J. Neurochem.* **6**:51–56.
76. Olsson, J. E., and Petterson, B., 1976, *Acta Neurol. Scand.* **53**:308–322.
77. Thompson, E. J., 1977, *Br. Med. Bull.* **33**:28–33.
78. Vandvik, B., and Skrede, S., 1973, *Eur. Neurol.* **9**:224–241.
79. Olsson, J. E., and Leik, H., 1973, *Arch. Neurol.* **28**:392–399.
80. Laterre, E. C., Callewaert, A., Heremans, J. F., and Staello, Z., 1970, *Neurology (Minneap.)* **20**:982–990.
81. Link, H., 1975, *Acta Neurol. Scand.* **52**:111–120.
82. Vindvik, B., Norrby, E., and Nordal, H. J., 1979, *Acta Neurol. Scand.* **60**:204–213.
83. Wolinsky, J. S., Berg, B. O., and Maitland, C. J., 1976, *Arch. Neurol.* **33**:722–723.
84. Norrby, E., and Vindvik, B., 1975, *Med. Microbiol. Immunol. (Berl.)* **162**:63–72.
85. Vindvik, B., Norrby, E., Nordal, H. J., and Degré, M., 1976, *Scand. J. Immunol.* **5**:979–992.
86. Johnson, K. P., and Nelson, B. J., 1977, *Ann. Neurol.* **2**:425–431.
87. McPherson, T. A., Marchalonis, J. J., and Lennon, V., 1970, *Immunology* **19**:929–933.
88. Sindic, C. J., Cambiaso, C. L., Masson, P. L., and Laterre, E. C., 1980, *Clin. Exp. Immunol.* **41**:1–7.
89. Sindic, C. J., Cambiaso, C. L., Masson, P. L., and Laterre, E. C., 1980, *Clin. Exp. Immunol.* **41**:8–12.
90. Hashim, G. A., and Sharpe, R. D., 1974, *Immunochemistry.* **11**:633–640.
91. Hashim, G. A., 1981, *J. Immunol.* **126**:419–423.
92. Kardys, E., and Hashim, G. A., 1981, *J. Immunol.* **127**:862–866.
93. Hashim, G. A., Hwang, F., and Schilling, F. J., 1973, *Arch. Biochem. Biophys.* **156**:298–309.
94. Hashim, G. A., and Sharpe, R. D., 1975, *Proc. Soc. Exp. Biol. Med.* **149**:646–651.
95. Hashim, G. A., and Schilling, F. J., 1973, *Biochem. Biophys. Res. Commun.* **50**:589–596.
96. Brody, J. A., Harlem, M. M., Kurtzke, J. F., and White, L. R., 1968, *N. Engl. J. Med.* **279**:202–204.
97. Hughes, D., Caspary, E. A., and Field, E. J., 1968, *Lancet* **2**:1205–1207.
98. Rocklin, R. E., Reardon, G., Sheffer, A., Churchill, W. H., and David, J. R., 1970, *Proceedings of the 5th Leukocyte Culture Conference* (J. E. Harris, ed.), Academic Press, New York, pp. 639–648.
99. Dau, P. C., and Peterson, R. D. A., 1970, *Arch. Neurol.* **23**:32–40.
100. Thor, D. E., Jureziz, R. E., Veach, S. R., Miller, E., and Dray, S., 1968, *Nature*, **219**:755–757.
101. Rocklin, R. E., Meyers, O. L., and David, J. R., 1970, *J. Immunol.* **104**:95–102.
102. Hashim, N., and Barr, M. L., 1963, *Lancet* **2**:1029–1030.
103. David, J. R., and Paterson, P. Y., 1965, *J. Exp. Med.* **122**:1161–1171.
104. David, J. R., al-Askali, S., Lawrence, H. S., and Thomas, L., 1964, *J. Immunol.* **93**:264–273.
105. David, J. R., and Schlossman, S. F., 1968, *J. Exp. Med.* **128**:1451–1459.
106. Rocklin, R. E., Shermata, W. A., Feldman, R. G., Kies, M. W., and David, J. R., 1971, *N. Engl. J. Med.* **284**:803–808.
107. Sheremata, W., Cosgrove, J. B. R., and Eylar, E. H., 1974, *N. Engl. J. Med.* **291**:14–17.
108. Hashim, G. A., Lee, D. H., and Pierce, J. C., 1977, *Neurochem. Res.* **2**:99–109.
109. Lee, D. H., Hashim, G. A., and Pierce, J. C., 1977, *Neurochem. Res.* **2**:293–302.
110. Hashim, G. A., Lee, D. H., Pierce, J. C., Braun, C. W., and Fitzpatrick, H. F., 1977, *Neurochem. Res.* **3**:37–48.
111. Hashim, G. A., 1980, *Basic Protein From Central Nervous System Myelin in Myelin: Chemistry and Biology* (G. A. Hashim, ed.), Alan R. Liss, New York, pp. 79–122.

112. Kies, M. W., and Alvord, E. C., Jr., 1958, *Nature* **183:**1106.
113. Shaw, C. M., Fahlberg, W. J., Kies, M. W., and Alvord, E. C., Jr., 1960, *J. Neuropathol. Exp. Neurol.* **19:**166–168.
114. Field, E. J., and Caspary, E. A., 1964, *Nature* **201:**936.
115. Alvord, E. C., Jr. Shaw, C. M., Hruby, S., and Kies, M. W., 1965, *Ann. N.Y. Acad. Sci.* **122:**333–345.
116. Roboz-Einstein, E., Cejtey, J., Davis, W. J., and Rauch, H. C., 1968, *Immunochemistry* **5:**567–575.
117. Rauch, H. C., and Roboz-Einstein, E., 1974, *J. Neurol. Sci.* **23:**99–116.
118. Hashim, G. A., Sharpe, R. D., Carvalho, E., and Stevens, L. E., 1975, *Proc. Soc. Exp. Biol. Med.* **149:**646–651.
119. Lennon, V. A., Whittingham, S., Carnegie, P. R., McPherson, T. A., and Mackay, I. R., 1971, *J. Immunol.* **107:**56–62.
120. Kibler, R. F., and Barnes, A. E., 1962, *J. Exp. Med.* **116:**807–825.
121. Hashim, G. A., Sharpe, R. D., Carvalho, E. F., and Stevens, L. E., 1976, *J. Immunol.* **116:**126–130.
122. Hashim, G. A., 1980, *Neurochem. Res.* **5:**101–112.
123. Gonsette, R. E., Delmotti, P., and Demorty, L., 1977, *J. Neurol.* **216:**27–31.
124. Campbell, B., Vogel, P. L. J., Fisher, E., and Lorenz, R., 1973, *Arch. Neurol.* **29:**10–15.
125. Lizak, R. P., Falk, G. A., Heinzi, R. G., Kies, M. W., and Alvord, E. C., Jr., 1970, *J. Immunol.* **104:**1435–1446.
126. Levine, S., and Hoenig, E. M., 1968, *Science* **161:**1155–1157.
127. Hashim, G. A., Sharpe, R. D., and Carvalho, E. F., 1979, *J. Neurochem.* **32:**73–77.
128. Hashim, G. A., Carvalho, E. F., Sharpe, R. D., 1978, *J. Immunol.* **121:**665–670.
129. Kardys, E., and Hashim, G. A., 1981, *J. Immunol.* **127:**862–866.
130. Hashim, G. A., and Eylar, E. H., 1969, *Arch. Biochem. Biophys.* **129:**635–644.
131. Hashim, G. A., and Schilling, F. J., 1973, *Arch. Biochem. Biophys.* **156:**287–297.
132. Antel, J. P., Arnason, B. G. W., and Medof, M. E., 1979, *Ann. Neurol.* **5:**338–342.
133. Welch, A. M., Holda, J. H., and Swanborg, R. H., 1980, *J. Immunol.* **125:**186–189.
134. Willenborg, D. O., 1979, *J. Immunol.* **123:**1145–1150.
135. Holda, J. H., Welch, A. M., and Swanborg, R. H., 1980, *Eur. J. Immunol.* **10:**657–659.
136. Waxman, F. J., Fritz, R. B., and Hinrichs, D. J., 1980, *Cell. Immunol.* **49:**34–42.
137. Holda, J. H., and Swanborg, R. H., 1981, *Eur. J. Immunol.* **11:**338–340.
138. Holda, J. H., and Swanborg, R. H., 1982, *Eur. J. Immunol.* **12:**453–455.
139. Antel, J. P., Weinrich, M., and Arnason, B. G. W., 1978, *Neurology (Minneap.)* **28:**999.
140. Hashim, G. A., Leo, D. H., Pierce, J. C., Braun, C. W., and Fitzpatrick, H. F., 1978, *Neurochem. Res.* **3:**37–48.
141. Hashim, G. A., and Brewen, M., 1985, *J. Neurosci. Res.* **13.**

12

The Biology of Huntington's Chorea

Ara Tourian

1. INTRODUCTION

This chapter is an overview of the analysis provided by the diverse disciplines concerned with the biology of Huntington's chorea, which is documented in a vast literature. It is not a comprehensive treatment, because excellent reviews (some of which are cited) cover many of the individual areas. This review has two goals: (1) to bring into sharp focus those observations that have endured the test of time and shed light on the more general features of the correlations among the clinical, neuropathological, and neurochemical–neuropathological observations of Huntington's chorea, and (2) to call attention to a newer approach—that of cell biology and nutrition of peripheral tissues such as skin fibroblasts—to the analysis of dominantly inherited brain disease. The insights generated by such a strategy regarding potential metabolic pathway abnormalities should be directly testable in HC brain. Ultimately, this type of approach may help solve the fundamental biochemical genetic problem of HC and serve as a model for unraveling the primary metabolic abnormalities of other dominantly inherited diseases.

Huntington's chorea (HC) is a devastating hereditary disorder of the central nervous system characterized by the appearance of progressive mental, motor, and emotional deterioration. The clinical onset of dysfunction is usually in adult life; however, childhood and juvenile varieties within the same family underscore the importance of the interaction of biochemical environment with the mutant gene in a given individual. The disease is inherited in an autosomal dominant fashion with an estimated gene frequency of 10^{-4} and has a worldwide distribution.[1] The childhood variety is characterized by deterioration of developmental, cognitive, and motor abilities ataxia and by a higher incidence of seizures. Whereas juvenile HC typically results in parkinsonianlike rigid and akinetic (nonchoreic) body tone, the adult-onset variety, which comprises more

Ara Tourian • Neurogenetics and Cell Biology Laboratory, Department of Medicine, Division of Neurology, Duke University Medical Center, Durham, North Carolina 27710.

than 80% of the clinical cases, may present and persist predominantly in mental, motor, or emotional dysfunction but more commonly in combinations of two or all three together with varying degrees of severity of each of the behavioral abnormalities.[2]

The clinical subtypes generated by an identical mutation emphasize the localization of the predominant neuronal net dysfunction and brain pathology that may be present in a given affected individual in this disease. Thus, the pathophysiology of rigidity in HC points to basal ganglia abnormalities but in an opposite direction with respect to neurotransmitter receptor mechanisms. Rigidity results from impaired dopamine–receptor interactions, whereas chorea results from overactivity as a result of dopamine excess.[3] Similarly, memory impairment and cognitive dysfunction imply temporal lobe, limbic, and cortical mantle dysfunction, in contrast to emotional deterioration and loss of drive, which localize to hypothalamic and limbic structures. The idea that the hypothalamic and pituitary regions are involved in HC is further supported by the hormonal abnormalities to be described.

It is clear that the primary target of dysfunction in this mutation is the brain and, more specifically, neuronal net function. However, this clinical and pathological localization of neuronal dysfunction in HC does not argue for exclusive involvement of neuronal cells. Precedents from the literature may illustrate this point. Acute intermittent porphyria is an autosomal dominant disease with clinical manifestations of brain and autonomic nervous system dysfunction as a result of a mutation of uroporphyrinogen synthetase I in the liver.[4] Both gray and white matter may be the targets of damage during an acute episode of acute intermittent porphyria.

The lack of accumulated product in neural or glial cells in HC brain suggests a synthetic pathway abnormality. Alternatively, the delayed clinical expression of HC may be caused by a defective temporal gene and may be analogous but not identical to the phenomenon of cellular aging, whereby temporal genes switch on the destruction of a cell after a finite life-span. Thus, in HC a normal time-dependent expression of phenotype may be altered by mutation to premature cell death.[5] This concept has also been formulated in terms of programmed cell death of the brain.

In the search for understanding of the varied clinical and biological manifestations of HC, there is the possibility of discovering a link between temporal genes, emotions, and the intellect and neural network functional organization and brain cell nutritional malfunction because of mutation.

2. *NEUROPATHOLOGY*

In HC, there is diffuse involvement of both neuronal and glial cell populations in the cortex, thalamus, hypothalamus, cerebellum, brainstem, and spinal cord, resulting in symmetrical atrophy of all gray and white matter. Certain selected areas suffer greater cell loss than others. The volume loss in the cortex, neostriatum, and globus pallidum has been recorded as 20, 60, and 50%, respectively. The diffuse loss of myelin, notably in the corpus callosum, the occipital lobes in the brain, and the lateral and anterior tracts of the spinal cord, argues for a common pathogenetic mechanism of cell death of both neural and glial origin.

In juvenile cases of HC, a diffuse loss of Purkinje cells has been observed in the cerebellum. Fibrillary gliosis and demyelination of central tegmental tracts and cerebelloolivary and olivocerebellar fibers have also been observed.[6]

3. CHEMICAL PATHOLOGY OF HUNTINGTON'S CHOREA BRAIN

The regional distribution of neurochemical abnormalities with respect to neurotransmitter receptor binding activity or immunoreactivity of neuropeptides closely correlates with the regional pathology of HC brain. All the neurochemical changes are region specific, and no universal changes throughout the brain are found (Table I). Thus, a greater than 80% drop in substance P immunoreactivity of globus pallidus and substantia nigra is found, and a moderate drop in Met-enkephalin activity of the same regions, but no changes in frontal cortex, thalamus, or hypothalamus.[7,8]

The changes in the regional distribution of receptor binding activity are listed in Table I.[9–15] These changes indicate reductions in total number of binding sites (which range from 40 to 60%) rather than decreases in affinity. An exception is the observation of increased affinity but unaltered maximum capacity for [^{3}H]GABA binding in cerebellum from HC brains compared with tissue from control patients. A similar pattern can be produced in control membranes by treatment with Triton X-100, phospholipase C, or glycerophosphate, which suggests that this observation represents an abnormality in phospholipid.[10]

4. NEUROTRANSMITTER AND HORMONAL ALTERATIONS IN HUNTINGTON'S CHOREA

Alterations in neurotransmitters and relevant synthetic enzymes in HC have been reviewed by Bird.[16] More recently, they have been determined in well-controlled autopsy brain studies by Spokes.[3] In HC, consistent increases in dopamine concentration were found in corpus striatum, nucleus accumbens, and pars compacta of the substantia nigra. Norepinephrine levels were raised in caudate nucleus, lateral pallidum, and pars reticulata of the substantia nigra. The dopamine and norepinephrine findings indicate a relative preservation of dopaminergic and noradrenergic pathways. Loss of choline acetyltransferase activity has been recorded in striatum, nucleus accumbens, septal nuclei, and hippocampus. Glutamic acid decarboxylase activity is predominantly reduced in striatum and lateral pallidum.

Hypothalamic dysfunction is the presumed basis of the clinical symptoms of cachexia, hyperphagia, increased libido, excessive sweating, and a variety of autonomic symptoms.[2,15]

Table I
Neurochemical Changes in Huntington's Chorea Brain[a]

Receptor/ligand	Caudate	Putamen	Globus pallidus	Substantia nigra	Thalamus	Hypothalamus	Cortex	Cerebellum
Ligand binding								
DA spiroperidol[9]	↓	↓						
[^{3}H]GABA[10]							N (Parietal)	↓
ACh[11] (muscarinic)	↓	↓						
Kainic acid[12]	↓	↓					N (Temp)	N
							N (Insula)	
							N (Hippocampus)	
Benzodiazepine[13]		↓						
Immunoreactivity								
Substance P[7,8]			↓	↓	N	N	N (Frontal)	
							N (Frontal)	
Met-Enkephalin[7,8]			↓	↓	N	N		
Cholecystokinin[14]		N	↓		N (VL nuc.)		N (Frontal)	
Vasoactive intestinal peptide[15]	N	N	N	N			N (Frontal)	

[a] Abbreviations: ↓, reduced; ↑, increased; N, normal; VL, ventrolateral.

A number of hypothalamic function tests have been done on HC patients, such as tests for control of the secretion of growth hormone, corticotropin, gonadotropin, thyroid-stimulating hormone (TSH), prolactin, and antidiuretic hormone (ADH).[17–19] The results showed that the normal rise in growth hormone in response to arginine/insulin, or bromocriptine is exaggerated in HC patients, whereas the normal suppression in response to a glucose load is reversed so that the level is increased.[17] In contrast to these results, Chalmers *et al.*[20] observed a blunted response to bromocriptine, and Levy *et al.*[21] found no response to apomorphine, a dopamine agonist, in HC patients. Also, Martinez-Campos *et al.*[18] found a normal growth hormone response to bromocriptine in three of their 14 patients. Gonadotropin-releasing hormone (GnRH) is modulated by dopaminergic neurons, and it has been shown to be significantly increased in the median eminence of choreic brain in females but not in males.[17] The best conclusion that can be derived from these studies is that there is measurable hypothalamic dysfunction in some HC patients. The variations in the test results can be best explained by differences in the stage and severity of damage to the neural networks in a given patient.

The critical questions to be answered are: (1) Are the behavioral motor abnormalities of chorea, or rigidity–akinesia and neuroendocrine abnormalities, in fact caused by neurotransmitter–receptor alterations? (2) Can the observed differences in neurotransmitter–receptor abnormalities be accounted for by methodological variations between different studies, or are they inherent in the variable severity of the effects in the involved area, which depends on the stage and extent of pathological change, and which results in a supervening predominant behavioral effect through the interplay of more than one neural net malfunction with a variety of neurotransmitter alterations? It is generally agreed that chorea results from unopposed dopaminergic overactivity. However, no differences between classical chorea and the rigid form have been found regarding neurotransmitter–receptor abnormalities to explain the behavioral difference.[3] Drugs that facilitate dopaminergic function, such as *d*-amphetamine, exacerbate chorea, and drugs that impede dopaminergic function, such as phenothiazines and butyrophenones (postsynaptic dopamine receptor blockers), are useful in ameliorating chorea. An elevation is seen in dopaminergically mediated GnRF gonadotropic-releasing factor concentration in the hypothalamus of female patients, which presumably is related to increased female fertility and libido.[17]

Pharmacological, neurochemical, anatomic, and behavioral evidence clearly indicates dopaminergic overactivity in Huntington's chorea. However, other neurotransmitter-mediated networks, such as the cholinergic and GABA-ergic, which are known to interact with dopaminergic systems, are also found to be abnormal in HC. The clinically observed behavioral responses to pharmacological treatment underscore the importance of a predominant effect in the presence of multisystem involvement. Thus, medications and nutrients that enhance cholinergic function such as choline, diethylaminoethanol, or physostigmine do not significantly improve chorea or memory. Similarly, muscimol or *p*-cholorophenylbutyric acid, a GABA analogue, L-glutamate, or imidizole-4-acetic acid (which stimulates postsynaptic GABA receptors) do not improve choreoathetosis.[22]

The success of dopamine receptor blockade in HC and precursor replenishment in Parkinson's disease have not been matched by any dramatic improvement of memory in Alzheimer's disease or Huntington's chorea. One possible reason for the failure of significant pharmacological modulation of a higher cortical function such as memory may be the complexity of such behavior with respect to neural organization and neurotransmitter mechanisms. Even a unitary cerebral behavioral function such as memory has been operationally divided into consolidation, storage, and retrieval, which implies multiple neurotransmitter and complex anatomic localization in temporal–limbic brain and thus a mechanism not subject to single neurotransmitter modulation. Neurotransmitter-replenishing maneuvers to improve memory may have to be tested against multiple precursor or receptor hypotheses if advances are to be made in the correction of complex cerebral cognitive malfunction.

5. *THE "KAINIC ACID MODEL" OF HUNTINGTON'S CHOREA*

The result of intrastriatal injection of kainic acid (KA), a rigid analogue of glutamic acid, into animals mimics some of the regional chemical pathology of HC. The neuropathological changes involve the dendrosomatic regions of neurons. The changes are rapid[23–25]: all neurons destined to die do so in hours or days—but they spare the axons of passage and require the integrity of corticostriatal glutamate pathways. The resulting lesion has decreased GABAergic and cholinergic indices, normal serotoninergic and adrenergic indices, and normal or elevated dopaminergic indices in the neostriatum. In the substantia nigra, the GABAergic indices and substance P are markedly decreased, and the tyrosine hydroxylase levels are normal. The animals' behavior does not mimic chorea. Five to 9 months after KA injection, the striatum is reduced to a gliovascular matrix interspersed with long axon tracts. An important finding is that neurons of hippocampal formation, olfactory cortex, lateral septum, lateral hypothalamus, and some amygdaloid and thalamic nuclei are more sensitive to the neurotoxic action of KA than are striatal neurons. This pattern of selective sensitivity does not correlate with the most intense neuronal loss in HC. Thus, the lesioning of the striatum with microinjection of KA serves as an interesting approach to studying HC but does not provide a model of HC.[23]

6. *NEUROPHARMACOLOGICAL MANAGEMENT OF HUNTINGTON'S CHOREA*

The goals of pharmacological treatment of HC are to control and attenuate the choreoathetotic movements and manic–depressive symptoms if present. Psychotropic drugs neither retard the progress of brain malfunction nor ameliorate higher cortical cognitive deficits. Excessive medication exacerbates the equilibratory deficit and further blunts cognitive abilities. Therefore, judicious use of psychotropic drugs, striking a balance between benefits and side effects, is critical.

Butyrophenones (such as haloperidol) and phenothiazines (such as thioridazine and fluphenazine) have been reported to be most consistently beneficial in attenuating choreoathetosis.[26,27] We find perphenazine to be superior to any of the above. The presumed mode of action of both butyrophenone and phenothiazine is dopamine receptor blockade, resulting in reduction of postsynaptic dopaminergic activity.

The emotional depression can be treated with tricyclic antidepressants in a significant number of patients.[28]

7. CELL BIOLOGY OF PERIPHERAL TISSUE IN HUNTINGTON'S CHOREA

More than 1000 well-substantiated hereditary disorders are known. The abnormal gene product has been identified in about 200 of these, and it is recessive in more than 95% of them. Dominant disorders account for half of the clinically recognized traits, and, in a large number of these mutations, the primary phenotype is expressed as brain damage.[29] It has been postulated that defects in structural, transport, cell-wall, or receptor proteins are the targets of mutation in autosomal dominant disease. Strategies of somatic cell genetics developed for investigation of autosomal recessive disorders have usually not been useful in unraveling biochemical abnormalities in dominantly inherited diseases because, in a dominant disorder, one of the pair of alleles is normal, and any biochemical phenotypic abnormality will probably express as a quantitative difference rather than a qualitative one.

The theoretical and practical reasons for using cultures of peripheral tissues such as fibroblasts to investigate the biochemical basis of dominant mutations in which the primary clinical target of dysfunction is the brain are as follows. (1) The nervous system and skin are of ectodermal origin in embryonal development. (2) The mutated locus is universally present on the chromosome in each cell of the body. (3) The dynamic aspects of the synthesis and degradation of carbodhyrate, peptide, and lipid components of membrane can be studied by incorporating tracer radioisotopes under nutritional stress or enrichment. (4) Membrane structure–function correlations can be made to test nutritional hypotheses for the basis of the defect and strategies for its correction. (5) When a specific metabolic abnormality is delineated, carrier and affected individuals should segregate into biochemical groups distant from normal.

A number of investigators using nuclear magnetic resonance (NMR), fluorescence, and concanavalin A as tools in studies on RBC, lymphocytes, and fibroblasts have detected differences between HC and normal cells.[30,31] Other inestigators have not observed the differences noted with the NMR and fluorescence procedures.[32–34] The disagreements among results remain largely unexplained. However, as shown below, at least for human fibroblasts, the nutritional components in the culture medium have a critical role in determining whether a phenotypic difference between cell lines can become manifest. One method for unmasking a nutritional requirement of mutant cells would be to stress the cells nutritionally and to test for nonessential amino acids, amino sugars, and other small molecules present in the dialyzable fractions of serum by first dialyzing the serum (to obtain dialyzed fetal calf serum, dFCS) and then filtering it through Sephadex G-50 (to obtain dfFCS) and heat inactivating the serum glycosidase activities, which can hydrolyze the sugars from serum glycoproteins.[35]

Thus, if fibroblasts from HC were tested with culture media containing whole serum and nonessential amino acids and amino sugars, the genetic metabolic block could be masked, since the culture media would contain all the components for cell multiplication and survival that normal fibroblasts do not need but that mutant cells might not survive without. The finding that [^{14}C] glucosamine incorporated into fibroblasts' oligosaccharide polypeptide is severely inhibited (by more than 90%, as evaluated by SDS gel electrophoresis) when the macromolecular fraction of serum (dfFCS) is used instead of whole fetal calf serum suggests that Sephadex G-50 removes an obligatory factor promoting protein glycosylation.[36]

Proliferating fibroblasts from HC grown in stock culture for 3 weeks in Eagle's minimal essential medium (MEM) plus the macromolecular fraction of serum (dfFCS) and then replated in plastic flasks express a defect in attachment to the plastic substratum of culture flasks measured at the G_0 phase of growth (2 weeks after plating). This defect of attachment can be corrected and is not observed when fibroblasts are grown in commercially dialyzed, not heat-inactivated, fetal calf serum (dFCS)[37] (Fig. 1).

In recent studies, Cassiman *et al.*[39] and Archer and Mancall,[40] purporting to use identical methods to test HC fibroblast attachment and the role of glucosamine, found no difference between HC and normal. Such studies are plagued with technical and conceptual problems. In fact, the test conditions under which HC fibroblasts were shown not to attach any differently from normal ones are the same as those under which we reported no difference in HC fibroblast attachment.[36,37] The critical conditions for HC attachment difference are (1) the propagation of stock cultures in 20% macromolecular fetal calf serum (dfFCS) for 2–4 weeks, (2) the use of logarithmically proliferating cells from these stock cultures (whose growth rate is equal to that in untreated FCS and is normal) and the further quantitative plating of these cells for 2 weeks in 20% dfFCS prior to studies of cell attachment to Falcon® flasks in the same culture medium, and (3) a specific quality and timing of trypsinization. Since none of these specified conditions was followed by these investigators,[39,40] any conclusions arrived at with respect to cell attachment or the role of glucosamine in HC metabolism depend on an assumption about the nature of autosomal dominant mutations that has no known basis, i.e., that HC fibroblasts should express a behavioral abnormality in response to a given nutritional variable under conditions other than the ones specified. The response of mammalian cells in culture to a variety of serum types and nutrients is very complex and cannot be expected to follow such logic.

A number of lines of evidence from our laboratory suggest that there is faulty hexosamine metabolism in HC fibroblasts in culture. These are as follows.

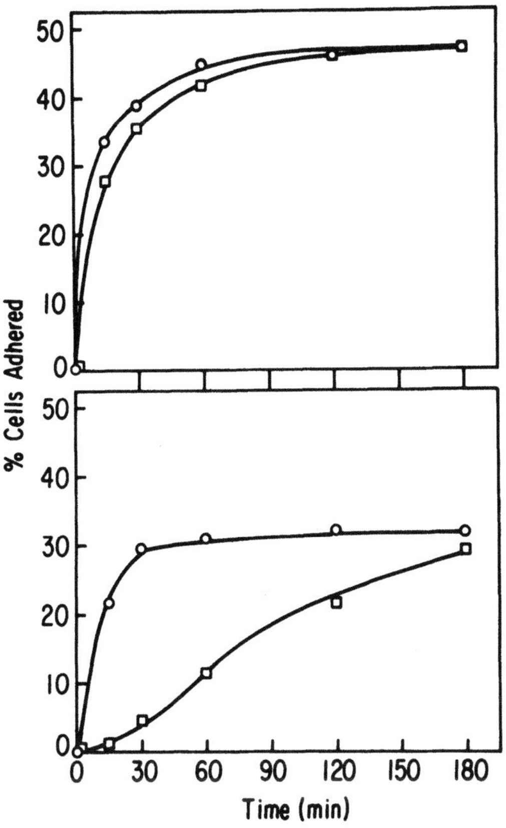

Fig. 1. Attachment of cells to surface substratum. Stock cultures were maintained in either MEM + 20% dFCS or MEM + 20% dfFCS for at least 2 to 4 weeks. Logarithmically growing fibroblasts from three HC and three normal patients matched for age, sex, passage number, and growth rate were replated and allow to grow for 2 weeks in the indicated medium. The cells were washed twice with PBS followed by a 5-min treatment with trypsin (0.04% Puck's saline A with 0.02% EDTA). The trypsin was neutralized with 5 ml of the initially specified culture medium, and the cells were counted; 250,000 cells were plated per flask in triplicate either in MEM + 20% dFCS or MEM + 20% dfFCS. The nonadherent cells were poured off at the indicated times and counted. Upper panel: Normal medium, MEM + 20% dfFCS. Lower panel: MEM + macromolecular fraction of serum, 20% dfFCS. ○——○, normal fibroblasts; □——□, HD fibroblasts. Each point is the mean of triplicate determinations (the S.E.M. is less than 5%). (From Tourian and Hung, 1977).

7.1. The Uniqueness of the Differential Labeling of Cellular Glycoprotein in HC Fibroblasts

To standardize our work on fibroblast culture and to reduce the sources of variables, cells were matched for growth phase and rate, age, sex, and passage number and were grown in a standard commercially dialyzed fetal calf serum (GIBCO) at 20% in Eagle's mimimal essential medium (MEM). Five pairs of HC fibroblasts—but not fibroblasts from normal individuals or from patients with autosomal dominant Alzheimer's disease, cerebellar degeneration, myotonic dystrophy, or recessive Charcot–Marie–Tooth disease—ex-

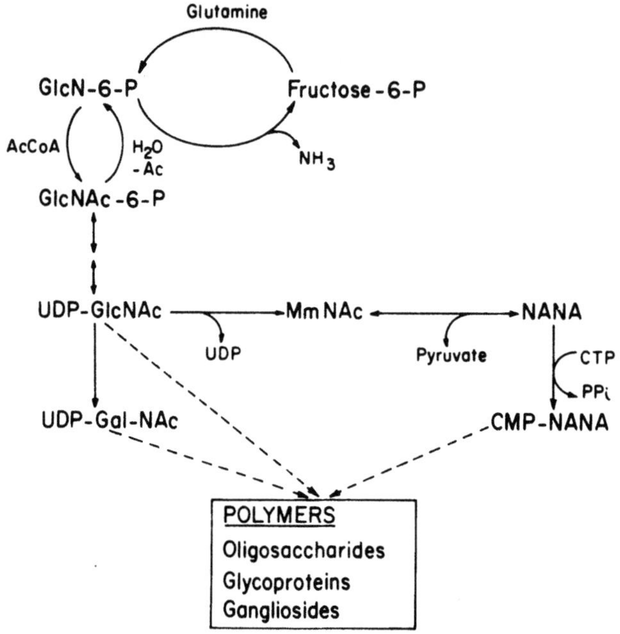

Fig. 2. Metabolic pathways from fructose-6-P to complex carbohydrates. Abbreviations are as follows: GlcN, glucosamine; NAc, N-acetyl; Mm, mannosamine; CMP, 5′ cytidylic acid; NANA, N-acetylneuraminic acid.

hibited strikingly greater than normal protein and lipid glycosylation when labeled with [^{14}C]glucosamine ([^{14}C]GLcN). The difference in protein glycosylation is not accompanied by any apparent difference in general cellular protein synthesis or by a differential rate of glucosamine uptake or decreased degradation of [^{14}C]glycosylated macromolecules. Additionally, [^{14}C]glucosamine exclusively labels hexosamine and sialic acid of cellular macromolecules[37,38] (Figs. 2,3).

Mammalian cells in culture maintain a very tightly regulated concentration of UDP-N-acetylglucosamine (UDP-GLcNAc) in the range of 0.1 to 2 nmol per 10^6 cells or per mg protein.[41] In interpreting radioisotope study results, it is important to bear in mind that cells in culture utilize preformed hexosamines generated by degradation of serum glycoproteins,[35] exogenously supplied [^{14}C]GLcN, and GLcN synthetized *de novo* from fructose-6-P (F-6-P). Therefore, the ultimate cellular specific radioactivity of UDP-N-acetylhexosamine will depend on the ratio of the utilization of exogenously supplied [^{14}C]GLcN to that of GLcN in the serum, which is affected by the relative contribution of cellular F-6-P and UDP-GLcNAc, which, in turn, is regulated in part by the concentration of glucose in the culture medium.[42]

7.2. The Differential Labeling of Hexosamines and UDP-N-Acetylglucosamine in HC Fibroblasts

Matched pairs of fibroblasts in culture were labeled with carrier-free $H_3[^{32}P]O_4$ and [^{14}C]GLcN. No quantitative difference was found in [^{32}P], but

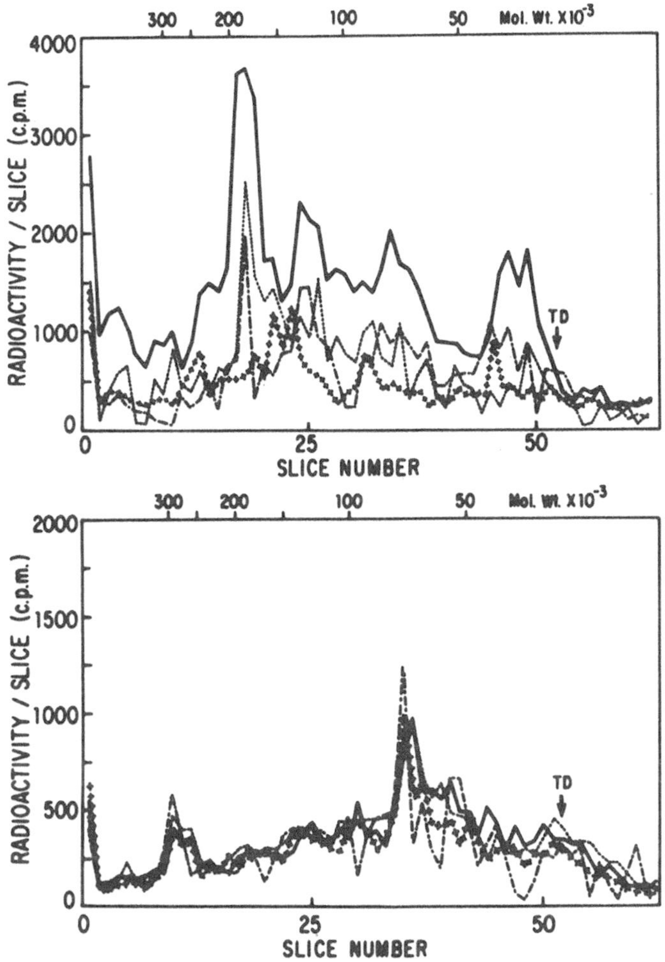

Fig. 3. The uniqueness of the differential labeling of cellular glycoprotein in HC fibroblasts. Top panel: SDS profile of [^{14}C]GLcN-labeled proteins. Bottom panel: [^{14}C]Leu-labeled proteins. HC ———— . Normal·········· . Others –·–, ++, such as familial Alzheimer's, myotonic dystrophy, spinocerebellar degeneration, and Charcot–Marie–Tooth disease. Only HC fibroblasts show strikingly greater than normal protein glycosylation.[38] (From Hung *et al.*[38] with permission of the Biochemical Society.)

a significant difference was found in [^{14}C] in UD[^{32}P]-[^{14}C]GLcNAc and [^{14}C]glucosamine-6-[^{32}P]phosphate, the ratio being approximately 1.5 to 1 for HC fibroblasts compared with normal (Fig. 4), suggesting that endogenous synthesis of UDP-N-acetylglucosamine is impaired in Huntington's chorea.[43] The true magnitude of the abnormality of *de novo* hexosamine synthesis in the fibroblasts cannot be evaluated from these experiments since the [^{32}P]-labeled hexosamines are derived from at least three independent sources: (1) preformed hexosamines from serum glycoprotein degradation, which are utilized by all cells,[35] (2) the [^{14}C]GLcN, and (3) *de novo* synthesized hexosamines from F-

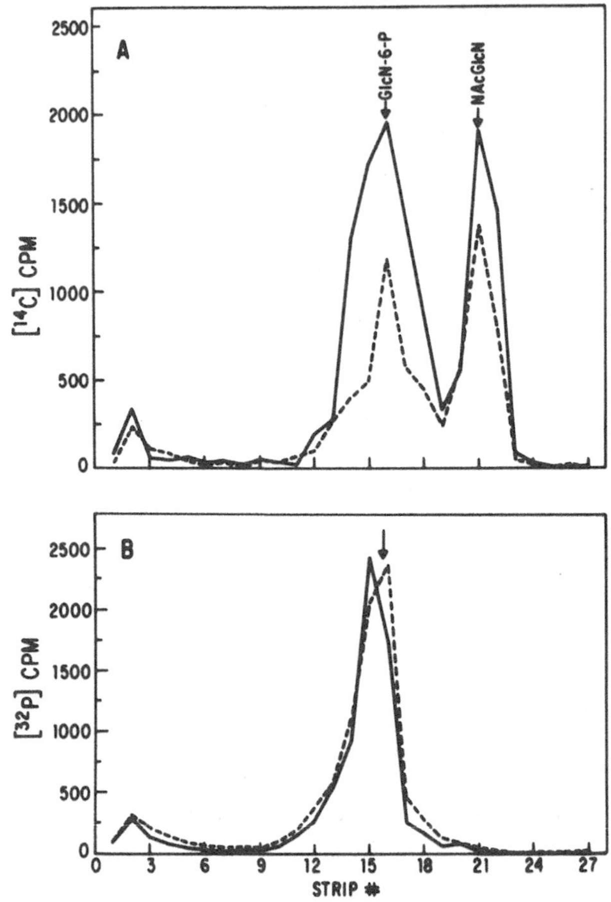

Fig. 4. The soluble cellular hexosamines and UDP-hexosamines of HC fibroblasts are labeled to a greater extent than normal by [^{14}C] when labeled by [^{14}C]GLcN and $H_3[^{32}P]O_4$. HC —— Normal -----. (From Hung and Tourian[43] with permission of the Biochemical Society.)

6-P. Thus, the difference in the ratio of [^{14}C]/[^{32}P] hexosamines between HC and normal fibroblasts is to be taken only as an indication of an abnormality and not as a measure of its true dimension. The paradox of equivalent hexosamine pool size (i.e., quantitative identity of [^{32}P] in hexosamine between HC and normal) in the presence of increased [^{14}C]hexosamine label in HC very strongly suggests that HC fibroblasts under the test conditions of growth correct the metabolic defect by utilizing preformed hexosamines from the culture medium.

7.3. The Conversion of [^{14}C]Glucose to [^{14}C]Hexosamines by HC Fibroblasts

Impaired conversion of [^{14}C]glucose to [^{14}C]UDP-hexosamines is observed in the soluble compartment of HC cells when a short (4-hr) pulse of

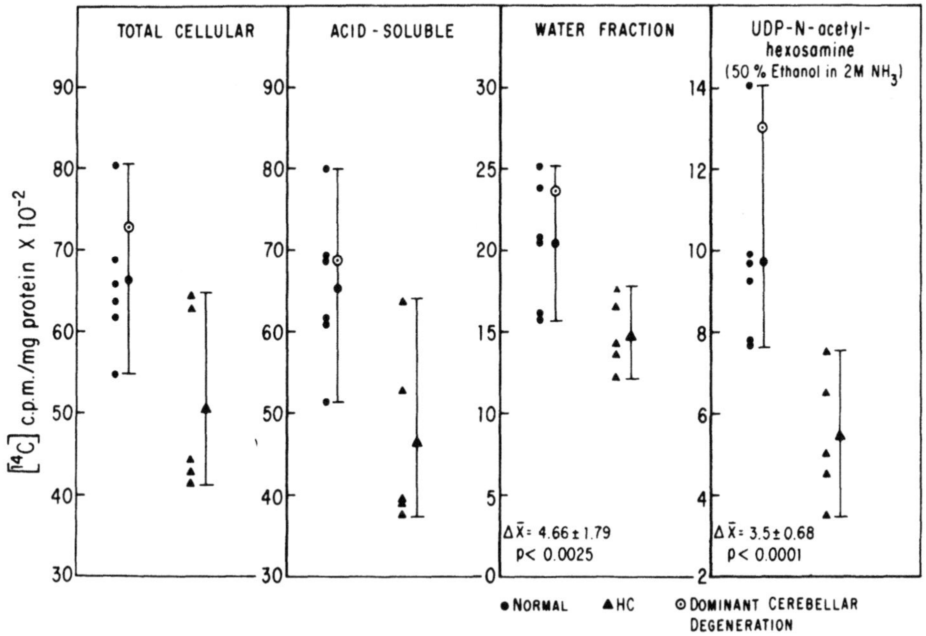

Fig. 5. Decreased conversion of [^{14}C]glucose to UDP-[^{14}C]hexosamines by HC fibroblasts. The third panel, *water fraction,* includes GLcN-6-P and glucose-6-P, which cannot be resolved by this method. The uptake of 3-O-CH_3[^{14}C]glucose and deoxyglucose was identical for HC and normal during the test period of this experiment. $\Delta\overline{X}$ represents the difference between the means of the two groups. Normal, ●. HC, ▲. Domninant cerebellar degeneration, ⊙.

[^{14}C]glucose is added to serum-free insulin-supplement culture medium.[44] This was found in studies in which five HC cell lines, a dominant cerebellar degeneration line, and six normal cell lines were coded, and the determinations were made without knowledge of the cells of origin by the investigator (Fig. 5).

This observation, in addition to the considerations listed below, makes it likely that the site of hexosamine synthesis impairment in HC fibroblasts is prior to the step of GLcN synthesis in HC fibroblasts. The [^{14}C]glucose and [^{14}C]glucosamine experiments circumscribe the metabolic abnormalities to the pathway

$$\text{Glucose} \xrightarrow{\text{Hexokinase}} \text{Glucose-6-P} \xrightarrow{\text{Isomerase}} \text{F-6-P} \xrightarrow{\text{Aminotransferase}}$$
$$\text{GLcN-6-P} \xrightarrow{\text{Acetylase}} \text{GLcNAc-6-P} \xrightarrow{\text{UTP}} \text{UDP-GLcNAc}$$

The amount of hexokinase in HC fibroblasts has not been measured, but it is unlikely to be abnormal, since the uptake of 3-0-CH_3-[^{14}C]glucose (not metabolized), [^{14}C]deoxyglucose (phosphorylated but not metabolized), [^{14}C]glucose, and [^{14}C]GLcN are all normal[45] (A. Tourian and W. Y. Hung, unpublished data) (cellular uptake is followed by phorphorylation of hexoses), and the conversion of [^{14}C]GLcN to UDP-GLcNAc, which depends on the initial phosphorylation of GLcN to GLcN-6-P, is greater in HC fibroblasts[45] (A. Tourian and W. Y. Hung, unpublished data).

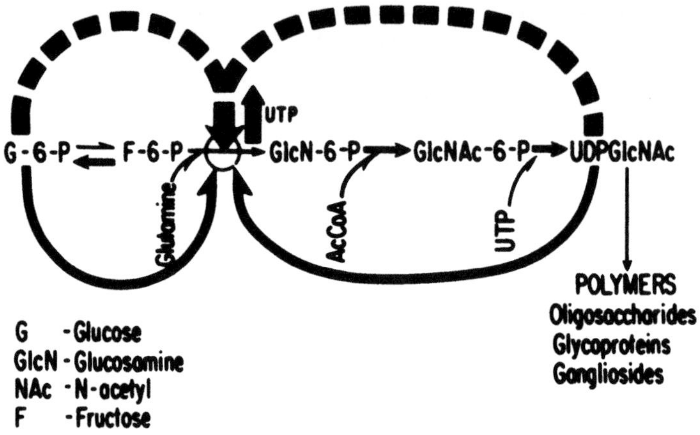

Fig. 6. Control L-glutamine F-6-P aminotransferse by G-6-P and UDP-GLcNAc.

The specific activity of glucose-6-phosphate isomerase is normal in HC fibroblasts. It was measured in the direction of F-6-P synthesis under V_{max} conditions (A. Tourian, M. Callahan, and W. Y. Hung, unpublished data) with the following results: HC, 1.29 ± 1.23 (S.E.M.) nmol/μg protein per min (n = 5); control, 1.07 ± 0.18 nmol/μg protein per min (n = 4).

A recessive somatic mutant cell in which the metabolic block is in the acetylation of glucosamine has the following characteristics: decreased protein glycosylation and decreased synthesis of GLcNAc and UDP-GLcNAc when the substrate utilized is either [^{14}C]glucose or [^{14}C]glucosamine.[46] A mutation at the activation step of GLcNAc (GLcNAc + UTP → UDP-GLcNAc) would result in a similar observation, i.e., decreased synthesis of UDP-GLcNAc when the substrate is either [^{14}C]glucose or [^{14}C]glucosamine.

The aminotransferase is situated at a branch point in carbohydrate metabolism and will compete for F-6-P with other pathways utilizing hexose monophosphate (glycolysis, pentose phosphate pathways, etc.). The fraction committed to hexosamine synthesis is only 0.5 to 2% for cellular (hepatic) F-6-P, even though it could be as high as 15 to 20% for F-6-P in skin (rat). The intracellular glutamine concentration is approximately 10 mM. The K_m of glutamine for the aminotransferase reaction is 0.67 mM and is not limiting in hexosamine synthesis.[47]

Since the only metabolic site at which hexosamine synthesis is exquisitely regulated in mammalian cells is at L-glutamine F-6-P aminotransferase, the obvious question is, have there been any enzyme activity abnormalities observed in HC fibroblasts?

L-Glutamine F-6-P aminotransferase is an extremely complex and labile enzyme. Besides the substrates L-glutamine and F-6-P, the end product of this pathway, which is UDP-N-acetylglucosamine (UDP-GLcNAc), in concert with G-6-P and AMP maintains the enzyme under constant inhibition (~90%), allowing for the function of a very small fraction of its potential activity (Fig. 6). This inhibition is reversed by UTP when the metabolic demand of the cell calls for increased synthesis of UDP-GLcNAc (1) during logarithmic growth of cell number, (2) when the ratio of F-6-P/G-6-P is increased, or (3) when the concentrations of UDP-GLcNAc drops.[47] In spite of a 1300-fold affinity purification of fibroblast aminotransferase and an extremely sensitive and specific radioisotope assay developed by us,[48] we are unable as yet to answer the question above. There seems to be a selective instability of the UDP-GLcNAc/[G-6-P] plus AMP functional sites, or, alternatively the regulatory sites are modified during affinity purification (A. Tourian, M. Callahan, and W. Y. Hung, unpublished data).

It is unlikely that the hexosamine metabolism abnormalities observed under a variety of extreme nutritional conditions such as enrichment or stress are secondary to a more fundamental metabolic problem (epiphenomenon) for the following reasons. Secondary metabolic abnormalities in human mutations can be corrected; for example, in phenylketonuria, in which tissue phenylpyruvic acid is elevated, this secondary phenomenon is corrected by reducing the load of phenylalanine in the diet.[49] The hexosamine abnormalities of HC fibroblasts, however, persist (1) under nutritional enrichment, i.e., 20% serum (a rich source of preformed hexosamines) or 5.5 mM glucose, (2) under starvation and stress, i.e., no serum, 5 U/ml insulin, and 0.5 mM glucose, (3) under growth conditions such that protein glycosylation is reduced by more than 90% (i.e., macromolecular fraction of serum growth condition), (4) after the cells have been passaged multiple times in tissue culture. All of these observations support the idea that hexosamine abnormalities in HC fibroblasts are stable and are unlikely to be secondary to a more fundamental metabolic abnormality.

REFERENCES

1. Myrianthopoulos, N., 1966, *J. Med. Genet.* **3**:298–314.
2. Bruyn, G. W., 1968, *Handbook of Clinical Neurology,* Volume 6 (P. J. Vinken and G. W. Bruyn, eds.), North-Holland, Amsterdam, pp. 298–378.
3. Spokes, E. G. S., 1980, *Brain* **103**:17–210.
4. Sassa S., Zalar, G. L., and Kappas, A., 1978, *J. Clin. Invest.* **61**:499–508.
5. Tourian, A., 1977, *Report: Commission for Control of Huntington's Disease and Its Consequences,* Volume III, Part 1, Public Health Service, Washington, pp. 301–310.
6. Bruyn, G. W., Bots, G. T. A. M., and Dom, R., 1979, *Adv. Neurol.* **23**:83–93.
7. Buck, S. H., Burks, T. F., Brown, M. R., and Yamamura, H. I., 1981, *Brain Res.* **209**:464–469.
8. Emson, P. C., Arregui, A., Clement-Jones, V., Sandburg, B., and Rossor, M., 1980, *Brain Res.* **199**:147–160.
9. Reisine, T. D., Fields, J. Z., Bird, E. D., Spokes, E. G., and Yamamura, H. I., 1978, *Commun. Psychopharmacol.* **2**:79–84.
10. Lloyd, K. G., and Davidson, L., 1979, *Science* **205**:1147–1149.
11. Olsen, R. W., Reisine, T. D., and Yamamura, H. I., 1980, *Life Sci.* **27**:801–808.
12. London, E. D., Yamamura, H. I., Bird, E. D., and Coyle, J. T., 1981, *Biol. Psychol.* **16**:155–162.
13. Reisine, T. D., Overstreet, D., Gale, K., Rossor, M., Iversen, L., and Yamamura, H. I., 1980, *Brain Res.* **199**:79–88.
14. Emson, P. C., Rehfeld, J. F., Langevin, H., and Rossor, M., 1980, *Brain Res.* **198**:497–500.
15. Emson, P. C., Fahrenkrug, J., and Spokes, E. G. S., 1979, *Brain Res.* **173**:174–178.
16. Bird, E. D., 1980, *Annu. Rev. Pharmacol. Toxicol.* **20**:533–551.
17. Bird, E. D., 1979, *Adv. Neurol.* **23**:291–297.

18. Martinez-Campos, A., Giovannini, P., Cocchi, D., Zanardi, P., Parati, E. A., Caraceni, T., and Muller, E. E., 1981, *Neurosecretion and Brain Peptides* (J. B. Martin, S. Reichlin, and K. L. Bick, eds.), Raven Press, New York, pp. 521–540.
19. Lavin, P. J. M., Bone, I., and Sheridan, P., 1981, *J. Neurol. Neurosurg. Psychiatry* **44**:414–418.
20. Chalmers, R. J., Johnson, R. H., Keogh, H. J., and Nanda, R. N., 1978, *J. Neurosurg. Psychiatry* **41**:135–139.
21. Levy, C. L., Carlson, H. E., Sowers, J. R., Goodlett, R. E., Tourtellotte, W. W., and Hershman, J. M., 1979, *Life Sci.* **24**:743–750.
22. Editorial, 1980, *Lancet* **1**:1119–1120.
23. Olney, J. W., 1979, *Adv. Neurol.* **23**:609–624.
24. McGeer, E. G., McGeer, P. L., Hattori, T., and Vincent, S. R., 1979, *Adv. Neurol.* **23**:577–591.
25. Coyle, J. T., London, E. D., Biziere, K., and Zaczek, R., 1979, *Adv. Neurol.* **23**:593–608.
26. Shoulson, I., and Chase, T. N., 1970, *Ann. Rev. Med.* **26**:419–426.
27. Shoulson, I., 1979, *Adv. Neurol.* **23**:751–757.
28. Folstein, S. E., Folstein, M. F., and McHugh, P. R., 1979, *Adv. Neurol.* **23**:281–289.
29. McKusick, V. A., 1978, *Mendelian Inheritance in Man*, 5th ed., Johns Hopkins University Press, Baltimore.
30. Butterfield, A., and Markesbery, W. R., 1981, *Life Sci.* **28**:1117–1131.
31. Pettegrew, J. W., Nichols, J. S., and Stewart, R. M., 1979, *J. Neurochem.* **33**:905–911.
32. Lakowicz, J. R., and Sheppard, J. R., 1981, *Am. J. Hum. Genet.* **33**:155–165.
33. Comings, D. E., Pekkala, A., Schuh, J. R., Kuo, P. C., and Chan, S. I., 1981, *Am. J. Hum. Genet.* **33**:166–174.
34. Fung, L. M. W., and Ostrowski, M. S., 1982, *Am. J. Hum. Genet.* **34**:469–480.
35. Block, R., Betschart, B., and Burger, M. M., 1977, *Exp. Cell. Res.* **104**:143–152.
36. Tourian, A., and Hung, W., 1980, *Prog. Clin. Biol. Res. Neurochem. Clin. Neurol.* **39**:195–204.
37. Tourian, A., and Hung, W., 1979, *Adv. Neurol.* **23**:371–386.
37a. Tourian, A., and Hung, W., 1977, *Biochem. Biophys. Res. Commun.* **78**:1296–1303.
38. Hung, W. Y., Mold, D., and Tourian, A., 1980, *Biochem. J.* **190**:711–719.
39. Cassiman, J. J., Verlinden, J., Vlietinck, R. F., Bellemans, J., VanLeuven, F., Deroover, J., Baro, F., and Vanden Berghe, H., 1979, *Hum. Genet.* **53**:75–86.
40. Archer, J. F., and Mancall, L. E., 1983, *Arch. Neurol.* **40**:24–27.
41. Yurchenco, P. D., Ceccarini, C., and Atkinson, A. H., 1978, *Methods Enzymol.* **50**:175–206.
42. Kim, J. J., and Conrad, E. H., 1976, *J. Biol. Chem.* **251**:6210–6217.
43. Hung, W. Y., and Tourian, A., 1981, *Biochem. J.* **196**:495–498.
44. Hung, W. Y., and Tourian, A., 1980, Poster presentation at meeting of the Society for the Study of Inborn Errors of Metabolism, Southampton, England.
45. Bachelard, H., 1981, *Chemisms of the Brain, Basic and Applied Neurochemistry* (R. Rodnight, H. S. Bachelard, and W. L. Stahl, eds.), Churchill Livingstone, Edinburgh, pp. 3–11.
46. Pouyssegur, J., Shiu, R. P. C., and Pastan, I., 1977, *Cell* **11**:941–947.
47. Winterburn, P. J., and Phelps, C. F., 1970, *Biochem. J.* **121**:711–720.
48. Callahan, M., Tourian, A., and Hung, W. Y., 1981, *Anal. Biochem.* **115**:347–352.
49. Tourian, A., and Sidbury, J. B., 1983, *The Metabolic Basis of Inherited Disease*, 5th ed. (J. B. Stanbury, J. B. Wyngaarden, D. S. Frederickson, J. L. Goldstein, and M. S. Brown, eds.), McGraw-Hill, New York, pp. 270–286.

13

Diseases and Induced Lesions of the Neuromuscular Endplate

Edith Heilbronn

1. INTRODUCTION

This chapter specifically addresses the issue of disorders of the neuromuscular junction at the mammalian skeletal muscle. Such disorders may be part of a disease or may be the result of the action of chemical agents, immunogenic compounds, or even transferred antibodies to proteins of the nerve–muscle synapse. Compounds of the latter category have been used to mimic diseases and have been used as tools in the elucidation of the function of the neuromuscular junction and the molecular events behind transmission there.

In this chapter, the autoimmune diseases or models of such diseases discussed have been divided into presynaptic and postsynaptic ones. Today, at least two disorders of the neuromuscular junction are known that may be classified as autoimmune diseases, i.e., the Eaton–Lambert syndrome and myasthenia gravis. Subclasses of these diseases, at least of myasthenia gravis, are known, but not the causes of the differences between them. Perhaps there are, after all, important differences in the pathogenesis of the various disease subclasses.

By necessity, the present chapter cannot discuss all known aspects of the various disorders. The reader is referred to the reviews and the literature quoted in the various chapters. A subject that would have merited a presentation within the framework of this chapter is the action of toxins that have been and are being used for the elucidation of neuromuscular transmission. It was, however, decided to exclude these because of limited space. A number of articles on this subject are available.

2. NEUROMUSCULAR TRANSMISSION

At the motor endplate of vertebrate skeletal muscle the transmitter is, according to current knowledge, acetylcholine (ACh). On nerve stimulation,

---

Edith Heilbronn • Unit of Neurochemistry and Neurotoxicology, University of Stockholm, S-172 46 Sundbyberg, Sweden.

the transmitter is released in multiples of "quanta" from special sites, the "active zones" of the nerve terminal membrane. The transmitter representing the quanta seems to be stored in synaptic vesicles within the motor neuron endings. Animal experiments suggest that the number of ACh molecules in a single quantum (a single vesicle?) is about 10,000. The size of a quantum for human muscle is not known. Vesicular release most probably occurs by exocytosis, after which the transmitter diffuses across the synaptic cleft, passes the junctional basement membrane, and reacts with the nicotinic acetylcholine receptor (nAChR) present at the tips of the foldings formed by the junctional part of the skeletal muscle cell membrane. The released transmitter is removed by diffusion and by hydrolysis, mediated by the enzyme acetylcholinesterase (AChE) present at the basement membrane (and also in the pre- and postsynapses). Formed choline diffuses away or is taken up presynaptically by a high-affinity uptake system and reused.

Nerve terminal depolarization produced by a nerve action potential opens voltage-sensitive calcium channels,[1] probably located in the "active zones."[2] The resulting influx of Ca^{2+} along its electrochemical gradient allows a synchronous exocytosis of a large number of synaptic vesicles and thereby produces the postsynaptic endplate potential. The amount of Ca^{2+} entering the nerve terminal is determined by the time the preterminal Ca^{2+} channel remains open, which is, among other factors, determined by the kinetics of Na^+ and K^+ channels. A cyclic-AMP-dependent protein kinase also seems to be involved in the modulation of Ca^{2+} channels. Impulse-evoked release is correlated to the amount of Ca^{2+} entering the nerve terminal, and the interior Ca^{2+} concentration is regulated by interactions with the mitochondria of the nerve endings and by uptake and extrusion systems.[2] Blockade of outward K^+ currents by drugs such as TEA and the aminopyridines prolongs the duration of the nerve terminal action potential and thereby greatly increases the inward Ca^{2+} flux and thus the amount of transmitter liberated.[7] Botulinum toxin (BTx) selectively blocks this mechanism for transmitter release by reducing its Ca^{2+} sensitivity.[3,4]

Quantal release is thus voltage dependent. A nerve impulse results in a near-synchronized release of 50–100 quanta, which evokes an endplate potential (epp) that depolarizes the muscle cell to the critical firing threshold, triggering the action potential. Single quanta are also released spontaneously. They generate a miniature endplate potential (mepp), which is about 1 mV in amplitude and is not able to trigger an action potential in the muscle fiber. In human muscle, the frequency of mepps is about 0.2 per sec.

There is also a substantial passive molecular leakage of ACh from the nerve terminal cytoplasm in resting muscle.[5–7,9] This is insufficient to significantly depolarize the postsynaptic membrane but may be a mechanism for neurotrophy at the target organ.[9] Quantitatively, it accounts for more than 90% of total transmitter release under resting conditions.

A further, probably vesicular, quantal release of ACh seems to occur from dispersed sites outside the "active zones."[4,9] This release is unaffected by transmembrane Ca^{2+} fluxes and nerve terminal depolarization and normally

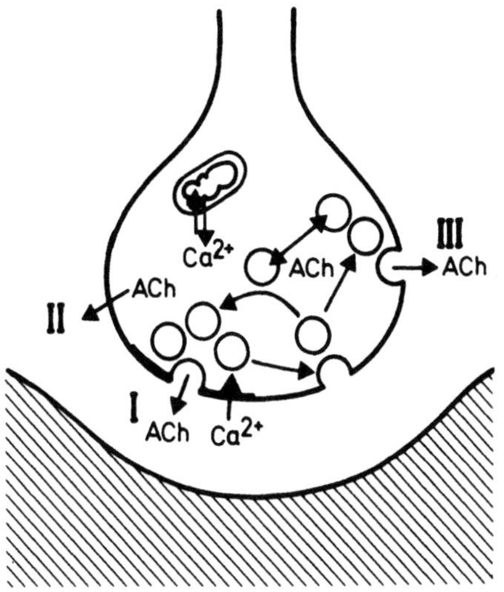

Fig. 1. Schematic drawing representing ACh release at the neuromuscular junction (From Thesleff[282]). I: Evoked vesicular quantal release. II: Leakage of ACh. III: Vesicular release independent of transmembrane Ca^{2+} fluxes.

accounts for the presence of only a small population of spontaneous giant miniature endplate potentials. This type of transmitter release is not affected by BTx.[4] Aminopyridines and similar compounds may enhance this release[7,9] depending on experimental conditions. Figure 1 is a schematic drawing representing the types of release seen at the neuromuscular junction.

To date, there is little information about a possible role for other transmitter systems or modulators at the motor endplate of mammals. The presence of enzymes synthesizing dopamine, taurine, and GABA at the vertebrate motor nerve ending was recently demonstrated immunocytochemically.[10] Their exact location and their function are not known. The occurrence of GABA at the electric organ of *Torpedo*, a model of the motor endplate, was shown. ATP has been found in synaptic vesicles together with ACh and has also been shown, in the *Torpedo*, to be released from the postsynaptic site in connection with neurotransmission[11] and from rat phrenic nerve–diaphragm preparations stimulated at physiological frequencies.[12] A direct adenosine nucleotide-induced potentiation of receptor activation by ACh was suggested.[13] Its role is not clear, but recent biochemical studies (ion translocation) performed in our laboratory on chicken embryo myotubes (Fig. 2) and studies with the membrane patch technique[14] suggest that there exists an ATP-mediated cation channel in addition to the ACh-mediated channel in these cells.

The search for modulating peptides has so far been relatively negative. There are some indications that substance P may be present at or close to the motor endplate.[15] In the nerve terminal of a slow skeletal motor neuron of cockroach, a neuropeptide cotransmitter, proctolin, was recently found.[16] Its release was shown to be caused by electrical stimulation or high K^+. Adrenocorticotropic hormone was suggested to act presynaptically as a modulator at the frog motor nerve terminal.[17]

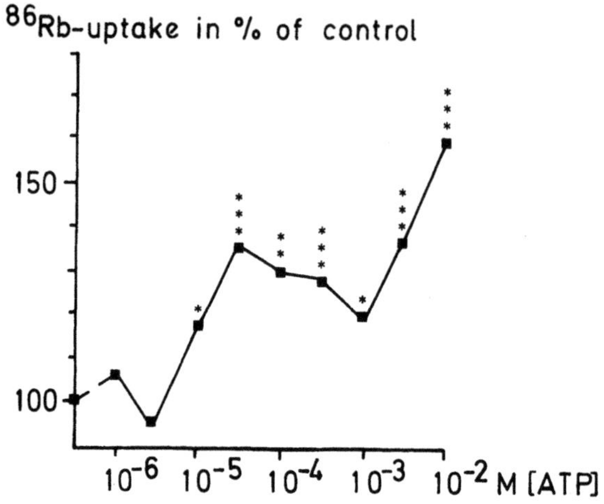

Fig. 2. Uptake of ^{86}Rb in cultured chick myotubes induced by ATP. Uptake of ^{86}Rb was studied during 30 sec periods at room-temperature in a medium containing Mg-ATP, 86 Rb (15 μCi/ml), and ouabain (5mM). Subsequent to uptake the cells were washed with ice-cold medium and radioactivity determined. *, $p < 0.05$; **, $p < 0.025$; ***, $p < 0.005$. Determined by a two-tailed t-test, $n = 3$. (Unpublished data from the author's laboratory).

3. *THE MOTOR NERVE TERMINAL IN RELATION TO TRANSMITTER RELEASE*

The motor nerve terminal, like other nerve terminals, is a specialized region free from myelin and thus open to contacts from the surrounding medium and from other neurons. Nerve and muscle fiber make contact at specialized areas where the very final part of the neuron has branched. This contact area is the motor endplate. Before reaching the muscle fiber, the myelinated neuron branches several times.

Present knowledge concerning transmitter release mechanisms, in particular, quantal release, originates largely from studies of the neuromuscular junction (see, e.g., ref. 18). New types of transmitter release are currently being discussed (see above).

Vesicular release is also studied in electric organs of *Torpedo* sp. The mechanism of vesicle fusion is not understood. In particular, nothing is known about the compounds causing membrane fusion or their activation by Ca^{2+} ions. Calmodulin has been suggested to mediate Ca^{2+} action in exocytosis at nerve terminals. Vesicle fusion and recycling have been discussed in a great number of papers (see, e.g., refs. 19,20). However, other authors[21] maintain that a gating mechanism for free cytoplasmic ACh may be responsible for evoked release.

New information concerning the molecular details of the quantal transmitter release may come from the study of a human condition only recently defined as a presumably autoimmune disease, the Eaton–Lambert syndrome (ELS, see Section 6.2). The antibodies involved in the disease may be directed to the calcium channel of the presynapse. It has been found that the rows of release sites normally observed on the inner membrane of the junctional part of the nerve terminal membrane are partly missing in ELS. Concomitantly, a reduction of transmitter release is noted, and thus neurotransmission becomes faulty.

Regulation of motor neuron transmitter release is still poorly understood. From other cholinergic systems such as brain[22] and smooth muscle,[23] evidence has accumulated that a negative feedback system involving a muscarinic receptor plays a role. Some authors postulate on pharmacological grounds the involvement of a presynaptic muscarinic acetylcholine receptor at vertebrate skeletal muscle junctions (see refs. in ref. 24) in such regulation. In recent work involving autoradiographic, pharmacological, and biochemical studies, we[24] found no signs of the involvement of such a receptor. A similar conclusion was drawn by Gundersen and Jenden.[25] In contrast, some of our data and also data by Miledi *et al.*[26] suggest that a presynaptic nicotinic mechanism may be involved at the neuromuscular junction; α-bungarotoxin (α-bgt), which binds to nicotinic ACh receptors, was found to augment evoked ACh release, suggesting a negative feedback control of transmitter release involving such a mechanism. More data are, however, needed before any such conclusion can be drawn.

4. *THE ACETYLCHOLINE RECEPTOR OF MAMMALIAN SKELETAL MUSCLE*

Acetylcholine released by a nerve impulse arriving at the motor nerve terminal diffuses across the synaptic cleft and binds transiently to nicotinic acetylcholine receptors (nAChR) located on the tip of the secondary foldings of the postsynaptic muscle membrane,[28] just opposite the transmitter release sites. As a consequence of this event, a conformational change occurs in the formed ACh–nAChR complex, resulting within a fraction of a millisecond in a transient alteration of the cation permeability of the muscle membrane. A local depolarization of the endplate occurs[29] and is terminated by the removal of the transmitter because of diffusion and hydrolysis.

The acetylcholine receptor molecule spans the membrane, extending extra- and intracellularly beyond it (*Torpedo* electric organ: 55Å and 15Å) (Fig. 3). It is able to serve both agonist and antagonist binding and a cation channel function. This has been thoroughly studied both *in vivo* and *in vitro* on microsac preparations (receptor-bearing membranes from *Torpedo* electric tissue) and on artificial membranes containing previously purified receptor (cf. refs. 30–33). Under the latter conditions, it was found that the receptor macromolecule itself is sufficient for physiological function: i.e., it is able to achieve the depolarization of the muscle membrane that eventually results in muscle contraction.

4.1. Structure

The receptor macromolecule consists of a number of subunits. One type (the 42 K subunits, see below) binds the transmitter and is thus an essential part of the signal-recognizing unit. The molecular nature of the other functional part, the cation channel (possibly channels), is not yet altogether understood.

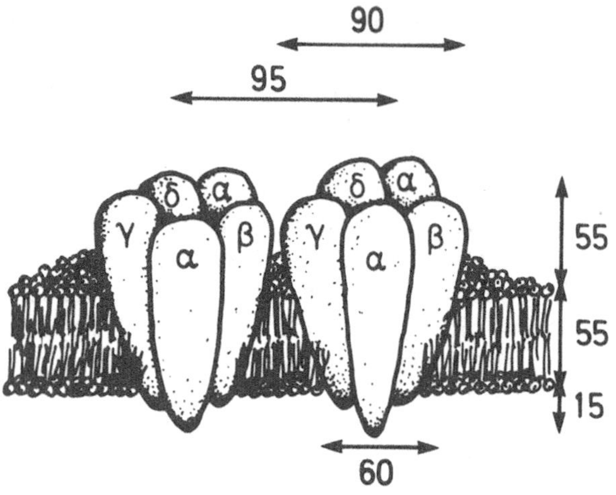

Fig. 3. Model of the nicotinic acetylcholine receptor from *Torpedo* electric organ. (Data from the literature, mostly Stroud.)

Immunologic and biochemical studies have shown that all subunits are transmembrane glycopeptides, and thus, all of them in fact could participate in the formation of the channel. (For discussions on the nature of the ion channel of nAChR, see ref. 34 and reviews 30–33; for references on receptor topography see also ref. 40.)

Information about the molecular nature of nicotinic acetylcholine receptors has accumulated for over 10 years. Functional aspects have been studied for much longer. The nAChRs from *Torpedo* spp. (marine elasmobranchs) or from *Electrophorus* (freshwater teleost) electric organ are the neuroreceptor models of preference; they are structurally closely related, and their subunits may derive from a common ancestral gene. Their structure is well described. Four homologous subunit types, α, β, γ, δ, in a stoichiometric ratio of 2:1:1:1, comprise the macromolecule. Recently, their primary structures were deduced from cDNA sequences.[35] The predominant native form of the *Torpedo* receptor seems to be a dimer[36] of 500 K cross linked by a disulfide bond between the δ subunits.

The structure of the mammalian muscle nAChR is less well understood. Such receptors have also been purified (cf. ref. 32), including, partly, human muscle receptor.[37] The subunit pattern is not clear; from one to six polypeptides have been reported for mammalian receptors.[38,39] Probably, the three nAChRs mentioned above are similar in structure and functions; differences in structure may largely be artifacts of purification that are caused by the presence in mammalian muscle of considerable amounts of proteases. Four homologous peptides in nAChR from fetal calf muscle copurifying with a fifth polypeptide suggested to be actin were described[39] and found to have molecular weights of 42, 44, 49, 55, and 58 K. Sometimes a sixth polypeptide is seen. Carbohydrate (tested with concanavalin A) is indicated on the 42 K, 49 K, 55 K, and 53 K peptide.[39] On the basis of its amino acid sequence, the presence of 3-methyl-histidine, and the binding of actin antibodies, the 44 K peptide is suggested to be actin. It is not certain but probable that the closeness of actin to the receptor has structural[32] rather than functional implication.[40] The 42 K peptide binds the transmitter derivative [^{3}H] bromoacetylcholine and thus corresponds to the α subunit of the electric organ nAChR.[32,33] As found for electric organ, four polypeptides of the muscle nAChR seem to be structurally related. A stoichiometry of 2:1:1:1 was found, and a pentameric complex with a molecular weight of 245 K is suggested.[39]

4.2. *Junctional and Extrajunctional Receptors*

At the adult synapse (endplate), nAChR is mainly found at the junction. Extrajunctional receptors exist, particularly on noninnervated and functionally or surgically denervated muscle fibers. The extrajunctional and junctional receptors differ not only in their location but also in some of their properties.[33,41] The ion channel open time of the junctional receptor is reduced to about one-third of that of the extrajunctional receptor,[41,42] and the turnover time of the junctional receptor is slower than that of the other.[41] Changes in the degree of phosphorylation,[33,43,44] methylation,[45] and glycosylation have been discussed. Further, more immunologic determinants seem available on the extrajunctional receptors than on the junctional ones.[46]

Innervation of the muscle fiber causes extrajunctional receptors, spread diffusely over the muscle membrane, to disappear. Muscle fiber contraction, even when locally induced, reduces nAChR levels in normal muscle. As shown in dysgenic mouse, skeletal muscle action potential generation without concomitant contraction also decreases nAChR levels. Denervation of muscle, on the other hand, causes supersensitivity to ACh through the appearance of new extrajunctional receptors and changes in electrical properties of the muscle membrane resulting in altered ion permeabilities.[8] The role of trophic factors in these processes, particularly in the receptor location at the junctions simultaneously with innervation, is not clear (see below).

4.3. *Biosynthesis, Localization, and Stabilization*

The skeletal muscle nAChR is synthesized within the muscle cell (cf. refs. 41, 47). Studies on cultured myotubes have given much information. The site of nAChR biosynthesis is apparently the rough endoplasmic reticulum.[28] The newly synthesized receptor is intracellular and is probably membrane-associated in the Golgi apparatus before transport to the plasma membrane.[49] On young myotubes in culture, the receptor appears rather evenly distributed. On more mature myotubes, nAChR clusters of low density are found. Despite a relatively rapid turnover, the positional stability of these receptors is rather high, and the clusters seem to attract new, arriving receptors.[50] *In vivo*, nAChR clusters are seen at developing neuromuscular junctions. The basal lamina may be involved in the organization of nAChR at the endplate[48,51] (for a thorough discussion and refs., see ref. 41). The receptors may be anchored to their place on the tip of the foldings of the muscle membrane by a network of filaments belonging to the cytoskeleton of the muscle cell.[42] The receptor concentration[38,43] on the tip of the postsynaptic membrane foldings is 10–20,000 sites/μm^2. Several laboratories[33] have described factors from nervous tissue that influence the number and distribution of nAChR *in vitro*, i.e., both overall receptor density and clustering.

Neurotrophic factors from the motor neurons have long-term effects on physiological, pharmacological, and biochemical properties of muscle fibers.[52] Denervation causes a fall in resting membrane potential, an appearance of TTX-resistant action potentials, an increase in extrajunctional ACh sensitivity (increase of nAChR synthesis), appearance of fibrillation, and decrease of AChE activity. Reinnervation restores membrane properties to normal. In other words, denervation changes the structural and functional properties of the muscle membrane. The motor neuron probably normally controls this partly by controlling muscle activity and partly by trophic substances. Many examples of the latter come from studies on the development of muscle cells in culture or from maintenance studies on matured muscle fibers, which led to the observation that a polypeptide of 84 K, sciatin, was necessary for myogenesis.

4.4. Modulation

The receptor concentration and receptor activity at the junction seem to be regulated by the transmitter concentration in the cleft. A gradual decrease of the permeability response to ACh, observed when ACh remains in contact with the muscle membrane for several minutes, i.e., "desensitization," was suggested by Katz and Thesleff[53] to be caused by two stages of the receptor existing, one of them "refractory." A more complicated model has also been discussed[33] and involves a four-stage model of the receptor.

Some recent results bear on receptor desensitization. An example is the neuropeptide substance P, which has been shown to block ACh-induced excitation in a number of systems. A function as a cholinergic antagonist, an ion channel blocker, or a modulator of the activity of the receptor–ionophore complex has been suggested.[54] In clonal cell line PC12, inhibition of carbamylcholine-induced $^{22}Na^+$ uptake was demonstrated. Both substance P and octahydrohistrionicotoxin (H_8-HTX), but not local anesthetics, were shown to enhance agonist-induced receptor desensitization. Substance P seemed able to trap the receptor in its desensitized state even when added after the actual desensitization by the transmitter. It thus may act by binding to a special regulatory site on the receptor–ionophore complex. High concentrations of Na^+ prevent the effect of substance P, which suggests that substance P may also block the ion channel and thus increase the rate of receptor desensitization.

Removal of the receptor from the junctional membrane occurs by an endocytotic process.[41] The neurotransmitter concentration at the receptor was found to regulate receptor concentration by "down-regulation."[55] A sustained increase in the level of transmitter may either decrease the normal median receptor lifetime or reduce the rate of synthesis (or insertion into the junctional membrane?) of the receptor. The stable ACh analogue carbachol was found to decrease the total number of α-bgt-binding sites present on muscle cells in culture.

The results suggest that the receptor level was reduced as a result of reduced synthesis.

The dependence of receptor activity and receptor concentration on transmitter levels is of considerable importance in medicine. Drugs used in the treatment of diseases may change transmitter levels or the duration of the presence of the transmitter within the cleft and may thus have an indirect effect on receptor function or concentration. A typical example are anticholinesterases used, e.g., in myasthenia gravis (MG, see Section 9).

A pathological down-regulation of junctional nAChR levels or, rather, an accelerated degradation has also been suggested to occur in MG. Experimental results suggest that the degradation increase is induced by the binding of specific antibodies to the receptors.[56] The receptor–antibody complex is probably recognized by the cell as nonfunctional and removed. Only divalent antibodies are able to form such complexes; monovalent Fab fragments are inactive and, in fact, are able to prevent divalent antibodies from cross linking receptors and accelerating their degradation.[57]

4.5. The "Myasthenic" Receptor

Changed receptor structures or surroundings may be responsible for neuromuscular diseases. Attack by virus or by foreign (or endogenous) pathogenic substances might be responsible for such changes.

The structure and properties of nAChR from myasthenic patients or animals with experimental MG are unknown. These receptors may be quite normal though reduced in number and complexed with antibodies *in situ*, as recognized from both neurophysiological measurements and morphological studies. On the other hand, the autoimmune disease can be caused experimentally by immunization of animals with purified receptor[58–60] from another or from the same animal species. Purified receptor differs in certain properties from receptor *in situ*; e.g., it carries mostly low-affinity sites for agonist binding instead of both low- and high-affinity binding sites,[33] suggesting a changed structural flexibility. These findings suggest that changes in the receptor molecule or perhaps its close surroundings could be recognized by the organism as foreign and become responsible for the induction of pathological antibody synthesis. This is also confirmed by the induction of severe experimental myasthenia with isolated nAChR α-subunit[61]; other subunits give rise to antibody titers with varying degrees of limited pathogenicity.

The explanation for this probably lies in the importance of the determinants on the developed antibodies for neuromuscular transmission or in their antigenicity. The antigenic determinants on the α-subunit of *Torpedo* nAChR have been discussed.[62] Such regions may be present on several of the subunits. Recently, it was confirmed by single-particle image averaging[63] that two different receptor sites bind α-toxin. One of these (α_1) is adjacent to the δ subunit; the second region (α_2) is diametrically across the molecule, about 50 Å away. In membrane-bound nAChR, the two recognition sites are indistinguishable in their ability to bind snake α-toxins and in the kinetics of this binding. Differences seem, however, to exist with respect to affinity labeling, and the environments of the chains differ.

Polyclonal antibodies (Abs) against nAChR from a number of sources have been prepared. When transferred to other animals, they cause experimental myasthenia gravis (see Section 9). Certain of these antibodies are probably not pathogenic and could even be protective. During the last few years monoclonal antibodies (mAbs) with high specificity for certain receptor sites have been prepared and are now used to study structural and functional properties of nAChR[64,65] and to study aspects of myasthenia gravis.

5. *BOTULISM*

Botulism is primarily a disturbance of the transmitter release at cholinergic nerve terminals caused by exotoxins from *Clostridium botulinum*, an anaerobic, spore-forming bacterium. Eight immunologically distinct toxins are known (A, B, C_1, C_2, D, E, F, and G) and are probably produced by different strains of bacteria. Types A, B, E, and G are known to cause human botulism. The protoxins consist of a hemagglutinating factor and a neurotoxin, the latter having molecular weights between 140,000 and 170,000. The neurotoxins contain two subunits held together by S-S bonds. The toxins are synthesized during cell growth and are released by cell lysis. The final toxic compound usually, but not always, appears after protease action (for refs. see ref. 66).

Three clinical syndromes are presently known; one is caused by ingestion of the toxin, the two others by organisms growing in a wound or in the gastrointestinal tract and producing small amounts of toxin over a period of time. A recently recognized and much studied form is infantile botulism[67,68] (for reviews see refs. 66,69). Clinical symptoms include diplopia, dysphagia, and dysarthria. Death is usually caused by respiratory failure. The average lethal dose of neurotoxin is 5–50 mg/kg body weight, different toxin types having different toxicities. Prophylaxis and treatment of botulism include immunization (toxoid injection) and the use of chemical agents that increase the influx of Ca^{2+} into the cholinergic nerve terminal (see below).

Most studies on the mode of action of botulinum toxins have been made with toxin A at the motor endplate. Morphologically, no presynaptic and generally no immediate changes are seen. Local application of toxin in nonlethal concentrations may cause sprouting of the motor nerve,[70] which occurs at different rates at "fast twitch" and "slow twitch" muscles. The sprouting apparently is a compensatory process caused by the interrupted neurotransmission at the original junctions. Eventually, new endplates are formed. Postsynaptically, the toxin-caused transmission failure results in muscle atrophy[71] of the type seen in denervation. Acetylcholine supersensitivity[72] and changes in receptor channel kinetics[66] are observed and most probably are caused by the formation of extrajunctional ACh receptors.

Electrophysiologically, mepp frequency and amplitude are decreased, and, later, evoked quantal release is impaired.[3,73] The evoked compound muscle action potential amplitude is reduced, but the quantal nature of the release remains. An incremental response occurs with high rates of repetitive nerve stimulation because of enhanced ACh release as a consequence of increased intracellular Ca^{2+} concentration,[74] but this may not be observed in poisoning from ingestion because of the very large amounts of toxin. Molecular ACh release is also reduced. Nerve conduction velocities are unaltered by the toxin,[75] and the muscle answers normally to iontophoretically applied ACh.[71,75] Endplate potentials, however, are too low to trigger muscle contraction.

The toxin is assumed to bind irreversibly to an unknown site on the external surface of the nerve terminal. Suggestions about the identity of this site include gangliosides, perhaps gangliosides at a vesicle fusion site. The toxin and its receptor are assumed to be transported into the interior of the nerve terminal, possibly by an endocytotic process.[76] This step is suggested[77] to be dependent on nerve activity and may thus be coupled to transmitter vesicle recycling. Nerve activity also shortens the time required for the onset of paralysis. The translocated complex is finally suggested to bind to a new site within the nerve terminal, where it blocks transmitter release.[77] Synthesis and storage of ACh and the number of ACh vesicles are not changed. The Ca^{2+} entry into the nerve terminal is not blocked.[73,78] The conclusion must be that some step in the release process itself is affected by the toxin.[73,76,78] A decreased Ca^{2+} sensitivity of the release process is suggested to be the cause of botulism.[73] In fact, increase of intracellular Ca^{2+} levels by the use of Ca^{2+} ionophore A23187 with high external Ca^{2+} concentration or prolongation of the nerve terminal action potential with tetraethylammonium (TEA) or 4-aminopyridine (4-AP) as well as 3,4-diaminopyridine (3,4-DAP) can be used to relieve botulism.

Hirokawa and Heuser[79] have, however, on the basis of studies on Ca^{2+} transport into nerve terminal mitochondria, suggested that botulinum toxin blocks Ca^{2+} channels.

In summary, though treatment of botulism has improved as a consequence of studies on the mode of action of the responsible toxins, there is still a long way to go before a rational therapy is available. A better understanding of the molecular events involved in the intraterminal coupling of Ca^{2+} to ACh secretion seems to be essential.

6. *PRESYNAPTIC AUTOIMMUNE DISEASES AND EXPERIMENTAL LESIONS*

Several human diseases involving the neuromuscular junction are of autoimmune origin. In some of these diseases patients have been shown to carry circulating antibodies directed against functionally important synaptic proteins. These observations have caused a considerable interest in the mechanisms of autoimmunity and also in animal models of diseases caused by the immunologic effects of synaptic proteins. For reviews on such pre- and postsynaptic diseases, sometimes excluding MG, see refs. 80,81.

6.1. *Autoimmune Effects of Choline Acetyltransferase*

Choline acetyltransferase (ChAT) (EC 2.3.1.6) catalyzes the last step in the synthesis of the neurotransmitter acetylcholine. A fatal autoimmune disease can be induced in guinea pigs by the injection of this enzyme (in complete Freund's adjuvant) purified from bovine caudate nuclei. The experimental disease is characterized by a progressive weakness, weight loss, and reduction of the respiratory rate. Sometimes diarrhea and incontinence occur. Cyanosis has been observed. The animals die after about 2 weeks from respiratory failure as evidenced by clinical, histological, and histochemical observations. Histologically and histochemically, numerous foci of inflammation are present at the neuromuscular junctions, particularly those of the diaphragm and the leg muscles. A tendency for muscle atrophy is seen in type 1 fibers as judged by ATPase histochemistry. Numerous type 1 and type 2 fibers demonstrate target and targetoid formations. Cellular reactions were occasionally found in the lungs, in liver, and in intestines. No significant abnormalities in the CNS of the diseased animals were observed. The main lesion observed is at the neuromuscular junctions, close to their ChAT-containing neurons.[82]

During purification of ChAT, the enzyme dissociated into three peaks simultaneously with a large loss of activity. All three peaks caused the formation of circulating antibodies in the injected animal, but only one fraction ("B"), with a very small protein content, caused an apparently autoimmune disease. Thus, injection of all three fractions caused formation of antibodies, but the two other antibody groups seem to be nonpathogenic.

Fluorescein-conjugated goat anti-guinea pig IgG revealed IgG bound to diaphragms from the diseased animals but not to those from controls. Electromyography (EMG) revealed slight electrophysiological changes, which were explained as evidence of a mild partial denervation. Repetitive nerve stimulation gave a slight decrement at 2/sec and a slight early incremental response at a higher frequency.

There seems to be no information concerning the ChAT activity or the amount of ACh present in the motor nerve endings of the immunized animals. There is also no information on possibly disturbed release mechanisms. Therefore, it is difficult to judge if the immunization with ChAT peak "B" caused an allergic neuropathy or a strictly neuromuscular disease.

An experimental, probably autoimmune disease also developed after injection of a soluble fraction of rat sciatic nerve mixed with complete Freund's adjuvant into female sheep.[83] The fraction used contained a high ChAT activity and a major protein band of 67,00 daltons that is suggested to be derived from the ChAT extract. Antibodies were produced to "soluble nerve proteins," and the animals became weak. Electrophysiology revealed a neuromuscular block. Lymphocytic and plasma cell infiltrations were observed in the ventral roots and around intramuscular nerve twigs. The authors discuss the possibility of a new important model of a neuromuscular disease; the data, however, suggest that neurotransmission was disturbed as a consequence of an experimental allergic neuropathy.

6.2. *The Eaton–Lambert Syndrome*

In 1957, Eaton and Lambert[84] described the generalized muscle weakness and fatiguability of a disease thereafter called the Eaton–Lambert syndrome

(ELS; myasthenic syndrome, recently sometimes Lambert–Eaton syndrome). Decrease of force,[85] proximal muscle weakness in lower extremities, weak or absent tendon reflexes, leg pain, paresthesias, and even autonomic disturbances such as dry mouth, constipation, difficult micturation, and impotence have been described. Seventy percent of cases have been found to have malignant neoplasm, mostly oat cell carcinoma of the lung.[81,86] Pernicious anemia and other autoimmune diseases have also been seen together with ELS. Another, but postsynaptic, disease, myasthenia gravis, may coexist with ELS.[87–89] The HLA antigens A3, B8, and DR3 occur with increased frequency in most of the autoimmune diseases associated with ELS.[81] For a recent review see ref. 89a.

6.2.1. Morphological, Electrophysiological, and Biochemical Evidence of Impaired Acetylcholine Release

Clinical and experimental evidence indicates that in ELS autoantibodies bind to the motor nerve terminal.[90,91] There are some postsynaptic morphological changes, but these do not impair neurotransmission. The sensitivity of the postsynaptic membrane to the released ACh[92,93] is normal, and so is the number of nAChR in the postsynaptic membrane.[94] Instead, degeneration and atrophy of terminal motor axons and demyelination and remyelination of the preterminal axons are seen ultrastructurally.[95] Postsynaptical folds show a secondary proliferation, and the ratio of postsynaptic to presynaptic length is increased.[96] This could be a compensatory development caused by degeneration and regeneration of the terminal axons.[95] Freeze–fracture electron microscopy of presynaptic membranes from ELS patients further reveals a reduced number of well-organized presynaptic ACh release zones and typical zone particles per unit area.[91] Often no release zones are present at all. Clusters of particles are seen. These observations could explain an observed reduced quantal release of transmitter (see below) provided the particles present at the "active zones" are involved in transmitter release, e.g., are voltage-sensitive Ca^{2+} channels that control transmitter release. Amounts and activity of acetylcholinesterase at endplates seem normal,[97] and anti-AChEs have few beneficial effects.

Biopsied intercostal muscle from patients with ELS reveals that both spontaneous (resting) and KCl-evoked release of ACh are abnormally low.[98,99] In contrast, the mepp frequency caused by the spontaneous release of packets of ACh ("quantal release") is unchanged in ELS and may be slightly increased,[93,99] but this type of release contributes only slightly to the total resting release of ACh from skeletal muscle.[100,101] Thus, the nonquantal leakage of transmitter seems decreased by a mechanism not yet understood. Chemical and electrophysiological[98,99,102] analyses confirm the low evoked quantal release. The impaired release produces endplate potentials too small to trigger muscle action potentials. The ACh content and level of ChAT activity are, however, the same as in healthy human intercostal muscle.[98]

Eaton–Lambert syndrome is thus caused by a presynaptic lesion that results in impaired ACh release from the nerve terminal at the neuromuscular junction.[103] Marked facilitation of endplate potentials (epps) occurs during a train of high-frequency stimuli.[102]

A single-fiber EMG shows increased jitter and blocking, which improve at higher firing rates.[89] Electromyographic studies show a low muscle action potential amplitude at nerve stimulation, which declines at low rates and augments greatly at higher rates (over 10/sec) of stimulation. Marked facilitation follows muscular exercise.[92] The motor unit potential varies in shape as a sign of disturbed neuromuscular transmission.

Elevation of external Ca^{2+} concentration or addition of guanidinium[104,105] augments transmitter release, suggesting disturbances in normal Ca^{2+} action at the presynapse. Drugs that increase the cyclic AMP content in the nerve terminal also augment ACh release.[106] Further, germine acetate improves muscle strength.[105] 4-Aminopyridine, which augments the Ca^{2+} concentration in the nerve terminal by blocking K^+ channels, is useful in ELS patients; for long-term treatment, it may, however, be too toxic.[107,108] Steroid treatment as well as other forms of immunosuppression (e.g., plasmapheresis) have positive effects.[109,110]

6.2.2. *Evidence of Autoimmunity*

Recent evidence suggests that ELS may be an autoimmune disease with circulating autoantibodies directed to the presynaptic ACh release mechanism, perhaps the prejunctional Ca^{2+} channels. Early search for a humoral factor was not very successful, but recently, a decrease in quantal content in rat muscle treated with ELS serum in the presence of 10 mM Mg^{2+} was reported.[111] Plasma exchange used on ELS patients resulted in improvement in muscle strength.[90] Significant increases in initial muscle action potential amplitude in abducti digiti minimi was described and subsided after 3 weeks.[112] The time lag was much greater than in the case of MG, where improvement may be seen within 2 days in contrast to about 15 days in ELS. The presynaptic structures that may be attacked in ELS probably have a low turnover rate.

Eaton–Lambert syndrome has been passively transferred to mice[90,113] by daily i.p. injections of 10 mg of ELS patient plasma IgG and normal IgG for control. After 37–77 days, diaphragms of mice showed no significant reduction of the number of α-bgt binding sites, i.e., of nAChR. During a train of nerve stimuli (40 Hz), epp amplitude decreased rapidly in controls to a plateau value of 32%, whereas ELS mice decreased only to 49–77%. The depression was sometimes preceeded by facilitation of epps. Patients IgG significantly reduced mean quantal content (range 31–5% to 58–9% of control).[113] The observations coincide with those made on ELS patients. Immunoglobulin G fractions from both noncarcinomatous and carcinomatous forms of ELS can transfer the disease. Small-cell carcinoma is believed to be of neuroectodermal origin and may deliver a neoantigen leading to the production of autoantibodies that cross react with nerve terminal determinants.[83] Muscle strength has occasionally been found to improve after tumor chemotherapy[58] or surgical removal.

A cerebellar disorder sometimes associated with these carcinomas may have a similar immunologic basis.

6.2.3. *Other Possible Causes of ELS*

The common occurrence of ELS together with malignant neoplasms, in particular oat-cell carcinoma of the lung, draws suspicion to these tumors not only as a possible source of autoimmune reactions directed against structures in the nervous system but also as a source of neuromuscular blocking agents or ACh-release blockers.[113a] Symptoms of ELS seem, however, largely to remain after tumor removal.[113b]

6.2.4. *Eaton–Lambert Syndrome in Animals*

Diseases that may belong to the ELS group have been described in animals. These include a neuromuscular disease resembling MG or congenital MG in man and dog, but also with features of ELS in man and with an autosomal recessive inheritance.[114]

7. *CONGENITAL MYASTHENIC SYNDROMES*

Congenital myasthenic syndromes[115,116] may have a nonautoimmune basis. They are associated with several different deficiencies of mixed location. Some disorders are inherited. Apart from these syndromes, a focal degeneration of the junctional folds has also been described in Duchenne dystrophy and other dystrophies. The clinical features of these diseases, however, differ totally from those of Myasthenia gravis, and no receptor deficiency is observed.

Congenital myasthenia starts in early infancy. The mother does not have myasthenia. Antibodies to nAChR are not seen in congenital myasthenia of the human[115,116] or the dog,[117] though some authors[118–120] have described HRPO-revealed immune complexes at the motor endplates in congenital canine myasthenia and circulating antibodies. It is questionable if the canine disease is comparable to human congenital myasthenia.

7.1. *Endplate Acetylcholinesterase Deficiency*

Inheritance of this disease is recessive.[115] Symptoms appear at birth or soon after and include fluctuating ptosis, intermittent strabismus, delayed motor development, generalized weakness increased by exertion, easy fatigability, and refractoriness to anticholinesterase drugs. Serum enzymes of muscle origin were analyzed in one case and found to be normal. No circulating antibodies to nAChR were found.

Morphological studies revealed no abnormalities of muscle fibers but a total absence of AChE from every endplate. The latter was confirmed with cytochemical studies. Nerve terminals are often small. Degeneration of junctional folds with concomitant loss of nAChR is sometimes seen.

There is a repetitive muscle action potential response to a single nerve stimulus, and the endplate current is prolonged. The number of ACh quanta released by a nerve impulse and the number of quanta readily released are reduced. Mepps are normal or slightly reduced.

The total lack of AChE explains the prolonged endplate current.

7.2. Slow-Channel Syndrome

The disorder is dominantly inherited.[97,115] Symptoms start in infancy or later. In all muscles there is a repetitive muscle action potential response to a single nerve stimulus, and endplate currents are markedly prolonged. As the AChE content at the endplate is normal, the prolonged endplate current is attributed to a prolonged open time of the ACh-induced ion channel. In clinically affected muscles, a degeneration of junctional folds with concomitant loss of nAChR is seen, and the amplitude of the mepps is reduced. Complement does not participate in the lysis of the folds; rather, the lesions observed are attributed to an altered ionic milieu caused by the prolonged openings of endplate ion channels. Down-regulated nAChR synthesis or insertion has, however, been discussed. Another possibility might be that the receptor, for genetic or other reasons, resembles an extrajunctional receptor more than the normal junctional receptor.

7.3. Congenital Endplate nAChR Deficiency

This disorder is observed early in life.[115] No autoantibodies to nAChR are seen; endplate AChE is normal, and so are endplate currents. The number of α-bgt sites per endplate is reduced. Decreased nAChR synthesis or membrane insertion or accelerated degradation of nAChR are considered as are possibly structural abnormalities including decreased affinity for ACh. A related recessively inherited syndrome in dogs has been discussed.

8. FAILURE IN NEUROMUSCULAR TRANSMISSION CAUSED BY CHANGES IN ACETYLCHOLINESTERASE ACTIVITY OR BY OTHER LOCALIZED ACTIONS OF ANTICHOLINESTERASES

At the neuromuscular junction of mammalian skeletal muscle, the enzyme acetylcholinesterase (acetylcholine acylhydrolase, E.C. 3.1.1.7, AChE) is found in the presynaptic membrane, in the synaptic cleft, and in the muscle fiber. The enzyme of the synaptic cleft occurs in different forms.[121,121a] One is a soluble form, whereas the other if found attached to the basal lamina (by a collagen tail) that ensheathes every muscle fiber and finally joins the membrane of the Schwann cell that covers the terminal. The lamina passes between the nerve terminal and muscle fiber at the endplate region and projects down into the secondary postsynaptic clefts formed by the enlargement of the muscle membrane at the endplate. Apart from AChE, the lamina seems composed of collagens, glycosaminoglycans, and glycoproteins. Its function is not well understood but is currently the subject of very active research.

Junctional AChE controls the ACh content of the synaptic cleft by hydrolyzing the transmitter. Failing this, a sudden accumulation of the transmitter occurs and has severe consequences. At worst, a cholinergic hyperactivity is produced, initially causing muscular fibrillations and cramps and eventually resulting in muscular paralysis and death. Sublethal consequences have been described, such as increase in mepp frequency and amplitude,[122] lengthening of the rising phase of the endplate potential and of the duration of underlying conductance changes,[123] and desensitization of the nicotinic acetylcholine receptor.[53] Prolonged increase in ACh concentration in the cleft sites[55] and muscle fiber degeneration starting at the motor endplate may occur.[124] Presynaptic consequences of AChE inhibition have been described[122] but could be attributable to other actions of the inhibitor rather than to its inactivation of AChE. Some carbamate inhibitors block the high-affinity choline uptake system at the nerve endings.

Diseases caused by changes in the amount of AChE present at the junction or by changes in the properties of the enzyme molecule seem to be extremely rare. A disorder caused by lack of AChE at the neuromuscular junction has, however, been described.[103,115]

Several different pathological conditions are caused by inhibitors of AChE. The type of lesion depends on the chemical nature of the inhibitor and also on the type of AChE inhibition, i.e., acute or chronic. Inhibitors of AChE are usually either quaternary ammonium compounds, carbamates, or organophosphates.

8.1. Carbamates

All carbamates have a common basic structure based on carbamic acid. The substitutions made on this structure determine the chemical nature and biological action of the final compound. Carbamates may thus have insecticidal, fungicidal, or herbicidal activities.[125] Some are used in medicine. Methyl- and dimethylcarbamates inhibit AChE, whereas dithiocarbamates have little AChE-inhibiting ability. Quaternary carbamates penetrate membranes slightly and therefore inhibit peripheral AChE. The carbamate Tensilon® (edrophonium) is used for diagnostic tests in patients with suspected myasthenia gravis (see Section 9) to increase the ACh concentration at the acetylcholine receptors of the muscle endplate area. In a situation with an overdose of AChE inhibitors (pyridostigmine, ambenium), a so-called cholinergic crisis is observed in the patient, a condition that may lead to paralysis of the respiratory system and is treated by withdrawal of the anti-AChE and by artificial respiration. As a result of AChE inhibition, two phenomena are seen in sequence at the neuromuscular junction, a muscle facilitation, followed, at continued enzyme inhibition, by muscle flaccidity and weakness. Generally, intoxidation by carbamate AChE inhibitors results in salivation, lacrimation, miosis, urination, defecation, muscle spasm, muscle weakness, prostration, and convulsions.[126] Carbamates have considerable neurobehavioral effects involving muscarinic acetylcholine receptors and other neurosystems in the CNS.

The carbamates inhibit AChE (and other ChEs) by carbamylation of the esteratic site of the enzyme,[121a] thus preventing the enzyme from catalyzing the hydrolysis of ACh. Carbamates thus replace the natural substrate at the active site of the enzyme. Inhibition is generally transient, though a temporary covalent binding to the enzyme occurs.[127] Some carbamates form anti-AChEs after biotransformation.

Dithiocarbamates produce peripheral neuropathies in man and animals. Hindlimb paralysis and ataxia have been seen in chronically exposed rats. Behavioral effects are noted.[128] Neuropathological and electrophysiological data from ataxic rats suggest demyelination and degeneration of peripheral nerve, decreased nerve conduction velocity, and abnormal electromyograms. This is also seen in the human, together with a great number of behavioral symptoms.

Carbamates may cause a Ca^{2+}-mediated myopathy initially affecting the motor endplate.[129] Transient Ca^{2+} deposits are observed at sites of carbamate-induced focal lesions and can be prevented by EGTA. With chronic drug treatment, the lesions and the Ca^{2+} deposits decrease, whereas alterations in the motor nerve terminal, the synaptic space, and junctional folds persist. Rod bodies and ribosomal clusters may be seen clumped with sarcoplasmic reticulum, T-systems, or mitochondria. The persistent endplate alterations are suggested to be related to Ca^{2+} flux through the postsynaptic membrane after prolonged nAChR interaction and prolonged damage to the postsynaptic membrane, particularly at the receptor sites.

8.2. Organophosphates

Organophosphates are used as insecticides and lethal chemical warfare agents. They inhibit AChE acutely and more or less irreversibly, thus causing severe acute effects and death by respiratory paralysis. Symptoms are as in carbamate poisoning (see above and ref. 130) and are caused largely by accumulation of ACh. Other, indirect toxic effects involve myopathies[129,131] and seem to be caused by Ca^{2+} accumulation. Organophosphates may cause a necrosis of the neuromuscular junction, of both the motor endplate and the nerve terminal,[132,133] when a critical loss of AChE activity has occurred. This neuropathic syndrome is typical for, e.g., a subacute diisopropylfluorophosfate (DFP) lesion. Presynaptically, mitochondrial changes are observed; otherwise a flattening of the primary cleft, increased basement membrane material, and accumulation of electron-dense debris are seen. The normal synaptic relationship is disturbed. Extensive axonal sprouting is observed after a week, emerging from a distal node of Ranvier.[134] Some of these effects may be caused by the organophosphate compounds themselves.

Organophosphates inhibit the enzyme by forming a covalent bond with it; i.e., they phosphorylate the enzyme at a serine group in its active center. Certain organophosphate residues, when bound to the enzyme, may split off another group ("aging"), after which the phosphorylated enzyme is refractory to reactivators, i.e., nucleophilic compounds able to restore enzyme activity by dephosphorylation.[135,136] The aging reaction seems to be triggered by the enzyme itself.

A certain group of organophosphates, e.g., triorthocresylphosphate (TOCP), do not inhibit AChE at all or not only (depending on structure) inhibit AChE but cause lesions known as delayed neurotoxic effects.[130,137,138] They produce a polyneuropathy. Their main targets are the large-diameter myelinated axons of both the PNS and the CNS. The polyneuropathy results in functional impairment of the nerve terminals. Symptoms[139] include nausea, vomiting, and diarrhea for a few hours or days, then ataxia and paralysis beginning in the distal portion of the hindlimb and advancing proximally. A latent period of 6–14 days is seen. Recovery is very slow and often incomplete. A massive outbreak of paralysis caused by TOCP is known as Ginger Jake disease.[140]

The molecular mechanism is not completely understood. The inhibition (phosphorylation) of a 155–180,000-dalton polypeptide on a so-called neurotoxic esterase, recently renamed neuropathy target esterase (NTE), seems to initiate the delayed neuropathy. For review see refs. 141 and 142.

There are nonneurotoxic compounds, including certain phosphinates, sulfonates, and carbamates, that are able to inhibit neurotoxic esterase.[143] It was found that they protect the enzyme from phosphorylation by neurotoxic inhibitors. Thus, inhibition of the enzyme alone is not the reason for the neurotoxic effect, but a second reaction, involving splitting off of the second group (R) from the phosphorus-containing moiety attached to the enzyme active site, needs to occur. A charged substituent is then left on the enzyme, which makes restoration of the enzyme activity no longer possible. In contrast to aging of phosphorylated AChE, the group cleaved off seems to be transferred to a second site on the abovementioned polypeptide.[143] The mechanism of the disease was mostly studied in hen, chicken, or rat. Certain aspects may be studied in human lymphocytes or platelets that carry the NTE.

9. *POSTSYNAPTIC AUTOIMMUNE DISEASES: MYASTHENIA GRAVIS*

It is impossible to summarize all information available on myasthenia gravis (MG) and experimental myasthenia gravis (EAMG) within the framework of this chapter. Most genetic, clinical, diagnostic, and treatment aspects are therefore excluded. There are many reviews on these subjects and on MG in general. For recent ones, see refs. 144–155.

An MG-like disease was first described in 1672, when Willis observed nerves "obstructed by a vicious humor." Erb, in 1877, recognized MG as a disease in itself. Relevant treatment, however, was not started until 1934, when MG, observed to be "like curare poisoning," was treated by Walker with physostigmine. After 1950, one began to understand the nature of the disease, and after 1960 evidence for its autoimmune nature began to accumulate. At present, MG is one of the main examples of an autoimmune disease. It occurs in only about 0.01–0.05% of the population, yet it is very well studied, satisfactory animal models are available, and several of its features are studied in cell cultures and other *in vitro* preparations.

The disease is characterized by weakness and extreme fatigability of voluntary muscles, often associated with thymic disorders. Certain genetic factors seem to predispose for MG. The symptoms of true MG depend largely on an antibody-induced reduction of the functional number of postsynaptic motor endplate acetylcholine receptors.[156,157,200] Aspects of cellular immunology are discussed and are probably most important, but these are as yet not well understood. As shown earlier in this chapter, a number of other diseases with characteristics not unlike those of MG are known, but these diseases are clinically, immunologically, therapeutically, and genetically different. A somewhat floating limit between the various diseases makes thorough diagnosis very important. Myasthenia gravis occurs in several different forms,[158] which relate to genetic factors, sex, or age.

Generalized myasthenia is often (>75%) associated with HLA-B8 and DRW3 and thymic hyperplasia when the disease starts before the age of 35 (female-to-male ratio 3 : 1). When it starts later, thymic tumor (usually thymoma of lymphoepithelial type) is common but without a clear-cut association to the HLA system. High titers of acetylcholine receptor antibodies are present, particularly in thymoma cases, and striated muscle antibodies are also found. Other autoimmune diseases may exist simultaneously. A number of different autoantibodies against the thyroid gland, thymus, gastric mucosa, striated and smooth muscle tissue, and subcellular elements such as nuclear factor or mitochondria, may exist. The occurrence of familial MG seems rare, but "myasthenic" EMGs in apparently healthy relatives occur, though their cause and significance are not yet understood. Possibly nAChR antibodies with low receptor affinity are present, or antibodies are directed against determinants of minor importance for neuromuscular transmission. Spontaneous MG has been observed in animals, e.g., dog and cat.[117,157]

Ocular MG, a localized form of MG in which weakness is limited to the extraocular muscles, is seldom helped by thymectomy but is usually helped by corticosteroids. Few antibodies are present, and these may be unique for the afflicted individual.[157a,158] Localized MG may be caused by antibodies bound in localized complexes. Electrophysiologically, weakness in other muscles may be detected.

Neonatal myasthenia is a transient form of MG seen in about 12% of babies born to myasthenic mothers. Placental transfer of maternal nAChR antibodies, probably via the umbilical cord,[159] seems to be the cause. Symptoms disappear gradually as the antibody titer falls. Recently, it was discussed whether children with neonatal MG may transiently synthesize nAChR antibodies.[160] Thus, in a study involving 17 myasthenic mothers and 19 neonates, only three neonates developed MG. No correlation was found between the severity or the treatment of the maternal disease and the development of neonatal MG. However, nAChR antibodies of the sick children survived considerably longer than those of the healthy neonates. Synthesis of IgG in a fetus can occur from week 20 of pregnancy on. Failure of exchange transfusion further supports the theory that antibodies are synthesized in the child. Synthesis may be triggered by material from endplates damaged by transferred maternal antibodies or may be from a transferred maternal cell clone. Antiidiotypic antibody showed differences, increasing after some weeks, between the idiotypes found in the mother and those in sick children but not in the healthy ones. The disease disappears within 45 days, and the antibody titer of the child declines to zero.

Drug-induced forms of MG have been described. Such lesions are mostly reversible. They are discussed in Section 10.

9.1. Diagnosis

Diagnosis of MG[152,154] includes muscle fatigability tests (EMG) with repetitive nerve stimulation, single-fiber EMG, drug tests, and nAChR antibody determinations by radioimmune assay (RIA) or enzyme-linked immunosorbent assay (ELISA).[161] For RIA, radioiodinated α-bungarotoxin (α-bgt) bound to nAChR in crude detergent extracts from human muscle[162] is routinely used. After incubation with patient sera, precipitation of immune complexes is performed with anti-IgG. Antibodies directed towards α-bgt binding sites are found by competition experiments. The range of antibody titers in MG may go from 0 to 900 nmol α-bgt binding sites per liter of serum.[163] Individual and organ specific differences in determinants may exist; e.g., in ocular MG antibodies may be missed with leg muscle receptor as an antigen but be found with ocular muscle receptor. Sometimes only the patient's own muscle nAChR finds the antibody. Generally, extrajunctional myasthenic human muscle nAChRs bind antibodies stronger than junctional nAChRs.[164] In about 5% of patients with generalized MG, no nAChR antibodies are found. Possibly antibody titers are low, and most antibodies are bound to receptor *in situ* or are circulating as complexes.

It is not advisable to use other than human muscle nAChR in testing myasthenic serum for antibodies. Cross reactivity between animal species varies[251] from about 80% with mammalian muscle receptor to very low with electric organ receptor.

9.2. Evidence of the Autoimmune Nature of MG and Nature of the nAChR Antibody

The similarity of the histological appearance of the thymus of a myasthenic patient and that of the thyroid in Hashimoto's disease[165] raised suspicions about an autoimmune cause of MG. Simpson observed the high frequency of other autoimmune phenomena in patients with MG and suggested that the disease was caused by the presence of antibodies at the motor endplate.[166] Fluctuations of complement levels with disease activity were reported,[167] as were circulating antibodies to striated muscle that cross react with thymic myoid cells in about one-third of patients with myasthenia[168] and in almost all myasthenic patients with a tumor of the thymus. These antibodies are, however, connected with the tumor and are not pathogenic at the muscle receptor. Lymph drainage was observed to relieve MG symptoms, whereas reinfusion of cell-free lymph aggravated the clinical symptoms.[169] Antibodies to skeletal muscle nAChR were found in 1973, after a cross-reacting nAChR had been purified. Antibodies blocking α-toxin binding to the receptor were demonstrated in the serum of patients with MG.[170] Almost 90% of MG cases have circulating polyclonal antibodies[171,172] directed to several determinants of the nAChR in the skeletal muscle membrane of varying pathogenic importance. The antibodies consist of populations belonging to the IgG fraction. In early stages of MG and experimental myasthenia gravis (EAMG), antibodies belonging to the IgM type may be seen.[153]

The occurrence of neonatal MG seems to confirm the autoimmune nature of MG, though, as discussed above, some neonates may transiently synthesize nAChR antibodies. Plasmapheresis, which probably improves the clinical status of patients because of an achieved reduction in circulating nAChR antibodies,[173,174] further confirms the autoimmune nature of the disease, and certainly so does the therapeutic effect of multiple immunosuppressive treatment. There are, however, MG cases described in which doubt is thrown on nAChR antibodies as the only source of MG symptoms. An example is an asymptomatic thymectomized mother with a baby with neonatal MG, with antibodies in both, where the antibodies do not cause symptoms in the mother.[175,176] Also, in postthymectomy patients with residual ocular signs, plasmapheresis has failed to achieve further clinical improvement despite a reduction in antibody titers of greater than 90%.[177] These patients respond to the anticholinesterase edrophonium and to low doses of prednisone.

The heterogeneity of nAChR antibody varies among individuals, and the polyclonal nature of the antibodies in MG is most probably the cause of a well-known low correlation between clinical status and serum nAChR antibody titer in MG or experimental MG in general. The correlation is better for the individual patient or experimental animal in different stages of the disease.[177] Various antibody clones seem directed towards receptor parts of varying pathogenic importance. Immunization of animals with isolated nAChR subunits (see below) has demonstrated considerable differences in the immunogenicity of these subunits and in their ability to cause EAMG.[178,178a,179] Further, certain antibodies not found with conventional RIA are directed against the α-bgt binding site of the receptor and may be related to the severity of the disease. However, at least two distinct subpopulations of antibodies block α-bgt binding to nAChR, and one of them seems not to be directed against the toxin binding site itself.[180]

Another unexplained observation concerns the degree of muscle weakness, which may vary in a patient, although all muscle groups are exposed to the same population of antibodies. This was suggested to reflect muscle factors rather than the antibodies, i.e., endplate accessibility, the safety margin of neuromuscular transmission,[181] or the ability of the muscle to compensate for the antibody-induced receptor deficit by increasing the synthesis of new receptors.[55]

It is not clear whether improvement in MG is correlated to the functional activities of antibodies. Age, sex, race, and duration of disease may have little effect on serum nAChR antibody levels,[182] whereas certain genetic factors seem to be related to a disposition for MG.

It is unknown what triggers and sustains the formation of antibodies in MG. Suggestions include existence of a primary antigen on the cell surface of thymic lymphocytes[183] and the presence of nAChR during development on myoid cells of the thymus. These antigens cross react with serum nAChR antibodies from MG patients and EAMG animals. Currently, an imbalance between suppressor and helper T lymphocytes as a cause of the disease (see 9.9.2f) is under investigation.

9.3. Role of Complement

From animal experiments, it seems clear that the cytotoxic reaction between nAChR and its antibody, which is partly responsible for the symptoms of MG and EAMG (see Section 9.6), needs complement C3 for the characteristic reduction of the postsynaptic membrane of the endplate[184] in these diseases. Both complement C3 and C9 have been shown to be present at the endplate in MG.[183a]

9.4. Immune Complexes

Immune complexes form at the motor endplate and have been suggested to be circulating.[185] Clq-binding activity in patient sera has been described.[185a] No correlation was found between the existence of immune complexes on one hand and the duration and severity of MG or the antibody titers on the other.[186] Thymectomy has been suggested to reduce the concentration of immune complexes. Immune complexes may prove to be of interest in protection or treatment: EAMG of rabbits using *Torpedo* nAChR for immunization (see below) is suppressed by pretreatment with immune complexes containing receptor and antireceptor antibodies. Antibody levels are reduced, and synthesis of antibody is terminated.[187] The composition and pathogenicity of various immune complexes are currently under investigation. The mechanism of pretreatment with immune complexes is not completely clear. The acid-treated receptor in the complex might act by inhibiting a subsequent response to untreated receptor as demonstrated for reduced carboxymethylated receptor.[188] It is more likely that the complexes stimulate[189] the generation of antiidiotypic antibodies or of antibodies directed against the complex and different from those directed against each of its components.

9.5. Role of Thymus

The association between MG and thymic tumor was recognized in 1901 but it still not understood, nor is the role of the cellular immune regulation. Thymectomy was first performed in the 1940s as a treatment of MG. The thymus, however, is responsible for the maturation of T lymphocytes, which serve as helper cells, suppressor cells, and cytotoxic T cells and may thus play an important role in causing an imbalance in these T-cell subclasses. B cells, precursors of the antibody-producing plasma cells, seldom occur in normal thymus.[190]

In young MG patients, thymus shows hyperplasia in about 80% of cases and is a site of anti-nAChR antibody production. Ten to 15% of MG patients, most of whom are around or over 40 years old, have a tumor of the thymus. In some older patients, the thymus may be involuted, and thymus removal may be of little help. However, some patients, particularly those without thymoma, improve after thymectomy.[152] Figures seem to vary (60–80%)[191] and may depend on complete removal of the thymus.[190a] In a number of patients, thymitis, i.e., infiltration of the thymic medulla with lymphocytes forming follicles with germinal centers in which B cells occur, is seen.[190b]

Histological observations of follicular hyperplasia and the recovery of IgG-positive cells from myasthenic thymus suggest an abnormal B-cell proliferation (see, e.g., refs. 192–194). Thymectomy reduces the number of circulating lymphocytes without causing serious immunologic deficiencies. It is worthwhile mentioning that the presence of both muscarinic and nicotinic[195] AChR on human lymphocytes has been reported, and a possible role for these receptors in MG pathogenesis has been discussed.

As mentioned above, the thymus may be an active site of nAChR antibody[196,197] synthesis. The ratio of nAChR antibodies to total IgG is higher in thymic tissue than in peripheral blood, suggesting synthesis or accumulation.[198] Irradiated thymic cells that are viable but incapable of antibody synthesis augment the production of antibodies to nAChR by autologous peripheral blood lymphocytes, suggesting that the thymus is the site of some factor, possibly a helper T cell, that enhances synthesis of nAChR antibodies.[199]

Thymus may thus be able to modulate autoimmunity to self-nAChR, perhaps after a viral infection (see below). The nAChR-bearing myoid cells[200] or other thymic cells may proliferate in myasthenics. T cells might then become sensitized to the thymic cells and cause B cells to synthesize antibodies to nAChR that cross react with skeletal muscle.[201] Both stages could be under genetic control. Another suggested possibility is that deficient suppressor cell mechanisms may fail to regulate the production or activity of the nAChR antibody. Other thymic factors have been considered to be involved, such as thymosin[203] and thymin (thymopoietin),[202,204] a thymic polypeptide.

Most normal as well as myasthenic thymocytes have receptors for sheep erythrocytes, the Fc part of IgG, and a thymocyte antigen predominantly present in cortical areas. Only a small part of thymus cells are IgG positive, with a slight increase in thymus biopsies from MG patients.[192] In most patients with thymoma (80%), there are no signs of IgG production. Two different subpopulations of thymocytes are found with respect to HLA and a thymocyte antigen, both in normal and myasthenic thymus. The reactivity of thymic cells to a large number of polyclonal lymphocyte activators seems equal in thymic cells from patients and normals.[196] An association between nodular areas and the follicles usually seen in myasthenic thymus within the medulla and an association between the number of follicles and C3- and IgG- positive nodules is suggested.

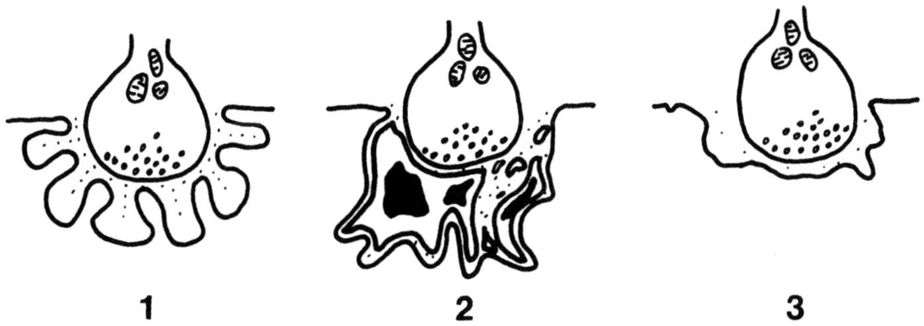

Fig. 4. Schematic drawing of the nAChR antibody- and complement-triggered muscle endplate destruction and simplification in MG and EAMG. 1, normal endplate. 2, endplate during inflammatory disease stage. Note debris caused by invasion of killer cells and macrophages, antibody binding and complement fixation. EAMG, perhaps MG. 3, during chronic disease stage. Note normalized synaptic cleft but reduction of postsynaptic membrane.

9.6. The Nature of the Antibody-Caused Lesion at the Neuromuscular Endplate

The breakdown in tolerance to self-nAChR seen in MG causes, for unknown reasons, the production of receptor antibodies, which results in loss of muscle nAChR and a destruction and simplification of the muscle endplate area. In EAMG, a similar final result is achieved (Fig. 4) after immunization of animals with solubilized (and purified) nAChR or subunits of it, which results in the formation of a polyclonal antibody population with varying pathogenicity. Mapping of immunogenic regions on nAChR with monoclonal nAChR antibodies[208] shows that over 50% of monoclonals bind, mutually exclusively, to a "main immunogenic region" on the same nAChR subunit to which most of the polyclonal nAChR antibodies in EAMG sera also bind.

Certain EAMG receptor antibodies of the IgG type block ACh receptor-induced $^{22}Na^{+}$ flux and are suggested to inhibit ACh receptor function by interfering with the agonist-induced conformational changes in the receptor–channel complex.[209]

Endplate channels in EAMG rats with high levels of nAChR antibodies have reduced mepps, although the mepp decay time constant and the endplate current decay time are unaffected as compared to controls. Noise analysis indicates unchanged single-channel conductance and mean channel lifetime. Mean quantal content and Ca^{2+} dependence of the endplate potential are also unchanged. The conclusion is that at endplates of EAMG animals the number of activated receptor–channel complexes is reduced without modification of single-channel properties.[209] This resembles the endplate of MG patients.

Extensive studies, particularly of EAMG but also of antibody effects on myotubes and of antibody populations and their effects in MG, have led to the following conclusions concerning the pathogenesis of MG.[150]

A direct and, probably depending on the affinity between nAChR and antibody species, more or less reversible immunopharmacological blockade of nAChR function by antibody binding occurs.[153] It is not clear which parts of the receptor are involved (ACh-binding site, ion channel, or other).

A complement-dependent lysis of the postsynaptic membrane occurs as a consequence of antibody binding.

Intermolecular cross linking of nAChR by their antibodies, resulting in increased degradation rates of nAChR, has been suggested.[210]

The two events mentioned first seem to be the important ones in MG and in EAMG, particularly the second. Recent studies suggest that nAChR synthesis may increase in compensation of the increased degradation,[55] but not as a consequence of cross-linking by antibody.[211]

9.7. Is the CNS Involved in MG and Experimental Autoimmune MG?

Epileptic patients without clinical signs of MG (electrophysiology not tested) occasionally (3/27 children) have low titers of muscle nAChR antibodies.[205] Deficiency of IGA, probably caused by antidepressive drugs, and HLA-A1 and B8 antigens as well as primary generalized seizures occur in these cases. These patients also have brain nAChR antibodies (measured with rat brain nAChR as antigen). A MG patient with a high titer of muscle nAChR antibodies and a low titer of brain nAChR antibody, lower than that of the epileptic children, is known. Some epileptics may primarily have an immune reaction directed against brain nAChR.

The occurrence of nAChR antibodies in the CNS has been discussed in several papers. Effects of nAChR IgG injected into the brain of rabbits are described,[206] and human myasthenic IgG, which has been found to cross react with rat brain nAChR *in vitro*, may react with a presynaptic protein and seems to decrease ACh release from rat hippocampal nerve endings.[207] The observation is difficult to interpret, since the cholinergic autoreceptor involved in ACh release from central cholinergic neurons is suggested to be of the muscarinic and not of the nicotinic type. On the other hand, cross reactivity between such brain receptors may yet be found. Biosynthesis of receptor antibodies in the brain would most probably be local, as passage of muscle nAChR antibodies into the brain has not been demonstrated.

9.8. Virus Induction: Possible Etiology of MG?

Simpson[166] was the first to point out that MG may be the consequence of a virus infection, and the involvement of virus has since been suggested by others.[192,212–216] Studies on patients after viral infections have given contradictory results. In one particular study,[216] five out of 50 patients suffering from MG with symptoms starting a few weeks after a proven viral infection (brainstem, encephalitis, measles, herpes zoster, upper respiratory tract infection) or immunization (measles vaccine) were described. A viral etiology as triggering factor in the appearance of autoimmune disease has been reported in a number of diseases (for refs. see ref. 216). In another study,[214] patients with MG were tested for antibodies against ornithosis, mycoplasma pneumonia, and 16 viral antigens, and no correlation was found. Cytomegalovirus was found in elevated titers in untreated MG patients.[213] The existence of new antigenic determinants on the surface of myasthenic thymocytes was suggested.[192] It has also been suggested that HLA antigen B8, often linked to MG, is associated with a greater degree of susceptibility to certain viral infections.[215] The status of the thymus in many MG patients also suggests the presence of viral antigen. Mice injected with measles viruses developed both antibodies to nAChR and clinical signs of muscle weakness.

It was suggested that a viral infection could release sequestered antigens from virus-infected cells or tissues (cf. ref. 216). Perhaps the virus could induce alterations of host cell membrane antigens, which then would appear as foreign. Rabies virus has quite recently been shown to enter muscle cells by way of nAChR.[217] Another possibility is cross reactivity between viral and host antigenic determinants. Virus could also alter the immune system by alteration of T–B or T–T cell interactions or by direct proliferative stimulation of autoaggressive cells or by acting as an adjuvant.[216]

9.9. *Experimental Studies on MG*

9.9.1. *Experimental Autoimmune MG*

A real breakthrough in the understanding of the autoimmune nature of MG came when experimental models of MG (EAMG) were obtained (see reviews 149, 150, 153, 155). As a consequence of the purification of the nicotinic acetylcholine receptor,[218–220] immunization of rabbits with the purified nAChR plus Freund's adjuvant was performed,[58–60] and it produced circulating receptor antibodies. These antibodies, which belong to the class of IgM in early stages and to IgG in later states of the disease, cause muscle weakness and a flaccid paralysis in the animals.[58–60] Electrophysiological analysis[221] revealed decreased mepp amplitudes, abnormal jitter, normal quantum content, and normal number of epps. Muscle decrement on repetitive nerve stimulation was observed.

Morphological[221a] and morphometric analysis[222] revealed two stages of the disease in the rabbit. The first is an inflammatory stage with complement- and antibody-mediated lysis and invasion of the endplate region by phagocytic cells, a swollen and debris-filled synaptic cleft, and destruction of the postsynaptic membrane from the tips. This stage is followed by a chronic stage characterized by a reduced number of receptors[221a] and with a normalized synaptic cleft and simplified postsynaptic membrane. Presynaptically, the nerve terminals do not seem broken, though changes in size may occur. Sprouting of the motor nerve is occasionally observed, and so are signs of new endplate formation. Experimental autoimmune MG in the rabbit responds positively to anti-ChE treatment in early stages of the disease; later little effect is seen. The reduced ACh sensitivity of the muscle[221,223] is accompanied by a reduced α-bgt binding capacity of the muscle. The chronic stage of EAMG, particularly in the rat,[224] reminds one more of human MG than the acute stage does, as inflammatory stages of MG in the human are seldom seen.

Experimental autoimmune MG may be obtained with nAChR purified from the electric organ of *Torpedo* or *Electrophorus* but equally well with skeletal

Fig. 5. Lewis rat immunized with purified nAChR. Lymphocyte-promoting factor from *B. pertussis* was used as an adjuvant. (From the author's laboratory.)

muscle nAChR[225a] including syngeneic muscle receptor. Antibodies are first directed against the injected nAChR, but these antibodies eventually decline while others directed to the animal's own endplate receptor appear. Antigen is probably provided by endplate receptor removed from the endplate by cytotoxic action. Another possibility would be that the thymus receptor shown[200] to be present in thymus, perhaps on myoid cells (cf. ref. 150), and to cross react with it is the source of the later antigen. The antibody production is T-cell dependent; thymectomized rats (E. Heilbronn, unpublished data) develop few antibodies and no symptoms of EAMG. In neonatally thymectomized rabbits no electrophysiological signs of EAMG were seen, though nAChR antibodies were present.[227] The reason is probably not a thymic factor necessary for EAMG but the reduced development of nAChR antibodies in the absence of thymus.

Experimental autoimmune MG by immunization with receptor has been caused in rats (Fig. 5), mice, monkeys, goats, etc. (see refs. in ref. 149). In mice only chronic EAMG is normally seen. Symptoms are weak, and only decreased mepp amplitudes may be observed. The choice of adjuvant could be important (cf. ref. 150); thus, EAMG accompanied by severe symptoms may be caused in rat when lymphocyte-promoting factor purified from *Bordetella pertussis* vaccine is used as an adjuvant[226] (cf. ref. 150).

Transferred cells from lymph nodes of EAMG guinea pigs induce a mild transient muscle weakness in 50% of tested animals. The same was seen in rats.[231] Total-body irradiation combined with thymectomy prevented EAMG,

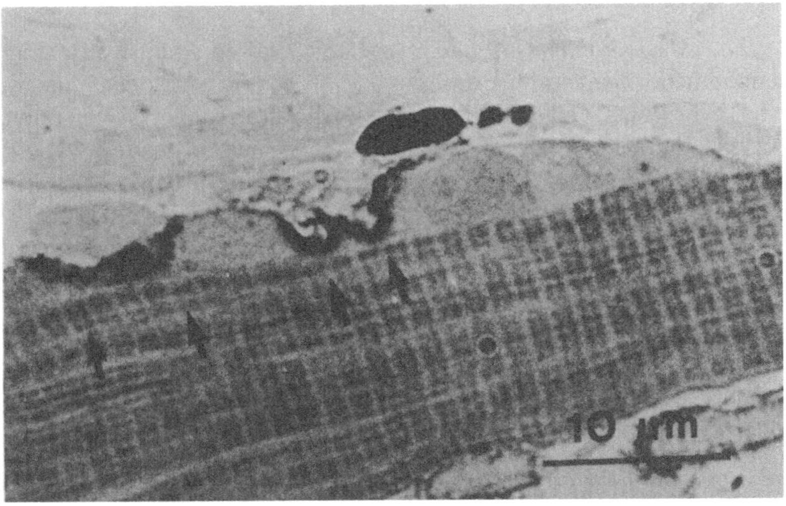

Fig. 6. Mouse diaphragm muscle fiber showing myasthenic IgG bound to the muscle endplate. Staining with protein A-conjugated peroxidase. (From the author's laboratory.)

experiments that indicate the importance of cellular mechanisms in the development of EAMG. T-Helper cells are suggested to be (one of) the crucial factor(s). Antibodies can cause EAMG in mice following passive transfer of human myasthenic IgG or IgG from EAMG animals provided an immunosuppressive (cyclophosphamide) is given.[184,228–230] Depletion of complement C3 in recipient mice diminishes the effect. Passive-transfer animals show fewer symptoms (see, however, ref. 230); their disease is generally only detected by mepp measurements. Antibodies to nAChR at the muscle endplate have been demonstrated[57] (Fig. 6). Even monoclonal antibodies directed against specific determinants of nAChR induce EAMG in passive-transfer experiments characterized by a reduction in miniature endplate potentials, binding of antibody to skeletal muscle nAChR, and a reduction in the availability of functional nAChR. It was shown that neuromuscular transmission can be impaired by monospecific nAChR antibodies that do not inhibit the binding of neurotransmitter to its receptor site.[208,239]

The amount of ACh released from endplates of MG patients[99] and the phrenic nerve–diaphragm preparation of EAMG rats is higher than normal (J. Häggblad, unpublished data). Biopsies from human MG patients contain more ACh than normal.[233] An increased level of choline acetyltransferase in MG muscle was also found.[234] The findings confirm the interpretation that increased ACh synthesis and release might be a compensatory mechanism for the reduced postsynaptic ACh sensitivity.

In EAMG as in human MG, the correlation between antibody titer and clinical severity of the disease is only rough because of the polyclonal and heterogeneous antibody population.

9.9.2. *Experimental Therapy*

9.9.2a. Monoclonal Antibody. Antibodies raised against any single determinant may be of several Ig classes and idiotypes.[236,238] Cloned lymphocyte–myeloma hybridoma cell lines have allowed detailed analysis of polyclonal antisera and the assessment of the relative contribution made by each component to the overall immune reaction *in vivo*. Certain monoclonal *Torpedo* nAChR antibodies, when injected *in vivo* into rats, induce EAMG.[238,239] Thus, a molecular antibody species reacting with a single antigenic determinant may produce the disease. Even monoclonal antibodies binding postsynaptically but remote from the binding site for cholinergic ligands in the muscle cause neuromuscular transmission defects.[237–239] The use of monoclonal antibodies will provide a better understanding of the induction of EAMG and thus may provide a background for understanding MG pathogenesis and therapy.

9.9.2b. Immune Complexes. Experimental autoimmune MG induced in rabbits by immunization with purified nAChR from *Torpedo* is suppressed by pretreatment of experimental animals with immune complexes consisting of nAChR and its antibodies. No severe paralysis is seen, antibody levels are reduced, and the synthesis of antibody is terminated. The mechanism is not altogether understood.[187,240]

9.9.2c. Antiidiotypic Antibodies. Immune response gene complexes. The antigen-combining site of antibodies, the idiotope, is structurally unique and is also antigenic in itself. Antibodies directed against these sites are produced in the body and may down-regulate antibody production by lymphocytes.[240a]

Experimentally, cross-reacting monoclonal antiidiotypic antibody directed against nAChR antibodies has been prepared[241] to the idiotype(s) in a subpopulation of antibodies directed against the toxin-binding region(s) of the receptor. These monoclonal antibodies cross reacted with antibodies from other myasthenic sera and thus probably shared idiotypic specificities. It was hoped that antiidiotypes of this kind might stop nAChR autoantibody production. A difficulty seems, however, to be lack of interanimal cross reactivity[242] and also the heterogeneity of the nAChR antibodies, which would demand the induction of antiidiotypes to all pathogenically important nAChR idiotypes. In tests on rabbits[242] or rats,[242a] so far, no protection against EAMG was found by prior immunization with purified sheep antibody.

Experimental autoimmune MG responsive to neostigmine has also been caused by an antiidiotypic route that led to the development of nAChR antibodies.[243] An agonist of nAChR, BisQ, constrained in its structure and probably structurally complementary to the combining site of nAChR, was conjugated with serum albumin and used as an antigen in rabbits. Formed antibodies cross reacted with *Torpedo* and rat receptors. The developed rabbit EAMG was probably caused by a set of antiidiotypic responses directed at determinants near or part of the binding site of the first antibody. The experiments suggest that an autoimmune disease involving antibodies to receptors may result from an antiidiotypic response to an antibody specific for a ligand of the receptor.

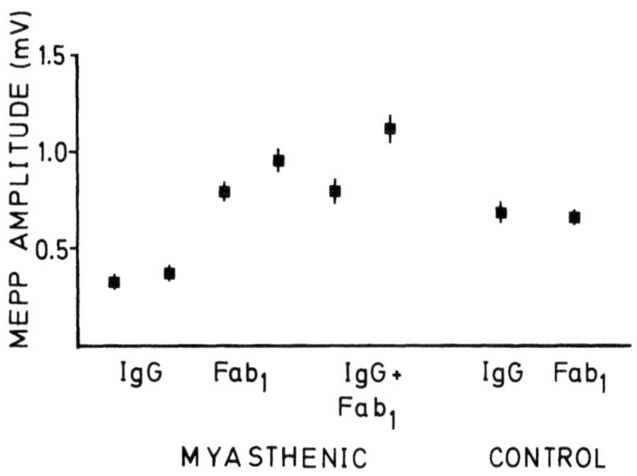

Fig. 7. Protection of nAChR from the blocking effect of myasthenic IgG by Fab_1 as demonstrated by mepp measurements. (From the author's laboratory.)

The role of immune response gene complexes in the manifestations of EAMG is currently being investigated with antibodies. *In vivo*, a monoclonal antibody to I-A gene products was found to suppress immune responses to nAChR.[244] In another study on MHC congenic and recombinant mice inoculated with *Torpedo* nAChR with adjuvant, the autoantibody responses to muscle nAChR and the defect of neuromuscular transmission closely parallel *in vitro* lymphocyte proliferative responses to *Torpedo* nAChR. The responses are controlled by a gene(s) at the I-A subregion of the H-2 complex.[245] The T-lymphocyte proliferative responses to nAChR may be controlled by a dominant gene linked to H-2 at the I-A subregion. Lymphocyte responses were eliminated by blocking Ia antigen on lymph node cell surfaces with appropriate anti-Ia alloantisera and by removal of adherent cells. A spontaneous mutation at the I-A subregion in the B6 strain converted high responsiveness to nAChR to a state of low responsiveness. A macrophage-associated Ia molecule may be implicated in the induction of EAMG, probably in the presentation of nAChR to helper T lymphocytes that thereby help B lymphocytes to differentiate into anti-AChR antibody-forming cells. The conclusion was drawn[245] that human susceptibility to the autoimmune form of MG might be determined by certain halotypes of DRW (the human analogue of murina Ia). Association of autologous nAChR (complexed with viral antigens) with macrophage-associated DREW could trigger "spontaneous" MG. It appears probable that non-MHC genes also influence susceptibility to MG in man, as has been demonstrated in non-B6 or -B10 congenic mice.

9.9.2d. Fab_1 Fragment. Fab_1 fragments prepared from myasthenic patient IgG and used in the study of antibody action were shown in mice to protect against the action of myasthenic IgG in transfer experiments (Fig. 7). With the protein A technique, the binding of Fab_1 to the motor endplate was shown. Fab_1 acted protectively when given either before or together with myasthenic

IgG. In contrast, control Fab_I could not prevent mepp amplitudes from decreasing.[57]

In the presence of Fab_I, binding of α-bgt to nAChR is decreased only 10–15%. Saturating Fab_I fragment concentrations (molar ratio maximum 4/nAChR) seem not to affect carbamylcholine (Carb)-induced increases in $^{22}Na^+$ influx or efflux at agonist concentrations sufficient to elicit maximum flux responses but shift the dose–response curves for $^{22}Na^+$ efflux to lower concentrations of carbachol, which suggests that bound Fab_I fragments may decrease the magnitude of desensitization.[248] Stable nAChR–Fab_I complexes have been isolated.

The protective effect of Fab_I most probably resides in its ability to bind to the receptor and to prevent myasthenic IgG from binding there. It is assumed that myasthenic IgG cross links receptors, causing the formed receptor–IgG complex to be removed from the membrane. The univalent Fab_I is unable to cross link receptors.

Fab_I fragments are univalent, and they are about one-third the size of intact IgG; thus, steric factors are smaller. The use of Fab_I in the treatment of MG may seem tempting, but it is contraindicated by its rapid excretion, which would demand waste amount of Fab_I fragments for clinical use.

9.9.2e. α-Fetoprotein. It was shown[246] *in vitro* that binding of antibodies from MG patients to nAChR is inhibited by human amniotic fluid from healthy women during the second trimester and by both umbilical cord serum and maternal serum during the second half of gestation. The inhibitory effect was shown to reside in α-fetoprotein (AFP), a glycoprotein with immunosuppressive activities on the humoral and cell-mediated immune response (for references, see ref. 247). It was also shown that AFP has an immunosuppressive effect on EAMG in rabbits if it is repeatedly injected during the induction period.[249] Further, clinical signs disappeared when the animals were treated with AFP after the development of the disease. α-Fetoprotein partly inhibited the binding of EAMG antibody to the *Torpedo* nAChR.

It was suggested[247] that AFP may be responsible for the clinical remissions that are seen in MG during pregnancy and that are also seen in other putative autoimmune diseases.

9.9.2f. Specific Suppressor Cells. Immune responses seem to be regulated by a balance between populations of T-helper cells and T-suppressor cells and their subpopulations. In accordance with this theory, an autoimmune disease may develop when helper cells dominate suppressor cells, and induction of specific suppressor cell might be a useful tool in blocking the development of an autoimmune disease. Methods are currently being developed to induce the production of specific T-suppressor cells. In the treatment of EAMG, a method using a changed form of antigen has caused some interest.

10. POSTSYNAPTIC AUTOIMMUNE DISEASE: MYASTHENIA GRAVIS BY FOREIGN SUBSTANCES

Several chemical compounds are known to induce reversible MG, e.g., D-penicillamine and chloroquine. These substances may act in a number of different ways, and, consequently, the study of their actions may provide us with important clues to the etiology of neuromuscular diseases. It is possible that the primary target of several or all of the substances is not nAChR but an endogenous compound with an antigenic region resembling one in the receptor. Antibodies created may then cross react with nAChR (molecular mimicry).

10.1. D-Penicillamine

D-Penicillamine (D-pen), used as a therapeutic agent in Wilson's disease, cystinurea, and rheumatoid arthritis, in a small proportion of patients causes a reversible MG indistinguishable from naturally occurring MG in its clinical features and sometimes also in its electrophysiology[249,250] yet probably a syndrome of its own. Receptor antibodies showing cross reactivity mainly with human nAChR develop in 90% of cases[251–253] and seem much like those seen in idiopathic MG. Thymic hyperplasia, characteristic of myasthenia in young adults, has also been described.[252–254] Other autoimmune diseases develop in some patients, suggesting a more general predisposition for autoimmune disease related to D-pen introduction into the body. There have been cases in which patients with rheumatoid arthritis have developed severe side effects (cutaneous, renal, and hematologic) after D-pen treatment.

Chemical modulation of crucial bioactive compounds may be the cause of this drug-induced MG. D-Penicillamine is capable of a number of reactions that may lead to new or altered antigenic determinants on proteins, producing loss of tolerance in D-pen-treated patients.[255,256] Thus, reactive sulfhydryl groups of D-pen may reduce S-S bonds in IgM and IgG[257,258] and in albumin[257,259] or may react with cystine[259] or with aldehydes.[260] Noncovalent interactions could involve heavy metal chelation[261] and complex formation with DNA.[262] It should be remembered that nAChR from electric fish and from mammalian muscle has at least two reducible disfulfide bonds. One is suggested to be near the cholinergic ligand binding site (fish[263,264]; human muscle[265]). The second disulfide bond is thought to be responsible for the formation of the nAChR dimer by linking two 67,000-dalton subunits in adjacent monomers.[266] It is still uncertain if such dimers exist in mammalian nAChR.[265,267] In any case, alteration of agonist binding (equilibrium dialysis) after D-pen binding to *Torpedo* nAChR was shown[268] and was suggested to lead to a loss of tolerance and to the induction of antibody formation as shown with arsenyl or sulfanyl modifications of thyroglobulin.[269]

D-Penicillamine has also been suggested to modify immunoinduction of the thymus with consequent development of MG.[254,270,271] Removal of the thymus reverses the symptoms.[254] In D-pen-treated animals, thymic hyperplasia has been observed. Thus, penicillamine might trigger "intrathymic autosensitization" to the myogenic cell, resulting in an autoimmune reaction against striated muscle and nAChR. Plasma exchange therapy may help[272] when symptoms remain after D-pen withdrawel and use of anti-ChE. D-Penicillamine reduces lymphocyte responsiveness to PHA *in vitro*.[250]

In guinea pigs,[278] electrophysiological, histological, and serological changes are reminiscent of both experimental allergic myositis and experimental allergic myasthenia gravis. These include electromyographic decrement following repetitive stimulation that is reversible with edrophonium and augmentation of twitch tension with edrophonium, nonspecific inflammatory and necrotic changes in muscle, and significant elevation of nAChR antibodies. The guinea pig may be a model of D-penicillamine-induced neuromuscular diseases in humans.

D-Penicillamine has also been suggested to act as a chemical hapten by binding cysteine-containing proteins, which may then act as a hapten–protein complex to produce autosensitization.[279]

At the rat neuromuscular junction, D-pen, applied in the bath, has little effect on mepp amplitude or action potential amplitude. The epp amplitude and spontaneous mepp frequency decrease significantly at concentrations about 40 times the human therapeutic level. Daily injections for 33–37 days cause no differences from controls, and EM shows no evidence of degeneration or simplification. Thymus histology by light microscopy is normal. No antireceptor antibodies are found. It is suggested that D-pen in rats has a weak presynaptic effect at high concentrations. This effect, however, is too small to explain the myasthenialike syndrome observed in humans. D-Penicillamine does not act by direct blockade of neuromuscular transmission in rats. The possibility remains that D-pen depletes stores of ACh.[280]

10.2. DL-Carnitine

DL-Carnitine causes a myasthenialike syndrome in some (one out of ten) patients with alterations in the EMG and decreased muscle action potentials. L-Carnitine does not produce this syndrome, whereas the D-isomer at high blood concentrations is said to block neuromuscular transmission.[274] A competitive action between DL-carnitine and ACh at nAChR is suggested,[273] similar to that of a hemicholiniumlike block.

10.3. Chloroquine and Quinidine Sulfate

Chloroquine treatment used for 2 months for rheumatoid arthritis in a patient resulted in increased nAChR antibody titers and a partial neuromuscular block. Withdrawal of the drug resulted in complete recovery, and antibody titers returned to normal. It is uncertain if this was a drug-induced MG or a latent MG manifested by the drug.[276] Chloroquine, however, is also known to act axonally or presynaptically[277] or to cause a myopathy.

Quinidine sulfate and its stereoisomer quinine decrease skeletal muscle endplate excitability. Quinine is used in MG diagnosis. In patients with, e.g., Graves' disease, it has been known to precipitate MG-like symptoms,[281] which may occur simultaneously.

11. CONCLUDING REMARKS AND PERSPECTIVES

The human diseases and the model diseases described in this chapter all result in changes of structure and/or function of the neuromuscular junction.

Defects may be either pre- or postsynaptic or both, and they may involve the synaptic cleft and its basement membrane. A number of the disorders described have a recognized or suspected autoimmune background, they are characterized by an abnormal, i.e., an anti-self, immune response.

In spite of the rapid progress that has been made in the understanding of autoimmunity and in the understanding of the consequences of uncontrolled antibody production against functionally important synaptic proteins—in particular, through studies on myasthenia gravis—basic information on the pathogenesis of such diseases is still lacking. Several theories are being pursued, from that of a single infectious agent as the cause of disease to the reasonable idea, sustained by knowledge about the complexity of some of these diseases, of a multifactorial etiology of some (all?) of them. Significant progress can be expected from the rapidly increasing information on the complex regulation of the immune system, which seems to involve cells and antibodies, and also on amplification systems such as complement on one hand and genetic factors governing them on the other. Alterations of the normal functioning of the immune regulatory systems obviously may result in disease.

Animal models of autoimmune diseases have greatly helped to clarify points of etiology, such as the production of the characteristic lesions, and have provided us with new tools for diagnosis and treatment. They are, however, only partly relevant in the search for the basic causes of autoimmune disease. In the study of MG, the passive transfer model may come closer to human disease though it bypasses the early steps in the etiology, i.e., the cause of the abnormal antibody production to self-antigen. Models using immunization with antigen may tend to overemphasize certain aspects of the autoimmune disease such as the cytotoxic process represented by the formation of antigen–antibody complexes able to bind complement and the phagocytic invasion of the myasthenic endplate, which seems not to be a characteristic of human MG.

The nAChR antibodies are hardly the primary cause of MG, the prototype of autoimmune disease, or, most probably, of any other autoimmune disease. Some of them are, however, responsible for the functional lesions obtained. Others may even be protective. Analysis of the polyclonal antibody population present in MG and EAMG serum reveals antibodies belonging to different subgroups and subclasses of immunoglobulins with different properties, including different receptor-binding characteristics. The presence of antibodies in patients in complete remission and in healthy children of myasthenic mothers demonstrates the presence of nonpathogenic antibodies and possibly of anti-idiotypic antibodies that may have protective properties. A multiplicity of antibody attack is demonstrated by the use of monoclonal antibodies and suggests that the induced immunologic attack may vary from patient to patient, a fact that may make treatment based on antiidiotypi difficult. The discovery that antibodies directed against a ligand may have properties reminiscent of those of the ligand receptor or that antibodies to a receptor may either belong to a ligand-binding, blocking category or even mimic ligand effects (insulin) opens interesting possibilities.

Among current approaches to understanding autoimmune disease is the study of changes in T-cell subpopulations connected with disease. The existence of an immunoregulatory antibody has been discussed. Another and as yet less studied aspect is that of "molecular mimicry," which suggests that an antibody directed against a certain compound may produce a second antibody that is in certain immunologically interesting ways related to the first compound. This implies that an antibody able to attack a certain biologically important antigen need not have been produced by this antigen. A consequence of this might be that with the ever-increasing use of chemical compounds, some of which may find their way into the body, the number of autoimmune diseases, at the neuromuscular junction and elsewhere, can be expected to increase. Such a development must be controlled not only by careful use of chemicals but also by a thorough pretesting of them. In this respect, the development and study of animal models and of *in vitro* test systems is of considerable importance, even in immunology.

Recent work on chemical neurotransmission has shown that our previous ideas represented a considerable oversimplification. Much of our knowledge about neuromuscular transmission is derived from electrophysiological experiments. Changes in "normal" activity as a consequence of desensitization, augmentation, super- and subsensitivity, and short- or long-term potentiation are not really understood on the molecular level. With increasing knowledge, however, our ability to understand the mechanisms behind diseases affecting neurotransmission will increase, and our ability to design working regimens for their treatment will improve considerably as we learn more about structure and function of the pre- and postsynaptic parts of synaptic junctions.

The current revolution in our view on neurotransmission in general makes it desirable to retrace our steps even where neuromuscular transmission is concerned and to start again, not from the beginning but perhaps from the concept of "one neuron, one transmitter," which now should at least be phrased as "one neuron, possibly one transmitter, and probably some modulators." The regulation of transmitter release from the neuron to skeletal muscle, apart from the role of Ca^{2+}, is not known. We do not know the nature of a possibly existing presynaptic autoreceptor or that of modulators. Once such information is available, the door will be open for the production of new drugs regulating transmitter release.

Another widely accepted working hypothesis for the nerve terminal is that of the quantal release of vesicle-packaged transmitter by fusion of the vesicle with the presynaptic membrane. Other ways of releasing vesicle-packaged transmitter and the release of cytoplasmic transmitter through a gating mechanism are discussed. Again, new information can be expected and may open new possibilities for drug-induced transmitter release control and thus probably for the treatment of neuromuscular diseases. Among these is probably the Eaton–Lambert syndrome. There, the so-far-unknown nature of the active zones and the function of their intramembrane particles can perhaps be studied by using antibodies developed to presynaptic components or obtained from the serum of ELS patients.

Finally, it may be appropriate to mention the need for more information concerning innervation and reinnervation of skeletal muscle and the role and nature of both anterogradely and retrogradely acting trophic factors. The importance of such information for the understanding and, when necessary, control of development or of changes with aging, accidents, diseases causing denervation, etc., is obvious.

ACKNOWLEDGMENTS. The author wishes to express her gratitude to her colleagues who critically read the manuscript.

REFERENCES

1. Hagiwara, S., and Byerly, L., 1981, *Annu. Rev. Neurosci.* **4**:69–125.
2. Åkerman, K. E. O., and Nicholls, D. G., 1983, *Trends Biochem. Sci.* **8**:63–64.
3. Cull-Candy, S. G., Lundh, H., and Thesleff, S., 1976, *J. Physiol.* (*Lond.*) **260**:177–203.
4. Thesleff, S., 1981, *Biochemical Aspects of Botulism* (G. E. Lewis, ed.), Academic Press, New York, pp. 65–80.
5. Katz, B., and Miledi, R., 1977, *Proc. R. Soc. Lond.* [*Biol.*] **196**:59–72.
6. Vyskocil, F., and Illes, P., 1978, *Physiol. Bohemoslov.* **27**:449–455.
7. Thesleff, S., 1982, *Aminopyridines and Similarly Acting Drugs: Effects on Nerves, Muscles and Synapses* (P. Lechat, S. Thesleff, and W. C. Bowman, eds.), Pergamon Press, New York, pp. 87–93.
8. Thesleff, S., and Sellin, L. C., 1980, *Trends Neurosci.* **3**:122–126.
9. Thesleff, S., and Molgo, J., 1983, *Neuroscience* **9**:1–8.
10. Chan-Palay, V., Engel, A. G., Palay, S. L., and Wu, J. Y., 1982, *Proc. Natl. Acad. Sci. U.S.A.* **79**:6717–6721.
11. Meunier, F. M., Israël, M., and Lesbats, B., 1975, *Nature* **257**:407–408.
12. Silinsky, E. M., and Hubbard, J. I., 1973, *Nature* **243**:404–405.
13. Ewald, D. A., 1976, *J. Membr. Biol.* **29**:47–65.
14. Kolb, H. A., and Wakelam, M. J. O., 1983, *Nature* **303**:621–623.
15. Ryall, R. W., 1982, *Substance P in the Nervous System. Ciba Foundation Symposium 91* (R. Porter and M. O'Connor, eds.), Pitman Medical, London, pp. 267–277.
16. Adams, M. E., and O'Shea, M., 1983, *Science* **221**:286–289.
17. Johnston, M. F., Kravitz, E. A., Meiri, H., and Rahamimoff, R., 1983, *Science* **220**:1071–1072.
18. Katz, B., 1969, *The Release of Neural Transmitter Substances*, Liverpool University Press, Liverpool.
19. Llinas, R. R., and Heuser, J. E., 1977, *Neurosci. Res. Progr. Bull.* **15**:557–687.
20. Reichardt, L. F., and Kelly, R. B., 1983, *Annu. Rev. Biochem.* **52**:871–926.
21. Tauc, L., 1982, *Physiol. Rev.* **62**:847–896.
22. Nordström, Ö., and Bartfai, T., 1981, *Brain Res.* **213**:467–471.
23. Kilbinger, H., 1977, *Naunyn Schmiedebergs Arch. Pharmacol.* **300**:145–151.
24. Häggblad, J., and Heilbronn, E., 1983, *Br. J. Pharmacol.* **80**:471–476.
25. Gundersen, C. B., and Jenden, D. J., 1980, *Br. J. Pharmacol.* **70**:8–10.
26. Miledi, R., Molenaar, P. C., and Polak, R. L., 1978, *Nature* **272**:641–643.
27. Couteaux, R., and Pécot-Dechavassine, M., 1970, *C.R. Acad. Sci.* [*D*] (*Paris*) **271**:2346–2349.
28. Fertuck, H. C., and Salpeter, M. M., 1974, *Proc. Natl. Acad. Sci. U.S.A.* **71**:1376–1378.
29. Gage, P. W., 1976, *Physiol. Rev.* **56**:177–247.
30. Karlin, A., 1980, *The Cell Surface and Neuronal Function* (G. Poste, G. L. Nicolson, and C. W. Cotman, eds.), Elsevier/North-Holland Biomedical Press, New York, pp. 191–260.
31. Adams, P. R., 1981, *J. Membr. Biol.* **58**:161–174.
32. Conti-Tronconi, B. M., and Raftery, M. A., 1982, *Annu. Rev. Biochem.* **51**:491–530.
33. Changeux, J. P., 1981, *Harvey Lect.* **75**:85–254.
34. Stevens, C. F., 1981, *Chemical Neurotransmission 75 Years* (L. Stjärne, P., Hedqvist, H. Lagercrantz, and Å. Wennmalm, eds.), Academic Press, London, pp. 361–370.
35. Noda, M., Takahashi, H., Tanabe, T., Toyosato, M., Kikyotani, S., Furutani, Y., Hirose, T., Takashima, H., Inayama, S., Miyata, T., and Numa, S., 1983, *Nature* **302**:528–532.

36. Wise, D. S., Schoenborn, B. P., and Karlin, A., 1981, *J. Biol. Chem.* **256**:4124–4126.
37. Stephenson, F. A., Harrison, R., and Lunt, G. G., 1981, *Eur. J. Biochem.* **115**:91–97.
38. Einarsson, B., Gullick, W., Conti-Tronconi, B., Ellisman, M., and Lindstrom, J., 1982, *Biochemistry* **21**:5295–5302.
39. Conti-Tronconi, B. M., Gotti, C. M., Hunkapiller, M. W., and Raftery, M. A., 1982, *Science* **218**:1227–1229.
40. Schlessinger, J., 1983, *Trends Neurosci.* **6**:360–363.
41. Pumplin, D. W., and Fambrough, D. M., 1982, *Annu. Rev. Physiol.* **44**:319–335.
42. Dennis, M. J., 1981, *Annu. Rev. Neurosci.* **4**:43–68.
43. Hugenir, R. L., and Greengard, P., 1983, *Proc. Natl. Acad. Sci. U.S.A.* **80**:1130–1134.
44. Eriksson, H., Salmonsson, R., Liljeqvist, G., and Heilbronn, E., 1984, *Proceedings of the Conference on the Dynamics of Cholinergic Function* (J. Thanin, ed.), Plenum Press, New York.
45. Flynn, D. D., Kloog, Y., Potter, L. T., and Axelrod, J., 1982, *J. Biol. Chem.* **257**:9513–9517.
46. Reiness, C. G., and Hall, Z. W., 1981, *Dev. Biol.* **81**:324–331.
47. Anderson, D. J., 1983, *Trends Neurosci.* **6**:169–171.
48. Hirokawa, N., and Heuser, J. E., 1982, *J. Neurocytol.* **11**:487–510.
49. Fambrough, D. M., and Devreotes, P. N., 1978, *J. Cell. Biol.* **76**:237–244.
50. Axelrod, D. P., Ravdin, P. M., and Podleski, T. R., 1978, *Biochim. Biophys. Acta* **511**:23–38.
51. Burden, S. Y., Sargent, P. B., and McMahan, U. J., 1979, *J. Cell Biol.* **82**:412–425.
52. Hasegawa, S., Kuromi, H., and Hagihara, Y., 1982, *Trends Pharmacol. Sci.* **3**:340–342.
53. Katz, B., and Thesleff, S., 1957, *J. Physiol.* (*Lond.*) **138**:63–80.
54. Stallcup, W. B., and Patrick, J., 1980, *Proc. Natl. Acad. Sci. U.S.A.* **77**:634–638.
55. Gardner, J. M., and Fambrough, D. M., 1979, *Cell* **16**:661–674.
56. Drachman, D. B., Angus, C. W., Adams, R. N., Michelson, J. D., and Hoffman, G. J., 1978, *N. Engl. J. Med.* **298**:1116–1122.
57. Toyka, K. V., Löwenadler, B., Heininger, K., Besinger, U. A., Birnberger, K. L., Fateh-Moghadam, A., and Heilbronn, E., 1980, *J. Neurol. Neurosurg. Psychiatry* **43**:836–842.
58. Patrick, J., and Lindstrom, J., 1973, *Science* **180**:871–872.
59. Sugiyama, H., Brenda, P., Meunier, J. C., and Changeux, J. P., 1973, *FEBS Lett.* **35**:124–128.
60. Heilbronn, E., and Mattsson, C., 1974, *J. Neurochem.* **22**:315–317.
61. Elfman, L., Thornell, L.-E., and Heilbronn, E., 1983, *J. Neurol. Sci.* **59**:111–121.
62. Noda, M., Takahashi, H., Tanabe, T., Toyosato, M., Furutani, Y., Hirose, T. , Asai, M., Inayama, S., Miyata, T., and Numa, S., 1982, *Nature* **299**:793–797.
63. Zingsheim, H. P., Barrantes, F. J., Frank, J., Hänicke, W., and Neugebauer, D.-C., 1982, *Nature* **299**:81–84.
64. Tzartos, S., Rand, D. E., Einarsson, B. L., and Lindstrom, J. M., 1981, *J. Biol. Chem.* **256**:8635–8645.
65. Tzartos, S., Langeberg, L., Hochschwender, S., and Lindstrom, J., 1983, *FEBS Lett.* **158**:116–118.
66. Sellin, L. C., 1981, *Med. Biol.* **59**:11–20.
67. Pickett, J., Berg, B., Chaplin, E., and Brunstetter-Shafer, M. A., 1976, *N. Engl. J. Med.* **295**:770–772.
68. Johnson, R. O., Clay, S. A., and Arnon, S. S., 1979, *Am. J. Dis. Child.* **133**:586–593.
69. Sugiyama, H., 1980, *Microbiol. Rev.* **44**:419–448.
70. Duchen, L. W., 1971, *J. Neurol. Sci.* **14**:47–60.
71. Guyton, A. C., and MacDonald, M. A., 1947, *Arch. Neurol. Psychiatry* **57**:579–592.
72. Thesleff, S., 1960, *J. Physiol.* (*Lond.*) **151**:598–607.
73. Lundh, H., Cull-Candy, S. G., Leander, S., and Thesleff, S., 1976, *Brain Res.* **110**:194–198.
74. Pickett, J. B., 1980, *The Physiology of Peripheral Nerve Disease* (A. J. Summer, ed.), W. B. Saunders, Philadelphia, pp. 238–264.
75. Harris, A. J., and Miledi, R., 1971, *J. Physiol.* (*Lond.*) **217**:497–515.
76. Thesleff, S., 1977, *Naunyn Schmiedebergs Arch. Pharmacol.* **297**:5–7.
77. Simpson, L. L., 1980, *J. Pharmacol. Exp. Ther.* **212**:16–21.

78. Kao, I., Drachman, D. B., and Price, D. L., 1976, *Science* **193:**1256–1258.
79. Hirokawa, N., and Heuser, J. E., 1981, *J. Cell Biol.* **88:**160–171.
80. Swift, T. R., 1981, *Muscle Nerve* **4:**334–353.
81. Newsom-Davies, J., 1982, *Clin. Immunol. Allergy* **2:**405–424.
82. Chao, L.-P., Kan, K. S., Angelini, C., and Keesey, J., 1982, *Exp. Neurol.* **75:**23–35.
83. Festoff, B. W., and Israel, R. S., Engel, W. K., and Rosenbaum, R. B., 1977, *Neurology (Minneap.)* **27:**963–970.
84. Eaton, L. M., and Lambert, E. H., 1957, *J.A.M.A.* **163:**1117–1124.
85. Henriksson, K. G., Nilsson, O., Rosén, I., and Schiller, H. H., 1977, *Acta Neurol. Scand.* **56:**117–140.
86. Hawley, R. J., Cohen, M. H., Saini, N., and Armbrustmacher, V. W., 1980, *Ann. Neurol.* **7:**65–72.
87. Takamori, M., and Gutmann, L., 1971, *Neurology (Minneap.)* **21:**47–54.
88. Dahl, D. S., and Sato, S., 1974, *Neurology (Minneap.)* **24:**897–901.
89. Schwartz, M. S., and Stålberg, E., 1975, *Neurology (Minneap.)* **25:**80–84.
89a. Schütte, J., Niederle, N., and Seeber, S., 1983, *Med. Welt* **34:**19–22.
90. Lang, B., Newsom-Davis, J., Wray, D., Vincent, A., and Murray, N., 1981, *Lancet* **2:**224–226.
91. Fukunaga, H., Engel, A. G., Osame, M., and Lambert, E. H., 1982, *Muscle Nerve* **5:**686–697.
92. Lambert, E. H., Rooke, E. D., Eaton, L. M., and Hodgson, C. H., 1961, *Myasthenia Gravis*, (H. R. Viets, ed.), Charles C. Thomas, Springfield, IL, pp. 362–410.
93. Elmqvist, D., and Lambert, E. H., 1968, *Mayo Clin. Proc.* **43:**689–713.
94. Lindstrom, J. M., and Lambert, E. H., 1978, *Neurology (Minneap.)* **28:**130–138.
95. Fukuhara, N., Takamori, M., Gutmann, L., and Chou, S.-M., 1972, *Arch. Neurol.* **27:**67–78.
96. Engel, A. G., and Santa, T., 1971, *Ann. N.Y. Acad. Sci.* **183:**46–63.
97. Engel, A. G., Lambert, E. H., Mulder, D. M., Torres, C. F., Sahashi, K., Bertorini, T. E., and Whitaker, J. N., 1982, *Ann. Neurol.* **11:**553–569.
98. Molenaar, P. C., Newsom-Davies, J., Polak, R. L., and Vincent, A., 1982, *Neurology (N.Y.)* **32:**1061–1065.
99. Cull-Candy, S. G., Miledi, R., Trautmann, A., and Uchitel, O. D., 1980, *J. Physiol. (Lond.)* **299:**621–638.
100. Fletcher, P., and Forrester, T., 1975, *J. Physiol. (Lond.)* **251:**131–144.
101. Miledi, R., Molenaar, P. C., and Polak, R. L., 1981, *International Symposium on Cholinergic Mechanisms* (G. L. Pepeu and H. Lasinsky, eds.), Plenum Press, New York, pp. 205–214.
102. Lambert, E. H., and Elmqvist, D., 1971, *Ann. N.Y. Acad. Sci.* **183:**183–199.
103. Engel, A. G., Lambert, E. H., and Gomez, M. R., 1977, *Ann. Neurol.* **1:**315–330.
104. Kamenskaya, M. A., Elmqvist, D., and Thesleff, S., 1977, *Arch. Neurol.* **32:**510–518.
105. Cherington, M., 1976, *Neurology (Minneap.)* **26:**944–946.
106. Takamori, M., Ishii, N., and Mori, M., 1973, *Arch. Neurol.* **29:**420–424.
107. Lundh, H., Nilsson, O., and Rosén, I., 1977, *J. Neurol. Neurosurg. Psychiatry* **40:**1109–1112.
108. Sanders, D. B., Kim, Y. I., Howard, J. F., and Goetsch, C. A., 1980, *J. Neurol. Neurosurg. Psychiatry* **43:**978–985.
109. Vroom, F. O., and Engel, W. K., 1969, *Neurology (Minneap.)* **19:**281.
110. Streib, E. W., and Rothner, A. D., 1981, *Ann. Neurol.* **10:**448–453.
111. Kim, Y. I., Sanders, D. B., and Johns, T. R., 1980, *Soc. Neurosci. Abstr.* **6:**403.
112. Newsom-Davis, J., Lang, B., Wray, D., Murray, N., Vincent, A., and Gwilt, M., 1982, *Neurology (N.Y.)* **32:**A221–222.
113. Gwilt, M., Lang, B., Newsom-Davis, J., and Wray, D., 1982, *J. Physiol. (Lond.)* **324:**29–30P.
113a. Ishikawa, K., Engelhardt, J. K., Fujisawa, T., Okamoto, T., and Katsuki, H., 1977, *Neurology (Minneap.)* **27:**140–143.
113b. Kennedy, W. R., and Jimenez-Pabow, E., 1962, *Neurology (Minneap.)* **18:**757.
114. Flagstad, A., 1982, *Hereditas* **96:**211–214.
115. Engel, A. G., 1980, *J. Neurol. Neurosurg. Psychiatry* **43:**577–589.

116. Vincent, A., Cull-Candy, S. G., Newsom-Davis, J., Trautmann, A., Molenaar, P. C., and Polak, R. L., 1981, *Muscle Nerve* **4**:306–318.
117. Palmer, A. C., 1980, *Neurol.* **10**:213–221.
118. Palmer, A. C., and Barker, J., 1974, *Vet. Rec.* **95**:452–454.
119. Johnson, R. P., Watson, A. D. J., Smith, J., and Cooper, B. J., 1975, *J. Small Animal Pract.* **16**:641–647.
120. Jenkins, W. L., Van Dyk, E., and McDonald, C. B., 1976, *J. Afr. Vet. Assoc.* **47**:59–62.
121. Rosenberry, T. L., and Richardson, J. M., 1977, *Biochemistry* **16**:3550–3558.
121a. Massoulié, J., 1980, *Trends Biochem. Sci.* **5**:160–164.
122. Bois, R. T., Hummel, R. G., Dettbarn, W.-D., and Laskowski, M., 1980, *J. Pharmacol. Exp. Ther.* **215**:53–59.
123. Katz, B., and Miledi, R., 1973, *J. Physiol.* (*Lond.*) **231**:549–574.
124. Salpeter, M. M., Kasprazak, H., Feng, H., and Fertuck, H., 1979, *J. Neurocytol.* **8**:95–115.
125. Miller, D. B., 1982, *Neurobehav. Toxicol. Teratol.* **4**:779–787.
126. O'Brien, R. D., 1967, *Insecticides: Action and Metabolism,* Academic Press, New York.
127. Hayes, W. J., 1982, *Pesticides Studied in Man*, Williams and Wilkins, Baltimore, pp. 436–462.
128. Lee, C.-C., and Peters, P. J., 1976, *Environ. Health Perspect.* **17**:35–43.
129. Kawabuchi, M., 1982, *J. Neuropathol. Exp. Neurol.* **41**:298–314.
130. Davis, C. S., and Richardson, R. J., 1980, *Experiments in Clinical Neurotoxicity* (P. S. Spencer and H. H. Schaumburg, eds.), Williams and Wilkins, Baltimore, London pp. 527–544.
131. Wecker, L., and Dettbarn, W. D., 1976, *Exp. Neurol.* **51**:281–291.
132. Laskowski, M. B., Olson, W. H., and Dettbarn, W. D., 1975, *Exp. Neurol.* **47**:290–306.
133. De Reuck, J., and Willems, J., 1975, *J. Neurol.* **208**:309–313.
134. Baker, T., Glazer, E., and Lowndes, H. E., 1977, *Neuropathol. Appl. Neurobiol.* **3**:377–390.
135. Heilbronn-Wikström, E., 1965, *Svensk Kem. Tidskr.* **77**:11–48.
136. Clothier, B., Johnson, M. K., and Reiner, E., 1981, *Biochim. Biophys. Acta* **660**:306–316.
137. Johnson, M. K., 1980, *Advances in Neurotoxicology* (L. Manzo, ed.), Pergamon Press, Oxford, pp. 225–226.
138. Abou-Donia, M. B., 1981, *Annu. Rev. Pharmacol. Toxicol.* **21**:511–548.
139. Cavanagh, J. B., 1973, *CRC Crit. Rev. Toxicol.* **2**:365–417.
140. Smith, M. I., Elvove, E., Valaer, P. J., Frazier, W. H., and Mallory, G. E., 1930, *U.S. Public Health Rep.* **45**:1703–1716.
141. Johnsson, M. K., 1982, *Rev. Biochem. Toxicol.* **4**:141–212.
142. Williams, D. G., and Johnson, M. K., 1981, *Biochem. J.* **199**:323–333.
143. Johnson, M. K., 1974, *J. Neurochem.* **23**:785–789.
144. Evered, D., and Whelan, J. (eds.), 1982, *CIBA Foundation Symposium 90*, Pitman Medical, London.
145. Toyka, K. V., Besinger, U. A., and Fateh-Moghadam, A., 1981, *Therapeutic Plasma Exchange* (H. J. Gurland, V. Heinze, and H. A. Lee, eds.), Springer-Verlag, Berlin, Heidelberg, New York.
146. Barkas, T., 1979, *Int. J. Immunopharmacol.* **1**:263–271.
147. Patten, B. M., 1978, *Muscle Nerve* **1**:190–205.
148. Brumback, R. A., 1981, *Am. Fam. Physician* **23**:126–133.
149. Lindstrom, J., 1979, *Adv. Immunol.* **27**:1–44.
150. Vincent, A., 1980, *Physiol. Rev.* **60**:756–824.
151. Scadding, G. K., and Havard, C. W. H., 1981, *Br. Med. J.* **283**:1008–1012.
152. Oosterhuis, H. J., 1981, *Clin. Neurol. Neurosurg.* **83**:105–135.
153. Heilbronn, E., Lefvert, A.-K., and Stålberg, E., 1978, *Muscle Nerve* **1**:427–431.
154. Hammarström, L., and Smith, C. I. E., 1981, *Afr. J. Clin. Exp. Immunol.* **2**:113–121.
155. Grob, D. (ed.), 1976, 1981, *Myasthenia Gravis. Annals of the New York Academy of Sciences*, Volumes 274, 377, New York Academy of Sciences, New York.
156. Fambrough, D. M., Drachman, D. B., and Satyamurti, S., 1973, *Science* **182**:293–295.
157. Lennon, V. A., Lambert, E. H., Palmer, A. C., 1981, *Myasthenia Gravis—Pathogenesis and Treatment* (E. Satoyoshi, ed.), Japan Medical Research Foundation Press, Toyko, pp. 41–45.

157a. Vincent, A., and Newsom-Davis, J., 1982, *Clin. Exp. Immunol.* **49**:257–265.
158. Compston, D. A. S., Vincent, A., Newsom-Davis, J., and Batchelor, J. R., 1980, *Brain* **103**:579–601.
159. Keesey, J., Lindstrom, J., Cokely, H., and Herrmann, C., 1977, *N. Engl. J. Med.* **296**:55.
160. Lefvert, A. K., and Osterman, P. O., 1983, *Neurology (N.Y.)* **33**:133–138.
161. Norcross, N. L., Griffith, I. J., and Lettieri, J. A., 1980, *Muscle Nerve* **3**:345–349.
162. Lindstrom, J., 1977, *Clin. Immunol. Immunopathol.* **7**:36–43.
163. Lindstrom, J. M., Seybold, M. E., Lennon, V. A., Whittingham, S., and Duane, D., 1976, *Neurology (Minneap.)* **26**:1054–1059.
164. Lefvert, A.-K., 1982, *J. Neurol. Neurosurg. Psychiatry* **45**:70–73.
165. Smithers, D. W., 1959, *J. Fac. Radiol.* **10**:3–16.
166. Simpson, J. A., 1960, *Scott. Med. J.* **5**:419–436.
167. Nastuk, W. L., Plescia, O. J., and Osserman, K. E., 1960, *Proc. Soc. Exp. Biol. Med.* **105**:177–184.
168. Strauss, A. J. L., Seegal, B. C., Hsu, K. C., Burkholder, P. M., Nastuk, W. L., and Osserman, K. E., 1960, *Proc. Soc. Exp. Biol. Med.* **105**:184–191.
169. Bergström, K., Franksson, C., Matell, G., and von Reis, G., 1973, *Eur. Neurol.* **9**:157–167.
170. Almon, R. R., Andrews, C. G., and Appel, S. H., 1974, *Science* **186**:55–57.
171. Lindstrom, J. M., and Lambert, E. H., 1978, *Neurology (Minneap.)* **28**:130–138.
172. Lefvert, A. K., Bergström, K., Matell, G., Osterman, P. O., and Pirskanen, R., 1978, *J. Neurol. Neurosurg. Psychiatry* **41**:394–403.
173. Vincent, A., Pinching, A. J., and Newsom-Davis, J., 1977, *Neurology (Minneap.)* **27**:364.
174. Dau, P. C., Lindstrom, J. M., Cassel, J. K., Denys, E. H., Shev, E. E., and Spitter, L. E., 1977, *N. Engl. J. Med.* **297**:1134–1140.
175. Namba, T., Brown, S. B., and Grob, D., 1970, *Pediatrics* **45**:488–504.
176. Olanow, C. W., Lane, R. J. M., Hull, K. L., Jr., and Roses, A. D., 1982, *J. Can. Sci. Neurol.* **9**:85–87.
177. Newsom-Davis, J., Pinching, A. J., Vincent, A., and Wilson, S. G., 1978, *Neurology (Minneap.)* **28**:266–272.
178. Elfman, L., 1981, Thesis, University of Stockholm, Stockholm.
178a. Elfman, L., Thornell, L.-E., and Heilbronn, E., 1983, *J. Neurol. Sci.* **59**:111–121.
179. Lindstrom, J. M., 1979, *Plasmapheresis and the Immunobiology of Myasthenia Gravis* (P. C. Dau, ed.), Houghton Mifflin, Boston, pp. 199–208.
180. Barkas, T., McPhail Gairns, J., Jackson Kerr, H., Coggins, J. R., and Simpson, J. A., 1982, *Eur. J. Immunol.* **12**:757–761.
181. Waud, D. R., 1971, *Ann. N.Y. Acad. Sci.* **183**:147–157.
182. Kornfeld, P., Nall, J., Smith, H., Mittag, T. W., Bender, A. N., Ambinder, E. P., Horowitz, S. H., Papatestas, A. E., Gross, H., and Genkins, G., 1981, *Muscle Nerve* **4**:413–419.
183. Appel, S. H., Blosser, J. C., McManaman, J. L., and Ashizawa, T., 1982, *Am. J. Physiol.* **243**:E31–E36.
183a. Engel, A. G., Lambert, E., and Howard, G., 1977, *Mayo Clin. Proc.* **52**:267–280.
184. Engel, A. G., Sahashi, K., Lambert, E. H., and Howard, F. M., 1979, *Excerpta Medica Int. Cong. Ser.* **455**:111–125.
185. Behan, W. H. M., and Behan, P. O., 1979, *J. Neurol. Neurosurg. Psychiatry* **42**:595–599.
185a. Casali, P., Borzini, P., and Zanussi, C., 1976, *Lancet* **2**:378.
186. Barkas, T., Boyle, R. S., and Behan, P. O., 1981, *J. Clin. Lab. Immunol.* **5**:27–30.
187. Barkas, T., and Simpson, J. A., 1982, *J. Clin. Lab. Immunol.* **7**:223–227.
188. Bartfeld, D., and Fuchs, S., 1977, *FEBS Lett.* **77**:214–218.
189. Klaus, G. G. B., 1978, *Nature* **272**:265–266.
190. Reinherz, E. L., and Schlossman, S. F., 1980, *Cell* **19**:821–827.
190a. Henze, A., Biberfeld, P., Christensson, B., Matell, G., and Pirskanen, R., 1983, *Scand. Assoc. Thorac. Cardiovasc. Surg.* Abstract 31.
190b. Castleman, B., 1966, *Ann. N.Y. Acad. Sci.* **135**:496–505.
191. Scadding, G. K., Thomas, H. C., and Havard, C. H. W., 1979, *Clin. Exp. Immunol.* **36**:205–213.
192. Abdou, N. I., Lisak, R. P., Zweiman, B., Abrahamsohn, I., and Penn, A. S., 1974, *N. Engl. J. Med.* **291**:1271–1275.

193. Opelz, G., Keesey, J., Glovsky, M. M., and Gale, R. P., 1978, *Arch. Neurol.* **35**:413–415.
194. Lisak, R. P., Zweiman, B., and Phillips, S. M., 1978, *Neurology (Minneap.)* **28**:1298–1301.
195. Richman, D. P., and Arnason, B. G. W., 1979, *Proc. Natl. Acad. Sci. U.S.A.* **76**:4632–4635.
196. Christensson, B., Biberfeld, P., Hammarström, L., Matell, G., and Smith, C. I., 1981, *Myasthenia Gravis. Pathogenesis and Treatment* (E. Satoyoshi, ed.), University of Tokyo Press, Tokyo, pp. 275–292.
197. Vincent, A., Scadding, G. K., Thomas, H. C., and Newsom-Davis, J., 1978, *Lancet* **1**:305–307.
198. Scadding, G. K., Newsom-Davis, J., Vincent, A., and Henry, K., 1981, *Neurology (N.Y.)* **31**:935–943.
199. Newsom-Davis, J., Willcox, N., Scadding, G. K., Calder, L., and Vincent, A., 1980, *Ann. N.Y. Acad. Sci.* **377**:393–402.
200. Aharonov, A., Tarrab-Hazdai, R., Abramsky, O., and Fuchs, S., 1975, *Proc. Natl. Acad. Sci. U.S.A.* **72**:1456–1459.
201. Wekerle, H., and Ketelson, U. P., 1977, *Lancet* **1**:678–680.
202. Goldstein, G., and Schlesinger, D. G., 1975, *Lancet* **2**:256–259.
203. Dalakas, M. C., Engel, W. K., McClure, J. E., and Goldstein, A. L., 1980, *N. Engl. J. Med.* **302**:1092–1093.
204. Goldstein, G., and Manganoro, A., 1971, *Ann. N.Y. Acad. Sci.* **183**:230–240.
205. Fontana, A., Fulpius, B. W., and Cuenod, S., 1978, *Schweiz. Med. Woschenschr.* **108**:1307–1310.
206. Tarrab-Hazdai, R., and Habib, E., 1979, *Abstracts, Seventh meeting of the ISN. Food Intake and Motor Disturbances after Intracerebroventricular Injection of Antiacetylcholine Receptor Antibodies.*
207. Heilbronn, E., Häggblad, J., and Kubat, B., 1981, *Ann. N.Y. Acad. Sci.* **377**:198–207.
208. Tzartos, S., and Lindstrom, J., 1980, *Proc. Natl. Acad. Sci. U.S.A.* **77**:755–759.
209. Alema, S., Cull-Candy, S. G., Miledi, R., and Trautman, A., 1981, *J. Physiol. (Lond.)* **311**:251–266.
210. Conti-Tronconi, B., Tzartos, S., and Lindstrom, J., 1981, *Biochemistry* **20**:2181–2191.
211. Drachman, D. B., Angus, C. W., Adams, R. N., and Kao, I., 1978, *Proc. Natl. Acad. Sci. U.S.A.* **75**:3422–3426.
212. Datta, S. K., and Schwartz, R. S., 1974, *N. Engl. J. Med.* **291**:1304–1305.
213. Tindall, R. S. B., Cloud, R., and Luby, J., 1978, *Neurology (Minneap.)* **28**:273–277.
214. Smith, C. L. E., Hammarstrom, L., and Berg, J. V. R., 1978, *Eur. Neurol.* **17**:181–187.
215. Feltkamp, T. E. W., Van den Berg Loonen, P. M., Nijenthis, L. E., Engelfriet, C. P., Van Rossum, A. L., Van Longhem, J. J., and Oosterhuis, H. J. G. H., 1974, *Br. Med. J.* **1**:131–133.
216. Lubetzkkorn, I., and Abramsky, O., 1981, *Eur. Neurol.* **20**:435–439.
217. Lentz, T. L., Burrage, T. G., Smith, A. L., Crick, J., and Tignor, G. H., 1982, *Science* **215**:182–184.
218. Karlsson, E., Heilbronn, E., and Widlund, L., 1972, *FEBS Lett.* **28**:107–111.
219. Olsen, R., Meunier, J. C., and Changeux, J. P., 1972, *FEBS Lett.* **28**:96–100.
220. Klett, R. P., Fulpius, B. W., Cooper, D., Smith, M., Reich, E., and Possani, L. D., 1973, *J. Biol. Chem.* **248**:6841–6853.
221. Heilbronn, E., Mattson, C., Thornell, L. E., Sjöström, M., Stålberg, E., Hilton-Brown, P., and Elmqvist, D., 1976, *Ann. N.Y. Acad. Sci.* **274**:337–353.
221a. Thornell, L.-E., Sjöström, M., Mattsson, C. H., and Heilbronn, E., 1976, *J. Neurol. Sci.* **29**:839–410.
222. Engel, A. G., Tsujihata, M., Lambert, E. H., Lindstrom, J. M., and Lennon, V. A., 1976, *J. Neuropathol. Exp. Neurol.* **35**:569–587.
223. Bevan, S., Heinemann, S., Lennon, V. A., and Lindstrom, J., 1976, *Nature* **260**:438–439.
224. Lennon, V. A., Lindstrom, J. M., and Seybold, M. E., 1975, *J. Exp. Med.* **141**:1365–1375.
225. Lambert, E. H., Lindstrom, J. M., and Lennon, V. A., 1976, *Ann. N.Y. Acad. Sci.* **274**:300–318.
225a. Lindstrom, J. M., Einarsson, B. L., Lennon, V. A., and Seybold, M. E., 1976, *J. Exp. Med.* **144**:726–738.

226. Elfman, L., Thornell, L. E., and Heilbronn, E., unpublished observations.
227. Penn, A. S., Lovelace, R. E., Lange, D. J., Toufexis, G., and Brockbank, K., 1977, *Neurology (Minneap.)* **27**:365.
228. Toyka, K. V., Drachman, D. B., Griffin, D. E., Pestronk, A., Winkelstein, J. A., Fishbeck, K. H., and Kao, I., 1977, *N. Engl. J. Med.* **296**:125–131.
229. Lindstrom, J., Engel, A. G., Seybold, M. E., Lennon, V. A., and Lambert, E. H., 1976, *J. Exp. Med.* **144**:739–753.
230. Oda, K., Korenaga, S., and Ito, Y., 1981, *Neurology (N.Y.)* **31**:282–287.
231. Lennon, V. A., 1976, *Immunol. Commun.* **5**:324–344.
232. Tzartos, S. J., and Lindstrom, Y. J., 1980, *Proc. Natl. Acad. Sci. U.S.A.* **77**:755–759.
233. Molenaar, P. C., Polak, R. L., Miledi, R., Alema, S., Vincent, A., and Newsom-Davis, J., *Prog. Brain Res.* **49**:449–458.
234. Molenaar, P. C., Newsom-Davis, J., Polak, R. L., and Vincent, A., 1981, *J. Neurochem.* **37**:1081–1088.
235. James, R. W., Kato, A. C., Rey, M.-J., and Fulpius, B. W., 1980, *FEBS Lett.* **120**:145–148.
236. Schwartz, M., Novick, D., Givol, D., and Fuchs, S., 1978, *Nature* **273**:543–545.
237. James, R. W., Lefvert, A.-K., Alliod, C., and Fulpius, B. W., 1982, *Neurochem. Int.* **4**:79–84.
238. Richman, D. P., Gomez, C. M., Berman, P. W., Burres, St. A., Fitch, F. W., and Arnason, B. G. W., 1980, *Nature* **286**:738–841.
239. Lennon, V. A., and Lambert, E. H., 1980, *Nature* **285**:238–240.
240. Barkas, T., and Simpson, J. A., 1981, *Experimental Myasthenia*, Teviot Scientific Publications, Edinburgh, pp. 1–5.
241. Lefvert, A.-K., James, R. W., Alliod, C., and Fulpius, B. W., 1982, *Eur. J. Immunol.* **12**:790–792.
242. Barkas, T., and Simpson, J. A., 1982, *Clin. Exp. Immunol.* **47**:119–126.
242a. Lennon, V. A., and Lambert, E. H., 1981, *Ann. N.Y. Acad. Sci.* **377**:77–96.
243. Wasserman, N. H., Penn, A. S., Freimuth, P. I., Treptow, N., Wentzel, S., Cleveland, W. L., and Erlanger, B. F., 1982, *Proc. Natl. Acad. Sci. U.S.A.* **79**:4810–4814.
244. Waldor, M. K., Shiram, S., McDevitt, H. O., and Steinman, L., 1983, *Proc. Natl. Acad. Sci. U.S.A.* **80**:2713–2717.
245. Christadoss, P., Lennon, V. A., Krco, C. J., and David, C. S., 1982, *J. Immunol.* **128**:1141–1144.
246. Abramsky, O., Brenner, T., Lisak, R. P., Zeidman, A., and Beyth, Y., 1979, *Lancet* **2**:1333–1335.
247. Fuchs, S., 1979, *Plasmapheresis and the Immunobiology of Myasthenia Gravis* (P. C. Dau, ed.), Houghton Mifflin, Boston, p. 20.
248. Farach, M. C., Mihovilovic, M., Paraschos, A., and Martinez-Carion, M., 1982, *Arch. Biochem. Biophys.* **214**:140–154.
249. Brenner, T., and Abramsky, O., 1981, *Immunol. Lett.* **3**:163–167.
250. Albers, J. W., Hodach, R. J., Kimmel, D. W., and Treacy, W. L., 1980, *Neurology (N.Y.)* **30**:1246–1250.
251. Garlepp, M. Y. M., Kay, P. H., Dawkins, R. L., Bucknall, R. C., and Kemp, A., 1981, *Muscle Nerve* **4**:282–288.
252. Masters, C. L., Dawkins, R. L., Zilko, P. J., Simpson, J. A., Leedman, R. J., and Lindstrom, J. M., 1977, *Am. J. Med.* **63**:689–694.
253. Vincent, A., and Newsom-Davis, J., 1982, *Clin. Exp. Immunol.* **49**:266–272.
254. Wysocka, K., Fabian, F., and Listewnik, M., 1981, *Z. Rheumatol.* **40**:135–137.
255. Dawkins, R. L., Zilko, P. J., Carrano, J., Garlepp, M. J., and McDonald, B. L., 1981, *J. Rheumatol.* **7**:56–61.
256. Naor, D., and Galili, N., 1977, *Prog. Allergy* **22**:107–146.
257. Ruiz-Torres, A., 1974, *Arzneim. Forsch.* **24**:914 917.
258. Virella, G., and Lopez-Virella, M., 1970, *Clin. Exp. Immunol.* **7**:85–108.
259. Tabachnick, M., Eisen, H. N., and Levine, B., 1954, *Nature* **174**:701–702.
260. Nimni, M., 1977, *Proc. R. Soc. Med.* **70**:65–72.
261. Aaseth, J., Forre, O., Munthe, E., Froland, S. S., and Abrahamsen, T., 1977, *Penicillamine Research in Rheumatoid Disease* (E. Munthe, ed.), Fabricius and Sonner, Oslo, pp. 80–85.

262. Jellum, E., 1965, *Int. J. Radiat. Biol.* **9:**185–200.
263. Karlin, A., 1969, *J. Gen. Physiol.* **54:**245–264.
264. Reiter, M. J., Cowburn, D. A., Prives, J. M., and Karlin, A., 1972, *Proc. Natl. Acad. Sci. U.S.A.* **69:**1168–1172.
265. Stephenson, F. A., Harrison, R., and Lunt, G. G., 1981, *Eur. J. Biochem.* **115:**91–97.
266. Chang, H. W., and Bock, E., 1977, *Biochemistry* **16:**4513–4520.
267. Dolly, J. O., and Barnard, E. A., 1977, *Biochemistry* **16:**5053–5061.
268. Bever, C. T., Chang, H. W., Penn, A. S., Jaffe, I. A., and Bock, E., 1982, *Neurology (N.Y.)* **32:**1077–1082.
269. Weigle, W. O., 1965, *J. Exp. Med.* **122:**1049–1063.
270. Froelich, C. J., Hashimoto, F., Searles, R. P., and Bankhurst, A. D., 1979, *J. Rheumatol.* **6:**237–239.
271. Wysocka, K., Kiczak, J., and Horylak, E., 1980, *Wiad. Lek.* **8:**607.
272. Lang, A. E., Humphrey, J. G., and Gordon, D. A., 1981, *J. Rheumatol.* **8:**303–307.
273. Philipott, E., Dallemagne, M. J., and Charlier, R., 1959, *Protides of the Biological Fluids* (H. Peters, ed.), Elsevier, Amsterdam, pp. 270–274.
274. Bazzato, G., Coli, U., Landini, S., Mezzina, C., and Ciman, M., 1981, *Lancet* **1:**1209.
275. Jacobus, D., Bokelman, D. L., and Mayka, J. A., 1976, *Penicillamine Research in Rheumatic Disease* (E. Munthe, ed.), Fabricius and Sonner, Merck, Sharp and Dohme, Oslo, pp. 25–38.
276. Schumm, F., Wiethölter, H., and Fateh-Moghadam, A., 1981, *Dtsch. Med. Woschenschr.* **106:**1745–1747.
277. Vartanian, G. A., and Chinyanga, H. M., 1972, *Can. J. Physiol. Pharmacol.* **50:**1099–1103.
278. Burres, S. A., Richman, D. P., Crayton, J. W., and Arnason, B. G. W., 1980, *Muscle Nerve* **2:**186–190.
279. Harpey, J. P., Caille, B., Moulias, R., and Goust, J. M., 1971, *Lancet* **1:**292.
280. Aldrich, M. S., Young, I. K., and Sanders, D. B., 1979, *Muscle Nerve* **2:**180–185.
281. Stoffer, S. S., and Chandler, J. H., 1980, *Arch. Intern. Med.* **140:**283–284.
282. Thesleff, S., 1982, *Advances in the Biosciences*, Volume 35 (P. Lechat, S. Thesleff, and W. C. Bowmann, eds.), Pergamon Press, Oxford, pp. 87–93.

14

Biochemistry of the Spinal Cord

Edward L. Hogan and Naren L. Banik

1. INTRODUCTION

The spinal cord is a particularly important part of the central nervous system as a conduit of motor, sensory, and sphincter function. Damage to it causes loss of motor function and, if sufficiently extensive, paralysis. The morphological architecture of spinal cord and its reflex physiology have been described in considerable detail, but the molecular mechanisms involved in the maintenance and loss of functions are relatively little understood. In the past, the biochemical and metabolic properties of spinal cord received little attention in comparison to the effort expended on brain. In the last decade, however, the situation has changed, and a substantial number of studies have been carried out on proteins, enzymes, amino acids, synaptology, and transmitters in the spinal cord. Studies on the metabolism of spinal cord proteins and lipids and of the enzymes mediating their breakdown provide insight into processes of axon and myelin degeneration. Damage to the spinal cord causes structural alterations and accompanying biochemical changes, some of which, e.g., proteinase release and/or activation, have been implicated in the breakdown of the axon–myelin unit. Relatively little is known at the molecular level of injury or diseases of spinal cord, and less concerning the factor(s) regulating spinal cord degeneration and their source. In order to understand spinal cord function, a detailed study of its biochemical properties (i.e. molecular events) is essential, both in the normal situation and in disease.

In this chapter, we aim to compile available data concerning spinal cord components with particular reference to proteins, lipids, enzymes, minerals, transmitter substances, and the effects of pharmacological agents on spinal cord following injury. Some areas are treated more thoroughly than others, and some, such as spinal cord in culture, are not covered. We have attempted to emphasize those findings that seem most likely to be the basis of further research directed at regulatory mechanisms of spinal cord function in the normal situation and following trauma.

Edward L. Hogan and Naren L. Banik • Department of Neurology, Medical University of South Carolina, Charleston, South Carolina 29425.

2. COMPOSITION OF NORMAL SPINAL CORD

The protein and mineral compositions of mammalian spinal cord from different species are remarkably similar. The most obvious regional difference, observed by comparison of spinal cord proteins to those of brain white matter, is a changed relative proportion of the major myelin proteins with less proteolipid protein (PLP) than basic protein in spinal cord. There is more cholesterol in spinal cord of mammals than of lower vertebrates, probably reflecting the increase in myelin that accompanies phylogenetic ascent.

2.1. Lipids

The first systematic studies of spinal cord lipids were those of Rossiter[1] and Brante.[2] Total lipid amounts to approximately 70% of spinal cord dry weight. In adult mammalian spinal cord, cholesterol accounts for approximately 21%, total phospholipids 47%, and galactolipids (cerebrosides and sulfatides) 27% of total lipid.[3] The composition of spinal cord lipid is remarkably similar in all species with the exception that the cholesterol content is significantly greater in mammalian spinal cord than in submammals (turtles, reptiles).[4] The phosphatidylcholine content of spinal cord is at least twice that of phosphatidylethanolamine. The plasmalogen contains ethanolamine for approximately 38% of total phospholipid with only 1.4% choline plasmalogen.[5] The other spinal chord phospholipids include phosphatidylinositol and sphingomyelin (approximately 10%) and phosphatidic acid (1.2%).[5] The cholesterol in adult spinal cord of the several species studied exists in free form, and none is esterified. The content in mammalian spinal cord is greater than that in lower vertebrates. Cholesterol synthesis in spinal cord explant culture shows a precipitous increase coincident with incipient myelination,[6] as would be expected from other studies in CNS. No cholesterol presursors such as desmosterol or 7-dehydrocholesterol have been found in adult spinal cord.

The galactolipid content of spinal cord is slightly greater than that of brain. Galactocerebroside (galactosylceramide) constitutes 18% of total spinal cord lipid compared to 13% in brain, whereas, sulfatide is similar (4–5%) in brain and spinal cord. Cerebroside accounts for 80% and sulfatide 18% of total galactolipid of spinal cord. Overall, spinal cord from various species has a higher percentage of lipid as cerebroside, cholesterol, and sphingomyelin than does brain.[4] On the other hand, brain ganglioside content is much higher than that in spinal cord.[7] In human and cat, the ganglioside content of spinal cord is about one-tenth of that in the cerebral gray matter. The ganglioside concentration of human spinal cord is considerably lower than that of the spinal cord of cat and rabbit.[8] Gangliosides represent only 1.8% of total galactolipid in spinal cord compared to 10% for brain. This may reflect the enrichment of gangliosides in neurons and low ganglioside content of myelin. However, most of the individual types of gangliosides found in brain are also found in spinal cord.[5]

There may be some species differences in the distribution of individual gangliosides (Table I). G_{M1}, G_{D1b}, and G_{T1b} are the most prominent gangliosides

Table I
Distribution of Gangliosides in Spinal Cord[a]

Ganglioside	Human	Cat	Rabbit
G_{M4}	12.8	1.1	—
G_{M3}	14.0	3.5	2.9
G_{M2}	3.7	3.2	1.3
G_{M1}	16.8	27.8	29.0
G_{D3}	16.4	19.5	10.5
G_{D1a}	4.2	6.1	14.7
G_{T1a}	0.5	0.3	2.0
G_{D2}	0.9	—	1.2
G_{D1b}	18.7	21.1	14.3
G_{D1b}-Fuc	—	—	2.3
G_{T1b}	9.1	13.6	15.6
G_{Q1b}	2.9	3.8	6.1

[a] Data from Ueno *et al.*[9] with permission. Results are expressed as percent total ganglioside sialic acid.

present in rat and human spinal cord,[5,8] and G_{D3} is more abundant in human than in rat. The ganglioside distribution in spinal cord differs from that of cerebrum with more G_{D3} and G_{M3} and less G_{D1a} in spinal cord.[9] Gray matter contains more ganglioside than white matter. The major gangliosides in gray matter are G_{M1}, G_{D3}, G_{D1a}, and G_{T1b}, whereas white matter has G_{M4}, G_{M3}, G_{M1}, G_{D3}, G_{D1b}, and G_{T1b}.[10] G_{M4} is one of the major gangliosides in human but not in rat, cat, and rabbit spinal cord. [5,8,11] Another minor ganglioside, G_{D1b}-FUC, has been detected in rat and cat spinal cord and cerebral white matter. An unknown ganglioside that comigrates on separation by TLC between human G_{D1b} and G_{T1b} has been observed in rat spinal cord.[5] These are probably species differences, although methodological differences may also contribute.

Small amounts of triglycerides and free fatty acids have been found in spinal cord myelin.[11] Phosphorylated precursors of phospholipids such as glycerylphosphorylcholine and glycerylphosphorylethanolamine have been reported in spinal cord of several species as in brain.[12] Other minor phosphorylated compounds found in fish spinal cord and brain include L-serine ethanolamine phosphate and L-threonine ethanolaminephosphate. The content of L-threonine ethanolaminephosphate is much greater in spinal cord and brain than the L-serine analogue. Both compounds are cleaved by microsomal enzymes. Their absence in mammalian CNS suggests a phylogenetic difference in phospholipid metabolism in spinal cord[13] that warrants further study.

Prostaglandins (PG) have been identified in the spinal cord (as in the brain) of numerous mammalian and nonmammalian species[14–17] and have been found as well in CSF.[18] Early studies revealed prostaglandins E_2 (400 ng/g wet wt.) and $F_{2\alpha}$ (10 ng/g wet wt. tissue) in chicken spinal cord. The amount of PGF in rat spinal cord (8.05 ng/g wet wt.) resembles the level in chicken.[17] Thromboxane A_2 and prostacyclin (PGI_2) have been detected in spinal cord. Throm-

Table II
Ganglioside Content of Spinal Cord Myelin[a]

Species	Salic acid (μg Neu-Ac per 100 mg myelin)	Ganglioside (μg per 100 mg myelin)	Ganglioside (% myelin lipid)
Human	32.2	137	0.20
Rabbit	46.7	199	0.29
Pigeon	196	743	1.1
Chicken	245	942	1.4
Cat	44.2[b]	—	—

[a] Data from Cochran *et al.*[28] with permission.
[b] From Ueno *et al.*[9]

baxane B_2 (TxB_2), a metabolite of thromboxane A_2, and 6-keto-$PGF_{1\alpha}$ (6-Keto), a metabolite of prostacyclin, have also been demonstrated in spinal cords of different mammlian species. The levels of TxB_2 (145 pg/mg dry wt. tissue) and 6-Keto (127 pg) are similar in spinal cord.[19]

2.1.1. *Lipids of Spinal Cord Myelin*

The lipid composition of adult spinal cord myelin isolated from several species is remarkably similar and is also comparable to that obtained from different regions of whole brain.[20–23] The proportion of the myelin lipids is the same as that in whole cord, namely, total phospholipid 42%, total galactolipid 35%, and cholesterol 22%.[21] Spinal cord myelin contains more ethanolamine plasmalogen, galactolipid, cholesterol, and phosphatidylcholine (lecithin) than brain.[20–22,24] It has been shown that vitamin B_{12} deficiency affects the composition of spinal cord myelin in the fruit bat. There is a reduction in the content of phosphatidylcholine and an increase in the level of C15:1 fatty acid. However, striking differences characterize all lipid classes except for serine and inositol phosphoglycerides when PNS and spinal cord myelin are compared.[20] Myelin of the PNS contains more sphingomyelin, whereas the contents of other lipids, e.g., galactolipid, ethanolamine plasmalogen, and cholesterol, are strikingly lower than in spinal cord. This suggests that oligodendroglia and Schwann cells in subhuman species synthesize myelin of different lipid composition.

The myelin prepared from spinal cord contains gangliosides[9,8,26,27] in about half the amount found in the myelin of brain white matter. The proportions of gangliosides in spinal cord myelin are similar to those in myelin of the same species (Table II). G_{M1} is the most abundant ganglioside in spinal cord myelin in all species, and G_{M4} is highly enriched in human cord myelin (Table III). As in mammals, G_{M4} is also localized in avian CNS myelin.[28] G_{D1a} is decreased in human and cat cord myelin relative to brain white matter, but the proportion is the same in rabbit. There is considerable G_{D1a} in both chicken and pigeon cord myelin. The fatty acids of G_{M4} ganglioside of avian myelin contain both the hydroxy and unsubstituted varieties, and the major long-chain base is C_{18} spingosine. Avian G_{M1} is resolved by TLC into two bands,

Table III
Ganglioside Composition of Spinal Cord Myelin[a]

Ganglioside	Human	Cat	Rabbit	Chicken	Pigeon
G_{M4}	25.3	1.7	—	31.4	32.2
G_{M3}	3.5	—	—	2.9	2.6
G_{M2}	6.6	4.3	—	3.4	3.4
G_{M1}	38.2	52.1	44.6	34.0	34.4
G_{D3}	5.0	—	5.3	7.3	7.2
G_{D1a}	0.7	2.1	7.8	11.1	8.5
G_{T1a}	0.1	1.1	2.4	—	—
G_{D2}	0.4	2.7	1.0	—	—
G_{D1b}	11.1	18.2	15.3	5.4	5.4
G_{D1b}-Fuc	—	2.4	2.4	—	—
G_{T1b}	7.1	—	12.5	3.0	6.5
G_{Q1b}	2.2	4.4	5.3	1.4	1.8

[a] Data from Ueno *et al.*[9] and Cochran *et al.*[28] with permission. Data expressed as percent total ganglioside sialic acid.

the slower one containing C_{18} and the faster predominantly longer-chain fatty acids.[28] These studies are all consistent with the first reports describing myelin gangliosides.[22, 29]

2.1.2. Changes during Development

As a rule, phylogenetically older tracts and regions of spinal cord myelinate earlier. The weight gain of spinal cord during development is primarily attributable to an accumulation of myelin and its constituent lipids and proteins. From the early studies of the lipid composition of whole spinal cord, a gradual accumulation of all lipid classes with age has been recognized. Increases occur in cholesterol, total phospholipid, and galactolipid in rat spinal cord during development.[30] The human spinal cord examined during the 17th to the 34th weeks of gestation shows an increase in cholesterol ester (5.8% to 49.0% of total cholesterol), which declines to 7.6% at birth and disappears in the adult.[31] Cholesterol ester has been found in human fetal and chick brain,[32] and desmosterol is also found in human fetal brain and early stages of developing rat brains.[31,33,34] Changes in ganglioside content and molecular species have been reported during development of rat, mouse, chicken, and human brain[35–37] but not in developing spinal cord.

2.1.3. Changes in Myelin

The increases in cholesterol, plasmalogen, sphingomyelin, and cerebroside coincide with the peak of myelination, consistent with the fact that these lipids are myelin constituents. Changes in these myelin lipids during development of the brain have been examined in detail[21,37–40] and surveyed as well in spinal cord.[21] Immature myelin contains more phospholipid, particularly phosphatidylcholine (PC), and less galactolipid than mature myelin, and the proportion

Table IV
Composition of Myelin Lipids of Spinal Cord during Development[a]

Lipids	Age (days) 20	25	30	60	90	180
Cholesterol	23.7	21.1	20.2	21.6	22.5	22.3
Total phospholipid	51.1	47.8	45.4	45.5	42.7	42.0
PS + sphing + P_I	12.9	13.1	12.6	13.1	13.8	13.1
Phosphatidylcholine	16.6	13.7	13.1	12.3	10.9	9.8
Phosphatidylethanolamine	20.1	20.7	19.6	20.2	18.5	19.3
Galactolipid	26.3	31.0	35.5	33.1	33.6	34.5

[a] PS, phosphatidylserine; sph, Sphingomyelin; P_I, inositide phosphoglycerides. Data from Smith[21] with permission. Data as percent total lipid.

of PC decreases with age as galactolipid increases (Table IV). Decreases in triglycerides and free fatty acids in myelin from human spinal cord occur prenatally and reach a plateau at 3 months post-natal.

Maturation of spinal cord is believed, therefore, to manifest decreases of triglycerides and free fatty acids and increases of myelin cholesterol and cerebroside. The fatty acids of brain myelin progress from short-chain fatty acids (C_{16} and below) to those of increased chain length. There is more palmitic acid and less stearic and oleic acid in immature myelin.[41] The gangliosides of myelin increase in mouse but not in rat and chicken,[34–36] although detailed studies of their developmental variation in spinal cord myelin have yet to be done.

2.2. *Proteins*

The total protein content of spinal cord is less than that of cerebral white matter.[8] The protein composition of whole spinal cord from different species is similar except that, like brain, rat and mouse spinal cord contains two major basic proteins rather than one as in nonrodent species.[42–45] Cerebral white matter contains, as does spinal cord, the characteristic myelin proteins: myelin basic protein (MBP), proteolipid protein (PLP), and Wolfgram protein (WP).[45] There is a distinct difference in the high-molecular-weight (HMW) proteins since additional protein bands are evident in this region in spinal cord. The proteins are resolved by sodium dodecyl sulfate polyacrylamide gel electrophoresis (SDS–PAGE).[8,46] The molar proportion of PLP in rat, rabbit, bovine, guinea pig, and calf spinal cord is higher than MBP,[42,43,45–47] although they are approximately equal in the human.[8] This comparison is notable because human white matter contains slightly lower amounts of PLP than MBP.

Myelin basic protein, which is encephalitogenic, is usually purified from spinal cord. Injection into animals of this protein or of cord myelin together with complete Freund's adjuvant causes experimental allergic encephalomyelitis (EAE).[48] A chronic progressive form of EAE has also been produced in rabbits by injection of PLP apoprotein.[49] The clinical signs and myelin damage found resemble the EAE induced by MBP. Experimental allergic encephalomyelitis serum causes demyelination and inhibits myelin formation in tissue culture, although the mechanism by which serum produces these effects is not known.[50] Antisera raised against MBP have been studied for their capacity to provoke complement-dependent demyelination and to release myelination inhibitor factor(s), but none have been found. The effect of antisera to PLP on the myelin of cultured rat spinal cord has also been examined, and no effects found here either.[51,52] A high-molecular-weight glycoprotein, the myelin-associated component (MAG), has also been shown in rat and human spinal cord by both biochemical and immunocytochemical means.[53–56]

2.2.1. *Cytoskeletal Proteins*

Spinal cord contains characteristic cytoskeletal proteins including the neurofilament (NFP) triplet, vimentin, glial filament (GFA), microtubular proteins, and others.[42,57–59] The electrophoretic migration of these proteins resembles those found in white matter,[42] but their proportions differ. Glial filament protein, the most abundant of the cytoskeletal proteins, can be synthesized *in vitro* employing messenger RNA purified from Jimpy mouse spinal cord, which has a prominent astrogliosis.[60] There is more of the 70K NFP than of the other two neurofilament proteins (160K and 210K) comprising the triplet. The crude cytoskeletal protein isolated from rat spinal cord has different proportions of the GFA and NFP triplet proteins.[61] There is more GFA than NFP proteins. The crude NFP triplet proteins were separated by gel filtration to obtain homogeneous preparations of the 210K, 160K, and 70K NF proteins. All three proteins showed similar dye-binding properties with fast green, but the 210K protein was underrepresented when stained with Coomassie blue. All three proteins contain high acidic amino residues, and the 210K protein contains more proline (8.5 mol%) than the other two. The large microtubular (MT) proteins (56K) are usually less than the small (54K) MT protein in spinal cord.[42]

The protein pattern of rat spinal cord at molecular weights in the range of 30–38K is notably different from that of bovine and rabbit cords. Two additional polypeptides are evident in rat that are negligible in rabbit and bovine samples.[42]

2.2.2. *Myelin Proteins*

The total protein of spinal cord myelin increases during development, with the largest increase in myelin protein at about 15 days in rat.[43,62] This undoubtedly reflects the early ontogenetic deposition of myelin in spinal cord, for the myelin protein accumulation starts after 15 days in other regions of CNS.[43]

The protein composition of spinal cord myelin isolated from a number of species has been investigated.[8,43,45,46,63,64] The spinal cord myelin proteins are similar to the proteins of brain myelin. As in brain, there are two major MBPs (large and small) in mouse and rat spinal cord myelin. Myelin isolated from brain contains a myelin-associated glycoprotein (MAG), which also increases in myelin with development. A similar increase in MAG is also seen in myelin isolated from spinal cord, and it is noteworthy that the apparent molecular weight of MAG by SDS–PAGE in immature spinal cord myelin is slightly greater than that of adult. There is no difference in the molecular weight of MAG isolated from the spinal cord of 22-day-old and mature rats.[55]

The relative proportions of the MBPs change during development of the rat spinal cord. During early myelination, the amount of the main large MBP (mol. wt. 18,500 or 18.5K), as estimated by dye binding, is greater than that of the small one (14K). With maturation, the amount of 18.5K protein decreases, and the 14K basic protein increases.[43,62] The HMW and Wolfgram proteins also decrease with increasing age, with a stable adult pattern attained at approximately 40 days of age. There is a slow increase in PLP content with maturation. Similar changes in the pattern of basic and Wolfgram proteins occur in myelin isolated from chicken, human, rat, rabbit, and mouse[35,65–69] except that the proportion of proteolipid protein in spinal cord myelin is less than that in myelin from other CNS regions.[46,70] The appearance of myelin containing MBP in rat spinal cord at 2 days post-natal corresponds to the earlier onset of myelination and the earlier appearance of myelin-specific components in the spinal cord than in other CNS regions.[43,62,71]

The protein composition of myelin isolated from explants of fetal spinal cord in culture has been compared with the myelin of rat spinal cords at corresponding developmental ages. The myelin of explant spinal cord contains a greater proportion of HMW proteins (63% of the total protein) than that of spinal cord of postnatal rats.[72] Yet the proportion of both large and small MBPs is the same in the explant myelin.[43,62] The higher proportion of HMW proteins in spinal cord culture suggests that the myelin is immature and/or may contain a membrane fraction (e.g., myelin-like) enriched in HMW proteins.[73]

The P_2 protein, a myelin basic protein characteristic of PNS myelin, induces experimental allergic neuritis (EAN).[74–79] This protein has been isolated from spinal cord and shown to have an amino acid composition and electrophoretic mobility similar to the P_2 protein of PNS myelin, suggesting that it is present in spinal cord and maybe in cord myelin.[80–82] Although its presence in the spinal cord could be attributed to inclusion of PNS myelin at cord–spinal root junctional regions, immunocytochemical studies indicate that it is in the myelin within the spinal cord.[83]

The protein patterns of other subcelluar fractions of mouse spinal cord have been examined by Stodieck and Luttges.[47] The nuclear fraction contains proteins of approximate molecular weight 150K and 68K (presumably NF proteins), an unidentified protein at 40K, and myelin PLP and MBP. This suggests that their nuclear fraction is contaminated by myelin and axonal components. The microsomal and soluble fractions contain HMW proteins and 40–50K proteins, and the mitochondrial fraction has typical myelin proteins together with HMW proteins. Judging from the protein profiles, there is either considerable heterogeneity of membrane fractions or disperison of proteins among cell membranes.

2.3. *Enzymes*

Few studies of the enzymatic activities of spinal cord have been published. Most of the enzymes found in brain are also present in the spinal cord. These

Table V
Activities of G6PDH, LDH, Carbonic Anhydrase, and Enolase in Spinal Cord and Purified Myelin of Rat[a]

	Spinal cord		Spinal cord myelin	
Enzyme	20 days	120 days	20 days	120 days
G-6-PDH	18.8	22.8	2.2	2.6
Glyc-P-DH	23.4	39.3	3.7	6.9
LDH	580	571	113	46.8
Carbonic anhydrase[b]	705	1,010	300[c]	475
Enolase[b]	—	16,500[d]		

[a] Data from Cammer and Zimmerman[84] with permission; activities expressed as nmol/min per mg protein.
[b] Expressed as ng/mg protein.
[c] Mean of two determinations.
[d] Adult rat, data from Marangos *et al.*[231]

include the enzymes mediating energy metabolism (Krebs cycle), transmitter and amino acid metabolism, and several proteinases, peptidases, and hydrolases.

There is little information available concerning protein catabolism. Spinal cord proteins can induce EAE in which the degradation of proteins is accompanied by increased activities of proteinases. A similar proteolysis by proteinases and peptidases may play an important role in the loss of spinal cord integrity in human disease.

2.3.1. *Glycolytic and Other Energy-Yielding Enzymes*

The enzymes mediating glycolysis and mitochondrial oxidation have been demonstrated both histochemically and biochemically in spinal cord.[84] In some, the levels of activity differ somewhat from brain, and the distribution of activity between gray and white matter also differs. The activity of glycerol phosphate dehydrogenase (GPDH) is higher in the rat brain than cord, whereas glucose-6-phosphate dehydrogenase (G6PDH) is higher in cord. The activities of both GPDH and G6PDH are higher in rat cord during development, as has also been found in brain (Table V). Hexokinase is more active in cord gray matter, whereas GPDH and G6PDH are enriched in white matter.[84–86] The other enzymes of the pentose phosphate shunt (transketolase and phosphogluconate dehydrogenase) are also found mainly in white matter, suggesting that these enzymes may be mainly in glia.[85,86] This is consistent with the enrichment of GDPH and G6PDH found in isolated oligodendrocytes of bovine and rat CNS.[84] These enzymes, however, are absent in purified myelin, suggesting that they are in the cell perikaryon.[84]

Glucose-6-phosphate isomerase (GPI), which catalyzes the reversible reaction glucose-6-phosphate $\rightleftharpoons$ fructose-6-phosphate, is found in the Purkinje cells, the molecule layer, and the granular layer in the cerebellar cortex[87] and is also found in spinal cord but is less active here than in brain. The ventral horn neurons contain more activity than those of the dorsal horn. Glucose-6-phosphate isomerase is more active in axons than in myelin.[88]

Hexokinase, a rate-limiting control of glucose metabolism in brain,[89] is found mainly in the gray matter and is low in tracts. Studies of subcellular fractions localize hexokinase in the synaptic ending fraction.[85,90–92]

Pyruvate decarboxylase and transketolase are thiamine-dependent enzymes with demonstrable activity in spinal cord.[93,94] The activity of pyruvate decarboxylase is lower in spinal cord than in brain. Activity in spinal cord is reduced in some human hereditary ataxias. Transketolase activity is present in myelinated nerves, suggesting that it is involved in oligodendroglial metabolism.[93] Lactic dehydrogenase activity and its five isozymes have been shown in spinal cord, with isozymes 1 and 2 predominating.[95–97] Glycogen phosphorylase and other enzymes of glycogen metabolism have been shown histochemically in spinal cord. Glycogen phosphorylase is in the white matter of spinal cord.[98]

Enzymes associated with the Krebs cycle and oxidative metabolism have also been demonstrated in spinal cord by both biochemical and histochemical means. Succinic dehydrogenase, cytochrome oxidase, isocitric dehydrogenase, NAD diaphorase, fumarase, and related enzymes are found in spinal cord.[96,99–102] These enzymes, particularly succinic dehydrogenase and cytochrome oxidase, are localized in the neuropil and are associated with mitochondria. The level of oxidative enzymes is lower overall in spinal cord than in brain. A similar relationship holds for respiratory rate at these two sites.[103]

2.3.2. Enzymes of Transmitters, Amino Acids, and Protein Metabolism

The enzymes of neurotransmitter synthesis are essential and characteristic for neuronal activities and brain function. The regulatory mechanisms for their synthesis have been extensively studied in brain, but the synthesis of the different neurotransmitters in cord has not been studied in detail.

2.3.2a. Choline Acetyltransferase (ChAT; EC 2.3.1.6) and Acetylcholinesterase (AChE; *EC* 3.1.1.7). Choline acetyltransferase (ChAT), which catalyzes the synthesis of the neurotransmitter acetylcholine, and acetylcholinesterase (AChE), which degrades acetylcholine, have been assayed in spinal cord and localized immunohistochemically. The distribution of ChAT activity in different regions of the nervous system has been studied by a number of investigators.[104–107] In spinal cord the enzyme activity is localized in the cell bodies of the ventral horn motor neurons, whereas in the cerebellum it is present mainly in mossy fibers.[105] In humans, ChAT is found in dorsal horn. The activity is low in bovine and negligible in rat and cat. The enzyme is enriched in gray matter, and subellular studies have shown its enrichment in the nerve terminal. The level of ChAT activity at different levels of rat spinal cord is similar to that in the cat.[108]

Acetylcholinesterase (AChE) activity in spinal cord has been studied for segmental distribution[109] and by histochemistry with delineation of ascending and descending cholinergic fibers in cat.[110] Acetylcholinesterase activity increases in spinal cord of rats and other species during fetal development.[111–113] Postnatally, the activity in rat cord peaks at 2 days and declines at 44 days. By comparison, the AChE activity in brain is lower than that in spinal cord and peaks at 24 days.[111] The AChE activity in brain increases with age,[41] although it decreases in spinal cord, so that activity in brain is greater than in spinal cord in the adult. This chronology of cholinergic enzymes with particular reference to fetal cord suggests that the functional development of synapses occurs earlier in cord ontogeny than elsewhere in CNS.

The AChE activity is substantially greater than that of ChAT in spinal cord, but their distribution is similar.[108] Acetycholinesterase activity is greater in the gray matter,[114] and activity is greater in neurons, cytoplasm, axons, dendrites, and synaptic endings than in nucleoplasm.[115–117] On subcellular fractionation of brain, AChE activity is enriched in synaptic endings, microsomes, and cytoplasm.[41,118,119] The AChE activity is greater in motor than sensory neurons.[120,121] Histochemical staining differs in the different laminae (strongest in laminae III and IV) of the dorsal horn. An intense reaction for AChE is found in cells of the dorsal part of the ventral horn between T13 and L6.[122] The similar pattern for ChAT indicates a substantial cholinergic function in spinal cord.[122,123]

The multiple forms of AChE found in rat brain[124–129] are also found in spinal cord.[130] Two of the four forms are of low molecular weight and are easily soluble in buffer, whereas the other two are larger, membrane bound, and soluble only in detergent (e.g., Triton X-100). The smaller ones may be derived from the larger in cleavage steps that presumably occur in the endoplasmic reticulum.[130]

Acetylcholinesterase has been studied in organotypic cultures of embryonic spinal cord and is localized in the cytoplasm of ventral horn neurons, in dorsal horn neuropil, and in neuroglia membranes.[131] A similar distribution has been reported in adult rat spinal cord.[122,132] The presence of AChE in neuroglia suggests a role for acetylcholine in these cells.

In contrast to AChE, butyrylcholinesterase activity is found mainly in glia and in endothelial and other cells in blood vessel walls in rat spinal cord.[102,116,120] In the cat, however, the vessels did not react, but the white matter and neuropil of the ventral horn stained strongly.[122] Other species differences in cholinesterase activity have been reported as well.[133]

2.3.2b. Glutamic Acid Decarboxylase (GAD; E.C. 4.1.1.15). The product of glutamic acid decarboxylase action is the neurotransmitter γ-aminobutyric acid (GABA). Glutamic acid decarboxylase activity is found in all areas of brain and spinal cord.[134,135] Activity is greater in brain than spinal cord and varies with species: there is more activity in rabbit than monkey.[134,135] The GAD activity is greater in the dorsal than ventral gray matter and is virtually absent in the corresponding areas of white matter (Table VI). The GAD activity and GABA concentration in gray matter correlate well.[136] The presence of GABA in white matter in the absence of GAD activity suggests that GABA is synthesized in the cell soma and transported in axons. Morphine administration increases the content of GABA and the activity of GAD.[137]

Table VI
Regional Enzyme Activities in Cat Spinal Cord[a]

Enzyme	Cord region: Doral white	Dorsal gray	Ventral gray	Ventral white
Glutaminase	67.0 ± 6.61	397 ± 34.9	331 ± 40.4	66.3 ± 8.04
Glutamine synthetase	20.1 ± 2.44	67.4 ± 5.05	66.8 ± 8.05	18.7 ± 1.64
GABA aminotransferase	3.02 ± 0.19	13.7 ± 1.04	12.8 ± 1.28	3.24 ± 0.14
Glutamate decarboxylase	0	13.3 ± 2.99	7.02 ± 1.59	0
Aspartate aminotransferase	2454 ± 205	5412 ± 107	5418 ± 151	2480 ± 186
Glutamate dehydrogenase	149 ± 24.4	446 ± 25.0	448 ± 13.1	139 ± 8.9

[a] Data from Graham and Aprison[135] with permission, expressed in μmol/g per hr.

The enzymes mediating metabolism of glutamine, aspartate, and GABA in spinal cord include glutamine synthetase, glutaminase, aspartate aminotransferase, GABA amino transferase, glutamate dehydrogenase, and succinic semialdehyde dehydrogenase. All are present in the spinal cord. The activity of glutamine synthetase is less than that of glutaminase. Both are enriched in gray matter. The activities of glutaminase and glutamine synthetase in dorsal and ventral gray matter are about five- and fourfold greater than in the corresponding white matter (Table VI). The synthetic enzyme activity and glutamine concentration are higher in the gray matter than the white matter.[137]

Glutamate dehydrogenase catalyzes conversion of glutamic acid to α-ketoglutarate. It is found mainly in the dorsal and ventral gray matter. The activity in the white matter is about one-third that of gray. The glutamate concentration is higher in the dorsal than ventral spinal cord.

GABA aminotransferase (E.C. 2.6.1.19) activity is also greater in the dorsal and ventral gray than white.[135,138] The GABA transaminase activity found in spinal cord culture is irreversibly inactivated by gabanilin.[139] The enzyme has been localized in neurons as well as in glial cells in cultures of developing rat spinal cord.[131]

Activity of succinic semialdehyde dehydrogenase (E.C. 1.2.1.16) is higher in gray than white matter and is enriched in dorsal over ventral gray.[140] The acitivity of aspartate aminotransferase (E.C. 2.6.1.1) is twofold greater in dorsal and ventral gray matter than in the corresponding white. Although activities in dorsal and ventral gray are equal,[135] the concentration of aspartate is higher in the ventral gray.

The distribution of enzymes for the metabolism of glutamate, GABA, glutamine, and aspartate does not consistently correspond to the distribtuion of these amino acids. There is no *a priori* reason for a simple correlation to exist, since assay conditions, intracellular enzyme regulation, and intracellular compartmentation are a few of the variables that could affect the functional expression of these catalysts. The preponderance of the data indicate that glutamate, aspartate, and GABA are neurotransmitters in cat spinal cord.

D-Amino acid oxidase (E.C. 1.4.3.3), which catalyzes the oxidative deamination of certain D-α-amino acids (e.g., glycine to glyoxylate), has been studied in different CNS regions. Its activity is higher in cerebellum than spinal cord and higher in the gray matter of spinal cord than the white.[141–143]

2.3.2c. Dihydroxyphenylalanine decarboxylase (DOPA Decarboxylase; E.C. 1.14.2.1). DOPA decarboxylase catalyzes a step necessary for the formation of catecholamines and 5-hydroxytryptamine (5-HT). Its activity is the same at different cord levels, and this is consistent with the even distribution of norepinephrine (NE) and 5-HT.[144] The activity is greater in gray than white matter of spinal cord.[145] Disappearance of NE and 5-HT correlates with loss of DOPA decarboxylase activity in the distal spinal cord after transection, suggesting that the enzyme is largely in the descending tracts of the cord.[146] The administration of 5,6-dihydroxytryptamine or 6-hydroxydopamine to rats produces a similar loss of DOPA decarboxylase and 5,6-dihydroxytryptophan decarboxylase activity in brain and in spinal cord. It appears then that a single decarboxylase regulates the synthesis of catecholamines and serotonin in CNS.[147] The activity of tryptophan hydroxylase is very low in the spinal cord in comparison to other CNS areas.

2.3.2d. Dopamine β-Hydroxylase (DBH; E.C. 1.14.2.1). Dopamine β-hydroxylase mediates the conversion of dopamine to norepinephrine and is a useful marker for noradrenergic neurons.[148,149] Flourescent histochemical studies have shown NE terminals in thoracic spinal cord, and immunocytochemical studies place DBH in neuronal processes.[150,151] The DBH is localized mainly in the intermediolateral cell column of the spinal cord.[151] It has also been found in synaptic vesicles.[152,153]

2.3.3. Proteinases and Peptidases

There have been only a few studies assaying the proteolytic activity and peptidases of normal spinal cord. The activities of proteolytic enzymes and peptidases in spinal cord are lower than in brain[154,155] and increase following organophosphorus intoxication[156] and in Wallerian degeneration.[157] An elegant and comprehensive study of proteinases and peptidases in spinal cord and its subcellular organelles was done by Marks and colleagues.[155] All enzymes found in brain are also found in spinal cord, and overall, the activities expressed in terms of tissue weight or protein are lower in spinal cord than brain. The activities of acid proteinase (cathepsin D), aminopeptidase, and arylamidase in spinal cord are half those of brain, whereas neutral proteinase activity is similar. In spinal cord, the acid proteinase activity is greater than the neutral proteinase, tripeptidase activity is greater than dipeptidase, and arylamidase activites are higher with monoacyl substrates than dipeptidyl derivatives (Table VII). Endogenous acid (cathepsin B, D) and neutral proteinase activities have been studied in many laboratories.[158–162] Cathepsin D in rat CNS is found predom-

Table VII
Proteolytic Enzymes in Spinal Cord and Brain of Rabbit[a]

Enzyme group and substrate	Total activity (μmol/g fresh tissue per hr)		Specific activity (μmol/g protein per hr)	
	Spinal cord	Brain	Spinal cord	Brain
Proteinase[b]				
Acid[c]	25	49	320	430
Acid[d]	13	22	180	210
Neutral	14	15	130	130
Peptide hydrolase				
Leu-Gly-Gly	280	490	3200	4100
Lys-Ala	79	100	880	860
Arylamidase				
Leu-β-NA	32	62	370	530
Arg-β-NA	35	96	390	790
Glu-β-NA	1.2	2.2	14	18
Ser-Tyr-β-NA	1.7	2.8	19	24
Lys-Ala-β-NA	2.6	3.8	28	31
Arg-Arg-β-NA	6.5	9.0	72	77

[a] Data from Serra *et al.*[155] with permission.
[b] Activity expressed as μmol amino group or tyrosine or naphthylamine produced.
[c] Ninhydrin method.
[d] Tyrosine equivalent.

inantly in neuronal perikarya and processes,[161] as are acid proteinases and other lysosomal hydrolases found in neurons.[163,164] A low activity of cathepsin D is also present in oligodendrocytes in corpus callosum and centrum semiovale, but the immunostaining is less prominent and convincing in the spinal cord and brainstem. The enzyme is enriched in spinal cord motor neurons and in large neurons in the caudate nucleus.[161] It may be that cathepsin D in neurons mediates important neural function and/or degeneration in disease. For example, its activity in demyelinating diseases is increased along with the reduction of MBP,[165–167] which is degraded by purified cathepsin D into antigenically active peptides.[74,168] An increased activity of cathepsin D is also found in the spinal cord of wobbler mice (a genetic motor neuron disease).[140,142]

Purified MBP is degraded into smaller peptides on incubation at several pHs (pH 3.6, 6.0, and 7.6) in a homogenate of spinal cord, indicating that there is neutral as well as acid proteinase activity in spinal cord. A Ca^{2+}-activated neutral proteinase whose natural substrates include MBP and NFP[169–172] also occurs in spinal cord. Its activity is inhibited by both EGTA and leupeptin, a neutral proteinase inhibitor. The GFAP of mouse spinal cord is degraded by such a Ca^{2+}-dependent proteinase into low-molecular-weight polypeptides, which have been identified by immunoelectrophoretic transfer.[173] Short polypeptides derived from GFAP are found in postmortem tissue[175] and resemble the peptide fragments generated by GFAP autolysis.[173] These studies suggest that catabolism of neurofilament and glial filament proteins depends on Ca^{2+}-activated neutral proteinases. The degradation of cytoskeletal proteins in dis-

Table VIII
Distribution of Enzyme Activities in Subcellular Fractions of Spinal Cord[a]

Enzyme	Enzyme activity (μmol/g protein per hr)			
	Nuclear	Mitochondrial	Myelin	Supernatant
Proteinase				
Acid	180	320	51	520
Neutral	200	170	340	190
Peptide hydrolase				
Leu-Gly-Gly	1300	1600	360	6400
Lys-Ala	640	600	—	980
Arylamidase				
Leu-β-NA	250	250	270	630
Arg-β-NA	350	240	52	600
Glu-β-NA	8	9	—	16
Ser-Tyr-β-NA	6	34	—	20
Lys-Ala-β-NA	1	26	—	47
Arg-Arg-β-NA	48	53	—	120

[a] Data from Serra *et al.*[155] with permission.

ease is found in the degeneration of axons in spinal cord injury and Wallerian degeneration.[175–177] It should be noted, however, that the turnover of GFAP in jimpy mice is not altered.[178] It is clear that spinal cord, like other regions of CNS, contains enzymes capable of degrading proteins and that changes in proteolytic activity accompany structural and biochemical alterations culminating in spinal cord degeneration.[176,177,179,180] Subcellular fractionation of rabbit spinal cord[155] reveals the bulk of acid proteinase activity in the supernatant, whereas in brain it is in the crude mitochondrial fraction, suggesting that spinal cord contains fewer or less hardy lysosomes. Approximately half of the total neutral proteinase activity of spinal cord, as in brain, is in the supernatant. Similarly, all-N-terminal peptidase activity (40–60% of the total) is in the supernatant. By specific activity (activity per milligram protein), the acid proteinase activity in the spinal cord supernatant is greater, and the particulate bound neutral proteinase activity is higher in the spinal cord than in brain (Table VIII).

Partially purified spinal cord myelin contains 13% and 1% of the total tissue neutral and acid proteinase activities, respectively.[155] The neutral proteinase activity observed in crude preparations of brain myelin had at first been attributed to contamination,[181–183] but a neutral proteinase in purified myelin prepared from spinal cord has now been found.[184,185] Myelin basic protein is degraded by this enzyme when myelin is incubated at neutral pH, but PLP is not. Proteolipid protein, which is trypsin resistant, is degraded when myelin is incubated with trypsin in the presence of Triton X-100.[185–187] The best possible explanation is either that the detergent releases enzyme from the myelin membrane or it alters the molecular structure of the protein and makes it accessible to the enzyme. Purified spinal cord myelin also contains a Ca^{2+}-activated neutral proteinase.[184,185] The proteinases in myelin have potential for regulating its protein turnover as well as for controlling myelin breakdown in demyelination.

A protein kinase activity is associated with a neurofilament-enriched preparation from CNS,[188] and a similar activity is found in neurofilaments isolated from rat spinal cord. It requires 20 μM ATP, is active at pH 8.0, and phosphorylates both serine and threonine residues in the NFP. The neurofilament-associated activity is distinct from microtubule-associated protein kinase activity.[188] Phosphorylation may play an important role in neurofilament function, and the protein kinase may regulate maintenance of NFPs. Myelin isolated from brain contains a protein kinase that phophorylates MBP.[189] Such an MBP kinase has not yet been sought in spinal cord. The activity of a myelin basic protein methyltransferase has been found in brain as well as in spinal cord. The enzyme activity has been shown to parallel myelination in spinal cord.[190,191]

2.3.4. Lysosomal Enzymes

2.3.4a. Acid Phosphatase and Alkaline Phosphatase (E.C. 3.1.3.1). Acid and alkaline phosphatase activity in spinal cord has been studied extensively.[192–198] Alkaline phosphatase activity is higher in the white matter of guinea pig than in rat spinal cord and is greater in the cervical than the thoracic tracts.[192,194] Isozymes of alkaline phosphatase are suspected on histochemical grounds.[194] Activity is high in differentiating CNS and decreases in spinal cord before other regions.[195]

Acid phosphatase activity, on the other hand, is not detectable by histochemical techniques in the dorsal horn of fetal spinal cord. It is demonstrable and has increased activity in spinal ganglion and motor neurons during early fetal development (between 13 and 17 days prenatally), prompting attempts to correlate the enzyme with the onset of movement and reflexes.[198] Enzyme activity is not detectable in substantia gelatinosa of the prenatal rat but appears at the first postnatal day, increasing to the adult level at approximately 6 days.[196] The rapid rise in activity may correlate with synaptogenesis.

Acid phosphatase activity is greater in gray than white matter of spinal cord.[197] The enzyme in motor neurons and spinal ganglion cells is localized in lysosomes in the neuronal cell body and is inhibited by sodium fluoride whereas that in the substantia gelatinosa is non lysosomal, unaffected by fluoride, and present in synapses.[196,199,200] These differences suggest that the function of these isozymes of acid phosphatase may be different.[196] The localization at the nerve ending is somewhat surprising and warrants further investigation. It is possible, for example, that the enzyme in the substantia gelatinosa may be involved in the functional maturation of synapses.

2.3.4b. β-Glucuronidase (E.C. 3.2.1.31), Glucosaminidase (E.C. 3.2.1.30), and β-Galactosidase (E.C. 3.2.1.23). These lysosomal enzymes, active at acid pH, are widely distributed in tissues including CNS. The activities of β-glucuronidase and glucosaminidase are higher in the gray than the white matter of dog spinal cord.[197] The levels of both in whole spinal cord approximate those in the white matter. However, the greater activity in the gray matter suggests that both are localized in lysosomes of the neuronal cell body, and the white matter activity may be in glial cells or axons.

β-Galactosidase activity has been demonstrated histochemically in the myelinated axons of turtle spinal cord and may be localized in the myelin sheath.[201]

2.3.5. *Miscellaneous Enzymes*

2.3.5a. Carbonic Anhydrase (C A; E.C. 4.2.1.1). Carbonic anhydrase, an enzyme associated with ion transport, is found in spinal cord[84,202–204] and appears along with incipient neural function.[202] Activity is lower in spinal cord than in brain and increases in spinal cord with maturation.[84,203,205,206] Carbonic anhydrase exists in soluble and membrane-bound forms. In spinal cord, the membrane-bound enzyme accounts for approximately 60% of total activity at all ages.[203] Carbonic anhydrase is present in myelin of brain and spinal cord.[207–211] It is in neuroglia and astrocytes,[211–213] serving as a glial marker, [214] and has been placed immunohistochemically in oligodendrocytes.[84,208,215] Its occurrence in myelin-forming cells suggests a role in the function of oligodendroglial cells, perhaps in the regulation of ion and fluid homeostasis in myelinated axons. A relationship of CA to demyelinations has been discussed, since its activity is decreased in isolated myelin from the hypomyelinated CNS of the quaking and shiverer mutants.[216]

2.3.5b. 5′-Mononucleotidase (E.C.3.1.3.5) 5′-Nucleotidase activity is accepted as a plasma membrane marker. Its activity is greater in gray than in white matter (guinea pig and mouse) of spinal cord.[101,103,200,217] The activity in dorsal gray is much higher (approximately double) than that in ventral gray in mouse, though no difference was found in different gray regions in cat. Substantia gelatinosa also has strong activity by histochemical staining.[101,200,217] In the white matter, the enzyme is associated with the myelin sheath.[218]

A low 5′-nucleotidase activity characterizes myelin isolated from brain[41,219] and myelin isolated from normal and shiverer mouse spinal cord.[219]

2.3.5c. Adenosine Triphosphatase (ATPase; E.C. 3.6.1.3) ATPase activity is clearly present in spinal cord.[43,101,193,197,220] It is activated by Mg^{2+} and further stimulated by Na^+ and K^+. Activity is greater in gray matter, and the enzyme has been found in neurons and neuroglial cells. Blood vessels of both gray and white matter show strong activity. Activity is highest in the substantia gelatinosa and near the central canal.[101,217] The Na^+,K^+,Mg^{2+}-ATPase activity increases in spinal cord with age. Activity is greater in rat spinal cord than in other CNS regions during early development[13] but decreases in the adult. Approximately 18–20% of the activity in the adult spinal cord is microsomal, and less than 1% is in myelin.[63] Although the activity present in myelin is very small, the activity in this fraction undoubtedly reflects the evolution of myelin as a modified glial plasma membrane.

2.3.5d. Adenosine-2',3'-Cyclic Nucleotide 3'-Phosphohydrolase (CNPase; E.C.3.1.4.37) 2',3'-Cyclic nucleotide phosphohydrolase has long been considered a myelin or oligodendroglial marker.[43,62,221,222] Its activity is greater in rat spinal cord at 3 days post-natal than in any other region of the CNS, and this developmental appearance correlates with that of MBP and of myelin lamellae.[43,62,71] Activity progressively increases in the rat spinal cord until 25 days after birth and then declines, consistent with the accumulated data indicating that the maximum rate of myelination in spinal cord occurs between 15 and 20 days post-natal.[43,62,223–225] The CNPase is localized in the nerve fibers and immature glial cells of 15-days-old rat spinal cord.[226]

In subcellular fractions, 40–60% of the enzyme activity is found in myelin isolated from adults, with activity lower in spinal cord than brain myelin.[28,43,46] During development, more of the activity is recovered in the microsomal fraction (e.g., 45–60% in rat spinal cord of 3- to 5-day-old and only 16–17% in myelin). This is consistent with the occurrence in the microsomes in early development of a myelin precursor membrane that evolves eventually to myelin.[227, 228]

2.3.5e. Enolase (E.C. 4.2.1.11) The enolase isozyme in CNS is functionally and structurally distinct from that in other tissues.[229,230] Two isozymes have been isolated: one is localized in neurons and neuroendocrine cells (neuron-specific enolase; NSE) and has two identical subunits (α) of 39,000 mol. wt., wherease the other is in glial cells (nonneuronal enolase; NNE) and has two identical subunits (α) of mol. wt. 43,500.[231] Activity has been shown immunohistochemically and by assay in spinal cord of rat, human, and monkey. The NSE activity is greater in the gray matter, wherease NNE is greater in the white. Both activities are lower in spinal cord than in other regions of rat and monkey CNS, and the NSE activity in spinal cord is substantially lower than NNE activity by approximately two- to fourfold.[231] Activity does not vary with segmental cord level. By immunohistochemistry, NSE is prominent in ventral horn motor neurons and their processes and is also seen in dorsal horn neurons *in vivo* and in tissue culture.[232,233] The appearance of NSE may correlate with development of functional activity of these cells.

2.3.5f. Nonspecific Esterase (E.C. 3.1.1.2) Nonspecific esterases have been demonstrated in mammalian and avian CNS tissues[41,234,235] but have not been examined to any extent in spinal cord. The enzyme activity is lower in spinal cord than in brain of chicken.[285] No studies on the distribution of nonspecific esterase in cord subcellular fractions have been carried out, but there is histochemical evidence that these enzymes are localized in rat cord motor neurons.[234] Recent studies on spinal cord injury have shown esterase activity in the nonneuronal cells in the lesion.[236]

2.4. Amino Acids, Transmitters, Neuropeptides, Amines, and Polyamines

2.4.1. Amino Acids

Amino acids and particularly the dicarboxylic amino acids of CNS are synthesized from glucose via the TCA cycle. They are modified by several

Table IX
Levels of Amino Acid in Spinal Cord[a]

Amino acid	Dorsal white	Dorsal gray	Ventral white	Ventral gray
Alanine	0.83	1.34	1.23	0.92
Aspartic acid	1.88	3.93	4.78	2.53
Cystathionine	1.36	1.62	1.29	1.70
GABA	0.43	2.18	1.04	0.44
Glutamic acid	3.74	5.91	4.41	3.25
Glutamine	4.00	5.48	4.75	3.68
Glycine	2.92	5.42	6.51	4.55
Serine	0.58	0.92	0.92	0.61

[a] Data from Johnston *et al.*[240] with permission.

transaminases, decarboxylases, and dehydrogenases. Other amino acids are transported from blood or synthesized by amination and deamination. Amino acid levels have been thoroughly described in brain[237,239] and spinal cord of different species.[136,240]

Glutamic acid formed by transamination or reducing amination of α-ketoglutarate is abundant in spinal cord. The content of glutamic acid is greater in hen than rat, which is also the case for GABA and glutamine.[241,242] Amino acids synthesized from glutamic acid, including alanine, aspartate, and glutamine, are all present in spinal cord. In spinal cord, as in brain, aspartate and glutamine levels exceed those of alanine.[238] There are appreciable amounts of depressant amino acids—alanine, GABA, cystathionine, glycine, and serine—in cat spinal cord. The level of glycine in spinal cord is greater than that in brain in many mammalian species.[238,243] It is elevated in spinal cord compared to brain in EAE, and this increase coincides with the appearance of clinical disease.[244] Other amino acids found in spinal cord are leucine, valine, tyrosine, threonine, and lysine.[155,241,242] Amino acid levels have been examined in the dorsal and ventral gray and white matter of cat spinal cord (Table IX). The levels of aspartate, glutamic acid, glutamine, GABA, and glycine are greater in the gray than white matter of spinal cord. There are others reported, but lesser differences in the distribution of amino acids in cervical, lumber, and thoracic areas of spinal cord (Table X).

Differences exist in the free amino acid pools of cervical and thoracic areas of spinal cord such that glutamic acid and glycine are higher and GABA and arginine are lower in the cervical cord. Levels of glycine and cystathionine in spinal cord are higher than in brainstem or cerebrum, whereas those of the other free amino acid are lower in cord than brain. There is more lysine in spinal cord than brain, which could reflect its content in the MBPs of spinal cord. More detailed accounts of metabolism, turnover, subcellular distribution, and functions of amino acids can be found in the reviews in Volume 3 of the *Handbook of Neurochemistry*.

Amino acids modified by acetylation and phosphorylation are also found in spinal cord.[245,263] The administration to mice of an amino acid that is not metabolized, 1-aminocyclopentane-1-carboxylic acid, reduces the weight and protein content of spinal cord and causes axonal degeneration.[246] Amino acid

Table X
Composition of Bound and Free Amino Acids in Spinal Cord[a]

	Bound amino acids (mol %)			Free amino acids (mol %)	
Amino acids	Cervical	Thoracic	Lumbar	Cervical	Thoracic
Glutamic acid	11.0	9.7	12.1	22.0	20.6
Alanine	10.4	10.2	10.0	5.2	4.5
Leucine	9.4	9.5	9.6	1.3	1.4
Glycine	8.5	9.4	8.7	16.0	9.5
Aspartic acid	8.0	7.4	8.7	10.7	10.3
Valine	6.5	5.9	6.8	0.5	0.6
Threonine	5.3	7.1	5.7	1.2	1.6
Lysine	6.5	6.2	5.2	1.5	2.0
Phenylalanine	5.2	5.2	5.3	0.5	0.8
Arginine	4.9	5.2	5.2	1.7	3.2
Serine	5.7	5.0	4.3	2.9	3.9
Isoleucine	4.5	5.9	4.6	0.3	0.4
Proline	4.3	4.5	3.4	0.67	0.6
Tyrosine	3.2	3.0	3.7	0.4	0.9
Histidine	3.6	2.9	2.9	0.5	0.6
Methionine	1.5	1.2	1.7	0.1	0.3
Cysteine	1.3	1.3	1.5		
Cystathionine				2.1	3.1
Phosphoserine				0.08	0.5
Phosphoethanolamine				1.7	2.7
Ethanolamine				1.5	2.5
Ornithine				0.3	0.3
Homocarnosine				0.5	0.5
GABA				2.2	5.6
Taurine				3.3	3.6
Glutamine				12.7	19.8

[a] Data from Serra *et al.*[155] with permission.

composition during development has been described in detail in brain (for detail see ref. 236) but not in spinal cord.

2.4.2. Transmitters

Glycine, glutamic acid, and GABA are thought to be the important amino acids involved in synaptic transmission in spinal cord. Others of interest include aspartic acid, taurine, and proline, but their role in neurotransmission is not as well established. Amino acids with inhibitory effects on nerve cells have been divided into groups of "glycinelike," which are blocked by strychnine, and "GABA-like,"which are inhibited by bicuculline.[247] The glycinelike amino acids in spinal cord include alanine, serine, taurine, and cystathionine. Much evidence has been provided supporting the postulate that glycine is a major spinal inhibitory neurotransmitter[136,240,247,248] since the depressant action of glycine on neurons was reported in 1960.[249] The inhibition of spinal motoneurons and interneurons by glycine has been shown,[248,250] and the release of glycine from cord (and also its release from and depressant action on tectal neurons) is consistent with neurotransmitter function.[251] The neurotransmitter action of glycine has recently been reviewed.[252]

Glutamic acid was first proposed as an excitatory neurotransmitter by Aprison and his colleagues.[135,136] After some controversy, glutamate is now accepted by most as a neurotransmitter. The distribution of glutamate content in discrete regions of spinal cord and in the dorsal and ventral roots suggests that it is a neurotransmitter released by primary afferent neurons. A loss of spinal interneurons in parallel with a reduction of glutamate suggested that these neurons may be glutamatergic.[136,253] Glutamatergic neurons have been identified by the stimulus-evoked tissue release of glutamate.[254–256] For further discussion of the transmitter role of glutamate, there is a recent review.[257]

GABA, synthesized from glutamate by GAD, has been accepted as a major inhibitory neurotransmitter as first proposed in 1963 by Eccles *et al.*[258] Bicuculline and picrotoxin, which are antagonists of the postsynaptic inhibitory action of GABA in spinal cord, block presynaptic inhibition as well,[247,258] suggesting a role for GABA in presynaptic inhibition. GABAergic terminals and neurons have been located in spinal cord and other CNS regions by immunocytochemical methods.[259,260] Detailed account of the uptake and depolarization-induced release of GABA and its role as a neurotransmitter have been published.[259,261,262]

Acetylcholine (ACh) is, of course, a classical and important neurotransmitter. Its presence in spinal cord together with its synthesizing (ChAT) and hydrolyzing (AChE) enzymes was shown long ago. The amount of ACh is much greater in spinal cord gray than white matter. The distribution of the enzymes ChAT and AChE is similar to that of ACh (see Section 2.3.2a). Several excellent reviews on cholinergic transmission in vertabrate CNS are available,[263] and this transmitter is discussed in detail in Volume 6 of the *Handbook*.

The cholinergic system appears to be involved in motor and sensory spinal cord pathways. Administration of ACh electrophoretically to spinal cord excites certain spinal interneurons and depresses others.[264,265] The immunocytochemical localization of ChAT in motor neurons and dendrites supports a role for cholinergic synapses on certain of these spinal neurons (e.g., Renshaw cells).[110,266]

The presence and distribution of catecholmines have been studied in spinal cord of several different mammalian species (Table XI). The levels of dihydroxyphenylalanine (DOPA), 5-hydroxytryptamine (5-HT), and norepinephrine (NE) in spinal cord of all mammalian species are approximately the same.[3] They are localized mainly in the gray matter of spinal cord.[267] The levels of 5-HT and monoamine oxidase increase in spinal cord during development and reach a peak at about 14 days after birth.[268]. The distribution of NE is greater in lateral horns than in the dorsal or ventral horns.[145,269] Tyramine is also present in the mammalian spinal cord.[270] The distribution of enzymes responsbile for the synthesis of catecholamines follows the pattern for NE, DOPA, and 5-HT (see Sections 2.3.2c.d).

The presence of NE in nerve fibers and terminals and the location of 5-HT in descending nerve fibers have been established. Their release by electrical

Table XI
Concentration of Catecholamines in Spinal Cord[a]

Species	DOPA	Dopamine	NE	5-HT
Human	350	320	130	—
Bovine	180	200	100	—
Cat	50	50	50	—
Rabbit	50	20	30	330
Rat	50	20	110	900[b]
Guinea pig	50	60	170	—
Hen	—	60	157	440
Frog	—	—	—	2530

[a] Data from Levy[3] with permission, expressed in ng/g wet weight.
[b] From Suzuki *et al.*[268]

stimulation of spinal cord suggests that they are neurotransmitters.[271–273] There is evidence that NE and 5-HT contribute to the control of spinal neurons. Monoamine-containing fibers are present in anterior horns with terminals mainly on autonomic preganglionic neurons.[274] Dorsal horn interneurons contact abundant monoamine-containing descending fibers, and NE inhibits spinal interneurons in thoracic and lumbar regions of spinal cord,[273,275] though excitatory effects are seldom seen.[276] Their role in spinal function and relevant pharmacological effects are summarized in a short review by Ryall.[262]

2.4.3. *Neuropeptides*

Several neuropeptides have been localized in spinal cord. These include substance P, somatostatin, methionine-enkephalin, neurotensin, and oxytocin. The postulated function of these peptides is as mediators and modulators of interneuronal communication. The dorsal horn of the spinal cord of many species is rich in such peptides.[277]

Substance P is an undecapeptide that was discovered by Von Euler and Gaddum in 1931[278] and is widely distributed in CNS. It has been localized in spinal cord by immunohistochemistry and is present throughout the spinal cord, being localized mainly in neuronal processes in the dorsal horn.[279–286] Substance P has also been identified in lumbosacral pia mater.[298] In dorsal horn, it is found in sensory axons terminating in the substantia gelatinosa, suggesting a role in primary sensory transmission.[288–291] Lesser amounts are found in ventral horn in motor neurons.[282] An excitatory action on motoneurons has been reported.[292] A similar distribution is reported in frog spinal cord.[293] This localization suggests to some that substance P is involved in vasodilatation and in transmission of pain.

β-Endorphin and β-lipotropin have been detected in CNS and immunocytochemically localized in spinal cord to ventral and dorsal horn cells. They are found in neurons and in some other cells in spinal cord explants in culture. β-Endorphin is synthesized transiently in the spinal cord during development.

Table XII
Distribution of Neuropeptides in Spinal Cord[a]

	Spinal cord	Cortex
Met-Enk-Arg-Phe	111.0	36.1
Met-Enk	752.0	302.0
Leu-Enk	86.0	47.0
α-Neoendorphin	22.9	15.0
$Dynorphin_{1-13}$	8.1	4.0

[a] Data from Giraud *et al.*[283] with permission, expressed in pmol/g wet tissue.

This biosynthetic activity disappears by 4 weeks after birth in the rat.[294] β-Lipotropin has also been found in the developing spinal cord in tissue culture.[295] It has been suggested that β-endorphin acts as a neurotransmitter but may also possess a neurotrophic (exerting specific effects on nervous system) rather than a neurotransmitter role.[296,297] It is not clear whether methionine-enkephalin is synthesized from β-endorphin or if individual precursors exist. (Table XII).

Peptides derived from the degradation of MBP in spinal cord can induce EAE. The nature of these peptides, their encephalitogenicity, amino acid sequence, and other properties have been reviewed.[298]

2.4.4. Trace Amines and Polyamines

The trace amines include β-phenylethylamine, tyramine, phenylethanolamine, octopamine, synephrine, tryptamine, and their derivatives. They have been studied in brain for the most part, and data concerning their occurrence level, distribution, etc. in spinal cord are sparse. The level of octopamine is higher in rat spinal cord (24 ng/g) than brain (3–14 ng/g).[299] Trace amines may act as synaptic activators[300] and as such could be implicated in mental and neurological disorders.[301,302] The metabolism of these amines, their storage and subcellular distribution in brain, and their potential clinical implications have been reviewed.[303]

Polyamines occur in all living organisms and apparently in all kinds of cells including those of CNS. Only three polyamines, putrescine, spermidine, and spermine, occur in as much as nanomolar concentration. Their distribution, synthesis, catabolism, and functional correlates have been studied in brains of several species, and this information has been summarized by Seiler.[304] Little work has been done on spinal cord. The level of spermine is higher in cortex and highest in the olfactory bulb (640 nmol/g fresh wt.), whereas it is low in spinal cord (71 nmol/g).[305] The concentrations of putrescine and spermidine are high in white and low in gray matter, and putrescine (24.4 nmol/g) and spermidine (1420 nmol/g) are higher in the spinal cord than in other areas of CNS (Table XIII). Spermidine concentration increases with increased white matter.[306] It is possible that the higher concentrations of spermidine and putrescine found in the white matter may have some roles in "clamping" (i.e., bringing together) adjacent myelin lamallae into myelin sheaths.[305,307,308] These

Table XIII
Content of Major Polyamines and Nucleic Acids in Rat Spinal Cord and Other Areas of CNS[a]

Region	Putrescine[b]	Spermidine[b]	Spermine[b]	RNA[c]	DNA[c]
Spinal cord	24.4	1420.0	71.0	2.52	1.16
Cortex	9.4	343.0	270.0	5.62	2.42
Cerebellum	13.0	593.0	352.0	6.81	12.5
Olfactory bulb	5.5	410.0	6.40	5.59	5.71

[a] Data from Seiler and Schmidt-Glenwinkel[305] with permission.
[b] Expressed in nmol/g wet wt.
[c] Expressed in μmol P/g wet wt.

polyamines are localized in the myelin sheath in PNS.[309] Their concentrations are lower in spinal cord of the myelin-deficient quaking and jimpy mutants.[310] Spermidine has a slow turnover rate in myelin in comparison to other subcellular organelles.[307] Study of the subcellular distribution of polyamines has been carried out in brain, and they have been found enriched in myelin, synaptic endings, and synaptic plasma membranes.[311]

Polyamines are thought to play a functional role in membrane structure, ion transport,[312] and regulation of GABA degradation, though their precise roles in CNS function are not known.

2.5. *Inorganic Elements in Spinal Cord*

Trace elements may have important roles in CNS development.[313] The spinal cord content of inorganic elements has been extensively studied. The water content of adult spinal cord (70% of the fresh wet weight) is lower than that of brain (about 80% of the fresh weight), and the water content of the gray matter is greater than that of the white. The lower water content in the spinal cord reflects its increased white matter compared to other CNS regions. The water content of brain gray matter (83%) is substantially higher than that of the white (70%).

The sodium and potassium levels in spinal cord are similar in different species. The potassium content slightly exceeds the sodium, and the levels of both are higher in gray than in white matter. In the spinal cord of fetal and newborn guinea pig, the sodium content is higher than in the adult. This has been attributed to a decrease in water content accompanying the accumulation of myelin.[314]

The total calcium concentration in mammalian spinal cord varies from 4.5 to 8.0 mmol/kg wet wt. Little is known of the levels of bound and free calcium in spinal cord because their accurate determination is only now becoming feasible. Levels of free intracellular Ca^{2+} have been measured in other tissues employing ion-selective electrodes. Hyperpolarization of the motorneurons of frog spinal cord provides both fast and slow action potentials; the latter is dependent on calcium.[315] Calcium is required for both transmitter release and fast axonal transport.

Table XIV
Distribution of Free Sugars in Spinal Cord[a]

Sugar	Amount (mmol/kg wet wt.)
Gucose	1.84
Fructose	0.04
Sorbitol	0.02
Galacticol	0.02
myo-Inositol	5.22

[a] Data from Stewart *et al.*[326] with permission.

In spinal cord, a calcium-activated enzyme(s) has been implicated in the turnover of neurofilament proteins. The calcium level increases in traumatized spinal cord and in the lesion of multiple sclerosis plaque.[316]

The levels of zinc and copper in spinal cord increase during development. This begins at 10 days post-natal in rat and reaches the adult level at 30 days. The zinc content is lower in spinal cord than in cerebrum or cerebellum. Zinc deficiency causes defective CNS development with degeneration of optic nerve and loss of myelin sheath staining,[318] which has been attributed to an effect on lipid metabolism.[319] Zinc provokes perivascular cuffing and lymphocyte activation resembling the histological changes observed in both EAE and MS.[320] An increase in the level of copper begins in rat prior to 10 days post-natal and rises again after 15 days, reaching the adult level at 30 days.

Most of these elements are localized in axon terminals, lysosomes, neuronal surface membranes, and glial cytoplasm.[321,322] Both Ca^{2+} and Zn^{2+} are found in motor neurons.[323]

The iron content of spinal cord is lower than that of the other CNS regions. Its level in spinal cord disease is not known, though elevation is reported in the demyelinating areas of MS brain.[317]

Magnesium, chloride, and phosphorus are also found in spinal cord. The chloride content of spinal cord does not vary in different species. [3,324] The manganese level in spinal cord has not yet been reported but is elevated in MS brain.[317]

2.6. Carbohydrates

The levels of glucose, other monosaccharides, and glycogen have been determined in the spinal cord of several different species.[3,324] The content of glycogen is lower in spinal cord than in other areas of CNS (e.g., cerebrum).[325] The level of glycogen decreases in spinal cord during development whereas it increases in brain. The content of glucose and other sugars and polyols is given in Table XIV. A small amount of sorbitol and galacticol has been found in spinal cord. The content of *myo*-inositol, on the other hand, is greater in spinal cord than in brain. Rats fed with galactose show a marked increase in the content of galacticol in spinal cord.[326] Glycogen is only slowly metabolized in chick spinal cord, whereas glucose is metabolized relatively rapidly, with insulin stimulating glucose uptake.[327] Insulin has much less effect on fructose and galactose. The enzymes mediating sugar metabolism have been discussed in Section 2.3.1.

2.7. Vitamins

Deficiencies of certain vitamins, especially thiamin and nicotinic acid and also B_6, B_{12}, pantothenic acid, and riboflavin, prominently affect the CNS. Thiamin lack has long been known to affect cerebral structures (e.g., endothelial proliferation, necrosis, and degeneration of neurons), lipid composition, and function.[328,329] Thiamin deficiency has a significant effect on the weight and the lipid composition of rat spinal cord,[330] indicating a growth retardation. The levels of thiamin in spinal cord of several mammalian species are uniform and are lower than those of brain. The morphological lesions seen in pellegra, a disease caused by lack of nicotinic acid, include degeneration of the large neurons of the anterior horn of the spinal cord.[331]

Vitamin B_6 or pyridoxine (e.g., pyridoxal phosphate) is required for many amino transfer reactions in the CNS. The lack of vitamin B_6 affects CNS in many ways, one of them being epileptic seizures.[332] Although the topographical CNS distribution of vitamin B_6 has been studied, its content in and the effect of deficiency on spinal cord have not.[333]

Lack of vitamin B_{12} (cobalamin) causes anemia and vacuolar degeneration of certain myelinated tracts in human spinal cord.[331] The latter lesion is mainly demyelinative and has been attributed to both abnormal fatty acid metabolism[334] and methyl group deficiency.[335] Lack of pantothenic acid in the rat can cause paralysis and convulsions. Morphological studies have shown demyelination in peripheral nerves and spinal cord.[328] Riboflavin deficiency in dogs produces demyelination in PNS and in the posterior columns of the spinal cord.[336] The level of this vitamin does not vary in different spinal cord regions.[335] Deficiency of vitamin A in developing animals causes degeneration of cells and fiber tracts mainly in the spinal cord.[337] Vitamin C levels (ascorbic acid) have been determined in brain but not in spinal cord.[338]

2.8. Nucleic Acids and Nucleotides

Little concerning the nucleic acids and nucleotides in spinal cord has been published. The RNA content of spinal cord exceeds the DNA by about twofold. The RNA and DNA contents of spinal cord are lower than in other areas of CNS.[305,339,340] The nuclear DNA content of large spinal neurons has been reported to be approximately 5.3 pg DNA/nucleus.[341]

The CNS is rich in nucleotides and phosphates. A detailed study of the levels of various nucleotides, phosphocreatine, and inorganic phosphate has been carried out in brain[328] but not in spinal cord. The levels of acid-soluble and lipid phosphorus in spinal cord of guinea pig and rat are similar. As in brain, ATP in the spinal cord preponderates over ADP and other nucleotides. Although mono-, di-, and triphosphates of guanosine, uridine, cytidine, and inosine are found in brain, no data are available concerning their spinal cord content. The uridine diphosphate derivatives of glucose and galactose that are necessary for synthesis of glycogen and galactolipids in brain[342,343] are undoubtedly present in spinal cord, as are the other nucleotides, such as ADP and NADPH, involved in essential catalytic reactions of the CNS.

2.9. Isolation of Cells and Subcellular Organelles from Spinal Cord

Spinal cord is composed mainly of white matter (approximately 80%) and contains, among other cell types, predominantly myelin-forming oligodendrocytes and astroglial cells. In order to study the biochemistry and metabolism of different cell types in CNS, oligodendrocytes, astrocytes, and neurons have been isolated and purified from brain.[344] Isolation of these cells from spinal cord is more difficult because of the inaccessibility and small size of the cord. In spite of this, different cell populations have been isolated from the spinal cord of the adult rat.[345] The oligodendroglial cells isolated from spinal cord by the method of Poduslo and Norton[344] comprise at least 80% small, round oligodendrocytes, with the remainder consisting mostly of small neurons and astrocytes. The fraction containing neurons is purer than those of astrocytes and oligodendrogcytes.

The isolated oligodendrocytes are enriched in adenosine 2′, 3′-cyclic nucleotide 3′-phosphohydrolase (CNPase), whose activity (160–225, μmol/mg protein per hr) is comparable to that of cells isolated from rat brain.[346] Oligodendrocytes isolated from spinal cord incorporate leucine into protein,[345] and these along with neurons and astrocytes also synthesize fatty acids and sulfatide.[347,348]

Subcellular organelles of CNS have been examined in spinal cord in only a few instances. These have been concerned mainly with the preparation of myelin, as in the isolation from squirrel monkey by Horrocks.[20] Purified myelin and other subcellular organelles have been prepared from spinal cord of several mammalian species.[43,45,46,62,155,349,350] The myelin preparation is usually from the mitochondrial fraction, though the crude microsomal fraction has also been used.[349, 351] The composition of spinal cord myelin approximates that of brain myelin, but the yield of cord myelin is slightly lower[426] (W. Cammer, personal communication). The myelin has a compact multilamellar structure and contains approximately 40–60% of the CNPase activity of the whole spinal cord.[43,46]

Other organelles have been prepared from developing spinal cord.[43,62] The fraction collecting at the 0.32 M and 0.88 M sucrose interface from the spinal cord of 5-day-old rats contains compact myelin unlike the cerebral hemisphere preparation, which shows amorphous membranous structures lacking compaction. Studies on the characterization of the microsomal fraction from early developing CNS suggest that it contains myelin precursor membranes,[43] supporting the earlier report of Sabri *et al.*,[73,227, 352] which has recently been confirmed.[353,354]

2.10. Metabolism of Spinal Cord

There are few studies of the metabolism of spinal cord. Respiration of spinal cord continues for a long time during incubation in an oxygenated medium, and electrical stimulation produces a marked increase in respiration, a change associated with increased glycolysis.[328] The metabolism of transmitters in spinal cord has been studied and reviewed.[262] More recent studies of spinal cord metabolism have included investigation of the synthesis and degradation of proteins and lipids of spinal cord myelin in adult and developing animals.[155,350] It is beyound the scope of this chapter to discuss the general metabolism of myelin lipids and proteins of other CNS regions and of PNS, and these can be found in an excellent review.[355] In spinal cord, labeled acetate and leucine have been used as precursors for the synthesis of lipids and proteins of spinal cord. [^{14}C]Acetate and [^{14}C]leucine are actively incorporated into myelin lipids and proteins, respectively, and the uptake of these precursors decreases with increasing age. A similar trend has been observed in brain.[350,355] A steady adult level is reached at about 6 months post-natal in rats. At earlier ages, the incorporation of acetate is mainly into cholesterol and phosphatidylcholine and then into cerebroside and sphingomyelin. This uptake into individual lipids is reduced in the adult. The incorporation of [^{14}C]acetate into myelin lipid of spinal cord is at least twofold greater than that of brain at early ages, whereas total incorporation of leucine into proteins is greater in brains at 45 days post-natal. The results in study of the metabolism of myelin then indicate that spinal cord is metabolically more active in early development.

The incorporation of leucine into spinal cord myelin proteins is similar to that in other CNS areas. The rate of synthesis is decreased with increasing age. A stable rate is reached at about 60 days of age.[350] Individual myelin proteins of spinal cord turn over at different rates. There are no consistent data, however, for the turnover rates of CNS myelin proteins.[355] Breakdown of MBP (which is encephalitogenic) may cause pathological alterations in spinal cord and in other CNS regions. The enzymes catalyzing the breakdown of MBP and/or other cord proteins occur in spinal cord (see Section 2.3.3.). The MBP is digested by cathepsin D and CANP.[74,168 185,356] Peptide hydrolases present in spinal cord mediate hydrolysis of proteins by sequential breakdown to peptides and amino acids, some of which are reutilized for the synthesis of macromolecules.

3. BIOCHEMISTRY OF DAMAGED SPINAL CORD

Injury to the spinal cord, causes tissue damage and necrosis with loss of neurological function. The extent of neurological deficit depends mainly on the degree of severity of damage. Severe injury leads to paralysis below the lesion with loss of control of bladder and bowel. A useful approach to understand the consequence of damage to the spinal cord is to study the effects of experimental cord injury in animals. Such studies on animals provide information on the sequence of changes that take place following injury.

Several animal models of experimental spinal cord trauma have been employed.[357,358] The model of inflicting injury by weight drop was developed by Allen in 1914[358] and has become the standard model in spinal cord injury research. The apparatus by which the injury is induced consists of a cylinder placed perpendicular to the exposed surface of the spinal cord. The cylinder is loosely fitted with a plastic impounder resting on the intact surface of the dura. A lead weight is dropped through the cylinder from a determined distance.

Acute damage to the spinal cord in experimental animals produces paralysis, edema, and a characteristic sequence of histological and ultrastructural changes.[358,363] The alterations in biochemical composition of the lesion have been correlated with morphologcal changes.[349,364–368]

Although no single factor or process responsible for the pathophysiological events and biochemical changes seen in spinal cord following injury has been established, it is axiomatic that understanding of the molecular events involved in production of the functional deficit is essential in developing research directed at a therapeutic approach.

Studies on the biochemical composition of spinal cord in damage and/or in disease have shown changes in lipids, proteins, enzymes, and transmitters.[19,180,197,349,364,368,369] Myelin isolated from the normal and lesioned areas of spinal cord was examined for yield, which was unaltered during the first hour of trauma in rats or monkeys[349,368] and progressively decreased from 2 hr onwards. A marked loss (approximately 30–60%) of myelin from the lesioned area at 4 to 24 hr following trauma has been reported.[349,368] The loss of myelin is probably caused by tissue degeneration as well as edema. Incubation of spinal cord *in vitro* and ischemia have been shown to result in the dissociation of and decrease in the yield of myelin.[370,371]

3.1. Changes in Lipids

The lipid composition in the lesion and its myelin has been studied by Horrocks and colleagues.[19,349,372] No significant decreases in the major lipids were found with the exception of ethanolamine plasmalogen.[349] A moderate decrease (15%) in its proportion in myelin from injured spinal cord has been observed at an hour following trauma. Dramatic changes in the content of ethanolamine plasmalogen are also noted in Wallerian degeneration in PNS. This decrease has been attributed to an increased plasmalogenase activity producing the myelinolytic compound lysoethanolamine phosphoglyceride.[373,374] Ganglioside content and species in spinal cord injury have not yet been studied. This would be of interest because peripheral nerve degeneration manifests a small decrease in the ganglioside content at 3 to 7 days following nerve transection and then increases markedly.[375]

It is important to examine levels of prostaglandin (PG) and related metabolites since activation of arachidonic acid metabolism is a common cellular response to injury. Marked increases (about 100%) in PG levels were found in the lesion in the first report.[16] Thromboxane A_2 and prostacyclin have been examined via levels of thromboxane B_2 (TxB_2), a metabolite of thromboxane A_2 and 6-keto-PGF_1 (6-Keto), a metabolite of prostacyclin.[19] The level of TxB_2

(380 pg/mg dry wt.) is significantly increased in the lesion compared to control (145 pg/mg dry wt.) 1 hr after injury, whereas that of 6-Keto (149 pg/mg) remained unchanged compared to control (127 pg/mg). These observations suggest that alterations of thromboxane/prostacyclin balance following spinal cord trauma may play a role in the progressive necrosis.[19]

3.2. Changes in Proteins

Morphological alterations of myelinated axons in the lesion led to study of the proteins of spinal cord, particularly those of myelin and axons.[175,368,369] The major proteins of spinal cord were analyzed by SDS–PAGE, and the extent of protein loss in the lesion was compared to normal. There is degradation of the neurofilament triplet proteins (NFP) (200K, 150K, 69K) as early as 15 min following injury, and the degradation is progressive. Losses are substantial at 30 min and much greater at 6 hr and beyound. At 6 hr, losses amounted to 85 and 100% for the 200K and 150K NFP, and losses of myelin PLP and MBP were extensive (Fig. 1). These proteins were completely degraded at 24–72 hr after injury. The degradation of the 69K NFP could not be reliably measured since a protein of 70–71K molecular weight comigrated with the NFP in the lesion. Whether this protein is a degradation product of a larger NFP or a contaminant is not known. A similar-size protein has been seen in Wallerian degeneration in optic nerve, and, interestingly, in tissue exposed to heat shock, synthesis of a 71K protein is induced.[376,377] Substantial losses of GFAP (51K) and microtubular proteins (MTP) (54K, 56K) are also encountered at 6 hr, and they are further reduced by 24–72 hr. Decreases in these proteins correlate well with the ultrastructural alterations in axons and are attributed to a degenerative process.[368,369] Chemical injury to the spinal cord produced by the application of exogenous $CaCl_2$ causes progressive and extensive losses of neurofilament, glial filament, microtubular, and myelin proteins (MBP and PLP) similar to those observed in spinal cord trauma. Ultrastructural alterations in myelinated axons have been correlated with these protein losses.[379] Injury of the spinal cord by gentamicin causes axonolysis and myelinolysis, but no analysis of proteins has been carried out.[380] It is likely that losses of structural proteins occur in this model of injury.

The proteins of purified myelin following injury have also been examined. The changes in Wolfgram proteins (WP), other high-molecular-weight proteins, MBP, PLP, and MAG are not proportional following spinal cord injury. The WP and PLP are less susceptible at 1–2 hr, whereas the MBPs are extensively degraded by 4 hr.[175,368] This MBP loss is very extensive (80–90% for the large MBP and 70–80% for the small MBP) at 24 hr after trauma (Table XV), by which time degradation of WP and PLP is also evident.[368,369] The reductions of these latter two are less dramatic than that of MBP. Other proteins found in the lesion include contaminating P_0 protein from PNS (peripheral nervous system myelin) and/or proteins from erythrocytes.[368,369] The progressive degradation of myelin proteins does correspond to the observed vesiculation, sloughing, and dissolution of myelin and axons following spinal cord trauma, consistent with an ultimate collapse of the structural unit of myelin and axon

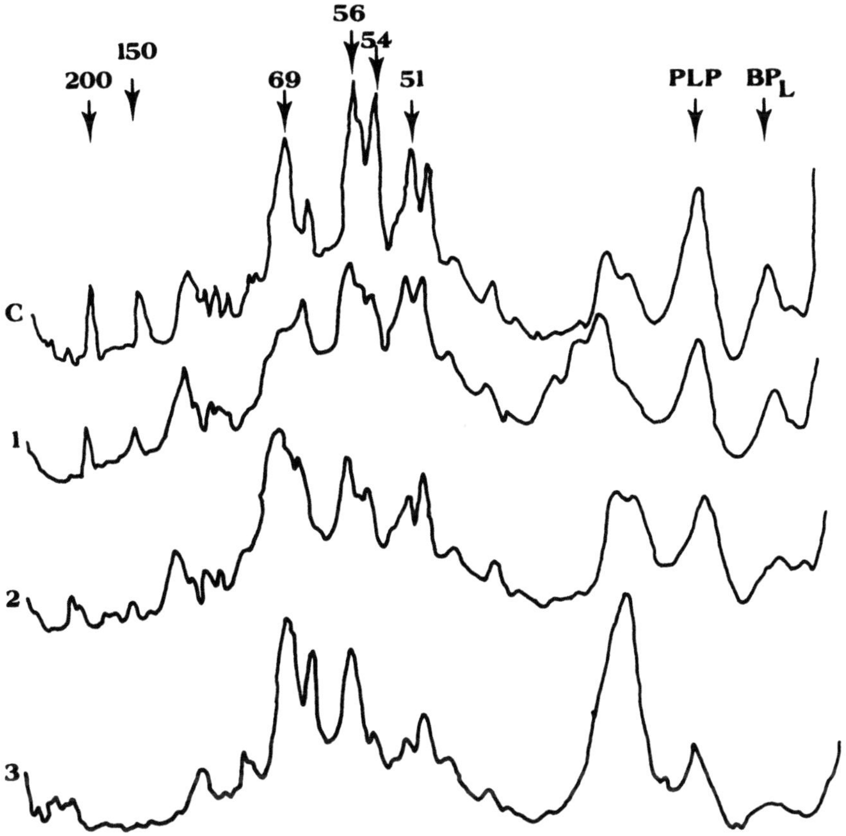

Fig. 1. Tracings of proteins separated by SDS–PAGE. C, control; 1, 2, and 3 are protein patterns at 15 min, 30 min, and 6 hr after trauma, respectively. Numbers indicate molecular weights (× 10) of proteins: 200, 150, and 69, NFPs; 56 and 54, microtubular proteins; 51, GFAP; PLP, proteolipid protein; BP_L, basic protein, large. Data from Banik *et al.*[175] and reproduced with permission.

Table XV
The Effect of Spinal Cord Trauma on Myelin and the Major Myelin Proteins in Rat[a]

Hours following trauma	Percent loss compared to control					
	Myelin	CNPase	WP	PLP	BP_L	BP_S
1	0	0	0	0	9 ± 0.67	6 ± 1.1
2	5	5	5 ± 1.0	5 ± 0.93	19 ± 3.0	12 ± 2.0
4	15 ± 2.1	15 ± 3.0	15 ± 2.0	10 ± 1.7	64 ± 5.2	46 ± 2.9
8	30 ± 3.6	24 ± 4.2	26 ± 2.9	20 ± 3.8	83 ± 7.1	68 ± 5.3
24	45 ± 3.9	32 ± 4.8	45 ± 4.6	35 ± 3.1	92 ± 6.8	75 ± 5.0
72	60 ± 5.3	55	70 ± 5.0	50 ± 6.8	97 ± 5.1	81 ± 3.9

[a] CNPase, 2,′3′-cyclic nucleotide-3′-phosphohydrolase; WP, Wolfgram protein; PLP, proteolipid protein; BP_L, basic protein, large; BP_S, basic protein, small. Data from Banik *et al.*[368] with permission.

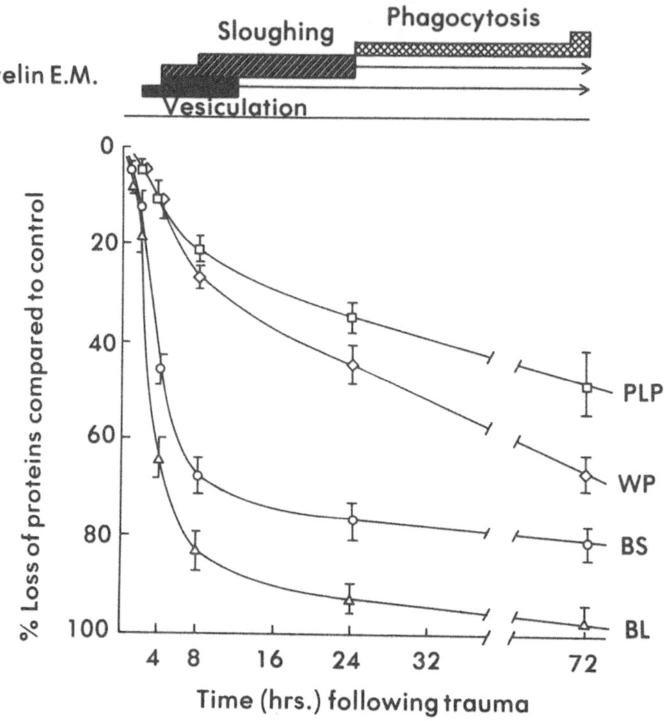

Fig. 2. Alteration of myelin proteins and of myelin ultrastructure following spinal cord trauma. PLP, proteolipid protein; WP, Wolfgram protein; BS, basic protein, small; BL, basic protein, large. Data from Banik *et al.*[368] and reproduced with permission.

(Fig. 2). The degradation of these proteins, whether axonal or myelin, suggests that proteolytic enzymes are involved. It is noteworthy that in demyelination (e.g., EAE and MS), the loss of proteins has been concomitant with increased activities of proteinases.[381,382] It is clear that such a correlation also exists in spinal cord injury.

3.3. Changes in Enzymes

The activity of CNPase, a marker enzyme of myelin, is decreased by about 15% in spinal cord as well as in isolated myelin at 3–4 hr following injury. This loss of enzyme activity in myelin amounts to approximately 60% at 72 hr after trauma.[368] The decreased enzyme activity corresponds with the myelin loss.[368,369] However, the extent of loss of CNPase later appears to be dependent on the severity of the injury.[368] Thus, trauma to the spinal cord of moderate severity is accompanied by less change in enzyme activity.

ATPase (adenosine triphosphatase) has been assayed in spinal cord injury.[220,383] No change in Na^+, K^+-activated ATPase or in acetylcholinesterase (AChE) activity was found in traumatized rat cord,[220] but Clendenon *et al.*[383] and Krikorian *et al.*[384] reported a significant decrease in APTase activity as early as 5 min following injury. This decrease is evident in both gray and white matter. This early change was interpreted as reflecting disruption of cell membranes and loss of the ionic pump in the lesion. Studies of oxygen pressure (Po_2) in the traumatized cord have implicated changes in blood flow and oxidative metabolism of cord following injury.[385,386] A marked decrease in cytochrome oxidase activity has been reported in the lesion 4 hr after injury.[387] The concentration of energy metabolites, ATP, and phosphocreatine is markedly decreased in both gray and white matter of spinal cord immediately after trauma. At about 2 hr, their levels return to the control level.[388] Lactic acid remains elevated in both gray and white matter of spinal cord after injury, and this suggests disordered regulation of glycolytic metabolism.[349,388] Attempts have been made to correlate the loss of ascorbic acid from spinal cord following injury with the loss of CNS function,[389] but the mechanism of the loss of ascorbic acid is not really clear.

Lysosomal hydrolases that mediate cellular digestion, phagocytosis, and autolysis are prime candidates for release and activation of processes of tissue degeneration, necrosis, and breakdown of myelinated axons.[390, 391] The reports of the levels of lysosomal enzymes in spinal cord trauma are inconsistent. No increase in the levels of β-glucuronidase, glucosaminidase, and acid phosphatase were found 2 hr after injury except for early decreases in activities of acid phosphatase (30 min) and β-glucuronidase (1 hr). Such a decrease in enzyme activity may be attributable to free radicals, for these can inactivate acid hydrolases.[197,392] However, an increased acid phosphatase activity at 24 and 68 hr after trauma was reported by Kakari *et al.*[390] suggesting that lysosomal enzymes are involved in the spinal cord damage, particularly at a later stage. No notable changes in lysosomes or histochemically demonstrable acid phosphatase have been found in acute spinal cord injury.[393] These conflicting observations suggest that lysosomal enzymes are probably involved in the late degeneration of tissue but not in the immediately acute injury.

The progressive degradation of cytoskeletal and myelin proteins in experimental spinal cord injury has been explored further by assay of proteolytic enzymes. Tissue extracts from spinal cord were incubated with purified MBP as substrate at pH 3.6 (cathepsin D), pH 5.5 (cathepsin B), and pH 7.0 (neutral proteinases), and the activities were determined by losses observed using SDS–PAGE. The levels of these enzymes progressively increase in the lesion following injury. The elevation of neutral proteinase activity is greater than that of acid proteinase in the lesion.[159] The neutral proteinase activity is significantly increased at 2 hr and by twofold at 12–24 hr. This activity was inhibited by leupeptin, a neutral proteinase inhibitor. Myelin basic proteins were degraded into smaller peptides by these enzymes. The degradation of neurofilament proteins in the lesions indicates a role in the process for Ca^{2+}-activated neutral proteinase(s) (CANP). It is now known that the degradation of neurofilament proteins in spinal cord *in vitro* as well as in Wallerian degeneration is mediated by CANP.[42] Both increased calcium and neutral proteinase activity are found in the spinal cord lesion.[316]

The source of these enzymes is not clearly understood. The degradation of myelin and axonal proteins occurs before cellular infiltration, and it therefore seems unlikely that circulating leukocytes are mainly responsible in the early stages. The early increase in enzymes may be derived from serum, axons, endogenous cells, (e.g., activated glia), and/or myelin, since myelin and axons may contain proteinase activity.[184,185,394,395] Further degradation of myelin at the later stage, however, is probably effected by lysosomal enzymes released by invading macrophages (see Section 3.6).

3.4. Changes in Elements

The importance of electrolytes and other elements in the maintenance of cellular and integrative function in excitable tissues has long been known. Any change of these resulting from injury by edema and/or mechanical or chemical means could impair function and even lead to cell necrosis. Necrosis of cells proportional in increased intracellular calcium has been shown in tissue culture.[396]

The levels of sodium and potassium have been determined in normal cord and in traumatic lesions of cat spinal cord at intervals following injury.[397] A decrease in the level of potassium and an increase in sodium level occurs in the lesion 1 day after injury. There is a net loss of potassium in tissue at 3–6 days, with the increased level of sodium remaining stable. Treatment of animals with dexamethasone prevents loss of potassium and is said to accompany functional improvement, but the level of sodium remains high.[397]

Changes in the calcium content are of real interest. The occurrence of calcification in myelinated axons in the severe model of spinal cord trauma in rat was reported first by Balentine and colleagues.[398] An increased calcium in the lesion is found in experimental animals and in human lesions.[316,363,368] Deposits of calcium are seen 30 min after injury, and large masses of crystalline calcium are found in degenerated axons at about 72 hr (see Section 3.6).[368] It is noteworthy that the administration of calcium in the presence of an ionophore causes granular changes in axoplasm and vesiculation of myelin.[394] This kind of degeneration of myelinated axons is similar to that in spinal cord injury as well as in $CaCl_2$-induced myelopathy in rat.[379] The increase in Ca^{2+} could activate enzymes and mediate the degradation of proteins and membranes.

The levels of total calcium have been determined by atomic absorption spectroscopy. The calcium level is significantly increased in the lesion within 45 min, reaches a maximum level at 8 hr post-trauma, and remains stable up to 72 hr following injury (Fig. 3).[316] The accumulation of calcium in the lesion site following trauma has been confirmed by Hsu *et al.*[399] A decline of extracellular calcium in spinal cord injury has been found by Young, *et al.*[400] and Stokes *et al.*[401] At this point, it is attributed to the influx of calcium into the cell and is consistent with the axonal calcification and increased total calcium levels in the lesions.[316,368]

An increased level of Ca^{2+} has also been found in the lesion created in spinal cord by application of exogenous calcium.[402] The mechanism of entry of calcium into the tissue is not known. A determination of levels of minerals in other chemically injured (e.g., gentamicin) spinal cord has not yet been done.[380]

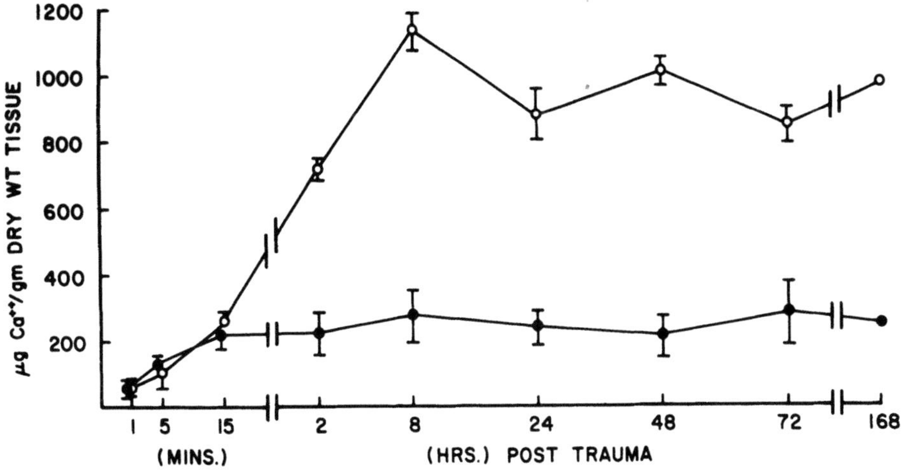

Fig. 3. Levels of spinal cord calcium in lumbar cord lesion (○) and cervical cord control (●). Calcium was determined by atomic absorption spectroscopy. Data from Happel *et al.*[316] and reproduced with permission.

3.5. Changes in Other Constituents

Increased amounts of catecholamines in the traumatized cord have been reported.[403] Others, however, have not confirmed the increase of norepinephrine (NE) in the injured cord.[404–408] The levels of norepinephrine (NE), dopamine (DA), histamine, and 5-hydroxytryptamine (5-HT) have been determined, and a significant elevation in the levels of DA found approximately 15 min following injury.[405–406] Norepinephrine levels remained unchanged or were reduced in the traumatized cord.[405,406,408] Histamine increased, whereas 5-HT remained unchanged.[405] The source of the trauma-related elevation of DA in the lesion is unknown, but it may be a result of inhibition of dopamine β-hydroxylase, the enzyme that converts DA to NE. No similar studies on the levels of catecholamines in the chemically injured spinal cord have been done. The role(s) of the elevated levels of amines (DA or histamine) in trauma must also be considered unknown at this point.

Alterations in the levels of tissue amino acids in traumatic cord injury have not been studied, but it is known that damage to the spinal cord produced by occlusion of thoracic aorta results in the loss of glycine and glutamate from spinal cord, although GABA and glutamine are not altered.[253]

3.6. Alterations in Morphology

Morphological changes in traumatized spinal cord have been studied in several species by light and electron microscopy. Light microscopic study of the lesion (irrespective of methods used for inducing trauma) has shown ischemic and hemorrhagic necrosis, edema, and inflammation. These changes occur first in the gray matter.[359,364,410–412] Later, there is progressive edema, hemorrhage, and necrosis extending from gray to peripheral white matter over a period of hours. White matter abnormalities consisting of edema and necrosis begin at 4–8 hr. Their extent depends on the force of impact and time after injury.

The extent and onset of these changes vary with the injury severity. Edema with an increase in the tissue water content may start within a few minutes following severe injury. An approximately linear relationship between edema and the degree of injury has been attained.[399] Tissue necrosis includes hemorrhage, lysis of cells, and formation of glial scar in the lesion area.

In the model of moderate spinal cord injury, the progressive ultrastructural changes occur several hours after injury, and no alterations in myelin and axon are found immediately,[413] whereas in the acute severe spinal cord trauma (300–400 gcf), ultrastructural alterations of myelin and axons are evident. Thus, at 15 min, the axons of myelinated fibers display swollen, hydropic axoplasm with reduction in the number of neurofilaments, although myelin sheaths are less affected. At 30 min, axons undergo axoplasmic granular changes and loosening of myelin lamellae (Fig. 4). Adaxonal spaces, ruffling and splitting of myelin, and breakage of axons are also reported.[363,368,413,414] The changes observed in the myelinated axons include granular axoplasm, axonal spheroids, tubulovesicular profiles, and vesicular degeneration of myelin with intramyelinic vacuoles.[368] Studies of the details of myelin structure reveal vesicular degeneration beginning at the innermost and outermost lamellae within the first 4 hr. This is followed by sloughing of the vesiculated myelin into the extracellular space by 12 hr following injury. Cavitated and swollen axons with numerous axonal organelles have also been seen in transected spinal cord.[391] A most striking observation is that of hydroxyapatite crystals of calcium within axons of the lesioned tissue (Fig. 5). Intraaxonal calcification has been found in mitochondria.

An increased level of calcium in injured spinal cord has been found,[316] and as a result Ca^{2+} is implicated in the degeneration of axons and myelin (see Section 3.4). Cytoplasmic debris, probably of axonal origin, as well as leaked blood elements, e.g., red cells, platelets, and blood vessels damaged by mechanical force, have been found widely scattered. Inflammatory cells, monocytes, and neurotophils have been found in the lesion within 12 hr of insult. These cells contain phagocytosed myelin debris, and their appearance usually follows the disintegration of myelinated axons (Fig. 6).

Administration of calcium in the presence of an ionophore causes granular changes in axoplasm and vesiculation of myelin. Damage to the spinal cord of rat *in vivo* has been produced chemically, i.e., by the application of $CaCl_2$ by dripping a solution of calcium chloride (10%) on the exposed spinal cord. The morphological alterations in the Ca^{2+}-exposed area of the spinal cord[379,415] include spongiosus of increasing severity in white matter and adaxonal and intramyelinic swelling that closely resemble changes in the traumatic lesion. Granular degeneration of axons, intraaxonal calcification, and vesicular demyelination are evident. These changes provoked by calcium are indistinguishable from those of experimental spinal cord injury. Chemical injury of the spinal cord induced by gentamicin in rabbits produces multiple lesions in the cord white matter. There are edema, swollen axons, and hypertrophied oligoden-

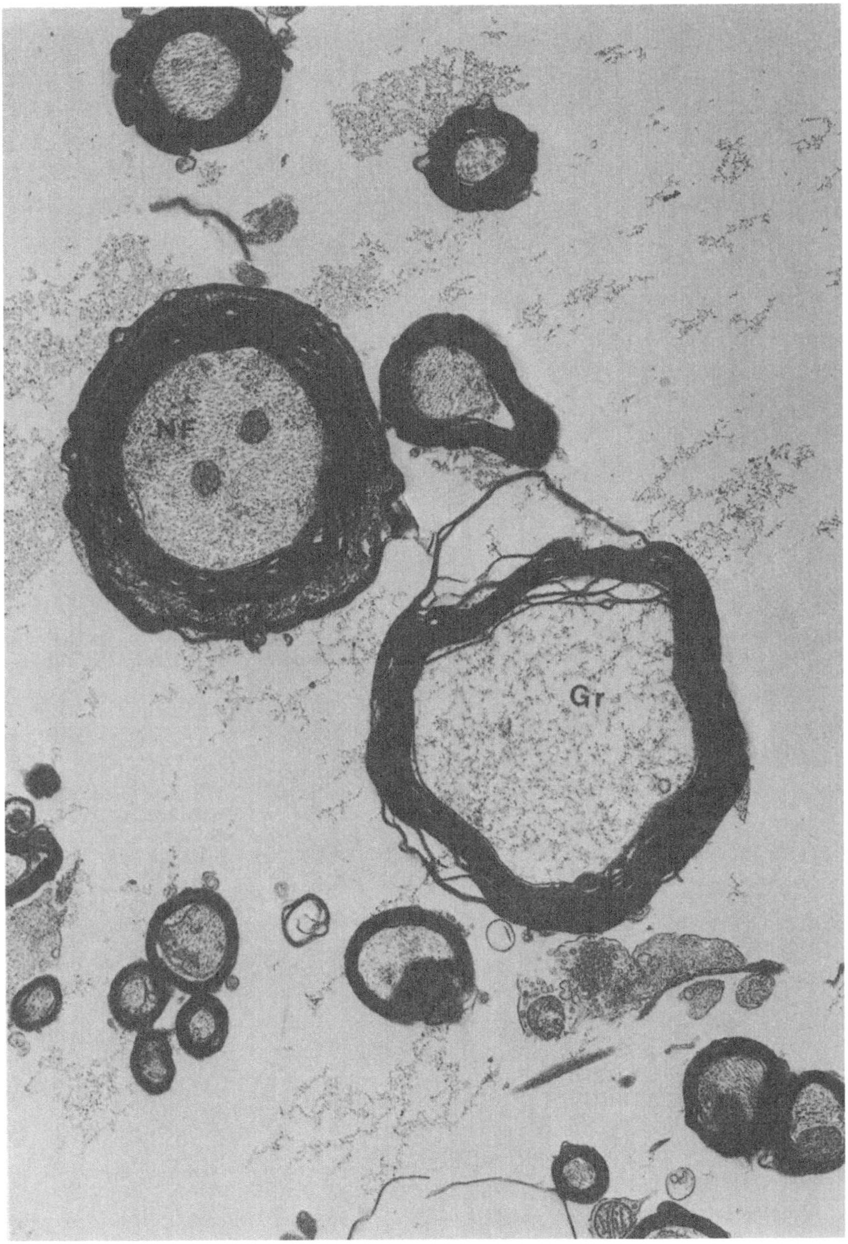

Fig. 4. Several myelinated axons have normal appearing axoplasm with intact neurofilaments (NF); another exhibits hydropic axoplasm with some granular change in neurofilaments (Gr); loosening of myelin lamellae is evident. ×14,000. Data from Banik *et al.*[175] and reproduced with permission.

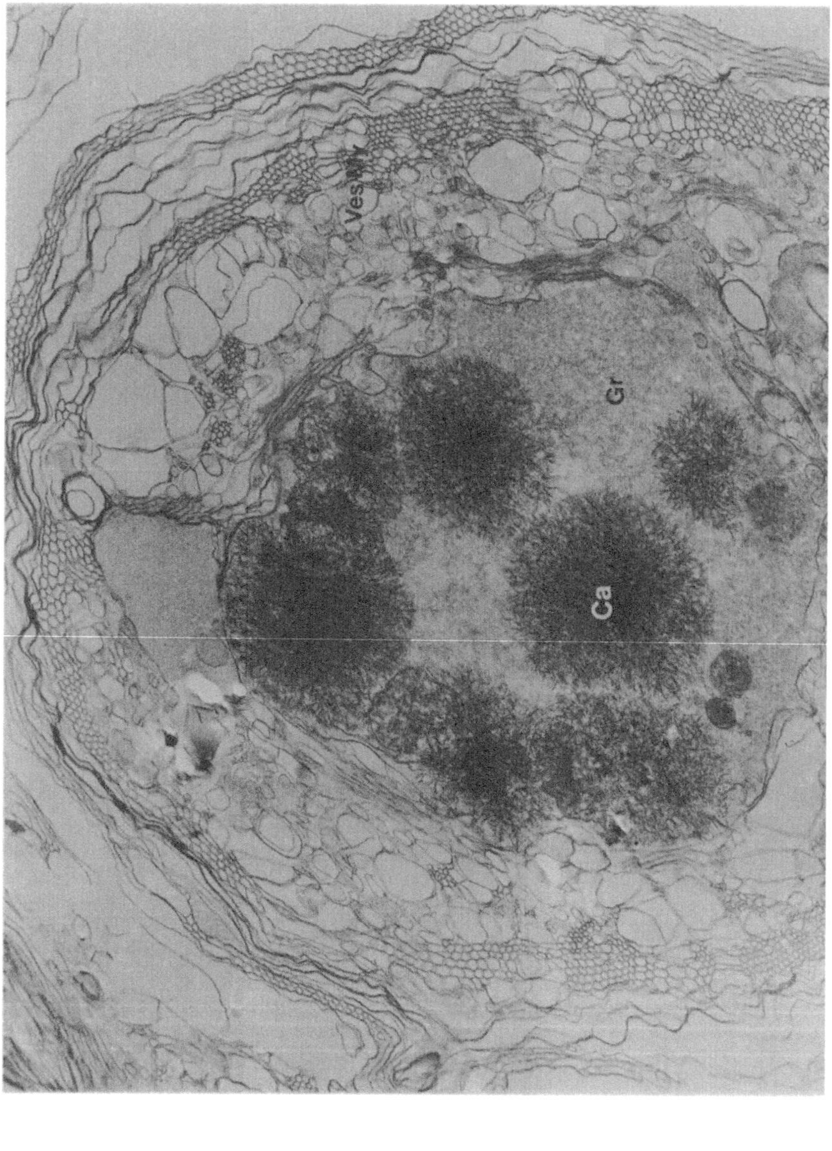

Fig. 5. Seventy-two hours following trauma. Vesiculated myelin (Ves My) surrounds a granular axon (Gr) containing large masses of crystalline calcium (Ca). ×13,000. Data from Banik *et al.*[368] and reproduced with permission.

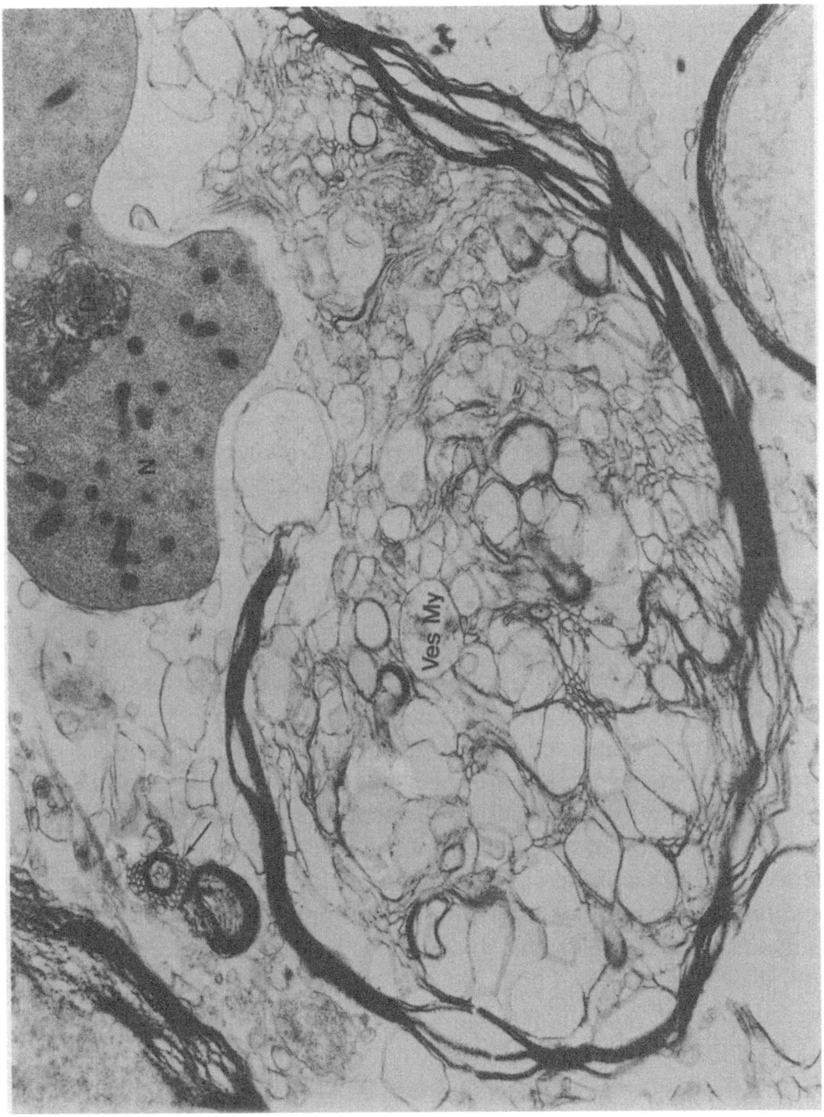

Fig. 6. Twelve hours following trauma. A myelin sheath demonstrating more advanced vesicular degeneration (Ves My) lies adjacent to neutrophil (N), which already contains some cytoplasmic debris(De). A neighboring small axon also demonstrates vesicular degeneration of its myelin sheath (arrow). × 10,500. Data from Banik *et al.*[366] and reproduced with permission.

droglia an hour after injection of gentamicin.[380] Astrocytes are severely edematous, and axonolysis and dissociation of myelin lamellae are evident at 24 hr. At a longer interval (48 hr) neuroaxonal end bulbs form at the lysed end. Wallerian degeneration is evident, and regeneration of axons occur later.

3.7. Mechanisms of Degradation of Myelin in Spinal Cord Trauma and Demyelinating Diseases

Destruction of myelin is common to a variety of pathological conditions, including edema, trauma, infection, and immune attack. The ultrastructural details of myelin dissolution suggest that several different modes of myelin breakdown occur. The process of breakdown of myelin in such different diseases as viral infections [e.g., subacute sclerosing panencephalitis (SSPE)] differs from that seen in the immunologic attack on myelin in human acute disseminated or allergic encephalitis or experimental allergic encephalomyelitis (EAE) in animals.

Our current understanding of cellular and molecular mechanisms of myelin dissolution is incomplete but is increasing with the surge of investigations of human tissues afflicted with demyelinating diseases, of animal disorders such as EAE, and of spinal cord injury. The available evidence implicates several factors in myelinolysis, which include the following: (1) increased activities of proteolytic and lipolytic enzymes,[165,381,415–418] (2) an increase of tissue calcium,[316] (3) myelinotoxic and phagocytic agencies of lymphocytes, macrophages, and perhaps other cells resident in myelinated tissue,[416,418] (4) serum factor,[419] and (5) myelin enzymes.[185]

Spinal cord trauma is an appropriate model to examine myelin breakdown for a number of reasons: (1) the chronological course of events can be accurately determined; (2) injury to the myelinated fiber may have a more damaging effect on the myelin-forming cell and its myelin sheath than on its axon, particularly in sites where the damage is partial and does not cause extensive tissue necrosis; and, finally, (3) functional impairment after injury parallels the site and extent of destruction of myelinated tracts.

In order to understand the chronology of myelin breakdown, the morphological and biochemical effects after physical trauma to spinal cord in experimental animals have been examined. There is progression of edema in the lesion with tissue necrosis and alterations in myelinated axons including vesicular degeneration of myelin, splitting of myelin lamellae, and axonal degeneration. The appearance of crystals of calcium phosphate precedes the infiltration of inflammatory cells with phagocytosis of myelin.[368] The pattern of breakdown and removal of myelin by invading macrophages resembles demyelinating conditions.[420] Minerals may play an important role in the demyelinative process in experimental animals,[320] Zinc, for example, elicits perivascular cuffing and activation of lymphocytes, features characteristic of EAE.[421] There is an increase in calcium and zinc in MS tissue.[317] The increased level of calcium in the lesion of spinal cord injury may contribute to the degeneration of the axon–myelin unit.[180,316]

One important remaining question concerns the source of calcium and its precise role in myelin breakdown. Likely sources of the lesion calcium influx include plasma and nerve cells, since calcium is localized predominantly in blood and is also present in axons and their terminals.[422] Calcium causes toxic cell death in tissue culture,[396] an observation supporting the hypothesis that damage to the cell membrane permits an influx of extracellular calcium, leading to cell destruction including processes. Relevant *in vitro* studies have shown that calcium in the presence of an ionophore causes granular changes in axoplasm and vesiculation of myelin.[394] Furthermore, administration of calcium to exposed spinal cord *in vivo* causes degeneration of myelinated axons.[379] Since these changes resemble those of the degeneration of myelin and axon in spinal cord injury, it is postulated that calcium activates enzymes (e.g., proteinases and lipases) determining the breakdown of myelin and axon. Such a neutral proteinase is present in CNS.[149–152]

The progressive loss of myelin and axonal proteins in trauma implicates proteolytic enzymes. Degradation of MBP was put forth as the initial event in myelinolysis[165] on the basis of the disappearance of MBP from the edge of MS plaques concomitant with increased proteinase activity.[166] Elevated proteolytic activities have been found in lesions of spinal cord, MS, and other demyelinating disease.[166] The activity of lysosomal acid proteinases is increased in the traumatic lesion, and that of neutral proteinases increased even more (see Section 3.3). Since vesiculation of myelin and degradation of proteins occur before cellular infiltration by cells, it is unlikely that circulating leukocytes are primarily responsible for the increased proteolytic activity in the early stages. The early release of protease activity may derive from serum or endogenous cells (e.g., glia), whereas later myelin degradation is carried out by enzymes released by macrophages. The increased activities in macrophages, proliferating astrocytes, and serum in MS and EAE and other demyelinating diseases are consistent with this.[416, 419, 423] Although myelin was once considered as a source for degradative enzymes in demyelinating diseases, recent findings revealing the presence of CANP in purified myelin suggest that myelin itself has autolytic capacity.[184,185] Important roles for CANP and for Ca^{2+} in the degeneration of axons and the breakdown of myelin must be considered. One interaction is the activation of proteinase by Ca^{2+} and its regulatory potential. Another is Ca^{2+} activation of phospholipase, for this activity is increased in the lesions of MS, EAE, and spinal cord injury.[349,424] The increased phospholipase releases arachidonic acid and its metabolites, which provoke edema and produce lysolipids (e.g., lysolecithin), which are myelinolytic.[373]

Macrophages contain phospholipase that may be activated by myelin contact, with the resultant release of proteinases and lipases and eventual myelin dissolution.[425] Dissolution of myelin entails destruction of its structural proteins (MBP, PLP). Since PLP is embedded and masked in the membrane and is resistant to proteolysis, phospholipases are necessary to remove lipids, or lysolecithin (with its detergentlike action) is needed to alter the conformation of the protein molecule and permit proteolysis. Such a degradation of PLP and/or myelin in the presence of both enzyme and detergent (not with enzyme alone) is known.[185,187,426] Hence, we think Ca^{2+} and CANP are involved in the degradation of axon and myelin in spinal cord injury.

If calcium and proteinases play a critical role in the degeneration of the structural unit of axon–myelin, calcium-entry blockers and/or proteinase inhibitors may be used as therapeutic means of minimizing tissue destruction and preserving motor function following spinal cord injury.

The effects of trauma to the spinal cord are complex and could well include primary biochemical changes (enzymes, proteins, etc.) and those secondary to vascular (ischemic, hemorrhagic) and metabolic alterations. Thus far, there is no treatment for spinal cord injury that provides arrest of disability or restoration of motor function following trauma. The major current efforts include (1) numerous investigators examining the effects of various drugs for treating animals with experimental spinal cord injury and (2) work directed at clarifying and facilitating CNS regeneration. Those seeking specific details can refer to reviews by De La Torre[427] on the approaches taken for studying therapeutic effects of drugs and by Kierman[428] and Puchala and Windle[429] on CNS regeneration following spinal cord injury.

An incomplete listing of the drugs proposed and/or used for experimental therapy include Ca^{2+}-entry blockers and enzyme inhibitors,[430] corticosteroids,[397,405,406,431,435] sympathetic blockers,[403,405,408] ϵ-aminocaproic acid,[405,436] naloxone,[437,438] dimethyl sulfoxide (DMSO[434,439] hyperbaric oxygen,[440–442] indomethacin, [443] and cyclophosphamide.[444] Hypothermia and surgical decompression also have advocates. The use of these drugs for treatment of spinal cord injury is without objective basis, and no general agreement regarding efficacy of any of them has yet been realized.

The dogma that the CNS is incapable of functional regeneration is no longer secure as the factors favoring and inhibiting it are coming under scrutiny. At this point, axonal sprouting has been observed at the site of spinal cord transection, but axonal regeneration of the severed axon is very limited (a few millimeters). Regeneratioin of cut axons and recovery of motor function occur in immature but not adult animals.[445,446] One proposal has been that the loss of axon capacity for regeneration after injury is a result of release of chemicals from axons and/or oligodendrocytes or the degradation of myelin products (e.g., MBP) which inhibit axonal growth.[447,448] Another factor to consider is the vascular supply, which is subject to trophic and inhibitory factors as well.

4. CONCLUSIONS

We have compiled available data concerning the biochemistry of normal spinal cord in disease states and after injury. Although complete data are unavailable and are probably unnecessary since different CNS regions must share many molecular features (it is the difference that may be important), there is no doubt that the composition of whole spinal cord is comparable to that of brain. The enzymes mediating metabolism of proteins, lipids, carbohydrates, amino acids, and transmitters of brain are present in the spinal cord, and their levels are either approximately the same as those of brain, or the differences are consistent with morphological preponderance of myelin or some other tissue constitutent in the spinal cord. Cholinergic, glutamatergic, and GABAergic neurons have been described in the spinal cord. Future studies of specific areas of spinal cord and/or of cell types and subcellular components may delineate significant regional variation. This will probably be most important in relation to neurotransmitters and peptide neuromodulators, which have evident implications in relation to the specificity of neuronal connections and functional modality specificities within the neuroaxis. At this time, it is noteworthy that the majority of the work on characterization of proteins and lipids has been done with purified spinal cord myelin.

The architectural simplicity of spinal cord underlies major body functions as is evident in the paralysis following the transecting lesion. The changes in degeneration of myelinated axons with losses of axonal and myelin proteins that occur in spinal cord following injury implicate increases of proteinase activity and of calcium. We propose that calcium-activated neutral proteinase is a critical mediator of spinal cord degeneration. Future studies will include investigation of the source of this enzyme (CANP) and other factors involved in spinal cord injury. An understanding of the molecular mechanism involved not only in the degeneration of the spinal cord but also in regeneration will stimulate further research concerning therapeutic drug use (e.g., current agents of interest include naloxone, verapamil, and clonidine) for restoring or minimizing the loss of motor function.

ACKNOWLEDGMENT. This work was supported in part by PHS grant No. NS-11066 from the NIH-NINCDS. We thank Dr. Wendy Cammer for providing data on spinal cord enzymes before publication.

REFERENCES

1. Rossiter, R. J., 1962, *Neurochemistry* (K. A. C. Elliot, I. H. Page, and J. H. Quastel, eds.), Charles C. Thomas, Springfield, Illinois, pp. 10–54.
2. Brante, G., 1949, *Acta Physiol Scand.[Suppl.]* **63:**13–189.
3. Levy, G., 1969, *Handbook of Neurochemistry,* Volume 2 (A. Lajtha, ed.) Plenum Press, New York, pp. 71–101.
4. McColl, J. D., and Rossiter, R. J., 1952, *J. Exp. Biol.* **29:**203–210.
5. Torello, L. A., Yates, A. J., and Horrocks, L. A., 1981, *Virchows Arch. [Cell Pathol]* **38:**219–228.
6. Pleasure, D., and Kim, G., 1976, *Brain Res.* **103:**117–126.
7. Maggio, B., Cumar, F. A., and Maccioni, H. J., 1972, *J. Neurochem.* **19:**1031–1037.
8. Yu, R. K., Ueno, K., Glaser, G. H., and Tourtellotte, W. W., 1982, *J. Neurochem.* **39:**464–477.
9. Ueno, K., Ando, S., and Yu, R. K., 1978, *J. Lipid Res.* **19:**863–871.
10. Yu, R. K., Endo, R., Takahashi, A., and Miyatake, T., 1980, *Proc. Jpn. Conf. Biochem.* **22:**332–334.
11. Fredman, P., Noren, R., Mansson, J. E., and Svennerholm, L., 1982 *Biochem. Biophys. Acta* **713:**410–418.
12. Porcellati, G., 1964, *Biochim. Biophys. Acta* **90:**183–186.
13. Phillis, I. W., and Tebecis, A. K., 1968, *Nature* **217:**1076–1077.
14. Ramwell, P. W., Shaw, J. E., and Jessop, R., 1966, *Am. J. Physiol.* **211:**998–1004.
15. Horton, E. W., and Main, I. H. M., 1967, *Br. J. Pharmacol. Chemother.* **30:**582–602.
16. Jonsson, H. T., and Daniell, H. B., 1976, *Prostaglandins* **11:**51–61.
17. Wolf, L. S., and Mamer, O. A., 1975, *Prostaglandins* **9:**183–192.

18. Veale, W. L., and Cooper, K. E., 1976, *J. Appl. Physiol.* **37**:942–945.
19. Hsu, C. Y., Halushka, P. V., Hogan E. L., Banik, N. L., Lee, W. L., and Perot, P. L., 1984, *Neurology (N.Y.)* **33(2)**:146.
20. Horrocks, L. A., 1967, *J. Lipid Res.* **8**:569–576.
21. Smith, M. E., 1973, *J. Lipid Res.* **14**:541–551.
22. Norton, W. T., and Autilio, L. A., 1966, *J. Neurochem.* **13**:213–222.
23. O'Brien, J. S., Sampson, E. L., and Stern, M. B., 1967, *J. Neurochem.* **14**:357–365.
24. Norton, W. T., Poduslo, S. E., and Suzuki, K., 1966, *J. Neuropathol. Exp. Neurol.* **25**:582–597.
25. Van Der Westhuyzen, J., Cantrill, R. C., Fernandez-Costa, F., and Metz, J., 1983, *J. Nutr.* **113**:531–537.
26. Ledeen, R. W., Yu, R. K., and Eng, L. F., 1973, *J. Neuropathol.* **21**:829–839.
27. Ledeen, R. W., Cochran, F. B., Yu, R. K., Samuels, F. G., and Haley, J. E., 1980, *Adv. Exp. Med. Biol.* **125**:167–176.
28. Cochran, F. B., Yu, R. K., Ando, S., and Ledeen, R. W., 1981, *J. Neurochem.* **36**:696–702.
29. Suzuki, K., Poduslo, J. F., and Poduslo, S. E., 1968, *Biochim. Biophys. Acta* **152**:576–586.
30. Adams, C. W. M., and Davison, A. N., 1965, *Neurohistochemistry* (C. W. M. Adams, ed.), Elsevier, New York, pp. 332–400.
31. Adams, C. W. M., and Davison, A. N., 1959, *J. Neurochem.* **4**:282–289.
32. Fumagalli, R., and Paoletti, R., 1963, *Life Sci.* **2**:291–295.
33. Banik, N. L., and Davison, A. N., 1967, *J. Neurochem.* **14**:594–596.
34. Yu, R. K., and Yen, S. I., 1975, *J. Neurochem.* **25**:229–232.
35. Cochran, F. B., Ledeen, R. W., and Yu, R. K., 1983, *Dev. Brain Res.* **6**:27–32.
36. Suzuki, K., Poduslo, S. E., and Norton, W. T., 1967, *Biochim. Biophys. Acta* **144**:375–381.
37. Horrocks, L. A., 1968, *J. Neurochem.* **15**:483–488.
38. Cuzner, M. L., and Davison, A. N., 1968, *Biochem. J.,* **106**:29–34.
39. Norton, W. T., and Poduslo, S. E., 1973, *J. Neurochem.* **21**:749–758.
40. Eng, L. F., and Noble, E. P., 1968, *Lipids* **3**:157–162.
41. Banik, N. L., and Davison, A. N., 1969, *Biochem. J.,* **115**:1051–1062.
42. Hogan, E. L., Banik, N. L., Happel, R. D., and Sostek, M., 1982, *Trans, Am. Soc. Neurochem.* **13**:P107.
43. Banik, N. L., and Smith, M. E., 1977, *Biochem. J.,* **162**:247–255.
44. Amaducci, L., 1962, *J. Neurochem.* **9**:153–160.
45. Morell, P., Lipkind, R., and Greenfield, S., 1973, *Brain Res.* **58**:510–514.
46. Lees, M. B., and Paxman, S. A., 1974, *J. Neurochem.* **23**:825–851.
47. Stodieck, L. S., and Luttges, M. W., 1983, *Neurochem. Res.* **8**:599–619.
48. Laatsch, R. H., Kies, M. W., Gordon, F., and Alvord, E. C., 1972, *J. Exp. Med.* **115**:777–788.
49. Williams, R. M., Lees, M. B., Cambi, F., and Macklin, W. B., 1982, *J. Neuropath. Exp. Neurol.* **41**:508–521.
50. Bornstein, M. B., and Raine, C. S., 1970, *Lab. Invest.* **23**:536–542.
51. Mithen, F., Bunge, R., and Agrawal, H., 1980, *Brain Res.* **197**:477–483.
52. Seil, F. J., 1977, *Ann. Neurol.* **2**:345–355.
53. Quarles, R. H., Everly, J. L., and Brady, R. D., 1973, *Brain Res.* **58**:506–509.
54. Sternberger, N. H., Quarles, R. H., Itoyama, Y., 1979, *Proc. Natl. Acad. Sci. U.S.A.* **76**:1510–1514.
55. McIntyre, L. J., Quarles, R. H., and Brady, R. O., 1978, *Brain Res.* **149**:251–256.
56. Itoyama, Y., Sternberger, N. H., Webster, H. de F., Quarles, R. H., Cohen, S. R., and Richardson, E. P., Jr., 1980, *Ann. Neurol.* **7**:167–177.
57. Chiu, F. C., Norton, W. T., and Fields, K., 1981, *J. Neurochem.* **37**:147–155.
58. Schlaepfer, W. W., and Zimmerman, U. J. P., 1981, *Neurochem. Res.* **6**:243–255.
59. Smith, M. E., Trotter, J., and Eng, L. F., 1982 *Trans. Am. Soc. Neurochem.* **13**:P108.
60. Bigbee, J., and Eng, L. F., 1982, *Brain Res.* **249**:383–386.
61. Chiu, F. -C, and Norton, W. T., 1982, *J. Neurochem.* **39**:1252–1260.
62. Zgorzalewicz, B., Neuhoff, V., and Waehneldt, T. V., 1974, *Neurobiology* **4**:265–267.
63. Amaducci, L., Pazzagli, A., and Pessina, G., 1962, *J. Neurochem.* **9**:509–518.

64. Wolfgram, F., and Kotorii, K., 1968, *J. Neurochem.* **15**:1281–1290.
65. Greenfield, S., Norton, W. T., and Morell, P., 1971, *J. Neurochem.* **18**:2119–2128.
66. Banik, N. L., Davison, A. N., Ramsey, R. B., and Scott, T., 1974, *Dev. Psychobiol.* **7**:539–549.
67. Fishman, M. A., Agrawal, H. C., and Alexander, A., Yolterman, J., Martenson, R. E., and Mitchell, R. F., 1975, *J. Neurochem.* **24**:689–694.
68. Adams, D. H., and Osborne, J., 1973, *Neurobiology* **3**:91–112.
69. Einstein, E. R., Dalal, K. B., and Csejtey, J., 1970, *Brain Res.* **18**:35–49.
70. Smith, M. E., and Sedgewick, L. M., 1975, *J. Neurochem.* **24**:763–770.
71. Kornguth, S. E., Anderson, J. W., and Scott, G., 1966, *J. Comp. Neurol.* **127**:1–17.
72. Fagg, G. E., Schipper, H. I., Neuhoff, V., 1979, *Brain Res.* **167**:251–258.
73. Agrawal, H. C., Banik, N. L., Bone, A. H., Davison, A. N., Mitchell, R. F., and Spohn, M., 1971, *Biochem. J.* **120**:635–662.
74. Brostoff, S. W., Reuter, W., Hichens, M., and Eylar, E. H., 1974, *J. Biol. Chem.* **249**:559–567.
75. Eylar, E. H., Szymanska, I., Ishaque, A., Ramwani, J., and Dubeski, S., 1980, *J. Immunol.* **124**:1086–1092.
76. Smith, M. E., Forno, L. S., and Hoffman, W. W., 1979, *J. Neuropathol. Exp. Neurol.* **38**:377.
77. Kadlubowski, M., and Hughes, R. A. C., 1979, *Nature* **277**:140–141.
78. Hoffman, P. M., Powers, J. M., Weise, M. J., and Brostoff, S. W., 1980, *Brain Res.* **195**:355–362.
79. Ishaque, A., Hofmann, T., and Eylar, E. H., 1979, *Fed. Proc.* **38**:514.
80. Uyemura, K., Kato-Yamenaka, T., and Kitamura, K., 1977, *J. Neurochem.* **29**:61–68.
81. Yo, S., and McPherson, C. F. C., 1972, *J. Immunol.* **109**:1009–1016.
82. Diebler, G. E., Driscoll, B. F., and Kies, M. W., 1978, *J. Neurochem.* **30**:401–412.
83. DeArmond, S. J., Diebler, G. E., Bacon, M., Kies, M. W., and Eng, L. F., 1980, *J. Histochem. Cytochem.* **28**:1275–1285.
84. Cammer, W., and Zimmerman, T. R., 1983, *Dev. Brain Res.* **6**:21–26.
85. McDougal, D. B., Schultz, D. W., Passonneau, J. V., Clark, J. R., Reynolds, M. A., and Lowry, O. H., 1961, *J. Gen. Physiol.* **44**:487–498.
86. Robinson, N., and Philips, B. M., 1964, *Biochem. J.* **92**:254–259.
87. Buell, M. V., Lowry, O. H., Roberts, N. H., Chang, M. L. W., and Kapphah, J. I., 1958, *J. Biol. Chem.* **232**:979–993.
88. Oster-Granite, M. L., and Gearhart, J., 1980, *J. Histochem. Cytochem.* **28**:250–254.
89. Bachelard, H., 1981, *Adv. Biochem. Psychopharmacol.* **29**:475–497.
90. Wilkin, G. P., and Wilson, J. E., 1977, *J. Neurochem.* **29**:1039–1051.
91. Wilson, J. E., 1972, *Arch. Biochem. Biophys.* **150**:96–104.
92. Bigl, V., Biesold, D., Dowedowa, E. L., and Pigarewa, S. D., 1971, *Acta. Biol. Med. Germ.* **26**:27–33.
93. Dreyfus, P. M., 1965, *J. Neurpathol. Exp. Neurol.* **24**:119–129.
94. Dreyfus, P. M., and Hauser, G., 1965, *Biochim. Biophys. Acta* **104**:78–84.
95. Bonavita, S., and Guarnieri, R., 1963, *J. Neurochem.* **10**:755–764.
96. Friede, R. L., Fleming, L. M., and Knoller, M., 1963, *J. Neurochem.* **10**:755–764.
97. Lowenthal, A., Karcher, D., and Van Sande, M., 1954, *J. Neurochem.* **10**:247–250.
98. Breckenridge, B. M., and Crawford, E. J., 1961, *J. Neurochem.* **7**:234–240.
99. McDougal, D. B., 1958, *Neurology (Minneap.)* **8**:58–59.
100. Manocha, S. L., and Bourne, G. H., 1966, *Exp. Brain Res* **2**:216–299.
101. Nandy, K., and Bourne, G. H., 1964, *J. Histochem. Cytochem.* **12**:188–193.
102. Nandy, K., and Bourne, G. H., 1964, *J. Anat. (Lond.)* **98**:647–653.
103. Himwich, H. E., and Fazekas, J. F., 1941, *Am. J. Physiol.* **132**:454–459.
104. Chao, L.-P., and Wolfgram, F., 1973, *J. Neurochem.* **20**:1075–1081.
105. Karenkan, K.-S., Chao, L. P., and Eng, L. F., 1978, *Brain Res.* **146**:221–229.
106. Motavkin, P. A., and Okhotin, V. E., 1980, *Neurosci. Behav. Physiol.* **10**:307–310.
107. Aquilonius, S.-M, Eckernas, S.-A., and Gilberg, T.-G., 1981, *Brain Res.* **211**:329–340.
108. Hetnarski, B., Wisniewski, H. M., Iqbal, K., Dziedzic, J. D., and Lajtha, A., 1980, *Ann. Neurol.* **7**:489–490.

109. Rosemann, U., Friede, R. L., 1967, *J. Anat. (Lond.)* **101**:27–32.
110. Gwyn, D. G., Silver, A., and Wolstencroft, J. H., 1969, *J. Physiol. (Lond.)* **201**:23–24.
111. Maletta, G. J., Vernadakis, A., and Timiras, P. S., 1966, *Proc. Soc. Exp. Biol. Med.* **121**:1210–1211.
112. Nachmansohn, D. M., 1940, *J. Neurophysiol.* **3**:396–402.
113. Wenger, D. S., 1961, *Fed. Proc.* **10**:268–269.
114. Bungen, A. S. V., and Chipman, L. M., 1951, *J. Physiol. (Lond.)* **114**:296–305.
115. Giacobini, E., and Holmstead, B., 1958, *Acta Physiol. Scand.* **42**:12–27.
116. Chacko, L. W., and Cerf, S. A., 1960, *J. Anat. (Lond.)* **94**:74–81.
117. Soderholm, U., 1965, *Acta Physiol. Scand.* [*Suppl.*] **256**:3–60.
118. Aldridge, W. N., and Johnson, M. K., 1959, *Biochem. J.* **73**:270–276.
119. Mahadik, S., and Rapport, M. M., 1979, *Advances in Neurochemistry,* Volume 3 (B. W. Agranoff and M. H. Aprison, eds.), Raven Press, New York, pp. 99–163.
120. Koelle, G. B., 1972, *J. Comp. Neurol.* **100**:211–228.
121. Potter, L. T., 1972, *The Structure and Function of Nervous System,* Volume 4 (G. H. Bourne, ed.), Academic Press, New York, pp. 105–128.
122. Silver, A., and Wolstencroft, J. H., 1970 *J. Physiol. (Lond.)* **210**:92–93.
123. Tewari, H. B., and Bhatnager, M., 1980, *Cell. Mol. Biol.* **26**:353–371.
124. Chan, S. L., Gordon, M. A., and Trevor, A. J., 1977, *Life Sci.* **21**:1611–1615.
125. Hollunger, E. G., and Niklasson, B. H., 1973, *J. Neurochem.* **20**:821–836.
126. Huther, G., and Luppa, H., 1977, *Histochemistry* **51**:245–251.
127. Koelle, W. A., Hossaini, K. S., Akbarzodeh, P., and Koelle, G. B., 1970, *J. Histochem. Cytochem.* **18**:812–819.
128. Reiger, F., and Vigny, M., 1976, *J. Neurochem.* **24**:121–129.
129. Vijayan, V. K., and Oschowka, J. A., 1977, *J. Neurochem.* **28**:1141–1143.
130. Huther, G., and Luppa, H., 1979, *Histochemistry* **63**:115–121.
131. Haynes, L. W., 1983, *Experientia* **39**:223–225.
132. Navaratham, V., and Lewis, R. R., 1970, *Brain Res.* **18**:411–425.
133. Brightman, M. W., and Albers, R. W., 1959, *J. Neurochem.* **4**:244–250.
134. Albers, R. W., and Brady, R. O., 1959, *J. Biol. Chem.* **234**:926–928.
135. Graham, L. T., and Aprison, M. H., 1969, *J. Neurochem.* **16**:559–566.
136. Graham, L. T., Shank, R. P., Werman, R., and Aprison, M. H., 1967, *J. Neurochem.* **14**:465–472.
137. Kuriyama, K., and Yoneda, Y, 1978, *Brain Res.* **148**:163–179.
138. Salvadar, R. A., and Albers, R. W., 1959, *J. Biol. Chem.* **234**:922–925.
139. Rando, R. R., Bangerter, F. W., and Farb, D. H., 1981, *J. Neurochem.* **36**:985–990.
140. Pitts, F. N., Jr., and Quick, C., 1965, *J. Neurochem.* **12**:893–900.
141. Goldstein, D. B., 1966, *J. Neurochem.* **13**:1011–1016
142. Neims, A. H., Zieverink, W. D., and Smilack, J. D., 1966, *J. Neurochem.* **13**:163–168.
143. DeMarchi, W. J., and Johnston, G. A. R., 1969, *J. Neurochem.* **16**:355–361.
144. Pscheidt, G. R., and Haber, B., 1965, *J. Neurochem.* **12**:613–618.
145. Anden, N. E., 1965, *Acta Physiol. Scand.* **64**:197–203.
146. Anden, N. E., 1964, *Life Sci.* **3**:473–478.
147. McLennan, I. S., and Lees, G. J., 1978, *J. Neurochem.* **30**.429–436.
148. Hartman, B. K., Zeid, D., and Udenfriend, S., 1972, *Proc. Natl. Acad. Sci. U.S.A.* **69**:2722–2726.
149. Lundberg, J., Bylock. A., Goldstein, M., Hansson, H., and Dahlstrom. A., 1977, *Brain Res.* **120**:549–552.
150. Dahlstrom, A., and Fuxe, K., 1965, *Acta Physiol. Scand.* [*Suppl.*] **247**:1–36.
151. Glazer, E., and Ross, L. L., 1980, *Brain Res.* **185**:39–49.
152. Axelrod, J., 1972, *Pharmacol. Rev.* **24**:233–243.
153. Grzanza, R., Morrison, J. H., Coyle, J. T., and Molliver, M. E., 1977, *Neurosci. Lett.* **4**:127–134.
154. Lajtha, A., 1961, *Regional Neurochemistry* (S. S. Kety and J. Elkes, eds.), pp. 25–36.
155. Serra, S., Grynbaum, A., Lajtha, A., and Marks, N., 1972, *Brain Res.* **44**:579–592.
156. Porcellati, G., Millo, A., and Monocchio, I., 1961, *J. Neurochem.* **7**:317–320

157. Adams, C. W. M., and Taquan, N. A., 1961, *J. Neurochem.* **6**:334–341.
158. Buletza, G., and Smith, M. E., 1975, *Biochem. J.* **156**:627–633.
159. Banik, N .L., Smith, K., Powers, J. M., and Hogan, E. L., 1979, *Trans. Int. Soc. Neurochem.* **7**:206.
160. Hirsch, H. E., Andrews, J. M., and Parks, M. E., 1974, *J. Neurochem.* **23**:935–941.
161. Nixon, R. A., 1983, *Neurofilaments* (C. A. Marotta, ed.), University of Minnesota Press, Minneapolis, pp. 117–154.
162. Chelmicka-Schorr, E., Sportiello, M., Antel, J. P., and Arnason, B. W. G., 1982, *J. Neurol. Sci.* **56**:141–145.
163. Hirsch, H. E., and Parks, M. E., 1973, *J. Neurochem.* **21**:453–458.
164. Freysz, L., Farooqui, A. A., Adamczewska-Goncerzewicz, Z., and Mandel, P., 1979, *J. Lipid Res.* **20**:503–508.
165. Hallpike, J. F., and Adams, C. W. M., 1969, *Histochem. J.* **1**:559–578.
166. Einstein, E. R., Csejtey, J., Dalal, K. B., Adams, C. W. M., Bayliss, D. B., and Hallpike, J. F., 1972, *J. Neurochem.* **19**:652–662.
167. Rauch, H., Einstein, E. R., and Csejtey, J., 1973, *Neurobiology* **3**:195–205.
168. Whitaker, J. N., and Seyer, J. M., 1979, *J. Neurochem.* **32**:325–333.
169. Guroff, G. E., 1964, *J. Biol. Chem.* **239**:149–155.
170. Banik, N. L., Hogan, E. L., Jenkins, M., McDonald, J. K., McAlhaney, W. M., and Sostek, M., 1983, *Neurochem. Res.* **11**:1389–1405.
171. Malik, M. N., Fenko, M. D., Iqbal, K., and Wisniewski, H. M., 1983, *J. Biol. Chem.* **258**:8955–8962.
172. Zimmerman, U. J. P., and Schlaepfer, W. W., 1982, *Biochemistry* **21**:3977–3983.
173. DeArmond, S. J., Fajardo, M., Naughton, S. A., and Eng, L. F., 1983, *Brain Res.* **262**:275–282.
174. Dahl, D., and Bignami, A., 1975, *Biochim. Biophys. Acta* **386**:41–51.
175. Banik, N. L., Hogan, E. L., Powers, J. M., and Whetstine, L. J., 1982, *Neurochem. Res.* **7**:1465–1475.
176. Soifer, D., Iqbal, K., Czosnek, H., DeMantini, J., Stwlman, J. A., and Wisniewski, H., 1981, *J. Neurosci.* **1**:461–470.
177. Dahl, D., Crosby, C. J., and Bignami, A., 1981, *Exp. Neurol.* **71**:421–430.
178. DeArmond, S. J., Lee, Y. L., and Eng, L. F., 1982, *J. Neuropathol. Exp. Neurol.* **41**:359.
179. Boehme, D. H., Fordice, M. W., and Marks, N., 1974, *Brain Res.* **75**:153–162.
180. Banik, N. L., Powers, J. M., and Hogan, E. L., 1981, *Chemisms of the Brain: Basic and Applied Neurochemistry* (R. Rodnight, H. S. Bachelard, and W. L. Stahl, eds.), Churchill Livingston, Edinburgh, London, pp. 296–306.
181. Palladin, A. V., and Belik, Y. V., 1970, *Protein Metabolism of the Nervous System* (A. Lajtha, ed.), Plenum Press, New York, pp. 77–91.
182. Riekkinen, P. J., Clausen, J., and Arstilla, A. U., 1970, *Brain Res.* **19**:213–227.
183. Demonte, B., Mela, P., and Marks, N., 1971, *Eur. J. Biochem.* **23**:355–365.
184. Sato, S., and Miyatake, T., 1983, *J. Neurochem.* **41**(Suppl.):546A.
185. Banik, N. L., McAlhaney, W. M., and Hogan, E. L., 1983, *J. Neurochem.* **41**(Suppl.):130A.
186. Lees, M. B., Messenger, B., and Burnham, J., 1967, *Biochem. Biophys. Res. Commun.* **28**:185–190.
187. Lees, M. B., and Chan, D. B., 1975, *J. Neurochem.* **25**:595–600.
188. Julien, J. P., Smoluk, G. D., and Mushynski, W. E., 1983, *Biochim. Biophys. Acta* **755**:25–31.
189. Miyamoto, E., and Kakiuchi, S., 1974, *J. Biol. Chem.* **149**:2769–2777.
190. Jones, G. M., and Carnegie, P. R., 1976, *J. Neurochem.* **23**:1231–1237.
191. Miyake, M., 1975, *J. Neurochem.* **24**:909–915.
192. Abood, L. G., Gerard, R. W., Banks, J., and Tschingi, R. D., 1952, *Am. J. Physiol.* **168**:728–738.
193. Fieschi, C., and Soriani, S., 1959, *J. Neurochem.* **4**:71–77.
194. Nandy, K., and Bourne, G. H., 1963, *Nature* **200**:1216–1217.
195. Chiquoine, A. D., 1954, *J. Comp. Neurol.* **100**:415–439.
196. Mattio, T. G., Rolenquist, T. H., and Kirby, M. L., 1981, *Exp. Brain Res.* **41**:411–413.

197. Clendenon, N. R., 1978, *Arch. Neurol.* **25**:432–448.
198. Shoenen, J., 1978, *Neuropathol. Appl. Neurobiol.* **4**:37–66.
199. Hirsch, H. E., 1968, *Neurochemistry and Clinical Neurology. Progress in Clinical and Biological Research,* Volume 39 (L. Battista, G. Hashim, and A. Lajtha, eds.), Alan R. Liss, New York, pp. 11–19.
200. Suran, A. A., 1974, *J. Histochem. Cytochem.* **22**:802–811, 812–818.
201. Sood, P. P., 1980, *Ind. J. Exp. Biol.* **18**:586–589.
202. Ashbey, W., and Schuster, E. M., 1950, *J. Biol. Chem.* **184**:109–116.
203. Sapirstein, V. S., Lees, M. B., and Trachtenberg, M. C., 1978, *J. Neurochem.* **31**:283–287.
204. Trachtenberg, M. C., and Sapirstein, V. S., 1980, *Neurochem. Res.* **5**:573–581.
205. Koul, O., and Kanungo, M. S., 1975, *Exp. Gerontol.* **10**:273–278.
206. Nair, V., and Bau, D., 1971, *Brain Res.* **31**:185–193.
207. Cammer, W., Fredman, T., Rose, A. L., and Norton, W. T., 1976, *J. Neurochem.* **27**:165–171.
208. Kumpulainen, T., and Korhonen, L. K., 1982, *J. Histochem. Cytochem.* **30**:283–292.
209. Yandrasitz, J. R., Ernst, S. A., and Salganicoff, L., 1976, *J. Neurochem.* **27**:707–715.
210. Korhonen, L. K., and Hyyppa, M., 1967, *Acta Histochem.* **76**:75–79.
211. Parthe, V., 1981, *J. Neurosci. Res.* **6**:119–131.
212. Giacobini, E., 1962, *J. Neurochem.* **9**:169–177.
213. Bhattacharjee, J., 1976, *Histochem. J.* **83**:63–70.
214. Tower, D. B., and Young, O. M., 1983, *J. Neurochem.* **20**:269–278.
215. Ghandour, M. S., Langley, O. K., Vincendon, G., Gombos, G., Fillipi, D., Limozin, N., Dalmasso, D., and Lahrcent, G., 1980, *Neuroscience* **5**:559–571.
216. Lees, M. B., Sapirstein, V. S., Reiss, D. S., and Kolodny, E. H., 1980, *Neurology (N.Y.)* **30**:719–725.
217. Sukumaran, M., and Sood, P. P., 1982, *Acta Morphol. Neurol. Scand.* **20**:43–55.
218. Naidoo, D., and Pratt, O., 1962, *J. Neurol. Neurosurg. Psychiatry* **14**:287–294.
219. Cammer, W., 1984, *J. Neurochem.* **42**:1372–1378.
220. Fried, R., 1965, *J. Neurochem.* **12**:815–832.
221. Kurihara, T., and Tsukada, Y., 1967, *J. Neurochem.* **14**:1167–1174.
222. Kurihara, T., and Tsukada, Y., 1968, *J. Neurochem.* **15**:827–832.
223. Sprinkle, R. J., Zaruba, M. E., and McKhann, G. M., 1978, *J. Neurochem.* **30**:309–316.
224. Guarnieri, M., Cohen, S. R., and Ginns, E., 1976, *J. Neurochem.* **26**:41–44.
225. Delassale, A., Zale, B., LaChapelle, F., Roul, M., Collier, P., and Jacque, C., 1981, *J. Neurosci. Res.* **6**:303–313.
226. Nishizawa, Y., Kurihara, T., and Takahashi, Y., 1981, *Brain Res.* **212**:219–222.
227. Sabri, M. I., Tremblay, C., Banik, N. L., Scott, T., Gohil, K., and Davison, A. N., 1975, *Biochem. Soc. Trans.* **3**:275–276.
228. Waehneldt, T. V., 1975, *Biochem J.* **151**:435–437.
229. Bock, E., and Dissing, J., 1975, *Scand. J. Immunol.* **6**(Suppl. 2):31–36.
230. Marangos, P. J., and Zomzely-Neurath, C., 1976, *Biochem. Biophys. Res. Commun.* **68**:1309–1316.
231. Marangos, P. J., Schmechel, D., Parma, A. M., Clark, R. L., and Goodwin, F. K., 1979, *J. Neurochem.* **33**:319–329.
232. Maxwell, G. D., Whitehead, M. E., Connothy, S. M., and Marangos, P. J., 1982, *Brain Res.* **255**:401–418.
233. Schmechel, D. E., Bryhtman, M. W., and Barker, J. L., 1980, *Brain Res.* **181**:391–400.
234. Poulsen, E., and Aldridge, W. N., 1964, *Biochem. J.* **90**:182–189.
235. Rumsby, M. G., Getliffe, H. M., and Reikkinen, P. J., 1973, *J. Neurochem.* **21**:959–968.
236. Schelfer, R., and Adrian, E. K., 1980, *J. Comp. Neurol.* **194**:829–844.
237. Lajtha, A., and Toth, J., 1974, *Brain Res.* **76**:546–551.
238. Agrawal, H. C., and Himwich, W. H., 1970, *Developmental Neurobiology* (W. A. Himwich, ed.), Charles C Thomas, Springfield, Illinois, pp. 287–310.
239. Perry, T. L., 1982, *Handbook of Neurochemistry*, Volume 1 (A. Lajtha, ed.), Plenum Press, New York, pp. 151–180.
240. Johnston, G. A. R., DeGroat, W. C., and Curtis, D. R., 1969, *J. Neurochem.* **16**:797–800.

241. Kandera, J., Levi, G., and Lajtha, A., 1968, *Arch. Biochem. Biophys.* **126**:249–260.
242. Nagata, Y., Yokoi, Y., and Tsukada, Y., 1966, *J. Neurochem.* **13**:1421–1431.
243. Tallan, H. H., 1962, *Amino Acid Pools* (J. T. Holden, ed.), Elsevier, Amsterdam, pp. 471–485.
244. Turecky, L., Liska, B., and Pechan, I., 1980, *J. Neurochem.* **35**:735–738.
245. Curatolo, A., D'Arcangelo, P., Lino, A., and Braneati, A., 1965, *J. Neurochem.* **12**:339–342.
246. Nixon, R. A., 1976, *J. Neurochem.* **27**:237–244.
247. Curtis, D. R., Duggan, A. W., and Johnston, G. A. R., 1971, *Exp. Brain Res.* **12**:547–565.
248. Werman, R., Davidoff, R. A., and Aprison, M. H., 1968, *J. Neurophysiol.* **31**:81–95.
249. Curtis, D. R., and Watkins, J. C., 1960, *J. Neurochem.* **6**:117–141.
250. Curtis, D. R., Hosli, L., and Johnston, G. A. R., 1968, *Exp. Brain Res.* **6**:1–18.
251. Fagg, G. E., Jones, I. M., and Jordan, C. C., 1978, *Neurosci. Lett.* **9**:71–75.
252. Daly, E. C., and Aprison, M. H., 1983, *Handbook of Neurochemistry*, Volume 3 (A. Lajtha, ed.), Plenum Press, New York, pp. 467–499.
253. Davidoff, R. A., Graham, L. T., Shank, R. P., Werman, R., and Aprison, M. H., 1967, *J. Neurochem.* **14**:1025–1031.
254. Baughman, R. W., and Gilbert, C. D., 1981, *J. Neurosci.* **1**:427–439.
255. Cotman, C. W., and Nadler, J. V., 1981, *Glutamate: Transmitters in the CNS* (P. J. Roberts, J. Storm-Mathisen, and G. A. R. Johnston, eds.), John Wiley and Sons, New York, Chichester, pp. 117–154.
256. Bradford, H. F., and Richard, C. D., 1976, *Brain Res.* **105**:168–172.
257. Shank, R. P., and Campbell, G. LeM., 1982, *Handbook of Neurochemistry*, Volume 3 (A. Lajtha, ed.), Plenum Press, New York, pp. 381–404.
258. Eccles, J. C., Schmidt, R. F., and Willis, W. D., 1963, *J. Physiol. (Lond.).* **168**:500–530.
259. Roberts, E., 1978, *Psychopharmacology: A Generation of Progress* (M. A. Lipton, A. DiMascio, and K. F. Killam, eds.), Raven Press, New York, pp. 95–102.
260. Fahn, S., 1976, *GABA in Nervous System Function* (E. Roberts, T. N. Chase, and D. B. Tower, eds.), Raven Press, New York, pp. 169–186.
261. Tapia, R., 1983, *Handbook of Neurochemistry*, Volume 3 (A. Lajtha, ed.), Plenum Press, New York, pp. 423–466.
262. Ryall, R. W., 1983, *Spinal Cord* (G. A. Austin, ed.), Igaku-Shoin, New York, pp. 95–137.
263. Krnjevic, K., 1974, *Physiol. Rev.* **54**:418–540.
264. Curtis, D. R., Ryall, R. W., and Watkins, J. C., 1966, *Exp. Brain Res.* **2**:97–106.
265. Zieglgansberger, W., and Reiter, C. H., 1974, *Neuropharmacology* **13**:518–527.
266. Eng, L. F., Yueda, C. T., Chao, L. P., and Wolfgram, F., 1974, *Nature* **250**:243–245.
267. Anderson, E. G., and Holgerson, L. O., 1966, *J. Neurochem.* **13**:479–485.
268. Suzuki, O., Noguchi, E., and Yagi, K., 1976, *J. Neurochem.* **27**:319–320.
269. Vogt, M. 1954, *J. Physiol. (Lond.)* **123**:451–481.
270. Spector, S., Melmon, K., Lovenberg, W., and Sjoerdsma, A., 1963, *J. Pharmacol. Exp. Ther.* **140**:229–235.
271. Dahlstrom, A., and Fuxe, K., 1965, *Life Sci.* **4**:129–132.
272. Anden, N. E., Carlsson, A., and Haggendal, J., 1969, *Annu. Rev. Pharmacol.* **9**:119–134.
273. Belcher, G., Ryall, R. W., and Schaffner, R., 1978, *Brain Res.* **15**:307–321.
274. Hongo, T., and Ryall, R. W., 1966, *Acta Physiol. Scand.* **68**:96–104.
275. Headley, P. M., Duggan, A. W., and Smith-Grier, B. T., 1978, *Brain Res.* **145**:185–189.
276. Engberg, I., and Ryall, R. W., 1966, *J. Physiol. (Lond.)* **185**:298–322.
277. Ljungdahl, A., Hokfelt, T., and Nilsson, G., 1978, *Neuroscience* **3**:861–863.
278. Von Euler, U. S., and Gaddum, J. H., 1931, *J. Physiol. (Lond.)* **72**:74–87.
279. Cuello, A. C., Polak, J. M., and Pearse, A. G. E., 1976, *Lancet* **2**:1054–1056.
280. Hockfelt, T., Kellereth, J. O., Nilsson, G., and Pernow, B., 1975, *Science* **190**:889–890.
281. Naftchi, N. E., Abrahams, S. J., St. Paul, H. M., Lowman, E. W., and Schlosser, W., 1978, *Brain Res.* **153**:507–513.
282. Vacca, L. L., Hobbs, J., Abrahams, S., and Naftchi, E., 1982, *Histochemistry* **76**:33–49.
283. Giraud, P., Castanas, E., Patey, G., Oliver, C., and Rossier, J., 1983, *J. Neurochem.* **41**:154–160.
284. Helme, R. D., and White, D. M., 1983, *J. Neuropathol. Exp. Neurol.* **42**:99–105.

285. Burt, A. M., 1975, *Exp. Neurol.* **47**:173–180.
286. Pickel, V. M., Reis, D. J., and Leeman, S. E., 1977, *Brain Res.* **122**:534–540.
287. Dalsgaard, C. J., Risling, M., and Cuello, C., 1982, *Brain Res.* **246**:168–171.
288. Henry, J. L., Krnjevic, K., and Morris, M. E., 1975, *Can. J. Pharmacol.* **53**:423–432.
289. Krnjevic, K., 1977, *Substance P* (U. S. Von Euler, and B. Pernow, eds.), Raven Press, New York, pp. 217–230.
290. Takahashi, T., and Otsuka, M., 1975, *Brain Res.* **87**:1–11.
291. Chan-Palay, V., and Palay, S. L., 1977, *Proc. Natl. Acad. Sci. U.S.A.* **74**:4050–4054.
292. Otsuka, M., and Konishi, S., 1976, *Cold Spring Harbor Symp. Quant. Biol.* **40**:135–143.
293. Inagaki, S., Senba, E., Shiosaka, S., Takagi, H., Kawai, Y., Takatsuki, K., Sakanaka, M., Takeshi, M., and Tohyama, M., 1981, *J. Comp. Neurol.* **201**:243–254.
294. Haynes, L. W., Smyth, D. G., and Zakariau, S., 1982, *Brain Res.* **232**:115–254.
295. Kim, S. U., 1980, *Neurosci. Lett.* **19**:179–184.
296. Bloom, F. E., Battenburg, E. F., Rossier, J. R., Ling, N., and Guillemin, R., 1978, *Proc. Natl. Acad. Sci. U.S.A.* **75**:1591–1595.
297. Haynes, L. W., 1980, *Neurosci. Lett.* **19**:185–190.
298. Brostoff, S. W., 1973, *Myelin* (P. Morrel, ed.), Plenum Press, New York, pp. 415–446.
299. Saavedra, J. M., 1974, *J. Neurochem.* **22**:211–216.
300. Boulton, A., 1976, *Trace Amines and the Brain* (E. Usdin and M. Sandler, eds.), Marcel Dekker, New York, pp. 21–39.
301. Sandler, M., and Reynolds, G. P., 1976, *Lancet* **1**:70–71.
302. Wyatt, R. J., and Gillis, J. C., Stoff, A. M., Majo, E. A., and Tinklenborg, J. R., 1977, *Neuroregulators and Psychiatric Disorders* (E. Usdin, D. A. Hamburg, and J. D. Barchas, eds.), Oxford University Press, New York, pp. 31–45.
303. Boulton, A., and Juorio, A. V., 1982, *Handbook of Neurochemistry,* Volume 1 (A. Lajtha, ed.), Plenum Press, New York, pp. 189–222.
304. Seiler, N., 1982, *Handbook of Neurochemistry,* Volume 1 (A. Lajtha, ed.), Plenum Press, New York, pp. 223–255.
305. Seiler, N., and Schmidt-Glenwinkel, T., 1975, *J. Neurochem.* **24**:791–795.
306. Kremzner, L. T., 1974, *Polyamines in Normal and Neoplastic Growth* (D. H. Russel, ed.), Raven Press, New York, pp. 27–40.
307. Georgi, P. P., 1978, *Neurosci. Lett.* **10**:335–340.
308. Mugnaini, E., 1978, *Proc. Eur. Soc. Neurochem.* **1**:3–31.
309. Fischer, H. A., and Schmatolla, E., 1972, *Science* **176**:1327–1329.
310. Russel, D. H., and Meier, H., 1975, *J. Neurobiol.* **6**:267–275.
311. Seiler, N., and Deckardt, K., 1976, *Neurochem. Res.* **1**:469–499.
312. Wedgwood, M. A., and Wolstencroft, J. H., 1977, *Neuropharmacology* **16**:445–446.
313. Warkamy, J., and Petering, H. G., 1972, *Teratology* **5**:319–334.
314. Wender, M., and Hierowski, M., 1960, *J. Neurochem* **5**:105–108.
315. Erulkar, S. D., and Fine, A., 1979, *Reviews of Neuroscience,* Volume 4 (D. Schneider, ed.), Raven Press, New York, pp. 179–232.
316. Happel, R. D., Smith, K., Powers, J. M., Banik, N. L., Hogan, E. L., and Balentine, J. D., 1981, *Brain Res.* **211**:476–479.
317. Craelius, W., Jacobs, R. M., and Lee Jones, A. O., 1980, *Proc. Soc. Exp. Biol. Med.* **165**:327–329.
318. Kozma, M., and Szerdaheylyi, P., 1982, *Acta Histochem.* **70**:54–61.
319. Prensky, A., and Hillman, I., 1977, *Brain, Fetal and Infant* (S. R. Bernberg, ed.), Martman Nijhoff, The Hague, pp. 124–136.
320. Levine, S., and Sowiniski, R., 1978, *J. Neuropathol. Exp. Neurol.* **37**:471–478.
321. Donaldson, J., Pierre, T. St., Minnich, J. L., and Barbeau, A., 1973, *Can. J. Biochem.* **51**:87–92.
322. Kozma, M., Ferke, A., and Kasa, P., 1978, *Acta Histochem.* **62**:142–154.
323. Kozma, M., and Ferke, A., 1979, *Acta Histochem.* **65**:219–227.
324. Dickerson, J. W. T., 1968, *Applied Neurochemistry* (A. N. Davison and J. Dobbing, ed.), Blackwell, London, pp. 48–115.
325. Chesler, A., and Himwich, H. E., 1943, *Arch. Biochem.* **2**:175–181.

326. Stewart, M. A., Sherman, W. R., and Kurien, M. M., 1967, *J. Neurochem.* **14:**1057–1066.
327. Smith, M. E., 1966, *Nature* **209:**1031–1032.
328. McIlwain, H., 1966, *Biochemistry of the Central Nervous System,* 3rd ed., J. and A. Churchill, London, pp. 167–196.
329. Geel, S. E., and Dreyfus, P. M., 1975, *J. Neurochem.* **24:**353–360.
330. Reddy, T. S., and Ramakrishnan, C. V., 1983, *J. Neurosci. Res.* **9:**111–114.
331. Blackwood, W., McMenemey, W. H., Meyer, A., Norman, R. M., and Russel, D. S., 1963, *Greenfield's Neuropathology,* Edward Arnold, London.
332. Chick, H., Macrae, T. F., Martin, A. J. P., and Martin, C. J., 1938, *Biochem. J.* **32:**2207–2224.
333. Horton, R. W., Chapman, A. G., and Meldrum, B. S., 1979, *J. Neurochem.* **33:**745–749.
334. Bailey, F. N., Sato, G. H., and Beles, A., 1972, *J. Biol. Chem.* **247:**4270–4276.
335. Dinn, J. J., Weir, D. G., McCann, S., Reed, B., Wilson, P., and Scott, J. M., 1980, *J. Med. Sci.* **149:**1–4.
336. Street, H. R., Cowgill, G. R., and Zimmerman, H. M., 1941, *J. Nutr.* **22:**7–24.
337. Coetzee, W. H. K., 1949, *Biochem. J.* **45:**628–637.
338. Damron, C. M., Monier, M. M., and Roe, J. H., 1952, *J. Biol. Chem.* **195:**599–606.
339. Edstrom, J. E., 1981, *J. Neurochem.* **1:**159–165.
340. Landholt, R., Hess, H. H., and Thalheimer, C., 1966, *J. Neurochem.* **13:**1441–1452.
341. McIlwain, D. L., and Copps-Covey, P., 1976, *J. Neurochem.* **27:**109–112.
342. Schmitz, H., Potter, V. R., Hurlbert, R. B., and White, D. M., 1954, *Cancer Res.* **14:**66–74.
343. Tarr, M., Brada, D., and Samson, F. E., 1962, *Am. J. Physiol.* **203:**690–692.
344. Poduslo, S. E., and Norton, W. T., 1972, *J. Neurochem.* **19:**727–736.
345. Banik, N. L., and Smith, M. E., 1976, *Neurosci. Lett.* **2:**235–238.
346. Deshmukh, D., Flynn, T., and Pieringer, R., 1974, *J. Neurochem.* **22:**479–485.
347. Benjamins, J. A., Guarnieri, M., Miller, K., Sonneborn, M., and McKhann, G. M., 1974, *J. Neurochem.* **23:**751–757.
348. Cohen, S. R., and Bernsohn, J., 1973, *Brain Res.* **60:**521–525.
349. Horrocks, L. A., Toews, A. D., Yashon, D., and Locke, G. E., 1973, *Neurobiology* **3:**256–263.
350. Smith, M. E., 1973, *J. Neurochem.* **21:**357–372.
351. Toews, A. D., Horrocks, L. A., and King, J. S., 1976, *J. Neurochem.* **27:**25–31.
352. Davison, A. N., 1972, *Lipids, Malnutrition and the Developing Brain, A Ciba Foundation Symposium,* Elsevier, New York, pp. 73–90.
353. Pereyra, P. M., and Braun, P. E., 1983, *J. Neurochem.* **41:**957–973.
354. Pereyra, P. M., Braun, P. E., Greenfield, S., and Hogan, E. L., 1983, *J. Neurochem.* **41:**974–988.
355. Benjamins, J. A., and Smith, M. E., 1976, *Myelin* (P. Morell, ed.), Plenum Press, New York, pp. 233–270.
356. Singh, I., and Singh, A., 1983, *Neurosci. Lett.* **39:**77–82.
357. Khan, M., and Griebel, R., 1983, *Can. J. Neurol. Sci.* **10:**161–165.
358. Allen, A. R., 1914, *J. Nerv. Ment. Dis.* **41:**141–147.
359. White, R. J., Albin, M. S., and Harris, L.-S., 1969, *Surg. Forum* **20:**432–434.
360. Ducker, T. B., Kindt, G. W, and Kempe, L. G., 1971, *J. Neurosurg.* **35:**700–708.
361. Wagner, F. C., Jr., Dohrmann, G., and Bucy, P. C., 1971, *J. Neurosurg.* **35:**272–276.
362. Yeo, J. D., Stabback, S., and McKenzie, B., 1977, *Paraplegia* **14:**276–285.
363. Balentine, J. D., 1978, *Lab. Invest.* **39:**254–266.
364. Osterholm, J. L., 1974, *J. Neurosurg.* **40:**5–33.
365. Albin, M. S., Bunegin, L., and Jannetta, P. J., 1974, *Exp. Neurol.* **53:**274–280.
366. Bingham, W. G., Ruffolo, R., and Friedman, S. J., 1975, *J. Neurosurg.* **42:**174–178.
367. Yashon, D., Bingham, W. C., Jr., Freedman, S. S., and Faddoul, E.-M., 1975, *Surg. Neurol.* **4:**43–51.
368. Banik, N. L., Powers, J. M., and Hogan, E. L., 1980, *J. Neuropathol. Exp. Neurol.* **39:**232–244.
369. Wells, M. R., 1978, *Exp. Neurol.* **62:**708–719.
370. Smith, M. E., 1979, *Neurochem. Res.* **4:**689–702.

371. Locke, G. E., Yashon, D., Feldman, R. A., and Hunt, W. E., 1971, *J. Neurosurg.* **34**:614–617.
372. Toews, A. D., King, J. S., Yashon, D., and Horrocks, L. A., 1979, *Neurol. Res.* **1**:271–279.
373. Fu, S. C., Goracci-Mozzi, R., Krakowa, S., and Horrocks, L. A., *Trans. Am. Soc. Neurochem.* **8**:205.
374. Hall, S., 1972, *J. Cell Sci.* **10**:535–546.
375. Yates, A. J., and Thompson, D. K., 1978, *J. Neurochem.* **30**:1649–1651.
376. Schläepfer, W. W., and Freeman, L. A., 1981, *Neuroscience* **5**:2305–2314.
377. Currie, R. W., and White, F. P., 1981, *Science* **214**:72–73.
378. Banik, N. L., Hogan, E. L., Balentine, J. D., and Whetstine, 1981, *Trans. Soc. Neurosci.* **7**:292.
379. Balentine, J. D., and Green, W., 1984, *J. Neuropathol. Exp. Neurol* **43**:500–510.
380. Hodges, G. R., and Watanabe, I., 1980, *J. Neuropathol. Exp. Neurol.* **39**:452–475.
381. Marks, N., Grynbaum, A., and Lajtha, A., 1976, *Neurochem. Res.* **1**:93–111.
382. Smith, M. E., 1976, *J. Neurochem.* **27**:1077–1082.
383. Clendenon, N. R., Allen, N., Gordon, W. A., and Bingham, W. G., 1978, *J. Neurosurg.* **49**:563–568.
384. Krikorian, J., Guth, L., Barrett, C. P., and Donati, E. J., 1982, *Exp. Neurol.* **76**:623–643.
385. Collman, H. R., Wullenweber, R., Spring, C., and Dinsberg, R., 1978, *Adv. Neurol.* **20**:443–450.
386. Stokes, B. T., and Garwood, M., 1982, *Exp. Neurol.* **75**:665–667.
387. Clendenon, N. R., and Allen, N., 1979, *Neural Trauma* (A. Pope, ed.), Raven Press, New York, pp. 115–129.
388. Walker, J. G., Yates, R. R., Yashon, D., and O'Neill, J. J., 1977, *J. Neurochem.* **29**:929–932.
389. Pietronigro, D. D., Hovespian, M., Demopoulos, H. B., and Flamm, E. S., 1983, *J. Neurochem.* **41**:1072–1076.
390. Kakari, S., De Crescito, V., Tomasula, J. J., Flamm, J. B., Campbell, J. B., and Ransohoff, J., 1976, *J. Neuropathol. Exp. Neurol.* **35**:109.
391. Kao, C. C., and Chang, L. W., 1977, *J. Neurosurg.* **46**:197–209.
392. Watkins, D. K., 1970, *Br. J. Radiol.* **43**:152–153.
393. Greenberg, J., McKeever, P. E., and Balentine, J. D., 1978, *Surg. Neurol.* **9**:361–364.
394. Schlaepfer, W. W., 1977, *Nature* **265**:734–736.
395. Sato, S., Quarles, R. H., and Brady, R. O., 1982, *J. Neurochem.* **39**:97–105.
396. Schanne, F. A. X., Kane, A. B., Young, E. E., and Farber, J. L., 1979, *Science* **206**:700–702.
397. Lewin, M. G., Hansebout, R. R., and Pappius, H. M., 1974, *J. Neurosurg.* **40**:65–75.
398. Balentine, J. D., and Spector, M., 1977, *Ann. Neurol.* **2**:520–523.
399. Hsu, C. Y., Hogan, E. L., Gadsden, R. H., Spicer, K. M, and Shi, M. P., 1983, *Excerpta Med.* **22**:181.
400. Young, W., Yen, V., and Blight, A., 1982, *Brain Res.* **253**:105–113.
401. Stokes, B. T., Fox, P., and Hollinden, G., 1983, *Exp. Neurol.* **80**:561–572.
402. Happel, R. D., Banik, N. L., Balentine, J. D., and Hogan, E. L., 1984, *Neurosci. Lett.* **49**:279–283.
403. Osterholm, J. L., and Mathews, G. J., 1972, *J. Neurosurg.* **36**:386–394.
404. Zivin, J. A., Doffman, J. L., Reid, J. L., Tappaz, M. L., Saavedra, J. M., Kopin, I. J., and Jacobowitz, D. M, 1976, *Neurology (Minneap.)* **26**:99–107.
405. Naftchi, N. E., Demeny, M., and Decrescito, V., 1974, *J. Neurosurg.* **40**:52–57.
406. Hedeman, L. S., Shellenberger, M. K., and Gordon, J. H., 1974, *J. Neurosurg.* **40**:37–43.
407. Felten, D. L., Hall, P. H., Campbell, R. L., and Kalsbeck, J. E., 1976, *J. Neural Transm.* **39**:209–221.
408. de la Torre, J. C., Johnson, C. M., Harris, L. H., Kajahara, K., and Mullan, S., 1974, *Surg. Neurol.* **2**:5–11.
409. Bulat, M., Lackovic, Z., Jakupeevic, M., and Damjanov, I., 1974, *Science* **185**:527–528.
410. Assenmacher, R. R., and Ducker, T. D., 1971, *J. Bone Joint Surg.* **53**:671–680.
411. Freeman, L. W., and Wright, T. W., 1953, *Ann. Surg.* **137**:433–443.

412. Wagner, F. C., Jr., Vangelder, J. C., and Dohrmann, G., 1978, *J. Neurosurg.* **48**:92–98.
413. Dohrmann, G. J., Wagner, F. C., Jr., and Bucy, P., 1972, *J. Neurosurg.* **36**:407–417.
414. Bresnahan, J. C., 1978, *J. Neurol. Sci.* **37**:59–82.
415. Balentine, J. D., and Dean, D., Jr., 1982, *Lab. Invest.* **47**:286–295.
416. Cuzner, M. L., Davison, A. N., and Rudge, P., 1978, *Ann. Neurol.* **4**:337–344.
417. Smith, M. E., Sedgewick, L. M., and Tagg, J. S., 1974, *J. Neurochem.* **23**:965–971.
418. Hirsch, H. E., 1981, *J. Histochem. Cytochem.* **29**:425–430.
419. Banik, N. L., 1979, *Neurosci. Lett.* **11**:307–312.
420. Lampert, P. W., 1978, *Am. J. Pathol.* **91**:176–208.
421. Andrews, J. M., 1972, *Multiple Sclerosis: Immunology, Virology and Ultrastructure* (E. Wolfgram, G. W. Ellison, J. G. Stevens, and J. M. Andrews, eds.), Academic Press, New York, pp. 23–52.
422. Baker, P. F., Hodgkin, A. L., and Ridgway, E. B., 1971, *J. Physiol.* (*Lond.*) **218**:709–755.
423. Alvord, E. C., Hruby, S., and Sires, L. R., 1979, *Ann. Neurol.* **6**:474–482.
424. Woelk, H., and Kanig, K., 1974, *J. Neurochem.* **23**:739–744.
425. Trotter, J., and Steinman, L., 1984, *Experimental Allergic Encephalomyelitis: A Good Model for Multiple Sclerosis* (E. C. Alvord, M. W. Kies, and A. Suckling, eds.), Alan R. Liss, New York, pp. 105–110.
426. Banik, N. L., Gohil, K., and Davison, A. N., 1976, *Biochem. J.* **159**:273–277.
427. De La Torre, J. C., 1981, *Spine* **6**:315–335.
428. Kierman, J. A., 1979, *Biol. Rev.* **54**:155–197.
429. Puchala, E., and Windle, W. F., 1977, *Exp. Neurol.* **55**:1–42.
430. Balentine, J. D., Hogan, E. L., Banik, N. L., and Perot, P. L., Jr., 1984, *Recent Advances in Neural Trauma* (H.R. Winn, R. Rimel, and J. A. Jane eds.), Raven Press, New York (in press).
431. Anderson, D. K., Means, E. D., and Waters, T. R., 1979, *Soc. Neurosci.* **5**:719.
432. Ducker, T. B., and Assenmacher, D. R., 1969, *Surg. Forum* **20**:428–430.
433. Nacimiento, A. C., Bartels, M., and Herrmann, H. D., 1979, *Soc. Neurosci.* **5**:727.
434. Eidelberg, E., Stalin, E., Watkins, C. J., and Smith, J. S., 1976, *Surg. Neurol.* **6**:243–246.
435. De La Torre, J. C., Johnson, C. M., Goode, D. J., and Mullan, S., 1975, *Neurology* (*Minneap.*) **25**:508–514.
436. Campbell, J. B., De Crescito, V., and Tomasula, J., 1973, *Surg. Neurol.* **1**:102–106.
437. Faden, A. I., Jacobs, T. P., and Holaday, J. W., 1981, *Science* **211**:493–494.
438. Young, W., Flamm, E. S., Demopoulos, H., Tomasula, J. J., and DeCrescito, V., 1981, *J. Neurosurg.* **55**:209–219.
439. Hill, P. K., De La Torre, J. C., Thompson, S. M., Rosenfield-Wessels, S., and Beckett, M. L., 1981, *Soc. Neurosci.* **7**:537.
440. Higgins, A. C., Pearlstein, R. D., Mullen, J. B., and Nashold, S., 1981, *J. Neurosurg.* **55**:501–510.
441. Kelly, D., Lassiter, K. R. L., and Calogero, J. S., and Alexander, E., 1970, *J. Neurosurg.* **33**:554–563.
442. Yeo, J. D., McKenzie, B., Hindwood, B., and Kidman, A., 1976, *Med. J. Aust.* **1**:538–540.
443. Hallenbeck, J. M., Jacobs, T. P., and Faden, A. I., 1983, *J. Neurosurg.* **58**:749–754.
444. Feringa, E. R., Vahlsing, H. L., Frye, R. A., and Kowalski, T. F., 1978, *Neurology* (*Minneap.*) **4**:338–339.
445. Kalil, K., and Reh, T., 1979, *Science* **205**:1158–1161.
446. Borklund, A., and Stenevi, U., 1979, *Brain Res.* **177**:555–560.
447. Berry, M., 1976, *Neuropathol. App. Neurobiol.* **2**:166.
448. McConnel, P., and Berry, M., 1982, *Bibl. Anat.* **23**:26–37.

15

The Epilepsies

Claude G. Wasterlain, Anne M. Morin, and Barney E. Dwyer

1. INTRODUCTION

1.1. History and Definition

The words *epilepsy*, of Greek origin, and *seizure* were originally used to describe the phenomenon in which a patient was seized by spirits or forces from outside, who took possession of him and temporarily changed his behavior. Epilepsy was recognized as a disease as early as 2080 B.C. in the Code of Hammurabi.[1] Hippocrates (460–357 B.C.) recognized epilepsy as a disease of the brain. Charles Le Pois (1563–1636) was the first to clearly state that all epilepsies are of cerebral origin.[2] The first experimental seizures were observed by Boyle, who placed mice in a jar in which he created a vacuum. The abbé Fontana (1760) and Caldanni (1760) were the first to induce convulsions in frogs by stimulation with frictional electricity. The modern understanding of epilepsy dates back to John Hughlings Jackson, who in 1870 introduced the concept of a discharging epileptic focus and recognized that seizures were hypersynchronous "occasional, excessive, and disorderly discharges of neurons." This view was confirmed by the recording of excessive hypersynchronous electrical discharges from the brain of seizing animals,[3] and later by recording of epileptiform discharges from the human brain.[4,5] The clinical manifestation of this discharge is the paroxysmal appearance of abnormal behavior. It is now well established that many causes can trigger such abnormal discharges, and that the symptomatology of seizures depends on the type and location of the neurons involved.

Epileptic seizures are a symptom of brain dysfunction, and not truly a disease. Five to seven percent of Americans will have a seizure at some time in their life.[6] The word epilepsy is used to define an illness only manifested by

---

Claude G. Wasterlain, Anne M. Morin, and Barney E. Dwyer • Epilepsy and Neurology Research Laboratories, Veterans Administration Medical Center, Sepulveda, California 91343; and Department of Neurology and The Brain Research Institute, University of California School of Medicine, Los Angeles, California 90024.

Table I
Summary of International Classification of Epileptic Seizures

I. Partial (focal, local) seizures
  A. Simple partial seizures (consciousness not impaired)
    1. With motor signs
    2. With sensory symptoms
    3. With autonomic symptoms or signs
    4. With psychic symptoms
  B. Complex partial seizures (temporal lobe or psychomotor seizures; consciousness impaired)
    1. Simple partial onset, followed by impairment of consciousness
      a. With simple partial features (A.1–A.4), followed by impaired consciousness
      b. With automatisms
    2. With impairment of consciousness at onset
      a. With impairment of consciousness only
      b. With automatisms
  C. Partial seizures, evolving to secondarily generalized seizures (tonic-clonic, tonic, or clonic)
    1. Simple partial seizures (A.), evolving to generalized seizures
    2. Complex partial seizures (B.), evolving to generalized seizures
    3. Simple partial seizures, evolving to complex partial seizures, evolving to generalized seizures
II. Generalized seizures (convulsive or nonconvulsive)
  A. Absence (petit mal) seizures
  B. Myoclonic seizures
  C. Clonic seizures
  D. Tonic seizures
  E. Tonic-clonic (grand mal) seizures
  F. Atonic seizures
III. Unclassified epileptic seizures (due to incomplete data)

the intermittent occurrence of epileptic seizures. That illness probably affects 0.5–1% of the American population. The incidence in developing countries with a higher frequency of birth injury and trauma may be even higher.[7] Epilepsy occurs in all races and in all geographic locations. It is more common in males than in females.[8] In view of the multiplicity of causes, mechanisms, and manifestations, most specialists feel that it does not represent a single illness but a group of highly diverse illnesses, the epilepsies.

Epilepsy is not unique to man and is seen in many vertebrate species including all mammalian species tested, birds, amphibians, and perhaps fish as well.

The International League Against Epilepsy has published a clinical and electroencephalographic classification of the human epilepsies (Table I).

1.2. Electroencephalographic Manifestations of Epileptic Seizures

The electroencephalogram (EEG) reflects the electrical activity of the brain. When recorded from the scalp, it is dominated by changes in membrane potential of the dendrites in the superficial cortex. During seizures, the EEG records rapid sequential increases and decreases in voltage manifested on paper

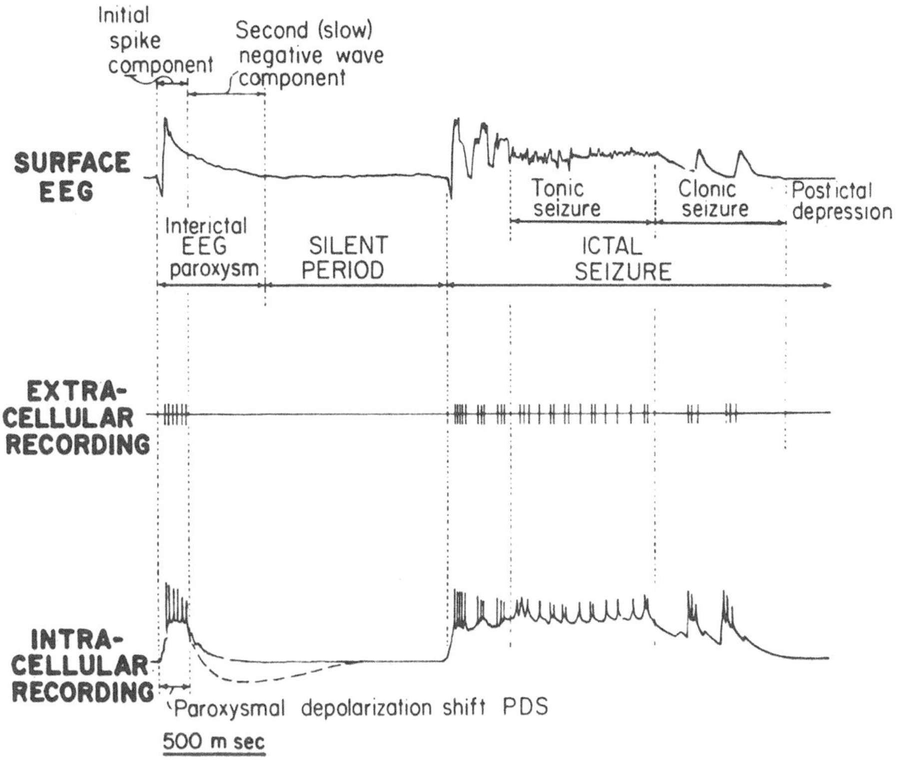

Fig. 1. Cellular events in spike generation. Reproduced from ref. 11 with permission from Elsevier Biomedical Press.

records as sharply angulated waves of high amplitude called spikes. When a spike is recorded on a scalp EEG, extracellular recordings in the underlying cortex reveal repetitive paroxysmal firing of neuronal populations. At the same time, intracellular records (Fig. 1) reveal a large depolarization shift of the resting membrane potential of neurons, and when this shift has depolarized the membrane potential sufficiently (e.g., 20 μV), the neurons tends to fire repeatedly. This paroxysmal depolarization shift (PDS)[9–13] is the hallmark of epileptiform activity and has been found in many seizure types, both in experimental animals and in man. Membrane events triggered by penicillin *in vivo* are very similar to those recorded from epileptic foci *in vitro* and have been studied extensively in the hippocampal slice preparation. These depolarization shifts have many characteristics in common with excitatory postsynaptic potentials (EPSPs), and may in fact be giant EPSPs. In these neurons, exemplified by CA3 in the hippocampus, large inward sodium and calcium currents activated by synaptic depolarization may serve to amplify and sustain membrane activity.[14–16] It has been shown that release from GABA-mediated presynaptic inhibition can generate PDSs and probably does so in the penicillin model.

These mechanisms underlie the generation of single spikes but not of seizures. During seizures, bursts of spikes are generated so that increasing syn-

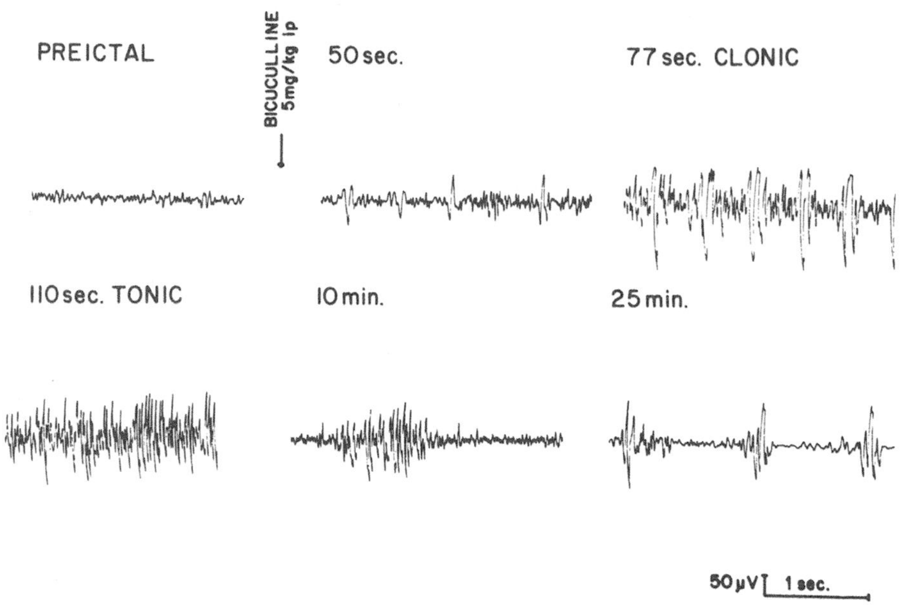

Fig. 2. EEG recorded from the scalp in a newborn marmoset monkey. Seizures were induced by the GABA antagonist bicuculline. High-voltage spikes synchronous with clinical seizure activity are seen.

chronicity and intensity of discharge lead to paroxysmal activity. Intracellular recording shows that the membrane potential of key neurons does not quite come back to baseline between PDSs and depolarization increases progressively. During those events, extracellular calcium falls and extracellular potassium rises, and both may participate in the large and progressive shift in membrane potential. The development of prolonged burst discharges in synapses may release massive amounts of neurotransmitter and may facilitate the development of an avalanche of excitation that characterizes the ictal episode. Electrical coupling seems to play a less important role than chemical coupling in this process.

1.3. Epileptic Phenomena in Man

1.3.1. Clinical Manifestations

The international classification of epileptic seizures is based on clinical, behavioral, and EEG criteria. The vast majority of epileptic seizures are focal in origin. The EEG manifestations of a focal seizure reflect a focal, paroxysmal, synchronous discharge of neuronal populations (Fig. 2). The clinical expression of this synchronous firing of neurons depends on the location and nature of the neurons involved in it. For example, if the focus is located in the part of the motor cortex that controls the arm, the first manifestation of the seizure will be clonic activity of the arm. If the focus is located in a part of the emotional brain where vigorous neuronal firing results in the sensation of fear, or of "déjà vu," these feelings will be the first manifestation of the seizure. Since these focal manifestations precede the spread of seizure activity, they are frequently perceived by the patient as announcing the seizure, but in fact they are the behavioral manifestations of the focal seizure. They are often called "aura," the latin word for "wind," since a patient seen by Galen's teacher, who described the first aura ever recorded, had a feeling of a wind blowing over his body before each seizure.

The manifestations of focal seizures are usually unilateral. However, in some locations, seizure discharges spread rapidly to both sides of the brain. They can go on for quite some time while involving only parts of the brain, most commonly both temporal lobes. During those partial bilateral seizures, awareness is often partially or completely lost, and later amnesia for the event is the rule. These seizures frequently have complex behavioral manifestation, and are often called psychomotor or partial complex seizures. For example, the patient may be unaware of his or her surroundings and unable to communicate, but still able to walk around, and carry out semipurposeful movements. Occasionally, the patient may even be able to drive a car or to perform rather complex actions while in that state.

If the seizure spreads to all parts of the brain, i.e., becomes generalized, the usual manifestation is a major motor convulsion characterized initially by contraction of all the muscles in the body with tonic rigidity, arrest of respiration, and severe muscle contractions to the point of fracturing long bones. Usually, after many seconds in that state, progressively longer bursts of inhibition develop in which the fully contracted muscles relax, resulting in alternating contraction and relaxation, called clonic seizure activity. Finally, inhibitory processes take over and the seizure is terminated. After major motor seizures, the patient is usually in a state of complete relaxation, unconscious, with a very low cerebral metabolic rate, and a slow EEG reflecting the hyperpolarization and lack of activity of many neurons.

1.3.2. Causes of Epileptic Seizures

Table I only gives a very partial idea of the variety of clinical seizure manifestations, which reflects the extraordinary complexity of cerebral anatomy. The diversity of etiologies (Table II) is equally formidable. As a result, the epilepsies form not one illness but a group of many separate illnesses, each with their own prognosis, natural history, and clinical manifestations. It is highly unlikely that a single biochemical mechanism could explain the pathogenesis of such a diverse group of illnesses. Different seizure types probably involve widely divergent pathophysiological mechanisms. The biochemical mechanisms by which seizures are generated have been identified in a few instances (Table III). Most of these result from transient metabolic disturbances. They show a diversity of etiology as great as that of chronic epilepsies. Any disorder that compromises the stability of the membrane potential of neurons seems to be able to produce seizures (Table III).

Table II
Causes of Seizures in Man

Inborn errors of metabolism: Many aminoacidurias, disorders of lipid metabolism, etc.
Developmental defects: Include Sturge–Weber, tuberous sclerosis, toxic effects of drugs, radiation, fetal infection such as toxoplasmosis, rubella, cytomegalovirus, malformation, etc.
Cerebrovascular disease: Anoxia-ischemia, cerebral hemorrhage, embolism, infarction, A-V malformation, vasculitis, disseminated intravascular coagulation
Trauma: Resulting from birth injury, closed and open head injuries
Infection: Viral, bacterial, and fungal meningitis or encephalitis, brain abscess, slow virus infections, etc.
Tumors, foreign bodies: Primary and metastatic brain tumors, cysticercosis, hydatid cysts
Nutritional disorders: Pyridoxine deficiency and dependency
Metabolic derangement: Hepatic, uremic, hypoxic, hypoglycemic, hyperosmolar encephalopathies; dialysis encephalopathy; acid–base imbalance; hyponatremia, hypocalcemia, and electrolyte imbalance
Degenerative diseases: Alzheimer's, Pick's, Huntington's diseases (occasionally)
Drug toxicity: Alcohol and depressant drug withdrawal; intoxication with many drugs such as penicillin, major tranquillizers, lead, mercury; nerve gas and organophosphate insecticides

1.3.3. Febrile Seizures

Febrile convulsions are by far the most common type of seizures. They are triggered by a sudden elevation in temperature, usually due to a common cold or other intercurrent infection. They are most common between the ages of 6 months and 3 years, and cease by the age of 6. There is a genetic predisposition to febrile convulsions. Clinically, they are generalized seizures of short duration. Similar seizures can be induced in animals by subjecting neonates to

Table III
Some Biochemical Mechanisms of Seizures in Man

Type	Cause	Mechanism	Treatment
Electrolyte balance	Hyponatremia	Fall in [Na^+] reduces membrane potential and triggers neuronal firing	Sodium
	Hypocalcemia	Unknown	Calcium
Oxidative metabolism	Anoxia, cyanide, fluoroacetate, fluorocitrate, etc.; hypoglycemia	Fall in ATP decreases the capacity of the Na^+,K^+-ATPase to maintain electrolyte gradients across membrane	Oxygen, glucose, antagonists
Neurotransmitter metabolism	Nerve gas, organophosphate insecticide poisoning	Cholinesterase inhibition increases excitatory muscarinic output	Atropine
	Pyridoxine deficiency	Lack of coenzyme for glutamate decarboxylase reduces GABA synthesis	Pyridoxine

a rapid rise in temperature. The vast majority of children who suffer febrile convulsions do not develop epilepsy or any other type of brain disease.

1.3.4. Reflex Epilepsy

In a small percentage of patients, epilepsy can be triggered by specific stimuli, such as flashing lights,[17] watching television,[18] reading,[19] or by more complex stimuli such as listening to a certain voice,[20] listening to a specific musical theme or tone, doing arithmetic,[21] or playing chess. Many of those forms of epilepsy triggered by complex stimuli are not hereditary.

1.3.5. Status Epilepticus

Occasionally, generalized seizures follow each other in such rapid succession that the patient remains comatose between seizures. This is a major medical emergency that frequently results in brain damage or death.[22]

1.3.6. Seizure Termination

The mechanisms involved in seizure termination are not clear. It is, however, certain that the seizure does not stop because of exhaustion of neuronal energy reserves. It is an active inhibitory process that terminates a seizure, and it is associated with hyperpolarization of neurons, perhaps associated with elevated extracellular potassium and intracellular sodium in the face of a persisting increased membrane conductance. With return of the resting membrane potential toward normal polarization, there is a slow recovery of normal EPSP and IPSP activity, and a progressive disappearance of the postictal electrical silence.[23]

1.4. Inheritance of Epilepsy

It is well documented that several rare types of epilepsy follow Mendelian inheritance. Much more commonly, in about a third of epileptics, no such hereditary factors are present, but a family history of epilepsy increases the chances of developing epilepsy following various kinds of cerebral insult, e.g., trauma or infection. Of course, since many types of cerebral pathology can cause seizures, many inherited brain diseases have seizures as one of their manifestations. Of course they differ from idiopathic epilepsy by the presence of other signs of brain disease.

1.5. Pathological Substrate of Human Epilepsy

In 1880, Sommer described a loss of Ammon's horn, or the CA1 sector of the hippocampus, in epileptics, and interpreted this finding as a causal factor in their disease.[24] However, the same change can result from the anoxic-hypermetabolic insult that occurs frequently during status epilepticus.[25–27] A similar change of neuronal loss and sclerosis in the area now called the Sommer sector has been found in nearly every patient subjected to temporal lobectomy for intractable epilepsy, even if those patients have not experienced status epilepticus. However, the possibility remains that single seizures could cause similar changes[28] (Fig. 3). Few data are available on patients with milder epilepsy who do not come to surgery. The question of whether these pathological changes are the cause or the consequences of epilepsy has remained extremely controversial. Evidence of misdirected regeneration has been produced in surgical samples of the hippocampus, from epileptic patients, and it has been suggested that such misdirected regeneration could transform inhibitory pathways into excitatory pathways and therefore generate seizures in the course of normal brain activity.[29] Golgi preparations of surgical samples have disclosed extensive and apparently progressive changes in dendritic trees with loss of dendritic spines, dendritic nodulation, swelling, and deformities, which might suggest that the pathological lesions are different from those induced by ischemia in the same distribution and may be of a progressive nature over a period of years.[30] After 100 years of controversy, the nature and significance of pathological changes in the brains of epileptics remains an area of active investigation.

1.6. Mirror Foci

It is a common observation that when an active epileptic focus is present on the cortical surface, the contralateral homotopic cortex tends to show epileptiform activity resembling that of the primary focus. Occasionally, if the primary focus is removed, the homotopic cortex continues to discharge in epileptiform fashion and may cause seizures. This phenomenon has been termed by Morrell the mirror focus. Since those two areas project to each other, and the end feet of the axons of the contralateral homotopic cortex are located in the primary lesion, it has been difficult to prove unequivocally that truly independent mirror foci occur and represent a plastic phenomenon in man. However, in view of the large amount of indirect evidence accumulated[31] and of the demonstration in the chemical kindling model in animals that synaptic activation is sufficient, under certain conditions, to generate a chronic epileptic focus, it is likely that the original interpretation of the mirror focus is correct.

1.7. Posttraumatic Epilepsy

Cerebral trauma is often followed by the development of chronic epilepsy. This is true of brain injury at birth due to difficult delivery as well as of trauma to the adult brain resulting from car accidents, missile injuries, etc. Some types of trauma, e.g., open head injuries by missiles in wartime, are followed by epilepsy in more than 50% of victims. What makes posttraumatic epilepsy particularly interesting for the investigator is the fact that there is frequently a free interval between trauma and development of seizures. In man for example, it is common to see the first seizures appear 6 months to 2 years after the trauma.

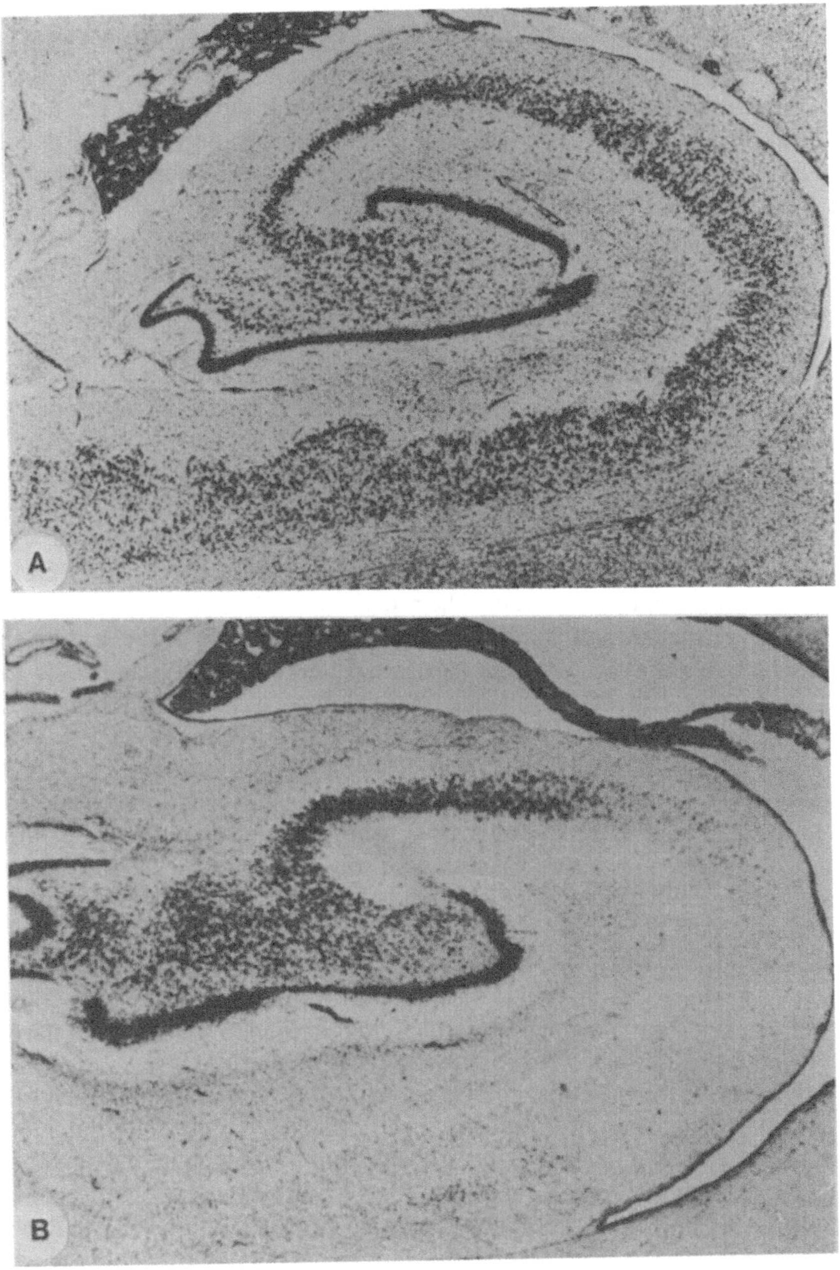

Fig. 3. Sclerosis of Sommer sector. Compared to a normal hippocampus (A), the hippocampus of an epileptic patient (B) shows marked loss of pyramidal cells in CAl. Reproduced from ref. 25 with permission from Raven Press.

1.8. Endocrine Aspects

Many endocrine factors modify epilepsy, which in turn can alter hormonal secretion patterns. Seizures induce increases in serum prolactin,[32,33] cortisol,[34] catecholamines,[35–37] insulin, and other hormones.[38]

The effect of hormones on epilepsy is too complex to review here. Cortisol, testosterone, and estrogens in physiological concentrations tend to decrease seizure threshold and increase the likelihood of seizures, whereas progesterone has the opposite effect.[39]

2. INTERMEDIATE METABOLISM

Epileptic seizures place the largest metabolic demands known upon brain tissue and severely challenge its ability to maintain homeostatic mechanisms. It is clear that brain energy metabolism in chronically epileptic animals is normal interictally, and that the disturbances that take place are the consequences, not the cause of seizures. However, seizures with attendant complications such as anoxemia, acidosis, partial breakdown of the blood–brain barrier,[40,41] and major changes in cerebral blood flow, represent one of the most severe metabolic stresses faced by the brain. The repetitive, high-frequency, synchronous firing of large numbers of neurons, which is characteristic of the seizure state, results in very large ionic fluxes across the neuronal membrane. It is likely that the energy expended to pump ions back across the membrane and to reestablish the resting membrane potential accounts for the bulk of the energy expended during seizures. In sympathetic ganglia, glycolytic rates have indeed been shown to increase linearly with the frequency of neuronal discharge. When a sympathetic ganglion *in vitro* is driven by periodic depolarization through electrical current, the increase in metabolic rate is proportional to the frequency of electrical stimulation as long as each stimulation results in neuronal firing. When the frequency of electrical stimulation becomes so high that the frequency of tissue depolarization reaches a plateau, the metabolic rate also reaches a plateau. If ionic fluxes are blocked by agents such as tetrodotoxin that block specific ionic channels, stimulation no longer results in increases in metabolic rate. It is likely that similar rules apply in the CNS.

It is likely that all other energy expenditures combined represent a relatively small proportion of the energy used during seizures. These include the pumping of extracellular ions such as potassium by glial cells, the active reuptake of neurotransmitters released into the synaptic cleft, the synthesis of new neurotransmitters, the generation of second messengers, and the phosphorylation of proteins and lipids through which the message is ultimately expressed. However, while all those reactions utilize energy, the total amount of energy expended in the metabolism of known neurotransmitters, second messengers, and message, as far as can be judged from the maximal velocities of the enzymes involved, is likely to represent a very small proportion of the energy expended during seizures. At the same time that catabolic activities are greatly enhanced, anabolic processes are profoundly inhibited[42–46] presumably in order to save energy essential for cell survival. A variety of adaptive mechanisms increase the supply of glucose and oxygen to the brain. These processes are very efficient if seizures are of short duration, and account for the fact that in human epilepsy, single seizures do not usually result in progressive brain damage and, between seizures, the brain can function normally for a lifetime. However, homeostatic mechanisms appear unable to keep up with metabolic demands for very long, and prolonged seizures are known to damage neurons. The following pages will review the complex interactions and adaptations that take place in intermediate metabolism of the brain during epileptic seizures, and will analyze how failure of homeostatic mechanisms can result in brain damage.

2.1. Cerebral Blood Flow and the Cerebral Supply of Glucose and Oxygen

A review of the large literature concerned with the physiological responses of the cerebral circulation to seizures is beyond the scope of this chapter. However, the relationship between blood supply and energy demand is so important that a brief review of the major principles involved will serve as a useful introduction to a more detailed discussion of metabolism.

2.1.1. The Brain Depends on a Continuous Blood Supply

The brain shares a very high metabolic rate for both oxygen and glucose with very low reserves of both metabolic substrates and a nearly complete lack of alternatives. In man, loss of consciousness may take place within 10 sec of complete circulatory arrest,[47] whereas in mice the EEG becomes isoelectric within 10 sec of decapitation, indicating that cerebral oxygen reserves can sustain the normal cerebral metabolic rate for only a few seconds.[48] Carbohydrate reserves, while larger, would last only minutes, illustrating the total dependence of brain function on a continuous blood supply. Moreover, lack of oxygen and glucose, in addition to stopping the machine, destroys the machinery, as evidenced by the brain damage frequently seen in man following only minutes of cardiac arrest.

The average blood flow to the human brain is approximately 55 ml/100 g tissue per min. There are considerable local variations in blood flow, gray matter averaging flows 4–5 times higher than white matter[49,50] (Table IV).

2.1.2. Coupling between Cerebral Blood Flow and Cerebral Metabolism

In a healthy brain, the local variations in metabolic rate are closely coupled to the local variations in cerebral blood flow. Physiological changes in local metabolic rates are invariably accompanied by corresponding increases or decreases in local blood flow. This phenomenon, which has been termed the "flow–metabolism couple,"[51] is preserved in a variety of pathological circumstances, including focal seizures which greatly increase both metabolic rate and local blood flow. The basis for this coupling is unknown, but many in-

Table IV
Cerebral Blood Flow and Metabolic Rate in Normal Young Adult Man[a]

	Unit	Rate
Cerebral blood flow	ml/100 g/min	57
Cerebral O_2 consumption	ml/100 g/min	3.5
	μmol/100 g/min	156
Cerebral glucose utilization	mg/100 g/min	5.5
	μmol/100 g/min	31
O_2/glucose ratio	mol/mol	5.5
Glucose equivalent of O_2 consumption	μmol Glc/100 g/min	26
CO_2 production	μmol/100 g/min	156
Cerebral respiratory quotient		0.97

[a] Modified from Ref. 50.

vestigators have suspected that it results from local release of metabolites such as CO_2, a potent dilator of cerebral blood vessels, or H^+.

2.1.3. *Uncoupling of Local Blood Flow from Local Cerebral Metabolism*

Recent studies suggest that stimulation of the cerebellar fastigial nucleus markedly increases local cerebral blood flow in the cerebral cortex, without changing cortical glucose utilization,[52] whereas stimulation of other brain stem areas such as the dorsal medullary reticular formation induces increases in both cortical blood flow and cortical metabolism.[53] Since the cerebral vasculature is known to receive a rich innervation, not from the autonomic nervous system, but from central neurons located within the brain substance,[54,55] these results suggest the existence of intrinsic brain systems by which the brain controls its own flow independently of metabolism.

In recent studies on neonatal seizures in the monkey, we have obtained evidence of apparent uncoupling of local cerebral blood flow from deoxyglucose metabolism in inferior coliculi, cortex, and hippocampus. These results, however, will require confirmation due to the uncertainty of the 2-deoxyglucose method.[56] Uncoupling between flow and metabolism in the hippocampus has also been observed by a similar method during chronic kindled seizures in the rat, and it has been speculated that such uncoupling might be responsible for the development of hippocampal sclerosis in human epilepsy.[28]

2.1.4. *Autoregulation of Cerebral Blood Flow*

Within very wide limits, cerebral blood flow is independent of systemic blood pressure. A fall in blood pressure results in automatic vasodilation, and a rise in blood pressure in vasoconstriction in the cerebral territories so that cerebral blood flow remains constant. The phenomenon of cerebral autoregulation is immediately abolished by seizures, so that cerebral blood flow during epileptic seizures becomes dependent on blood pressure.[51] It has been shown that during single seizures, which are accompanied by a large increase in arterial blood pressure, the abolition of autoregulation results in a large increase in cerebral blood flow that protects the brain by delivering additional glucose and oxygen to cerebral tissue.[57] However, during sustained seizures or status epilepticus, we and others have shown that systemic arterial blood pressure frequently falls for a variety of reasons.[45,58,59] During convulsive status epilepticus, large amounts of lactic acid are produced from muscle and smaller amounts from brain. Severe lactic acidosis develops both in animal models and in man. This is associated with progressive arterial hypotension, which can be partially corrected by correcting the acidosis. When this happens, cerebral blood flow remains dependent on arterial blood pressure and falls with it.[60] In other words, the loss of autoregulation, which protected the brain by increasing its blood flow during a single seizure, can become a liability during multiple or sustained seizures with their attendant hypotension.

2.1.5. *Seizures Increase Cerebral Blood Flow*

Focal seizures increase cerebral blood flow in and around the focus, and generalized seizures uniformly increase total cerebral blood flow.[61,62] The mechanism of this increase is only partially understood. It may reveal the liberation of vasodilating metabolites such as CO_2 in the seizure focus. At least part of the response in generalized seizures is well understood: seizures abolish autoregulation of cerebral blood flow. This appears to protect the brain, since seizures also induce marked arterial hypertension, and blood flow, which is now pressure dependent, increases markedly, improving delivery of glucose and oxygen to the brain.[63,51] Spinal cord transection, which prevents the peripheral release of catecholamines by seizures, and abolishes the hypertensive response to seizures, also abolishes the accompanying increase in cerebral blood flow.

2.2. *Glycolysis*

Glycolytic rates increase severalfold during seizures.[64–67] This elevation has been observed with both deoxyglucose and glucose itself as tracers, and therefore cannot be a methodological artifact. However, neither method has been adequately validated for use during seizure states, and the precise extent of the increase in glycolytic flux varies with the method used. In spite of these methodological problems, there is no doubt that the increase of glycolytic flux during seizures is large, and its existence is also supported by abundant indirect evidence. In convulsing animals, brain glucose and glycogen fall during seizures, while lactate rises rapidly (Table V). Paralysis and ventilation with oxygen prevent the convulsions and the reduction of glucose and glycogen by single seizures, but do not prevent the elevation of lactate.[48] The extent of depletion of glucose, glycogen, ATP, and phosphocreatine varies with the intensity and the duration of the seizures. The existence of a fall in glucose, while blood glucose is within the normal range or higher, suggests that glycolytic fluxes can be more rapid than glucose transport from blood to brain. This mechanism may become quite important in neonates, in which the transport

Table V
Brain Glycolytic Metabolism during Status Epilepticus[a]

Seizure model	Time	Area	Species	[Glucose]	[Glycogen]	[Lactate]	[Pyruvate]	L/P ratio	Reference
Bicuculline	30 sec	ctx	rat	−17%	NS	+186%	+39%	+103%	78
	60 min			−50%	−46%	+474%	+84%	+210%	
	120 min			−68%	NS	+618%	+84%	+286%	
Metrazol	60 min			−73%	—[b]	+166%	+3%	+159%	88
	120 min			−82%	—[b]	+119%	−36%	+241%	
	60 min		cat	NS	—[b]	+521%	+40%	+317%	
	120 min			−40%	—[b]	+470%	+125%	+242%	
Bicuculline	20 min		rat	−40%	−82%	+298%	+79%	+122%	70
	120 min			−73%	−71%	+183%	+57%	+70%	
	20 min	hpe		−49%	−85%	+384%	+71%	+190%	
	120 min			−73%	−46%	+150%	+35%	+94%	
	20 min	cbl		NS	−82%	+348%	+106%	+120%	
	120 min			−44%	−25%	+55%	+26%	+25%	

[a] All animals were paralyzed and ventilated with 100% O_2; all were adult.
[b] Not measured.

Table VI
Glucose Transport and Glucose Utilization during Neonatal Seizures

Blood (glucose)	27 mg/100 ml	90 mg/100 ml	200 mg/100 ml
Max. rate of glycolysis	0.69	0.69	0.69
Max. rate of glycolysis from glucose transported across BBB	0.144	0.336	0.48
Max. rate of reserve mobilization	0.37	0.37	0.37
Energy balance			
Calculated	−0.176	+0.016	+0.160
Measured	negative	−0.028	positive

[a] Data are expressed in nmol ~P/kg per min; they compare the amounts of high-energy phosphate bonds generated or consumed in a 4-day-old rat during anoxic status epilepticus. Since glycogen reserves in the immature brain total approximately 24 nmol ~P/kg, an animal could maintain relatively good energy balance for up to an hour during status epilepticus if it could maintain a blood glucose concentration of at least 200 mg/100 ml, and mobilize all its glycogen. However, if its blood level was 27 mg/100 ml, it would rapidly deplete its brain glucose and be in negative energy balance. After depletion of glycogen reserves (seizures lasting over 1 hr), energy balance would be inadequate even at the higher blood glucose levels, and could only be maintained by restoring aerobic metabolism.

capacity across the blood–brain barrier is considerably smaller than that in adults.[59,68,69] and in status epilepticus, where massive demands for glucose over long periods of time may outstrip transport capacity and result in brain glucose depletion.[70] Table V shows the large decreases in brain glucose and glycogen concentrations that occur during status epilepticus, even in animals protected against anoxemia and convulsive activity by paralysis and ventilation with oxygen. The severe decreases in glucose concentrations recorded reflect the fact that in normoglycemic animals glucose transport cannot match the rate of glucose utilization, so that if seizures are prolonged brain glucose and glycogen reserves fall. At the same time, the moderate rise in pyruvate concentration and the greater rise in lactate, with increase in lactate/pyruvate ratio, are expected in the presence of very high glycolytic rates and of a shift in redox potential. These changes are seen in several species and in several brain regions. Table VI, which compares the maximal rates of glycolysis, of glucose transport across the blood–brain barrier, and of glycogen mobilization in the neonate, confirms that a similar situation applies to the newborn rat. Maximal rates of glycolysis are much higher than the sum of reserve mobilization and glucose transport across the barrier. As a result, balance can only be maintained if blood glucose is high, if glycogen reserves are available, or if the efficiency of energy-generating systems is increased by the use of aerobic metabolism through the Krebs cycle.

2.2.1. Regulation of Hexokinase Activity

The form of hexokinase that is bound in mitochondria is more active than that that is free in the cytoplasm.[71] During the times when the cell needs a lot of energy, and as a result shows a reduced ratio of intracellular ATP over ADP, the bound state is favored.[72] The bound form of the enzyme in calf brain has a lower K_m for ATP[73,74] and is less sensitive to product inhibition by G6P.[75]

It is possible that inorganic phosphate partially releases the enzyme from inhibition by G6P. Seizures reduce ATP and G6P and increase inorganic phosphate, and as a result they activate hexokinase and increase glycolytic flux when hexokinase is rate-limiting. However, under most normal circumstances hexokinase is not rate-limiting. The normal glucose concentration in brain is 20 times the K_m of hexokinase, and stronger regulatory control is exerted at the phosphofructokinase (PFK) level.

2.2.2. Regulation of PFK

Under normal circumstances, the transport of glucose from blood to brain is not rate-limiting for glucose utilization. Regulation of glycolytic flux is exerted primarily at the level of PFK, while hexokinase and possibly pyruvate kinase play a lesser role.[76] Seizures reduce several glycolytic intermediates, including G6P and F6P, suggesting activation of PFK, even when hypoxemia is prevented by paralysis and oxygenation (Fig. 4). After several minutes, brain levels of G6P, F6P, and FDP return to normal, suggesting that hexokinase and pyruvate kinase also get activated. Seizures increase 5′-AMP, cAMP, inorganic phosphate, FDP, and ammonia, which stimulate the enzyme, and reduce levels of ATP, which releases the enzyme from inhibition.[77]

2.2.3. Regulation of Pyruvate Kinase

Pyruvate kinase catalyzes the formation of pyruvate from phosphoenolpyruvate and generates ATP in the process. It may be a glycolytic control point, although its role has not been defined as well as that of PFK. It is usually assumed that the flux of triosephosphates through these enzymes is increased during seizures since overall glycolytic flux is increased.[78] The enzyme may be activated by FDP, which releases inhibition by ATP[79] although this would not be the case after prolonged seizures when levels of FDP approach control values.

The pyruvate form can then diffuse into the mitochondrial compartment where it is converted to acetyl-CoA, which enters the citric acid cycle, or to lactate.

2.3. The Krebs Cycle

When oxygen is available, the Krebs cycle is the main source of energy of the cell. It provides a means of oxidizing acetyl-CoA derived from the glycolytic pathway, from fatty acids, or from amino acids to CO_2 and water through NAD^+-linked dehydrogenase enzymes in the mitochondria (Fig. 5). In the process, metabolism of 1 molecule of glucose, which only produced 2 ATPs through glycolysis, generates 36 ATPs through the Krebs cycle and electron transport chain. During seizures in rats that are paralyzed and ventilated with oxygen, most Krebs cycle intermediates increase in concentration, with the exception of α-ketoglutarate and oxaloacetate, which decrease slightly immediately after seizure onset.[78,80] This increase in Krebs cycle intermediates

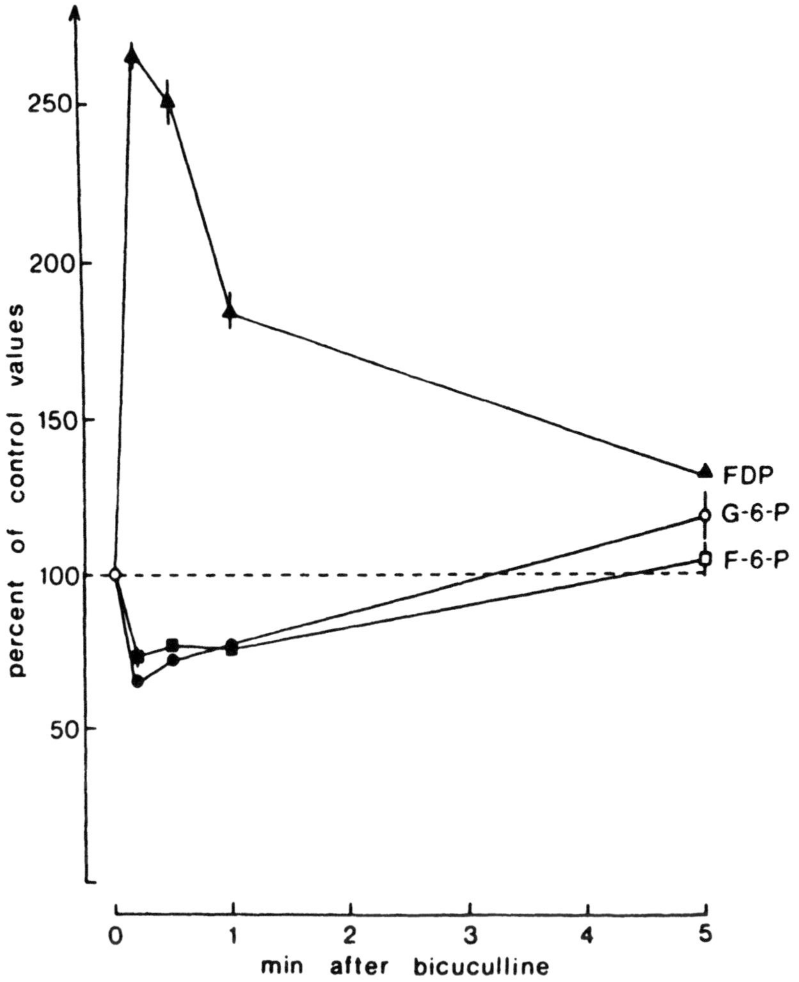

Fig. 4. Changes in glycolytic intermediates during seizures. Reproduced from ref. 78 with permission from Raven Press.

suggests increased flux through the cycle, but an exact rate of increase has not been measured.

2.3.1. *Cerebral Energy Use Rates*

Seizures are highly energy-intensive. ATP utilization increases severalfold during seizures.[81–84] A variety of seizures have similar effects on brain energy reserves and energy use rate (Table VII). Even single convulsive seizures invariably reduce the concentration of brain high-energy phosphates. This is true after ECS,[82–85] audiogenic seizures,[84] and homocysteine seizures, but not after seizures induced by methionine sulfoximine.[86] A large proportion of this depletion of high-energy phosphates could be prevented by paralyzing animals to prevent convulsion activity and ventilating them with pure oxygen (Table

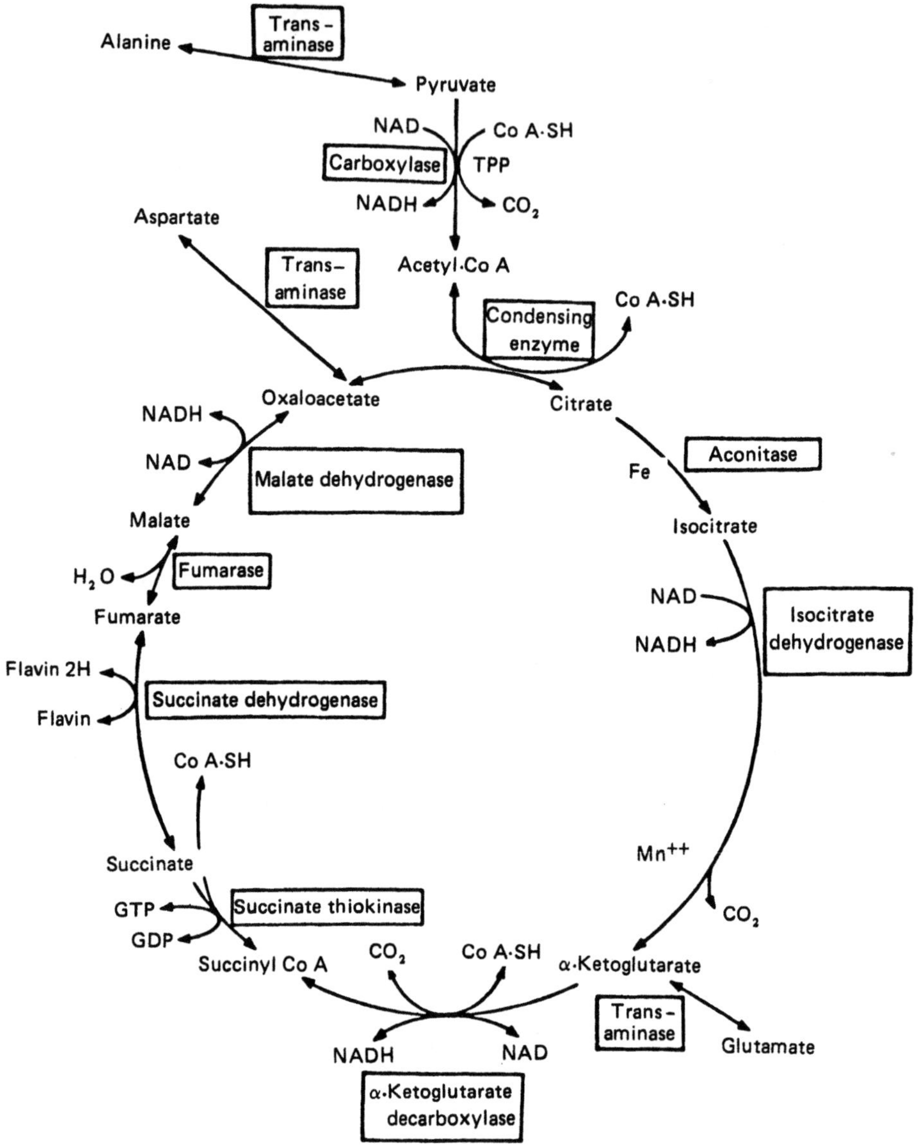

Fig. 5. The Krebs cycle. Involvement of several amino acids. Reproduced with permission from Little, Brown and Co.

VIII). Even prolonged seizures on a respirator do not produce profound depletion of energy reserves (Table IX). The energy charge potential of the cell, defined by Atkinson,[87] is a measure of the availability of high-energy phosphate bonds for performing energy-intensive work in the cell. This energy charge potential of the cell is reduced mildly but significantly after ECS, pentylenetetrazole, flurothyl, or bicuculline seizures.[48,70,78,88] ATP reserves in the right cerebral cortex fall within the first minute of seizures[78] and then remain at a plateau close to control values until seizures become self-sustaining in the ab-

Table VII
Brain Energy Reserves during Convulsive Seizures

Seizure model	Time (sec)	Area	Species	Age	[ATP]	[Phosphocreatine]	Reference
Flurothyl	30	ctx	mouse	10 days	−15%	−50%	80
	60				−20%	−53%	
ECS	3			adult	−29%	−76%	82
	20				−86%	−94%	83
	4				−41%	−74%	
	10				−57%	−78%	
DL-Methionine					NS	NS	86
Sulfoximine		cbl			NS	NS	
		subctx			NS	NS	
ECS	5	ctx			−22%	−70%	84
	17				−56%	−88%	
	5	thal			−30%	−59%	
	17				−61%	−82%	
	5	cbl			−25%	−62%	
	17				−49%	−92%	
Audiogenic	3.5	ctx			NS	NS	
	3.5	thal			−18%	−45%	
	3.5	cbl			NS	−20%	
	20				−9%	−49%	
DL-Homocysteine		ctx			−86%	−40%	338

sence of convulsion stimuli, while a further fall in ATP concentration takes place. In other words, brain energy failure is greatly minimized or prevented by oxygenation and prevention of convulsive activity (Tables VII and VIII). It is clear from the comparison of Tables VII and VIII that convulsive seizures reduce both ATP and phosphocreatine severely across a number of seizure models. However, in several species the same single seizures fail to reduce ATP concentrations significantly in paralyzed, O_2-ventilated animals. Even

Table VIII
Brain Energy Reserves during Seizures on the Respirator

Seizure model	Time (sec)	Area	Species	[ATP]	[Phosphocreatine]	Reference
ECS		ctx	mouse	NS	NS	83
Flurothyl	10			−15%	−49%	48
	10	cbl		−22%	−59%	
	10	BS		NS	NS	
ECS	10	ctx		NS	NS	
Metrazol	20			NS	NS	
	45			−4%	−19%	
Flurothyl	10		cat	−12%	−14%	88
	30			−8%	−24%	
ECS	10		rat	−5%	−16%	48
Metrazol	60			−10%	−30%	

Table IX
Brain Energy Reserves during Status Epilepticus on the Respirator

Seizure model	Time	Area	Species	[ATP]	[Phosphocreatine]	Reference
5 ECS	10 sec	ctx	mouse	−5%	NS	48
10	10 sec			−3%	NS	
25	10 sec			−3%	−36%	
ECS	10 sec	cerebellum: molecular layer		−59%	87	85
	10 sec	Purkinje		−52%	83	
	10 sec	granular		−58%	83	
	10 sec	white matter		−33%	NS	
5 flurothyl	10 sec	ctx		−9%	−34%	48
10	10 sec			−16%	−57%	
20	10 sec			−35%	−59%	
5	10 sec	cbl		−10%	−33%	
10	10 sec			−10%	−33%	
20	10 sec			−10%	NS	
Metrazol	5 min	ctx		−11%	−41%	
Bicuculline	30 sec		rat	−9%	−50%	78
	5 min			−9%	−35%	
	120 min			NS	−45%	
Metrazol	30 min			−12%	−33%	88
	60 min			−6%	−41%	
	120 min			−13%	−37%	
	60 min		cat	−9%	−52%	
	120 min			−9%	−54%	
	20 min		rat	−7%	−29%	70
	120 min			−11%	−42%	
	60 min	hpc		−10%	−24%	
	120 min			−11%	−23%	
	60 min	cbl		−5%	−15%	
	120 min			NS	NS	

paralyzed and ventilated animals in status epilepticus fare better than those subjected to single seizures and left to convulse freely (Table IX). Substantial decreases in concentrations of phosphocreatine are observed, but the decreases in ATP concentrations are small to moderate. They are seen in several species and seizure models. In other words, it appears that in the adult brain in well-oxygenated animals, ATP production by Krebs cycle and glycolytic metabolism can prevent severe energy depletion for a considerable period of time.

2.3.2. Amino Acid Metabolism

As shown in Fig. 5, intracellular pools of some nonessential amino acids and Krebs cycle intermediates are linked enzymatically so that the Krebs cycle is also involved in regulating amino acid concentrations and ammonia metabolism in brain. Falls in brain glutamate[80,89,90] have been reported during convulsive seizures, a fate shared by α-ketoglutarate.[78,91] It is possible that ictal energy demands increase flux through the Krebs cycle, reducing levels of α-ketoglutarate, which is then replaced by conversion of glutamate through amino acid transaminases. The equilibrium constants for such reactions are close to 1, so that they are freely reversible. The relatively large brain content of glutamate[78] of over 12 mM, the relatively low concentration of α-ketoglutarate (0.16 mM), and the reduction of α-ketoglutarate in the first minute after seizures are compatible with this interpretation. A similar argument can be made for aspartic acid. Oxaloacetate falls immediately after the onset of seizure activity.[78,80]

2.3.3. Ammonia Metabolism

Even in animals that are paralyzed and oxygenated, brain ammonia increases early in the course of bicuculline seizures, and reaches a plateau at approximately 5 min.[78] After 5–20 min of seizures, brain glutamate and aspartate start falling while glutamine and alanine increase. Glutamate is a substrate for glutamine synthetase, which catalyzes the ATP-dependent formation of glutamine, which in turn is used for GABA production. Glutamate can be replenished through the Krebs cycle, by transamination of aspartate with α-ketoglutarate. Formation of asparagine to detoxify ammonia may also occur. The increase in alanine probably reflects increased transamination of pyruvate (Fig. 5).

It has been suggested[92] that when inadequate amounts of glucose, pyruvate, and acetyl-CoA are available to the Krebs cycle, glutamate stores may be used as a substrate to keep the Krebs cycle going. Glutamate would fall as a result of transamination to α-ketoglutarate, and aspartate would rise as a result of transamination of oxaloacetate. This would provide a shunt mechanism by which glutamate could be introduced into the Krebs cycle at the level of α-ketoglutarate and exit at the level of oxaloacetate. From this shunt, 1 molecule of glutamate would generate 2 molecules of NADH, capable of producing 6 molecules of ATP, and 1 molecule of GTP would be produced during the hydrolysis of succinyl-CoA to succinate. This mechanism might provide an additional source of ATP when blood and brain glucose levels can only marginally support ATP production through the Krebs cycle. The evidence for this attractive mechanism is entirely indirect, and consists mainly of the observation that in starved rats, which have lower energy reserves and lower blood glucose and glucose transport into brain, seizures result in a much slower ammonia fixation than in fed animals. With sustained seizures, ammonia levels continue to rise, glutamate falls, and aspartate increases.[92] Alanine formation is reduced, probably because reduced glucose transport and reduced glycolytic rates provided less pyruvate for its generation. All of these results are compatible with the proposed shunt, and might be yet another example of a way by which the brain cannibalizes its own transmitters and proteins to sustain its energy metabolism.

2.4. Electron Transport Chains

Little attention has been paid to this aspect of oxidative metabolism in the epileptic brain. During single seizures on a respirator, the partial pressure of oxygen in venous blood coming from the brain increases, in spite of the fact that the lactate/pyruvate ratio increases and that the cortical redox potential goes down. This might suggest that plenty of oxygen is available, but that the brain is unable to utilize it, so that phosphocreatine and brain energy reserves fall during the seizure. The presumed site of the block would be in the electron transport chains, since there is abundant evidence of a massive increase in both glycolytic rates and Krebs cycle metabolism. However, several lines of evidence suggest that the proposed block is either of minimal importance or the result of an erroneous interpretation of the data. It is likely that the fall in phosphocreatine is the result of a pH change and does not truly reflect depletion of energy reserves. The change in lactate/pyruvate ratio might be of a similar nature, and probably does not reflect tissue hypoxia. During prolonged status epilepticus, a transition phase occurs when energy reserves begin to be compromised. However, the work of Kreisman *et al.*[93] clearly shows that at the same time, the partial pressure of oxygen within the cortex falls, the reactivity of blood volume to seizures disappears, and cytochrome A3 fluorescence shifts toward a reduced state, in contrast to single or earlier seizures on a respirator, which were accompanied by a shift of cytochrome A3 in the direction of a more oxidized state. This suggests that brain damage in status epilepticus is associated with reduced oxygen availability to neurons and not with a reduced ability to utilize available oxygen.

2.5. Lipid Metabolism

A variety of changes occur in the lipid constituents of membrane bilayers during convulsive seizures.[94] Their role is not yet fully understood, but important clues have begun to emerge. Excitable membranes in the CNS contain a wide variety of phospholipids enriched in arachidonoyl acyl chains, including phosphatidylcholine, phosphatidylethanolamine, and a phosphatidylinositol. The relative proportions of these phospholipids differ in the various layers of the neuronal membrane. The proportions of acidic (e.g., phosphatidylinositol, phosphatidylserine) and zwitterionic (e.g., phosphatidylcholine) phospholipids vary, even within different areas of a given membrane. Specific phospholipids may form an annulus around specific proteins, such as receptors, giving rise to marked regional heterogeneity of membrane composition. In addition, the polar head of phosphatidylethanolamine may undergo methylation and membrane translocation, and these modifications may result in changes in membrane fluidity and thus the function of receptors and other membrane-bound proteins. In the case of muscarinic cholinergic receptors, a change in receptor conformation through ligand binding may in turn result in marked modifications of phospholipid turnover.

The hydrophobic moiety of phospholipids is made of fatty acids varying in number of unsaturated bonds and in chain length. They contain a high pro-

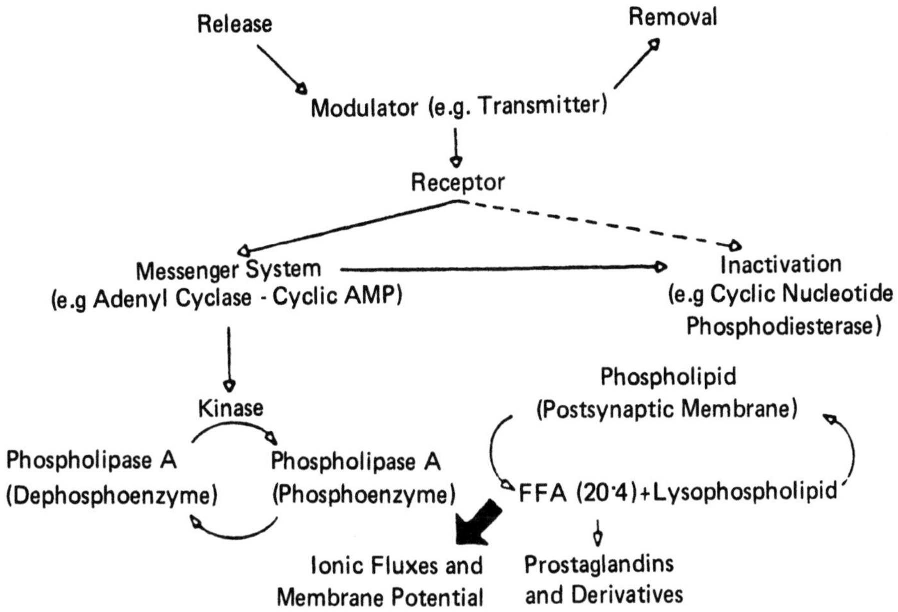

Fig. 6. Hypothesis on the involvement of lipid acyl chains in the permeability of synaptic membranes. Activation systems of arachidonic acid release in the CNS after electroconvulsive shock. Reproduced with permission of N. Bazan.

portion of long-chain highly unsaturated fatty acids such as arachidonic acid (20:4 omega 6) and docosahexaenoic acid (26:6 ω 3). The distribution of these fatty acids in the membrane results in microdomains of different fluidity and properties, and may result in differences in interactions with ions such as calcium and other metabolites. It has been shown[94–96] that marked changes in brain free fatty acids and particularly in arachidonic acid occur during seizures. Free arachidonic acid increases 10-fold in the forebrain during bicuculline seizures in rats; stearic, oleic, and palmitic acids increase to a lesser extent. Changes in the cerebellum are smaller than in the forebrain. Docosahexaenoic acid also increases but more so in the cerebellum than in the cerebrum. These results illustrate the striking diversity in the regional response of membrane phospholipid metabolism within different parts of the nervous system. Unsaturated fatty acids are released at faster rates than saturated fatty acids, but at the same time degradation of triglycerides is not enhanced by seizures. Bazan has proposed that seizures activate phospholipase A by a phosphorylation mechanism, and that this activation in turn has a variety of effects on the CNS (Fig. 6).

Arachidonic acid release, in part, provides precursors from stored membrane glycerolipids, for the synthesis of cyclooxygenase and lipooxygenase reaction products with potent biological effects: prostaglandins, thromboxane, leukotrienes, and a variety of peroxidation products that could generate free radicals and could have major effects on the integrity of the neuronal membrane. It is also likely to provide major changes in microdomains in membrane fluidity.

Table X
Effect of Single Seizures on the Incorporation of Radioactive Amino Acids into Brain Protein, in Vivo

Species	ECS[a]	Precursor amino acid	Protein synthesis, % of control (T[b])	Reference
Mouse	12 mA, 0.2 sec	L-[1-^{14}C]-Leucine, i.v.	76%	43[c]
	40 V a.c., 0.1 sec	L-[4,5-^{3}H]-Leucine, i.p.	20% (immediately after ECS)	146[d]
	40 V, 0.1 sec	L-[4,5-^{3}H]-Leucine, i.p.	No difference (20 min)	
	17 mA, 0.1 sec	L-[U-^{14}C]-Leucine, i.p.	50% (immediately after ECS)	339[e]
	17 mA, 0.1 sec	L-[U-^{14}C]-Leucine, i.p.	No difference (20 min)	
	12.5–15 mA, 0.2 sec	L-[4,5-^{3}H]-Lysine, i.p.	47% (immediately after ECS)	340[f]
	12.5–15 mA, 0.2 sec	L-[1-^{14}C]-Leucine, i.p.	35% (immediately after ECS)	

[a] Current was applied through transcorneal electrodes.
[b] T indicates the time at which isotope was injected following administration of ECS.
[c] Male Swiss–Webster mice were injected with 4 μCi of [1-^{14}C]-leucine via the tail vein. Two minutes after ECS, animals were killed.
[d] Male Swiss white mice were injected i.p. with 30 μCi of [^{3}H]-leucine at various times after ECS or sham treatment. Animals were killed 5 min after isotope injection.
[e] Conditions for footnote *d* except that 5 μCi of [^{14}C]-leucine was injected.
[f] Male albino mice were injected with 100 μCi of [^{3}H]-lysine or 5 μCi of [^{14}C]-leucine i.p. and killed 5 min later. The relative radioactivity of brain proteins from ECS-treated and control mice measured in brain homogenate is given.

A postictal decrease in accumulated free fatty acids as a function of time after seizures is also observed, suggesting that the free fatty acid activating and acylating enzymes remove the released free fatty acids, and that seizures activate lipases and inhibit reacylation reactions, terminating the initial increases in free fatty acid levels. The rise of free fatty acids in the brain following seizures is minimized by pretreatment with α-methyl-*p*-tyrosine, suggesting that neurotransmitters are involved in the effect of seizures on membrane lipids.

The lipid effects of epileptic seizures are not due to an anoxic-ischemic component during the convulsive activity, since they have been shown to occur in paralyzed, oxygen-ventilated animals. Since polyunsaturated free fatty acids that are released are highly susceptible to peroxidation, there has been speculation that this mechanism may be important in generating membrane damage during status epilepticus.

A completely separate but synchronous effect is an increase in the rate of production of diacylglycerols, possibly from phosphatidylinositol. Similar efects are observed in cerebral ischemia. The simultaneous production of diacylglycerols and of free fatty acids such as arachidonic acid may be due to the activation of phospholipase C, followed by the activation of diglyceride lipase. The fact that these lipid effects of seizures are minimized in neonates, which are more resistant to seizures and to anoxia, supports the speculation that they are involved in the mechanism of brain damage. Whatever the precise molecular mechanism, the role of these lipid changes in membrane function and brain damage from seizures appears important.

2.6. Cerebral Energy Reserves

In seizures as in ischemia, sustained depletion of cerebral energy reserves is closely associated with brain damage or death. Yet we have no idea of the mechanism by which depletion of energy reserves might not only stop the machine but also wreck the machinery. In fact, some lines of evidence would even suggest that the association, albeit close, does not imply a causal relationship. Recent reviews discuss these issues in depth.[97]

2.6.1. ATP and Phosphocreatine

ATP turnover may increase from three- to fivefold during seizures.[81–84] Convulsive seizures induced by a variety of convulsants rapidly reduce brain ATP and phosphocreatine concentration (Table VII). This decrease is nearly immediate after ECS,[82–85] audiogenic,[84] and homocysteine seizures, but not after methionine sulfoximine.[86] The reduction in ATP concentration and most of the fall in phosphocreatine are prevented by paralyzing the animals and ventilating them with oxygen during the seizures (Table VIII).

2.6.2. Energy Charge Potential

The energy charge potential of the brain[87] is a measure of the availability of high-energy phosphate bonds for performing energy-intensive activities in the cell. It is defined as:

$$\frac{[\mathrm{ATP}] + \frac{1}{2}[\mathrm{ADP}]}{[\mathrm{ATP}] + [\mathrm{ADP}] + [\mathrm{AMP}]}$$

The energy charge potential is severely reduced by convulsive seizures. A mild but significant reduction occurs during the first seizure, even in animals that are paralyzed and ventilated with oxygen, but subsequent seizures result in a plateau that is maintained until late in status epilepticus, when seizures become self-sustaining in the absence of convulsive stimuli.

While convulsive seizures rapidly and grossly compromise the brain's energy balance, the situation is entirely different on the respirator, where anoxemia and competition for the available oxygen between brain, heart, and muscle are prevented or minimized. In the absence of those noncerebral complications, the brain's energy reserves are only minimally compromised, even by multiple seizures. Even prolonged status epilepticus, when appropriate precautions are observed to ensure rapid inactivation of metabolism, results in rather mild depletion of ATP concentrations (Table IX). Phosphocreatine is affected more severely, but as far as we know its only role is that of a reserve for ATP, and its depletion alone has no adverse effects on the brain. The inevitable conclusion is that under appropriate circumstances the adult brain can maintain its ATP production by oxidative phosphorylation to a level high enough to prevent severe energy depletion even during status epilepticus. This is obviously a precarious balance, which may eventually fail in prolonged status and during convulsions.

2.7. *Redox Potential of the Cerebral Cortex*

Seizures increase brain lactate and brain pyruvate, and the ratio of lactate to pyruvate is elevated despite adequate oxygenation (Table V). This occurs in the face of increased oxygen tension in cerebral venous blood, suggesting a nonhypoxic lactic acidosis in brain. Together with the lactate/pyruvate ratio, the cytoplasmic redox potential $NADH/NAD^+$ shifts toward a more reduced state. Howse and Duffy[80] suggested that the redox potential is regulated by the energy charge potential in brain as it is in liver.[98,99] However, they could not exclude the possibility that this increase in $NADH/NAD^+$ could reflect impaired transport of NADH reducing equivalents into mitochondria through the malate–aspartate shuttle (Fig. 7).

A shift toward a more reduced cell cytoplasmic during seizures would presumably ensure that NADH was available to support maximal rates of oxidative phosphorylation through the mitochondrial electron transport chain. It is obvious that this change in redox potential would shift the equilibrium of the lactate dehydrogenase reaction in favor of lactate production:

$$\text{Pyruvate} + \text{NADH} + \text{H}^+ \rightarrow \text{lactate} + \text{NAD}^+$$

This reaction consumes H^+ and decreases the intracellular acidosis produced by anaerobic glycolysis. However, the overall stoichiometry of the formation of lactate from glucose still produces considerable intracellular lactic acidosis, since 1 molecule of glucose produces 2 of lactate and 2 H^+. High levels of intracellular lactic acid are thought to cause neuronal damage[100,101] when concentrations reach the 20–30 mM range. This is never reached in the newborn during status epilepticus, but may occasionally be reached in the adult and might then play an accessory role in the mechanism of cell damage from status. This cerebral intracellular lactic acidosis is, of course, totally separate from the severe arterial lactic acidosis observed in freely convulsing animals including man. Circulating lactic acid is released predominantly from muscle during convulsions, and may be a major factor in the circulatory complications of status epilepticus.

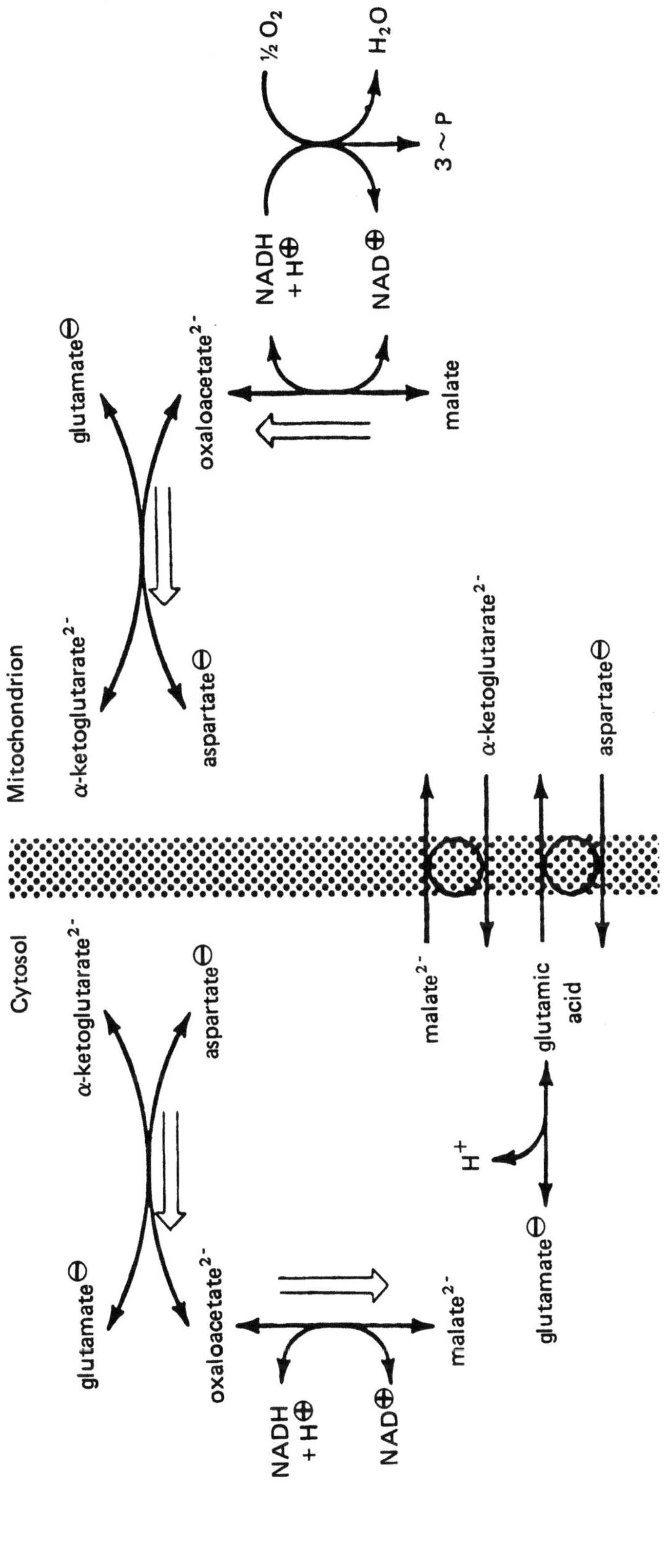

Fig. 7. Redox state in cytosol and mitochondria and the malate/aspartate shuttle. Reproduced from ref. 79 with permission from WB Saunders and Co.

2.8. Mechanisms of Cell Damage in the Adult

Brain damage is a common outcome of status epilepticus in man[25,26,102,103] and in experimental animals.[104–106] It is clear that the presence of systemic complications such as anoxemia vastly increases mortality and brain damage from seizures. It would seem logical to assume that cerebral anoxia is a key component of neuronal damage, especially since the histological appearance and distribution of cell damage induced by status epilepticus have a close resemblance to those induced by ischemia. Paradoxically, the precise role of cerebral anoxia and the molecular mechanism by which it might damage neurons are completely unknown. A similar situation, of course, applies in ischemic brain damage. Subcellular components isolated from brain-damaged animals, including mitochondria, appear to function well *in vitro* after a period of recovery, and most biochemical evidence of damage is reversible even when neurons are irreversibly injured.

2.8.1. Mismatch between Supply and Demand

The discovery of an association between loss of pyramidal cells in the hippocampus and epilepsy, over 100 years ago,[24,107] led German neuropathologists to postulate that a relative insufficiency of blood supply was responsible for brain damage during seizures.[27,102,108] This hypothesis is supported by the many similarities between neuronal damage induced by ischemia and that induced by status epilepticus,[25,38,109] although some reports emphasize that differences do exist between the two types of neuronal damage.[106,110] With a single exception dealing with a very particular type of seizure,[28] all recent studies in the adult[111] report that cerebral blood flow increases during seizures. However, this increase may not be as large as the increase in metabolism, and it may not be as sustained.[38,59,109] As a result, during prolonged seizures even in paralyzed, ventilated animals, a mismatch between metabolic demands and blood supply may develop. Several reports[77,112,155] describe the extent of failure of circulation late in status, in paralyzed, O_2-ventilated animals. The occurrence of a relative failure of oxygenation in the hippocampus late in the course of status epilepticus[93] also supports the hypothesis that substrate depletion is responsible for neuronal damage during status. In newborn marmoset monkeys cerebral blood flow was studied by the method of Lear *et al.*,[113] metabolism by the method of Sokoloff *et al.*,[114] and protein synthesis by that of Dwyer *et al.*[46] Seizures were induced by bicuculline (10 mg/kg i.p.) under anesthesia. As shown in Fig. 8, in some brain regions 2-deoxyglucose accumulation increased much more than blood flow, while in others no such mismatch developed. If one accepts the rate of protein synthesis as indicative of the intensity of the metabolic insult to a particular brain region, there was a significant association between existence of mismatch between flow and metabolism and inhibition of protein synthesis in the same region. While all methods comparing metabolism measured by the 2-deoxyglucose method technique to blood flow are subject to major pitfalls due first to the lack of a well-validated, precise lumped constant, and second to the difference in timing of the experiment between the

Table XI
The Effect of Repetitive Seizures on the Incorporation of Amino Acids into Brain Protein, in Vivo

Species	Convulsant	Precursor	Protein synthesis inhibition during seizures: Brain region	% of control	p	Reference
Rat[a] (80–120 g)	electroshock	[^{3}H]leucine	forebrain (0–5 min)[b]	72	< 0.001	44
			forebrain (25–30 min)	61	< 0.001	
			forebrain (55–60 min)	51	< 0.001	
Rat[c] 4 days old	bicuculline (8 mg/kg, i.p.)	[^{3}H]lysine	forebrain (1 hr)	42	< 0.001	160b
			hindbrain (1 hr)	45	< 0.001	
12 days old	bicuculline (8 mg/kg, i.p.)	[^{3}H]lysine	forebrain (1 hr)	25	< 0.001	
			hindbrain (1 hr)	34	< 0.001	
Marmoset monkey 5–7 days old[d]	bicuculline (5 mg/kg, i.m.)	[^{14}C]valine	frontal pole (30 min–1 hr)	48	< 0.01	56

[a] Rats were paralyzed by injection of succinylcholine, tracheotomized, and ventilated with 100% oxygen. ECS was administered every 30 sec for the duration of the experiment. [^{3}H]-Leucine was injected intravenously and animals were killed 5 min later. See Ref. 44 for further details.

[b] Time interval after onset of seizures during which animals were pulsed with radioactive amino acids.

[c] Seizures were induced in 4- and 12-day-old rats with bicuculline (8 mg/kg). Protein synthesis was estimated using the method of Dunlop *et al.*[345] Rats were injected with [^{3}H]lysine (10 μmol/g; 0.1 μCi/μmol) and killed 1 hr later. Forebrain was defined as the part of the brain situated below the olfactory lobes and above the inferior colliculi, and hindbrain as the part between the intercollicular cut and the foramen magnum.

[d] Marmoset monkeys were injected with [1-^{14}C]valine (0.1 μCi/μmol; 10 μmol/g, i.p.) followed immediately with bicuculline (5 mg/kg, i.m.). Seizures began within 2–3 min and animals were killed between 30 and 60 min later. Right frontal pole, defined as the forebrain region to the right of the midline and extending back to the optic chiasm, was dissected out and used to measure rates of protein synthesis.

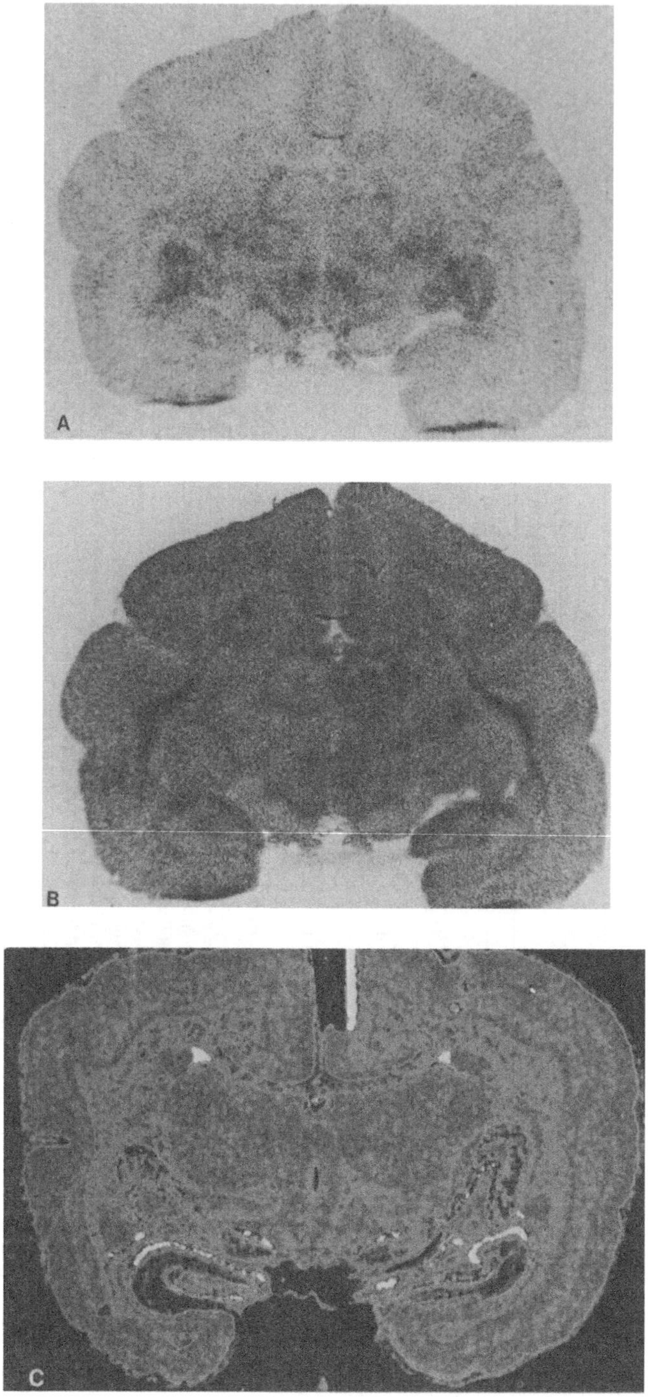

Fig. 8. Imaging of cerebral blood flow with [^{131}I]iodoamphetamine (A, D), glycolytic metabolism with [^{14}C]-2-deoxyglucose (B, E), and protein synthesis with [1-^{14}C]valine (C, F) in newborn marmoset monkeys. A–C are from control animals, D–F from animals receiving bicuculline seizures (5 mg/kg IP) for 30 min. Note that in the cortex and hippocampus, seizures induce a mild increase

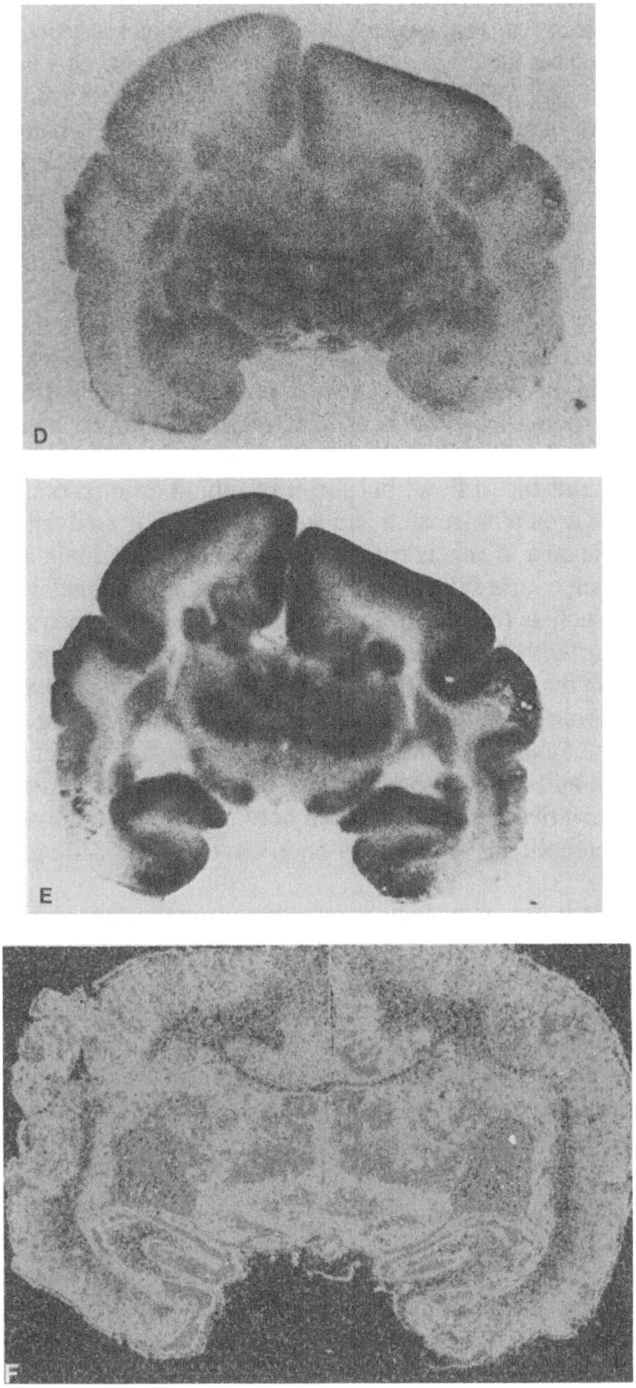

in iodoamphetamine uptake and a severe increase in deoxyglucose uptake, suggesting a mismatch between flow and metabolism. In those same areas, protein synthesis is profoundly inhibited. By contrast, in the lateral geniculate, during seizures, flow is high and metabolism is low, and protein synthesis is maintained throughout the seizure.

1 min necessary to measure flow and the 45 min necessary to measure metabolism, the existence of this negative correlation suggests a possible etiology for mismatch and for the brain damage associated with repetitive seizures. This does not explain why ATP and energy reserves decreased only minimally during status epilepticus in rats on O_2 at a time when brain damage is known to occur.[78] Presumably, where tissue metabolism and subsequently energy-demanding processes outmatch blood flow, local changes in energy reserves result in inhibition of protein synthesis, which can be taken as a rough measure of the intensity of the cerebral insult in those regions. In some areas, large increases in 2-deoxyglucose metabolism are unmatched by any increase whatsoever in blood flow. This is illustrated by the inferior colliculi, where the outer rim of the colliculus, a region rich in synaptic connections and poor in cell bodies, shows a striking mismatch between flow and metabolism. An element of uncertainty is introduced by the lack of a validated lumped constant for this seizure model, and by the difference in timing between the studies of metabolism and those of blood flow; but a rim of inhibition of protein synthesis is observed in the area of mismatch and suggests that it is an event of real physiological significance. If this is true, it may indicate a total failure of the blood flow–metabolism couple in certain brain regions during neonatal seizures. Vulnerable areas such as the cortex and hippocampus, where the increase in flow did not keep up with the increase in deoxyglucose metabolism, showed profound inhibition of protein synthesis, while areas where no mismatch between flow and metabolism occurred during seizures, such as the lateral geniculate nuclei, optic tract and cranial nerve nuclei, showed a fully intact rate of protein synthesis in the midst of seizure activity. This suggests that, during bicuculline seizures in young marmosets, mismatch between flow and metabolism results in a severe metabolic insult and may be involved in the pathogenesis of local brain damage.

Systemic, extracerebral factors may play an important role in brain damage, particularly during repetitive seizures. Anoxemia during convulsions may make oxygen reserves even more precarious.[115] Since autoregulation of cerebral blood flow is abolished, any fall in systemic blood pressure because of acidosis, desensitization of peripheral catecholamine receptors, or for other reasons[59] will result in a fall in cerebral blood flow proportional to the fall in blood pressure. As a result, the cerebral supply of O_2 or glucose may become inadequate to meet increased needs.

2.8.2. *Role of Sustained Cell Activity*

In recent years, evidence has accumulated that, in key neuronal populations, long-lasting seizures can result in neuronal necrosis, even when the systemic effects of seizure activity are prevented and adequate cerebral blood flow and oxygenation are maintained.[106,115,116] It is possible that a relative mismatch develops between moderately increased blood flow and vastly increased metabolic needs; it is also possible that the process of sustaining maximal firing rates of some neurons is so energy-intensive that blood supply and delivery of oxygen and glucose cannot follow maximal metabolic needs. This was already suggested by the work of Meldrum *et al.*[115] and by the apparently transynaptic effects of kainic acid.[117,118]

2.8.3. *Lactic Acidosis*

In ischemia, accumulation of lactic acid within cell bodies can be an important factor in generating brain damage.[100] Blood glucose elevation increases mortality and morbidity from cerebral ischemia in rats[119] and histological recovery is poor in cat brain after a period of ischemia if blood glucose was elevated by glucose pretreatment.[119a] Levels of lactic acid in brain associated with ischemic cell change were well in excess of 20 mM.[120–122] Similar concentrations can be seen in convulsing animals, particularly if there is an ischemic component to the seizures, but are almost never found in status epilepticus in neonates. The probable explanation is that, in the neonate, lactate is easily transported from blood to brain by the lactate carrier, which is abundant in the blood–brain barrier of suckling animals. As long as cerebral perfusion and oxygenation are maintained, and transport across the blood–brain barrier is effective, such high levels of lactic acid are not likely to accumulate.

2.8.4. *Calcium Cytotoxicity*

Calcium may have a role in heart and liver cytotoxicity.[123–125] Both Meldrum[38] and Siesjo[97] recently proposed mechanisms by which abnormally large calcium entry into vulnerable cells during seizures could lead to selective cell deaths. Sustained entry of calcium into neurons can occur as a result of burst firing in neurons associated with a PDS in membrane potential during epileptic seizures.[126] When the ability of the cell to maintain low intracellular calcium levels is exceeded, accumulation of calcium may result in swollen and possibly damaged mitochondria (although irreversible damage to mitochondria has never been demonstrated) and may result in activation of intracellular lipases and proteases, leading to ischemic cell change and neuronal death. It is not clear whether or how energy failure intervenes in this process as a result of mitochondrial calcium accumulation and concomitant swelling. Siesjo[97,110] postulated that increased cellular calcium resulting from a repetitive seizure discharge could overwhelm cell mechanisms that buffer calcium in the cytoplasm or extrude it into the extracellular space. These processes might divert respiratory energy that would otherwise be utilized for ATP production. Increased intracellular calcium might activate hydrolases, including phospholipase A2, resulting in large increases in free fatty acids including arachidonic acid, while release potassium from neurons might be taken up by glia. This and the free fatty acid accumulation may contribute to the swelling of astrocytic processes.[106] Since potassium uptake into astrocytes is oxygen-dependent, this and astrocytic swelling may further impede delivery of energy substrates to neurons. The accumulation of arachidonic acid in the presence of oxygen may result in the formation of prostaglandins and leukotrienes (via cyclooxygenase and lipooxygenase, respectively), possibly contributing to microcirculatory problems and to formation of cell-damaging free radicals.

Meldrum[127] made the point that Purkinje cells and many other vulnerable cells have a high calcium conductance, and that accumulation of large amounts of calcium into mitochondria has been observed during status epilepticus.[38,128]

Our recent studies of the calcium and calmodulin-dependent synaptic enzyme calmodulin kinase 2 suggest that this enzyme is extremely sensitive to ischemic conditions, which result in enzyme activation within minutes. This is a result of inhibition of kinase activity and of the inhibition of the ability of the kinase to bind calmodulin, while phosphatase activity is unchanged. Preliminary results suggest that similar changes occur in synaptic membranes during status epilepticus (Bronstein and Wasterlain, in preparation), and might suggest a critical role for damage to nerve terminals from status epilepticus. The exact significance of those molecular changes remains to be established.

2.9. Mechanisms of Neuronal Damage in the Neonate

2.9.1. Cerebral Energy Use Rates Vary with Age

Immature rats have a metabolic rate in brain regions that is 1/10th to 1/20th of that of adults, and the latency to last gasp after decapitation is at least 20 times longer in the neonate than in the adult. We have already reviewed (Fig. 8) the evidence in favor of presumed mismatch between metabolic needs and blood supply of essential nutrients during neonatal seizures. Epileptic seizures can elevate cerebral metabolic rate nearly fourfold in the neonate's brain.[129] This relative increase is similar to that found in the adult brain[64,78,130,131] even if the absolute metabolic rates are relatively low. However, forebrain blood flow may not be able to match those increases in neonates.[111]

A protective factor in the newborn is the presence of large amounts of lactate carrier in the blood—brain barrier,[132,133] so that lactate produced in brain escapes easily into blood as long as circulation is maintained (Fig. 9). As a result, brain lactate concentrations never reach a toxic level during neonatal seizures in the experimental models that we use.[68,129]

Several factors, however, make neonatal seizures a period of great risk from metabolically induced damage to the immature brain. These factors are the possible mismatch during seizures between flow and metabolism; the need for protein synthesis and DNA synthesis, two energy-intensive processes that are easily compromised by metabolic stress; the slow mobilization of glycogen reserves in the immature brain; and the limited transport capacity of the blood–brain barrier of the neonate.

2.9.2. Mobilization of Energy Reserves

The neonate possesses fairly sizable reserves of glycogen.[134] As a result, the brain can satisfy its energy demands by drawing to some extent upon its glycogen reserves, but the ability to mobilize those in the neonate is less efficient than in the adult, due to lower levels of the enzymes phosphorylase and phosphoglucomutase[134,135] and to a lesser accumulation of cAMP, needed to

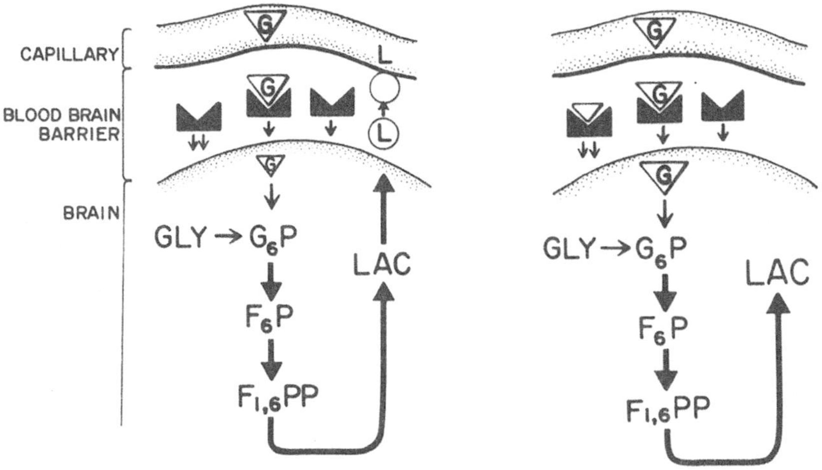

Fig. 9. Transport of glucose and lactate across the blood–brain barrier as a function of age. In neonates (left), when glycolytic rates are high (e.g., in seizures and ischemia), transport of glucose cannot keep up with utilization, resulting in decreased brain glucose concentration. Lactate, on the other hand, easily escapes the brain of the neonate but not of the adult (right).

activate a protein kinase that transforms phosphorylase B (inactive) to phosphorylase A (active) (Fig. 10).

2.9.3. Limited Transport Capacity of the Blood–Brain Barrier

The maximal rate of glucose transport across the immature blood–brain barrier has been estimated to be one-fifth of the rate of the adult, probably as the result of lower capillary density in the neonate.[136] In spite of normal or elevated blood glucose levels, brain glucose falls rapidly after onset of seizures in newborn rats, rabbits, and marmosets,[68,129] possibly reaching concentrations of glucose that are rate-limiting for brain hexokinase and hence for energy production.

Since the K_m for glucose transport into brain is about 7 mM, the carrier is less than half-saturated at physiological blood glucose concentrations (5 mM). Increasing blood glucose levels can increase transport from blood to brain and prevent brain glucose depletion in newborn rats, rabbits, and marmosets during seizures (Fig. 11). Thus, glucose supplements may offer a strategy to protect the brain of human neonates during seizures.

2.9.4. Utilization of Ketone Bodies

The healthy suckling neonate's brain may obtain as much as one-third of its energy from ketone bodies, which cannot be utilized in anoxic states such as convulsive seizures, because molecular oxygen is needed to regenerate NAD^+ from NADH. This increases the brain's requirements for glucose during seizures and anoxia, and makes glucose supply even more critical.

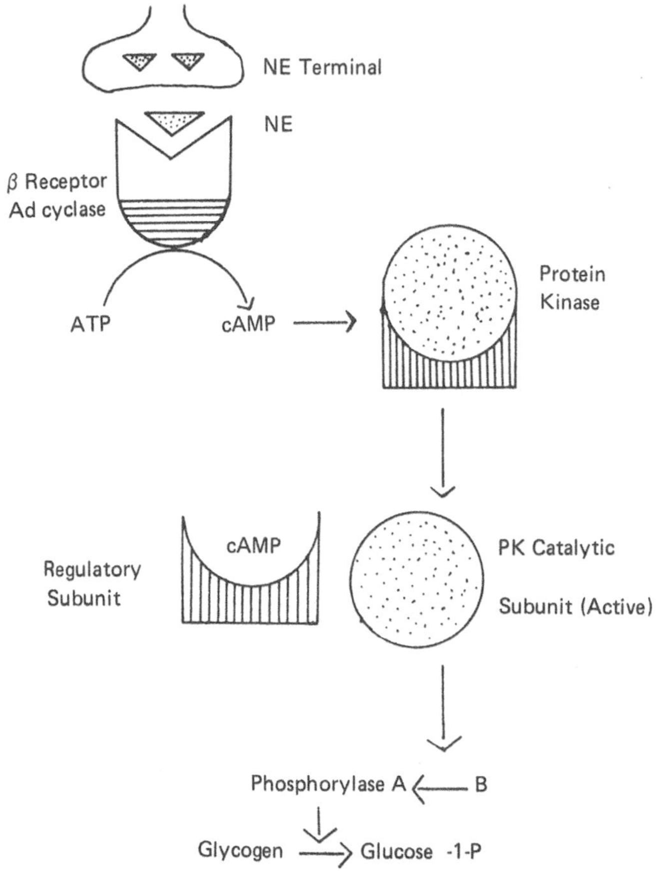

Fig. 10. Mechanism of glycogen mobilization.

2.10. Cerebral Metabolism in Human Epilepsy

Earlier methods based on A–V differences and on the Fick principle only permitted measurements of overall cerebral metabolism and blood flow during seizures. Isotopic methods also based on the Fick principle, using xenon or other radioactive inert gases, allowed studies of regional blood flow but not of regional metabolism, and were limited to superficial regions. Recently, the development of positron emission tomography has permitted the study of cerebral blood flow using $H_2\,^{15}O$, of oxidative metabolism with $^{15}O_2$, and of glycolytic metabolism with [^{18}F]fluorodeoxyglucose.[137–141] Regional blood flow in healthy controls has been found by these methods to range from 18 to 25 ml/100 g per min in white matter, and from 55 to 90 ml/100 g per min in gray matter. The regional cerebral metabolic rate for oxygen ranges from 4 to 8 ml/100 g per min in gray matter and from 1.5 to 2 ml/100 g per min in white matter.

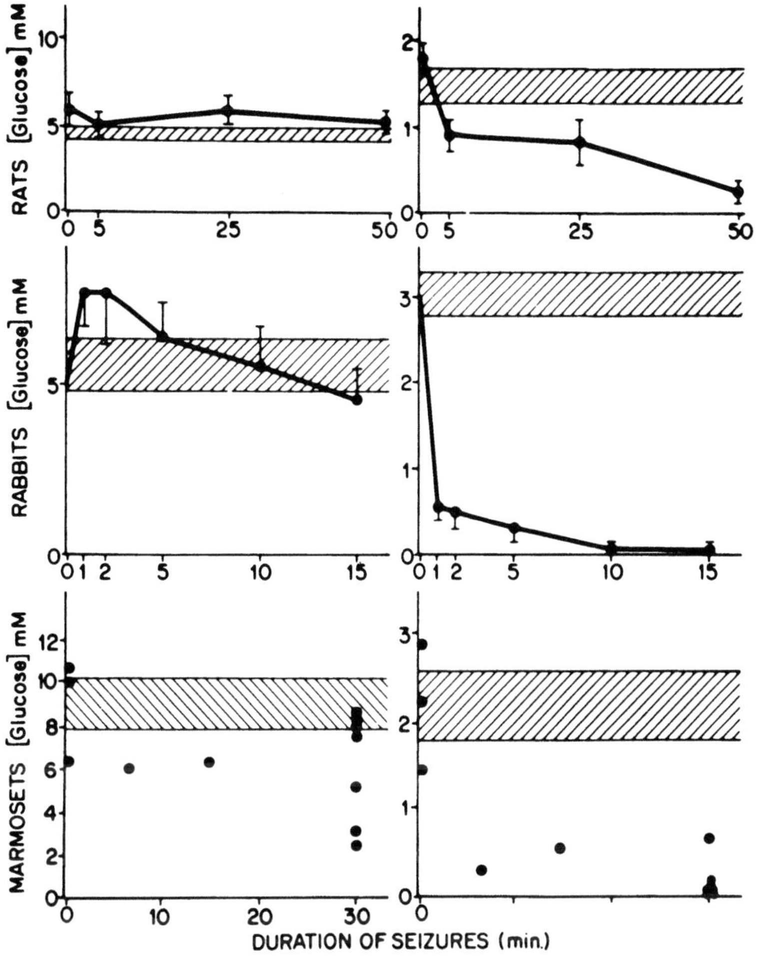

Fig. 11. Blood (left) and brain (right) glucose concentration during neonatal seizures. Reprinted from Ref. 59, with permission from Raven Press.

The cerebral metabolic rate for glucose has been in the range of 8 ml/100 g per min with considerable variations.

Between seizures, the epileptic focus in patients suffering from chronic epilepsy is almost universally hypometabolic. It might seem surprising that a region of frequent cell discharge and EEG spiking would have a decreased metabolic rate, averaging 33% in one series.[67] A number of possible explanations have been put forth. It is clear that this decreased metabolic rate does not represent only cell loss, although some degree of cell loss is present in the majority of patients.[142] In spite of the cell loss, the hypometabolic zone of the focus is quite capable of sustaining very high metabolic rate at times, and such increases in deoxyglucose uptake are recorded whenever a seizure occurs during a study.[144] The suggestion that the hypometabolic zone in focal epilepsy represents an inhibitory surround is appealing, but this must be reconciled with data demonstrating increased glucose metabolism in zones of decreased cell firing and neuronal membrane hyperpolarization.[145] On the other hand, generalized seizures such as those induced by ECS in psychotic patients have uniformly been found to be associated with a markedly increased rate of [^{18}F]fluorodeoxyglucose accumulation, and postictal states have been associated with decreased uptake of fluorodeoxyglucose. The precise extent of ictal hypermetabolism is difficult to determine, although the direction of change is undoubtedly toward a marked increase, because of two uncertainties associated with the fluorodeoxyglucose method: seizures could modify the lumped constant, which has not been measured under those circumstances, and the method is designed to deal with steady states and not well suited to the study of seizures where the 30 to 45 min of incorporation integrates periods of intense ictal activity and periods of decreased postictal activity. In spite of those limitations, this methodology offers the unique attraction of permitting the noninvasive study of biochemistry in the human brain during disease states, and its use will undoubtedly expand vastly in the future.

*3. PROTEIN SYNTHESIS AND OTHER BIOSYNTHETIC PROCESSES**

The large expenditures of energy incurred by the brain during epileptic seizures represent a major metabolic stress. Energy used to pump ions across cell membranes, to synthesize and take up neurotransmitters, and to restore membrane potential, is no longer available for biosynthetic processes. Experimental data uniformly indicate that seizures inhibit brain protein synthesis. The effect has been found across several species including subhuman primates in animals ranging in age from newborn to adult. Recent autoradiographic analysis of brain protein synthesis has revealed striking heterogeneity in response in different brain structures.

3.1. Rate of Brain Protein Synthesis

Inhibition of amino acid incorporation in brain protein ranging from 24% to 80% has been found immediately after ECS in mice (Table XII). The variability of the results likely derives in part from the choice of precursor and strains of mice and from the use of tracer quantities of amino acid that were injected either intravenously or intraperitoneally so that absorption, metabolism, and the extent of their cerebral uptake may vary during seizures altering brain precursor specific activity in unforeseen ways. Initially depressed by seizures, amino acid incorporation into protein had recovered to pre-ECS levels by 15 min after treatment.[146,147] When adult rats were paralyzed and oxygen-ventilated to prevent the effects of hypoxia and muscle convulsions, the incorporation of [^{3}H]leucine into cerebral protein was reduced by half during a 5-min period in which 10 seizures were delivered by electroshock (every 30 sec) (Table X). The specific activity of leucine in brain was greater in seizing

* Written by B. E. Dwyer.

Table XII
Rates of Local Brain Protein Synthesis (LBPS) in Newborn Marmoset Monkeys during Bicuculline-Induced Status Epilepticus[a]

Brain region	LBPS (%/hr) Control	Seizure	% of control
Cortex: Striate	1.69 ± 0.16	0.59 ± 0.17	35*
Parietal	1.60 ± 0.07	0.52 ± 0.11	33*
Temporal	1.53 ± 0.05	0.54 ± 0.07	35*
Entorhinal	1.38 ± 0.06	0.68 ± 0.20	49*
Hippocampus: CA1	1.68 ± 0.15	0.65 ± 0.08	39*
CA3	2.26 ± 0.11	0.90 ± 0.13	40*
Dentate	1.87 ± 0.04	0.77 ± 0.04	41*
Thalamus	1.31 ± 0.08	1.02 ± 0.28	78 (NS)
Lateral geniculate	2.04 ± 0.12	1.89 ± 0.46	93 (NS)

[a] Marmoset monkeys (5–7 days old) were injected with 150 mM L-[1-^{14}C]valine (10 μmol/g, 0.1 μCi/μmol, i.p.) followed immediately by bicuculline (5 mg/kg, i.m.) or 0.9% NaCl in controls. Animals were killed up to 1 hr after seizures had started. Each measurement is the mean ± S.D. for three animals.
* $p < 0.001$; NS, not significant.

animals than in controls, suggesting that the results may have underestimated the extent of inhibition of protein synthesis. Rates of brain protein synthesis measured in grossly dissected brain regions during generalized convulsions in immature rats and newborn marmoset monkeys were reduced to a similar extent (Table XI). These latter studies utilized the injection of large amounts of radioactive precursor amino acid essentially as described by Dunlop *et al.*[148] This method "floods" precursor amino acid pools thereby maintaining precursor specific activity constant over the interval in which incorporation is measured. This minimizes the problem of accurately estimating precursor specific activity under pathological conditions. Thus, it is reasonably certain that the inhibition of amino acid incorporation seen during seizures reflects true inhibition of protein synthesis.

3.2. Local Variations

Autoradiography of brain protein synthesis has revealed a striking regional diversity of effects. Kiessling and Kleihues[149] followed the incorporation of [^{3}H]tyrosine into brain protein during bicuculline-induced seizures lasting 30 min in freely convulsing adult rats. Bilateral reduction of tyrosine incorporation during seizures was shown in the cerebral cortex, hippocampus, and thalamus (Fig. 12). Several regions including the cingulate cortex, lower brain stem nuclei, and cerebellum appeared largely unaffected. In the cortex, inhibition was described as "pseudolaminar" in appearance being most severe in layers 2 and 5. In the hippocampus, patchy decrease of tyrosine incorporation was described affecting both granule cells of the dentate gyrus and pyramidal cells in sections CA1 and CA3. The mediodorsal and lateral thalamic nuclei also incorporated less tyrosine during seizures. Furthermore, in certain cell groups in the hippocampus and in the dorsal parts of the ventral thalamic nuclei, increased

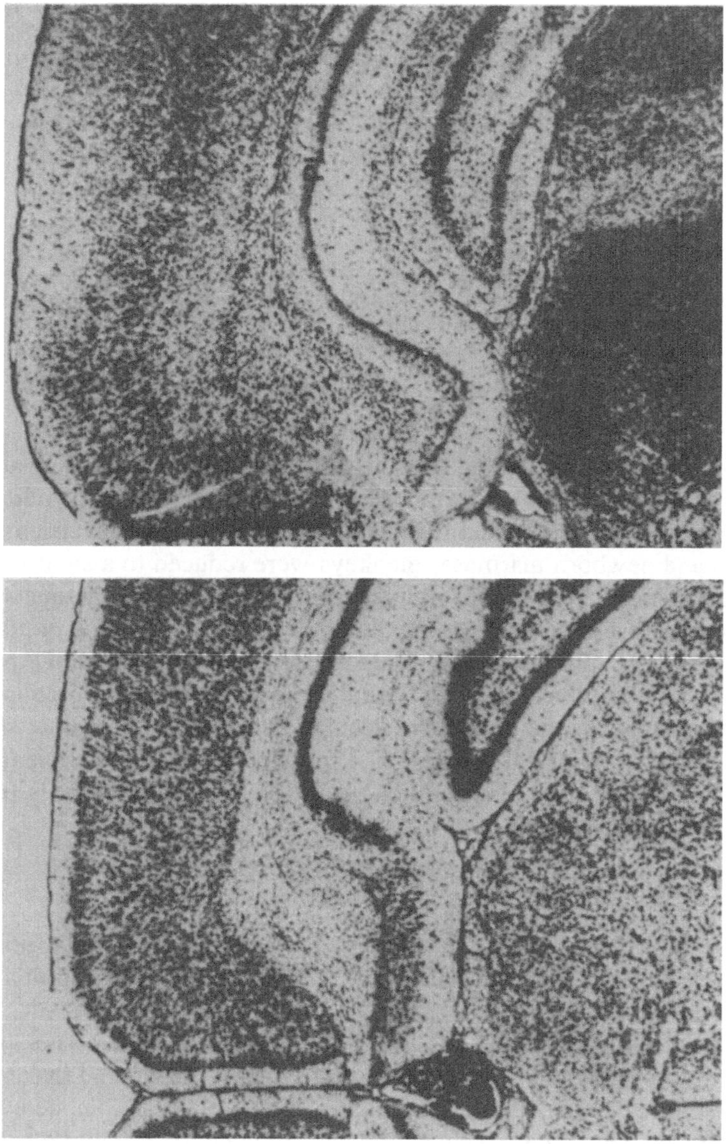

Fig. 12. Unstained autoradiograms ($\times$ 33) demonstrating marked focal reduction of [^{3}H]tyrosine incorporation into neurons of the cortex, of the pyramidal layer (PL) and granular layer (GL) of the hippocampus. In addition, note the irregular extent of incorporation within the adjacent midbrain structures. Left: control rat; right: status epilepticus. For further details and discussion see Kiessling and Kleihues.[149] Reprinted with permission of Springer-Verlag.

amounts of tyrosine incorporation were reported. This study utilized tracer quantities of amino acid and the specific activity of tyrosine was not measured. Thus, the pictured autoradiogram may not represent only protein synthesis but also seizure-related effects on amino acid transport, metabolism, or precursor pools.

Focal inhibition of brain protein synthesis by generalized seizures was also described in freely convulsing newborn marmoset monkeys.[46,56] Valine incorporation into protein was measured in 5- to 7-day-old marmosets during bicuculline-induced seizures lasting 30–60 min (Fig. 8). Measured rates of protein synthesis in brain regions are shown in Table XII. Cells in the cerebral cortex and hippocampus were most severely affected where measured rates were only 30–40% of control levels. Unlike the adult rat cortex where laminar inhibition was described, the cortex of the immature marmoset showed a columnar pattern where regions of severe protein synthesis inhibition alternated with regions in which protein synthesis was relatively spared. A similar patchy inhibition was found in the hippocampus. In the cerebellum the external granular layer, composed of germinal cells without synaptic connections, appeared largely unaffected while the internal granular layer, which shares the same blood supply, showed severe inhibition suggesting that inhibition was related to participation in seizure activity rather than to anoxemia or other systemic factors. The high dose of [^{14}C]valine injected in this study expands brain valine pools 10- to 20-fold and maintains cerebral valine specific activity constant for the duration of the seizures.[56] These concentrations are well below the concentrations that interfere with amino acid or protein synthesis. Thus, it is likely that the ^{14}C autoradiograms and our measured rates of brain protein synthesis accurately reflect *in situ* conditions.

The striking focality of protein synthesis inhibition in these two studies argues against a major role for the systemic effects of seizures (e.g., acidosis, hypoxemia, and cerebral hypoxia) in the observed inhibition. It is possible that certain cells may be more sensitive to these effects than others. More likely, inhibition reflects the degree of involvement of certain brain structures in seizure activity. This hypothesis is supported by the autoradiographic study of Collins and Vandi,[150] which showed that inhibition of leucine incorporation into protein was most severe in the seizure focus and in synaptically connected thalamic sites where retrograde burst firing is as intense as firing in the cortical seizure focus. In the nucleus reticularis thalami, which receives orthodromic input from the cortical focus but does not project axons to the cortex, glucose utilization was increased during focal seizures (consistent with increased neuronal activity and synaptic transmission) but the incorporation of leucine was not affected.

This study also used tracer quantities of radioactive amino acid and its results are subject to the same caution regarding the possible effects of seizures on local amino acid uptake and metabolism.

3.3. Translational Regulation during Seizures

Analysis of cerebral tissue following a single electroshock in rabbits[151,152] or in paralyzed and oxygen-ventilated rats after a series of electroshocks (status epilepticus)[153] suggests that inhibition of amino acid incorporation into protein does not result from damage to ribosomes or from the loss of mRNA, from the activation of ribonucleases, or from inactivation of soluble factors necessary for protein synthesis. The evidence for this is summarized below:

1. The integrity of ribosomes appeared unaffected after status epilepticus or electroshock. Total ribosomal RNA content was unchanged, and the ability to incorporate [^{3}H]phenylalanine into acid-precipitable material by brain PMS systems was not impaired after status epilepticus. Furthermore, analysis of monoribosomes that accumulated after electroshock suggested that the rate of chain elongation and termination was the same as in controls.
2. Hybridization analysis of poly(A)-containing RNA in brain using [^{3}H]poly(U) suggested that this fraction (and by inference, mRNA) was not degraded after electroshock nor was the mRNA population different as judged by the size distribution following gel electrophoresis and zone sedimentation analysis, and by analysis of radioactive polypeptide synthesized in reconstituted cell-free systems utilizing ribosomes derived from electroshock or control brain. Further, it appears that the integrity of mRNA is preserved during status epilepticus. Polyribosomes that disaggregate during status epilepticus tend to reaggregate after the end of the seizures. When rats are pretreated with actinomycin D, an inhibitor of RNA synthesis, the incorporation of [^{3}H]uridine into acid-precipitable material is reduced by over 80%, but this does not prevent reaggregation of polyribosomes, which suggests that synthesis of new mRNA is not required and that mRNA is preserved during status epilepticus.
3. The decrease in polyribosomes and inhibition of protein synthesis could not be accounted for by increased ribonuclease activity. The integrity of mRNA seemed unaffected as described in the preceding paragraph. More direct measurements using degradation of [^{3}H]poly(U) in postmitochondrial supernatants from electroshock or status epilepticus animals revealed no differences compared with controls. Furthermore, the monoribosomes isolated after electroshock or status epilepticus were dissociable into subunits in buffers containing high salt concentrations. Monoribosomes formed by ribonuclease action are still bound to small fragments of mRNA and do not dissociate in buffers containing high salt concentrations.
4. No evidence for depletion of necessary components (e.g., tRNA, amino acyl-tRNA synthetases) could be found in cell sap from electroshock animals. Assay for elongation factor activity using a poly(U)-programmed *in vitro* system revealed that cell sap from electroshock animals could support polypeptide elongation as well as that from controls. On the basis of these observations, it was concluded that the initiation of new polypeptides was the major regulatory site for protein synthesis during seizures. However, it must be noted that reduced rates of [^{3}H]leucine incorporation into status epilepticus rats was found by Wasterlain[44] within 5 min after electroshock seizures were begun. At this time there was no evidence for polyribosomal disaggregation in a parallel group of rats treated the same way[153] so that in this model polypeptide elongation (or termination) as well as initiation may also be impaired. As a footnote to the conclusions reached above, *in vitro* protein-synthesizing systems are derived from whole brain or grossly dissected brain regions, while the results of

Table XIII
Effect of Added GDP on Ternary Complex Formation[a]

	[GTP] (mM)				
GDP/GTP	0.1	0.3	0.5	1.5	3.0
0	100	100	100	100	100
1:10	62	63	57	57	52
1:2	31	31	27	23	22

[a] Ternary complex formation was assayed by retention of radioactivity on nitrocellulose filters at the GTP concentrations indicated. GDP was added in molar ratios of 0 (complete system without added GDP), 1:10, and 1:2 with GTP over the range of concentrations employed. GTP-independent binding was subtracted from all determinations. The results are expressed as percent of control.

autoradiographic analysis just reviewed suggest this approach may not always be adequate.

It is clear then that protein synthesis is inhibited during seizures, most likely by an initiation block; the mechanisms involved are largely unknown, but some of the most important steps are beginning to be understood (see below).

3.4. Role of GDP and of the Energy Charge Potential

The translation of mRNA into protein involves a series of steps by which mRNA is bound to ribosomes forming a competent 80 S ribosome initiation complex (initiation), synthesis of the polypeptide chain by the energy-dependent polymerization of amino acids coded for by the mRNA (elongation), and release of the completed polypeptide chain (termination). One of the first steps in brain protein synthesis initiation is the formation of a ternary complex between initiation factor eIF-2, the initiator species of $tRNA^{Met}$, and CTP, which subsequently binds to a 40 S ribosomal subunit in one of the several steps in the initiation process. Ternary complex formation *in vitro* is inhibited when the GDP/GTP ratio increases.[154] This holds true over at least a 30-fold concentration range of GTP (Table XIII); GMP has little effect (Fig. 13), nor do ADP, AMP, UDP, UMP, CDP, and CMP. The GDP/GTP ratio is directly linked to the ADP/ATP ratio via the enzyme nucleoside diphosphate kinase (EC 2.7.4.6), which catalyzes the reversible reaction

$$\text{XTP} + \text{YDP} \rightleftharpoons \text{XDP} + \text{YTP}$$

and thus to the adenylate energy charge potential of the cell. The adenylate charge potential is defined[87] as:

$$\frac{[\text{ATP}] + \frac{1}{2}[\text{ADP}]}{[\text{ATP}] + [\text{ADP}] + [\text{AMP}]}$$

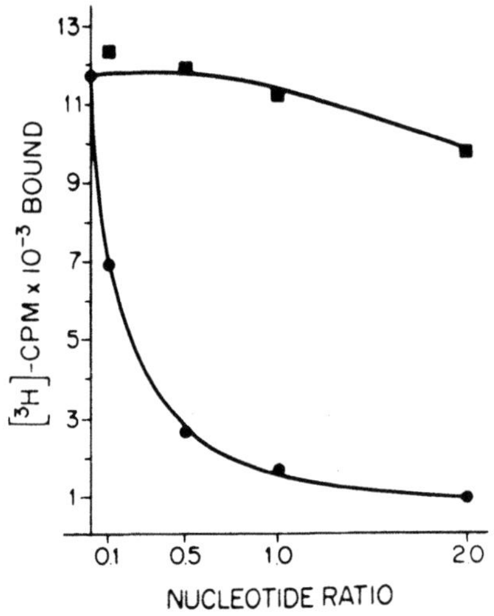

Fig. 13. Effect of GDP and GMP on GTP-dependent ternary complex formation. Ternary complex formation was assayed by retention of radioactivity on nitrocellulose filters.[154] GTP was present in the assay mixture at 1.5 mM and various amounts of GDP (●—●) or GMP (■—■) were added to give the desired concentration ratio. GTP-independent binding was subtracted from all measurements. Reproduced from ref. 154 with permission from Raven Press.

and is a measure of the availability of high-energy phosphate bonds for performing cell work.

Recently we have described a mechanism by which the adenylate energy charge of the cell may regulate brain protein synthesis during seizures.[45,154] Epileptic seizures place severe demands on cerebral energy metabolism. Rates of cerebral energy utilization have been estimated to increase three- to fivefold during the seizure discharge,[81–84] and significant reductions of brain high-energy phosphate compounds are found shortly after convulsive onset of seizure. While most of the reduction in high-energy phosphates can be prevented by paralyzing and oxygen-ventilating animals,[83] there remains a small but significant fall in the brain adenylate energy charge potential.[48,70,78,88] A small increase in the [ADP]/[ATP] ratio presumably results in an increased [GDP]/[GTP] ratio sufficient to inhibit ternary complex formation. A similar mechanism regulates polypeptide synthesis in the reticulocyte[155] and may also play a role in the inhibition of brain protein synthesis during seizures. Furthermore, the restoration of protein synthesis in the brain after ECS lags behind the recovery of ATP. Protein synthesis initiation is a highly regulated process[156–158] and in brain during seizures, regulation at several sites is likely. One prominent mechanism for the regulation of protein synthesis initiation involving the phosphorylation of eIF-2 is well established in the reticulocyte and is a distinct possibility in the brain.

3.5. *Role of Ionic Equilibrium*

Massive membrane depolarization during a seizure discharge can disturb ionic equilibrium in brain. Moody *et al.*[159] found that the extracellular potassium concentration rose to 10 mM in penicillin epileptogenic foci in cat cortex (baseline value 3.4 mM). Disturbed ionic equilibrium may lead to altered rates of protein synthesis. Lipton and Heimbach[160] showed that *in vitro* lysine incorporation into hippocampal slice protein was sensitive to the intracellular

$[K^+]/[Na^+]$ ratio. Since the restoration of normal extracellular $[K^+]$ occurs very quickly in the interictal period,[159] ionic imbalance is unlikely to explain protein synthesis inhibition after the termination of seizures, but may contribute in part to the inhibition of protein synthesis during seizures, possibly via regulation of ternary complex formation during the initiation process[154] or through a direct effect on the elongation of polypeptide chains.[160a,160b]

3.6. Role of Extracerebral Physiological Changes

Status epilepticus in freely convulsing adult animals produces dramatic physiological effects including hypoxemia and cerebral hypoxia, metabolic and respiratory acidosis, and hyperthermia.[45,58] In the neonate, the brain may also become depleted of glucose in spite of normal blood glucose concentrations.[68,129] Several studies suggest that all of these factors are associated with reduced brain protein synthesis. Protein synthesis was significantly reduced *in vivo* in rat brain during hypoxia[161] and *in vitro* in rabbit cortex slices obtained from animals breathing 8% oxygen.[162] Lactic acidosis during seizures may be partially responsible for protein synthesis inhibition.[161,163] Hypoglycemia also results in reduced brain protein synthesis *in vivo*.[161] In hypoglycemia, lactic acid does not accumulate in the brain. More likely, inhibition is related to energy failure and falling high-energy phosphate concentrations. Hyperthermia is also known to inhibit brain protein synthesis in some species. Little is known about mechanisms that regulate protein synthesis under these pathological conditions, and specific local effects on rates of brain protein synthesis have not been investigated.

3.7. RNA Metabolism

Inhibition of protein synthesis during a seizure discharge or in the immediate postictal period may result from short-acting regulatory mechanisms responding to the disturbed cellular homeostasis. However, inhibition of protein synthesis several hours after seizure termination is not easily explainable by these mechanisms, although the possibility cannot be entirely discounted. Long-term inhibition of protein synthesis may be due to impaired RNA metabolism. Early studies suggested that RNA processing was inhibited in rat brain following a single flurothyl-induced seizure[164] or in cat cortex during status epilepticus induced by the topical application of cobalt powder.[165] This effect was found in rabbit brain in the poly(A)-containing RNA species immediately following electroshock[166] and persisted 30 min postictally at a time when the EEG pattern had returned to normal.[167] Later studies showed that abnormal RNA metabolism could persist in the postictal recovery period for at least 12 hr.[164,168] The meaning of altered RNA metabolism to normal cell function is not clear. Protein synthesis returns to control levels shortly after a single ECS at a time when abnormal RNA metabolism is reported in the brain. It is possible that the effects of altered RNA metabolism in the cell are delayed.

4. SYNAPTIC NEUROCHEMISTRY

This aspect was covered extensively in the first edition of this handbook, and has been reviewed frequently. Only selected aspects of this vast field will be addressed.

4.1. Putative Neurotransmitter Amino Acids*

4.1.1. Amino Acid Concentrations in Epileptic Foci

Amino acids have long been implicated in the mechanisms of epileptogenicity. They are key cellular components that serve as precursors for proteins; they play a prominent role in intermediary metabolism, and several, including glutamate, aspartate, GABA, and glycine, have been suggested as neurotransmitter candidates in the mammalian CNS. In particular, strong lines of evidence suggest that in brain, glutamate and GABA are major excitatory and inhibitory neurotransmitters, respectively. Aspartate is a possible excitatory neurotransmitter and glycine has been established as a major inhibitory neurotransmitter in the spinal cord. Taurine, a sulfur-containing amino acid, has been postulated to play an inhibitory role in brain, possibly as a neurotransmitter, but more likely as a neuromodulator. See McGeer *et al.*[169] for a more detailed discussion of the role of amino acids in neurotransmission. Evidence for a specific role of amino acids in cerebral hyperexcitability has been sought by several laboratories. Amino acid concentrations in epileptogenic foci measured in surgically excised portions of the cerebral cortex obtained from intractable epileptics are shown in Table XIV. In the study of van Gelder *et al.*,[170] in tissue peripheral to the center of the focus, GABA and aspartate were reduced compared to control values extrapolated from the literature. In the center of the focus itself, reduced levels of glutamate and taurine were found in addition to those of aspartate and GABA. In contrast, glycine was elevated in the focus only and little change was reported for the other amino acids measured including serine and alanine. These results were not confirmed by Perry and Hansen[171] (Table XIV). As reported in an expanded study of human epileptogenic foci,[171] glutamate, glycine, and GABA levels were elevated in the focus compared to levels in "normal" tissue excised from nonepileptic patients undergoing neurosurgery. Aspartate and taurine levels were unchanged. Tissue sampling, the nature of control tissue, the type of seizure disorder, and the role of long-term anticonvulsant therapy may contribute to the discrepant data. Furthermore, structural changes including neuronal necrosis and glial scarring are prominent in chronic human epileptogenic foci[172] so that differences may also represent altered neuron/glia ratios or the effects of cell damage. Abnormal amino acid patterns have been reported in several experimental models of chronic and acute focal epilepsy and during chemically induced generalized seizures (Table XV).

Amino acid changes are often quite variable and time-dependent. Cobalt and alumina cream applied to the cerebral cortex produce a seizure focus in

* Written by B. E. Dwyer.

Table XIV
Amino Acid Abnormalities in Surgically Removed Human Epileptogenic Foci[a]

	Taurine	GABA	Glycine	Glutamate	Aspartate	Glutamine	Serine	Alanine
Focus[b] (Ref. 170)	↓	↓	↑	↓	↓	=	=	=
Peripheral tissue[b] (Ref. 171)	=	↓	=	=	↓	=	=	=
Focus[c]	NS	↑	↑	↑	NS	NS	NM	NM

[a] ↑, increased; ↓, decreased; =, not changed; NM, not measured; NS not significant.
[b] This series was obtained from 11 temporal and 5 frontal lobe epileptogenic foci. Focal tissue was sampled in the most abnormal region. Peripheral tissue was sampled in tissue furthest removed from the focus. Control values were extrapolated from literature sources.
[c] This series consisted of 54 epileptogenic foci removed from frontal (12) and temporal (42) lobes of 35 patients. Control tissue was from biopsies from 14 nonepileptic patients undergoing neurosurgery.

Table XV
Amino Acid Alterations in Epileptic Tissue

Experimental study	Tissue sampling	Amino acid changes								
		Tau	GABA	Gly	Glu	Gln	Asp	Ser	Thr	Ala
Cat, cobalt powder (Ref. 177)[a]	Primary focus									
	Preconvulsive (60–360 min)	=	=	=	↓	=	=	NM	NM	NM
	Convulsive (24 hr)	=	↓	=	↓	=	=	NM	NM	NM
	Postconvulsive (30–70 days)	=	=	=	=	=	=	NM	NM	NM
	Secondary focus									
	Preconvulsive	=	=	=	=	=	=	NM	NM	NM
	Convulsive	=	=	=	=	=	=	NM	NM	NM
	Postconvulsive	↑	↓	=	=	=	=	NM	NM	NM
Rat, cobalt implant (Ref. 178)[d]	Primary focus									
	Convulsive	NM	↓	=	↓	=	↓	NM	NM	NM
	Secondary focus									
	Convulsive	NM	=	=	=	=	=	NM	NM	NM
Rat, cobalt implant (Ref. 184)[e]	Primary focus									
	Convulsive	↓								
	Secondary focus									
	Convulsive	↓								
Cat, cobalt implant (Ref. 341)[f]	Primary focus									
	Preconvulsive (30 min)	↓	=	=	=	=	=	=	=	=
	Convulsive (12–48 hr)	↓	↓	↑	↓	↓	↓	↑	↑	↑
	Postconvulsive (14–30 days)	=	=	=	=	↓	=	=	=	=
	Secondary focus									
	Convulsive	↓	↓	↑	↓	↓	=	=	↑	=
Cat, penicillin (Ref. 341)[g]	Primary focus									
	Preconvulsive (1 min)	↓	=	=	=	=	=	=	=	=

	(Convulsive (3–120 min)	↓	↓	=	=	=	=	=	=	↑
	Postconvulsive (240 min)	=	=	↓	↓	↓	↓	↓	=	=
	Secondary focus Convulsive	↓	↓	=	=	=	=	=	=	=
Cat, cobalt powder (Ref. 342)[h]	Primary focus (7 days)	=	↓	↑	↓	↑	↓	↑	NM	↑
	Seconday focus (7 days)	NM	=	=	=	=	=	NM	NM	NM
Mouse, cobalt powder (Ref. 343)[i]	Primary focus Postconvulsive (7 days)	↓	↓	↑	↓	=	=	↑	NM	↑
Rat, cobalt implant (Ref. 344)[j]	Primary focus Postconvulsive (7 days)	↓	NM	↑	↓	=	↓	=	↑	↑
	Secondary focus Postconvulsive (19 days)	=	NM	=	=	=	=	=	=	=

[a] Six or seven cats were used for preconvulsive study, 7 or 8 during convulsions, and 7 after convulsions. The statistical significance of results was not indicated in the text. Increased or decreased concentrations are reported here for mean value differences whose standard deviations did not overlap. Cobalt powder was applied to the left anterior sigmoid gyrus. Sampling for focal tissue was from the posterior sigmoid gyrus.

[b] Time at which tissue was sampled after cobalt or penicillin applications.

[c] Cortex that is contralateral and homotopic to the primary focus.

[d,e] Cobalt, in a gelatin pellet, was implanted into the frontal cortex of male Glaxo rats. Amino acid concentrations were measured in the primary and secondary focus between 5 and 25 days (*d*) or 3 and 30 days (*e*) after cobalt was implanted.

[f] Cobalt, in a gelatin pellet, was implanted into both anterior and posterior sigmoid gyrus, and removed 30 hr later. Seven cats were used as controls. For tissue sampling at 30 min, 2 and 4 hr, and 14 and 30 days after cobalt implant, 3 animals were used. For sampling at 12 hr, 7 animals were used.

[g] Sodium penicillin G was applied to the pial surface over the left precruciate and removed 3 min later to avoid convulsions. Seven cats were used as controls. For tissue sampling, 3 cats were taken at 1 and 3 min and 2 and 4 hr after penicillin was applied. Seven cats were used at 20 min.

[h] Cats were made epileptic by exposing an area of the anterior motor cortex and applying 50 mg of cobalt powder directly to the cortical surface. The severity of epilepsy was graded from none to severe in the series of 11 cats. Amino acid concentrations were measured in the primary focus on tissue obtained from cortex immediately posterior to the site of cobalt application 7 days later. At this time most cats were reported to be seizure free. Amino acid changes in the table reflect trends shown by the authors.

[i] Mice were made epileptic by application of cobalt powder to the cerebral cortex. Tissue was sampled 7 days later, ipsilateral and posterior to the site of cobalt application.

[j] Rats were made epileptic by implanting a cobalt wire in the right parietal cortex. Tissue was sampled from the primary lesion site 7 days after implant and in the homotopic contralateral cortex 19 days after cobalt implant.

which high-amplitude spikes appear in the EEG. One advantage of these models is that epilepsy develops over a period of time so that amino acid changes in the focus can be followed and correlated with the appearance of epileptic activity. A major drawback is that severe cortical necrosis takes place in the focus, so that neuronal fallout and attendant changes in amino acid neurotransmitters could be expected to occur in addition to changes that might play a possible role in epileptogenicity. The most significant alterations were found during the convulsive period when spike frequency in the seizure focus was greatest. Some general trends were suggested. Prominent at the time of maximal epileptogenicity after cobalt application were reduced levels of glutamate and GABA in the primary focus while glycine was elevated. Mixed results were obtained for the other amino acids including taurine.

Similar findings were reported in active epileptogenic foci produced by freeze lesioning the cortex[173] and penicillin[174] and following chronic deafferentation of cortical tissue (cat undercut-cortex model).[173,175,176] During the preconvulsive period, amino acid changes in the cobalt focus were less severe although reduced levels of glutamate and taurine have been reported (Table XV).

4.1.2. Methodological Considerations

While amino acid concentrations clearly are found to change, it is not clear what relationship this has to epileptogenicity and several confounding factors must be considered. Glutamate, which is directly involved in intermediary metabolism, and GABA are easily affected by changes in Krebs cycle activity and ammonia metabolism. Ischemia is unavoidable during sampling of human epileptic tissue, but its duration and intensity vary enormously with the surgical method. Hypoxia and ischemia alone have been reported to alter brain amino acid concentrations. In addition, ischemia is difficult to avoid during sampling of experimental epileptic foci and many investigators have not taken necessary precautions to do so. Another major problem relates to interpretation of the results found in both human and animal studies and the uncertain relationship between concentration, turnover, and release of substances acting as neurotransmitters. Studies of biogenic amines have abundantly documented instances when concentrations were increased but transmitter release was decreased, and vice versa. Neuronal necrosis, gliosis in the focus, and in the case of the cobalt model, the presence of heavy metal ions must also be considered. The fact that the severity of amino acid changes in brain were roughly proportional to the severity of epilepsy and were most prominent in the area of the focus[174,177,178] suggests that changes may possibly reflect the degree of involvement of cells in epileptic activity. This is not surprising in light of the central role of amino acids, particularly of glutamate, in cell metabolism. Glutamate, aspartate, and alanine are directly convertible by transamination reactions to α-ketoglutarate, oxaloacetate, and pyruvate, respectively, major intermediates in the citric acid cycle and glycolytic pathways. GABA is formed from glutamate and it may enter the citric acid cycle at the level of succinyl-CoA. Serine and glycine are interconvertible and can be made from glucose.

Furthermore, glutamate serves to detoxify ammonia in the cell and ammonia is produced during seizure. Those agents that perturb cerebral energy metabolism and hence glycolytic flux and citric acid cycle activity are likely to affect steady-state amino acid levels in brain. Thus, glutamate was reduced during the tonic phase of a pentylenetetrazol-induced convulsion but not during the preconvulsive and clonic phase.[179] Nor was glutamate reduced in paralyzed and oxygenated rats after a single electroshock.[80] This treatment also prevents anoxemia and any fall in brain ATP resulting from electroshock. Sacktor *et al.*[81] did not find a fall in brain glutamate in 10-day-old rats with flurothyl-induced seizures. However, cerebral metabolic rate is lower in young rats than adults and ATP depletion was not as severe as that found in animals suffering seizures.

The sensitivity of brain amino acid concentrations to the state of the animal at the time of the experiment is also demonstrated by two studies[78,92] where bicuculline seizures were induced in rats either fed *ad libitum* or starved for 24 hr prior to the experiment. Under these conditions, striking differences in cerebral amino acid concentrations are found during seizures (Fig. 14). It is conceivable that cerebral hyperexcitability resulting in epilepsy could derive from an amino acid imbalance in the focus. This in turn could result from altered amino acid metabolism, trauma, or abnormal release and reuptake mechanisms, which could produce locally high concentrations of excitatory amino acids (glutamate, aspartate) or conversely lower than normal levels of inhibitory amino acids (GABA). In fact, Tower[180] had found excessive release of glutamate from brain slices prepared from excised human epileptic foci, and an increase in glutamate, among other amino acids, has been found in the cortical superfusate around cobalt-induced foci[177,181,182] and in the cat undercut-cortex model.[174] Koyama[177] had speculated that increased glutamate release may play a role in neuronal hyperexcitability in the cobalt focus; furthermore, glutamate (and aspartate) administered systemically in subconvulsant doses can activate the cobalt focus in rats.[183] How it does this is not clear since circulating glutamate penetrates normal brain tissue very poorly.

4.1.3. Secondary Foci

In an effort to avoid tissue trauma and neuronal necrosis characteristic of the primary epileptogenic focus, for example, in the cobalt model, amino acid concentrations have been measured in the secondary focus, the region in the contralateral homotopic cortex that receives projections from the site of the primary focus. No change in glutamate, aspartate, GABA, or glycine was found in the secondary focus,[178] although a significant reduction in taurine in the secondary focus during the period when epileptic spiking was most pronounced was reported.[184] Since spike frequency in the secondary focus is even higher than in the primary focus 8–14 days after cobalt implantation,[185] these results do not support a direct relationship between amino acid concentration (possibly with the exception of taurine) and epileptogenicity. They suggest that differences in amino acid concentrations result from tissue damage or secondary metabolic changes in establishing the primary focus. Furthermore, normal

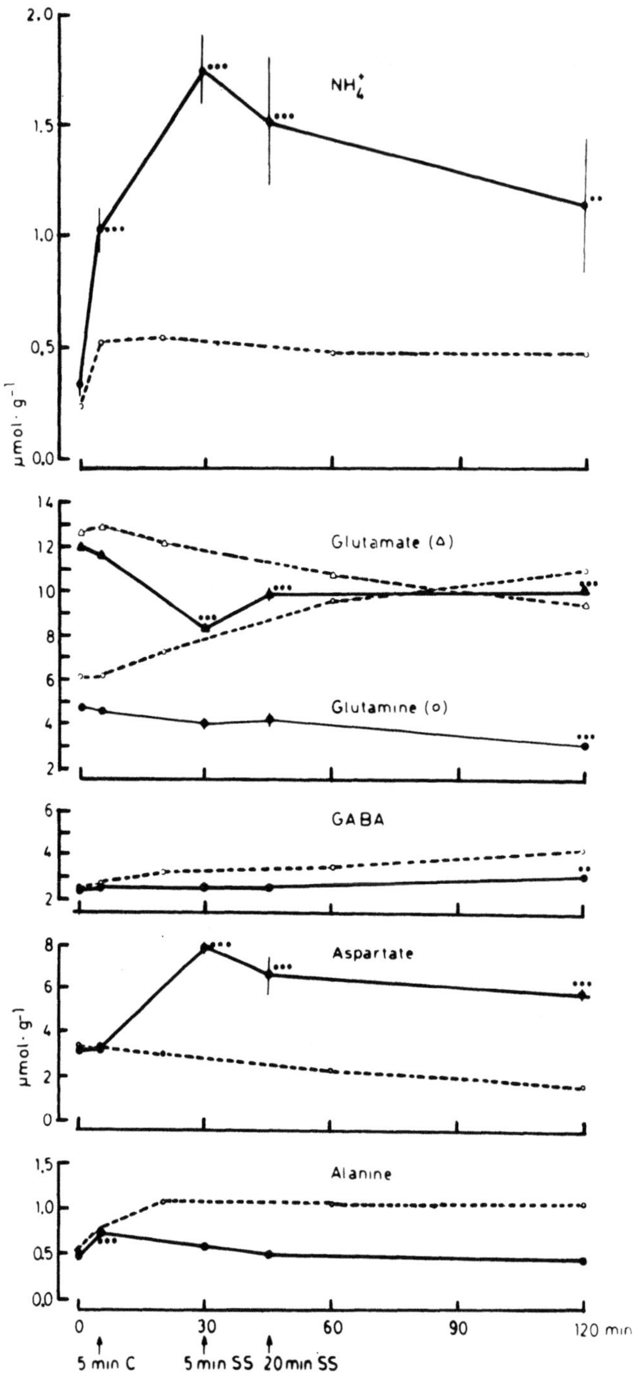

Fig. 14. Changes in tissue concentrations of ammonia and amino acids during bicuculline-induced seizures in starved rats. Results for starved rats (closed symbols) were compared with those previously obtained[78] for fed rats (open symbols). Group 5 min C denotes animals studied after 5 min of continuous seizure activity; groups 5 min SS and 20 min SS are those that showed a single-spike pattern in the EEG for 5 and 20 min, respectively. Values are means ± S.E.M. (not shown if smaller than the size of the symbol; S.E.M. for fed animals not given). ($*p < 0.05$, $**p < 0.01$; experimental group compared to control group.) See Blennow *et al.*[92] for further details. Reprinted with permission of Little, Brown.

amino acid concentrations are found in brain of rats,[186] baboons,[187] and beagle dogs[188] with a genetic predisposition for generalized seizures, and in the brain of fully kindled rats despite the permanent reduction in seizure threshold characterizing this model of epileptogenesis.[189]

4.1.4. Conclusions

There are many speculations as to the role in epileptogenesis of glutamate, a putative excitatory neurotransmitter, and taurine, which may act as an inhibitory neurotransmitter or neuromodulator. The metabolism of taurine in brain and its anticonvulsant activity have recently been reviewed.[190,191] In particular, the work of van Gelder and collaborators suggests that taurine may play a role in regulating glutamate levels in tissue and that local taurine loss in brain combined with disturbances in the metabolism of glutamate (influenced by taurine) may be a precipitating factor in some forms of epilepsy. The role of taurine and glutamate in epileptogenesis has been recently discussed.[174,192–195]

In conclusion, the hypothesis that an imbalance of excitation and inhibition in brain may be a causative factor in epilepsy[196] and that this may result from an imbalance between transmitter amino acids remains attractive, but has not yet been established and will be difficult to prove by measuring static levels of amino acids in brain tissue. Nonetheless, changes in neurotransmitter amino acids may be of relevance in some types of epilepsy.

*4.2. Biogenic Amines**

The involvement of norepinephrine (NE), dopamine (DA), and serotonin (5-HT) in epilepsy can be discussed in two categories: (1) changes in NE, DA, and 5-HT *in situ*, which may directly lead to altered cerebral excitability, and (2) long-term changes in biogenic amine function following seizures, which can alter behavior. For the purposes of this review, only the former aspect of neurotransmitter changes will be considered.

Evidence that an underlying defect in biogenic amine metabolism may be a contributing factor in epilepsy has been accumulating over the past 20 years. In general, the preponderance of data suggest that reduced biogenic amine function is associated with reduced threshold to electrically and chemically induced seizures and increased severity of convulsions.[197,198] The role of abnormal neurotransmitter metabolism in several species of genetically seizure-prone animals has recently been reviewed.[199] It is clear that the biochemical results from similar genetic models are often in disagreement. For example, in the seizure-susceptible rat,[200] data suggest a modulatory role for NE. Abnormally low levels of NE in this rat strain were reported by Laird *et al.*[207] The work of Jobe *et al.*[201] suggested that these epilepsy-prone rats carry another genetic factor and that a reduced level of monoamines was not itself a sufficient cause. Thus, they depleted central monoamines with Ro4-1284 in non-epilepsy-prone control rats, and in the nonseizure-susceptible progeny of seizure-prone

* Written by B. E. Dwyer.

parents. Only in the latter animals did central monoamine depletion result in increased seizure susceptibility. DA does not seem to play a major role. Only a slight effect on seizure susceptibility or intensity was found after brain DA levels were selectively altered.[202–205] Elevation of NE levels resulted in decreased seizure intensity in this model.[202,205,206] Serotoninergic transmission was also implicated. 5-HT levels were reduced below normal levels.[207] Depletion of brain 5-HT in these rats after treatment with *p*-chlorophenylalanine[202] and *p*-chloromethamphetamine[208] resulted in an increase in seizure susceptibility whereas elevation of brain 5-HT reduced seizure intensity.[202]

In mice susceptible to similar audiogenic seizures, a more prominent role for DA was suggested but available data have failed to establish a definite role for 5-HT.[199] This underscores the possibility that several animal models may share common characteristics (e.g., audiogenic seizure susceptibility), and display similar behavioral convulsions, but may involve quite different neurotransmitter systems.

In the kindling model of epileptogenesis,[209] a series of brief subconvulsive electrical stimuli in time result in a permanently reduced seizure threshold. Altered function of catecholaminergic neurons has been implicated in the mechanism. The role of neurotransmitters in the kindling phenomenon has recently been reviewed.[210] Depletion of brain monoamine stores was found to facilitate amygdala kindling.[211,212] Measured levels of catecholamines in the brain of kindled animals do not provide conclusive evidence of a role for either NE or DA.[213–216] However, one pharmacological study supports a role for NE (particularly acting at the β-adrenergic receptor) over DA in the facilitation of kindling.[212] Thus, α-methyl-*p*-tyrosine, which depletes brain NE and DA, facilitates kindling. Pimozide, a DA receptor blocker, and apomorphine, a DA agonist, had no effect on the rate of kindling, nor did phenoxybenzamine or clonidine, which block α-adrenergic receptors. In contrast, disulfiram treatment, which depletes central NE, and the β-adrenergic receptor blocker propranolol facilitated kindling.

The role of serotoninergic transmission in kindling is not clear. Raphe nucleus stimulation was found to increase seizure thresholds in amygdala-kindled cats,[217] suggesting that serotoninergic function could influence seizure susceptibility. Fluoxetine, an inhibitor of 5-HT uptake, also increased seizure threshold, but the acute administration of *p*-chlorophenylalanine (10 mg/kg), which inhibits 5-HT synthesis, had no effect on seizure threshold. In amygdala-kindled rabbits, enhanced serotoninergic action had little effect on the seizure intensity and its behavioral or electrical correlates. Nor did chronic 5-HT administration alter the course of kindling.[218] However, chronic administration of *p*-chlorophenylalanine (80 mg/kg daily) could delay the development of kindled seizures.

Biogenic amine metabolism may also be abnormal in focal epilepsy. There is some evidence that DA and NE act antagonistically in the cobalt focus. Thus, apomorphine, a DA receptor agonist, and amphetamine, which stimulates release of endogenous DA, both suppress spike activity in cobalt-treated rats.[219] Spike activity was increased in cobalt-treated animals with the DA receptor blocker spiroperidol and by clonidine. The latter effect can be counteracted by either apomorphine or amphetamine. The possible relevance of altered biogenic amine function to the development of the cobalt focus has been briefly discussed by Emson.[185]

Clearly, the role of neurotransmitters in epilepsy is not understood, and compiling and comparing data is made difficult by the variety of animal models used, the various treatment regimens, the time at which neurotransmitter levels and the effects of their alterations are measured with respect to the development of seizures, and quite probably the way animals are housed and handled by various investigators. Furthermore, it is very difficult to directly link change in one neurotransmitter with a change in seizure susceptibility, as there is ample experimental evidence that individual neurotransmitter systems may serve to modulate other neurotransmitter systems.

Many pitfalls are encountered in exploring the role of biogenic amines in epilepsy. Their function may be compromised by abnormal synthesis, storage, release, or reuptake in the presynaptic nerve ending or by abnormal receptor function in the postsynaptic terminal. In order to study these particular aspects of biogenic amine function and their role in epilepsy, pharmacological agents are routinely used. These drugs often alter whole brain levels of biogenic amines and may obscure changes in particular neurotransmitter systems that may be of significance. Furthermore, these drugs are often not specific for a single neurotransmitter. Lastly, most studies to date have reported static levels of biogenic amines and thus present only part of the picture. More detailed studies, including turnover studies, will be useful in interpreting their meaning.

However, in spite of the many impediments, progress toward the goal of understanding how neurotransmitter function is related to epilepsy is rapid, not only with respect to 5-HT, NE, and DA, but also with respect to the role of cholinergic and GABAergic mechanisms in regulating cerebral excitability.

*4.3. Receptors for Putative Neurotransmitters**

The concept that long-term or permanent modification of neuronal activity in seizure disorders could be explained in terms of receptor modification is a fascinating idea indeed. With this possibility in mind, the binding properties of a number of receptors have been examined in a variety of seizure models. Initial studies have focused on the affinity of various receptors for radiolabeled ligands and the concentration of receptors present in brain areas associated with seizure activity, i.e., cortex, hippocampus. As it becomes more apparent which receptors are involved with ictal events, it will be necessary to correlate receptor modifications with receptor function in order to fully understand the neurochemical process of seizure propagation. Studies to date suggest that measurement of receptor density and affinity does not in itself give sufficient information for understanding seizure models, since observed changes have been both small and transient. Evaluation of the significance of such changes in terms of expression of receptor function will lead to a greater understanding of the role of various neurotransmitter receptors in seizure expression.

* Written by A. M. Morin.

Table XVI
Acetylcholine Muscarinic Receptors and Seizures

Types of seizures	Area	Change	Time	Reference
Kindling				
Electrical-amygdala	Amygdala	↓	15 hr	226
	Hippocampus	↓	15 hr	227
	Cortex	N.C.	15 hr	
	Amygdala, hippocampus, striatum, cortex, cerebellum	N.C.	1 week	228
Electrical-perforant path	Hippocampus	↑	18 hr	225
Carbachol-amygdala	Amygdala, cortex, hippocampus, striatum, cerebellum	N.C.	1 week	
Electrical-medial septum	Cortex, hippocampus	N.C.	1 sec, 30 min	232
Freeze-induced focal epilepsy	Cortex	↓	12 hr	241
	Adjacent cortex	N.C.	12 hr	
ECS				
Multiple	Cortex, hippocampus	N.C.	24 hr	233
	Forebrain	N.C.	30 min	234
Single	Rat neonate forebrain	N.C.	30 min	[a]

[a] Morin and Wasterlain (unpublished).

4.3.1. *Muscarinic Receptor*

It is well known that *in situ* application of acetylcholine (ACh) to the cerebral cortex causes an increase in cerebral excitability sometimes manifested by seizures, as well as evidence of epileptic activity in brain regions distant to the cortex.[220] Rats made epileptic by local cortical freezing have been shown to have a higher spontaneous release of ACh from synaptosomes as well as having a higher endogenous level of ACh.[221] Thus, it would seem likely that ACh along with its receptor may play a role in increasing neuronal excitability and in producing seizures in some experimental models of epilepsy (Table XVI).

4.3.1a. Kindling. During kindling an animal is exposed to periodic electrical stimuli or subconvulsive doses of drugs producing seizures.[222] The animals progress through a series of behavioral responses to the stimuli, which culminate in a generalized seizure.[223] The model allows for study of the development of seizures and of the biochemical phenomena that accompany the gradual lowering of seizure threshold. Because the kindling effect is permanent, the biochemical nature of this long-lasting effect is the subject of much study.

That kindled seizures could develop through gradual modification of muscarinic or other receptors is an interesting proposal. Burchfield *et al.*[224] found that rats kindled through stimulation of the hippocampus were more responsive to *in situ* stimulation by ACh. Neuronal sensitivity to ACh was correlated with response of the hippocampal cells to additional kindling stimuli. These authors suggested that alteration of synaptic sensitivity to ACh may contribute to epileptogenesis and that increased ACh responsiveness could result from (1) increased postsynaptic receptors, (2) increase in sensitivity of receptors, (3) a decrease in AChE activity, or (4) alterations in other neurotransmitter systems affected by ACh. Results from several laboratories have been variable when [^{3}H]-QNB binding to receptors was measured. Morrell *et al.*[225] reported that the number of muscarinic receptors increased 2–3 times in hippocampi of animals kindled through hippocampal stimulation and killed shortly (18 hr) after their last stimulation. There was no change in receptor affinity. Measurement of receptor number in these animals after long resting periods as an indication of the permanency of this effect was not done. Some studies of the muscarinic receptor indicate that the cerebral excitability associated with seizure activity may exert a "down-regulation" of muscarinic receptors but that this effect is transient and may not be directly related to receptor occupancy. McNamara[226] reported a slight but significant decrease in muscarinic receptors in the electrode-implanted amygdala of rats that received hourly stimulations until fully kindled. The animals were then examined for receptor binding shortly after their last full seizures. Further studies established that a similar decline in muscarinic receptors occurred in the dentate gyrus of the hippocampus[227] but not in the cortex. In both the stimulated amygdala and the hippocampus, the decrease observed was due to fewer [^{3}H]-QNB-binding sites and not to a change in receptor affinity for this radiolabeled antagonist.

While kindling is permanent, the loss of muscarinic receptors is transient. Seeking to relate the long-lasting effect of kindling with permanent receptor loss, we measured [^{3}H]-QNB binding in a number of brain areas in animals electrically kindled in the amygdala, but rested 7–14 days after their last seizure. We were not able to observe long-term loss of binding in the amygdala, hippocampus, striatum, cortex, or cerebellum of these animals, suggesting that receptor loss occurred as a result of ictal events.[228] Dashieff *et al.*[229] studied the time course of the observed muscarinic receptor decrease and found that (1) animals having stage 4 seizures[223] showed a significant decrease in [^{3}H]-QNB binding; (2) animals having a single stage 5 seizure exhibited the greatest decrease in binding; and (3) as few as 3 days later, the receptor number was back to control levels. Dashieff and McNamara[230] suggested that the repeated neuronal depolarization occurring in kindling induces the temporary loss of muscarinic receptors, but the mechanism of this action remains unknown. Several lines of study suggested that the receptor loss was independent of agonist interaction. For example, in kindled rats that had previously been lesioned in the medial septum, the dentate gyrus of the hippocampus was examined and found to have a small decrease in [^{3}H]-QNB binding consistent with previous observations in kindled but nonlesioned animals, and occurring in spite of interruption of cholinergic input to the hippocampus from the medial septum.[230] These results suggest that other input to the hippocampus may be responsible for the decrease in receptors in amygdaloid kindling.

We conducted a study involving direct chemical kindling of the cholinergic system with carbachol given in a cannula implanted in the amygdala. Animals developed kindled seizures rapidly but did not do so when atropine, a muscarinic receptor blocker, was added to the carbachol given *in situ*. The chemical kindling was long-lasting. Various brain areas (amygdala, cortex, striatum, hippocampus, and cerbellum) were examined for cholinergic parameters after the animals were rested for 1 week. There were no significant differences in [^{3}H]-QNB binding, choline acetyltransferase or cholinesterase activities in carbachol-kindled animals when compared to controls.[231] While it is known that this type of kindling is initiated through the muscarinic receptors in the amygdala, the nature of the amygdala–hippocampus connection is not clear and presents problems for interpretation of the receptor events. In an attempt to clarify this problem, kindling was carried out in the medial septum, which is the source of a large monosynaptic muscarinic projection passing through the fimbria to the hippocampus and supplying over 90% of muscarinic synapses in the hippocampus. After animals were kindled, they were either killed 30 min after their last seizures (acute) or rested for 1 week. There was no differences in [^{3}H]-QNB binding in either group in either the hippocampus or the cortex.[232] Because no acute changes were noted in this system, this observation suggests that the transient decreases in binding previously observed in amygdaloid kindling may not be present in all types of kindling (i.e., medial septal vs. amygdaloid).

The muscarinic system has been studied in other seizure models. Coutinho-Netto *et al.*[221] studied endogenous levels of ACh and muscarinic receptors in the cortex made epileptic by topical freezing. Synaptosomes were prepared from focal and adjacent cortex at 12 hr post-initiation of the lesion. Elevated ACh levels (3- to 4-fold) in epileptic areas were found and there was also an increase in spontaneous release. There was a 12% decrease in muscarinic receptors occurring only in the focal area and not evident in adjacent cortical areas. These authors suggest that the observed decrease in receptors could be due to prolonged exposure to muscarinic agonist (down-regulation) while noting that no changes were observed in adjacent epileptic cortex where high ACh concentrations were also observed.

If epileptic activity produces a sustained increase in ACh release, a subsequent decrease in muscarinic receptors, and an increase in muscarinic-mediated hyperexcitability, one could expect that carbachol-kindled animals would be more susceptible to electrical kindling and vice versa. We have studied the interaction between carbachol and electrical amygdaloid kindled rats as a means of discovering a "transfer effect" due to cholinergic-induced hyperexcitability. We found that there was no change in the kindling rates of animals that had been previously kindled with either carbachol or electrical stimulation and retested in the alternate mode. The conclusion drawn from this study is that the involvement of muscarinic receptors in various types of seizure models may not be uniform or even necessary.[231]

4.3.1b. ECS. Muscarinic receptors do not appear to be reduced in animals that have received a series of daily ECS.[233] Examining the cortex and hippocampus of rats receiving a series of 14 ECS and killed 1 day after the last shock, no changes in muscarinic receptor binding were found. We also measured [^{3}H]-

QNB binding in the forebrains of rat neonates immediately after a single ECS and did not observe any change in receptor density in these animals. Muscarinic receptors have been measured in mice receiving repetitive electroshock in a single test period[234] as a model of status epilepticus. We found no changes in receptors in forebrain from animals killed 15 min after their last seizure. We also measured cGMP levels in a group of similarly treated animals, since desensitization of muscarinic receptors has been shown to decrease carbamylcholine-stimulated cGMP formation. cGMP levels increased sharply after the first seizures and remained elevated throughout the procedure. These data support a lack of desensitization of muscarinic receptors during repetitive seizures.

It appears that muscarinic receptors are either not affected by seizure activity or are transiently and minimally decreased in some seizure models. It is not known if decreased muscarinic receptor concentrations could contribute to the development of seizures. It may well be that receptors down-regulate as a result of depolarization of exposure to excess amounts of ACh released during seizures. If comparisons to *in vitro* experiments with synaptosomes can be made, then depolarization and exposure to agonists may cause down-regulation of muscarinic receptors[235] with the process being reversible and requiring protein synthesis.[235–237] Alternatively, measurement of receptor concentration and affinity by use of an antagonist may not give enough information to allow for a complete explanation of receptor activity.

4.3.2. *Catecholamine Receptors*

Early studies on the adrenergic system indicated that diminished functional activity of this system may be implicated in the etiology of depression.[238] The possibility that ECS may exert its antidepressant effect through the catecholamine systems has provided the basis for studies of this receptor in various seizure models. In a number of cases, β-adrenergic receptors as well as dopaminergic receptors have decreased in number in response to seizures (Table XVII). While the mechanism for these changes in not known, it does seem probable that the alteration in receptor number produces a change in functional aspects of this as well as other related systems.

4.3.2a. ECS. Bergstrom and Kellar[239] reported that a single ECS in rats produces no discernible changes in β- or α-adrenergic or serotoninergic receptors. However, repeated daily exposure to ECS had a down-regulatory effect (−25%) on β receptors in the cortex, while having no effect on the other receptors in the cortex and on dopaminergic receptors in the striatum. Thus, the effect is rather specific. In a further study, these authors found that receptors significantly decreased after 4 days of exposure to ECS (1 shock/day) and that the decrease was still present 7 days after the seizures. Receptors also decreased by 24% in the hippocampus.[233] In a related study, Gillespie *et al.*[240] reported a rapid subsensitivity of the adenylate cyclase system linked to β receptors in rats that experienced a single ECS. These animals had a lowered response to adrenergic stimulation of cAMP formation in the cortex (−37%)

Table XVII
β-Adrenergic and Dopamine Receptors and Seizures

Type of seizures	Area	Change	Time	Reference
	β-Adrenergic			
ECS				
Single	Cortex	↓	18 hr	240
Single	Cortex	N.C.	24 hr	239
Multiple	Cortex, hippocampus	↓	24 hr	233
Kindling				
Electrical-amygdala	Amygdala	↓	3 days	344
Electrical-medial septum	Cortex	N.C.	1 week	232
β-Carboline[a]	Cortex	N.C.	1 week	241
	Dopamine			
ECS-multiple	Striatum	N.C.	24 hr	239
Kindling				
Electrical-amygdala	Cortex	↑	24 hr	244
	Amygdala	↓		
Pentylenetetrazol	Amygdala, cortex	↓	30 hr	245

[a] Drug-induced kindling occurred after daily i.p. injections.

and in the limbic forebrain (−58%). In this study the authors also observed a 17% decrease in dihydroalprenolol binding in the frontal cortex.

4.3.2b. Kindling. Electrical kindling of the amygdala[222] is a gradual process that culminates in full clonic seizures. Examination of β receptors in the site of the subconvulsant electrical stimulation, the basolateral amygdaloid nucleus, showed a significant decrease in receptor number. There were no receptor changes in adjacent areas or in animals receiving electrical current with no subsequent kindling.[226]

Down-regulation of receptors occurs in different neuroanatomical sites in response to various stimuli, implying that very specific circuitry is turned on as a result of the stimuli. In some types of drug-induced seizures, this circuitry is not easily discerned and the seizure itself does not appear to down-regulate the receptors. One such type of kindling has been through the use of β-carboline,[241] a benzodiazepine receptor antagonist. We observed no changes in [^{3}H]-DHA binding in β-carboline-kindled animals. While the mechanism of this type of kindling is not known, β-carboline has been shown to bind to the benzodiazepine receptor and to antagonize *in vivo* the action of benzodiazepines, a major class of anticonvulsants.[242,243] The motor seizures resulting from this kind of stimulation apparently do not produce the appropriate stimuli for down-regulation of the β receptor.

Amygdaloid kindling produces down-regulation of high-affinity [^{3}H]spiperone binding in both the stimulated and the contralateral amygdala. No changes in high-affinity binding were observed in the striatum or cortex. Low-affinity binding of this ligand was increased in the cortex of the kindled animals; however, it is not known if the binding of this ligand in the cortex is to DA or 5-HT receptors.[244] These same authors reported a decrease in high- and low-affinity receptors in pentylenetetrazol kindling induced with i.p. injections.[245]

It is quite apparent that the role of catecholamines in seizures is complex. While depletion of catecholamines by reserpine or 6-hydroxydopamine facilitates amygdaloid kindling,[211] kindling itself causes depletion of catecholamine content[214] and down-regulation of the β receptor.

The role of such alterations in the etiology of seizure disorders is enigmatic.

4.3.3. GABA

The association of GABA with seizure disorders has a long history. Early observations that dietary deficiencies in vitamin B_6 in humans resulted in seizures were subsequently explained by the requirement of pyridoxal phosphate for glutamate decarboxylase activity.[246,247] B_6-deficient animals may develop spontaneous seizures.[248]

Some convulsants (e.g., hydrazine) are believed to act through inactivation of glutamate decarboxylase[249] and subsequently lower the endogenous GABA levels. Interference with GABA synthesis has been shown to lower threshold to experimentally induced seizures.[250] Also, compounds shown to block the postsynaptic inhibitory action of GABA have also been shown to cause convulsions, e.g., bicuculline and picrotoxin.[251] GABA itself has been shown to stop seizures when given *in situ*.[252] The concept that seizures could be mediated through decreased function of the GABA receptor has provided the basis for study of this system in a number of experimentally induced seizure models.

4.3.3a. Kindling. One extensively studied model is kindling. It has been found that agents such as 3-mercaptopropionic acid (a GABA synthesis inhibitor) and bicuculline (a GABA receptor antagonist) actually facilitate kindling,[253] presumably by lowering the level of GABA-mediated inhibition. These authors did not observe any suppression of seizure development by pretreatment with the GABA metabolism inhibitor γ-vinyl GABA or with imidazole acetic acid, a GABA agonist. Alternatively, blockade of GABA-transaminase by pretreatment with α-acetylenic GABA significantly reduced or eliminated behavioral seizures associated with amygdaloid kindling while having a less pronounced effect of EEG afterdischarges.[254]

The results imply that levels of endogenous GABA may be maximally involved in suppression of seizure activity and that kindling may occur in spite of pharmacologically elevated GABA levels induced by GABA-transaminase inhibitors.

In an effort to determine if the kindling process itself may decrease the overall function of the GABA system, we have carried out a number of different kinds of kindling and studied the binding of [^{3}H]muscimol to the postsynaptic GABA receptor. In an extensive study, rats were kindled by stimulation of the basolateral amygdaloid nucleus. The GABA receptors were measured both immediately after a kindled seizure and after a resting period of 7 days. No changes were observed in [^{3}H]muscimol binding in either case in the cortex and hippocampus and no changes were observed in the striatum or cerebellum in acutely seized kindled animals. A parallel experiment carried out on animals kindled by stimulation of the medial septum gave similar results in the cortex and in both left and right hippocampi. We carried out chemical kindling with bicuculline given by cannula in the motor cortex. Examination of GABA receptors in the cortex in both focal and contralateral cortex by measuring [^{3}H]muscimol binding showed a small but insignificant decrease in the cortex from kindled rats. Chemical kindling with β-carboline showed no difference in [^{3}H]muscimol binding in the cortex either. The kindling process is a developmental phenomenon requiring repeated exposure to stimuli. It would seem logical that the lowered seizure threshold that is integral to the kindling process would be accompanied by a lessening in the efficiency of the GABA system. However, this remains to be shown. It is quite possible that kindling lowers GABA levels and/or receptor efficacy in neuroanatomically distant but physiologically relevant areas that have not as yet been identified.

4.3.3b. Genetic Models. A second type of seizure model in which the GABA system has been studied is the genetically audiogenic seizure-prone mouse (DBA/2). Drug studies suggest an impairment of GABA-mediated neurotransmission in these DBA/2 mice. These mice do not appear to have impaired synthetic enzyme, glutamate decarboxylase, or lower GABA concentrations[255] in whole brain. Systemic treatment with γ-acetylenic GABA or γ-vinyl GABA, inhibitors of GABA-transaminase, protects the DBA/2 mice from audiogenically induced seizures. Seizure incidence is reduced to 50% when whole brain GABA levels are increased to about 320% of controls.[256]

Examination of high-affinity GABA receptors by [^{3}H]GABA binding shows significant differences in whole forebrain membranes. Highly seizure-prone animals have fewer high-affinity binding sites (about 10%) but with higher affinity than the controls.

Only at the earliest stages of development is there a significant decrease in the B_{max} for the low-affinity receptor in the seizure-prone animals. While the decrease in receptor numbers may lead to a decrease in GABA-mediated inhibition, no direct correlation can be made between decreased inhibition and seizure susceptibility since these two events do not follow the same age-dependent time course in these animals.[257]

4.3.3c. ECS. A third approach to the study of seizures has been through the induction of seizures by ECS. We have given single ECS to both adult and neonate rats and adult mice and given repetitive ECS to mice. In none of these cases did we observe any change in [^{3}H]muscimol binding to extensively washed membranes prepared shortly after the seizure.

We have also examined the binding of [^{3}H]muscimol to the membranes of the forebrain and cerebellum in mice given convulsant doses of bicuculline. No change in GABA receptor binding was noted in the forebrain, but a significant increase was noted in the cerebellum. Examination of cortical membranes from mice given convulsant doses of pentylenetetrazol showed a slight but significant decrease in receptor binding after these seizures. The inconsistent and small changes observed in these two examples stand out as the sole exceptions for changes in GABA receptors in various seizure paradigms. Evidence to date suggests that seizure activity does not cause changes in GABA receptors in spite of changes in GABA metabolism.

What appears to be the more consistent observation is that reinforcement of GABAergic transmission has an anticonvulsant effect and that seizure activity may in fact lower the overall components of the GABA system.

Repeated audiogenic seizures produce a significant decrease in endogenous GABA levels in specific areas of the brain, i.e., nucleus caudatus, substantia nigra, hippocampus, and cortex.[258] Examination of epileptic foci after alumina gel-induced seizures showed a decreased number of glutamate decarboxylase-containing terminals in the epileptic foci.[259] When the GABA agonist muscimol was given *in vivo*, it delayed the onset of isoniazid- and picrotoxin-induced seizures, and eliminated the tonic forelimb extension associated with bicuculline and metrazole.[260]

When GABA-transaminase blockers are given *in situ* in the substantia nigra, the maximal subsequent buildup of nerve terminal GABA concentrations correlates directly with the maximal protection against bicuculline-, pentylenetetrazol-, and ECS-induced seizures.[261] It has been further established that not all areas in which GABA levels can be elevated offer the anticonvulsant effect. Severalfold increases in GABA in the cortex, hippocampus, striatum, thalamus, and hypothalamus do not protect against ECS, whereas increase in the midbrain area containing the substantia nigra was quite effective.[262] The GABA receptors, which are activated by large focal concentrations of GABA in the nigra, exert their anticonvulsant effect because of their unique location. This area appears to be the conduit for neuronal pathways involved in motor manifestations of seizures.

4.3.4. Benzodiazepine Receptors

The recent demonstration that benzodiazepines, a major class of anticonvulsants, bind with high specificity and affinity to receptors within the nervous system has indicated a number of experiments to determine the biochemical mode of action of these drugs.[263] There is excellent correlation between the affinity of various benzodiazepines for receptor binding and their clinical potency as anticonvulsants and antianxiety agents. Braestrup and Squires[264] strongly suggest that these receptors are involved in the mechanism of seizure suppression. In understanding the mechanism of action of these drugs, it may be possible to determine the nature of seizure propagation both in animal models and in human epilepsy.

4.3.4a. ECS. The first report of modification of benzodiazepine receptors in response to seizures was by Paul and Skolnick.[265] Rats having a single ECS were shown to increase the number of receptors in cortex by 20% within 15 min of the seizure. (Table XVIII) The concentration remained elevated for an additional 15 min and by 1 hr postictally had returned to preseizure levels. This

Table XVIII
Benzodiazepine Receptors and Seizures

Type of seizures	Area	Change	Time	Reference
ECS-single	Cortex		15 and 30 min	265
	Cortex	N.C.	30 min	266
	Forebrain, 13 day old	N.C.	30 min	[a]
ECS-multiple	Cortex forebrain, hippocampus	N.C.	30 min	234
Genetic Models				
DBA/2J	Forebrain		22 days old	268
DBA/2J	Forebrain		28–29 days old	257
DBA/2J	Forebrain	N.C.	48 days old	269
Mongolian gerbil	Cortex, hippocampus, brainstem, striatum		5 min after seizure	267
Kindling—electrical				
Medial Septal	Cortex, hippocampus	N.C.	30 min and 1 week	[a]
Amygdala	Hippocampus		24 hr	270
	Amygdala, cortex		24 hr	
Kindling—drug				
Beta carboline	Cortex, hippocampus	N.D.	1 week	241
Pentylenetetrazol	Forebrain		18 hr	272

[a] Morin and Wasterlain (unpublished).

effect was not confirmed by Bowdler and Green,[266] who after a single ECS failed to find changes in benzodiazepine receptor number in cortical membranes. Mice given repeated seizures in a model of status epilepticus did not show any change in benzodiazepine receptors in the forebrain, hippocampus, or cortex.[234] Benzodiazepine binding sites have been measured in two genetically seizure-prone animals, the Mongolian gerbil[267] and the DBA/2J mouse.[257,268] Asano and Mizutani[267] found that [^{3}H]diazepam binding increased rapidly (within 10 min) in the cortex, hippocampus, brain stem, and striatum after a seizure with the elevation due to an increase in receptor affinity. It is possible that a postictal increase in GABA may have increased the binding affinity since the membranes were not washed prior to assay to reduce the endogenous level of GABA. This increase differs from that observed by Paul and Skolnick[265] both in its nature (higher affinity) versus concentration, and in its time course (return to control levels within 20 min).

Inconsistent results have been found in the audiogenic seizure-prone DBA/2 mouse. Robertson[268] found that increased [^{3}H]flunitrazepam binding in the seizure-prone mice was age-related and independent of their seizure state (interictal vs. ictal). The increase in receptor number was significant at 22–24 days of age but by age 35 days was no longer significant. The strain to which the DBA/2 mice were compared was a seizure-resistant strain, C57BL/6J. Horton *et al.*[257] carried out a study on the binding of [^{3}H]flunitrazepam in DBA/2 mice of different ages. They found that only at days 28–29 was there a decrease in the B_{max} in the DBA/2 mice. This is in contrast to the work of Robertson in which an increased number of receptors was present at a slightly younger age. Horton *et al.*[257] were comparing their seizure-prone animals to TO mice, which have a lower audiogenic seizure susceptibility. This may account for the incompatibility of the two sets of data. However, examination of DBA/2 and C57 mice at 21 and 43 days of age showed no differences in ^{3}H-benzodiazepine-binding sites in a separate series of experiments.[269]

4.3.4b. Kindling. We have examined [^{3}H]diazepam binding in membranes prepared from animals kindled through stimulation of the medial septum. Hippocampal membranes were prepared from animals killed either 30 min or 1 week after a seizure. In neither group of animals did we find any change in binding of [^{3}H]diazepam in the cortex or hippocampus. There were no significant differences in the ability of GABA (10 μM) to stimulate benzodiazepine binding[232] between control and kindled tissues. In contrast, in amygdaloid kindled animals, McNamara *et al.*[270] found a 35% increase in [^{3}H]diazepam binding in left and right hippocampi. No change in receptor number was seen in the cortex or amygdala. Further examination showed that the number of seizures was critical. Animals requiring a greater number of stimulations (16.3) had significantly greater increases in receptor number than animals requiring fewer (10.3) stimulations to achieve kindled status. Animals given repetitive ECS treatment also increased receptors in response to a greater number of seizures (experiencing only an 8% increase after 7 ECS, and a 19% increase after 17 seizures). Further examination of the site of increase within the hippocampus indicated that the change occurred within the granule cell and molecular layers of the fascia dentata.[271]

Kindling with drugs has not given uniform responses to the binding assays for benzodiazepine receptors. Pentylenetetrazol-injected mice experiencing multiple seizures in a kindling paradigm showed a 34% increase in B_{max} in forebrain that Syapin and Rickman[272] suggest may be a response to the drug and not seizure activity. We have kindled rats with norharman, a β-carboline shown to antagonize the benzodiazepine receptor both through competition with diazepam for binding sites and through behavioral effects opposite to diazepam.[241–243] We did not find any lasting receptor changes in binding of [^{3}H]diazepam or [^{3}H]-β-carboline ethyl-ester, previously shown to bind to receptors with high affinity,[273] or of [^{3}H]muscimol in various brain areas in these animals, even though their response to the drug had increased and this increase persisted for at least 7 days.[241]

The nature of receptor modifications appears to be quite variable. The observations that seizure activity can cause an immediate increase in receptor density would mechanistically fit into a model of increased inhibition and enhanced anticonvulsant activity. However, this increase is not always seen and may in fact be very specific with regard to seizure origin. It has been shown that there is a direct correlation between benzodiazepine receptor occupation and some anticonvulsant effects of diazepam.[274] Perhaps the role of these receptors in seizures will be elucidated when their endogenous ligand(s) is identified.

4.3.5. *Opiate Receptors*

4.3.5a. Opioid-Induced Seizures. An observation that the endogenous opioid compounds could induce seizure activity was made coincidentally during a study on the analgesic properties of morphine and enkephalin. Intracerebroventricular (i.c.v.) administration of methionine-enkephalin and morphine produced electrographic and behavioral epileptic activity in rats within the first minute of exposure.[275] The behavioral component included twitching and "wet dog shakes" while the EEG gave evidence of repetitive abnormal rapid spiking. Because the analgesia and epileptic seizures did not always occur in the same animals in the same time frames, it appeared unlikely that both effects would have a common mechanism. Naloxone, an opiate receptor antagonist, blocked the EEG effects elicited by i.c.v. morphine and enkephalin. A subsequent study[276] showed that doses as low as 10 μg of methionine- or leucine-enkephalin produced no analgesia but persistently gave EEG abnormalities. Systemic naloxone blocked these effects, suggesting that seizures resulted from an interaction of these compounds with opiate receptors. Leucine-enkephalin had greater epileptic potency than methionine-enkephalin and both were more potent than morphine. In this same study, repeated *in situ* exposure to morphine indicated that the epileptogenic effect may be subject to both tolerance and potentiation.

The potency of the enkephalins in inducing seizures as opposed to analgesia suggested to Frenk and co-workers that the site of application and relative concentration of receptors in these sites was critical to the epileptic effect. Application in the lateral ventricle would allow more ready access to forebrain structures known to be involved in epileptogenesis, and provide less access to mid- and hindbrain areas mediating analgesia.

Frenk *et al.*[276] showed that injection of methionine-enkephalin near or into the forebrain dorsomedial nucleus of the thalamus induced seizure activity, while analgesia was induced by application near the ventral caudal midbrain and periaqueductal gray matter. These authors also suggested that perhaps not only are the opiate receptors mediating epileptogenic and analgesic activity located in different brain areas, but that they may in fact be pharmacologically distinct, i.e., the sigma receptors in the dorsomedial thalamus mediate seizures while the mu receptors of the periaqueductal gray mediate analgesia. Evidence has also been gathered indicating there may be two opiate-sensitive systems, one with excitatory-epileptogenic effects and a second possessing inhibitory anticonvulsant properties. While i.c.v. injections of enkephalin and morphine cause naloxone-sensitive EEG seizures, pretreatment with systemic injection of morphine is potently anticonvulsant against these seizures.[277] Animals that were given 520 nmol morphine i.c.v. experienced wet dog shakes, myoclonic twitching, and rapid EEG spiking. Pretreatment with systemic morphine of doses greater than 50 mg/kg abolished i.c.v. morphine-induced seizure activity while a higher dose of systemic morphine (100 mg/kg) was required to suppress enkephalin-induced seizures. Naloxone given systemically at 10 mg/kg blocked morphine-induced seizures and at 1 mg/kg suppressed enkephalin-induced seizures.

It is interesting that, when given systemically, both opiate agonists and antagonists, depending on concentration, block the epileptogenic activity of *in situ* administration of both opioids and opiates. One interpretation for this phenomenon is that systemically administered opiates reach anatomically diverse regions and may in fact activate anticonvulsant activity in areas quite distant from the injection site. Urca and Frenk[277] have found that preinjection of opiates into the nucleus accumbens but not the periaqueductal gray prevented leucine-enkephalin (i.c.v.)-induced seizures.

Henrickson *et al.*[278] have produced seizure activity through i.c.v. injection of β-endorphin, and found that lesions in the medial dorsal thalamus or amygdalectomy had no effect on β-endorphin-induced epileptogenesis. Henrickson *et al.*[278] suggested the hippocampus as a possible site for the epileptogenic activity of i.c.v. injections since neuronal discharge rates increase in the area in response to iontophoresis of opiates and endorphins.[279] This epileptogenic excitation may be due to a naloxone-reversible disinhibition of inhibitory basket cell interneurons. These observations and the fact that opioids and morphine differ in their rank potencies in eliciting epileptic and analgesic effects suggest the two processes involve anatomically and pharmacologically distinct receptors. The fact that i.c.v. levorphanol has no epileptic effect even at high concentration and yet is a potent anticonvulsant and analgesic supports this view.[280]

4.3.5b. Kindling. The role of endogenous opioids in amygdaloid kindled seizures appears to be in mediating postictal events while their participation in the seizure itself is doubtful or controversial.

Pretreatment with naloxone does not consistently alter the rate of amygdaloid kindling or alter the threshold for afterdischarge.[281] Only at a single dose (20 mg/kg) was kindling facilitated. Hardy *et al.*[282] reported that lower doses of naloxone (1 mg/kg) given prior to the kindling stimulus facilitated kindling such that at day 9 of stimulation, 83% of naloxone-treated animals reached stage 5 seizures while no untreated animals had stage 5 seizures. Naloxone at this dose had no effect on afterdischarge duration. Morphine significantly increased postictal depression time. They suggested that endogenous opioids may function to suppress seizures. Le Gal La Salle *et al.*[283] reported that pretreatment with morphine (5–10 mg/kg) increased the duration of afterdischarge and that this effect could be abolished by naloxone. Alternatively, Post *et al.*[284] examined the effect of pretreatment with morphine (10 mg/kg) and naloxone (10 mg/kg) on amygdaloid kindling and found no change in kindling rate or afterdischarge duration.

Frenk *et al.*[285] found that morphine did not alter kindling rate but prolonged postictal behavioral depression (from 102.5 sec to 566.9 sec) and increased frequency of interictal spiking (IISs). Naloxone shortened the postictal depression period from 114 sec to 1.3 sec and reversed the appearance of interictal spiking. Engel *et al.*[286] suggested that endogenous opioids may be released during seizures and mediate postictal depression and suppress further seizure activity. These authors pointed out that opioid-induced seizures resemble the postictal behavioral aspects of amygdaloid seizures, e.g., wet dog shakes, catatonic posturing, and myoclonic jerks. The increased uptake of 2-deoxyglucose into the hippocampus seen in opioid seizures is also seen in amygdaloid seizures.[278,287] Important recent developments in this area are the observations that hippocampal blood flow fails progressively during the kindling process[28] and that a similar failure follows hippocampal injection of very large doses of opiate agonists, an effect that is reversed by naloxone.[288]

Opiate receptors have not been reported in the cerebral vasculature; however, the role of endogenous opioids in vascular adaptations during seizures may represent an important area for future investigations.

4.3.5c. ECS. ECS produces postictal respiratory and behavioral depression. It has also been shown to activate an endogenous opioid system. Tortella *et al.*[289] found that pretreatment of animals with naloxone antagonized postictal depression, and suggested that this was due to an antagonism of endogenous opioids released during the seizure. Sarne *et al.*[290] and Haladay *et al.*[291] have reported that there is a regional increase in specific opioids in response to repeated ECS. Enkephalin was shown to increase in the hindbrain, mesencephalon, striatum, and hypothalamus, while β-endorphin levels increased in the striatum and hypothalamus. Hong *et al.*[292] and Green *et al.*[293] found increases in hypothalamic methionine-enkephalin but observed no change in levels of β-endorphin. A single ECS did not produce any observable differences in opioid levels. Tortella *et al.*[289] and Elazar *et al.*[294] observed that tolerance developed to the epileptogenic effects of enkephalins; however, the nature of this effect is not known. It has not been shown that the ECS-induced changes in the endogenous opioids are subject to tolerance. In fact, it appears that repeated ECS sensitized rats to opiate challenge.[295] It is not known if an increase in release of endogenous opioids or a change in opiate receptor number of affinity is the cause of this enhanced sensitivity.

4.3.5d. Conclusion. It appears likely that opioid seizures are induced through opiate receptors distinct from those that exert anticonvulsant and analgesic effects. There is evidence to suggest that these seizures may be elicited through limbic structures that are known to be susceptible to kindling, i.e., the hippocampus.

Lastly, it is likely that the behavioral depression experienced by animals having electrically induced seizures may be mediated in part through opiate receptors, since this depression is alleviated by opiate receptor blockade. The observed release of endogenous opioids during and after seizures may support the idea that they mediate such depression and the anticonvulsant state associated with it. However, little biochemical evidence is available to support these conclusions based mostly on pharmacological data, and the molecular mechanism involved remains largely unknown.

4.4. Second Messengers

This topic has been reviewed a number of times,[296–300] and is treated more extensively elsewhere in this handbook. It has been suggested that cAMP is a second messenger for several neurotransmitters, such as the biogenic amines, which have a predominantly inhibitory role in the mammalian brain. Indeed, cAMP analogs such as 8-bromo-cAMP (0.1–10 μM) consistently depressed evoked potentials and spontaneous activity in slices of guinea pig hippocampus, suggesting an anticonvulsant action.[298] By contrast, the cGMP analog 8-bromo-cGMP increased the duration of evoked potentials and enhanced spontaneous activity in slices.[298] Iontophoretic application of cGMP onto cortical neurons in the cat increased their firing rates.[300] A similar effect was observed in rat pyramidal tract neurons.[301] Some evidence has linked cGMP with the action of predominantly excitatory neurotransmitters such as ACh acting on muscarinic systems. However, definite proof that cGMP mediates muscarinic depolarization in the CNS is still lacking.

4.4.1. cAMP

ECS and a number of convulsants markedly elevate the brain concentrations of cAMP.[302–312] The increase in cerebral cAMP concentration becomes apparent only after the onset of EEG seizures or of clinical convulsive activity.[308,312] It would therefore appear to be a result rather than a cause of the epileptiform activity. However, it is not simply a consequence of anoxia or other systemic changes in blood pressure, acid–base balance, etc. since the elevation in cAMP in brain is observed in the appropriate brain regions during focal seizures induced by penicillin[313] and freezing,[314] It is also seen in animals that are paralyzed and ventilated with oxygen during their seizures, so that convulsive activity is prevented and no anoxemia takes place. Anticonvulsants block or reduce the elevation of cerebral cAMP induced by seizures.[298,304,305,307–309,311] A number of other drugs modify cAMP accumulation in brain in response to seizures. Some may act as adenosine blockers, others have no known mechanism of action.[298,305,307,311]

Depletion of cerebral stores of catecholamines by reserpine reduces or prevents the seizure-induced elevation of cAMP. Furthermore, depletion of NE by 6-hydroxydopamine decreases seizure latency and threshold and reduces seizure-induced accumulation of cAMP in brain. It also facilitates kindling. The β-adrenergic receptor antagonist propranolol reduces cAMP accumulation in response to seizures, an effect mimicked by the α_2-adrenergic receptor antagonist yohimbine and opposed by the α_1-adrenergic antagonist phentolamine. The adenosine receptor blocker aminophylline also facilitates seizures and inhibits cAMP accumulation in brain.[298] Therefore, it appears at least part of the action of cAMP reflects a second messenger function resulting from the liberation of biogenic amines playing a predominantly inhibitory role, perhaps in response to epileptic activity in brain. It is possible that the increase in cAMP activity plays a role in the inhibitory response of the brain to seizures. A direct role in seizure termination is not obvious since the time courses of seizure termination and cAMP elevation are quite different. However, the facilitation of seizures and marked reduction of cAMP elevation by locus coeruleus lesions or by chemical amputation of noradrenergic systems favors such a hypothesis. The mechanism by which cAMP effects its actions during seizure activity has not been studied extensively. The activation of cAMP-dependent protein kinases is bound to play a role, in many different adaptive reactions, but only one is relatively well understood. Activation of local serotoninergic-dependent noradrenergic mechanisms resulting in cAMP elevation in the cortex, plays an important role in the activation of phosphorylase through cAMP-dependent phosphorylase kinase, and therefore plays an essential role in the mobilization of glycogen during sustained seizures such as status epilepticus. Elimination of those noradrenergic pathways ablates the ability of the brain to mobilize glycogen in response to seizure activity.[315]

4.4.2. *cGMP*

Epileptic seizures elevate cGMP levels in brain.[302–304,306,307,312,313,316] The elevation of cGMP is observed in the epileptic focus during partial seizures.[313,314] It is also seen in animals in which anoxia is prevented by paralysis and ventilation with oxygen. Furthermore, many convulsants induce an increase in cGMP activity before clinical or electrographic seizures are evident.[297,308,312] Furthermore, dibutyryl cGMP is epileptogenic when injected into the cerebral cortex.[312] and cGMP analogs have powerful excitatory actions resulting in epileptiform activity when applied iontophoretically to hippocampal transplants maintained in the anterior chamber of the eye.[316a,b] Anticonvulsants block the accumulation of cGMP in brain, as they do that of cAMP.[304,305,308] In the kindling model of epilepsy, however, we have been unable to find any change in cGMP in the hippocampal target cells before clinical seizures are evident, and it appears that the time course for cGMP elevation is too slow to account for the rapid depolarization of neurons in response to kindled stimulation. Therefore, it seems that cGMP plays an important regulatory role in epilepsy, but that role has yet to be defined.

4.5. *The Expressed Message: Synaptic Phosphoproteins*

The cell regulates the activity of its components by modifying the shape and charge of important molecules, using varied means. In addition to allosteric or transcriptional regulation of enzymes, these encompass methylation of phospholipids, carboxymethylation of key proteins such as calmodulin, and most commonly it involves the addition or subtraction of phosphate groups to membrane proteins. The effectors of the latter reaction are protein kinases. Generally speaking, some protein kinases are stimulated by cyclic nucleotides, while others are stimulated by calcium. The cyclic nucleotide-dependent reactions tend to be rather slow and involve adaptive processes over a relatively long time course, while the calcium-dependent processes are rapid and account for cellular adaptations that may be over in a matter of milliseconds.[317–320] Both of those processes appear to play important roles in cellular adaptation during seizures, but the very limited amount of work done in this area has led to fragmentary information that at present is not easily integrated into a conceptual framework.

4.5.1. Effects of Generalized Seizures

Ehrlich *et al.*[321] reported transient changes in the phosphorylation of a cortical membrane protein of 18,000 daltons following electroconvulsive shock. Delgado-Escueta and Horan[322] found similar changes in proteins of 18,000 to 20,000 daltons following metrazol or electroconvulsive seizures. In animals receiving repeated ECS, these changes were associated with modifications of the ionic permeability of synaptosomes *in vitro*, suggesting that the change in phosphorylation resulted in a modification of ionic channel permeability. Metrazol and picrotoxin have also been found to increase the state of phosphorylation of synapsin I and of other proteins after *in vivo* administration of the convulsive agents.[323]

This work has recently been extended to two genetic models of epilepsy. Tuchek *et al.*[324] reported modification of the baseline phosphorylation of a 16,000-dalton protein in the brains of homozygous epileptic fowl; heterozygotes and noncarriers did not show this abnormality. Bajorek *et al.*[325] reported similar changes in seizure-prone gerbils. There has been little work on cAMP- or cGMP-dependent protein kinases during seizures, but in view of the large increases in neurotransmitter release and in cyclic nucleotide synthesis during epileptic activity, it is likely that these enzymes are involved in the regulation of many cellular processes associated wtih epilepsy.

4.5.2. Calcium- and Calmodulin-Dependent Processes

In the resting cell, calcium ions are present in the cytosol in very low concentration (0.1 μM). In mitochondria, calcium ions exist in a separate pool highly segregated from the cytoplasm, which plays an important buffering role in case of calcium overload during excitation. The calcium concentration in the extracellular space ranges from 1 to 2 mM, and is therefore over a thousand times higher than that in the cytosol. Additionally, intracellular stores of calcium exist in the endoplasmic reticulum, and in specialized cells such as muscle, they can reach very high concentrations (20 mM). The extremely sharp gradients between those pools are maintained by energy-dependent mechanisms. The calcium pump (Ca^{2+}, Mg^{2+}-ATPase) present in both the plasma membrane and the endoplasmic reticulum, hydrolyzes ATP in order to transport calcium outside the cytosol against a concentration gradient. The sodium gradient, itself resulting from ATP hydrolysis, is also used to maintain a calcium gradient, by calcium-sodium exchange through specific channels. This extremely steep calcium gradient allows calcium to be both a rapid trigger and an amplification mechanism when calcium channels open in the membrane. At the same time, precise buffering in specific domains probably closely related to the calcium channels and involving calcium-binding proteins is likely to restrict the extent of calcium diffusion and to precisely regulate the effects of channel opening.

Both cAMP-dependent protein kinases[326–330] and calcium plus calmodulin-dependent protein kinases catalyze phosphorylation of membrane protein or proteolipids that control calcium influx and efflux. For example, calcium and calmodulin protein kinase in muscle activates phospholamban, which stimulates the calcium pump in sarcoplasmic reticulum.[331] This helps terminate the action of calcium by pumping it out of the cytosol. Calcium-dependent phosphorylation can also modify the influx of calcium into cytosol. For example, in muscle, a 60,000-dalton protein that is phosphorylated in a calcium- and calmodulin-dependent manner is involved in closing a gate in the calcium release channel, previously open in the absence of a proton gradient.[332] In certain neurons, the conductances of potassium channels[333] and calcium channels[334] in the neuronal membrane are regulated by intracellular calcium ion concentrations. These processes, which have not been studied extensively in specific neuronal populations, are probably of great importance in regulating the expression of seizure activity.

The state of phosphorylation of synaptic vesicle membrane proteins modulates the amount of several neurotransmitters released from these vesicles *in vitro*.[335] It is possible that the effects of some anticonvulsants, and the long-term plastic regulation of synaptic excitability may both involve the calcium- and calmodulin-dependent phosphorylation of synaptic proteins.[336] Kindling is a model of epilepsy in which repeated stimulation of some brain sites with small amounts of electrical current[222] or with agonists of excitatory neurotransmitters[337] results in a progressive buildup of response culminating in full-fledged epileptic seizures upon every stimulation. Once established, it may last a lifetime. It is highly selective for some brain sites, and there is good evidence that the key changes are transsynaptic.[337] In recent studies, we found that kindling by stimulation of the medial septal nuclei was associated with an inhibition of the post hoc phosphorylation of several synaptic plasma membrane proteins of the kindled hippocampus. In control rats, the ^{32}P incorporation into proteins of 50,000, 58,000, and 60,000 daltons was markedly stimulated by calcium plus calmodulin, whereas in kindled animals, the response to combined calcium and calmodulin was reduced. Calcium alone, cAMP, or cGMP modulated ^{32}P incorporation into several synaptic plasma membrane proteins but did not differentiate control from kindled tissues. Both control and kindled rats showed nonspecific inhibition of calcium plus calmodulin-stimulated phosphorylation in the post hoc assay by ACTH and by leucine-enkephalin. The differences between control and kindled animals were most striking in the hippocampus and in the amygdaloid–pyriform–entorhinal area, less pronounced in the cortex, basal ganglia, and brain stem, and were not present in the cerebellum, a region where kindling cannot be elicited. They were not found in stimualted controls. An 8-week period of rest after kindling did not reduce these changes, suggesting that they may be as persistent as the kindling behavior itself. Benzodiazepines (e.g., diazepam, 10 μM in brain) blocked the development of kindling and that of the associated changes in hippocampal synaptic protein phosphorylation. *In vitro*, diazepam inhibited the calcium plus calmodulin-stimulated phosphorylation of synaptic plasma membrane proteins with an IC_{50} of 20–50 μM, in both control and kindled animals (Farber and Wasterlain, in preparation). Control and kindled hippocampi had similar levels of protein phosphatase activities, while calcium plus calmodulin-dependent kinase activities were reduced by about half in membranes from kindled animals, suggesting either substrate limitation or kinase down-regulation.

These results suggest an important role of calmodulin kinase and of calcium plus calmodulin-dependent phosphorylation in the regulation of synaptic excitability, and possibly in the pathogenesis of the kindling phenomenon. They also suggest mechanisms by which the amount of activity through a synaptic apparatus could, by modulating calcium influx and calcium-dependent phosphorylation, regulate the excitability of that synaptic apparatus; and for the first time a molecular mechanism involved in epileptogenesis appears to also be the site of action of several major anticonvulsants.

ACKNOWLEDGMENTS. The authors' research was supported by the Research Service of the Veterans Administration and by Research Grant NS-13515 from NINCDS. We are indebted to Eva Csiszar and Sheryl Weems for technical assistance, and to Joan Semasinghe, Phyllis Oblander, and Kim De Lisi for help in preparing the manuscript.

REFERENCES

1. Temkin, O., 1945, *Of the Falling Sickness*, Johns Hopkins Press, Baltimore.
2. Penfield, W., and Jasper, H., 1954, *Epilepsy and Functional Anatomy of the Human Brain*, Little, Brown, Boston.
3. Kaufman, P. Y., 1912, *Obz. Psikhiat. Nevrol. Eksp. Psikhol. (St. Petersburg)* **7–8**:3–24, 513–535.
4. Berger, H., 1831, *Arch. Psychiatr. Nervenkr.* **94**:16–60.
5. Gibbs, R. A., David, H., and Lennox, W. G., 1935, *Arch. Neurol. Psychiatry* **34**:1133–1148.
6. Kurland, L. T., 1960, *Epilepsia* **1**:143–161.
7. Messell, S., Torres, J. M., and Kurland, L. T., 1962, *Arch. Neurol.* **7**:37–44.
8. Leibowitz, U., and Alter, M., 1968, *Epilepsia* **9**:87–105.
9. Matsumoto, H., and Ajmone-Marsan, C., 1964, *Exp. Neurol.* **6**:305–326.
10. Goldensohn, E. S., and Purpura, T. P., 1963, *Science* **139**:840–842.
11. Ayala, G. F., Dichter, M., Gumnit, R. J., Matsumoto, H., and Spencer, W. A., 1972, *Brain Res.* **62**:1–17.
12. Ayala, G. F., Matsumoto, H., and Gumnit, R. J., 1970, *J. Neurophysiol.* **33**:73–85.
13. Prince, D. A., 1978, *Annu. Rev. Neurosci.* **1**:395–415.
14. Johnston, D., Hablitz, J. J., and Winston, W. A., 1980, *Nature* **286**:391–393.
15. Wong, R. K. S., and Prince, D. A., 1978, *Brain Res.* **159**:385–390.
16. Wong, R. K. S., and Prince, D. A., 1981, *J. Neurophysiol.* **45**:86–97.
17. Gastaut, H., and Tassinari, C. A., 1966, *Epilepsia* **7**:85–138.
18. Binnie, C. D., Darby, C. E., and Hindley, A. D., 1973, *Br. Med. J.* **4**:378–379.
19. Lasater, G. M., 1962, *Arch. Neurol.* **6**:492–495.
20. Foster, F. M., Hansotia, P., Cleeland, C. S., and Ludwig, A., 1969, *Neurology* **19**:325–331.
21. Ingvar, D. H., and Nyman, G. E., 1962, *Neurology* **12**:282–287.
22. Delgado-Escueta, A. V., Wasterlain, C. G., Treiman, D. M., and Porter, R. J. (eds.), 1983, *Status Epilepticus*, Raven Press, New York.
23. Oakley, J. C., Sypert, G. W., and Ward, A. A., Jr., 1972, *Exp. Neurol.* **37**:300–311.
24. Sommer, W., 1880, *Arch. Psychiatr. Nervenkr.* **10**:631–675.
25. Corsellis, J. A. N., and Bruton, C. J., 1983, *Status Epilepticus* (A. V. Delgado-Escueta, C. G. Wasterlain, D. Treiman, and R. Porter, eds.), Raven Press, New York, pp. 129–140.
26. Norman, R. M., 1964, *Med. Sci. Law* **4**:46–51.
27. Peiffer, J., 1963, *Mongr. Neurol. Psychiatry* **100**:1–185.
28. Ackermann, R., Chugani, H. T., Finch, D. M., Babb, T. L., Lear, J. L., and Engel, J., Jr., 1983, *J. Cerebr. Blood Flow Metab.* **3**(Suppl.).

29. DeLorenzo, R., 1983, *Status Epilepticus* (A. V. Delgado-Escueta, C. G., Wasterlain, D. M., Trieman, and R. J. Porter, eds.), Raven Press, New York, pp. 325–338.
30. Scheibel, M. E., and Scheibel, A. M., 1973, *Epilepsy: Its Phenomena in Man* (N. Brazier, ed.), 1959/1960, Academic Press, New York, pp. 311–337.
31. Morrell, F., 1959/1960, *Epilepsia* **1**:538–560.
32. Abbott, R. J., Browning, M. C. K., and Davidson, D. L. W., 1980, *J. Neurol. Neurosurg. Psychiatry* **43**:163–167.
33. Arato, M., Erdos, A., Kurex, M., Vermes, I., and Feketerm, M., 1980, *Acta Psychiatr. Scand.* **61**:239–244.
34. Meldrum, B. S., Horton, R. W., Bloom, S. R., Butler, J., and Keenan, J., 1979, *Epilepsia* **20**:527–534.
35. Havens, L. L., Zileli, N. S., DiMascio, L. A., Boling, L., and Goldfien, A., 1959, *J. Ment. Sci.* **105**:821–828.
36. Weil-Malherbe, H., 1955, *J. Ment. Sci.* **101**:156–162.
37. Ohman, R., Walinder, J., Balldin, J., Wallin, L., and Abrahamson, L., 1976, *Lancet* **2**:936–937.
38. Meldrum, B. S., 1980, *Status Epilepticus* (A. V. Delgado-Escueta, C. G. Wasterlain, D. M. Treiman, and R. J. Porter, eds.), Raven Press, New York, pp. 399–403.
39. Jonec, V., Freckelton, W., and Wasterlain, C. G., 1980, *Neurology* **29**:597.
40. Nitsch, C., and Klatzo, I., 1983, *Neurol. Sci.* **59**:305–322.
41. Petito, C., Schaefer, J. A., and Plum, F., 1977, *Brain Res.* **127**:251–267.
42. Vesco, C., and Guiditta, J., 1968, *J. Neurochem.* **15**:81–85.
43. Cotman, C. W., Banker, G., Zornetze, S. V., and McGaugh, J. L., 1971, *Science* **173**:454–456.
44. Wasterlain, C. G., 1974, *Neurology* **24**:175–180.
45. Wasterlain, C. G., 1974, *Epilepsia* **15**:155–176.
46. Dwyer, B. E., Donatoni, P., and Wasterlain, C. G., 1982, *Trans. Am. Soc. Neurochem.* **12**:115.
47. Rosen, R., Kabat, H., and Anderson, J. P., 1943, *Arch. Neurol. Psychiatry* **50**:510–528.
48. Duffy, T. E., Howse, D. C., and Plum, F., 1975, *J. Neurochem.* **24**:925–934.
49. Kety, S. S., and Schmidt, C. F., 1943, *Trans. Assoc. Am. Physicians* **60**:52–58.
50. Sokoloff, L., and Kety, S., 1960, *Physiol. Rev.* **40**(Suppl. 4):38–44.
51. Plum, F., and Duffy, T. E., 1975, *Brain Work: The Coupling of Function Metabolism and Blood Flow in the Brain* (D. H. Ingvar and N. A. Tassen, eds.), Munksgaard, Copenhagen, pp. 197–214.
52. Nakai, M., Iadecola, C., Ruggiero, D. A., Tucker, L. W., and Reis, D. J., 1983, *Brain Res.* **260**:35–49.
53. Iadecola, C., Nakai, M., Arbit, E., and Reis, D. J., 1983, *J. Cereb. Blood Flow Metab.* **3**:270–279.
54. Pinard, E., Purves, M. J., Seylaz, J., and Vasquez, J. V., 1979 *Pfluegers Arch.* **379**:165–172.
55. Vasquez, J., and Purves, M. J., 1979, *Pfluegers Arch.* **379**:157–163.
56. Dwyer, B. E., and Wasterlain, C. G., 1983, *Neurology* **33**:213.
57. Posner, J. B., Brodersen, P., Paulson, O. B., Bolwig, E. G., Rogon, Z. E., Rafaelsen, O. J., and Lassen, N. A., 1973, *Arch. Neurol.* **28**:334–338.
58. Meldrum, B., and Horton, R. W., 1973, *Arch. Neurol.* **28**:1–9.
59. Wasterlain, C. G., and Dwyer, B., 1983, *Status Epilepticus* (A. V. Delgado-Escueta, C. G. Wasterlain, D. M., Treiman, and R. J. Porter, eds.), Raven Press, New York, pp. 241–260.
60. Wasterlain, C. G., and Graham, S. L., 1980, *Ann. Neurol.* **8**:94.
61. Posner, J. B., Plum, F., and Van Poznak, A., 1969, *Arch. Neurol.* **20**:388–395.
62. Kety, S. S., Woodford, R. B., Harmel, M. H., Freyhan, F. A., Appel, K. E., and Schmidt, C. F., 1947/1948, *Am. J. Psychiatry* **104**:765–770.
63. Plum, F., Posner, J. B., and Troy, B., 1968, *Arch. Neurol.* **18**:1–13.
64. Borgstrom, L., Chapman, A. G., and Siesjo, B. K., 1976, *J. Neurochem.* **27**:971–973.
65. Hawkins, R., Hass, W. K., and Ransohoff, J., 1979, *Stroke* **10**:690–703.
66. Kuhl, D. E., Engel, J., Jr., Phelps, M. E., and Selin, C., 1980, *Ann. Neurol.* **8**:348–360.
67. Engel, J., Jr., Kuhl, and Phelps, M. E., 1983, *Status Epilepticus* (A. V. Delgado-Escueta, C. G. Wasterlain, D. Treiman, and R. Porter, eds.), Raven Press, New York, pp. 141–148.

68. Wasterlain, C. G., and Duffy, T. E., 1976, *Arch. Neurol.* **33**:821–827.
69. Dwyer, B., and Wasterlain, C. G., 1981, *Neurology* **31**:162.
70. Folbergrova, J., Ingvar, M., and Siesjo, B. K., 1981, *J. Neurochem.* **37**:1228–1238.
71. Wilson, J. E., 1968, *J. Biol. Chem.* **243**:3640–3647.
72. Knull, H. R., Taylor, W. F., and Wells, W. W., 1973, *J. Biol. Chem.* **248**:5414–5417.
73. Fromm, H. J., and Zewe, V., 1962, *J. Biol. Chem.* **237**:1661–1667. J. P. Fabisiak and Schwark, W. S., 1981, *Neuropharmacology* **21**:179–182.
74. Copley, M., and Fromm, H. J., 1967, *Biochemistry* **6**:3503–3509.
75. Tuttle, J. B., and Wilson, J. E., 1970, *Biochim. Biophys. Acta* **212**:185–188.
76. Lowry, O. H., and Passonneau, J. V., 1964, *J. Biol. Chem.* **239**:31–42.
77. Maker, H. S., Clarke, D. D., and Lajtha, A., 1976, *Basic Neurochemistry* (G. J. Siegel, R. W. L. Albers, R. Katzman, and B. W. Agranoff, eds.), Little, Brown, Boston, pp. 279–307.
78. Chapman, A. G., Meldrum, B. S., and Siesjo, B. K., 1977, *J. Neurochem.* **28**:1025–1035.
79. McGilvery, R. W., (ed.), 1979, *Biochemistry: A Functional Approach*, Saunders, Philadelphia, p. 719.
80. Howse, D. C., and Duffy, T. E., 1975, *J. Neurochem.* **24**:935–940.
81. Sacktor, B., Wilson, J. E., and Tiekert, C. G., 1966, *J. Biol. Chem.* **241**:5071–5075.
82. King, L. J., Lowry, O. H., Passonneau, J., and Venson, V., 1967, *J. Neurochem.* **14**:599–611.
83. Collins, R. C., Posner, J. B., and Plum, F., 1970, *Am. J. Physiol.* **218**:943–950.
84. Ferrendelli, J. A., and McDougall, D. B., Jr., 1971, *J. Neurochem.* **18**:1197–1205.
85. McCandless, D. W., Feussner, G. K., Lust, W. D., and Passonneau, J. V., 1979, *J. Neurochem.* **32**:743–753.
86. Folbergrova, J., Passonneau, J. V., Lowry, O. H., and Schulz, D. W., 1969, *J. Neurochem.* **16**:191–203.
87. Atkinson, D. E., 1968, *Biochemistry* **7**:4030–4034.
88. Howse, D. C., 1979, *Can. J. Physiol. Pharmacol.* **57**:205–212.
89. Whisler, K. E., Tews, J. K., and Stone, W. E., 1968, *J. Neurochem.* **15**:315–320.
90. Nahorski, S. R., Roberts, D. J., and Stewart, C. G., 1970, *J. Neurochem.* **17**:621–631.
91. King, L. J., Carl, J. L., and Lao, L., 1973, *J. Neurochem.* **20**:477–485.
92. Blennow, G., Folbergrova, J., Nilsson, B., and Siesjo, B. K., 1979, *Ann. Neurol.* **5**:139–151.
93. Kreisman, N. R., Rosenthal, M., LaManna, J. C., and Seck, T. J., 1983, *Status Epilepticus* (A. V. Delgado-Escueta, C. Wasterlain, D. Treiman, and R. Porter, eds.), Raven Press, New York, pp. 231–239.
94. Bazan, N. G., 1970, *Biochim. Biophys. Acta* **218**:1–10.
95. Bazan, N. G., Jr., 1971, *J. Neurochem.* **18**:1379–1385.
96. Bazan, N. G., Morelli de Liberti, S., and Rodriguez de Turco, E. G., 1982, *Neurochem. Res.* **7**:839–843.
97. Siesjo, B. K., 1981, *J. Cerebr. Blood Flow Metab.* **1**:155–185.
98. Krebs, H. A., and Veech, R. L., 1970, *Pyridine Nucleotide-dependent Dehydrogenases* (H. Sund, ed.), Springer-Verlag, Berlin, pp. 413–434.
99. Stubbs, M., Veech, R. L., and Krebs, H. A., 1972, *Biochem. J.* **126**:59–65.
100. Myers, R. E., 1977, *Intrauterine Asphyxia in the Developing Fetal Brain* (L. Gluck, ed.), Year Book, Chicago, pp. 37–97.
101. DeCourten, G. M., Myers, R. E., and Yamaguchi, S., 1981, *Neurology* **31**:131.
102. Scholtz, W., 1959, *Epilepsia* **1**:36–55.
103. Aicardi, J., and Baraton, J., 1971, *Dev. Med. Child Neurol.* **13**:660.
104. Meldrum, B. S., and Brierley, J. B., 1973, *Arch. Neurol.* **28**:10–17.
105. Soderfeldt, B., Kalimo, H., Olsson, Y., and Siesjo, B. K., 1981, *Acta Neuropathol.* **54**:219–231.
106. Soderfeldt, B., Blennow, G., Kalimo, H., Olsson, Y., and Siesjo, B. K., 1983, *Acta Neuropathol.* **60**:81–91.
107. Pfleger, L., 1880, *Allg. Z. Psychiatr.* **36**:359–365.
108. Spielmeyer, W., 1927, *Z. Neurol. Psychiatr.* **109**:501–520.
109. Siesjo, B. K., Ingvar, M., J. Folbergrova, and Chapman, A. G., 1983, *Status Epilepticus* (A. V. Delgado-Escueta, C. G. Wasterlain, D. Treiman, and R. Porter, eds.), Raven Press, New York, pp. 217–230.

110. Siesjo, B. K., 1981, *J. Cereb. Blood Flow Metab.* **1**:155–185.
111. Fujikawa, D. G., Dwyer, B. E., and Wasterlain, C. G., 1984, *Neurology* **34**(Suppl. 1):220.
112. Ingvar, M., and Siesjo, B. K., 1983, *Acta Neurol. Scand.* **68**:129–144.
113. Lear, J. L., Ackermann, R. F., Kameyama, M., and Kuhl, D. E., 1982, *J. Cereb. Blood Flow Metab.* **2**:179–185.
114. Sokoloff, L., Reivich, M., Kennedy, C., Des Rosiers, M. H., Patlak, C. S., Pettigrew, K. D., Sakurada, O., and Shinohara, M., 1977, *J. Neurochem.* **28**:897–916.
115. Meldrum, B. S., Vigoroux, R. A., and Brierley, J. B., 1973, *Arch. Neurol.* **29**:82–87.
116. Blennow, G., Brierley, J. B., Meldrum, B. S., and Siesjo, B. K., 1978, *Brain* **101**:687–700.
117. Ben-Ari, Y., Tremblay, E., Ottersen, O. P., and Meldrum, B. S., 1980, *Brain Res.* **191**:79–97.
118. Collins, R. C., Lothman, E. W., and Olney, J. W., 1983, *Status Epilepticus* (A. V. Delgado-Escueta, C. G. Wasterlain, D. Treiman, and R. Porter, eds.), Raven Press, New York, pp. 277–288.
119. Siemkowicz, E., 1981, *Acta Neurol. Scand.* **64**:207–216.
119a. Welsh, F. A., Greenberg, J. H., Jones, S. C., Ginsberg, M.D., and Reivich, M., 1980, *Stroke* **11**:79–84.
120. Rehncrona, S., Siesjo, B. K., and Westerberg, E., 1978, *Acta Physiol. Scand.* **104**:453–463.
121. Kalimo, H., Rehncrona, S., Soderfeldt, B., Olsson, Y., and Siesjo, B. K., 1981, *J. Cerebr. Blood Flow Metab.* **1**:313–327.
122. Kalimo, H., Rehncrona, S., and Soderfeldt, B., 1981, *Acta Neuropathol. Suppl.* **7**:20–22.
123. Wrogemann, K., and Pena, S. D. J., 1976, *Lancet* **1**:672–674.
124. Schanne, O. F., Kane, A. B., Young, E. E., and Farber, J. L., 1979, *Science* **206**:700–702.
125. Schanne, O. F., Ruiz-Ceretti, E., Payet, M. D., and Deslauriers, Y., 1979, *J. Mol. Cell. Cardiol.* **11**:47–84.
126. Schwartzkroin, P. A., and Wyler, A. R., 1979, *Ann. Neurol.* **7**:95–107.
127. Meldrum, B., 1981, *Metabolic Disorders of the Nervous System* (F. C. Rose, ed.), Pitman, London, pp. 175–187.
128. Griffiths, T., Evans, M. C., and Meldrum, B. S., 1982, *Neurosci. Lett.* **30**:329–334.
129. Wasterlain, C. G., and Dwyer, B. E., 1981, Abstracts 12th World Congress of Neurology, Kyoto, p. 76.
130. Meldrum, B. S., and Nilsson, B., 1976, *Brain* **99**:523–542.
131. Abdul-Rahman, A., Dahlgren, N., Ingvar, M., Rehncrona, S., and Siesjo, B. K., 1979, *Acta Physiol. Scand.* **106**:53–60.
132. Cremer, J. E., Braun, L. D., and Oldendorf, W. H., 1976, *Biochim. Biophys. Acta* **448**:633–637.
133. Cremer, J. E., Cunningham, V. J., Pardridge, W. M., Braun, L. D., and Oldendorf, W. H., 1979, *J. Neurochem.* **33**:439–445.
134. Vannucci, S. J., and Vannucci, R. C., 1980, *J. Neurochem.* **34**:1100–1105.
135. Shapiro, B., and Wertheimer, E., 1943, *Biochem. J.* **37**:397–403.
136. Moore, T. J., Lione, A. P., Regen, D. M., Tarpley, H. L., and Raines, P. L., 1971, *Am. J. Physiol.* **221**:1746–1753.
137. Ter-Pogossian, M. M., Phelps, M. E., and Hoffman, E. J., 1975, *Radiology* **114**:89.
138. Phelps, M. E., Huang, S. C., Hoffman, E. J., Selin, C. E., and Kuhl, D. E., 1979, *Ann. Neurol.* **6**:371–388.
139. Jones, T., Chesler, D. A., and Ter-Pogossian, M. M., 1976, *Br. J. Radiol.* **49**:339–343.
140. Barron, J. C., Bousser, M. G., Comar, D., Soussaline, F., and Castonigne, P., 1981, *Eur. Neurol.* **20**:273–284.
141. Frackowiak, R. S. J., Pozzelli, C., and Legg, N. J., 1981, *Brain* **104**:753–778.
142. Engel, J., Jr., Kuhl, D. E., Phelps, M. E., and Mazziotta, J. C., 1982, *Ann. Neurol.* **12**:510–517.
143. Engel, J., Jr., Brown, W. J., Kuhl, D. E., Phelps, M. E., Mazziota, J., and Crandall, P. H., 1982, *Ann. Neurol.* **12**:518–528.
144. Engel, J., Jr. Kuhl, D. E., Phelps, M. E., Rausch, R., and Nuwar, M., 1983, *Neurology* **33**:400–413.
145. Ackerman, R. H., Correia, J. A., Alpert, N. M., Baron, J. C., Gouliamos, A., Grotta, J. C., Brownell, G. L., and Taveras, J. M., 1981, *Arch. Neurol.* **38**:537–543.

146. Dunn, A., Giuditta, A., and Pagliucad, N., 1971, *J. Neurochem.* **18**:2093–2099.
147. Dunn, A., 1971, *Brain Res.* **35**:254–259.
148. Dunlop, D. S., van Elden, W., and Lajtha, A., 1975, *J. Neurochem.* **24**:337–344.
149. Kiessling, M., and Kleihues, P., 1980, *Acta Neuropathol.* **55**:157–162.
150. Collins, R. C., and Vandi, N., 1980, *Trans. Am. Soc. Neurochem.* **11**:231.
151. Metafora, S., Persico, M., Felsani, A., Ferraiuolo, R., and Giuditta, A., 1977, *J. Neurochem.* **28**:1335–1346.
152. Giuditta, A., Metafora, S., Popoli, M., and Perrone-Capano, C., 1980, *Status Epilepticus* (A. V. Delgado-Escueta, C. G. Wasterlain, D. M. Treiman, and R. J. Porter, eds.), Raven Press, New York, pp. 289–295.
153. Wasterlain, C. G., 1977, *J. Neurochem.* **29**:707–716.
154. Dwyer, B., and Wasterlain, C. G., 1980, *J. Neurochem.* **34**:1639–1647.
155. Walton, G. M., and Gill, G. N., 1976, *Biochim. Biophys. Acta* **447**:11–19.
156. Austin, S. A., and Clemens, J. J., 1980, *FEBS Lett.* **110**:1–7.
157. Hunt, T., 1980, *Trends Biochem. Sci.* **5**:178–181.
158. Gross, M., 1980, *Mol. Cell. Biochem.* **31**:25–36.
159. Moody, W. J., Futamachi, K. J., and Prince, D. A., 1974, *Exp. Neurol.* **42**:248–263.
160. Lipton, P., and Heimbach, C. J., 1978, *J. Neurochem.* **31**:1299–1307.
160a. Zomzely, C. E., Roberts, S., and Rapaport, D., 1964, *J. Neurochem.* **11**:567–582.
160b. Fando, J. L., and Wasterlain, C. G., 1980, *Neurochem. Res.* **5**:197–207.
161. Metter, E. J., and Yanagihara, T., 1979, *Brain Res.* **161**:481–492.
162. Blomstrand, C., 1970, *Exp. Neurol.* **29**:175–188.
163. Morimoto, K., Brengman, J., and Yanagihara, T., 1978, *J. Neurochem.* **31**:1277–1282.
164. Wynter, C. V. A., Ioannow, P., and Mathias, A. P., 1975, *Biochem. J.* **152**:449–467.
165. Cupello, A., Ferillo, F., Lazzarini, G., and Rosadini, G., 1976, *Exp. Neurol.* **53**:601–609.
166. Cupello, A., Ferrillo, F., and Rosadini, G., 1979, *Exp. Neurol.* **67**:451–457.
167. Cupello, A., Ferrillo, F., and Rosadini, G., 1980, *Exp. Neurol.* **69**:247–252.
168. Cupello, A., Ferrillo, F., and Rosadini, G., 1981, *Neurochem. Res.* **6**:175–182.
169. McGeer, P. L., Eccles, J. C., and McGeer, E. G., (eds.), 1983, *Molecular Neurobiology of the Mammalian Brain*, Plenum Press, New York.
170. van Gelder, N. M., Sherwin, A., and Rasmussen, T., 1972, *Brain Res.* **40**:385–393.
171. Perry, T. L., and Hansen, S., 1981, *Neurology* **31**:872–876.
172. Corselles, J. A. N., 1974, *Epilepsy* (P. Harris and C. Mawdsley, eds.), Churchill Livingstone, London, pp. 111–114.
173. Berl, S., and McMurthy, J. C., 1967, *Arch. Biochem. Biophys.* **118**:645–648.
174. Mutani, R., Bergamini, L., and Durelli, L., 1978, *Taurine and Neurological Disorders* (A. Barbeau and R. J. Huxtable, eds), Raven Press, New York, pp. 359–374.
175. Reiffenstein, R. J., and Neal, M. J., 1974, *Can. J. Physiol. Pharmacol.* **52**:286–290.
176. Koyama, I., and Jasper, H., 1977, *Can. J. Physiol. Pharmacol.* **55**:523–553.
177. Koyama, I., 1972, *Can. J. Physiol. Pharmacol.* **50**:740–752.
178. Emson, P. C., and Joseph, M. H., 1975, *Brain Res.* **93**:91–110.
179. Nahorski, S. R., Roberts, D. J., and Stewart, G. G., 1970, *J. Neurochem.* **17**:621–631.
180. Tower, D. B., 1960, *Neurochemistry of Epilepsy*, Charles C Thomas, Springfield, Illinois.
181. Dodd, P. R., Bradford, H. F., Abdul-Ghani, A. S., Cox, W. G., and Coutinho-Netto, J., 1980, *Brain Res.* **193**:505–517.
182. Dodd, P. R., and Bradford, H. F., 1976, *Brain Res.* **111**:377–388.
183. Bradford, H. F., and Dodd, P. R., 1973, *Biochem. Pharmacol.* **26**:253–254.
184. Joseph, M. H., and Emson, P. C., 1976, *J. Neurochem.* **27**:1495–1501.
185. Emson, P. C., 1976, *Biochemistry and Neurology* (H. F. Bradford and D. D. Marsden, eds.), Academic Press, New York, pp. 163–173.
186. Huxtable, R. J., and Laird, M. E., 1978, *Neurosci. Lett.* **10**:341–345.
187. Hansen, S., Perry, T. L., Wada, J. H., and Sokol, M., 1973, *Brain Res.* **50**:480–483.
188. van Gelder, N. M., Edmonds, H. L., Hegreberg, G., Chatburn, C. C., Clemmons, R., and Sylvester, D. M., 1980, *J. Neurochem.* **35**:1087–1091.
189. Fabisiak, J. P., and Schwark, W. S., 1981, *Neuropharmacology* **21**:179–182.
190. Huxtable, R. J., 1981, *Prog. Clin. Biol. Res.* **68**:53–97.

191. Durelli, L., and Mutani, R., 1983, *Clin. Neuropharmacol.* **6**:37–48.
192. van Gelder, N. M., 1978, *Can. J. Physiol. Pharmacol.* **56**:362–374.
193. Emson, P. C., 1978, *Taurine and Neurological Disorders* (A. Barbeau and R. J. Huxtable, eds.), Raven Press, New York, pp. 319–338.
194. Laird, H. E., and Huxtable, R. J., 1978, *Taurine and Neurological Disorders* (A. Barbeau and R. J. Huxtable, eds.), Raven Press, New York, pp. 339–358.
195. Takahashi, R., and Nakane, Y., 1976, *Taurine and Neurological Disorders* (A. Barbeau and R. J. Huxtable, eds.), Raven Press, New York, pp. 375–386.
196. Brazier, M. A. B., 1974, *Neurology* **24**:903–911.
197. Carlsson, A., 1974, *Epilepsy* (P. Harris and C. Mawdsley, eds.), Churchill Livingstone, London, pp. 1–4.
198. Maynert, E. W., Marczynski, T. J., and Browning, R. A., 1975, *Adv. Neurol.* **13**:79–147.
199. Jobe, P. C., and Laird, H. E., 1981, *Biochem. Pharmacol.* **30**:3137–3144.
200. Consroe, P., Picchioni, A., and Chin, L., 1979, *Fed. Proc.* **38**:2411–2416.
201. Jobe, P. C., Brown, R. D., and Dailey, J. W., 1981, *Life Sci.* **28**:2031–2038.
202. Jobe, P. C., and Picchioni, A. L., and Chin, L., 1973, *Life Sci.* **13**:1–13.
203. Bourn, W. M., Chin, L., and Picchioni, A. L., 1977, *Life Sci.* **21**:701–706.
204. Bourn, W. M., Chin, L., and Picchioni, A. L., 1978, *J. Pharm. Pharmacol.* **30**:800–801.
205. Ko, K. H., Dailey, J. W., and Jobe, P. C., 1952, *J. Pharmacol. Exp. Ther.* **222**:662–669.
206. Jobe, P. C., Picchioni, A. L., and Chin, L., 1979, *Pharmacologist* **21**:182.
207. Laird, H. E., Chin, L., and Picchioni, A. L., 1980, *Soc. Neurosci. Abstr.* **6**:724.
208. Laird, H. E., Chin, L., and Picchioni, A. L., 1974, *Proc. West. Pharmacol. Soc.* **17**:46–50.
209. Goddard, G. V., McIntyre, D. B., and Leech, C. K., 1969, *Exp. Neurol.* **25**:243–330.
210. Wada, J. A. (ed.), 1981, *Kindling* 2, Raven Press, New York.
211. Arnold, P. S., Racine, R. J., and Wise, R. A., 1973, *Exp. Neurol.* **40**:457–470.
212. Corcoran, M., Fibiger, H. C., McCaughran, J., and Wada, J. A., 1974, *Exp. Neurol.* **45**:118–133.
213. Sato, M., and Nakashima, T., 1976, *Kindling* (J. A. Wada, ed.), Raven Press, New York, pp. 10, 80, 103–116.
214. Engel, J., Jr., and Sharpless, N. S., 1977, *Brain Res.* **136**:381–386.
215. Callaghan, D. A., and Schwark, W. S., 1979, *Neuropharmacology* **18**:541–545.
216. Burnham, W., King, G. A., and Lloyd, K. G., 1981, *Prog. Neuro-Psychopharmacol.* **5**:537–541.
217. Siegel, J., and Murphy, G. J., 1979, *Brain Res.* **174**:337–340.
218. Stach, R., Lazarova, M. B., and Kacz, D., 1981, *Nauyn-Schmiedebergs Arch. Pharmacol.* **316**:56–58.
219. Ashcroft, G. W., Dow, R. C., Emson, P. C., Harris, P., Ingleby, J., Joseph, M. H., and McQueen, J. K., 1974, *Epilepsy* (P. Harris and C. Mawdsley, eds.), Churchill Livingstone, London, pp. 115–123.
220. Arduini, A., and Machna, X., 1949, *Arch. Fisiol.* **13**:152–167.
221. Coutinho-Netto, J., Boyar, M., Bradford, H. F., Birdsall, N. J. M., and Hulme, E. C., 1981, *Exp. Neurol.* **74**:837–846.
222. Goddard, G. V., 1969, *Nature* **214**:1020–1021.
223. Racine, R., 1972, *Electroencephalogr. Clin. Neurophysiol.* **32**:281–294.
224. Burchfield, J., Duchowny, M., and Duffy, K., 1979, *Science* **204**:1096–1098.
225. Morrell, F., Hitri, A., Hoeppner, T. J., Bergen, D., Kessler, E., and Flemins, S., 1980, *Soc. Neurosci. Abstr.* **6**:11.
226. McNamara, J. O., 1978, *Brain Res.* **154**:415–420.
227. Byrne, M. C., Gottlieb, R., and McNamara, J., 1980, *Exp. Neurol.* **69**:85–98.
228. Wasterlain, C. G., Morin, A. M., and Jonec, V., 1982, *Brain Res.* **247**:341–346.
229. Dashieff, R. M., Byrne, M. C., Patrone, V., and McNamara, J. O., 1981, *Brain Res.* **206**:233–238.
230. Dashieff, R. M., and McNamara, J. O., 1980, *Brain Res.* **195**:345–353.
231. Wasterlain, C. G., Morin, A. M., and Jonec, V., 1982, *Electroencephalogr. Clin. Neurophysiol. Suppl.* **36**:264–273.
232. Morin, A. M., and Wasterlain, C. G., 1982, *Soc. Neurosci. Abstr.* **8**:456.

233. Kellar, K. J., Cascio, C. S., and Bergstrom, D. A., 1981, *J. Neurochem.* **37**:830–836.
234. Morin, A. M., and Wasterlain, C. G., 1983, *Status Epilepticus* (A. V., Delgado-Escueta, C. G. Wasterlain, D. M., Treiman, and R. J. Porter, eds.), Raven Press, New York, pp. 369–374.
235. Luqmani, Y. A., Bradford, H. G., Birdsall, N. J., and Hulme, E. C., 1979, *Nature* **277**:481–483.
236. Klein, W. L., Nathanson, N., and Nirenberg, M., 1979, *Biochem. Biophys. Res. Commun.* **90**:506–516.
237. Klein, W., 1980, *Proc. West. Pharmacol. Soc.* **23**:449–458.
238. Schildkraut, J. J., and Kety, S., 1967, *Science* **156**:21–37.
239. Bergstrom, D. A., and Kellar, K. J., 1979, *Nature* **278**:464–466.
240. Gillespie, D. D., Manier, D. H., and Sulser, F., 1979, *Commun. Psychopharmacol.* **3**;191–195.
241. Morin, A. M., Watson, A. L., and Wasterlain, C. G., 1983, *Eur. J. Pharmacol.* **88**:131–134.
242. Morin, A. M., Tanaka, I. A., and Wasterlain, C. G., 1981, *Soc. Neurosci. Abstr.* **7**:426.
243. Morin, A. M., Tanaka, I. A., and Wasterlain, C. G., 1981, *Life Sci.* **28**:2257–2263.
244. Gee, K. W., Hollinger, M. A., Bowyer, J., and Killam, E., 1979, *Exp. Neurol.* **66**:771–777.
245. Gee, K. W., Hollinger, M. A., and Killam, E. K., 1981, *Exp. Neurol.* **74**:265–275.
246. Coursin, D. B., 1954, *J. Am. Med. Assoc.* **154**:406–408.
247. Holtz, P., and Palm, D., 1964, *Pharmacol. Rev* **16**:113–178.
248. Tower, D. B., 1956, *Am. J. Clin. Nutr.* **4**:329–345.
249. Bain, J. A., and Williams, H. L., 1960, *Inhibition in the Nervous System and Gamma amino butyric acid* (E. Roberts, C. F. Baxter, A. Van Harreveld, C. A. G. Wiersman, W. R. Adey, and K. F. Killam, eds.), pp. 275–293.
250. Davenport, V. D., and Davenport, H. W., 1948, *J. Nutr.* **36**:263.
251. Curtis, D. R., Ruggan, A. W., Felix, D., Johnston, G. A. R., 1971, *Brain Res.* **32**:69–96.
252. Hayashi, T., 1959, *J. Physiol.* (*London*) **145**:570–578.
253. Kalichman, M. W., Livingston, K. E., and Burnham, W. M., 1981, *Exp. Neurol.* **74**:829–836.
254. Myslobodsky, M. S., and Valenstein, E. S., 1980, *Epilepsia* **21**:163–175.
255. Sykes, C. C., and Horton, R. W., 1982, *Neurochemistry* **39**:1489–1491.
256. Schechter, P. J., Tranier, Y., Jung, M. J., and Bohlen, P., 1977, *Eur. J. Pharmacol.* **45**:319–328.
257. Horton, R. W., Prestwick, S. A., and Meldrum, B. S., 1982, *J. Neurochem.* **39**:864–870.
258. Ciesielski, L., Simler, S., and Mandel, P., 1981, *Neurochem. Res.* **6**:267–272.
259. Ribak, C., Harris, A. B., Vaughn, J., and Roberts, E., 1979, *Science* **205**:211–213.
260. Matthews, W. D., and McCafferty, G. P., 1979, *Neuropharmacology* **18**:885–889.
261. Gale, K., and M. J. Iadarola, J. J., 1980, *Science* **208**:288–289.
262. Iadarola, M. J., and Gale, K., 1982, *Science* **218**:1237–1240.
263. Squires, R., and Braestrup, C., 1977, *Nature* **266**:732–734.
264. Braestrup, C., and Squires, R., 1978, *Br. J. Psychiatry* **133**:249–260.
265. Paul, S., and Skolnick, P. 1978, *Science* **202**:892–894.
266. Bowdler, J. M., and Green, A. R., 1982, *Br. J. Pharmacol.* **76**:291–298.
267. Asano, T., and Mizutani, A., 1980, *Jpn. J. Pharmacol.* **30**:783–788.
268. Robertson, H. A., 1979, *Eur. J. Pharmacol.* **66**:249–252.
269. Kellogg, C., and Syapin, P., 1981, *Soc. Neurosci. Abstr.* **7**:847.
270. McNamara, J. O., Peper, A. M., and Patrone, V., 1980, *Proc. Natl. Acad. Sci. USA* **77**:3029–3032.
271. Valdes, F., Dashieff, R. M., Brimingham, F., Crutchen, K. A., and McNamara, J. O., 1982, *Proc. Natl. Acad. Sci. USA* **79**:193–197.
272. Syapin, P. J., and Rickman, D. W., 1981, *Eur. J. Pharmacol.* **72**:117–120.
273. Braestrup, C., Nielsen, M., and Olsen, C., 1980, *Proc. Natl. Acad. Sci. USA* **77**:2288–2292.
274. Paul, S., Syapin, P. J., Paugh, B. A., Moncada, V., and Skolnick, P., 1979, *Nature* **281**:688–689.
275. Urca, G., Frenk, H., Liebeskind, J., and Taylor, A., 1977, *Science* **197**:83–86.
276. Frenk, H., McCarty, B., and Liebeskind, J., 1978, *Science* **200**:335–337.

277. Urca, G., and Frenk, H., 1982, *Brain Res.* **246**:121–126.
278. Henrickson, S., Floom, F., McCoy, S., Ling, N., and Guillemin, R., 1978, *Proc. Natl. Acad. Sci. USA* **75**:5221–5225.
279. French, E. D., Siggins, G., Henricksen, S., and Ling, N., 1977, *Soc. Neurosci. Abstr.* **3**:926.
280. Lewis, J. W., Caldecott-Hazard, S., Cannon, J. T., and Liebeskind, J. C., 1981, *Neurosecretion and Brain Peptides* (J. B. Martin, S. Reicklin, and K. L. Bick, eds.), Raven Press, New York, pp. 213–224.
281. Corcoran, M., and Wada, J., 1979, *Life Sci.* **24**:791–796.
282. Hardy, C., Panksepp, J., Rossi, J., and Zolonick, A., 1980, *Brain Res.* **194**:293–297.
283. Le Gal La Salle, G., Calvine, B., and Ben-Ari, Y., 1977, *Neurosci. Lett.* **194**:293–297.
284. Post, R. M., Davenport, S., Pert, A., and Squillace, K., 1979, *Commun. Psychopharmacol.* **3**:185–191.
285. Frenk, H., Engel, J., Ackermann, R., Shavit, Y., and Liebeskind, J., 1979, *Brain Res.* **167**:435–440.
286. Engel, J., Ackermann, R., Caldecott-Hazard, S., and Kuhl, D., 1981, *Kindling 2* (J. Wada, ed.), Raven Press, New York.
287. Engel, J., Jr., Wolfson, L., and Brown, L., 1978, *Ann. Neurol.* **3**:538–544.
288. Chugani, H. T., Ackermann, R. F., Chugani, D. C., and Engel, J., Jr., 1983, International Symposium on Cerebral Blood Flow, Metabolism and Epilepsy, Montpellier.
289. Tortella, F., Cowan, A., Belenky, G., and Haladay, J., 1981, *Eur. J. Pharmacol.* **76**:121–128.
290. Sarne, Y., Weissman, B. A., and Urca, G., 1982, *J. Neurochem.* **39**:1478–1480.
291. Haladay, J. W., Belensky, G. L., Tok, H. H., and Meyerhoff, J. L., 1978, *Soc. Neurosci. Abstr.* **4**:409.
292. Hong, J. S., Gillin, H., Yang, Y. T., and Costa, E., 1979, *Brain Res.* **177**:273.
293. Green, A. R., Peralta, E., Hong, J. S., Mao, C. C., Atterwill, C. K., and Costa, E., 1978, *J. Neurochem.* **31**:607–613.
294. Elazar, Z., Simantov, R., Motles, E., and Ely, Y., 1978, *Israel J. Med. Sci.* **14**:996.
295. Belensky, G. L., and Haladay, J. W., 1979, *Soc. Neurosci. Abstr.* **5**:549.
296. Bloom, F. E., 1975, *Rev. Physiol. Biochem. Pharmacol.* **74**:1–103.
297. Daly, D., 1977, *Cyclic Nucleotides in the Nervous System*, Plenum Press, New York.
298. Ferrendelli, J. A., Blank, A. C., and Gross, R. A., 1983, *Status Epilepticus* (A. V. Delgado-Escueta, C. G. Wasterlain, D. M. Treiman, and R. J. Porter, eds.), Raven Press, New York, pp. 353–357.
299. Nathanson, J. A., 1977, *Physiol. Rev.* **57**:157–256.
300. Phillis, J. W., 1977, *Can. J. Neurol. Sci.* **4**:151–195.
301. Stone, T. W., Taylor, D. A., and Bloom, F. E., 1975, *Science* **187**:845–847.
302. Berti, F., Bernareggi, V., Folco, G. C., Fumagalli, R., and Paoletti, R., 1976, *Adv. Biochem. Psychopharmacol.* **15**:367–377.
303. Ferrendelli, J. A., Blank, A. C., and Gross, R. A., 1980, *Brain Res.* **200**:93–103.
304. Ferrendelli, J. A., and Kinscherf, D. A., 1977, *Epilepsia* **18**:525–531.
305. Folbegrovo, J., 1977, *Brain Res.* **135**:337–346.
306. Folco, G. W., Longiave, D., and Besisio, E., 1977, *Prostaglandins* **13**:893–900.
307. Lust, W. D., Goldberg, N. D., and Passonneau, J. V., 1976, *J. Neurochem.* **26**:5–10.
308. Lust, W. D., Kupferberg, H. J., Yonekawa, W. D., Passonneau, J., and Wheaton, A. B., 1978, *Mol. Pharmacol.* **14**:347–356.
309. Palmer, C. G., Jones, D. J., Medina, M. A., and Stavinhova, W. B., 1979, *Epilepsia* **29**:95–104.
310. Rehncrona, S., Seisjo, B. K., and Westerberg, E., 1978, *Acta Physiol. Scand.* **104**:453–463.
311. Sattin, A., 1971, *J. Neurochem.* **18**:1087–1096.
312. Wasterlain, C. G., and Csiszar, E., 1980, *Exp. Neurol.* **70**:260–268.
313. Raabe, W., Nichol, S., Gumnit, R. J., and Goldberg, N. D., 1978, *Brain Res.* **144**:185–188.
314. Walker, J. E., Lewin, E., Sheppard, T. R., and Cromwell, R., 1973, *J. Neurochem.* **21**:79–85.
315. Harik, S. I., LaManna, J. C., Light, A. I., and Rosenthal, M., 1979, *Science* **206**:69–71.
316. Gross, R. A., and Ferrendelli, J. A., 1982, *Neuropharmacology* **21**:655–661.

316a. Hoffer, B., Seiger, A., and Friedman, R., 1977, *Brain Res.* **119**:107–132.
316b. Hoffer, B., Seiger, A., and Taylor, D., 1977, *Exp. Neurol.* **54**:233–250.
317. Greengard, P., 1979, *Fed. Proc.* **38**:2208–2217.
318. Greengard, P., 1981, *Harvey Lect.* **75**:277–331.
319. Demaille, J. G., and Pecherre, J. F., 1983, *Adv. Cyclic Nucleotide Res.* **15**:337–371.
320. Haiech, J., and Demaille, J. G., 1983, *Phil. Trans. R. Soc. London Ser. B* **302**:91–98.
321. Ehrlich, Y. H., Reedy, M. V., Keen, P., Davis, L. G., Daugherty, Y., and Brunngraber, E. G., 1980, *J. Neurochem.* **34**:1327–1330.
322. Delgado-Escueta, A. V., and Horan, M., 1980, *Status Epilepticus* (A. V. Delgado-Escueta, C. G. Wasterlain, D. M. Treiman, and R. J. Porter, eds.), Raven Press, New York, pp. 311–323.
323. Strombom, U., Forn, J., Dolphin, A. C., and Greengard, P., 1979, *Proc. Natl. Acad. Sci. USA* **76**:4687–4690.
324. Tuchek, J. M., Johnson, D. D., and Crawford, R. D., 1980, *Soc. Neurosci. Abstr.* **6**:444.
325. Bajorek, J. G., 1983, A Syllabus for an International Symposium on Basic Mechanisms of the Epilepsies, UCLA Extension, 13a-c.
326. Kirchberger, M. A., and Antonetz, T., 1982, *Biochem. Biophys. Res. Commun.* **105**:152–156.
327. Kirchberger, M. A., and Tada, M., 1976, *J. Biol. Chem.* **251**:725–729.
328. Kirchberger, M. A., Tada, M., and Katz, A. M., 1974, *J. Biol. Chem.* **249**:6166–6173.
329. Rinaldi, M. L., LePeuch, C. J., and Demaille, J. G., 1981, *FEBS Lett.* **129**:277–281.
330. Rinaldi, M. L., Capeny, J. P., and Demaille, J. G., 1982, *J. Mol. Cell. Cardiol.* **14**:279–289.
331. Le Peuch, E. J., Le Peuch, D. A. M., and Demaille, J. G., 1980, *Biochemistry* **19**:3368–3373.
332. Campbell, K. P., and MacLennan, D. H., 1982, *J. Biol. Chem.* **257**:1238–1246.
333. Meech, R. W., 1978, *Annu. Rev. Biophys. Bioeng.* **7**:1–18.
334. Hagiwara, S., and Byerly, L., 1981, *Annu. Rev. Neurosci.* **4**:69–125.
335. DeLorenzo, R., 1980, *Ann. N.Y. Acad. Sci.* **356**:92–109.
336. DeLorenzo, R. J., and Freedman, S. D., 1977, *Biochem. Biophys. Res. Commun.* **71**:590–597.
337. Wasterlain, C. G., and Jonec, V., 1983, *Brain Res.* **271**:311–323.
338. Folbergrova, J., 1974, *Brain Res.* **81**:443–454.
339. Dunn, A., and Bergert, B. J., 1976, *J. Neurochem.* **26**:369–375.
340. Mutani, R., Durelli, L., Mazzarine, M., Valentini, C., Monaco, F., Fumero, S., and Mondino, A., 1977, *Brain Res.* **122**:513–521.
341. van Gelder, N. M., and Courtois, A., 1972, *Brain Res.* **43**:477–484.
342. van Gelder, N. M., 1972, *Brain Res.* **47**:477–484.
343. Craig, C. R., and Hartman, E. R., 1973, *Epilepsia* **14**:409–414.
344. McNamara, J. O., 1978, *Exp. Neurol.* **61**:582–591.
345. Dwyer, B. E., and Wasterlain, C. G., 1984, *Brain Res.* (in press).

16

Pathological Neurochemistry of Cerebrospinal Fluid
Neurotransmitters and Neuropeptides

Theodore A. Hare and James H. Wood

1. INTRODUCTION

Measurement of neuroactive substances in human cerebrospinal fluid (CSF) offers a unique opportunity for elucidating the pathology of neuropsychiatric disorders as well as for evaluating therapeutic approaches to these disorders. Human CSF is readily available and, in fact, offers the only source of such information from living patients that does not require brain biopsy with its associated morbidity. Studies of CSF effectively circumvent vagaries associated with extrapolation of data from autopsy tissue or from imperfect animal models.

In the previous edition of the *Handbook of Neurochemistry*, Davson ended his chapter,[1] "The Cerebrospinal Fluid," in Volume 2 with the conclusion that:

> The significance of the cerebrospinal fluid is now emerging. Originally considered by the physiologist as little more than a fluid cushion that supported the brain and protected it from stress, and by the pathologist as a useful fluid whose closeness to the central nervous parenchyma allowed it to reflect some of the diseases that afflicted it, the cerebrospinal fluid has now been revealed as a specific secretion that serves to maintain the stable and peculiar environment that is required for the adequate functioning of the central neurons.

In Volume 7 of the same series, Lowenthal concluded his chapter,[2] "Chemical Physiopathology of the Cerebrospinal Fluid," with the statement that:

> We can therefore conclude that although the biochemical composition of normal CSF may be relatively well known, in the pathological field little accurate information has been obtained, except perhaps in the study of proteins. For the rest, we believe that an enormous field still remains open.

Theodore A. Hare • Department of Pharmacology, Jefferson Medical College, Thomas Jefferson University, Philadelphia, Pennsylvania 19107. *James H. Wood* • Cerebral Blood Flow Laboratories, Division of Neurosurgery, Emory University Clinic, Atlanta, Georgia 30322.

Both observations reflected the realization that study of CSF offers significant potential for evaluating the pathology of central nervous system (CNS) disorders but also that this potential had not been developed.

During the intervening decade, exponential progress has confirmed the conclusion that significant knowledge of pathology can result from study of CSF. A prime factor limiting this approach has also been uncovered in that it has been documented that careful consideration of both *in vivo* and *in vitro* variables is essential if valid conclusions are to be drawn.[3–7] Inadequate regard for these variables, such as assay specificity and precision, sample storage, patient age and sex, etc., leads to much less useful data and in some cases obfuscation in the literature.

A coherent rationale has evolved, as described recently by several authors,[4,5,8–13] to support the supposition that CSF measurements constitute an appropriate source for meaningful information about CNS function. For example, CSF is thought to be essentially continuous with brain and spinal interstitial extracellular fluid and therefore is in intimate contact with the entire CNS.[13]

The composition of the CSF is finely regulated. For example, the blood–brain barrier (BBB) effectively insulates the brain and CSF from peripheral constituents in many instances, and, therefore, CSF is much more a reflection of central activity than of peripheral function. This conclusion is supported by studies documenting correlations between CSF and brain responses to pharmacological perturbation in the absence of correlations with blood responses.[14–18] In addition, rostrocaudal concentration gradients have been demonstrated for several key neurochemicals,[9,19–22] and certain CSF neurotransmitters have been shown to be altered[3,23–33] in disorders for which the same neurotransmitters are known to be altered in CNS tissue.[34–41]

Thus, at least for certain key neurochemicals, the emerging picture is that CSF constitutes, if not a window, at least a mirror offering a reflection of CNS activity. Various technical and methodological difficulties, however, have been encountered in attempts to focus this reflection.

To maintain the integrity of CNS function, various mechanisms (e.g., reuptake and the "sink" effect of CSF) precisely regulate the concentrations of neuroactive substances in CSF to very low (picomole or subpicomole) levels. The resulting minute quantities of neurotransmitters in CSF have constituted the chief difficulty in utilizing CSF measurements to elucidate pathology associated with neuropsychiatric disorders because adequate analytical procedures have not been available. For this reason, initial studies were carried out using metabolites of neurotransmitters, as these are present at somewhat higher concentrations in CSF. Extrapolation from metabolite data to CNS neurotransmitter function, however, suffers from the drawback that the metabolites are not subject to the same regulation through barriers or via reuptake as are the neurotransmitters themselves. Central mechanisms generally function to maintain neurotransmitters proximate to their site of action, whereas the same mechanisms facilitate transport of metabolites away from the site of neurotransmitter action and to eventual excretion. Another complicating factor is that some of the metabolites are present at low concentrations and therefore have also been difficult to detect. A useful but perhaps complicating technique for achieving more readily detectable metabolite levels has been through use of a transport-blocking drug such as probenecid[42] to reduce transport of acidic metabolites out of the CSF. Thus, study of metabolites has provided preliminary information, but extrapolation from such data is risky and certainly much less precise than would be that from direct measurement of neurotransmitters themselves.

The primary impetus underlying the current interest in the neurochemical pathology of CSF has been the relatively recent development of several assay procedures with adequate sensitivity, selectivity, and precision for reliably measuring the neurotransmitters. These procedures have included the radioimmune,[43–54] radioenzymatic,[55–60] and radioreceptor[20,54,61–66] assays and also chromatographic procedures such as ion-exchange or reverse-phase high-performance liquid chromatography with fluorescence[19,67,68] or amperometric detection[69–75] and gas chromatography with mass spectrometric detection.[76–80] These procedures differ somewhat with respect to selectivity and precision as well as sensitivity, but all have proven adequate for neurotransmitter measurement in CSF and have provided useful information about the pathology of CNS neurotransmitter function in various neuropsychiatric disorders.

For certain, the base-line concentrations of some neurotransmitters have been established, and investigators have begun to undertake clinical evaluations. Physiological studies to date, however, have demonstrated the significant influence of several *in vivo* and *in vitro* variables.[3–7] Unfortunately, some of the early studies of CSF neurochemical pathology were carried out before the significance of these variables was elucidated, producing some initial confusion in the literature. It is important that these studies be critically evaluated. For other neurotransmitters, systematic evaluation of basic parameters has not been adequately carried out, and, in those instances, clinical studies must still be regarded as preliminary. Future patient and reference group samples should be stringently controlled in light of the relevant variables, and only appropriate homogeneous subgroups should be compared during clinical studies.

In some cases, adequate assay procedures are just being developed (e.g., dopamine, serotonin), and in other cases, concern has been expressed about the precision and/or specificity of assay procedures (e.g., radioimmune, radioreceptor).

A previous chapter[22] in Volume 1 of this edition of the *Handbook of Neurochemistry* has reviewed the recent extensive literature that has now provided rather sophisticated knowledge of the physiological neurochemistry of CSF. The purpose of this chapter is to critically discuss and evaluate the studies of the pathology and therapeutics of neuropsychiatric disorders and thereby provide a more solid foundation of data to facilitate the design of future investigations.

2. γ-*AMINOBUTYRIC ACID* (*GABA*)

GABA is uniquely present at high concentrations throughout brain and spinal cord tissue. Extensive investigations have led to the conclusion that it functions as an inhibitory neurotransmitter by inducing membrane hyperpolarization secondary to elevation of chloride ion permeability.[80,81] It is generally thought to be the principal inhibitory neurotransmitter in the CNS and evidently is the most ubiquitous of all the neurotransmitters in light of the estimate[82] that 40% of all the synapses in the brain are GABAergic.

Thus, GABA appears to occupy a key position with regard to neurochemical and pathological considerations, especially if the supposition is accepted that neuropsychiatric disorders are the consequence of altered balances of opposing neurotransmitter systems. An inhibitory–excitatory balance would be the most basic form of balance between opposing neurotransmitter systems, and for this type of balance, GABA would evidently most often provide the drive on the inhibitory side of the fulcrum. Thus, GABA can logically be visualized as providing a general inhibitory tone to the CNS in opposition to various other neurotransmitters that produce excitatory influences. In this manner of viewing neurotransmitter interaction, GABA could be pictured as providing a more or less constant background drive essential for normal CNS function, whereas the more numerous excitatory neurotransmitters would more logically be pictured as providing the more specific regulations required for normal function. A chief requirement of such a scheme is that regulation of the GABA system would generally be directed toward maintenance of a relatively constant base-line level of GABAergic function.

It would be consistent with the above consideration that alterations of GABAergic activity be associated not with a limited number of neuropsychiatric disorders but that such alterations would be a factor common to many of these disorders. In other words, deficient GABAergic activity may provide a propensity toward various neuropsychiatric disorders, which would then be expressed depending on the more specific anomalies on the other sides of balance fulcrums.

Because of its unique central role, GABA has been the most definitively evaluated[3,5,7,29,30,33,83–92] among the neurotransmitters in CSF, and, therefore, the sequence of studies evaluating GABA in CSF can be regarded as a prototype for other neurotransmitter systems.

Human CSF GABA has been measured by ion-exchange/fluorometric (I-E/F),[19,67,68] radioreceptor (RR),[20,61,62] gas chromatography/mass spectrometric (GC/MS),[76–78] and enzymatic methods.[93,94] The I-E/F, RR, and GS/MS procedures have been compared[76–78,95] and have provided generally consistent data, whereas the enzymatic procedure has been criticized,[3,68] evidently lacks specificity and sensitivity, and has provided data inconsistent with those of the other three methods. Of the three available procedures for reliably measuring GABA in CSF, the I-E/F and RR methods have been most widely used for clinical investigations, and of these two, the I-E/F method is the more precise.[5] This difference of precision could be highly significant in view of the numerous reports consistently suggesting differences between controls and patients even when statistical comparisons do not reveal significance of the difference. For example, comparison of GABA in CSF from patients with Parkinson's disease to that in CSF from controls has not always revealed a significant difference; however, in most studies the concentration in CSF from Parkinson's disease patients is lower than that from controls.[27,31,78,86,96,97] It must be recognized that failure to demonstrate a significant difference cannot lead to the conclusion that there is no difference.

Early during the evaluation of GABA in CSF, GABA itself was found to be relatively stable in CSF.[78] Later investigators found that concentrations of GABA become artifactually elevated under various *in vitro* conditions via several mechanisms.[10,68,83,98] In untreated CSF, these increases are the consequence of enzyme action,[10] and there are indications in the literature that the sources of these elevations are glutamate via glutamic acid decarboxylase (GAD) activity[5,99] as well as hydrolysis of GABA peptides such as homocarnosine.[100–102] The GABA level is more stable following acid deproteinization[98]; however, even under these conditions, the level can be artifactually elevated during storage following breakdown of GABA-containing peptides.[5,83,98,102]

Recently, a pool of readily releasable GABA has also been found to be present in CSF by virtue of the fact that the amount of GABA determined is dependent on the concentration of acid used for deproteinization.[97,102] The lowest values are observed when no acid is utilized,[97] i.e., following short-term exposure of the CSF to 100°C, a technique that does not alter GABA concentration[10] but halts GABA-elevating enzyme action. The amount of GABA determined was observed to be directly proportional to the concentration of sulfosalicylic acid (SSA) utilized up to about 1% SSA, but values determined using SSA concentrations of 5, 7.5, and 10% were the same. Thus, the acid does not appear to be producing hydrolysis of peptides in general, because these peptides are present in CSF at an order of magnitude higher concentrations.[15,100–103] Rather, the data indicate either a pool of readily releasable GABA or an intrinsic property of GABA *per se*. In either event, patient data have been found to be proportional when comparison is made between results from heat-in activated (no acid) CSF and from CSF exposed to 5% SSA for deproteinization.[97]

Recently, freeze/thaw cycles were reported to produce significant increases of results obtained by a muscimol RR method,[104] whereas results obtained by the I-E/F procedure are not influenced by freeze/thaw cycles.[97] Apparently, a substance is released by freeze/thaw that interferes with the muscimol RR assay.

Measurement of GABA in CSF from normal volunteer human subjects[8] has documented the significant influence of several *in vivo* variables including age and sex and has confirmed the existence of a rostrocaudal concentration gradient[19–21,84] for GABA in CSF.

The above consideration of *in vitro* and *in vivo* variables is quite relevant to a discussion of the pathological neurochemistry of CSF because several of the early patient studies, and even some of the more current studies, were carried out without adequate regard for the influence of these variables. These earlier studies need not be discussed here because evaluations have been presented previously.[3–10,22,83] Nevertheless, patient studies to elucidate neurochemical pathology through evaluation of CSF must utilize properly matched control and patient groups as well as storage and assay conditions that have been demonstrated to provide valid data.

Several lines of evidence indicate that CSF concentrations of GABA reflect GABA content of the brain. Correlations have been observed between ventricular or cisternal CSF levels of GABA and GABA concentrations in brain following administration of GABA-elevating drugs.[14,15] The presence of a rostrocaudal concentration gradient of GABA in CSF obtained at the lumbar level[8,9,19–21] implies that GABA in lumbar CSF originates in brain and therefore reflects brain GABA content. Preliminary clinical investigations support this concept, documenting low GABA levels in lumbar CSF[3,23–30,78] from patients with disorders known to be associated with deficiencies of brain GABA content.[34–39] Increases of lumbar CSF GABA content have been observed during therapeutic trials using drugs that elevate brain GABA content.[16,17,105–107]

Recent reports have presented results of general studies involving GABA measurement in CSF of patients with various neurological[27,30,78,105] and psychiatric[32,33,92,108–113] disorders. Although these reports do not all uncover significant differences between values from specific disease groups and control groups, results are generally in agreement and lead to the conclusion that reduced GABA in CSF is a characteristic of a variety of neuropsychiatric disorders. A larger number of more specific studies also support this conclusion.

2.1. Neurological Disorders

In concert with demonstrations of reduced GABA concentrations and GAD activity[34,36,114–116] in brain tissue obtained post-mortem from individuals with Huntington's disease (HD), several studies have shown GABA also to be reduced in CSF from patients with HD.[23–27,30,78] In two other studies,[102,108] GABA values in CSF from HD patients were lower than but not significantly different from those of controls. Measurement of GABA in CSF from children of parents with known HD has indicated that about half of these at-risk individuals also have reduced CSF GABA content.[24,111] Thus, although definitive conclusions cannot be drawn, the data are consistent with the possibility that the at-risk individuals with the low CSF GABA levels are those who are presymptomatic for HD, even though they were clinically normal when the CSF was obtained.

The possibility that GABA levels may be altered prior to the appearance of symptoms adds a possibly complicating consideration. Altered GABA activity has been associated for various reasons with a variety of both neurological and psychiatric disorders. If GABA alteration is necessary but not by itself sufficient for manifestation of the symptoms of these disorders, then the GABA alteration would precede the appearance of clinical symptoms. Thus, even a well-screened population of normal volunteers may not be an ideal reference group, since some of these individuals will develop various neurological and psychiatric symptoms at a later age. In this connection, CSF GABA values obtained during studies of normal volunteers have been found to vary more within groups of younger individuals than within groups of older individuals.[5,8]

Deficient GABA has long been associated with epilepsy. Hypothetically, epilepsy can be readily pictured to be the consequence of a loss of inhibition. Experimental associations stem from studies of interactions between convulsant drugs and vitamin B_6 with resulting influence on GAD activity in animal models as well as direct studies of the neurotransmitter in human brain taken at autopsy.[37,38] Several studies have shown that GABA is also reduced in CSF of patients with epilepsy,[27,28,91,117–119] most prominently in those with grand mal and psychomotor seizures.[28] Study of dogs has revealed a highly significant positive correlation between pentylenetetrazole seizure threshold and GABA levels in canine CSF but not in the plasma.[120]

A number of reports[27,30,31,78,90,108,121,122] have presented values for GABA in CSF from patients with Parkinson's disease (PD). Mean values have consistently been lower than those from respective control groups, although the reductions have not always been statistically significant. Generally, values from PD patients have been reported to be higher in L-DOPA-treated patients[27,30,31,108] than in untreated patients[27,30,31,78,90] although lower levels in CSF from L-DOPA-treated patients have also been reported.[121] Data have been presented[122] suggesting that differences in CSF GABA levels in L-DOPA-treated patients may be correlated with the patients' responses. In that report, those who responded well to L-DOPA had CSF GABA levels similar to controls, whereas those who responded poorly or who exhibited the "on–off" effect had significantly lower values.

GABA has been measured in a limited number of patients with dementias of various origin.[63,78,108,123,124] Again, these reports generally indicate that demented patients have reduced concentrations of GABA in their CSF. It has been pointed[123] out, however, that at least for Alzheimer's disease, the reductions may be more a reflection of the fact that CSF GABA levels are reduced as a function of age even in control populations.[5,18,123,124]

Reduced GABA values have been reported in CSF from various other patient groups such as multiple sclerosis,[25,108] cerebellar cortical atrophy,[25,108,125] and essential tremor.[30] Significantly elevated GABA values have been reported in CSF from patients with meningitis[108] and acute hypoxic encephalitis.[25]

2.2. *Psychiatric Disorders*

Many neurological and psychiatric disorders tend to be age and/or sex specific, and, therefore, invalid conclusions could be drawn when comparisons are made between patient and reference populations unmatched with respect to age and sex distribution. This difficulty can be illustrated by considering conflicting conclusions that have recently been published based on evaluations of GABA in CSF from psychiatric patient groups.[32,33,86,92,109–112] In some of these studies,[110–112] the results are biased by major disparities existing among patient and/or control groups with respect to age and sex distribution.

In an initial study,[92] GABA levels in the second 13-ml CSF fraction from both male and female depressed patients had been shown to be lower than but not significantly different from those of corresponding normal volunteers of similar age range. In a similar study also utilizing CSF from normal volunteers of matching age range,[109] GABA levels in the first 13-ml CSF fraction were found to be significantly reduced in both male and female depressed patients, and, as in the previous study, GABA values from the second CSF fraction were observed to be lower than but not significantly different from those of the normal individuals.

Another group of investigators[111] reported a seemingly more dramatic reduction of GABA in CSF from depressed patients in comparison to that of patients with various neurological disorders. In this study, however, questions can be appropriately raised as to whether the GABA differences were biased by differences in the age and sex distributions between the two groups, since only one of the 17 depressed patients (mean age 45) but four of the neurological patients (mean age 52) were male. Thus, in view of the propensity of older females to have low CSF GABA, it is possible, depending on the precise age distribution (not specified), that the age and sex discrepancies could have produced an exaggeration of the GABA differences.

In the same report[111] and a subsequent expanded report,[112] conclusions were drawn that GABA in the CSF from a group of psychotic patients was not significantly different from that of either the depressed or the neurological control group. However, for these comparisons, the age and sex distributions were especially inappropriate because the psychotic group consisted of nine men and five women (subsequently 11 men and eight women) of considerably younger age (mean 25 years) than either of the other two groups. Similar disparities of age and sex distribution exist between psychiatric and nonpsychiatric control groups in a report[110] that concluded that GABA in the CSF of chronic schizophrenic patients is elevated relative to that of nonpsychiatric controls. Contrarily, GABA in CSF from schizophrenic patients has been found to be significantly reduced in comparison to that of normal volunteers.[33] Subsequent evaluation of the data[32] following division into male and female groups of matching age ranges revealed that the significant reduction was limited to the female schizophrenic patients. This result is compatable with data[109] that also compared GABA levels in CSF from schizophrenic patients and normal volunteers. In this study, the values from male patients were lower than but not significantly different from age-matched normal individuals, whereas the values (mean $\pm$ SD) from female patients were substantially lower (173 $\pm$ 25 pmol/ml *vs.* 224 $\pm$ 20 pmol/ml) than age-matched normal subjects, although the low number (n = 3) of female patients precluded statistical comparison.

Data concerning GABA in CSF from schizophrenic and depressed individuals have also been presented in another report.[113] In this report, values for untreated schizophrenic patients were higher but not significantly different from a control group, whereas values from a short-term-treated schizophrenic group were lower but not significantly different from another control group, and evidence was presented that neuroleptic treatment produces an increase of CSF GABA levels. This report seems internally inconsistent and is difficult to evaluate because there were age and sex disparities among the various groups, which were not precisely delineated. In addition, values were determined either by an ion-exchange/fluorometric procedure or a gas-chromatographic/mass spectrometric procedure, and it is not clear in some cases which procedure was utilized. Finally, the CSF GABA values, in general, were substantially higher than those presented in the other reports discussed, possibly because samples may have been subjected to acid under conditions not precisely described. As discussed above, CSF samples exposed to acid conditions are subject to artifactual elevation of GABA levels secondary to hydrolysis of GABA-containing peptides.[83,98]

Thus, critical review of the literature to date suggests that GABAergic activity as reflected in CSF may be reduced in both depressed and psychotic patients and that this reduction is more prominent in women. In order to evaluate GABA alterations, great care must be exercised (1) to avoid conditions that would produce artifactual alterations and (2) in drawing comparisons with reference populations, because influences based on age and sex even in normal individuals are of similar magnitude to that resulting from the disorder.[8]

In summary, GABA has been shown to be deficient in CSF from patients with a variety of neuropsychiatric disorders. Genetic factors have been implicated in the etiology of several of these disorders, and in at least one, Huntington's disease, GABA may be reduced prior to the appearance of symptoms. Presumably, some of the normal individuals, especially those of younger age from whom reference data have been derived, will develop a neuropsychiatric disorder at some later time in their life: i.e., they are "presymptomatic" for these disorders. In one sense, therefore, they could not be considered to be definitively normal except retroactively, after they have lived a "normal" life.

Thus, the data available to date seem compatible with the possibility that GABA deficiency may be a necessary but not sufficient etiologic factor in many neuropsychiatric disorders. Only continued evaluation of these and other individuals can prove or disprove this hypothesis. In conclusion, ample justification exists for continuing this area of investigation and expanding such studies to include careful evaluation and comparison with other neurotransmitters.

2.3. *Cerebrospinal Fluid Studies to Evaluate Pharmacological Manipulation*

Following the realization that GABA is deficient in a variety of degenerative neurological and psychiatric disorders and that the GABA content of CSF offers a reflection of CNS GABA activity, several studies have been carried out attempting to therapeutically manipulate CNS GABA activity. Such studies combine clinical evaluation of patient response with neurochemical evaluation of neurotransmitter response through measurement of GABA in CSF.

Attempts to augment GABA activity in patients have utilized two approaches: GABA agonists and GABA-elevating agents. Treatment of patients with GABA agonists such as imidazole-4-acetic acid,[126] muscimol,[127,128] and 4,5,6,7-tetrahydroisoxazolo-(5,4-C)pyridin-3-ol (THIP)[129,130] have not provided clinical benefit, although during a trial of THIP for treating epilepsy,[130] a trend toward lower seizure frequency was noted only during a period on submaximal dose. Several pharmacological agents produce elevation of GABA concentrations through inhibition of GABA transaminase (GABA-T), the chief catabolic enzyme for GABA. Compounds such as isoniazid (INH) and aminooxyacetic acid inhibit GABA-T by reacting with the coenzyme pyridoxal phosphate. More specific GABA-T inhibitors include the enzyme-activated agents ethanolamine-O-sulfate (EOS), γ-acetylenic GABA (GAG), γ-vinyl-GABA (GVG), and the neurotoxin gabaculline. Valproic acid (VPA) also functions to elevate brain GABA levels, although its mechanism of action remains equivocal (see refs. 131–135 for reviews). These compounds, except EOS, which does not cross into the brain, are known to produce increases of brain GABA levels in various species, and, in general, it has been shown that brain increases are also reflected in CSF.[14,17,26,105]

Clinical studies of certain GABA-elevating agents have been carried out, offering the advantage that CSF GABA measurement can be utilized to verify the biochemical consequences of treatment. Isoniazid was initially reported to provide clinical benefit for individuals with HD.[136] Subsequent studies[16,105,137–139] have not confirmed the initial clinical observation even though monitoring of CSF GABA levels confirmed that the drug produced the desired biochemical effect.[16,105]

Clinical trial with GAG has similarly indicated a lack of clinical benefit for HD patients during treatment, although again CNS GABA levels have been demonstrated to be elevated as monitored through measurement in CSF.[106] In a study of tardive dyskinesia (TD), GAG was reported[140] to be moderately effective in suppressing TD symptoms, although side effects were seen such as sedation and confusion plus myoclonic jerks in one patient. γ-Acetylenic GABA also produced consistent decreases of SGPT levels, evidently through its nonspecific effects on other transaminase enzymes.

γ-Vinyl-GABA is a specific GABA-T inhibitor[132] and therefore would presumably be less likely to produce side effects mediated through effects on other enzymes. Recent clinical studies evidently have confirmed this presumption. GVG has been shown to produce dose-dependent increases of free and conjugated GABA when administered to patients with miscellaneous disorders.[17] Administration of GVG to patients with TD has recently been shown in three studies to significantly reduce dyskinesia. In one of these studies,[107] patients were withdrawn from their neuroleptic regimen and were drug-free for at least 4 weeks prior to the beginning of the study. All seven patients exhibited antidyskinetic responses while receiving GVG, and CSF GABA levels were elevated in the patients monitored. No clinically significant side effects occurred during the study, and side effects reported during previous GABA agonist studies were suggested to be secondary to drug–drug interactions with neuroleptics.

In the other two trials of GVG in patients with TD, GVG was administered along with the patients' neuroleptic regimen. In one of these investigations,[141] hyperkinesia was reduced in seven of the nine subjects but increased in two subjects, both of whom had senile dementia, whereas parkinsonism was unchanged in six subjects and changed in a mixed fashion in the others. Sedation was reported to be the most prominent side effect. In the other of these reports,[142] mean TD scores were significantly reduced during GVG therapy, but the reductions were correlated with exaggeration of parkinsonian symptoms. Sedation was reported to be the most common side effect; one patient exhibited reversible mental deterioration during treatment, and SGPT values were decreased.

γ-Vinyl-GABA has also been reported to provide benefit during a trial in patients with epilepsy.[143] During this study, significant reductions in seizure frequency were observed along with only mild and transient side effects.

3. BIOGENIC AMINE NEUROTRANSMITTERS

Of the biogenic amine neurotransmitters, only norepinephrine (NE) has been extensively studied in CSF. Unfortunately, studies of epinephrine (E), dopamine (DA), and serotonin (HT) have been limited by lack of assay procedures suitable for CSF analysis[22,144,145] and therefore data concerning these compounds have been extrapolated from studies of metabolites.

Recent work has established liquid chromatography as a procedure for measuring catechol- and other biogenic amines.[69] This advance resulted from the development of the electrochemical (EC) detector, which is based on the principle that compounds that can be oxidized (or reduced) will donate (or accept) electrons to a detector electrode in a cell maintained under proper potential. These donated electrons can be detected amperometrically and readily amplified to provide sensitive detection in the subpicomole range. This detector is somewhat selective by depending on the particular electronic configuration of the specific molecule as well as the potential and polarity of the detector cell. The primary specificity of the procedure, however, results from the high resolving capability of high-performance liquid chromatography (HPLC). The detector is particularly compatible with ion-exchange and reverse-phase HPLC in view of the fact that aqueous solvents are utilized for both of these forms of HPLC. Catecholamines and their metabolites are readily oxidized and, therefore, are particularly suited for EC detection. Despite the potential for ultrasensitivity, utilization of these procedures for CSF measurements has been hampered because of various difficulties having to do, for example, with column and detector consistency.

3.1. Norepinephrine

Utilization of fluorometric assays had demonstrated NE in CSF at levels near the lower limits of detection.[57,146,147] More recently, two radioenzymatic procedures[58,148] have been developed that are more sensitive and therefore more suitable for study of CSF. Both radioenzymatic procedures depend on an enzyme-mediated reaction utilizing the methyl donor [^{3}H]methyl-S-adenosylmethionine to produce a tritiated derivative of norepinephrine. The catechol-O-methyltransferase (COMT)-based assay utilizes purified COMT to produce [^{3}H]normetanephrine from NE, whereas the phenylethanolamine-N-methyltransferase (PNMT)-based assay utilizes purified PNMT to produce [^{3}H]epinephrine from NE. The PNMT-based procedure is more sensitive and more precise than the COMT-based procedure.[55] The PNMT-based procedure has been verified through comparison with a GC/MS procedure[148] and has provided most of the useful information in the literature concerning pathological alterations of NE in CSF.

Studies have been carried out characterizing several of the basic parameters required for the proper evaluation of pathological alteration of NE function associated with human disorders. Rostrocaudal concentration gradients of NE in human CSF obtained by lumbar puncture have been documented, reflecting a rate of increase of about 2%/ml within the first 16 ml and less or no gradient reflected by aliquots from the 16th to 40th ml.[149] These data have been interpreted to indicate that lumbar CSF NE is primarily a reflection of spinal tissue NE, probably influenced by noradrenergic nerves from the brainstem terminating in the spinal cord.

Correlations have been demonstrated between plasma and CSF NE concentrations; however, CSF NE is primarily of central origin, and there is little communication of NE between plasma and blood across the blood–CSF and blood–brain barriers.[150] For these reasons, the plasma–CSF correlations have been attributed to similarity of function in the peripheral and central nervous systems.[55]

Circadian rhythms have been documented for NE in CSF with concentrations during normal wakeful periods almost twice those during sleep periods,[149] apparently paralleling the functional effects of NE on activity.[55] Drug effects also support the concept that CSF NE levels provide a reflection of central activity; accordingly, amphetamine facilitates release of norepinephrine and produces striking increases of CSF NE concentrations, whereas bromocriptine inhibits release of NE and produces significant reductions of NE in CSF.[55]

Norepinephrine in the 14th to 26th milliliter fraction of lumbar CSF from a population ($n = 37$) of normal humans was found to be 93 ± 35 ($\pm$SD) pg/ml (range 36–173 pg/ml). Differences between males (90 ± 35 pg/ml; $n = 25$) and females (98 ± 34 pg/ml; $n = 12$) were not statistically significant. The specimens were chilled immediately and then divided into aliquots, acidified with ascorbic acid, and frozen at $-70°C$ within 30 min.[12] Presumably, these values were artifactually reduced during this period of sample preparation because, as noted in the report, more rapid acidification and freezing results in much higher normal values for NE in CSF.[151]

In CSF from another population ($n = 240$) consisting of patients with various neurological disorders not known to be associated with altered NE activity, a much wider distribution of NE values ranging from 40 to 2000 pg/ml was found with a mean value of about 175 pg/ml. The differences between these two populations may reflect artifactual alterations resulting from differences in specimen sampling; however, the wider range may also be indicative of a deviation from normal associated with some of the neurological disorders included.

Altered NE in lumbar CSF has been associated with several neuropsychiatric disorders. For example, mean values ($\pm$SE) of NE in CSF from nine patients with Huntington's disease were shown to be 171 ± 18 pg/ml, which was significantly lower than the 239 ± 17 pg/ml determined for CSF from age- and sex-matched control patients.[152,153] Intermittent electrical stimulation of the caudate nucleus has been demonstrated to produce significant reductions of CSF NE levels. Similarly, intermittent cerebral cortical stimulation significantly depressed NE levels in CSF from epileptic patients, whereas levels were significantly elevated by chronic cerebellar stimulation.[154]

Study of CSF from 34 amyotrophic lateral sclerosis (ALS) patients revealed a mean (±SE) value of 365 ± 44 pg/ml, significantly higher than that from the control group (251 ± 14 pg/ml).[155] Significantly elevated NE has been reported in CSF from male alcoholic patients during alcohol withdrawal.[156] The mean (±SE) value of 193 ± 22 pg/ml from the patient group was significantly higher than the 129 ± 14 observed for a control group and significantly reduced toward the control values following cessation of withdrawal symptoms. A highly significant correlation of CSF NE with blood pressure has been described and discussed with respect to alterations of blood pressure associated with disorders having known CSF NE anomalies.[55] Propranolol has also been reported to produce dose-dependent increases in CSF NE accompanied by reductions in blood pressure in dogs.[157]

In a recent report,[158] NE was studied in cisternal, ventricular, and lumbar CSF from patients with cerebral vasospasm following subarachnoid hemorrhage using an automated chromatographic/fluorometric assay procedure. The data indicated that NE increased significantly in CSF from those with vasospasm in comparison to those without vasospasm.

Several lines of evidence point toward involvement of NE in the mood alterations symptomatic of manic–depressive illnesses. In one study,[159] NE levels in CSF from manic patients were found to be more than twice as high as those in depressed patients or in neurological control patients. During treatment, the elevated CSF NE concentrations tended to decrease toward normal. Depressed patients with a high degree of anxiety had significantly higher levels of CSF NE than did patients who were depressed but not anxious. Other studies have documented elevated levels of NE in CSF from schizophrenic patients, particularly those with paranoid features, when compared to those in age-matched normal volunteers[160,161] or other control groups.[162] Furthermore, chronic treatment with the neuroleptic pimozide produced a reduction of CSF NE toward normal, which correlated with the patients' decrease in global psychosis.[161]

Another approach for estimating central NE activity has utilized measurements of noradrenergic metabolites in CSF. Norepinephrine is degraded by monoamine oxidase and catechol-*O*-methyltransferase to produce the metabolites 3-methoxy-4-hydroxphenylethylene glycol (MHPG) and vanillylmandelic acid (VMA). Of the two principal metabolites, MHPG is the more prevalent, being present in CSF at concentrations nearly two orders of magnitude higher than NE, whereas the concentration of VMA is approximately one order of magnitude greater than that of NE.[149] Measurements of MHPG in CSF do not provide as precise a reflection of CNS NE activity as does measurement of the neurotransmitter *per se*. Even though MHPG levels correlate with those of NE in CSF, several parameters such as circadian rhythms and age correlations are detectable only through determination of NE concentrations.[163,164]

A third approach for estimating central NE activity is through measurement of associated enzyme activities in CSF. Dopamine-β-hydroxylase (DBH), the catalyst for NE formation from dopamine, is the principal enzyme that may provide an index in CSF of central NE activity. Dopamine-β-hydroxylase is measureable in CSF,[165–174] evidently being released from noradrenergic neurons by exocytosis during neuronal function. Both pharmacological[165] and physiological[166,167] perturbations in central NE activity are positively associated with alterations of DBH activity in CSF.

Clinical study of DBH in CSF has primarily been directed toward evaluation of psychiatric disorders. Several studies[169,171–174] have not demonstrated significant differences of CSF DBH activities among depressive, schizophrenic, alcoholic, and/or personality disorders or controls. However, data have indicated that CSF DBH activity may be reduced during the manic phase in bipolar affective patients when compared to the euthymic or depressed phases.[171] Similarly, studies have indicated correlations between CSF DBH activities and degree of psychopathology or vulnerability to psychopathology.[172–174] In the most recent of the cited reports,[173] the psychotic patients who became nonpsychotic during neuroleptic treatment had pretreatment CSF DBH activities significantly lower than did those who remained psychotic during neuroleptic therapy. Low pretreatment CSF DBH activity also correlates with adverse reactions to disulfiram.[172]

3.2. Dopamine and Serotonin

Direct measurement of dopamine (DA) and serotonin (HT) in CSF has not been possible because adequate procedures for quantification of these two neurotransmitters are still under development. We can hope that currently promising analytical techniques such as the use of amperometric (EC) or fluorescent detection coupled with high-performance liquid chromatography will soon overcome this deficiencey. In the meantime, it has been necessary to rely primarily on metabolite measurements to gain insight into central DA and HT activities through CSF measurements. In view of the apparent lack of precise reflection of noradrenergic activity offered by MHPG measurements in comparison to that offered by NE measurement *per se,* DA and HT metabolite studies would also be expected to be of limited value. This expectation seems even more reasonable when considered in light of the relative functions of the neurotransmitters and their metabolites. The metabolites are generally weakly acidic compounds synthesized for the purpose of excretion into and then out of the CSF. On the other hand, the neurotransmitters *per se* are specifically removed from the CSF through reuptake mechanisms to maintain the functional integrity of the CNS. Thus, neurotransmitter content in CSF represents a "spillover" from normal regulated function, a circumstance more likely to reflect functional activity than that of the more dynamic excretory process.

Recently, a sensitive radioimmunoassay for serotonin (HT) was described and utilized to measure HT in lumbar CSF from 21 patients.[175] Seven of the specimens showed HT in the range of 2–3 pmol/ml, and the remainder did not reveal HT (i.e., <2 pmol/ml). The authors concluded that <4 pmol/ml constitutes the upper limit for HT in CSF and that a more sensitive procedure is required for detailed examination of HT in CSF. Nevertheless, this procedure was utilized to study ventricular CSF from recent stroke patients and indicated values over the range of <2 to 5 pmol/ml.[176] Subsequently, a sensitive GC/MS procedure was presented[161] and utilized to evaluate diurnal variation in monkey CSF.[177,178] Results indicated mean (±SE) values to be 378 ± 146 and 46 ± 10 pg/ml (i.e., 2.14 ± 0.83 and 0.26 ± 0.06 pmol/ml) during light and dark periods, respectively.

A GC/MS procedure has also been described recently for measuring dopamine (DA) in CSF.[79] This procedure, when applied to lumbar CSF samples from six monkeys, did not detect DA in four of the samples, and the other two samples showed mean values of 1.7 and 4.5 pmol/ml. After being subjected to acid hydrolysis, all of the specimens contained detectable DA, the mean ± SD values being 13.3 ± 7.0 pmol/ml, indicating the mean presence of conjugated DA at the concentration of 12.3 ± 6.5 pmol/ml. An HPLC procedure with EC detection was also recently utilized to measure DA as well as NE in CSF from schizophrenic and control patients.[179] Mean (±SD) values of DA were reported to be 207 ± 263 pg/ml (1.35 ± 1.72 pmol/ml) for control patients, which were not significantly different from those from the schizophrenic subjects off drugs, i.e., 200 ± 124 pg/ml (1.31 ± 0.81 pmol/ml). The values from schizophrenic subjects receiving neuroleptic drugs (i.e., 524 ± 549 pg/ml or 3.42 ± 3.58 pmol/ml) were significantly higher than those from control subjects or schizophrenic patients off drugs. Mean values for NE were 133 ± 37, 163 ± 44, and 210 ± 73 pg/ml (0.79 ± 0.22, 0.96 ± 0.26, and 1.24 ± 0.43 pmol/ml) for the control individuals, schizophrenic subjects off drugs, and schizophrenic patients on drugs, respectively. As was reported by others,[160,161] the schizophrenic subjects off drugs showed CSF NE values higher than control individuals, but in this instance the difference was not statistically significant. Values from the schizophrenic patients on drugs were significantly higher than those of control individuals. Significant positive correlation was noted between DA and NE values in these subjects.

4. PEPTIDE NEUROTRANSMITTERS

Recent studies by many investigators have provided rather conclusive evidence that at least 30 peptides are present in the CNS and function as neurotransmitters.[180–184] Certain peptides such as the endorphins, substance P, neurotensin, and some of the hypothalamic releasing factors have been isolated, characterized, and unequivocally demonstrated to be present in the brain.[185] However, without concentration through elaborate isolation procedures, their presence in such low levels can only be measured through various bio- or radioassay procedures. Many peptides have not been directly demonstrated in brain but are felt to exist there primarily because CNS tissue exhibits immune reaction to antibodies prepared for peptides that are known to function outside the blood–brain barrier. These immunoassays are based on competition between tissue constituents and the labeled peptide for interactive sites on the antibody.[185] These immunointeractive sites are not necessarily similar to sites important for biological activity, and, thus, categorical conclusions that immunoassay results indicate concentrations of specific compounds or reflect levels of neurochemical activity are not justified. Radioreceptor and other bioassay methods are also sufficiently sensitive for peptide measurements; however, these assay procedures may also lack specificity. Comparison of data from these various bioassay methods can aid in uncovering instances of nonspecificity; however, more specific procedures will be necessary to elucidate definitive peptide neurochemical function. During the interim, data must be expressed with qualification, using terms such as "peptide-like" immunoreactivity.

Reviews in a previous volume of this *Handbook*[22] and elsewhere[44,183,186–191] have summarized recent studies documenting the existence in CSF of immunoreactivity similar to that of nearly all of the putative neurotransmitter peptides that have been studied in the CNS parenchyma. In general, these studies have shown that neurotransmitter peptide-like reactivities are present at minute (femtomole/milliliter) concentrations, and, in many case, normal or control values have been assigned. More detailed studies have served to illustrate that the situation is much more complicated than perhaps had been generally recognized.

Among the peptide neurotransmitters, the opioid class has been most widely studied. These substances have been assayed by bioassay, receptor-binding assay, and radioimmunoassay.[192] The latter two assay techniques have been most widely used for measurements in CSF. Receptor- and immunoglobulin-binding studies have proven complementary in some circumstances; however, the natures of both the radioreceptor and radioimmunoassays present problems in elucidating the pathological neurochemistry in CSF. For example, various studies[192–195] have demonstrated a multiplicity of opioid receptors in brain tissue. Morphine, its congeners, and various peptides not only bind differentially to these sites but also bind according to the regional distribution of these sites. Thus, the specific radioligand or membrane preparation could influence results, as could various competing substances, possibly not at all related to opioid action, contained in the CSF to be analyzed.[181] These considerations make formulation of conclusions about specific ligands difficult, especially during comparison of various studies presented in the literature.

Similar considerations limit the present use of radioimmunoassay procedures. A chief drawback stems from the fact that antisera from various sources demonstrate various specificities, again impeding comparison of studies presented in the literature. These specificities can be characterized in each case with respect to known constituents; however, current evidence suggests that not all the interactive constituents are known.

The significance of the limitations of the binding-type assay procedures for peptide studies in CSF can be illustrated by considering recent CSF studies of opioid-type binding. Opioid-, adrenocorticotropin- (ACTH-), β-lipotropin-, dynorphin-, and enkephalin-like immunoreactivities (and receptoreactivities) have been demonstrated and extensively studied in CSF.[44,49–54,63,65,66,190,196,197] These reactivities represent the net results of various cross-reactive ligands, not all of which are even necessarily peptides.[198,199] Use of antibodies of differing specificities has characterized the ligands to some extent, demonstrating the probable presence of β-endorphin[49–51,65,190] and

Met-enkephalin[52,63] in CSF. Studies that have utilized immunoassays have been characterized for cross reactivity only with respect to known related ligands.

Utilizing separation procedures prior to immunoassay, various investigators have found that only a portion of the immunoreactivity in unfractionated CSF can be attributed to the actual intended ligand. For example, one study[50] revealed β-endorphin-like immunoreactivity in CSF from four control patients at a mean (± SE) concentration of 91.6 ± 11.3 pg/ml. Following gel chromatography, however, only 20% of the reactivity was found to have been eluted with authentic β-endorphin; the remainder cochromatographed with β-lipotropin and as a larger unknown molecule.

Several other studies have been reported in which CSF was fractionated using gel or reverse-phase chromatography followed by detection of opioid-like reactivity using radioimmunoassay procedures "selective" for endorphins[47,51] or enkephalins[53] as well as opioid receptor-binding assays[65,66,75,190,200] or bioassay.[200–203] These studies have also documented the multiplicty of opioid-like reactive components in CSF and have served to illustrate that a significant portion of the reactivity cannot yet be attributed to specific known components.

Fractionation studies have yet to be carried out for many of the other peptide neurotransmitters that have been measured in CSF. In those cases in which attempts have been made to chromatographically characterize CSF immunoreactivity, a multiplicity of reactivity has also sometimes[46,47,188,204] but not always[45,48] been observed.

The apparent complexity of the data generated by the respective binding procedures makes difficult the comparison of data or the explanation of discrepancies between studies. This difficulty is compounded by variations of antisera or other procedures utilized as well as the realization that a portion of the binding observed may not even be related to physiological action. In addition, many reports, by presenting data either on a weight or molar basis, using equivalents to one particular ligand, obviate comparison by not allowing a clear proper means for interconverting these values. Moreover, as recently reviewed,[205,206] only a limited number of studies have been carried out to evaluate the impact of *in vivo* and *in vitro* variables on control or patient data (summarized in Table I). These factors must be thoroughly characterized to provide us with basis for avoiding artifacts or inappropriate comparison, which could lead to invalid conclusions.

Thus, among the peptide neurotransmitter systems, base-line data have not yet been adequately established, and only the presence of large numbers of "binding" constituents in CSF, probably including β-endorphin and Met-enkephalin *per se,* can be concluded. We can hope that continued refinement of existing procedures, along with additional, more specific analytical procedures, will soon provide more definitive information.

From several standpoints, then, attempts to extensively evaluate the pathological peptide neurochemistry in CSF appear premature. Given the above limitations, however, preliminary studies of CSF from patient populations have provided some possible leads, but a certain probability exists that some of these leads may be "blind" tangents.

Table I
Physiological and Dynamic Characteristics of CSF Peptide and Steroid Hormones[a]

Peptide	Rostrocaudal gradient	Circadian rhythm	Sex variation	Age variation
TRH	+	↑ Afternoon	---	
TSH				
T_4				
T_3				
LH				
Testosterone		↑ Night	↑ Male	
Estradiol			---	
Progesterone			---	
Prolactin	---			
Somatostatin		↑ Night		
Growth hormone				
ACTH				
Cortisol		↑ Morning		
Vasopressin	+?			+
Calcitonin				
VIP				
Insulin				
Substance P	--			+
MSH				
Opiods		↑ Morning		
Melatonin	--	↑ Night		+

[a] Summary of *in vivo* variations in CSF peptide and steroid hormones. The obvious voids in the table document areas requiring clarification. (From Kee and Wood,[205] with permission.)

Pathological neurochemistry of peptides in CSF has primarily involved study of the opioid class of compounds. These studies have included two prominent general areas—pain and psychiatric disorders. Of these, correlations have been more consistent during the studies of pain. Evaluation of patients with chronic pain syndrome has indicated that low endogenous levels of CSF opioid receptor reactivities are directly associated with low pain threshold and tolerance levels.[207] Similar studies of patients with organic pain syndromes have found significantly lower CSF opioid reactivities when compared with those having predominantly psychogenic pain syndromes.[208] Opioid-like activity, measured by bioassay, has been found to be significantly lower in CSF of patients with acute pain than in CSF of patients with chronic pain, the values from both groups being significantly lower than those from control patients with no history of pain.[209] Decreased levels of enkephalin-like binding activity have been reported in CSF from patients during migraine headaches as well as in cluster headache sufferers.[210] On the other hand, no statistically significant differences of CSF β-endorphin-like immunoreactivity were seen in patients with trigeminal neuralgia.[211] During experiments with cats,[212] cisternal CSF Met-enkephalin-like material (attributed to high-molecular-weight compounds) was found to increase in response to tooth-pulp stimulation or direct electrical stimulation of the trigeminal nerve.

Evaluation of the analgesic effects of various forms of electrical stimulation has consistently found associated significant increases of β-endorphin- and/or Met-enkephalin-like immunoreactivity in CSF.[213–217] These data suggest that release of endogenous opioid activity may mediate the analgesic effect. Peripheral administration of β-endorphin to human subjects has been reported to result in substantial elevation of CSF β-endorphin-like immunoreactivity.[218]

A series of studies of opioid receptor reactivities in CSF from schizophrenic individuals has led to the conclusions that concentrations in acute schizophrenic patients are significantly elevated with respect to those of healthy volunteers, that levels in chronic schizophrenic subjects are lower than those in acute schizophrenic patients, and that neuroleptic treatment significantly lowers opioid receptor reactivity in patients with acute schizophrenia.[64,219,220] Another evaluation of a larger population, however, found significantly lower levels of CSF opioid receptor reactivity in comparison to that of normal individuals, the difference being accounted for primarily by the values from the male patients.[54] Data in this report also suggested possible elevation of activity in the CSF of manic–depressive patients during the manic phase. β-Endorphin-like immunoreactivity was also measured in some of the same CSF samples. Comparison of the data did not reveal correlation between opioid receptor reactivity and β-endorphin-like immunoreactivity, suggesting that immunoreactivity does not provide a major reflection of the total CSF opioid activity.

Measurement of somatostatin-like immunoreactivity in CSF from patient and control individuals has documented low immunoreactivity associated with multiple sclerosis during relapse,[221] Parkinson's disease,[222], Alzheimer's disease and other dementias,[223], Huntington's disease (both affected patients and symptom-free offspring,[224]), and depressed patients.[225]

Substance-P-like immunoreactivity (SP-LI) has been reported to be not significantly altered in CSF from patients with Huntington's disease,[223] parkinsonism, dyskinesias, progressive supranuclear palsy, myopathy, or amyotrophic lateral sclerosis,[226] but significantly reduced in CSF from patients with neuropathy and the multiple-system atrophy of Shy–Drager syndrome.[188,226] This pattern along with other indications suggests that SP-LI in lumbar CSF may originate, and therefore reflect, activity, in the spinal cord.[188,226]

Thyrotropin-releasing hormone-like immunoreactivity has been reported to be significantly elevated in CSF from patients with endogenous depression relative to that from a reference propulation both when the patients were depressed and after recovery.[227]

Prolactin-like immunoreactivity (PRL-LI) measurement in CSF has indicated elevated immunoreactivity in pregnant women and some patients with pituitary tumors relative to that of control subjects.[228] Study of CSF PRL-LI in psychotic patients has shown elevation of immunoreactivity following chlorpromazine therapy.[229] During this study, significant differences of PRL-LI were noted between the pretreatment values for male and female patients. Indications have been presented that elevated PRL-LI in CSF correlates with PRL-LI in plasma.[227,228]

Elevations of the concentrations of pituitary peptide hormone-like reactivities in the CSF of patients with pituitary tumors have been associated with suprasellar extension of the tumor.[230] In addition, CSF vasopressin-like immunoreactivity has been found to be elevated in CSF and reduced in plasma of patients with diabetes insipidus.[231]

5. CONCLUDING REMARKS

A large and rapidly expanding body of literature now exists describing studies of CSF. These investigations have provided more or less definitive knowledge about the physiology and pharmacology of CSF in several important fields (reviewed in this *Handbook*[22] and other major references texts[232–234]), leading to the conclusion that the study of CSF is an appropriate source of information about CNS dysfunction. Attention is increasingly being directed toward elucidation of CNS pathology through neurochemical measurements in CSF. Current efforts will be most fruitful when investigators seek, inasmuch as possible, to define and utilize standardized protocols. Such standardization will provide a basis for comparing data between laboratories and, therefore, a broader foundation of knowledge on which to build.

Certain portions of such a foundation are currently being formed, although a continuing need exists for improvements of technique in some areas, especially neuropeptides. The future now seems clear, however, supported by indications that further CSF investigations will produce significant knowledge of the pathophysiology of CNS disorders.

REFERENCES

1. Davson, H., 1969, *Handbook of Neurochemistry,* Volume 2 (A. Lajtha, ed.), Plenum Press, New York, p. 43.
2. Lowenthal, A., 1972, *Handbook of Neurochemistry,* Volume 7 (A. Lajtha, ed.), Plenum Press, New York, p.448.
3. Hare, T. A., Manyam, N. V. B., and Glaeser., B. S., 1980, *Neurobiology of Cerebrospinal Fluid I* (J. H. Wood, ed.), Plenum Press, New York, pp. 171–187.
4. Wood, J. H., 1980, *Neurology (Minneap.)* **30:**645–651.
5. Hare, T. A., 1981, *Mol. Cell. Biochem.* **39:**297–304.
6. Wood, J. H., 1980, *Neurobiology of Cerebrospinal Fluid I* (J. H. Wood, ed.), Plenum Press, New York, pp. 71–96.
7. Manyam, B. V., and Hare, T. A., 1983, *Clin. Neuropharmacol.* **6:**25–36.
8. Hare, T. A., Wood, J. H., Manyam, B. V., Gerner, R. H., Ballenger, J. C., and Post, R. M., 1982, *Arch. Neurol.* **39:**247–249.
9. Wood, J. H., 1980, *Neurobiology of Cerebrospinal Fluid I* (J. H. Wood, ed.), Plenum Press, New York, pp. 53–62.
10. Hare, T. A., Wood, J. H., and Manyam, B. V., 1981, *Arch. Neurol.* **38:**491–494.
11. Cutler, R. W. P., and Spertel, R. B., 1982, *Ann. Neurol.* **11:**1–10.
12. Ballenger, J. C., Post, R. M., and Goodwin, F. K., 1983, *Neurobiology of Cerebrospinal Fluid II* (J. H. Wood, ed.), Plenum Press, New York, pp. 143–155.
13. Oldendorf, W. H., 1980, *Neurobiology of Ceresppronal Fluid I* (J. H. Wood, ed.), Plenum Press, New York, pp. xiii–xiv.
14. Loscher, W., 1979, *J. Neurochem.* **32:**1587–1591.
15. Bohlen, P., Huot, S., and Palfreyman, M. G., 1979, *Brain Res.* **167:**297–305.
16. Manyam, B. V., Katz, L., Hare, T. A., Kaniefski, K., and Tremblay, R. D., 1981, *Ann. Neurol.* **10:**35–37.

17. Grove, J., Schechter, P. J., Tell, G., Koch-Weser, J., Sjoerdsma, A., Warter, J. -M., Marescaux, C., and Rumbach, L., 1981, *Life Sci.* **28**:2431–2439.
18. Kuriyama, K., and Sze, P. Y., 1971, *Neuropharmacology* **10**:103–108.
19. Hare, T. A., and Manyam, N. V. B., 1980, *Anal. Biochem.* **101**:349–355.
20. Enna, S. J., Wood, J. H., and Snyder, S. H., 1977, *J. Neurochem.* **28**:1121–1124.
21. Grove, J., Schechter, P. J., Hanke, N. F. J., deSmet, Y., Agid, Y., Tell, G., and Koch-Weser, J., 1982, *J. Neurochem.* **39**:1618–1622.
22. Wood, J. H., 1982, *Handbook of Neurochemistry*, 2nd ed., Volume 1 (A. Lajtha, ed.), Plenum Press, New York, pp 415–487.
23. Glaeser, B. S., Vogel, W. H., Oleweiler, D. B., and Hare, T. A., 1975, *Biochem. Med.* **12**:380–385.
24. Manyam, N. V. B., Hare, T. A., Katz, L., and Glaeser, B. S., 1978, *Arch. Neurol* **35**:728–730.
25. Manyam, N. V. B., Hare, T. A., and Katz, L., 1979, *Huntington's Disease: Advances in Neurology*, Volume 23 (T. N. Chase, N. S. Wexler, and A. Barbeau, eds.) Raven Press, New York, pp. 547–555.
26. Manyam, N. V. B., Hare, T. A., and Katz, L., 1980, *Life Sci.* **26**:1303–1308.
27. Manyam, N. V. B., Katz, L., Hare, T. A., Gerber, J. C. III, and Grossman, M. H., 1980, *Arch. Neurol.* **37**:352–355.
28. Wood, J. H., Hare, T. A., Glaeser, B. S., Ballenger, J. C., and Post, R. M., 1979, *Neurology (Minneap.)* **29**:1203–1208.
29. Enna, S. J., Ziegler, M. C., Lake, C. R., Wood, J. H., Brooks, B. R., and Butler, I. J., 1980, *Neurobiology of Cerebrospinal Fluid I* (J. H. Wood, ed.), Plenum Press, New York, pp. 189–196.
30. Manyam, N. V. B., Hare, T. A., and Katz, L., 1980, *Brain Res. Bull.* **5**(Suppl. 2):741–745.
31. Manyam, B. V., 1982, *Arch. Neurol.* **39**:391–392.
32. Van Kammen, D. P., Sternberg, D. E., Hare, T. A., Waters, R. N., and Bunney, W. E., Jr., 1982, *Arch. Gen. Psychiatry* **39**:91–97.
33. Van Kammen, D. P., Sternberg, D. E., Hare, T. A., Ballenger, J. C., and Post, R. M., 1980, *Brain Res. Bull.* **5**(Suppl. 2):731–736.
34. Bird, E. D., and Iversen, L. L., 1974, *Brain* **97**:457–472.
35. Perry, T. L., Hansen, S., and Kloster, M., 1973, *N. Engl. J. Med.* **288**:337–342.
36. Urquhart, N., Perry, T. L., Hansen, S., and Kennedy, J., 1975, *J. Neurochem.* **24**:1071–1075.
37. Van Gelder, N. M., Sherwin, A. L., and Rassmussen, T., 1972, *Brain Res.* **40**:385–393.
38. Ribak, C. E., Harris, A. B., Vaughn, J. E., and Roberts, E., 1979, *Science* **205**:211–214.
39. Spokes, E. G. S., 1980, *Brain* **103**:179–210.
40. Perry, T. L., Buchanan, J., Kish, S. J., and Hansen, S., 1979, *Lancet* **1**:237–239.
41. Hornykiewicz, O., Lloyd, K. G., and Davidson, L., 1976, *GABA in Nervous System Function* (E. Roberts, T. N. Chase, and D. B. Tower, eds.), Raven Press, New York, pp. 479–485.
42. Ebert, M. H., Kartzinel, R., Cowdry, R. W., and Goodwin, F. W., 1980, *Neurobiology of Cerebrospinal Fluid I* (J. H. Wood, ed.), Plenum Press, New York, pp. 97–112.
43. Wilkes, M. M., Stewart, R. D., Bruni, J. F., Quigley, M. E., Yen, S. S. C., Ling, N., and Chretien, M., 1979, *J. Clin. Endocrinol. Metab.* **50**:309–315.
44. Post, R. M., Gold, P. W., Rubinow, D. R., Bunney, W. E., Jr., Ballenger, J. C., and Goodwin, F. K., 1983, *Neurobiology of Cerebrospinal Fluid II* (J. H. Wood, ed.), Plenum Press, New York, pp. 107–141.
45. Yamada, T., Takami, M. S., and Gerner, R. H., 1981, *Brain Res.* **223**:214–217.
46. Rehfeld, J. F., and Kruse-Larsen, C., 1978, *Brain Res.* **155**:19–26.
47. Kruse-Larsen, C., and Rehfeld, J. F., 1979, *Brain Res.* **176**:189–191.
48. Fahrenkrug, J., Schaffalitzky de Muckadell, O. B., and Fahrenkrug, A., 1977, *Brain Res.* **124**:581–584.
49. Jeffcoate, W. J., Rees, L. H., McLoughlin, L., Ratter, S. J., Hope, J., Lowry, P. J., and Besser, G. M., 1978, *Lancet* **2**:119–121.
50. Nakao, K., Nakai, Y., Oki, S., Matsubara, S., Konishi, T., Nishitani, H., and Imura, H., 1980, *J. Clin. Endocrinol. Metab.* **50**:230–233.

51. Nakao, K., Oki, S., Tanaka, I., Horii, K., Nakai, Y., Furui, T., Fukushima, M., Kuwayama, A., Kageyana, N., and Imura, H., 1980, *J. Clin. Invest.* **66**:1383–1390.
52. Clement-Jones, V., Lowry, P. J., Rees, L. H., and Besser, G. M., 1980, *J. Endocrinol.* **86**:231–243.
53. Sarne, Y., Azov, R., and Weissman, B. A., 1978, *Brain Res.* **151**:399–403.
54. Picker, D., Naber, D., Post, R. M., VanKammen, D. P., Ballenger, J., Rubinow, D., Waters, R., Kaye, W. H., Ebert, M. H., and Bunney, W. E., Jr., 1982, *Brain Peptides and Hormones* (R. Collu, J. R. Ducharme, A. Barbeau, and G. Tolis, eds.), Raven Press, New York, pp. 207–219.
55. Ziegler, M. H., Lake, C. R., Wood, J. H., and Ebert, M. H., 1980, *Neurobiology of Cerebrospinal Fluid I* (J. H. Wood, ed.), Plenum Press, New York, pp. 141–152.
56. Zurcher, G., and DaPrada, M., 1979, *J. Neurochem.* **33**:631–639.
57. Ziegler, M. G., Lake, C. R., Foppen, F. H., Shoulson, I., and Kopin, I. J., 1976, *Brain Res.* **108**:436–440.
58. Passon, P. G., and Peuler, J. D., 1973, *Anal. Biochem.* **51**:618–631.
59. Peuler, J. D., and Johnson, G. A., 1975, *Clin. Res.* **23**:474A.
60. Michaels, R. R., Huber, M. J., and McCann, D. S., 1976, *Science* **192**:1242–1244.
61. Bernasconi, R., Bittiger, H., Heid, J., and Martin, P., 1980, *J. Neurochem.* **34**:614–618.
62. Abbott, R., and Nahorski, S. R., 1981, *Br. J. Pharmacol.* **72**:164P.
63. Akil, H., Watson, S. J., Sullivan, S., and Barchas, J. D., 1978, *Life Sci.* **23**:121–126.
64. Rimon, R., Terenius, L., and Kampman, R., 1980, *Acta Psychiatr. Scand.* **61**:395–403.
65. Terenius, L., and Walstrom, A., 1975, *Life Sci.* **16**:1759–1764.
66. Walstrom, A., and Terenius, L., 1980, *Acta Physiol. Scand.* **110**:427–429.
67. Glaeser, B. S., and Hare, T. A., 1975 *Biochem. Med.* **12**:274–282.
68. Bohlen, P., Schechter, P. J., VanDamme, W., Coquillat, G., Dorsch, J. -C., and Koch-Weser, J., 1978, *Clin. Chem.* **24**:256–260.
69. Sasa, S., and Blank, L., 1979, *Anal. Clin. Acta* **104**:29–45.
70. Marsden, C. A., 1981, *J. Neurochem.* **36**:1621–1626.
71. Tyce, G. M., and Creagan, E. T., 1981, *Anal. Biochem.* **112**:143–150.
72. Davis, G. C., Kissinger, P. T., and Shoup, R. E., 1981, *Anal. Chem.* **53**:156–159.
73. Causon, R. C., Carruthers, M. E., and Rodnight, R., 1981, *Anal. Biochem.* **116**:223–226.
74. Kissinger, P. T., Bruntlett, C. S., and Shoup, R. E., 1981, *Life Sci.* **28**:455–465.
75. Bennett, G. W., Brazell, M. P., and Marsden, C. A., 1981, *Life Sci.* **29**:1001–1007.
76. Huizinga, J. D., Teelken, A. W., Muskiet, F., Jeuring, H., and Wolthers, B., 1978, *J. Neurochem.* **30**:911–913.
77. Faull, K. P., DoAmaral, J. R., Berger, P. A., and Barchas, J. D., 1978, *J. Neurochem.* **31**:1119–1122.
78. Enna, S. J., Stern, L. Z., Wastek, G. J., and Yamamura, H. I., 1977, *Arch. Neurol.* **34**:683–685.
79. Elchisak, M. A., Powers, K. H., and Ebert, M. H., 1982, *J. Neurochem.* **39**:726–728.
80. Dreifuss, J. J., Kelly, J. S., and Krnjevic, K., 1969, *Exp. Brain Res.* **9**:137–154.
81. Krnjevic, K., and Schwartz, S., 1967, *Exp. Brain Res.* **3**:320–336.
82. Lader, M., 1980, *Introduction to Psychoparmacology*, Upjohn, Kalamazoo, Michigan, p.26.
83. Hare, T. A. Grossman, M. H., Wood, J. H., Glaeser, B. S., and Manyam, N. V. B., 1980, *Brain Res. Bull.* **5**(Suppl. 2):725–729.
84. Wood, J. H., Hare, T. A., Enna, S. J., and Manyam, N. V. B., 1980, *Brain Res. Bull.* **5**(Suppl. 2):111–114.
85. Bohlen, P., Huot, S., Mellet, M., and Palfreyman, M. G., 1980, *Brain Res. Bull.* **5**(Suppl. 2):905–908.
86. Hare, T. A., Wood, J. H., Manyam, N. V. B., Ballenger, J. C., Post, R. M., and Gerner, R. H., 1980, *Brain Res. Bull.* **5**(Suppl. 2):721–724.
87. Brooks, B. R., Ziegler, M. G., Lake, C. R., Wood, J. H., Enna, S. J., and Engel, W. K., 1980, *Brain Res. Bull.* **5**(Suppl. 2):765–768.
88. Bohlen, P., Tell, G., Schechter, P. J., Koch-Weser, J., Agid, Y., Coquillat, G., Chazot, G., and Fischer, C., 1980, *Brain Res. Bull.* **5**(Suppl. 2):761–764.
89. Prasad, A. L. N., and Fahn, S., 1980, *Brain Res. Bull.* **5**(Suppl. 2):737–740.

90. Teychenne, P. F., Ziegler, M. G., Lake, C. R., and Enna, S. J., 1980, *Brain Res. Bull.* **5**(Suppl. 2):769–772.
91. Wood, J. H., Hare, T. A., Glaeser, B. S., Brooks, B. R., Ballenger, J. C., and Post, R. M., 1980, *Brain Res. Bull.* **5**(Suppl. 2):747–753.
92. Post, R. M., Ballenger, J. C., Hare, T. A., Goodwin, F. K., Lake, C. R., Jimerson, D. C., and Bunney, W. E., Jr., 1980, *Brain Res. Bull.* **5**(Suppl. 2):755–760.
93. Achar, V. S., Welch, K. M. A., Chabi, E., Bartosh, K., and Meyers, J. S., 1976, *Neurology (Minneap.)* **26**:777–780.
94. Lichtshtein, D., Dobkin, J., Ebstein, R. P., Biederman, J., Rimon, R., and Balmaker, R. H., 1978, *Br. J. Psychiatry* **132**:145–148.
95. Wood, J. H., Glaeser, B. S., Enna, S. J., and Hare, T. A., 1978, *J. Neurochem.* **30**:291–293.
96. Teychenne, P. F., Lake, C. R., and Ziegler, M. G., 1980, *Neurobiology of Cerebrospinal Fluid I* (J. H. Wood, ed.), Plenum Press, New York, pp. 197–206.
97. Ferraro, T. N., Manyam, B. V., and Hare, T. A., 1983, *J. Neurochem.* **41**:1057–1064.
98. Grossman, M. H., Hare, T. A., Manyam, N. V. B., Glaeser, B. S., and Wood, J. H., 1980, *Brain Res.* **182**:99–106.
99. Ferraro, T. N., Tremblay, R., Manyam, B. V., and Hare, T. A., 1983, *Fed. Proc.* **42**:2008.
100. Grove, J., Schechter, P. J., Rumbach, L., Marescaux, C., Warter, J. -M., and Koch-Weser, J., 1983, *J. Neurochem.* **39**:1061–1065.
101. Loscher, W., and Alnfelt-Ronne, I., 1982, *J. Neurochem.* **39**:251–254.
102. Perry, T. L., Hansen, S., Wall, R. A., and Gauthier, S. G., 1982, *J. Neurochem.* **38**:766–773.
103. Perry, T. L., Hansen, S., Stedman, D., and Love, D., 1968, *J. Neurochem.* **15**:1203–1206.
104. Abbott, R. J., Keidan, J., Pye, I. F., and Nahorski, S. R., 1981, *J. Neurochem.* **37**:1042–1044.
105. Yamamoto, M., Otsuki, S., Kuroda, H., and Ogawa, N., 1983, *Acta Neurol. Scand.* **67**:124–127.
106. Tell, G., Bohlen, P., Schechter, P. J., Koch-Weser, J., Agid, Y., Bonnet, A. M., Coquillat, G., Chazot, G., and Fischer, C., 1981, *Neurology (N.Y.)* **31**:207–211.
107. Tamminga, C. A., Thaker, G. K., Ferraro, T. N., and Hare, T. A., 1983, *Lancet* **2**:97–98.
108. Kuroda, H., Ogawa, N., Yamawaki, Y., Nukina, I., Ofuji, T., Yamamoto, M., and Otsuki, S., 1982, *J. Neurol. Neurosurg. Psychiatry* **45**:257–260.
109. Gerner, R. H., and Hare, T. A., 1981, *Am. J. Psychiatry* **138**:1098–1101.
110. McCarthy, B. W., Gomes, U. R., Neethling, A. C., Shanley, B. C., Taljaard, J. J. F., Potgieter, L., and Roux, J. T., 1981, *J. Neurochem.* **36**:1406–1408.
111. Gold, B. I., Bowers, M. B., Jr., Roth, R. H., and Sweeney, D. W., 1980, *Am. J. Psychiatry* **137**:362–364.
112. Bowers, M. B., Jr., Gold, B. I., and Roth, R. H., 1980, *Psycopharmacology* **70**:279–282.
113. Zimmer, R., Teelken, A. W., Meier, K. D., Alken-Heil, M., and Zander, K. J., 1981, *Prog. Neuropsychopharmacol.* **4**:613–620.
114. Bird, E. D., Mackay, A. V. P., Rayner, C. N., and Iversen, L. L., 1973, *Lancet* **1**:1090–1092.
115. McGeer, P. L., McGeer, E. G., and Fibiger, H. C., 1973, *Neurology (Minneap.)* **23**:912–917.
116. Stahl, W. L., and Swanson, P. D., 1974, *Neurology (Minneap.)* **24**:813–819.
117. Airaksinen, E. M., and Leino, E., 1982, *Acta Neurol. Scand.* **66**:666–672.
118. Loscher, W., Dietz, R., and Siemes, H., 1981, *Epilepsia* **22**:697–702.
119. Ohisalo, J. J., Murros, K., Fredham, B. B., and Hare, T. A., 1983, *Arch. Neurol.* **40**:623–625.
120. Loscher, W., 1982, *J. Neurochem.* **38**:293–295.
121. Abbott, R. J., Pye, I. F., and Nahorski, S. R., 1982, *J. Neurol. Neurosurg. Psychiatry* **45**:253–256.
122. Teychenne, P. F., Ziegler, M. G., Lake, C. R., and Enna, S. J., 1982, *Ann. Neurol.* **11**:76–79.
123. Bareggi, S. R., Franceschi, M., Bonini, L., Zecca, L., and Smirne, S., 1982, *Arch. Neurol.* **39**:709–712.
124. Foster, N. L., Hare, T. A., and Chase, T. N., 1983, *Neurology (N.Y.)* **33**:68.

125. Ogawa, N., Kurona, H., Ota, Z., Yamamoto, M., and Otsuki, S., 1982, *Lancet* **2**:215.
126. Shoulson, I., Chase, T. N., Roberts, E., and Van Baloogy, J. N. A., 1975, *N. Engl. J. Med.* **293**:504–505.
127. Shoulson, I., Goldblatt, D., Charlton, M., and Joynt, R. J., 1978, *Ann. Neurol.* **4**:279–284.
128. Tamminga, C. A., Crayton, J. W., and Chase, T. N., 1978, *Am. J. Psychiatry* **135**:746–747.
129. Foster, N. L., Chase, T. N., Denaro, A., Hare, T. A., and Tamminga, C. A., 1983, *Neurology (N.Y.)* **33**:637–639.
130. Petersen, H. R., Jensen, I., and Dam, M., 1983, *Acta Neurol. Scand.* **67**:114–117.
131. Enna, S. J., 1980, *J. Can. Sci. Neurol.* **7**:257–259.
132. Seiler, N., and Sarhan, S., 1980, *Neurochemistry and Clinical Neurology* (L. Battistin, G. Hashim, and A. Lajtha, eds.), Alan R. Liss, New York, pp. 425–439.
133. Enna, S. J., 1981, *Neuropharmacology of Central Nervous System and Behavioral Disorders* (G. C. Palmer, ed., pp. 507–537.
134. DeFeudis, F. V., 1981, *Trends Pharmacol. Sci.* **4**:1–4.
135. Palfreyman, M. G., Schechter, P. J., Buckett, W. R., Tell, G. P., and Koch-Weser, J., 1981, *Biochem. Pharmacol.* **30**:817–824.
136. Perry, T. L., Wright, J. M., Hansen, S., and Macleod, P. M., 1979, *Neurology (Minneap.)* **29**:370–375.
137. Manyam, B. V., Katz, L., Hare, T. A., Kaniefski, K., and Tremblay, R. D., 1980, *Trans. Am. Neurol. Assoc.* **105**:484–486.
138. Enna, S. J., Ferkany, J. W., Van Woert, M., and Butler, I. J., 1979, *Huntingtons Disease, Advances in Neurology,* Volume 23 (T. N. Chase, N. S. Wexler, and A. Barbeau, eds.), Raven Press, New York, pp. 741–750.
139. Paulson, G. W., Malarkey, W. B., and Shaw, G., 1979, *Huntingtons Disease, Advances in Neurology,* Volume 23 (T. N. Chase, N. S. Wexler, and A. Barbeau, eds.), Raven Press, New York, pp. 797–801.
140. Casey, D. E., Gerlach, J., Magelund, G., and Christensen, T. R., 1980, *Arch, Gen. Psychiatry* **37**:1376–1379.
141. Hosoya, R., Miwa, M., Nishamura, K., and Yamamoto, T., 1981, *N. Engl. J. Med.* **305**:581–582.
142. Korsgaard, S., Casey, D. E., and Gerlach, J., 1983, *Psychiatry Res.* **8**:261–269.
143. Gram, L., Lyon, B. B., and Dam, M., 1983, *Acta Neurol. Scand.* **68**:34–39.
144. Sharpless, N. S., Tyce, G. M., Thal, L. J., Waltz, J. M., Tabaddor, K., and Wolfson, L. I., 1981, *Brain Res.* **217**:107–118.
145. Tyce, G. M., Sharpless, N. S., Kerr, F. W. L., and Muenter, M. D., 1980, *J. Neurochem.* **34**:210–212.
146. Cummins, B. H., and Lothian, D., 1973, *Br. J. Surg.* **60**:910.
147. Dencker, S. J., Haggendal, J., and Ilves-Haggendal, M., 1967, *Acta Physiol. Scand.* **69**:140–146.
148. Meyer, J. S., Chase, T. N., and Engel, W. K., 1972, *Arch. Neurol.* **25**:320–325.
149. Ziegler, M. G., Wood, J. H., Lake, C. R., and Kopin, I. J., 1977, *Am. J. Psychiatry* **134**:565–568.
150. Ziegler, M. G., Lake, C. R., Wood, J. H., Brooks, B. R., and Ebert, M. H., 1977, *J. Neurochem.* **28**:677–679.
151. Lake, C. R., Ballenger, J. C., Ziegler, M. G., Post, R. M., Polinsky, R. J., Wood, J. H., Williams, A. C., and Ebert, M. H., 1984, *Psychiatr. Res.* (in press).
152. Wood, J. H., Ziegler, M. G., Lake, C. R., Shoulson, I., Brooks, B. R., and VanBuren, J. M., 1976, *Trans. Am. Neurol. Assoc.* **101**:68–72.
153. Wood, J. H., Ziegler, M. G., Lake, C. R., Shoulson, I., Brooks, B. R., and VanBuren, J. M., 1977, *Ann. Neurol.* **1**:94–99.
154. Wood, J. H., Lake, C. R., Ziegler, M. G., Sode, J., Brooks, B. R., and Van Buren, J. M., 1977, *Neurology (Minneap.)* **27**:716–724.
155. Ziegler, M. G., Brooks, B. R., Lake, C. R., Wood, J. H., and Enna, S. J., 1980, *Neurology (N.Y.)* **30**:98–101.
156. Hawley, R. J., Major, L. F., Schulman, E. A., and Lake, C. R., 1981, *Arch. Neurol.* **38**:289–292.

157. Tackett, R. L., Webb, J. G., and Privitera, P. J., 1981, *Science* **213**:911–913.
158. Shigeno, T., 1982, *J. Neurosurg.* **56**:344–349.
159. Post, R. M., Lake, C. R., Jimerson, D. C., Bunney, W. E., Jr., Wood, J. H., Ziegler, M. G., and Goodwin, F. K., 1978, *Am. J. Psychiatry* **135**:907–912.
160. Lake, C. R., Sternberg, D. E., Van Kammen, D. P., Ballenger, J. C., Ziegler, M. G., Post, R. M., Kopin, I. J., and Bunney, W. E., 1980, *Science* **207**:331–333.
161. Sternberg, D. E., VanKammen, D. P., Lake, C. R., Ballenger, J. C., Marder, S. R., and Bunney, W. E., Jr. 1981, *Am. J. Psychiatry* **138**:1045–1051.
162. Gomes, U. C. R., Shanley, B. C., Potgieter, L., and Roux, J. T., 1980, *Br. J. Psychiatry* **137**:346–351.
163. Perlow, M., Ebert, M. H., Gordon, E. K., Ziegler, M. G., Lake, C. R., and Chase, T. N., 1978, *Brain Res.* **139**:101–113.
164. Ziegler, M. G., Lake, C. R., Wood, J. H., and Ebert, M. H., 1976, *Nature* **264**:656–658.
165. De Potter, W. P., De Potter, R. W., De Smet, F. H., and Fraeyman, N. N., 1978, *Arch. Int. Pharmacodyn.* **232**:334–335.
166. De Potter, W. P., 1976, *J. Physiol.* (*Lond.*) **258**:26P–27P.
167. De Potter, W. P., Chanh, C. P. -H., De Smet, F., and De Schaepdryver, A. F., 1976, *Neuroscience* **1**:523–524.
168. Goldstein, D. J., and Cubeddu, L. X., 1976, *J. Neurochem.* **26**:193–195.
169. Okada, T., Ohta, T., Shinoda, T., Kato, T., Ikutu, K., and Nagatsu, T., 1976, *Neuropsychobiology* **2**:139–144.
170. Fujita, K., Maruta, K., Teradaira, R., Beppu, H., Shinpo, K., Maeno, Y., Ito, T., Nagatsu, T., and Kato, T., 1977, *J. Neurochem.* **29**:1141–1142.
171. Lerner, P., Goodwin, F. K., VanKammen, D. P., Post, R. M., Major, L. F., Ballenger, J. C., and Lovenburg, W., 1978, *Biol. Psychiatry* **13**:685–695.
172. Major, L. F., Lerner, P., Ballenger, J. C., Brown, G. L., Goodwin, F. K., and Lovenberg, W., 1979, *Biol. Psychiatry* **14**:337–345.
173. Sternberg, D. E., Van Kammen, D. P., Lerner, P., Ballenger, J. C., Marder, S. R., Post, R. M., and Bunney, W. E., Jr., 1983, *Arch. Gen. Psychiatry* **40**:743–747.
174. Major, L. F., Lerner, P., Dendel, P. S., and Post, R. M., 1983, *Neurobiology of Cerebrospinal Fluid II* (J. H. Wood, ed.), Plenum Press, New York, pp. 179–196.
175. Engbaek, F., and Voldby, B., 1982, *Clin. Chem.* **28**:624–628.
176. Voldby, B., Engbaek, F., and Enevoldsen, E. M., 1982, *Stroke* **13**:184–189.
177. Taylor, P. L., Garrick, N. A., Burns, R. S., Tamarkin, L., Murphy, D. L., and Markey, S. P., 1982, *Life Sci.* **31**:1993–1999.
178. Garrick, N. A., Tamarkin, L., Taylor, P. L., Markey, S. P., and Murphy, D. L., 1983, *Science* **221**:474–476.
179. Gattaz, W. F., Riederer, P., Reynolds, G. P., Gattaz, D., and Beckmann, H., 1983, *Psychiatry Res.* **8**:243-n250.
180. Hokfelt, T., Johansson, O., Ljungdahl, A., Lundberg, J. M., and Shultzberg, M., 1980, *Nature* **284**:515–521.
181. Snyder, S. H., 1980, *Science* **209**:976–982.
182. Hakanson, R., and Sundler, F., 1983, *Trends Pharmacol. Sci.* **4**:41–44.
183. Wood, J. H., 1983, *Neurobiology of Cerebrospinal Fluid II* (J. H. Wood, ed.), Plenum Press, New York, pp. 43–65.
184. Wood, J. H., 1982, *Neurosurgery* **11**:293–305.
185. Klee, W. A., 1979, *Adv. Protein Chem.* **33**:243–285.
186. Jackson, I. M. D., 1980, *Neurobiology of Cerebrospinal Fluid I* (J. H. Wood, ed.), Plenum Press, New York, pp. 625–650.
187. Bloom, F. E., and Segal, D. S., 1980, *Neurobiology of Cerebrospinal Fluid I* (J. H. Wood, ed.), Plenum Press, New York, pp. 651–664.
188. Nutt, J. G., 1983, *Neurobiology of Cerebrospinal Fluid II* (J. H. Wood, ed.), Plenum Press, New York, pp. 67–75.
189. Firemark, H. M., 1983, *Neurobiology of Cerebrospinal Fluid II* (J. H. Wood, ed.), Plenum Press, New York, pp. 77–81.
190. Von Knorring, L., Terenius, L., and Wahlstrom, A., 1983, *Neurobiology of Cerebrospinal Fluid II* (J. H. Wood, ed.), Plenum Press, New York, pp. 83–96.

191. Hosobuchi, Y., and Bloom, F. E., 1983, *Neurobiology of Cerebrospinal Fluid II* (J. H. Wood, ed.), Plenum Press, New York, pp. 97–105.
192. Lord, J. A. H., Rance, M. J., and Smith, C. F. C., 1982, *Endorphins* (J. B. Malick, and R. M. S. Bell, eds.), Marcel Dekker, New York, pp. 57–88.
193. Frederickson, R. C. A., 1980, *The Endocrine Functions of the Brain* (M. Motta, ed.), Raven Press, New York, pp. 233–270.
194. Hiller, J. M., Angel, L. M., and Simon, E. J., 1981, *Science* **214**:468–469.
195. Chavkin, C., James, I. F., and Goldstein, A., 1982 *Science* **215**:413–415.
196. Wahlstrom, A., and Terenius, L., 1980, *FEBS Lett.* **118**:241–244.
197. Allen, J. P., Kendall, J. W., McGilvra, R., and Vancura, C., 1974, *J. Clin. Endocrinol. Metab.* **38**:586–593.
198. Gintzler, A. R., Levy, A., and Spector, S., 1976, *Proc. Natl. Acad. Sci. U.S.A.*, **73**:2132–2136.
199. Gintzler, A. R., Gershon, M. D., and Spector, S., 1978, *Science* **199**:447–448.
200. Nyberg, F., Wahlstrom, A., Sjolund, B., and Terenius, L., 1983, *Brain Res.* **259**:267–274.
201. Bowman, E. R., Welch, S., Rosemond, R., Martin, B. R., and Dewey, W. L., 1982, *Fed. Proc.* **41**:1075.
202. Dewey, W. L., Natsuki, R., Bowman, E., Kachur, J., Welch, G., Rosemond, R., and Martin, B. R., 1983, *Pharmacologist* **25**:149.
203. Natsuki, R., Bowman, E., Kachur, J., Welch, S., Rosemond, R., Martin, B. R., and Dewey, W. L., 1983, *Pharmacologist* **25**:261.
204. Pavlinac, D. M., Lenhard, L. W., Parthemore, J. G., and Deftos, L. J., 1980, *J. Clin. Endocrinol. Metab.* **50**:717–720.
205. Kee, D. B., Jr., and Wood, J. H., 1983, *Psychopharmacol. Bull.* **19**:364–368.
206. Kee, D. B., Jr., and Wood, J. H., 1984, *Prog. Neuropsychopharmacol. Biol. Psychiatry* (in press).
207. Von Knorring, L., Almay, B. G. L., Johansson, F., and Terenius, L., 1978, *Pain* **5**:359–365.
208. Almay, B. G. L., Johansson, F., Von Knorring, L., Terenius, L., and Wahlstrom, A., 1978, *Pain* **5**:153–162.
209. Puig, M. M., Laorden, M. L., Miralles, F. S., and Olaso, M. J., 1982, *Anesthesiology* **57**:1–4.
210. Anselmi, B., Baldi, E., Casacci, F., and Salmon, S., 1980, *Headache 20*:294–299.
211. Salar, G., Mingrino, S., Trabucchi, M., Bosio, A., and Semenza, C., 1981, *J. Neurosurg.* **55**:935–937.
212. Cesselin, F., Oliveras, J. L., Bourgoin, S., Sierralta, F., Michelot, R., Besson, J. M., and Hamon, M., 1982, *Brain Res.* **237**:325–338.
213. Hosobuchi, Y., Rossier, J., Bloom, F. E., and Guillemin, R., 1979, *Science* **203**:279–281.
214. Clement-Jones, V., McLoughlin, L., Lowry, P. J., Besser, G. M., Rees, L. H., and Wen, H. L., 1979, *Lancet* **2**:380–383.
215. Clement-Jones, V., McLoughlin, L., Tomlin, S., Besser, G. M., Rees, L. H., and Wen, H. L., 1980, *Lancet* **2**:946–949.
216. Akil, H., Richardson, D. E., Barchas, J. D., and Li, C. H., 1978, *Proc. Natl. Acad. Sci. U.S.A.* **75**:5170–5172.
217. Akil, H., Richardson, D. E., Hughes, J., and Barchas, J. D., 1978, *Science* **201**:463–465.
218. Gerner, R. H., Sharp, B., and Catlin, D. H., 1982, *J. Clin. Endocrinol. Metab.* **55**:358–360.
219. Terenius, L., Wahlstrom, A., Lindstrom, L., and Widerlov, E., 1976, *Neurosci. Lett.* **3**:157–162.
220. Lindstrom, L. H., Widerlov, E., Gunne, L. -M., Wahlstrom, A., and Terenius, L., 1978, *Acta Psychiatr. Scand.* **57**:153–164.
221. Sorensen, K. V., Christensen, S. E., Dupont, E., Hansen, A. P., Pedersen, E., and Orskov, H., 1980, *Acta. Neurol. Scand.* **61**:186–191.
222. Dupont, E., Christensen, S. E., Hansen, A. P., Olivarius, B. deF., and Orskov, H., 1982, *Neurology (N.Y.)* **32**:312–314.
223. Wood, P. L., Etienne, P., Lal, S., Gauthier, S., Cajal, S., and Nair, N. P. V., 1982, *Life Sci.* **31**:2073–2079.
224. Cramer, H., Kohler, J., Oepen, G., Schomburg, G., and Schroter, E., 1981, *J. Neurol.* **225**:183–187.

225. Rubinow, D. R., Gold, P. W., Post, R. M., Ballenger, J. C., Cowdry, R., Bollinger, J., and Reichlin, S., 1983, *Arch. Gen. Psychiatry* **40**:409–412.
226. Nutt, J. G., Mroz, E. A., Leeman, S. E., Williams, A. C., Engel, W. K., and Chase, T. N., 1980, *Neurology (N.Y.)* **30**:1280–1285.
227. Kirkegaard, C., Faber, J., Hummer, L., and Rogowski, P., 1979, *Psychoneuroendocrinology* **4**:227–235.
228. Schroeder, L. L., Johnson, J. C., and Malarkey, W. B., 1976, *J. Clin. Endocrinol. Metab.* **43**:1255–1260.
229. Wode-Helgodt, B., Eneroth, P., Fyro, B., Gullberg, B., and Sedvall, G., 1977, *Acta Psychiatr. Scand.* **56**:280–293.
230. Post, K. D., Biller, B. J., and Jackson, I. M. D., 1980, *Neurobiology of Cerebrospinal Fluid I* (J. H. Wood, ed.), Plenum Press, New York, pp. 591–604.
231. Luerssen, T. G., and Robertson, G. L., 1980, *Neurobiology of Cerebrospinal Fluid I* (J. H. Wood, ed.), Plenum Press, New York, pp. 613–623.
232. Wood, J. H., ed., 1980, *Neurobiology of Cerebrospinal Fluid I,* Plenum Press, New York.
233. Wood, J. H., ed., 1983, *Neurobiology of Cerebrospinal Fluid II,* Plenum Press, New York.
234. Fishman, R. A., 1980, *Cerebrospinal Fluid in Diseases of the Nervous System,* W. B. Saunders, Philadelphia.

17

Cerebrospinal Fluid Proteins in Pathology

A. Lowenthal and D. Karcher

1. INTRODUCTION

Clinicians are currently showing increasing interest in the study of the CSF. This interest can be estimated from the great number of recent publications devoted to the subject.[1] Human CSF proteins have been examined quantitatively and qualitatively. The quantitative examinations concern the determination of the total protein content, determinations of isolated, characterized proteins, or enzymatic determinations. Qualitative analysis is obtained in most cases by electrophoretic studies, often combined with immunologic assays. With these two approaches to the study of neurological disease, conclusions can be drawn that lead to diagnosis and to therapy as well. Many methodological applications, however, still need to be questioned. We shall discuss the problems that arise from the study of human CSF proteins in normal and pathological subjects in terms of (1) methodology and (2) known anomalies occurring in various neurological diseases.

2. METHODOLOGICAL ASPECTS

2.1. Quantitative Studies

2.1.1. Total Protein Determinations

These determinations have been discussed at length, and reviews have dealt with the matter in detail, but no real progress has been made.[1] Several methods have been considered, but the various authors could not agree on the validity of any given method or on the values to be referred to as normal. Among the biochemical methods, the most satisfactory appears to be that of Lowry. However, physical methods that do not alter the samples and that use

A. Lowenthal and D. Karcher • Laboratory of Neurochemistry, Born-Bunge Foundation, Universitaire Instelling Antwerpen, B-2610 Antwerp, Belgium.

a mimimum of CSF are likely to be the preferred methods in the future. Ultraviolet or laser[2] readings seem most promising. They are becoming routine methods in neurology. With the present technical array, no doubt, a really precise and reproducible method could be set up. Some laser determinations are coupled to immunologic measurements, which make them all the more advantageous.

2.1.2. Study of Characterized and Well-Defined Proteins

It is necessary to differentiate proteins found in normal serum from those that arise in pathological syndromes such as tumor proteins, paraproteins, specific proteins of nervous tissue (α-albumin or GFA, myelin basic protein, enolase, S-100, etc.), and enzymes.

2.1.2a. Normal Serum Proteins. The techniques used for the quantitative determination of these proteins in the CSF are the same as those used for serum (Ouchterlony, Mancini, Laurell, etc.). They always require numerous antisera, but these antisera are specific for serum proteins and are generally not raised against CSF proteins. These methods are relatively expensive. Combinations with physical methods, such as laser methods, made it possible to work on smaller quantities.

Almost all serum proteins are also present in the CSF. There was some doubt about some proteins, such as fibrinogen, but it has been found in the CSF in pathological syndromes. It may also be asked whether all CSF proteins are present in the serum too. In the early works devoted to the study of CSF proteins by electrophoresis, it was stated that prealbumin, which is always seen in normal CSF but seldom in serum, and the slow β-globulin of the CSF, later called τ-globulin, which, as a rule, is not seen in the serum, are specific for the CSF. In later studies, prealbumin could be shown in the serum. τ-globulin is also found in the serum, and it may be asked to what degree it could be a product of β_2 serum globulin degradation. During storage of CSF or serum, there is indeed a loss of this β_2-globulin. Up to now it has not been possible to demonstrate the presence of specific proteins in the CSF except for specific proteins of the nervous tissue. It seems that real qualitative differences do not normally exist between serum and CSF proteins, but there are considerable differences in the absolute and relative concentrations. These relative concentrations are not parallel in serum and CSF in development, aging, or pathological states.

2.1.2b. Tumor-Specific Proteins. In recent years serum proteins have been identified that appear specific for the formation of tumors. The diagnostic interest of these determinations has been emphasized several times. It is possible that some of them are found in pathological CSF, e.g., carcinoma embryonic antigen (CEA).[3] Attempts were made to use CEA determinations in the CSF for the differentiation of primary from secondary and benign from malignant nervous system tumors.[4] This study is only in its earliest stages.

2.1.2c. Nervous-System-Specific Proteins. The results obtained in the study of specific proteins of nervous tissue are more promising. It was possible to measure S-100, myelin basic protein (MBP), α-albumin (GFA), enolase (14.3.2), and α_2-glycoprotein in the CSF. These proteins are generally seen in normal CSF in concentrations of a few nanograms per milliliter. Authors more or less agree that these nervous-tissue-specific proteins are not found in tissues or fluids other than CSF. This distinction, however, is not accepted by all. The enolase observed in the liver seems to be γγ enolase, which differs from the nervous tissue αα enolase. The presence of S-100 in other than nervous tissues is not to be ruled out either, and one wonders, taking into account these results, whether the specificity of some proteins for nervous tissue is more than a quantitative difference. The presence of these specific proteins of nervous tissue in the CSF may point to modifications in the metabolism of certain structures of the nervous system. S-100 is indeed principally located in the glia,[5] MBP is specific for myelin, α-albumin (GFA) for glia also, and enolase for neurons. But 14.3.3.[6] and α_2-glycoproteins[7] seem to be glial proteins too. The liberation of these proteins from the cells where they are normally found and their appearance in the CSF might allow localization of pathological processes at the cellular level in nervous tissue. This implies that quantitative determinations of these proteins could be of major clinical importance. The titration methods do not always allow us to obtain precise results easily. Some proteins are better immunogens than others: S-100 is a poor immunogen, α-albumin (GFA) a better one. Some proteins are more stable than others. Isolated α-albumin (GFA) becomes insoluble and cannot be solubilized again without loss. Thus, each protein needs a specific titration method. Many aspects of CSF have to be studied before it would be possible to draw valid conclusions.

S-100[8–10] titrations have never been done in large series. An increase of S-100 in the CSF was seen in some glial lesions. Studies on CSF MBP titers[11–14] have been much more elaborate. At present, it is accepted that MBP levels increase each time the myelin is acutely injured, for instance, in multiple sclerosis (MS) bouts.[15–18] These increases, however, are not typical of this disease. Titration of MBP often presents technical problems[19] because this protein is easily attacked by proteolytic enzymes.

α-Albumin (GFA)[20,21] has been studied extensively. Many hundreds of cases have been investigated.[20] Normal values are less than 9 ng/ml. α-Albumin (GFA) levels in the CSF increase in many neurological syndromes, e.g., in acute meningitis, the acute phase of encephalitis, syringomyelia, malignant astrocytic tumors that are relatively poor in α-albumin (GFA), the acute phase of intoxications including ethanol intoxication. α-Albumin (GFA) levels increase with age. This is probably related to specific pathologies of elderly people. High values of α-albumin (GFA) are found in the nervous tissues of patients suffering from senile dementia,[22,23] and similar increases are observed in the CSF. In gangliosidosis G_{M2},[24] the α-albumin (GFA) values are also increased in the CSF and in nervous tissue. On the whole, increases of more than 9 ng/ml are found in about 20% of examined cases.

Research on enolase[8,25,26] is still in its early stages. It seems possible that increases of CSF enolase are observed in patients affected with astrocytic tumors.[26,27]

2.1.2d. Concentration Ratios. More or less complicated formulae have been used in an attempt to calculate the quotients that should be able to give information on the permeability of the blood–brain barrier or indicate the possibility of synthesis of certain proteins in nervous tissue. This has been done especially for IgGs. The authors who elaborated these formulae started from the idea that serum and CSF protein levels do not necessarily follow a parallel course. In some cases, they tend to do so, and we can conclude that there is an alteration of the permeability of the blood–brain barrier. In other cases, however, differences become more pronounced, and this divergence has been interpreted as a consequence of an intrathecal synthesis of some proteins.

2.1.2e. Enzyme Activity. The last field where quantitative measurements have been made is that of enzymatic activities. For that purpose, many authors have only adapted the techniques used for serum to the study of the CSF. Original methods have seldom been proposed except perhaps for enolase. The adjustment of the serum techniques to the CSF posed technical problems, which investigators have tried to solve by proposing unusual methods such as incubation times of 24 to 48 hr.

At present, methods are improving. In the beginning, only enzymes such as the transaminases GOT and GPT, lactate dehydrogenase and its isoenzymes,[28] and creatine phosphokinase and its isoenzymes[29,30] were measured. Now, more specific enzymes can be determined, such as adenylate kinase[30,31] or enolase.[8,25–27] Adenylate kinase[30,31] could be a good marker of even slight neuronal lesions.[31] It is not impossible that slight lesions of nervous tissue cause increases in some enzymatic activities in the CSF. Titrations of dopamine-β-hydroxylase[32–35] may give indications of the psychiatric condition of the patient. On the other hand, titrations of transaminases, lactate dehydrogenase and its isoenzymes, and creatine phosphokinase and its isoenzymes do not yet give very specific results. Titrations of esterase, muramidase,[36] β-glycuronidase,[37,38] and lysosomal enzymes[39] have not given valuable information up to now. By these titrations the authors have tried to find out whether they could differentiate benign from malignant tumors or viral from bacterial meningitis. Results have been disappointing and not specific at all. The authors rarely take into account the variations of the CSF total protein content and the CSF white cells. Variations of these parameters can be considerable and should be taken into consideration for the interpretation of results. Inhibiting factors of some enzymic activities such as the proteolytic activity have also been reported.[40,41]

2.2. Qualitative Studies

Numerous techniques have been proposed to identify the CSF proteins qualitatively. The so-called colloidal methods should be completely abandoned. In fact, the techniques that are used lead to the identification of proteins on the basis of their electrophoretic migration or immunologic characteristics.

The CSF is poor in proteins, and most of the proposed methods require prior concentration. This can be a source of artifacts, and in recent years, new techniques have been proposed to avoid this concentration.

Many different proteins have been shown, but they have not always been identified, for instance, by means of isoelectric focusing and bidimensional electrophoresis.[42] We shall not review in detail all methods proposed and used.

According to our personal experience and confirmed by recent literature, agar gel electrophoresis is the best method for revealing the most important phenomenon for clinical neurology: fractionation of the γ-globulins or formation of IgGs with restricted heterogeneity. We must keep in mind that this phenomenon is of the greatest consequence for clinicians and that any new method should take this into consideration.

Electrophoretic methods were supposed to reveal specific proteins of the CSF. Prealbumin and τ-globulin were considered to be specific proteins of the CSF. We have already said that these proteins were also found in the serum. One should be very careful, however, when drawing conclusions about the specificity of proteins revealed only by their electrophoretic mobility. The detection of these proteins is a question of sensitivity of the method used: more fractions can be found with other methods than agar gel electrophoresis, and more with bidimensional electrophoresis. Immunoelectrophoresis[43] and enzymoelectrophoresis[44,45] enabled us to show that so-called homogeneous fractions of agar gel electrophoresis can be further fractionated and that in some parts of the pherogram, where classic staining does not show anything, staining by enzymatic or immunologic reactions shows proteins. It is at the present time difficult to establish the existence of CSF-specific proteins by relying on electrophoretic studies. The levels of the nervous-tissue-specific proteins in the CSF are probably not high enough for detection by electrophoretic, immunoelectrophoretic, or enzymoelectrophoretic methods.

The future in this field will most probably be electrophoretic techniques with silver staining.[46] They already enable us to work with CSF volumes of 5 to 10 μl, do not require prior concentration, and permit staining of bidimensional electropherograms[42] to reveal many more proteins. The major problem remains the identification of the stained proteins.

3. ANOMALIES OF CSF PROTEINS IN NEUROLOGICAL DISEASES

It should be noted that findings cannot be expected to be uniform. The stage of the disease, the rate of development, the age and sex of the patient, etc., should be taken into account. Some assays of nervous-tissue-specific proteins demonstrate this very clearly: α-albumin (GFA) values reached during viral encephalitis are very high but become normal again less than 3 weeks later, and in the sequelae of encephalitis, results are normal.[20]

To present the anomalies of the CSF proteins in neurological diseases, we shall differentiate neurological diseases in which CSF proteins remain normal, diseases in which total protein contents are increased, diseases in which restricted heterogeneity of the IgGs is observed, and finally, as in any good classification, we shall present a series of cases that do not fit in any of the abovementioned categories.

3.1. Neurological Diseases with CSF Proteins That Are Quantitatively and Qualitatively Normal

We can list[47] here heredodegenerative and metabolic diseases, e.g., amyotrophic lateral sclerosis, spongiform encephalopathies (Creutzfeldt–Jacob disease, Kuru, etc.), parkinsonism, and most of the vascular diseases. One can also add the sequelae of acute or subacute neurological diseases, i.e., nonevolutive neurological syndromes. However, some qualifications have to be given. In Refsum's disease, which is a heredoataxia, one can observe increases of total CSF protein content and obtain a serumlike electropherogram.

In a few cases of spongiform encephalitis, either spontaneous or experimental, European or Oceanic, one can observe a slight increase of IgG with a tendency to show a pherogram of the serum type. However, in the very large series that we have been able to examine in this field, our pherograms were all normal. In vascular diseases, depending on the stage of the disease, one can find either blood or a slight increase of the total proteins in the acute phase and, over the following days, a serumlike pherogram. In these same diseases, in the acute phase, one notes increases of enzymatic activities such as transaminase,[48] lactate dehydrogenase and its fifth isoenzyme, creatine phosphokinase and its cerebral isoenzyme,[49] and adenylate kinase.[50,51] The latter results, however, are controversial, probably because they are transient. In amyotrophic lateral sclerosis, no anomalies of the CSF proteins have ever been found except an increase of β-glycuronidase in some cases.

3.2. Neurological Diseases with Increase of CSF Total Protein Content

These are diseases with greatly divergent physiopathology. One can explain these increases of total protein either by a modification of the permeability of the blood–brain barrier or by CSF circulation problems with localized stasis. These two explanations, however, do not always seem convincing. The increased permeability of the blood–brain barrier is evoked to explain the increase of total protein content observed in meningitis and some encephalitis. Metabolic modifications might explain the observed increases in case of diabetes or myxedema. During difficulties of the CSF circulation under medullary compression or in the whole arachnoid space in cases of expansive lesions of the brain, the electropherogram becomes a serumlike type with reduced relative content of albumin and τ-globulin, whereas the relative content of polyclonal γ-globulins increases. In CSF, one can normally observe regional variations in electropherograms of the proteins from top to bottom, i.e., from the ventricular to the lumbar region. In the ventricular region, total protein content is lower, and prealbumin and τ-globulin contents are higher. In the lumbar region, protein content increases, and relative concentrations of prealbumin and τ-globulin decrease. The pherogram looks more like a serum-type pherogram.

In general, levels of nervous-tissue-specific proteins do not vary with the total protein content except the 14.3.3 levels, which increase together with the total protein levels. It is probable that some enzymatic activities such as those of phosphorylases increase identically. In fluids with a high protein content, as in different forms of meningitis or in patients with cerebral tumors, enzymatic determinations were done in an attempt to distinguish benign from malignant cerebral tumors, primary from metastatic tumors, or viral from bacterial meningitis.[52–54] Nonspecific increases, mainly in bacterial meningitis and malignant tumors, were observed. The question is now to what extent they represent increases of total protein or cell content.

3.3. Neurological Diseases with IgG Restricted Heterogeneity

We are dealing here with an extremely important phenomenon, which we described for the first time in 1959.[55] This phenomenon is more easily revealed in the CSF than in other biological fluids because the γ-globulin content is normally very low in the CSF. The same phenomenon is likely to be observed in the serum but can be hidden by the relatively large quantities of polyclonal IgGs appearing there.

The γ-globulins or IgG with restricted heterogeneity are also observed in the CSF of patients presenting either an acute disease such as herpetic encephalitis[56] or chronic diseases such as MS, subacute sclerosing panencephalitis (SSPE), trypanosomiasis, cysticercosis, syphilis, filariosis, or meningitis that has become chronic.[47] In the serum, one can observe anomalies on a par with those of the CSF in SSPE, trypanosomiasis, cysticercosis, and herpetic encephalitis.[56] The same may well be true for MS.[57]

This restricted heterogeneity poses different problems: whether some of the IgGs are synthesized in the nervous tissue (intrathecal synthesis), and the meaning of the restricted heterogeneity of the IgGs.

3.3.1. Intrathecal Synthesis

Many authors believe that the restricted heterogeneity is caused by intrathecal synthesis of the IgGs. We need, first of all, a precise definition of intrathecal synthesis. It may be asked whether this synthesis occurs in neurons, glial, hematogenous cells that have migrated in the central nervous system, or possibly in CSF cells. We believe that at the present time, nothing proves that IgGs can be synthesized in neurons or in glial cells. One has, therefore, to accept the fact that IgGs that are found in the CSF originate either from the blood or from hematogenous cells migrated into the CNS or from CSF cells. It could be asked whether an intrathecal synthesis could explain the finding that in some diseases the modifications of the serum IgGs are similar to those of the CSF, e.g., in SSPE. Is it possible that intrathecal cells synthesize IgGs, which then are poured into the CSF and from there into the serum? An alternate possibility is that the observed modifications are generalized modifications that can be revealed in the serum as well as in the CSF. The latter hypothesis is the one we defend, and for three reasons:

1. In myeloma, without any neurological lesion, one can observe identical protein anomalies in the serum and in the CSF after electrophoresis.

2. In patients affected with ataxia–telangiectasia,[58] CSF and serum modifications similar to those observed in SSPE are seen after long continued treatment with γ-globulins.
3. A restricted heterogeneity of the IgGs is also seen both in the serum and the CSF in chronic relapsing experimental allergic encephalomyelitis, and this is caused by a hyperimmunization to MBP.[59]

We believe that intrathecal synthesis can occur either in cells found in the CSF (but they are rare) or in cells of hematogenous origin that have migrated into the nervous tissue. One should also take into account, however, the possibility that CSF modifications are nothing else than a consequence of the serum modifications.

3.3.2. Significance of the Restricted Heterogeneity

The second question is: What is the significance of this restricted heterogeneity? It is certainly not specific for a disease but may be specific for a certain immunologic reaction and, more precisely, for an immunologic dysregulation, specific or not.

In SSPE, it leads to the formation of antibodies against all the measles virus polypeptides although the SSPE virus is not identical to the measles virus and is missing one of the measles virus polypeptides, the matrix protein. Thus, one can observe in the CSF and the serum of patients suffering from SSPE antibodies against all the polypeptides of the measles virus.[60]

The oligoclonal bands reveal an antiherpetic activity in the CSF of herpetic encephalitis.[61] In acute EAE, the CSF oligoclonal bands do not react specifically as MBP antibodies, although MBP antibodies are found in the entire IgG region. On the other hand, MBP does not absorb the serum IgG bands.[62] In chronic relapsing EAE, similar conclusions were drawn.[63] One has the impression that this work should be repeated before the antibody activity of the IgG with restricted heterogeneity, at least in chronic relapsing EAE, can be defined.

In MS it has never been demonstrated that the formed antibodies or the IgGs with restricted heterogeneity are directed against a determined or specific antigen. It was shown that antibody titers against a whole series of antigens (measles, herpes, rabies, etc.) are increased. It has never been possible to evaluate precisely the very small portion of CSF antibodies that has today been identified in MS. This induced some authors to set forth the hypothesis that the immunologic reaction in the CSF of patients affected with MS is a nonsense reaction.[64]

We believe that the restricted heterogeneity is the consequence of the persistence of one or several antigens that stimulate the humoral immune reaction, i.e., the synthesis of immunoglobulins by B lymphocytes. This stimulation is not necessarily specific, and the B cells, while producing specific antibodies in large quantities, also produce other, nonspecific antibodies. The increases of antibody content against measles, rabies, herpes, etc. in the CSF of patients affected with MS can thus be explained. The restricted heterogeneity is therefore a humoral hyperimmunization caused by the persistence of an antigen. It may be asked what the reason is for the persistence of the antigen or, in other words, why this antigen is not neutralized by the produced antibodies, or else, why the humoral and perhaps cellular immune reaction does not permit the organism to get rid of the antigen causing the disease. The intervention of a factor inhibiting the cellular immune reaction and the antigen–antibody reaction can be evoked. Many arguments plead in favor of this hypothesis, but no real evidence has been given.[65] We have good reasons to believe that this inhibiting factor would be bound to the α_2-globulins and that it would be thermostable. It is distinct from the complement.

3.4. Neurological Diseases with Anomalies of the CSF Proteins That Are Difficult to Classify

In many diseases increases of specific proteins are observed. Increases of α-albumin (GFA) can be seen in syringomyelia, acute encephalitis, senile dementia, and some cerebral tumors. α-Albumin also increases in some acute intoxications, in ethanol intoxication, and in gangliosidosis G_{M2}.[24]

Increases of MBP are observed in cases in which the myelin is injured, e.g., in MS bouts. This is also observed in other diseases with severe lesions of the myelin.

Increases of adenylate kinase have been reported after long surgical anesthesia, and this might be the proof of nervous cell deterioration. This question remains controversial.[31]

The content of dopamine-β-hydroxylase may increase in schizophrenia. This is still to be confirmed.[66]

In some diseases antibodies are formed against specific proteins of the CNS. Antibodies are found, indeed, against S-100, MBP, and α-albumin (GFA) in serum. Increased antibodies levels against α-albumin (GFA) are seen, for example, in the blood of patients affected with polyneuritis.[67]

4. CONCLUSIONS

We have tried to show the importance and the complexity of the study of normal and pathological human CSF proteins. We wish to emphasize the fact that although these studies are of the greatest interest for clinicians, they are not usually mentioned by those involved in fundamental research in the CSF. We believe that better contacts between clinical and fundamental research should be established in order to achieve a better understanding of the mechanisms causing the observed changes in the CSF and their utilization in diagnosis and therapy.

ACKNOWLEDGMENT. We wish to thank all our colleagues, technicians, and secretaries at the Laboratory of Neurochemistry who helped us with the preparation of this chapter. We wish to thank for financial support the National Fonds voor Geneeskundig Wetenschappelijk Onderzoek, the Universitaire

Instelling Antwerpen, the Ministerie van Nationale Opvoeding, and the Born-Bunge Foundation.

REFERENCES

1. Einstein, E. R., 1982, *Proteins of the Brain and Cerebrospinal Fluid in Health and Disease*, Charles C Thomas, Springfield, Illinois, pp. 88–92.
2. Schliep, G., and Felgenhauer, K., 1978, *J. Clin. Chem. Clin. Biochem.* **16**:631–635.
3. Schold, S. C., Wasserstrom, W. R., Fleisher, M., Schwartz, M. K., Posner, J. B., 1980, *Ann. Neurol.* **8**:597–604.
4. Wasserstrom, W. R., Schwartz, M. K., Fleisher, M., and Posner, J. B., 1981, *Ann. Clin. Lab. Sci.* **11**:239–251.
5. Hyden, H., and McEwen, B. S., 1966, *Proc. Natl. Acad. Sci. U.S.A.* **55**:354–358.
6. Boston, P. F., Jackson, P., and Thompson, R. J., 1982, *J. Neurochem.* **38**:1475–1482.
7. Leonhardt, K. F., Renschler, H., and Warecka, K., 1976, *Munch. Med. Wochenschr.* **117**:1113–1116.
8. Kato, K., Nakajima, T., Ishiguro, Y., and Matsutani, T., 1982, *Biomed. Res.* **9**:24–28.
9. Michetti, F., Massaro, A., and Murasio, M., 1979, *Neurosci. Lett.* **11**:171–175.
10. Michetti, F., Massaro, A., Russo, G., and Rigon, G., 1980, *J. Neurol. Sci.* **44**:259–263.
11. Jacque, C. M., Delassalle, A., Rancurel, G., Raoul, M., Lesourd, B., and Legrand, J. C., 1982, *Arch. Neurol.* **39**:557–560.
12. Kohlschutter, A., 1978, *Eur. J. Pediatr.* **127**:155–161.
13. Prange, H., Kohlschutter, A., and Ritter, G., 1980, *Deut. Med. Wochenschr.* **105**:1119–1121.
14. Alling, C., Karlsson, B., and Vällfors, B., 1980, *J. Neurol.* **223**:225–230.
15. Cohen, S. R., Brooks, B. R., Herndon, R. M., and McKhann, G. M., 1980, *Ann. Neurol.* **8**:25–31.
16. Biber, A., Englert, D., Dommasch, D., and Hempel, K., 1981, *J. Neurol.* **225**:231–236.
17. Ohta, M., Nishitani, H., Matsubara, F., and Inaba, G., 1980, *N. Engl. J. Med.* **302**:1093.
18. Gerson, B., Cohen, S. R., Gerson, I. M., and Guest, G. H., 1981, *Clin. Chem.* **27**:1974–1977.
19. Alvord, E. C., Hruby, S., and Sires, L. R., 1979, *Ann. Neurol.* **6**:474–482.
20. Crols, R., Noppe, M., Caers, J., and Lowenthal, A., 1983, *Neurochem. Pathol.* **1**:91–102.
21. Klaes, R., Crols, R., and Lowenthal, A., 1984, *Protides of the Biological Fluids*, 30th Colloquium Brussels, Abstract 05-06 (H. Peeters, ed.) Pergamon Press, Oxford, New York (in press).
22. Yen, S. H., Dahl, D., Schachner, M., and Shelanski, M. L., 1976, *Proc. Natl. Acad. Sci. U.S.A.* **73**:529–733.
23. Shelanski, M. L., Yen, S. H., and Lee, V. M., 1976, *Cell Mobility* (R. Goldman, T. Pollard and J. Rosenbaum, eds.) Cold Spring Harbor Laboratory, Cold Spring Harbor, New York, pp. 1007–1020.
24. Karcher, D., Lowenthal, A., and Zeman, W., 1972, *Advances in Experimental Medicine and Biology*, Volume 19 (B. W. Volk and S. M. Aronson, eds.), Plenum Press, New York, pp. 151–162.
25. Brown, K. W., Kynoch, P. A. M., and Thompson, R. J., 1980, *Clin. Chim. Acta* **101**:257–264.
26. Royds, J. A., Timperley, W. R., and Taylor, C. B., 1981, *J. Neurol. Neurosurg. Psychiatry* **44**:1129–1135.
27. Parma, A. M., Marangos, P. J., and Goodwin F. K., 1981, *J. Neurochem.* **36**:1093–1096.
28. Hall, R. T., Kulkarni, P. B., Sheehan, M. B., and Rhodes, P. G., 1980, *Dev. Med. Child. Neurol.* **22**:300–307.
29. Briem, H., and Rombo, L., 1981, *Clin. Chim. Acta.* **112**:361–364.
30. Fishman, R. A., 1982, *Lancet* **2**:37.
31. Aberg, T., Ronquist, G., Tyden, H., Ahlund, P., and Bergström, K., 1982, *Lancet* **2**:1139–1141.
32. Fujita, K., Maruta, K., Teradaira, R., Beppu, H., Ikegama, M., Kawai, K., and Nagatsu, T., 1982, *Clin. Chem.* **28**:1403–1404.

33. Major, L. F., Lerner, P., and Goodwin, F. K., 1980, *Arch. Genet. Psychiatry* **37**:308–310.
34. Nagatsu, T., Wakui, Y., Kato, T., Fujita, K., Kondo, T., Yokochi, F., and Narabayashi, N., 1982, *Biomed. Res.* **3**:95–98.
35. Cubeddu, L. X., Davila, J., Zschaeck, D., Barbella, Y. R., Ordaz, P., and Dominguez, J., 1981, *Hypertension (Dallas)* **3**:448–455.
36. Corberand, J., and Péré, P., 1980, *Rev. Med. Toulouse* **15**:386.
37. Kilpeläinen, H., Halonen, T., Pitkänen, A., and Riekkinen, P. J., 1982, *Acta Neurol. Scand.* [*Suppl.*] **90**:272.
38. Shuttleworth, E., and Allen, N., 1980, *Arch. Neurol.* **37**:684–687.
39. Cherian, R., and Balasubramanian, A. S., 1978, *Clin. Chim. Acta* **89**:411–416.
40. Bieth, J., Miesch, F., and Metais, M., 1969, *Clin. Chim. Acta* **24**:203–209.
41. Aroor, A. R., Krishnan, K. A., and Pattabiraman, T. N., 1981, *Indian J. Med. Res.* **73**:941–945.
42. Merril, R. C., Goldman, D., Sedman, S. A., and Ebert, M. H., 1981, *Science* **211**:1437–1438.
43. Dencker, S. J., 1966, *Symposium über die Zerebrospinalflüssigkeit,* Volume 31 (J. Sayk, ed.), Gustav Fischer-Verlag, Jena, pp. 135–216.
44. Lowenthal, A., Van Sande, M., and Karcher, D., 1961, *J. Neurochem.* **7**:135–140.
45. Lowenthal, A., Karcher, D., and Van Sande, M., 1966, *Symposium über die Zerebrospinalflüssigkeit,* Volume 31 (J. Sayk, ed.), Gustav Fisher-Verlag Jena, pp. 129–137.
46. Karcher, D., Lowenthal, A., and Van Soom, G., 1979, *J. Clin. Chem. Clin. Biochem.* **17**:505–511.
47. Lowenthal, A., 1964, *Agar Gel Electrophoresis in Neurology*, Elsevier, Amsterdam.
48. Engelhardt, P., and Avenarius, H. J., 1976, *Med. Klin.* **71**:699–702.
49. Wolintz, A. H., Jacobs, L. D., Christoff, N., Solomon, M., and Chernik, N., 1969, *Arch. Neurol.* **20**:54–61.
50. Frithz, G., Ericsson, P., and Ronquist, G., 1977, *Upsala J. Med. Sci.* **82**:11–14.
51. Terent, A., and Ronquist, G., 1980, *Acta Neurol. Scand.* **62**:327–335.
52. Stiffel, M., Dittmann, J., Faulhauer, K., and Loew, F., 1973, *Wien. Z. Nervenheilkd.* **31**:325–333.
53. Seidenfeld, J., and Marton, L. J., 1979, *J. Natl. Cancer Inst.* **63**:919–931.
54. Vivekanadan, A. P., Rao, K., Selvam, R., and Kanaka, T. S., 1982, *Acta Neurol. Scand.* **66**:347–354.
55. Karcher, D., Van Sande, M., and Lowenthal, A., 1959, *J. Neurochem.* **4**:135–140.
56. Rappel, M., Dubois-Dalcq, M., Sprecher, S., Thiry, L., Lowenthal, A., Pelc, S., and Thys, J. P., 1971, *J. Neurol. Sci.* **12**:443–458.
57. Lowenthal, A., 1979, *Humoral Immunity in Neurological Diseases* (D. Karcher, A. Lowenthal, and A. D. Strosberg, eds.), Plenum Press, New York, pp. 281–288.
58. Lowenthal, A., Adriaenssens, K., Colfs, B., Karcher, D., and Van Heule, R., 1972, *Z. Neurol.* **202**:58–63.
59. Karcher, D., Lassmann, H., and Lowenthal, A., 1982, *J. Neuroimmunol.* **2**:93–106.
60. Karcher, D., Thormar, H., Lowenthal, A., and Noppe, M., 1982, *J. Neurol.* **227**:29–34.
61. Vandvik, B. Vartdal, F., and Norrby, E., 1982, *J. Neurol.* 25–38.
62. Whitacre, C. C., Mattson, D. H., Day, E. D., Peterson, D. J., Paterson, P. Y., Roos, R. P., and Arnason, B. G. W., 1982, *Neurochem. Res.* **7**:1209–1222.
63. Glynn, R., Weedon, D., Edwards, J., Suckling, A. J., and Cuzner, M. L., 1982, *J. Neurol. Sci.* **57**:369–384.
64. Paterson, P. Y., and Whitacre, C. C., 1981, *Immunol. Today* **2**:111–117.
65. Karcher, D., 1979, *Humoral Immunity in Neurological Diseases* (D. Karcher, A. Lowenthal, and A. D. Strosberg, eds.), Plenum Press, New York, pp. 541–551.
66. Sternberg, D. E., Van Kammen, D. P., Lerner, P., and Bunney, W. E., 1982, *Science* **216**:1423–1425.
67. Melse, J., Noppe, M., Crols, R., Gheuens, J., and Lowenthal, A., 1982, *Acta Neurol. Belg.* **83**:17–22.

18

Retinal Degenerations

Susan Y. Schmidt

1. INTRODUCTION

This chapter focuses exclusively on retinal degenerations involving the degeneration of photoreceptor cells. The most prevalent types of hereditary retinal degenerations in man and animals fall within this category. Retinitis pigmentosa in man comprises a number of different hereditary retinal diseases having in common the degeneration of photoreceptor cells generally followed by the appearance of pigmentation within the retina that becomes visible with the ophthalmoscope. Hereditary retinal degenerations involving the photoreceptor cells also occur in a wide number of animal species, and these may serve as models of human diseases in which early stages of the disease processes can be studied. The following animal models are considered: Royal College of Surgeons (RCS) rats, retinal degeneration (*rds*) mice, mice with a slow form of retinal degeneration (*rds*), Purkinje cell degeneration (*pcd*) mice, and several affected dog breeds, including Irish setters with rod–cone dysplasia and collie and miniature poodle dogs with progressive rod–cone degeneration. Consideration is given to nutritionally induced photoreceptor cell degenerations such as vitamin A deficiency in man, rat, and ground squirrel, taurine deficiency in cat and rat, and vitamin E deficiency in rat, dog, and monkey. Consideration is also given to certain agents that can cause selective photoreceptor cell degeneration, including toxic effects of light, free radical formation (i.e., peroxidation), iodoacetate, tunicamycin, and hemicholinium. It is believed that understanding the effect(s) of these agents on photoreceptor cell structure and function may reveal certain fundamental aspects of photoreceptor cell metabolism and/or function that, when altered, result in photoreceptor cell death.

In recent years, large numbers of books and articles have been published on retinal degenerations, and no attempt is made here to refer to all of them or to present a historical perspective. For the sake of brevity, in some instances, review articles are cited in preference to numerous original publications. Emphasis is on general principles rather than experimental details. A preceding

Susan Y. Schmidt • Berman-Gund Laboratory for the Study of Retinal Degenerations, Harvard Medical School, Massachusetts Eye and Ear Infirmary, Boston, Massachusetts 02114.

Table I
Retinitis Pigmentosa and Allied Retinal Diseases[a]

1. Dominant retinitis pigmentosa with complete penetrance
2. Dominant retinitis pigmentosa with reduced penetrance
3. Autosomal recessive retinitis pigmentosa
4. Sex-linked (X-chromosome-linked) retinitis pigmentosa
5. Progressive cone–rod degeneration
6. Sector retinitis pigmentosa
7. Atypical forms of retinitis pigmentosa
   a. Paravenous
   b. Pericentral
   c. Unclassified
8. Retinitis pigmentosa sine pigmento
9. Syndromes or diseases of which retinitis pigmentosa is a part
   a. Usher's syndrome
   b. Laurence–Moon–Bardet–Biedl syndrome
   c. Bassen–Kornzweig disease
   d. Refsum disease
   e. Kearns–Sayre syndrome
   f. Osteopetrosis
   g. Hereditary cerebroretinal degenerations
   h. Vitreoretinal degeneration of Goldmann–Favre
10. Congenital amaurosis of Leber
11. Sex-linked choroideremia
12. Generalized choroidal sclerosis
13. Gyrate atrophy of the choroid and retina
14. Progressive albipunctate dystrophies
15. Cone degenerations
16. Hereditary macular degenerations including Stargardt's disease, fundus flavimaculatus, central areolar choroidal dystrophy, Best's disease, etc.
17. Stationary forms of night blindness including dominant (Nougaret) nyctalopia, autosomal recessive and sex-linked nyctalopia, Oguchi's disease, and fundus albipunctatus
18. Congenital rod monochromacy, blue cone monochromacy, etc.

[a] From the *Annual Report of The Berman-Gund Laboratory for the Study of Retinal Degenerations 1977–1978.*

volume of the *Handbook of Neurochemistry* contains several chapters on the biochemistry of the retina and visual excitation, which provide additional details. A list of a few recent books and issue of journals dealing with retinal degenerations is included in the Appendix.

2. *RETINITIS PIGMENTOSA AND ALLIED DISEASES*

Retinitis pigmentosa and allied diseases have been classified on the basis of genetic patterns of inheritance and, in some cases, on associated abnormalities (Table I).An estimated 50,000 to 100,000 people are affected in the United States alone.[1–7] Patients characteristically report problems in adaptation, night blindness, loss of peripheral vision, and eventually loss of central vision. The symptoms usually become evident between the ages of 10 and 35, and about half are legally blind by age 40. At more advanced stages of the

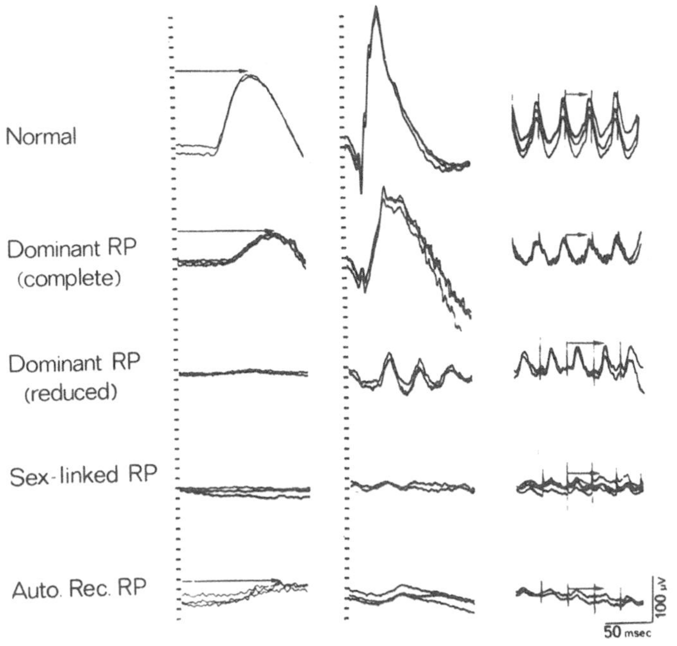

Fig. 1. Full-field ERG responses of one normal subject and four patients (ages 13, 14, 14, and 9) with retinitis pigmentosa. Responses were obtained after 45 min of dark adaptation to single flashes (10 μsec in duration) of dim blue ($\lambda < 470$ nm) light (left column) and suprathreshold white (8 foot lamberts) light (middle column). Right column responses were obtained with white (8 foot lamberts) flicker presented at 30 cps. Calibration symbol (lower right corner) signifies 50 msec horizontally and 100 μV vertically. Rod b-wave implicit times in column 1 and cone implicit times in column 3 are designated with arrows. (From Berson.[12])

disease, patients have extensive visual field loss, attenuation of retinal arterioles, and widespread changes in retina and pigment epithelium and intraretinal pigmentation, typically in a bone spicule configuration in peripheral regions of the fundus.

Over the last 20 years, the electroretinogram (ERG) has served as a diagnostic aid in evaluating photoreceptor cell function in these patients. Although at advanced stages of retinitis pigmentosa the ERG is undetectable (i.e., <5 μV in amplitude), abnormalities in the ERG can be detected in young patients (under 20 years of age), often several years before any other signs of the disease become evident.[8–14] With careful attention to dark adaptation, stimulus wavelength, stimulus intensity, and frequency of stimulation, the full-field (*Ganzfeld*) ERG has been resolved into rod and cone components. Patients with progressive forms of retinitis pigmentosa have delays in their ERG cone and/or rod b-wave implicit times (i.e., time interval between stimulus onset and the major cornea-positive peak of the cone or rod b-wave response—see Fig. 1), whereas patients with self-limited sector retinitis pigmentosa or stationary forms of night blindness have normal or reduced amplitudes with normal b-wave implicit times.[6,12] Percentages of normal and abnormal ERGs in young

(6- to 20-year-old) patients in families with autosomal dominant and autosomal recessive disease have been observed to correspond closely with percentages that would be predicted from Mendelian laws governing these patterns of inheritance. The ERGs are useful not only for determining which children are affected but also for identifying early in life those that have normal retinal function in families with retinitis pigmentosa, as patients with normal ERGs have not been observed to develop retinitis pigmentosa at a later time.[13,14]

Although patients can be identified early in life, the rates of progression in the different types of retinitis pigmentosa are difficult to predict. Dominantly inherited forms characteristically proceed at a slower rate than the autosomal and sex-linked forms. Patients with dominant retinitis pigmentosa can often retain central vision beyond age 60 or 70, but patients with autosomal recessive types characteristically lose vision by age 45–55 (although there may be exceptions), and those with sex-linked forms characteristically lose central vision by age 30–40.

The ERG has been used as an aid in identifying carriers of sex-linked (i.e., X-chromosome-linked) retinitis pigmentosa within a 95% confidence limit.[15] These female carriers of X-linked retinitis pigmentosa may sometimes show patches of bone spicule pigment in the peripheral fundus and other features of retinitis pigmentosa [such as reduced early receptor potential (ERP)[16] and reduced amounts of visual pigment[17,18]] consistent with the idea that the X-chromosome with the defective gene is randomly expressed within each photoreceptor cell. Thus, retinas of carrier females may represent a mosaic of normal diseased photoreceptor cells. Females identified as carriers of X-chromosome-linked retinitis pigmentosa can be made aware that they have a 50% chance with each male childbirth of having an affected son and a 50% chance with each female childbirth of having a carrier daughter. Affected male relatives would know that all their sons would be normal and that all their daughters would be carriers. Detection and effective genetic counseling of both carrier females and affected males in families with X-chromosome-linked disease could potentially reduce the incidence of retinitis pigmentosa in the United States by 6 to 10%.

Histopathological studies of donor eyes from patients with retinitis pigmentosa have revealed widespread degeneration of photoreceptor cells, loss of pigment epithelial cells, and abnormalities in the surviving cells of both types.[19–28] In donor eyes from a number of patients with advanced retinitis pigmentosa (one case identified as dominant[21] and several donor eyes in which the type of retinitis pigmentosa was not defined[20,22,24]), the only remaining photoreceptors in the retina were cones. The stages of cone degeneration were more advanced in the perifovea and less advanced in the fovea. Remaining foveal cones showed a conspicuous loss or absence of outer segments with remarkable preservation of cone inner segments, cell bodies, and synaptic pedicles. Large autophagic vacuoles (or secondary lysosomes) were observed within the perinuclear cytoplasm of cones. Pigment epithelial cells apposed to these remaining cones were larger than normal their nuclei were apically displaced, and they contained many melanolysosomes, lysosomes, and occasional phagosomes. An ultrastructural study of a postmortem donor eye from a 24-

year-old male patient with sex-linked retinitis pigmentosa showed abnormalities in all remaining cones and rods.[23] Central foveal cones were reduced in number by about 50% and had shortened and severely distorted outer segment. From the parafovea through the midperiphery, there was a gradual reduction in the number of remaining cones, which had no organized outer segments. The pigment epithelium contained large numbers of melanolysosomes and few free melanin granules from the fovea through the midperiphery and few melanolysosomes and many free melanin granules in the far periphery. Because of the close functional relationship between the pigment epithelium and the photoreceptor cells, it has not been possible to ascertain whether the defect is primarily within the photoreceptors or within the pigment epithelium in these donor eyes.

In a few rare types of hereditary retinal diseases allied to retinitis pigmentosa, the ophthalmoscopic appearance of the fundus is suggestive of a primary involvement of the pigment epithelium and/or the choroid in the disease process: these diseases include gyrate atrophy of choroid and retina, generalized choroidal sclerosis, sex-linked choroideremia, central areolar choroidal dystrophy, and Best's disease (Table I). The earliest fundus abnormalities in these diseases appear to be associated with the choroid and/or pigment epithelium and not the retina.

Abnormalities in the pigment epithelium have also been associated with central progressive retinal atrophy (CPRA), described to occur in several dog breeds.[29] At the earliest stages of CPRA, the first ophthalmoscopically visible abnormality was the presence of irregular pigment foci near the area centralis. At the level of the light microscope, focal hypertrophy of pigment epithelial cells was noted at an early stage, while the overlying photoreceptor cells were normal. As the disease progressed, the hypertrophy became widespread, and the pigment epithelial cells accumulated increasing amounts of coarse light-brown brilliantly autofluorescent lipopigments. At later stages, the pigment epithelial abnormalities were associated with progressively more severe photoreceptor cell degeneration, although the inner (i.e., neural) retina remained intact. The cause of the excessive accumulation of lipopigments, either lipofuscin or ceroid, within the pigment epithelium in dogs affected with CPRA is not known.

Lipofuscin has been shown to accumulate with age within the pigment epithelium in normal human eyes,[30,31] whereas accumulation of ceroid within mammalian tissues has been associated with disease processes (neuronal ceroid lipofuscinosis, i.e., Batten's disease[32–34]). Increased amounts of lipopigments, lipofuscinlike material, or melanolysomes within the pigment epithelium have also been observed in donor eyes from one patient with dominant,[21] another with sex-linked,[23] and a number of other eyes with atypical or undefined-type retinitis pigmentosa,[24,27,35] Best's disease, and fundus flavimaculatus[36,37] and in rats,[38,39] dogs,[40] and monkeys[41] with experimentally induced vitamin E deficiency.

The photoreceptor cells appear to be dependent on the normal functioning of the pigment epithelium for the transport of nutrients, vitamins A and E, and taurine, removal of waste products such as retinol or oxidized tocopherol, and the daily phagocytosis of the outer segment tips that have been shed from rod and cone photoreceptor outer segments.[42–46] Disruption of photoreceptor–pigment epithelial interaction as in retinal detachment in man or experimentally induced retinal detachment in the monkey has been shown to lead to photoreceptor cell degeneration.[47,48] Through the process of shedding followed by phagocytosis, the pigment epithelium participates in the renewal of outer segment membranes. These membranes are continuously renewed; new membrane constituents are either inserted at the base of the outer segments (in rods) via a process of new membrane assembly or are randomly distributed throughout the outer segment membranes via molecular replacement (in both rods and cones).[49–51] The outer segments are maintained at a constant length through daily rhythmic shedding of the outermost portions of the outer segment tips: in rats or frogs maintained on regular diurnal light–dark cycles, shedding of rod outer segment tips has been shown to occur daily within 1–2 hr after light onset.[52–55] The idea that rod outer segment renewal occurs similarly in human photoreceptor cells is supported by the recent demonstration of a diurnal rhythm in the human rod ERG.[56] One eye in normal subjects was entrained for 3–4 days to regular daily 14-hr light and 10-hr dark cycles through wearing of an eyepatch during the defined 10-hr dark period; in these eyes, ERG amplitudes and sensitivity of rod photoreceptor cells were reduced within 1½ hr after light onset, consistent with a shortening of outer segments through shedding of about 10% of the distal portion of the rod outer segment tips. Shedding of cone outer segment tips has been observed in some species (lizard, goldfish, chicken, and rabbit) during the dark phase of the light/dark cycle,[57–60] but such a rhythm has not been observed in all species.[55,61,62]

The biochemical abnormalities within photoreceptor cells and/or the photoreceptor–pigment epithelium complex that lead to the degeneration of photoreceptor cells in retinitis pigmentosa remain to be defined. Studies of donor eyes from these patients must be interpreted with caution, as it is often difficult or impossible to differentiate a primary abnormality from secondary changes resulting from advanced degeneration. For instance, reduced amounts of photoreceptor retinol-binding protein[35] or interphotoreceptor matrix glycoprotein[63,64] reported in donor eyes from patients with retinitis pigmentosa may be secondary to the loss of photoreceptor cells.

Other than in the case of a few rare diseases, nothing is known of the biochemical defect in most types of retinitis pigmentosa. Efforts in several laboratories have been directed at detecting some abnormalities in plasma or urine of these patients. Substances that have been measured include vitamin A,[65–67] retinol-binding protein,[68–70] amino acids,[71–73] serum lipids,[74,75] and serum immunoglobulins and immunoreactivity to photoreceptor-specific proteins.[76–78] So far these studies yielded mainly negative results.

Studies of patients in India have raised the possibility that patients have very low plasma ceruloplasmin concentrations and high urinary copper excretion.[79] Abnormal serum copper-to-zinc ratios were reported in a family with sex-linked retinitis pigmentosa in New Jersey.[80] Elevated levels of arachidonic acid were reported in plasma of some patients with retinitis pigmentosa.[81] These latter reports prompted evaluation of patients on file in the Berman–Gund

Laboratory for the Study of Retinal Degenerations. Serum copper, zinc, and plasma ceruloplasmin levels as well as plasma arachidonic acid levels and serum phospholipid and fatty acid concentrations were found to be within the normal range.[74,75] Except for the observed elevations in plasma ornithine in patients with gyrate atrophy of the choroid and retina, the plasma levels of amino acids including taurine appeared to be within the normal range in patients with different types of retinitis pigmentosa.[72,73]

Some types of mucopolysaccharidoses (characterized by mucopolysaccharidurla) have also been associated with retinitis pigmentosa. The biochemical abnormalities have been defined in some of these diseases, including sulfoiduronate sulfatase deficiency in Hunter's disease,[82] heparin sulfaminidase deficiency in Sanfilippo A disease,[83] N-acetyl-α-glucosaminidase deficiency in Sanfillippo B disease,[84] and α-L-iduronidase deficiency in Hurler and Scheie syndromes.[85] The involvement of the eye has been characterized by a "cherry red spot in the macula" in some of the lipid storage diseases including Tay–Sachs disease, generalized G_{M1} gangliosidosis, and Niemann–Pick disease.[86] The spot in the macula is caused by infiltration of lipid into parafoveal ganglion cells. The defective enzymes in these diseases are, respectively, hexoseaminidase A, β-galactosidase, and sphingomyelinase.[86] Some of the diseases, grouped under the heading of neuronal ceroid lipofuscinosis, are associated with retinitis pigmentosa (Batten's disease). Recently, abnormal accumulations of dolichols (polyisoprenols) were described within postmortem brain specimens[32,33] (probably in association with lipofuscin), and urinary excretion of dolichols has also been shown to be higher than normal in patients with neuronal ceroid lipofuscinosis.[32] Since dolichols have been associated with glycoprotein synthesis, a possible abnormality in glycoprotein synthesis is currently being explored in these patients.

In patients with Friedreich's ataxia, both retinal degeneration and cerebellar degeneration have been known to occur. The nature of the retinal abnormality in these patients has not been clearly defined and has been described as typical or atypical pigmentary degeneration of the retina.[1–3] Recent studies have suggested an abnormality in the transport of taurine and β-alanine into tissues. Plasma levels of taurine and β-alanine were within the normal range, but their levels were reduced in spinal fluid, and urinary excretion of these amino compounds was shown to be higher than normal in affected patients.[87,88]

A recent report indicates abnormalities in folate metabolism in a patient with Kearns–Sayre syndrome.[89] This disease, characterized by neurological abnormalities, ophthalmoplegia, conduction heart block, and atypical pigmentary degeneration of the retina, becomes manifest during childhood. Affected patients develop seizures during the first decade, muscle biopsies show evidence of a mitochondrial disorder, and computerized tomographic brain scans show calcification of the basal ganglia or white matter hypodensities. Histopathological examination of postmortem donor eyes from a 52-year-old man and sections from the postmortem donor eyes of a 14-year-old patient with this disease suggest that the primary defect is associated with the pigment epithelium. The degenerative changes within the photoreceptor cells appeared to follow degenerative changes in the pigment epithelium.[90] Recently, low levels of folate were reported in plasma and cerebrospinal fluid over a 15-year period in a young woman who was also receiving phenytoin for seizures. Additionally, muscle biopsies showed that carnitine levels and mitochondrial enzyme activities were reduced from normal. Administration of folate to the patient resulted in a rapid increase in plasma and cerebrospinal fluid folate levels, but the plasma-to-cerebrospinal fluid ratio remained high (~2.5), whereas in normal subjects this ratio was shown to be low (0.25–0.35). However, improvement of neurological abnormalities was observed, and muscle carnitine levels increased following folate administration. Visual function was not tested in this patient.[89]

Systemic biochemical abnormalities have been clearly associated with photoreceptor cell degeneration in some rare forms of retinitis pigmentosa and allied diseases: Bassen–Kornzweig syndrome (abetalipoproteinemia), Refsum's disease, and gyrate atrophy of the retina and choroid. The findings in these patients are briefly reviewed.

2.1. Bassen–Kernzweig Syndrome

Bassen–Kornzweig syndrome (abetalipoproteinemia) has been associated with abnormally low levels of β-lipoproteins and other lipoproteins in serum, including retinol-binding protein, which is important in the transport of vitamin A to the eye.[6,91–94] These patients become effectively vitamin A deficient, and the photoreceptor and pigment epithelial cells undergo degenerative changes. The role of vitamin A deprivation in causing the retinal degeneration in this syndrome is supported by the observation that large doses of vitamin A have led to return of dark adaptation thresholds and ERG responses to normal in two patients in the early stages.[94,95] More advanced cases have not responded, but in one of these cases postmortem examination of the retina showed advanced photoreceptor cell degeneration.[96] Whether or not vitamin A therapy can maintain retinal function over the long term has not been resolved, as in some patients the retinal degeneration appeared to progress even though vitamin A levels had been restored to normal.[97] More recently, plasma levels of vitamin E have also been reported to be low in untreated patients,[98,99] and eight patients maintained on diets supplemented with both vitamin A (50,000 I.U. per day) and vitamin E (100 mg/kg of body weight per day up to 3000 mg per day) showed no loss of retinal function over 2–6 years.[100]

2.2. Refsum's Disease

Refsum's disease is an inborn error of metabolism in which large amounts of phytanic acid, 3,7,11,15-tetramethylhexadeconoic acid, accumulate in the tissues and serum of affected patients.[6,101,102] The findings in these patients include retinitis pigmentosa, peripheral neuropathy, and ataxia. Anosmia or lack of smell, neurogenic impairment of hearing, cardiac abnormalities, lens opacities, and skeletal malformations can also be present. The fundus can be granular around the periphery with a subnormal ERG in the early stages or show more typical retinitis pigmentosa with undetectable ERG in more advanced stages.[6] The enzyme defect in these patients (as demonstrated in culture skin fibroblasts) is a deficiency of phytanic acid β-hydroxylase.[103,104] Phytanic acid is the oxidation product of phytol and is present in animal fat, cow's milk, and foods derived from milk. The phytol presumably originates from plant sources (green leafy vegetables). With careful diet and reduced consumption of foods containing both phytanic acid (dairy products, meat, and fish oils) and phytol (green leafy vegetables), serum phytanic acid levels have been lowered to normal, and improvements in neurological deficits have beeen reported. In two patients maintained for many years on this diet, the retinitis pigmentosa and hearing impairment have not been observed to progress.[105]

Two different processes may help to explain some of the neurological and retinal abnormalities in patients with Refsum's disease: (1) phytanic acid may replace long-chain fatty acids in membrane phospholipids and triglycerides with consequent malfunction; or (2) accumulation of phytanic acid in membranes may disrupt the packing of myelin because of its branched methyl groups. In one autopsy specimen, a large amount of fat-staining substance was seen in the pigment epithelium,[106] which may have compromised the pigment epithelial cells and eventually the photoreceptors. Although the mechanism that links phytanic acid to the disease process remains to be defined, successful reversal of at least some of the abnormalities after dietary treatment supports the idea that the phytanic acid itself is responsible for some, if not all, of the clinical manifestations.

2.3. Gyrate Atrophy of the Choroid and Retina

Gyrate atrophy of the choroid and retina is a recessively inherited chorioretinal degeneration.[6,107,108] Patients experience night blindness and loss of peripheral vision between the ages of 10 to 20 years. Ocular findings include myopia, constricted visual fields, elevated dark adaptation thresholds, very small or undetectable ERG responses, and chorioretinal atrophy distributed around the peripheral fundus and sometimes near the optic disk. Abnormalities in electroencephalogram, muscle and hair morphology, and mitochondrial structure in the liver have also been reported.[109,110] Plasma ornithine levels have been shown to be elevated 10- to 20-fold above normal,[72,73,111,112] and a near complete deficiency of ornithine–ketoacid aminotransferase (OKT) activity (Fig. 2) has been demonstrated in cultured skin fibroblasts[113–115] or transformed lymphocytes[116] of affecte patients. Carrier parents showed OKT activity that was intermediate between normal subjects and affected patients.[6,114] In addition to abnormally high ornithine levels, other findings in these patients include reduced levels of plasma lysine, glutamate, and glutamine as well as serum and urine creatine.[72,73,117]

The deficiency in OKT activity in these patients results in an inability to convert ornithine to pyrroline-5-carboxylic acid (PCA) (Fig. 2). A genetic heterogeneity has been demonstrated in gyrate atrophy, since in fibroblasts of some patients OKT could be activated by increasing the concentration of the cofactor, pyridoxal phosphate (i.e., vitamin B_6) in the reaction mixture, and

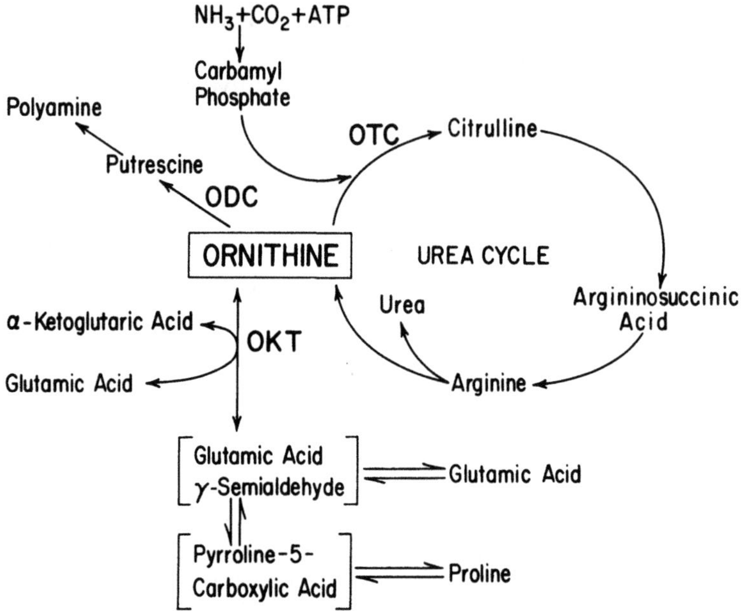

Fig. 2. Pathways of ornithine metabolism. A deficiency in ornithine ketoacid transaminase (OKT) activity has been demonstrated in patients with gyrate atrophy of the choroid and retina.

in these patients, ornithine levels could be lowered *in vivo* by administration of vitamin B_6. In other patients, vitamin B_6 did not affect OKT activity in extracts of fibroblasts; neither did it reduce plasma ornithine concentrations.[6,112,114,118,119] Plasma ornithine levels, however, have been lowered toward normal in all patients by restricting the dietary intake of arginine (direct precursor of ornithine) (Fig. 2) and total protein (source of arginine).[120–123] Studies are in progress in several laboratories to determine the long-term effects of these dietary restrictions on the progression of gyrate atrophy.

As a model for gyrate atrophy, the effects of intravitreal injections of ornithine have been investigated in the normal primate. Such injections lead to swelling of pigment epithelial cells with subsequent death of overlying photoreceptors.[124] A similar pathogenic process may also be taking place in patients with gyrate atrophy, particularly if OKT (known to be present in normal pigment epithelium[125]) is also lacking in their pigment epithelium.

3. *ANIMAL MODELS OF HEREDITARY RETINAL DEGENERATIONS*

The hereditary retinal degenerations in animals (i.e., those presented in Table II) are autosomal recessive diseases, and the abbreviations used (i.e., *rd*, *rds*, *pcd*, RCS, etc.) designate in each case homozygous affected animals. These diseases have been associated with photoreceptor cell degeneration and have been classified either as early or late onset depending on the stage of

Table II
Animal Models with Hereditary Retinal Degenerations

Animal models	Observations	Time course of photoreceptor cell death
Early-onset degenerations		
rd/rd Mice	p8[a] Deficiency in PDE activity Elevated cyclic GMP levels p10–11 Morphological abnormalities Arrested outer segment development	p11–18 Most photoreceptor cells degenerate p18 Only some conelike nuclei remain
rds/rds Mice	p8–21 Complete absence of outer segments Cilia protrude from inner segments PDE activity is reduced Cyclic GMP levels are low; cyclic AMP normal p21–163 Persistence of dark-light differences in cyclic nucleotides Cyclic GMP and cylic AMP levels are reduced with light exposure	p21–1 year Initially selective loss of rod photoreceptors 9 mos. Complete disappearance of photoreceptor cells in the peripheral retina 12 mos. Complete disappearance of photoreceptor cells in the central area
Irish setter (rod–cone dysplasia)	p14–26 Arrested outer segment development Deficiency in PDE activity Elevated cyclic GMP levels	p26–1 year Rod degeneration precedes cone degeneration Histological abnormalities more severe in periphery than in central macular area
Collie dogs (progressive retinal atrophy)	p15–35 Delayed and arrested outer segment development p15–58 Elevated levels of cyclic GMP, PDE activity reduced by about 25% PDE activity demonstrated in normal-appearing outer segments with EM histochemistry but is shown lacking in disorganized segments	p15–58 Loss of photoreceptors; only a few outer segments remain p58–6 to 8 mos. Degeneration of all photoreceptor cells
Late-onset degenerations		
pcd/pcd Mice	p13 Inner segment abnormalities p18–25 Apparently normal outer segment development Pyknotic nuclei noted p53 Reduced rates of synthesis and shedding of outer segments Cyclic GMP not elevated	p18–15 to 16 mos. Rods degenerate faster than cones 6.5 mos. Rod density ~20% of control, cone density ~50% of control Degeneration most severe in the inferior retina

(continued)

Table II (Continued)

Animal models	Observations	Time course of photoreceptor cell death
Miniature poodle (progressive rod–cone degeneration)	p40 Apparently normal outer segment development 14.5 wk. Normal cone outer segments Disorientation, disorganization of some rod outer segments Reduced rate of assembly of outer segment membranes	30 wks. to 60 mos. Slow progression Rods more severely affected than cones 60 mos. Complete degeneration of rods Cones present only in central area, with broad inner segments and shortened outer segments
RCS rat	p10–22 Normal outer segment development p21–23 Normal adult function p18–60 Accumulation of debris as a result of phagocytic defect in pigment epithelium	p18–30 to 40 wks. Complete degeneration of photoreceptor cells. Onset and progression 3–5 days faster in albino than in pigmented strains. Rods degenerate more rapidly than cones

[a] p designates postnatal day.

photoreceptor cell development when the disease becomes manifest (Table II). Early-onset diseases occur in *rd* and *rds* mice and in affected Irish setters and collie dogs, whereas late-onset degenerations occur in RCS rats, miniature french poodles, and *pcd* mice (Table II). It has been proposed that the early-onset diseases are associated with abnormalities in the processes of photoreceptor cell differentiation, whereas the late-onset diseases are associated with abnormalities in the maintenance of the differentiated function(s) of photoreceptor cells.[126] The most studied animal models so far have been the RCS rat and the *rd* mouse, representing late-onset and early-onset diseases, respectively.

3.1. Royal College of Surgeons (RCS) Rats

It is now known that photoreceptor cell degeneration in RCS rats results from an abnormality in the pigment epithelium, which fails to phagocytize the shed tips of photoreceptor outer segments.[127,128] As a consequence of this abnormality, whorls of membrane debris accumulate in the subretinal space between the photoreceptors and the pigment epithelium. During early postnatal life, photoreceptor cells in the RCS rat retina develop normally, and until about 21–23 postnatal days, the RCS rat retina is similar to normal with respect to the thickness of the outer nuclear layer, retinal DNA and taurine concentrations, and ERG function[129] (Fig. 3). These latter observations were obtained with a pigmented strain of RCS (RCS p^+) rats[130]; in albino RCS rats, photoreceptor cell degeneration has been shown to progress more rapidly.[131–133] Although photoreceptor development occurs normally, small amounts of mem-

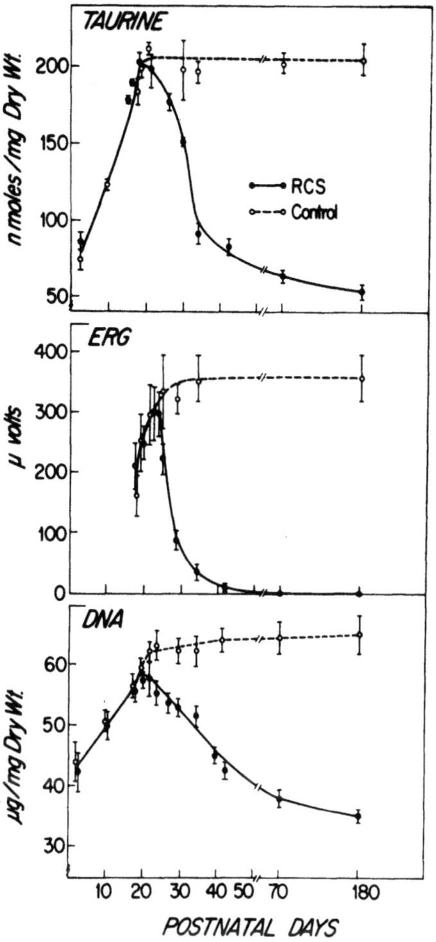

Fig. 3. Retinal taurine concentrations, ERG amplitudes, and retinal DNA concentrations at various postnatal ages in normal control Long–Evans and pigmented RCS p/+ rats with hereditary retinal degeneration. Each data point and vertical bar represents the mean ± S.E.M. for measurements from three to five rats. (From Schmidt and Berson.[129])

brane debris begin to accumulate in the subretinal space between the photoreceptor cells and the pigment epithelium after ~12–16 postnatal days in parallel with rapid outer segment growth. This layer of debris increases in thickness until 30–50 days of age and then decreases gradually as the photoreceptor cells degenerate.[134,135]

The pigment epithelium has been shown to be the primary site of expression of the mutant RCS gene both with studies of chimeras[136] and cultured pigment epithelium.[137,138] In chimeras (formed by combining embryos of normal pigmented and albino RCS rats), accumulation of membranous debris and loss of photoreceptor cells were seen only opposite the RCS pigment epithelium; photoreceptor cells with outer segments in contact with normal pigment epithelium were unaffected.[136] Cultured pigment epithelial cells from normal rats phagocytized both RCS and normal outer segments, whereas cultured RCS pigment epithelial cells failed to phagocytize rod outer segments isolated from normal or RCS rats.[137] Selective phagocytosis of liposomes made up of phosphatidylinositol plus phosphatidylserine was observed in cultured explants of RCS pigment epithelium. Liposomes made up of combinations of these two phospholipids and other phospholipids or of phospholipids other than phosphatidylinositol or phosphosphatidylserine were not phagocytized.[139] In contrast, explants of normal pigment epithelium phagocytized all preparations of liposomes. These studies suggest that membrane phospholipids may participate in the phagocytic mechanism.[139] Recently, the defect in outer segment phagocytosis has been shown to be associated with the ingestion phase of phagocytosis.[140] These *in vitro* findings have provided direct evidence that the mutant gene in RCS rats is expressed in the pigment epithelium as a greatly reduced capacity for outer segment phagocytosis and is not expressed in the photoreceptor cells. A deficiency in the esterification of retinal (a known function in normal pigment epithelium) has also been associated with the pigment epithelium in RCS rats.[141] These studies with RCS rats clearly show that photoreceptor cell degeneration can occur secondary to an abnormality of the pigment epithelium.

Recent studies of RCS rat retinas have shown early abnormalities in the distribution of histochemically staining interphotoreceptor matrix (IPM) glycoconjugates.[135] In normal rat retinas after the 12th postnatal day, an intense band of IPM staining was seen at the apical surface of the PE and a less intense band between the outer segments. In contrast, in RCS rat retinas, the intense band of IPM staining was incompletely formed at the apical surface of the pigment epithelium at 12 days of age and almost completely disappeared as the thickness of the outer segment debris zone increased. In addition, the basal outer segment region stained more heavily in dystrophic retinas than in normal retinas, which suggests an abnormal accumulation of IPM in this region of mutant retinas. The role of IPM has not been defined, but its abnormal distribution in the RCS rat (6–8 days prior to any other abnormality) suggests that it may be related either to the phagocytosis defect or to the pathogenesis of photoreceptor cell death.

Isolation of the components of the interphotoreceptor matrix in rat and bovine retinas has provided evidence that it consists of glycoproteins and/or proteoglycans, some of which are derived from the pigment epithelium, although others are derived from the retina.[142] It is not yet known which specific component of the interphotoreceptor matrix is different from normal in the retina of the RCS rat.

Several hypotheses have been proposed that relate photoreceptor cell death in RCS rats to the accumulation of debris: (1) the debris, by separating the photoreceptors from the pigment epithelium, may provide physical barrier for the diffusion of nutrients and metabolites from the choriocapillaris (the vascular layer beneath the pigment epithelium) to the photoreceptors; (2) the excessive amounts of visual pigments within the membranous debris might result in an accumulation of retinol in the pigment epithelium, which then may disrupt lysosomal membranes, leading to the release of hydrolytic enzymes; and (3) a heat-denaturable and nondialyzable substance within the membranous debris has been shown to induce an abnormality in the activity and kinetics of cyclic GMP PDE in the RCS photoreceptor cells such that the activity of PDE and its affinity for cyclic GMP were both enhanced.[143] In agreement with this observation, biochemical studies showed that in retinas of RCS rats both light-

and dark-adapted cyclic GMP levels were slightly (~40%) below respective normal values.[144]

In contrast to the retinal degeneration in the RCS rat, the pigment epithelium in postmortem donor eyes from patients with retinitis pigmentosa so far studied did not appear to have an abnormality in outer segment phagocytosis. Histological examination of postmortem donor eyes of these patients revealed that pigment epithelial cells apposed to photoreceptor cells with outer segments contained phagososomes, indicative of phagocytic activity. Studies of cultured pigment epithelium from postmortem donor eyes of patients with retinitis pigmentosa have not indicated an impairment in the *in vitro* phagocytic capacity of these cells (R. B. Edwards, personal communication).

3.2. Retinal Degeneration (rd) Mice

In mice homozygous for the retinal degeneration (*rd*) gene, the photoreceptor outer segments never attain normal adult structure or function, and degenerative changes in photoreceptor cells are evident by the tenth postnatal day, during early stages of photoreceptor outer segment development.[145–149] In contrast with RCS rats, the pigment epithelium has been shown to have phagocytic capabilities, and studies of chimeras have clearly demonstrated that the genetic mutation in *rd* mice is expressed in the photoreceptor cells.[150] Photoreceptor cell degeneration has a very rapid progression in these animals, and initially rod photoreceptor cells are affected more severely, such that by the 18th postnatal day, only a single layer of conelike nuclei remains within the photoreceptor cell layer.[148,149,151]

Biochemical studies have demonstrated a deficiency in cyclic nucleotide phosphodiesterase (PDE) activity in the photoreceptor cells of *rd* mice from the onset of their differentiation.[152–155] This PDE enzyme within photoreceptor cells has been shown in several species to have a characteristically high Michaelis constant (i.e., K_m value) for cyclic nucleotides, and on the basis of this property, it can be distinguished from other phosphodiesterases in the retina that have much lower K_m values for either cyclic AMP or cyclic GMP. In contrast to normal mouse retinas, retinas of *rd* mice fail to show the normal postnatal increase in PDE activity at the time of outer segment development between 6 and 20 days (Fig. 4). The substrate of the PDE enzyme *in situ* in outer segments has been shown to be cyclic GMP, and it has become evident from studies of normal outer segment preparations obtained from bovine or frog retinas that the outer segment PDE enzyme becomes activated in light,[156–160] and rapid light-dependent reductions in the concentration of cyclic GMP have been reported in isolated outer segments.[161–164] Apparently, in darkness, because of close interaction with an inhibitor, PDE activity is low, and cyclic GMP levels increase within the outer segments; in several species, dark-adapted cyclic GMP levels have been shown to be higher than after light adaptation *in vivo*.[144,165–167] The process whereby PDE becomes activated in light has been shown to involve a cascade of events initiated by the bleaching of rhodopsin and the light-dependent activation of an outer segment guanosine triphosphatase.[168–174] The process may also involve dissociation of an inhibitor

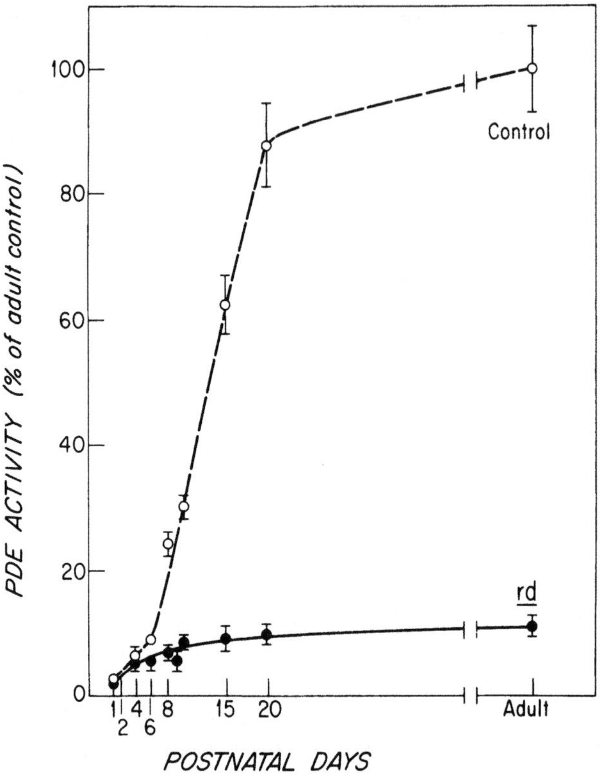

Fig. 4. Cyclic nucleotide phosphodiesterase (PDE) activity in retinas of control (○) and *rd* (●) mice with hereditary retinal degeneration as a function of postnatal age. The data at each age are expressed as percentages of the PDE activity in control adult mouse retinas (16.5 nmol of cyclic AMP hydrolyzed/retina per min in the resence of 2 mM cyclic AMP). In the normal mouse retina, the postnatal increase in PDE activity reflects the increase in the photoreceptor-specific enzyme activity (with a high K_m value for cyclic nucleotides) occurring at the time of outer segment development between 6 and 20 days. The developing photoreceptor cells in *rd* mice are shown to be deficient in the activity of the high-K_m PDE, and homogenates of their retinas only show a low-K_m PDE activity (localized within the inner retina).[152] The substrate of the PDE enzyme *in situ* has been shown to be cyclic GMP, and the patterns of postnatal changes in PDE activity are essentially identical when cyclic GMP is used as the substrate.[154,155] (From Schmidt.[153])

and interactions with additional activator(s).[159,168,170,173,174] Recent studies suggest that a PDE enzyme is present in the *rd* photoreceptor cells,[175] but the question as to whether the deficient activity is caused by an abnormality in the structure of PDE or by an abnormality in the regulatory components or in their interactions with the PDE enzyme remains to be defined.

In retinas of *rd* mice, the deficiency of PDE results in the accumulation of cyclic GMP[154,155,176] (with levels that are three- to fivefold higher than normal) in or around the photoreceptor cells, which may disrupt the metabolism and/or function of these cells and initiate their degeneration. Indeed, selective photoreceptor cell degeneration was induced in developing eye rudiments of *Xenopus laevis* in culture by the addition of dibutyryl cyclic GMP and isobutylmethylxanthine (IBMX), an inhibitor of phosphodiesterase; under these conditions, the extent of the photoreceptor cell degeneration could be correlated with the accumulation of cyclic GMP in these cultured eye rudiments.[177] Elevated levels of cyclic GMP and IBMX have also been shown to be toxic to photoreceptor cells in human retinas.[178] Isolated human retinas incubated in the presence of dibutyrylcyclic GMP and IBMX for 4 to 8 hr showed reduced rates of protein synthesis within the rod photoreceptor cells, resulting in rod photoreceptor degeneration, although cone photoreceptor cells appeared unaffected.

3.3. *Retinal Degeneration Slow (rds) Mice*

In mice homozygous for the retinal degeneration slow (*rds*) gene, a complete failure of outer segment development has been described.[179,180] The degeneration of photoreceptor cells after the second postnatal week progresses slowly and results in a gradual reduction in the thickness of the outer nuclear layer over a period of 8–12 months. In contrast to the *rd* mouse, in which an increase in rhodopsin content has been observed between 6 and 11 postnatal days,[181] in *rds* retinas, rhodopsin content has not been detectable.[180] Similar to *rd* mice, the developing photoreceptor cells in *rds* mice show very low levels of cyclic GMP PDE activity; however, cyclic GMP levels remained low throughout 163 days of study.[180] Despite the lack of outer segments and measurable amounts of rhodopsin, light-dependent reductions in cyclic GMP have been observed up to 163 days of age in *rds* mice. This indicates that the photoreceptor cells retained light sensitivity. Physiological studies to prove this point, however, have not yet been done.

A genetic interaction has been demonstrated between the *rd* and *rds* mutations even though it is known that the genes involved in these diseases are located on different chromosomes (chromosome 5 for *rd*[182] and chromosome 17 for *rds*)[183]. In double homozygous mutants (obtained from matings of *rd*/*rd* and *rds*/*rds* mice), the photoreceptor cells remained lacking in outer segments (as in *rds*), the inner segments showed abnormal vacuoles by 10–11 postnatal days (as in *rd*), and the rate of photoreceptor cell degeneration was intermediate between the rapid degeneration in *rd* and the slowly progressive degeneration in *rds* mice.[183]

3.4. *Photoreceptor Cell Degeneration in Irish Setters and Collie Dogs*

The retinal degenerations in Irish setters (rod–cone dysplasia) and in collie dogs (progressive rod–cone degeneration) are autosomal recessive diseases with early-onset photoreceptor cell degeneration. In both species, reduced rates of outer segment growth become evident by the 15th postnatal day, at early stages of photoreceptor cell development.[184–187] In affected Irish setter dogs, as in *rd* mice, a near complete deficiency was observed in the activity of photoreceptor-cell-specific cyclic GMP PDE enzyme,[184] whereas in affected collie dogs, the activity of this enzyme was reduced by about 25% below control during early stages of the disease.[185]

In homozygous affected dogs of both strains, the concentrations of cyclic GMP within the retina were shown to be 10- to 20-fold higher than in control animals by the 15th postnatal day. As in *rd* mice, in affected dogs of both breeds, rod photoreceptor cells initially appeared more severely affected than the cones; however, by the end of the first year, both rod and cone photoreceptor cells have degenerated. Whereas in Irish setters, PDE activity in retinal homogenates at low substrate concentrations (i.e., low-K_m enzyme) appeared to be enhanced in the presence of calmodulin, in collie dogs no such interaction was noted.[188] In affected Irish setters (as in *rd* mice), the pigment epithelium remained normal, whereas in affected collie dogs, the pigment epithelium appeared hypertrophic and vacuolated by 15 days of age. The diseases in these dogs were shown to be genetically distinct entities as evidenced by the normal progeny that resulted from mating homozygous affected animals of the two different breeds.[189]

3.5. Progressive Rod–Cone Degeneration in Miniature French Poodles

This is a late-onset, slowly progressive disease of both rod and cone photoreceptor cells. Unlike the disease in affected Irish setters or collie dogs, photoreceptor cell development appears to be normal as determined structurally and by electrophysiological tests.[190,191] Ultrastructural studies, however, show abnormalities within the distal portions of photoreceptor outer segments and reduced rates of outer segment assembly by 6–9 weeks of age, although the pigment epithelium appeared normal throughout this period.[190] Cyclic GMP levels and PDE activity in the retina remained within normal limits throughout the extended period (5–8 years) of gradual photoreceptor cell degeneration in affected poodles.

3.6. Purkinje Cell Degeneration (pcd) Mice

Photoreceptor cell degeneration in the Purkinje cell degeneration (pcd) mutant mouse occurs gradually during the first year of life.[192,193] The first sign of an abnormality in photoreceptors was observed by the 13th day within the region of photoreceptor cell inner segments.[194] With the electron microscope, the abnormality was characterized by tubular membranous extensions from the cell membrane of inner segments that appear on cross sections as vesicles in the extracellular space adjacent to photoreceptor inner segments. By day 25, numerous pyknotic photoreceptor nuclei are noted, outer segments become shorter than normal, and the rate of outer segment renewal has become reduced compared with littermate controls.[193] The pigment epithelium appears to have normal phagocytic capacity, and cyclic nucleotide metabolism within the retina also appears to be normal. In contrast to the gradual loss of photoreceptor cells, cerebellar Purkinje cell loss progresses rapidly between 3 and 5 postnatal weeks in homozygous affected animals.

The relationship between Purkinje cell and photoreceptor cell degeneration in these animals is not understood. Hereditary neurological diseases also occur in man in which cerebellar degeneration and ataxia have been associated with typical or atypical pigmentary retinal degenerations, including Friedreich's ataxia and Pierre–Marie diseases.[1–3] The affected cell types in retina or cerebellum, however, have not been identified in the human diseases.

4. *ROD VERSUS CONE INVOLVEMENT IN HEREDITARY RETINAL DEGENERATIONS IN MAN AND ANIMALS*

The animal models as well as the different types of retinitis pigmentosa in human may all represent different genetic diseases with different biochemical etiologies. Although none of the animal models so far studied replicates the exact features of the human diseases, collectively they may contribute to our overall understanding of disease processes associated with photoreceptor cell degeneration. In most of these animals, as in most types of retinitis pigmentosa, regional differences in the progression of the disease have been noted: photoreceptor cells in the peripheral or midperipheral regions of the retina appear more severely affected, whereas those in the macular area in man or the central areas in animals appear less severely affected. Exceptions to this have been observed in miniature French poodles with rod–cone dystrophy and in diseases that affect the macula.

In man as in animals, in most of these diseases the cone photoreceptor cells appear to survive for longer periods than do rod photoreceptor cells. The reason for this better survival of cones is not understood, but it is suggestive of some basic difference(s) between rods and cones. Some of the differences that may hypothetically protect cones in comparison with rods include: (1) structural differences between their outer segments (in rods the disks pinch off and become surrounded by a plasma membrane; in cones the outer segment lamellae remain continuous with the plasma membrane), which may possibly allow for a larger surface area in cones for the uptake of some needed nutrient or for the outward diffusion of a toxic metabolite; (2) differences in the distribution of chromatin in rod and cone nuclei (nuclei of rods being smaller and more dense than those of cones), although how this may render cone nuclei more resistant to injury is not known; (3) large autophagic vacuoles seen within cone photoreceptor cells in donor eyes with retinitis pigmentosa but not within rods, which suggest that cones may possess a better capacity to eliminate defective organelles than do rods; (4) differences in the outer segment renewal process, which may provide for more effective utilization of retinol in cones than in rods (membrane biogenesis in rods, molecular replacement in both rods and cones); and (5) differences in the apoprotein (i.e., opsin) component of the visual pigment, which may allow for faster regeneration of visual pigment in cones than in rods. The extent to which these or other differences contribute to the differential survival of rods and cones remains to be defined.

A common disease process or end-stage pathology in retinitis pigmentosa and in animal models of hereditary retinal degenerations may be associated with reduced renewal of outer segment membranes. This has been clearly demonstrated in the animal models (Table II). In postmortem donor eyes of patients with retinitis pigmentosa, the remaining photoreceptor cells either had no organized outer segments or their outer segments were reduced in length. Reduced outer segment synthesis in these patients was also suggested by the reduced amounts of visual pigments measured by retinal densitometry[17,18] and by the reductions in early receptor potential amplitudes[16] recorded in young patients with the early stages of these diseases. Whether reduced renewal is a primary abnormality or a secondary expression of disease processes remains a difficult problem to resolve.

5. *NUTRITIONAL DEFICIENCIES*

Nutritional factors have been shown to play a role in the maintenance of normally functioning photoreceptor cells. Deficiencies that have been specifically associated with photoreceptor cell degeneration include those of vitamins A and E and taurine.

5.1. *Vitamin A Deficiency*

Vitamin A (i.e., retinol) is specifically required for the synthesis and regeneration of the visual pigments (rhodopsin in rods and less well characterized cone-opsins in cones.[195,196] Vitamin A is transported to the eye via a retinol-binding protein in the plasma[197, 198] and taken up by the pigment epithelium, where it may become associated with intracellular retinol-binding protein(s).[199–202] In the pigment epithelium, vitamin A is esterified and stored primarily as the palmitate.[203] Several additional retinol-binding proteins may be involved in the transport of retinol to the photoreceptor cells. Visual pigment synthesis or regeneration after a bleach requires the formation of a Schiff base linkage between opsin (i.e., the ε-amino group of a lysine) and 11-*cis*-retinaldehyde.[204–208] The process whereby all-*trans*-retinol is oxidized to retinaldehyde and convereted to the 11-*cis* isomer prior to its insertion into visual pigment has not yet been completely elucidated.[208–210] Recent studies with postmortem human donor eyes revealed that Vitamin A is stored within the pigment epithelium mainly in the 11-*cis*-retinyl ester form: in a donor eye from a patient with a chorioretinal degeneration resembling sector retinitis pigmentosa, total vitamin A levels within the pigment epithelium were within the normal range, whereas selective depletion of 11-*cis*-retinyl esters was noted.[211] Since isomerization of retinol is thought to occur within the photoreceptor cells, these observations suggest either that too few photoreceptor cells were present to contribute 11-*cis* isomer to the pigment epithelium or that the remaining photoreceptor cells had an impaired capacity to isomerize retinol or its derivatives from the all-*trans* to the 11-*cis* form.

Absorption of light by the visual pigments results in (1) isomerization of 11-*cis*-retinaldehyde to all-*trans*-retinaldehyde; (2) conformational changes within the opsin moiety; (3) the generation of series of intermediate visual pigments; and, (4) finally, the hydrolysis of the Schiff base linkage between all-*trans*-retinaldehyde and opsin.[212–214] These events constitute the first steps in the visual transduction mechanism leading to light-dependent hyperpolarization of photoreceptor cells.[215–219]

The specific effects of vitamin A in visual pigment metabolism have been studied in vitamin-A-deficient animals using retinoic acid as a vitamin A substitute. Retinoic acid has been shown to relieve the systemic effects of vitamin A deficiency, but it cannot be utilized for visual pigment synthesis.[208,209,220–222] The albino rat, in which storage of retinol in the pigment epithelium does not occur, has been particularly favored for the study of the effects of vitamin A deficiency on the retina. Vitamin A deficiency in the albino rat was associated with a rise of visual threshold that was related to the loss of retinol from the visual pigment.[221]

Weanling rats raised on a vitamin-A-free diet supplemented with retinoic acid showed changes in the outer segments at about 2 months and loss of outer segments, inner segments, and about half the photoreceptor nuclei at about 6 months. At 10 months, the photoreceptors have disappeared except for one row of nuclei. The level of rhodopsin declined in these rats to 5–10% of normal after 2 months, whereas the concentration of the visual pigment protein opsin decreased more slowly, with 50% remaining after 2 months. The ERG function declined in these rats in parallel with the disruption of outer segment structure at a time preceding photoreceptor cell loss. After 6 months of depletion, ERG function and outer segment structure in surviving receptor cells could be restored to normal within 16 days of vitamin A supplementation. In the vitamin-A-deficient rat, rod degeneration was shown to precede cone degeneration.[223]

Recently, it has been shown that light is required to produce this sequence of events, as vitamin-A-deficient rats maintained in darkness did not show photoreceptor cell degeneration until they were placed in cyclic light.[224,225] In the vitamin-A-deficient 13-lined ground squirrel supplemented with retinoic acid, photoreceptor cell structure also appeared relatively intact in squirrels maintained in very dim cyclic light for about 1 year.[226] Exposure of some of these squirrels to brighter cyclic light for a few weeks resulted in scattered white lesions at the photoreceptor–pigment epithelial cell interface, some of which were visible with the opthalmoscope as scattered white deposits. These fundus abnormalities resembled the scattered white deposits reported in vitamin A deficiency in man.

In a manner similar to vitamin-A-deficient animals, reduced ERG amplitudes, elevated thresholds, and shortened outer segments have also been observed in patients with retinitis pigmentosa. Several lines of evidence, however, suggest that vitamin A deficiency is not the problem in these diseases. Serum levels of vitamin A appear to be within normal limits, and large doses of vitamin A given orally or intramuscularly to patients over a relatively short period have had no therapeutic effect.[227,228] Furthermore, the retinal malfunction in early retinitis pigmentosa differs from that observed in vitamin-A-deficient patients with chronic alcoholism, cirrhosis of the liver, and nondetectable levels of serum vitamin A or retinol-binding protein. In all types of progressive, widespread retinitis pigmentosa, the ERGs were shown to be not only reduced in amplitude but also substantially delayed in their temporal aspects (Fig. 1). In contrast, patients with chronic alcoholism, vitamin A deficiency, and night blindness showed only reduced amplitudes and did not show delays in their ERGs either prior to treatment or during stages of recovery on vitamin A therapy.[6,229]

5.2. Vitamin E Deficiency

Several lines of evidence indicate that vitamin E (α-tocopherol), a naturally occurring lipid antioxidant, has a role in maintaining the structural and functional integrity of photoreceptor outer segment membranes. High levels of vitamin E were measured within photoreceptor outer segments in bovine and rat retinas.[230–232] Bovine outer segments have shown a seasonal variation in vitamin E content; summer outer segments with the lowest levels showed the greatest susceptibility to peroxidative injury. Addition of vitamin E to outer segments exerted a protective effect but did not completely eliminate peroxidation, as vitamin E was rapidly destroyed in the process (i.e., inactivated through oxidation). Disappearance (i.e., oxidation) of vitamin E from outer segment membranes has been correlated with an increased rate of fatty acid peroxidation.[231,233] Recent eletrophysiological studies have shown that lipid peroxidation alters the permeability characteristics of the light-dependent sodium channels in frog rod outer segment membranes.[234] In other studies, peroxidation of polyunsaturated fatty acids (measured by the formation of malonyldialdehyde) in isolated bovine and frog outer segments on exposure to air or light was associated with disruption, swelling, and vesiculation of disk membranes.[231,235–237] These changes occurred spontaneously and could be accelerated by the addition of ferrous (Fe^{2+}) ions (known to have potent oxidizing characteristics) and prevented by the addition of vitamin E or EDTA (a chelator of divalent cations) or by replacing air with an inert gas (i.e., argon or nitrogen).[231,234]

Outer segment membranes appear to be highly susceptible to peroxidative injury because of (1) high levels of long-chain polyunsaturated fatty acids[238–240] (which comprise more than 50% of the total fatty acid content), mainly docosahexaenoic (22:6, i.e., 22 carbons, six double bonds) and arachidonic (20:4) acids; (2) the presence of a high oxygen flux from the choriocapillaries and pigment epithelium towards the outer layers of the retina[241,242]; (3) high concentrations of retinaldehyde (either complexed with opsin or released with bleaching) contributing to the generation of singlet oxygen[243]; (4) the presence of electron transport and enzyme systems[244–248] (retinol and NADPH oxidases) known to generate superoxide radicals ($O_2^{-}\cdot$) and/or hydroxyl (HO·) radicals that can initiate fatty acid autooxidation. In the rat retina, the major site of lipoperoxide formation was associated with the photoreceptor cells.[249]

The function of vitamin E as an antioxidant is schematically illustrated (Fig. 5); vitamin E can prevent the initiation of free radical formation and the propagation of free radicals.[250–253] Both of these steps involve the abstraction of hydrogen at a double bond in fatty acids. Peroxidation can be initiated by oxygen, highly reactive singlet oxygen, superoxide radical, hydroxyl radical, and metal ions, and it is propagated by peroxy radicals within fatty acids abstracting hydrogen from neighboring polyunsaturated fatty acids. The process,

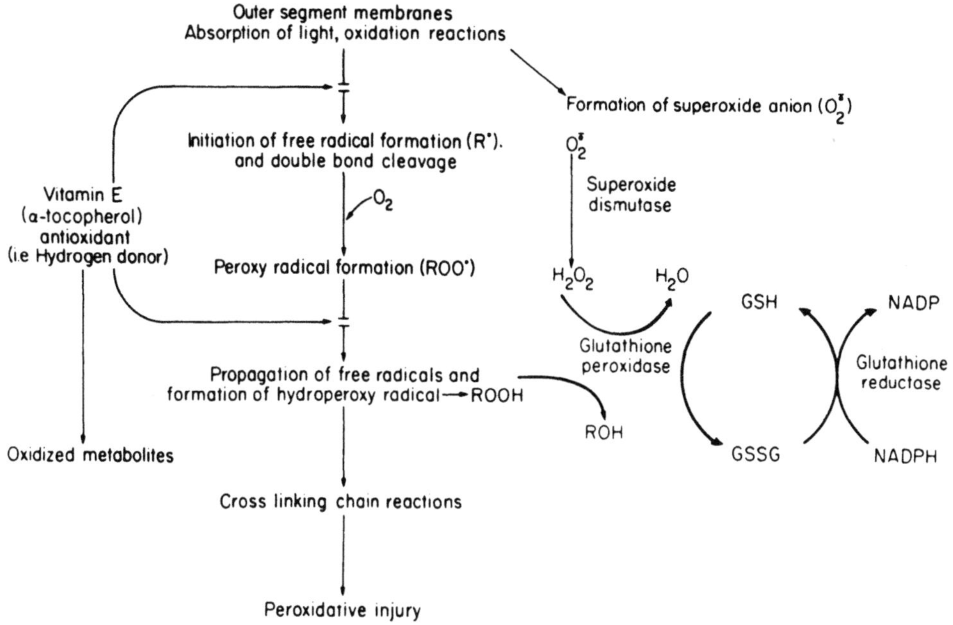

Fig. 5. Proposed scheme for lipid peroxidation reactions and for the role of vitamin E in preventing the initiation of free radical formation and the propagation of free radicals in outer segment polyunsaturated fatty acids.[250–253] In addition to vitamin E, other cellular defense mechanisms in photoreceptors and/or pigment epithelium include superoxide dismutase,[254,255] glutathione peroxidase, [256–259] catalase,[260] and a mechanism for maintaining glutathione in the reduced form (GSH).[261] Oxidized glutathione (GSSG) is converted to GSH via glutathione reductase in the presence of reduced nicotinamide adenine dinucleotide phosphate (NADPH) cofactor. Pathways to generate NADPH have also been shown to be present in the retina.[244–248]

if unchecked, is self-perpetuating and leads to production of fluorescent breakdown products (including malonyldialdehyde) and cross-linking chain reactions, which can involve any and all cellular components. This process of autooxidation (i.e., peroxidation) can lead to various levels of cellular malfunction as enzymes and macromolecules become inactivated and as autofluorescent lipopigments (e.g., lipofuscin) accumulate within lysosomes. The lipofuscin granules, which contain lipids complexed with protein, are thought to represent undigestible end products of lipid peroxidation.

In view of the above considerations, it has not been surprising to find that vitamin E deficiency leads to photoreceptor degeneration as shown in monkeys,[41] dogs,[40] and rats.[38,39] The early abnormalities observed in these animals have included extensive disruption and degeneration of outer segment membranes. Increased accumulation of lipofuscin within the pigment epithelium (up to fivefold) has been observed simultaneously with the disruption and disappearance of outer segment membranes, suggesting that the breakdown products of these outer segments contribute to the formation of the lipofuscin granules.[38,39,232,262,263] The initial site of injury in vitamin E deficiency has been associated with the outer portion of the outer segments; then the entire outer segment becomes disrupted, which is followed by photoreceptor cell death. The inner retina retains its normal structure.

Species differences have been noted in the progression of photoreceptor cell degeneration in vitamin-E-deficient animals; in monkeys, photoreceptor cell degeneration has occurred mainly within the macular region after 2 years[41]; in pigmented rats, degenerative changes were evident within 6 months; in albino rats, disruption of outer segments and massive accumulation of lipofuscin within the pigment epithelium has been noted within 20–34 weeks; and in dogs retinal lesions had become ophthalmoscopically visible within 3 months of deprivation.[40] Recovery of photoreceptor cells from vitamin E deficiency has been shown to be possible if vitamin E is given prior to irreversible injury to photoreceptor cells (~6 months in the rat).[264] Histopathological and electron microscopic studies of tissues other than the eye have lent further support to the concept that the earliest tissue damage in vitamin-E-deficient animals involves cellular membrane systems.[232]

Additional nutritional factors that can affect the animals' antioxidant status include selenium (component of glutathione peroxidase), sulfur amino acids (cysteine and methionine, precursors of glutathione), and chromium (required for insulin-dependent glucose uptake into cells and generation of NADPH via the pentose phosphate pathway). Albino rats deficient in vitamin E and selenium were compared to those deficient in vitamin E, selenium, sulfur amino acids, and chromium and to control animals receiving all four nutrients after 24–26 weeks on their respective diets.[265,266] In rats deficient in vitamin E and selenium, a large increase in lipofuscin content was noted in the pigment epithelium accompanied by an irregularity in size and shape and an increase in the height of the cells. The photoreceptor cells were also shown to be adversely affected by these nutritional deficiencies: the outer segment membranes appeared swollen, vesiculated, and disoriented; there was significant loss of photoreceptor cells, particularly within the central retina (20–34% reduction in cell numbers). These changes appeared more severe in rats deficient in vitamin E, selenium, sulfur amino acids, and chromium than in rats deficient in vitamin E and selenium only.

An interrelationship has recently been demonstrated between deficiencies in vitamins E and A. On the one hand, photoreceptor cell loss was shown to be accelerated in weanling rats maintained for 35 weeks on a diet deficient in both vitamins A and E compared with rats maintained on diets deficient in either vitamin A or vitamin E alone.[232,263] On the other hand, lipofuscin accumulation within the pigment epithelium was highest in rats deficient in vitamin E alone (~5-fold higher than in control) and was lower in rats deficient in both vitamins A and E, (only ~2-fold higher than in rats on control diets containing both vitamins A and E). Increased lipofuscin accumulation in vitamin-E-deficient animals supplied with vitamin A also suggests that vitamin A may undergo peroxidative changes within outer segment membranes either prior to or subsequent to their phagocytosis and that this process is accelerated in vitamin E deficiency. The better preservation of outer segment structure despite the increased lipofuscin in pigment epithelium on diets deficient in vitamin E but containing vitamin A compared with diets deficient in both vitamins E and A suggests that peroxidation of vitamin A occurs within the pigment epithelium rather than the outer segments.

One explanation for these observations is that vitamin A may contribute to the accumulation of lipofuscin within the pigment epithelium. Retinoyl complexes have been associated with autofluorescent lipofuscinlike granules isolated from postmortem human brain specimens from patients with Batten disease.[267] Another possibility is that the requirement for vitamin A is increased in vitamin-E-deficient animals (as storage of vitamin A is known to be reduced with vitamin E deficiency), and, therefore, reduced lipofuscin accumulation in animals deficient in both vitamins E and A compared with animals deficient in vitamin E alone may reflect reduced rates of outer segment renewal secondary to reduced availability of retinol.

5.3. Taurine Deficiency

Taurine deficiency has provided a model for studying the role of taurine in the retina. In the cat, as in all other species so far studied including man, taurine (an aminosulfonic acid) is the predominant free amino acid in the retina.[268–271] With miscrodissection, the highest concentrations of taurine (40–80 mM) have been shown to be within the outer nuclear layer where photoreceptor cell soma and Muller cell processes are located.[272–275] In the cat, exogenous taurine has been shown to be essential to maintain normal retinal taurine concentrations and normal structure and function of photoreceptor cells. Cats fed taurine-free diets (with purified casein as the source of protein) have been shown to develop retinal taurine deficiency associated initially with photoreceptor cell malfunction (20–50% reductions in retinal taurine concentration) and subsequently with photoreceptor cell death (>50% reductions in retinal taurine concentration).[276–280] In taurine-deficient cats, reductions in retinal taurine concentration have been lineraly related to reductions in ERG amplitudes even before photoreceptor cell death (Fig. 6). As retinal taurine concentration became reduced below 50% of normal, granularity was seen in the area centralis, retinal DNA concentration became reduced, and ultrastructural studies showed photoreceptor cell death. Although at early stages of taurine deficiency ultrastructural abnormalities in photoreceptor cells were most pronounced in the area centralis, with continued taurine depletion for up to 18–24 months, eventually all the photoreceptor cells have been shown to degenerate. Disruption of the orderly arrangement of tapetal rods within cells of the tapetum has also been observed in taurine-deficient cats at about the same time when degenerative changes within photoreceptor cells became manifest.[281]

Biochemical and autoradiographic studies suggest that taurine is concentrated in the outer retina (photoreceptor cell layer) by uptake from the plasma (possibly via transport across the pigment epithelium) and not by synthesis *in situ*.[12,274,282–284] Injections of tracer doses of radioactive taurine showed that in normal cats the retina had a capacity to concentrate taurine against a 400-fold gradient (retina ~40 mM; plasma ~0.1 mM); in slightly taurine-deficient cats (preceding photoreceptor loss), this ratio was increased ~6,000 to ~10,000 such that near-normal retinal taurine levels were maintained during taurine

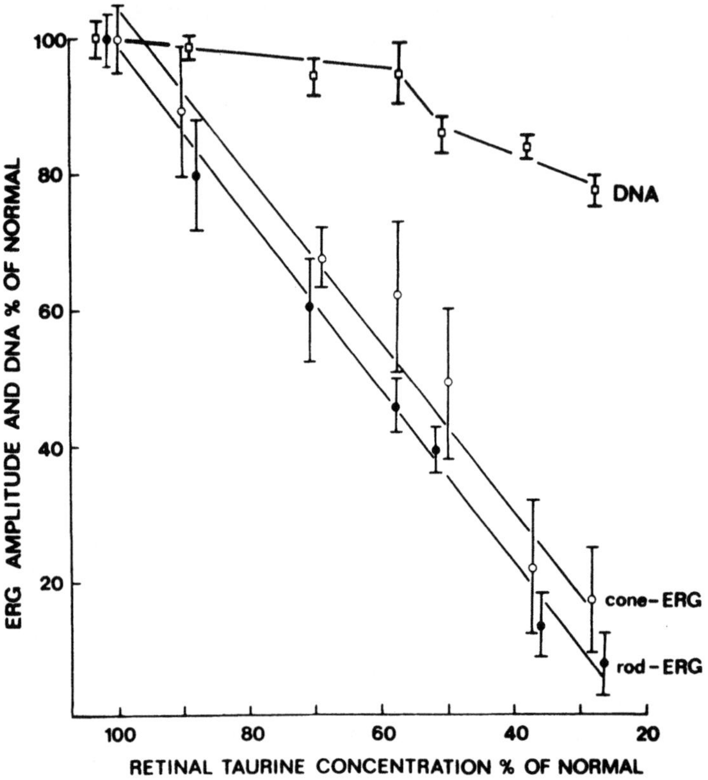

Fig. 6. Peak-to-peak ERG amplitudes and retinal DNA concentrations related to retinal taurine concentrations in ten control and 38 taurine-deficient cats. Controls were considered to have 100% retinal taurine concentration, ERG amplitude, and retinal DNA concentration. Taurine-deficient cats were divided into six groups of four to eight according to the amount of taurine in their retinas. For each group, average amplitudes (mean ± S.E.M.) for rod (●) and cone (○) responses and average DNA values (□) are presented. The coefficients of correlation for rod ERG amplitude and cone ERG amplitude to retinal taurine concentration were, respectively, 0.90 and 0.84. (From Schmidt *et al.*[279])

deprivation until plasma values had fallen below 1–5% of normal.[279] These ratios become even higher if one considers that taurine uptake may have involved mainly the photoreceptor cells. *In vitro* studies have shown that a high-affinity uptake mechanism for taurine is associated with the photoreceptor cell layer and that the affinity for taurine is increased in slightly taurine-deficient retinas compared with normal cat retinas.[285] Figure 7 illustrates that in the isolated cat retina uptake of labeled taurine occurred to the greatest extent within the photoreceptor cell layer.[283] This is also in agreement with the finding that endogenous levels of taurine are highest within the photoreceptor cell layer in the cat and in other species as well.

Feeding taurine-free diets to rats failed to produce taurine deficiency since rat liver (in contrast to that in cats) has a high capacity for biosynthesis of taurine.[286] Recently, however, taurine deficiency has also been produced in rats by adding to their drinking water guanidinoethane sulfonate (GES), a taur-

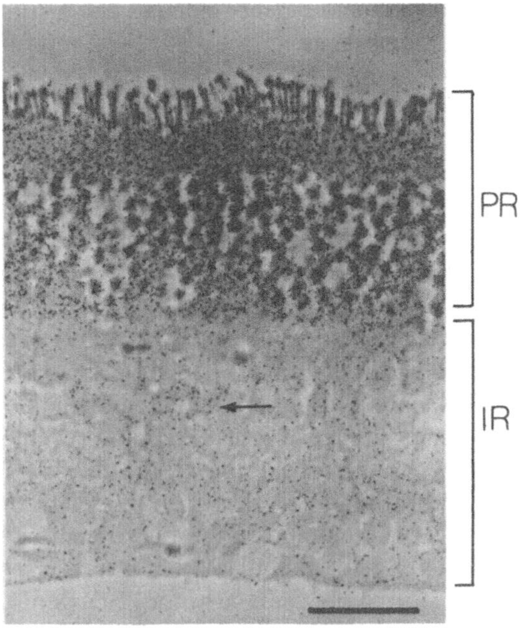

Fig. 7. Phase-contrast autoradiogram of an isolated cat retina (from a slightly taurine-deficient cat with ~20% reduction in retinal taurine content) incubated in medium containing [^{3}H]-taurine (23.1 Ci/mmol; 250 μCi/ml). The radiolabeled grains appear to be concentrated throughout the photoreceptor cell layer (PR) with the highest density over the nuclei and inner segments of photoreceptor cells. Some grains are found in inner retina (IR), particularly over Muller cells and some amacrine cells (the arrow points to one such amacrine cell). Bar, lower right, indicates 50 μm. (From Schmidt and Szamier.[283])

ine analogue that has been shown to antagonize taurine transport into retina, brain, and cardiac tissue *in vivo* and *in vitro* as well.[287–289] In rats given GES, rapid depletion of retinal taurine to about 60% of normal occurred within 9 days and to 40% after 22 days accompanied by progressive reductions in ERG amplitudes and photoreceptor cell degeneration. These findings indicate that in both rat and cat and possibly other species as well, high levels of taurine are required for normal photoreceptor function. The rapid reduction in retinal taurine content in the presence of GES indicates that continuous uptake of taurine plays an important role in the maintenance of high concentrations of retinal taurine. Light-dependent taurine fluxes have been demonstrated in isolated rat and cat retinas.[290] Light onset and offset were associated with release of taurine followed by reuptake; if such a process were to occur *in vivo*, then the presence of GES in the extracellular spaces may interfere with the reuptake of taurine, and this could explain the rapid loss of taurine from the retina in rats given GES.

The mechanism by which retinal taurine deficiency leads to photoreceptor cell death remains to be defined. The observed reductions in the ERG amplitudes in taurine-deficient cats and in rats given GES suggest that taurine deficiency may have some effect on the fluxes of Na^+ and K^+ involved in the generation of the ERG a- and b-waves or that taurine deficiency has led to abnormal concentrations of Na^+ and K^+ in or around photoreceptor and Muller cells. Taurine may also affect the distribution of calcium (Ca^{2+}) in photoreceptor cell outer segments: in isolated frog rod outer segments, taurine was shown to inhibit Ca^{2+} accumulation.[291] More recently, addition of taurine has been shown to prevent peroxidation-induced permeability changes and disruption of isolated frog rod outer segment membranes.[292]

6. LIGHT-INDUCED INJURY TO PHOTORECEPTOR CELLS

The photoreceptor cells, specialized for capturing light energy, have been shown to be highly susceptible to light damage.[232,293–299] The detrimental effects of bright, excessive, continuous or low-wavelength light on the morphology of photoreceptor cells have been demonstrated in a number of species including man, mice, rats, rabbits, and monkeys. Albino rats or mice have been shown to be particularly susceptible to damaging effects of light; albino rats exposed to low levels (18 foot candles, ~200 lux) of continuous illumination for as little as 4 days showed extensive morphological damage to photoreceptor cells and within 30 days complete degeneration of photoreceptor cells.[232] In albino mice, the extent of injury to photoreceptor cells has been related to the intensity of illumination and length of exposure: exposure to fluorescent light (~45 foot candles) for 12 hr was associated with mild disruption of outer segment membranes, and continuous exposure to the same light for 15 days or to 100 foot candles for only 12 hr resulted in severe disruption of the outer segments.[299] Continuous exposure to 100–150 foot candles resulted in 90% loss of photoreceptor nuclei within 6 days.

Numerous studies on the toxic effects of visible, near-visible, low-wavelength light, laser injury, and sun gazing or eclipse blindness suggest that at least two different phototoxic effects can be distinguished: (1) photic effects related to excessive bleaching of visual pigments and (2) photochemical effects caused by exposure to short-wavelength (440–460 nm) light apparently associated with photosensitized oxidations.[300] The deleterious effects of continuous light in albino rats (or mice) can be explained in terms of bleach-related phototoxic effects: in the rat, the extent of light damage, outer segment deterioration, and loss of ERG function has been related to the bleaching of rhodopsin, and the threshold for injury had an action spectrum that corresponded to the absorption maximum of rhodopsin (500 nm).[296] In the rat vitamin A deficiency exerted a protective effect, possibly through reduced levels of visual pigment in the photoreceptor cells.[224]

Phototoxic effects have also been observed in the primate[300–308] and in the rabbit.[309,310] The threshold for damage for 12-hr exposures in these diurnal animals was about 1–2 log units higher than in albino rats. In the monkey, the threshold for damage to cones was lower than for rods, and the action spectrum appeared shifted towards the lower wavelengths (440 nm or less). In partially restrained unanesthetized monkeys with fully dilated pupils, the threshold intensity for damage to cones in the macula for a single 12-hr exposure to fluorescent, broad-spectrum light (400–700 nm) was between 5,900 and 10,800 lux (550–1000 foot candles), and for rods it was 10,800–19,400 lux (1000–1800 foot candles). At near-threshold intensities of exposure, the light damage was restricted to the outer segments with swelling and vesiculation of rod outer segments and disorganization and disorientation of cone outer segments.[307]

These changes closely paralleled the earliest observable abnormalities in photoreceptor cells of albino rats exposed to constant light and are suggestive of a similar pathology probably related to sustained and excessive bleaching of visual pigments. In the monkey, involvement of visual pigments is supported by the demonstration of wavelength-specific phototoxic damage to cones produced by additive effects of repeated exposures to low-intensity intermittent light (40 min/day, 6–7 days); 460-nm light produced irreversible blue cone loss, 520-nm light produced long-term loss of green cone responses.[311,312]

Photochemical (i.e., photosensitized oxidative) effects can be differentiated from bleach-dependent phototoxic effects in that in the latter type of damage the photoreceptor cells do not appear to be primarily involved. Such photochemical lesions have been produced by relatively short (1000 sec, 4 hr) exposures to low-wavelength light (441–458 nm) of moderate intensity focused on discrete areas of the retina in anesthetized immobilized monkeys and rabbits.[301,308] Such exposures resulted in extensive photoreceptor cell degeneration, although initially the pigment epithelium appeared more severely affected in the monkey, whereas all cells appeared similarly affected in the rabbit retina. The earliest abnormalities in monkey involved the pigment epithelium with swelling, disruption, and cell death within 2 days after 1000-sec exposure to 441-nm light. Macrophages appeared in the subretinal space and were filled with melanin granules and cellular debris from degenerating pigment epithelium. Abnormalities in photoreceptors included disruption and disorientation of outer segments and became a prominent feature 5–6 days after exposure. Photoreceptor cell degeneration appeared secondary to abnormalities in the pigment epithelium.

These observations led to the proposal that photosensitized oxidations form the basis of such low-wavelength light injury. This idea is supported by the recent demonstration that similar pathological changes could be produced in cultured bovine pigment epithelium following exposure to 435-nm light in the absence of photoreceptor cells. It was also shown that these effects were oxygen dependent: the threshold to produce injury was much higher in the absence of oxygen, whereas cell death was increased tenfold when exposures were done in 95% oxygen instead of 20% oxygen.[313] It is interesting to note that under slightly different conditions (4-hr exposures to 458-nm light) in the monkey, the initial injury appeared to be associated with the distension of outer segments followed by a similar reaction in the pigment epithelium as above.[303] Furthermore, under these latter conditions, all cells in the retina showed pathological edematous changes, which suggests that photic injury to photoreceptor cells and photochemical injury to all cells of the retina may have occurred simultaneously. The macula in the monkey and possibly in man as well appear to be protected from low-wavelength light by the lens and the macular yellow pigments; although the macula was the most sensitive area for phototoxic effects of broad-spectrum light, it seemed less susceptible to the phototoxic effects of short-wavelength light than the perimacular area. Low-wavelength light has also been shown to be toxic to photoreceptor cells in albino mice, but the mechanism of the effects have not been defined.[314,315]

In addition to phototoxic effects, thermal and mechanical injury may also contribute to light damage, particularly after laser treatment.[306,316,317] Thermal effects may potentiate light damage, although with most exposures, temperature changes within the retina are small. In the rat, increases in body temperature were shown to greatly increase susceptibility to light damage.[296] The mild laser lesion in the monkey suggests a combination of phototoxic and thermal effects with subtle changes in photoreceptors and pigment epithelium. At higher intensities, significant destruction of the pigment epithelium occurs at the site of the lesion followed by discrete foci of photoreceptor cell degeneration.

Melanin granules within the pigment epithelium may enhance or alleviate light damage or have no effect depending on the intensity and wavelength of the light stimulus. Thermal injury induced by laser photocoagulation was shown to be increased in areas with increased density of melanin in the pigment epithelium, apparently because of absorption of heat energy by the melanin granules.[317] In contrast to thermal injury, in phototoxic injury (constant exposure to moderate fluorescent light in rodents), the presence of melanin seemed to slow considerably the rate of photoreceptor cell destruction.[232,296,297] The threshold for injury and the time required to produce the same effects were up to tenfold higher in pigmented than in albino mice or rats. The protective effect of melanin can, at least in part, be explained by the absorption of light since in pigmented rats with dilated pupils the light damage produced by a given intensity of light was similar to that in albino rats.[318] The density of melanin did not seem to affect the extent of phototoxic injury in rabbits with regional variations in fundus pigmentation or differences between rabbits with different degrees of pigmentation. This may appear to be different from rodents; however, even in the lightly pigmented regions of the rabbit fundus, a sufficient number of melanin granules may have been present to exert a protective effect. Therefore, differences attributable to melanin may not be detectable. In support of this idea, even the very light pigmentation of the pigment epithelium was sufficient to exert a protective effect in pink-eyed rats compared with albino rats under conditions of constant light.[297] Photosensitized oxidations caused by low-wavelength light apparently occur independently of the melanin content, and even under culture conditions, no differences were found in the threshold for damage in cultured albino or pigmented pigment epithelium.[313]

In addition to phototoxic and thermal effects, some other processes may also contribute to the pathology of light damage; these include (1) peroxidation reactions, (2) metabolic disruption leading to permeability and electrolyte changes within photoreceptor and pigment epithelial cells, (3) injurious effects of toxic photoproducts (i.e., retinol), and (4) changes in the conformation of opsin molecule (on sustained bleaching) such as to prevent the regeneration of visual pigment(s).

Peroxidation of lipids has been implicated in light damage and outer segment disruption in albino rats and isolated outer segments.[319] Reductions in docosahexaenoic acid and concurrent generation of malonydialdehyde have been reported. Vitamin E levels were also shown to be ~50% reduced from normal in albino rats exposed to 2 days of constant light.[320,321] In view of the high content of polyunsaturated fatty acids within photoreceptor outer segment membranes, it is possible that peroxidation reactions are associated with phototoxic injury. Despite a relationship between light damage and peroxidation of fatty acids, there is as yet no clear-cut evidence that dietary supplementation with vitamin E exerts a protective effect in the rat.[232] However, further studies may be needed to establish optimal doses. Vitamin E deficiency, on the other hand, was associated with decreased ERG amplitudes and increased accumulation of malonyldialdehyde following light stress in rats.[322]

Structural and functional recovery of photoreceptor cells and pigment epithelium following light damage has been observed in albino rats following bright light exposure and in monkeys after exposure to light from an indirect ophthalmoscope or low-wavelength light. Photoreceptor cell recovery occurs by the regeneration of outer segments provided the cells have not been irreversibly damaged. In contrast, recovery of the pigment epithelium involves cell division with the appearance of mitotic figures and hypopigmented cells. Infiltration of macrophages with phagocytosis of cellular debris has also been documented. In the monkey, 10–11 days after exposure, remarkable recovery of rod and cone outer segments was noted, although outer segment length was slightly reduced in the area of the lesion. Outer segment regeneration was complete within ~30 days, although macrophages persisted in the subretinal space until ~60 days, and even 3–5 months after exposure, the pigment epithelium remained hypopigmented.[301,305,306] Gradual recovery has also been noted in human subjects blinded by eclipse gazing or solar retinitis; about 50% of the subjects recovered normal visual acuity within 6 months.[323,324]

Studies of light damage in the primate retina have shown a low threshold (i.e., high sensitivity) to damaging effects of low-wavelength light and have revealed that light damage may occur at levels of illumination potentially encountered in the natural environment. The intensity of light provided by a bright fluorescent lamp (~100 foot candles) or tungsten filament (~1000 foot candles) or bright sunlight (~10,000 foot candles) approach levels shown to be damaging in the primate. In establishing safety guidelines for man, both exposure time and light intensity are under careful consideration, and particular attention is paid to low-wavelength light; for instance, sungazing at bright midnoon for 100 sec could produce a threshold lesion, which can be explained by the low-wavelength light component of the solar spectrum (unattenuated sunlight has a 470-nm peak).

It is not clear to what extent one can extrapolate to man from studies of the primate, and, furthermore, very little information exists on the possible damaging effects of chronic exposures to nearly damaging light levels. Studies in the primate, however, have led to the recommendation of the use of protective sunglasses, to the reexamination of the potentially harmful effects of ophthalmic instrumentation and to the institution of precautionary measures during surgical procedures or laser treatment to minimize the occurrence of light damage.[326] Patients with retinitis pigmentosa may be particularly sensitive to light damage. However, a study of two patients who wore an opaque scleral occluding lens over one eye during a 7-year period showed that the progression of retinitis pigmentosa in these patients was similar in both eyes.[327] Since these eyes were not totally protected from light, the possibility that light sensitivity contributes to the progression of retinitis pigmentosa cannot be excluded. In addition, the sensitivity of the retina to short-wavelength light may have possible clinical significance for certain age-related retinal pathologies as well as retinitis pigmentosa.

7. PEROXIDATION REACTIONS

There is evidence that suggests that a relationship exists among lipid peroxidation, destruction of polyunsaturated fatty acids, increased production of malonyldialdehyde, and degenerative changes within photoreceptor cells that initially appear to involve the outer segment membranes and subsequently culminate in photoreceptor cell death. A direct association between peroxidation reactions and photoreceptor cell degeneration has been shown secondary to X-ray irradiation,[328] exposure to high levels of oxygen,[329,330] intravitreal injection of linoleic acid peroxide,[328,331] ferrous form of iron (an oxidizing agent),[332–336] or surgical implantation of an iron nail into the vitreous.[337] Exposure to high levels of oxygen (95% oxygen at 2 atmospheres) was associated with photoreceptor cell degeneration and increased lipoperoxides in retinas in 14-day-old chick embryos and in mature rabbits.[329,330] In rabbits, increases in lipoperoxide production paralleled the reductions in ERG amplitudes.

Increased lipid peroxidation and disruption of outer segment structure have also been observed in isolated bovine and frog rod outer segments exposed to air and could be counteracted by addition of vitamin E and by replacing air with argon or nitrogen gas. Intravitreal injections of $FeCl_2$ (ferrous form of iron ~0.01 mg) in the squirrel monkey was associated with disruption of photoreceptor outer segments and pigment epithelium within 4 hr; in contrast, similar injections of $FeCl_3$ (ferric, nonoxidizing form of iron) had no effects[332] (slight effects observed at a later time could be explained by the conversion of the iron to the ferrous form). Intravitreal injection of $FeSO_4$ to frogs resulted in progressive reduction of docosahexaenoic acid within 4 and 24 hr; concurrently, lipid peroxides (measured as malonyldialdehyde and related autofluorescent dienes) increased in the retinas compared with Na_2SO_4-injected controls. In contrast, the content of linoleic acid (a saturated fatty acid not subject to peroxidation) did not change in retinas of $FeSO_4$-injected frog eyes.[333]

Although some of these studies may point to lipid peroxidation as a common pathogenic mechanism in oxygen toxicity, light damage, and vitamin E deficiency, additional factors might require consideration. A recent report suggests that light damage in albino rat eyes was associated with significant reductions in vitamin E levels, but lipoperoxide levels remained relatively low when extractions were done under stringently controlled conditions. These studies suggest that in some experiments high levels of lipoperoxides might be artifactual and might be caused by oxidative changes during tissue preparation, particularly if, under different conditions, endogenous levels of vitamin E are variable.[231,320,321]

The high polyunsaturated fatty acid content of outer segment lipids renders them susceptible to peroxidation, particularly in the presence of light and oxygen. Several antioxidant mechanisms are present in the retina and pigment epithelium, presumably to minimize the occurrence of peroxidative injury (Fig. 5). Recent studies have provided evidence that the redox state of the photoreceptor cells can modify their electrophysiological function and that peroxidation of membrane lipids can alter the function of sodium channels[234] in rod outer segments as well as the regenerability of visual pigments.[338] Within the pigment epithelium, the melanin granules appear to be part of an antioxidant defense mechanism by absorbing light or heat energies, by binding free radicals and metal ions, and by their capacity to undergo electron-transfer reactions.[339,340]

Deficiency in any of the components of the antioxidant defense mechanism including vitamin E, superoxide dismutanse, glutathione peroxidase, or the amount of reudced glutathione could lead to excessive peroxidation in outer segment polyunsaturated fatty acids, peroxidative decomposition of membrane lipids with cross-linking chain reactions, and enzyme inactivation as well as increased accumulation of lipofuscin in the pigment epithelium. It is tempting to speculate that failure in some aspect of the antioxidant mechanisms may play a role in photoreceptor cell degeneration in some hereditary diseases, in light damage, or in aging. Such a failure may occur as a deficiency or as a reduction in the activity of a particular enzyme in photoreceptors or pigment epithelium (i.e., superoxide dismutase, glutathione peroxidase, glutathione reductase, or catalase) or in the transport of vitamin E or selenium or may represent a secondary consequence of some other disease process that results in excessive generation of peroxides (this may result from altered lipid-to-protein ratio within outer segments, lability and/or unregenerability of visual pigments, or inhibition of protective enzymes because of accumulation of toxic metabolites) or in an inadequate supply of flavine adenine nucleotide cofactors (needed for the reduction of glutathione).

8. SUBSTANCES WITH TOXICITY TO PHOTORECEPTOR CELLS: IODOACETATE, TUNICAMYCIN, HEMICHOLINIUM-3

Iodoacetate has been known to have selective toxicity towards photoreceptor cells. In the rabbit retina, intracardiac injection of a single dose of iodoacetate resulted in selective and rapid degeneration of rod photoreceptor cells within 10 hr and a slower degeneration of some cones after 2–3 weeks.[341,342] Cone degeneration has recently been demonstrated in cone-dominant retina of 13-line ground squirrels following injections of iodoacetate.[343] The specific metabolic sequence(s) that iodoacetate interferes with and that is crucial for the survival of rod and cone photoreceptor cells is not known. Because iodoacetate is an inhibitor of glycosis, recent studies suggest that it also inhibits the formation of glycerol phosphate and phosopholipid synthesis. In isolated rat retinas incubated in the presence of iodoacetate (1 or 2×10^{-3} M), 90% reductions were noted in the uptake and incorporation of [^{14}C]glucose into phospholipid and in the incorporation of [^{3}H]inositol into phosphatidylinositol.[344,345] In the ground squirrel retina, concomitant with loss of cone photoreceptor cells, significant reductions were observed in the concentration of cyclic AMP. In this particular case, the disappearance of cyclic AMP supported the idea that cyclic AMP represents the nucleotide that is normally associated with cone photoreceptor cells.[343] In contrast, cyclic GMP appears to be the nucleotide associated with rod photoreceptor cells.

Tunicamycin is known to be an inhibitor of dolichol-mediated protein glycosylation. Isolated human or toad (*X. laevis*) retinas incubated in the presence of [^{3}H]mannose plus tunicamycin followed by a 4-hr chase incubation showed great reductions in the radioactivity within photoreceptor cell inner segments and decreased or even complete absence of new outer segment membrane assembly at the base of the outer segments. In contrast, control retinas incubated with [^{3}H]mannose (using the same protocol) showed a striking concentration of radioactive grains throughout the photoreceptor layer and discrete bands of radioactivity at the base of the outer segments indicative of assembly of new outer segment membranes.[346] Intravitreal injection of tunicamycin in the rabbit was associated with degeneration of photoreceptor cells.[347] Biochemical studies have provided evidence that in the presence of tunicamycin glycosylation of proteins and particularly that of rhodopsin was reduced by ~80%, whereas protein synthesis was only slightly reduced.[346] These results suggest that opsin glycosylation occurs via a tunicamycin-sensitive (i.e., dolichol-mediated) pathway and that glycosylation reactions play a role in the synthesis and assembly of outer segment membranes. One possible interpretation of these results is that opsin glycosylation is a prerequisite for its insertion into outer segment membranes and the subsequent assembly of new outer segment membranes.

Hemicholinium-3, an analogue of choline, has been shown to interfere with choline metabolism by competitively inhibiting choline uptake into cells. A high-affinity mechanism for choline uptake has been associated with the photoreceptor cells, in which choline was shown to be utilized for synthesis of phosphatidylcholine but not acetylcholine.[348,349] In the rabbit retina, intravitreal injection of hemicholinium-3 (final concentration estimated at 20 μM) resulted within 2–5 days in shedding of most outer segments and their disappearance by 7–10 days. Cone photoreceptor cells appeared more sensitive to hemicholinium-3 than were rods, and by 14 days, most cone photoreceptor cells had completely degenerated.[349] In contrast, rod photoreceptor cells remained viable and could regenerate their outer segments once hemicholinium-3 was no longer present in the eye (~20–30 days after the intraocular injection of hemicholinium-3).

Intravitreal injection of hemicholinium-3 (final concentration ~50–100 μM) was also associated with rapid loss of photoreceptor outer segments in the cone-dominant retina of the 13-lined ground squirrel.[350] Outer segments were phagocytized by pigment epithelial cells and macrophages that appeared in the subretinal space. At the lower dose, the entire superior retina was denuded of outer segments, whereas the inferior retina was preserved selectively. At the higher dose, outer segments were lost throughout the retina. Recent *in vitro* studies extend these findings and show that in isolated rabbit retinas hemicholinium-3 is an effective inhibitor of choline incorporation into phosphatidylcholine and provide evidence that phosphatidylcholine synthesis is important in maintaining outer segment structure.[351] This is not surprising since phosphatidycholine comprises ~45% of outer segment phospholipids.

9. AGE-RELATED CHANGES IN PHOTORECEPTOR AND PIGMENT EPITHELIAL CELLS

With age, changes occur within the photoreceptor and pigment epithelial cells as well as other structures of the eye, including the vasculature, choroid, and Bruch's membrane, which may directly affect the function of these cells. In addition, characteristic aging changes in the lens (reduced transparency), abnormalities in the vitreous, and constriction of the pupil reduce the amount of light that is transmitted to the retina.[352,353] Aging changes within photoreceptor cells observed to increase in frequency after age 40 include thinning of the outer nuclear layer, displacement and pyknosis of photoreceptor cell nuclei, and various degrees of structural disorganization in the arrangement of disk within the rod outer segments.[354–356] The pigment epithelial cells become irregular in size and shape[357,358]; their content of lipofuscin granules increases dramatically,[31] as does the frequency of melanolysomes and/or melanolipofuscin granules (which arise from the fusion of melanin and lipofuscin granules),[30] and there are reductions in the numbers of free melanin granules and melanin content.[359,360] It is interesting to note that topographical variations in the distribution of these cellular inclusions are such that lipofuscin is highest while melanin is lowest within the macular pigment epithelium.[31,359,360]

The basement membrane underlying the pigment epithelium (Bruch's membrane) becomes irregular and thickened. Areas of drusen accumulation appear as discrete yellowish spots most concentrated in the macula. Drusen represents granular substance between Bruch's membrane and the pigment epithelium, probably produced by the degeneration of pigment epithelial cells and/or material extruded from lipofuscin-laden pigment epithelial cells. The pigment epithelial layer may become very thin with preserved gap junctions between adjacent cells or discontinuous as the cells become stretched over drusen.[361] Frequent aging changes within the choroidal and retinal vasculature involve narrowing of the arteries, disappearance of capillaries, with ensuing ischemia of the entire retina.[361] These aging changes in the pigment epithelium, Bruch's membrane, and the vasculature undoubtedly reduce the efficacy of the transport of nutrients and waste products to and from the photoreceptor cells and the choriocapillaries.

Age-related changes within pigment epithelial cells may directly affect their function: the capacity for phagocytosis may be reduced (as a consequence of excessive lipofuscin accumulation), and their function as a defense against peroxidative injury may be reduced (because of reduced melanin content). In view of all these changes, the photoreceptor cells and pigment epithelium may become increasingly susceptible to light damage with age. In support of this idea, both the absence of melanin and ischemia have been associated with greatly increased susceptibility to light damage in experimental animals.[362]

Reduced visual function in older people reflects a combination of aging changes within the structures of the eye and within the entire visual system. Decrease in visual acuity and rise in threshold with age are thought to result in part from optical changes (decreased pupil size, reduced transparency of lens and vitreous) and in part from cell death within retina and brain. In the retina, photoreceptor cell death with age exceeds by far that of any other cell type in the retina and is more pronounced than could be accounted for by uniform and random cell death throughout the visual system. Rod function appears to deteriorate more rapidly than cone function. The rise in visual threshold (based on changes per decade) has been shown to be two- to threefold higher for rods than for cones.[363] Cumulative effects of numerous "mild" phototoxic injuries may also contribute to photoreceptor cell loss with age. In this context, photoreceptor cells in hereditary retinal diseases may show accelerated effects of aging either as a direct consequence of the inherited abnormality or as a consequence of failure of some protective and/or detoxification mechanism in the photoreceptor cells and/or the pigment epithelium.

Age-related macular degenerations (senile macular degenerations) represent a major cause of loss of vision in the elderly. The pathogenesis of these conditions is not understood at present. In view of the discrete localization in the region surrounding the optical axis of the eye where the flux of light falling on the retina and the density of the photoreceptor cells are the greatest, one may speculate that these macular changes are related to light damage. Although a parallel can be drawn between age-related pathological changes and light damage (particularly in the primate models), a clear-cut relationship has not been established. Many different types of age-related macular degenerations fall in either of two broad categories: hemorrhagic or dry.[1–3,355] A pathogenic mechanism for the former could involve breakdown of tight junctions between adjacent pigment epithelial cells, seepage of fluid into the subretinal space, formation of small areas of retinal detachments, ingrowth of blood vessels, and bleeding. In the dry form of age-related macular degenerations, the pathological changes may initiate within the photoreceptor cells as a gradual failure in their function and in their ability to repair themselves, which is followed by degenerative changes and cell death. Rather than representing discrete diseases, these degenerations probably result from the interactions of many factors, including (1) culmination of normal or accelerated aging changes in photoreceptors and/or pigment epithelium (2) cumulative effects of a life-long history of chronic light exposure, (3) long-term nutritional factors, (4) genetic factors, (5) vascular and/or systemic diseases, (6) abnormalities in Bruch's membrane, (7) pupil size, (8) the state of the lens and its presence or absence, (9) melanin content in pigment epithelium and iris, and (10) current light exposures and measures taken for the protection of the retina from low-wavelength and ultraviolet light.

Aging changes and age-related photoreceptor cell degeneration have also been observed in several strains of rats, including Osborn–Mendel rats and hypertensive rats.[364,365] Extensive loss of photoreceptor cells was noted by 10–20 months of age in these rats. The rate of photoreceptor cell loss, monitored by the reduction in the thickness of the outer nuclear layer, appeared quite variable between rats of the same strain and suggested that some extrinsic factor may have contributed to the progressive cell loss with age. Since these studies were done with albino rats, increased light sensitivity with age cannot be excluded from consideration. In albino rats but not in pigmented Long–Evans rats, retinal taurine levels decreased with age and were about 35% lower at 1 year than at 60 days.[129,366] Since taurine is known to be concentrated within the photoreceptor cells, this observation suggests that significant loss of photoreceptor cells occurs in albino rats compared with pigmented rats during the first year of life. More studies are needed to systematically define aging changes in photoreceptor cells and pigment epithelium in experimental animals.

10. CONCLUSION

Over the last 15–20 years, some of the disease processes in retinitis pigmentosa and in animal models with hereditary retinal degenerations have become better defined, and the affected cell types have been identified. Progress has also been made in understanding some of the biochemical properties of the photoreceptor cells and the photoreceptor–pigment epithelium complex. These include (1) the regulation of cyclic nucleotide metabolism, (2) the process of outer segment renewal and the role of the pigment epithelium, (3) transport of retinol and metabolism of visual pigments, (4) the role of vitamin E and other antioxidant mechanisms in the prevention of lipid peroxidation in outer segment membranes, (5) maintenance of high levels of taurine (up to 80 mM) within photoreceptor cell somata, and (6) rapid rates of phospholipid (particularly phosphatidylinositol) synthesis, which have been demonstrated in photoreceptor cells in the rat. It has also become evident that a number of agents can exert selective toxicity towards photoreceptor cells. These include light, oxygen toxicity, peroxidation of membrane lipids, hemicholinium-3 (through interference with the synthesis of phosphatidylcholine), isobutylmethylxanthine (an inhibitor of phosphodiesterase), and tunicamycin (which inhibits the glycosylation of rhodopsin and its insertion into outer segment membranes).

Recent studies have helped to delineate nutritional factors that can affect the photoreceptor cells and have provided evidence that through dietary manipulations certain specific requirements of the photoreceptor cells can be met (requirements for vitamins A and E, taurine, selenium, sulfur amino acids, chromium, etc.) and that in some specific diseases toxic substances can be avoided (Refsum's disease, gyrate atrophy of the retina and choroid). Better understanding and careful attention to nutritional and possibly other environmental factors may provide means of reducing the risk factors for age-related degenerative changes in photoreceptor and pigment epithelial cells and for altering the course of some types of hereditary retinal degenerations.

The integrity the photoreceptor–pigment epithelium complex has been shown to play an important role in maintaining photoreceptor cell viability. This may be related to the process(es) of outer segment renewal, transport and diffusion of nutrients including glucose, phospholipid precursors, vitamins A and E, and taurine into the photoreceptors, and also diffusion of metabolites (i.e., retinol, oxidized tocopherol) away from the photoreceptor cells. Disruption of normal photoreceptor–pigment epithelium interactions (in RCS rats as a result of a genetic mutation or in man due to detachment of the retina) has also been associated with photoreceptor cell degeneration.

Recent advances have made possible not only the early diagnosis of progressive forms of retinitis pigmentosa and allied diseases but also the detection of carriers of sex-linked retinitis pigmentosa. Through awareness and genetic counseling, these women may decide to have fewer children, which may help to reduce the incidence of this disease entity. Furthermore, with new advances in molecular genetics, it may become possible to identify known marker genes on the X-chromosome closely linked to the gene(s) responsible for sex-linked retinitis pigmentosa. It may even be feasible in the not too distant future to obtain the DNA sequence of this gene (or genes) and to characterize the mRNA(s) and the final gene product(s) (i.e., proteins) derived from this gene (or genes).

In summary, our understanding of hereditary retinal degenerations has been expanding during the last 15–20 years, and we are approaching a period during which it will be possible to elucidate some of the disease processes not only at the cellular but also at the subcellular and molecular levels.

ACKNOWLEDGMENTS. This work was supported in part by Research Career Development Award EY00148 from the National Eye Institute and by grants from the National Retinitis Pigmentosa Foundation, Baltimore, and the George Gund Foundation, Cleveland.

APPENDIX

Bergsma, D., Bron, A. J., and Cotlier, E., eds., 1976, *The Eye and Inborn Errors of Metabolism, Birth Defects: Original Article Series*, Volume 12, Alan R. Liss, New York.

Clayton, R. M., Haywood, J., Reading, H. W. and Wright, A., eds., 1982, *Problems of Normal and Genetically Abnormal Retinas*, Academic Press, New York.

Cotlier, E., Maumenee, I. H., and Berman, E. R., eds., 1982, *Genetic Eye Diseases, Birth Defects: Original Article Series*, Volume 18, Alan R. Liss, New York.

Daentl, D. L., ed., 1982, *Progress in Clinical and Biological Research*, Volume 82, Alan R. Liss, New York.

Dowling, J. E., Proenza, L. M., and Atwell, C. W., eds., 1982, *J. Retinal Vitreous Dis.* **2**:231–359.

Landers, B. III, Wolbarsht, L., Dowling, J. E., and Laties, A. M., eds., 1977, *Advances in Experimental Medicine and Biology*, Volume 77: *Retinitis Pigmentosa: Clinical Implications of Current Research*, Plenum Press, New York.

Newell, F. W., ed., 1980, *Hereditary Disorders of the Eye and Ocular Adnexa*, Ophthalmic Publishing, Chicago.

Osborn, N., and Chader, G., eds., 1983, *Progress in Retinal Research*, Volume 1, Pergamon Press, Oxford.

Sears, M. L., ed., 1981, *New Directions in Ophthalmic Research*, Yale University Press, New Haven.

Vinken, P. J., and Bruyn, G. W., eds., 1972, *Handbook of Clinical Neurology*, Volume 13, American Elsevier/North-Holland, New York.

Zinn, K. M., and Marmor, M. F., eds., 1979, *The Retinal Pigment Epithelium*, Harvard University Press, Cambridge.

REFERENCES

1. Duke-Elder, S., and Dobree, J. H. 1967, *Systems of Ophthalmology*, Volume 10, *Diseases of the Retina*, C. V. Mosby, St. Louis.

2. Franceschetti, A., Francois, J., and Babel, J., eds., 1974, *Chorioretinal Heredodegenerations*, Charles C Thomas, Springfield, Illinois.
3. Krill, A. E., 1977, *Hereditary Retinal and Choroidal Diseases*, Volume 2, Harper and Row, New York.
4. Fishman, G. A., 1978, *Arch. Ophthalmol.* **96:**822–826.
5. Jay, M., 1982, *Br. J. Ophthalmol.* **66:**405–416.
6. Berson, E. L., 1982, *Retina* **2:**236–255.
7. Boughman, J. A., Conneally, P. M., and Nance, W. E., 1980, *Am. J. Hum. Genet.* **32:**223–235.
8. Goodman, G., and Gunkel, R. D., 1958, *Am. J. Ophthalmol.* **46:**142–178.
9. Gouras, P., and Carr, R. E., 1964, *Arch. Ophthalmol.***73:**104–110.
10. Berson, E. L., Gouras, P., and Gunkel, R. D., 1968, *Arch. Ophthalmol.* **80:**68–76.
11. Berson, E. L., 1981, *Int. Ophthalmol.* **4:**7–22.
12. Berson, E. L., 1976, *Trans. Am. Acad. Ophthalmol. Otolaryngol.* **81:**659–666.
13. Berson, E. L., and Simonoff, E. A., 1979, *Arch. Ophthalmol.* **97:**1286–1291.
14. Berson, E. L., Rosner, B., and Simonoff, E. A., 1980, *Am. J. Ophthalmol.* **89:**763–775.
15. Berson, E. L., Rosen, J. B., and Simonoff, E. A., 1979, *Am. J. Ophthalmol.* **87:**460–468.
16. Berson, E. L., and Goldstein, E. B., 1970, *Invest. Ophthalmol. Vis Sci.* **9:**58–63.
17. Highman, V. N., and Weale, R. A., 1973, *Am. J. Ophthalmol.* **75:**822–832.
18. Gieser, D. K., Fishman, G. A., and Cunha-Vaz, J., 1980, *Arch. Ophthalmol.* **98:**307–310.
19. Cogan, D., 1949, *Trans. Am. Acad. Ophthalmol.* **54:**629–661.
20. Mizuno, K., and Nishida, S., 1967, *Am. J. Ophthalmol.* **63:**791–803.
21. Kolb, H., and Gouras, P., 1974, *Invest. Ophthalmol. Vis. Sci.* **13:**487–498.
22. Szamier, R. B., and Berson, E. L., 1977, *Invest. Ophthalmol. Vis. Sci.* **16:**947–962.
23. Szamier, R. B., Berson, E. L., Klein, R., and Meyers, S., 1979, *Invest. Ophthalmol. Vis. Sci.* **18:**145–160.
24. Szamier, R. B., and Berson, E. L., 1982, *Invest. Ophthalmol. Vis. Sci.* **22:**559–570.
25. Gartner, S., and Henkind, P., 1982, *Ophthalmology* **89:**1425–1432.
26. Santos-Anderson, R. M., Tso, M. O. M., and Fishman, G. A., 1982, *Ophthalmol. Paediatr. Genet.* **1:**151.
27. Bunt-Milam, A. H., and Kalina, R. E., and Pagon, R. A., 1983, *Invest. Ophthalmol. Vis. Sci.* **24:**458–469.
28. Meyer, K. T., Heckenlively, J. R., Spitznas, M., and Foos, R. Y., 1982, *Ophthalmology* **89:**1414–1424.
29. Aguirre, G. D., and Laties, A., 1976, *Exp. Eye Res.* **23:**247–256.
30. Feeney-Burns, L., 1980, *Curr. Top. Eye Res.* **2:**119–178.
31. Wing, L., Blanchard, C., and Weiter, J., 1978, *Invest. Ophthalmol. Vis. Sci.* **7:**601–607.
32. Wolfe, L. S., and Kim, N. M. K., Ng, 1982, *Birth Defects* **18:**234–239.
33. Ng Ying Kin, N. M. K., Palo, J., Haltia, M., and Wolfe, L. S., 1983, *J. Neurochem.* **40:**1465–1473.
34. Zeman, W., 1971, *Adv. Gerontol. Res.* **3:**147–170.
35. Bergsma, D. R., Wiggert, B., Funahashi, M., Kuwabara, T., and Chader, G., 1977, *Nature* **265:**66–67.
36. Miller, S. A., 1978, *Br. J. Ophthalmol.* **62:**256–258.
37. Eagle, R. C., Jr., Lucier, A. C., Bernardino, V. B., Jr., and Yanoff, M., 1980, *Ophthalmology* **87:**1189–1200.
38. Katz, M. L., Stone, W. L., and Dratz, E. A., 1978 *Invest. Ophthalmol. Vis. Sci.* **17:**1049–1058.
39. Robison, W. G., Jr., Kuwabara, T., and Bieri, J. G., 1979, *Invest. Ophthalmol. Vis. Sci.* **18:**683–690.
40. Riis, R. C., Sheffy, B. E., Loew, E., Kern, T. J., and Smith, J. S., 1981, *Am. J. Vet. Res.* **42:**74–86.
41. Hayes, K. C., 1974, *Invest. Ophthalmol.* **13:**499–510.
42. Young, R. W., 1969, *The Retina: Structure, Function and Clinical Characteristics* (B. Straatsma, R. Allen, M. Hall, and F. Crescitelli, eds.), University of California Press, Los Angeles, pp. 177–209.

43. Miller, S. S., and Steinberg, R. H., 1977, *Exp. Eye Res.* **25**:235–248.
44. Bridges, C. D. B., 1976, *Exp. Eye Res.* **22**:435–455.
45. Bok, D., and Young, R. W., 1979, *Retinal Pigment Epithelium*, Harvard University Press, Cambridge, pp. 148–174.
46. Young, R. W., and Bok, D., 1969, *J. Cell. Biol.* **42**:392–403.
47. Kroll, A. J., and Machemer, R., 1969, *Am. J. Ophthalmol.* **68**:58–77.
48. Aarberg, T. M., and Machemer, R., 1970, *Am. J. Ophthalmol.* **69**:640–650.
49. Young, R. W., and Droz, B., 1968, *J. Cell Biol.* **39**:169–184.
50. Young, R. W., 1976, *Invest. Ophthalmol.* **15**:700–725.
51. Young, R. W., 1971, *Vision Res.* **11**:1–5.
52. LaVail, M. M., 1976, *Science* **194**:1071–1074.
53. Basinger, S. F., Hoffman, R., and Matthes, M., 1976, *Science* **194**:1074–1976.
54. Basinger, S. F., and Hollyfield, J. G., 1980, *Neurochemistry of the Retina* (N. G. Bazan and R. N. Lolley, eds.) Pergamon Press, Oxford, pp. 81–92.
55. Besharse, J. C., 1982, *Retinal Research*, Volume 1 (N. Osborne and G. Chader, eds.), Pergamon Press, New York, pp. 82–118.
56. Birch, D. G., Berson, E. L., and Sandberg, M. A., 1984, *Invest. Ophthalmol. Vis. Sci.* **25**:236–238.
57. Young, R. W., 1977, *J. Ultrastruct. Res.* **61**:172–185.
58. Young, R. W., 1978, *Invest. Ophthalmol. Vis. Sci.* **17**:105–116.
59. O'Day, W. T., and Young, R. W., 1978, *J. Cell. Biol.* **76**:593–604.
60. Hollyfield, J. G., 1982, *The Structure of the Eye* (J. G. Hollyfield, ed.), Elsevier/North-Holland, Amsterdam, pp. 67–73.
61. Anderson, D. H., Fisher, S. K., Erickson, P. A., and Tabor, G. A., 1980, *Exp. Eye Res.* **30**:559–574.
62. Fisher, S. K., Pfeffer, B. A., and Anderson, D. H., 1983, *Invest. Ophthalmol. Vis. Sci.* **24**:844–856.
63. Bridges, C. D. B., Alvarez, A., and Fong, S. L., 1983, *Invest. Ophthalmol. Vis. Sci.* **24**(Suppl.):141.
64. Bridges, C. D. B., 1984, *The Retinoids* (M. B. Sporn, A. B. Roberts, D. S. Goodman, eds.), Academic Press, New York (in press).
65. Campbell, D. A., Harrison, R., and Tonks, E. L., 1964 *Exp. Eye. Res.* **3**:412–426.
66. Mehra, K. S., and Khare, B. A., 1965, *Oriental Arch. Ophthalmol.* **3**:80–81.
67. Krachmer, J. H., Smith, J. L., and Tocci, P. M., 1966, *Arch. Ophthalmol.* **75**:661–664.
68. Gouras, P., and Chader, G., 1974, *Invest. Ophthalmol.* **13**:239–242.
69. Futterman, S., Swanson, D., and Kalina, R. E., 1974, *Invest. Ophthalmol.* **13**:798–801.
70. Maraini, G., 1974, *Invest. Ophthalmol.* **13**:288–290.
71. Campbell, D. A., and Tonks, E. L., 1962, *Br. J. Ophthalmol.* **46**:151–164.
72. Berson, E. L., Schmidt, S. Y., and Rabin, A. R., 1976, *Br. J. Ophthalmol.* **60**:142–147.
73. Arshinoff, S. A., McCulloch, J. C., Matuk, Y., Phillips, M. J., Gordon, B. A., and Marliss, E. B., 1979, *Metabolism* **28**:979–988.
74. Cotlier, E., and Berson, E. L., 1978, *Metab. Ophthalmol.* **2**:251.
75. Berson, E. L., *Annual Review of the Berman Gund Laboratory for the Study of Retinal Degenerations*, 1976–1977, Harvard Medical School, p. 49.
76. Brinkman, C. J. J., Pinckers, A. J. L. G., and Broekhuyse, R. M., 1980, *Invest. Ophthalmol. Vis. Sci.* **19**:743–750.
77. Rahi, A. H. S., 1973, *Br. J. Ophthalmol.* **57**:904–909.
78. Spiro, R., Weleber, R., and Kimberling, W., 1978, *Clin. Genet.* **13**:295–304.
79. Gahlot, D. K., Khosla, P. K., Makashir, P. D., Vasuki, K., and Basu, N., 1976, *Br. J. Ophthalmol.* **60**:770–774.
80. Bastek, J., Bogden, J., Cinotti, A., TenHove, W., Stephens, G., Markopoulos, M., and Charles, J., 1977, *Retinitis Pigmentosa: Clinical Implications of Current Research* (M. B. Landers III, M. E. Wolbarsht, D. E. Dowling, and A. M. Laties, eds.), Plenum Press, New York, pp. 43–50.
81. Dawson, G., and Newell, F. W., 1974, *Lancet* **1**:1119.
82. Bach, G., Eisenberg, F., Cantz, M., and Neufeld, E., 1973, *Proc. Natl. Acad. Sci. U.S.A.* **70**:2134–2138.

83. Kresse, H., 1973, *Biochem. Biophys. Res. Commun.* **54:**1111–1118.
84. Von Figura, K., Logering, M., Mersmann, G., and Kresse, H., 1973, *J. Pediatr.* **83:**607–611.
85. Bach, G., Friedman, R., Weissman, B., and Neufeld, E., 1972, *Proc. Natl. Acad. Sci. U.S.A.* **69:**2048–2051.
86. Brady, R. O., 1978, *Trans. Am. Acad. Ophthalmol. Otolaryngol.* **85:**1007–1013.
87. Barbeau, A., 1978, *Taurine and Neurological Disorder* (A. Barbeau, and R. J. Huxtable, eds.), Raven Press, New York, pp. 429–440.
88. Lemieux, B., Barbeau, A., Beroniade, V., Shapcott, D., Breton, G., Geoffroy, G., and Melancon, S., 1976, *Can. J. Neurol. Sci.* **3:**373–378.
89. Allen, R. J., DiMauro, S., Coulter, D. L., Papadimitriou A., and Rothenberg S. P., 1983, *Ann. Neurol.* **13:**679–682.
90. Eagle, R. C., Hedges, T. R., and Yanoff, M., 1982, *Ophthalmology* **89:**1433–1440.
91. Bassen, F. A., and Kornzweig, A. L., 1950, *Blood* **5:**381–387.
92. Salt, H. B., Wolff, O. N., Lloyd, J. K., Fosbrooke, A. S., Camaron, A. H., and Hubble, D. V., 1960, *Lancet* **2:**325–329.
93. Kayden, H. J., 1972, *Annu. Rev. Med.* **23:**285–296.
94. Carr, R. E., 1976, *Birth Defects* **12:**385–399.
95. Gouras, P., Carr, R. E., and Gunkel, R. D., 1971, *Invest. Ophthalmol. Vis. Sci.* **10:**784–793.
96. von Sallmann, L., Gelderman, A. H., and Laster, L., 1969, *Doc. Ophthalmol.* **26:**451–460.
97. Wolff, O. H., Lloyd, J. K., and Tonks, E. L., 1964, *Exp. Eye Res.* **3:**439–442.
98. Sperling, M. A., Hiles, D. A., and Kennerdell, J. S., 1972, *Am. J. Ophthalmol.* **74:**342–351.
99. Muller, D. P. R., Lloyd, J. K., and Bird, A. C., 1977, *Arch. Dis. Child.* **52:**209–214.
100. Bishara, S., Merin, S., Cooper, M., Azizi, E., Delpre, G., and Deckelbaum, R. J., 1982, *Br. J. Ophthalmol.* **66:**767–770.
101. White, A., Handler, P., and Smith, E. L., 1973, *Principles of Biochemistry*, Fifth edition, McGraw-Hill, New York.
102. Klenk, E., and Kahlke, W., 1963, *Hoppe Seylers Z. Physiol. Chem.* **333:**133–139.
103. Avigan, J., Steinberg, D., Gutman, A., Mize, C. E., and Milne, G. W. A., 1966, *Biochem. Biophys. Res. Commun.* **24:**838–844.
104. Eldjarn, L., Stokke, O., and Try, K., 1966, *Scand. J. Clin. Lab Invest.* **18:**694–695.
105. Refsum, S., 1981, *Arch. Neurol.* **38:**605–606.
106. Toussaint, D., and Danis, P., 1971, *Am. J. Ophthalmol.* **72:**342–347.
107. Fuchs, E., 1896, *Arch. Augenheilk.* **32:**111–116.
108. Takki, K., and Simell, O., 1974, *Br. J. Ophthalmol.* **58:**907–916.
109. McCullough, C., and Marliss, E. B., 1975, *Am. J. Ophthalmol.* **80:**1045–1057.
110. Kaiser-Kupfer, M. I., Kuwabara, T., Askanes, V., Brody, L., Takki, K., Dvoretzky, I., and Engel, W. K., 1981, *Ophthalmology* **88:**302–306.
111. Simell, O., and Takki, K., 1973, *Lancet* **1:**1030–1033.
112. Berson, E. L., Schmidt, S. Y., and Shih, V. E., 1978, *Ophthalmology* **85:**1018–1027.
113. Kennaway, N. G., Weleber, R. G., and Buist, N. R. M., 1977, *N. Engl. J. Med.* **297:**1180.
114. Shih, V. E., Berson, E. L., Mandell, R., and Schmidt, S. Y., 1978, *Am. J. Hum. Genet.* **30:**174–179.
115. Trijbels, J. B. F., Sengers, R. C. A., and Bakkeren, J. A. M. M., Dekort, A. F. M., and Deutman, A. F., 1977, *Clin. Chim. Acta* **79:**371–377.
116. Valle, D. L., Kaiser-Kupfer, M., and Del Valle, L. A., 1977, *Proc. Natl. Acad. Sci. U.S.A.* **74:**5159–5161.
117. Sipila, I., 1980, *Biochim. Biophys. Acta* **613:**79–84.
118. Kaiser-Kupfer, M. I., Valle, D., and Bron, A. J., 1980, *Am. J. Ophthalmol.* **89:**219–222.
119. Weleber, R. G., Kennaway, N. G., and Buist, N. R. M., 1978, *Lancet* **2:**1213.
120. Valle, D., Walser, M., Brusilow, S. W., and Kaiser-Kupfer, M., 1980, *J. Clin. Invest* **65:**371–378.
121. Stoppolini, G., Prisco, F., Santinelli, R., and Tolone, C., 1978, *Helv. Paediatr. Acta* **33:**429–433.
122. Berson, E. L., Shih, V. E., and Sullivan, P. L., 1981, *Ophthalmology* **88:**311–315.
123. Kaiser-Kupfer, M. I., deMonasterio, F. M., Valle, D., Walser, M., and Brusilow, S., 1981, *Ophthalmology* **88:**307–310.

124. Kuwabara, T., Ishikawa, Y., and Kaiser-Kupfer, M. I., 1981, *Ophthalmology* **88**:331–334.
125. Hayasaka, S., Shiono, T., Mizuno, K., Saito, T., and Tada, K., 1980, *Invest. Ophthalmol. Vis. Sci.* **19**(Suppl.):185.
126. Lolley, R. N., 1982, *Problems of Normal and Genetically Abnormal Retinas* (R. M. Clayton, J. Haywood, H. W. Reading, and A. Wright, eds.), Academic Press, New York, pp. 189–195.
127. Herron, W. L., Riegel, B. W., Myers, O. E., and Rubin, M. L., 1969, *Invest. Ophthalmol.* **8**:595–604.
128. Bok, D., and Hall, M. O., 1971, *J. Cell. Biol.* **49**:664–682.
129. Schmidt, S. Y., and Berson, E. L., 1978, *Exp. Eye Res.* **27**:191–198.
130. LaVail, M. M., Sidman, R. L., and Gerhardt, C. O., 1975, *J. Hered.* **66**:242–244.
131. Dowling, J. E., and Sidman, R. L., 1962, *J. Cell. Biol.* **14**:73–109.
132. LaVail, M. M., Sidman, R. L., and O'Neil, D., 1972, *J. Cell. Biol.* **53**:185–209.
133. LaVail, M. M., and Battelle, B. A., 1975, *Exp. Eye Res.* **21**:167–192.
134. LaVail, M. M., 1981, *Invest. Ophthalmol. Vis. Sci.* **21**:638–657.
135. LaVail, M. M., Pinto, L. H., and Yasumura, D., 1981, *Invest. Ophthalmol. Vis. Sci.* **21**:658–668.
136. Mullen, R. J., and LaVail, M. M., 1976, *Science* **192**:799–801.
137. Edwards, R. B., and Szamier, R. B., 1977, *Science* **197**:1001–1003.
138. Tamai, M., and O'Brien, P. J., 1979, *Exp. Eye Res.* **28**:399–411.
139. Effron, J. T., Szamier, R. B., and Edwards, R. B., 1981, *Invest. Ophthalmol. Vis. Sci.* **21**:611–616.
140. Chaitin, M. H., and Hall, M. O., 1983, *Invest. Ophthalmol. Vis. Sci.* **24**:812–820.
141. Berman, E. R., Merin, S., Segal, N., and Photiou, S., 1982, *Birth Defects* **18**:135–148.
142. Adler, A., and Severin, K. M., 1981, *Exp. Eye Res.* **32**:755–769.
143. Lolley, R. N., and Farber, D. B., 1975, *Exp. Eye Res.* **20**:585–589.
144. Farber, D. B., and Lolley, R. N., 1977, *J. Neurochem.* **28**:1089–1095.
145. Keeler, C. E., 1970, *Arch. Ophthalmol.* **84**:499–504.
146. Karli, P., 1963, *Prog. Ophthalmol.* **14**:51–89.
147. Caley, D. W., Johnson, C., and Liebelt, R. A., 1972, *Am. J. Anat.* **133**:179–212.
148. Sanyal, S., and Bal, A. K., 1973, *Z. Anat. Entwickl. Gesch.* **142**:219–238.
149. Carter-Dawson, L., LaVail, M. M., and Sidman, R. L., 1978, *Invest. Ophthalmol. Vis. Sci.* **17**:489–498.
150. LaVail, M. M., and Mullen, R. J., 1976, *Exp. Eye Res.* **23**:227–245.
151. Noell, W. K., 1958, *Arch. Ophthalmol.* **60**:702–733.
152. Schmidt, S. Y., and Lolley, R. N., 1973, *J. Cell Biol.* **57**:117–123.
153. Schmidt, S. Y., 1978, *Metab. Ophthalmol.* **2**:247–250.
154. Farber, D. B., and Lolley, R. N., 1974, *Science* **186**:449–451.
155. Farber, D. B., and Lolley, R. N., 1976, *J. Cyclic. Nucleotide Res.* **2**:139–148.
156. Bitensky, M. W., Wheeler, G. L., Aloni, B., Vetury, S., and Matuo, Y., 1978, *Adv. Cyclic Nucleotide Res.* **9**:553–672.
157. Yee, R., and Liebman, P. A., 1978, *J. Biol. Chem.* **253**:8902–8909.
158. Baehr, W., Devlin, M. J., and Applebury, M. L., 1979, *J. Biol. Chem.* **254**:11669–11677.
159. Stein, P. J., Rasenick, M. M., and Bitensky, M. W., 1982, *Progress in Retinal Research*, Volume 1 (N. Osborne, and G. Chader, eds.), Pergamon Press, New York, pp. 227–243.
160. Kohnken, R. E., Eadie, D. M., Revzin, A., and McConnell, D. G., 1981, *J. Biol. Chem.* **256**:12502–12509.
161. Kilbride, P., and Ebrey, T. G., 1979, *J. Gen. Physiol.* **74**:415–426.
162. Polans, A. S., Kawamura, S., and Bownds, D., 1981, *J. Gen. Physiol.* **77**:41–48.
163. Woodruff, M. L., and Bownds, M. D., 1979, *J. Gen. Physiol.* **73**:629–653.
164. Fletcher, R. T., and Chader, G. J., 1976, *Biochem. Biophys. Res. Commun.* **70**:1297–1302.
165. Orr, H. T., Lowry, O. H., Cohen, A. I., and Ferrendelli, J. A., 1976, *Proc. Natl. Acad. Sci. U.S.A.* **73**(12):4442–4445.
166. deAzeredo, F. A. M., Lust, W. D., and Passonneau, J. V., 1978, *Biochem. Biophys. Res. Commun.* **85**:293–300.
167. DeVries, G. W., Cohen, A. I., Hall, I. A., and Ferendelli, J. A., 1978, *J. Neurochem.* **31**:1345–1351.

168. Hurley, J. B., and Stryer, L., 1982, *J. Biol. Chem.* **257**:11094–11099.
169. Kuhn, H., 1980, *Nature* **283**:587–588.
170. Somers, R. L., and Shichi, H., 1979, *Biochem. Biophys. Res.* Commun. **89**:479–485.
171. Robinson, W. E., and Hagins, W. A., 1979, *Nature* **280**:398–400.
172. Godchaux, W. III, and Zimmerman, W. F., 1979, *J. Biol. Chem.* **254**:7874–7884.
173. Kuhn, H., and Hargrave, P. A., 1981, *Biochemistry* **20**:2410–2417.
174. Hurley, J. B., 1980, *Biochem. Biophys. Res. Commun.* **92**:505–510.
175. Lee, R. H., Lieberman, B. S., and Lolley, R. N., 1983, *Invest. Ophthalmol. Vis. Sci.* **24**(Suppl.):208.
176. Ferrendelli, J. A., and Cohen, A. I., 1976, *Biochem. Biophys. Res. Commun.* **73**:421–427.
177. Lolley, R. N., Farber, D. B., Rayborn, M. E., and Hollyfield, J. G., 1977, *Science* **196**:664–666.
178. Ulshafer, R. J., Garcia, C. A., and Hollyfield, J. G., 1980, *Invest. Ophthalmol. Vis. Sci.* **19**:1236–1241.
179. Sanyal, S., De Ruiter, A., and Hawkins, R. K., 1980, *J. Comp. Neurol.* **194**:193–207.
180. Cohen, A. I., 1983, *Invest. Ophthal. Vis. Sci.* **24**:832–843.
181. Caravaggio, L. L., and Bonting, S. L., 1963, *Exp. Eye Res.* **2**:12–19.
182. Sidman, R. L., and Green, M. C., 1965, *J. Hered.* **56**:23–29.
183. Sanyal, S., and Hawkins, R. K., 1981, *Exp. Eye Res.* **33**:213–222.
184. Aguirre, G., Farber, D., Lolley, R., O'Brien, P., Alligood, J., Fletcher, T., and Chader, G., 1982, *Exp. Eye Res.* **35**:625–642.
185. Woodford, B. J., Liu, Y., Fletcher, R. T., Chader, G. J., Farber, D. B., Santos-Anderson, R., and Tso, M. O. M., 1982, *Exp. Eye Res.* **34**:703–714.
186. Aguirre, G., O'Brien, P., Marshall, J., and Buyukmihci, N., 1982, *Problems of Normal and Genetically Abnormal Retinas* (R. M. Clayton, J. Haywood, H. W. Reading, and A. Wright, eds.), Academic Press, New York, pp. 199–200.
187. Buyukmihci, N., Aguirre, G., and Marshall, J., 1980, *Exp. Eye Res.* **30**:575–591.
188. Chader, G., Liu, Y., O'Brien, P., Fletcher, T., Krishna, G., Farber, D., and Lolley, R. N., 1980, *Neurochem. Int.* **1**:441–458.
189. Acland, G., Aguirre, G., Chader, G., Fletcher, R. T., and Farber, D., 1980, *Invest. Ophthalmol. Vis. Sci.* **19**(Suppl.):250.
190. Aquirre, G., Alligood, J., O'Brien, P., and Buyukmichi, N., 1982, *Invest. Ophthalmol. Vis. Sci.* **23**:610–630.
191. Aguirre, G., and Rubin, L. F., 1972, *J. Am. Vet. Assoc.* **160**:190–201.
192. LaVail, M. M., 1981, *Invest. Ophthalmol. Vis. Sci.* **21**:638–657.
193. LaVail, M. M., Blanks, J. C., and Mullen, R. J., 1982, *J. Comp. Neurol.* **212**:217–230.
194. Blanks, J. C., Mullen, J., and LaVail, M. M., 1982, *J. Comp. Neurol.* **212**:231–246.
195. Bridges, C. D. B., Alvarez, R. A., and Fong, S. L., 1982, *Invest. Ophthalmol. Vis. Sci.* **22**:706–714.
196. Akhtar, M., Blosse, P. T., and Dewhurst, P. B., 1967, *Chem. Commun.* pp. 631–632.
197. Kanai, M., Raz, A., and Goodman, D. S., 1968, *J. Clin. Invest.* **47**:2025–2044.
198. Smith, J. E., and Goodman, D. W., 1979, *Fed. Proc.* **38**:2504–2509.
199. Maraini, G., and Gozzoli, F., 1975, *Invest. Ophthalmol.* **14**:785–787.
200. Bok, D., and Heller, J., 1976, *Exp. Eye Res.* **22**:395–402.
201. Saari, J., Futterman, S., and Bredberg, L., 1978, *J. Biol. Chem.* **253**:6432–6436.
202. Chader, G., 1982, *Cell Biology of the Eye* (D. S. McDevitt, ed.), Academic Press, New York, pp. 377–433.
203. Bridges, C. D. B., 1976, *Exp. Eye Res.* **22**:435–455.
204. Bownds, D., 1967, *Nature* **216**:1178–1181.
205. Wang, J. K., McDowell, J. H., and Hargrave, P. A., 1980, *Biochemistry* **19**:5111–5117.
206. Akhtar, M., Blosse, P. T., and Dewhurst, P. B., 1968, *Biochem. J.* **110**:693–702.
207. Heller, J., 1968, *Biochemistry* **7**:2914–2920.
208. Hubbard, R., Bownds, D., and Yoshizawa, T., 1965, *Cold Spring Harbor Symp. Quant. Biol.* **30**:301–315.
209. Wald, G., 1968, *Science* **162**:230–239.
210. O'Brien, D. F., 1982, *Science* **218**:961–966.

211. Bridges, C. D. B., and Alvarez, R. A., 1982, *Retina* **2**:256–260.
212. Baumann, C., 1972, *J. Physiol.* (*Lond.*) **222**:643–663.
213. Abrahamson, E. W., and Wiesenfeld, J. R., 1972, *Handbook of Sensory Physiology*, Volume VII, Part 1 (H. J. A. Dartnall, ed.), Springer-Verlag, New York, pp. 69–121.
214. Morton, R. A., 1972, *Handbook of Sensory Physiology*, Volume VII, Part 1 (H. J. A. Dartnall, ed.), Springer-Verlag, New York, pp. 33–68.
215. Brown, K. T., 1968, *Vision Res.* **8**:633–677.
216. Sillman, A. J., Ito, H., and Tomita, T., 1969, *Vision Res.* **9**:1435–1442.
217. Toyoda, J. I., Nosaki, H., and Tomita, T., 1969, *Vision Res.* **9**:453–463.
218. Hagins, W. A., 1972, *Annu. Rev. Biophys. Bioeng.* **1**:131–158.
219. Yoshikami, S., and Hagins, W. A., 1973, *Biochemistry and Physiology of Visual Pigments*, (H. Langer, ed.), Springer-Verlag, New York, pp. 245–255.
220. Dowling, J. E., and Wald, G., 1960, *Proc. Natl. Acad. Sci. U.S.A.* **46**:587–608.
221. Dowling, J. E., and Gibbons, I. R., 1961, *The Structure of the Eye*, (G. Smelser, ed.), Adacemic Press, New York, pp. 85–99).
222. Robison, W. G., Kuwabara, T., and Bieri, J. G., 1980, *Invest. Ophthalmol. Vis. Sci.* **19**:1030–1037.
223. Carter-Dawson, L., Kuwabara, T., O'Brien, P. J., and Bieri, J. G., 1979, *Invest. Ophthalmol. Vis. Sci.* **18**:437–446.
224. Noell, W. K., DelMelle, M. C., and Albrecht, R., 1971, *Science* **172**:72–76.
225. Noell, W. K., and Albrecht, R., 1971, *Science* **172**;76–80.
226. Berson, E. L., 1973, *Invest. Ophthalmol.* **12**:35–44.
227. Bergsma, D. R., and Wolf, M. L., 1976, *Retinitis Pigmentosa* (M. A. Landers, M. L. Wolbarsht, A. M. Laties, and J. E. Dowling, eds.), Plenum Press, New York, pp. 197–209.
228. Chatzinoff, A., Nelson, E. Stahl, N., and Clahane, A., 1968, *Arch. Ophthalmol.* **80**:417–419.
229. Sandberg, M. A., Rosen, J. B., and Berson, E. L., 1977, *Am. J. Ophthalmol.* **84**:658–665.
230. Dilley, R. A., and McConnell, D. G., 1970, *J. Membr. Biol.* **2**:317–323.
231. Farnsworth, C. C., and Dratz, E. A., 1976, *Biochim. Biophys. Acta* **443**:556–570.
232. Robinson, W. G., Kuwabara, T., and Bieri, J. G., 1982, *Retina* **2**:263–281.
233. Kagan, V. E., Shvedova, A. A., Novikov, K. N., and Kozlow, Y. P., 1973, *Biochim. Biophys. Acta* **330**:76–79.
234. Wormington, C. N., and Cone, R. A., 1978, *J. Gen. Physiol.* **71**:657–681.
235. Pasantes-Morales, H., and Ademe, R. M., 1981, *Invest. Ophthalmol. Vis. Sci.* **20**(Suppl.):6.
236. Pasantes-Morales, H., and Cruz, C., 1983, *Abstracts, 13th Annual Meeting, Society for Neuroscience*, Society for Neuroscience, Bethesda, p. 807.
237. McConnell, D. G., 1975, *J. Biol. Chem.* **250**:1898–1906.
238. Poincelot, R. P., and Abrahamson, E. W., 1970, *Biochim, Biophys. Acta* **202**:382–385.
239. Anderson, R. E., Benolken, R. M., Dudley, P. A., Landis, D. J., and Wheeler, T. G., 1974, *Exp. Eye Res.* **18**:204–215.
240. Bazan, N. G., DeEscalante, M. S., Careaga, M. M., Bazan, H. E. P., and Giusto, N. M., 1982, *Biochim. Biophys. Acta.* **712**:702–706.
241. Sickel, W., 1972, *Handbook of Sensory Physiology* Volume VII, Part 2 (M. G. F. Fuortes, ed.), Springer-Verlag, New York, pp. 667–727.
242. Feke, G. T., Zuckerman, R., Green, G. J., and Weiter, J. J., 1983, *Invest. Ophthalmol. Vis. Sci.* **24**:136–141.
243. Delmelle, M., 1978, *Photochem. Photobiol.* **27**:731–734.
244. Futterman, S., Hendrickson, A., Bishop, P. E., Rollins, M. H., and Vacano, E., 1970, *J. Neurochem.* **17**:149–156.
245. Graymore, C., Power, J., and Kissun, R. D., 1974, *Exp. Eye Res.* **19**:163–166.
246. McConnell, D. G., Ozga, G. W., and Solze, D. A., 1969, *Biochim. Biophys. Acta* **184**:11–28.
247. Futterman, S. and Saari, J. C., 1981, *Adler's Physiology of the Eye* (R. A. Moses, ed.), C. V. Mosby, St. Louis, pp. 411–426.
248. Shichi, H., and Rafferty, C. N., 1980, *Photochem. Photobiol.* **31**:631–639.
249. Organisciak, D. T., Favreau, P., and Wang, H. M., 1983, *Exp. Eye Res.* **36**:337–349.
250. McCay, P. B., and King, M. M., 1980, *Vitamin E—A Comprehensive Treatise* (L. J. Machlin, ed.), Marcel Dekker, New York, pp. 289–317.

251. Draper, H. H., 1980, *Vitamin E—A Comprehensive Treatise* (L. J. Machlin, ed.), Marcel Dekker, New York, pp. 272–288.
252. Machlin, L. J., 1980, *Vitamin E—A Comprehensive Treatise* (L. J. Machlin, ed.), Marcel Dekker, New York, pp. 637–645.
253. Lubin, B., and Machlin, L. J., (eds.), 1982, *Vitamin E: Biochemical, Hematological and Clinical Aspects, Annals of the New York Academy of Sciences* (Volume 393, New York Academy of Sciences, New York.
254. Hall, M. O., and Hall, D. O., 1975, *Biochem. Biophys. Res. Commun.* **67**:1199–1204.
255. Crouch, R., Priest, D. G., and Duke, E. J., 1978, *Exp. Eye Res.* **27**:503–509.
256. Armstrong, D., Santangelo, G., and Connole, E., 1981, *Curr. Eye Res.* **1**:225–242.
257. Leuenberger, P. N., and Novikoff, A. B., 1975, *J. Cell Biol.* **65**:324–334.
258. Stone, W. L., and Dratz, E. A., 1982, *Exp. Eye Res.* **35**:405–412.
259. Reim, M., Heuvels, and Cattepoel, H., 1974, *Ophthalmol. Res.* **6**:228–234.
260. Robison, W. G., and Kuwabara, T., 1975, *Invest. Ophthalmol.* **14**:866–872.
261. Winkler, B. S., and Giblin, F. J., 1983, *Exp. Eye Res.* **36**:287–297.
262. Feeney-Burns, L., Berman, E. R., and Rothman, H., 1980, *Am. J. Ophthalmol.* **90**:783–791.
263. Robison, W. G., Jr., Kuwabara, T., and Bieri, J. G., 1980, *Invest. Ophthalmol. Vis. Sci.* **19**:1030–1037.
264. Amemiya, T., 1981, *Int. J. Vitam. Nutr. Res.* **51**:114–118.
265. Stone, W. L., Katz, M. L., Lurie, M., Marmor, M. F., and Dratz, E. A., 1979, *Photochem. Photobiol.* **29**:725–730.
266. Katz, M. L., Parker, K. R., Handelman, G. J., Bramel, T. L., and Dratz, E. A., 1982, *Exp. Eye Res.* **34**:339–369.
267. Wolfe, L. S., Ng Ying Kin, N. M. K., Baker, R. R., Carpenter, S., and Andermann, F., 1977, *Science* **195**:1360–1362.
268. Kubicek, R., and Dolenek, A., 1958, *J. Chromatogr.* **1**:266–268.
269. Pasantes-Morales, H., Klethi, J., Ledig, M., and Mandel, P., 1972, *Brain Res.* **41**:494–497.
270. Schmidt, S. Y., and Berson, E. L., *Taurine in Neurological Disorders* (R. Huxtable, and A. Barbeau, eds.), Raven Press, New York, pp. 281–287.
271. Cohen, A. I., McDaniel, M., and Orr, H., 1973, *Invest. Ophthalmol.* **12**:686–693.
272. Orr, H. T., Cohen, A. I., and Lowry, O. H., 1976, *J. Neurochem.* **26**:609–611.
273. Voaden, M. J., Lake, N., Marshall, N., and Morjaria, B., 1977, *Exp. Eye Res.* **25**:249–257.
274. Schmidt, S. Y., 1980, *Proceedings of 21st Annual A. N. Richards Symposium: Actions of Taurine in Excitable Tissues* (S. I. Baskin, S. W. Schaffer, and J. J. Kocsis, eds.), Spectrum, New York, pp. 177–185.
275. Kennedy, A. J., Neal, M. J., and Lolley, R. N., 1977, *J. Neurochem.* **29**:157–159.
276. Hayes, K. C., Carey, R. E., and Schmidt, S. Y., 1975, *Science* **188**:949–951.
277. Berson, E. L., Hayes, K. C., Rabin, A. R., Schmidt, S. Y., and Watson, G., 1976, *Invest. Ophthalmol.* **15**:52–58.
278. Schmidt, S. Y., Berson, E. L., and Hayes, K. C., 1976, *Invest. Ophthalmol.* **15**:47–52.
279. Schmidt, S. Y., Berson, E. L., Watson, G., and Huang, C., 1977, *Invest. Ophthalmol. Vis. Sci.* **16**:673–678.
280. Sturman, J. A., and Hayes, K. C., 1980, *Advances in Nutritional Research*, Volume 3 (H. H. Draper, ed.), Plenum Press, New York, pp. 231–299.
281. Wen, G. Y., Sturman, J. A., Wisniewski, H. M., Lidsky, A. A., Cornwell, A. C., and Hayes, K. C., 1979, *Invest. Ophthalmol. Vis. Sci.* **18**:1200–1206.
282. Miller, S. S., and Steinberg, R. H., 1979, *J. Gen. Physiol.* **74**:237–259.
283. Schmidt, S. Y., and Szamier, R. B., 1978, *Invest. Ophthalmol. Vis. Sci.* **17**(Suppl.):255.
284. Edwards, R. B., 1977, *Invest. Ophthalmol.* **16**:201–208.
285. Schmidt, S. Y., 1980, *Exp. Eye Res.* **31**:373–379.
286. Jacobsen, J. G., and Smith, L. H., 1968, *Physiol. Rev.* **48**:424–511.
287. Huxtable, R. J., and Lippincott, S. E., 1981, *Arch. Biochem. Biophys.* **210**:698–709.
288. Lake, N., 1982, *Retina* **2**:261–262.
289. Pasantes-Morales, H., Quesada, O., Cazabez, A., and Huxtable, R. J., 1983, *J. Neurosci. Res.* **9**:135–143.
290. Schmidt, S. Y., 1978, *Exp. Eye Res.* **26**:529–535.

291. Lopez-Colome, A. M., and Pasantes-Morales, H., 1981, *Exp. Eye Res.* **32**:771–780.
292. Pasantes-Morales, H., Ademe, R. M., and Quesada, O., 1981, *J. Neurosci. Res.* **6**:337–348.
293. O'Steen, W. K., and Anderson, K. V., 1971, *Exp. Neurol.* **30**:525-534.
294. Gorn, R. A., and Kuwabara, T., 1967, *Arch. Ophthalmol.* **77**:119–123.
295. Friedman, E., and Tso, M. O. M., 1968, *Arch. Ophthalmol.* **79**:315–320.
296. Noell, W. K., Walker, V. S., Kang, B. S., and Berman, S., 1966, *Invest. Ophthalmol.* **5**:450–473.
297. LaVail, M. M., 1980, *The Effects of Constant Light on Visual Processes*, (T. P. Williams, and B. N. Baker, eds.), Plenum Press, New York, pp. 357–387.
298. Dobson, V., 1976, *Invest. Ophthalmol.* **15**:595–598.
299. Robison, W. G., and Kuwabara, T., 1976, *Exp. Eye Res.* **22**:549–557.
300. Ham, W. T., Ruffolo, J. J., Mueller, H. A., and Guerry, D., 1980, *Vis. Res.* **20**:1105–1111.
301. Ham, W. T., Ruffolo, J. J., Mueller, H. A., Clarke, A. M., and Moon, M. E., 1978, *Invest. Ophthalmol Vis. Sci.* **17**:1029–1035.
302. Ham, W. T., and Mueller, H. A., 1976, *Nature* **260**:153–155.
303. Lawwill, T., Crockett, R. S., and Currier, G., 1977, *Doc. Ophthalmol.* **44**:379–402.
304. Lawwill, T., Crockett, R. S., Currier, G., and Rosenberg, R. B., 1980, *Vis. Res.* **20**:1113–1115.
305. Tso, M. O. M., Wallow, I. H. L., Powell, J. O., and Zimmerman, L. E., 1972, *Trans. Am. Acad. Ophthalmol. Otol.* **76**:1247–1262.
306. Tso, M. O. M., Wallow, I. H. L., and Powell, J. O., 1973, *Arch. Ophthalmol.* **89**:228–234.
307. Tso, M. O. M., 1973, *Invest. Ophthalmol.* **12**:17–34.
308. Sykes, S. M., Robinson, W. R., Waxler, M., and Kuwabara, T., 1981, *Invest. Ophthalmol, Vis. Sci.* **20**:425–434.
309. Lawwill, T., 1973, *Invest. Ophthalmol. Vis. Sci.* **12**:45–51.
310. McKechnie, N. M., Johnson, N. F., and Foulds, W. S., 1982, *Invest. Ophthalmol. Vis. Sci.* **22**:449–459.
311. Haverth, R. S., and Sperling, H. G., 1971, *Science* **174**:520–522.
312. Sperling, H. G., Johnson, C., and Harverth, R. S., 1980, *Vis. Res.* **20**:1117–1125.
313. Crockett, R. S., and Lawwill, T., 1984, *Curr. Eye Res.* **3**:209–215.
314. Zigman, S., and Vaughn, T., 1974, *Invest. Ophthalmol.* **13**:462–465.
315. Zigman, S., Groff, J., Yulo, T., and Vaughn, T., 1975, *Invest. Ophthalmol.* **14**:710–713.
316. Marshall, J., Hamilton, A. M., and Bird, A. C., 1975, *Br. J. Ophthalmol.* **59**:610–630.
317. Marshall, J., Grindle, J., Ansell, P., and Borwein, B., 1979, *Br. J. Ophthalmol.* **63**:181–187.
318. Howell, W. L., Rapp, L. M., and Williams, T. P., 1982, *Invest. Ophthalmol. Vis. Sci.* **22**:139–144.
319. Wiegand, R. D., Giusto, N. M., Rapp, L. M., and Anderson, R. E., 1983, *Invest. Ophthalmol. Vis. Sci.* **24**:1433–1435.
320. Joel, C., Briggs, S., Gaal, D., Hannan, J., Kahlow, M., Stein, M., Tarver, A., and Yip, A., 1981, *Invest. Ophthalmol. Vis. Sci.* **20**(Suppl.):166.
321. Joel, C. D., Stein, M. A., Thome, J. A., Tarver, A. P., Sievert, T. J., and Hannan, J. E., 1983, *J. Neurochem.* **41**(Suppl.):134.
322. Kagan, V. E., Kuliev, I. Y., Spirichev, V. B., Shvedova, A. A., and Kozlov, Y. P., 1981, *Bull. Exp. Biol. Med.* **91**:144–148.
323. Penner, R., and McNair, J. N., 1966, *Am. J. Ophthalmol.* **61**:1452–1457.
324. Hatfield, E. M., 1970, *Sight. Sav. Rev.* **40**:79–86.
325. Sliney, D. H., 1983, *Ophthalmology* **90**:937–944.
326. Sperling, H. G., and Ruiz, R. S., eds., 1980, *Proceedings of a Symposium on Intense Light Hazards in Ophthalmic Diagnosis and Treatment*, *Vis. Res.* **20**:1033–1203.
327. Berson, E. L., 1980, *Vis. Res.* **20**:1179–1184.
328. Hiramitsu, T., Majima, Y., Hasegawa, Y., and Hirata, K., 1974, *Acta. Soc. Ophthalmol. Jpn.* **78**:819–825.
329. Yagi, K., Matsuoka, S., Ohkawa, H., Ohishi, N., Takeuchi, Y. K., and Sakai, H., 1977, *Clin. Chim. Acta* **80**:355–360.
330. Hiramitsu, T., Hasegawa, Y., Hirata, K., Nishigatsi, I., and Yagi, K., 1975, *Experientia* **32**:622–623.

331. Armstrong, D., Hiramitsu, T., Gutteridge, J., and Nilsson, S. E., 1982, *Exp. Eye Res.* **35**:157–171.
332. Barber, A. N., Catsulis, C., and Cangelosi, R. J., 1971, *Br. J. Ophthalmol.* **55**:91–105.
333. Rapp, L. M., Wiegand, R. D., and Anderson, R. E., 1982, *Problems of Normal and Genetically Abnormal Retinas* (R. M. Clayton, J. Haywood, H. W. Reading, and A. Wright, eds.), Academic Press, London, pp. 109–119.
334. Kagan, V. E., Arkhipenko, Y. V., Belousova, L. V., Tyurin, V. A., Shvedova, A. A., Shukolyukov, S. A., and Kozlov, Y. P., 1981, *Vis. Res.* **21**:1029–1034.
335. Shvedova, A. A., Sidorov, A. S., Novikov, K. N., Galushchenko, I. V., and Kagan, V. E., 1979, *Vis. Res.* **19**:49–55.
336. Masciulli, L., Anderson, D. R., and Charles, S., 1972, *Am. J. Ophthalmol.* **74**:638–661.
337. Hiramitsu, T., Majima, Y., Hasegawa, Y., and Hirata, K., 1975, *Acta. Soc. Ophthalmol. Jpn.* **79**:1468–1473.
338. Crouch, R., 1980, *The Effects of Constant Light on Visual Processes* (T. Williams, and B. Baker, eds.), Plenum Press, New York, pp. 309–318.
339. Gan, E. V., Haberman, H. F., and Menon, I. A., 1976, *Arch. Biochem. Biophys.* **173**:666–672.
340. Proctor, P., McGinness, J., and Corry, P., 1974, *J. Theor. Biol.* **48**:19–41.
341. Noell, W. K., 1952, *J. Cell. Comp. Physiol.* **40**:25–28.
342. Noell, W. K., 1965, *Biochemistry of the Retina* (C. N. Graymore, ed.), Academic Press, London, pp. 51–72.
343. Farber, D. B., Souza, D. W., and Chase, D. G., 1983, *Invest. Ophthalmol. Vis. Sci* **24**:1236–1249.
344. Schmidt, S. Y., 1983, *J. Neurochem.* **40**:1630–1638.
345. Schmidt, S. Y., 1983, *J. Biol. Chem.* **258**:6863–6868.
346. Fliesler, S. J., Rayborn, M. E., and Hollyfield J. G., 1983, *Invest Ophthalmol. Vis. Sci.* **24**(Suppl.):279.
347. Cotlier, E., Fletcher, G., Mead, A., and Carella, R., 1983, *Invest. Ophthalmol. Vis. Sci.* **24**(Suppl.):293.
348. Masland, R. H., 1982, *Retina* **2**:282–287.
349. Pu, G. A., and Masland, R. H., 1981, *Invest. Ophthalmol. Vis. Sci.* **20**(Suppl.):40.
350. Flaherty, W. A., and Szamier, R. B., 1982, *Invest. Ophthalmol. Vis. Sci.* **22**(Suppl.):204.
351. Pu, G. A., and Anderson, R. E., 1983, *Invest. Ophthalmol. Vis. Sci.* **24**:288–293.
352. Marmor, M. F., 1982, *Aging and Human Visual Function* (K. Dismukes, D. Kline, and R. Sekular, eds.) Alan R. Liss, New York, pp. 59–78.
353. Balazs, E. A., and Denlinger, J. L., 1982, *Aging and Human Visual Function* (K. Dismukes, D. Kline, and R. Sekular, eds.), Alan R. Liss, New York, pp. 45–57.
354. Gartner, S., and Henkind, P., 1981, *Br. J. Ophthalmol.* **65**:23–28.
355. Sarks, S. H., 1976, *Br. J. Ophthalmol.* **60**:324–341.
356. Marshall, J., Grindle, J., Ansell, P., and Borwein, B., 1979, *Br. J. Ophthalmol.* **63**:181–187.
357. Friedman, E., and Tso, M. O. M., 1968, *Arch. Ophthalmol.* **79**:315–320.
358. Lerche, W., 1974, *Albrect v. Graefes Arch. Klin. Exp. Ophthalmol.* **189**:323–338.
359. Wing, L., Delori, F. C., Weiter, J. J., and Kunis, K., 1982, *Invest. Ophthalmol Vis. Sci.* **22**(Suppl.):173.
360. Peisch, R. D., and Schmidt, S. Y., 1979, *Invest. Ophthal. Vis. Sci.* **18**(Suppl.):20.
361. Hogan, M. J., 1972, *Trans. Am. Acad. Ophthalmol. Otol.* **76**:64–80.
362. McKechnie, N. M., Johnson, N. F., and Foulds, W. S., 1982, *Invest. Ophthalmol. Vis. Sci.* **22**:449–459.
363. Weale, R. A., 1982, *Aging and Human Visual Function*, Alan R. Liss, New York, pp. 161–171.
364. von Sallman, L., and Grimes, P., 1972, *Arch. Ophthalmol.* **88**:404–411.
365. von Sallman, L., and Grimes, P., 1974, *Invest. Ophthalmol. Vis. Sci.* **14**:1010–1015.
366. Baskin, S. I., Cohn, E. M., and Kocsis, J. J., 1976, *Taurine* (R. Huxtable, and A. Barbeau, eds.), Raven Press, New York, pp. 201–208.

19

Virus-Induced Changes in Neural Cells

Erik Lycke

1. INTRODUCTION

Virus infections of the CNS have attracted particular interest not only because of their medical importance but also because of the unique features that infections display within a compartmentalized and highly specialized organ system such as the CNS. To reach the CNS, the virus must overcome or circumvent the anatomical and physiological barriers of the CNS. Once inside the CNS, there seems to be no CNS intrinsic immune system activated but a recruitment of macrophages and an invasion of immune-competent and phagocytic cells. Immune reactions are probably crucial for clearance of virus from the infected CNS but they may also be involved in disease-associated reactions. Although most clinically manifested virus infections tend to become generalized, and virus infections of the CNS therefore may represent the result of a systemic spread of the infection, some viruses demonstrate a greater affinity for neural cells than others and exhibit neurotropic properties. The present chapter presents various aspects of virus–nerve cell interactions, but as information in particular at the molecular level is still very limited, it has been necessary for a more complete presentation to add some relevant observations on viral effects on nonneural cells.

2. NEUROTROPIC VIRUSES

The concept of tropisms, repeatedly challenged, has regained credibility with growing knowledge of viral pathogenesis. Neurotropism might be regarded as reflecting a summation of multifactorial influences of which the presence of receptors on neuronal cells for attachment of a virus appears to be one. Permissiveness of nerve cells to a virus infection and cytocidal effects of the in-

Erik Lycke • Department of Virology, Institute of Medical Microbiology, University of Göteborg, 413 46 Göteborg, Sweden.

fection, including those evoked by immune reactions against infected cells, are other factors decisive for the appearance of clinical manifestations. These might depend on degeneration and death of neurons or on metabolic and functional changes of infected but still vital neural cells. The variability in the outcome of the infection has generally been attributed to factors extrinsic to the nervous system, e.g., the dose of infecting virus, the virulence and other genetic properties of the virus, and the defense mechanisms of the host. However, the functional and metabolic state of the neural cells might be equally important in determining the course of infection. The medical implications of neural cell dysfunctions as consequences of latent or past virus infections have not been closely examined.

Neurotropism is difficult to define. To comply with conditions of neurotropism in very restricted terms, one may assume that the pathogenesis of infection should reveal the specific interactions of virus and neural cells that monitor the course of the infection. With this presumption it is doubtful that most viruses that reach the CNS by hematogenous spread and after infection of meningeal and ependymal cells might be referred to as neurotropic since the CNS involvement in many of these infections mainly reflects the generalization of the infection. Infection of the neuronal cells is preceded by infection of vascular endothelial cells and glia and by migration of infected white blood cells into the CNS. The CNS infection reveals that the immune defense has not been able to compete in the race against the infection. It would be preferable to consider those viruses that reach the CNS by neural routes and that are spread within the CNS by neuronal pathways as more strictly neurotropic. However, this kind of distinction is by no means unequivocal since it is obvious that, for example, transmission of infection within the CNS also might depend upon diffusion of virus into the intercellular space from degenerating infected neurons, or be mediated by the inflammatory response. Moreover, it is difficult to discriminate infections reaching the CNS by axonal transport from those that infect neurons via satellite cells and glial cells.

Most of the virus–cell interactions and virus-induced changes of metabolic and functional states of infected neurons described in the following refer to findings with three virus families, i.e., the herpes-, paramyxo-, and rhabdoviruses. Short descriptions of these viruses are presented below. Comprehensive reviews on pathogenesis of virus infections of the nervous system have recently appeared.[1–6]

2.1. Herpesviruses

Of the approximately 70 herpesviruses, 6 are pathogenic for man [herpes simplex virus (HSV) types 1 and 2, cytomegalovirus, varicella–zoster virus (VZV), Epstein–Barr virus (EBV), and herpes B virus]. Some herpesviruses display biological properties making them characteristic representatives of neurotropic viruses. The herpesviruses are ubiquitous, causing as a rule subclinical infections but may occasionally develop into life-threatening infections. Infections with herpesviruses regularly lead to establishment of latent reactivatable infections of particular target cells. Of these, not only sensory neurons of dorsal root ganglia but also other neuronal cells have been shown to harbor latent infections with HSV and VZV.

Most interactions of herpesviruses and neuronal cells have been studied experimentally as well as clinically with HSV. The virions of HSV have an icosahedral nucleocapsid made up of 162 capsomeres. The 100-nm nucleocapsid is surrounded by an envelope, giving the virion a total diameter of 150–200 nm. The genome is a linear double-stranded DNA molecule with a molecular weight of 10^8. Theoretically, the HSV genome might code for up to 100 different polypeptides. Of these, about 50 proteins have been identified and half of them are referred to as viral structural proteins. The 7 glycoproteins of the envelope have been studied in some detail. Glycoprotein B seems important for penetration and fusion processes between the virion and the plasma membrane. gC carries type-specific antigenic characteristics immunologically differentiating the two subtypes (HSV-1 and HSV-2) and is, like gD, probably involved in attachment reactions binding the virion to the plasma membrane of the cell. gE has the capacity to combine with the Fc fragment of IgG immune globulin (the Fc receptor). The occurrence of at least two other HSV-induced glycoproteins has been noted. The envelope glycoproteins are synthesized late during the infectious cycle and also appear to be inserted in the plasma membrane of the infected cell where they constitute major immunogenic proteins. The nucleocapsids are enveloped by a budding process through the inner layer of the nuclear membrane. This maturation process also occurs in neuronal cells and might be important for the mode of transport and release of virus synthesized.

Latent HSV infections have been demonstrated in neurons of peripheral ganglia (cranial, spinal, and autonomic nerves) and in the CNS. During the latent infection, viral replication is restricted. Presence of viral DNA in latently infected cells is demonstrable by nucleic acid hybridization, and transformation of the latent infection to a virus replicative phase has been demonstrated by reactivation of virus in cocultivation experiments *in vitro*, culturing latently infected ganglionic tissues with permissive indicator cells. Participation of both cellular and viral genes is involved in establishment and maintenance of latency. It is not known if parts or all of the viral genome are integrated with cellular genes. At least a minor part of the virus genome seems to be transcribed in latently infected neurons but the gene products have not been identified. One of the HSV-derived proteins discussed in this context is an immediate–early viral protein, VP175. However, establishment of latent herpesvirus infections seems to be dependent on the ability of the nerve cell to restrict the virus infection rather than due to specific genetic viral characteristics.

Three principally different sets of reactions leading *in vivo* to reactivation of latent infections in trigeminal ganglia are discernible: peripheral injuries directed against the skin and/or the nerve of the corresponding dermatome (surgical operations or damage caused by trauma, irradiation, chemicals, etc.); centrally induced stimuli often of emotional character (distress, anxiety, depression, etc.); and changes in hormonal status (associated with the menstrual cycle, medication, etc.). It is likely that reactivation frequently occurs in patients with recurrent herpesvirus infections but that intervening immune reactions as a rule prevent the appearance of clinically overt symptoms.

2.2. Paramyxoviruses

Several members of the paramyxoviruses are infectious for neural cells. Thus, infections with parainfluenza virus, mumps and measles virus have all been associated with acute and/or chronic diseases of the CNS in man, canine distemper virus and canine parainfluenza virus with neurological diseases in dogs, and Newcastle disease virus with CNS infections in birds. In experimental animals, acute measles virus infection of the CNS seems primarily to involve oligodendroglia and neurons, although there are marked differences in resistance to infection between neurons of different parts of the CNS. Replication of measles virus in the brain is restricted. Virus is probably spread by cell-to-cell transmission and through dendritic extensions. In certain experimental animals, chronic CNS infection with measles virus demonstrates features that are similar to what has been observed in chronic measles virus infection of the CNS in man, i.e., subacute sclerosing panencephalitis (SSPE). This fatal neurological disease, which occurs in 1 per million children, is probably acquired before the second year of life; the symptoms of disease may become detectable by the beginning of the school age period. The SSPE infection of the CNS seems restricted and is therefore asymptomatic for a long period. However, the infection persists and is influenced, but not cleared from the CNS, by immune reactions. When extensive pathological changes have been created, the disease becomes clinically apparent. Virions are not seen budding from plasma membranes of infected neural cells, and as signs of restriction of viral replication, infectious virus is demonstrable only by cocultivation of brain specimens with measles virus-permissive cells.

The paramyxoviruses demonstrate virions that are pleomorphic (100–300 nm) with an enveloped nucleocapsid of helical structure and a single-stranded linear RNA genome with a molecular weight of 6×10^6. Six viral proteins are recognized. The matrix (M) protein is present on the inside of the envelope and two glycoproteins are inserted in the lipid layer of the envelope. Of the glycoproteins, the HN protein carries hemagglutinin and neuraminidase activity and the F protein participates in fusion reactions with the cellular membrane to which the virion attaches via the NH protein. The protein of the nucleocapsid is referred to as NP. In addition, the nucleocapsid contains two other proteins designated L (large) and P for an RNA-dependent RNA polymerase. After attachment and fusion of the virion with the host cell, the RNA is transcribed by the polymerase of the nucleocapsid. The glycoproteins are glycosylated by cellular glycosyltransferases and are inserted in the plasma membrane of the infected cell where they are demonstrable as viral antigens. Virions are produced by budding from the modified plasma membrane. Measles virus lacks neuraminidase activity and does not—like the neuraminidase-supplied paramyxoviruses—exclude sialyl groups from the plasma membrane of infected cells.

2.3. Rhabdoviruses

Both rabies virus and VSV cause CNS disease and in rabies the selective involvement of neuronal cells is striking. Infection with rabies virus in man causes an acute lethal CNS disease and although virus also replicates in several different tissues, the course of the infection and the symptomatology are determined by the virus—nerve cell interactions. Rhabdoviruses are large (175 × 70 nm) negative-stranded RNA viruses. For RNA transcription, the virion carries an RNA-dependent RNA polymerase in the nucleocapsid. The helical nucleocapsid is covered with an envelope containing approximately 500 copies of a glycoprotein (G) that is essential for virion attachment to cells. As the G protein is glycosylated, the G protein and the M protein are located on the surface of the nucleocapsid to which the G protein is anchored. Like most other viral glycoproteins, the G protein appears to be inserted in the plasma membrane of infected cells.

In rabies-infected nerve cells, the morphogenesis of the virus occurs inside the cell in the endoplasmic reticulum, whereas in nonneuronal cells, rabies virus, and VSV, buds out from the plasma membrane. The possible influences on the axonal transport and the release of virions from infected neurons when the morphogenesis of rabies virus is intracellular have not been evaluated.

The relatively long incubation time of rabies seems largely dependent on slow production of virus in muscle cells preceding the nerve cell infection. In the CNS, virus is spread and replicated selectively by neuronal cells. Both clinical symptoms and histopathological findings suggest that certain areas of the brain (particularly structures of the limbic system and the cerebellum) are affected. Later, virus reappears in muscles, abdominal organs, skin, and other tissues. The relative independence of engagement of cellular elements other than nerve cells is reflected in the insignificant inflammatory reactions.

3. VIRAL RECEPTORS ON NEURAL CELLS

The perineurium, which acts as a diffusion barrier, does not totally encapsulate the motor and sensory terminals but terminates before it reaches the synaptic cleft or the nerve terminal, leaving the nerve open-ended.[7,8] Near the plasma membrane of sensory nerve terminals, the portion of the terminal area sometimes designated the receptorplasm, endocytotic activity is high with uptake of low-molecular-weight matter as well as exogenous macromolecules.[9] The endocytotic activity is assumed to be a part of the mechanism for plasma membrane retrieval[10] or to represent activity compensatory to the exocytosis.[11] The significance of macromolecular uptake in nerve terminals and the axonal transport have been discussed recently.[12,13]

The mechanisms of macromolecular uptake vary from the seemingly nonspecific (e.g., horseradish peroxidase) to the highly selective (e.g., nerve growth factor). It is not known if neuronal uptake of viruses always requires the presence of specific virus-binding receptors. It is reasonable to assume, however, that viruses and viral subunits may be incorporated by endocytotic activity of the nerve terminal even in the absence of viral receptors. Whether the processing of the virus and the outcome of the infection are different from infections that occur when virus internalization is preceded by interactions between virions and specific virus-binding receptors is not known.

Since the virus infection of the cell is initiated by attachment of the virion to the plasma membrane, presence, localization, and density of virus-binding receptors on plasma membranes might be decisive for the relative affinity between viruses and cells of a particular organ system. The fact that nerve terminals and synapses are amply supplied with glycoconjugates containing sialyloligosaccharides and that glycoproteins and gangliosides carry functions of biologically important surface membrane receptors suggest that interest should be focused on glycoconjugates selectively concentrated in terminals of neuritic extensions.

A multitude of various cells of most species studied exhibits receptors to which HSV adsorbs. Nevertheless, these receptor reactions reveal a considerable degree of specificity since they can differentiate between the HSV subtypes and specifically bind either HSV-1 or HSV-2.[14] Blocking of receptors selectively binding HSV-1 with homologous virus does not affect the attachment rate of HSV-2 but excludes the adsorption of HSV-1. Also, nerve and glial cells of brain tissue[15] as well as ependymal cells[16] of different species demonstrate HSV-binding receptors. In man, monkey, rabbit, and rat, neuronal receptors with affinity for HSV-1 predominate, but there seems to be an uneven distribution of the receptors along the nerve cell plasma membrane. Synaptosomal fractions of brain tissue contain HSV-adsorbing receptors in abundance whereas the viral receptors seem not at all or only sparsely scattered on the neuronal perikarya.[17] Therefore, the entry of HSV into the CNS might be monitored by the relative access to receptors present in sensory nerve endings and synapses. In agreement, *in vitro* cultured rat sensory neurons exposed to HSV become infected via their neuritic extensions without displaying signs of infection such as retraction.[18] Chemically, HSV-binding cellular receptors are considered to be of glycoprotein nature, and to react with one or more of the viral envelope glycoproteins.[19,20] Attachment of HSV is promoted by thyroid hormone[21] and inhibited by parathyroid hormone.[22]

The use of lectins has suggested that concanavalin A binds to HSV but does not interfere with the adsorption of virus to cellular receptors.[23–25] If the lectin is added together with the virus to rat sensory neurons in culture, HSV is rendered noninfectious,[25] presumably by destruction of virions. After treatment of cultures of rat sensory neurons with various lectins, wheat germ agglutinin (WGA) and *Ricinus communis* agglutinin consistently blocked the infection of the neurons apparently by interfering with attachment of HSV to neurites.[25] Preincubation of HSV with *N*-acetylneuraminic acid but not with *N*-acetyl-D-glucosamine or *N*-acetyl-D-galactosamine reduced the infectivity of HSV for neurites of cultured neurons. WGA is readily taken up by neuritic extensions,[26] and when labeled, for example, with fluorescein, WGA is recovered in the somas of the neurons. For these reasons, WGA is well suited for studies of the mechanisms of membrane binding and neuronal receptor turnover.

It should be pointed out that the experiments with HSV infection of cultured neurons were performed in a culture system allowing infections of neurites without the access of virus to neuronal perikarya.[18] The observations suggest that sialoglycoproteins on sensory neurites may constitute a receptor structure for HSV. However, it is at present not possible to exclude that the influences on HSV attachment by WGA are due to, for example, nonspecific steric hindrance of the virus–receptor binding or lectin-induced redistribution and internalization of receptors,[27] nor has the viral specificity of the uptake of virions by neurites in culture been ascertained.

Histopathological studies on animals experimentally infected with rabies virus have suggested that rabies virus enters the CNS via sensory nerve endings of muscle and tendon spindles.[28,29] Immunofluorescence in rabies virus-infected mice indicated in addition that sites of viral antigens were demonstrable in form and distribution similar to those of acetylcholinesterase (AChE), suggesting that rabies virus might well use motor endplates at neuromuscular junctions for entry into the nervous system.[30,31] In agreement, a similar pattern of coincidence of viral antigen-positive sites and ACh receptor sites has been observed with rabies virus-infected cultured myotubes from chicken embryos and with preparations of mouse diaphragms with attached phrenic nerves exposed to rabies virus.[31] Using electron microscopy, viral particles were observed in close apposition to plasma membrane sites associated with high-density clusters of ACh receptors. Pretreatment of cultured myotubes with α-bungarotoxin or *d*-tubocurarine reduced the attachment and infection rates drastically as revealed by immunofluorescence and assays of virus adsorption. In contrast, Sindbis virus, an alfa-togavirus lacking the neurotropic properties of rabies virus, demonstrated no affinity for ACh receptors.[32] As neuromuscular junctions reveal a high density of ACh-binding receptors, the affinity of rabies virus for the ACh receptor probably represents an important pathogenetic mechanism of rabies virus infection.

Most virus receptors are considered to be of glycoprotein nature.[20] However, Sendai virus, one of the paramyxoviruses, adsorbs well to liposomes containing ganglioside.[33,34] This finding was initially interpreted as if both glycoproteins and glycolipids naturally might serve as virus receptors. However, treatments with neuraminidase and trypsin, and observations that in glycoconjugate-containing liposomes virtually all of the Sendai virus-binding activity was associated with the glycoprotein rather than with the glycolipid fraction[35] have emphasized the importance of protein as carrier of the Sendai virus receptor. The identification of the molecular structure of the oligosaccharide receptor was carried further by studies on the binding of Sendai virus to plastic-adsorbed gangliosides. It was then demonstrated that the virus adsorbed with high affinity to gangliotetraosylceramides with a terminal disialosyl group.[36] A recognition structure, NeuAcα2,8NeuAα2,3Galβ1,3GalNAc, was proposed on basis of maximal binding capacity. In agreement, neuraminidase-treated bovine kidney cells, which by the treatment became resistant to infection with Sendai virus, regained susceptibility after a brief exposure to CMP-sialic acid and β-galactoside : α2,3-sialyltransferase.[37,38] Moreover, receptor activity and susceptibility to infection were also restored in cells treated with trypsin to remove the original glycoprotein receptor. Treatment of 3T6 cells with neuraminidase renders these cells resistant to polyoma virus but susceptibility to infection can be restored by reacting the cells with β-galactoside : α2,3-sialyltransferase and CMP-NeuAc, forming the sequence NeuAcα2,3Galβ1,3GalNAc, which seems to be serving as the specific receptor.[39] The relevance of these observations to viral attachment to neural cells has not been evaluated.

The oligosaccharide moiety of viral glycoproteins are considered responsible for formation and maintenance of the tertiary structure of the peptide constituent and necessary for the function, protection, and transport of the viral protein in the cell.[40] HSV-1 particles produced in the presence of glycosylation inhibitors (2-deoxy-D-glucose and tunicamycin), and thus lacking *N*-linked oligosaccharides, still demonstrate the capacity to adsorb to cellular receptors.[41] HSV adsorption apparently does not require the presence of *N*-linked oligosaccharides in envelope glycoproteins. The nonglycosylated G protein of VSV is not transported to the plasma membrane and no budding of virus is observed.[42] Measles virus, lacking neuraminidase in the virion, is reported to incorporate NeuNAc into the envelope when virus is budding out from the plasma membrane,[43] but NeuNAc on the envelope glycoprotein does not seem to be essential for attachment. The general impression is that inhibition of glycosylation of viral envelope glycoproteins results in defective production of virions[42,44,45] but that the carbohydrates of the viral envelopes are of no or insignificant importance in virus recognition of cellular receptors.

4. AXONAL TRANSPORT OF VIRUSES

The concept that virus infections may spread of the CNS by the axonal flow of nerves is based on numerous clinical, histopathological, and experimental observations. One of the earliest indications for axonal transport of a virus was the classical experiment of Bodian and Howe[46] demonstrating poliovirus retrograde progression along the sciatic nerve of the monkey at a rate estimated to 2.4 mm/hr and recovery of the virus from the spinal cord. It is now accepted that virus infections can be transferred axonally, probably in the same way as newly synthesized cellular proteins, i.e., by fast axonal flow.[47]

In herpesvirus infections, axonal transport of the virus is unequivocally demonstrated with HSV and VZV and is probable also in herpes B virus infections of man. In mice infected intradermally, the spread of HSV infection was prevented by ligating, freezing, or colchicine soaking of the sciatic nerve prior to infection.[48] Ultrastructurally, virions have been observed intraxonally in HSV-infected animals,[49,50] but it is questionable if the particles detected represent virus traveling from the periphery toward the CNS or virus transported in the opposite direction. If the virions observed were transported with the retrograde flow, there should have been no preceding fusion between the viral envelope and the membrane of the nerve terminal.

Nucleocapsids have also been seen in nonmyelinated axons,[49,51] but again it is uncertain if these nucleocapsids originate from, for example, leakage through breaks of the nuclear membrane of infected neurons. We have found that HSV-1 infection of neuritic extensions of rat sensory neurons in culture is initiated by fusion of the viral envelope of attached virions with the neuritic plasma membrane.[178] Subsequently, the infection is axonally transported from the periphery toward the cell soma with the viral nucleocapsids. In turn, virus produced *de novo* in infected nerve cell bodies is transported in the neuritic extensions from the cell soma toward the periphery but now as enveloped virus confined to vesicles.

The *in vitro* studies failed to demonstrate infectivity of HSV nucleocapsids taken up by neuritic extensions. Because fractions with nucleocapsids produced by heat treatment of HSV suspensions have been reported to maintain infectivity as revealed by transfection experiments,[52,53] it is unlikely that HSV nucleocapsids that are internalized passively into neurons are able to cause infection.

Rats infected with HSV-1 by microinjection of virus into the left neostriatum demonstrate infected neurons in the substantia nigra, cortex, and dorsal raphe nuclei.[54] This spread of the infection is compatible with a retrograde transport of the infection from the neostriatum. The massive degeneration of afferent nerve terminals observed throughout the neuropil probably reflects degeneration of nerve cells at distant sites. In fact, in mice infected in the snout, the HSV infection may be followed by the peroxidase–antiperoxidase (PAP) method to the third order of neurons.[55] Elongated deposits of PAP-reaction products, the length of which might exceed that of a Schwann cell segment, have been found in trigeminal nerve axons. Such accumulations of viral antigens presumably representing material transported with the axoplasmic flow toward sensory endings have also been observed in rabies-infected hamsters.[1] Of the brain stem, trigeminal sensory nuclei, and trigeminal spinal tract nuclei, and, moreover, the reticular formation including the raphe nuclei, the thalamic nuclei and the locus coeruleus demonstrated the presence of HSV antigens. Thus, the infection had passed two or more synapses on its way to the brain stem. In cultures of dorsal root ganglionic neurons[18] and autonomic neurons[56] of the rat, the HSV infection appears to be spread from cell to cell via neuritic extensions without being influenced by antiviral antibody added to the culture overlay medium.

Several reports emphasize the axonal transport of rabies virus.[28,29,57–59] It has been suggested that after fusion of viral envelope and the plasma membrane, the rabies virus nucleocapsid is transported in the axoplasm.[1] In accord with the hypothesis of axonal transport, treatment with colchicine or vinblastine of mouse sciatic nerve after virus inoculation prevented spread of the infection to the CNS.[58]

Other evidence, some of which is circumstantial, suggests that rabies infection is transmitted transsynaptically. In this way, nerve cell processes can transfer the virus infection without exposing the virus to neutralizing antibodies and thus facilitating persistence of the infection even in the immune animal.[60,61] It is possible that budding of virions on axonal membranes of infected neurons occurs, and an intraaxonal accumulation of viral nucleocapsids at nodes of Ranvier has also been noted.[1]

In cultures of dissociated mouse neurons, VSV infection is markedly influenced by addition of antiviral antibodies. In the presence of antibodies, viral nucleocapsids accumulate in dendritic extensions and budding sites are frequent on the side of the postsynaptic density. Viruses seem to enter directly into the lateral side of the presynaptic terminal to which the virus-delivering dendrite is connected.[61] These findings suggest somatofugal transport of nucleocapsids and are in agreement which *in vivo* experiments with rabies-infected animals.

Transsynaptic dissemination within the CNS of measles virus and viral antigens has been visualized by immunofluorescence, immunoperoxidase, and electron microscopy techniques. In measles virus-infected animals, antigens were traced to neuronal perikarya with somatofugal spread into dendritic and synaptic sites. Nucleocapsids and antigenic viral subunits were identified in the nucleus and cytoplasm of the neuron including dendrites at various distances from the cell body. A frequent finding was the presence of viral antigens in postsynaptic endings.[62]

In cases of SSPE, measles virus antigens are widely distributed within the brain. Infected cells demonstrate antigens in both nucleus and cytoplasm, and sometimes in the neuritic processes.[63,64] Of neurons involved, both cortical pyramidal cells and large brain-stem neurons have been observed.[64] It has been suggested that in SSPE, viral genetic information spreads from cell to cell along dendritic processes.[65] However, in measles as in rabies, the presence of antiviral antibodies seems important for restriction of the infection to the neuronal cells.

5. *VIRAL PROTEINS IN PLASMA MEMBRANES INFLUENCE NEURAL CELL ACTIVITIES*

Cells infected with enveloped viruses acquire new membrane-active components in association with the fusion process by which the viral envelope merges into the cellular membrane and when viruses become enveloped by budding out from cellular membranes and viral glycoproteins are inserted.[66,67] Changes in membrane-mediated cellular functions may also occur when viral antigens in membranes of infected cells react with immunoglobulins, causing redistribution of ligand–globulin complexes.[68–71]

Theoretically, virus infections affecting membranes of nerve cells may interfere with, for example, protein transport maintaining functional axonal regions and synaptic terminal membranes. This transport seems predominantly membrane-associated.[47] The vesicles that constitute the vesiculotubular structures acting as transport vectors within the axon and that become associated with synaptic vesicles at the terminal are, like the viral envelope glycoproteins, Golgi-derived and may be affected by viral glycoprotein synthesis. Insertion of viral glycoproteins in cell membranes may in addition imply exclusion of constituents with receptor functions normally present.

An influence of virus infections on nerve cell receptor functions has been observed with cell line 108-CC-15 (NG-108-15), a mouse neuroblastoma × rat glioma hybrid, demonstrating a variety of neuronal receptor activities.[72] Intracellular cAMP levels were likely reduced by membrane receptor changes in cells persistently infected with rabies virus.[73] Incubation with prostaglandin E_1 (PGE_1), which in the infected hybrid cells is followed by a rapid rise in cAMP levels, was in the rabies-infected cultures followed by an increase that was only 50% of that of the uninfected control. In comparison with uninfected cells, rabies-infected hybrid cells exposed to L-isoproterenol, an adrenergic agonist, and PGE_1 revealed impaired α-adrenergic receptor functions. In contrast, the ACh receptor functions remained unaffected when cells were exposed to PGE_1 and ACh. The latter finding may be at some variance with observations on binding to homogenates of rabies-infected rat brain of [^{3}H]quinuclidinyl benzylate, an antagonist to the muscarinic ACh receptors.[74]

The binding of the antagonist declines at the time of the appearance of clinical symptoms; whether this is a result of a reduced number of receptors available or due to more complex mechanisms of impaired neuronal function is unknown. Recent observations suggest that the number of opiate receptors on neuroblastoma–glioma hybrid cells persistently infected with rabies virus is not reduced but that instead their binding affinity is decreased.[75] The reported findings were all obtained with fixed rabies virus strains. In street rabies virus infections, the importance of viral glycoproteins of infected cells is doubtful, since in contrast to infections with fixed virus, only small amounts of viral glycoproteins are present in the plasma membrane of infected neurons.[76]

Established mouse neuroblastoma cells (clone N 115) persistently infected with lymphocytic choriomeningitis virus (LCMV) display maintained receptor functions for ACh as revealed by their affinity for neurotoxin A and α-bungarotoxin. On the other hand, the virus infection causes significantly reduced cellular levels of AChE and choline acetyltransferase.[77] A similar reduction of AChE activity has been noted in neuroblastoma cells (clone 41 A3) acutely infected with measles virus.[78]

Failure of another mouse neuroblastoma cell line (clone N_2A) to manifest signs of neural differentiation when persistently infected with measles virus was considered to be related to impairment of receptor-mediated cAMP synthesis.[79] The reduction of AChE formation in N_2A cells persistently infected with measles virus was associated with a general decrease of RNA and protein synthesis. It seems evident from several studies that intracellular cAMP levels are lowered in established cell lines of neural origin when cells become persistently infected with measles virus and that these effects presumably are caused by alterations of plasma membrane receptors. Thus, a rat glioma cell line (C-6) persistently infected with an SSPE-derived measles virus strain demonstrated only about half the maximum values of intracellular cAMP of the uninfected cells and drastically reduced adenylate cyclase activity.[80] It was stated that the measles virus infection caused impairment of the membrane receptor signal transfer. Whether the virus infection also affected the β receptor itself was not demonstrated, although this cannot be excluded. In addition, studies on persistently infected C-6 cells have revealed that closely related viruses (measles and canine distemper viruses) may influence cellular membrane functions in different ways. With both viruses a reduction of adenylate cyclase activity is seen, but a decrease in the number or affinity of the β-adrenergic receptor is clearly observed only after the canine distemper virus infection.[81] Some of the virus-induced effects on neural cell receptor functions are summarized in Table I.

Table I
Virus-Induced Influences on Neural Cell Receptor Functions

Virus	Cell/tissue	Receptor	Effect	Author
Rabies	Mouse neuroblastoma × rat glioma hybrid, 108-CC-15 (NG-108-15)	α-Adrenergic	Impairment of receptor-mediated cAMP response	Koschel and Halbach[73]
Rabies	Mouse neuroblastoma × rat glioma hybrid, 108-CC-15 (NG-108-15)	Opiate receptor	Impairment of agonist binding	Münzel and Halbach[75]
Rabies	Mouse neuroblastoma × rat glioma hybrid, 108-CC-15 (NG-108-15)	Muscarinic ACh receptor	Reduced receptor-mediated cAMP levels	Koschel and Halbach[73]
Rabies	Rat brain	Muscarinic ACh receptor	Reduced binding of quinuclidinyl benzylate	Tsiang[74]
CDV	Rat glioma, C-6	β-Adrenergic	Reduction in number of receptors	Koschel and Münzel[81]
Measles (SSPE)	Rat glioma, C-6	β-Adrenergic	Reduced cAMP and adenylate cyclase activities	Halbach and Koschel[80]
Murine picornavirus	Mouse brain	Catecholamine-binding brain protein	Reduction of binding capacity in neonatally infected mice	Lee[117]
Scrapie agent	Hamster brain	5-HT receptor	Activation	Goudsmit *et al.*[114]

6. VIRUS-INDUCED CHANGES IN CELLULAR PERMEABILITY AND ELECTROPHYSIOLOGY

Permeability changes might be secondary to viral effects on the cytoskeleton of the cell and have consequences for various cell functions and, ultimately, cell vitality. Viral influences on microfilament structures of cells are manifested in alterations of the distribution of actin and myosin.[82–84] Viral modulation of nerve cell microfilament structures has been studied in rat (B 103) neuroblastoma cells where the HSV-1-induced disorganization of contractile proteins[85] was demonstrated by the random alignment of actin cables and loss of the fibrous appearance of the myosin. Obviously, such changes might result in reduced stability and leakage through the plasma membrane. A ^{51}Cr-permeability test has been used as an approach to characterize virus-induced plasma membrane changes.[86] Although the results reported varied with respect to which viruses and cells were studied, there seemed to be a reduced plasma membrane permeability to ^{51}Cr in most cell types infected with HSV-1. Other viruses associated with induction of dysfunctions in ion transport through membranes of infected cells were poliovirus, other picornaviruses,[87–89] and Sindbis virus.[90] However, of greater interest are the more specific changes in permeability induced by some virus infections.

Experiments with paramyxoviruses, the Sendai and Newcastle disease viruses, have revealed that the fusion of viral envelope and cell plasma membrane introduces hydrophilic channels into the cell membrane.[91,92] As a result there are currents of low-molecular-weight compounds including ions (Na^+ and K^+) into and out of the cells. Divalent cations (Ca^{2+}) can specifically and reversibly prevent these effects[93,94] and the virally induced channels have been considered to resemble intercellular communicating junctions. The channels are opened by removal of Ca^{2+} and closed when Ca^{2+} is added.[94] The permeability changes induced by adsorption of Sendai virus were observed with both brain cells and nonneuronal cells.

A decline in cellular K^+ has been noted with VSV-infected L cells at 7 hr postinfection. This alteration was not associated with virus-induced inhibition of cellular protein synthesis,[95] suggesting that blocking of cell metabolism by the virus and impaired cellular stability occurred independently of each other.

Specific changes of electrical parameters related to virus-induced alterations in plasma membrane permeability have previously been suggested for HSV-infected heart cells.[96,97] The transmembrane ionic currents, action potentials, and electrical excitability of nerve cells are functions presumably sensitive to virus-induced membrane changes. Only a few studies of the electrophysiology of virus-infected neurons have been reported. Dissociated dorsal root ganglionic neurons of the rat responded after 4 hr of HSV-1 infection with a decreased spike rate and lowered amplitude and overshoot.[98] The full width of spikes at half-maximum and the resting membrane potential were increased. At 16 hr postinfection, when virus-induced morphological changes first appeared in the culture, less than half of the neurons in the culture demonstrated an action potential. A similar study[99] with HSV-2-infected guinea pig dorsal root ganglionic cells cultured *in vitro* also demonstrated a decrease in membrane excitability. The changes observed were explained in both studies as resulting from specifically altered ionic currents leading to reduction of Na^+ conductance at an early stage of the infection, although changed depolarization conditions were among the other possibilities discussed.[98] One further type of electrophysiological alteration has been detected in HSV-2-infected neurons.[100] Rat dorsal root ganglionic neurons generate action potentials spontaneously upon HSV-2 infection. This activity was initially of low amplitude and with low-frequency discharges but increased with time postinfection. The waveform of the spontaneous potentials seemed identical to that of the evoked action potentials from the same cells. HSV-2 infection thus seems able to cause two Na^+ conductance-related effects, one of which suggested a reduced Na^+ channel activity and the other an increased movement presumably in another part of the infected cell.

7. *TURNOVER OF NEUROTRANSMITTERS IN VIRUS-INFECTED NEURAL CELLS*

Apparently, *in vitro* virus infections can provoke alterations of cholinergic neurotransmission and alter the catabolism of ACh. There are also indications that in animals, defects of ACh metabolism might appear subsequent to virus infections.[77,78]

In mice intracerebrally inoculated with HSV-1, an increased turnover of brain monoamines has been observed during the acute stage of the infection by demonstration of high levels of homovanillic acid (HVA) and 5-hydroxyindoleacetic acid (5-HIAA).[101] The rises of HVA and 5-HIAA brain concentrations were associated with unchanged levels of dopamine (DA) and 5-hydroxytryptamine (5-HT), but dose–response relationships, normalization of the brain concentrations by use of inhibitors of virus replication, as well as results with hydroxylation inhibitors [α-methyltyrosine, *p*-chlorophenylalanine, 2-(4-methyl-1-homopiperazinylthiocarbonyl)disulfide] all suggested a causality between the virus infection and the increase in monoamine turnover. Effects observed after administration of probenecid, pargyline, or reserpine to HSV-infected animals indicated that neither impaired elimination of metabolites nor reduced capacity of uptake of amines was of significant influence on the concentrations of the acids.[102] On the other hand, there seemed to be an increased release and turnover of catecholamines and 5-HT in the virus-infected mouse brain. This assumption was corroborated by results obtained with precursors (levodopa and 5-HTP), receptor agonists (apomorphine, clonidine), and antagonists (pimozide, phenoxybenzamine).[103]

Enzymatic activities related to synthesis of putative transmitter substances have also been studied in HSV-infected rats injected stereotactically into the left neostriatum.[104] Choline acetyltransferase, glutamic acid decarboxylase (GAD), and tyrosine hydroxylase (TH) activities decreased in the ipsilateral neostriatum, GAD and TH decreased in the ipsilateral pallidum and/or substantia nigra regions, but levels in the cerebral cortex were unchanged. It is noteworthy that in rats displaying diminished enzymatic activities upon infection, the reduced biosynthesis is correlated to severe nerve terminal destruction and to other morphological changes in cholinergic and catecholaminergic regions.[104] However, in the mouse model with signs of an increased turnover of transmitters, the histopathological picture demonstrates neuronal destruction in other parts of the brain excluding the lower brain-stem area. Histofluorometry has indicated that in the infected mice, the monoaminergic neurons are essentially intact.[101]

Similar results were achieved with Venezuelan equine encephalomyelitis virus (VEEV), an alfa-togavirus that stimulates DA and 5-HT brain metabolism in the mouse. This influence was manifested by increased brain DA, 5-HT, and HVA levels.[105] In the VEEV-infected rat, on the other hand, decreased TH[106] and GAD[107] activities were found.

Infection of neonatal rats with LCMV results in immunologically mediated necrotic lesions of the cerebellar cortex. Concomitant elevation of the activity of catechol-O-methyltransferase and production of normetanephrine but not that of monoamine oxidase are demonstrable. These alterations were interpreted as consequences of immune-mediated cellular infiltrate rather than changes in activity of brain parenchymal cells.[108] Intracerebral inoculation of Newcastle disease virus into mice that are in a depressed metabolic state with low body temperature produces uptake and release characteristics of putative neurotransmitters that are not different from those of matched uninfected controls.[109]

It has been suggested that some of the virally induced behavioral consequences may be secondary to altered turnover of brain monoamines.[103,110,111] Among behavioral changes of HSV-infected mice, hyperactivity and aggressiveness appearing shortly after infection have been attributed to an increased turnover of catecholamines,[103,111] whereas decreased locomotor activity and low responsiveness to catecholaminergic drugs in mice having passed the acute stage of the CNS infection were considered to reflect a reduced turnover.[111] These latter observations are in agreement with a higher sensitivity to *d*-amphetamine sulfate of intracerebrally inoculated mice by day 2 postinfection and a reduced responsiveness to the drug in infected and partly immune mice examined at a later stage postinfection.[112]

Reduction of 5-HT brain concentrations has also been noted in scrapie-infected animals during the late clinical phase.[113] Infected animals exhibited hypersensitivity reactions to *l*-5-hydroxytryptophan and to a 5-HT agonist, quipazine maleate.[114] As seen in infections with conventional viruses, the behavioral effects of scrapie encephalopathy seemed dependent on genetic properties of the agent as well as genetic qualities of the host.[115]

These and other observations[116,117] suggest varying effects of viral CNS infections on neurotransmitter synthesis dependent not only on the virus but also on the species and age of the experimental animals studied. Generally, viral CNS infections occurring in the neonatal period of the mouse seem to lead to sequelae with impairment of neurotransmitter synthesis, whereas increased turnover is seen in acutely infected adult animals.

Virus infections have been used to modify animal CNS models. A consistent and reproducible granuloprival hypoplasia of newborn hamster cerebellum is produced by intracerebral inoculation of feline or rat parvoviruses.[118–120] Studies on these animals have provided evidence for the presence of glutamic acid as a neurotransmitter in granule cells of the cerebellum.

8. VIRUS-INDUCED DEMYELINATION

There are principally two kinds of reactions by which virus infections may induce demyelination: (1) viruses may infect oligodendrocytes and Schwann cells and cause demyelination by cytocidal effects on myelinating cells, and (2) virus infections may initiate reactions resulting in an immunopathologically based demyelination. Several experimental models have been elaborated for evaluation of underlying mechanisms,[121] and for some of the systems pathogenetically important factors have been elucidated although the medical importance of virus-induced demyelination is still obscure.

The JHM virus is an endogenous coronavirus of the mouse causing lesions with demyelination if intracerebrally inoculated in low doses into 4-week-old mice,[122] the dosage of virus and the age of the animals being decisive for the induction of demyelination. Oligodendrocytes are apparently the main target cells and demonstrate morphological and functional abnormalities associated with myelin degradation and membrane vesiculation.[123–126] The demyelination caused by the infection seems fundamentally dependent on the cytocidal effects of the virus.

In infections with measles, canine distemper virus (CDV), or other paramyxoviruses, development of a demyelinating disease as a direct consequence of cytopathogenic viral effects seems unlikely. Viral components in measles[2,127] or CDV infections[128] are rarely seen in oligodendrocytes, but the presence of inflammatory cells in areas of demyelination, in both acute and chronic types of infection, suggests immunopathological reactions. Intriguing is the late-appearing demyelination demonstrable in mice experimentally infected with murine picornavirus (Theiler virus). The acute infection resembling poliomyelitis is followed 2–3 weeks later by a disease characterized by spinal cord myelin destruction and mononuclear cell infiltration.[129] Nerve cell infection predominates during the acute poliomyelitis-like phase and, as with other picornaviruses, the infected cells undergo a cytolytic degeneration. The demyelination appearing at later stages can be prevented by immunosuppression, i.e., by treatments with cyclophosphamide and rabbit anti-mouse thymocyte serum.[130] No viral antigens are detectable in Schwann cells but are abundant in neurons, astrocytes, and macrophages.[131] Remyelinating Schwann cells may be seen invading the demyelinated lesions.

A similar process of remyelination can be observed in HSV-infected mice suffering from a demyelinating process in the transitional area of the trigeminal nerve where the central and peripheral nervous systems meet.[132,133] Both oligodendrocytes and astrocytes[133,134] display signs of infection. It has been suggested that the astrocytes, which are amply supplied with HSV receptors,[15] may be the cells responsible for transferring the axoplasmically transported infection from the axolemmal membrane of Ranvier nodes to the myelinating cells.[134] Myelin destruction shortly after the inoculation of HSV might indicate that cytocidal effects of the virus are of importance for the initiation of the demyelinating process,[133,135] whereas host inflammatory responses seem essential for its maintenance.[136,137] Thus, the reduction of the mononuclear infiltrate appears concomitantly with reduced demyelination in cyclophosphamide-treated[137] and in T-cell-deficient mice.[138]

In Semliki forest virus-infected animals demonstrating destruction of oligodendrocytes[139] as well as an immunologically monitored demyelination, the remyelination of initially relatively mild lesions seems to be inhibited.[140,141] Possibly, the immunopathological reactions may be maintaining the demyelination by interfering with the processes of remyelination.

The immunological mechanisms by which various virus infections create myelin-destructive reactions are still unknown. Viral antigens might be absent in plaques of demyelination. Some antigenic relatedness has been noted between measles virus protein and myelin basic protein,[142] but such virus–myelin antigenic similarities can hardly account for demyelination with all viruses involved. It has been shown that recurrent exposures of the CNS to viral antigens can exacerbate experimental allergic encephalomyelitis in guinea pigs immunized with spinal cord myelin.[143] One theory has suggested that macrophage-derived substances such as proteases of inflammatory immune reactions against viral antigens of infected CNS cells may be responsible for the injuries of the myelinated tissue.[144,145] A pattern for possible collaboration of various reactions in development and maintenance of virus-induced demyelination is presented in Fig. 1.

9. *INFLUENCES OF THE CNS AND NEURAL CELLS ON VIRUS INFECTIONS*

The impacts of virus infections on CNS structures and functions have been discussed above. However, the progress and outcome of the virus infections themselves are also subject to modification by factors and functions intrinsic to the nervous system. Aspects of these influences include restriction of viral replication in neural cells, host-modified spread and expression of virulence, and development of persistent and latent virus infections.

HSV synthesizes its own thymidine kinase, perhaps an essential ability for a virus that might have to replicate in neuronal cells with low DNA-synthesizing activity. The replication of thymidine kinase-negative HSV mutants is restricted in sensory ganglia of infected animals[146–148] and in ganglia cultured *in vitro*.[149] Whether this phenomenon is due to insufficient capacity of neurons to provide phosphorylated thymidine derivatives is not known. Restriction of HSV replication in neural cells has been attributed to lack of synthesis of a necessary gene product.[150–152] and also to the presence in neural cells of virus-inhibiting non-interferon factors.[153] Undoubtedly, the HSV infection can vary

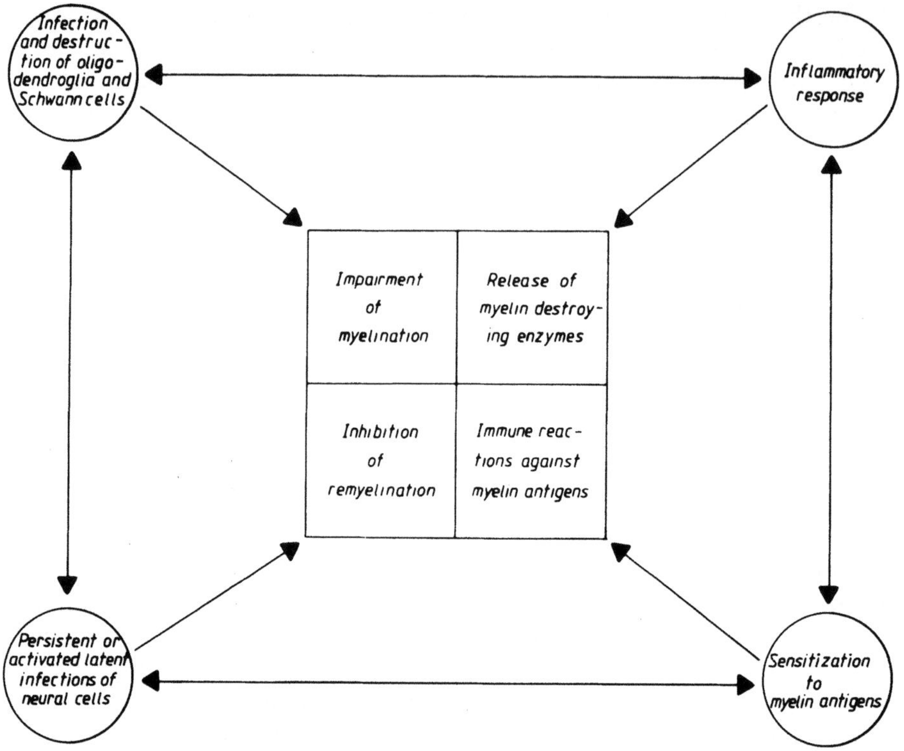

Fig. 1. Virus-induced reactions associated with development and maintenance of demyelination.

considerably in neural cells from permissive infections to a relatively low degree of permissiveness, depending on the species origin of the cells studied.[150]

It is well recognized that the integrity of postganglionic axons influences the HSV ganglionic infection. Neurectomy in mice with a ganglionic infection reactivates latent HSV infection[154,155] and favors the viral replicative phase over that of latency. Both mechanical[156] and electrical stimulations[157] of the nerve induce reactivation of latent infections with shedding of virus. 6-Hydroxydopamine-induced destruction of adrenergic nerve terminals enhances HSV replication in ganglion cells.[158] Thus, virus produced in the superior cervical ganglion of mice was augmented when the drug was administrated 2 days after infection. It was suggested that injury to the terminal part of the axon directly influenced viral replication in the neuronal soma.

In cultures of rat dorsal ganglion neurons, the time course of HSV infection is altered in the presence of antiviral antibody[56] but despite antibody concentration adequate to neutralize the virus, spread of the infection from neuron to neuron cannot be prevented. Also, HSV infection in mice is greatly influenced by immune reactions shunting the infection from an acute productive to a nonproductive stage.[159,160] The precise role of the immune reactions in establishment of latency is not known but in the immune individual it is probably only within the neuronal network of the CNS that the infection may be spread and maintained. In no other cells but neurons has establishment of latent HSV infections been unequivocally shown. In peripheral tissues, persistent infections are maintained by virus-productive infections.[161]

Reduction of HSV replication occurred when intracellular levels of cAMP were raised, whereas enhanced viral production was the rule when cellular cGMP concentrations were increased.[162] Possibly, changes in intracellular cyclic nucleotides represent parts of a cellular control mechanism of HSV replication. Other observations suggest that cyclic nucleotides might also be involved in cellular control of the latent infection.[163] The existence of control mechanisms of cyclic nucleotides for selection of gene expression leading to induction of viral replication in somatic cell hybrids carrying but not expressing EBV genomes has been described.[164]

Conversion of productive measles virus infection in neural cells to low or nonproductive infections may be achieved by treatment with drugs capable of modifying cyclic nucleotide metabolism.[165,166] Thus, inhibition of cAMP phosphodiesterase via papaverine-increased cAMP levels reduces viral replication concomitantly with the abolishment of viral matrix (M) protein. Absence of M protein in measles virus-infected cells is considered an important pathogenetic feature of SSPE[167,168] and is also demonstrable in experimental models of measles virus persistency.[169] Effects on measles virus replication by changes in cyclic nucleotide metabolism have been observed not only with neural cells but also in a nonneural cell system, i.e., in human amnionic cells.[165] As for HSV-infected cells, addition of cGMP to measles virus-infected neural cells counteracts the influence of raised cAMP levels. The observations reported suggest that evolution of an acute measles virus infection to a persistent type of infection might be influenced if not regulated via the cyclic nucleotides.

In contrast to nonneural cells, neurons sustain replication of VSV with relatively maintained integrity.[170] This morphological stability seems to be associated with cellular maturity and increases gradually with the stage of neuronal differentiation. However, the preservation of neuronal cells seems in addition dependent on the presence in the inoculum of defective interfering virus particles.[170,171] It is not known if in infection of neural cells the production of defective virus particles is enhanced relative to that of virus infections of cells of nonneural origin, although this has been suggested.[170] Interference with VSV replication by defective virus occurs in the neuron as in the nonneural cell at the level of viral replication. In this context, it should be recalled that virus-induced interferon production seems to parallel and stimulate cellular prostaglandin secretion.[172–174] Prostaglandins restore the antiviral reactivity of interferon in hyporeactive animals[175] and regulate cAMP enhancement of interferon.[176,177]

10. CONCLUDING REMARKS

Morphological studies on the virus-infected CNS have revealed the particular affinity of some viruses for neural cells and emphasize the predilection of neurotropic viruses for different CNS structures. Viral neurotropism can be discussed in terms of occurrence and density of viral receptors on neural cells and axonal transport of the infection. In recent years, progress in the delineation of molecular events involved in virus-induced functional changes of infected nerve cells is discernible although many of the virus–nerve cell interactions cannot as yet be explained neurochemically. There is also a growing interest in the various mechanisms by which the nervous system is able to modify the outcome of virus infections. The findings discussed in this chapter have mainly been gathered by studies on experimental animals or neural cell cultures, but their medical implications are essentially unexplored. There are reasons to believe, however, that virus-induced CNS dysfunctions might be of greater medical importance in neurology and psychiatry than is recognized today. Finally, certain functions of the brain may be advantageously studied on the virus-infected animal since virus infections specifically infect particular cell populations or compartments of the CNS, providing a model otherwise difficult to achieve.

ACKNOWLEDGMENT. The author acknowledges support by the Swedish Medical Research Council (Grant 4514).

REFERENCES

1. Murphy, F. A., 1977, *Arch. Virol.* **54**:279–297.
2. Dubois-Dalcq, M., 1979, *Int. Rev. Exp. Pathol.* **19**:101–135.
3. Fields, B. N., and Weiner, H. L., 1982, *Life Sciences Research*, Report 20 (T. A. Sears, ed.), Springer-Verlag, Berlin, pp. 217–228.
4. Johnson, R. T., 1982, *Viral Infections of the Nervous System*, Raven Press, New York.
5. Klein, R. J., 1982, *Arch. Virol.* **72**:143–168.
6. Stroop, W. G., And Baringer, J. R., 1982, *Progress of Medical Virology*, Volume 28 (J. L. Melnick, ed.), Karger, Basel, pp. 1–43.
7. Bannister, L H., 1976, *The Peripheral Nerve* (D. N. London, ed.), Chapman and Hall, London, pp. 396–454.
8. Halata, Z., 1977, *J. Anat.* **124**:717–729.
9. Persson, L. A., and Kristensson, K., 1979, *Acta Neuropathol.* **46**:191–196.
10. Broadwell, R. D., and Brightman, M. W., 1979, *J. Comp. Neurol.* **185**:31–74.
11. Holtzman, E., 1977, *Neuroscience* **2**:327–355.
12. Kristensson, K., 1978, *Annu. Rev. Pharmacol. Toxicol.* **18**:97–110.
13. Ellisman, M. H., and Lindsey, J. D., 1982, *Axoplasmic Transport* (D. G. Weiss, ed.), Raven Press, New York, pp. 55–63.
14. Vahlne, A., Svennerholm, B., and Lycke, E., 1979, *J. Gen. Virol.* **44**:217–225.
15. Vahlne, A., Nyström, B., Sandberg, M., Hamberger, A., and Lycke, E., 1978, *J. Gen. Virol.* **40**:359–371.
16. Tardieu, M., and Weiner, H. L., 1982, *Science* **215**:419–421.
17. Vahlne, A., Svennerholm, B., Sandberg, M., Hamberger, A., and Lycke, E., 1980, *Infect. Immun.* **28**:675–680.
18. Ziegler, R. J., and Herman, R. E., 1980, *Infect. Immun.* **28**:620–623.
19. Spear, P. G., 1980, *Cell Membrane and Viral Envelopes*, Volume 2 (H. A. Blough and J. M. Tiffany, eds.), Academic Press, New York, pp. 709–750.
20. Lonberg-Holm, K., and Philipson, L., eds., 1981, *Virus Receptors*, Part 2, *Animal Viruses*, Chapman and Hall, London.
21. Roizman, B., 1962, *Proc. Natl. Acad. Sci. U.S.A.* **48**:973–977.
22. Roizman, B., 1962, *Proc. Natl. Acad. Sci. U.S.A.* **48**:795–803.
23. Okada, Y., and Kim, J., 1972, *Virology* **50**:507–515.

24. Ito, M., and Barron, A. L., 1974, *J. Virol.* **13:**1312–1318.
25. Ziegler, R. J., and Pozos, R. S., 1981, *Infect. Immun.* **34:**588–595.
26. Steindler, D. A., 1981, *Brain Res.* **223:**367–373.
27. Carbonetto, S., and Argon, Y., 1980, *Dev. Biol.* **80:**364–378.
28. Murphy, F. A., Bauer, S. P., Harrison, A. K., and Winn, W. C., 1973, *Lab. Invest.* **28:**261–376.
29. Harrison, A. K., and Murphy, F. A., 1978, *Arch. Virol.* **57:**167–175.
30. Watson, H. D., Tignor, G. H., and Smith, A. L., 1981, *J. Gen. Virol.* **56:**371–382.
31. Lentz, T. L., Burrage, T. G., Smith, A. L., Crick, J., and Tignor, G. H., 1982, *Science* **215:**182–184.
32. Smith, A. L., and Tignor, G. H., 1980, *Arch. Virol.* **66:**11–26.
33. Haywood, A. M., 1974, *J. Mol. Biol.* **83:**427–436.
34. Haywood, A. M., 1975, *Negative, Strand Viruses*, Volume 2 (B. W. J. Mahy and R. D. Berry, eds.), Academic Press, New York, pp. 923–928.
35. Wu, P.-S., Ledeen, R. W., Udem, S., and Isaacson, Y. A., 1980, *J. Virol.* **33:**304–310.
36. Holmgren, J., Svennerholm, L., Elwing, H., Fredman, P., and Strannegård, Ö., 1980, *Proc. Natl. Acad. Sci. U.S.A.* **77:**1947–1950.
37. Markwell, M. A. K., and Paulson, J. C., 1980, *Proc. Natl. Acad. Sci. U.S.A.* **77:**5693–5697.
38. Markwell, M. A. K., Svennerholm, L., and Paulson, J. C., 1981, *Proc. Natl. Acad. Sci. U.S.A.* **78:**5406–5410.
39. Fried, H. T., Cahan, L. D., and Paulson, J. C., 1981, *Virology* **109:**188–192.
40. Gibson, R., Kornfeld, S., and Schlesinger, S., 1980, *Trends Biochem. Sci.* **5:**290–293.
41. Svennerholm, B., Olofsson, S., Lundén, R., Vahlne, A., and Lycke, E., 1982, *J. Gen. Virol.* **63:**343–349.
42. Gibson, R., Leavitt, R., Kornfeld, S., and Schlesinger, S., 1978, *Cell* **13:**671–679.
43. Dore-Duffy, P., and Howe, C., 1978, *Proc. Soc. Exp. Biol. Med.* **157:**622–625.
44. Leavitt, R., Schlesinger, S., and Kornfeld, S., 1977, *J. Virol.* **21:**375–385.
45. Nakamura, K., and Compans, R. W., 1978, *Virology* **84:**303–319.
46. Bodian, D., and Howe, H. A., 1941, *Bull. Johns Hopkins Hosp.* **69:**79–85.
47. Hammerschlag, R., and Stone, G. C., 1982, *Trends Neurosci.* **5:**12–15.
48. Kristensson, K., Lycke, E., and Sjöstrand, J., 1971, *Acta Neuropathol.* **17:**44–53.
49. Hill, T. J., Field, H. J., and Roome, A. P. C., 1972, *J. Gen. Virol.* **15:**253–255.
50. Cook, M. L., and Stevens, G., 1973, *Infect. Immun.* **7:**272–288.
51. Schwartz, J., and Elizan, T. S., 1973, *J. Neuropathol. Exp. Neurol.* **32:**303–312.
52. Schwartz, J., and Elizan, T. S., 1975, *J. Neuropathol. Exp. Neurol.* **34:**359–368.
53. Fenyves, A., and Strupp, L., 1982, *Intervirology* **17:**228–239.
54. Bak, I. J., Markham, C. H., Cook, M. L., and Stevens, J. G., 1977, *Brain Res.* **136:**415–429.
55. Kristensson, K., Nennesmo, I., Persson, L., and Lycke, E., 1982, *J. Neurol. Sci.* **54:**149–156.
56. Price, R. W., Rubenstein, R., and Khan, A., 1982, *Arch. Virol.* **71:**127–140.
57. Murphy, F. A., Harrison, A. K., Winn, W. C., and Bauer, S. P., 1973, *Lab. Invest.* **29:**1–16.
58. Bijlenga, G., and Heaney, T., 1978, *J. Gen. Virol.* **39:**381–385.
59. Tsiang, H., 1979, *J. Neuropathol. Exp. Neurol.* **38:**286–296.
60. Iwasaki, Y., and Clark, H., 1975, *Lab. Invest.* **33:**391–399.
61. Dubois-Dalcq, M., Hooghe-Peters, E. L., and Lazzarini, R. A., 1980, *J. Neuropathol. Exp. Neurol.* **39:**507–522.
62. Van Pottelsberghe, C., Rammohan, K. W., McFarland, H. F., and Dubois-Dalcq, M., 1979, *Lab. Invest.* **40:**99–108.
63. Esiri, M. M., Oppenheimer, D. R., Bradnell, B., and Haire, M., 1981, *J. Neurol. Sci.* **53:**29–43.
64. Budka, H., Lassman, H., and Popow-Kraupp, T., 1982, *Acta Neuropathol.* **56:**52–62.
65. Haase, A. T., Swoveland, P., Stowring, L., Ventura, P., Johnson, K. P., Norrby, E., and Gibbs, C. J., Jr., 1981, *J. Infect. Dis.* **144:**154–160.
66. Goldstein, J. L., Anderson, R. G. W., and Brown, M. S., 1979, *Nature* **279:**679–685.
67. Dubois-Dalcq, M., and Rentier, B., 1980, *Progress of Medical Virology*, Volume 26 (J. L. Melnick, ed.), Karger, Basel, pp. 158–213.

68. Oldstone, M. B. A., Fujinami, R. S., and Lampert, P. W., 1980, *Progress of Medical Virology*, Volume 26 (J. L. Melnick, ed.), Karger, Basel, pp. 45–93.
69. Raff, M., 1976, *Nature* **259**:265–266.
70. Heineman, S., Merlie, J., and Lindstrom, J., 1978, *Nature* **274**:65–67.
71. Schechter, Y., Hernsey, L., Schlessinger, J., and Cuatrecasas, P., 1979, *Nature* **278**:835–838.
72. Hamprecht, B., 1977, *Int. Rev. Cytol.* **49**:99–170.
73. Koschel, K., and Halbach, M., 1979, *J. Gen. Virol.* **42**:627–632.
74. Tsiang, H., 1982, *J. Gen. Virol.* **61**:277–281.
75. Münzel, P., and Koschel, K., 1981, *Biochem. Biophys. Res. Commun.* **101**:1241–1250.
76. Tsiang, H., and Guillon, J. C., 1981, *Acta Neuropathol.* **55**:263–267.
77. Oldstone, M. B., Holmstoen, J., and Welsh, R. M., Jr., 1977, *J. Cell. Physiol.* **91**:459–471.
78. Lundén, R., Vahlne, A., and Lycke, E., 1980, *Proc. Soc. Exp. Biol. Med.* **165**:55–62.
79. Miller, C. A., Erlich, S., and Raine, C. S., 1981, *Life Sci.* **29**:2473–2480.
80. Halbach, M., and Koschel, K., 1979, *J. Gen. Virol.* **42**:615–619.
81. Koschel, K., and Muenzel, P., 1980, *J. Gen. Virol.* **47**:513–517.
82. Fagraeus, A., Tyrell, D. L. J., Norberg, R., and Norrby, E., 1978, *Arch. Virol.* **57**:291–296.
83. Meyer, R. K., Burger, M. M., Tschannen, R., and Schäfer, R., 1981, *Arch. Virol.* **67**:11–18.
84. Lösse, D., Lauer, R., Veder, D., and Radsak, K., 1982, *Arch. Virol.* **71**:353–359.
85. Winkler, M., Dawson, G. J., Elizan, T. S., and Berl, S., 1982, *Arch. Virol.* **72**:95–103.
86. Schlehofer, J. R., Habermehl, K.-O., Diefenthal, W., and Hampl, H., 1979, *Intervirology* **11**:158–166.
87. Nair, N., 1981, *J. Virol.* **37**:268–273.
88. Carrasco, L., and Smith, A. E., 1976, *Nature* **264**:807–809.
89. Egberts, E., Hackett, P. B., and Traub, P., 1977, *J. Virol.* **22**:591–597.
90. Garry, R. F., Bishop, J. M., Parker, S., Westbrook, K., Lewis, G., and Waite, M. R. F., 1979, *Virology* **96**:108–120.
91. Foster, K. A., Gill, K., Micklem, K. J., and Pasternak, C. A., 1980, *Biochem. J.* **190**:639–646.
92. Poste, G., and Pasternak, C. A., 1978, *Cell Surface Rev.* **5**:305–367.
93. Impraim, C. C., Foster, K. A., Micklem, K. J., and Pasternak, C. A., 1980, *Biochem. J.* **186**:847–860.
94. Micklem, K. J., and Pasternak, C. A., 1982, *J. Physiol.* (*London*) **326**:11P.
95. Francoeur, A. M., and Stanners, C. P., 1978, *J. Gen. Virol.* **39**:551–554.
96. Batra, G. K., Nahmias, A. J., and De Haan, R. L., 1976, *Nature* **259**:677–678.
97. Shrier, A., Nahmias, A. J., and De Haan, R. L., 1978, *Am. J. Physiol.* **234**:170–176.
98. Oakes, S. G., Petry, R. W., Ziegler, R. J., and Pozos, R. S., 1981, *J. Neuropathol. Exp. Neurol.* **40**:380–389.
99. Fukuda, J., and Kurata, T., 1981, *Brain Res.* **211**:235–241.
100. Lima, P. H., Oakes, S. G., Pozos, R. S., and Ziegler, R. J., 1981; International Workshop on Herpesviruses, Esculapio, Bologna, p. 158 (abstract).
101. Lycke, E., Modigh, K., and Roos, B.-E., 1970, *Brain Res.* **23**:235–246.
102. Lycke, E., and Roos, B.-E., 1972, *Brain Res.* **44**:603–613.
103. Lycke, E., and Roos, B.-E., 1974, *J. Neurol. Sci.* **22**:277–289.
104. Kataoka, K., Bak, I. J., and Markham, C. H., 1979, *Brain Res.* **169**:401–405.
105. Bonilla, E., Ryder, S., and Hernandez, H., 1975, *J. Neurochem.* **25**:529–530.
106. Levine, S., Bonilla, E., Ryder, S., Salazar, M., and Rangel, P., 1981, *Neurochem. Res.* **6**:691–697.
107. Bonilla, E., Ryder, E., and Ryder, S., 1980, *Neurochem. Res.* **5**:209–215.
108. Guchhait, R. B., and Monjan, A. A., 1980, *Neuroscience* **5**:1105–1111.
109. Bondy, S. C., Burks, J. S., and Harrington, M. E., 1979, *Arch. Neurol.* **36**:540–543.
110. Lycke, E., Modigh, K., and Roos, B.-E., 1969, *Experientia* **25**:951–953.
111. Seegal, R., Sikora, E., and Hotchin, J., 1980, *Pharmacol. Biochem. Behav.* **12**:61–66.
112. Seegal, R., Hotchin, J., and Sikora, E., 1981, *Life Sci.* **29**:777–782.
113. Rohwer, R. G., Neckers, L. M., Trepel, J. B., Gajdusek, D. C., and Wyatt, R. J., 1981, *Brain Res.* **220**:367–371.

114. Goudsmit, J., Rohwer, R. G., Silbergeld, E. K., and Gajdusek, D. C., 1981, *Brain Res.* **220**:372–377.
115. McFarland, D. J., Baker, F. D., and Hotchin, J., 1980, *Physiol. Behav.* **24**:911–914.
116. Lycke, E., and Roos, B.-E., 1975, *J. Neurol. Sci.* **26**:49–60.
117. Lee, C. J., 1975, *Pediatr. Res.* **9**:645–652.
118. Herndon, R. M., Margolis, G., and Kilham, L., 1971, *J. Neuropathol. Exp. Neurol.* **30**:196–205.
119. Young, A. B., Oster-Granite, M. L., Herndon, R. M., and Snyder, S. H., 1974, *Brain Res.* **73**:1–13.
120. Snyder, S. H., Young, A. B., Oster-Granite, M. L., and Herndon, R. M., 1975, *Metabolic Compartmentation and Neurotransmission: Relation to Brain Structure and Function* (S. Berl, D. D. Clarke, and D. Schneider, eds.), Plenum Press, New York, pp. 1–10.
121. Dal Canto, M. C., and Rabinowitz, S. G., 1982, *Ann. Neurol.* **11**:109–127.
122. Weiner, L. P., 1973, *Arch. Neurol.* **28**:298–303.
123. Powell, H. C., and Lampert, P. W., 1975, *Lab. Invest.* **33**:440–445.
124. Fleury, H. J. A., Sheppard, R. D., Bornstein, M. B., and Raine, C. S., 1980, *Neuropathol. Appl. Neurobiol.* **6**:165–179.
125. Knobler, R. L., Dubois-Dalcq, M., Haspel, M. V., Claysmith, A. P., Lampert, P. W., and Oldstone, M. B. A., 1981, *J. Neuroimmunol.* **1**:81–92.
126. Takahashi, K., Goto, N., Ishida, T., Katami, K., and Fujiwara, K., 1981, *Jpn. J. Exp. Med.* **51**:323–330.
127. Raine, C. S., Prineas, J. W., Sheppard, R. D., Bornstein, M. B., and Dubois-Dalcq, M., 1977, *J. Neurol. Sci.* **33**:13–20.
128. Wisniewski, H., Raine, C. S., and Kay, W. J., 1972, *Lab. Invest.* **26**:589–599.
129. Lipton, H. L., 1975, *Infect. Immun.* **11**:1147–1155.
130. Lipton, H. L., and Dal Canto, M. C., 1976, *Science* **192**:62–64.
131. Dal Canto, M. C., 1982, *Infect. Immun.* **35**:1133–1138.
132. Townsend, J. J., and Baringer, J. R., 1976, *J. Neuropathol. Exp. Neurol.* **35**:100.
133. Kristensson, K., Vahlne, A., Persson, L. A., and Lycke, E., 1978, *J. Neurol. Sci.* **35**:331–340.
134. Townsend, J. J., 1981, *J. Neuropathol. Exp. Neurol.* **40**:369–379.
135. Kristensson, K., Svennerholm, B., Persson, L., Vahlne, A., and Lycke, E., 1979, *J. Neurol. Sci.* **43**:253–264.
136. Townsend, J. J., and Baringer, J. R., 1979, *Lab. Invest.* **40**:178–182.
137. Kristensson, K., Svennerholm, B., Vahlne, A., Nilheden, E., Persson, L., and Lycke, E., 1982, *J. Neurol. Sci.* **53**:205–216.
138. Townsend, J. J., 1981, *J. Neurol. Sci.* **50**:435–441.
139. Sheahan, B. J., Barett, P. N., and Atkins, G. J., 1981, *Acta Neuropathol.* **53**:129–136.
140. Illavia, S. J., Webb, H. E., and Pathak, S., 1982, *Neuropathol. Appl. Neurobiol.* **8**:35–42.
141. Kelly, W. R., Blakemore, W. F., Jagelman, S., and Webb, H.E., 1982, *Neuropathol. Appl. Neurobiol.* **8**:43–53.
142. Panitch, H. S., Swoveland, P., and Johnson, K. P., 1979, *Neurology* **29**:548–549.
143. Hochberg, F. H., Lehrich, J. R., and Arnason, B. G. W., 1977, *Neurology* **27**:584–587.
144. Wisniewski, H. M., 1977, *Br. Med. Bull.* **33**:54–59.
145. Cammer, W., Bloom, B. R., Norton, W. T., and Gordon, S., 1978, *Proc. Natl. Acad. Sci. U.S.A.* **75**:1554–1558.
146. Field, H. J., and Wildy, P., 1978, *J. Hyg.* **81**:267–277.
147. Tenser, R. B., Miller, R. L., and Rapp, F., 1979, *Science* **205**:915–917.
148. Tenser, R. B., and Dunstan, M. E., 1979, *Virology* **99**:417–422.
149. Price, R. W., and Khan, A., 1981, *Infect. Immun.* **34**:571–580.
150. Adler, R., Glorioso, J. C., and Levine, M., 1978, *J. Gen. Virol.* **39**:9–20.
151. Rice, M., Holland, L., and Wagner, E. K., 1979, *Arch. Virol.* **59**:345–355.
152. Levine, M., Goldin, A. L., and Glorioso, J. C., 1980, *J. Virol.* **35**:203–210.
153. Vahlne, A., and Lycke, E., 1978, *J. Gen. Virol.* **39**:321–332.
154. Walz, M. A., Price, R. W., and Notkins, A. L., 1974, *Science* **184**:1185–1187.
155. Price, R. W., and Schmitz, J., 1979, *Infect. Immun.* **23**:373–383.

156. Nesburn, A. B., Dickinson, R., and Radnoti, M., 1976, *Invest. Ophthalmol.* **15**:726–731.
157. Green, M. T., Rosborough, J. P., and Dunkel, E. C., 1981, *Infect. Immun.* **34**:69–74.
158. Price, R. W., 1979, *Science* **205**:518–520.
159. Openshaw, H., Puga, A., and Notkins, A. L., 1979, *Fed. Proc.* **38**:2660–2664.
160. Openshaw, H., Shavrina Asher, L. V., Wohlenberg, C., Sekizawa, T., and Notkins, A. L., 1979, *J. Gen. Virol.* **44**:205–215.
161. Scriba, M., and Tatzber, F., 1981, *Infect. Immun.* **34**:655–661.
162. Stanwick, T. L., Anderson, R. W., and Nahmias, A. J., 1977, *Infect. Immun.* **18**:342–347.
163. Blue, W. T., Winland, R. D., Stobbs, D. G., Kirksey, D. F., and Savage, R. E., 1981, *Antimicrob. Agents Chemother.* **20**:547–548.
164. Zimmerman, J. E., Jr., Glazer, R., and Rapp, F., 1973, *J. Virol.* **12**:1442–1445.
165. Robbins, S. J., and Rapp, F., 1980, *Virology* **106**:317–326.
166. Miller, C. A., and Carrigan, D. R., 1982, *Proc. Natl. Acad. Sci. U.S.A.* **79**:1629–1633.
167. Hall, W. W., and Choppin, P. W., 1979, *Virology* **99**:443–447.
168. Stephensson, J. R., Siddell, S. G., and ter Meulen, V., 1981, *J. Gen. Virol.* **57**:191–197.
169. Johnson, K. P., Norrby, E., Swoveland, P., and Carrigan, D.R., 1981, *J. Infect. Dis.* **144**:161–169.
170. Faulkner, G., Dubois-Dalcq, M., Hooghe-Peters, E., McFarland, H. F., and Kazzarini, R. A., 1979, *Cell* **17**:979–991.
171. Clark, H. F., 1980, *Infect. Immun.* **27**:1012–1022.
172. Yaron, M., Yaron, I., Gurani-Rotman, D., Revel, M., Linder, H., and Zor, U., 1977, *Nature* **267**:457.
173. Fitzpatrick, F. A., and Stringfellow, D. A., 1980, *J. Immunol.* **125**:431–437.
174. Schultz, R. M., Stoychkov, J. N., Pavlidis, N., Chirigos, M.A., and Okowski, Z. L., 1979, *J. Reticuloendothel. Soc.* **26**:93–102.
175. Stringfellow, D., 1978, *Science* **201**:376–378.
176. Gorman, R., 1975, *J. Cyclic Nucleotide Res.* **1**:1–9.
177. Weber, J., and Stewart, R., 1975, *J. Gen. Virol.* **28**:363–372.
178. Lycke, E., Kristensson, K., Svennerholm, B., Vahlne, A., and Ziegler, R., 1984, *J. Gen. Virol.* **65**:55–64.

20

Slow Viruses and Prions

Stanley B. Prusiner and Ashley T. Haase

1. SLOW VIRUSES AND PRIONS

Almost two decades have passed since the discovery of the first human slow infection, kuru.[1] Several additional slow infections of humans have been discovered in recent years, but all of these diseases are rather rare. Despite an intensive effort, it has not been possible to show that a common degenerative disorder possesses a slow infectious etiology. Either slow infections are indeed rare and not the cause of such disorders, or our present techniques are inadequate to detect most slow infectious agents.

Eleven years before the discovery that kuru was caused by a slow transmissible agent, Sigurdsson introduced the concept of slow infections.[2] He characterized slow infections as having four cardinal features: (1) a prolonged incubation period ranging from several months to decades, (2) a brief progressive clinical course leading to death, (3) a pathology confined to a single organ, and (4) a natural host usually confined to a single species. It appears that slow infections are caused by at least two different classes of infectious agents: viruses and prions.

Many different types of viruses have been found to cause slow infections in humans and animals (Table I). In contrast, the prions causing scrapie, transmissible mink encephalopathy (TME), Creutzfeldt–Jakob disease (CJD), and kuru are probably quite similar (Table II).

Although considerable progress has been made in elucidating the macromolecular structures of slow viruses, relatively little biochemical information is available about how they destroy their target tissue. One of the well-studied slow viruses is visna virus, which attacks the central nervous system (CNS) of sheep and goats. Not only is the structure of this slow virus well documented, significant progress has been made toward understanding how the virus causes a chronic, progressive demyelinating disorder of the CNS.

Stanley B. Prusiner • Departments of Neurology and Biochemistry and Biophysics, University of California, San Francisco, California 94143. *Ashley T. Haase* • Departments of Medicine and Microbiology and Immunology, University of California School of Medicine, San Francisco, California 94143; and Veterans Administration Hospital, San Francisco, California 94121.

Table I
Slow Virus Diseases

Disease	Virus class	Natural host
Visna	Retrovirus	Sheep
Distemper	Paramyxovirus	Dogs
Subacute sclerosing panencephalitis (SSPE)	Paramyxovirus	Humans
Progressive multifocal leukoencephalopathy (PML)	Papovavirus	Humans
Progressive rubella panencephalitis (PRP)	Togavirus	Humans

In contrast to the progress made in studying slow viruses, the acquisition of knowledge about the structure of prions has been slow, tedious, and frustrating. The enigmatic properties of the prototypic prion causing scrapie have fueled the imagination of many investigators. Until recently, so little was known about the structure of the scrapie agent that no fewer than 16 hypotheses about its composition seemed viable. Progress in purification of the agent from hamster brain has shown that a protein component is required for infectivity and that the dominant biophysical characteristics of the infectious particle are similar to those of a hydrophobic protein. On this background, the term "prion" was introduced to distinguish these novel particles from viruses and viroids.

In this chapter two slow infectious agents will be reviewed in detail: the slow virus causing visna and the prion causing scrapie.

2. *VISNA VIRUS*

The lentiviruses (for reviews, see refs. 3–5), of which visna is the prototype, comprise a subfamily of retroviruses that includes visna–maedi,[6,7] zwoegerziekte,[8] progressive pneumonia virus,[9] an arthritis-inducing virus (O. Narayan and J. Gorham, unpublished data cited in ref. 10) from sheep, and caprine encephalitis–arthritis virus (CAEV)[11] from goats. Visna, maedi, zwoegerziekte, and progressive pneumonia virus isolates are most likely variants of a single virus. Although various maedi isolates appear more closely related to

Table II
Prion Diseases[a]

Disease	Natural host
Scrapie	Sheep and goats
Transmissible encephalopathy (TME)[b]	Mink
Kuru[b]	Humans—Fore
Creutzfeldt–Jakob Disease (CJD)[b]	Humans

[a] Alternative terminology includes unconventional slow virus diseases and subacute transmissible spongiform encephalopathies.
[b] Proteinaceous infectious particles causing these diseases not yet demonstrated.

one another than to visna isolates, and *vice versa*, there is as yet no correlation of the different disease syndromes (see below) with particular molecular variations. The CAEV does exhibit some differences from the visna/maedi group. These viruses comprise a subfamily with common morphological features, related structural polypeptides and antigens, and shared nucleotide sequences. In tissue-culture cells derived from their natural host, the viruses replicate productively and lytically, causing formation of polykaryocytes. The viruses are nontransforming and do not cause neoplasms. In their natural hosts *in vivo*, they cause slowly progressive inflammatory destructive lesions.

2.1. Lentivirus-Induced Diseases

Visna–maedi first came to attention in Iceland, where introduction of the virus into a susceptible population of sheep led to particularly widespread disease, but similar conditions are endemic throughout the world.[12] Although the disease occurs primarily in sheep, both experimental and natural transmission to goats also occurs. Natural infection of goats has been reported in India, Holland, and Germany. In these cases, sheep seem to be the source of infection.[13,14] Attempts to transmit the disease to small animals have been unsuccessful.[12,15]

2.1.1. Visna

Visna (wasting) is primarily a neurological disease of the CNS. Visna virus affects animals of both sexes and at any age, although the disease is infrequent in sheep less than 2 years old; about 10% of infected animals may not show clinical symptoms at 8 years post-infection. The first symptoms include an aberration in gait, trembling of the lips, unnatural tilting of the head, and, infrequently, blindness. Progressive paralytic changes follow, especially of the hind limbs. The survival of the animal can be extended, and the appearance of emaciation can be avoided for long periods of time, if the animals are helped to obtain food and water. The only effective eradication program has proved to be slaughter of all animals from flocks showing infection.

The principal histopathological change in visna is destruction of tissue within inflammatory foci, which are located in the periventricular areas, choroid plexus, and meninges.[16,17] The inflammatory infiltrates consist of lymphocytes (at various stages of maturation, including mature plasma cells) and macrophages. These cells accumulate in the CNS as perivascular cuffs and as discrete foci; the cells may also infiltrate the interalveolar septa of the lung. The CNS tissue is destroyed within the inflammatory lesions; when damage becomes extensive, paralysis ensues. In animals with advanced lesions, demyelination may occur, but this is not a universal finding. Although visna virus induces syncytia formation *in vitro*, no syncytia have been observed *in vivo*,[18] suggesting that this phenomenon is not important in the generation or progression of the lesions.

In nature, infection is transmitted horizontally from ewe to lamb via the milk (in which virus is excreted) or, in older animals, by respiratory aerosol or infected saliva.[19] There is no evidence for transplacental infection.[19]

2.1.1. Maedi

Maedi (shortness of breath) is the name given to the slowly progressive pulmonary manifestations in virus-infected sheep. In addition to respiratory difficulties and a dry cough, the animals often die from acute bacterial pneumonia. The most striking postmortem change is a two- to threefold increase in the weight of the lung. The principal histopathological alteration consists of a thickening of the interalveolar septa as a result of infiltration of inflammatory cells; eventually, the alveolar spaces may become obliterated. Foci of inflammatory lesions occur throughout the lung parenchyma with peribronchial and perivascular hyperplasia and fibrosis.[12] As the lesions in the lung enlarge, the infiltration leads to impairment of gaseous exchange and results in shortness of breath.

Isolation of virus is most frequent from choroid plexus and lungs, although many other tissues are also productive. The disease occurs in sheep of both sexes, but rarely before 3 years of age. As with visna, transmission occurs through direct contact or by ingesting colostrum from infected ewes.

2.1.3. Caprine Encephalitis–Arthritis Virus

Goats have been found to develop an arthritic syndrome caused by CAEV that is distinct from visna and maedi. The encephalomyelitis–arthritis syndrome in goats has been recognized in Germany,[20] Australia,[21] and the United States[22–24] Interestingly, visna–maedi is not found in Australia, providing further evidence for considering CAEV a distinct virus.

The clinical course of CAEV infection shares several features with visna–maedi: institutional pneumonia with shortness of breath, leukoencephalitis, paralysis, and the diagnostic inflammatory lesions in the CNS and lung.[11,22] The principal hallmark of the CAEV syndrome that distinguishes it from visna–maedi is arthritis in the joints.[11,25] Although arthritis may occur in visna/maedi-infected sheep, it is most prevalent in goats that develop a CAEV infection. Focal necrosis and hyperplasia of synovial membrane lining cells occur; as the necrotic process progresses, it may result in extreme fibrosis and destruction of surrounding cartilage. Some of these changes are reminiscent of rheumatoid arthritis.

Following experimental inoculation of goats, inflammatory lesions appear in the brain, joints, and lungs as early as 1 week post-infection and may persist for several years. Adult and newborn goats are equally susceptible. Virus can be recovered from many tissues, including choroid plexus, brain, synovial membrane, spleen, lymph nodes, thymus, and peripheral blood leukocytes.[26] In nature, transmission occurs horizontally, either from ewe to kid or via other methods of direct contact.

2.2. Structure of Visna Virus

Visna virions are enveloped spheres. They generally measure 80 to 120 nm in diameter with a 40-nm electron-dense core. The virions have an $S_{20,w}$

Table III
Characteristics and Properties of Visna Virus

Morphology[a]
Spherical enveloped virions 80–120 nm in diameter with a 40-nm central electron-dense core; surface bears 8-nm knoblike projections
Physical and chemical properties
Isopycnic density: 1.15–1.16 g/cm^3 in sucrose; isoelectric point: 3.8[b]
Sedimentation: 600 S in sucrose[c]
Probable composition: 60% protein, 35% lipid, 3% carbohydrate, 2% RNA[d]
Inactivation: relatively resistant to ultraviolet irradiation; infectivity abolished by lipid solvents, periodate, phenol, trypsin, ribonuclease, formaldehyde, and low pH (less than 4.2)
Thermal stability: infectivity preserved for months in the presence of serum at −50°C; relatively stable at 0–4°C; infectivity destroyed at 56°C; at 37°C one-half the infectivity is lost in 6–8 hr, 90% in 20–30 hr[e]

[a] Refs. 27–30.
[b] Ref. 31.
[c] Ref. 32.
[d] Refs. 3, 33.
[e] Refs. 34, 35.

of 600 S and a density of approximately 1.15 g/cm^3 like other enveloped viruses (Table III).

Visna virus resembles other retroviruses in its structure and composition. Like all retroviruses, it possesses a virion-associated reverse transcriptase that transfers information from the RNA genome of the virus to a DNA intermediate in the cell[36,37] (Table IV).

Visna virus enters the cell by fusion,[29] and its RNA is released into the cytoplasm within the first hour of infection.[64] The completed viral DNA in the nucleus consists of linear duplex molecules equivalent to a transcript of a subunit of viral RNA and to rare circular forms. Many of the linear molecules have a gap in the plus strand.[65] A variable proportion of viral DNA is associated with high-molecular-weight cellular DNA,[66–68] but there is no evidence as yet that establishes covalent linkage of viral and host cell sequences.

Viral DNA synthesis continues throughout the life cycle of the virus to reach 200–300 copies per cell,[3,65,69] representing about a tenfold amplification over input RNA. The mechanism involved in this amplification seems to be superinfection and is not necessary for the transcription of viral RNA and subsequent production of progeny virus. However, viral DNA synthesis to the level of about 40 copies per cell does seem to be required for the productive life cycle of the virus.[70]

Viral RNA synthesis is initiated asynchronously in the nucleus of infected cells as early as 5–7 hr after infection[70] and proceeds exponentially to reach levels of several thousand copies of RNA per cell.[64] The aggregation of subunits to form the 70 S complex begins during virus maturation and is completed extracellularly.[71] There are three classes of mRNA with sedimentation coefficients of 35 S, 28 S, and 21 S.[72] By analogy to other retroviruses, the 35 S and 28 S mRNAs may code for the major core polypeptides and reverse transcriptase and envelope glycoprotein, respectively[73]; as is the case with other

Table IV
Molecular Structure of Visna Virus

Proteins: structural organization and antigenicity[31,38–40]
Four structural polypeptides designated gp135, p30, p16, and p14 (see ref. 41 for nomenclature) comprise 90% of the protein mass of the virus[31,38–40]
gp135 is the glycoprotein of the knoblike projections emanating from the virion surface and elicits type-specific neutralizing antibody[38,42–44]
p30 and p14 are located in the virion core[40]; p 30 is the major group-specific antigen, with antigenic determinants shared by visna, maedi, PPV, and zwogerziekte virus[45]

Enzymes[46–52]
RNA-directed DNA polymerase or reverse transcriptase; dimer with subunits of molecular weight 68,000 located in the virion core

RNAs
Virion nucleic acid is single-stranded RNA with extensive secondary structure; sediments at 60–70 S in sucrose gradients. On denaturation, two or three subunits sedimenting at 35 S are released[53–59]
Base composition (%): C, 16; A, 36; G, 26; U, 22[60]
Each subunit contains essentially identical information; i.e., the genome is polyploid[60–62]
Subunits have a molecular weight of 3.6×10^6, are of plus strand polarity, and have a poly(A) tract at the 3′ end[60,63]
Virions also contain ribosomal and low-molecular-weight RNAs derived from the cell. One species of tRNA binds to the 70 S genome and functions as the primer for DNA synthesis[59] (A. J. Faras, A. T. Haase, and J. M. Bishop, unpublished)

retroviruses, synthesis of polypeptides takes place in the cytoplasm,[74,75] and virion polypeptides are generated by proteolytic cleavage of larger precursors.[76] In the final step in the life cycle, virions are formed by budding from the plasma membrane of the infected cell.[29,77,78]

The lentivirus type- and group-specific antigens do not share determinants with the group-specific antigens of other retroviruses, including isolated examples[45] of retroviruses that are associated with slow infections (equine infectious anemia virus) or cause cell fusion (bovine syncytial virus) or CPE (retriculoendotheliosis virus). The polypeptides of lentiviruses also differ somewhat in size and number from oncornaviruses. The glycopeptide gp135 is apparently larger, although this may reflect only the extent of glycosylation, and there is one report that the glycopeptide has a molecular weight of 70,000.[44] There are three core polypeptides in lentiviruses and four in oncornaviruses. Lentiviruses do not have a second internal membrane either, since the nucleocapsid is immediately apposed to the virion envelope in the formation of the virion bud.

2.3. Visna Pathogenesis

Slow infections pose three central questions in viral pathogenesis: (1) How does a virus escape from the immunologic surveillance mechanisms of its host over a period of years? (2) How is a virus disseminated in the animal in the face of these defense mechanisms? (3) How does the virus cause the pathological lesions?

Viruses ordinarily are eliminated by the concerted action of the host's macrophage/granulocyte phagocytic system, humoral and cellular immunity, and interferon. The persistence of the lentiviruses for periods of years implies that there are mechanisms that frustrate these defensive measures. In visna, a cellular immune response is demonstrable in the first weeks of infection,[79] and readily detectable neutralizing antibodies appear later. Subsequently, high levels of antibodies are sustained throughout the course of the disease,[17,80] and there are nonimmunoglobulin components in serum and cerebrospinal fluid that neutralize virus as well.[81,82] These defense mechanisms undoubtedly suppress overall virus production without clearing the infection. In most animals, the continued production of virus with attendant pathology eventually leads to patent disease and death. The incubation period from the time of exposure to the overt appearance of symptoms is variable, ranging from months to several years; later, the clinically recognizable phase may also be extremely protracted, generally extending over several years.

2.3.1. Virus Gene Expression and Persistence

Considerable insight into the mechanism of persistence of lentiviruses has been gained by analysis of virus replication in individual cells by *in situ* hybridization refined to the point of detection of single copies of the viral genome.[83] In the tissues of animals infected with visna virus, only a small proportion of the cells containing proviral DNA synthesize antigens or viral particles detectable by immunofluorescence and electron microscopy, respectively.[84] This restriction *in vivo* is imposed, at least in part, at a transcriptional level.[85] Some cells in the CNS of infected sheep contain virus-specific RNA corresponding quantitatively to levels in tissue culture of cells producing 0.1–1 plaque-forming units per cell or 0.1–1% of the full yield in permissive infections *in vitro*. These low levels of RNA synthesis may be responsible for the small amounts of virus always present in tissues that perpetuate and extend the infectious process.

The restriction of viral gene expression and the small proportion of cells synthesizing viral antigen provide a satisfying explanation for virus persistence at the molecular level, since the immunologic surveillance system of the host will neither detect nor destroy most of the infected but antigenically "silent" cells. Such cells are also not likely to be eliminated by other mechanisms such as interferon, since the lentiviruses are exceptionally resistant to its antiviral activity.[86] Thus, the animal is left with a burden of infected cells that can perpetuate the infection.

2.3.2. Virus Dissemination

The spread of virus to distant sites in the presence of antibody can also be explained by latent dissemination in motile cell populations (peripheral blood leukocytes) with comparable limitations on gene expression in the cells that carry the virus. This hypothesis is substantiated by the observation that all of the cells in the hematopoietic system can be chronically infected with visna virus, although virus can be recovered from peripheral blood leukocytes at only a very low frequency (10^{-6}), presumably because of cell association of viral particles.[17] Conceivably, peripheral blood leukocytes, their precursors in the marrow, and other elements of the reticuloendothelial system might harbor the viral genome with intermittent or continuous dissemination to distant sites. As long as antigen expression remained restricted in the blood leukocytes, circulating neutralizing antibody would have little effect in clearing the infection.

Antigenic variation is a second plausible explanation for the spread of virus. In visna-virus-infected sheep, variants arise that are not neutralized by antibody to the inoculated strain and therefore could replicate and spread temporarily unchecked.[87,88] Antigenic variants of visna virus probably arise by mutation and show alterations in the *env* gene that codes for the surface glycoprotein antigen(s) to which neutralizing antibody is directed.[43,89] However the role of the variants in dissemination is unclear for two reasons. First, parental and variant strains are not found extracellularly but rather are always isolated from peripheral blood leukocytes. Second, despite equal replicative ability and the postulated selective advantage to the variant of temporary escape from immunolysis, the variants do not replace the parental strain or other variants in successive episodes. Instead, parental and variant strains are isolated contemporaneously.[88]

2.3.3. Slowness

Another striking feature of lentivirus infections is the protracted course of the disease. Slowness clearly refers to the pace of progression of disease in animals, not to the replication of these viruses, whose life cycles are completed *in vitro* in 3 or 4 days. Presumably, slowness reflects the interacting effects of the restricted viral gene expression and diminished production of infectious virus. Continuous reinfections enhance the pace of disease progression, as do conditions of stress (e.g., breeding, parasitic infections, and harsh climate).

2.3.4. Immunopathological Processes

In lentivirus infections, the basic pathological changes likely result from deleterious effects of the interaction between virus and the host defensive apparatus. Destruction of tissue in inflammatory foci provides evidence that the host's immune response is contributory to disease manifestiations. Immunosuppressive treatment in these infections has the sparing effect predicted for immunologically mediated disease,[18,90] whereas immunopotentiating treatment increases the severity of the lesions.[91]

3. SCRAPIE PRION

In 1959, Hadlow suggested that kuru, a CNS degenerative disease of New Guinea highlanders, might be similar to scrapie because the pathologies of these disorders share many features.[92] The transmission of kuru to chimpanzees in 1965 by Gajdusek, Gibbs, and Alpers forced a major reconsideration of the etiology of all degenerative disorders and made scrapie a subject of intense medical interest.[1] Subsequently, CJD, a progressive presenile dementia, was shown by Gibbs, Gajdusek, and co-workers to be caused by a transmissible agent.[93,94]

A recent study suggests that there may be similarities between the agents causing scrapie and CJD.[95] Goats inoculated with brain tissue from demented patients dying of CJD developed a neurological disorder 3 to 4 years after inoculation. Five out of ten CJD inocula have produced disease in goats (W. J. Hadlow and S. B. Prusiner, unpublished observations). Experimental CJD in goats is indistinguishable both clinically and neuropathologically from natural scrapie. Monkeys have been used as a common experimental host for scrapie and CJD; curiously, chimpanzees are susceptible to CJD but not scrapie.[98] Numerous attempts to link scrapie epidemiologically to CJD have been unsuccessful.[99] At present, there is no direct evidence that the scrapie agent causes disease in humans.

In contrast to CJD, which occurs worldwide, kuru is found only in a small mountainous region of Papua–New Guinea. Epidemiological studies of kuru provide evidence for incubation periods of 20 to 30 years.[100,101] Although considerable evidence implicates cannibalism in the spread of kuru, no direct observations of cannibalistic acts in the "endemic" region have been recorded. Attempts to transmit kuru by feeding infected brain tissue to chimpanzees have been unsuccessful, although one monkey developed a kurulike illness 36 months after oral ingestion of the kuru agent.[102] In contrast, goats fed scrapie-infected tissue frequently develop disease.[103,104] Recently, we have taken advantage of the natural cannibalistic activities of hamsters to develop an experimental model of scrapie transmitted by cannibalism (S. B. Prusiner and S. P. Cochran, unpublished data). Oral transmission of the scrapie agent appears to be extremely inefficient. Cannibalism requires a dose of agent 10^9 times greater than that needed to produce scrapie by intracerebral injection. These results provide compelling evidence for oral transmission of the scrapie agent and may offer new insights into the spread of kuru by cannibalism among the Fore people and their neighboring tribes.

3.1. Hypothetical Structures for the Scrapie Agent

Investigators have been aware of the unusual properties of the scrapie agent for more than three decades. Hypotheses on the chemical structure of the scrapie agent have included *Sarcosporidia* parasite,[105,106] "filterable" virus,[107–110] small DNA virus,[111] replicating protein,[112–115] replicating abnormal polysaccharide within membranes,[116,117] DNA subvirus controlled by a transmissible linkage substance,[118,119] provirus consisting of recessive genes generating RNA particles,[120] naked nucleic acid similar to plant viroids,[121,122] unconventional virus,[100,123–128] aggregated conventional virus with unusual properties,[129] replicating polysaccharide,[130,131] nucleoprotein complex,[132] nucleic acid surrounded by a polysaccharide coat,[133–135] spiroplasmalike organism,[136–138] multicomponent system with one component quite small,[139,140] and membrane-bound DNA.[139–141]

3.2. Prions

Recent progress in the purification and characterization of the scrapie agent suggests that it is probably not a virus.[142] The novel properties of the scrapie agent clearly distinguish it from other small infectious agents such as viroids and plasmids and have prompted introduction of the term, "prion."[142] Prions are defined as small proteinaceous infectious particles that resist inactivation by most procedures that modify nucleic acids. Current knowledge does not allow exclusion of a small nucleic acid within the interior of the prion. Neuropathological observations and transmission studies on the kuru, CJD, and TME agents indicate that they are similar to the scrapie agent in many respects[95,100]; however, direct evidence for a proteinaceous structure, such as reduction of infectivity by protease digestion, has only been demonstrated for the scrapie agent.

The definition of a prion must remain operational until the detailed molecular structure of these particles is known. The prototypic scrapie agent is clearly a novel infectious entity because it is resistant to procedures attacking nucleic acids, resistant to inactivation by heat,and has an apparent small size.[142] Although six lines of evidence show that the scrapie agent contains a protein required for infectivity, all attempts to demonstrate a nucleic acid within the agent have been unsuccessful.[142–145]Assuming a viruslike substance, the apparent small size of the agent makes it unlikely that a hypothetical nucleic acid buried within the prion is of sufficient size to act as a gene coding for the protein(s) of the surrounding shell. Nevertheless, replication of the scrapie agent in natural and experimental hosts is well documented. For example, one infectious dose of the agent inoculated into the brain of a hamster appears to induce the production of $>10^9$ ID_{50} units during the ensuing 120-day period.[146,147]

3.3. Protein Required for Prion Infectivity

The purification of prions allowed application of techniques possessing a high degree of specificity for the hydrolysis and modification of proteins. We have been able to show convincingly that scrapie infectivity depends on an intact protein component, since protease degradation inactivates the agent.[142,145] Inactivation of the agent was also observed after chemical modification with diethylpyrocarbonate (DEP).[148] The infectivity of the inactive, modified agent was restored by subsequent treatment with hydroxylamine. Like proteins, single-stranded nucleic acids can also be chemically modified by DEP, but in contrast to proteins, these biologically inactive, modified nucleic acids cannot be reactivated by hydroxylamine treatment. In fact, hydroxylamine alone modifies and inactivates nucleic acids.

Neither protease digestion nor chemical modification inactivated the scrapie agent in crude homogenates. Substantial purification of the agent was required for these techniques to diminish scrapie infectivity. Four lines of evidence using denaturation procedures, including detergents such as sodium dodecyl sulfate (SDS), chaotropic ions such as guanidinium thiocyanate, compounds such as urea, and organic solvents such as phenol, provided additional information about the proteinaceous structure of the infectious agent.[142,144,149] In all, six lines of evidence provide compelling data for a protein requirement in the expression of scrapie agent infectivity (Table II).[142]

3.4. Search for a Prion Nucleic Acid

Having identified a macromolecular component of the scrapie agent, we began to probe for the putative genome.[142] Like other investigators,[125,150] we found that the agent was extremely resistant to treatment by nucleases. However, virtually all viruses resist inactivation by nucleases because the viral protein coat prevents access of the nuclease to the genome buried within its core. In addition, we extended the earlier studies of others[151,152] that probed for a genome within the scrapie agent using UV irradiation. Even after substantial purification, the agent remained highly resistant to inactivation by irradiation at 254 nm.

In our search for the putative scrapie genome, we have employed three additional methods involving selective modification or hydrolysis of nucleic acids.[142,153] Psoralens, which readily penetrate the coats of most viruses and form covalent adducts on photoactivation, have been used. Scrapie agent infectivity was unaltered by exposure to five different psoralens exhibiting varying degrees of hydrophobicity. In addition, we subjected the scrapie agent to exposure to zinc ions at 65° for up to 24 hr. Under these conditions, RNA is reduced to mononucleotides, and DNA is extensively cleaved, but the infectivity of the scrapie agent remained unaltered. The third method employed hydroxylamine, a chemical that modifies pyrimidines. Not only is the scrapie agent resistant to inactivation by hydroxylamine, but, in fact, hydroxylamine can be used to reactivate the DEP-inactivated agent as described above.[148]

3.5. Molecular Size of the Prion

One possible explanation for the extreme resistance of the scrapie agent to inactivation by irradiation at 254 nm is that the putative nucleic acid within the agent is quite small.[152] The resistance of the agent to inactivation by irradiation at 254 nm by pyrimidine dimer formation could be explained by a putative nucleic acid of 50 bases or fewer. This estimate assumes that the pyrimidines are randomly distributed within the nucleic acid and that one dimer is sufficient to inactivate the agent. Alternatively, the protein(s) of the scrapie agent might be modified by irradiation at 254 nm. Interestingly, a similar size for the putative nucleic acid of the scrapie agent can be calculated from the psoralen experiments. Assuming that the psoralen was able to freely penetrate the protein exterior of the agent, then only a scrapie nucleic acid of 40 bases or fewer could have eluded psoralen photoadduct formation under the conditions of our experiments. These data are consistent with four other lines of evidence indicating that the scrapie particle is quite small, as described below.

Almost two decades ago, Alper and colleagues found that the scrapie agent was extremely resistant to ionizing radiation.[154,155] They chose to interpret those data as indicating that the agent has a molecular weight of 64,000 to 150,000.[152] Target size estimates can be greatly influenced by aggregation of the infectious agent and the efficiency of the cellular repair processes. Even though we know that the scrapie agent is hydrophobic and has a propensity to aggregate, those target size estimates may prove to be reasonably accurate.

Using HPLC gel permeation chromatography with zwitterionic detergents, we found that the agent eluted as a uniform peak between bovine serum albumin (mol. wt. 69,000) and ovalbumin (mol. wt. 45,000), indicating that its molecular weight was approximately 50,000.[142] More recently, rate-zonal gradient centrifugation studies have indicated that the scrapie agent may have an observed sedimentation coefficient as low as ~2 S (S. B. Prusiner and D. F. Groth, unpublished observations). Since we do not know the density of the scrapie agent in detergent solutions, it is possible that the agent was floating in these rate-zonal gradients and that this value is artifactually low. We have also found that the scrapie agent, after disaggregation in zwitterionic detergents, passes through Nucleopore® filters with 15-nm pores (S. B. Prusiner and D. F. Groth, unpublished observations).

Thus, ionizing radiation, gel filtration, rate-zonal sedimentation, and membrane filtration data all suggest that the scrapie agent might be considerably smaller than any known virus. If the apparent diameter of the scrapie agent is as small as 5 nm, as suggested by gel permeation chromatography experiments, then we can calculate the size of a nucleic acid that could fit within such a particle. Let us assume that the agent has a protective protein that is 1 nm thick (10 Å); then the volume of the core will be 14.1 nm^3. From measurements of DNA packing in crystals and bacteriophage, there is space for a 12-nucleotide polymer consisting of six base pairs within the interior core.[142] Dehydration of the nucleic acid would allow 32 nucleotides to be encapsulated. If such oligonucleotides exist within the agent, they must operate differently from the traditional function of coding for the synthesis of a coat protein—they are simply too small to encode a protein product. If we enlarge the hypothetical scrapie particle to 11 nm in diameter, there would be space for a 346-nucleotide polymer consisting of 173 base pairs. Viroids contain 300 to 400 nucleotides, but even they appear to be of insufficient size to code for a protein.[156]

It seems doubtful that a protein shell only 10 Å thick could provide a coat impervious to UV irradiation, psoralens, Zn^{2+} ions, and NH_2OH. The protein shells of many small spherical viruses are 30 to 40 Å thick, but even these readily admit UV irradiation, psoralens, and hydroxylamine. Comparative studies on viroids and prions show that the scrapie agent is stable under conditions in which the potato spindle tuber viroid is readily inactivated by UV irradiation at 254 nm, psoralen photoadduct formation, Zn^{2+}-catalyzed hydrolysis, and chemical modifications by NH_2OH.[153]

None of our data allow us to exclude a nucleic acid from the scrapie agent, but we are able to conclude that it is unlikely that the scrapie agent contains a nucleic acid of sufficient size to code for the protein or proteins comprising its exterior.

3.6. Improved Purification Protocol—Further Advances

Recently, an improved purification protocol for the scrapie agent has been developed.[157] This protocol has several advantages over the earlier ones noted above. First, in the earlier protocols, the sedimentation of the agent in a microsomal membrane fraction required prolonged ultracentrifugation and thus severely limited the size of the preparations. Second, the preparative Sarkosyl electrophoresis of the agent was slow, tedious, and of limited capacity. In our new purification protocol, differential ultracentrifugation has been supplanted by polyethylene glycol-8000 (PEG-8000) precipitation, and preparative gel electrophoresis has been replaced by rate-zonal gradient centrifugation. This new protocol has allowed us to purify the agent from 100- to 1,000-fold with respect to protein. The protocol includes Triton X-100/sodium deoxycholate extraction and PEG-8000 precipitation, nuclease and protease digestion, cholate and Sarkosyl extraction, ammonium sulfate precipitation, Triton X-100/SDS extraction, and rate-zonal sedimentation through a discontinuous sucrose gradient. The highest degree of purification was found in a fraction from the 25%/60% sucrose interface near the bottom of the gradient. When Triton X-100 or octylglucoside extractions were used in place of the Triton X-100/SDS extraction, the distribution of the scrapie agent in the discontinuous sucrose gradients was altered, and no substantial purification was obtained.

Electron microscopic examination of rotary-shadowed samples from the gradient interface, which contained greater than $10^{9.5}$ ID_{50} units/ml of the scrapie agent, revealed aggregates composed of an amorphous material and numerous flattened rods measuring 25 nm in diameter by 100–200 nm in length. These are the same rods that we observed in the fractions prepared by Sarkosyl gel electrophoresis. Recent studies have shown that the rods are aggregates of prions and are composed of PrP 27-30 protein molecules. The prion rods are indistinguishable from amyloid and form long filaments within amyloid plaques that are found in scrapie-infected hamster brain tissue.

3.7. Radiolabeling Studies: Identification of a Prion Protein (PrP 27-30)

The apparent small size of the scrapie agent suggests that the monomer of the agent is too small to be identified by electron microscopy.[157] For this reason, we turned to radiolabeling procedures. We began with radiolabeling probes for proteins, since we had accumulated compelling evident for a protein within the agent. In our initial radiolabeling studies, both experimental and control samples generated by preparative Sarkosyl gel electrophoresis were labeled with [^{125}I]Bolton–Hunter reagent. The radioiodinated samples were then analyzed by polyacrylamide gel electrophoresis. On occasion, we found a protein of approximately 29,000 daltons that was present in scrapie and absent in control preparations; however, the protein was barely detectable in most of the scrapie preparations.

Table V
Evidence that PrP 27-30 is a Component of the Scrapie Agent

1. The protein is found only in scrapie preparations; not in controls.
2. The protein has been found in preparations purified from both scrapie-infected hamster and mouse brain.
3. One-dimensional peptide maps have distinguished the protein from others of similar mol. wt. that are found in control preparations.
4. The protein and agent copurify by two different procedures—properties of each must be similar.
5. The protein is a major component of purified preparations with high specific infectivities.
6. Amount of the protein correlates with agent titers.
7. [^{14}C]DEP labels the protein, and DEP inactivates the agent.
8. After protease digestion for 30 min at 25° under nondenaturing conditions, the protein and the agent remain intact. All other proteins are degraded.
9. Prolonged protease digestion of 3 hr or more at 37° under nondenaturing conditions degrades the protein and diminishes the infectivity of the agent.
10. Denaturation by boiling in 1.25% SDS for 2 min diminishes the infectivity of the agent and renders the protein susceptible to proteolysis as well as facilitating its entry into polyacrylamide gels during electrophoresis.
11. The protein (30,000 mol. wt.) is small enough to be part or all of the agent (~50,000 mol. wt.).
12. By animal infectivity assay, there are ~10^4 protein molecules per ID_{50} unit.

Subsequent radioiodination of the sucrose gradient interface fractions with substantially higher titers obtained by the improved purification protocol lead to identification of a protein that is unique to fractions purified from scrapie-infected brains.[158] This protein migrates as a diffuse band during electrophoresis through polyacrylamide gels (Fig. 1). The apparent molecular weight of the protein using polyacrylamide gels of three different concentrations is between 27,000–30,000. The protein was also labeled with [^{14}C]DEP. Fortuitously, we found that the protein is resistant to protease digestion for 30 min at 25° under nondenaturing conditions. Under these same conditions, scrapie infectivity is unaltered, whereas virtually all proteins except for this unique protein undergo proteolytic cleavage. Thus, the only protein consistently found in these protease-treated preparations that retains its scrapie infectivity is the unique protein. Denaturation in SDS and β-mercaptoethanol followed by boiling for 2 min renders this protein susceptible to protease-catalyzed degradation. Other studies indicate that the amount of this protein correlates with the titer of the scrapie agent over a 100-fold range. All of these studies (including our ability to label the protein with [^{14}C]DEP, which, as discussed above, has been shown to reversibly inactivate the scrapie agent) suggest, but do not prove, that the protein is a structural component of the scrapie agent. Perhaps the strongest evidence we have is that the protein purifies with scrapie infectivity by two different protocols. Indeed, the properties of this protein and the scrapie agent must be quite similar for them to copurify. Recently, we have found that this same protein is also present in purified fractions of the agent prepared by Sarkosyl gel electrophoresis, as described above.

Data supporting our hypothesis that PrP 27-30 is a component of the scrapie agent are summarized in Table V. Although all observations to date are consistent with this view, further studies are needed to eliminate the pos-

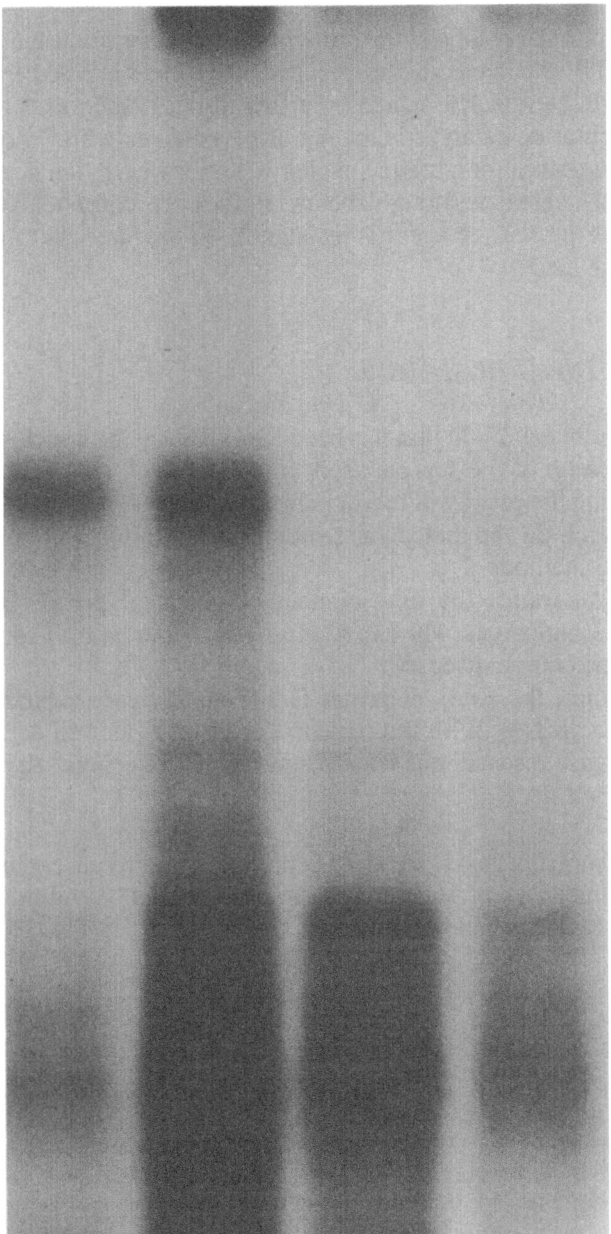

Fig. 1. Radioiodination of purified fractions from scrapie-infected and control hamster brains. Following radiolabeling with N-succinimidyl-3-(4-hydroxy, 5-[^{125}I]iodophenyl)proprionate,[96] the purified samples were treated with 100 μg/ml of proteinase K for 30′ at 25°. The digestion was terminated by the addition of an equal volume of twofold concentrated SDS electrophoresis sample buffer and heating to 100° for 2 min. Electrophoresis was performed in 15% polyacrylamide gels by the Laemmli method.[97] Lanes 1 and 2 (left to right), discontinuous sucrose gradient fraction 2 from scrapie agent preparations; lanes 3 and 4, from control preparations.

sibility, although unlikely, that this protein is a pathological product of infection.

In contrast to the results of the protein-radiolabeling studies, 5′-endlabeling with γ-[^{32}P]ATP of nucleic acid molecules in the same purified fraction failed to identify a unique nucleic acid.[157] Further studies, such as 3′-end labeling and thallium-chloride-catalyzed labeling of pyrimidines with [^{125}I] are needed to continue the search for a unique nucleic acid. However, the negative results of these initial nucleic-acid-radiolabeling studies are consistent with the observations that the scrapie agent is resistant to procedures that modify or hydrolyze nucleic acids.

4. CONCLUDING REMARKS

If the protein PrP 27-30 that has been identified can be conclusively shown to be a component of the scrapie agent, then identification of all macromolecules comprising the prion will follow. The slow, imprecise bioassays that have impeded research on the molecular structure of the prion will be replaced by rapid chemical methods.

Equally important, the development of chemical probes for detecting prions such as antibodies will allow us to renew our search for prions in a variety of degenerative diseases.

At this point, the study of prions is still many years behind the investigations of slow viruses. With the structure of visna virus well defined, we ask how does it cause disease, but we are still trying to decipher the structure of prions.

ACKNOWLEDGMENT. This work was supported in part by research grants from the National Institutes of Health NS14069 and AG02132. Additional support was given by R. J. Reynolds Industries, Inc.

REFERENCES

1. Gajdusek, D. C., Gibbs, C. J., Jr., and Alpers, M., 1966, *Nature* **209:**794–796.
2. Sigurdsson, B., 1954, *Br. Vet. J.* **110:**341–354.
3. Haase, A. T., 1975, *Curr. Top. Microbiol. Immunol.* **72:**101–156.
4. Nathanson, N., and Robinson, W., eds., 1979, *Persistent Viral Infections*, Volume 3, DHEW Publication No. NIH 79-1833, Department of Health, Education and Welfare, Washington, pp. 1–213.
5. Brahic, M., and Haase, A. T., 1981, *Comparative Diagnosis of Viral Diseases*, Volume 3 (E. Kurstak, ed.), Academic Press, New York, pp. 627–647.
6. Sigurdsson, B., Palsson, P. A., and Grimmson, H., 1957, *J. Neuropathol. Exp. Neurol.* **16:**389–403.
7. Gudnadottir, M., and Palsson, P. A., 1967, *J. Infect. Dis.* **117:**1–6.
8. DeBoer, G. F., 1975, *Res. Vet. Sci.* **18:**15–25.
9. Kennedy, R. C., Eklund, C. M., Lopez, C., and Hadlow, W. J., 1968, *Virology* **35:**483–484.
10. Narayan, O., Clements, J. E., Strandberg, J. D., Cork, L. C., and Griffin, D. E., 1980, *J. Gen. Virol.* **50:**69–79.

11. Cork, L. C., Hadlow, W. J., Crawford, T. B., Gorham, J. R., and Piper, R. C., 1974, *J. Infect. Dis.* **129:**134–141.
12. Palsson, P. A., 1976, *Slow Virus Diseases of Animals and Man* (R. H. Kimberlin, ed.), North-Holland, Amsterdam, pp. 17–43.
13. Weinhold, E. 1974, *Zentralbl. Veterinarmed.* [*B*] **21:**32–36.
14. Weinhold, E., and Triemer, B., 1978, *Zentralbl. Veterinarmed.* [*B*] **25:**525–538.
15. Thormar, H., 1976, *Slow Virus Diseases of Animals and Man* (R. H. Kimberlin, ed.), North-Holland, Amsterdam, pp. 97–114.
16. Sigurdsson, B., Palsson, P. A., and van Bogaert, L., 1962, *Acta Neuropathol.* (*Berl.*) **1:**343–362.
17. Petursson, G., Nathanson, H., Georgsson, G., Panitch, H., and Palsson, P. A., 1976, *Lab. Invest.* **35:**402–412.
18. Georgsson, G., Palsson, P. A., Panitch, H., Nathanson, N., and Petursson, G., 1977, *Acta Neuropathol.* (*Berl.*) **37:**127–135.
19. DeBoer, G. F., Terfstra, C., and Houwers, D. J., 1978, *Bull. Off. Int. Epizoot.* **89:**487–506.
20. Stavrou, D., Deutschlander, N., and Dahme, E., 1969, *J. Comp. Pathol.* **79:**393–396.
21. O'Sullivan, B. M., Eaves, F. W., Baxendell, S. A., and Rowan, K. J., 1978, *Aust. Vet. J.* **54:**479–483.
22. Cork, L. C., Hadlow, W. J., Gorham, J. R., Piper, R. C., and Crawford, T. B., 1974, *Acta Neuropathol.* (*Berl.*) **29:**281–292.
23. Sherman, D. N., 1978, *Vet. Med. Small Anim. Clin.* **73:**1439–1440.
24. Williams, C. S., 1979, *Vet. Med. Small Anim. Clin.* **74:**9.
25. Crawford, T. B., Adams, D. A., Cheevers, W. P., and Cork, L. C., 1980, *Science* **207:**997–999.
26. Cork, L. C., and Narayan, O., 1980, *Lab. Invest.* **42:**596–602.
27. Thormar, H., and Cruickshank, J. G., 1965, *Virology* **25:**145–148.
28. Coward, J. E., Harter, D. H., and Morgan, C., 1970, *Virology* **40:**1030–1038.
29. Chippaux-Hyppolite, C., Taranger, C., Tamalet, J., Pautrat, G., and Brahic, M., 1972, *Ann. Inst. Pasteur* (*Paris*) **123:**409–420.
30. Takemoto, K. K., Aoki, T., Garon, C., and Sturm, M. M., 1973, *J. Natl. Cancer Inst.* **50:**543–547.
31. Haase, A. T., and Baringer, J. R., 1974, *Virology* **57:**238–250.
32. Stone, L. B., Takemoto, K. K., and Matin, M. A., 1971, *J. Virol.* **8:**573–578.
33. August, M. J., Harter, D. H., and Compans, R. W., 1977, *J. Virol.* **22:**832–834.
34. Thormar, H., 1961, *Arch. Ges. Virusforsch.* **10:**501–509.
35. Thormar, H., 1965, *Res. Vet. Sci.* **6:**117–129.
36. Baltimore, D., 1970, *Nature* **226:**1209–1211.
37. Temin, H. M., and Mizutani, S., 1970, *Nature* **226:**1211–1213.
38. Mountcastle, W. E., Harter, D. H., and Choppin, P. W., 1972, *Virology* **47:**542–545.
39. Lin, F. H., and Thormar, H., 1974, *J. Virol.* **14:**782–790.
40. Lin, F. H., 1978, *J. Virol.* **25:**207–214.
41. August, J. T., Bolognesi, D. P., Fleissner, E., Gilden, R. V., and Nowinski, R. C., 1974, *Virology* **60:**595–601.
42. Lin, F. H., and Thormar, H., 1979, *J. Virol.* **29:**536–539.
43. Scott, J. V., Stowring, L., Haase, A. T., Narayan, O., and Vigne, R., 1979, *Cell* **18:**321–327.
44. Bruns, M., and Frenzel, B., 1979, *Virology* **97:**207–211.
45. Stowring, L., Haase, A. T., and Charman, H. P., 1979, *J. Virol.* **29:**523–528.
46. Lin, F. H., and Thormar, H. A., 1970, *J. Virol.* **6:**702–704.
47. Lin, F. H., and Thormar, H., 1972, *J. Virol.* **10:**228–233.
48. Stone, L. B., Scolnick, E., Takemoto, K. K., and Aaronson, S. A., 1971, *Nature* **229:**257–258.
49. Lin, F. H., Genovese, M., and Thormar, H., 1973, *Prep. Biochem.* **3:**525–539.
50. Haase, A. T., Garapin, A. C., Faras, A. J., Varmus, H. E., and Bishop, J. M., 1974, *Virology* **57:**251–258.
51. Scolnick, E. M., and Parks, W. P., 1974, *Virology* **59:**168–178.
52. Lin, F. H., and Papini, M., 1979, *Biochim. Biophys. Acta* **561:**383–395.

53. Harter, D. H., Rosenkranz, H. S., and Rose, H. M., 1969, *Proc. Soc. Exp. Biol. Med.* **131**:927–933.
54. Harter, D. H., Schlom, J., and Spiegelman, H., 1971, *Biochim. Biophys. Acta* **240**:435–441.
55. Brahic, H., Tamalet, J., and Chippaux-Hyppolite, C., 1971, *C. R. Acad. Sci.* [*D*] (*Paris*) **272**:2115–2118.
56. Brahic, M., Tamalet, J., Filippi, P., and Delbecchi, L., 1973, *Biochimie* **55**:885–898.
57. Lin, F. H., and Thormar, H. A., 1971, *J. Virol.* **7**:582–587.
58. Friedmann, A., Coward, J. E., Harter, D. H., Lipset, J. S., and Morgan, C., 1974, *J. Gen. Virol.* **25**:93–104.
59. Haase, A. T., Garapin, A. C., Faras, A. J., Taylor, J. M., and Bishop, J. M., 1974, *Virology* **57**:259–270.
60. Vigne, R., Brahic, M., Filippi , P., and Tamalet, J., 1977, *J. Virol.* **21**:386–395.
61. Beemon, K. L., Faras, A. J., Haase, A. T., Duesberg, P. H., and Maisel, J. E., 1976, *J. Virol.* **17**:525–537.
62. Vigne, R., Filippi, P., Brahic, M., and Tamalet, J., 1978, *J. Virol.* **28**:543–550.
63. Gillespie, D., Takemoto, K., Robert, M., and Gallo, R. C., 1973, *Science* **179**:1328–1330.
64. Brahic, M., Filippi, P., Vigne, R., and Haase, A. T., 1977, *J. Virol.* **24**:74–81.
65. Harris, J. D., Scott, J. V., Traynor, B., Brahic, M., Stowring, L., Ventura, P., Haase, A. T., and Peluso, R., 1981, *Virology* **113**:573–583.
66. Haase, A. T., and Varmus, H. E., 1973, *Nature* (*New Biol.*) **245**:237–239.
67. Haase, A. T., Traynor, B. L., and Ventura, P. E., 1976, *Virology* **70**:65–79.
68. Clements, J. E., Narayan, O., Griffin, D. E., and Johnson, R. T., 1979, *Virology* **93**:377–386.
69. Traynor, B. L., and Haase, A. T., 1977, *Abstracts Annual Meeting American Society for Microbiology*, American Society for Microbiology, Washington, Abstract S362, p. 339.
70. Haase, A. T., Stowring, L., Harris, J. D., Traynor, B., Ventura, P., Peluso, R., and Brahic, M., 1982, *Virology* **119**:399–410.
71. Brahic, M., and Vigne, R., 1975, *J. Virol.* **15**:1222–1230.
72. Filippi, P., Brahic, M., Vigne, R., and Tamalet, J., 1979, *J. Virol.* **31**:25–30.
73. Weiss, M. J., Zeelong, E. P., Sweet, R. W., Harter, D. H., and Spiegelman, S., 1977, *Virology* **76**:851–854.
74. Harter, D. H., Hsu, K. C., and Rose, H. M., 1967, *J. Virol.* **1**:1265–1270.
75. Thormar, H., 1969, *Acta Pathol. Microbiol. Immunol. Scand.* **75**:296–302.
76. Vigne, R., Filippi, P., Quérat, G., Sauze, N., Vitu, C., Russo, P., and Delori, P., *J. Virol.* **42**:1046–1056.
77. Thormar, H., 1961, *Virology* **14**:463–475.
78. Dubois-Dalcq, M., Reese, T. S., and Narayan, O., 1976, *Virology* **74**:520–530.
79. Griffin, D. E., Narayan, O., and Adams, R. J., 1978, *J. Infect. Dis.* **138**:340–350.
80. Gudnadottir, M., and Palsson, P. A., 1965, *J. Immunol.* **95**:1116–1128.
81. Griffin, D. E., Narayan, O., Bukowski, J. F., Adams, R. J., and Cohen, S. R., 1978, *Ann. Neurol.* **4**:212–218.
82. Thormar, H., Wisniewski, H. M., and Lin, F. H., 1979, *Nature* **279**:245–246.
83. Brahic, M., and Haase, A. T., 1978, *Proc. Natl. Acad. Sci. U.S.A.* **75**:6125–6129.
84. Haase, A. T., Stowring, L., Narayan, O., Griffin, D., and Price, D., 1977, *Science* **195**:175–177.
85. Brahic, M., Stowring, L., Ventura, P., and Haase, A. T., 1981, *Nature* **292**:240–242.
86. Carroll, D., Ventura, P., Haase, A., Rinaldo, C. R., Jr., Overall, J. C., Jr., and Glasgow, L. A., 1978, *J. Infect. Dis.* **138**:614–617.
87. Narayan, O., Griffin, D. E., and Chase, J., 1977, *Science* **197**:376–378.
88. Narayan, O., Griffin, D. E., and Clements, J. E., 1978, *J. Gen. Virol.* **41**:343–352.
89. Clements, J. E., Pedersen, F. S., Narayan, O., and Haseltine, W. A., 1980, *Proc. Natl. Acad. Sci. U.S.A.* **77**:4454–4458.
90. Nathanson, N., Panitch, H., Palsson, P. A., Petursson, G., and Georgsson, G., 1976, *Lab. Invest.* **35**:444–451.
91. Petursson, G., Martin, J. R., Georgsson, G., Nathanson, N., and Palsson, P. A., 1979, *Aspects of Slow and Persistent Virus Infections* (D. A. J. Tyrrell, ed.), Martinus Nijhoff, The Hague, pp. 165–197.

92. Hadlow, W. J., 1959, *Lancet* **2**:289–290.
93. Gibbs, C. J., Jr., Gajdusek, D. C., Asher, D. M., Alpers, M. P., Beck, E., Daniel, P. M., and Matthews, W. B., 1968, *Science* **161**:388–389.
94. Gibbs, C. J., Jr., and Gajdusek, D. C., 1973, *Science* **182**:67–68.
95. Hadlow, W. J., Prusiner, S. B., Kennedy, R. C., and Race, R. E., 1980, *Ann. Neurol.* **8**:628–631.
96. Bolton, E. A., and Hunter, W. M., 1973, *Biochem. J.* **133**:529–539.
97. Laemmli, U. K., 1970, *Nature* **227**:680–685.
98. Gibbs, C. J., Jr., and Gajdusek, D. C., 1972, *Nature* **236**:73–74.
99. Masters, C. L., Harris, J. O., Gajdusek, D. C., Gibbs, C. J., Jr., Bernoulli, C., and Asher, D. M., 1979, *Ann. Neurol.* **5**:177–188.
100. Gajdusek, D. C., 1977, *Science* **197**:943–960.
101. Alpers, M. P., 1979, *Slow Transmissible Diseases of the Nervous System*, Volume 1 (S. B. Prusiner and W. J. Hadlow, eds.), Academic Press, New York, pp. 67–92.
102. Gibbs, C. J., Jr., Amyx, H. L., Bacote, A., Masters, C. L., and Gajdusek, D. C., 1980, *J. Infect. Dis.* **142**:205–208.
103. Pattison, I. H., and Millson, G. C., 1961, *J. Comp. Pathol.* **71**:171–176.
104. Pattison, I. H., Hoare, M. N., Jebbett, J. N., and Watson, W. A., 1972, *Vet. Rec.* **90**:465–468.
105. M'Gowan, J. P., 1914, *Investigation into the Disease of Sheep Called "Scrapie,"* William Blackwood and Sons, Edinburgh, pp. 1–114.
106. M'Fadyean, J., 1918, *J. Comp. Pathol.* **31**:102–131.
107. Cuille, J., and Chelle, P. L., 1939, *C.R. Acad. Sci. [D] (Paris)* **208**:1058–1060.
108. Wilson, D. R., Anderson, R. D., and Smith, W., 1950, *J. Comp. Pathol.* **60**:267–282.
109. Eklund, C. M., Hadlow, W. J., and Kennedy, R. C., 1963, *Proc. Soc. Exp. Biol. Med.* **112**:974–979.
110. Cho, H. J., 1976, *Nature* **262**:411–412.
111. Kimberlin, R. H., and Hunter, G. D., 1967, *J. Gen. Virol.* **1**:115–124.
112. Griffith, J. S., 1967, *Nature* **215**:1043–1044.
113. Pattison, I. H., and Jones, K. M., 1967, *Vet. Rec.* **80**:1–8.
114. Lewin, P., 1972, *Lancet* **1**:748.
115. Lewin, P., 1981, *Can. Med. Assoc. J.* **124**:1436–1437.
116. Gibbons, R. A., and Hunter, G. D. 1967, *Nature* **215**:1041–1043.
117. Hunter, G. D., Kimberlin, R. H., and Gibbons, R. A., 1968, *J. Theor. Biol.* **20**:355–357.
118. Adams, D. H., and Field, E. J., 1968, *Lancet* **2**:714–716.
119. Adams, D. H., 1970, *Pathol. Biol. (Paris)* **18**:559–577.
120. Parry, H. B., 1962, *Heredity* **17**:75–105.
121. Diener, T. O., 1972, *Nature (New Biol.)* **235**:218–219.
122. Diener, T. O., 1973, *Ann. Clin. Res.* **5**:268–278.
123. Pattison, I. H., 1965, *J. Comp. Pathol.* **75**:159–164.
124. Stamp, J. T., 1967, *Br. Med. Bull.* **23**:133–137.
125. Hunter, G. D., 1972, *J. Infect. Dis.* **125**:427–440.
126. Adams, D. H., 1973, *Biochem. Soc. Trans.* **1**:1061–1064.
127. Gajdusek, D. C., 1978, *Human Diseases Caused by Viruses* (H. Rothschild, F. Alison, Jr., and C. Howe, eds.), Oxford University Press, New York, pp. 231–258.
128. Gajdusek, D. C., and Gibbs, C. J., Jr., 1978, *Viruses and Environment* (E. Kurstak and K. Maramorosch, eds.), Academic Press, New York, pp. 79–98.
129. Rohwer, R. G., and Gajdusek, D. C., 1980, *Search for the Cause of Multiple Sclerosis and Other Chronic Diseases of the Central Nervous System* (A. Boese, ed.), Verlag Chemie, Weinheim, pp. 333–355.
130. Field, E. J., 1966, *Br. Med. J.* **2**:564–565.
131. Field, E. J., 1967, *Dtsch. Z. Nervenheilkd.* **192**:265–274.
132. Latarjet, R., Muel, B., Haig, D. A., Clarke, M. C., and Alper, T., 1970, *Nature* **227**:1341–1343.
133. Adams, D. H., and Caspary, E. A., 1967, *Br. Med. J.* **3**:173.
134. Narang, H. K., 1974, *Acta Neuropathol. (Berl.)* **29**:37–43.

135. Siakotos, A. N., Raveed, D., and Longa, G., 1979, *J. Gen. Virol.* **43:**417–422.
136. Bastian, F. O., 1979, *Arch. Pathol. Lab. Med.* **103:**665–669.
137. Gray, A., Francis, R. J., and Scholtz, C. L., 1980, *Lancet* **2:**152.
138. Bastian, F. O., Hart, M. N., and Cancilla, P. A., 1981, *Lancet* **1:**660.
139. Hunter, G. D., Kimberlin, R. H., Collis, S., and Millson, G. C., 1973, *Ann. Clin. Res.* **5:**262–267.
140. Somerville, R. A., Millson, G. C., and Hunter, G. D., 1976, *Biochem. Soc. Trans.* **4:**1112–1114.
141. Marsh, R. F., Malone, T. G., Semancik, J. S., Lancaster, W. D., and Hanson, R. P., 1978, *Nature* **275:**146–147.
142. Prusiner, S. B., 1982, *Science* **216:**136–144.
143. Prusiner, S. B., Hadlow, W. J., Garfin, D. E., Cochran, S. P., Baringer, R. J., Race, R. E., and Eklund, C. M., 1978, *Biochemistry* **17:**4993–4999.
144. Prusiner, S. B., Groth, D. F., Cochran, S. P., Masiarz, F. R., McKinley, M. P., and Martinez, H. M., 1980, *Biochemistry* **19:**4883–4891.
145. Prusiner, S. B., McKinley, M. P., Groth, D. F., Bowman, K. A., Mock, N. I., Cochran, S. P., and Masiarz, F. R., 1981, *Proc. Natl. Acad. Sci. U.S.A.* **78:**6675–6679.
146. Kimberlin, R., and Walker, C., 1977, *J. Gen. Virol.* **34:**295–304.
147. Prusiner, S. B., Cochran, S. P., Groth, D. F., Downey, D. E., Bowman, K. A., and Martinez, H. M., 1982, *Ann. Neurol.* **11:**353–358.
148. McKinley, M. P., Masiarz, F. R., and Prusiner, S. B., 1981, *Science* **214:**1259–1261.
149. Prusiner, S. B., Groth, D. F., McKinley, M. P., Cochran, S. P., Bowman, K. A., and Kasper, K. C., 1981, *Proc. Natl. Acad. Sci. U.S.A.* **78:**4606–4610.
150. Hunter, G. D., 1979, *Slow Transmissible Diseases of the Central Nervous System*, Volume 2 (S. B. Prusiner and W. J. Hadlow, eds.), Academic Press, New York, pp. 365–385.
151. Alper, T., Cramp, W. A., Haig, D. A., and Clarke, M. C., 1967, *Nature* **214:**764–766.
152. Latarjet, R., 1979, *Slow Transmissible Diseases of the Central Nervous System*, Volume 2 (S. B. Prusiner and W. J. Hadlow, eds.), Academic Press, New York, pp. 387–408.
153. Diener, T. O., McKinley, M. P., and Prusiner, S. B., 1982, *Proc. Natl. Acad. Sci. U.S.A.* **79:**5220–5224.
154. Alper, T., Haig, D. A., and Clarke, M. C., 1966, *Biochem. Biophys. Res. Commun.* **22:**278–284.
155. Alper, T., Haig, D. A., and Clarke, M. C., 1978, *J. Gen. Virol.* **41:**503–516.
156. Diener, T. O., 1979, *Viroids and Viroid Diseases*, John Wiley and Sons, New York, pp. 1–252.
157. Prusiner, S. B., Bolton, D. C., Groth, D. F., Bowman, K. A., Cochran, S. P., and McKinley, M. P., 1982, *Biochemistry* **21:**6942–6950.
158. Bolton, D. C., McKinley, M. P., and Prusiner, S. B., 1982, *Science* **218:**1309–1311.

21

Neurochemical Aspects of Schizophrenia

Lynn Eleanor DeLisi and Richard Jed Wyatt

1. INTRODUCTION

Schizophrenia, a syndrome characterized by delusions, hallucinations, and disturbances of thought, is one of the most pervasive and socially disabling of all psychiatric disorders. Approximately 27% of annual psychiatric hospital admissions (340,000) in the United States are patients with this disorder,[1] and most individuals with a chronic schizophrenic illness, if not institutionalized, are unemployable and spend years in community care facilities. Despite the severe morbidity and relatively high prevalence of this disorder in the general population (approximately 1%),[2] its etiology remains unknown. In the postwar years, schizophrenia research focused on putative psychodynamic and socioenvironmental vulnerabilities for development of this disorder. Severely disordered mother–infant relationships as well as pathological communication patterns among family members were, in some psychiatric circles, considered important environmental inducers of schizophrenia. Terms such as "schizophrenigenic mother,"[3] "double bind," and "schism"[4,5] were used to describe these pathological relationships. There is no doubt that disturbed relationships within some, but no all, families of schizophrenics are accurate observations, although how these stressors are associated with the etiology of schizophrenia remains obscure. Nevertheless, decreasing familial and other social stresses has been demonstrated to be an extremely important treatment strategy.[6] Since neither the biochemical nor psychosocial factors that create schizophrenia are understood, their interactions remain even more obscure.

The idea that psychotic-like states were based on a chemical imbalance predates the establishment of present-day psychiatry and is implied by the practices of blood letting as well as the ingestion of iron filings, quinine so-

Lynn Eleanor DeLisi • Neurogenetics Branch, Intramural Research Program, National Institute of Mental Health, Bethesda, Maryland 20205. *Richard Jed Wyatt* • Neuropsychiatry Branch, Intramural Research Program, National Institute of Mental Health, Saint Elizabeths Hospital, Washington, D.C. 20032.

lutions, and other special chemical preparations as remedies for "madness" in the 18th and 19th centuries.[7] In the early 20th century, fever induced by typhoid vaccine, foreign protein, sulfur injections, and induction of malaria were used with varied success.[8] After the discovery of insulin, hypoglycemia with induction of insulin subcoma was claimed to be beneficial in schizophrenia and is still used in some parts of the world.[9] The newest era in biochemical investigations of schizophrenia was opened, however, with the introduction of neuroleptic medication and the knowledge that widespread pharmacological intervention dramatically reduced the number of hospitalized psychiatric patients. Neuroleptics (the word derived from the Greek, *neuro* for nerve, *lepsis*, a taking hold),[10] drugs that produce neurological symptoms, were found to be potent blockers of postsynaptic dopamine receptors and are the basis for the hypothesis that schizophrenia may be related to an excess of dopamine or dopaminergic function in the brain.[11]

In addition, a biological basis for schizophrenia was greatly supported by the publication of results of several carefully executed important genetic studies revealing a probable genetic component to schizophrenia.[12,13] Although the conclusions of the genetic studies are still controversial, most investigators would agree that the etiology of schizophrenia is based, at least in part, on a genetic vulnerability. Unfortunately, the degree of genetic vulnerability is not known, nor is it clear that genetic vulnerability is present in the majority of patients. Furthermore, a specific genetic mechanism or mode of inheritance has not been elucidated.

Although an anatomic structure, or even an organ system, that is directly involved in the development of schizophrenia has not been defined, it is assumed that the primary lesion must be in the central nervous system since the major symptoms relate to higher cerebral functions. Nevertheless, other disorders principally affecting the brain are known to be influenced by pathology in peripheral organ systems. An example is phenylketonuria, a disease characterized by mental retardation and other central nervous system disturbances, which classically is caused by an inherited deficiency of the hepatic enzyme phenylalanine hydroxylase.[14]

It is thought that the primary positive symptoms of schizophrenia result from changes in the brain mesolimbic and related cerebral cortical projections. This concept is supported by several lines of evidence including studies of limbic lesions and stimulation in humans.[15–17] Some electrophysiological (EEG) studies point to temporal lobe dysfunction in schizophrenia,[18–21] whereas others implicate more generalized dysfunction.[22–24] The therapeutic rationale behind prefrontal lobotomies for schizophrenic patients, popularized in the 1950s,[25] resulted from the belief that frontal lobe dysfunction was the basis for the emotional aspects of this disorder. More recently, cerebral blood flow studies and measurements of regional cerebral glucose metabolism by positron emission tomography (PET) have shown relatively low frontal compared with occipital lobe activity in schizophrenic patients versus non-psychiatrically-ill controls.[26,27] These studies also indicate increased temporal lobe blood flow and metabolism.[27,28] Although computed tomography (CT) studies do not show focal frontal lobe deficits, there is evidence of both cortical and subcortical

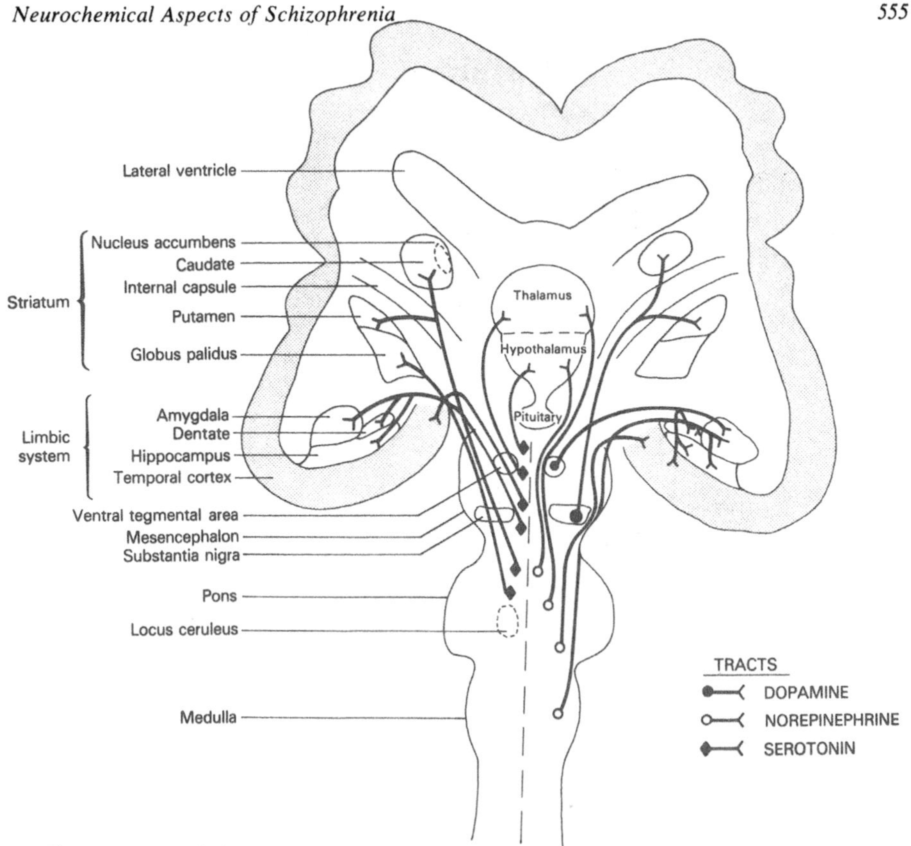

Fig. 1. Anatomic illustration of the major neurotransmitter pathways in the human brain.

cerebral atrophy in some schizophrenic patients.[29,30] In addition, histopathological studies have shown lesions such as subcortical islands of gliosis, particularly in periventricular regions.[31,32] Attempts to trace biochemical abnormalities to specific anatomic structures in the brain have not, however, revealed consistent correlations.

There are also a number of biochemical hypotheses defining subgroups of schizophrenia, with varying amounts of supportive data available for each. Most of these remain unresolved at this writing. Altered activity of a number of neurotransmitters or neuromodulators in addition to dopamine (i.e., serotonin, norepinephrine, GABA) have been proposed (see Figs. 1, 2). Furthermore, decreased activity of major enzymes important to neurotransmitter and neuromodulator metabolism (i.e., monoamine oxidase, MAO; Dopamine β-hydroxylase, DBH) has also been hypothesized. Endogenous hallucinogen production through altered kinetics of both catecholamine and indoleamine metabolism may also explain some of the reported biochemical findings (see Figs. 3, 4). Other active theories include an autoimmune mechanism and a viral etiology to schizophrenia.

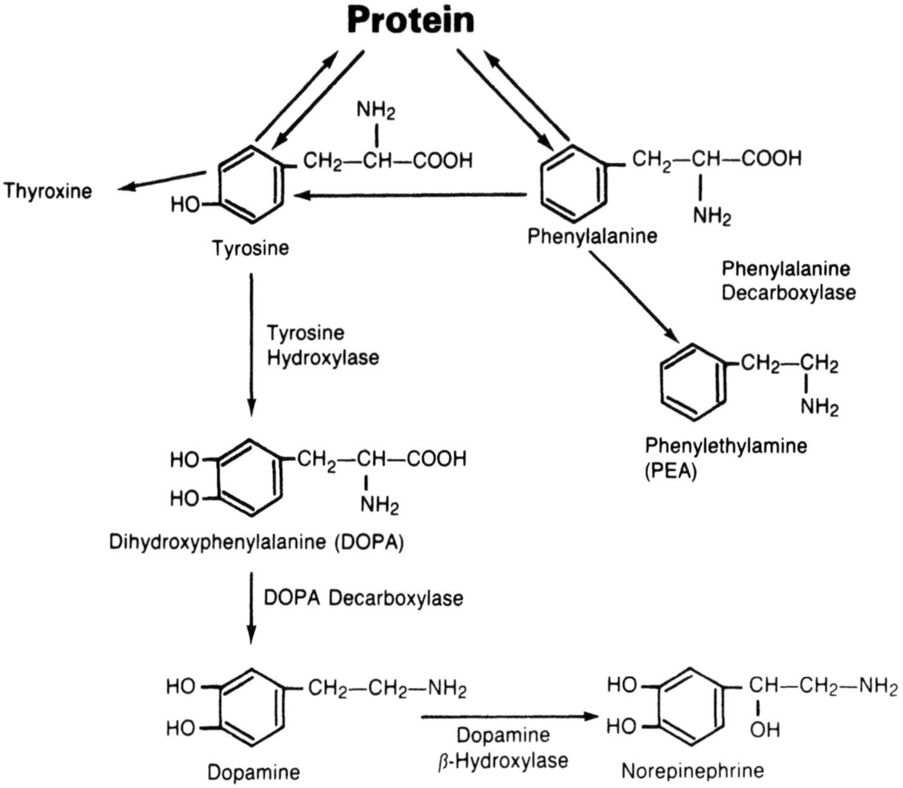

Fig. 2. Schematic pathways for catecholamine synthesis.

It is possible that all these are interrelated and are relevant to the development of schizophrenia. For example, a virus could cause structural brain damage, induce an autoimmune reaction, and interact with the host genome to produce alterations in key proteins involved in multiple neurotransmitter metabolism. Similarly, the inheritance of a gene defect that alters enzyme activity could produce all or some of the above. Alternatively, it is possible that all, or a few, of these are a consequence of schizophrenia and are secondary to the illness and its treatment rather than being etiologically related. Finally, there may be several independent etiologies to a common schizophrenic-like syndrome as, for example, there are for pneumonia. Thus, each hypothesis could relate to a subgroup of schizophrenic patients. Biochemical subgrouping of schizophrenic patients, regardless of etiologic considerations, may lead to new and more appropriate future treatment strategies.

2. *THE DOPAMINE HYPOTHESIS*

As previously mentioned, the dopamine hypothesis gained prominence among researchers concerned with the biochemical aspects of schizophrenia

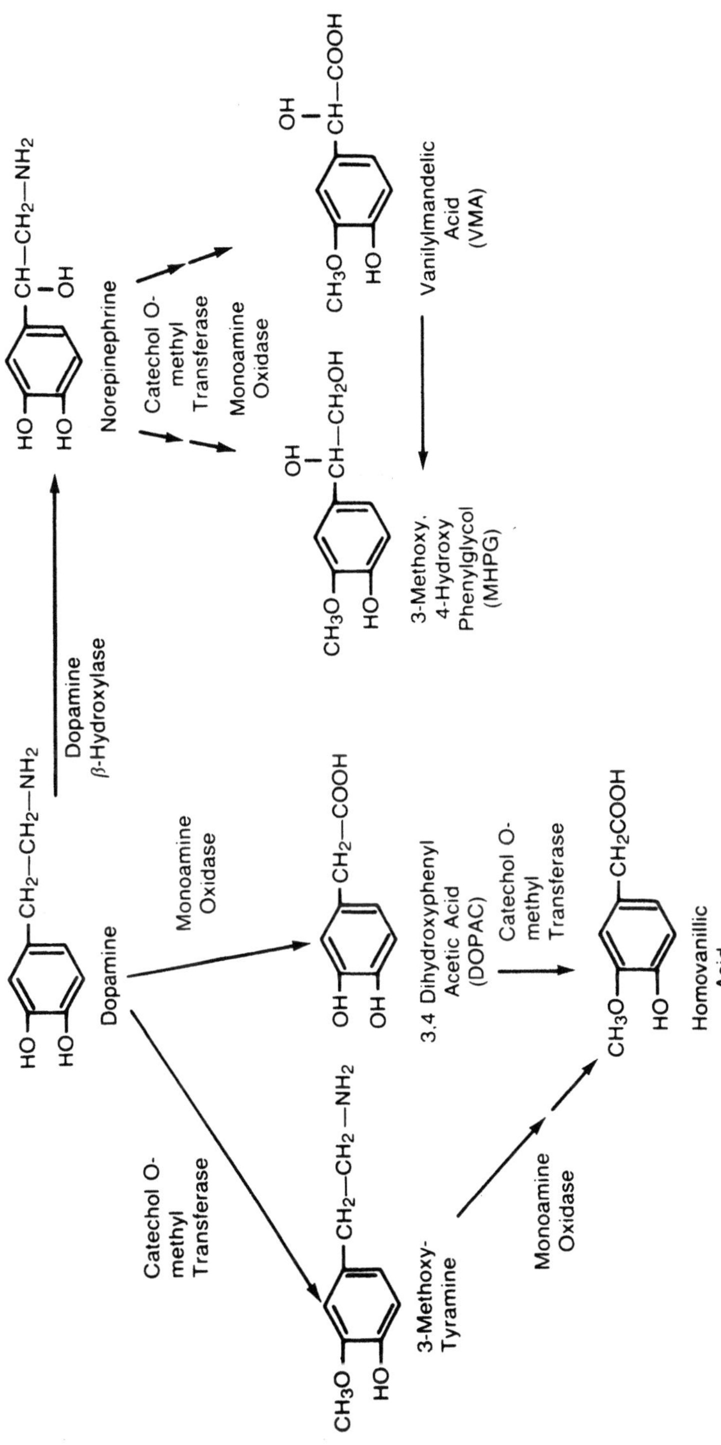

Fig. 3. Schematic pathways for the metabolism of dopamine and norepinephrine.

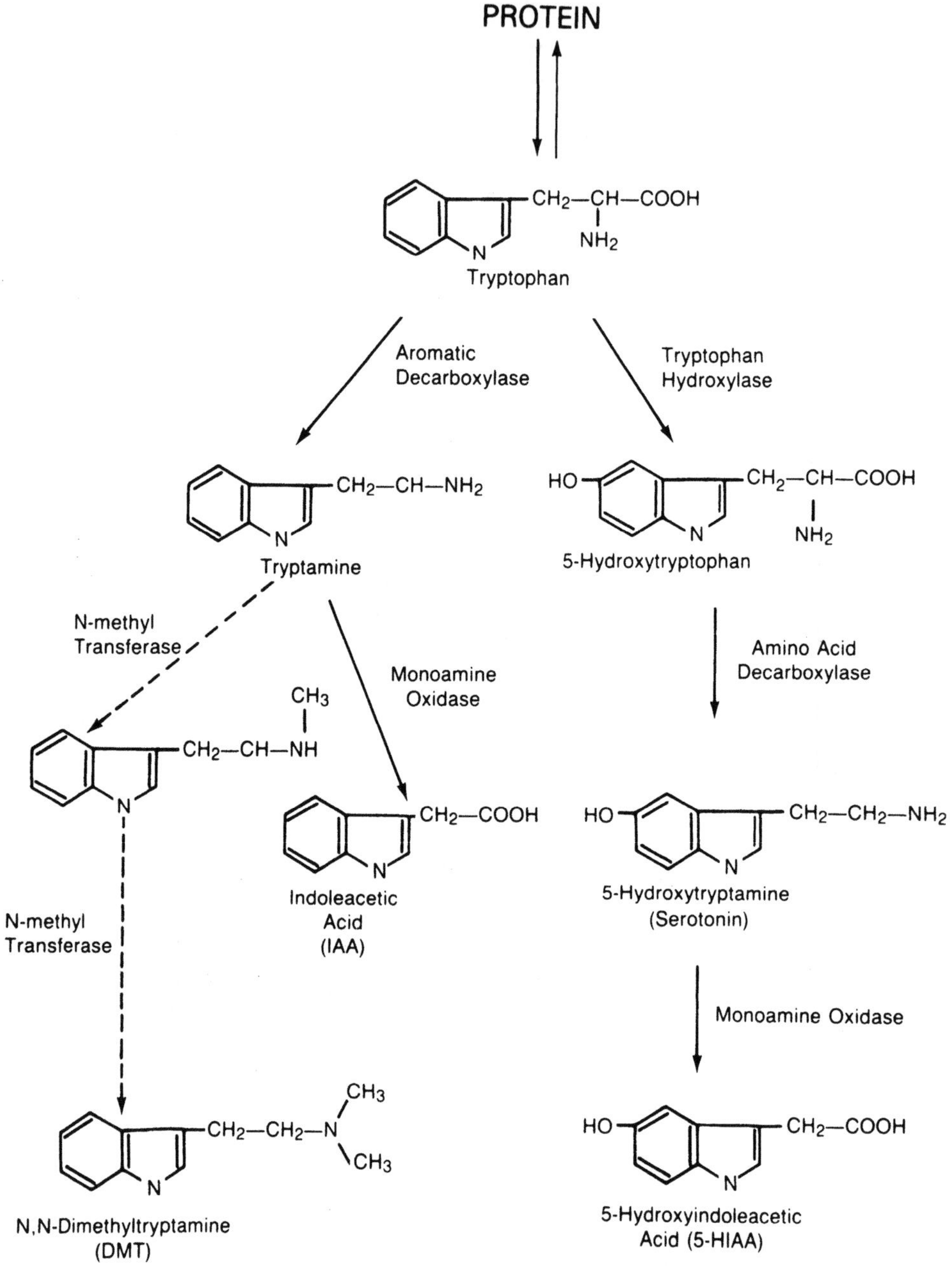

Fig. 4. Indoleamine metabolism.

as a result of the finding that neuroleptics block dopamine receptors and that their potency generally correlates with their ability to block these receptors.[11]

Neurons containing dopamine are located primarily in three regions of the brain: (1) the substantia nigra within the midbrain, with their axons projecting to the striatum (the nigrostriatal tract); (2) the ventromedial tegmental nuclei with projections to the limbic system and areas of the temporal and frontal cortex; and (3) the hypothalamus, where they project from the arcuate nucleus to the pituitary stalk[33] (see Fig. 1). In addition, there are short dopaminergic neurons within the retina, the olfactory bulb, the periaqueductal gray region, and the zona incerta connecting the hypothalamus and septum. The synthesis of dopmaine occurs predominantly intraneuronally in the brain from the amino acids tyrosine and phenylalanine (see Fig. 2). It can be either converted to the other catecholamines by DBH or metabolized to an inactive product, 3-methoxytyramine, dihydroxyphenylacetic acid (DOPAC), or homovanillic acid (HVA). Homovanillic acid, however, is the major metabolite[34] (see Fig. 3).

Drugs that increase brain dopamine, such as L-DOPA or amphetamine, induce a schizophrenic-like psychosis and can also exacerbate preexisting schizophrenia.[35–39] Nevertheless, the dose of amphetamine required to release dopamine is substantially lower than that required to produce a paranoid psychosis, and the time lag between amphetamine injection and the development of psychotic symptoms does not correlate with dopamine release. This suggests that another process is also required for the production of psychoses.[40] Furthermore, a subgroup of schizophrenic patients actually show clinical improvement with acute doses of amphetamine.[41] Dopamine receptor agonists such as bromocriptine and liseride have been reported to produce psychotic symptoms,[42] and, conversely, decreasing dopamine synthesis with drugs such as α-methyl-*para*-tyrosine (AMPT) in some studies appears to reduce the amount of neuroleptic needed for clinical efficacy.[11,43] Apomorphine also has an antipsychotic effect in some patients, presumably by stimulating presynaptic receptors to reduce dopamine synthesis and release.[44–46] These pharmacological effects, although significant, are not produced in all patients and have not been replicated in all published studies.

Furthermore, studies of dopamine and its metabolites (see Fig. 3) in brain and body fluids have failed to find consistent elevations. No overall elevation in dopamine concentrations or turnover has been found in postmortem brains of schizophrenic patients.[47,48] Several recent postmortem investigations, however, have found discrete increases in dopamine concentrations and dopamine receptor numbers in structures of the basal ganglia and limbic forebrain.[49–52] Although most patients in these studies were medicated with neuroleptics at the time of death,[50,51,53–55] a small number who were medication-free for a least 1 year prior to death also had increases. Nevertheless, the effect of chronic neuroleptic treatment on these findings has not been resolved.

Recently, Mackay *et al.*[52] indicated that the results of receptor binding studies in postmortem schizophrenic brains were probably neuroleptic related, since in their studies patients not on neuroleptics the month prior to death failed to have receptor changes. Patients on neuroleptics did have increased receptor numbers. In addition, animal studies indicate there are increases in brain dopamine metabolites and dopamine receptors after long-term treatment with neuroleptic drugs.[56–58] If increased dopamine receptor numbers is a neuroleptic-related finding, then patients having other disorders treated with neuroleptics would also be expected to have similar findings. At least one study of dopamine receptors in Huntington's disease, a neurological disorder often treated with neuroleptics, found decreased rather than increased dopamine receptor numbers.[59] Perhaps postmortem brain specimens of acute schizophrenic patients early in their illness, for example, from suicides, might clarify the significance of these findings.

Several CSF studies have failed to find increases in dopamine or its metabolites in schizophrenic patients.[60] Patients in these studies have also been on chronic neuroleptic medication, and even if they had been withdrawn from medication, the time from withdrawal may not have been long enough for baseline concentrations to be achieved.

An easily obtainable indirect measure of central dopamine activity is serum prolactin concentration. Dopamine stimulates the production of the hypothalamic prolactin inhibitory factor (PIF), and dopamine may in fact be PIF.[61] Thus, the blockade of postsynaptic dopamine receptors with neuroleptic drugs, if effective, can be measured by a rise in serum prolactin concentration. This prolactin change appears to be related to clinical response in schizophrenic patients[62] and can also be a predictor of relapse.[63,64] No studies, however, have found decreased serum prolactin concentrations in schizophrenic patients,[65] although this would be expected if dopamine in the hypothalamus were increased.

Finally, although dopamine receptor blockade with neuroleptic medication has been of substantial benefit for the majority of schizophrenic patients, it may be only indirectly treating the basic defect, with its site of action at a major common pathway in the production of psychosis. Alternatively, assuming a basic disturbance in dopaminergic transmission in a subgroup of the schizophrenias, if multiple etiologies exist, looking for dopaminergic abnormalities in a heterogeneous population of schizophrenics may be most apt to produce falsely negative results.

Several investigators have found it useful to subdivide patients into those with active or positive symptoms, such as delusions and hallucinations, and those with negative symptoms, such as social withdrawal and blunted affect, for investigating the dopamine hypothesis.[66–68] Some investigators postulate that the positive symptoms represent a dopamine excess or receptor hypersensitivity state that is responsive to neuroleptic medication, but the pathology of negative symptoms has been less well thought out.

Some investigators have proposed that a dopamine deficiency is specifically related to negative symptoms[66–68] and that positive and negative symptoms are at opposite ends of a dopamine spectrum. Crow,[69] on the other hand, hypothesizes two distinct syndromes: type I, predominantly positive symptoms, and type II, predominantly negative symptoms. Whereas type I is postulated to be related to dopamine excess, type II may be an illness unrelated to dopamine and might be caused by an encephalitic process with associated cell loss, intellectual impairment, and irreversibility. These separate hypotheses may not actually be contradictory, however, if a dopamine excess is related to only specific dopaminergic pathways (i.e., limbic and hypothalmic) whereas a dopamine deficiency is present in areas of cell loss and atrophy and is a result of the independent pathological process Crow describes. Consistent with this notion are the series of CT studies of schizophrenic patients suggesting that lateral ventricular enlargement may delineate a specific schizophrenic subgroup comprised of patients with predominantly negative symptoms, whose responsiveness to neuroleptic medication is reduced, whose symptoms are uncorrelated with serum prolactin concentrations, and who fail to react to dopamine agonists.[70–73] Whether these patients comprise a separate dopamine-deficient subgroup or a syndrome independent of a dopamine etiology remains an important research issue.

There is an increasing body of literature, particularly with animal models, providing evidence for a dopamine deficiency hypothesis for certain characteristic schizophrenic symptoms. For example, rats depleted of dopamine by neuroleptic treatment were found to have deficient arousal responses associated with goal-directed behavior. This behavior was suggestive of the anhedonia often characteristic of chronic schizophrenic patients.[74] In another study, loss of dopamine-containing neurons in rats following administration of the neurotoxin 6-hydroxydopamine (6-OHDA) resulted in the loss of ability to orient towards sensory stimuli, sensory neglect.[75] Finally, destruction of the mesolimbic cortical dopamine projection system, sparing the striatal system, produced deficits in exploratory behavior in rats that were restored by apomorphine administration.[76] These are novel animal models for behavior reminiscent of several of the negative symptoms ascribed to schizophrenia. Subgrouping patients by presence of predominantly positive or negative symptoms might then be one valid way of further testing the dopamine hypothesis of schizophrenia and the evaluation of pharmacological treatment response. The dopamine hypothesis, explaining a subgroup of psychoses, thus remains a viable and important research issue.

3. *THE SEROTONIN HYPOTHESIS*

5-Hydroxytryptophan (serotonin), a derivative of the indole amino acid tryptophan (see Fig. 4), is an important brain neurotransmitter with cell bodies in the midline raphe system. Wooley and Shaw[77] first hypothesized a major role for serotonin in mental processes and suggested that the suppression of its action could result in a psychiatric disorder. They based this hypothesis on the striking similarities between serotonin and the hallucinogen LSD. Since LSD is thought to block serotonin receptor sites, it could produce a relative decrease in serotonergic activity. Later reports that serotonergic depleting agents, such as reserpine, appear to alleviate some of the symptomatology of schizophrenia led Wooley and Shaw to shift their position and postulate that an actual increase in serotonin may be etiologically related to schizophrenia.[78] Even if schizophrenia is characterized by a deficiency in serotonin metabolism, it does not necessarily follow that these patients have low brain serotonin concentrations. Any substance, such as LSD, that blocks serotonin receptors might lead to an absolute increase in serotonin concentrations via a feedback loop while maintaining decreased serotonergic function. Numerous clinical studies measuring serotonin and its metabolites have since been pursued.

Peripheral measurements of serotonin and its principal metabolite, 5-hydroxyindoleacetic acid (5-HIAA), have yielded inconsistent results. Urinary excretion studies have, for the most part, been negative[79,80] and are of questionable value since most of urinary 5-HIAA excretion is derived from the gut, and very little represents brain metabolism.[81] Whole-blood and platelet concentrations of serotonin have also been measured in schizophrenic patients as well as in patients with a variety of other psychiatric and neurological disturbances.[82–84] Although several earlier studies either found low blood serotonin concentrations or showed no abnormalities in schizophrenic patients,[85,86] more recent research has revealed a group of these disorders, including schizophrenia, that have elevated blood serotonin concentrations. Whether this finding relates to brain metabolism is far from clear, since blood serotonin is also predominantly derived from gut metabolism of tryptophan.[81] In fact, there is some suggestion in the literature of an inverse relationship between blood and brain serotonin.[87] The derivation of CSF serotonin and 5-HIAA is controversial[88]; nevertheless, they are thought to be closer indicators of the status of brain serotonin. Whole-blood serotonin was found to be significantly higher in schizophrenic patients with cerebral atrophy detected by CT than in those with normal CT scans and controls.[83] In another study of many of the same patients, lower CSF 5-HIAA was correlated with the presence of cerebral atrophy by CT scanning and was inversely correlated with blood serotonin concentrations.[83,89]

Postmortem brain studies unfortunately do not add any clarity to this issue. The cell bodies of serotonergic neurons are located in the midline raphe nuclei in lower midbrain and upper pons, although their axons are widely distributed throughout the brain (see Fig. 1). Although one extensive study[90] by Joseph and colleagues reported no differences in serotonin and metabolites in several brain regions from schizophrenics compared with controls, another study by the same group of investigators determined a significant increase in serotonin, as well as 5-HIAA, in the putamen of schizophrenics.[91] They did not, however, assay other related brain regions. Farley *et al.*[92] found serotonin and 5-HIAA elevations in several forebrain regions (hypothalamus, medial olfactory area, nucleus accumbens, and globus pallidus) in a small group of schizophrenic patients, although they did not specifically examine the putamen. E. Korpi, L. E. DeLisi, J. E. Kleinman, and R. J. Wyatt (unpublished data) more recently found serotonin elevations in several brain areas including putamen, although there was considerable variability in the data. Winblad *et al.*,[93] on the other hand, reported reduced serotonin and 5-HIAA concentrations in several areas of brains from chronic schizophrenic patients. The inconsistencies in these findings may simply represent the difficulties inherent in obtaining reliable data from human postmortem studies. Cause of death, time from death to autopsy, medication status, and reliability of postmortem diagnoses are only a few of these problems.

Nevertheless, taken together, these several approaches appear to indicate abnormal serotonin metabolism in some chronic schizophrenic patients. The relationship of this finding to the illness, its etiology, and treatment warrants further investigation. The effect of neuroleptic medication on serotonergic activity is also an open issue. Most studies report an inhibitory effect of neuroleptics on platelet serotonin uptake.[94–96] With decreased uptake, whole-blood serotonin concentrations probably would be decreased rather than increased, as has been found in some studies. Bacopoulos *et al.*[97] found increased CSF 5-HIAA after fluphenazine administration in primates, although no change was noted in brain serotonin metabolites.

Pharmacological studies have added little in support of this hypothesis. Lauer *et al.*[98] administered tryptophan to both schizophrenic patients and normal control subjects. Although they found no increase in 5-HIAA excretion in the urine from the schizophrenic patients, there was a 100% increase of 5-HIAA in the urine from normal controls. This study suggests decreased serotonin production in schizophrenics, although others have been unable to confirm this finding.[99,100]

Assuming a serotonin deficiency hypothesis, treatment of schizophrenia with serotonin or its precursor, 5-hydroxytryptophan (5-HTP), has been attempted. These trials have been generally unsuccessful,[101–104] and in one study, 5-HTP significantly worsened the psychosis of at least three of 15 chronic schizophrenic patients.[105] Sherwood[106] and Ljungberg,[107] however, both reported significant improvement in patients given 5-HTP. Alternatively, if a serotonin excess is related to schizophrenia, then agents aimed at reducing serotonergic activity should have clinical efficacy. *para*-Chlorophenylalanine (PCPA), an inhibitor of serotonin synthesis, however, had no beneficial effect when given in a double-blind trial to seven chronic schizophrenic patients.[108] Four out of the seven studied had elevated blood serotonin concentrations, and two of these became clinically worse after 5-HTP administration. It is still possible, however, that specific serotonergic inhibiting agents could be of benefit to a subgroup of schizophrenic patients not clearly defined. There are also many other aspects of tryptophan metabolism in nervous tissue, as well as the physiological interaction of indoleamines with catecholamines at nerve synapses, that may be crucial to the disordered system and require further investigation.

4. THE NOREPINEPHRINE HYPOTHESIS

Noradrenergic axons are distributed throughout the brain, with their cell bodies prominent in the locus ceruleus, a nucleus of the brainstem (see Fig. 1). Animal behavioral studies support the concept that the norepinephrine neurons of the locus ceruleus function as a "reward system," since activation of this system with implanted electrodes produces behavior such as "self-stimulation" lever pressing.[109] On the basis of these and other studies, Wise and Stein[110–112] hypothesized that schizophrenia may be related to a defect in the "noradrenergic reward system" or selective norepinephrine neuron degeneration, producing the anhedonia characteristic of many chronic schizophrenics. This was supported by postmortem findings that dopamine β-hydroxylase (DBH), the enzyme that converts dopamine to norepinephrine and is found in noradrenergic neurons in the brain, was decreased in brains from schizophrenic patients. Although these studies could not be replicated,[113,114] they led to other hypotheses of noradrenergic dysfunction in schizophrenia. Even though the therapeutic effect of neuroleptic medication has been attributed largely to dopamine receptor blockade,[115–117] noradrenergic transmission is also inhibited by administration of these drugs[118] and should not be overlooked. In recent years extensive studies of both peripheral and central noradrenergic metabolism have been completed.

Farley *et al.*[119] reported increased norepinerphrine concentrations in the limbic regions of brains from paranoid schizophrenics. This finding was not confirmed by two other groups,[120,121] both of whom did not look at paranoid schizophrenia as a subclassification. Carlsson[122] found norepinephrine increased in the mesencephalon of paranoid schizophrenics, and Kleinman *et al.*[48] found both norepinephrine and its metabolite MHPG increased in the nucleus accumbens of paranoid schizophrenics. The CSF and plasma concentrations of norepinephrine also have been found to be elevated in some chronic schizophrenic patients compared with controls.[123–125]

In addition, investigations of noradrenergic receptor function have been conducted. Receptors for norepinephrine in the brain are located primarily in areas of the limbic system, hypothalamus, thalamus, and cerebellum. In some areas, their juxtaposition to dopaminergic neurons has led to interesting hypotheses of feedback loops and inhibitory–excitatory mechanisms. α-Adrenergic receptors are found in cortical and limbic structures, whereas β receptors are found in cortical projections from the locus ceruleus and in the cerebellum. Although studies of β-receptor function have not yet been examined, α-receptor function on platelets, as a model for α-receptor function throughout the body, has been studied in medication-free schizophrenic patients. Elevated α-receptor numbers have been found in these patients associated with a deficiency of PGE_1-stimulated cyclic AMP production, a function known to be inhibited by α-receptor stimulation.[126] It is not known, however, whether these peripheral findings parallel brain metabolism. In addition, the effect of long-term neuroleptic administration on these measures has not been tested adequately. One study reported an elevation of plasma norepinephrine associated with neuroleptics,[127] whereas another[128] reported decreased rather than increased norepinephrine concentrations after administration of pimozide.

The specificity of the noradrenergic alterations are also in question. Anxiety present at the time of the blood drawing and CSF sampling may be related to these findings, and studies designed to control for this nonspecific state are needed. Medication-free schizophrenic patients could be expected to display a higher level of anxiety than those "tranquilized" with neuroleptics. In addition, they may be more prone to anxiety about these procedures, particularly if they relate to their delusional systems. Physiological data also support this notion. Significantly higher pulse rates in schizophrenics compared with controls concomitant with elevations in CSF and plasma norepinephrine concentrations have been noted.[125]

Pharmacological receptor manipulation has been another interesting approach to investigate noradrenergic mechanisms. Clonidine, a predominantly centrally acting α_2-receptor agonist, has been given acutely and in short trials

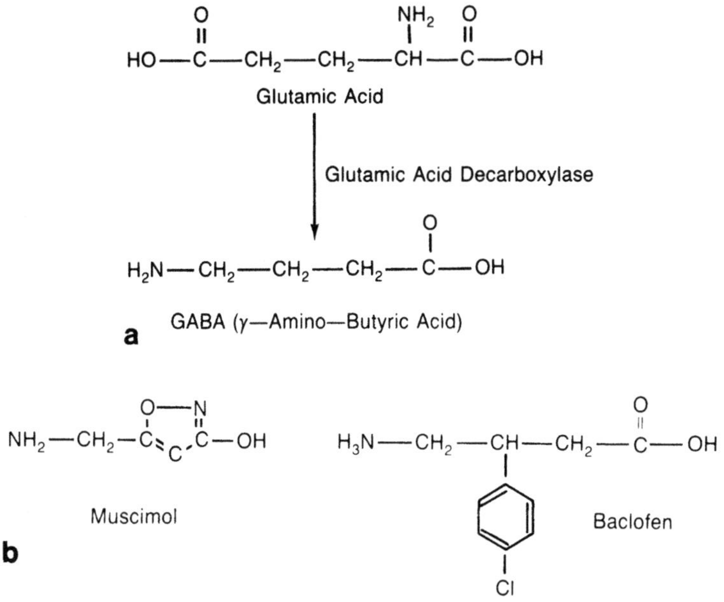

Fig. 5. (a) GABA synthesis. (b) GABAergic compounds.

to schizophrenics, producing a transient decrease in symptoms in some patients while concurrently reducing plasma norepinephrine and MHPG concentrations.[129,130] Propranolol, a β-receptor blocking agent, has been used in several therapeutic trials for schizophrenia. Most studies find propranolol not to be as efficacious as neuroleptics and not to produce obvious improvement as was initially thought (reviewed by Davis *et al.*[131]). When improvement occurs, it may be a result of a propranolol-induced increase in neuroleptic plasma concentrations, thus potentiating their action.[132] In addition, the high doses required to produce any decrease in symptoms cast doubt on the premise that the drug's only effect is through β-receptor blockade. Nevertheless, the improvement reported in some patients warrants further trials, perhaps using specific criteria for patient selection such as a predetermined marker of β-adrenergic receptor function or norepinephrine concentrations.

5. *THE γ-AMINOBUTYRIC ACID (GABA) HYPOTHESIS*

GABA (Fig. 5a) is thought to be an inhibitory neurotransmitter, with the highest concentrations found in the substantia nigra, globus pallidus, cortex, hippocampus, and hypothalamus. It is the most widely distributed neurotransmitter in the brain; 30% of all brain synapses are believed to be GABAergic.[133–136] GABAergic and dopaminergic pathways in both the nigrostriatal and mesolimbic systems closely interact, and inhibition of dopamine activity by GABA has been postulated.[137,138]

The observations of behavioral abnormalities in animals following mesolimbic injections of GABA antagonists led to a GABA deficiency hypothesis of schizophrenia.[139] If decreased GABA results in increased dopaminergic activity, then this notion would also be consistent with the dopamine hypothesis of schizophrenia.

Postmortem brain GABA and GABA receptor binding studies of schizophrenics compared with controls have in most cases not detected any differences.[140–143] Although most CSF GABA studies have also been negative,[144,145] a decrease in GABA concentrations has been reported in a small proportion of schizophrenics patients[146] in one study, and an increase in another.[147] In addition, most therapeutic trials of GABAergic drugs (Fig. 5b) have found no beneficial response or significant exacerbation of psychotic symptoms.[148–150] The knowledge that benzodiazepines bind to GABA receptors and appear to potentiate GABA activity[151] renewed interest in the use of this class of tranquilizers for schizophrenia. Although they generally appear not to be of major benefit,[152] there are some reports of marked improvement in psychotic symptoms with some benzodiazepine analogues[153,154] in small subgroups of patients.

The inconsistencies in these reports further support the necessity for biological subgrouping of schizophrenia. It is possible that one subgroup may be characterized by a high-dopamine–low-GABA syndrome, and another low dopamine and high GABA. Pharmacological studies might then appear to be contradictory depending on the nature of the patient group studied. On the basis of their CSF GABA studies, Van Kammen *et al.*[146] have suggested that a subgroup of acute, actively psychotic patients compose the decreased-GABA subgroup, whereas the more chronic patients, who have signs of emotional blunting and withdrawal, appear not to fit into this group.

6. *ENDOGENOUS OPIOIDS (ENDORPHINS)*

The literature on endogenous opioids, the endorphins, and schizophrenia is another confusing area. The notion that opioids may play a role in psychiatric disorders dates back to early reports of the use of morphine for schizophrenia.[155] Although, in general, these studies produced negative results, the idea gained prominence again with the discovery of endogenous opiates and their receptors. Various hypotheses have since been generated relating opioid activity to schizophrenia as well as other psychiatric disorders.[156,157]

Two major proposals have been experimentally explored with respect to the role of opioids and schizophrenia. Bloom *et al.*[158] proposed that an excess in central opioids, which induced catalepsy in rats, might have a role in the pathology of schizophrenia. At the same time, Jacquet and Marks[159] speculated that β-endorphins may be of therapeutic benefit as neuroleptics in view of their production of extrapyramidal-like effects in animals. Most subsequent clinical studies have been approached in an attempt to support or refute one of these opposing hypotheses.

Elevations of a variety of endogenous opioids have been detected in CSF and dialysate fluid from schizophrenic patients, although these have not been confirmed in other studies[157,160–162] and have been found to be decreased in at least one study.[163] If an excess of opioids is related to schizophrenia, than an opioid antagonist such as naloxone or naltrexone should alleviate some of the symptoms. In general, however, most acute double-blind administrations of naloxone to schizophrenic patients have not been beneficial. Although there are some reports of improvement in subgroups of patients with auditory hallucinations,[164–166] it is unclear whether the improvements are related to a therapeutic potentiation of concurrent neuroleptic treatment in some of these patients.[167,168]

β-Endorphins administered to schizophrenics in open trials were claimed to have a therapeutic effect by Kline *et al.*[169] Only one double-blind trial, however, produced positive effects[157]; others had no effect or even exacerbated symptoms.[170,171] More hopeful results have been obtained with another endorphin, des-tyrosine-γ-endorphin, in some but not all investigative trials,[172–174] and the Met-enkephalin analogue Fk 33-824 was also reported to reduce psychotic symptoms.[175] These initial trials provide some hope for the future usefulness of this class of agents for certain schizophrenias and may, in additional trials, help to clarify the role of opioids in the development of psychotic symptoms. Since opioid pathways in the brain are closely intertwined with the pathways of the major neurotransmitters already discussed, particularly in the limbic and striatal systems, it is not surprising that abnormalities in this system also have been associated with schizophrenia.

7. CHOLECYSTOKININ

Only very recently has the gastrointestinal peptide CCK been thought to play a significant role as a neurotransmitter. In specific areas of the brain it has been found to coexist with dopamine in neurons and is thought to act as a modulator of dopamine receptor activity.[176] A deficiency in CCK could lead to an increase in dopaminergic function in much the same way that GABA is thought to act. Preliminary measurements of CCK in postmortem schizophrenic brains have found limbic system decreases in subgroups of schizophrenic patients, although these are not consistent findings.[177,178] Nevertheless, ceruletide, a CCK derivative, may have an effect if given to acutely psychotic schizophrenic patients; two recent reports of dramatic improvement with this agent warrant further trials.[179,180]

8. ENDOGENOUS HALLUCINOGEN HYPOTHESIS

The idea that schizophrenia is caused by the abnormal production of a substance with hallucinogenic properties has been considered for several years. Osmond and Smythies first proposed in 1952[181] that abnormal methylation of neurotransmitter-like substances could produce endogeneous hallucinogens. From the knowledge that mescaline, a hallucinogenic drug, is a methylated catecholamine derivative (see Fig. 6), they hypothesized that it might be possible, *in vivo*, for an O-methylating enzyme (such as is involved in the synthesis of epinephrine) to produce a similar abnormally methylated substance such as

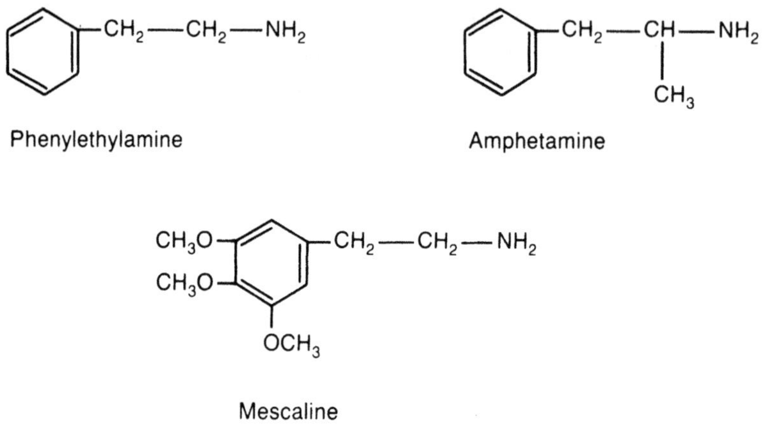

Fig. 6. Structural similaries among phenylethylamine, amphetamine, and mescaline.

dimethoxyphenylethylamine (DMPEA). With the subsequent discovery of the presence of an N-methylating enzyme in human brain having the potential to produce N-methylated compounds similar to the hallucinogen LSD, this hypothesis was expanded into the more general "transmethylation hypothesis" of schizophrenia.[182,183] It states that the abnormal (N- or O-) methylation of biogenic amines may give rise to schizophrenic symptoms. Studies in the last two decades have accumulated in support of this hypothesis, and the properties of the catechol derivative DMPEA and the indole derivative dimethyltryptamine (DMT) have been studied extensively.

Major support for this hypothesis came from reports suggestive of abnormal response to L-methionine loading in schizophrenic patients. L-Methionine, the main source of methyl groups in the body, can precipitate an acute psychosis as well as exacerbate the symptoms of chronic schizophrenia,[184–188] which may result from altered dimethylation of the compounds formed from methionine methyl transfer. Other studies of the reverse, attempting to decrease methylation reactions by the use of nicotinamide or low-methionine diets, have not alleviated the psychosis.[189–191]

9. *DIMETHOXYPHENYLETHYLAMINE*

Dimethoxyphenylethylamine was reported in one early study by Friedhoff and Van Winkle to be present in the urine from the majority of schizophrenic patients sampled and not from controls.[192] Referred to as the "pink spot" because of its staining properties when appropriate reagents are added to urine, DMPEA was extensively investigated in further studies. When given to humans in doses up to 1000 mg, it failed to produce schizophrenic-like symptoms in normal volunteers. Furthermore, other experiments failed to demonstrate *in vivo* synthesis of DMPEA. Numerous studies subsequent to Friedhoff and Van Winkle's initial report generally failed to confirm differences between schizophrenics and controls. Several studies provide support for the notion that DMPEA is of dietary origin rather than related to a psychotic state. Other investigations suggest that the pink spot may actually be *p*-tyramine, that it contains monoacetylcadaverine, that it is the product of intestinal bacterial metabolism, or that it may be a metabolite of drug treatment (reviewed by Wyatt *et al.*[183]).

10. DIMETHYLTRYPTAMINE

Although a psychotomimetic effect of DMPEA has never been demonstrated, DMT, when administered to normal volunteers, causes an acute transient psychotic-like state.[193] Dimethyltryptamine (Fig. 4) has been measured in blood and urine from schizophrenic patients and normal controls.[194,195] Consistent differences, however, between chronic schizophrenics or acute schizophrenics and controls with respect to both urine and plasma concentrations have not been reported.[196] The origin of DMT in these fluids has also been questioned. Although its presence could be explained by peculiarities in diet, the enzyme N-methyltransferase, thought to be required for DMT production, has been demonstrated in human lung and brain.[197,198]

In vivo conversion of [^{14}C]NMT to [^{14}C]DMT by the addition of a methyl group has been demonstrated in rabbit lung.[199] In a further study, [^{15}N,^{13}C]tryptamine was administered to seven chronic hallucinating schizophrenic patients in order to demonstrate actual *in vivo* conversion of tryptamine to NMT. No NMT was extracted from urine collected for 24 hours subsequent to the infusion, although virtually all of the label was detected in the urine specimens.[200] It is possible that urinary excretion was not representative of brain NMT or that the NMT was further metabolized and thus no longer detectable. Nevertheless, this experiment lends doubt to the significance of endogenous DMT production to schizophrenia. Difficulties in development of sensitive and specific assays for DMT have delayed further pursuit of this hypothesis.

Frohman *et al.*[201–203] isolated from plasma of schizophrenic patients a protein that they hypothesize is related to abnormal indole metabolism and the accumulation of DMT. Their factor, which they named the S-protein, is an α_2-globulin. When isolated from plasma of schizophrenic patients, they hypothesized, the protein is in an active form, an α-helical configuration, and causes more tryptophan to accumulate in cells than the inactive form, a β-helix, isolated from controls. They also found a high percentage of the α-helical molecule in plasma from normal volunteers under psychological stress.

11. PHENYLETHYLAMINE

Phenylethylamine (PEA), an endogenously produced decarboxylated metabolite of phenylalanine, is structurally and pharmacologically similar to amphetamine (Fig. 6) and has also been hypothesized to be an endogenous psychotomimetic compound relevant to the development of schizophrenia.[204,205] Indirect support for this notion comes from the behavioral effects of these agent in animals. Both PEA- and amphetamine-induced stereotypies in animals are blocked by neuroleptics. Furthermore, although there are several exceptions (i.e., clozapine and thioridizine), the relative doses required for blocking approximate the relative clinical potency of neuroleptics for treating schizophrenia.[206]

Fischer *et al.*[207,208] were the first to report elevated urinary PEA excretion in psychiatric patients, and several subsequent studies, although not all, have confirmed this finding. Potkin *et al.*,[209] using a gas-chromatography mass-spectrometric procedure, measured PEA in 24-hr urine collections from 31 chronic paranoid schizophrenic patients and 32 controls. The urinary PEA excretion was significantly higher in the paranoid schizophrenics than in the nonparanoid schizophrenics and controls. This finding was confirmed in an independent population of schizophrenics in India.[210] The diagnostic specificity of this finding, however, is not known. Fischer's original study, as well as others, also found elevations in manic patients, and another group recently reported elevations related to stress.[211] Linnoila *et al.*[212] and DeLisi *et al.*[213] have found a subgroup of atypically depressed patients also to have elevations in PEA excretion, although most PEA studies in depressed patients show marked deficits in urinary PEA concentrations.[214] The question of whether these findings are more specific to level of anxiety than to a particular illness etiology needs to be further explored.

Finally, carbidopa, a peripheral decarboxylation inhibitor that results in substantial decreases in urinary PEA excretion, has failed to be of benefit to three paranoid schizophrenics who had base-line elevations in PEA excretion (S. Potkin, L. E. DeLisi, C. N. Carson, and R. J. Wyatt, unpublished data), although it did alleviate several symptoms in multiple trials given to one delusionally depressed woman.[215]

12. IMMUNE DYSFUNCTION HYPOTHESIS

Heath and associates[216] reported the presence in blood from schizophrenic patients of a substance they named "taraxein," which means "confusion" or "disorder of the mind." Within 20 min of injection into normal humans, as well as primates, it produced a variety of psychotic-like symptoms from catatonia with waxy flexibility to delusions, auditory hallucinations, and a thought disturbance. In addition, it produced septal spikes on EEGs in monkeys, not unlike a similar abnormality seen in schizophrenic patients.[217] Since taraxein activity was found to be present in a particular subfraction of IgG, they theorized that schizophrenic symptoms may result from the formation of autoantibodies to brain tissue, particularly in the limbic system. Antibody deposits on neural membranes might then interfere with certain brain functions, resulting in schizophrenia. In support of this, Heath and Krupp[218] were able to demonstrate antibody deposition on neural cell nuclei from postmortem schizophrenic brain tissue, particularly from the septal region and basal caudate nuclei, which were the most intense when first reacted with schizophrenic sera. Although some investigators have been unable to confirm this work,[219,220] others have published supportive data.[221–224] Boehme *et al.*,[225] while demonstrating antibody staining in brain tissue of schizophrenics, in contrast to Heath's group, found it to be cytoplasmic only and present in schizophrenic as well as nonschizophrenic brain tissue.

Other evidence of immune dysfunction, such as abnormal lymphocyte histology,[226] abnormal lymphocyte function tests,[227,228] and abnormal percentages of lymphocyte subtypes,[229–231] have also been described in schizophrenic patients.

In addition, studies of histocompatibility antigens (HLA), the genetic component to the immune response, have shown increases of several types in schizophrenic populations, although each study finds elevated frequencies of different types.[232–234] Racial and other population factors unrelated to schizophrenia may be determining these differences. Nevertheless, it is significant that response to neuroleptic medication may be correlated with specific HLA types.[235] If future studies confirm this, the inheritance of particular immune capabilities may be of use in subtyping the schizophrenic syndrome[235] and predicting treatment response.

The array of signs of immune dysfunction among schizophrenic patients is, on the other hand, suggestive of nonspecific changes that may be related to the chronicity of the disorder and its treatment or, indirectly, to the harboring of an infectious agent. They may be manifestations of a generally disabled physical state in these patients.

13. VITAMIN AND MINERAL DEFICIENCIES

Reports[236] of several vitamins as remedies for schizophrenia (combinations of niacin, ascorbic acid, vitamin E, and others) have not been substantiated in controlled scientific trials.[237] Pauling[238] and others reported low excretion of ascorbic acid in groups of schizophrenics, and improvement in some symptoms has been observed with massive doses of ascorbic acid.[239] In the only double-blind trial of this vitamin,[240] however, the majority of patients participating had clinical signs of scurvy. In addition, the early reports of improvement with open trials of niacin were followed by several studies failing to confirm this. There is no doubt that, particularly in the past, the nutritional status of institutionalized patients may have been poor and that vitamin deficiencies may have been exacerbating the symptoms of schizophrenia as well as other psychiatric disorders.

Pauling, however, has hypothesized the presence of a genetic trait for schizophrenia characterized by a greater than normal requirement for cofactors, such as vitamins, to activate enzymes crucial to neurotransmitter metabolism. There is extremely little experimental evidence, however, to support this. Despite the widespread use of large doses of vitamins to treat schizophrenia by some clinicians, the few controlled studies that exist suggest that vitamins are not beneficial to specific psychotic symptoms other than in cases of vitamin deficiencies, that their effects are not specific to schizophrenia, nor that a basic deficiency in vitamins can lead to the development of schizophrenia.

14. ENDOCRINE DYSFUNCTION HYPOTHESIS

The classical post-pubertal development of schizophrenia, coupled with a peak age of incidence between 18 and 30, coincides with physiological fluctuations in hormonal levels.

Kraepelin[241] first postulated that schizophrenia was related to a disease of the endocrine glands. In the early 20th century, atrophy of the gonads was reported in postmortem examination of patients with dementia praecox.[242–244] Several reports also appeared of deficient secondary sexual characteristics in schizophrenic patients.[245,246] In addition, Mott[243] observed signs typical of thyroid deficiency in schizophrenics and reported improvement with thyroid hormone. It is now thought, however, that these early reports were complicated by other concurrent disease states in these patients (reviewed by Richter[247]). Furthermore, no clinical response to either testosterone[248] or gonadotrophin[249] has been noted in schizophrenic patients.

More recently, Shader *et al.*[250] found low follicle-stimulating hormone (FSH) concentrations in a subgroup of schizophrenic patients. Johnstone *et al.*[251] found a subgroup of patients with low luteinizing hormone (LH) and FSH and an inverse relationship between FSH and delusions in these patients. In a later study from the same group, absence of the normal episodic LH release or only small increments of change in LH concentrations were observed in 14 of 20 patients.[252] Clearly, dysfunctional hormonal regulation may play an important role in schizophrenia, and further research in this area warrants pursuit.

15. PROSTAGLANDIN HYPOTHESES

Horrobin[253] has been a proponent of a relationship between prostaglandin E deficiency and schizophrenia, whereas Feldberg[254] presents arguments in favor of an association with prostaglandin excess.

Prostaglandins are cyclic unsaturated fatty acids. Some prostaglandins are synthesized in the brain and have been found to be responsible for modulating neuronal release of catecholamines as well as having their formation mediated by a variety of neurotransmitters.[255] Thus, their association with psychiatric disorders has been of interest. The most direct evidence for prostaglandin alteration was reported by Mathe *et al.*,[256] who found increased total PGE in CSF of schizophrenic patients. This study, however, has yet to be confirmed in the literature. In addition, it should be noted that an increase in CNS prostaglandins occurs in a variety of other neurological disorders such as subarachnoid hemorrhage, transient ischemic attacks, meningoencephalitis, and multiple sclerosis.[257]

16. DISCUSSION

If not artifactual, the group of biochemical alterations associated with schizophrenia, separately or together, may be the result of any one of the following mechanisms:

1. Heredity. An inherited trait could lead to the production of a schizophrenic disorder or a vulnerability to its development.
2. Congenital. An environmental insult to the developing fetus arising during gestation and up to the moment of birth may predispose individuals to the late development of schizophrenia.
3. Trauma. A traumatic insult to the brain in prenatal or perinatal periods or childhood (physical, toxic, metabolic, or even emotionally initiated) may cause disruptions in neuronal growth and neurotransmitter mechanisms that could have permanent effects on the developing brain. Johnstone *et al.*[29] and more recently others[30] have reported structural brain changes including ventricular enlargement and cortical atrophy in some schizophrenic patients. These could be caused by such traumatic insults.
4. Infection. An infectious agent, such as a virus, may lay dormant for several years, cause an acute infectious reaction, or both. Segments of the viral DNA may fuse with the host DNA and alter the message communicated for production of specific brain proteins, thus altering neurotransmitter metabolism by this route. Viral infections have been known to cause changes in dopamine and other neurotransmitter activities in the brain and may also produce the previously described measures of cerebral atrophy.[258,259].
5. Environmental toxins. Exposure to lead, other metals, or multiple industrial pollutants may produce central nervous system damage, particularly to the developing brain in childhood. The finding of higher rates of schizophrenia in urban versus rural areas is well known and is consistent with this as well as the infectious hypotheses.

16.1. Support for the Genetic Hypothesis of Schizophrenia

The role heredity plays in the development of schizophrenia has been disputed for many years, although consanguinity and twin studies as well as the series of Danish adoption studies strongly suggest an inherited component to schizophrenia.[260–263] Although the concordance rate for schizophrenia within twin pairs varies from study to study, monozygotic twins consistently have greater concordance rates than dizygotic twins. Gottesman and Shields,[264] in a review of a series of these studies, found that the monzygotic concordance rate ranged between 34 and 58% while the dizygotic rate vaired between 9 and 26%. Although twins are reported to have emotional disturbances related to their twinship itself, such as retardation in ego development as well as identity,[265,266] they do not have an increased incidence of schizophrenia.[267] The most compelling argument against environmental induction of these high concordance rates comes from studies of schizophrenic and nonschizophrenic adoptees and their relatives. These studies showed a high prevalence of schizophrenic spectrum disorders among biological relatives of index cases. Schizophrenia in the biological parent, not the adoptive parent, appeared to be the crucial determinant.

There are several hypotheses for the inheritance of one or more enzyme defects or the existence of structural protein defects. One of the genetic models that has been suggested for schizophrenia is a single autosomal locus with two alleles.[268] Recent application of segregation analyses, however, to a large cohort of families failed to support this hypothesis.[269] X-chromosome-linked inheritance is unlikely as well, since there is no support for vertical transmission from one sex parent in greater frequency than another, and affected males are just as likely to have affected sons as daughters.[269] Multiple gene interaction through more than one locus, a polygenic hypothesis, also remains a possibility but is difficult to explore in population studies, particularly when the extent of heterogeneity within schizophrenic populations is unknown. A genetic–environmental interactional hypothesis is also tested only with great difficulty but may provide a substrate for further thought. Common to many medical illnesses are inherited gene traits that remain dormant until stimulated by environmental factors (i.e., galactosemia induced by milk exposure; favism, an inherited hemolytic anemia produced by exposure to fava beans; diabetes and exposure to high carbohydrates; and inherited susceptibility to certain infectious agents).

The inheritance of low activity of an enzyme or enzymes responsible for neurotransmitter metabolism has been proposed to be associated with schizophrenia. The two enzymes most extensively studied are monoamine oxidase (MAO) and dopamine β-hydroxylase (DBH). Monoamine oxidase catalyzes the oxidative deamination of several of the biogenic amines reviewed in this chapter, including dopamine, norepinephrine, serotonin, dimethyltryptamine, and phenylethylamine. It is found in brain as well as other tissues throughout the body and is most conveniently studied in platelets.

Although initial studies suggested that low platelet MAO activity was a marker for vulnerability for schizophrenia, more extensive research revealed many complicating environmental factors that may have contributed to the differences between patients and controls (reviewed by DeLisi *et al.*[270]). It is now thought that although MAO activity has an inherited component, chronic neuroleptic medication causes sufficient lowering of the enzyme activity to lend doubt to its significance as a marker for illness vulnerability.

Dopamine β-hydroxylase (DBH) is the enzyme that converts dopamine to norepinephrine. It is also a major enzyme in the metabolism of phenylethylamine. Family studies reveal a strong inheritance for the activity of this enzyme,[271] although most studies do not appear to link the inherited pattern to schizophrenia.[272] Studies of plasma activity of this enzyme in schizophrenic patients vary. Some studies show decreases in schizophrenic populations or subpopulations compared with controls.[273,274] but the majority of studies find no differences.[275] In one study, schizophrenics were found to have higher than normal plasma dopamine β-hydroxylase activity.[276] Dopamine β-hydroxylase activity has also been studied in CSF of schizophrenic patients. Sternberg *et al.*[277] report decreased CSF activity in medication-free patients and recently have shown that the patients with cerebral atrophy are the ones who have the lowest dopamine β-hydroxylase activity.[278] Postmortem brain dopamine β-hydroxylase activity was only found to be decreased in one early study.[112] Other investigators now attribute this initial finding to the difficulties inherent in obtaining accurate postmortem enzyme data.[113] It is also possible that neuroleptic medication causes alterations in dopamine β-hydroxylase activity as well as MAO activity and may be contributing to some of the positive findings, as has been seen in some preliminary studies.[113,279] Patients who have developed tardive dyskinesia appear to be a group with higher dopamine β-hydroxylase activity than a comparable schizophrenic group without tardive dyskinesia.[280] Although the reason for this increase is unknown, trials of dopamine β-hydroxylase-inhibiting drugs such as disulfiram or fusaric acid in initial reports were of benefit to patients with dyskinesias,[281,282] and this finding may help clarify the pathogeneses of tardive dyskinesia.

It is thus far unlikely that these two enzymes alone or together are responsible for a genetic predisposition to schizophrenia. There are, however, many other enzymes involved in neurotransmitter metabolism in the brain that have been less well studied clinically and may be worth further investigation.

16.2. Evidence in Support of a Viral Hypothesis

Several historical associations of viral epidemics with acute psychoses as well as more chronic sequelae have been reported. Karl Menninger[283] identified a psychosis associated with the influenza outbreaks of 1889–1892 and 1918–1919 as well as similar syndromes occurring during outbreaks of other major infections (i.e., tuberculosis, typhus, typhoid fever, and cholera). In addition, several documented CNS viral infections such as herpes encephalitis may, at the onset, present as a schizophrenia-like illness. The rare Russian tick-borne encephalitis, endemic to the Yakut Republic of the USSR, is said to be, in its chronic form, indistinguishable from classical schizophrenia.[284]

The epidemiology of schizophrenia, although controversial, appears to be consistent with the known epidemiology of viral illnesses in general. Schizophrenia appears to be increased in urban industrialized areas of the world and appears rarely, or in a more benign form, in rural areas and underdeveloped countries. In certain geographical areas, such as in particular counties of Ireland, it occurs in clusters, almost in epidemic form; moreover, the centers of these geographical areas change slowly with time.[285]

Also consistent with a viral etiology is the finding of a seasonal distribution in the birth rates of schizophrenics; a disproportionate number are born in the late winter and early spring. A seasonal peak in illness onset (early summer) is present as well. Some viral infections also show similar seasonality.[286,287]

A few studies examining the crucial factor of history of contact with a schizophrenic suggest increased incidences of schizophrenia among individuals in the age range of risk[288,289] who reside in close proximity to an index case of schizophrenia. Even the genetic adoption studies of Rosenthal and Kety, previously mentioned, do not present a conflict with this notion. Although they found an increased prevalence of schizophrenia among the biological and not adoptive relatives of schizophrenics, they did notice a clustering of schizophrenia in only a small portion of the biological families, which could be consistent with an infectious spread of the disease (S. S. Kety, personal communication).

Further support for the viral hypothesis comes from the extensive series of independent reports of immune system abnormalities in schizophrenic patients. Most of these could be consistent with immune reponse to a viral infection, although genetic control of the immune response, as well as the effect of chronic neuroleptic medication on immunity, may play a principal role in these findings.

Serum and CSF immunoglobulins, often elevated in viral diseases, have been found in some studies to be increased in schizophrenic patients,[290–292] although this finding may depend on the population of patients studied. Acute patients may show transient rises in immunoglobulin levels coinciding with the exacerbation of the illness. Chronically ill schizophrenics, on the other hand, may have decreased antibody production as a result of chronic insult to the central nervous system secondary to the disorder itself or to its treatment with neuroleptic medication.[293] It is interesting, however, that Amkraut *et al.*[294] found that patients with IgA and IgM below the median had a better prognosis than those who had levels above the median.

Serum antinuclear antibodies occur in increased frequency in schizophrenic populations, and increased antithymic antibody titers have been reported as well.[295–297] Although some investigators may use these studies to support a primary autoimmune mechanism for schizophrenia, they also are consistent with viral induction of autoantibody production.

Other investigations have focused on the presence of antibodies specific to certain viruses. Increased serum antibody titers to herpes simplex type I virus were reported in one study,[298] although this was not subsequently confirmed.[299] In addition, an elevation of the ratio of CSF to serum IgG antibody titers has been observed for both measles[291] and cytomegalovirus (CMV),[300] although an actual increase in CSF antibodies was not found, but rather a decrease in serum antibody titers. More significant is the further recent study by the same investigators reporting absolute increases in CSF CMV IgM antibodies in a small subgroup of schizophrenic patients.[301] Since IgM elevation is specifically suggestive of active viral infection, this last finding warrants more extensive exploration.

Another approach has been the attempt to isolate the unknown viral agent and further characterize it. Tyrrell *et al.*[302] demonstrated degenerative cytopathic changes in tissue cultures inoculated with CSF from schizophrenic patients and those with other serious neurological diseases. This effect, however, was not serially transferred from culture to culture, and no viral particles were detected by electron microscopy. The agent appears similar in size and in chemical characteristics to RNA-type viruses, although the likelihood of it being an active virus appeared diminished with the finding that RNA and protein synthesis inhibitors had little effect on its cytotoxic properties.[303] Nevertheless, despite the above initial negative reports, a more recent finding of behavioral changes in primates approximately 3 years after injection of the viral-like agent does suggest transfer of an active virus.[309]

Additional findings have emerged from several studies of postmortem brains of schizophrenics. Fisman[32] described "encephalitic-like" lesions in seven of ten schizophrenic brains. These lesions were concentrated in the area of the trigeminal nucleus, for which the herpes virus has a predilection. Herpes simplex type I nucleic acid sequences have also been demonstrated by another group in postmortem brain specimens from psychotic individuals.[304] One attempt to isolate the CMV viral genome using a hybridization technique was unable to detect CMV-related genetic information in a small group of schizophrenic brains. Stevens[31] recently reported evidence of patchy fibrillary gliosis most marked in the periventricular and periaqueductal structures as well as the basal forebrain. Although evidence of inflammation was not present in the brain tissue, gliosis in the absence of inflammation has been found previously in some specific viral disorders of the CNS.[305]

Finally, there is a relationship between viral CNS infection and brain neurotransmitter concentrations. Decreased brain serotonin has been reported in brain tissue from patients with Creutzfeldt–Jakob disease,[306] and brains of mice acutely ill with herpes simplex encephalitis have increases in serotonin as well as catecholamine concentrations.[258,259]

Taken together, these studies remain significant preliminary findings that may prove fruitful in further more extensive studies. Future emphasis might be to isolate specific viral genome material from DNA in postmortem brain areas hypothesized to be relevant to schizophrenic symptoms.

17. CONCLUSION

The etiology of schizophrenia remains unknown. In this chapter we attempted to provide an overview of some of what is known of the biochemistry associated with schizophrenic-like disorders. Research, however, has been hampered by several considerations. First, there has been little consensus over the years on how to clinically define the disorder called schizophrenia or even if it consists of one or many disorders with common clusters of symptoms. Consequently, studies of schizophrenics compared with controls have in many instances been misleading and contradictory. Moreover, if there are multiple biochemical subtypes of this disorder, the study of selected chemically homogeneous subgroups for evaluation of etiologic hypotheses is of considerable importance.

Aside from heterogeneity, the issue of where in the course of the illness scientific studies should focus is also important. Most studies thus far have focused on the chronically ill, deteriorated patients. It may be necessary to focus some of the research away from these patients who have been hospitalized for several years and refocus on acutely ill patients at the onset of the disorder or those who are established to be at increased risk for schizophrenia. New treatments already abandoned for lack of response may be false negatives if only examined in trials given to the chronically ill, since it has long been recognized that the chronically ill schizophrenics are less responsive even to conventional neuroleptic medications.[307,308] It is only by studying the younger patients that the effects of chronic mental deterioration and pharmacological treatment will be separated from the etiology of schizophrenia.

Additionally, methods used to study this disorder have been limited by multiple technical problems. Most of the research thus far has focused on the use of body fluids and postmortem tissue for the determination of various chemical substances. Aside from problems of biochemical assay sensitivity and reliability, technical aspects of postmortem, blood, CSF, and urine collections, if not carefully considered, can lead to artifactual findings.

Perhaps of considerably more importance, however, is that the ability to study the neurochemical aspects of the schizophrenic state is hampered by the limitations in our general knowledge of human brain neurotransmitter and neuromodulator interactions in normal individuals. Most of our knowledge of neurochemistry, in fact, is derived from studies in animals; and although the details of the major neurotransmitter pathways have been well described, little has been confirmed in postmorten human brains. It is most noteworthy also that although most of the hypotheses mentioned in this chapter are derived from animal experimentation, there are no good animal models for schizophrenia. Disorders of "the mind," after all, are human entities.

Finally, it is evident from the array of different chemical, physiological, and anatomic findings associated with schizophrenia that no system is operating in isolation from the others. Unfortunately, most reports appearing in the literature do not address this issue. It is rare, for example, that one can find reported correlations of dopamine, serotonin, norepinephrine, GABA, and other substances within one patient or one postmortem brain specimen. Investigators generally study one substance or hypothesis of interest in isolation from others. The data reviewed in this chapter must also be seen in the context of the limitations of our present knowledge of basic genetics, virology, immunology and neurochemistry and of how each field interrelates. What remains clear is that all the biochemical studies thus far have provided limited insight into the dynamic interactions of brain neurotransmitters and neuromodulators with gene and environmental controls of the pathogenesis of schizophrenia. We hope, however, that they have laid the groundwork for future biochemical classification of psychoses and may provide the basis for new pharmacological approaches to treatment.

REFERENCES

1. Rosenstein, M. J., and Milazzo-Sayre, L. J., 1975, Mental Health Service System Reports, U.S. Department of Health and Human Services, NIMH, Division of Biometry and Epidemiology, Survey and Reports Branch, Rockville, Maryland, p. 45.
2. World Health Organization, 1975, *Public Health Papers* No. 63, *Schizophrenia: A Multinational Study*, World Health Organization, Geneva, pp. 35–38.
3. Fromm-Reichmann, F., 1948, *Psychiatry* **11**:263–273.
4. Lidz, T., Cornelison, A. R., Fleck, S., and Terry, D., 1957, *Psychiatry* **20**:329–342.
5. Singer, M. T., and Wynne, L. C., 1966, *Psychiatr. Res. Rep.* **20**:25–38.
6. Keith, S. J., and Matthews, S. M., 1982, *Psychiatry: 1982, The American Psychiatric Association Annual Review* (L. Grinspoon, ed.), American Psychiatric Press, Washington, pp. 166–178.
7. Faucault, M., 1967, *Madness and Civilization*, Tavistock Publications, London.
8. Tusques, J., 1937, *Sem. Hop. Paris* **13**:109–111.

9. Kalinowsky, L. B., and Hippius, H., 1969, *Pharmacological, Convulsive, and Other Somatic Treatments in Psychiatry*, Grunne & Stratton, New York.
10. *Dorland's Illustrated Medical Dictionary*, 1965, (J. P. Friel, ed.), W. B. Saunders, Philadelphia.
11. Carlsson, A., 1978, *Am. J. Psychiatry* **135:**164–173.
12. Rosenthal, D. and Kety, S. S., eds., 1968, *The Transmission of Schizophrenia*, Pergamon Press, Oxford.
13. Gottesman, I. I., and Shields, J., 1982, *Schizophrenia: The Epigenetic Puzzle*, Cambridge University Press, Cambridge.
14. Knox, W. E., 1972, *The Metabolic Basis of Inherited Disease*, 3rd edition (J. B. Stanbury, J. B. Wyngaarden, and D. S. Fredrickson, eds.), McGraw-Hill, New York, p. 266.
15. Teuber, H. L., 1972, *Neurosci. Res. Prog. Bull.* **10**(4):381–388.
16. Penfield, W., and Jasper, G., 1954, *Epilepsy and the Functional Anatomy of the Human Brain*, Little, Brown, Boston.
17. Mullan, S., and Penfield, W., 1959, *Arch. Neurol. Psychiatry* **81:**269–284.
18. Bazhin, E. F., Wasserman, L. I., and Tonkonogii, I. M., 1975, *Neuropsychologia* **13:**481–487.
19. Abrams, R., and Taylor, M. A., 1979, *Arch. Gen. Psychiatry* **36:**1355–1358.
20. Flor-Henry, P., 1976, *Ann. N.Y. Acad. Sci.* **280:**777–797.
21. Stevens, J. R., Bigelow, L., Denney, D., Lipkin, J., Livermore, A. H., Rauscher, F., Wyatt, R. J., 1979, *Arch. Gen. Psychiatry* **36:**251–262.
22. Itil, T. M., 1977, *Schizophren. Bull.* **3:**61–79.
23. Fenton, G. W., Fenwick, P. B. C., Dollimore, J., Dunn, T. L., and Hirsch, S. R., 1980, *Br. J. Psychiatry* **136:**445–455.
24. Buchsbaum, M. S., King, A. C., Cappelletti, J., Coppola, R., van Kammen, D. P., 1982, *Adv. Biol. Psychiatry* **9:**50–56.
25. Malitz, S., Lozzi, V., and Kanzler, M., 1969, *Schizophrenia: Current Concepts and Research* (D. V. Siva Sankar, ed.), P. J. D. Publications, Hicksville, New York, pp. 273–302.
26. Buchsbaum, M. S., Ingvar, D. H., Kessler, R., Waters, R. N., Cappelletti, J., van Kammen, D. P., King, A. C., Johnson, J. L., Manning, R. G., Flynn, R. W., Mann, L. S., Bunney, W. E., and Sokoloff, L., 1982, *Arch. Gen. Psychiatry* **39:**251–259.
27. Ingvar, D. H., and Frazen, G., 1974, *Acta Psychiatr. Scand.* **50:**425–462.
28. DeLisi, L. E., and Buchsbaum, M. S. *Biol. Psychiatry* (in press).
29. Johnstone, E. C., Crow, T. J., Frith, C. D., Husband, J., and Kreel, L., 1976, *Lancet* **2:**924–926.
30. Weinberger, D. R., and Wyatt, R. J., 1982, *Biological Markers in Psychiatry and Neurology* (I. Hanen and E. Usdin, eds.), Pergamon Press, New York, pp. 505–512.
31. Stevens, J. R., 1982, *Arch. Gen. Psychiatry* **39:**1131–1139.
32. Fisman, M., 1975, *Br. J. Psychiatry* **126:**414–422.
33. Kandel, E. R., and Schwartz, J. H., 1981, *Principles of Neural Science*, Elsevier/North-Holland, New York.
34. Cooper, J. R., Bloom, F. E., and Roth, R. H., 1982, *The Biochemical Basis of Neuropharmacology*, Oxford University Press, New York.
35. Young, D., and Scoville, W. B., 1938, *Med. Clin. North Am.* **22:**637–646.
36. Connell, P. H., 1958, *Maudsley Monographs No. 5.*, Oxford University Press, London.
37. Griffith, J. D., Cavanaugh, J. H., and Oates, J. A., 1970, *Psychotomimetic Drugs* (D. H. Efron, ed.), Raven Press, New York.
38. Janowsky, D., and Davies, J., 1976, *Arch. Gen. Psychiatry* **33:**304–308.
39. Woodrow, K. M., Reifman, A., and Wyatt, R. J., 1978, *Neuropharmacology and Behavior* (B. Haber and M. H. Aprison, eds.), Plenum Press, New York, pp. 1–22.
40. Wyatt, R. J., Potkin, S. G., Kleinman, J. E., Weinberger, D. R., Luchins, D. L., and Jeste, D. V., 1981, *J. Nerv. Ment. Dis.* **169:**100–112.
41. van Kammen, D. P., Bunney, W. E., Docherty, J. P., Marder, S. R., Ebert, M. H., Rosenblatt, J. E., and Rayner, J. N., 1982, *Am. J. Psychiatry* **139:**991–997.
42. Frye, P. E., Pariser, S. F., Kim, M., and O'Shaughnessy, R. W., 1982, *J. Clin. Psychiatry* **43:**252–253.

43. Carlsson, A., 1974, *J. Psychiatr. Res.* **11**:57–64.
44. Corsini, G. U., Del Zompo, M., Manconi, S., Piccardi, M. P., Onali, P. L., and Mangoni, A., 1977, *Adv. Biochem. Psychopharmacol.* **16**:645–648.
45. Smith, R. C., Tamminga, C., and Davis, J. M., 1977, *J. Neural Transm.* **40**(2):171–176.
46. Tamminga, C. A., Schaffer, M. H., Smith, R. C., and Davis, J. M., 1978, *Science* **200**(4341):567–568.
47. Bird, E. D., Crow, T. J., Iverson, L. L., Longden, A., Mackay, A. V. P., Riley, G. J., and Spokes, E. C., 1979 *J. Physiol.* (*Lond.*) **293**:36–37P.
48. Kleinman, J. E., Karoum, F., Rosenblatt, J. E., Gillin, J. C., Hong, J., Bridge, T. P., Zalcman, S., Storch, F., del Carmen, R., and Wyatt, R. J., 1982, *Biological Markers in Psychiatry and Neurology* (I. Hanin and E. Usdin, eds.), Pergamon Press, Oxford, pp. 67–78.
49. Bird, E. D., Spokes, E. G., Barnes, J., Mackay, A., Iversen, L., and Shepherd, M., 1977, *Lancet* **2**:1157–1159.
50. Owen, F., Cross, A. J., Crow, T. J., Longden, A., Poulter, M., and Riley, G. J., 1978, *Lancet* **2**:223–225.
51. Bird, E. D., Spokes, E. G., and Iversen, L. L., 1979, *Science* **204**:93–94.
52. Mackay, A. V. P., Iverson, L. L., Rossor, M., Spokes, E., Bird, E., Arregui, A., Creese, I., and Snyder, S. H., 1982, *Arch. Gen. Psychiatry* **39**:991–997.
53. Lee, T., and Seeman, P., 1980, *Am. J. Psychiatry* **137**:191–197.
54. Schwartz, J. C., Baudry, M., Markes, M. P., Costentin, J., and Protais, P., 1978, *Life Sci.* **23**:1785–1790.
55. Crow, T. J., Cross, A. J., Johnstone, E. C., and Owen, F., 1982, *Schizophrenia As a Brain Disease* (F. A. Henn and H. A. Nasrallah, eds.), Oxford University Press, New York, pp. 196–234.
56. Burt, D. R., Creese, I., and Snyder, S. H., 1977, *Science* **197**:326–328.
57. Clow, A., Theodorou, A., Jenner, P., and Marsden, C. D., 1980, *Eur. J. Pharmacol.* **63**:135–144.
58. Owen, F., Cross, A. J., Waddington, J. L., Poulter, M., Gamble, S. J., and Crow, T. J., 1980, *Life Sci.* **26**:55–59.
59. Reisine, T. D., Fields, J. Z., Bird, E. D., Spokes, E., and Yamamura, H. I., 1978, *Commun. Psychopharmacol.* **2**:79–84.
60. Haracz, J. L., 1982, *Schizophren. Bull.* **8**:438–469.
61. Muller, E. E., Nistico, G., and Scapagini, E., 1977, *Neurotransmitters and Anterior Pituitary Function*, Academic Press, New York.
62. Meltzer, H. Y., and Fang, V. S., 1976, *Arch. Gen. Psychiatry* **33**:279–286.
63. Laughren, T. P., Brown, W. A., and Williams, B. W., 1979, *Am. J. Psychiatry* **136**:108–110.
64. Brown, W. A., and Laughren, T. P., 1981, *Am. J. Psychiatry* **138**:237–239.
65. Meltzer, H. Y., Sachar, E. J., and Frantz, A. G., 1974, *Arch. Gen. Psychiatry* **31**:564–569.
66. Mackay, A. V. P., 1980, *Br. J. Psychiatry* **137**:379–383.
67. Chouinard, G., and Jones, B. D., 1978, *Lancet* **2**:99–100.
68. Wyatt, R. J., 1984, The Stanley R. Dean Research Aware Lecture (in press).
69. Crow, T. J., 1980, *Br. Med. J.* **280**:66–68.
70. Crow, T. J., 1981, *Br. J. Psychiatry* **139**:251–254.
71. Kleinman, J. E., Weinberger, D. R., Rogol, D. A., Bigelow, L. B., Klein, S. T., and Wyatt, R. J., 1982, *Arch. Gen. Psychiatry* **39**:655–657.
72. Andreasen, N. C., Olsen, S. A., Dennert, J. W., and Smith, M. R., 1982, *Am. J. Psychiatry* **132**:297–302.
73. Jeste, D. V., Zalcman, S., Weinberger, D. R., Bigelow, L. B., Kleinman, J. E., Rogol, A., Gillin, J. C., and Wyatt, R. J., 1983, *Prog. Neuropsychopharmacol. Biol. Psychiatry* **7**:83–88.
74. Wise, R. A., 1982, *Behav. Brain Sci.* **5**:39–87.
75. Siegfried, B., and Bures, J., 1979, *Brain Res.* **167**:139–155.
76. Fink, S. J., and Smith, G. P., 1980, *Brain Res.* **199**:359–384.
77. Wooley, D. W., and Shaw, E., 1953, *Proc. Natl. Acad. Sci. U.S.A.* **40**(4):228–231.
78. Wooley, D. W., 1962, *The Biochemical Bases of Psychoses or the Serotonin Hypotheses about Mental Illness*, John Wiley and Sons, New York, pp. 168–179.

79. Haverback, B. J., Sjoerdsma, A., and Terry, L. L., 1956, *N. Engl. J. Med.* **255**:270–274.
80. Feldstein, A., Hoagland, H., and Freeman, H., 1958, *Science* **128**:358.
81. Toh, C. C., 1954, *J. Physiol. (Lond.)* **126**:248–254.
82. Garelis, E., Gillin, J. C., Wyatt, R. J., and Neff, N., 1975, *Am. J. Psychiatry* **132**:184–186.
83. DeLisi, L. E., Neckers, L. M., Weinberger, D. R., and Wyatt, R. J., 1981, *Arch. Gen. Psychiatry* **38**:647–654.
84. Freedman, D. X., Belenduik, K., Belenduik, G. W., and Crayton, J. W., 1981, *Arch. Gen. Pscyhiatry* **38**:655–659.
85. Feldstein, A., Hoagland, H., and Freeman, H., 1959, *J. Nerv. Ment. Dis.* **129**:62–68.
86. Jus, A., Laskowska, D., and Zimny, S., 1958, *Ann. Med. Psychol. (Paris)* **116**:898–913.
87. Coleman, M., Hart, P. N., and Randall, J., 1977, *Neuropaediatrie* **8**:459–466.
88. Post, R. M., Ballenger, J. C., and Goodwin, F. K., 1980, *Neurobiology of Cerebrospinal Fluid* (J. H. Wood, ed.), Plenum Press, New York, pp. 685–717.
89. Potkin, S. G., Weinberger, D. R., Linnoila, M., and Wyatt, R. J., 1983, *Am. J. Psychiatry* **140**:21–25.
90. Joseph, M. H., Baker, H. F., Crow, T. J., Riley, G. J., and Risby, D., 1979, *Psychopharmacology* **62**:279–275.
91. Crow, T. J., Baker, H. F., Cross, A. J., Joseph, M. H., Lofthouse, R., Longden, A., Owen, F., Riley, G. J., Glover, V., and Killpack, W. S., 1979, *Br. J. Psychiatry* **134**:249–256.
92. Farley, I. J., Shannak, K. S., and Hornykiewicz, O., 1980, *Receptors for Neurotransmitters and Peptides Hormones* (G. Pepeu, M. J., Kuhar, and S. J. Enna, eds.), Raven Press, New York, pp. 427–433.
93. Winblad, B., Bucht, G., Gottfries, C. G., and Roos, B. E., 1979, *Acta Psychiatr. Scand.* **60**:17–28.
94. Ahtee, L., and Paasonen, M. K., 1968, *Ann. Mod. Exp. Biol. Fenn.* **46**:416–422.
95. Todrick, A., and Tait, A. C., 1969, *J. Pharm. Pharmacol.* **21**:751–762.
96. Oxenkrug, G. F., 1978, *J. Pharm. Pharmacol.* **30**:740.
97. Bacopoulos, N. G., Redmond, D. E., and Roth, R. H., 1979, *J. Neurochem.* **32**(4):1215–1218.
98. Lauer, J. W., Inskip, W. M., Bernsohn, J., and Zeller, E. A., 1958, *Arch. Neurol. Psychiatry* **80**(1):122–130.
99. Kopin, I. J., 1959, *Science* **129**:(3352):835–836.
100. Shaw, C. R., Lucas, J., and Rabinovitch, R. D., 1959, *Arch. Gen. Psychiatry* **1**(4):366–371.
101. Wooley, D. W., 1962, *The Biochemical Bases of Psychoses or the Serotonin Hypothesis about Mental Illness*, New York, John Wiley and Sons, pp. 168–170.
102. Klee, G. D., Bertino, J., Goodman, A., and Aronson, H., 1961, *J. Ment. Sci.* **106**(442):309–316.
103. Hoagland, H., 1958, *J. Nerv. Ment. Dis.* **126**(3):211–220.
104. Pollin, W., Cardon, P. V., Jr., and Kety, S. S., 1961, *Science* **133**(3446):104–105.
105. Bigelow, L. B., Walls, P., Gillin, J. C., and Wyatt, R. J., 1979, *Biol. Psychiatry* **14**:53–67.
106. Sherwood, S. L., 1955, *Proc. R. Soc. Med.* **48**(10):855–863.
107. Ljungberg, E., 1963, *West Eur. Symp. Clin. Chem.* **2**:169–170.
108. DeLisi, L. E., Freed, W. J., Gillin, J. C., Kleinman, J. E., and Wyatt, R. J., 1982, *Biol. Psychiatry***17**:471–477.
109. Crow, T. J., Spear, P. J., and Arbuthnott, G. W., 1972, *Brain Res.* **36**:275–287.
110. Stein, L., and Wise, C. D., 1971, *Science* **171**(3975):1032–1036.
111. Wise, C. D., and Stein, L., 1973, *Science* **181**:344–347.
112. Wise, C. D., Baden, M. M., and Stein, L., 1974, *J. Psychiatr. Res.* **11**:185–198.
113. Wyatt, R. J., Schwartz, M. D., Erdelyi, E., and Barchas, J. D., 1975, *Science* **187**:368–369.
114. Cross, A. J., Crow, T. J., Killpack, W. S., London, A., Owen, F., and Riley, G. L., 1978, *Psychopharmacology* **59**:117–121.
115. Creese, I., Burt, D. R., and Snyder, S. H., 1976, *Science* **192**:481–483.
116. Seeman, P., Lee, T., Chan-Wong, M., and Wong, K., 1976, *Nature* **261**:717–719.
117. Snyder, S. H., 1976, *Am. J. Psychiatry* **133**:197–202.
118. Blumberg, J. B., Vetulani, J., Stawarz, R. J., and Sulser, F., 1976, *Eur. J. Pharmacol.* **37**(2):357–366.
119. Farley, I. J., Price, K. S., McCullough, E., Deck, J. H. N., Hordynski, W., and Hornykiewicz, O., 1978, *Science* **200**:456–458.

120. Crow, T. J., Owen, F., Cross, A. J., Lofthourse, R., and Longden, A., 1978, *Lancet* **1**:36–37.
121. Bird, E. D., Spokes, E. G., and Iverson, L. L., 1979, *Science* **204**:93–94.
122. Carlsson, A., 1979, *Catecholamines: Basic and Clinical Frontiers*, Volume 1 (E. Usdin, I. J. Kopin, and J. D. Barchas, eds.), Pergamon Press, New York, pp. 4–19.
123. Gomes, V. C. R., Shanley, B. C., Potgieter, L., and Roux, J. T., 1980, *Br. J. Psychiatry* **137**:346–351.
124. Lake, C. R., Sternberg, D. E., van Kammen, D. P., Ballenger, J. C., Ziegler, M. G., Post, R. M., Kopin, I. J., and Bunney, W. E., 1980, *Science* **207**:331–333.
125. Kemali, D., Vecchio, M. D., and Maj, M., 1982, *Biol. Psychiatry* **17**:711–717.
126. Kafka, M. S. vanKammen, D. B., Kleinman, J. E., Nurenberger, J. I., Siever, L. J., Uhde, T. W., and Polinsky, R. J., 1982, *Commun. Psychopharmacol.* **4**(6):477–486.
127. Naber, D., Finkbeiner, C., Fischer, B., Zander, K. J., and Ackenheil, M., 1980, *Neuropsychobiology* **6**:181–189.
128. Sternberg, D. E., van Kammen, D. P., Lake, C. R., Ballenger, J. C., Marder, S. R., and Bunney, W. E., 1981, *Am. J. Psychiatry* **138**:1045–1050.
129. Freedman, R., Bell, J., and Kirch, D., 1980, *Am. J. Psychiatry* **137**(5):629–630.
130. Ko, G., 1983, Abstr. Am. Psychiatr. Assoc.
131. Davis, J. M., Janicak, P., Chang, S., and Klerman, K., 1982, *Psychiatry 1982: Annual Review*, American Psychiatric Press, Washington, pp. 178–228.
132. Peet, M., Middlemiss, D. N., and Yates, R. A., 1980, *Lancet* **2**(8201):978.
133. Awapara, J., 1950, *Fed. Proc.* **9**:148.
134. Roberts, E., and Frankel, S., 1950, *Fed. Proc.* **9**:219.
135. Snyder, S. H., 1975, *Biochem. Pharmacol.* **24**:1371–1374.
136. Baxter, C., 1976, *GABA in Nervous System Function* (E. Roberts, T. N. Chase, and D. B. Tower, eds.), Raven Press, New York, pp. 61–87.
137. Yoshida, M., and Precht, W., 1971, *Brain Res.* **32**:225–228.
138. Fuxe, K., Perez de la Mora, M., Hokfelt, T., Agnati, L., Ljungdahl, A., and Johanson, O., 1977, *Psychopathology and Brain Dysfunction* (C. Shagass, S. Gershon, and A. J. Friedhoff, eds.), Raven Press, New York, pp. 97–111.
139. Roberts, E., 1972, *Neurosci. Res. Prog. Bull.* **10**:468–483.
140. Cross, A. J., Crow, T. J., and Owen, F., 1979, *Lancet* **1**:560–561.
141. Bennett, J. P., Enna, S. J., Bylund, D. B., Gillin, J. C., Wyatt, R. J., and Synder, S. H., 1979, *Arch. Gen. Psychiatry* **36**:927–934.
142. Iversen, L. L., Bird, E., and Spokes, E., 1979, *GABA-Neurotransmitters: Pharmacochemical, Biochemical and Pharmacological Aspects* (P. Krogsgaard-Larsen, J. Scheel-Kruger, H. Kofod, eds.), Academic Press, New York, pp. 177–190.
143. Petty, T. L., Kish, S. J., Buchanan, J., and Hansen, S., 1979, *Lancet* **1**:237.
144. Gold, B. I., Bowers, M. B., Jr., Roth, R. H., and Sweeney, D. W., 1980, *Am. J. Psychiatry* **137**:362–364.
145. Lichtshtein, D., Dobkin, J., Ebstein, R. P., Biederman, J., Rimon, R., and Belmaker, R. H., 1978, *Br. J. Psychiatry* **132**:145–148.
146. van Kammen, D. P., Steinberg, D. E., Hare, T. A., Waters, R. N., Bunney, W. E., 1982, *Arch. Gen. Psychiatry* **39**:91–97.
147. McCarthy, B. W., Gomes, U. R., Neethling, A. C., Shanley, B. C., Talgaard, J. J., Potgieter, L., and Roux, S. T., 1981, *J. Neurochem.* **36**(4):1406–1408.
148. Tamminga, C. A., Crayton, J. W., and Chase, T. N., 1978, *Am. J. Psychiatry* **135**:746–747.
149. Schulz, S. C., van Kammen, D. P., Buchsbaum, M. S., Roth, R. H., Alexander, P., and Bunney, W. E., Jr., 1981, *Pharmacopsychiatria* **14**:129–134.
150. Simpson, G. M., Branchex, M. M., and Shrivastava, R. K., 1975, *Lancet* **1**:966–967.
151. Tallman, J. F., Paul, S. M., Skolnick, P., and Gallager, D. W., 1980, *Science* **207**:274–281.
152. Jimerson, D. C., van Kammen, D. P., Post, R. M., Docherty, J. P., and Bunney, W. E., 1982, *Am. J. Psychiatry* **139**:489–491.
153. Beckmann, H., and Haas, S., 1980, *Psychopharmacology* **71**:79–82.
154. Lingjaerde, O., 1982, *Acta Psychiatr. Scand.* **65**:339–354.
155. Wikler, A., 1952, Public Health Monograph No. 52, U. S. Dept. of Health, Education and Welfare, Washington.

156. Watson, S. J., Akil, H., Berger, P. A., and Barchas, J. D., 1979, *Arch. Gen. Psychiatry* **36**:35–41.
157. Berger, P. A., Watson, S. J., Akil, H., Elliot, G. R., Rubin, R. T., Pfefferbaum, A., Davis, K. L., Barchas, J. D., and Li, C. H., 1980, *Arch. Gen. Psychiatry* **37**:635–640.
158. Bloom, F., Segal, D., Ling, N., and Guillemin, R., 1976, *Science* **194**:630.
159. Jacquet, Y., and Marks, N., 1976, *Science* **194**:632.
160. Lindstrom, L. H., Widerlow, E., Gunne, L. M., Wahlstrom, A., and Terenius, L., 1978, *Acta Psychiatr. Scand.* **57**:153–164.
161. Domschke, W., Dickschas, A., and Mitznegg, P., 1979, *Lancet* **1**:1024.
162. Terenius, L., Wahlstrom, A., Lindstrom, L., and Widerlou, E., 1976, *Neurosci. Lett.* **3**:157–162.
163. Pickar, D., Cohen, M. R., Naber, D., and Cohen, R. M., 1982, *Biol. Psychiatry* **17**:1243–1276.
164. Gunne, L. M., Lindstrom, L., and Terenius, L., 1977, *J. Neural Transm.* **40**:13–19.
165. Berger, P., Watson, S., Akil, H., and Barchas, J., 1981, *Am. J. Psychiatry* **138**:913–915.
166. Lehman, H., Nair, N. P. V., and Kline, N. S., 1979, *Am. J. Psychiatry* **136**:762–766.
167. Pickar, D., Vartanian, F., Bunney, W. E., Maier, H. P., Gastpar, M. T., Prakash, R., Sethi, B. B., Lideman, R., Belyaer, B. S., Tsutsulkovskaja, M. V. A., Jungkunz, G., Nedopil, N., Verhoeven, W., and van Praag, H., 1982, *Arch. Gen. Psychiatry* **39**:313–319.
168. Davis, G. L., and Bunney, W. E., Jr., 1980, *Neural Peptides and Neuronal Communication* (E. Costa and M. Trabucchi, eds.), Raven Press, New York, pp. 455–464.
169. Kline, N. S., Li, C. H., Lehmann, H. E., Lajtha, A., Laski, E., and Cooper, T., 1977, *Arch. Gen. Psychiatry* **34**:1111–1113.
170. Gerner, R. H., Catlin, D. H., Gorelick, D. A., Hui, K. K., and Li, C. H., 1980, *Arch. Gen. Psychiatry* **73**:642–647.
171. Pickar, D., Davis, G. C., Schulz, C., Exstein, I., Wagner, R., Naber, D., Gold, P. W., van Kammen, D. P., Goodwin, F. K., Wyatt, R. J., Li, C. H., and Bunney, W. E., 1981, *Am. J. Psychiatry* **138**:160–166.
172. Verhoeven, W. M. A. van Praag, H. M., van Ree, J. M., and de Wied, D., 1979, *Arch. Gen. Psychiatry* **36**:294–298.
173. Meltzer, H. Y., Busch, D. A., Tricou, B. J., and Robertson, A., 1982, *Psychiatr. Res.* **6**(3):313–326.
174. Tamminga, C. A., Tighe, P. J., Chase, T. N., De Fraites, E. G., and Shaffer, M. H., 1981, *Arch. Gen. Psychiatry* **38**:167–174.
175. Jorgenson, A., Fog, R., and Veilisi, B., 1979, *Lancet* **1**:935.
176. Hökfelt, T., Skirboll, L., Rehfeld, J. F., Goldstein, M., Markey, K., and Dann, O., 1980, *Neuroscience* **5**(12):2093–2124.
177. Crow, T. J., 1982, *Abstracts of the Annual meeting of the American College of Psychopharmacology*, ACNP, Inc., Vanderbilt University, Nashville, p. 28.
178. Kleinman, J. E., Hong, J., Gellin, J. C., Govoni, S., and Wyatt, R. J., 1982, *Abstract of the Annual Meeting of the American College of Neuropsychopharmacology*, p. 28.
179. Nair, N. P., Bloom, D. M., and Nestoros, J. N., 1982, *Prog. Neuropsychopharmacol. Biol. Psychiatry* **6**:509–512.
180. Morogi, T., Watanabe, N., Aoki, N., and Itoh, S., 1982, *Arch. Gen. Psychiatry* **39**:485–486.
181. Osmond, H., and Smythies, J., 1952, *J. Ment. Sci.* **98**:309–315.
182. Kety, S. S., 1967, *Amines and Schizophrenia* (H. E. Himwich, S. S. Kety, and J. R. Smythies, eds.), Pergamon Press, Oxford, pp. 271–277.
183. Wyatt, R. J., Termini, B. A., and Davis, J., 1971, *Schizophren. Bull* **4**:10–44.
184. Pollin, W., Cardon, P. V., Jr., and Kety, S. S., 1961, *Science* **133**:104–105.
185. Brune, G. C., and Himwvich, H. E., 1962, *J. Nerv. Ment. Dis.* **137**:447–450.
186. Alexander, F., Curtis, G. C., Sprince, H., and Crosley, A. P. J., 1963, *J. Nerv. Ment. Dis.* **137**:135–142.
187. Antun, F. T., Burnett, G. B., Cooper, A. J., Daly, R. J., Smythies, J. R., and Zealley, A. K., 1971, *J. Psychiatr. Res.* **8**:63–71.
188. Antun, F. T., and Kurkjian, R., 1982, *Br. J. Psychiatry* **140**:611–614.
189. Pscheidt, G. R., Berlet, H. H., Spaide, J., and Himwich, H. E., 1961, *Clin. Chim. Acta* **13**:229–334.

190. Kline, M. S., 1967, *Br. J. Psychiatry* **113:**731–742.
191. Meltzer, H., Shader, R., and Grinspoon, L., 1969, *Psychopharmacologia* **15:**144–152.
192. Friedhoff, A. J., and Van Winkle, E., 1962, *Nature* **194:**897–898.
193. Szara, S., 1956, *Experientia* **12:**441–442.
194. Franzen, F., and Gross, H., 1965, *Nature* **206:**1052.
195. Checkley, S. A., Murray, R. M., Oon, M. C. H., Rodnight, R., and Birley, J. L. T., 1980, *Br. J. Psychiatry* **137:**236–239.
196. Gillin, J. C., Kaplan, J., Stillman, R., and Wyatt, R. J., 1976, *Am. J. Psychiatry* **133:**203–208.
197. Wyatt, R. J., Saavedra, J. M., and Axelrod, J., 1973, *Am. J. Psychiatry* **130:**754–760.
198. Mandel, L. R., Ahn, H. S., and Vandenheuvel, W. J., 1972, *Biochem. Pharmacol.* **21:**1197–1200.
199. Mandel, L. R., Prasad, R., Lopez-Ramos, B., and Walker, R. W., 1977, *Res. Commun. Chem. Pathol. Pharmacol.* **16:**47–58.
200. Walker, R. W., Mandel, L. R., DeLisi, L. E., Wyatt, R. J., and Vandenheuvel, W. J. A., 1983, *Abstracts of the International Symposium on Advances in Chromatography*, Amsterdam.
201. Frohman, C. E., Latham, K. L., Beckett, P. G. S., and Gottlieb, J. S., 1960, *Arch. Gen. Psychiatry* **2:**255–262.
202. Frohman, C. E., Harmison, C. R., Arthur, R. E., and Gottlieb, J. S., 1971, *Biol. Psychiatry* **3:**113–121.
203. Frohman, C. E., Arthur, R. E., and Gottlieb, J. S., 1973, *Biol. Psychiatry* **7:**53–61.
204. Sandler, M., and Reynolds, G. P., 1976, *Lancet* **1:**70–71.
205. Wyatt, R. J., Gillin, J. C., Stoff, D. M., Moja, E. A., and Tinklenberg, J. H., 1977, *Neuroregulators and Psychiatric Disorders* (E. Usdin, J. Barchas, and D. Hamburg, eds.), Oxford University Press, New York, pp. 31–45.
206. Moja, E. A., Stoff, D. M., Gillin, J. C., and Wyatt, R. J., 1978, *Biol. Psychiatry* **13:**291–295.
207. Fischer, E., Spatz, H., Saavedra, J. M., Reggiani, H., Mira, A., and Heller, B., 1972, *Biol. Psychiatry* **5:**139–147.
208. Fischer, E., Heller, B., and Miro, A. H., 1968, *Arzneim. Forsch.* **18:**1486.
209. Potkin, S., Karoum, F., Chuang, L. W., Cannon-Spoor, E., Philips, I., and Wyatt, R. J., 1979, *Science* **206:**470–471.
210. Jeste, D. V., Doongaji, D. R., Panjwani, D., Datta, M., Potkin, S. G., Karoum, F., Thatle, S., Sheth, A. S., Apte, J. S., and Wyatt, R. J., 1981, *Psychiatr. Res.* **5:**341–352.
211. Paulos, M. A., and Tessel, R. E., 1982, *Science* **215:**1127–1129.
212. Karoum, F., Linnoila, M., Potter, W. Z., Chuang, L. W., Goodwin, F. K., and Wyatt, R. J., 1982, *Psychiatry Res.* **6:**215–222.
213. DeLisi, L. E., Murphy, D., Karoum, F., Targum, S., and Wyatt, R. J., 1984, *Psychiatry Res.* (in press).
214. Sabelli, H. C., and Borison, R. L., 1976, *Advances in Biochemical Psychopharmacology*, Volume 15 (E. Costa, E. Ciacobini, and R. Paoletti, eds.), Raven Press, New York, pp. 75–86.
215. Linnoila, M., Potter, W., Potkin, S. G., and Wyatt, R. J., 1984, *Br. J. Psychiatry* **144:**428–431.
216. Heath, R. G., Martens, S., Leach, B. E., Cohen, M., and Angel, C., 1957, *Am. J. Psychiatry* **114:**14–24.
217. Heath, R. G., Guschwan, A. F., and Coffey, J. W., 1970, *Dis. Nerv. Syst.* **31:**391–395.
218. Heath, R. G., and Krupp, I. M., 1967, *Arch. Gen. Psychiatry* **16:**1–9.
219. Robins, E., 1957, *Neuropharmacology* (H. S. Abramson, ed.), Josiah Macy, Jr., Foundation, New York, p. 123.
220. Siegal, M., Niswander, C. D., Sachs, E., Jr., and Stavros, D., 1959, *Am. J. Psychiatry* **115:**819.
221. Bergen, J. R., Grinspoon, L., Pyle, H. M., Martinez, J. L., and Pennell, R. B., 1980, *Biol. Psychiatry* **15:**369–379.
222. Martens, S., Vallbo, S., and Melander, B., 1959, *Acta Psychiat. Neurol. Scand. Suppl.* **136:**361.
223. Meckler L. B., Lapteva, N. N., Lozovskii, D. V., and Balezinc, T. I., 1960, *Proc. Acad. Sci. SSSR* **130:**1148.

224. Nelson, J. W., Daniels, E. G., Mann, K. M., and Kinman, J. W., 1963, *Serological Fractions in Schizophrenia* (R. G. Heath, ed.), Hoeber Medical Division, Harper and Row, New York, pp. 43–56.
225. Boehme, D. H., Cottrell, J. C., Dohan, F. Ç., and Hillegass, L. M., 1974, *Biol. Psychiatry* **8**(1):89–94.
226. Hirata-Hibi, M., Higashi, S., Tachibana, T., and Watanabe, N., *Arch. Gen. Psychiatry* **39**:82–87.
227. Vartanian, M. E., Koliaskina, G. I., Lozovsky, D. V., Burbaeva, G. S., and Ignator, S. A., 1978, *Birth Defects* **14**:339–364.
228. Liedeman, R. R., and Prilipko, L. L., 1978, *Birth Defects* **14**:365–377.
229. Zarrabi, M. H., Zucker, S., Miller, F., Derman, R. M., Romeno, G. S., Hartnett, J. A., and Varma, A. O., 1979, *Ann. Intern. Med.* **91**:194–199.
230. DeLisi, L. E., Goodman, S., Neckers, L. M., and Wyatt, R. J., 1982, *Biol. Psychiatry* **17**(9):1003–1007.
231. Coffey, C. E., Sullivan, J. L., and Rice, J. R., 1983, *Biol. Psychiatry* **18**:113–119.
232. McGuffin, P., 1979, *Psychol. Med.* **9**:721–728.
233. McGuffin, P., Farmer, A. E., and Yonace, A. H., 1981, *Psychol. Res.* **5**:115–122.
234. Luchins, D. J., Weinberger, D. R., Zalcman, S., DeLisi, L. E., Rogentine, N., and Wyatt, R. J., 1980, *Br. J. Psychiatry* **136**:243–248.
235. Smeraldi, E., Sacchetti, E., and Cazzullo, C. L., 1976, *Br. J. Psychiatry* **129**:486–489.
236. APA Task Force Report 7, 1973, *Megavitamin and Orthomolecular Therapy in Psychiatry*, American Psychiatric Association, Washington, D.C.
237. Wyatt, R. J., 1974, *Am. J. Psychiatry* **131**:1258–1261.
238. Pauling, L., 1974, *Am. J. Psychiatry* **131**:1251–1257.
239. Herjanic, M., 1973, *Orthomolecular Psychiatry: Treatment of Schizophrenia* (D. Hawkins and L. Pauling, eds.), W. H. Freeman, San Francisco, pp. 303–315.
240. Milner, G., 1963, *Br. J. Psychiatry* **109**:294–299.
241. Kraepelin, E., 1910, *Geschlachtliche Verirrungen und Volksvermehrungpsychiatric*, edition 8, J. Barth, Leipzig.
242. Lewis, N. D. C., and Davies, G. R., 1921, *J. Nerv. Ment. Dis.* **54**:385–493.
243. Mott, F. W., 1922, *Br. Med. J.* **1**:463–467.
244. Morse, M. E., 1923, *J. Neurol. Psychopathol.* **4**:1–26.
245. Geller, F. C., 1923 *Arch. Gynakol.* **120**:237–248.
246. Rey, J. H., and Coppen, A. J., 1959, *Br. Med. J.* **11**:1445–1447.
247. Richter, D., 1970, *Biol. Psychiatry* **2**:153–164.
248. Hoskins, R. G., 1943, *Psychosom. Med.* **5**:3–9.
249. Hemphill, R. E., and Reiss, M., 1945, *J. Ment. Sci.* **91**:1–7.
250. Shader, R. I., Taymoor, M. I., and Grinspoon, L., 1968, *Proceedings IV World Congress of Psychiatry*, pp. 2988–2991.
251. Johnstone, E. C., Crow, T. J., and Mashiter, K., 1977, *Psychol. Med.* **7**:223–228.
252. Ferrier, I. N., 1982, *Biological Aspects of Schizophrenia and Addiction*, (G. Hemmings, ed.), John Wiley and Sons, New York, pp. 35–47.
253. Horrobin, D. F., 1977, *Lancet* **1**:936–937.
254. Feldberg, W., 1976, *Psychol. Med.* **6**:359–369.
255. Price, E. J., and Rowe, C. E., 1972, *Biochem. J.* **126**:575.
256. Mathe, A. A., Sedvall, C., Wiesel, F. A., and Nyback, H., 1980, *Lancet* **1**:16–17.
257. Wolfe, L. S., and Coceani, F., 1979, *Annu. Rev. Physiol.* **41**:669–684.
258. Lycke, E., and Roos, B. E., 1968, *Experientia* **24**:687–689.
259. Lycke, E., and Roos, B. E., 1974, *J. Neurol. Sci.* **22**:277–289.
260. Rosenthal, D., Wender, P. H., Kety, S. S., Schulsinger, F., Welner, J., and Oskergaard, J., *The Transmission of Schizophrenia* (D. Rosenthal and S. S. Kety, eds.), Pergamon Press, New York.
261. Slater, E., 1968, *The Transmission of Schizophrenia* (D. Rosenthal and S. S. Kety eds.), Pergamon Press, New York.
262. Kety, S. S., Rosenthal, D., Wender, P. H., Schulsinger, F., and Jacobsen, B., 1975, *Genetic Research in Psychiatry* (R. R. Fieve, D. Rosenthal, and H. Brill, eds.), Johns Hopkins University Press, Baltimore.

263. Cancro, R., 1982, *Psychiatry 1982: The American Psychiatric Association Annual Review* (L. Grinspoon, ed.), American Psychiatric Press, Washington, pp. 92–96.
264. Gottesman, I. I., and Shields, J., 1976, *Schizophren. Bull.* **2**:360–401.
265. Lidz, T., Schafer, S., Fleck, S., Cornelison, A., and Terry, D., 1962, *J. Am. Psychoanal. Assoc.* **10**:74–90.
266. Jackson, D. D., 1959, *The Study of Schizophrenia* (D. D. Jackson, ed.), Basic Books, New York, pp. 37–90.
267. Allen, M. G., Cohen, S., and Pollin, W., 1971, *Am. J. Psychiatry* **128**:939–945.
268. Elston, R. C., and Cambell, M. A., 1970, *Behav. Genet.* **1**:3–10.
269. Tsuang, M. T., Bucher, K. D., and Fleming, J. A., 1982, *Br. J. Psychiatry* **140**:595–599.
270. DeLisi, L. E., Wise, C. D., Bridge, T. P., Potkin, S., Phelps, B., and Wyatt, R. J., 1981, *Biological Markers for Mental Illness*, Pergammon Press, New York, pp. 79–96.
271. Weinshilboum, R. M., Raymond, F. A., Elveback, L. R., and Weidman, W. H., 1973, *Science* **181**:943.
272. Ross, S. B., Book, J. A., and Wetterberg, L., 1981, *Clin. Genet.* **19**(5):415–426.
273. Fujita, D., Ito, T., Maruta, K., Teradaera, R., Beppu, H., Nakagama, Y., Kato, Y., Nagatsu, T., and Kato, T., 1978, *J. Neurochem.* **30**:1569–1572.
274. Böök, J. A., Wetterberg, L., and Modrzewska, S., 1978, *Clin. Genet.* **14**(6):373–394.
275. DeLisi, L. E., Wise, C. D., Phelps, B. H., Zalcman, S., Potkin, S. G., and Wyatt, R. J., 1980, *Biol. Psychiatry* **15**(6):899–905.
276. Markianos, E. S., Nystrom, I., Reichel, H., and Matussek, N., 1976, *Psychopharmacology* **50**:259–267.
277. Sternberg, D. E., van Kammen, D. P., Lerner, P., and Bunney, W. E., 1980 *Abstracts of the Annual Meeting of the Society of Biological Psychiatry*, San Francisco.
278. van Kammen, D. P., Mann, L., Sternberg, D. E., Scheinan, M., Ninan, P., Marder, S. R., van Kammen, W. B., Reider, R., and Linnoila, M., 1983, *Science* **220**:974–977.
279. DeLisi, L. E., Phelps, B., Wise, C. D., Apostoles, P. S., and Wyatt, R. J., 1981, *Biol. Psychiatry* **16**:873–878.
280. Jeste, D. V., DeLisi, L. E., Zalcman, S., Wise, C. D., Phelps, B. H., Rosenblatt, J. E., Potkin, S. G., Bridge, T. P., and Wyatt, R. J., 1981, *Psychiatry Res.* **4**:327–332.
281. Viukari, M., and Linnoila, M., 1977, *Acta. Psychiatr. Scand.* **56**:57–61.
282. Birket-Smith, E., and Anderson, J. V., 1973, *Lancet* **1**:431.
283. Menninger, K. A., 1925, *Assoc. Res. Nerv. Ment. Dis. Proc.* **5**:182–203.
284. Petrov, P. A., 1970, *Am. J. Trop. Med. Hyg.* **19**:146–150.
285. Torrey, E. F., 1980, *Civilization and Schizophrenia*, Jason Aronson, New York.
286. Torrey, E. F., Torrey, B., and Peterson, M., 1977, *Arch. Gen. Psychiatry* **34**:1065–1070.
287. Hare, E. H., and Walter, S. D., 1978, *J. Epidemiol. Commun. Health* **32**:47–52.
288. Abe, K., 1969, *Br. J. Psychiatry* **115**:519–531.
289. Crow, T. J., 1983, *Lancet* **1**:173–175.
290. Strahilevitz, M., Fleishman, J., Fischer, G., Harris, R., and Narasimhachari, N., 1976, *Am. J. Psychiatry* **133**:772–777.
291. Torrey, E. F., Peterson, M. R., Brannon, W. L., Carpenter, W. T., Post, R. M., and van Kammen, D. P., 1978, *Br. J. Psychiatry* **132**:342–348.
292. Gowdy, J. M., 1980, *Psychosomatics* **21**:751–756.
293. DeLisi, L. E., Weinberger, D. R., Neckers, L. M., Potkin, S. G., Shiling, D., and Wyatt, R. J., 1981, *Br. J. Psychiatry* **139**:513–519.
294. Amkraut, A., Solomon, G. F., Allansmith, M., McClellan, B., and Rappaport, M., 1973, *Arch. Gen. Psychiatry* **28**:673–677.
295. Johnstone, E. C., and Whaley, K., 1975, *Br. Med. J.* **2**:724–725.
296. Koliaskina, G., Tsutsulkovskaya, M., Domashneva, I., Maznina, T., Kielholz, P., Gastpar, M., Bunney, W., Rafaelson, O., Heltberg, J., Coppen, A., Hippius, H., Hoechesl, B., and Vartanian, F., 1980, *Neuropsychobiology* **6**:349–355.
297. Watanabe, M., Funahashi, T., Suzuki, T., Nomura, S., Nakazawa, T., Noguchi, T., and Tsukada, Y., 1982, *Biol. Psychiatry* **17**(6):699–710.
298. Halonen, P., Rimon, R., Arochonka, R., and Jantti, V., 1974, *Br. J. Psychiatry* **125**:461–465.
299. Rimon, R., Halonen, P., Pukakka, P., Laitinen, L., Marttila, R., and Salmela, L., 1979, *J. Clin. Psychiatry* **40**:241–243.

300. Albrecht, P., Torrey, E. F., Boone, E., Hicks, J., and Daniel, N., 1980, *Lancet* **2**:769–772.
301. Torrey, E. F., Yolken, R. H., Winfrey, C. J., 1982, *Science* **216**:892–894.
302. Tyrrell, D. A. J., Parry, R. P., Crow, T. J., Johnstone, E., and Ferrier, I. N., 1979, *Lancet* **1**:839–841.
303. Taylor, G. R., Crow, T. J. Ferrier, I. N., Johnstone, E. C., Parry, R. P., and Tyrrell, D. A. J., 1982, *Lancet* **2**:1166–1167.
304. Sequiera, L. W., Carrasco, L. H., Curry, A., Jennings, L. C., Lord, M. A., and Sutton, R. N. P., 1979, *Lancet* **2**:609–612.
305. Johnson, R. T., 1982, *Viral Infections of the Nervous System*, Raven Press, New York.
306. Nyberg, P., Almay, B. G. L., Carlsson, A., Masters, C., and Winblad, B., 1982, *Acta Neurol. Scand.* **66**:16–24.
307. Hughes, J. J., and Little, J. C., 1967, *Br. J. Psychiatry* **113**:867–873.
308. Letemendia, F. J. J., and Harris, A. D., 1967, *Br. J. Psychiatry* **113**:950–958.
309. Baker, H. F., Bloxham, C., Crow, T. J., Davies, H., Ferrier, I. N., Johnstone, E. C., Parry, R. P., Ridley, R. M., Taylor, G. R., Tyrrell, D. A. J., 1983, *Adv. Biol. Psychiat.* **12**:1–19.

22

Endorphins and Mental Disease

Karl Verebey and Mark S. Gold

1. INTRODUCTION

It was recognized many years ago that endogenous regulation of physiological processes is often performed by chemicals synthesized in the organism. Many of the chemicals are peptide hormones. A certain portion of each peptide chain has a specific amino acid sequence representing the "keys" that turn on or off the "locks" of the receptors, which regulate specific physiological functions. Recently, it has been discovered that a very-well-studied group of drugs, the opioids, have endogenous peptide analogs called the endorphins. Various sizes have been found, the smallest being the pentapeptide enkephalins. Somewhat larger are the dynorphins, containing up to 21 amino acids. The largest oligopeptides (e.g., β-endorphin) contain 31 amino acids. One characteristic common to other endorphins is that somewhere in the polypeptide molecule, the amino acid sequence of Met-enkephalin is present. This indicates that the active principle for opioid activity is coded by 4 or 5 amino acids in a specific sequence.

The endorphins and the exogenous opioids (i.e., morphine, heroin, methadone, and meperidine) have very different primary and secondary chemical structures. However, significant similarities were discovered when the tertiary, three-dimensional, configurations of morphine and Met-enkephalin were reconstructed. The functional groups were in similar orientation in the tertiary model of morphine and Met-enkephalin, indicating the basis for their similar pharmacological activity. The behavioral effects of opiates have been observed for centuries but specifically studied only during the past few decades. These studies are an excellent source of information for the construction of hypotheses for the function of the newly discovered endogenous opioid peptides.

In this chapter the pharmacology and behavioral effects of the opiate agonists, antagonists, and opioid peptides will be reviewed. Possibilities will be explored for the involvement of the opioid system in the pathogenesis of emo-

Karl Verebey • New York State Division of Substance Abuse Services Testing and Research Laboratory, and Department of Psychiatry, Downstate Medical School, Brooklyn, New York 11217. *Mark S. Gold* • Fair Oaks Hospital and Psychiatric Diagnostic Laboratory of America, Summit, New Jersey 07901.

tional disorders and also the therapeutic potentials for opioid drugs in some forms of mental illness.

2. HISTORICAL ACCOUNT OF OPIATE EFFECTS ON BEHAVIOR

The behavior-modifying effects of opiates have been widely described throughout written history. In Greek mythology are found descriptions of the familiar effects of morphine. From Homer's *Odyssey*: ". . . Helen, the daughter of Zeus, turned her thoughts elsewhere. Straightway, she cast into the wine of which they drank a drug which quenches pain and brings forgetfulness to every ill."[1] Although opium or morphine was not specifically mentioned, the quote clearly describes the psychotropic-analgesic action of a morphinelike drug. In 1680, Sydenham wrote: "Among the remedies which it has pleased almighty God to give to man to relieve his sufferings, none is so universal and so efficacious as opium."[2] Although the word "universal" is mentioned, little did Sydenham know that 300 years later opiate receptors and endogenous brain opioids would be discovered. Nevertheless, "universality" of opiates is now the key word suggesting the presence of opiate receptors in every human.

The multitude of opiate effects were observed early and were utilized for the treatment of various conditions. Some of these included cough suppression, sedative-euphoriant for social and ceremonial events, and the relief of physical pain and anxiety of various etiology. For depression, opiates were used widely during the 19th and early 20th century. In Germany, Kraepelin and Weygandt described the use of opioids especially in the condition of melancholia.[3] Interestingly, the early literature was not overly concerned about the addiction liability of the opiates. Perhaps in emotional diseases the observed anhedonia results from defects in neuronal sensitivity to opiates and consequently dependence does not occur. These neurons are normally responsible for the transmission or regulation of pleasure and reward in the CNS. Berridge[4] and Way[5] explain that addiction to opioids became a public health problem after the invention of the hypodermic syringe and the availability of purified parenteral preparations. The policies toward narcotics (in England) have been influenced by a small group of doctors and administrators.[4] In the United States, opiate regulatory policies became political. Way describes it in the following way: "Over a span of 200 years narcotic usage in this country was widely extolled, strongly advocated, fiercely rejected, excessively regulated, and reasonably tolerated." These reactions to opiates were, according to Way, the result of subjectivity, emotionality, and overreactivity.[5] Future research needs a prejudice-free environment, and regulations should be based on solid, objective, scientific data rather than the continuation of the emotional subjectivity of the past.

3. ALTERED PAIN MECHANISMS IN MENTAL ILLNESS

There are two major ways the pharmacology of a drug can be studied. Both have advantages and shortcomings. The first is observing the pharmacological response in the whole organism. The second involves studying the effects of the drug in isolated organ or cellular systems. The advantage of the first approach is that an integrated outcome can be observed at once, although the quantitative criteria are rather subjective and general, in most cases. The second approach is more objective in its measurements, but their implications to the whole organism are questionable, because the response of other cells and organs are not considered. Information from both approaches is available on the pharmacological effects of opioids on pain and behavior. Our review of the literature will hopefully unify the information into a reasonable hypothesis and explain the mechanism of opioid action on behavior.

Analgesia is the most important pharmacological effect of the morphinelike alkaloids. Possibly opioid peptides will be used for the same purpose in the future. Currently, opiates are the only class of drugs that can effectively eliminate severe intractable pain. The precise mechanism of opiate-related analgesia is not known. It is, however, acknowledged that a major part of pain relief is a result of the calming, soothing, and fear-relieving effect of opiates.[2] Surgical patients treated with morphine or other opiates relate that their pain is still felt, but they do not care about it or that it does not bother them. It appears that the opiates' direct analgesic effect is not as important as their ability to alter pain sensory perception. Consequently, the most potent pain relievers known are effective in relieving intractable physical pain by controlling the emotional components of pain at the level of perception.[6] Psychotic episodes have common behavioral components with severe, intractable pain. These symptoms include: fear, anxiety, rage, and confusion. Strangely, the simple association between severe pain and psychosis has not been commonly made.

Following the discovery of the endorphin system, one of the functions of endogenous opioids was hypothesized to be as antistress neurohormones. Kelly reported that the analgesia achieved following stress was partly antagonized by naloxone, an opiate antagonist.[7] This indicated that some nonopioid mechanism of analgesia also exists. Another interesting observation was that certain stresses may cause maximal elevation of plasma endorphin, ACTH, and adrenal corticosteroids without causing measurable increase in analgesia.[7] This may indicate the action of an endogenous opiate antagonist, although the existence of such substances has not been firmly established.

Stress in animal studies suggested a linkage of the hypothalamus–pituitary–adrenal axis. Thus, increases of circulating β-endorphin along with ACTH and cortisol may also prove to be a biochemical marker of human stress or arousal.[8] For this reason, Cohen *et al.* studied the effect of amphetamine and surgical stress on plasma β-endorphin levels.[8] Amphetamine has been shown to potentiate the behavioral effects of stress noted in other studies, and surgery itself is a known stress-producing stimulus. The findings indicate that amphetamine and surgery-related stress both independently increased plasma β-endorphin. The authors suggest that circulating β-endorphin may be a marker of CNS arousal. Thus, surgery patients with low β-endorphin levels after surgery may require larger postsurgery doses of opioid analgesics. This action may be necessary in order to activate the poorly "stress-activated" CNS opioid system.[8]

Ginzler showed that pregnant rats approaching parturition have a significant increase in their blood levels of β-endorphin. The author considers that the opioid system is an important intrinsic mechanism that modulates responsiveness to aversive stimuli during pregnancy.[9] It was considered earlier that pain responses are similar to symptoms of acute psychosis. Thus, it is possible, as the author suggests, that this internal mechanism prevents untoward behavioral changes in the mother to protect both mother and fetus. This mechanism is not functional in nonpregnant rats, indicating that a special protective system is operational only during parturition in rats.[9] The extensive increase in β-endorphin was noted also in human subjects during delivery.[10] The similar symptom components in intractable physical pain and acute psychosis suggest a common neuronal pathway in the two conditions. In physical pain, the signal is transmitted through the spinal cord, through the reticular formation leading to the cortex where the signal is recognized, interpreted, and a typical response is generated, characterized by fear, confusion, and rage. In case of acute psychosis, a "false signal" travels to the cortex utilizing the same pathway. Although physical pain is not present in psychosis, a similar reaction is triggered as a response to severe pain. We know that physical pain responds to opiate analgesics. Whether or not acute psychosis responds to opiate treatment is a question for future research to answer. A similar phenomenon is observed between grand mal epilepsy and psychomotor epilepsy. In this situation, the indiscriminate firing of CNS neurons are the common mechanism in both conditions but different areas of the brain are affected. One condition results in grand mal convulsions, while the other in atypical behavior. However, the therapeutic approach to both diseases is the same—the use of an anticonvulsant drug. This example suggests that if the model of the atypical pain pathway is a correct hypothesis for psychosis, opiate analgesics may be effective by decreasing the perception of the atypical signals in the cortex.

Davis *et al.*[11] found that schizophrenic subjects are generally pain insensitive. In such subjects following the administration of naltrexone (a narcotic antagonist), pain sensitivity was restored. The authors suggest that excess endorphins may be responsible for the pain insensitivity in schizophrenics. However, the use of narcotic antagonists in schizophrenia has yielded mostly negative results in relieving major schizophrenic symptoms. The authors suggest that the disturbance in pain appreciation in schizophrenics may be a reflection of a wider-scale disturbance in perceptual functioning related to defective selective attention.[11] These effects on attention were studied and discussed in detail by Buchsbaum *et al.*[12] They point out that attentional disorders are the most prominent and consistent psychopathological findings in schizophrenia. Various evidence suggests that this is true: (1) pain perception is dependent on endorphinergic neuronal mechanisms and is closely related to attention; (2) schizophrenics are relatively pain insensitive; (3) opiates and their antagonists affect attentional performance; and (4) schizophrenic thought disorders may be modified by opiates and their antagonists.[12]

In human studies, evoked potentials were utilized for the experimental technique.[12] ACTH and naloxone increased selective attention whereas opiate agonists such as morphine and endorphins had the opposite effect. Stimulated attention had been closely tied to pharmacological agents (e.g., amphetamines and caffeine) that are known to stimulate the adrenergic nervous system, especially at sites such as the locus coeruleus (LC). It was postulated that the stabilizing tranquilizing effects of opiates are due to the modulating effects of opiate agonists at sites of the sympathetic nervous system. Electrophysiological studies showed that morphine inhibits LC activity,[13,14] and high opiate-receptor binding was also found in the LC.[15] Coincidentally, these basic studies made it possible for Gold *et al.* to propose an LC theory for opiate withdrawal and human panic and discover the efficacy of clonidine and clonidinelike drugs.[16–19] A final point of evidence is that naloxone reverses opioid suppression of LC activity. Thus, the attention-enhancing effects of naloxone are possibly mediated by reversal of endorphin suppression of noradrenergic activity in the LC. The LC apparently plays an important modulating role in the transmission of pain signals and it also regulates levels of attention, arousal, and anxiety by acting on the cortex and the limbic system. Buchsbaum *et al.* suggest that alteration in pain and attention promotes complex changes in norepinephrine activity, which may not alter norepinephrine metabolite (MHPG) levels in body fluids because the small localized effect of the LC would be too little when diluted in total body water.[12]

4. ANIMAL STUDIES ON OPIOIDS WITH POSSIBLE RELEVANCE TO MENTAL ILLNESS

Behavioral studies on endorphins have been evaluated in living animals. Endorphins and both enkephalins have been shown to produce transient analgesia and catatonia in animals. Enkephalin-tolerant animals showed cross-tolerance to morphine, and withdrawal signs can be precipitated by the injection of naloxone to enkephalin-tolerant animals. Thus, the opioid peptides possess most of the morphinelike characteristics.

Belluzzi and Stein's work evaluated the effects of endorphins on memory formation in animals. A biphasic effect was observed, i.e., low doses of the peptides were inhibitory whereas high doses stimulated memory facilitation in appetitive learnings.[20]

Henriksen *et al.* found that the action especially of dynorphin resulted in a disinhibitory process within the hippocampus. The authors suggest possible behavioral effects of the opioid peptides especially through their regulatory influence on limbic excitability.[21]

Katz showed that the opiate-related activation in certain strains of mice was an excellent preclinical analog of human mania.[22] The activation is mediated by enkephalinlike neuropeptides in the CNS. Certain strains of mice genetically transfer greater or lesser capacity to be activated by opioids. This finding indicates that symptoms of mania may be passed on within families by genetic transmission from one generation to the next.

The fact that morphine can be either excitatory or inhibitory in different species was studied by Jacquet.[23] Morphine is overtly excitatory in animals such as the cat, horse, and tiger whereas it is sedative in the monkey and man.

Both excitatory and inhibitory effects of morphine were found in the periaqueductal gray neurons of the brain; in the midbrain reticular formation, only excitatory action was observed. It was hypothesized that depending on which receptor is in excess, the stimulant or sedative responses will be observed. Atypical cases have been seen in rare human subjects in whom opiates are excitatory. These characteristics may be genetically transferred and/or pathological and should be further studied in detail.

Differential reaction to opiates and amphetamines can be also diagnostic. The administration of even low doses of amphetamines often precipitates an acute psychosis in susceptible individuals.[24] Hyperreactivity to a specific drug or stimulus is likely mirrored by neurochemical change, which may be measured and become a marker of a specific disease. Thus, differential reaction to opiate drugs can also be diagnostic. Such drug-induced responses, although currently not in common use, may become useful diagnostic tools in the future, like the dexamethasone suppression test. Such test provides a picture of the neurochemical state of the individual. It is possible that at steady state the neurochemical-neurohormone levels are similar in normal controls and psychiatric patients; however, when stimulated or inhibited by psychoactive substances, the different neurochemical responses may be diagnostic. The proposed name for such testing methods is "DINAR" (*drug-induced neurochemical adoptive response*).

Jacquet noticed that the CNS action of morphine was paralleled by β-endorphin and ACTH. Although ACTH caused fearful hyperactivity just as morphine and β-endorphin, it did not produce analgesia.[23] The author suggests that morphine exerts a dual action by stimulating different receptors, one for β-endorphin and another for ACTH. Behaviorally, it is possible that β-endorphin is responsible for sedation and immobility and that ACTH causes fearful hyperactivity. The relative number of cells responding to either stimulant or depressive effects may determine the observed response in a given individual. Thus, the functional balance of the endorphin system in the CNS is diverse among species and among individuals. Holtzman used the technique of stimulus discrimination in rats and squirrel monkeys for various opioid drugs.[25] The results indicated that different opioid drugs produce variable effects owing to their differential affinity to the various types of opiate receptors.

The scientific and clinical significance of the multiple opiate receptors and their pharmacological profile were recently reviewed by Adler.[26] The heterogeneity of opiate receptors complicates the understanding of the pharmacological and behavioral effects of opiates, but provides an opportunity to synthesize substances that are more specific at one or another opiate receptor.

5. *SATISFACTION, REWARD, AND EUPHORIA*

Soon after the discovery of endogenous opioids, it was hypothesized that they may regulate euphoria among other behavioral events.[27] Because the opposite of euphoria (anhedonia) is commonly observed in depression and schizophrenia, a possible defect in the endorphin system was considered. Several mechanisms such as endorphin deficiency or receptor dysfunction may be responsible for the individual's inability to sense well-being, satisfaction, and reward. If endorphin deficiency produces anhedonia, opiate supplementation may be effective, whereas if neuronal dysfunction is the cause, it would not respond to opioids. Gold and Byck postulated that substances that increase binding of agonists may cause euphoria, and increased binding of antagonists may result in dysphoria. Sodium and lithium are known to increase antagonist binding. This may be the reason for the effectiveness of lithium in manic-depressive illness.[27] Thus, sufficient quantities of endorphins at the receptors are only one of the requirements for euphoria. The sodium ion content of the interstitial fluid is another factor influencing preferential binding of agonists or antagonists at the opiate receptors. Thus, ion concentration may be the ultimate determinant of the particular mood exhibited.

Stein and Belluzzi tested whether release of enkephalins in animals causes release of catecholamines and thus activates of the "reward" receptors in the brain. Their data indicate that enkephalins are the natural euphoria and reward transmitters.[28] Similar conclusions were found by Kornetsky and Bain studying the biobehavioral basis of the reinforcing properties of opiate drugs.[29] Their experiments indicate that morphine increases the sensitivity of animals to rewarding brain stimulation. The authors suggest that important interaction occurs between the catecholaminergic and endorphinergic systems. Nitrous oxide (N_2O) is an analgesic-anesthetic gas. It is well known that this substance causes euphoria, for which the common name "laughing gas" was coined to describe its effects. Naloxone was able to eliminate the N_2O-related analgesia in mice. This indicates that N_2O may exert its euphorigenic activity through endorphin release and, like the analgesia, it can also be blocked by naloxone.[30]

The experimental studies of opioid peptides in animals indicate that endorphins are intimately involved in analgesia, memory formation, and in the mediation of satisfaction, reward, and euphoria. Some of these effects are motivating and thus provide the reinforcing properties of opioid drugs. Depending on the animal species, opiates can be excitatory or depressant and opiates can act on receptors with different activity profiles. In fact, in mice the activating effect of opiates provides an excellent experimental profile for the human condition of manic-depressive psychosis. Investigation in this exciting area of research is continuing to provide more information in this ever-expanding field.

6. *THE POSSIBLE LINK BETWEEN OPIATE AND ALCOHOL DEPENDENCE*

Along with opiates, the other class of addictive substance most intensively studied is alcohol. This is a subject especially interesting theoretically, in considering a universal theory of addiction, and clinically in treating alcohol abuse. Former opiate addicts are notoriously liable to develop overuse of and even addiction to alcohol, especially when opiates are not available.[31–33] Alcohol in high doses may have significant analgesic and antiwithdrawal properties.[34,35] Even while on methadone many addicts develop a pattern of alcohol abuse, at times resulting in discontinuation of maintenance treatment in as many as 20% of methadone patients.[34,36] Alcohol addiction might occur secondary to the generation of addictive alkaloid, which is an agonist at brain opiate receptors. The effects of alcohol have often been compared to those of opiates, especially because both substances cause analgesia, euphoria, tranquilization,[2,37] as well as similar neurochemical and physiological effects.[38] Wikler *et al.* reported that a 60-ml dose of 95% ethanol in human subjects raises the pain threshold approximately 35–40% while not altering other sensory perceptions.[39] Euphorigenic activity might be the principal property of ethanol's dependence liability. The euphoria is described as a "high" manifested by a decrease of anxiety. This leads to a positively reinforcing, relaxed, and physically active state. It has long been debated whether alcohol is a stimulant or depressant. Alcohol is not a stimulant according to a respected textbook of pharmacology, but like other general anesthetics, it is a primary and continuous depressant of the CNS.[37] The activity of certain inhibitory centers in the CNS are depressed by low doses of alcohol and this inhibition results in a "false sense" of stimulation. This simple explanation, although logical and possibly partially true, has certain shortcomings. First, it presumed that there is greater sensitivity of inhibitory neurons to alcohol than that of other neurons. Second, it is not clear how the inhibition of the inhibitory centers caused (in addition to stimulation) euphoria, analgesia, and antianxiety effects. It would be easier to understand the euphorigenic effects of alcohol if alcohol could be shown to release endogenous, opiatelike substances thus providing the known opiate pharmacological effects, i.e., analgesia, euphoria, and anxiety relief.[2] Perhaps the stimulatory effects of alcohol could be viewed as a counteraction of the anxiety-related immobilization seen in stressed and depressed individuals.

Some interesting and provocative reports are supportive of both direct and indirect interactions between alcohol and endorphins at the opiate receptors. Lorens and Sainati present evidence for the ability of naloxone to block the excitatory effects of ethanol in the rat.[40] Naloxone injected 1 hr after ethanol administration immediately reversed the excitatory effects of ethanol. The authors suggest the possibility that ethanol releases endogenous opioids that act at the opiate receptors and enhance locomotor activity similarly as seen after the administration of an opiate agonist or after direct electrical stimulation of lateral hypothalamic centers. The implication of the study is that endorphin release might underlie the positive reinforcing properties of ethanol.[40] If the hypothesis is tenable, opiate antagonists (naloxone or naltrexone) should be able to block alcohol-induced analgesia, euphoria, and catatonia and possibly improve alcoholic stupor by displacing endorphins from opiate receptor sites. A preliminary report provided encouraging results. Schenk *et al.*[41] selected four subjects who were severely intoxicated with alcohol; their blood alcohol levels ranged from 0.15 to 0.32 g/100 ml. Following the administration of naloxone at various doses from 4 to 28 mg i.v., the authors observed that nearly unconscious alcoholic subjects spontaneously awoke and were able to talk and perform fairly well on motor tests and coordination tasks. In addition, the previously unresponsive corneal reflex could be stimulated. Changes in an objective measure, blood gas analysis, were also striking after naloxone injection. It showed a marked increase in pO_2 levels 15 min after naloxone from 58 to 104 mm Hg while the pCO_2 levels dropped from 61 to 48 mm Hg. It is interesting to note that in another study, β-endorphin was shown to produce respiratory depression in dogs, and this effect was also promptly reversible by naloxone administration paralleled by pO_2 increase and pCO_2 decrease.[42] Jeffcoate *et al.* were able to prevent the impairment of psychomotor performance in human volunteers induced by alcohol when they injected 0.4 mg naloxone i.v. before alcohol administration.[43] The performance scores measured by the "four-choice serial reaction-time test" were the same after baseline measurements and after naloxone plus alcohol administration. The authors postulated that naloxone may act by blocking endogenous opioids that have been released by alcohol.

Using psychometric testing, another interesting hypothesis has been presented by Senter *et al.*[44] These authors found that the euphoria experienced by alcoholics, induced by the consumption of large amounts of alcohol, is far greater than that experienced by social drinkers. The authors suggest that the sustained drinking behavior of alcoholics might arise from the intense alcohol-induced euphoria, providing the positive reinforcing effects. This hypothesis is an alternative to the theory that alcohol dependence is due to the tension reduction of alcohol as a negative reinforcement. Based on opiate pharmacology, both theories may be partly correct. However, it would be interesting to know whether or not the higher euphoria observed in the work of Senter *et al.* would be paralleled by higher levels of endorphins in the plasma or increased number of opiate receptors of alcoholics.

The general implication of these research data is that if narcotic antagonists can block certain alcohol-related effects, there is a likelihood for a common mechanism of action by alcohol and opiates. Although alcohol dependence and opiate dependence are different because of other effects on various organs, e.g., the liver in alcoholics, the primary reinforcing effects in both may be related to their ability to stimulate directly or indirectly the opiate receptor system in the brain.

7. *INTERACTION OF OPIOIDS WITH ADDICTION AND MENTAL ILLNESS*

Reports and case studies of the clinical literature have long suggested that opiates may have therapeutic effects on mental illness. The majority of observations are uncontrolled, but the large number of similar findings collectively provide more convincing evidence for their validity.

Khantzian[45] suggested at a recent international conference on "*Opioids in Mental Illness*" that the special psychological appeal of narcotics resides in their antiaggression effects. Individuals most vulnerable to be dependent on opioid drugs had lifelong experiences and problems with rage and aggression. Such behavioral trends were often tied to environmental influences of physical abuse, brutality, violence, and sadism. These subjects dependent on opioids did not crave euphoria; their main interest was the relief of dysphoria associated with anger, rage, violence, and related anxiety.[45]

After the introduction of methadone for the maintenance treatment of heroin addicts, many professionals working in the field recognized the psychotropic effects of methadone.[46–52] There was a clear indication that many subjects in the methadone programs had had prior psychological problems.[53]

McKenna, among others, described the areas of psychopathology that seemed to be helped by opiates.[49] Interestingly, as varied symptoms as mania, depressive psychosis, agitated depression, and paranoid schizophrenia were all among the conditions that were improved by methadone. The subjects experienced periods of intense anxiety, depression, and self-doubt, and subsequently found symptomatic relief with the use of an opiate. Methadone motivated them to pursue relationships, gave them a new energy for life, and blunted their previous feelings of fear, doubt, and imminent failure.

Some subjects repeatedly attempted to become drug free; but as the doses of methadone became low, they also became severely depressed or psychotic, and returned to opiate use to abolish or attenuate their psychotic disturbances. None of the patients were hedonistic and their dependence did not result from being victims of peer pressure. It was obvious that in their frustration they were actively seeking any substance to relieve their painful psychological problems. These subjects were specific for opiates and they were not deterred by threats of criminal prosecution.[48,49]

The psychopharmacological action of methadone is conceptualized by its stabilizing, antiaggression effects. While providing stabilization, it counteracts the disorganizing influences of rage and aggression and thereby reverses regressed states and makes a healthier adaptation possible.[51] In general, a shift is observed from liability to stability and from egocentricity to concerned interest in others. "Craving" is usually the sign of physical dependence, and in these subjects Khantzian[45] and Wurmser[48] found that it resulted more from a desire for relief of threatening and dysphoric feelings associated with unmitigated aggression.

Resnick *et al.* as early as 1970 recognized that addicts are a heterogeneous group.[46] Through a questionnaire, they rated individuals based on the importance opiates played in their lives. Many of their patients viewed heroin use as an effective measure for reducing anger, tension, and disappointments and increasing their ability to work, study, and relate to others. Resnick *et al.* suggest that this group of subjects may fit the Dole and Nyswander[54] paradigm in which opiate dependence was related to a state of metabolic deficiency (endorphin deficiency). Recent data summarized by Gold and Rea[55] support the hypothesis that narcotic addicts have a preexisting or acquired functional endorphin deficiency.

Methadone patients who are successful in a slow detoxification schedule belong to another group. These subjects use heroin for their social interactions, while the drug has no major role in improving their daily functioning. These subjects may fit Wikler's description, which stipulates opiate addiction as a result of conditioning by environmental reinforcing stimuli.[56] These subjects can live opiate free without major psychological problems after slow, gradual detoxification from methadone.

The major difference between the groups is that the "metabolic deficiency" group becomes psychotic when the doses of methadone are reduced below 10–15 mg while the other group does not. An excellent example of the group that can be easily detoxified are the heroin-using Vietnam veterans described by Robins.[57] Soldiers, while under stress in the combat zone, used heroin to help cope with the horrible conditions in Vietnam. However, after the stess was removed by their return to the U.S., most stopped using heroin. Less than 1% of the addicted veterans continued the use of heroin. These data would support the notion that while a healthy endorphin system is functional in such individuals, they would not become permanently dependent on opioids. But it also indicates that when the environmental stress is high and opiates are easily available, opiate use can become popular.

Still another group of subjects is described by Powell, called the "chipers".[58] They are a very elusive group to study because they use opiates only intermittently and do not take on the stereotypic addict life-style. Because of the pattern of their opiate use, they do not develop tolerance or physical dependence, and their habits go undetected. This group uses opiates as self-medication of various psychological problems, but somehow they are able to control the dose and the frequency of their opiate use so that only a few of their intimate friends are aware of it. Although this group of subjects does depend psychologically on opiates, they are able to conform to the social mores by holding responsible jobs and by having socially acceptable personal lives. The "chipers" existence negates the myth that everyone who has tried heroin or morphine is liable to become addicted to opiates. This idea was further denounced by Pittel: "dependence is most likely to occur among individuals who lack the psychological resources needed to deal with inner conflicts and/or environmental frustrations. The opiates are desired by such individuals to compensate for the absence of inner structure."[59] Although it is provocative, the suggestion may be reasonable that a functional endorphin system is necessary for what Pittel calls "an inner structure."

The personalities of addicts and speculation over the reasons for their dependence on opiates were also hypothesized. Wikler and Rasor describe former heroin addicts in general as noncompetitive individuals who prefer to handle their anxieties of pain, sexuality, and aggression by avoiding situations that provoke them.[60] Opiate addiction often begins in adolescence and is related to the intensity of conflicts in the area of sexuality and assumption of aggressive masculine roles. The almost instantaneous relief for such emotional suffering serves to heighten the addicts' esteem for opiates.[60] Not only opiate abuse but also alcohol abuse, schizophrenia, and affective disorders often begin during the mid or late adolescent years. There is a possibility that during these neuroendocrinologically explosive years, some juveniles develop a temporary deficiency of endorphins that results in the particular psychopathology. If this can be substantiated, temporary exogenous opiate treatment may abridge the deficiency period and perhaps prevent the concomitant development of lifelong psychosis or substance abuse.[61]

That opiates provoke aggressive, assaultive behavior is more of a myth than reality. Nyswander describes that the choice of drug reveals what makes the user feel at his best. The narcotic addict is more comfortable when his hostility and aggressive feelings are blunted as with opiates. The significance of the opiate is to make the addict feel completely at peace with himself and the world. He has succeeded in killing two birds with one stone; in his passivity he experiences the feelings of satisfaction that come to an aggressor.[62]

Even though the antipsychotic effects of opiates were observed by many investigators, in most methadone maintenance programs an effort has been made to exclude patients with mental illness. Despite this general policy, Salzman and Frosch admitted only patients with a psychiatric history.[63] In general, the disturbed patients seemed to benefit from the methadone maintenance program indicated by the fact that 50% of the patients were able to hold jobs compared to only 10% before methadone. Weissman believes that in certain individuals, treatment with methadone should be lifelong. The use of methadone dampens physiologically some of the unchanneled tension, and individuals can gradually learn new and more efficient ways of binding tension in a socially acceptable fashion. Methadone serves the tension-reducing role previously played by the subject's acting-out behavior.[64] On a similar note, Winkelstein proposed that in some cases of opiate dependence it is useless to attempt further psychiatric treatment to achieve abstinence. One such patient almost killed his father in a rage. He then took opiates to avoid acting out his overt aggressive behavior.[65] Another case is of a paranoid schizophrenic who, by using opiates, has made a reasonably good work adjustment. He is isolated in a file room and does excellent work. He received several promotions with increase in salary, but he needs his daily dose of opiates.[65] These subjects indefinitely require maintenance doses of methadone or other opiates to keep them in some state of stability.

Evaluation of 6 years of records of a drug-abusing population provided evidence for the prophylactic effects of opiates against the development of psychiatric illness.[66] McLellan *et al.* reported that at the beginning of the study, the psychiatric examinations showed low symptom levels in various drug-abusing groups. By the end of 6 years a large percentage of stimulant and depressant users had serious psychoses while the opiate users showed no deterioration from their control values. These investigators also recognized the specific psychopharmacological effects of opiates. Their clinical impression was that opiates may function to medicate underlying psychological problems by reducing the symptoms of anxiety, depression, and paranoia.[66]

Levinson in 1971 proposed outright the use of methadone for the treatment of schizophrenic patients. He stressed that the organizing and antianxiety effects of methadone should help psychotic patients in mental institutions become functional. He believes that methadone maintenance would provide a much more humane and effective control of psychotic symptoms than some of the currently used neuroleptics in controlling rage, paranoia, and anxiety. Thus, methadone would likely improve the physical and mental conditions of many institutionalized patients.[67]

8. THE EFFECTS OF METHADONE ON SUICIDAL TENDENCY

At present there is no reliable drug for the treatment of suicidal ideation. For this reason the observation of Litman is important. While working in a

California-based suicide control agency, he observed that methadone was very effective in the control of suicidal tendency. When detoxification from methadone was attempted, suicidal ideation returned. The author concluded that "depressive suicidal components are often present in drug addicts and that, indeed, narcotics are used as self-medication to prevent depressive pain and as an alternative to suicide."[68] The report by Fredrick *et al.* on both the depression scale and the assessment of morbid attitudes reveals fewer expressions of anxiety, resentment, and depression for methadone users compared to other addicts (abstinent or on heroin). However, the greatest difference was in the expectation of violent death, where the percentage of subjects on methadone was less than half of those abstaining from methadone treatment. The author finds that methadone may be helpful in bringing about a happier and better-adjusted attitude by decreasing the level of depression and attitudes toward violence, aggression, and morbidity.[69] A more recent case study by Berken *et al.* described a 19-year-old woman with psychotic rage and a compulsion toward self-mutilation and suicidal attempts among other serious psychopathology.[70] Conventional, medically accepted psychopharmacological agents were ineffective in controlling her symptoms. As a last resort, the investigators obtained permission from the FDA to use methadone in this patient. The authors describe that "the 'normalizing' effect of methadone proved more satisfactory for this patient than any other treatment she had received." These investigators, similarly as others in this section, made their observations at a methadone maintenance clinic. They noticed that addicts with histories of crimes of rage, current aggressive rage, or repressed rage who received methadone could recognize the onset of rage and could verbalize their needs rather than acting out in violence.[70]

9. *SIMILARITIES AND DIFFERENCES BETWEEN OPIOID AGONISTS AND ANTIPSYCHOTIC AGENTS*

Some of the biochemical and neurophysiological data on opiates and neuroleptics indicate similarity of actions between the two classes of drugs. There are theories depicting dopamine hyperactivity as a possible cause of schizophrenia.[71–73] Hartman described schizophrenia as a single-deficit condition and a multiplicity of causes or aggravating factors. The single basic deficit is vulnerability to stress, described as a constant pressure that can be conceptualized as unneutralized psychic energy. For some relatively health individuals, it takes a major stressful event to precipitate a psychotic episode while others are on the brink of psychosis chronically.[71] The author suggests that increased levels of dopamine produces the buildup of what he calls "unneutralized psychic energy." Hartman described an experiment in which he administered to himself a precursor of dopamine (dopa) and fusaric acid, an inhibitor of the enzyme dopamine β-hydroxylase, which is necessary for the conversion of dopamine to norepinephrine. This combination theoretically would increase the levels of dopamine in the brain. The temporary psychosis created by these chemicals was described by the author.[71]

Fitting this theory, the mechanism of antipsychotic action of various neuroleptics, such as chlorpromazine and haloperidol, is dopamine receptor blockade.[74,75] This action was also demonstrated for morphine and methadone.[76–78]

Haloperidol (a meperidine derivative quintessential antipsychotic) and morphine are equally effective in blocking morphine-withdrawal body shakes in rats while increasing the spontaneous electrical activity in the nigrostriatal nerve endings.[79–84] Neuroleptic drugs directly compete with dopamine for the postsynaptic dopamine receptor, by inhibiting the actions of dopamine on dopamine-sensitive adenylate cyclase,[74,75] whereas opiates do not inhibit this enzyme.[76,77] However, the similarity of the behavioral and neurochemical responses of opiates and neuroleptics suggests a common final pathway in their mechanisms of action in various brain regions. Another biochemical correlation is that both neuroleptics and methadone are able to increase the levels of serum prolactin, possibly by acting through the hypothalamic–adrenergic system.[85]

EEG studies of Fink[86] and Martin *et al.*[87] in human subjects evaluated the changes in electrical activity of the brain after the administration of neuroleptics and methadone. Although the methods of EEG evaluations were somewhat different in the two studies, the electrical changes were quite similar after phenothiazines and methadone.

10. THE THERAPEUTIC POSSIBILITIES FOR OPIOIDS AND THEIR ANTAGONISTS IN MENTAL DISEASE

The potential involvement of the endorphin system in mental disease was promptly hypothesized by several investigators. Their differences in approach were mainly on whether an excess or a deficiency of endorphins is responsible for the observed pathophysiology of psychosis.

In order to evaluate the possible use of narcotic agonists in mental disease, it is helpful to review the psychopharmacology of these agents. Mixed narcotic agonist-antagonists such as cyclazocine and nalorphine have been shown to have agonist activity[88,89] as well as dysphoria and psychotomimetic effects.[87] Naloxone and naltrexone have generally been considered to be relatively pure antagonists, but some agonistlike physiological effects were also described by Martin *et al.*[87] and Verebey *et al.*[90] after naltrexone administration in ex-addicts: a decrease in respiratory rate, body temperature, and pupillary constriction. Jones administered 20 mg naloxone i.v. in volunteers who have never been addicted. This also caused some effects: tremors, EEG changes, increase in body temperature, and changes in mood. Those changes showed considerable interindividual variability, but naloxone had clearly different effects than placebo in this double-blind experiment. Many of the effects were those observed during mild withdrawal syndromes in addicts.[91] Buchsbaum *et al.* have observed an increased sensitivity to pain (in pain-sensitive normal subjects) after administration of 2 mg naloxone of i.v. Pain-insensitive subjects reported the opposite effect after the same dose. Similar group differences were observed in somatosensory-evoked potentials using the same subjects.[92] Naltrexone produced minor changes in ex-addicts: an increase in attention and perception and some dysphoria.[93] Naltrexone was shown to be extremely safe in acute and chronic treatments, an important consideration for use in chronic psychiatric conditions.[90]

11. THE USE OF NARCOTIC ANTAGONISTS TO COUNTERACT ENDORPHIN EXCESS

Terenius *et al.* found increased CSF levels of endorphins in chronic psychotic patients; these levels decreased after successful treatment with neuroleptics.[94] Therefore, the logical therapeutic measure implicated by these observations was the administration of a narcotic antagonist. Naloxone (0.4 mg i.m.) was administered to six chronic schizophrenic patients. Four patients reported a dramatic reduction of hallucinations for several hours after naloxone.[95] Attempts of replication and extension of similar studies using naloxone to treat schizophrenic patients yielded negative results.[96–98] Various alterations in experimental design were implemented because of questions about the size of the doses, patient selection, and even the drugs used. A review by Volavka describes in detail all of the past and recent clinical findings on the narcotic antagonists naloxone and naltrexone in schizophrenia, mania, and depression.[99] In essence, his findings were that low doses of naloxone (up to 6.0 mg) had no effect on mania, schizophrenia, and depression whereas doses of 10–20 mg yielded some promising results in occasional patients. Small doses of 0.4–1.2 mg are effective in reversing serious opiate overdose-related respiratory depression, indicating high potency and specificity of naloxone in displacing opioids from the opiate receptors. The fact that in mental illness 20 to 50 times more naloxone is necessary for a limited success than its normal narcotic antagonist dose indicates another mechanism than exclusive effects at the opiate receptors.

12. THE USE OF OPIOID AGONISTS FOR THE SUPPLEMENTATION OF ENDORPHIN DEFICIENCY

In contradistinction to the endorphin excess theory, Jacquet and Marks suggested that endorphin deficiency may result in some mental disease.[100] Their suggestion was based on the apparent similarity between the extrapyramidal rigidity seen in patients taking neuroleptic drugs and the rigidity elicited in rats by β-endorphin. Along with the extrapyramidal side effects, the neuroleptics are effective antipsychotic agents. Thus, they suggested that β-endorphin may have similar pharmacological effects. Testing this hypothesis, Kline *et al.* administered 6–10 mg β-endorphin to a total of 15 patients with various psychiatric diagnoses, including schizophrenia, depression, personality disorders, and mental deficiency.[101,102] Some of the patients became more talkative and less anxious followed by a period of sedation. Improvement of depressive and schizophrenic symptoms was apparent after these injections of β-endorphin. Several subsequent studies were undertaken by other investigators using β-endorphin and synthetic analogs in mental diseases instituting several changes in the experimental conditions. Berger *et al.* studied the effects of β-endorphin in nine chronic schizophrenic patients using a double-blind cross-over design. A 20-mg injection of β-endorphin produced beneficial EEG changes by increasing α waves and also by releasing prolactin similarly as other opiate agonists. Consequently, the treatment showed a statistically significant improvement of the patients' symptoms but these benefits were not apparent clinically.[103] Gerner *et al.* investigated the effects of β-endorphin in both schizophrenic and depressed patients. Temporary but significant improvement was seen in the depressed patients, but there was no significant change in the schizophrenic patients. In fact, six of the eight schizophrenic subjects had worse symptoms than before β-endorphin.[104]

The results of other studies are discussed in a review by Olson *et al.*[105] The most interesting data emerged from the studies utilizing some of the synthetic endorphin analogs, destyrosine-γ-endorphin (DTγE) and FK-33-824, a potent analog of Met-enkephalin. Verhoven *et al.* administered DTγE to schizophrenic patients and found improvement in all six patients.[106] This study, however, was not double-blind, but their second study was. Using eight patients, 1 mg DTγE was injected daily for 8 days. After 4 days, in most patients the psychotic symptoms were greatly reduced.[107] Studies were also conducted with FK-33-824. Krebs and Roubicek reported short-term improvement in the patients' psychotic symptoms[108] as did Nedapil and Ruther.[109] The comparative evaluation of these studies is difficult because of the differences in patient selection and the variations in psychiatric evaluation by the different groups. Also, the clinical trials involving endorphins and their synthetic analogs are hindered by lack of availability and the very high cost of these substances. Nevertheless, there is considerable promise in these preliminary results to warrant further studies.

β-Endorphin, although the naturally occurring substance in brain, may not be the optimal substance for therapeutic use. Its effectiveness in treating mental disease would suffer from the same difficulty as would the short-acting narcotic agonists such as heroin and morphine: they would have to be administered systematically at short intervals. And since β-endorphin is a peptide, it would be rapidly degraded by various peptidases limiting the bioavailability of the active molecules in the circulatory system and at their sites of action. If opioid agonists and/or antagonists prove to be useful therapeutically in mental disease, the longer-acting, orally effective methadone and 1α-acetylmethadol provide more suitable characteristics for chronic treatment.

13. *RECENT STUDIES OF THE ENDORPHIN SYSTEM IN HUMAN SUBJECTS*

The effects of opioid peptides in neuroendocrine and neurochemical systems were reviewed by Simon.[110] Like opiates, the endorphins and enkephalins interact with the endocrine system, producing *increases* in the release of growth hormone, prolactin, and ACTH and *decreases* in the release of luteinizing hormone and follicle-stimulating hormone. On the neurotransmitter system, opioid peptides were found to be mostly *inhibitory* just as was reported for opiate drugs. Thus, Met-enkephalin has been shown to inhibit both electrically and high K^+-induced Ca^{2+}-dependent norepinephrine release *in vitro* from the neurons of the cerebral cortex. Enkephalin, furthermore, inhibited the release of acetylcholine and dopamine in a similar test system. Substance P release was also inhibited by β-endorphin. These results indicate an inhibitory neuromodulatory role for opioid peptides and suggest a presynaptic site for their mode of action. In behavioral terms, some symptoms of psychosis are related to sympathetic nervous system hyperactivity. The endorphin system seems to negatively modulate such a sympathetic neurochemical discharge. Thus, their behavioral role may be the promotion of psychological homeostatis.[111]

Certain depressive disorders are known to respond to electroconvulsive treatment (ECT). The study of Inturrisi *et al.* concentrated on the effects of single and repetitive ECT on the plasma β-endorphin levels and the therapeutic outcome.[112] Nine of the twelve subjects positively responded to ECT and had substantial increases in β-endorphin levels. Two patients had incomplete remission with increased β-endorphin levels. If the increased endorphin levels are responsible for the improvement, the two subjects with increased endorphin levels and incomplete response must have had problems with their opiate receptors. Such a mechanism was postulated by Extein *et al.* to explain the blunted or lack of prolactin response of depressed patients receiving i.v. morphine infusion.[113] The authors explain that opioids may inhibit dopamine release and thus the dopamine-dependent prolactin release is blunted or blocked completely. This would also manifest if the opiate receptors on dopamine neurons are not functional.

Terenius and Wahlstrom described fractions I and II of opioid peptides in the CSF[114] and O'Brien *et al.* studied the levels of these peptides in various phases of addiction.[115] The CSF samples were collected from addicts while on methadone maintenance, early and late withdrawal, and during antagonist therapy with naltrexone. The most striking were the elevated levels of fraction II in methadone-maintained addicts, however, both fraction I and fraction II levels were higher in methadone-maintained addicts as compared to normals, indicating that a feedback inhibition in endorphin biosynthesis is not operational during opiate dependence. There was a drop in the endorphin levels in early withdrawal followed by a rise over the methadone-maintenance endorphin levels in late withdrawal. This pattern of changes in endorphin levels is similar to the one reported by Picker *et al.* in manic-depressive psychosis, i.e., low during depression and high during the manic period.[116] An even further increase in endorphin levels was observed following naltrexone therapy. The authors suggest that a short-acting antagonist like naloxone may release enough endorphin that the observed effects may be caused by the endorphins rather than the antagonist. This observation is important especially in the interpretation of the recent positive naloxone effects in Alzheimer's syndrome.[117] An opiate agonist (deodorized opium tincture) was found effective in senile mental dementia by Abse *et al.*[118] Thus, it is possible that the agonists and not the antagonists are beneficial in that disorder.

Agren *et al.* used naloxone in clinically depressed patients. Two of six became worse, and their clinical symptoms were paralleled by an increase in their CSF endorphin level.[119] It is difficult to evaluate what the increases and decreases in peripheral endorphin levels mean. Such measurements may not always indicate the true picture occurring in the CNS at the opiate receptors. Such discrepancy is reflected in an early study of Agren *et al.* They found CSF endorphin levels sometimes elevated and occasionally low in depression. This phenomenon was explained by the differences between the types of endorphins measured. Fractions I and II have not yet been identified but they appear to be distinct from circulating β-endorphin and the enkephalins. Hypercortisolism and nonresponse to dexamethasone suppression became an aid in diagnosing certain types of depressions. The studies of Agren *et al.* found no connection between the levels of fraction I or II and the levels of monoamine metabolites in the CSF.[119]

An earlier work of Pickar *et al.* indicated that the CSF opioid activity was higher in mania than during depression in manic-depressive subjects.[116] This finding seems easy to accept based on the well-described pharmacological effect of opiates, producing a state of well-being and presumably depression in its absence.[120] Although other studies describe excess endorphins, these investigators reported lower endorphin levels in some schizophrenic patients. The anecdotal clinical observations found opioids effective in relieving a variety of schizophrenic symptoms in opiate addicts. Thus, simple logic would also anticipate lowered endorphin levels in schizophrenics as described by Pickar *et al.*[121]

Further development in this area of research was the testing of opiate agonists' effects in schizophrenic subjects who were never addicted to opioids. At the time of this review, the first pilot study was completed by Breizer and Millman, using methadone in nonaddicted schizophrenic patients.[122] Despite the poor selection criteria of the protocol, assembling patients who did not respond to any known therapy and conventional medication, methadone provided significant improvement in the four subjects studied. There were lowered Brief Psychiatric Rating Scale scores mainly affecting the positive symptomatology of schizophrenia (which also are found in mania) such as conceptual disorganization, grandiosity, delusions, and hallucinations. There was less effect on paranoid ideation and no significant effects were observed on the negative schizophrenic symptoms of emotional withdrawal and blunted effect.

The partial improvement reported in this study indicates that it is not likely to find a single agent that will be totally effective in all symptoms of schizophrenia. The variation of symptoms in different patients indicates a multilocus defect in schizophrenia, and for this reason successful treatment will probably need numerous agents to balance the deficit or excess of various neurochemicals responsible for the particular symptoms. Another interesting future project is the trial of methadone in subjects who respond to antipsychotic medication. Is there a possibility for efficacy without continued degeneration or return to premorbid function? Even the combination of the two drugs, neuroleptics and opiates, may allow reduction of the doses of phenothiazines and thus reduce the likelihood of tardive dyskinesia.

In an interesting recent case report, 4–500 mg of L-tryptophan and 100 mg of carbidopa were effective in reversing the hallucinogenic effects caused by LSD.[123] It may be possible that future therapy of schizophrenia may include a number of pharmacological and neurochemical precursers for a successful combination therapy.

14. THE USE OF β-ENDORPHIN IN SCHIZOPHRENIC AND DEPRESSED PATIENTS

The first trial of β-endorphin in human subjects was reported by Kline *et al.*[101,102] They used 1.5 to 9 mg β-endorphin in single i.v. doses. There were 15 psychotic patients who received one or more injections. The summary of their results on the type and on the duration of effects were the following. The autonomous effects were present directly after injection between 30 and 120 sec. The euphoric effects were observed between 1 and 6 hr, the sedative effects between 2 and 3 hr, and the therapeutic effects for 1 to 10 days. The favorable therapeutic effects reported were better mood and relief from anxiety.

Such wide-ranging pharmacological effects and long-time action from a peptide cannot be accepted as *direct* effect. The pharmacokinetic works of Catlin *et al.*[124] and Foley *et al.*[125] report a very short half-life for β-endorphin. In fact, Catlin *et al.* reports a two-to-fivefold increase of β-endorphin-immunoreactive substances for 17 hr after injection, the plasma half-life reported is 14 min (β phase). Such information implies that if the observed long action of infused β-endorphin is related to β-endorphin, the effects must be secondary or tertiary ones.

The resting, mean, baseline CSF β-endorphin-like immunoreactivity level in nine normal subjects was 100 ng/ml. There was no difference found from this value in depressed or schizophrenic patients.[124] It would be interesting to see whether or not a stress or pain stimulus-related response in the β-endorphin levels would be the same in the three groups. Similarly as with evoked potentials, abnormality can be seen only in the stimulated state.

Catlin *et al.* report that β-endorphin injection produced significant physiological and hormonal effects such as decrease in systolic and mean arterial blood pressures, oral temperature, and increased prolactin levels. Serum cortisol and growth hormone levels were unaffected and most unexpectedly no pupillary constriction and respiratory depression were observed.[124] These differences from exogenous opioids may be explained in at least two ways. It is possible that the injected β-endorphin, due to its short half-life and large size, did not reach opiate receptors that mediate these functions. A second possibility is that the β-endorphin molecule selectively does not act at opiate receptors that mediate the pupillary and respiratory effects.

Statistically significant but not clinically obvious lessening in schizophrenic symptoms was found after a 20-mg i.v. injection of β-endorphin as compared with saline injection in a double-blind cross-over study by Berger and Barchas.[126] The authors suggested that possibly the i.v. administration of β-endorphin limits the effectiveness of their treatment. For more recent studies, intrathecally administered β-endorphin is promising and such studies permit the use of lower doses of β-endorphin (which is still very expensive). Also, the effectiveness of multiple-doses versus single-dose trials is important.[126]

Six acute schizophrenic patients off all other medication in a double-blind design were studied by Petho *et al.*[127] All six patients showed a certain improvement for a variable period. Improvement was only in the area of mood and activity. The nature of the improvement "in all cases indicated a tendency to restore the homeostasis." There was no relief of the delusions and hallucinations. The authors suggest that the more intense the psychosis, the less the homeostatic influence of β-endorphin. No antipsychotic effects of β-endorphin were seen in any patient in this study. The authors also suggest that β-endorphin in nonpsychotic subjects may even upset the equilibrium. This was indicated by the symptoms seen in certain normal individuals such as retardation, confusion, and bewilderment.[128,129]

Emrich *et al.* reported on the possible antidepressant effects of the opiate agonist/antagonist buprenorphine.[130] The subjects in the study were unresponsive to various conventional psychotropic medications. The authors suggest this therapy as an alternative physiologically safer approach to ECT. As shown by Inturrisi *et al.*[112] and others, ECT does increase endogenous opioid levels. This finding suggests that a combination treatment of cyclazocine and naltrexone may also be effective in certain depressions.

15. EFFECTS OF CHRONIC OPIATE USE: SAFETY AND TOXICITY

Addiction liability is the major safety problem of long-term opiate treatment. It is interesting to note that apparently no addiction resulted from the old opiate treatments of depression.[3] The schizophrenic patients of Wikler *et al.* treated with morphine have developed physical dependence reflected by the withdrawal syndrome when the treatment was stopped. However, none of their three patients exhibited any interest in morphine at any time during or after the study.[131] Thus, it may be suggested that psychotic patients can develop physical dependence on opiates without taking on the life-style of street addicts. Similar observations were made in "normal" individuals treated for pain. When opiates are used for medical reasons, the chances of developing drug-seeking behavior are very slight.[59,132,133]

The gross fluctuations in performance normally associated with a single dose of short-acting narcotics such as heroin were eliminated by maintenance on methadone.[134,135] Measurements using the Eysenck Personality Inventory showed that subjects undergoing methadone maintenance therapy were not more anxious or unable to cope with stress than a group of apparent normals.[136] Some opiate users stated that methadone did not impair functions, but in fact improved their ability to do useful work. Observations made under experimental conditions showed that as long as adequate amounts of opiates were available to the subjects, their aggressive and antisocial behavior was practically eliminated, personal hygiene was maintained, and assigned responsibilities were discharged satisfactorily.[60] Psychological tests of performance revealed little or no impairment, and the sensorium remained clear, while anxiety was reduced.[60,137]

The advent of methadone maintenance in the treatment of opiate addiction introduced the concept of eliminating dysfunctional drug use rather than drug use itself. Methadone clinics have provided the addict with a stable source of drug, eliminating the disruptive need to procure supplies. Substitution of heroin with methadone enables the avoidance of illegality, and the exposure to the criminal life-style and multiple medical illnesses so often associated with drug-seeking behavior. The psychological craving is reduced and the individual can function more effectively as the cost and frequency of drug-taking decrease.

Good health and productive work are compatible with stable (maintenance) dependence on opiates. Addicts receiving high daily oral doses of methadone carried out their daily tasks and held responsible positions. Nyswander stated that opiate addicts were not unusual in appearance when maintained at the proper dose of opiates. Their retention and recall of simple mathematics and conceptual thinking were intact, with no evidence of mental deterioration.[62] A study of Gearing and Schweitzer of 17,500 methadone maintenance patients found increased social productivity and decreased antisocial behavior among patients in methadone maintenance treatment versus before treatment.[138] Cutting described a physician who was functioning well both physically and mentally after 62 years of addiction.[139] Thus, moderate morphine use does not necessarily interfere with regular activities during a normal life span.

Although clinical experience suggests that chronic intake of controlled amounts of opiates is not necessarily detrimental to the general health, there are some undesirable side effects of chronic opiate consumption. The most common symptoms and signs in patients receiving high maintenance doses of methadone are sweating, constipation, minor sleep disorders, decreased testosterone levels, decreased libido and fertility, and abnormal menses.[140–144] Tolerance develops to many of these effects during the repeated intake of methadone.[143] The premier social problems are medical diversion, continued use of illicit substances, and society's concern over the failure of the patient to return to a totally narcotic-free life.

In comparison with the existing antipsychotic drugs, opiates have a number of clinical advantages and disadvantages.[145] Opiates overdose is very dangerous; however, it is easier to treat than an overdose of tricyclic antidepressants or of phenothiazines. Since specific narcotic antagonists are available, opiate overdoses can and are readily managed. Withdrawal syndrome following the discontinuation of opiates is certainly much more unpleasant than that occurring after stopping neuroleptics. However, the recent introduction of clonidine seems to eliminate many of the unpleasant symptoms of withdrawal,[146] and the effects of chronic opiate treatment subside at some point after the treatment is stopped. This is not true of neuroleptics; tardive dyskinesia is irreversible in many cases. Recent research indicates that tardive dyskinesia may be associated with significant mortality.[147] Thus, the dangers of opiates should be weighed against the serious toxic effects of currently used psychoactive drugs.[148,149] Treating nonaddicts with opiates will raise ethical problems. The medical profession as well as the public have shown increasing acceptance of opiates as analgesics for patients with cancer as well as for the treatment of other diseases.[150] It is perhaps possible to view some psychoses as mentally painful psychological states, in a sense similar to painful physical states, that require an analgesic. In addition to the heuristic value of therapeutic studies using opiates, there is an obvious and practical need for more effective treatment of this type of mental disease. For this disease there are truly no antipsychotics. The state of the art is generally to reduce length of hospitalization and ability to function at some level in the community. Most clinicians assume that the more overt symptoms are treated by antipsychotics but the disease continues. Special subjects for initial trials of opiates would include particularly those showing aggressive behavior, taking advantage of the potent antiaggression effects of opiates.[70] Another group would be patients showing first signs of tardive dyskinesia, who cannot discontinue neuroleptic therapy without serious deterioration of psychiatric symptoms. Such patients could be effectively managed with methadone, especially because chronic methadone use in humans does not result in extrapyramidal side effects. Another group of prospective patients are psychotic depressives who are considered for ECT, but who cannot receive it for medical or other reasons.

16. PSYCHOLOGICAL HOMEOSTASIS

Verebey *et al.* suggested that one of the physiological roles of endorphins is the preservation of "psychological homeostasis."[111] This is based on the quasi-equivalency of endorphins with opiate agonists and on the opioids' clinically observed psychopharmacological effects, which include the promotion of stress tolerance, tranquilization, and fear relief among others. A crude but descriptive term for the role of endorphins would be "brain shock absorbers." Following stressful stimuli, the autonomic response is usually a major sympathetic discharge that is likely to result in dopamine and norepinephrine hyperactivity in the limbic system of the brain. This would possibly trigger some signs and symptoms of psychoses. In "normal" individuals simultaneously with the sympathetic discharge, a stress-induced release of endorphins would promptly antagonize dopamine and would also produce inhibitory modulation to most stimulant neuronal activities.

Evidence by Pickar *et al.* indicates that a rise and fall of endorphin levels is a possible mechanism of manic-depressive psychosis.[116] Plasma levels of β-endorphin were high when the patient exhibited mania. This was followed by some decrease in plasma β-endorphin levels, paralleled by a mixed psychological state and further decrease in β-endorphin levels leading to a motorically retarded depression. These data are consistent with the data of Extein *et al.*[113] mentioned earlier. The same patient at a later date while in the manic state had increased opioid activity in her plasma again. This mechanism stipulates that certain levels of endorphins are needed for a psychological equilibrium and that appropriate adjustment of these levels may be therapeutically effective at least in manic-depressive illness.

Apparent endorphin deficiency can result through various mechanisms. Deficiency may be realized by slow biosynthesis or too rapid enzymatic deactivation of endorphins. Or if the endorphins are present in sufficient amounts in brain storage, then the release mechanisms may be defective or the receptors may be nonfunctional. The concomitant endorphin deficiency symptoms would include anxiety, irritability, lability, supersensitivity to stress, rage, aggression, depression, paranoia, suicidal ideation, and low self-esteem. These symptoms are commonly observed in various psychiatric disorders. The healthiest approach would be the reactivation of the endorphin system by stimulation of production, facilitation of release, or inhibition of enzymatic decay of endorphins. The second approach would be the supplementation of long-acting synthetic endorphins, and the third route the use of long-acting, orally effective opiate agonists. Supplementation of endorphins or opiates would not be effective, however, if the problem lies with the receptors.[111]

A more likely possibility is that the endorphin system is only one of the behaviorally active peptide and neurotransmitter systems in the CNS and that all systems must interact to produce the integrated responses manifested as "normal" behavior.

17. CONCLUSION

This review brings into focus the fact that opiates are not entirely foreign molecules to the human organism based on the similar pharmacology of the naturally occurring brain peptides and the exogenous opioids. This indicates that normal human subjects are constantly under the influence of morphinelike substances. An acceptance of these facts may liberate opiates from the prevailing social stigma and legal attitudes that identify them entirely with addiction. Further research on endorphins may elucidate their precise role in physiology and behavior, and in this way their excess or deficiency states may be properly treated.

The clinically observed antipsychotic effects of opiates seem very reasonable based on the anatomical distribution of opiate receptors and endorphins and on the biochemical and neurophysiological similarity between the effects of opiates and neuroleptic drugs. The clinical observations, although stimulating, also exposed many important questions for future research to answer. The anecdotal clinical evidence seems to imply the effectiveness of opiates in many diverse psychopathological states. It would be necessary to identify the diagnostic category of responding conditions.

The timeliness of investigation in this area is exemplified by the numerous reports of tardive dyskinesia in patients chronically treated with antipsychotic drugs. These patients would be left without any medication after the discontinuation of antipsychotic agents. Effective psychoactive drugs without extrapyramidal side effects are needed, and the orally effective long-lasting opiate agonists might be excellent candidates for this purpose.

REFERENCES

1. Gay, G. R., and Way, E. L., 1971, *J. Psychedelic Drugs* **4**:31–39.
2. Jaffee, J. H., and Martin, W. R., 1980, *Pharmacological Basis of Therapeutics*, 6th ed. (A. G. Gilman, L. S. Goodman, and A. Gilman, eds.), Macmillan Co., New York, pp. 494–502.
3. Weygandt, W., 1935, *Lehrbuch der Nerven und Geistes-Krankheiten,* Marhold Verlagsbuchhandlung, Halle, West Germany, p. 507.
4. Berridge, V., 1982, *Ann. N.Y. Acad. Sci.* **398**:1–11.
5. Way, E. L., 1982, *Ann. N.Y. Acad. Sci.* **398**:12–23.
6. Verebey, K., 1981, *Adv. Alcohol Substance Abuse* **1**:101–123.
7. Kelly, D., 1982, *Ann. N.Y. Acad. Sci.* **398**:260–271.
8. Cohen, M. R., Pickar, D., Dubois, M., Nurnberger, J., Roth, Y., Cohen, R. M., Gershon, E., and Bunney, W. E., Jr., 1982, *Ann. N.Y. Acad. Sci.* **398**:424–431.
9. Ginzler, A. R., 1982, *Ann. N.Y. Acad. Sci.* **398**:302–306.
10. Csontos, K., Rust, M., Holt, V., Mahr, W., Kromer, U., and Teschemacher, H. J., 1979, *Life Sci.* **25**:835–844.
11. Davis, G. C., Buchsbaum, M. S., Naber, D., Pickar, D., Post, R., Van Kammen, D., and Bunney, W. E., Jr., 1982, *Ann. N.Y. Acad. Sci.* **398**:366–373.
12. Buchsbaum, M. S., Reus, V. I., Davis, G. C., Holcomb, H. H., Cappelletti, J., and Silberman, E., 1982, *Ann. N.Y. Acad. Sci.* **398**:352–365.
13. Bird, S. J., and Kuhar, M. J., 1977, *Brain Res.* **122**:523–533.
14. Korf, J., Bunney, B. S., and Aghajanian, G. K., 1974, *Eur. J. Pharmacol.* **25**:165–169.
15. Pert, C. B., Kuhar, M. J., and Snyder, S. H., 1976, *Physiology* **73**:3729–3733.
16. Gold, M. S., Pottash, A. L. C., Sweeney, D. R., and Kleber, H. D., 1980, *Drug Alcohol Depend.* **6**:201–208.
17. Gold, M. S., Byck, R., Sweeney, D. R., and Kleber, H. D., 1979, *Biomedicine* **30**:1–4.
18. Gold, M. S., Pottash, A. L. C., Extein, I., and Kleber, H. D., 1980, *Lancet* **2**:1078–1079.
19. Gold, M. S., Pottash, A. L. C., Annitto, W. J., Extein, I., and Kleber, H. D., 1981, *Lancet* **1**:992–993.
20. Belluzzi, J. D., and Stein, L., 1982, *Ann. N.Y. Acad. Sci.* **398**:221–229.
21. Henriksen, S. J., Chouvet, G., McGinty, J., and Bloom, F., 1982, *Ann. N.Y. Acad. Sci.* **398**:207–219.
22. Katz, R. J., 1982, *Ann. N.Y. Acad. Sci.* **398**:291–299.
23. Jacquet, Y. F., 1982, *Ann. N.Y. Acad. Sci.* **398**:272–290.
24. Gold, M. S., and Bowers, M. B., Jr., 1978, *Am. J. Psychiatry* **135**:1546–1548.
25. Holtzman, S. G., 1982, *Ann. N.Y. Acad. Sci.* **398**:230–239.
26. Adler, M. W., 1982, *Ann. N.Y. Acad. Sci.* **398**:340–350.
27. Gold, M. S., and Byck, R., 1978, *Natl. Inst. Drug Abuse Res. Monogr. Ser.* **19**:192–209.
28. Stein, L., and Belluzzi, J. D., 1978, *Adv. Biochem. Pharmacol.* **18**:299–311.
29. Kornetsky, C., and Bain, G., 1982, *Ann. N.Y. Acad. Sci.* **398**:241–259.
30. Berkowitz, B. A., Finck, A. D., and Ngai, H., 1977, *J. Pharmacol. Exp. Ther.* **203**:539–547.
31. Brown, B. S., Kozel, N. J., Meyers, M. A., and DuPont, R. L., 1973, *Am. J. Psychiatry* **130**:599–601.
32. Perkins, M. E., and Bloch, H. I., 1971, *Am. J. Psychiatry* **128**:47–51.
33. Stimmel, B., Sturiano, U., Cohen, M., Korts, D., Hanbury, R., and Jackson, G., 1982, *Alcoholism* **6**:362–368.
34. Jones, M. A., and Sparatto, G. R., 1977, *Life Sci.* **20**:1549–1555.
35. Smith, S. G., Werner, T. E., and Davis, W. M., 1975, *Psychol. Rec.* **25**:17–20.
36. Gearing, F. R., 1971, Public Health Service Publication No. 2171, U.S. Government Printing Office, Washington, D.C.
37. Ritchie, J. M., 1975, *Pharmacological Basis of Therapeutics*, 5th ed. (L. A. Goodman and A. Gilman, eds.), Macmillan Co., New York, pp. 137–151.
38. Blum, K., Hamilton, M. L., and Wallace, J. E., 1977, *Alcohol and Opiates* (K. Blum, ed.), Academic Press, New York.
39. Wikler, A., Goodell, H., and Wolf, H. G., 1945, *J. Pharmacol. Exp. Ther.* **83**:294–299.

40. Lorens, S. A., and Sainati, S. M., 1978, *Life Sci.* **23**:1359–1364.
41. Schenk, G. K., Enders, P., Engelmeier, M. P., Ewert, T., Hendermerten, S., Kohler, K. H., Lodenmann, E., Matz, D., and Pach, J., 1983, *Arzneim. Forsch. Drug Res.* **28**:1274–1277.
42. Moss, I. R., and Friedman, E., 1978, *Life Sci.* **23**:1271–1276.
43. Jeffcoate, W. J., Herbert, M., Cullen, M. H., Hastings, A. G., and Walder, C. P., 1979, *Lancet* **2**:1157–1159.
44. Senter, R. J., Heintzelman, M., Dorfmeuller, M., and Hinkle, H. A., 1979, *Psychol. Rec.* **29**:49–56.
45. Khantzian, E. J., 1982, *Ann. N.Y. Acad. Sci.* **398**:24–30.
46. Resnick, R. B., Fink, M., and Freedman, A. M., 1970, *Am. J. Psychiatry* **126**:1256–1260.
47. Resnick, R. B., Schuyten-Resnick, E., and Washton, A., 1979, *Compr. Psychiatry* **20**:116–125.
48. Wurmser, L., 1982, *Ann. N.Y. Acad. Sci.* **398**:33–43.
49. McKenna, G., 1982, *Ann. N.Y. Acad. Sci.* **398**:44–55.
50. Khantzian, E. J., Mack, J. E., and Schatzberg, E. F., 1974, *Am. J. Psychiatry* **131**:160–164.
51. Khantzian, E. J., 1974, *Am. J. Psychother.* **28**:59–70.
52. Gold, M. S., Pottash, A. L. C., Sweeney, D. R., Kleber, H. D., and Redmond, D. E., 1979, *Am. J. Psychiatry* **136**:982–983.
53. Kleber, H. D., and Gold, M. S., 1978, *Ann. N.Y. Acad. Sci.* **311**:81–98.
54. Dole, V. P., and Nyswander, M., 1965, *J. Am. Med. Assoc.* **193**:646–650.
55. Gold, M. S., and Rea, W. S., 1983, *Psychiatric Clinics of North America*, Vol. 6, No. 3 (S. C. Risch and D. Pickar, eds.), Saunders, Philadelphia, pp. 489–520.
56. Wikler, A., 1965, *Narcotics* (D. M. Wilner and G. G. Kassenbaum, eds.), McGraw–Hill, New York.
57. Robins, L., 1974, *Special Action Office Monograph*, Ser. A, No. 2, U.S. Government Printing Office, Washington, D.C.
58. Powell, D. H., 1973, *Arch. Gen. Psychiatry* **28**:586–594.
59. Pittell, S. M., 1971, *J. Psychedelic Drugs* **4**:40–45.
60. Wikler, A., and Rasor, R. W., 1953, *Am. J. Med.* **14**:566–570.
61. Verebey, K., and Blum, K., 1979, *J. Psychedelic Drugs* **11**:305–311.
62. Nyswander, M., 1956, *The Drug Addict as a Patient*, Grune and Stratton, New York, pp. 57–82.
63. Salzman, B., and Frosch, W. A., 1972, *Natl. Conf. Methadone Treat. Proc.* **4**:117–118.
64. Weissman, A., 1970, *Int. J. Addict.*, **5**:717–730.
65. Winkelstein, C., 1956, *J. Hillside Hosp.* **5**:78–90.
66. McLellan, T. A., Woody, G. F., and O'Brien, C. P., 1979, *N. Engl. J. Med.* **310**:1310–1314.
67. Levinson, P., 1971, *Perspect. Biol. Med.* **14**:671–674.
68. Litman, R. E., Shaffer, M., and Peck, M. L., 1972, *Natl. Conf. Methadone Treat. Proc.* **4**:482–485.
69. Fredrick, C. J., Resnick, H. L. P., and Wiltin, B. J., 1973, *Arch. Gen. Psychiatry* **28**:579–585.
70. Berken, G. H., Stone, M. M., and Stone, S. K., 1978, *Am. J. Psychiatry* **135**:248–249.
71. Hartman, H., 1976, *Am. J. Psychiatry* **135**:248–249.
72. Snyder, S. H., Taylor, J. J., Coyle, J. L., and Myerhoff, J. L., 1970, *Am. J. Psychiatry* **127**:199–207.
73. Snyder, S. H., Banerjee, S. P., Yamamura, H. J., and Greenberg, D., 1974, *Science* **184**:1243–1253.
74. Kebabian, J. W., Petzold, G. J., and Greengard, P., 1972, *Proc. Natl. Acad. Sci. U.S.A.* **69**:2145–2149.
75. Clement-Cormier, Y. D., Kebabian, J. W., and Petzold, G. J., 1974, *Proc. Natl. Acad. Sci. U.S.A.* **71**:1113–1117.
76. Iwatsubo, K., and Clouet, D. H., 1975, *Biochem. Pharmacol.* **24**:1499–1503.
77. Carenzi, A., Guidotti, A., and Reunelta, A., 1975, *J. Pharmacol. Exp. Ther.* **194**:311–318.
78. Sesame, H. A., and Perez Cruet, J., 1972, *J. Neurochem.* **19**:1953–1957.
79. Lal, H., and Numan, R., 1976, *Life Sci.* **18**:163–168.
80. Bunney, B. S., Walters, J. R., and Roth, R. H., 1973, *J. Pharmacol. Exp. Ther.* **185**:560–571.

81. Iwatsubo, K., and Clouet, D. H., 1977, *J. Pharmacol. Exp. Ther.* **202**:429–436.
82. Ahtee, L., 1973, *J. Pharm. Pharmacol.* **25**:649–651.
83. Clouet, D. H., and Ratner, M., 1970, *Science* **168**:854–856.
84. Gauchy, C., Agid, Y., and Glowinski, J., 1973, *Eur. J. Pharmacol.* **22**:311–319.
85. Gold, M. S., Donabedian, B. K., and Dillard, M., 1977, *Lancet* **2**:398–399.
86. Fink, M., 1953, *Clinical Psychiatry*, Wiley, New York.
87. Martin, W. R., Jasinski, D. R., and Mansky, P. A., 1973, *Arch. Gen. Psychiatry* **28**:784–791.
88. Haertzen, C. A., 1973, *Narcotic Antagonists* (M. C. Brande, L. S. Harris, and E. L. May, eds.), Raven Press, New York.
89. Martin, W. R., and Gorodetzki, C. S., 1965, *J. Pharmacol. Exp. Ther.* **150**:437–442.
90. Verebey, K., Volavka, J., Mule, S. J., and Resnick, R., 1976, *Clin. Pharmacol. Ther.* **20**:315–328.
91. Jones, R., 1978, *Endorphins in Mental Illness* (E. Usdin and W. E. Bunney, Jr., eds.), MacMillan Press, London.
92. Buchsbaum, M. S., Davis, G. C., and Bunney, W. E., Jr., 1977, *Nature* **270**:620–622.
93. Gold, M. S., Dackis, C. A., Pottash, A. L. C., Sternbach, H., Annitto, W. J., Martin, D. M., and Dackis, M., 1982, *Med. Res. Rev.* **2**:211–246.
94. Terenius, L., Wahlstrom, A., and Agren, H., 1977, *Psychopharmacologia* **54**:31–33.
95. Gunne, L. M., Lindstrom, L., and Terenius, L., 1977, *J. Neural Transm.* **40**:13–19.
96. Volavka, J., Mallya, A., and Baing, S., 1977, *Science* **196**:1227–1228.
97. Davis, G. C., Bunney, W. E., Jr., and DeFreites, E. G., 1977, *Science* **197**:74–77.
98. Janowsky, D. S., Segal, D. S., and Bloom, F., 1977, *Am. J. Psychiatry* **134**:926–927.
99. Volavka, J., 1981, *Modern Problems of Pharmacopsychiatry* (H. M. Emrich, ed.), Karger, Basel.
100. Jacquet, Y. F., and Marks, N., 1976, *Science* **194**:632–635.
101. Kline, N. S., Li, C. H., and Lehman, H. E., 1977, *Arch. Gen. Psychiatry* **34**:111–113.
102. Kline, N. S., and Lehman, H. E., 1978, *Endorphins in Mental Illness* (E. Usdin and W. E. Bunney, Jr., eds.), MacMillan Press, London.
103. Berger, P. A., Watson, S. J., Akil, H., Elliott, G. R., Rubin, R. T., Pietterbaum, A., Davis, K. L., Barchas, J. D., and Li, C. H., 1980, *Arch. Gen. Psychiatry* **37**:635–640.
104. Gerner, R. H., Catlin, D. H., Gorelick, D. A., Lui, K. K., and Li, C. H., 1980, *Arch. Gen. Psychiatry* **37**:642–647.
105. Olson, R. D., Kastin, A. J., Olson, G. H., and Coy, D. H., 1980, *Psychoneuroendocrinology* **5**:47–52.
106. Verhoven, W. M., Van Praag, H. M., Botter, P. A., Sunier, A., Van Ree, J. M., and DeWied, D., 1978, *Lancet* **2**:1046–1047.
107. Verhoven, W. M., Van Praag, H. M., Van Ree, J. M., and DeWied, D., 1979, *Arch. Gen. Psychiatry* **36**:294–298.
108. Krebs, E., and Roubicek, J., 1979, *Pharmacopsychiatry* **12**:86–93.
109. Nedapil, N., and Ruther, E., 1979, *Pharmacopsychiatry* **12**:277–280.
110. Simon, E. J., 1982, *Ann. N.Y. Acad. Sci.* **398**:327–338.
111. Verebey, K., Volavka, J., and Clouet, D., 1978, *Arch. Gen. Psychiatry* **35**:877–886.
112. Inturrisi, C. E., Alexopoulos, G., Lipman, R., Foley, K., and Ressier, J., 1982, *Ann. N.Y. Acad. Sci.* **398**:413–422.
113. Extein, I., Pottash, A. L. C., and Gold, M. S., 1982, *Ann. N.Y. Acad. Sci.* **398**:113–119.
114. Terenius, L., and Wahlstrom, A., 1974, *Acta Pharmacol. Toxicol.* **33**(Suppl. 1):55.
115. O'Brien, C. P., Terenius, L., Wahlstrom, A., McLellan, T., and Krivoy, W., 1982, *Ann. N.Y. Acad. Sci.* **398**:377–383.
116. Pickar, D., Cutler, N. R., Naber, D., Post, R. M., and Bunney, W. E., Jr., 1980, *Lancet* **1**:937.
117. Reisberg, B., Ferris, S. H., Anand, R., Mir, P., Geibel, V., and DeLeon, M., 1983, *N. Engl. J. Med.* **308**:721–722.
118. Abse, D. W., Rheuban, W. J., and Akhtar, S., 1982, *Ann. N.Y. Acad. Sci.* **398**:79–82.
119. Agren, H., Terenius, L., and Wahlstrom, A., 1982, *Ann. N.Y. Acad. Sci.* **398**:388–398.
120. Gold, M. S., Pottash, A. L. C., Sweeney, D. R., Martin, D., and Extein, I., 1982, *Ann. N.Y. Acad. Sci.* **398**:140–150.

121. Pickar, D., Naber, D., Post, R. M., Van Kammen, D. P., Kaye, W., Rubinow, D. R., Ballenger, J. C., and Bunney, W. E., Jr., 1982, *Ann. N.Y. Acad. Sci.* **398**:399–411.
122. Breizer, D., and Millman, R., Clinical Effects of Methadone HCl on Neuroleptic-Refractory Chromic Schizophrenic Patients: A Pilot Study, Payne Whitney Psychiatric Clinic.
123. Abraham, H. D., 1983, *Am. J. Psychiatry* **140**:456–458.
124. Catlin, D. H., Gorelick, D. A., and Gerner, R. H., 1982, *Ann. N.Y. Acad. Sci.* **398**:434–446.
125. Foley, K. M., Kourides, I. A., Inturrisi, C. E., Kaiks, R. F., Zaorulis, C. G., Posner, J. B., Houde, R. W., and Li, C. H., 1979, *Proc. Natl. Acad. Sci. USA* **76**:5377–5381.
126. Berger, P. A., and Barchas, J. D., 1982, *Ann. N.Y. Acad. Sci.* **398**:448–458.
127. Petho, B., Grof, L., Karczag, I., Borvendez, J., Bitter, I., Barna, I., Harmann, I., Tolna, J., and Baraczka, K., 1982, *Ann. N.Y. Acad. Sci.* **398**:460–469.
128. Kline, N. S., and Lehman, H. E., 1979, *Endorphins in Mental Health Research* (E. Usdin, W. E. Bunney, Jr., and N.S. Kline, eds.), MacMillan Press, London, pp. 500–517.
129. Oyama, T., Finn, T., Yamaya, R., Ling, N., and Guillemin, R., 1980, *Lancet* **1**:122.
130. Emrich, H. M., Zandig, M., Kissling, W., Dirlich, G., von Zerssen, D., and Herz, A., 1982, *Ann. N.Y. Acad. Sci.* **398**:108–112.
131. Wickler, A., Pescor, M. J., and Kalbaugh, E. P., 1952, *Arch. Neurol. Psychiatry* **67**:510–521.
132. Jaffe, J. H., 1968, *Mod. Treat.* **5**:1121–1135.
133. Beecher, H. K., 1959, *Quantitative Effects of Drugs*, Oxford University Press, London.
134. Rothenberg, S., Schottenfeld, S., and Meyer, R. E., 1977, *Psychopharmacologia* **52**:299–306.
135. Gordon, N. B., and Appel, P. W., 1972, *Natl. Conf. Methadone Treat. Proc.* **4**:425–427.
136. Gasser, E. S., Langrod, J., and Valdes, K., 1974, *Br. J. Addict.* **69**:85–88.
137. Grevert, P., Masover, B., and Goldstein, A., 1977, *Arch. Gen. Psychiatry* **34**:849–853.
138. Gearing, F. R., and Schweitzer, M. D., 1974, *Am. J. Epidemiol.* **100**:101–112.
139. Cutting, W. C., 1942, *Stanford Med. Bull.* **1**:39–41.
140. Cicero, T. J., Bell, R. D., and Wiest, W. G., 1975, *N. Engl. J. Med.* **292**:882–887.
141. Kay, D. C., 1975, *Electroencephalogr. Clin. Neurophysiol.* **38**:35–39.
142. Yaffe, J., Strelinger, R. W., and Parwatikar, S., 1973, *Natl. Conf. Methadone Treat. Proc.* **5**:507–522.
143. Kreek, M. J., 1973, *JAMA* **223**:665–668.
144. Leen, N. J., Senay, E. C., and Renault, P. F., 1976, *Drug Alcohol Depend.* **1**:305–311.
145. Kreek, M. J., and Hartman, N., 1982, *Ann. N.Y. Acad. Sci.* **398**:151–172.
146. Gold, M. S., Pottash, A. L. C., Sweeney, D. R., and Kleber, H. D., 1980, *JAMA* **243**:343–346.
147. Mehta, D., Mallya, A., and Volavka, J., 1978, *Am. J. Psychiatry* **135**:371–372.
148. Leestma, J. E., and Koenig, K. L., 1968, *Arch. Gen. Psychiatry* **18**:137–148.
149. Davis, J. M., and Casper, R., 1977, *Drugs* **14**:260–282.
150. Holden, C., 1977, *Science* **198**:807–809.

23

Biochemistry of Senile Dementia

Steven T. DeKosky and Norman H. Bass

1. INTRODUCTION

This review of senile dementia shall be confined to the major pathological entity, Alzheimer's disease, a devastating and increasingly recognized illness that has become the focus of intensifying research over the decade since the first edition of this handbook. Demographic statistics showing the increasing life span of the American population, coupled with an enhanced age-related risk of affliction with Alzheimer's disease, have heightened the urgency for research into what is and promises to become an increasingly common and debilitating health problem of the elderly.[1] Dementia is defined as a progressive loss of memory and cognitive function over an extended period of time, i.e., months to years. Although there are several diseases that result in the insidious onset of cognitive loss during the middle years, Alzheimer's disease represents the largest percentage of presently incurable senile dementias. However, major progress in identifying biochemical abnormalities of the brain in Alzheimer's disease has been made during the past decade, distinguishing these patients from those with other disorders of cognitive function and from the changes occurring in normal, age-matched elderly controls. Specific biochemical and structural alterations associated with "normal" or "accelerated" aging need to be further differentiated from the progressive biochemical changes of Alzheimer's disease.

Although *senile dementia* was the term applied to progressive cognitive decline with onset after age 65, there appear to be no reliable differences between senile and presenile forms of the disease, as initially described by Alzheimer in 1907.[2] Therefore, neurological and neuropathological studies now refer to Alzheimer's disease in the elderly as *senile dementia of the Alzheimer type*, suggesting a unitary pathological and biochemical change in the brains of patients with progressive cognitive and memory loss whether onset of symptoms occurs prior to or following age 65. However, there is some suggestion

---

Steven T. DeKosky and Norman H. Bass • Department of Neurology, Lexington Veterans Administration Medical Center and University of Kentucky College of Medicine, and Sanders Brown Research Center on Aging, Lexington, Kentucky 40536.

that the clinical presentation and biochemical findings in patients with Alzheimer's disease with onset in their 60s quantitatively differ from those in patients with onset in their 80s.[3–11] For purposes of this review, however, we will use the terms *Alzheimer's disease* and *senile dementia* interchangeably.

The diagnosis of senile dementia is one of exclusion, as there is no non-invasive diagnostic test for Alzheimer's disease. While cortical biopsy is highly likely to reveal the presence or absence of the disease process, the absence of therapy has mitigated against biopsy of the cerebral cortex in patients with cognitive decline. Hence, diagnosis of Alzheimer's disease during life is based upon a careful neurological examination, neuropsychological testing, blood studies, and radiological evaluation to exclude chronic infection, tumor, vascular disease, vitamin or other nutritional deficiencies, or alterations in intracranial pressure such as hydrocephalus. If all of such studies are normal, and the neurological and neuropsychological findings are compatible with the progressive profile of cognitive decline found in Alzheimer's disease, the patient is regarded as having senile dementia. Thus, studies of biochemical alterations in Alzheimer patients done prior to the establishment of such a battery of diagnostic tests or without brain biopsy or autopsy verification of Alzheimer's disease are associated with a small but real possibility that some of the cases may not represent the disease under investigation. Similarly, control groups of age-matched patients who have not been carefully assessed for memory function, may inadvertently include patients with early or mild cognitive decline, thereby increasing the variance of the control group and possibly obscuring differences between changes of Alzheimer's disease and those of normal aging. This problem is compounded by the fact that the progressive changes associated with Alzheimer's disease are qualitatively similar to those that occur with normal aging. Hence, a "threshold effect" for dementia may be postulated whereby clinical symptoms of memory loss and thought process disturbances occur when the amount of degenerative brain change exceeds the threshold and impairs the functional capabilities of the brain.[12]

Numerous excellent reviews of the clinical, biochemical, and neuropsychological aspects of senile dementia have been published and the reader is referred to Terry and Katzman,[13] Corsellis,[14] Schneck *et al.*,[15] McGeer,[16] Marchbanks,[17] Terry and Davies,[18] Ball,[19] Coyle *et al.*,[20] Beck *et al.*,[21] and Crook and Gershon[22] for additional information.

The quantitative differentiation of morphological changes found in senile dementia from those seen in normal aging constituted a major neuropathological achievement and stimulated investigators to examine biochemical changes that might be associated with this important disease entity. Since the mid-1970s the major focus has been on determining biochemical changes, specifically alterations in chemical markers of neurotransmission. The finding of a significant and consistent decrement in cholinergic neurotransmission, with its potential for therapeutic intervention, is at the forefront of these discoveries. Alterations in other neurotransmitter systems are also under intensive study. Second, the issue of neuronal cell loss as measured by neurochemical and histological means has extended beyond generalized loss in the cerebral cortex to specific subcortical nuclei such as the nucleus basalis of Meynert, nucleus of the diagonal band, and medial septal nuclei that supply cholinergic innervation to the hippocampus and cortex. Other source nuclei such as the locus coeruleus, which provides noradrenergic innervation to anterior regions of the brain, are under evaluation.

2. *BRAIN CHANGES ASSOCIATED WITH NORMAL AGING*

2.1. *Gross Structural Changes*

The brain decreases in weight and size during senescence, as indicated in numerous pathological studies. Reviews of such pathological changes are found in Tomlinson,[48] Corsellis,[14] and Katzman and Terry.[13a] These normative changes have been confirmed in recent years by the *in vivo* measurement of ventricular volume and brain size by radiographic imaging with computerized axial tomography.[23–28] Caution should be exercised in the interpretation of these and other gross anatomical studies of aging whole brain, as it has been suggesed that mean human brain weights may have increased over the past half-century.[29] Thus, some portion of the observed weight difference between young and old brains results from a mixture of aging brain weights with an increasing weight of the standard normal adult brain. Additionally, the ratio of gray to white matter in whole brain shows significant changes with aging. Interpretation of this ratio is complicated by the fact that early tissue loss occurs predominantly in gray matter, while later decrements in tissue volume appear to occur predominantly in white matter accompanied in some instances by ventricular dilation.[29]

2.2. *Loss of Neural Cells in Aging*

Early workers assumed that diffuse loss of neurons and/or other brain cells was responsible for the atrophic changes seen in normal aging. Recent studies have confirmed that neural cell loss indeed occurs in the aging brain, but that this process shows great variability in various brain regions. Thus, Brody,[30] Anderson *et al.*,[31] and Henderson *et al.*[32] all noted significant loss of neurons in the cortex, but while some areas lost as many as 50% of their cells, other areas did not seem to be affected.[32] Neuronal loss has also been reported in the hippocampus[32,33] and in the basal ganglia.[34] Other subcortical regions do not appear to be subject to cell loss during normal sensecence.[35] It should be emphasized that this age-related loss of neural cells in the cerebral cortex is unaccompanied by reactive neuropathological changes usually found in association with cell death.[13a]

2.3. *Neuropathological Alterations in Aging*

The light microscopic findings of senile plaques, neurofibrillary tangles, and granulovacuolar degeneration of the hippocampal pyramidal cells form the basis for the neuropathological definition of Alzheimer's senile dementia. However, Tomlinson *et al.*[36] clearly showed that these changes can all be found in the brains of normal elderly people, but they are significantly more abundant in age-matched patients with Alzheimer's disease.[37] Further discussion of these tissue changes associated with senescence and senile dementia will be found in the section on pathological findings in Alzheimer's disease.

2.4. Lipid Changes in Normal Aging

2.4.1. Lipofuscin in the Aging Brain

Lipofuscin is an autofluorescent lipopigment that accumulates progressively with normal aging.[161] The lipofuscin granules are thought to reflect an accumulation of peroxidated lipids sequestered in vacuoles and held in an inert form within the cell body. Early speculation suggested that intracellular lipofuscin accumulation might be directly or indirectly related to cell loss in the aging brain. However, there are regions (e.g., inferior cochlear nucleus) where lipofuscin is plentiful but no cell dropout is seen in aging.[38] Nevertheless, the biochemical accumulation of autofluorescent lipopigments in neural cells deserves additional study to further explore observations such as those suggesting that there are inverse relationships between the amount of lipofuscin in the cell body and the amount of RNA noted by histochemical staining.[39]

2.4.2. Membrane Lipid Alterations in Normal Aging

Gangliosides are found in relatively great abundance in neuronal membranes and have been shown to decrease in aging cortex.[40,41] Additionally, galactocerebroside and other lipids that enrich myelin membranes also decline during senescence[40,42] as does myelin basic protein.[43]

2.5. Neurotransmitter Alterations in Aging

Several reviews of changes in neurotransmitter metabolism in the brain during normal aging have been published.[13,16,44–47] It must be emphasized that agonal and postmortem changes are major factors that confound interpretation of data reporting changes in neurotransmitter-related enzymes, receptors, and both established and putative neurotransmitters.[14,215] Interpretation of neurotransmitter changes in senile dementia must therefore be made with reference to changes in brain tissue during normal aging. Several different profiles of change in neurotransmitter system constituents have been reported to occur during the period of young adult life through senescence. For example, McGeer and McGeer noted the progressive rapid decline over the first two or three decades of life of several neurochemical components followed by stabilization or very slow decline thereafter.[46] Alternatively, other components may remain relatively stable until age 60 and then begin a rapid decline. These varied lifespan changes in neurochemical components, especially those related to neurotransmitters, indicate the absolute necessity for using age-matched control tissues, if meaningful study of neurochemical changes in dementia is to be performed. They further emphasize the need for better understanding of the dynamic biochemistry of normal aging upon which the pathological changes of senile dementia are quantitatively superimposed. This aspect of research design in senile dementia is particularly emphasized as brain tissue specimens with Alzheimer's disease are usually obtained 2–10 years after onset of dementia and with variable postmortem time intervals prior to performance of biochemical analysis.

3. STRUCTURAL AND CHEMICAL PATHOLOGY OF SENILE DEMENTIA

3.1. Brain Weight and Ventricular Size

There is no clear gross anatomical differentiation between brains of demented and normal patients. Whereas patients with severe dementia who have long-standing disease may have marked cortical atrophy and ventricular dilation, variability among both "normal" elderly and dementia patients obscures a consistent difference. In one small well-defined series, a statistically significant difference in weights was found,[49] but with much overlap between the two groups. There are reports of correlation of decreased hemisphere size with dementia,[50] and significantly lower whole temporal lobe weights in senile dements.[4] Some cases with severe cognitive loss have minimal cortical atrophy, and while size of the third ventricle has been reported as a discriminative index of dementia,[24] other reports indicate no differences between normal and demented brain ventricular size on computerized axial tomograms.[3] Thus, brain weight and ventricular size do not accurately select cases or consistently indicate the severity of pathological change in senile dementia.

In so-called classical cases of Alzheimer's disease with cortical gyral atrophy and ventricular enlargement, Terry and Katzman[13] have pointed out that the bulk of the weight loss is in the subcortical white matter. However, even in such cases and with consideration of difficulties of measurement, it is notable that no significant loss of the gray matter thickness of the cerebral cortex has been demonstrated. This is in contrast to significant diminution of white matter volume as measured in the subcortical centrum semiovale. These white matter changes are presumably secondary to axonal degeneration with secondary loss of myelin.[13] The pattern of atrophy appears selective, sparing the basal ganglia and cerebellum and involving mainly the frontal and temporal cortices and hippocampus; the pericentral gyri and occipital cortices are less often affected.[13]

3.2. Histological Changes: Neuronal Cell Counts

3.2.1. Cortical Cell Counts

Quantitative loss of cortical neurons in specific brain regions has been shown to occur in Alzheimer's disease.[49,51] Schechter *et al.* reported a 25%

loss of small neurons in the frontal cortex and a 45% decline of larger neurons in both frontal and superior temporal cortices. Glial cell number appears to be maintained in most cortical areas of brains with Alzheimer's disease, although the number of fibrous astrocytes may specifically increase in the frontal lobe.[52] In another study, loss of cortical neurons was observed in frontal, temporal, and cingulate regions of demented patients as compared with age-matched controls, but such differences were not found in the parietal and occipital regions.[51] The frontal areas of the cerebral cortex did not show increased numbers of glia in association with obvious neuronal loss, while in the temporal lobe the number of glial cells increased. Although such quantitative studies of cell counts within brain tissue are both tedious and difficult to perform, it would appear that changes in the vulnerability of neurons and glia predominantly occur in the frontal and temporal regions of the cerebral cortex of patients with senile dementia as compared with elderly patients with neurological disease. A similar loss of neural cells in the cerebral cortex has been reported by Shefer[53] and Colon,[54] the latter study in presenile cases of Alzheimer's disease.

3.2.2. Cell Counts of Other Brain Regions

Pathological changes in cell composition of the hippocampus are of major importance in the study of senile dementia because of the role this brain structure plays in memory function, which is to a large degree lost in senile dementia. Ball has reported a loss of hippocampal neurons in senile dementia and correlated the cell loss with the density of neurofibrillary tangles found in that region.[33]

Large acetylcholinesterase (AChE)-positive neurons of the nucleus basalis of Meynert (NBM) in the basal forebrain have been shown to be markedly decreased in senile dementia of the Alzheimer type.[55–57d] This dramatic loss of cells in a subcortical region of the brain has been shown to relate directly to the diminution of the neurotransmitter synthetic enzyme, choline acetyltransferase (CAT), from the frontal cortex, since the cortex is supplied with its cholinergic input by a projection from the NBM. Bondareff *et al.*[10] reported losses in the nucleus of origin of forebrain noradrenergic innervation, the locus coeruleus of the midbrain, in cases of Alzheimer's disease. These cellular alterations will be discussed in the sections on specific neurotransmitter alterations. Neurotransmitter changes have been observed in the hypothalamus in cases of dementia,[58] and computerized axial tomography studies indicate increased width of the third ventricle[24] suggesting hypothalamic changes. To date, no quantitative studies of cellular composition in the hypothalamus in dementia have been reported and this may be a productive area for future investigation.

3.3. Microscopic Neuropathological Alterations in Dementia

Historical accounts of the discovery of the neuropathological hallmarks of senile dementia of the Alzheimer type are found in Bowen[59] and Torack.[60] Alzheimer's original case (a presenile dementia) had cortical atrophy, neurofibrillary tangles, and senile plaques in the cerebral cortex. Because these abnormalities were also found in normal aging brain, it was not until Corsellis[61] demonstrated excessive numbers of tangles and plaques in brains of patients with progressive senescent cognitive decline that more detailed work on these abnormal structures was initiated. Subsequently, a series of studies showed that the rate of acquisition of neurofibrillary tangles and senile plaques during normal aging was relatively slow and regionally limited. In contrast, in patients with senile dementia, plaques and tangles were more widely distributed and showed increased numbers per unit volume in the cerebral cortex.[36,37] This quantitative distinction between normative aging change and senile dementia focused attention on senile plaques and neurofibrillary tangles as important clues to understanding the pathophysiological mechanisms associated with Alzheimer's disease. These studies also deemphasized the relationship between dementia and atherosclerotic change, a process that previously had been considered to be a major cause of dementia but was clearly shown to be a significant correlate of dementia in only a small percentage of cases.[62]

3.3.1. Senile (Neuritic) Plaques

Neuritic plaques are silver-staining clumps of altered tissue that accumulate in high density in the cerebral cortex, hippocampus, and amygdala in patients with senile dementia[37,63–65]; they also accumulate to a lesser degree and at a slower rate in the cerebral cortex and limbic system during normal aging.[36] Electron microscopic examination of neuritic plaques has shown them to be composed of distended and morphologically altered neurites, i.e., neuronal processes thought to primarily represent presynaptic terminals.[66–68] A fully developed or "mature" neuritic plaque has swollen distended mitochondria, distorted presynaptic terminals, microglia, and a central core of amyloid. Some plaques do not contain amyloid, and based on light and electron microscopic criteria have been subdivided into three groups: primitive, mature, and burnt-out.[69] In addition, the degree of AChE staining differs in the three subgroups. This observation has been used to support a hypothesis relating cholinergic deafferentation to neuritic plaque formation.[70]

Chemical Pathology of the Neuritic Plaque. Density gradient enrichment of plaques in bulk indicated elevated silicon content.[71] Elevation of oxidative enzymes in the region of the plaque has been observed[72–73] as has nucleotide phosphatase.[74] These elevations are thought to be secondary to extension of processes of satellite glial cells surrounding the plaque.[59] Intense AChE staining is seen in young and mature plaques but is not seen as the plaques become older or burnt-out.[70]

The chemical makeup of the amyloid fibrils in the plaque or cerebral vessels has not been clearly defined.[13,60,75,91] It is not certain whether the amyloid is derived from central sources or peripheral (circulating) sources.[76] It was suggested that the amyloid was derived from local cells in or near the plaque.[67] Recent histochemical studies suggest the material is APUD-amyloid, which could be made by local cells.[77] A peripheral source for the amyloid was suggested by its histochemical identification as IgG,[78] and suggestions that it contains prealbumin.[79] Penetration of the blood–brain barrier by amyloid or amyloid precursors has also been proposed.[80]

Most recently, Prusiner and his colleagues have purified the protein associated with the scrapie agent, termed the prion. The prion-associated protein when purified associates in polymeric form and resembles amyloid.[80a] They have speculated that amyloid in neuritic plaques in Alzheimer's disease might be the prion protein.[80b]

3.3.2. Neurofibrillary Tangles

Alzheimer's neurofibrillary degeneration is the term given to the silver-staining structures in the cell bodies of neurons in the hippocampus and cortex.[14,48,60] Restricted in normal aged brains to the hippocampus,[36,62] they are present diffusely in the temporal and frontal regions in senile demented brains.[14,37,48] They are also present in much greater numbers in hippocampal neurons of demented patients compared to controls.[33] Electron microscopic studies indicate that the argyrophilic structures are helically wound paired filaments (PHF),[81–83] 100 Å wide with a periodicity of 800 Å.[13] Further identification of the chemical composition of the PHF has been hampered by their marked resistance to the usual methods of protein analysis (see next section).

Neurofibrillary tangles and neuritic plaques are also associated with the process of brain aging seen in Down's syndrome. These structures are present in all brains of patients dying of Down's syndrome after age 40.[84,85] Additionally, neurofibrillary tangles in the absence of neuritic plaques are found in a wide diversity of neurological conditions including dementia pugilistica, postencephalitic Parkinson's disease, progressive supranuclear palsy, and the Parkinsonian-dementia and amyotrophic lateral sclerosis syndromes of Guam.[14,86] Despite the presence of neurofibrillary tangles in diseases other than senile dementia, the quantitative relationship of tangles and the diagnosis of Alzheimer's disease[37] and the negative correlation of tangle density and cell loss in the hippocampus[33] and cortex[51] have led to extensive studies of the protein constituents of tangles and an extensive study of their relationships to neurotubules, neurofilaments, and other structural neuronal proteins.[60,75,87,88]

Chemical Pathology of the Neurofibrillary Tangle. Summaries of neurofibrillary tangle chemistry are found in Selkoe,[75] and Marotta.[89]

The chemical makeup of the PHF, which are found in large numbers in the perikaryon, and in the periphery of the neuritic plaque, is not known. Early studies reported a 50,000-dalton protein from fractions of cortical tissue, which was thought to be a purified component of the PHF.[90] Several other polypeptides have been described, and antisera raised to several of these proteins.[92] A major recent discovery indicates that the PHF fractions are so insoluble in conventional protein analytic reagents that they do not descend on SDS gels, and are remarkably resistant to agents that solubilize most proteins. Because the intact large polymers of PHF were found at the tops of the gels,[93] it was concluded that the insoluble polymerized PHF contained covalent bonds, leading to a rigid, highly stable structure. Selkoe speculated that such cross-links may be produced by transglutaminase, forming glutamyl-lysine covalent bonds between human neurofilaments, but emphasized that other bonding mechanisms must be considered.[75] Igbal *et al.*[93a] have recently found that SDS treatment does fractionate PHF, but that rapid reaggregation of the proteins may have led to the failure of all the protein to descend on the gel. Further, antisera raised to purified fractions of these PHF do not stain any normal cytoskeletal proteins, suggesting that PHF contain antigenic determinants not found on normal fibrous proteins of neurons.[94a,b] However, antibodies raised against normal neurofilament fractions can stain PHF.[94c]

Such massive insoluble structures within the neuron could well disrupt intraneuronal axonal transport and perikaryal metabolism, and might certainly compromise cytoskeletal integrity. Study of the as yet difficult to assay PHF, its biochemical and antigenic relationship to normal neuronal cytoskeletal proteins, and any relationship of PHF to the cholinergic deficit remain a major focus of research in senile dementia.[75,88]

3.3.3. Granulovacuolar Degeneration

These vacuoles (4–5 μm in diameter) are found in the cytoplasm of the pyramidal cells of the hippocampus, and have a central, argyrophilic core. Under the electron microscope they appear as membrane-bound inclusions, the central core of which is electron dense.[14,95] They are present in the hippocampus of young adults and their numbers significantly increase with aging.[96–98] The increase is quantitatively greater in the brains of patients with senile dementia.[96–99] Methods to obtain enriched fractions of granulovacuolar bodies have not been described and such work would certainly facilitate the neurochemical characterization of these structures.

3.3.4. Hirano Bodies

Hirano bodies are small, rod-shaped, strongly eosinophilic structures found in the hippocampus of normals and, it is said, in greater numbers in brains of senile dementia patients. Initially described by Hirano and Zimmerman in the Guamanian dementia and ALS syndromes,[100] recent evidence suggests that they may be actin aggregates[101] or derived from ribosomes.[102] Tomlinson regards them as nonspecific, since they are seen in many pathological conditions, and no specific role in senile dementia has been proposed for them.[48]

3.4. Histochemical Changes in Neural Cells and Their Processes

The limited amount of tissue available from biopsy samples and the infrequent use of cortical biopsy to confirm the diagnosis in suspected cases of Alzheimer's senile dementia have resulted in the majority of histochemical and neurochemical observations on tissue specimens deriving from autopsy examination. The impetus for neurochemical studies usually derives from histological observations, and a quantitative analysis of such microscopic changes is helpful, if not critical, for assessing primary versus secondary neurochemical changes.

3.4.1. Neuronal Membrane Changes

The Golgi stain delineates the perikaryon and processes of only 5% of all neurons in tissue. Nevertheless, regression of the dendritic arborization in aging and in senile dementia has been described by this method.[104]

Detailed Golgi studies of lamina II of the human parahippocampal gyrus have confirmed earlier qualitative impressions of neuronal membrane loss in senile dementia compared to normal aging.[107,108] Interestingly, Buell and Coleman[107,108] also found that in normal aged controls, neurons in this brain region continued to expand their dendritic lengths with increasing age. While that study measured dendritic length and branching patterns, Mehraein *et al.*[109] quantitated the loss of dendritic spines (outgrowths on the dendritic shaft on which the majority of synapses occur) and found greater loss of these structures in senile dementia than in normal age-matched brains. Thus, in addition to the description of aberrant neurites in senile plaques, evidence for decreased or altered neuronal connectivity emerged. The regional variations in this loss of connectivity have yet to be specified. The decline of cortical connectivity may certainly account for the loss of characteristic higher "cortical" functions in patients with senile dementia.

Recent systematic investigation of the reliability and consistency of the Golgi method as applied to human tissue that has been immersed in fixative for a long time (months to years) suggests that such histochemical techniques may be subject to artifact.[105,106] However, the decline of neurochemical neuronal membrane markers suggests that loss of connectivity is associated with the effects of the disease process.[41]

3.4.2. Histochemical Alterations in Glia

While glia have not been accorded a primary role in pathogenesis of senile dementia, secondary alteration of this cell population often occurs during most neuropathological processes. While early studies suggested that an increased proliferation of astrocytes occurs in Alzheimer's disease, quantitative cell counts have not confirmed this. However, immunocytochemical staining of astrocytes with antibody to glial fibrillary acidic protein indicates that an increase in "reactive astrocytes" may be found in neuritic plaques.[110] Moreover, fibrous astrocytes appear to be much more numerous in layers two through six of the cerebral cortex in Alzheimer's disease than in age-matched controls, despite the absence of any significant change in the total glial population. Accordingly, it has been suggested that this change may be a reactive pathological process involving transformation of protoplasmic to fibrous astrocytes.[52]

3.4.3. Lipofuscin

The rate of accumulation of lipofuscin pigments is not increased in the neurons of Alzheimer brains. Thus, the substance is considered to have little relevance for the neurochemist seeking to better define the chemical pathology of senile dementia.[111] Further information on lipofuscin may be found in Horrocks *et al.*[161] and in Nandy.[112]

3.5. Nucleic Acids

3.5.1. Changes in DNA

Alterations of DNA content have been measured by Bowen *et al.*[113,114] in homogenates of whole temporal lobe from Alzheimer brains and age-matched controls. There was no significant difference in total DNA per lobe between the two groups,[114] although other neurochemical markers of neurons or glia in those homogenates suggested that alterations of cell membranes had occurred that could be considered predominantly neuronal. The authors concluded that DNA assay of temporal lobe homogenates was not a good quantitative index for the total cell population, especially if the tissue had been subjected to hypoxia prior to death. It has been shown that cortical neuronal populations in different gyri of the temporal lobe respond differently to aging, some remaining stable and others declining,[32] and selected neuronal populations are lost in Alzheimer's disease.[49,51] Hence, the variability and the lack of consistent change in DNA in homogenates from postmortem specimens of whole temporal lobe is not totally unexpected (see also 18).

Microassays of DNA in the six cortical lamina of the frontal cortex (Brodmann's area 9) have also been performed and do not show a difference between senile dementia and similarly aged controls.[41,115] A loss of total DNA was found when these groups were compared to younger controls.[41] When the DNA assays were combined with differential cell counts of histologically processed cortical tissue taken adjacent to the cortical core samples used for chemistry, there was no statistically significant decline in the neuronal population. Small sample size and difficulty differentiating small neurons from glia may have contributed to the lack of detection of the neuronal loss that has been reported by extensive cell counting studies. Again, it must be emphasized that assay of nucleic acids in tissue specimens of brain subject to a prolonged hypoxic insult prior to death is potentially unreliable. This factor in addition to the heterogeneity of the subpopulations of neurons assayed after homogenization of such extensive brain areas as the whole frontal or temporal lobe demands caution in the interpretation of either positive or negative results.

3.5.2. Alterations in Chromatin

A series of studies by Crapper-McLachlan and colleagues has focused on potential alterations in DNA transcription. They found altered distribution of fractions of chromatin in both neuron-enriched and glia-enriched fractions of brains from senile dementia patients, compared to control brains similarly analyzed.[116] A reduction in euchromatin was accompanied by a reciprocal increase in the percentage of DNA content that fractionated as heterochromatin. An increase in the latter was regarded as indicating greater areas of the DNA molecule that, via conformational alteration, do not permit attachment of RNA polymerases. Such a restriction of transcription could thus result in a protein synthesis decrement and the metabolic alterations of Alzheimer's disease. While labeled nucleotides introduced into euchromatin-enriched fractions from normal and Alzheimer brains showed no differences in total uptake or chain length of mRNA produced, subsequent studies employing micrococcal nuclease indicated that senile dementia brain DNA was less susceptible to digestion than controls, suggesting at least an altered state of DNA.[117] Finally, the significant increase in a protein resembling histone H1^{O} was found in a digest of Alzheimer chromatin.[117]

A potential mechanism for such chromatin alterations in Alzheimer's disease has been proposed, involving excessive accumulation of aluminum within nuclear and perinuclear structures and subsequent abnormal cross-linking with DNA. This will be further discussed in Section 3.9 (Trace Metals).[118,119]

3.5.3. RNA

Biochemical assays of RNA have indicated either stability or loss of this nucleic acid in the cortex in Alzheimer's disease. In a single case, Pope *et al.* did not find differences from their normal controls in a laminar assay of the frontal cortex.[115] With similar techniques, total RNA was shown to decrease in a group of normal aging frontal cortices and in senile dementia cortex but the two aged groups did not differ from each other.[41] The RNA/DNA ratio was not changed, indicating that either cell loss did not involve predominately large cells, or that a hyperplastic response was occurring in other cell types, thus allowing the RNA/DNA ratio to remain stable. The hypertrophic astrocytic response seen in Alzheimer frontal cortex might account for such lack of change in the RNA/DNA ratio.[52]

In their series of neurochemical studies of the temporal lobe, Bowen *et al.*[113,114] showed that prolonged hypoxia prior to death increased RNA in homogenates of normal brain tissue. Since patients with senile dementia frequently die of pneumonia with attendant terminal hypoxia, RNA in the samples of dementia tissue might be artifactually elevated. Thus some senile dementia samples may have been subjected to an anoxic terminal event, artifactually elevating their RNA levels to the level of nonhypoxic normal aged-matched patients. Since the degree and duration of terminal hypoxia are often difficult to quantitate, the authors suggested that biochemical measurement of RNA and the RNA/DNA ratio were unreliable neurochemical markers if there was a question of terminal anoxic damage.[114] This careful examination and presentation of results using two different control groups explains why this study is occasionally cited as showing either a decrease or no decrease in RNA in senile dementia brains. Terminally hypoxic normal brains and Alzheimer brains had higher RNA levels than age-matched nonhypoxic controls.

Measurement of RNA in single dissected neurons in Brodmann's area 9 indicated an age-related decrease in single neuron RNA content but no differences from normals when neurons of three demented patients were examined.[120] Mann *et al.*[121] utilized histological stains to measure pyramidal cells of cortical layers III and V of the temporal cortex and reported losses of 50% in cytoplasmic RNA content. They also reported decreases of up to 40% in nucleolar volume and nuclear volume, suggesting that the cytoplasmic loss of RNA might constitute a secondary rather than a primary pathological event.

Most recently, decline in total cellular RNA and polyadenylated RNA in AD cortex has been reported, along with increased alkaline ribonuclease activity, indicating a post-transcriptional abnormality in AD brain (See section 3.8.1).[261]

3.6. Lipid Membrane Alterations

Early work on lipid alterations in senile dementia has been extensively reviewed by Embree *et al.*[122] Several studies have placed special emphasis on decrements on ganglioside, a sialic acid-containing glycolipid, shown to be enriched in neuronal plasma membranes and to have markedly diminished content in tissue specimens from Alzheimer brains.[4,41,42,114,123] Both age-related losses of ganglioside and the greater loss of ganglioside in senile dementia appear to affect the lower laminae of the cortex more than the upper laminae.[41] This may relate to loss of large neurons as reported by Terry *et al.*[49] and Mountjoy *et al.*[51] with attendant loss of the large dendritic arbors of such neurons. It may also reflect inability of metabolically impaired neurons to sustain their normal membrane expanse or axodendritic arborization to the extent possible under normal conditions (see Section 3.4).

Bowen *et al.*[114] have also shown a decrease in mitochondrial light fraction protein (a substance enriched in brain myelin subfractions). The myelin-specific enzyme 2′,3′-cyclic nucleotide 3-phosphohydrolase, and lipid galactocerebroside, a marker for myelin and microsomes, have been reported decreased in senile dementia.[41,42,124] This alteration of protein, enzyme, and lipid presumably reflects a loss of myelinated axons associated with progressively decreased axonal branching, a phenomenon well documented to occur during the course of the disease process.

Other lipid membrane components, such as phospholipids and cholesterol, have not shown major changes in reported composition in Alzheimer's disease (but see 125), and a primary role for lipids in the neurochemical pathology of senile dementia has not been proposed. Further clues concerning the selectivity of the degenerative process may come from studies that carefully examine well-defined anatomical regions of the nervous system in association with quantitative neuropathological observations. However, at present, abnormalities of lipid metabolism are not emphasized in current theories of causation of senile dementia.

3.7. Blood Flow and Metabolism

Cerebral blood flow is reduced in patients with senile dementia.[126–130] Some studies have correlated the decreased flow with the degree of cortical atrophy and ventricular enlargement,[129] while others have not found such a relationship.[131] The normal increments in regional blood flow during problem-solving mental exercises do not occur in the demented patient.[132] Since these are noninvasive studies, the diagnosis in most of these patients is by clinical examination and exclusion of treatable causes of dementia; thus, neuropathological confirmation in most of these cases is not available. Differences are apparent in both the history and regional cerebral blood flow between multi-infarct dementia (MID) and senile dementia of the Alzheimer type.[130,133]

Recent studies utilizing positron-emission tomography (PET scans) have enabled determination of regional glucose utilization by employing ^{18}F-2-deoxy-2-fluoro-D-glucose label. Alzheimer patients had a 35–45% lower glucose metabolic rate in several brain regions; the decrement was reported to correlate with the degree of cognitive impairment.[134] It is of interest that areas not traditionally associated with the pathological (or behavioral) abnormalities of Alzheimer's disease, e.g., the caudate, were also markedly affected. Other PET studies have not been consistent,[135] suggesting that the limited resolution of the current PET scanners may not allow definite conclusions yet. An excellent review of the PET scan studies in Alzheimer's Disease and normal aging is found in Cutler *et al.*[262]

The new imaging technology utilizing nuclear magnetic resonance (NMR) scanners indicate that both proton measurements and spin-lattice relaxation times (T_2) are different in Alzheimer white matter from controls.[135b] NMR in dementia is an active field of study.

In homogenates of whole temporal lobe, Bowen *et al.*[4] reported decrements of 20–45% in enzymes involved in glycolysis. Pyruvate dehydrogenase was also reported to be decreased in senile dementia and the degree of cognitive deficit correlated with the enzyme loss.[136] The possibility was raised that this loss related to pyruvate dehydrogenase compartments within cholinergic cells, which are known to be greatly decreased in senile dementia (see Section 3.2.2). Glucose metabolism has been measured in cortical biopsy specimens by Suzuki *et al.*[123] and by the Queen's Square group,[59,137] and in both cases was normal. An elevation in $^{14}CO_2$ production (from [U-^{14}C]glucose) was noted in the latter study, possibly secondary to a decrease in the utilization of the label in acetylcholine metabolism (see Section 3.10). Some of these data have been critically reviewed by Marchbanks.[17]

3.8. Protein Alterations

3.8.1. Protein Synthesis

In a biopsy sample of Alzheimer cortex, protein synthesis measured by radiolabeled lysine incorporation was no less than that of a similarly aged patient, whereas younger controls had higher rates of synthesis than did the older group.[138] Histochemical studies of biopsy and postmortem samples of the temporal cortex indicate significant reduction in cytoplasmic RNA content, nuclear and nucleolar volume in the presenile and senile forms of Alzheimer dementia.[121]

As noted above (Section 3.5.3), both RNA content and mRNA translational activity have been reported deficient in AD cortex compared to controls.[261]

Preliminary PET scan studies also suggest decreased methionine incorporation into brain tissue in AD.[263]

3.8.2. Total, Enzymatic, and Structural Proteins

Total brain protein decreases with aging and is thought to correlate with the degree of parenchymal loss or atrophy.[88] Differences in the total protein content of the cerebral cortex and temporal lobes of patients with senile dementia compared to normal aged controls have been demonstrated.[114,123] Certain proteins are discussed in sections where their chemistry is more relevant to a pathological evaluation: neurofibrillary proteins in Section 3.3.2, amyloid in Section 3.3.1, and specific neurotransmitter-related enzymes and receptors are discussed in those subsections. Glycolytic enzymes are discussed in Bowen *et al.*[4]

Carbonic anhydrase, a predominantly glial marker enzyme, appears lower in the temporal lobes in patients with senile dementia as compared with age-matched controls. Additionally, β-galactosidase, an enzyme thought by some to be enriched in neuronal membranes, was also lower in the temporal lobe of patients with Alzheimer's disease when compared with controls. Succinic dehydrogenase was reported to be decreased in senile dementia but terminal hypoxia probably confounds interpretation of this change. As previously discussed, similar cautions regarding hypoxic changes in brain constituents have been expressed for glutamic acid decarboxylase (GAD), lipid galactose, RNA, and possibly DNA.[113,114]

Membrane protein abnormalities as revealed by electron spin resonance have also been reported in circulating red cells in patients with Alzheimer's disease as compared with controls, raising the possibility of a more widespread involvement of membranes and protein metabolism.[140]

3.9. Trace Metals

Aluminum has received considerable attention as a possible causal factor for this nontreatable form of senile dementia. In 1973, Crapper *et al.* reported that levels of aluminum in brains of Alzheimer patients living in Toronto, Canada, contained significant elevations of aluminum compared with age-matched controls.[141] Markesbery *et al.*, utilizing neutron activation techniques, showed that brain aluminum content increased in normal aging, but did not find the aluminum in Alzheimer brains to be elevated above controls.[142] They suggested that geographic changes in availability of aluminum in environmental sources such as water supply might alter the brain content of this trace metal. However, the role of aluminum in this disease process is still unclear, since aluminum has been localized in neurofibrillary tangle-containing neurons but not in adjacent non-tangle-bearing cells by X-ray spectrophotometry.[143] This observation has refocused attention on the relationship of aluminum to the pathogenesis of Alzheimer's disease and the role of this trace metal in such basic processes as cross-linking with DNA and polymerization of proteins associated with the production of neurofibrillary tangles.[139]

Silicon and iron have both been found elevated within or near neuritic plaques.[71,144] The significance of their presence is uncertain.

3.10. Neurotransmitter Systems

3.10.1. Cholinergic Neurotransmission

3.10.1.a. Neurochemical Studies. Discussions of senile dementia of the Alzheimer type that specifically focus on cholinergic mechanisms or involvement are found in Bartus *et al.*,[145] Marchbanks,[17] McGeer and McGeer (Volume 6, this series), Coyle *et al.*,[20] Bowen,[59] and Davies.[146]

Although it is apparent that several neurotransmitter systems may be affected in Alzheimer's disease, the profound vulnerability of cholinergic neurotransmission in the disease process received considerable attention when reported and confirmed by several investigators.[113,147–151] Further findings of cholinergic enzyme alterations, especially CAT, have since been reported.[4–8,114,153–155] A review of cholinergic abnormalities in memory dysfunction and dementia is found in Bartus *et al.*[145] A review of ACh synthesis and CAT activity is found in Tucek (Volume 4, this series).

We have recently determined the laminar distribution of CAT in the frontal cortex (Brodmann's area 9) in normal age-matched controls and patients with senile dementia. The bulk of activity appears to reside in the upper laminae (largely lamina 2 and upper lamina 3). The alteration in Alzheimer's disease involves loss of activity in all layers; if any remnant of activity was left, it was usually in lamina 2.[156] Henke and Lang[155] reported a similar distribution in the frontal cortex.

In middle temporal gyrus (area 21), normal CAT activity has a similar peak in lamina II and III, with a second peak in lamina IV and a third elevation of activity in lamina VI.[156a] Loss of CAT was similar to that found in area 9: large losses everywhere with the highest remnant of activity in lamina II. While other markers of amino acid and LDH laminar distribution were measured, the most marked and consistent changes where cholinergic.

A cholinergic metabolic defect met ready acceptance, since memory loss, the sine qua non of senile dementia, can be produced by interfering with cholinergic function in normals.[145,157–159] Another strong suggestion of cholinergic involvement in senile dementia came from correlation of the degree of loss of CAT with the number of neuritic plaques and severity of dementia as assessed by behavioral testing procedures.[160] A subsequent study did not find such a correlation with plaques but did correlate neurofibrillary tangle density with CAT loss.[4]

Although it is unlikely that a defect of cholinergic neurotransmitter metabolism constitutes the primary or sole abnormality relating to the cause of Alzheimer's disease, the loss of CAT appears to represent a change found early in the course of the disease so that its quantitation in autopsy and biopsy specimens from brains of patients with dementia has been shown to be of diagnostic value.

While CAT has been reported decreased, there is still discussion about whether its loss is generalized or concentrated in the cortex and hippocampus.[6,8,160a] While the amount of cortical CAT loss has varied in different studies, and ranges from 10 to 50% of age-matched controls, regional differences, dissection methods, and duration of the patients' illness prior to death may partially affect such values. Muscarinic cholinergic receptor density in the cortex and hippocampus appears unchanged.[145,149,151,156,160,162–164] One study has reported decreases in hippocampal (though not cortical) binding with tritiated quinuclidinyl benzylate ([^{3}H]-QNB).[154]

Lange and Henke also found no differences in total muscarinic cholinergic binding between normal cortex or hippocampus and that of senile dementia brains.[162] In three of their seven dementia brains, however, they reported loss of a laminated pattern of QNB binding, and a more "diffuse" autoradiographic pattern of binding sites was seen. In our studies of normal and Alzheimer cortex, no change in the binding affinity for QNB is seen and a nonlaminated pattern of QNB binding is found in cortical specimens from demented patients and normal age-matched controls. Cortical regions appear to vary, however, in the degree of laminar distribution of the QNB binding.[156] The relative proportions of the high- and low-affinity muscarinic receptor subtypes[165,165a] in Alzheimer cortex have been reported unchanged from normal.[166] or altered[166b] in a small number of cases. No alterations were detected in nicotinic cholinergic binding in the hippocampus and cortex in one study[162]; in another, there was a decline in α-bungarotoxin binding to cerebral tissue but the suitability of the ligand for CNS tissue was not regarded as optimal.[167]

AChE was also low in the brains of dementia patients,[8,160] but was not found to be low in the lumbar cerebrospinal fluid (CSF) in Alzheimer's disease.[8,152] One study reported decreased levels of CSF AChE in senile dementia patients.[168] AChE is not confined to cholinergic cells, and some cholinergic systems are not affected by senile dementia. Lumbar CSF is far removed from the circulatory pathways of spinal fluid that bathe the cortex and hippocampus, and may not accurately reflect the status of the more rostral cholinergic systems. Many of the data are derived from live patients with "presumptive" Alzheimer's disease. The lack of pathological confirmation of senile dementia and the other factors cited above may add to variability of such results.

3.10.1b. Biopsy Results. The majority of clinically suspected cases of Alzheimer's disease that undergo biopsy (either frontal or temporal cortex) are found to have decreased CAT activity, diminished high-affinity choline uptake, and lowered ACh synthesis.[4,153,169,170] These deficits have tended to correlate with degree of cognitive impairment,[153] but were found in patients who have clinical symptoms for less than 1 year.[169] Thus, biopsy evidence indicates that loss of cholinergic synthetic activity is an early neurochemical alteration, which correlates to at least some degree with the degree of cognitive impairment.

At either of two concentrations of K^+ (to simulate alterations in ACh synthesis rates brought about by increased neuronal activity), Alzheimer tissue activity was below that of the biopsied controls. As in the normal samples, there was increased ACh production in the Alzheimer tissue in response to the increased K^+ concentration.[59,153] CAT is likely not the rate-limiting step in ACh synthesis (59,153, and Tucek, Volume 4, this series). There appeared to be many times more CAT available in the cortical samples of the demented group than was needed to account for the synthesis rate of the control group. Perhaps only a portion of the CAT, as measured by maximizing the *in vitro* enzyme activity, is available for ACh synthesis.[146,153] More biopsy studies would be of great value in determining the relationship of amount of CAT and ACh synthesis ability, the relationship of early cholinergic changes to behavioral changes, and the time sequence of cholinergic chemical alterations and classic histopathological changes. Biopsy findings exploring other neurotransmitters are discussed in those individual sections.

3.10.1c. Anatomic Systems Underlying the Cholinergic Decline. Several excellent reviews of the anatomy of cholinergic systems are available.[57,171,172] With the evidence that the cholinergic deficit in senile dementia appeared to be presynaptic (loss of CAT- and ACh-synthesizing ability without change in the cholinergic receptors), the focus shifted to the source cells of the cholinergic input to the cortex and hippocampus. The NBM in the substantia innominata was reported to have lost great numbers of the large AChE-positive cells that are the source of the cortical cholinergic input.[55–57d,173] For example, lesion of the homologous nucleus in rodents produces loss of CAT activity in the cerebral cortex.[174–176] The NBM in normal humans is a scattered group of neurons with extremely high CAT activity, which constitutes a challenging dissection for the anatomist who tries to segregate this region.[177] Nevertheless, studies have shown markedly decreased CAT activity in the NBM of patients with senile dementia.[178] Moreover, nucleolar volume and cytoplasmic RNA, both indicative of protein-synthesizing capacity of this brain region, are significantly altered in the neurons of the NBM of demented patients when compared with normal age-matched controls.[179] Finally NBM neuron number is reported to decrease in normal aging[57d], and greater numbers of NBM cells are lost in earlier onset cases.[57b,181] The less profound loss of NBM neurons in older Alzheimer patients is not explained by duration of the disease[57b] and is consistent with the relative sparing of frontal cortical CAT activity in older Alzheimer patients.[156,179a,]

While the loss of cortical and hippocampal CAT appeared related to the pathology of their source nuclei (no intrinsic neurons have yet been identified in either cortex or hippocampus by CAT antibody methods), some questions remain. Perry *et al* reported loss of both cortical and NBM CAT without massive neuronal loss in the basal nucleus.[180] They suggested that down-regulation of CAT synthesis was possible, with delayed cell loss resulting. The medial septal nucleus supplies the majority of hippocampal CAT via the fornix, yet one study was unable to find a correlation between the loss of CAT in the hippocampus and the loss of the same enzyme in the medial septum.[6] Henke and Lang[155] did not find significant decline in NBM CAT activity in Alzheimer's disease in their study. However, a different nucleus, that of the diagonal band of Broca (NDB), was severely affected. The possibility exists that there are several anatomic subunits coupled with various manifestations of Alzheimer's dementia. The uncertainty of accurate dissection of the difficult-to-isolate basal forebrain structures, and the varying amount of time after diagnosis that autopsy material becomes available may explain these inconsistencies.

Rossor's group[6] found several cases of dementia with CAT loss in both cortex and substantia innominata that did not have the characteristic plaques and tangles upon histological examination. They speculated that these cases might represent an early preneuropathological stage of the disease. Price and colleagues attempted to relate cholinergic denervation of the cerebral cortex of patients with Alzheimer's disease to the quantitative appearance of neuritic plaques. In aged simian brain, they found that AChE is present in large amounts in immature and mature plaques but less frequently seen by histochemical techniques in old plaques.[70] Relationships among the cholinergic source nuclei (medial septum, NBM, NDB) and their terminal fields in the cortex and hippocampus need further exploration in the brains of clinically well-characterized demented patients. Further details of these cholinergic pathways in the simian brain will also be of great value.[182–184]

Further delineation of neurotransmitter interconnectivity within the mammalian brain is necessary in conjuction with more detailed studies of the pathological chemistry and anatomy in Alzheimer's disease. Questions as to whether cell death is primary or secondary to dying back of processes, down-regulation of CAT activity, or to induced or acquired transcriptional or translational problems lie at the basis for understanding the disease and any attempt to devise rational future therapies. Therapeutic trials of agents directed toward a functional enhancement of the cholinergic system and cerebral cortex are under study and discussed in Section 3.12.

3.10.2. Noradrenergic Transmission

Norepinephrine is decreased in the cerebral cortex of Alzheimer brains,[185] and is markedly reduced in the hypothalamus of patients with senile dementia.[58,186] While dopamine β-hydroxylase, the enzyme that metabolizes dopamine to norepinephrine, is decreased in activity,[187,188] tyrosine hydroxylase, the rate-limiting enzyme of catecholamine synthesis, is not decreased.[147] Additionally, while β-adrenergic receptor density does not appear affected,[4] norepinephrine uptake is deficient in biopsy samples of the temporal cortex from patients with senile dementia.[189]

It has been well described that the locus coeruleus constitutes the major source for noradrenergic axonal projections to the forebrain,[190] and this brain stem nucleus shows progressive cell loss during normal aging.[191] Several investigators have reported cell loss in the locus coeruleus in Alzheimer's disease as compared with age-matched controls.[10,186,188,192] In two studies, younger Alzheimer patients suffered a more severe loss of noradrenergic neurons, while an older group (mean age 83) had a less severe decrement.[10,57b] Perry *et al.* showed that regardless of whether neuron counts in the locus coeruleus were markedly low or only moderately reduced, CAT activity was low in both groups.[188] Although the number of patients used in these studies was small, it was noted that CAT activity was slightly more reduced in the subjects who had decreased number of neurons in the locus coeruleus, and dopamine β-hydroxylase activity was equally depressed in both groups regardless of the degree of loss of noradrenergic neurons in the locus coeruleus.

Monoamine oxidase B (MAO-B) is elevated in both brain and peripheral platelets in Alzheimer's disease and MAO-A is unchanged.[193,160a]

3.10.3. Dopaminergic Neurotransmission

An early report found decreased homovanillic acid in the CSF of senile dementia patients.[194] Decrements in cortical homovanillic acid have also been reported.[44,185,195] There are conflicting reports for both CSF[196] and brain.[58] Because most Alzheimer patients do not exhibit the symptoms of Parkinson's disease, dopamine metabolism disturbances have not generally been regarded as a major contributor to the primary pathology. A multiple neurotransmitter deficit in senile dementia, which includes dopamine, has been proposed.[44,146,160a] The monoamine alterations are reviewed by Gottfries[197]; differing reports of alterations in dopaminergic transmission in senile dementia have been ascribed to regional variations in the areas of basal ganglia where samples were taken, or in the case of CSF studies, uncertainties as to a conclusive diagnosis of Alzheimer's disease.[13] A recent extensive study of dopamine found it reduced in caudate nucleus in Alzheimer's disease.[160a]

3.10.4. Serotonin Neurotransmission

Serotonin was not found to be reduced in senile dementia by one group[196] but was found to be lower in the caudate and hippocampus by other researchers. The major metabolite of serotonin, 5-hydroxyindoleacetic acid (5-HIAA), was decreased in senile dementia brains.[4,185] LSD binding is also decreased,[4] although there is uncertainty as to the specificity of this ligand.[59] Recent studies indicate that uptake of labeled serotonin was severely deficient in fresh temporal cortex biopsies in Alzheimer's disease.[189,210] Nucleolar volume and cytoplasmic RNA are also significantly lower in the serotoninergic dorsal tegmental nucleus of the brain stem of patients with senile dementia.[260] The role of serotonin metabolism in the pathology of Alzheimer's disease remains unclear.

3.10.5. GABA Metabolism

The postmortem instability of GAD and its sensitivity to terminal hypoxia[113] cast doubts on early reports that GAD was lower than normal in Alzheimer's disease. In a study correlating cognitive loss in dementia with postmortem neurochemical changes, a positive correlation was found with loss of CAT activity; however, no relationship was found between GAD activity and cognitive loss.[160] GAD had been reported normal in biopsy samples of Alzheimer brain[198] and in most areas of the brain subjected to postmortem study.[8] GABA, which is less sensitive to terminal hypoxia and postmortem alterations than is GAD, is normal in Alzheimer brain compared to controls.[6,156a] Biopsy tissue from the temporal cortex showed a normal efflux of GABA in response to depolarizarion, as did the excitatory amino acid neurotransmitters glutamate and aspartate.[199]

3.10.6. Peptides and Opiates

Evaluation of these putative peptide neurotransmitters or neuromodulators in Alzheimer's disease is still in an early stage. Substance P[200] and somatostatin[201,202] are deficient in regions of Alzheimer brain, though another group has found the regional losses of the latter peptide more restricted[203] The role of vasopressin in memory function is under continued examination.[204] Arginine vasopressin (AVP), measured by RIA in frozen brain extracts, did not show changes, except in the lateral portion of the globus pallidus where levels were lower in dementia brains. Levels of AVP did not correlate with CAT activity in any areas of the brain.[205] AVP levels were found to be lower than those of controls in CSF of senile dementia patients.[206]

Little or no data are available on endorphin or enkephalin content of Alzheimer versus normal brain tissue. Animal studies indicate a role for opiates in memory processes, and possible disruption of such a system has been speculated upon as a mechanism of senile dementia.[207,208] Preliminary studies of intravenous naloxone, a block of opiate receptor sites, has been reported to show mild short-term improvement of memory in a small number of patients with Alzheimer's disease.[209] Opiate receptors as measured by radiolabeled naloxone appear unaltered in the temporal lobe of Alzheimer victims.[4]

3.11. Down's Syndrome (Trisomy 21)

Down's syndrome is associated with an age-related dementia in which neurofibrillary tangles and neuritic plaques, the classic histopathological findings of Alzheimer's disease, are found in autopsy specimens of brain in almost all patients over 40 years of age.[84,85,211–213] Moreover, a defect in cholinergic function has been found,[214] and like Alzheimer's dementia, low norepinephrine content of the hypothalamus has been reported in patient's with Down's syndrome.[58] Other similarities between these two diseases have been noted, possibly indicating an underlying or associated abnormal gene associated with the well-described chromosomal abnormality of Down's syndrome. This observation suggests that patients with Alzheimer's disease may have a genetic predisposition to develop the illness. Down's syndrome is more frequent in families of patients with Alzheimer's disease and an increased frequency of myeloproliferative disorders (a known risk in Down's syndrome) is also reported in families that have a member affected by Alzheimer's disease.[216–218]

3.12. Pharmacological Interventions

In an effort to mitigate the deficit in central cholinergic neurotransmission in Alzheimer's disease, clinical trials of enhancers of central cholinergic metabolism were initiated in the late 1970s.[145,146,159] With the analogy of the successful use of L-dopa in Parkinson's disease to partially restore dopamine neurotransmission in the nigrostriatal pathway, it was postulated that nutritional supplementation of large quantities of a neurotransmitter precursor might be helpful in ameliorating the symptoms by increasing brain ACh. Alternatively, a cholinergic agonist active at the postsynaptic receptor (e.g., arecoline) might be analogous to the use of bromocriptine at the dopamine receptor. Like the dopaminergic system of the substantia nigra, the isodendritic core[219] of the cholinergic basal forebrain system appeared to be pharmacologically accessible to manipulation.

Initial animal studies suggested that elevated dietary administration of choline could raise brain ACh levels.[220,221] Several other groups did not find such increases[222–224] (see Marchbanks[17] for a review and discussion). Exogenous choline can "protect" brain ACh levels from the alterations produced by exogenous agents such as atropine,[225] implying that augmented peripheral choline may aid the stability of the central cholinergic system during times of increased demand. Finally, animal studies using choline administration or physostigmine infusion showed clear-cut effects on age-related memory deficits in rats[226] and nonhuman primates.[227,228] Bartus *et al.*[145] have reviewed these studies.

Administration of oral choline or phosphatidyl choline (lecithin), even in doses demonstrably elevating peripheral serum choline levels, has not produced significant stabilization or improvement of dementia in patients with Alzheimer's disease (for reviews see 1,17,145,229). Since the capability of the synthetic enzyme CAT to make ACh is impaired in senile dementia, loading of the precursor, perhaps understandably, has not proved helpful for most patients.

Inhibition of AChE by physotigmine has indicated at least measurable improvement in memory function in patients with presenile dementia.[158,230] Use of such an inhibitor has the advantage of enhancing the pharmacological effect of age at sites where the diseased tissue has relatively maintained its ability to synthesize the neurotransmitter.[17] A combination of intravenous physostigmine and oral lecithin has been reported to induce a mild, detectable improvement in some human cases of dementia.[231] Most recently, lecithin and oral physostigmine have resulted in a more stable but modest improvement in memory function of Alzheimer patients.[232–234]

Choline loading combined with the administration of piracetam (a drug that stimulates oxidative metabolism) resulted in improved cognitive function in 3 of 10 patients with Alzheimer's disease, and the responders had higher levels of choline within their circulating red blood cells than did the nonresponders.[235] Combined use of piracetam and choline has also been shown to have a much greater effect on passive avoidance retention in aged rats than when either substance is administered alone.[236] This effect may relate to a beneficial role of choline or lecithin in aiding a required increase in brain oxidative metabolism and cholinergic turnover during periods of heightened cholinergic demand. Wecker's studies with choline and atropine indicate that in rats, exogenous choline exerts a protective effect on the decrease in ACh normally seen after injection of the cholinergic blocker atropine. Thus, ACh levels are stabilized if choline is administered before or after atropine.[222,225] It is notable that clinically, Alzheimer patients may cope reasonably well if left alone, but decompensate cognitively if stressed or if requested to perform intellectual or memory functions of which they are incapable. It would appear, then, that the mild beneficial effects of dietary supplementation of choline may relate to the ability of the choline to stabilize central cholinergic neurotransmission.

Based on review of human and animal studies on the pharmacological effects of various dietary supplements and drugs on memory, Bartus and colleagues suggested that pharmacological strategy for improving memory should focus on agents that directly affect cholinergic neurotransmission at the central postsynaptic receptor sites.[145] To date, only one muscarinic agonist, arecoline, has been subjected to clinical trials, and initial reports have shown modest improvement.[230] Further trials with this agent and a search for a more effective muscarinic agonist may prove to be a useful strategy for future research. If the supposition is correct that the M_2 receptor is the presynaptic autoreceptor in human cortex and the M_1 is postsynaptic,[264] selective M_1 agonists and M_2 antogonists might be sought.

Reisberg *et al.*[209] have reported a small group of patients who exhibited brief improvements in cognitive function when administered naloxone, an opiate receptor antagonist. This effect may relate to the incompletely understood role of the opiate system in memory function,[207] or to effects on cholinergic metabolism, since opiates can decrease ACh turnover.[237]

3.13. Etiological Issues

The plethora of reported abnormalities (neurochemical and neuropathological) and the data suggesting subsets of patients within the group termed *senile dementia of the Alzheimer type*[11] have pointed to no specific etiology. "The proliferation of seemingly unrelated leads . . . make a unitary hypothesis for the etiology of Alzheimer's disease inadmissible."[87]

Price and co-workers have attempted to link the cholinergic denervation to appearance of the neuritic plaque, speculating that the latter is a tissue response to the loss of cholinergic afferents.[57] Correlations exist between the amount of CAT activity lost and the number of neuritic plaques[160,179a] or neurofibrillary tangles[4,179a] present in cortex. However, since cortical distribution of CAT involves predominantly upper lamina[155,156,156a] one might expect consistent reports of topographic distribution of plaques in this area of the cortex. Such observations have not yet been made, suggesting that while these neuropathological and neurochemical alterations correlate strongly with disease severity and with each other, causative relationships among them are not apparent. Heuristic attempts to relate the neurochemical and neuropathological changes, such as that proposed above[57] are sorely needed to provide a unitary theory of the mechanism of Alzheimer's disease.

Most researchers regard the genetic data (at least in terms of predisposition to development of the disease) as essential for the etiology. The genetic data derive from a wide variety of sources, including familial Alzheimer's disease (5–10% of all cases),[238–240] high concordance rates for monozygotic and dizygotic twins,[241] and increased frequency of the disease in relatives of affected patients.[242] Theories have proposed an autosomal dominant gene with increasing penetrance in senescence or polygenetic determinance.[243,244] Aneuploidy has been found in the peripheral white cells in some patients with senile dementia in some studies,[245,246] but this finding has not been confirmed in other cases of sporadic or familial Alzheimer's disease.[247] Consistent HLA type groupings have not been apparent.[13]

Superimposed upon a genotype that might predispose individuals to Alzheimer's disease, there exists an array of proposed etiological agents including aluminum, slow virus infection, autoimmune processes, and aberrant protein production. These agents has been postulated to lead to morphological changes (neuritic plaques, neurofibrillary tangles, granulovacuolar degeneration, neuronal loss) and/or biochemical alterations (loss of neurotransmitter, decrease in neurotransmitter turnover, loss of specific neurotransmitter enzymes or receptors).

Protein changes regarded as having implications for the etiology of senile dementia include altered neurofibrillary proteins[75] and the accumulated amyloid within neuritic plaques. Aluminum's role is supported by its relatively selective presence in tangle-bearing neurons,[143] and its potential for interfering with DNA transcription.[116–119]

Viral etiologies or other atypical transmissible agent etiologies are supported by the loss of CAT and formation of neuritic plaques in mouse cortex that has been infected with scrapie, the production and presence of amyloid with the plaque and along the injection site in such affected animals,[80] and the recent suggestion that plaque core amyloid might be the prion-associated protein of the scrapie agent.[80a,80b] An infectious etiology could explain the selective vulnerability of certain groups of neurons, analogous to the relatively specific effects of poliovirus on motor neurons.

Kuru and Creutzfeldt–Jakob disease both include dementia in their clinical presentation and are caused by subviral transmissible agents yet to be completely described. However, numerous inoculations of brain tissue from patients with Alzheimer's disease have not provided evidence for the "slow virus" hypothesis and no viral particles have been found in the brains of patients with Alzheimer's disease despite exhaustive search.[248] One report showing production of PHF in cultured human neurons when Alzheimer brain extract was added to the culture medium has not been replicated.[249]

Any theory designed to explain a cause of senile dementia must seek to answer the vexing question of the cause of death of neurons of a specific cell type within specific regions of brain. Although Gowers introduced the concept of abiotrophy over 70 years ago, the process that determines the selective vulnerability of certain cell types in specific brain regions needs to be identified.

Thus, cells of the NBM, medial septum, NDB, locus coeruleus, large neurons of the cortex, and possibly median raphe cells of the brain stem may die, but there is no indication of whether the process of neuritic plaque and neurofibrillary tangle formation is primary or secondary. Identifying Alzheimer's disease as a disorder of cortical (and hippocampal) cholinergic metabolism, though valuable and productive of palliative therapeutic ventures, does not address the question of why the cells and their processes, extending for great distances within the brain, disappear. To account for such changes the existence of trophic factors has been proposed.[250]

During the past several years, numerous "trophic factors" have been isolated from the CNS or peripheral nerve target tissues, which prolong survival and neuritic outgrowth of neural cells in culture.[251–253] Early work on such factors as nerve growth factor (NGF) was directed toward understanding neuritic development and programmed cell death during brain development. Most recent studies have assessed the role of such factors in enabling continued cell survival (neuronotrophic activity) and maintenance of the neuritic expanse (specifying factors).[251] The responses of cholinergic neurons to such factors have been demonstrated.[250,254] The survival in culture of cholinergic neurons of the chick ciliary ganglion is clearly enhanced when such cells are exposed to extracts containing such factors.[251] *In vivo* effects of other factors are seen in rat brain when such proteins are centrally administered following septohippocampal disconnection.[255]

Titers of a "trophic factor" that enhances neuron survival in culture increase after traumatic brain injury.[256] Survival of fetal striatal cell implants into rat brains correlates with the increase of trophic activity in the brains of such animals as a function of the time after brain injury or surgery.[257] Thus, *in vivo* survival may parallel the *in vitro* trophic activity.

The action of such trophic factors in prolonging survival *in vitro* and *in vivo* and enabling innervation of target tissue suggests that lack of such a factor or factors might be associated with etiology of several neurodegenerative diseases such as senile dementia.[250] Identification of such trophic factors in human brain represents an area of future research that may have considerable impact on the etiology and stabilization of numerous neurological diseases including senile dementia of the Alzheimer type.[255,258,259]

3.14. Treatment Perspectives

Several explanations have been offered for the therapeutic failure of high oral doses of choline or lecithin to improve cognition, or to explain the minimal increments in memory noted with the combined administration of physostigmine and lecithin.[17] Whereas large doses of L-Dopa can be shown to produce severalfold elevations in brain levels of the neurotransmitter dopamine, a similar increment in brain ACh does not occur following massive administration of oral choline. While dopa decarboxylase is present in many nondopaminergic cells, CAT is confined to the presynaptic terminals and cell bodies of cholinergic neurons. Hence, if the cholinergic cell population is lost, there is little or no remaining synthetic enzyme despite the availability of an abundance of substrate[16] (McGeer and McGeer, Volume 6, 2nd ed., this series). Deficiencies of other neurotransmitter systems (norepinephrine, serotonin, somatostatin) may be involved in the disease process, mitigating against any striking improvement in cognitive function when therapy is directed toward a single neurotransmitter system.

The mild cognitive or memory improvement seen with cholinergic therapies might be illustrative of the necessary, but not sufficient role of cholinergic metabolic alteration in the production of Alzheimer's disease. The search for better cholinergic system enhancers, postsynaptic agonists, and anticholinesterases continues, including assessment of the use of combined therapeutic modalities if indeed multiple neurotransmitter system defects are present.[17,146] The role of disordered structural proteins in production of the disease needs clear definition. Until the etiology of cell death and degeneration in senile dementia is determined and either prevented or halted, the best hope for palliative therapies likely lies with such multiple neurotransmitter-directed interventions. For the long term, a strategy involving the isolation and purification of "trophic proteins" directed toward increasing neuronal survival and stimulating restorative sprouting of neuronal processes may offer a rational innovative approach to future therapy.

REFERENCES

1. Terry, R. D., and Katzman, R., 1983, *Ann. Neurol.* **14**:497–506.
2. Alzheimer, A., 1907, *Allg. Z. Psychiatr. Ihre Grenzgeb.* **64**:146–148.
3. Hubbard, B. M., and Anderson, J. M., 1981, *J. Neurol. Neurosurg. Psychiatry* **44**:631–635.
4. Bowen, D. M., Spillane, J. A., Curzon, G., Meier-Ruge, W., White, P., Iwangoff, P., and Davison, A. N., 1979, *Lancet* **1**:11–14.
5. Rossor, M. N., Iversen, L. L., Johnson, A. L., Mountjoy, C. Q., and Roth, M., 1981, *Lancet* **2**:1422.
6. Rossor, M. N., Garrett, N. J., Johnson, A. L., Mountjoy, C. Q., Roth, M., and Iversen, L. L., 1982, *Brain* **105**:313–330.
7. Wilcock, G. K., Esiri, M. M., Bowen, D. M. J., and Smith, C. C. T., 1982, *J. Neurol. Sci.* **57**:407–417.
8. Davies, P., 1979, *Brain Res.* **171**:319–327.
9. Jacoby, R. J., and Levy, R., 1980, *Br. J. Psychiatry* **136**:256–259.
10. Bondareff, W., Mountjoy, C. Q., and Roth, M., 1982, *Neurology* **32**:164–168.
11. Bird, T. D., Stranahan, S., Sumi, S. M., and Raskind, M., 1983, *Ann. Neurol.* **14**:284–293.
12. Roth, M., 1972, *Biol. Psychiatry* **5**:103–125.
13. Terry, R., and Katzman, R., 1983, *The Neurology of Aging* (R. Katzman and R. D. Terry, eds.), Davis, Philadelphia, pp. 51–84.

13a. Katzman, R., and Terry, R., 1983, *The Neurology of Aging* (R. Katzman and R. D. Terry, eds.), Davis, Philadelphia, pp. 15–50.

14. Corsellis, J. A. N., 1976, *Greenfield's Neuropathology* (J. A. N. Corsellis and W. Blackwood, eds.), Yearbook, Chicago, pp. 796–848.
15. Schneck, M. K., Reisberg, B., and Ferris, S. H., 1982, *Am. J. Psychiatry* **139**:165–173.
16. McGeer, E. G., 1981, *Prog. Neuro-Psychopharmacol.* **5**:435–445.
17. Marchbanks, R. M., 1982, *J. Neurochem.* **39**:9–15.
18. Terry, R. D., and Davies, P., 1980, *Annu. Rev. Neurosci.* **3**:77–95.
19. Ball, M. J., 1982, *Arch. Pathol. Lab. Med.* **106**:157–162.
20. Coyle, J. T., Price, D. L., and DeLong, M. R., 1983, *Science* **219**:1184–1189.
21. Beck, J. C., Benson, F. D., Scheibel, A. B., Spar, J. F., and Rubenstein, L. Z., 1982, *Ann. Intern. Med.* **97**:231–241.
22. Crook, T., and Gershon, S. A. (eds.), 1981, *Strategies for the Development of an Effective Treatment for Senile Dementia*, Mark Pawley Associates, New Canaan, Connecticut.
23. Zatz, L. M., Jernigan, T. L., and Ahumada, A. J., 1982, *Am. J. Neuroradiol.* **3**:1–11.
24. Soinenen, G., Puranen, M., and Riekkinen, P. J., 1982, *J. Neurol. Neurosurg. Psychiatry* **45**:50–54.

25. Ford, C. V., and Winter, J., 1980, *J. Gerontol.* **36:**164–169.
26. Yamamura, H., Ito, M., Kubota, K., and Matsuzawa, T., 1980, *J. Gerontol.* **35:**492–498.
27. Jacoby, R. J., Levy, R., and Dawson, J. M., 1980, *Br. J. Psychiatry* **136:**249–255.
28. Gyldensted, C., 1977, *Neuroradiology* **14:**183–192.
29. Corsellis, J. A. N., 1976, *Neurology of Aging* (R. D. Terry and S. Gershon, eds.), Raven Press, New York, pp. 205–209.
30. Brody, H., 1955, *J. Comp. Neurol.* **102:**511–556.
31. Anderson, J. M., Hubbard, B. M., Coghill, G. R., and Slidders, W., 1983, *J. Neurol. Sci.* **58:**233–244.
32. Henderson, G., Tomlinson, B. E., and Gibson, P. H., 1980, *J. Neurol. Sci.* **46:**113–136.
33. Ball, M. J., 1977, *Acta Neuropathol.* **37:**111–118.
34. McGeer, E., 1978, *Alzheimer's Disease: Senile Dementia and Related Disorders* (R. Katzman, R. D. Terry, and K. Bick, eds.), Raven Press, New York, pp. 427–440.
35. Konigsmark, B. W., and Murphy, E. A., 1972, *J. Neuropathol. Exp. Neurol.* **31:**304–316.
36. Tomlinson, B. E., Blessed, G., and Roth, M., 1968, *J. Neurol. Sci.* **7:**331–356.
37. Tomlinson, B. E., Blessed, G., and Roth, M., 1970, *J. Neurol. Sci.* **11:**205–242.
38. Monagle, R. D., and Brody, H., 1974, *J. Comp. Neurol.* **155:**61–66.
39. Mann, D. M. A., and Yates, P. O., 1974, *Brain* **97:**489–498.
40. Rouser, G., and Yamamoto, A., 1969, *Handbook of Neurochemistry*, Volume 1 (A. Lajtha, ed.), Plenum Press, New York, pp. 121–169.
41. DeKosky, S. T., and Bass, N. H., 1982, *Neurology* **32:**1227–1233.
42. Cherayil, G. D., and Cyrus, A. C., 1966, *J. Neurochem.* **13:**579–590.
43. Ansari, A., and Loch, J., 1975, *Neurology* **25:**1045–1050.
44. Carlsson, A., 1981, *Strategies for the Development of an Effective Treatment for Senile Dementia* (T. Crook and S. A. Gershon, eds.), Mark Pawley Associates, New Canaan, Connecticut, pp. 93–104.
45. Carlsson, A., 1981, *Funktionsstorungen Des Gehirns in Alter* (D. Platt, ed.), Schattauer-Verlag, Stuttgart, pp. 67–81.
46. McGeer, E., and McGeer, P. L., 1976, *Neurobiology of Aging* (R. D. Terry and S. A. Gershon, eds.), Raven Press, New York, pp. 389–403.
47. McGeer, P. L., and McGeer, E. G., 1981, *The Molecular Basis of Neuropathology* (A. N. Davison and R. H. S. Thompson, eds.), Igaku-Shoin, New York, pp. 631–648.
48. Tomlinson, B. E., 1979, *Recent Advances in Neuropathology*, Volume 1 (W. T. Smith and J. B. Cavanaugh, eds.), Churchill Livingstone, London, pp. 129–159.
49. Terry, R. D., Peck, A., deTeresa, R., Schechter, R., and Horoupian, D. S., 1981, *Ann. Neurol.* **10:**184–192.
50. Miller, A. K. H., Alston, R. C., and Corsellis, J. A. N., 1980, *Neuropathol. Appl. Neurobiol.* **6:**119–132.
51. Mountjoy, C. Q., Roth, M., Evans, N. J. R., and Evans, H. M., 1983, *Neurobiol. Aging* **4:**1–11.
52. Schechter, R., Yen, S.-H.C., and Terry, R. D., 1981, *J. Neuropathol. Exp. Neurol.* **40:**95–101.
53. Shefer, V. F., 1972, *Zh. Nevropatol. Psikhiatr. im. S.S. Korsakova* **72:**1024–1029.
54. Colon, E. J., 1973, *Acta Neuropathol.* **23:**281–290.
55. Whitehouse, P. J., Price, D. L., Clark, A. W., Coyle, J. T., and DeLong, M., 1981, *Ann. Neurol.* **10:**122–126.
56. Whitehouse, P. J., Price, D. L., Struble, R. G., Clark, A. W., Coyle, J. T., and DeLong, M. R. J., 1982, *Science* **215:**1237–1239.
57. Price, D. L., Whitehouse, P. J., Struble, R. G., Clark, A. W., Coyle, J. T., DeLong, M. R., and Hedreen, J. C., 1982, *Neurosci. Comment.* **1:**84–92.
57a. Wilcock, G. K., Esiri, M. M., Bowen, D. M., and Smith, C. C. T., 1983, *Neuropath. Appl. Neurobiol.* **9:**175–179.
57b. Mann, D. M. A., Yates, P. O., and Marcyniuk, B., 1984, *J. Neurol. Neurosurg. Psychiat.* **47:**201–203.
57c. Arendt, T., Bigl, T., Arendt, A., and Tennstedt, A., 1983, *Acta Neuropathol.* **61:**101–108.
57d. McGeer, P. L., McGeer, E. G., Suzuki, J., Dolman, C. E., and Nagai, T., 1984, *Neurology* **34:**741–745.

58. Yates, C. M., Ritchie, I. M., Simpson, J., Maloney, A. F., and Gordon, A., 1981, *Lancet* **2**:39–40.
59. Bowen, D. M., 1981, *The Molecular Basis of Neuropathology* (A. N. Davison and R. H. S. Thompson, eds.), Igaku-Shoin, New York, pp. 649–665.
60. Torack, R. M., 1978, *The Pathologic Physiology of Dementia*, Springer-Verlag, Berlin.
61. Corsellis, J. A. N., 1962, *Mental Illness and the Aging Brain*, Oxford University Press, London.
62. Tomlinson, B. E., 1977, *Aging and Dementia* (W. L. Smith and M. Kinsbourne, eds.), Prentice–Hall, Englewood Cliffs, New Jersey, pp. 25–56.
63. Hooper, M. W., and Vogel, F. S., 1976, *Am. J. Pathol.* **85**:1–13.
64. Herzog, A. G., and Kemper, T. L., 1980, *Arch. Neurol.* **37**:625–629.
65. Muramatusu, M., Hirai, S., Muramatusu, A., and Yosikawa, M., 1975, *J. Amer. Geriatr. Soc.* **23**:390–406.
66. Gonatas, N. K., Anderson, W., and Evangelista, I., 1967, *J. Neuropathol. Exp. Neurol.* **26**:25–39.
67. Terry, R. D., Gonatas, N. K., and Weiss, M., 1964, *Am. J. Pathol.* **44**:269–297.
68. Kidd, M., 1964, *Brain* **87**:307–320.
69. Terry, R. D., and Wisniewski, H. M., 1970, *Ciba Foundation Symposium on Alzheimer's Disease and Related Conditions* (G. E. W. Wolstenholme and M. O'Connor, eds.), Churchill, London, pp. 145–165.
70. Struble, R. G., Cork, L. C., Whitehouse, P. J., and Price, D. L., 1982, *Science* **216**:413–415.
71. Nikaido, T., Austin, J., Trueb, L., and Rinehart, R., 1972, *Arch. Neurol.* **27**:549–554.
72. Ishii, T., 1969, *Acta Neuropathol.* **14**:250–260.
73. Friede, R. C., 1965, *J. Neuropathol. Exp. Neurol.* **24**:477–491.
74. Johnson, A. B., and Blum, N. R., 1970, *J. Neuropathol. Exp. Neurol.* **29**:463–478.
75. Selkoe, D. J., 1982, *Trends Neurosci.* **5**:332–339.
76. Glenner, G. G., 1978, *Alzheimer's Disease: Senile Dementia and Related Disorders* (R. Katzman, R. D. Terry, and K. Bick, eds.), Raven Press, New York, pp. 493–501.
77. Powers, J. M., and Spicer, S. S., 1978, *Virchows Arch.* **373**:107–115.
78. Ishii, T., and Haga, S., 1976, *Acta Neuropathol.* **36**:243–249.
79. Shirahama, T., Skinner, M., Westermark, P., Rubinow, A., Cohen, A. S., Brun, A., and Kemper, T. L., 1982, *Am. J. Pathol.* **107**:41–50.
80. Wisniewski, H. M., Moretz, R. C., and Lossinsky, A. S., 1981, *Ann. Neurol.* **10**:517–522.
80a. Prusiner, S., McKinley, M. P., Bowman, K. A., Bolton, D. C., Bendheim, P. E., Groth, D. F., and Glenner, G., 1983, *Cell* **35**:349–398.
80b. Prusiner, S. B., 1984, *New Engl. J. Med.* **310**:661–663.
81. Kidd, M., 1963, *Nature* **197**:192–193.
82. Terry, R. D., 1963, *J. Neuropathol. Exp. Neurol.* **22**:629–634.
83. Wisniewski, H. M., Narang, H. K., and Terry, R. D., 1976, *J. Neurol. Sci.* **27**:173–181.
84. Burger, P. C., and Vogel, F. S., 1973, *Am. J. Pathol.* **73**:457–468.
85. Owens, D., Dawson, J. C., and Lowsin, S., 1971, *Am. J. Ment. Defic.* **75**:606–612.
86. Wisniewski, K., Jervis, G. A., Moretz, R. C., and Wisniewski, H., 1979, *Ann. Neurol.* **5**:288–294.
87. Selkoe, D. J., 1978, *Current Neurology*, Volume 1 (H. R. Tyler and D. M. Dawson, eds.), Houghton Mifflin, Boston, pp. 360–387.
88. Shelanski, M. L., and Selkoe, D. J., 1981, *The Molecular Basis of Neuropathology* (A. N. Davison and R. H. S. Thompson, eds.), Igaku-Shoin, New York, pp. 591–600.
89. Marotta, C. A. (ed.), 1983, *Neurofilaments*, University of Minnesota Press, Minneapolis.
90. Iqbal, K., Wisniewski, H. M., Shelanski, M. L., Brostaff, S., Liwnicz, B. H., and Terry, R. D., 1974, *Brain Res.* **77**:337–343.
91. Allsop, D., Landon, M., and Kidd, M., 1983, *Brain Res.* **259**:348–352.
92. Gambetti, P., Autelio-Gambetti, L., Perry, G., Shecket, G., and Crane, R. C., *Lab. Invest.* **49**:430–435.
93. Selkoe, D. J., Ihara, Y., and Salazar, F. J., 1982, *Science* **215**:1243–1245.
93a. Iqbal, K., Zaidi, T., Thompson, C. H., Merz, P. A., and Wisniewski, H. M., 1984, *Acta Neuropathol.* **62**:167–177.

94. Ihara, Y., Abraham, C., and Selkoe, D., 1983, *Nature* **304**:727–730.
94a. Grundke-Iqbal, I., Iqbal, K., Tung, Y.-C., and Wisniewski, H., 1984, *Acta Neuropathol.* **62**:259–267.
94b. Wang, G. P., Grundke-Iqbal, I., Kascsak, R. J., Iqbal, K., and Wisniewski, H., 1984, *Acta Neuropathol.* **62**:268–275.
94c. Gambetti, P., Autilio-Gambetti, L., Perry, G., Shecket, G., and Crane, R. C., 1983, *Lab. Invest.* **49**:430–435.
95. Hirano, A., Dembitzer, H. M., Kurland, L. T., and Zimmerman, H. M., 1968, *J. Neuropathol. Exp. Neurol.* **26**:167–182.
96. Woodard, J. S., 1962, *J. Neuropathol. Exp. Neurol.* **21**:85–91.
97. Woodard, J. S., 1962, *Am. J. Pathol.* **49**:1157–1169.
98. Tomlinson, B. E., and Kitchener, D., 1972, *J. Pathol.* **106**:165–185.
99. Ball, M. J., and Lo, P., 1977, *J. Neuropathol. Exp. Neurol.* **56**:474–487.
100. Hirano, A., and Zimmerman, H. M., 1962, *Arch. Neurol.* **7**:227–242.
101. Goldman, E., and Suzuki, K., 1982, *J. Neuropathol. Exp. Neurol.* **41**:359.
102. O'Brien, L., Shelley, K., Towfighi, J., and McPherson, A., 1980, *Proc. Natl. Acad. Sci. U.S.A.* **77**:2260–2264.
103. Gibson, P. E., and Tomlinson, B. E., 1977, *J. Neurol. Sci.* **33**:199–206.
104. Scheibel, A., 1979, *Congenital and Acquired Cognitive Disorders* (R. Katzman, ed.), Raven Press, New York, pp. 107–122.
105. Williams, R. S., Ferrante, R. J., and Caviness, V. S., 1978, *J. Neuropathol. Exp. Neurol.* **37**:13–33.
106. Buell, S. J., 1982, *J. Neuropathol. Exp. Neurol.* **41**:500–507.
107. Buell, S. J., and Coleman, P. D., 1979, *Science* **206**:854–856.
108. Buell, S. J., and Coleman, P. D., 1981, *Brain Res.* **214**:23–41.
109. Mehraein, P., Yamada, M., and Tarnowska-Dziduszko, E., 1975, *Advances in Neurology*, Volume 1 (G. W. Kreutzerg, ed.), Raven Press, New York, pp. 453–458.
110. Duffy, P. E., Rapport, M., and Graf, L., 1980, *Neurology* **30**:778–782.
111. Mann, D. M. A., and Sinclair, K. G. A., 1978, *Neuropathol. Appl. Neurobiol.* **4**:129–135.
112. Nandy, K., 1981, *Strategies for the Development of an Effective Treatment for Senile Dementia* (T. Crook and S. A. Gershon, eds.), Mark Pawley Associates, New Canaan, Connecticut, pp. 231–245.
113. Bowen, D. M., Smith, C. B., White, P., and Davison, A., 1976, *Brain* **99**:459–476.
114. Bowen, D. M., Smith, C. B., White, P., Flack, R. H., Carrasco, L. H., Geyde, J. L., and Davison, A., 1977, *Brain* **100**:427–453.
115. Pope, A., Hess, H. H., and Lewin, E., 1964, *Morphological and Biochemical Correlates of Neural Activity* (M. M. Cohen and R. Snider, eds.), Harper and Row (Hoeber), New York, pp. 98–111.
116. Crapper, D. R., Quittkat, S., and DeBoni, U., 1979, *Brain* **103**:483–495.
117. Lewis, P. N., Lukiw, W. J., DeBoni, U., and Crapper-McLachlan, D., 1981, *J. Neurochem.* **37**:1193–1202.
118. Karlik, S. J., Eichorn, G. L., and Crapper-McLachlin, D. R., 1980, *Neurotoxicology* **1**:83–88.
119. DeBoni, U., and Crapper-McLachlan, D. R., 1981, *Strategies for the Development of an Effective Treatment for Senile Dementia* (T. Crook and S. A. Gershon, eds.), Mark Pawley Associates, New Canaan, Connecticut, pp. 215–228.
120. Uemura, E., and Hartman, H. A., 1978, *J. Neuropathol. Exp. Neurol.* **37**:487–496.
121. Mann, D. M. A., Neary, D., Yates, P. O., Lincoln, J., Snowden, J. S., and Stanworth, P., 1981, *J. Neurol. Neurosurg. Psychiatry* **44**:97–102.
122. Embree, L. J., Bass, N. H., and Pope, A., 1972, *Handbook of Neurochemistry*, Volume 7 (A. Lajtha, ed.), Plenum Press, New York, pp. 329–369.
123. Suzuki, K., Katzman, R., and Korey, S. R., 1965, *J. Neuropathol. Exp. Neurol.* **24**:211–224.
124. Pope, A., and Embree, L. J., 1976, *Handbook of Clinical Neurology*, Volume 27 (P. J. Vinken and G. W. Bruyn, eds.), North-Holland, Amsterdam, pp. 477–501.
125. Bowen, D. M., Smith, C. B., and Dawson, A. N., 1973, *Brain* **96**:849–856.
126. Freyhan, F. A., Woodford, R. B., and Kety, S. S., 1951, *J. Nerv. Ment. Dis.* **113**:449–456.

127. Obrist, W. D., Chirian, E., Cronquist, S., and Ingvar, D. H., 1970, *Neurology* **20**:315–322.
128. Lavy, S., Melamed, E., Benton, S., Cooper, G., and Rinof, Y., 1978, *Ann. Neurol.* **4**:445–450.
129. Raichle, M. E., Grubb, R. L., Gado, M.-H., Eichling, J. O., and Hughes, J. P., 1978, *Senile Dementia: A Biomedical Approach* (K. Nandy, ed.), Elsevier, Amsterdam, pp. 131–138.
130. Yamaguchi, F., Meyer, J. S., Yamamoto, M., Fumihiko, S., and Shaw, T., 1980, *Arch. Neurol.* **37**:410–418.
131. Melamed, E., Lavy, S., Siew, E., Benton, S., and Cooper, G., 1978, *J. Neurol. Neurosurg. Psychiatry* **41**:894–899.
132. Ingvar, D. W., Brun, A., Hagberg, B., and Gustafson, L., 1978, *Alzheimer's Disease: Senile Dementia and Related Disorders* (R. Katzman, R. Terry, and K. Bick, eds.), Raven Press, New York, pp. 441–451.
133. Hachinski, V. C., Iliff, L. D., Zilha, E., DuBoulay, G. H., McAllister, V. L., Marshall, J., Russell, R. W. R., and Symon, L., 1975, *Arch. Neurol.* **32**:632–637.
134. Farkas, T., Ferris, S., Wolff, A., Deleon, M. J., Christman, D. R., Reisberg, B., Alavi, A., Fowler, J. S., George, A. E., and Reivich, M., 1982, *Am. J. Psychiatry* **139**:352–353.
135. Frackowiak, R. S. J., Pozzilli, C., Legg, N. J., DuBoulay, G. H., Marshall, J., Lenzi, G. L., and Jones, T., 1981, *Brain* **104**:753–778.
135a. Besson, J. A. O., Corrigan, F. M., Foreman, E. I., Ashcroft, G. W., Eastwood, L. M., and Smith, F. W., 1983, *Lancet* **1**:789.
136. Perry, E., Perry, R. H., Tomlinson, B. E., Blessed, G., and Gibson, P. H., 1980, *Neurosci. Lett.* **18**:105–110.
137. Sims, N. R., Bowen, D. M., and Davison, A. N., 1981, *Biochem. J.* **196**:867–876.
138. Suzuki, K., Korey, S. R., and Terry, R. D., 1964, *J. Neurochem.* **11**:403–412.
139. Perl, D. P., Gajdusek, D. C., Garruto, R. M., Yanagihara, R. T., and Gibbs, C. J., 1982, *Science* **217**:1053–1055.
140. Markesbery, W. R., Leung, P. K., and Butterfield, D. A., 1980, *J. Neurol. Sci.* **45**:323–330.
141. Crapper, D. R., Krishnan, S. S. and Dalton, A. J., 1973, *Science* **180**:511–513.
142. Markesbery, W. R., Ehmann, W. D., and Hossain, T., 1981, *J. Neuropathol. Exp. Neurol.* **40**:359.
143. Perl, D. P., and Brody, A. R., 1980, *Science* **208**:297–299.
144. Austin, J., Rinehart, R., Williamson, T., Bucar, P., Russ, K., Nikaido, T., and Lafrance, M., 1973, *Prog. Brain Res.* **40**:485–495.
145. Bartus, R. Y., Dean, R. L., Beer, B., and Lippa, A. S., 1982, *Science* **217**:408–417.
146. Davies, P., 1981, *Strategies for the Development of an Effective Treatment for Senile Dementia* (T. Crook and S. Gershon, eds.), Mark Pawley Associates, New Canann, Connecticut, pp. 19–32.
147. Davies, P., and Maloney, A. J. F., 1976, *Lancet* **2**:1403.
148. Spillane, J. A., White, P., Goodhardt, M. J., Flack, R. H. A., Bowen, D. M., and Davison, A., 1977, *Nature* **266**:558–559.
149. White, P., Hiley, C. R., Goodhardt, M. J., Carrasco, L. H., Keet, J. P., Williams, I. E. I., and Bowen, D. M., 1977, *Lancet* **1**:668–671.
150. Perry, E. K., Gibson, P. H., Blessed, G., Perry, R. H., and Tomlinson, B. E., 1977, *J. Neurol. Sci.* **34**:247–265.
151. Perry, E. K., Perry, R. H., Blessed, G., and Tomlinson, B. E., 1977, *Lancet* **1**:189.
152. Johnson, S., and Domino, E. F., 1971, *Clin. Chim. Acta* **35**:421–428.
153. Sims, N. R., Bowen, D. M., Smith, C. C. T., Flack, R. H. A., Davison, A. N., Snowden, J. S., and Neary, D., 1980, *Lancet* **1**:333–336.
154. Reisine, T. D., Yamamura, H. I., Bird, E. D., Spokes, S., and Enna, S. J., 1978, *Brain Res.* **159**:477–481.
155. Henke, H., and Lang, W., 1983, *Brain Res.* **267**:281–291.
156. DeKosky, S. T., Scheff, S. W., and Markesbery, W. R., 1984, (submitted).
156a. Perry, E. K., Atack, J. R., Perry, R. H., Hardy, J. A., Dodd, P. R., Edwardson, J. A., Blessed, G., Tomlinson, B. E., and Fairbairn, A. F., 1984, *J. Neurochem.* **42**:1402–1410.
157. Drachman, D., and Leavitt, J., 1974, *Arch. Neurol.* **30**:113–121.
158. Smith, C. M., and Swash, M., 1978, *Ann. Neurol.* **3**:471–473.

159. Drachman, D., and Glosser, G., 1981, *Strategies for the Development of an Effective Treatment for Senile Dementia* (T. Crook and S. A. Gershon, eds.), Mark Pawley Associates, New Canaan, Connecticut, pp. 35–51.
160. Perry, E. K., Tomlinson, B. E., Blessed, G., Bergman, K., Gibson, P. H., and Perry, R. H., 1978, *Br. Med. J.* **2**:1457–1459.
160a. Gottfries, C.-G., Adolfsson, R., Acquilonius, S. M., Carlsson, A., Eckernas, S.-A., Nordberg, A., Oreland, L., Svennerholm, L., Wiberg, A., and Winblad, B., 1983, *Neurobiol. Aging* **4**:261–271.
161. Horrocks, L. A., VanRollins, M., and Yates, A. J., 1981, *The Molecular Basis of Neuropathology* (A. N. Davison and R. H. S. Thompson, eds.), Igaku-Shoin, New York, pp. 601–630.
162. Lang, W., and Henke, W., 1983, *Brain Res.* **267**:271–280.
163. Bowen, D. M., and Davison, A. N., 1980, *Psychol. Med.* **10**:315–319.
164. Davies, P., and Verth, A. H., 1978, *Brain Res.* **138**:385–392.
165. Hammer, R., Berrie, C. P., Birdsall, N. J. M., Burge, A. S., and Hulme, E. C., 1980, *Nature* **283**:90–92.
165a. Hirschowitz, B. I., Hammer, R., Giachetti, A., Keirns, J. J., and Levine, R. R., eds., 1983, Subtypes of muscarinic receptors, 1983, *Trends Pharmacol. Sci.* **4(Suppl)**:1–103.
166. Caulfield, M. P., Straughan, D. W., Cross, A. J., Crow, T., and Birdsall, N. J. M., 1982, *Lancet* **2**:1277.
166a. Mash, D. C., Sevush, S., Flynn, D. D., Norenberg, M. D., and Potter, L. T., 1984, *Neurology* **34(Suppl. 1)**:120.
167. Davies, P., and Feisullin, S., 1981, *Brain Res.* **216**:449–454.
168. Soininin, H., Halonen, T., and Reikkinen, P. J., 1981, *Acta Neurol. Scand.* **64**:217–224.
169. Bowen, D. M., Benton, J. S., Spillane, J. A., Smith, C. C. T., and Allen, S. J., 1982, *J. Neurol. Sci.* **57**:191–202.
170. Sims, N. R., Bowen, D. M., Allen, S. J., Smith, C. C. T., Neary, D., Thomas, D. J., Davison, A. N., 1983, *J. Neurochem.* **40**:503–509.
171. Kuhar, M. J., 1976, *Biology of Cholinergic Function* (A. M. Goldberg and I. Hanin, eds.), Raven Press, New York, pp. 3–27.
172. Emson, P. C., and Lindvall, O., 1979, *Neuroscience* **4**:1–30.
173. Averback, P., 1981, *Arch. Neurol.* **38**:230–235.
174. Lehmann, J., Nagy, J. I., Atmadja, S., and Fibiger, H. C., 1980, *Neuroscience* **5**:1161–1174.
175. Hartgraves, S. L., Mensah, P. L., and Kelly, P. H., 1982, *Neuroscience* **7**:2369–2376.
176. Johnston, M. V., McKinney, M., and Coyle, J. T., 1979, *Proc. Natl. Acad. Sci. USA* **76**:5329–5396.
177. Davies, P., and Feisullin, S., 1982, *J. Neurochem.* **39**:1743–1747.
178. Rosser, M. N., Svendsen, C., Hunt, S. P., Mountjoy, C. Q., Roth, M., and Iversen, L. L., 1982, *Neurosci. Lett.* **28**:217–222.
179. Mann, D. M., and Yates, P. O., 1982, *J. Neurol. Neurosurg. Psychiatry* **45**:936–943.
179a. Mountjoy, C. Q., Rossor, M. N., Iverson, L. L., and Roth, M., 1984, *Brain* **107**:507–518.
180. Perry, R. H., Candy, J. M., Perry, E. K., Irving, D., Blessed, G., Fairbairn, A. F., and Tomlinson, B. E., 1982, *Neurosci. Lett.* **33**:311–315.
181. Tagliavini, J., and Pilleri, G., 1983, *Lancet* **1**:469–470.
182. Jones, E. G., Burton, H., Saper, C. B., and Swanson, L. W., 1976, *J. Comp. Neurol.* **167**:385–420.
183. Armstrong, D. M., Saper, C. B., Levey, A. I., Wainer, B. H., and Terry, R. D., 1983, *J. Comp. Neurol.* **216**:53–68.
184. Pearson, R. C. A., Gatter, K. C., Brodal, P., and Powell, T. P. S., 1983, *Brain Res.* **259**:132–136.
185. Adolfson, R., Gottfics, C. G., Roos, B. E., and Winblad, B., 1979, *Br. J. Psychiatry* **135**:216–223.
186. Mann, D. M. A., Lincoln, J., Yates, P. O., Stamp, J. E., and Toper, S., 1980, *Br. J. Psychiatry* **136**:533–541.
187. Cross, A. J., Crow, T., Perry, E. K., Perry, R. H., Blessed, G., and Tomlinson, B. E., 1981, *Br. Med. J.* **282**:93–94.

188. Perry, E. K., Tomlinson, B. E., Blessed, G., Perry, R. H., Cross, A. J., and Crow, T., 1981, *Lancet* **2:**149.
189. Benton, J. S., Bowen, D. M., Allen, S. J., Haan, E. A., Davison, A. N., Neary, D., Murphy, R. P., and Snowden, J. S., 1982, *Lancet* **1:**456.
190. Moore, R. Y., 1982, *Ann. Neurol.* **12:**321–327.
191. Vijayashankar, N., and Brody, H., 1979, *J. Neuropathol. Exp. Neurol.* **38:**490–497.
192. Tomlinson, B. E., Irving, D., and Blessed, G., 1981, *J. Neurol. Sci.* **49:**419–428.
193. Adolfsson, R., Gottfries, C. G., Oreland, L., Wiberg, A., and Winblad, B., 1980, *Life Sci.* **27:**1029–1034.
194. Gottfries, C. G., Gottfries, I., and Roos, B. E., *J. Neurochem.* **16:**1341–1345.
195. Carlsson, A., Adolfsson, R., Aquilonius, S. M., Gottfries, C. G., Oreland, L., Svennerholm, L., and Bengt, W., 1980, *Ergot Compounds and Brain Function: Neuroendocrine and Neuropsychiatric Aspects* (M. Goldstein, ed.), Raven Press, New York, pp. 295–304.
196. Parkes, J. D., Marsden, C. D., Rees, J. E., Curzon, G., Kantamaneni, B. D., Knill-Jones, R., Akbar, A., Das, S., and Katari, M., 1974, *Q. J. Med.* **43:**49–61.
197. Gottfries, C. G., 1981, *Strategies for the Development of an Effective Treatment for Senile Dementia* (T. Crook and S. A. Gershon, eds.), Mark Pawley Associates, New Canaan, Connecticut, pp. 107–120.
198. Bowen, D., 1980, *Aging of the Brain and Dementia* (L. Amaducci, A. N. Davison, and P. Antuono, eds.), Raven Press, New York, pp. 127–138.
199. Smith, C. C. T., Bowen, D. M., Sims, N. R., Neary, D., and Davison, A. N., 1983, *Brain Res.* **264:**138–141.
200. Crystal, H. A., and Davies, P., 1982, *J. Neurochem.* **38:**1781–1784.
201. Davies, P., Katzman, R., and Terry, R. D., 1980, *Nature* **288:**279–280.
202. Davies, P., and Terry, R. D., 1981, *Neurobiol. Aging* **2:**9–14.
203. Rossor, M. N., Emson, P. C., Mountjoy, C. Q., Roth, M., and Iversen, L. L., 1980, *Neurosci. Lett.* **20:**373–377.
204. Gash, D. M., and Thomas, G. J., 1983, *Trends Neurosci.* **6:**197–198.
205. Rossor, M., Iversen, L., Mountjoy, C. Q., Roth, M., Hawthorn, J., Ang, V.Y., and Jenkins, J. S., 1980, *Lancet* **2:**1367–1368.
206. Sundquist, J., Forlsling, M. L., Olsson, J., and Akerlund, M., 1983, *J. Neurol..* **46:**14–17.
207. Kastin, A. J., Olson, G. A., Sandman, C. A., and Ehrensing, R. H., 1981, *Strategies for the Development of an Effective Treatment for Senile Dementia* (T. Crook and S. A. Gershon, eds.), Mark Pawley Associates, New Canaan, Connecticut, pp. 139–152.
208. Roberts, E., 1981, *Strategies for the Development of an Effective Treatment for Senile Dementia* (T. Crook and S. A. Gershon, eds.), Mark Pawley Associates, New Canaan, Connecticut, pp. 247–320.
209. Reisberg, B., Ferris, S., Anand, R., Mir, P., Geibel, V., and DeLeon, M. J., 1983, *N. Engl. J. Med.* **308:**721–722.
210. Bowen, D. M., Allen, S. J., Benton, J. S., Goodhardt, M. J., Haan, E. A., Palmer, A. M., Sims, N. R., Smith, C. C. T., Spillane, J. A., Esiri, M. M., Neary, D., Snowdon, S. S., Wilcock, G. K., and Davison, A. N., 1983, *J. Neurochem.* **41:**266–272.
211. Ellis, W. G., McCulloch, J. R., and Corley, C. L., 1974, *Neurology* **24:**101–106.
212. Wisniewski, K., Howe, J., Willians, D. G., and Wisniewski, H. M., 1978, *Biol. Psychiatry* **13:**619–627.
213. Ball, M. J., and Nuttall, K., 1981, *Neuropathol. Appl. Neurobiol.* **7:**13–20.
214. Yates, C. M., Simpson, J., Maloney, A. F. J., Gordon, A., and Reid, A. H., 1980, *Lancet* **2:**979.
215. Spokes, E. G., 1979, *Brain* **102:**333–346.
216. Heston, L. L., 1977, *Science* **196:**322–323.
217. Heston, L. L., 1979, *Congenital and Acquired Cognitive Disorders* (R. Katzman, ed.), Raven Press, New York, pp. 167–176.
218. Heston, L. L., Mastri, A. R., Anderson, V. E., Roth, V., and Matsuyama, S. S., 1981, *Arch. Gen. Psychiatry* **38:**1085–1090.
219. Rossor, M. N., 1981, *Br. Med. J.* **283:**1588–1590.
220. Haubrich, D. R., Wedeking, P. W., and Wang, P. F. L., 1974, *Life Sci.* **14:**921–927.

221. Cohen, E. L., and Wurtman, R. J., 1975, *Life Sci.* **16**:1095–1102.
222. Wecker, L., Dettbarn, W. D., and Schmidt, D. E., 1978, *Science* **199**:86–87.
223. London, E. D., and Coyle, J. T., 1978, *Biochem. Pharmacol.* **27**:2962–2965.
224. Flentge, F., and Vandenberg, C. J., 1979, *J. Neurochem.* **32**:1331–1333.
225. Tromner, B. A., Schmidt, D. E., and Wecker, L., 1982, *J. Neurochem.* **39**:1704–1709.
226. Bartus, R. T., Dean, R. L., Goas, J. A., and Lippa, A. S., 1980, *Science* **209**:301–303.
227. Bartus, R. T., 1979, *Science* **206**:1087–1089.
228. Bartus, R., Dean, R. L., and Beer, B., 1980, *Neurobiol. Aging* **1**:145–152.
229. Davis, K. L., Mohs, R. C., Davis, B. M., Levy, M., Rosenberg, G. S., Horvath, T. B., DeNigris, Y., Ross, A., Decker, P., and Rothpearl, M. A., 1981, *Strategies for the Development of an Effective Treatment for Senile Dementia* (T. Crook and S. A. Gershon, eds.), Mark Pawley Associates, New Canaan, Connecticut, pp. 53–69.
230. Christie, J. E., Shering, A., Ferguson, J. J., and Glen, A. I. M., 1981, *Br. J. Psychiatry* **138**:46–50.
231. Peters, B. H., and Levin, H. S., 1979, *Ann. Neurol.* **6**:219–221.
232. Davis, K. L., Mohs, R. C., Rosen, W. G., Greenwald, B. S., Levy, M. I., and Horvath, T. B., 1983, *N. Engl. J. Med.* **308**:721.
233. Thal, L. J., and Fuld, P. A., 1983, *N. Engl. J. Med.* **308**:720.
234. Thal, L. J., Fuld, P. A., Masur, D. M., and Sharpless, N. S., 1983, *Ann. Neurol.* **13**:491–496.
235. Friedman, E., Sherman, K. A., Ferris, S. H., Reisberg, B., Bartus, R. T., and Schneck, M. K., 1981, *N. Engl. J. Med.* **304**:1490–1491.
236. Bartus, R. T., Dean, R. L., Sherman, K. A., Friedman, E., and Beer, B., 1981, *Neurobiol. Aging* **2**:105–111.
237. Ansell, G. B., 1981, *Central Neurotransmitter Turnover* (C. J. Pycock and P. V. Taberner, eds.), University Park Press, Baltimore, pp. 81–104.
238. Feldman, R. D., Chandler, D. A., Levy, L. L., and Glasser, G., 1963, *Neurology* **13**:811–824.
239. Masters, C. L., Gajdusek, D. C., and Gibbs, C. J., Jr., 1981, *Brain* **104**:535–558.
240. Goudsmit, J., White, B. J., Weitkamp, L. R., Keats, B. J. B., Morrow, C. H., and Gajdusek, D. C., 1981, *J. Neurol. Sci.* **49**:79–89.
241. Jarvik, L. F., Ruth, V., and Matsuyama, S. S., 1980, *Arch. Gen. Psychiatry* **37**:280–286.
242. Heston, L. L., and White, J., 1980, *Proc. Am. Psychopathol. Assoc.* **69**:63.
243. Larsson, T., Sjogren, T., and Jacobsen, G., 1963, *Acta Psychiatr. Scand.* **39**(Suppl. 167):3–259.
244. Roth, M., 1978, *Alzheimer's Disease: Senile Dementia and Related Disorders* (R. Katzman, R. D. Terry, and K. Bick, eds.), Raven Press, New York, pp. 71–85.
245. Cook, R. H., Ward, B. E., and Austin, J. H., 1979, *Neurology* **29**:1402–1412.
246. Nordenson, I., Adolfsson, R., Beckman, G., Bucht, G., and Winblad, B., 1980, *Lancet* **1**:481–483.
247. White, B. J., Crandall, C., Goudsmit, J., Morrow, C. H., Alling, D. W., Gajdusek, D. C., and Tijio, J. H., 1981, *Am. J. Med. Genet.* **10**:77–89.
248. Goudsmit, J., Morrow, C. H., Asher, D. M., Yanagihara, R. T., Masters, G. L., Gibbs, C. J., and Gajdusek, D. C., 1980, *Neurology* **30**:945–950.
249. DeBoni, U., and Crapper, D. R., 1978, *Nature* **271**:566–568.
250. Appel, S. H., 1981, *Ann. Neurol.* **10**:499–505.
251. Varon, S., Adler, R., and Skaper, S. D., 1981, *Mechanisms of Growth Control* (R. O. Becker, ed.), Charles C Thomas, Springfield, Illinois, pp. 54–63.
252. Barde, Y.-A., Edgar, D., and Thoenen, H., 1983, *Annu. Rev. Physiol.* **45**:601–612.
253. Perez-Polo, R., deVellis, J., and Haber, B., (eds.), 1982, *Clinical and Biological Research Progress in Growth and Trophic Factors*, Volume 118, Liss, New York.
254. Smith, R. G., and Appel, S. H., 1983, *Science* **219**:1079–1081.
255. Rothman-Schonfeld, A., and Katzman, R., 1983, *Biological Aspects of Alzheimer's Disease* (R. Katzman, ed.), Cold Spring Harbor Laboratories, Cold Spring Harbor, New York, pp. 269–274.
256. Nieto-Sampedro, M., Lewis, E. R., Cotman, C. W., Manthorpe, M., Skaper, S. D., Barbin, G., Longo, F. M., and Varon, S., 1982, *Science* **217**:860–861.

257. Manthorpe, M., Luyten, W., Longo, F. M., and Varon, S., 1983, *Brain Res.* **267**:57–66.
258. Cotman, C. W., and Nieto-Sampedro, M., 1983, *Biological Aspects of Alzheimer's Disease* (R. Katzman, ed.), Cold Spring Harbor Laboratory, Cold Spring Harbor, N.Y., pp. 275–284.
259. Ojika, K., and Appel, S., 1983, *Biological Aspects of Alzheimer's Disease* (R. Katzman, ed.), Cold Spring Harbor Laboratories, Cold Spring Harbor, New York, pp. 285–295.
260. Mann, D. M., and Yates, P. O., 1983, *J. Neurol. Neurosurg. Psychiatry* **46**:96.
261. Sajdel-Sulkowska, E. M., and Marotta, C. A., 1984, *Science* **225**:947–949.
262. Cutler, N. R., Duara, R., Creasey, H., Grady, C. L., Haxby, J. V., Schapiro, M. B., and Rapoport, S. E., 1984, *Ann. Int. Med.* **101**:355–369.
263. Bustany, P., Henry, J. F., Soussaline, F., and Comar, D., 1983, *Functional Radionuclide Imaging of the Brain* (P. G. Mastretti, ed.), Raven Press, New York, pp 319–326.
264. Potter, L. T., Flynn, D. D., Hanchett, H. E., Kalinoski, D. L., Luber-Narod, J., and Mash, D. C., 1983, *Trends Pharmacol. Sci.* (Supplement: Subtypes of Muscarinic Receptors):22–31.

24

Depression

F. Petty, A. D. Sherman, and G. Winokur

1. INTRODUCTION

A variety of perhaps insurmountable problems confront the neuroscientist attempting to study the neurochemistry of depression. First, it is necessary to appreciate the difficulties involved in the classification of this disorder.

That everyone is depressed at one time or another in life is a normal part of human existence. However, since ancient times, it has been recognized that occasionally individuals sink into a low mood, often accompanied by psychological and somatic signs and symptoms, which is not appropriate to their surroundings or experience. Over the years, psychiatrists, most of whom do not have a neurochemical orientation, have attempted a variety of diagnostic schemes. As reviewed by Kendell,[1] these schemes in general are based on clinical observation and not on objective criteria, particularly not on biologically empirical criteria. They range from a one-category scheme, "depressive illness," considered to encompass the entire spectrum of depressive disorders; to a five-category scheme that subdivides depression into (1) physiological retardation depression, (2) tension depression, (3) schizoaffective depression, (4) depression secondary to a problem in life, and (5) depression as a prodrome of organic illness. Usually the diagnostic schemes proposed are not particularly helpful in determining appropriate therapy.

Within the last 20 years a more useful way of categorizing the affective disorders has been developed originally by Leonhard,[2] and later by Angst,[3] Perris,[4] and Winokur.[5] These schemes differentiate by polarity. Bipolar affective disorder is diagnosed in patients who have episodes not only of depression but also mania, whereas unipolar depression describes patients who have episodes of depression only. Bipolar illness differs in that it has an earlier age of onset and more frequent recurrence. In the depressive phase of bipolar illness, however, bipolar patients respond similarly to antidepressant treatment or electroconvulsive therapy as do unipolars. In the manic phase, however, they re-

F. Petty • Veterans Administration Medical Center and Department of Psychiatry, University of Texas, Dallas, Texas 75216. *A. D. Sherman and G. Winokur* • Department of Psychiatry, University of Iowa College of Medicine, Iowa City, Iowa 52242.

spond differentially to treatment with lithium salts, and there is evidence that lithium is also useful in these patients in prophylaxis for both mania and depression.

Another potentially useful classification is that of primary versus secondary depression.[6] Primary depression is judged to occur in the absence of preexistent psychiatric or serious physical illness, whereas secondary depression is preceded by, for example, anxiety neurosis, schizophrenia, alcoholism, etc. As we shall see there are neurochemical differences between primary and secondary depression, although again it is not clear whether response to antidepressant treatment is different in one or the other. More recently, Winokur has proposed subdividing primary unipolar depression according to family history.[7] Depression spectrum disease is considered to occur in patients who develop depression but whose family history is positive for alcoholism or sociopathy, pure depressive disease in patients whose family history is positive only for depression, and sporadic depressive disease in depressed patients in whom a family history for psychiatric illness cannot be demonstrated. Again, as will subsequently be reviewed, there are neurochemical differences that validate this as a useful classification at least for research purposes.

In 1980 the American Psychiatric Association accepted criteria similar to those originally developed by Feighner *et al.*[8] for the diagnosis of depression, hopefully at least standardizing the confusion. Briefly, these criteria require, for the diagnosis of depression, a dysphoric mood with feelings of depression, sadness, blueness, and hopelessness for a period of at least 2 weeks accompanied by four of eight symptoms: (1) poor appetite or weight loss or excess appetite and weight gain, (2) insomnia or hypersomnia, (3) psychomotor agitation or retardation, (4) loss of interest or pleasure in usual activities, (5) loss of energy or fatigue, (6) feelings of worthlessness or self-reproach, (7) complaints or evidence of diminished ability to think or concentrate, and (8) recurrent thoughts of death, suicidal ideations, or suicidal attempt. Although the criteria have demonstrated interrater reliability for diagnosis of the syndrome, from the standpoint of neurochemistry, a patient with poor appetite and weight loss, insomnia, and psychomotor retardation may be suffering a different process than one with weight gain, hypersomnia, or psychomotor agitation. The picture is additionally complicated by the fact that the diagnosis can be modified if there are psychotic features such as delusions and/or hallucinations, and that these features may be "mood congruent" or "mood incongruent." Also, the depression may additionally be described as "with melancholia," which means loss of all pleasure in all or most all activities, lack of reactivity to usually pleasurable stimuli, and at least three of the following: distinct quality of depressed mood, the depression being worse in the morning, early morning awakening, marked psychomotor retardation or agitation, significant anorexia or weight loss, and excessive or inappropriate guilt.

Thus, we are confronted with a situation in which psychiatrists often disagree about a particular diagnosis based on their clinical examination of the patient. We are faced with a situation in which standardized diagnostic criteria are of *very* recent adoption. These diagnostic problems apply to an illness whose systematic study is important for at least two reasons: first, there is a high incidence of depression in the general population (8–10%), making it easily one of the most common national health problems; and second, depression is a disease having a very high mortality—about 15% of patients with affective disorder will eventually take their own life.[9] This is comparable to the mortality for one round of Russian roulette or one myocardial infarction.

To complicate the problem beyond the diagnostic confusion, a case can be made that neurochemical studies on depression are impossible. Direct neurochemical studies on the brains of depressed people simply cannot be done. A few studies on brains of depressed people who committed suicide or were depressed and died of other causes will be reviewed but must be interpreted with great caution, as significant changes in brain metabolism occur within seconds after death; often it is not possible to obtain autopsy material for several hours after this event. Thus, we are restricted to measuring body fluids: urine, plasma, saliva, and cerebrospinal fluid—faced with the strong possibility that nothing found in the fluids necessarily reflects central metabolic activity in the brain. Even studies with CSF are difficult to interpret because concentrations of biologically active compounds in CSF do not necessarily reflect regional brain metabolism. Likewise, the contribution to CSF from the spinal cord (as opposed to the brain) has to be considered and is difficult to assess.

Another approach to the problem would be the development of an adequate animal model of depression, and some progress has been made in this respect.

We cannot attempt to pretend an exhaustive review of the field, nor should we, since many aspects have been recently and well reviewed by other authors. Instead, we will briefly discuss some of the traditional hypotheses of depression and try to focus on some of the more recent findings in the area, clinical and preclinical.

2. *MONOAMINES*

2.1. *Norepinephrine*

The catecholamine hypothesis of affective disorder has been well reviewed recently.[10–12] Briefly, it evolved from the findings that reserpine, which depletes monoamines, sometimes leads to a depressive syndrome, and that the prototypical tricyclic antidepressants and monoamine oxidase inhibitors appear to increase the concentration of monoamines at the synaptic cleft. Two hypotheses were formulated in the mid-1960s: the catecholamine hypothesis and the serotonin hypothesis. The catecholamine hypothesis[13,14] probably has less objective support than the serotoninergic hypothesis. Autopsy material from one patient with severe depression demonstrated higher levels of norepinephrine in the substantia nigra but no real differences from controls in eight other brain regions.[15] MHPG, on the other hand, was found to be low in the globus pallidus, hypothalamus, mammillary bodies, substantia nigra, and raphe in endogenous depressives as compared to controls, but there are confounding variables in that the patients had been on long-term therapy with antidepressant drugs with their final dose soon before death.

Studies of CSF MHPG, the principle metabolite of norepinephrine, and of HVA, another metabolite, have in general yielded contradictory and conflicting data.[11] Again this probably reflects differences in techniques of obtaining spinal fluid, possible differences in assay, and difficulty in finding appropriate controls for patients with depression.

Another approach to catecholamine studies has been measurement of MHPG in urine. Although it was originally thought that most MHPG in urine was of a central origin, this in fact has recently been demonstrated not to be the case[16] with most recent estimates indicating that 20% or less of total urinary MHPG is derived from brain.

Obviously, in addition to being a neurohumor, norepinephrine is also a peripheral hormone and its levels in the periphery, as well as those of its metabolite MHPG, would be enormously affected by factors such as diet, exercise, psychological or physical stress. MHPG was originally thought to be lower in the urine of patients suffering from depression.[17] These findings, however, have not always been replicated. For example, Hollister *et al.* recently found that "the calculated range of excretion of MHPG for depressed patients was similar to that for normal persons."[18] It has been suggested that depressives with low levels of urinary MHPG might preferentially respond to tricyclic antidepressants, which preferentially blockade norepinephrine reuptake. Data to support this hypothesis are still preliminary and it is unclear whether this will develop into a useful prognostic tool for clinical practice.[18]

In summary, in spite of its popularity for almost 20 years and the enormous amount of research undertaken in attempts to confirm this particular hypothesis, little compelling evidence has accumulated to implicate norepinephrine as *the* neurotransmitter responsible for depression.

2.2. Serotonin

The serotonin hypothesis of depression also originated in the mid-1960s and for much the same reasons. Evidence implicating it in depression may, however, rest on slightly firmer grounds than that for norepinephrine. Several studies of serotonin levels in postmortem brain samples have been undertaken.[11] Again, suicide data must be interpreted with great caution since not all suicides suffer a depressive syndrome. Alcoholism accounts for a large proportion of suicides in the U.S. and Great Britain. In any event, few differences were found in comparison with controls but there was some indication of decreased levels of serotonin in the nuclei raphe, dorsalis, and centralis. One study comparing 28 controls with 23 depressed suicides found a small but significant decrease in levels of 5-HIAA in the hindbrain. A significant problem in these studies is that the nature of death in suicide cannot be considered to be the same as that for controls and we are given little information as to the means used to effect the suicides.

Data and studies on CSF 5-HIAA from depressed patients are contradictory, with some groups reporting a significant decrease below controls, in some cases as much as 67%. Others report slight elevations in depressives. When the probenecid technique is used, again, a decreased 5-HIAA accumulation is sometimes found in depressives, but other studies report no significant difference.[19]

Precursor loading strategies with the serotonin precursors tryptophan and 5-hydroxytryptamine similarly have yielded inconsistent results. Some investigators have found tryptophan to have utility as an antidepressant, whereas others claim that 5-hydroxytryptamine has more antidepressant efficacy.[11,19] The simple fact that use of tryptophan in the treatment of depression has not achieved widespread popularity should speak for itself.

A recent addition to the serotonin story involves finding of a decreased uptake for serotonin in platelets of patients with depression.[20] This work needs replication, but again may tell us nothing about brain biochemistry.

In summary, in spite of many years and much labor, we are left with suggestive findings that serotonin may be involved in depression, but certainly with nothing that can be called overwhelming hard-core evidence.

2.3. Tyramine

Reports indicate that in response to an oral tyramine load, significantly lower amounts of conjugated tyramine are excreted by depressed patients compared to controls.[21] Of interest is the fact that this phenomenon does not revert to normal even after a patient is completely improved or recovered. There is also some suggestive evidence that this low tyramine excretion may be a predictor of vulnerability to depressive illness.[21]

3. AMINO ACIDS

3.1. Tryptophan

As a precursor of serotonin, tryptophan has logically been a focus of interest in the neurochemistry of depression. In a recent review of 13 studies in which blood levels of tryptophan was measured,[22] varying results were obtained, with some studies demonstrating no difference in plasma tryptophan between depressed and control subjects and some demonstrating a lowering of tryptophan. The authors wisely point out that factors that would affect plasma tryptophan such as diet, time of day, state of activity, medications, and other blood constituents such free fatty acids and competing amino acids were usually not controlled for, or, if this was done, it was done with questionable methods, probably accounting for many of the discrepancies in the reviewed results. Clearly the effect of diet is important to consider in any study with depressives as many of them have anorexia with weight loss. In their own study,[22] they did demonstrate that if the ratio of tryptophan to five other neutral amino acids was considered rather than the absolute level, depressed patients had significantly lower ratios than controls. This phenomenon reverted to normal as the depression resolved.

Similarly, another group[23] found a relative deficiency of tryptophan as measured by the ratio of total tryptophan to competing amino acids in some depressives, and claimed that these patients responded to tryptophan treatment; whereas patients with a normal pattern of plasma amino acids were resistant to oral tryptophan. The small number of patients in this study makes it difficult to appreciate without further replication. Of peripheral interest is the finding that patients with a diagnosis of primary depression, in one study that did not find abnormal patterns of tryptophan, demonstrate a rise in plasma unesterified fatty acid after physiological and psychiatric testing. These authors interpret this as suggesting that if tryptophan reflects anything, it reflects agitation and retardation, a not altogether surprising conclusion.[24]

3.2. GABA

Although abnormalities in GABA metabolism in several neurological diseases are fairly well established and some evidence has accumulated of abnormalities in schizophrenia, three recent reports suggest that inpatients suffering from depression have CSF GABA levels lower than controls.[25–27]

An alternative to measuring GABA in CSF consists of measuring it in plasma, which is more readily obtained. It is tempting to believe that the plasma levels reflect brain levels in that manipulations that elevate brain levels in CSF do so to a similar extent in plasma, but it has not been firmly established that plasma levels are a true reflection of central metabolism.

Nevertheless, in a preliminary study of 35 patients with affective disorder, Petty and Schlesser[28] found low levels of plasma GABA in patients with unipolar depression. Of particular interest was that patients with "sporadic depressive disease" had levels that although significantly lower than controls were significantly higher than those of patients with "familial pure depressive disease" and "depression spectrum disease," supporting the Winokur diagnostic classification. Patients with mania were found to have elevated plasma levels of both while clinically ill and on recovery, suggesting the possibility that elevated plasma GABA may reflect a trait marker for this illness. Three additional studies[29,30] have been completed with over 200 subjects, and the findings have consistently supported the original observation of low plasma GABA levels in depression. Schizophrenics have levels that are not significantly different from controls. This phenomenon is not unique to unipolar depression since a series of alcoholics also demonstrated levels that were below control values. Studies have not yet demonstrated whether the low levels found in depression are a state or trait variable, nor have medication effects been entirely excluded.

4. ENZYMES

4.1. Monoamine Oxidase

At least 15 studies have examined MAO activity in unipolar patients. A recent review[31] found no consistent differences between unipolar patients and controls. Although elevated MAO activity has been described in patients with secondary depression who had a primary diagnosis of chronic anxiety or borderline personality disorder, this is not necessarily of diagnostic or theoretical interest, since elevated levels of MAO have been found in manic patients as well. Again, most of these studies suffer from problems of selection of controls, inadequate control for medication, etc. MAO levels in alcoholism do appear to be significantly reduced in most studies that have examined them.[31]

4.2. Catechol-O-Methyltransferase

This enzyme was first reported to be decreased in activity in unipolar depressed women when compared to bipolar and normal controls.[32] Again, like so much else in the neurochemistry of depression, other workers were not able to replicate these findings. More recently, with a cohort of 88, no significant differences were found across several diagnostic categories for COMT activity, except for a significant increase in agitated depressed males.[33]

5. HORMONES

5.1. Dexamethasone Suppression Test

Recently the dexamethasone suppression test has been presented as a specific laboratory blood test for "melancholia."[34] If a normal person takes 1 mg of dexamethasone orally at about 11:00 p.m., adrenal function will be suppressed during the ensuing 24 or more hours. In about 40% of patients suffering from "melancholia," an early escape from this suppression will be observed, i.e., they will have serum cortisol values higher than 5 μg/dl when samples are obtained at 8:00 a.m., 4:00 p.m., and/or 11:00 p.m. on the following day. Although the original findings of Carroll *et al.* have been adequately replicated, there is a considerable range in the sensitivity reported (anywhere from 24 to 100%). Initial reports suggested this test to be specific for depression, but recent findings have called this into question. For example, 53% of patients with dementia,[35] 30% of patients with chronic schizophrenia,[36] and 33% of patients with alcoholism[37] also have been reported to have a positive dexamethasone suppression test.

Even though the test may not be as specific for depression as originally thought, it may be useful for research purposes, or for clinical management. Indeed, it has been reported to serve as a potentially useful index of severity in depression.[38] Also, depressives who fail to normalize the dexamethasone suppression test may have a greater tendency to relapse.[39]

Regardless, the dexamethasone suppression test has provided another validation of the familial classification scheme. Patients with pure depressive disease have been shown to have a significantly higher proportion of nonsuppressors than those with depression spectrum disease.[40–42] Furthermore, the dexamethasone suppression test supports the differentiation of primary from secondary depression on a biological as well as clinical basis in that patients with secondary depression have repeatedly been found to have dexamethasone suppression tests similar to normal controls.[40,43]

Another potential utility is in prediction of treatment response. In one study, patients with abnormal dexamethasone suppression tests had better response to electroconvulsive therapy[44] and to sleep deprivation.[45] Its utility in predicting suicidal behavior is still open to question,[46,47] although some preliminary data suggest that primary depressive patients with an abnormal suppression may be more likely to commit suicide. This would, of course, correlate with the test possibly predicting severity of depressive illness.

What does the dexamethasone suppression test mean neurochemically? Here the evidence is somewhat contradictory in that there are reports that suppression correlates with ACTH levels and that it does not.[48,49] The case has been made that dexamethasone nonsuppression indicates a central muscarinic cholinergic overactivity based on the fact that abnormal response to dexamethasone is found in normal subjects premedicated with physostigmine and scopolamine methylsulfate.[50] The case has also been made that noradrenergic systems exert a tonic inhibitory influence on the secretion of CRF and that hypofunction of hypothalamic noradrenergic neurons may explain the test.[51] In light of the extremely complex neurohumoral feedback mechanisms involved in adrenal regulation, such speculations are no doubt premature, but of theoretical interest, and useful for indicating future research directions.

5.2. Growth Hormone

We will discuss the studies of growth hormone in depression in some detail for two reasons. First, the subject has not been recently reviewed, and second, its literature represents a good example of the difficulties encountered in trying to compare studies into the neurochemistry of depression. The articles chosen are representative, if not exhaustive, and for each we will present (1) diagnostic criteria, (2) number of subjects studied, (3) method, and (4) result.

5.2.1. Langer et al.[52]

5.2.1a. Diagnostic Criteria. International Classification of Diseases, 8th Edition, to World Health Organization.

5.2.1b. Number of Subjects. Twenty-one normal controls, 20 depressives, 8 schizophrenics, 6 alcoholics. Depressives included 9 "endogenous," 4 "unipolar recurrent," 7 "reactive."

5.2.1c. Method. Subjects were challenged with 0.1 mg/kg of amphetamine. Blood samples were obtained at 15-min intervals after amphetamine infusion and samples were analyzed by RIA.

5.2.1d. Results. Endogenous depressives secreted significantly less growth hormone than controls. Reactive depressives secreted significantly more. Authors did not state what statistic was used for arriving at significance; however, their standard deviations are rather large, e.g., for normal controls, the peak change in hormone was 11.53 ± 10.09.

5.2.2. *Matussek*[53]

5.2.2a. Diagnostic Criteria. Not stated.

5.2.2b. Number of Subjects. Not stated.

5.2.2c. Method. Clonidine (0.15 mg) was injected i.v. and blood was collected every 15 min afterwards. Analytic procedure was not stated.

5.2.2d. Results. Endogenous depressives ($N = 9$) secreted significantly less human growth hormone than did neurotic reactive depressives ($N = 10$). Statistical comparison with control was not given.

5.2.3. *Checkley*[54]

5.2.3a. Diagnostic Criteria. Patients were accepted into the study if they fulfilled certain exclusion criteria and had a score of at least 6 on the Carney–Roth–Garside Questionnaire.

5.2.3b. Number of Subjects. Twenty patients with endogenous depression were studied, 6 with reactive depression, 8 with other functional psychosis, and 10 with other psychiatric illness.

5.2.3c. Method. Methamphetamine (15 mg/75 kg) was given i.v. and blood samples were collected at 15-min intervals for 1.5 hr. Samples were analyzed by RIA.

5.2.3d. Results. No difference in median plasma growth hormone concentrations was demonstrated for any of the sampling times between the four groups.

5.2.4. *Maeda et al.*[55]

5.2.4a. Diagnostic Criteria. Agreement between two psychiatrists that the patient was depressed.

5.2.4b. Number of Subjects. Eight unipolar and 5 bipolar patients were studied along with 16 controls.

5.2.4c. Method. Thyrotropin-releasing hormone (500 mg i.v.) was administered with blood samples obtained at 15-min intervals for 2 hr.

5.2.4d. Results. Response of bipolars was significantly lower than controls, and that of unipolars was significantly higher than controls.

5.2.5. *Sachar et al.*[56]

5.2.5a. Diagnostic Criteria. Not stated.

5.2.5b. Number of Subjects. Eight unipolar and 2 bipolar patients were studied along with 3 with secondary depression. No controls were investigated.

5.2.5c. Method. Insulin tolerance test.

5.2.5d. Results. The unipolars were significantly lower than the controls used from another study.

5.2.6. *Sachar et al.*[57]

5.2.6a. Diagnostic Criteria. Not stated.

5.2.6b. Number of Subjects. Twelve male and 18 female unipolars and 10 male bipolars were studied along with 19 male controls and 13 female controls.

5.2.6c. Method. L-Dopa was administered (500 mg p.o.) with blood samples obtained at hourly intervals for 5 hr.

5.2.6d. Results. There were no significant differences found for the unipolars once the effect of menopause was taken into account for the females. For the bipolars, no significant differences were found compared to controls.

5.2.7. *Maany et al.*[58]

5.2.7a. Diagnostic Criteria. Criteria of Feighner *et al.* and Research Diagnostic Criteria were used with patients being diagnosed as having primary or secondary affective illness.

5.2.7b. Number of Subjects. Eight normal, 10 unipolar, 8 bipolar I (depressed), 9 bipolar II, and 8 secondary or reactive depressed.

5.2.7c. Method. After obtaining baseline samples, 200 mg of L-dopa p.o. or 0.75 mg of apomorphine s.c. was administered with blood samples collected at 20-min intervals for 3.5 hr. Peak response to growth hormone after apomorphine was not significantly different from controls among unipolar, bipolar I, bipolar II, or secondary depressives.

5.2.7d. Results. No significant differences among any of the groups.

5.2.8. *Mendlewicz et al.*[59]

5.2.8a. Diagnostic Criteria. Feighner *et al.*

5.2.8b. Number of Subjects. Eighteen bipolar and 25 unipolar females were studied.

5.2.8c. Method. Patients were administered L-dopa by mouth and blood samples were collected every 30 min for 2 hr.

5.2.8d. Results. No differences between bipolars and unipolars were noted nor were there differences between pre- and postmenopausal women. There was no control group.

5.2.9. Linkowski et al.[60]

5.2.9a. Diagnostic Criteria. Feigner *et al.*

5.2.9b. Number of Subjects. Twenty menopausal and 30 postmenopausal female patients were studied. There were 27 unipolars and 23 bipolars. Twenty-eight premenopausal and 11 postmenopausal women served as controls.

5.2.9c. Method. Synthetic TRH was administered (200 mg i.v.). Blood samples were collected at 15-min intervals for 2 hr.

5.2.9d. Results. There were no significant differences in growth hormone response between patients and controls, although base levels of growth hormone were higher in the unipolar postmenopausal group compared to controls.

5.2.10. Catlin et al.[61]

5.2.10a. Diagnostic Criteria. Not stated.

5.2.10b. Number of Subjects. Three male and 3 female depressed patients were studied. Four had a diagnosis of depressive disorder, 1 of bipolar affective disorder, depressed, and 1 of schizoaffective disorder. Four methadone addicts served as controls.

5.2.10c. Method. β-Endorphin (10 mg/kg) was administered i.v. over a period of 30 min. Blood samples were collected at 15, 30, 60, and 120 min.

5.2.10d. Results. β-Endorphin did not increase levels of growth hormone in either group of subjects.

5.2.11. Puig-Antich et al.[62]

5.2.11a. Diagnostic Criteria. A combination of Research Diagnostic Criteria and DSM-III criteria was used.

5.2.11b. Number of Subjects. Twenty-seven drug-free prepubertal children: 10 with major depressive disorder endogenous type, 10 with nonendogenous major depressive disorder, and 7 with nondepressed neurotic disorder.

5.2.11c. Method. After obtaining baseline blood samples, regular insulin (0.1 U/kg) was infused i.v. as a bolus. Blood samples were collected every 15 min for 2 hr.

5.2.11d. Results. The endogenous depressed group hyposecreted growth hormone as compared to the other two groups.

5.2.12. Kirkegaard et al.[63]

5.2.12a. Diagnostic Criteria. Idiosyncratic.

5.2.12b Number of Subjects. Twenty patients were studied, all of whom had "endogenous depression." They were separated into two groups according to their TSH response to TRH, with group 1 demonstrating an increased TSH response to TRH and a good prognosis, and group 2 having an unaltered TSH response and relapse within 6 months of cessation of treatment.

5.2.12c. Method. After obtaining baseline samples, 200 mg of TRH was injected i.v. as a bolus with blood samples drawn at 20 and 60 min.

5.2.12d. Results. Significantly increased responses of growth hormone were found in group 1; those in group 2 were unchanged.

5.2.13. Checkley et al.[64]

5.2.13a. Diagnostic Criteria. Research Diagnostic Criteria were used. Additionally, patients were required to have a score higher than 6 on the Newcastle scale for depression.

5.2.13b. Number of Subjects. Ten drug-free patients were used and matched with 10 normal controls.

5.2.13c. Method. Clonidine (2 mg/kg) was injected i.v. slowly over 10 min. Blood samples were collected every 15 min for 2 hr.

5.2.13d. Results. Patients with depression had significantly lower growth hormone response to clonidine than did the normal subjects, but had the same hypotensive and sedative side-effect incidence.

5.2.14. Siever et al.[65]

5.2.14a. Diagnostic Criteria. Research Diagnostic Criteria were used.

5.2.14b. Number of Subjects. Nineteen patients with major depressive disorder and 20 normal controls were studied.

5.2.14c. Method. Clonidine (2 mg/kg) was administered i.v. over 5 min. Blood samples were collected at 15-min intervals for 1 hr and again at 90 min after infusion.

5.2.14d. Results. Growth hormone response was significantly lower for all patients than for controls. Maximum growth hormone response to clonidine was also significantly lower in postmenopausal patients than postmenopausal controls.

5.2.15. *Halbreich et al.*[66]

5.2.15a. Diagnostic Criteria. Research Diagnostic Criteria were used.

5.2.15b. Number of Subjects. Six normal men, 7 normal postmenopausal women, 11 endogenously depressed (including 6 postmenopausal women, 3 premenopausal women, and 2 men), and 6 patients with conditions diagnosed as atypical.

5.2.15c. Method. Dextro-amphetamine sulfate (0.15 mg/kg) was administered i.v. Blood samples were collected at 15-min intervals for 90 min. The procedure was performed at both 9:00 a.m. and 6:30 p.m.

5.2.15d. Results. The growth hormone response to amphetamine was reduced in endogenously depressed postmenopausal women but equally reduced in normal postmenopausal women and in patients with atypical depression. Depressed and normal men had larger growth hormone responses but no differences between depressed and controls.

What is to be made of this? First, that different groups at times arrive at different results even using comparable diagnostic criteria. Second, that much of the work in the literature cannot be adequately evaluated due to lack of description of patient selection. Third, that most studies do not adequately control for factors such as diet and weight and even psychological stress, which can profoundly influence growth hormone. Fourth, that sometimes differences between pre- and postmenopausal women are found and that this may be an important factor for which to control. Fifth, that using this technique to try to get a handle on neurotransmitter control of a neuroendocrine function is very problematic because compounds such as amphetamine are capable of potentiating both dopamine and norepinephrine. Compounds such as L-dopa are capable of potentiating both compounds as well. Additionally, clonidine, while originally thought to be an α-adrenergic postsynaptic agonist, has recently been demonstrated to have several other pharmacological actions. Sixth, that when you measure a compound that is difficult to assay under the best of circumstances and the levels of which in both normal and ill people fluctuate widely during the day in response to a host of factors, you are likely to come up with a significant result quite by statistical chance. For example, it is conceivable that depressed patients might react to having an intravenous catheter inserted in a different manner physiologically than nondepressed patients, which would obfuscate any possible results obtained. Thus, the studies on growth hormone in depression illustrate the many difficulties in the field.

5.3. *TRH*

In 1972 Prange and colleagues administered TRH to 10 women with depression. Two of the patients showed virtually no TSH response to this challenge.

Since then, over 41 studies involving more than 917 depressed patients have corroborated Prange's original findings while 5 studies with 36 patients have been negative. This subject has been well reviewed recently.[67] The finding of a blunted TRH-induced TSH response does not occur in all depressives. Generally, it is found in about a third of patients studied depending on a variety of factors such as those described above. Nor is it specific for depression, as blunting is also found in some patients with mania, anorexia nervosa, and alcoholism even when sober. The test is usually administered much as the growth hormone stimulation test in which TRH is infused i.v. and TSH is measured at intervals subsequent to that. The routine procedure is to give 0.5 mg of TRH over a course of 1 min; normally, a peak TSH response occurs about 30 min after TRH injection. Indeed, a single 30-min value yields as much information as obtaining sequential aliquots over a period of several hours. The test can be influenced by nutritional state (of obvious importance in studying depression), but increasing age, previous drug intake, and severity of illness do not seem to affect the test. Studies comparing the use of the TRH-induced TSH response do not clearly differentiate between secondary and primary depression, as some have found the fault to occur only in primary depression while others have found no significant differences. Also, this test is not useful in distinguishing unipolar versus bipolar patients, for 10 studies found no significant differences. There is some suggestion that elevated serum cortisol, found in some depressives, might contribute to the TSH blunting; however, other studies demonstrating a lack of association with the dexamethasone suppression test would contradict that possibility. Again, this test is of limited utility as a state/trait discriminator since 5 studies have found that it normalized upon recovery and 5 have found persistent blunting to be the rule. There are some preliminary suggestions that the TRH test may predict response to tricyclic antidepressant and electroconvulsive shock therapy but again this work is tentative and requires considerable replication before it can be assumed to be of clinical utility. As to its neurochemical significance, there are suggestions that norepinephrine, dopamine, and/or serotonin may be involved and as Prange wisely points out "at every level of the system there are complex interactions of multiple factors of opposing sign. Therefore its usefulness in understanding brain neurochemistry in an indirect fashion is limited." It is certainly of historical interest, however, since for more than 200 years, relationships between thyroid and behavior, particularly affective state, have been suspected and these have now been documented to occur in at least some patients.

Recently, it has been reported[68] that the TRH stimulation test does not distinguish between the subtypes of unipolar depression using the familial criteria of Winokur.

6. *IMIPRAMINE BINDING*

Specific high-affinity binding sites for [^{3}H]imipramine have been demonstrated in human brain.[69] Binding sites have also been demonstrated on human platelets whose kinetic characteristics are almost identical to those for brain.[70–72] Recently, the [^{3}H]imipramine-binding sites on platelets were demonstrated to be decreased in untreated human depressed patients. This was a decrease in maximal binding rather than the affinity constant.[73] The imipramine-binding site in human brain has been demonstrated to be located on serotoninergic neurons. In a similar fashion, patients demonstrate a significant decrease in the maximal uptake of serotonin into platelets, suggesting association with the serotonin transport system.[74,75] Also, a decrease in imipramine-binding sites in the frontal cortex of human suicides has been demonstrated.[76] Preliminary research with the learned helplessness animal model of depression described below has demonstrated a decrease in the density but not affinity of imipramine-binding sites in rat cortex of helpless animals.[77]

The clinical findings must be considered very preliminary because they have not been shown to be specific for depression and since imipramine and other tricyclic antidepressants themselves as agonists cause a down-regulation of the binding site. This raises a curious paradox since preliminary studies with electroconvulsive shock have also shown it to down-regulate imipramine-binding sites in rat cortex. We are presented with a situation in which the disease produces the same neurochemical abnormalities as does its cure.[78]

7. *ANIMAL MODELS*

Another approach to the problem involves studying the neurochemistry of depression with an animal model of the illness. There are many obvious difficulties at both the philosophical and the practical level in dealing with animal models of any psychiatric illness particularly since verbal communication is essential in the diagnosis. To put it another way, how can you tell if a rat is depressed? Nevertheless, several putative animal models of depression have been developed. Of these, the only one in which significant neurochemical research has been performed is the learned helplessness animal model of depression.[79]

Learned helplessness is a maladaptive behavioral state that can be induced in a variety of species by exposure to uncontrollable aversive stimuli.[80] It has been demonstrated to fulfill published criteria[81] for a valid animal model of depression. That is to say, there is a behavioral similarity between helplessness and psychomotor retardation seen in depressed humans. Helplessness is reversed with chronic but not acute administration of tricyclic antidepressants, atypical antidepressants, MAO inhibitors, or electroconvulsive shock, all agents that are useful in the treatment of human depression. Helplessness is not reversed by neuroleptics, anxiolytics, stimulants, or alcohol.[82] Of considerable interest is the fact that chronic drug treatment is required to effect a reversal of the helpless state. This situation is similar to that seen in the treatment of human depression, for several days or weeks of treatment is usually required before symptomatic improvement is noted. With a series of intracranial microinjections into nine brain regions, it has been demonstrated that desipramine reverses helplessness only when injected into the anterior neocortex.[83] This reversal can also be effected by injecting serotonin into this brain region, GABA into the hippocampus, or serotonin into the septum. Prevention of learned helplessness involves more complex, less well understood pathways, as desipramine will prevent helplessness in the anterior neocortex, lateral geniculate body, or hippocampus. If bicuculline is injected into the hippocampus, desipramine is not effective in the behavioral reversal of learned helplessness. When desipramine is injected into the cortex, rats are cured from helplessness immediately, that is to say, on recovery from ether anesthesia. This suggests that the explanation for the delayed onset of tricyclic antidepressant drug action may reflect the time required to achieve a pharmacologically active drug level in the cortex. This has been demonstrated by studies involving i.p. injection of a series of animals with imipramine at twice the ED_{50} for reversal of helplessness after helplessness induction. In the examination of behavior and cortical drug level at varying intervals of time, a strong correlation was found between cortical tricyclic drug level and behavioral reversal, regardless of the length of time involved in treatment with imipramine.[84]

That GABA and the hippocampus play a modulating role in helplessness was demonstrated by testing rats with an injection of bicuculline into the hippocampus with no behavioral manipulation. These rats were found to be as helpless on testing as rats receiving behavioral learned helplessness[85] induction with foot shock. Additionally, when flux through the "small" hippocampal pool with GABA was determined, chronic treatment with imipramine or iprindole but not lorazepam or chlorpromazine elevated GABA flux.[85]

Additional confirmation for the hypothesized neuronal pathway involved in learned helplessness was obtained by exposing rats to learned helplessness training and subsequently examining regional Ca^{2+}-specific neurotransmitter release. Learned helplessness induction caused a decrease in release of serotonin from slices of the neocortex in septum and of GABA from hippocampal slices when measured either 1 or 4 days later. Interestingly, chronic but not acute administration of imipramine produced opposite changes in control animals, and reversed the decreased release measured in helpless rats. Thus, rats receiving both helplessness induction and chronic imipramine had neurotransmitter release comparable to controls.[86]

Thus, the learned helplessness animal model of depression suggests that, insofar as it is an analog of human depression, septal and cortical serotonin play a key role, as does hippocampal GABA, in the development of the helpless state and in its reversal by both tricyclic antidepressant drugs and electroconvulsive shock.[87]

8. SUMMARY

What do we really know about the neurochemistry of depression that we did not know 10 or 20 years ago? Surprisingly little. But new findings and theories have been heuristically productive and we have some reason to expect that new findings will push forward the frontiers of the basic mechanisms involved in clinical depression.

ACKNOWLEDGMENT. This work was supported in part by the Veterans Administration.

REFERENCES

1. Kendell, R. E., 1976, *Br. J. Psychiatry* **129:**15–28.
2. Leonhard, K., 1959, *Aufteilung der Endogenen Psychosen*, 2nd ed., Akademie-Verlag, Berlin.
3. Angst, J., 1966, *Zur Atiologie und Nosologie endogener depressiver Psychosen*, Springer-Verlag, Berlin.
4. Perris, C., 1966, *Acta Psychiat. Scand. Suppl.* **194:**1–188.
5. Winokur, G., and Clayton, P., 1967, *Recent Advances in Biological Psychiatry*, Volume 9 (J. Wortis, ed.), Plenum Press, New York, pp. 35–50.
6. Robins, E., Munoz, R. A., Martin, S., and Gentry, K. A., 1972, *Disorders of Mood* (J. Zubin and F. A. Freyhan, eds.), Johns Hopkins Press, Baltimore, pp. 33–45.
7. Winokur, G., 1974, *Int. Pharmacopsychiatry* **9:**5–13.
8. Feighner, J. P., Robins, E., Guze, S. B., Woodruff, R. A., Winokur, G., and Munoz, R., 1972, *Arch. Gen. Psychiatry* **26:**57–63.
9. Winokur, G., 1981, *Depression: The Facts,* Oxford University Press, London.
10. Schildkraut, J. J., 1978, *Psychopharmacology: A Generation of Progress* (M. A. Lipton, A. DiMascio, and K. F. Killam, eds.), Raven Press, New York, pp. 1223–1234.
11. van Praag, H. M., 1982, *Compr. Psychiatry* **23:**124–135.
12. van Praag, H. M., 1981, *Compr. Psychiatry* **21:**44–54.
13. Schildkraut, J. J., 1965, *Am. J. Psychiatry* **122:**509–522.
14. Bunney, W. E., Jr., and Davis, J. M., 1965, *Arch. Gen. Psychiatry* **13:**483–494.
15. Birkmayer, W., and Riederer, P., 1975, *J. Neural. Transm.* **37:**95–109.
16. Blombery, P. A., Kopin, I. J., Gordon, E. K., Markey, S. P., and Ebert, M. H., 1980, *Arch. Gen. Psychiatry* **37:**1095–1098.
17. Schildkraut, J. J., Orsulak, P. J., Schatzberg, A. F., Gudeman, J. E., Cole, J. O., Rohde, W. A., and LaBrie, R. A., 1978, *Arch. Gen. Psychiatry* **35:**1427–1433.
18. Hollister, L. E., Davis, K. L., Kenneth, L. D., and Berger, P. A., 1980, *Arch. Gen. Psychiatry* **37:**1107–1110.
19. Murphy, D. L., Campbell, I., and Costa, J. L., 1978, *Psychopharmacology: A Generation of Progress* (M. A. Lipton, A. DiMascio, and K. F. Killam, eds.), Raven Press, New York, pp. 1235–1247.
20. Meltzer, H. Y., Arora, R. C., and Song, P., 1982, *Biological Markers in Psychiatry and Neurology* (E. Usdin and I. Hanin, eds.), Pergamon Press, Elmsford, New York, pp. 39–48.
21. Bonham Carter, S. M., Reveley, M. A., Sandler, M., Dewhurst, Sir J. Little, B. C., Hayworth, J., and Priest, R. G., 1980, *Psychiatr. Res.* **3:**13–21.
22. DeMyer, M. K., Shea, P. A., Hendrie, H. C., and Yoshimura, N. N., 1981, *Arch. Gen. Psychiatry* **38:**642–646.
23. Moller, S. E., Kirk, L., and Fremming, K. H., 1976, *Psychopharmacologia* **49:**205–213.
24. Curzon, G., Kantamaneni, B. D., Lader, M. H., and Greenwood, M. H., 1979, *Psychol. Med.* **9:**457–463.
25. Kasa, K., Otsuki, S., Yamamoto, M., Sato, M., Kuroda, H., and Ogawa, N., 1982, *Biol. Psychiatry* **17:**877–883.
26. Gold, B. I., Bowers, M. B., Roth, R. H., and Sweeney, D. W., 1980, *Am. J. Psychiatry* **137:**362–364.
27. Gerner, R. H., and Hare, T. A., 1981, *Am. J. Psychiatry* **138:**1098–1101.
28. Petty, F., and Schlesser, M. A., 1981, *J. Affect. Dis.* **3:**339–343.
29. Petty, F., and Sherman, A. D., 1982, *Res. Commun. Psychol. Psychiatry Behav.* **7:**431–440.
30. Petty, F., and Sherman, A. D., 1984, *J. Affective Disord.* **6:**131–138.
31. Murphy, D. L., Coursey, R. D., Haenel, T., Aloi, J., and Buchsbaum, M. S., 1982, *Biological Markers in Psychiatry and Neurology* (E. Usdin and I. Hanin, eds.), Pergamon Press, Elmsford, New York, pp. 123–134.

32. Dunner, D. L., Cohn, C. K., Gershon, E. S., and Goodwin, F. K., 1971, *Arch. Gen. Psychiatry* **25**:348–353.
33. Davidson, J. R. T., McLeod, M. N., Turnbull, C. D., White, H. L., and Feuer, E. J., 1979, *Biol. Psychiatry* **14**:937–942.
34. Carroll, B. J., 1982, *Br. J. Psychiatry* **140**:292–304.
35. Spar, J. E., and Gerner, R., 1982, *Am. J. Psychiatry* **139**:238–240.
36. Dewan, M. J., Pandurangi, A. K., Boucher, M. L., Levy, B. F., and Major, L. F., 1982, *Am. J. Psychiatry* **139**:1501–1503.
37. Schwartz, C. M., and Dunner, F. J., 1982, *Arch. Gen. Psychiatry* **39**:1309–1312.
38. Davis, K. L., Hollister, L. E., Mathe, A. A., Davis, B. M., Rothpearl, A. B., Faull, K. F., Hsieh, J. Y. K., Barchas, J. D., and Berger, P. A., 1981, *Am. J. Psychiatry* **138**:1555–1562.
39. Greden, J. F., Albala, A. A., Haskett, R. F., James, N. M., Goodman, L., Steiner, M., and Carroll, B. J., 1980, *Biol. Psychiatry* **15**:449–458.
40. Schlesser, M. A., Winokur, G., and Sherman, B. M., 1980, *Arch. Gen. Psychiatry* **37**:737–743.
41. Coryell, W., Gaffney, G., and Burkhardt, P. E., 1982, *Biol. Psychiatry* **17**:33–40.
42. Targum, S. D., Byrnes, S. M., and Sullivan, A. C., 1982, *J. Affect. Dis.* **4**:21–27.
43. Coryell, W., Gaffney, G., and Burkhardt, P. E., 1982, *Am. J. Psychiatry* **139**:120–122.
44. Coryell, W., 1982, *Psychiatr. Res.* **6**:283–291.
45. Nasrallah, H. A., and Coryell, W. H., 1982, *Psychiatr. Res.* **6**:61–64.
46. Coryell, W., and Schlesser, M. A., 1981, *Am. J. Psychiatry* **138**:1120–1121.
47. Coryell, W., 1982, *Am. J. Psychiatry* **139**:1214.
48. Kalin, N. H., Weiler, S. J., and Shelton, S. E., 1982, *Psychiatry Res.* **7**:87–92.
49. Fang, V. S., Tricou, B. J., Robertson, A., and Meltzer, H. Y., 1981, *Life Sci.* **29**:931–938.
50. Checkley, S. A., 1980, *Psychol. Med.* **10**:35–53.
51. van Praag, H. M., 1982, *Compr. Psychiatry* **23**:216–226.
52. Langer, G., Heinze, G., Reim, B., and Matussek, N., 1976, *Arch. Gen. Psychiatry* **33**:1471–1475.
53. Matussek, N., 1978, *Nervenarzt* **49**:431–435.
54. Checkley, S. A., 1979, *Psychol. Med.* **9**:107–115.
55. Maeda, K., Kato, V., Ohgo, S., Chihara, K., Yoshimoto, Y., Yamaguchi, N., Kuromaru, S., and Imura, H., 1975, *J. Clin. Endocrinol. Metab.* **40**:501–505.
56. Sachar, E. J., Finkelstein, J., and Helman, L., 1971, *Arch. Gen. Psychiatry* **25**:263–269.
57. Sachar, E. J., Altman, N. K., Gruen, P. H., Glassman, A., Halpern, F. S., and Sassin, J., 1975, *Arch. Gen. Psychiatry* **32**:502–503.
58. Maany, I., Mendels, J., Frazer, A., and Brunswick, D., 1979, *Neuropsychobiology* **5**:282–289.
59. Mendlewicz, J., Linkowski, P., and van Cauter, E., 1979, *J. Affect. Dis.* **1**:25–32.
60. Linkowski, P., Brauman, H., and Mendlewicz, J., 1980, *Br. J. Psychiatry* **137**:229–232.
61. Catlin, D. H., Poland, R. E., Gorelick, D. A., Gerner, R. H., Hui, K. K., Rubin, R. T., and Li, C. H., 1980, *J. Clin. Endocrinol. Metab.* **50**:1021–1025.
62. Puig-Antich, J., Tabrizi, M. A., Davies, M., Goetz, R., Chambers, W. J., Halpern, F., and Sachar, E. J., 1981, *Biol. Psychiatry* **16**:801–818.
63. Kirkegaard, C., Eskildsen, P. C., and Bjorum, N., 1981, *Psychoneuroendocrinology* **6**:253–259.
64. Checkley, S. A., Slade, A. P., and Shur, E., 1981, *Br. J. Psychiatry* **138**:51–55.
65. Siever, L. J., Uhde, T. W., Silberman, E. K., Jimerson, D. C., Aloi, J. A., Post, R. M., and Murphy, D. L., 1982, *Psychiatr. Res.* **6**:171–183.
66. Halbreich, U., Sachar, E. J., Asnis, G. M., Quitkin, F., Nathan, R. S., Halpern, F. S., and Klein, D. F., 1982, *Arch. Gen. Psychiatry* **39**:189–192.
67. Loosen, P. T., and Prange, A. J., 1982, *Am. J. Psychiatry* **139**:405–416.
68. Targum, S. D., Byrnes, S. M., and Sullivan, A. C., 1982, *J. Affect. Dis.* **4**:29–34.
69. Rehavi, M., Paul, S. M., Skolnick, P., and Goodwin, F. K., 1980, *Life Sci.* **26**:2273–2279.
70. Paul, S. M., Rehavi, M., Skolnick, P., and Goodwin, F. K., 1980, *Life Sci.* **26**:953–959.
71. Briley, M. S., Raisman, R., and Langer, S. Z., 1979, *Eur. J. Pharmacol.* **58**:347–348.
72. Langer, S. Z., Briley, M. S., Raisman, R., Henry, J.-F., and Morselli, P. L., 1980, *Naunyn-Schmiedebergs Arch. Pharmacol.* **313**:189–194.

73. Briley, M. S., Langer, S. Z., Raisman, R., Sechter, D., and Zarifian, E., 1980, *Science* **209**:303–305.
74. Paul, S. M., Rehavi, M., Skolnick, P., and Goodwin, F. K., 1982, *Biological Markers in Psychiatry and Neurology* (E. Usdin and I. Hanin, eds.), Pergamon Press, Elmsford, New York, pp. 193–204.
75. Paul, S. M., Rehavi, M., Skolnick, P., Ballenger, J. C., and Goodwin, F. K., 1981, *Arch. Gen. Psychiatry* **38**:1315–1317.
76. Stanley, M., Virgilio, J., and Gershon, S., 1982, *Science* **216**:1337–1339.
77. Petty, F., and Sherman, A. D., 1982, Society of Biological Psychiatry Annual Meeting, Toronto, Canada.
78. Langer, S. Z., Zarifian, E., Briley, M., Raisman, R., and Sechter, D., 1982, *Pharmacopsychiatry* **15**:4–10.
79. Petty, F., and Sherman, A. D., 1983, *Psychopharmacology* (D. G. Grahame-Smith, H. Hippius, and G. Winokur, eds.), Excerpta Medica, Amsterdam, pp. 444–459.
80. Seligman, M. E. P., 1978, *Research in Neurosis* (H. M. van Praag, ed.), Spectrum, New York, p. 72–107.
81. McKinney, W. T., and Bunney, W. E., 1979, *Arch. Gen. Psychiatry* **21**:240–248.
82. Sherman, A. D., Sacquitne, J. L., and Petty, F., 1982, *Pharmacol. Biochem. Behav.* **30**:1811–1815.
83. Sherman, A. D., and Petty, F., 1980, *Behav. Neural. Biol.* **30**:119–134.
84. Petty, F., Sacquitne, J. L., and Sherman, A. D., 1982, *Neuropharmacology* **21**:475–478.
85. Petty, F., and Sherman, A. D., 1981, *Pharmacol. Biochem. Behav.* **15**:567–570.
86. Sherman, A. D., and Petty, F., 1982, *Behav. Neur. Biol.* **35**:344–353.
87. Petty, F., and Sherman, A. D., 1980, American Psychiatric Association Annual Meeting.

25

Biochemical Studies on Autism

A. Yuwiler, E. Geller, and E. Ritvo

1. INTRODUCTION

Like many contemporary clinical designations, the term *early infantile autism* refers to a clinical syndrome and not an etiologically homogeneous disease. This is a distinction of some importance in evaluating the existing literature on the syndrome and in devising strategies for its biochemical elucidation. The term was coined by Kanner[1] in 1944 to identify a syndrome of four key symptoms: the early onset of profound disturbances in relating to others; great difficulty or inability in developing communicative speech; displays of stereotypic motor behavior; and insistence on a stable environment. While Kanner's clinical descriptions were clear, whether they defined a distinct clinical entity was open to debate.[2–6] Currently defined as a syndrome of disturbances in perception, developmental rate, relating, language, and motility, and sanctified by inclusion in the American Psychiatric Association's Diagnostic and Statistical Manual of Mental Disorders (DMS-III), the question of whether the syndrome should be considered a distinct "disease" is unresolved. What is clear, however, is that the syndrome can be associated with, and may be secondary to, a number of etiologies. Autistic behavior occurs in a significant number of phenylketonurics[7] and in about 8% of the cases of maternal rubella.[8] It has been found associated with tuberous sclerosis,[9] Hurler's disease, congenital hypothyroidism,[10] and some cases of fragile x syndrome.[11,12] Consequently, biological studies have the dual tasks of defining the substrates of the *symptoms* and of identifying the underlying diseases that can produce the syndrome of autism; a task analogous to identifying the metabolic and developmental etiologies that can present as mental retardation. In the case of autism, this task is additionally complicated by its low incidence (estimated as 0.04% of live births[13]), its developmental nature, and the uniquely human character of its identifying symptoms. Together, these complications limit the available sub-

---

A. Yuwiler and E. Geller • Neurobiochemistry Laboratory, Veterans Administration Medical Center, Los Angeles, California 90073; and Department of Psychiatry, University of California, Los Angeles, California 90024. *E. Ritvo* • Department of Psychiatry, University of California, Los Angeles, California 90024.

jects, tissues, and approaches. Current information on autism is limited to some meager genetic data, a few scattered biochemical measures, and the effects of a few pharmacological agents on symptomatology. This chapter will summarize the current status of data in these areas and then consider the general strategic problem of dealing with symptomatic classifications such as autism.

2. GENETICS

The role of genetic factors in childhood autism is still unclear but concordance for the syndrome is higher among identical twins than among either nonidentical twins or siblings.[14–17] This of course is expected if there are genetic contributions to the syndrome and, because aberrant DNA is usually translated into aberrant protein, it is compatible with the assumption that there is a biochemical basis for the syndrome or one of its underlying diseases. However, the concordance between identical twins has been previously reported as 30–50%. This suggests that nongenetic factors also contribute to the manifestation of the syndrome, that the syndrome is produced by several etiologically distinct diseases only some of which have strong genetic controls, that the criteria for establishing twins as identical are imperfect, or that the criteria for assessing autism are inadequate. Recent data show that of 31 sets of identical twins currently collected at the UCLA Registry for Genetic Studies of Autism, 30 (98%) are concordant for autism in contrast to only 2 (8%) of 24 sets of nonidentical twin siblings. The possibility that the syndrome of autism may encompass several distinct diseases, some of which are genetically controlled, is compatible with the early onset of symptoms and the occurrence of symptoms secondary to diseases of differing etiology and also with family studies. Investigations on relatively small populations failed to find evidence of an increased incidence of autism or schizophrenia among families with an autistic child.[18,19] More recent studies[20,21] (E. R. Ritvo, unpublished) have discovered a number of families in which there is a multiple incidence of autism or *other developmental disorders*. If the syndrome of autism is the common expression of several etiologically distinct diseases, it would be expected that the syndrome might occur among siblings in some families but not in others, and that disease expression might range over a number of developmental disorders. Indeed, as Spence[22] pointed out, several genetic mechanisms rather than a single one may be responsible for some or all of the cases of autism. A subgroup that seems to follow the autosomal recessive pattern has recently been found (E. R. Ritvo, unpublished).

3. BIOCHEMISTRY

3.1. Indoleamines

The most studied biochemical variable in infantile autism is blood serotonin. This is not because of some *a priori* link between autism and serotonin in general or blood serotonin in particular, but rather because Schain and Freedman[23] in the course of their studies on indoles in mental disease reported in 1961 a high incidence of hyperserotoninemia among autistics. The phenomenon was ignored for a number of years until independently rediscovered and confirmed.[24] Assessing its relevance to the syndrome, however, requires information on the generality, stability, specificity, and, above all, mechanisms for the hyperserotoninemia among autistics. The question of whether in this instance brain and blood serotonin are linked, is important but secondary.

The findings of Schain and Freedman[23] seem to be general to infantile autism and have been replicated by a number of investigators[25–31] on populations differing in age, inpatient/outpatient status, country of origin, diagnostic criteria, nature of control groups, and even differing in whether whole blood or platelet serotonin was measured. Indeed, one group replicated the finding on *serum* serotonin,[32] which presumably reflects that fraction of serotonin liberated from the platelets into serum and not yet destroyed by autoxidation or the actions of plasma amine oxidases and monoamine oxidase. The magnitude of the population differences in these studies varied from a high of a 2.3-fold[30] greater blood serotonin concentration among autistics to a low of a 1.2-fold elevation[27,33] with an average among studies of about 1.5-fold. Most of that difference was due to the 30% of the autistics who had blood serotonin concentrations 2 to 4 times greater than controls or other autistic children. This last comparison alone shows that hyperserotoninemia is not a global biological marker for autism but either reflects some attribute of the disorder, marks an etiologically distinct subgroup, is a fortuitous correlate, or indicates some experimental artifact. Information on the stability of hyperserotoninemia within the subpopulation would distinguish between these possibilities. If it is consistently high in a subpopulation of autistics, it may be related to etiology or risk; if it varies with clinical state, it may be related to disease expression; and if it varies seemingly at random, it is probably irrelevant to any fundamental feature of the syndrome.

In normal adults, blood serotonin shows large interindividual, but small intraindividual, variance,[34] and for an individual, concentrations may remain constant over decades. Generally, the coefficient of variation in blood serotonin for normal adults is less than 8%. In contrast, the coefficient of variation for the blood serotonin of four autistic children sampled 30 times over the period of a year ranged from 13% to 27%. Unfortunately, normative data on multiply-sampled children of like age are lacking, so that it is unclear if this increased variance reflects developmental oscillations or state-dependent influences. Some studies suggest that whole blood serotonin concentration and platelet number,[24] but perhaps not serotonin per milligram platelet protein,[29] decline with age until age 12–14. Thus, maturational delays might contribute to the increased levels and variance seen in the young autistic population. These data also indicate that age matching may be necessary in comparison studies.

Circadian variations in blood serotonin could contribute to variance in serotonin concentration and abnormal phasing could even account for population differences. Circadian and/or seasonal variations in blood or platelet serotonin have been reported by some workers[35–36] but have not been found

Table I
Population Differences in Blood Serotonin Concentrations

Increased concentration	Decreased concentration
Childhood autism	Down's syndrome
Mental retardates	Phenylketonuria
Carcinoid syndrome	Histindinemia
Maternal rubella	Hyperactivity in children
Kernicterus	De Lange syndrome
Infantile spasms	Pellagra with psychotic symptoms
Huntington's chorea	Duchenne's disease (childhood muscular dystrophy)
Bipolar manic-depressive illness	Migraine headaches
Chronic schizophrenia	Some depressive disorders
Dominant male vervet	Rheumatoid arthritis
Nontropical sprue	Chediak–Higashi syndrome

by others.[37] The only study investigating circadian differences in autism failed to detect any consistent or significant rhythm in young autistics, age matched controls, or normal adults.[27]

The hyperserotoninemia of hyperserotoninemic autistics appears to persist over the years for at least some individuals.[30,34] While indicating the measure is stable for autistics as well as normals, these subjects were still clinically ill and the data do not help determine if the elevations are linked to clinical status. The stability of blood serotonin in normal adults has been taken to mean it is trait characteristic, but recent data on groups of vervet monkeys indicate that even though intraindividual variance in concentration is low when social groups are stable,[38] blood serotonin concentration can be affected by manipulating social status[39,40] and therefore can at least be influenced by state as well as trait. Social influences on blood serotonin concentrations in man have not yet been demonstrated nor is there clear evidence for clinical correlates of hyperserotoninemia. Several reports suggest that blood serotonin is inversely related to IQ.[31,41] Hyperserotoninemia seems unrelated to activity level.[32,42] An association between hyperserotoninemia and enlargement of cerebral ventricles and/or cerebral atrophy among schizophrenics has been reported.[43] About 15% of autistics are claimed to have enlarged ventricles[44,45] and another 15–20% show other geometric abnormalities on computer tomographic scans,[45–48] but whether these subjects also show hyperserotoninemia is unknown. There are no reports as yet of neuropathological abnormalities characteristic of autistics or retardates with autistic features,[49] other than these brain scan anomalies.

While the initial report of Schain and Freedman[23] appears to have generality, and while blood serotonin concentrations show the stability required for clinical utility, hyperserotoninemia is certainly not a marker for autism. Depending upon how "hyperserotoninemia" is defined, it occurs in only 30–50% of autistics and in about 4% of normals. Further, as seen in Table I, elevated blood serotonin is associated with at least 10 different clinical conditions including severe mental retardation. However, a single biological variable, like a single behavioral one, is the resultant of many interactive processes and, to be medically useful, the *mechanism for the difference*, not the difference itself, must be unique to the disease. For example, hyperglycemia is found not only in diabetes but also in sepsis, meningitis, anoxia, hyperadrenalism, hyperpituitarism, hyperthyroidism, and encephalitis; hyperuricemia occurs in gout, renal failure, Lesh–Nyhan syndrome, toxemia in pregnancy, leukemia, polycythemia, and Down's syndrome. Despite multiple incidence, hyperglycemia is still useful in detecting diabetes and hyperuricemia in detecting gout. Again, it is the mechanism that is important. If common to many clinical states it is unlikely to be useful, but if unique it may provide an insight into the underlying pathology.

The serotonin found in blood is made primarily in the enterochromaffin cells of the gastrointestinal tract[50] and virtually disappears if the large and small intestine are resected.[51,52] A number of factors can influence serotonin formation. Substrate availability to the enterochromaffin cell could be altered by changes in tryptophan uptake during digestion; by changes in tryptophan metabolism in the liver and other tissues; by changes in tryptophan transport perhaps secondary to changes in albumin concentration or the number of tryptophan-binding sites; by amino acid competition with tryptophan for uptake into the enterochromaffin cell; by alterations in biosynthetic and degradative enzymes within the cell, etc. In blood, serotonin is almost exclusively confined within the platelet,[53] which takes up serotonin both by passive diffusion and by a sodium-[54] and chloride-sensitive[55] high-affinity uptake system with a K_m of about 0.3–0.5 μM[56] and an uptake rate of about 7 pmol/min per billion platelets.[57] Serotonin transport through the membrane seems dependent upon the membrane potential as well as the sodium gradient[58] and may involve sodium cotransport.[59] Uptake appears to occur in two steps, transport through the plasma membrane into platelet cytoplasm and transport through the vesicular membranes into "dense-core" vesicular granules.[56,60] These are 500–1000 Å in diameter,[61] each capable of holding some 500,000 molecules of serotonin[62] in association with ATP,[63] calcium, and magnesium.[64] Membrane potential and pH gradients generated by an ATP-dependent proton translocase serve as the driving mechanisms for serotonin uptake and entrapment in the vesicle. Once in the vesicle, there is little or no net movement back to the cytoplasm.[65] Serotonin uptake and storage could be affected by factors that alter platelet shape and thereby receptor exposure, ionic environment and receptor activity, vesicular pH and storage capacity, the binding of serotonin to cytoplasmic and vesicular compartments, efflux, and any intracellular metabolism by monoamine oxidase that might occur. Some processes, such as those controlling the ionic composition of blood, are affected by organ systems far removed from the platelet.

Platelets themselves are not uniform and featureless organelles but rather are heterogeneous in size,[66–68] density,[67,69–72] and biological properties.[69,73–75] The size differences have been attributed to platelet age[71] and to random growth and membrane demarcation of megakaryocytes.[76] Whatever their origin, size variations create a practical problem in platelet studies; slight changes in experimental procedure and even subtle factors like hematocrit, hemoviscosity, and tube dimensions relative to blood volume may alter the subpopulation of platelets isolated and thus lead to confusion.

Blood serotonin concentrations could reflect alterations in the number of platelets, their size distribution, their biological half-life, their transport time through regions of high serotonin formation, or their distribution in the spleen and other tissues.

A few of the many other factors that could affect blood serotonin are listed in Table II. It is not known if the hyperserotoninemia of autism results from changes in platelet physiology or indole metabolism; the related question of whether platelet number or platelet serotonin content is primary to the hyperserotoninemia among autistics has been investigated by determining the serotonin content of isolated platelets or by calculating this from whole blood serotonin and platelet counts. In most studies,[24,27–30,33,77] but not all,[78] autistics had greater platelet serotonin content than controls, although sometimes the differences were small. The largest population difference was 230%[30] from a study examining hyperserotoninemic autistics.

The hyperserotoninemia in autism thus appears to reflect an increased concentration of serotonin in platelets. In some studies, however, autistics also appear to have higher platelet counts than controls[24,78] although of a magnitude too small to account for the differences in blood serotonin. Methodological problems in platelet counting also contribute to variance between studies. Visual platelet counts sometimes overestimate true counts because of difficulties in platelet identification; machine counting sometimes underestimates depending upon the setting of the electronic "sensing window." Such differences may contribute to conflicting findings even by the same investigators.[24,27,33]

If elevated platelet serotonin content results from a general increase in serotonin biosynthesis, it should be accompanied by increased urinary 5-HIAA output. Some evidence for this exists. In two studies,[23,30] urinary 5-HIAA was 2–4 times greater for autistics than for controls. Neither obtained 24-hr collections; instead, samples were normalized to creatinine output and/or 24-hr values were calculated from estimates of creatinine clearance. However, it is possible that differences in the activity levels of controls and autistics could lead to differences in muscle mass and thereby in blood and urinary creatinine, producing an abnormal 5-HIAA/creatinine ratio. The results then are compatible with increased indoleamine metabolism but are not definitive. Some precursor-load studies indicate that both controls and autistics increase urinary 5-HIAA production in response to oral tryptophan to roughly the same extent.[23,30,79] The distribution of blood tryptophan between albumin-bound and free forms in hyperserotoninemia has not received much attention but there is no indication that this is abnormal for autistics as a whole.[28,32]

Serotonin could also be increased by decreased catabolism. A general inhibition of serotonin catabolism in autism seems ruled out both by the elevated basal excretion of 5-HIAA by autistics and by the increase in urinary 5-HIAA seen after tryptophan loading. A more specific decrease in the catabolism of platelet serotonin is possible but unlikely on several grounds. First, the bulk of platelet serotonin is vesicular and is protected from degradation, and second,

Table II
Factors That May Increase Blood Serotonin

- Indoleamine metabolism
  - Increased substrate
    - In blood
      - Increased consumption in diet
      - Increased absorption from GI tract
      - Decreased renal excretion
      - Decreased catabolism (TPO in liver or incorporation into protein)
    - In synthesizing cells
      - Increased free tryptophan in blood
        - Due to decreased albumin
        - Due to increased free fatty acids or other ligands for albumin sites
      - Decreased pool size
      - Decreased K_m or increased V_{max} for transport
      - Increased blood transit time through tissue (length, number of capillaries)
      - Increased number of ADP cells
  - Increased activities of biosynthetic enzymes
    - More enzyme protein
    - Higher enzyme activities
      - Increased halo-/apoenzyme
        - Increased cofactor production
        - Increased coupling to cofactor biosynthesis
        - Increased cofactor delivery
      - Altered cellular environment (pH, inhibitors)
      - Increased activity of enzyme due to phosphorylation, isozymic form, altered enzyme localization
  - Decreased serotonin degradation
    - Decreased activity of MAO
    - Altered ratio of isozymic MAO forms
    - Fewer mitochondria
    - Altered mitochondrial lipids
- Platelet
  - Number
    - Increased production
    - Decreased destruction
    - Decreased splenic storage
  - Properties
    - Increased ratio of large to small
    - Increased serotonin uptake through outer membrane
      - Altered ion concentration
      - Altered transport kinetics
      - Altered activity of Na^+,K^+-ATPase
    - Increased content in vesicles
      - More vesicles
      - Larger vesicles
      - Altered internal composition (pH, ATP, Ca^{2+})
      - Altered transport into or from vesicles

while the platelet does contain monoamine oxidase, it contains the B form of the enzyme, which has a relatively poor affinity for serotonin. Further, platelet MAO activity is the same for autistics as for normals.[80–84] Finally, the activity of the B form of the enzyme in cultured skin fibroblasts derived from autistic children did not differ from that in fibroblasts from children with Gilles de la Tourette syndrome.[85]

Platelet MAO is reported to correlate inversely with hematocrit among autistics but not other subjects.[82] Since MAO activity is greatest in large, young platelets, and large platelets are more easily lost into the red cell mass upon centrifugation when hematocrit is high, this finding might indicate that there is a high proportion of large platelets in autistic blood, perhaps mirroring increased platelet turnover. No direct evidence for this exists. In one study (E. Geller, unpublished), the median volumes for platelets from autistics and controls were not significantly different, nor was there a significant difference between the median volumes of platelets drawn from autistics with the highest blood serotonin concentration and from those with the lowest blood serotonin concentration.

Examinations of monoamine catabolite concentrations in CSF of autistics are limited, and because little normative data on children are available, comparisons have generally been between groups with different pathologies and mostly *after* probenecid treatment. Probenecid competitively inhibits the transport of organic acids, like 5-HIAA and HVA, from brain. Hence, accumulation of these compounds after pretreatment can yield a measure of monoamine turnover, provided that inhibition is complete and that the steady-state equilibrium between substrate, product, and catabolite is not perturbed. In animals the technique works quite well because massive doses of probenecid can be given, the timing between injection and obtaining tissue can be controlled, and regional turnover can be assessed because the brain itself is available. There are problems, however, in its application to man. Oral probenecid is noxious and is therefore administered in doses divided over time, each of which exerts some unknown degree of inhibition of catabolite transport and produces some unknown accumulation of catabolites in the CSF. More importantly, the total dose given is usually insufficient to cause complete blockade of transport. This appears as a correlation between acid catabolites and CSF probenecid concentrations. To adequately estimate turnover in a partially inhibited system requires estimates of the concentrations of probenecid and competing organic acids at transport sites, estimates of their relative affinities for those sites, and their individual transport rates. None of these are known. To further complicate interpretation, there is considerable individual variance in CSF probenecid concentrations after the same dosage regimen. For example, in one study,[86] probenecid concentrations ranged 10-fold, from 3 to 32 μg/ml. These problems are not much ameliorated by normalizing monoamine catabolite concentrations to the corresponding concentration of probenecid or to the log of its concentration. A final problem is that CSF 5-HIAA and HVA concentrations show a maturational decline[87–89] reminiscent of the decline in whole blood serotonin, thereby requiring age as well as group comparisons.

Within these constraints, existing data[86,87,90,91] fail to demonstrate abnormalities in central monoaminergic systems in autism; neither CSF 5-HIAA or

HVA appears to consistently differentiate autism or other diverse neurological, psychiatric, or medical diseases of children. The possible exception is low 5-HIAA among epileptics.[86] One study noted that three autistics showed no elevation in CSF 5-HIAA after probenecid.[91] Most probably this means only that the probenecid concentration at transport sites was insufficient to prevent 5-HIAA transport and cause it to accumulate. The alternate, that neither the cord nor the brain produces 5-HIAA in these children, seems unlikely.

In summary, then, there is reasonable evidence that whole blood serotonin concentration is elevated in some autistic children but no direct evidence that this reflects brain serotonin metabolism. There is also some evidence that peripheral serotonin biosynthesis may also be enhanced but catabolism by platelets clearly is not.

3.2. Platelets

As already discussed, the number and size of platelets could influence blood serotonin concentrations but the existing evidence suggests that neither accounts for the hyperserotoninemia of autism.

Serotonin uptake, efflux, and storage by platelets of autistics have been examined to some extent but often under conditions that confound one process with another. Most of the data represent serotonin accumulation by platelets incubated with solutions having relatively large serotonin concentrations (10 μM) after relatively long periods of exposure (30–90 min). In one series of studies,[92–95] heparinized platelets from autistics and/or childhood schizophrenics reportedly showed subnormal uptake; in another,[78] uptake was greater than that of controls: while in still other studies, platelets from autistics and normals did not differ.[33,96] Short-term uptake from dilute serotonin solutions was reported to be 30% greater for platelets from a small number of autistics than for hospitalized psychotic controls.[97]

Studies on serotonin efflux from platelets are in no better agreement. Boullin *et al.*[77,78] reported increased serotonin efflux by platelets from children diagnosed as autistic from a parental questionnaire[98] and were able to predict the diagnosis from the efflux values. Unfortunately, later studies were unable to replicate this with populations subdivided by parental questionnaire,[98] clinical criteria,[33] or both.[100]

These inconsistent results probably reflect methodological differences between studies. For example, estimates of uptake and efflux generally require a preliminary centrifugation of whole blood to obtain a platelet-rich plasma that is then used directly, or the platelets are further concentrated by recentrifugation and resuspended in some medium. Until the recent advent of procedures for the quantitative recovery of platelets from blood,[75,101] even this preliminary step was a source of variance between investigations because the centrifugation conditions of speed, time, blood volume, and tube dimensions were idiosyncratic. As a consequence, the relative recoveries of heavy and light platelets varied in an unspecified manner, and to the extent that the property to be measured varied with platelet size or age, at least quantitative discrepancies between studies were almost certain to occur. There are numerous other complications as well. Even substituting anticoagulants could affect replication of results. For example, heparin prevents clotting by a different mechanism than do EDTA and citrate and while the latter act by chelating external calcium, their effects are not identical. Citrated platelets tend to clump if rapidly cooled and warmed whereas EDTA-treated platelets do not, while EDTA-treated platelets have a greater tendency to swell and change shape than do citrated platelets and there are situations where these differences could be important to the results.

Platelet studies carried out on platelet-rich plasma or on platelets concentrated and resuspended in fresh platelet-free plasma could confuse the effects of differences in plasma composition, particularly ionic composition, with effects attributable to the platelets themselves. Separation of platelets from plasma and resuspension in artificial media involve additional manipulations. Because platelets are designed to self-destruct, increasing the number of manipulations increases the risk of damaging the structure and functional integrity of the platelet or platelet subpopulations.

Each measure also has its own inherent problems. Uptake studies generally examine the changing distribution of radiolabeled serotonin between platelet and medium. Experimentally, the uptake rate is multiphasic, with an initial fast linear uptake for about the first 2 min followed by a somewhat slower linear rate of uptake that gradually plateaus off to a steady state. The first phase presumably involves movement though the plasma membrane into the cytoplasm; the second, the first movement combined with transport from the cytoplasm into storage vesicles; and the last phase, the sum of true uptake, passive diffusion, efflux, and loss of label due to platelet destruction.[97,102,103] Low concentrations of ligand minimize the contribution of passive diffusion but measurements must be taken at short intervals, not only so as to delineate the first segment of the uptake curve (passage through the plasma membrane) but also to avoid changes in substrate concentration sufficient to affect rate. High concentrations buffer against concentration changes at the risk of increasing the contribution of passive diffusion to the apparent uptake rate.

Similar considerations apply to interpreting "efflux" measures. Efflux methods generally involve loading platelets with radioactive serotonin, isolating and resuspending the loaded platelets in isotope-free medium, then measuring the loss of radioactivity from the platelet and/or the gain in the medium. All of the above-mentioned problems associated with platelet isolation and separation are even more relevant here, where the effects of platelet destruction can easily pass for efflux. Indeed, under appropriate conditions, platelets do not lose serotonin from either storage vesicles or cytoplasm,[65,104] and unless reuptake inhibitors are added to the medium, apparent efflux is the sum of true efflux minus reuptake and, without platelet loss, is usually negligible.

In light of these factors, a closer look at the short-term serotonin uptake rates of platelets from autistic children, and especially hyperserotoninemic autistic children, may be warranted. Further, the contradictory efflux data may be explicable as a consequence of an increased fragility of platelets from autistics that appeared, under the experimental conditions employed, as increased efflux. Such a possibility could be tested by a combination of careful efflux and fragility studies. To date, however, the data on platelet properties do not differentiate autistics nor have they, as yet, contributed to understanding the mechanism for hyperserotoninemia.

3.3. Catecholamines

The catecholamines, like the indoleamines, have also been implicated in a variety of mental disorders; the noradrenergic system is often linked to affective states and the dopaminergic system to schizophrenia. A few studies have been carried out on autistics. Autistics are reported to have elevated blood norepinephrine concentrations compared to age-matched controls, and slightly elevated supine blood pressure and heart rate.[80] This was interpreted to indicate heightened peripheral sympathetic tone, a view compatible with a reported increased peripheral blood flow and blood pressure and decreased peripheral vascular resistance among autistics.[105] Dopamine β-hydroxylase (DBH) activity was examined on the assumption that this enzyme, which converts dopamine to norepinephrine in noradrenergic neurons and is partly released with norepinephrine upon sympathetic stimulation,[106] should increase in serum as sympathetic tone increases. The data are ambiguous. Blood DBH activity in autism was found to be elevated,[107] to be equal to that of children with organic psychosis but lower than that of childhood schizophrenics,[108] and to be lower than normal.[25,80] Apart from these inconsistent data, it now appears that the relationship between DBH activity and sympathetic tone is more complex than initially thought: the enzyme has a very long lifetime compared to that of circulating catecholamines, and while it may provide some measure of chronic sympathetic tone, it is a poor tool for assessing acute changes in noradrenergic activity. There are suggestions that DBH activity may be under genetic control but there has been no attempt as yet to determine if DBH activity follows any pattern among the members of families with a multiple incidence of autism.

There are two routes of degradation of the catecholamines in brain and in the periphery, oxidation catalyzed by MAO and *O*-methylation carried out by catechol-*O*-methyltransferase. MAO can be found in platelets and its activity in relationship to autism has already been discussed. Catechol-*O*-methyltransferase activity can be detected in red cells and in fibroblasts. The existing data uniformly indicate the activity of this enzyme is unaltered in autism.[108,109]

In contrast to the report of elevated blood norepinephrine in autism,[80] blood 3-methoxy-4-hydroxyphenylglycol (MHPG), the major degradation product of norepinephrine in brain, has been reported to be either lower in autistics[110] or the same as that of normals.[111] Urinary and CSF levels of MHPG also appear normal and 30–60% of MHPG in peripheral tissues is believed to derive from brain. These results thus suggest there is no gross noradrenergic involvement in the disease.

Similarly, studies measuring CSF HVA, the major degradation product of central dopamine metabolism, also fail to support gross involvement of this system in autism.[86,87,90] Autistics diagnosed according to responses to the parental inventory[98] are reported to have an elevated urinary HVA/creatinine ratio roughly correlating with symptom severity.[112] This was not seen in another study, which instead found that clinically diagnosed autistics and controls excreted equal amounts of HVA over a 24-hr period.[113] The apparent difference could result from a population difference in urinary creatinine, from differences in the populations selected by the two types of diagnostic procedures, or chance. In any event, the relationship between urinary and central HVA is still to be defined.

3.4. *Miscellaneous*

A sprinkling of other biochemical measures also appears in the literature. A general screening of blood and CSF from autistics for amino acid concentrations and for concentrations of some select organic acids that have been associated with inborn errors in metabolism, failed to reveal any significant abnormalities.[114]

Plasma cAMP concentrations, which likely reflect spillover of cAMP production by all the major visceral organs, have been reported[32,108] to be significantly higher in autistics than in normal controls and to correlate directly with both hyperkinesia and serum serotonin.[32] Concentrations of cAMP in the CSF of autistic children have also been estimated[91] but without control group data for comparison.

Hyperuricosuria in the absence of hyperuricemia has been reported for some children with the diagnosis of childhood autism[115] and hypouricosuria coupled to excessive hypoxanthine excretion in others.[116] Erythrocyte adenine deaminase, which converts adenosine to inosine, was found to be lower in autistics than in controls; erythrocyte nucleoside phosphorylase, hypoxanthine-guanine phosphoribosyltransferase, and adenosine phosphoribosyltransferase, as well as lymphocyte adenosine kinase and ecto-5′-nucleotidase were the same for both populations.[117] The significance or consistency of these findings has yet to be determined.

The mineral content of scalp hair has been examined to see if heavy metal exposure contributes to the onset of autism.[118] Analysis for calcium, magnesium, zinc, copper, lead, and cadmium showed only that autistics may have had a lesser exposure to cadmium than controls.

4. *PHARMACOLOGY*

Pharmacological response has been useful in developing biochemical models for conditions like schizophrenia and depression and could serve autism as well. Results must be interpreted with caution, however, since medical history has many examples of therapies that have worked or failed for unsuspected reasons. Pharmacological studies are especially difficult to design, perform, and interpret with syndromes like autism. The low incidence of the disease limits the size of treatment populations and increases the problem of statistically demonstrating efficacy. As with other childhood diseases, it is difficult to obtain appropriate controls, so that most studies are longitudinal in design rather than cross-sectional. Yet the spontaneous fluctuations in clinical state, so characteristic of the syndrome, can confound spontaneous change with therapeutic response in such longitudinal studies. Additionally, symptoms are often idiosyncratic and measuring instruments generally too insensitive, nonspecific, and subjective to adequately reflect clinical change. Finally, the possibility that the syndrome may involve early irreversible damage obscures the significance of a failure to respond.

The discussion below summarizes the pharmacological studies that appear relevant to proposed biochemical abnormalities in autism.

4.1. Indoleamines

Dopa, 5-HTP, imipramine, and fenfluramine have been used to examine the linkage between blood serotonin and clinical state. The efficacy of dopa in treating the motor disturbance of Parkinsonism was used to justify trials with other abnormalities, including autism. The major action of dopa administration is to elevate dopamine and, to a lesser extent, norepinephrine, but it also affects the serotonin system by competing with 5-HTP for decarboxylation and by acting as a false transmitter in serotoninergic neurons. As expected, treatment of autistics with dopa decreased excretion of urinary 5-HIAA and lowered blood serotonin although generally concentrations after treatment still exceeded those of matched controls. Platelet counts were unaffected. Clinical change was minor. One study, evaluating behavior by global clinical evaluations of perception, motility, relatedness, and language, failed to detect significant behavioral changes after dopa beyond those that could be attributed to spontaneous change.[119] Another study, with more discrete behavioral indices, reported some, statistically insignificant, improvement in half the subjects despite dopa-induced exacerbations of stereotypic behavior.[120]

Two studies[121,122] on a total of five autistic children found that 5-HTP administration alone or in combination with a decarboxylase inhibitor did not improve, and perhaps worsened, clinical symptoms. This indicates that the syndrome is probably not due to a general deficiency in brain 5-HTP and, indeed, may involve excessive serotonin. However, the pharmacological use of 5-HTP suffers from some of the same problems as the use of dopa since it too can act as a false transmitter. Interpretation of the results is therefore somewhat ambiguous. Similarly, reports that the serotonin reuptake inhibitor, imipramine, magnified autistic symptoms[123] also indicate a link with hyperserotoninemia but again interpretation is unclear. Imipramine is rapidly demethylated *in vivo* to the rather specific norepinephrine reuptake inhibitor, desmethylimipramine; the exacerbation of symptoms could as well be due to hypernoradrenergic stimulation as to hyperserotoninemic activity. The behavioral effects on autistic children of more specific serotonin reuptake inhibitors like fluoxetine, chlorimipramine, or zymilidine have yet to be examined.

The best pharmacological evidence that the serotoninergic system may be involved in autism comes from reports that fenfluramine ameliorates the symptoms of autism.[124,125] This compound, originally used as an anorectic, reversibly and rather specifically decreases serotonin and 5-HIAA in brains of experimental animals,[126,127] decreases serotonin vesicular storage,[128] and decreases serotonin reuptake.[129,130] Initial reports of long-term fenfluramine-induced injury to serotoninergic neurons[131] have been challenged[132] and the actions of the drug on the serotonin system seem reversible. The compound does not affect the noradrenergic or dopaminergic systems at doses sufficient to lower serotonin[133] although at high doses striatal HVA, but not dopamine, is increased by a mechanism thought to involve postsynaptic dopamine receptors.[134] Given to man, fenfluramine lowers circulating norepinephrine concentrations,[135] presumably by acting on the autonomic nervous system, and lowers blood serotonin concentration by mechanisms that may involve both serotonin storage in enterochromaffin cells and uptake into platelets. The slow fall in blood serotonin following continuous treatment with fenfluramine roughly corresponds with the appearance of therapeutic response, beginning with decreased stereotypic motor behaviors, followed by changes in the personality, motor, and attentional components of the syndrome, and finally improving performance on standard IQ tests. Upon return to placebo, symptoms reappear in the same order, abnormal stereotypic behavior reappearing first and impaired performance on IQ tests last. Only a few studies have been carried out, however, and the data are too limited to permit characterization of responders with regard to age at treatment, severity, particular clinical features, or blood serotonin concentrations; to define the generality, duration, and magnitude of its therapeutic effects; or to assess the possible risks in its chronic use with prepuberal children. It seems reasonable that some of the clinical effects of fenfluramine probably reflect actions on the serotonin system but, even if so, serotonin could be involved indirectly in symptom production rather than directly in disease etiology. Preliminary data indicating that clinical efficacy is not restricted to hyperserotoninemics could suggest that this is the case. Considerably more data are required to answer any of these questions.

4.2. *Catecholamines*

Presumed similarities between autistics and schizophrenics prompted clinical trials with the neuroleptics, especially the dopamine receptor blocker haloperidol,[136–139] and clinical responses to haloperidol have been summarized by Campbell *et al.*[123,140] In general, haloperidol appears to slightly reduce the frequency of stereotyped behaviors and withdrawal, decrease hyperactivity, and facilitate behavioral therapies. The effects do not appear to be dramatic or long-lasting. Amphetamine is reported to exacerbate preexisting stereotypic motor behavior.[129] These results indicate dopamine may be involved in at least the motor abnormalities in autism.

4.3. *Pyridoxine*

There is no direct evidence linking avitaminosis and autism and what little data exist indicate autistics have normal blood levels of ascorbate, folate, pyridoxine, and riboflavin.[141] Nonetheless, megavitamin therapy has become a widespread home treatment for many illnesses, including autism, largely because of Pauling's suggestion[142] that genetically directed structural changes in enzymes may decrease their cofactor affinity and that mental disease results from localized cerebral avitaminosis accompanying such changes. The possibility of disease resulting from genetic abnormalities leading to genetic-linked excessive cofactor binding is generally not considered; apparently it is assumed that all enzymatic activities should run at V_{max}. Megavitamin therapy for autism has largely focused on pyridoxine, seemingly because pyridoxal phosphate is the cofactor for aromatic amino acid decarboxylase and it was therefore assumed that increasing pyridoxine would increase monoamine biosynthesis. Why increasing monoamine biosynthesis *per se* would necessarily be beneficial is unclear. Of course, pyridoxal phosphate is not only the cofactor for aromatic aminoacid decarboxylase but also for all decarboxylations as well as for all transaminations, and for a variety of more specific reactions such as α-β and β-γ elimination reactions like serine dehydration and homocysteine desulfuration. Many metabolic systems then could theoretically be affected by manipulating pyridoxal cofactors. Further, decarboxylation is normally not the rate-limiting reaction in monoamine biosynthesis; aromatic aminoacid decarboxylase is usually in the form of the haloenzyme so that only small increases in rate would result from further cofactor saturation; and the activity of pyridoxal kinase is probably more important in regulating pyridoxal phosphate concentration under normal dietary conditions than is pyridoxine.

However weak the rationale, efficacy has been claimed. Thus, interpersonal responses of 40% of autistics are reported to improve when given combinations of 500–1300 mg pyridoxine/day (2–4 mg is considered the minimum daily requirement) and magnesium salts (usually 200–500 mg magnesium lactate). Subjects who had additional illnesses and/or other disturbed behaviors appeared to be better responders than subjects with uncomplicated autism.[143] Unfortunately, 75% of the subjects were on other medications as well throughout the study. Another study reported that subjects regarded as responding to doses of pyridoxine ranging from 75 to 3000 mg/day in previous uncontrolled studies, deteriorated upon being given placebo. Behavior was assessed from a parental questionnaire and concurrent medications were taken on an uncontrolled basis.[144] In contrast to the preceding study, response was reported to be best among those autistics without associated illnesses.

The massive doses of pyridoxine used in these studies suggest that, if effective, the vitamin probably acts by mechanisms other than serving as cofactor. For example, pyridoxal and catecholamines can combine to form isotetraquinolines[145] and at these massive doses the vitamin may interfere with catecholaminergic activity. Because pyridoxine abuse runs the possible risk of sensory neuropathies,[146] which could well go undetected in uncommunicative infants, it seems important to more carefully test its therapeutic use and to include careful neurological examinations throughout such studies.

5. *COMMENTS*

A high incidence of hyperserotoninemia among autistics is the only confirmed biochemical abnormality reported for this population. The mechanism for this hyperserotoninemia is unknown and because it occurs in only a fraction of the autistic population and is seen in other pathological populations as well, its significance is obscure. Other reports of abnormalities either have not been verified yet or have been contested. Indeed, there are a surprising number of opposite, statistically significant population differences reported in this relatively small literature. Statistically significant discrepancies can, of course, result by chance and must occur when large numbers of similar studies are conducted or large numbers of variables examined. The probability estimate of $p < 0.05$, which has become the criterion for asserting a difference between sets of numbers, does, after all, mean that such a difference will occur by chance on the average of one time in 20. When studies are limited, however, statistically significant results that are flatly contradictory generally mean significant differences in subjects or methods. If findings differ because of subjects, then diagnostic criteria are inadequate or inappropriate to the task of defining distinct populations. If findings are due to differences in methodology, then the relevant variable is unlikely to be the one thought to be assayed but is instead some other variable inadvertently measured by the experimental technique. The most effective way to decide between subjects and methods is independent replication on the same patient population. Obviously, if the finding can be replicated on the test population but cannot be generalized to other populations, the test population merits more detailed study. On the other hand, if the results cannot be independently replicated on the initial test population, the details of the methodologies used need examination for possibly contributory variables that can be tested. For example, the conflicting findings on serotonin efflux in autism may reflect differences in platelet survival under the different experimental conditions employed in these studies rather than platelet efflux properties.

At least some of the seeming contradictions in the literature probably reflect differences in the populations studied. Until recently, ther terms *autism* and *childhood schizophrenia* were sometimes used interchangeably and frequently the operational criterion for either classification was idiosyncratic. The advent of more standardized criteria today have at least provided a frame of reference although the medical validity of that referent is uncertain. Indeed, if terms like autism, schizophrenia, depression, or coronary heart disease identify populations with shared symptoms, but not shared etiologies, clear population differences on some variables reveal, at best, correlates of symptom expression and, at worst, consequences of shared environmental artifacts. A more appropriate strategy may be to more closely examine the clusters of patients who exhibit a common biological abnormality that may not be shared by others carrying the same diagnostic label or by the general population. At worst, this may provide information about biological expression and at best may define a disease hidden beneath the diagnosis. However, until the basis for such an abnormality is established and means for manipulating it are developed, there are few clear guides beyond shared pharmacological responsivity, medical history, or disease course to determine if the shared abnormality is medically relevant.

The magnitude of the abnormality used to define the population provides some clue for the direction of the search for mechanism. Magnitude usually increases with proximity to the primary metabolic lesion. Thus, the primary metabolic defect in a genetic disease like phenylketonuria is a *qualitative* difference in DNA sequence and in its protein translation product. Consequences of this qualitative change become quantitative and progressively smaller as they are increasingly influenced by other systems.

Most biochemical studies on disease begin with variables that are distant consequences of the prime defect and are most prone to provide conflicting results. Such conflicts are a serious problem for disease of low incidence like autism since the numbers of subjects, investigators, and studies are likely to remain insufficient to provide the kind of "preponderance of evidence" upon which most biomedical judgments are based. As indicated above, such conflicts are generally resolvable and could become clues for disease resolution rather than impediments to progress. All this will require more cooperation and coordination than is needed to resolve conflicting findings in diseases with greater incidence, but without such efforts progress is likely to remain slow.

ACKNOWLEDGMENTS. This work was supported in part by the Veterans Administration Research Service. The authors are indebted to Elaine Martin for excellent editorial assistance.

REFERENCES

1. Kanner, L., 1944, *J. Pediatr.* **25**:211–217.
2. Rutter, M., 1972, *J. Autism Child. Schizophr.* **2**:315–337.
3. Eisenberg, L., 1972, *J. Autism Child. Schizophr.* **2**:338–342.
4. Ornitz, E. M., and Ritvo, E. R., 1976, *Am. J. Psychiatry* **133**:609–621.
5. Fish, B., Schapiro, T., Campbell, M., and Wile, R. A., 1968, *Am. J. Psychiatry* **124**:1415–1423.
6. Bender, L., 1956, *Am. J. Orthopsychiatry* **26**:499–506.
7. Friedman, E., 1969, *Schizophrenia* **1**:249–261.
8. Chess, S., 1971, *J. Autism Child. Schizophr.* **1**:33–47.
9. *Rutter, M., 1978, Autism: A Reappraisal of Concepts and Treatment*, Plenum Press, New York, pp. 1–26.
10. Knoblock, H., and Pasamanick, B., 1975, *Pediatrics* **55**:182–191.
11. Meryash, D. L., Szymanski, L. S., and Gerald, P. S., 1982, *J. Autism Dev. Disord.* **12**:295–301.
12. Brown, W. T., Jenkins, E. C., Friedman, E., Brooks, J., Wisniewski, K., Raguthu, S., and French, J., 1982, *J. Autism Dev. Disord.* **12**:303–308.
13. Torrey, E. G., Hersh, S. P., and McCabe, K. D., 1975, *J. Autism Child. Schizophr.* **5**:287–297.
14. Chapman, A. H., 1957, *Arch. Neurol. Psychiatry* **78**:621–623.
15. Vaillant, G. E., 1963, *Arch. Gen. Psychiatry* **9**:163–167.
16. Folstein, S., and Rutter, M., 1977, *Nature* **265**:726–728.
17. Hanson, D. R., and Gottesman, I., 1976, *J. Autism Child. Schizophr.* **6**:209–234.
18. Kolvin, I., Ounsted, C., and Humphrey, M., 1971, *Br. J. Psychiatry* **118**:385–395.
19. Rutter, M., 1965, *Dev. Med. Child Neurol.* **7**:518–528.
20. Folstein, S., and Rutter, M., 1977, *J. Child Psychol. Psychiatry Allied Discip.* **18**:297–321.
21. Ritvo, E. R., Ritvo, E. C., and Brothers, A. M., 1982, *J. Autism Dev. Disord.* **12**:109–115.
22. Spence, M. A., 1976, *Autism: Diagnosis, Current Research, and Management* (E. R. Ritvo, B. J. Freeman, E. M. Ornitz, and P. Tanguay, eds.), Spectrum, New York, pp. 169–174.
23. Schain, R. J., and Freedman, D. X., 1961, *J. Pediatr.* **58**:315–318.

24. Ritvo, E., Yuwiler, A., Geller, E., Ornitz, E., Saeger, K., and Plotkin, S., 1970, *Arch. Gen. Psychiatry* **23**:566–572.
25. Goldstein, M., Mahanad, D., Lee, J., and Coleman, M., 1976 *The Autistic Syndromes* (M. Coleman, ed.), Elsevier/North-Holland, Amsterdam, pp. 57–64.
26. Yuwiler, A., Plotkin, S., Geller, E., and Ritvo, E., 1970, *Biochem. Med.* **3**:426–436.
27. Yuwiler, A., Ritvo, E., Bald, D., Kipper, D., and Koper, A., 1971, *J. Autism Child. Schizophr.* **1**:421–435.
28. Douay, O., Debray-Ritzen, P., and Kamoun, P., 1980, *Ann. Biol. Clin.* **38**:243.
29. Takahashi, S., Kanai, H., and Miyamoto, Y., 1976, *J. Autism Child. Schizophr.* **6**:317–326.
30. Hanley, H. G., Stahl, S. M., and Freedman, D. X., 1977, *Arch. Gen. Psychiatry* **34**:521–531.
31. Campbell, M., Friedman, E., Devito, E., Greenspan, L., and Collins, P. J., 1974, *J. Autism Child. Schizophr.* **4**:33–41.
32. Hoshino, Y., Kumashiro, H., Kaneko, M., Numata, Y., Honda, K., Yashima, Y., Tachibana, R., and Watanabe, M. O., 1979, *Fukushima J. Med. Sci.* **26**:79–91.
33. Yuwiler, A., Ritvo, E., Geller, E., Glousman, P., Schneiderman, G., and Matsuno, D., 1975, *J. Autism Child. Schizophr.* **5**:83–98.
34. Yuwiler, A., Brammer, G. L., Morley, J. E., Raleigh, M. J., Flannery, J. W., and Geller, E., 1981, *Arch. Gen. Psychiatry* **38**:619–626.
35. Wirz-Justice, A., Lichtsteiner, M., and Feer, H., 1977, *J. Neural Transm.* **41**:7–15.
36. Sauerbier, I., and Von Mayersbach, H., 1976, *Horm. Metab. Res.* **8**:157–158.
37. Sarai, K., and Kayano, M., 1968, *Folia Psychiatr. Neurol. Jpn.* **22**:138–140.
38. Raleigh, M. J., Yuwiler, A., Brammer, G. L., McGuire, M. T., Morley, J., and Flannery, J. W., 1981, *Psychopharmacologia* **72**:241–246.
39. Raleigh, M. J., Brammer, G. L., and McGuire, M. T., 1981, *Am. J. Primatol.* **1**:361.
40. Raleigh, M. J., McGuire, M. T., Brammer, G. L., and Yuwiler, A., 1984, *Arch. Gen. Psychiatry* **41**:405–410.
41. Halevy, A., Moos, R. H., and Solomon, E. F., 1965, *J. Psychiatr. Res.* **3**:1–10.
42. Partington, M. W., Tu, J. B., and Wong, C. Y., 1973, *Dev. Med. Child Neurol.* **15**:616–627.
43. DeLisi, L. E., Neckers, L. M., Weinberger, D. R., and Wyatt, R. J., 1981, *Arch. Gen. Psychiatry* **38**:647–650.
44. Hauser, S., DeLong, S., and Rosman, N., 1975, *Brain* **98**:667–688.
45. Gillberg, C., and Svendsen, P., 1983, *J. Autism Dev. Disord.* **13**:19–32.
46. Damasio, H., Maurer, R. G., Damasio, A. R., and Chiu, H. C., 1980, *Arch. Neurol.* **37**:504–510.
49. Hier, D. E., LeMay, M., and Rosenberg, P. B., 1978, *Neurology* **28**:348.
48. Hier, D. E., LeMay, M., and Rosenberg, P. B., 1979, *J. Autism Dev. Disord.* **9**:153–159.
49. Williams, R. S., Hauser, S. L., Purpura, D. P., Delong, G. R., and Swisher, C. N., 1980, *Arch. Neurol.* **37**:749–753.
50. Erspamer, V., 1966, *Handbuch Exp. Pharmakol.* **19**:132–181.
51. Haverback, B. J., and Davidson, J. D., 1958, *Gastroenterology* **35**:570–578.
52. Bertaccini, G., 1960, *J. Physiol. (London)* **153**:239–243.
53. Demet, E. M., Halaris, A. E., and Bhatarakamul, S., 1978, *Clin. Chim. Acta* **89**:285–292.
54. Sneddon J. M., 1969, *Br. J. Pharmacol.* **37**:680–688.
55. Lingjaerde, O., 1971, *Acta Physiol. Scand.* **81**:75–83.
56. Born, G. V. R., and Gillson, R. E., 1959, *J. Physiol. (London)* **46**:472–491.
57. Lingjaerde, O., 1970, *Eur. J. Pharmacol.* **13**:76–82.
58. Rudnick, G., 1977, *J. Biol. Chem.* **252**:2170–2174.
59. Crane, R. K., 1965, *Fed. Proc.* **24**:1000–1006.
60. Holmsen, H., Ostvold, A. C., and Day, H. J., 1973, *Biochem. Pharmacol.* **22**:2599–2608.
61. Tranzer, J. P., Da Prada, M., and Pletscher, A., 1967, *Adv. Pharmacol.* **6A**:125–128.
62. Costa, J. L., Reese, T. S., and Murphy, D. L., 1974, *Science* **183**:537–538.
63. Born, G. V. R., Ingram, G. I. C., and Stacey, R. S., 1958, *Br. J. Pharmacol.* **13**:62–64.
64. Pletscher, A., Da Prada, M., Berneis, K. H., and Tranzer, J. P., 1971, *Experientia* **27**:993–1002.
65. Costa, J. L., Murphy, D. L., and Kafka, M. S., 1977, *Biochem. Pharmacol.* **26**:517–521.
66. McDonald, T. P., Odell, T. T., Jr., and Gosslee, D. G., 1964, *Proc. Soc. Exp. Biol. Med.* **115**:684–689.

67. Karpatkin, S., 1969, *J. Clin. Invest.* **48**:1073–1082.
68. Von Behrens, W. E., 1972, *Thromb. Diath. Haemorrh.* **27**:159–172.
69. Booyse, R. M., Hohn, D. C., and Rafelson, M. E., 1969, *S. Afr. Med. J.* **43**:964-967.
70. Nakeff, A., and Ingram, M., 1970, *J. Appl. Physiol.* **28**:530–533.
71. Minter, F. M., and Ingram, M. J., 1971, *Br. J. Haematol.* **20**:55–68.
72. Pennington, D. G., Lee, N. L. Y., Roxburgh, A. E., and McCready, J. P., 1976, *Br. J. Haematol.* **34**:365–376.
73. Karpatkin, S., 1972, *Ann. N. Y. Acad. Sci.* **201**:262–279.
74. Corash, S., Shafer, B., and Perlow, M., 1978, *Clin Res.* **26**:345.
75. Murphy, D. L., Costa, J. L., Shafer, B., and Corash, L., 1978, *Psychopharmacologia* **59**:193–197.
76. Paulus, J. M., 1975, *Blood* **46**:321–336.
77. Boullin, D. J., Coleman, M., O'Brien, R. A., and Rimland, B., 1971, *J. Autism Child. Schizophr.* **1**:63–71.
78. Boullin, D. J., Coleman, M., and O'Brien, R. A., 1970, *Nature* **226**:371–372.
79. Shaw, C., Lucas, J., and Rabinovitch, R., 1959, *Arch. Gen. Psychiatry* **1**:366–370.
80. Lake, C. R., Ziegler, M. G., and Murphy, D. L., 1977, *Arch. Gen. Psychiatry* **34**:553–556.
81. Cohen, D. J., Young, J. G., and Roth, J. A., 1977, *Arch. Gen. Psychiatry* **34**:534–537.
82. Young, J. G., Cohen, D. J., and Roth, J. A., 1978, *Life Sci.* **23**:797–806.
83. Boullin, D. J., Bhagavan, H. N., Coleman, M., O'Brien, R. A., and Youdim, M. B. H., 1975, *Med. Biol.* **53**:210–213.
84. Boullin, D. J., Bhagavan, H. N., O'Brien, R. A., and Youdim M. B. H., 1976, *The Autistic Syndromes* (M. Coleman, ed.), North-Holland, Amsterdam, pp. 51–56.
85. Giller, E. L., Jr., Young, G., Breakefield, X. O., Carbonari, C., Braverman, M., and Cohen, D. J., 1980, *Psychiatr. Res.* **2**:187–197.
86. Cohen, D. J., Shaywitz, B. A., Johnson, W. T., and Bowers, M., 1974, *Arch. Gen. Psychiatry* **31**:845–853.
87. Leckman, J. F., Cohen, D. J., Shaywitz, B. A., Caparulo, B. K., Heninger, G. R., and Bowers, M. B., Jr., 1980, *Arch. Gen. Psychiatry* **37**:677–681.
88. Seifert, W. E., Jr., Foxx, J. L., and Butler, I. J., 1980, *Ann. Neurol.* **8**:38–42.
89. Shaywitz, B. A., Cohen, D. J., Leckman, J. F., Young, J. G., and Bowers, M. B., Jr., 1980, *Dev. Med. Child Neurol.* **22**:748–754.
90. Cohen, D. J., Caparulo, B. K., Shaywitz, B. A., and Bowers, M. B., Jr., 1977, *Arch. Gen. Psychiatry* **34**:545–556.
91. Winsberg, B. G., Sverd, J., Castells, S., Hurwic, M., and Perel, J. M., 1980, *Neuropaediatrie* **11**:250–255.
92. Sankar, D. V. S., Gold, E., Phipps, E., and Sankar, D. B., 1962, *Ann. N.Y. Acad. Sci.* **96**:392–398.
93. Sankar, D. V. S., Cates, N., Broer, H. B., and Sankar, D. B., 1963, *Recent Advances in Biological Psychiatry*, Volume 5 (J. Wortis, ed.), Plenum Press, New York, pp. 76–83.
94. Sankar, D. V. S., 1970, *Acta Paedopsychiat. Int J. Child Psychiatry* **37**:174–182.
95. Sankar, D. V. S., 1977, *Neuropsychobiology* **5**:234–239.
96. Lucas, A. R., Warner, K., and Gottlieb, J. S., 1971, *Biol. Psychiatry* **38**:39–48.
97. Rotman, A., Caplan, R., and Szekely, G. A., 1980, *Psychopharmacologia* **67**:245–248.
98. Rimland, B., 1979, *Serotonin and Mental Abnormalities* (D. J. Boullin, ed.), Wiley, New York, pp. 29–40.
99. Boullin, D. J., 1978, *Serotonin and Mental Abnormalities* (D. J. Boullin, ed.), Wiley, New York, p. 7.
100. Boullin, D., Freeman, B. J., Geller, E., Ritvo, E., Rutter, M., and Yuwiler, A., 1982, *J. Autism Dev. Disord.* **12**:97–98.
101. Corash, L., Tan, H., and Gralnick, H. R., 1977, *Blood* **49**:71–87.
102. Costa, J. L., Kirk, K. L., and Stark, H., 1981, *Res. Commun. Chem. Pathol. Pharmacol.* **33**:547–557.
103. Stahl, S. M., and Meltzer, H. Y., 1978, *J. Pharmacol. Exp. Ther.* **205**:118–132.
104. Costa, J. L., Kirk, K. L., Murphy, D. L., and Stark, H., 1981, *Br. J. Pharmacol.* **72**:449–459.

105. Cohen, D. J., and Johnson, W. T., 1977, *Arch. Gen. Psychiatry* **34**:561–567.
106. Weinshilboum, R. M., Raymond, F. A., Elveback, L. R., and Weidman, W. H., 1973, *Frontiers in Catecholamine Research* (E. Usdin and S. Snyder, eds.), Pergamon Press, Elmsford, New York, pp. 1115–1122.
107. Freedman, L. S., Roffman, M., and Goldstein, M., 1973, *Frontiers in Catecholamine Research* (E. Usdin and S. Snyder, eds.), Pergamon Press, Elmsford, New York, pp. 1109–1114.
108. Belmaker, R. H., Hattab, J., and Ebstein, R. P., 1978, *J. Autism Child. Schizophr.* **8**:293–298.
109. O'Brien, R. A., Semenuk, G., Coleman, M., and Spector, S., 1976, *J. Clin. Exp. Pharmacol. Physiol.* **3**:9–14.
110. Young, J. G., Cohen, D. J., Caparulo, B. K., Brown, S. L., and Maas, J. W., 1979, *Am. J. Psychiatry* **136**:1055–1057.
111. Young, J. G., Cohen, D. J., Kavanagh, M. E., Landis, H. D., Shaywitz, B. A., and Maas, J. W., 1981, *Life Sci.* **28**:2837–2845.
112. Garreau, B., Barthelemy, C., Domenechi, J., Sauvage, T., Muh, J. P., Lelord, G., and Callaway, E., 1980, *Acta Psychiatr. Belg.* **80**:249–265.
113. De Villard, R., and Dalery, J., 1979, *Ann. Med. Psychol.* **137**:109–114.
114. Perry, T. L., Hansen, S., and Christie, R. G., 1978, *Biol. Psychiatr.* **13**:575–586.
115. Coleman, M., Landgrebe, M. A., and Landgrebe, A. R., 1976, *The Autistic Syndromes* (M. Coleman, ed.), North-Holland, Amsterdam, pp. 183–195.
116. Lis, A. W., McLaughlin, D. I., McLaughlin, R. W., Lis, E., and Stubbs, E. G., 1976, *Clin. Chem.* **22**:1528–1532.
117. Stubbs, G., Litt, M., Lis, E., Jackson, R., Voth, W., Linburgg, A., and Litt, R., 1982, *J. Am. Acad. Child Psychiatry* **21**:71–74.
118. Shearer, T. R., Larson, K., Neuschwander, J., and Gedney, B., 1982, *J. Autism Dev. Disord.* **12**:25–34.
119. Ritvo, E., Yuwiler, A., Geller, E., Kales, A., Rashkis, S., Schicor, A., Plotkin, S., Axelrod, R., and Howard, C., 1971, *J. Autism Child. Schizophr.* **2**:190–205.
120. Campbell, M., Small, A. M., Colin, P. J., Friedman, E., David, R., and Genieser, N. B., 1976, *Curr. Ther. Res. Clin. Exp.* **18**:70–86.
121. Sverd, J., Kupietz, S. S., Winsberg, B. G., Hurwic, M. J., and Becker, L., 1978, *J. Autism Child. Schizophr.* **8**:173–180.
122. Zarcone, V., Kales, A., Scharf, M., Tan., T. L., Simmons, J. Q., and Dement, W. C., 1973, *Arch. Gen. Psychiatry* **28**:843–846.
123. Campbell, M., Cohen, I. L., and Anderson, L. T., 1981, *Can. J. Psychiatry* **26**:265–273.
124. Geller, E., Ritvo, E. R., Freeman, B. J., and Yuwiler, A., 1981, *N. Engl. J. Med.* **307**:165–169.
125. Ritvo, E. R., Freeman, B. J., Geller, E., and Yuwiler, A., 1983, *J. Am. Acad. Child Psychiatry* **22**:549–558.
126. Duhault, J., and Verdavainne, C., 1967, *Arch. Int. Pharmacodyn. Ther.* **170**:276–286.
127. Duhault, J., and Boulanger, M., 1977, *Eur. J. Pharmacol.* **43**:203–205.
128. Costa, E., Gropetti, A., and Revuelta, A., 1971, *Br. J. Pharmacol.* **41**:57–64.
129. Fuxe, K., Farnebo, L. O., Hamberger, B., and Ogren, S. O., 1975, *Postgrad. Med. J.* **51**(Suppl. 1):35–45.
130. Clineschmidt, B. V., Zacchei, A. G., Totaro, J. A., Pflueger, A. B., Guffin, J. C., and Wyshousky, T. I., 1978, *Ann. N.Y. Acad. Sci.* **305**:222–241.
131. Harvey, J. A., and McMasters, S. E., 1977, Psychopharmacol. Commun. **1**:3–17.
132. Sotelo, C., and Zamora, A., 1979, *Curr. Med. Res. Opin.* **6**(Suppl. 1):55–62.
133. Garattini, S., Bizzi, A., De Gaetano, G., Jori, A., and Samanin, R., 1975, *Recent Advances in Obesity Research* (A. Howard, ed.), Newman, London, pp. 354–367.
134. Garattini, S., Buczko, W., Jori, A., and Samanin, R., 1975, *Postgrad. Med. J.* **51**(Suppl. 1):27–35.
135. De La Vega, C. E., Slater, S., Ziegler, M. G., Lake, C. R., and Murphy, D. L., 1977, *Clin. Pharmacol. Ther.* **21**:216.
136. Faretra, G., Dooher, L., and Dowling, J., 1970, *Am. J. Psychiatry* **129**:1670–1673.

137. Engelhardt, D. M., Polizos, P., Waizer, J., and Hoffman, S. P., 1973, *J. Autism Child. Schizophr.* **3**:128–137.
138. Campbell, M., Anderson, L. T., Meier, M., Choen, I. L., Small, A. M., Samit, C., and Sachar, E. J., 1978, *J. Am. Acad. Child Psychiatry* **17**:640–655.
139. Cohen, I. L., Campbell, M., Posner, D., Small, A. M., Triebel, D., and Anderson, L. T., 1980, Behavioral effects of haloperidol in young autistic children, *J. Amer. Aca. Child Psychiat.* **19**:665–677.
140. Campbell, M., Anderson, L. T., Small, A. M., Perry, R., Green, W. H., and Caplan, R., 1982, *J. Autism Dev. Disord.* **12**:167–175.
141. Sankar, D. V. S., 1979, *J. Autism Dev. Disord.* **9**:73–82.
142. Pauling, L., 1968, *Science* **160**:265–271.
143. Lelord, G., Muh, J. P., Bvarthelemy, C., Martineau, J., Garreau, B., and Callaway, E., 1981, *J. Autism Dev. Disord.* **11**:219–230.
144. Rimland, B., Callaway, E., and Dreyfus, P., 1978, *Am. J. Psychiatry* **135**:472–475.
145. Schott, H. F., and Clark, W. C., 1952, *J. Biol. Chem.* **196**:449–462.
146. Schaumburg, H., Kaplan, J., Windebank, A., Vick, N., Rasmus, S., Pleasure, D., and Brown, M. J., 1983, *N. Engl. J. Med.* **309**:445–448.

26

Metabolic Encephalopathies

L. Battistin and M. Dam

1. INTRODUCTION

Metabolic encephalopathies (MEs) is a nonspecific term that includes a large group of neurological disorders characterized by metabolic derangement that affects the CNS and other organs as well, absence of mass lesions of the brain, aspecificity of neurological symptoms, and potential reversibility. MEs can be divided into inherited, i.e., secondary to an abnormality of metabolism, and acquired, usually produced by underlying diseases of organs or systems. Cerebral metabolic derangement need not depend upon an organic pathology, as it may represent an episodic accident during the life of healthy individuals (e.g., hypoxia from a foreign body in the trachea, CO poisoning). Most likely, there is a systemic disease that, if chronic, makes the ME a potential complication, always impending and life-threatening. In this case, the ME is the end result of multiple pathogenetic factors acting in concert to alter CNS functional activity. Nevertheless, ME may be caused by a single specific alteration; for instance, lack or excess of glucose is sufficient to produce cerebral disturbances.

From the clinical standpoint, MEs exhibit extreme polymorphism. The neurological symptoms and signs include altered states of consciousness (lethargy, stupor, coma), hyper- or hyporeflexia, paralysis, tremor, asterixis, myoclonic jerks, focal and generalized convulsions, frontal lobe signs, and a wide variety of psychiatric abnormalities. As a rule, MEs present a relatively common clinical pattern indicating an aspecific involvement of the CNS. Accordingly it has been suggested that at least some MEs (such as those related to hypoxia, hypoglycemia, nutritional deficits, hyperammonemia) may share the same pathophysiological mechanism in spite of diverse etiologies.[1,2] Since aspecificity of the neurological symptomatology is the hallmark of MEs, only the signs of the underlying disease and/or the patient's history can clarify the diagnosis. Hypoglycemic, hypoxic, or hepatic encephalopathies may start with a progressive or sudden impairment of consciousness; extrapyramidal symptoms may con-

---

L. Battistin and M. Dam • Clinica delle Malattie Nervose e Mentali, Università di Padova, 35100 Padova, Italy.

note hypoparathyroidism; dementia or severe psychiatric disturbances are common presenting symptoms of Addison's disease, Cushing's disease, and thyroid diseases.

MEs constitute a chapter of increasing interest in human pathology since they are dreaded complications of very common diseases such as hepatic, renal, or respiratory failure. At times, MEs may represent the factor limiting the survival of patients in spite of advances in the treatment of the underlying disease. This derives from the fact that, up to now, the pathogenetic mechanisms of most MEs have not been clearly delineated. Thus, the specific therapy and/or the prevention is far from satisfactory.

The aim of this chapter is to review some of the MEs in the following list. Particular emphasis will be placed on the pathogenetic mechanisms involved in producing the cerebral metabolic derangement. Obviously not all the MEs will be discussed: space limitations impose a selection. The reader therefore is referred to other chapters of this handbook for discussions of the MEs not included in this review.

- Encephalopathies from metabolic substrate abnormalities
  Hypoglycemia
  Anoxic-ischemic encephalopathy
  Hyperosmolality
  Hypoosmolality
  Hyper- and hypocalcemia
  Vitamin deficiencies*
- Encephalopathies from organ and system pathology
  Hepatic encephalopathies
  Uremic encephalopathies
  Diabetic ketoacidosis
  Respiratory encephalopathies
  Pancreatic encephalopathies
  Encephalopathies related to the digestive system
  Encephalopathies related to the endocrine system
- Encephalopathies from abnormalities of basic metabolism*
  Lipid metabolism
  Carbohydrate metabolism
  Protein metabolism

2. *ENCEPHALOPATHIES FROM METABOLIC SUBSTRATE ABNORMALITIES*

2.1. *Hypoglycemia*

It is well known that glucose is the primary energy source for the brain. An inadequate glucose supply produces CNS functional derangement and, if prolonged, irreversible damage. The cerebral tissue is particularly sensitive to

* This section will not be discussed since it is presented in other chapters.

the reduction of blood glucose because of its energy requirements and its low levels of endogenous stores.[3] Furthermore, other plasma compounds, such as ketone bodies, cannot, under normal circumstances, support brain metabolism.[4] Hypoglycemia may be consequent to many pathological conditions[5,6]; in adult life, common causes are overdose of insulin or oral antidiabetic agents; pancreatic or extrapancreatic insulin-secreting tumors; uremia; hepatic failure; endocrinopathies, such as adrenal insufficiency, hypothyroidism, or hypopituitarism; and drugs such as ethanol, salicylates, or propranolol.

The clinical picture is dominated by the symptoms and signs related to the cerebral metabolic derangement and to the increase in plasma of insulin counterregulatory hormones, particularly catecholamines: release of epinephrine accounts for the symptoms occurring as soon as hypoglycemia begins. The clinical manifestation of acute hypoglycemia depends upon the rapidity of onset, the rate of glucose lowering, and the duration.[5–8] Briefly, when the blood glucose is about 20–30 mg/100 ml, palpitation, sweating, trembling, hunger, lassitude, and nervousness appear. The next stage is characterized by confusion, amnesia, drowsiness, muscular spasms, focal symptoms, and with glucose blood levels about 10 mg/100 ml, by mydriasis, bradycardia, hypotonicity, and deep coma. Focal or generalized seizures can be observed. EEG abnormalities (slowing of the rhythms and burst of polyspikes in the case of seizures) parallel the progressive deterioration of brain function and the severity of the neurological signs and symptoms[9]; corresponding alterations of the sensory evoked potentials have been reported.[10]

With prolonged hypoglycemia, permanent lesions of the CNS may ensue, and concomitant cardiovascular and/or respiratory failure and convulsions may be contributory to the brain injury. Neuropathological findings from human and experimental animal brains indicate that the cortex, hippocampus, and basal ganglia are extensively damaged, whereas the cerebellum, brain stem, and spinal cord are the structures most likely to be spared.[11–13]

During prolonged hypoglycemia, the brain undergoes profound metabolic alterations. Concentrations of high-energy phosphates (ATP and phosphocreatine) are severely lowered during coma associated with an isoelectric EEG.[14,15] Cerebral blood flow is relatively unchanged, whereas the cerebral metabolic rate for oxygen and for glucose is reduced by 45 and 73%, respectively, suggesting oxidation of nonglucose substrates.[15] Thus, in the absence of glucose the cerebral tissue utilizes other fuels to produce energy[16–18]; that may explain the decreased brain content of glycolytic metabolites, citric acid cycle intermediates, total fatty acids, RNA, amino acids, and proteins.[3] Other metabolic abnormalities include enhanced brain ammonia concentration and electrolyte (mainly CA^{2+} and K^{+}) imbalance.[3,15] Brain ammonia, a by-product of amino acid metabolism, may reach levels during hypoglycemic coma sufficient to induce coma in normoglycemic animals.[15] Some regional differences indicate variable brain sensitivity to hypoglycemia; glucose-6-phosphate and fructose-1, 6-diphosphate fall 80% in all brain regions, lactate levels are low in all cerebral areas except the cerebellum, and pyruvate mainly decreases in the cerebral cortex and brain stem.[19] It is likely that all these changes particularly in brain energy levels account for cerebral functional derangement and brain damage occurring in the course of acute hypoglycemia. In terms of energy balance, the concentration of ATP in the brain is normal during lethargy and decreased by 6% during precoma and by 40% during coma associated with an isoelectric EEG.[15] ATP in the precoma stage was found depleted by 30%, and phosphocreatine by 55%, in cells of the reticular activating system,[20] whereas, even in the presence of an isoelectric EEG, ATP and phosphocreatine levels do not significantly decrease in the cerebellum.[19] The lowered levels of ATP and phosphocreatine in the reticular activating system may explain the impairment of consciousness in the absence of overt brain energy failure. It has been reported that hypoglycemia affects rostral brain regions more than caudal structures[21]: cortical activity may be absent (isoelectric EEG) in the presence of cardiovascular and respiratory regulation, which indicate a persistence of medullary function.[19] Since glucose and energy reserves are uniformly depleted except in the cerebellum, it is unlikely that energy failure accounts for this regionally different cerebral functioning. On the other hand, a mild to moderate reduction of oxygen or glucose supply to the brain is associated with deranged cerebral activity in spite of normal levels of energy metabolites[1]; accordingly, Ferrendelli and Chang[22] found that hypoglycemia inducing loss of righting reflex in mice had no effect on regional energy reserves. Changes of neurotransmitter metabolism, particularly of acetylcholine and monoamines, related to the altered oxidative processes, may explain the early cerebral functional alterations in the absence of brain energy failure.[1,3]

The duration and severity of the hypoglycemia necessary to produce neuronal damage are not known. Histopathological findings indicate that irreversible brain damage occurs after 30 min of hypoglycemic coma, and is exaggerated after 60 min.[3] However, Ghajar *et al.*[15] suggested that brain lesions may develop in a shorter period of time, which implies the need for prompt treatment once hypoglycemia is suspected in a comatose patient. As stated before, the neocortex is particularly damaged by prolonged hypoglycemia whereas the cerebellum is relatively spared. This selectivity of effects may partially reside in the fact that glucose delivery shows regional differences: the cerebellum may be more resistant since its metabolic rate for glucose during hypoglycemia is reduced less than that of other, more vulnerable structures.[3] Although the mechanisms involved in the neuronal damage are not completely elucidated, oxidative breakdown of cellular structures may be relevant. Substrates such as phospholipids or proteins, which derive from cellular components, could be used to produce energy in the absence of glucose.[23] The reported disintegration of ribosomes and endoplasmic membranes during hypoglycemia may account for this phenomenon.[24,25] It is also likely that the metabolic degradation of cellular components may lead to the formation and accumulation of products toxic for the brain. Free fatty acids and lysophospholipids, derived from phospholipid breakdown, greatly increase in the cortex of rats during hypoglycemic coma.[26] The damaging action of free fatty acids (particularly arachidonic acid) may depend upon their capacity to impair mitochondrial function, to induce brain edema, and to promote prostaglandin synthesis.[27–30] It is noticeable that free fatty acid concentration is more pronounced in the cortex than in the cerebellum. Accordingly the former is less "resistant" to hypoglycemia than the latter.[31]

Clinically, in addition to acute hypoglycemia, two other forms, subacute hypoglycemia and chronic hypoglycemia, have been identified.[5,7] Subacute hypoglycemia is characterized by progressive somnolence and lethargy, the chronic form by personality changes, psychotic behavior, gradual mental deterioration, and frank dementia. Islet cell hypertrophy or tumor and malignancy are alleged causes of these syndromes.

2.2. *Anoxic-Ischemic Encephalopathy*

Oxygen deficiency is a common pathogenetic mechanism by which many human diseases may functionally and/or anatomically involve the brain. Hypoxia may accompany cerebral hypoperfusion, which can result from inadequate blood supply to the brain, as in cardiac arrest; increased vascular resistance, as in severe hypertension; or increased blood viscosity, as in macroglobulinemia, polycythemia, or plasmacytoma. Small vessel occlusion, as in thrombotic thrombocytopenic purpura or in systemic lupus erythematosus, can also cause brain hypoxia and consequently neurological and/or psychiatric disturbances.

Hypoxia may occur without concomitant impairment of cerebral blood flow, as in severe pulmonary diseases, anemia, or hemoglobinopathies, where only oxygen supply to the brain is reduced. The various forms of hypoxia have been classified as follows:[32]

1. Anoxic anoxia. The arterial partial pressure of oxygen is reduced because oxygen is deficient in the environment (high altitudes) or oxygen exchange to the pulmonary capillaries is impaired (chronic pulmonary diseases, paralysis of the respiratory muscles).
2. Anemic anoxia. The capacity of the blood to carry an adequate quantity of oxygen is lowered because hemoglobin oxygenation or the amount of hemoglobin is insufficient: CO poisoning, anemia, gross bleeding.
3. Histotoxic anoxia. The tissues cannot utilize the oxygen because it cannot be released from hemoglobin or the tissue respiration is blocked: CO intoxication, cyanide poisoning.
4. Ischemic anoxia. The cerebral blood flow is partially or completely reduced so that the amount of oxygen available is insufficient to meet the metabolic requirement of the brain: thrombotic or embolic occlusion of cerebral arteries, systemic circulatory collapse.

One cannot easily separate the effects produced on the brain by hypoxia-anoxia and by ischemia since the two conditions almost always coexist.

Experimentally, various degrees of pure hypoxia may be obtained in man and in animals by breathing a limited amount of oxygen, a condition that mimics high-altitude exposure.

An impaired oxygen supply to the brain is accompanied by clinical disturbances, but it does not provoke gross alterations in cerebral energy levels. Experimental evidence in man indicates that, at Pa_{O_2} below 40 mm Hg, the cerebral metabolic rate for oxygen is unchanged in the presence of loss of consciousness and severe EEG alterations.[33] Even with a Pa_{O_2} less than 30 mm Hg, cerebral energy levels (i.e., ATP, ADP, AMP concentrations) are still normal.[34,35] In animals, hypoxia may produce an isoelectric encephalogram without affecting ATP concentrations.[36] Based on these observations it can be concluded that the clinical effects of hypoxia are not due to changes in brain energy metabolism, which implies that other processes must be altered. Since the synthesis of several amino acids and neurotransmitters depends upon oxygen availability, the attention of investigators has been focused on the study of these substances in various experimental models of hypoxia. The cholinergic system is severely impaired by hypoxia, even if mild, as demonstrated by the fact that ACh synthesis is reduced during anemic, histotoxic, and hypoxic hypoxia.[37,38] Furthermore, physostigmine, which reduces ACh degradation, may reverse the symptoms produced by oxygen deficiency,[37] whereas cholinergic antagonists mimic these symptoms.[39] Other neurotransmitters also may be altered in conditions of hypoxia: cerebral synthesis of catecholamines and serotonin is reduced when oxygen content in the inspired air is decreased from 21 to 9.4%.[40–42] It is interesting that hypoxia does not provoke these effects if the animals are stressed.[40,42] In addition, dopamine and noradrenaline levels were affected differently in several brain areas of guinea pigs adapted to simulated high altitude.[43]

Hypoxia may also affect the metabolism of amino acids such as alanine, aspartate, GABA, glutamine, glutamate, and serine.[44,45] All the above-mentioned alterations of amino acids and neurotransmitters may account for the functional derangement observed in the clinical setting of acute and pure hypoxia.

The effects of oxygen deprivation (hypoxic hypoxia) on the brain may be briefly summarized as follows.[32] During hypoxia, aerobic metabolism fails and the anaerobic degradation of glucose is increased, with production of lactate and pyruvate. The consequent parenchymal acidosis and/or increased adenosine concentration may produce an augmented cerebral blood flow (CBF),[46,47] which supplies more glucose to the brain in order to maintain an adequate energy balance. However, an altered metabolism of neurotransmitters ensues, which may account for the clinical evidence of deranged cerebral function. The blood flow can wash out the acidic metabolites formed as a consequence of the anaerobic glycolysis so that they cannot damage brain tissue. In experimental conditions, anoxia with Pa_{O_2} as low as 15 mm Hg can be sustained for at least 30 min without brain damage if cerebral circulation is not impaired.[48] However when Pa_{O_2} is between 45 and 15 mm Hg the cardiac function fails and cardiovascular collapse occurs.[32] The impaired CBF lowers the glucose supply to the brain and induces disturbances in the cerebral energy state. In addition, lactic acid accumulates at high concentrations (> 20 μmol/g tissue) and alters the cell membrane structure and the blood–brain barrier.[32]

In conclusion, severe brain damage may depend upon cerebral hypoperfusion, whereas hypoxia has only a permissive role. Even in experimental conditions of histotoxic hypoxia (cyanide or CO poisoning) circulatory complications leading to ischemia may be crucial in determining brain damage.[49,50]

Unlike in pure hypoxia, profound modifications occur in the brain during ischemia, along with clinical evidence of loss of consciousness, isoelectric electroencephalogram, and disappearance of cortical evoked potentials. The ischemic changes are stereotyped; they include glucose, glycogen, ATP, and phosphocreatine depletion, depolarization of cell membranes, anaerobic generation of intracellular osmotically active particles, shift of water into the cells, shrinking of the extracellular space, phospholipid catabolism, free fatty acid accumulation, and suppression of polypeptide synthesis.[51–53]

The great vulnerability of the brain to ischemia has been attributed to the interruption of an adequate delivery of oxygen and substrates to the brain and consequent energy failure.[53] Deprived of an energy source, neurons may catabolize themselves in order to preserve functional activity; therefore, irreversible cellular damage and death ensue. However, energy failure cannot, alone, explain the development of permanent brain lesions.[54]

Under experimental conditions it has been shown that complete ischemia (30–60 min duration) may be followed by noticeable recovery.[55–57] On the other hand, severe incomplete ischemia, i.e., with reduced and inadequate circulation, may damage the brain more severely.[58,59]

A possible explanation is that during severe incomplete ischemia, lactate progressively accumulates since the residual blood flow supplies the brain with substrate for anaerobic glycolysis. Conversely, during complete ischemia the amount of lactate is limited by the preexistent stores of glucose and glycogen in the tissue. Thus, the accumulation of lactate during incomplete ischemia may explain the apparent discrepancy.[54]

The deleterious effect of lactic acid accumulation is suggested by the fact that recirculation after 30 min of incomplete ischemia was not followed by recovery of energy balance and return of neuropsychological functions under conditions of severe lactic acidosis.[54] The exact mechanism whereby lactic acid leads to irreversible brain injury is not completely elucidated, but likely depends upon a severe reduction in intracellular pH.[60]

Accumulation of free fatty acids (particularly arachidonic acid) during ischemia has been indicated as a further cause of brain lesion.[61] Among the different mechanisms whereby free fatty acids could damage brain tissue, the fact that they promote prostaglandin synthesis when the ischemic tissue is reoxygenated is worthy of consideration. This phenomenon may account for the so-called delayed hypoperfusion. Once circulation is restored in a cerebral ischemic area, ischemia is followed by hyperemia and subsequent reduction of CBF (delayed hypoperfusion), which may aggravate brain damage.[62] Prostaglandins seem to be involved in this phenomenon through their vasoconstrictive effect, and pretreatment with indomethacin, an inhibitor of prostaglandin synthesis, prevents[61,63] postischemic hypoperfusion in dogs with transient global ischemia.

On the other hand the concentration of free fatty acids in the brain does not differ during complete and severe incomplete ischemia; consequently, they do not explain the more severe outcome after the latter insult.[61]

Other factors than free fatty acids and lactate may play a role in the pathogenesis of brain ischemic lesions, for instance, glutamate and lipid peroxidation, the latter producing damage as soon as reoxygenation of ischemic areas occurs.[64,65]

The role of glutamate in producing brain damage is suggested by the fact that, in experimental models of cerebral ischemia, the topography of regional injuries corresponds to areas (such as cerebellar Purkinje cells, hippocampal pyramidal neurons CA1, 3, 4) with pronounced concentration of glutamate receptors.[64] However, the role of the amino acids in the pathogenesis of ischemic brain damage needs to be further clarified.

Neuropathological studies have clearly demonstrated that the most common expression of cerebral ischemia is laminar necrosis. The more damaged brain areas are the cortex, pyramidal neurons in Sommer's sector of the hippocampus, and cerebellar Purkinje cells. Other regions such as basal ganglia and brain stem can also be involved. The most likely causes for this typical pattern of lesions seem to be the anatomical characteristics of the cerebral vasculature and the selective sensitivity of some neurons to anoxia-ischemia. Accordingly, the most vulnerable areas are the "border zones" between the regions supplied by the major cerebral arteries; thus, a severe reduction in CBF produces a dualistic insult from two impaired circulations. In addition, it has been recognized that some neurons, such as the pyramidal neurons of Sommer's sector of the hippocampus, have a particular sensitivity to anoxia-ischemia. These two factors may account for the resulting pathological picture of the postischemic brain.

Mild hypoxia "pure" syndrome. A typical example is high-altitude hypoxia. When the pO_2 of inspired air is reduced by 10% to that at sea level, night vision is impaired. At higher altitudes, headache, difficulty in mental concentration, deficits of short-term memory, impaired judgement, and uncoordination progressively appear. With pO_2 below 30–40% of that at sea level, unconsciousness occurs. Usually a "pure" mild hypoxia does not lead to permanent brain damage, and when normal pO_2 is restored the symptoms disappear.

Ischemic anoxic syndrome. A typical severe ischemic anoxic state occurs with cardiac arrest, in which consciousness is lost almost immediately. If the anoxia-ischemia persists for more than a few minutes, permanent damage occurs. Several factors can determine the degree of the damage: conditions with reduced cerebral energy requirement (anesthesia, hypothermia), starvation, and low cerebral carbohydrate are beneficial, whereas hyperosmolar and hypermetabolic states increase the severity of the brain damage.

When consciousness is restored, some patients fully recover, whereas others are left with permanent neurological sequelae such as Korsakoff's amnesia, dementia, cortical blindness, Parkinsonism, choreoathetosis, cerebellar ataxia, paresis, paralysis, visual agnosia, and intention or action myoclonus seizures. Some patients never regain consciousness and remain in a vegetative state.

Delayed postanoxic encephalopathy is an uncommon phenomenon, which occurs frequently after CO poisoning, but is not peculiar to this condition. The patients show an initial recovery after the hypoxic state, but later (usually in 2–3 weeks but even after 5–6 weeks) they may develop confusion, disorientation, aphasia, aphatia. Later coma, rigidity, and death may ensue.

Usually the neuropathological feature of this encephalopathy is an irregular necrosis of white matter. However, a few cases of delayed neurological deterioration without white matter lesions or marked damage of the basal ganglia have been reported.

2.3. Encephalopathy Related to Hyperosmolality

Under normal conditions there is no osmotic difference between brain cells, CSF, and plasma. Changes in plasma osmolality produce related modifications in the brain osmolality, leading to an impaired neuronal function. Plasma hyperosmolality may be associated with hypernatremia or hyperglycemia, both conditions presenting neurological manifestations such as confusion, stupor, and coma. Increasing osmolality parallels behavioral changes or severity of the symptoms, and above 435 mOsm/liter, death occurs in animals.[66–68]

The main cause of the cerebral derangement is ascribed to the fact that the brain loses water in response to plasma hyperosmolality: an experimentally hypertonic solution of urea, administered acutely, reduces brain volume when the blood–brain barrier is not altered.[69] However, the amount of the water lost (and the consequent brain shrinkage) is not linearly related to the degree of plasma osmolality, and it depends upon the substances used to produce hyperosmolality (NaCl, sucrose, mannitol, etc.). In addition, brain cells exposed to hypertonic solutions undergo profound metabolic adjustments, which also may vary in relation to the "type" of hyperosmolality. *In vitro,* slices of brain in hypertonic medium lose water and decrease in volume, but not to the degree expected in relation to the degree of hyperosmolality.[70] This phenomenon is explained by the fact that substances formed in the brain in response to hyperosmolality enhance intracellular osmolality, thus preserving cellular volume. The metabolic changes accounting for the increased osmolality include intracellular accumulation of glucose, lactate, myoinositol, sorbitol, free amino acids, and other unidentified osmotically active substances, so-called idiogenic osmoles.[67,70–72]

In vivo, hyperosmotic stress may induce metabolic alterations, increased cellular osmolality, and water loss.[67,73–75]

Brain and plasma amino acid levels and cerebral content of Na^+, K^+, lactate, and ammonia are augmented by sucrose; NaCl or glucose induces hyperosmolality. However, some differences are apparent[75]: GABA and glycine brain levels are increased more by NaCl than by sucrose or glucose; the salt increases the brain content of aspartate, taurine, and glutamate, whereas no effect on these amino acids is produced by sucrose or glucose. The enhanced concentrations of the above-described components, and eventually the formation of "idiogenic osmoles," account for the augmented brain osmolality.[75] However, the increased cerebral osmolality is mainly secondary to water loss when plasma osmolality is enhanced with sucrose or mannitol solutions.[67,74,75] Conversely, NaCl and glucose produce less brain water loss, whereas they induce the formation of idiogenic osmoles.[67,74,75] This may account for the increased brain osmolality and may contribute to preserve cellular volume from the effects of hypertonic stress.[75–77] The above-reported phenomena could be responsible for the alteration of brain function accompanying plasma hyperosmolality; since, in clinical practice, hyperglycemia is one of the most important causes of hyperosmolality and related encephalopathy, it will be described below.

2.3.1. Hyperosmolar Hyperglycemic Nonketotic (HHNK) Diabetic Coma

HHNK coma is associated with several pathological conditions, such as extensive burns, steroid or immunosuppressive therapy, and dialysis procedures, but diabetes is the most common cause. HHNK diabetic coma occurs in aged patients with mild or occult diabetes, whereas it rarely develops in juvenile diabetics.

At times the onset is related to particular events such as infections, enteritis, pancreatitis, or ingestion of drugs complicating diabetes (i.e., thiazides, diphenylhydantoin). An insidious appearance of polyuria, polydipsia, thirst, polyphagia, weakness, and vomiting may antecede the major clinical manifestations, which are neurological in nature.[78] They include hallucinations, confusion, alteration of muscle tone, focal or generalized convulsions, and lethargy progressing to coma. Frequently, focal symptoms or signs occur such as hemiparesis, unilateral hyperreflexia, or unilateral Babinski sign, leading to an erroneous diagnosis of acute stroke.[78]

The laboratory findings show average glucose levels from 400 to 1000 mg/100 ml, related plasma hyperosmolality (not always detectable), absence of ketoacidosis, and hemoconcentration. The absence of ketoacidosis is explained by the fact that there is a small amount of circulating insulin, which can prevent excessive lipolysis but not hyperglycemia.[79]

Early recognition of the metabolic derangement before the impairment of consciousness is vital, since HHNK coma results in a mortality of 30–60% of the patients affected.[79,80]

Plasma hyperosmolality and the related metabolic alterations, particularly brain dehydration, may account for the deranged brain function.[68] Although postmortem studies have failed to demonstrate severe cerebral dehydration,[81] nonketotic hyperglycemia is often associated with transient or permanent focal neurological deficits.[82] Concomitant hyponatremia may contribute to those alterations acting on cerebral areas where circulation is impaired by preexistent vascular lesions.[78] Borderline areas of cerebral ischemia, ascribed to atherosclerosis, may be present in these patients since they are usually aged and diabetic. Espinas and Poser[83] have experimentally confirmed such a possibility: hyponatremia rather than hyperosmolality produced contralateral hemiparesis in dogs if hemispheric circulation was impaired by middle cerebral artery ligation.

About 25% of all patients with nonketotic hyperglycemia of diabetes mellitus experience epilepsy, and focal motor seizures (sometimes as epilepsia partialis continua) frequently occur.[84–86] In some cases they present symptoms prior to the onset of the depression of the sensorium. Correction of the metabolic imbalance (and not specific anticonvulsive therapy) stops the epileptic discharges. Derangement of cerebral amino acid levels has also been implicated as a cause of epilepsy: GABA concentrations may be lowered during hyperglycemia. GABA can be used by the brain as a source of energy, since glucose utilization and the Krebs cycle are depressed in conditions of nonketotic hyperglycemia.[85] Decreased levels of the inhibitory amino acid may lower cerebral seizure threshold.

Experimental evidence supports the hypothesis that hyperosmolality may cause seizures,[87] although patients with epilepsia partialis continua may have only mild hyperosmolality and moderate hyperglycemia, hyponatremia, or normonatremia.[85] In addition, depression of consciousness, which parallels increasing hyperglycemia and hyperosmolality, is associated with cessation of epileptic discharges.[85] Hyponatremia (or electrolytic disturbances) may play a critical role, since it has been found to occur in patients with seizures.[88,89] It is likely that the above-mentioned alterations act, eventually in concert, on silent areas of cerebral damage, triggering the epileptic phenomena.[85,87,90] However, Venna and Sabin[86] suggested that, in some patients, seizures may be due to focal cortical venous sludging or thrombosis induced by hyperglycemia, hyperosmolality, and dehydration.

2.4. Encephalopathies Related to Hypoosmolality

Hyponatremia may be the consequence of loss of sodium or dilution of body fluids by water excess. Several clinical states are associated with hyponatremia such as adrenocortical insufficiency, chronic congestive heart failure, cirrhosis with ascites, nephrotic syndrome, iatrogenic water intoxication, or inappropriate secretion of antidiuretic hormone.

Neurological disturbances are often observed during hyponatremia; they depend upon the severity and rapidity of electrolyte lowering.[91] Thus, if hyponatremia develops slowly, the symptoms appear when blood sodium levels are as low as 110 mEq/liter, whereas a rapid fall of natremia to 130 mEq/liter is associated with brain functional derangement.

The hyponatremic syndrome includes protean clinical manifestations; nausea, vomiting, headache, weakness, asterixis, myoclonus, lethargy, stupor, and coma are usual features. Focal or generalized convulsions may occur along with EEG slowing and EEG focal abnormalities. The mechanisms producing brain disturbances are still controversial; cerebral edema and increased intracranial pressure, hyponatremia *per se,* and intracellular potassium depletion are the most usual causes. Acute hyponatremia may provoke plasma hypoosmolality and consequent brain edema and increase in intracranial pressure, these phenomena being responsible for the neurological derangement. However, the increase in intracranial pressure *per se* is not the cause of cerebral dysfunction,[92] whereas the clinical manifestations of hyponatremia correlate with brain water content.[93] Accordingly, the brain content of water-intoxicated animals parallels the alert state, and improvement after coma is related to the reduction in cerebral water.[70,93] Under conditions of isoosmotic hyponatremia, cerebral function can still be deranged as evidenced by a decrease of seizure threshold and by EEG changes, suggesting, in some experimental conditions, a direct effect of hyponatremia *per se*.[94,95] Lowered Na^+ brain content may impair cerebral function, interfering with neurotransmission: it is known that Na^+ is necessary for the uptake of amino acids such as glutamate, aspartate, and taurine.[96] Katzman and Pappius[97] and Dila and Pappius[98] have noticed that osmotically induced cell swelling is a transient phenomenon and does not explain the neurological disturbances associated with chronic hyponatremia. A net loss of K^+ from cerebral tissue accompanies chronic hyponatremia. This phenomenon, demonstrable within 2 or 3 hr of induction of hyponatremia, is related to hyponatremia and is independent of hypoosmolality.[97,98]

The fall in cerebral K^+ content has been considered an adaptive phenomenon: it may contribute to reduce intracellular osmolality and so limit cellular swelling.[70] However, it could be possible that hypokalemia *per se* induces or contributes to the brain function derangement.[97] At present the mechanism(s) whereby low K^+ levels lead to neurological disturbances is not clearly elucidated.[97]

2.5. *Hyper- and Hypocalcemia*

It is well known that neurologic disorders may accompany alterations of calcium balance. However, the mechanism(s) whereby hyper-hypocalcemia produces cerebral derangement is not known. The main reason resides in the fact that the Ca^{2+} ion is involved with the regulation of manifold physiological processes. Calcium regulates the activity of many enzymes, such as adenylate cyclase, guanylate cyclase, phosphodiesterase, protein kinase, tryptophan hydroxylase, tyrosine hydroxylase, pyruvate dehydrogenase.[99] Thus, calcium imbalance may interact with brain functional activity in a variety of ways. As an example, deranged regulation of tryptophan hydroxylase may imply a secondary disorder of brain serotonin synthesis.[100] In addition, calcium alterations may affect the passive permeability of membranes to cations, water, or nonelectrolytes.[101–103] Calcium is a stabilizing agent of the excitable membranes; its removal from the external medium makes the membrane more sensitive to stimulation.[104] It has been demonstrated that a decline in the Ca^{2+} concentration enhances the excitability of the peripheral and central nervous system.[97] Conversely, in hypercalcemic states, increased membrane stability may reduce cell excitability.

Accordingly, stupor may be produced in animals by intraventricular injection of calcium salts[105]; hypocalcemia is associated with tetany, along with Chvostek's and Trousseau's sign, whereas hypercalcemia may produce muscular weakness, fatigability, and atrophy.

Among the other effects, calcium is implicated in the secretion of hypophyseal hormones,[106] the liberation of catecholamines from the adrenal medulla,[107] and neurotransmitter release.[99] The last effect may be one of the main causes of cerebral derangement, associated with altered calcemic states.

It has been clearly established that Ca^{2+} causes release of transmitters once it has entered the cell: depolarization in the presence of Ca^{2+} is a prerequisite for the release of norepinephrine, ACh, dopamine, and serotonin from synaptosomes.[99] This observation gains further support from the fact that ACh secretion in the cerebral cortex, which is related to the cortical activity, is reduced by low levels of Ca^{2+}.[108] Besides the direct effects of Ca^{2+}, other abnormalities secondary to ion imbalance may contribute to functional derangement of the brain and to neurological manifestations. For instance, hypercalcemia may lead to dehydration, hypokalemia, intravascular coagulation, hypertension, heart arrhythmias; hypocalcemia may be complicated by alkalosis or hypomagnesemia.[109] Hypercalcemia can be caused by primary or secondary hyperparathyroidism, sarcoidosis, malignant diseases including lymphomas and myelomas.[109] Hypercalcemia associated with malignancy may depend upon bone demineralization related to bone metastasis or upon the action of parathyroid hormone-like peptides or parathyrotropic compounds, secreted by tumors; both cause a kind of "secondary hyperparathyroidism."[110] The clinical features include headache, nausea, vomiting, anorexia, muscle weakness and atrophy, ataxia, and sensory motor neuropathy. Progessive lethargy and coma may ensue. Mental symptoms are protean. Depression, confusion, loss of recent memory, disorientation, hallucinations, and frank psychosis have all been reported.[109] Aspecific EEG alterations may accompany hypercalcemia and hypocalcemia; also, abnormalities of CSF (i.e., elevation of protein content and pleocytosis) were observed during hyperparathyroidism. This phenomenon could be related to deposition of calcium precipitates in the meninges.[111]

Hypocalcemia may result from hypoparathyroidism, i.e., from parathyroid hormone deficiency, ineffectiveness, resistance.[112]

Hypoparathyroidism may be consequent to several pathological conditions, such as accidental removal of parathyroid glands during thyroid surgery, infiltration of the glands by iron in hemochromatosis or by metastasis, chronic renal failure, hypomagnesemia, acute pancreatitis, and vitamin D deficiency.[112]

Common clinical signs are carpopedal spasms, muscle cramps, paresthesias, seizures, stupor, and coma. Anxiety, hallucinations, and delirium connote mental status alterations.[109] In addition, hypoparathyroidism may cause dementia, which may progress unrecognized despite the possibility of reversal with adequate therapy.[113]

3. ENCEPHALOPATHIES FROM ORGAN AND SYSTEM PATHOLOGY

3.1. Hepatic Encephalopathy

Hepatic encephalopathy (HE) is a metabolic disorder that may appear in the course of liver failure or as a complication of portal-systemic venous shunting.[114,115] The hepatocellular disease may be acute (toxic hepatitis) or chronic (cirrhosis) and the portal-systemic shunt may be either spontaneous, as a consequence of illness itself, or surgical.

Personality changes, affective disorders, and impairment of mentation are the central features in the early clinical stages of the syndrome. The patients may appear irritable, dull, apathetic, disinterested, with inappropriate behavior.[114] Gradually the level of consciousness worsens and frank confusion, stupor, and finally coma supervene. As deep coma develops, decerebrate or decorticate posturing may occur. The clinical course of the disease is accompanied by a great variety of symptoms and signs; they include motor abnormalities: involuntary movements, tremor, asterixis, hyperactive reflexes, muscular rigidity, neuroopthalmological changes, focal or generalized seizures.[114] The stigmata of the underlying hepatic failure—jaundice, ascites, cutaneous angiomas, fetor hepaticus, palmar erythema, parotid gland enlargement, and Dupuytren's contractures—round out the clinical picture. The neurological manifestations may vary considerably from patient to patient, in relation to the nature and the course of the liver disease. Accordingly several subtypes of HE can be distinguished: acute hepatic coma, subacute or chronic HE, progressive HE, intermittent HE.

There is no predictable relationship between the onset of the brain metabolic involvement and the underlying hepatic disease: HE may occur spontaneously, without any evident precipitating factor, or in relation to complications such as intestinal bleeding, infections, increased dietary proteins, or profuse vomiting. The EEG is a useful technique to ascertain the degree of the brain metabolic involvement inasmuch as the EEG changes parallel the severity of the metabolic disease. They are distinctive, presenting first as slowness of the rhythms, followed by the appearance of the typical triphasic waves, and in the advanced stages, by the absence of electrical activity.[116]

In neuropathological studies, moderate and aspecific brain alterations have been reported.[117] A constant finding is the hypertrophy and hyperplasia of protoplasmic astrocytes (Alzheimer-type astrocytes), mainly increased in the cerebral cortex, basal ganglia, and cerebellum. However, more severe changes, such as spongy degeneration of the deep layers of the cerebral cortex and of white matter, may be observed in protracted cases of HE. The pathogenesis of HE has been the subject of numerous investigations, but the real cause is still unknown. It seems reasonable to suppose that several factors can act in concert to produce cerebral derangement; aspecific metabolic complications of hepatic failure such as hypoxia, hypokalemia, hypocapnia, or respiratory alkalosis may also be contributory. It is known that portal systemic shunting and/or hepatocellular failure are prerequisites for HE: experimental evidence indicates that portocaval-shunted animals are a model remarkably similar to human HE.[117] Thus, it is likely that various substances that are produced by intestinal bacteria, absorbed from the gut, but not metabolized since portal blood bypasses the liver and/or liver function is impaired, accumulate in the brain with toxic consequences.[118] This view gains further support from the fact that intestinal sterilization is effective in the correction of metabolic alterations in portocaval-shunted rats, and is a cornerstone in the therapy of human HE.[118,119] Various compounds have been claimed to be relevant for HE; among them, ammonia has been considered one of the main pathogenetic factors. This view is suggested by the evidence that blood ammonia levels are elevated in patients with severe chronic liver disease and associated HE,[120,121] and that EEG abnormalities similar to those of human HE can be induced in animals after chronic ammonia infusion.[122,123] In addition, therapeutic measures that reduce ammonia concentrations can be effective in the treatment of hepatic coma.[124]

Experimental findings indicate that ammonia can directly damage the brain. Ammonium ions may exert an inhibitory effect upon cerebral electrical activity: ammonia reduces the frequencies of spontaneous action potentials in guinea pig cerebellar slices, altering the permeability of the cells to Cl^-, possibly to K^+ and Na^+.[125] On the other hand, it has also been suggested that hyperammonemia increases neuronal excitability, which may explain convulsions during acute administration of ammonia at high dosage.[126] Ammonia may interfere with cerebral metabolic processes in a variety of ways.[127] Accordingly, several metabolic steps can be altered: the decarboxylation of pyruvate and α-ketoglutarate, the maltate–aspartate system, the citric acid cycle, the electron transport chain. The resulting impairment of the energy balance may be responsible for the morphological changes of astrocytes. As reported, Alzheimer type II astrocytes are a feature of human HE and of experimental hyperammonemia.[128–130] These pathological alterations run in parallel with serum ammonium concentrations and probably with the energy levels of the astrocytes.[131,132] Thus, diminished glial ATP may cause an impairment of the Na^+-K^+ pump, enhanced inflow of Na^+ and water, and consequent cell swelling.[127] The deranged brain function during HE may also result from ammonium induced modifications, of neurotransmitter synthesis and release. There is evidence that ammonium salts decrease cerebral ACh synthesis *in vitro,* and concentrations of ACh *in vivo.*[127] Ammonia may also interfere with other neurotransmitter systems, such as glutamate and aspartate. Ammonia may also disturb brain function indirectly through its metabolites. α-Ketoglutarate, which derives from glutamine, is related to ammonia metabolism. Since α-ketoglutarate concentrations were found to be increased in patients during hepatic coma, it has been implicated as a putative cause of HE.[133,134] It may exert a toxic effect by competing for glutamic acid receptors in the brain. However, at present, the role of α-ketoglutarate in the pathogenesis of HE remains uncertain. Admittedly, in chronic hepatopathic patients with prominent neurological symptoms, blood and CSF levels of ammonia can be unrelated to the severity of ME.[135,136] Because it is not always possible to correlate the development of HE with raised ammonia levels, other etiological factors have been proposed.

In humans and animals during hepatic failure or after portocaval shunting, alterations of plasma and brain amino acid levels have been found. In the blood the most striking pattern is the decrease of branched-chain amino acids (valine, leucine, and isoleucine) and the increase of the aromatic amino acids (phenylalanine, tyrosine, and tryptophan), whereas in the brain large increases of aromatics but also a slight increase of branched-chain amino acids have been observed.[137–140] Some experimental evidence indicates that the abnormal cerebral levels of amino acids may be unrelated to the changes found in the plasma, and altered transport mechanisms of the blood–brain barrier have been proposed.[141–143]

In *in vivo* studies, neutral amino acid permeability was greatly increased, whereas basic amino acids showed a net decrease in their rate of passage from blood to brain.[140]

Thus, in the brain, the large increase of aromatic amino acids and the slight increase of valine, leucine, and isoleucine fit well with the reported increased permeability to the neutral amino acids.[140] It is likely that modifications of the blood–brain barrier leading to the abnormal brain amino acid concentrations occur at the level of the capillary endothelial membrane, but the factor causing these changes is not yet known; the suggested role of ammonia or its metabolite glutamine is still controversial.[144]

Recently, Mass *et al.*[144] reported regional differences in the influx of phenylalanine, tryptophan, and leucine into the brain of portocaval-shunted animals. All brain regions showed an increased influx of these amino acids, which was greater in the reticular formation and hippocampus. These findings suggest that some brain areas would be more affected by the metabolic derangement consequent to the hepatic failure and they may be particularly responsible for the symptoms of the encephalopathy. The altered amino acid levels may have several consequences for the brain: they can be directly toxic, alter levels and distributions of neurotransmitters, cause "false neurotransmitter" synthesis, or modify protein synthesis.[140]

It has been suggested that increased levels of amino acids such a phenylalanine and tyrosine can interfere with neurotransmitter metabolism, particularly with monoamine synthesis.[145–147] Furthermore, high cerebral levels of phenylalanine and tyrosine seem to be associated with increased synthesis of the "false neurotransmitters" phenylethanolamine and octopamine.[148–150] Some experimental evidence suggests a possible connection between these "false neurotransmitters" and HE. For instance, the brain concentration of octopamine is enhanced in portocaval-shunted rats[150]; plasma and CSF octopamine and phenylethanolamine increase markedly as clinical features of HE that occur in portocaval-shunted dogs.[149] Octopamine and phenylethanolamine belong to the class of so-called "trace amines" which are proposed neuromodulators related metabolically and probably functionally to norepinephrine, dopamine, and serotonin.[151]

These compounds may produce brain functional alterations interfering with monoamine-mediated neurotransmission; their clinical relevance is further provided by the fact that these substances have been implicated in the etiology of various psychiatric disorders.[151]

Another "trace amine," tyramine, may contribute to the development of HE. Tyramine concentration is augmented in the plasma of patients with cirrhosis and hepatitis.[152,153] In addition, in portocaval-shunted dogs the levels of tyramine in plasma, CSF, and selected brain areas (i.e., hypothalamus, midbrain, and striatum) increase, mainly following the appearance of HE.[154,155] Tyramine (or its metabolite octopamine) may affect the synthesis of the catecholamines and it may displace them from effector neurons; this could produce concomitant brain depletion of norepinephrine and dopamine and their increase in the CSF.[154]

Other factors may play a pathogenetic role in HE, for instance short-chain fatty acids, mercaptans, methionine.

Short-chain fatty acids (butyrate, valerate, octanoate, caproate) have sometimes been found to be increased in plasma and CSF of patients during

HE.[114] In experimental conditions, infusions of fatty acids produced coma, with normal amounts in brain of ATP and phosphocreatine.[156,157] However, the concentrations of these substances necessary to provoke coma are higher than those encountered in clinical practice.[156] Methyl mercaptans may be elevated in plasma and urine during hepatic coma.[114] Experimentally, infusion of mercaptans induces coma in rats through unknown mechanisms.[158] Phear *et al.*[159] have reported that methionine can be toxic in hepatopathic patients. Since methyl mercaptan is a derivative of methionine, the metabolic link between the two compounds may account for the cerebral toxicity of the amino acid. Thus, there are only indications of the role played by each putative toxin in the pathogenesis of HE. In addition, up to now, it is unclear whether or not other substances can contribute to the brain metabolic derangement; some experimental evidence leads to the speculation that unknown medium-size molecules may be involved.[160–162]

Thus, HE and many MEs need further investigation in order to be clarified.

3.2. Uremia

Uremia is the clinical condition resulting from renal failure, which is consequent to any grave kidney disease. The uremic state produces profound metabolic alterations that can easily affect brain function, hence uremic encephalopathy is a common manifestation of renal insufficiency.[163,164] Clinically, the early stage of cerebral uremic dysfunction is characterized by mild disturbances in mentation (i.e., inability to concentrate, impairment of attention, forgetfulness), indicating depression of general cerebral activities. Gradually, as the renal failure worsens, sleep abnormalities, disorientation, confusion, irritability, and hallucinations appear. Depressive, paranoid, or schizophrenic states may also occur. Finally, the patient lapses into increasing lethargy and coma with decerebrate or, more frequently, decorticate posturing.

The clinical picture is not always "progressive," but may present cyclic fluctuations in intensity; coma may suddenly appear after a short and severe delirious state. Various other symptoms and signs can be associated: weakness, monoplegia, hemiplegia, fasciculations, asterixis, myoclonic jerks, cerebellar tremor, choreoathetosis, restless leg syndrome, aphasia, apraxia, amaurosis, nystagmus, vertigo, deafness, peripheral seventh nerve palsy, peripheral neuropathy.

Generalized tonic-clonic convulsions or focal motor seizures may occur, especially as late manifestations of the renal disease. EEG aspecific changes are commonly seen during uremia and may help to uncover encephalopathy before the clinical onset.[165] Diffuse slow-wave activities, in the β and/or δ range, depending on the severity of brain involvement, can be observed; paroxysmal discharges of spike waves appear in the case of seizures.

The atypical changes occurring in the brain during chronic uremia have been extensively described by Olsen[166]; they include neuronal degeneration of the brain stem, reticular formation, cerebral cortex, and cerebellar Purkinje cells. No correlation between the pathological and the clinical picture has been found.[165]

The pathophysiology of uremic encephalopathy is not well clarified. The metabolic abnormalities caused by renal failure are so complex that it is difficult to determine the exact role played by each factor. In addition, the many aspecific alterations associated with the uremic state, including hypertension, acidosis, alkalosis, hyponatremia, hypocalcemia, hyperkalemia, hypermagnesemia, hyperhydration, and dehydration may also play a role in the causation of the encephalopathy. Brain involvement is demonstrated by several lines of evidence. Decreased cerebral glucose utilization, defective oxygen consumption, increased cerebral permeability, disturbed nucleotide metabolism, elevated ATP and phosphocreatine concentrations with diminished high-energy phosphate utilization, a block in the phosphofructokinase step of the glycolytic cycle, abnormalities in cerebral Na^+,K^+-ATPase, and elevation of cerebral Na^+ content have all been reported.[167]

Cerebral amino acid and neurotransmitter metabolism is altered in renal insufficiency; serotoninergic and dopaminergic systems seem to be particularly affected. These observations go along with the reported changes of amino acid concentrations and enhanced content of 5-HIAA and homovanillic acids in the CSF of patients with chronic renal insufficiency,[168,169] with elevated serotonin and reduced dopamine levels in postmortem brain and samples from uremic patients.[170] In addition, chronic renal insufficiency is associated with altered cerebral amino acid uptake and arterial amino acid levels[171]: in the brain, glycine uptake increases, valine and isoleucine extraction decreases, glutamine uptake disappears, ammonia extraction becomes evident; in the plasma, glycine levels are enhanced, whereas valine is lowered.

Since CBF is slightly modified,[172,173] the alterations in arterial amino acid concentrations may account for the abnormal amino acid entry into the brain.[171]

Uremic encephalopathy can be reversed by dialysis treatment, indicating that substances of small molecular size are involved, at least in part, in producing the cerebral metabolic derangement.[167]

Brain oxygen uptake is depressed during uremic coma; this inhibition can be provoked with a mixture of those small molecules that accumulate in uremia, whereas each one individually does not give inhibition.[174] This observation again suggests the multifactorial pathogenesis of the encephalopathy.

Substances with toxic effects upon the brain may be produced and may accumulate, as a consequence of the impaired renal excretion and abnormal protein metabolism associated with developing uremia. Based on experimental evidence, many organic compounds have been candidates for "uremic toxins": urea, methylguanidine, AMP, polyamines, phenols, indoles, glucagon, parathyroid hormone, growth hormone, gastrin, renin, creatinine, "middle molecules", etc.[167,175] They may alter brain function through diverse mechanisms, for instance, by inhibiting cerebral enzyme activity.[167]

Historically, urea can be considered the first putative "uremic" toxin. Urea can produce cerebral metabolic derangement, since, for example, it inhibits oxygen uptake of rat brain at the concentrations found in the end stage of renal failure.[176] Furthermore, urea may be responsible for the myoclonic jerks that can occur in the course of renal failure.[167] This seems to be a rather peculiar effect of urea, probably mediated by the "activation" of the reticular formation. However, it is clear that urea does not cause all of the neurological alterations during uremia, since marked clinical improvement has been noted in patients dialyzed against baths with high concentrations of urea.[177]

One of the cardinal disturbances of the endocrine system in uremic patients is hyperplasia of the parathyroid glands and the consequent excess of circulating parathyroid hormone.[178,179] The effect upon the brain may be secondary to the related increased cerebral calcium content.[180] Parathyroid hormone may also affect the brain, increasing the rate of absorption of aluminum.[181] Moreover, Goldstein *et al.*[182] have found a correlation between EEG alterations and serum levels of the N-terminal fragments of parathyroid hormone. This observation suggests a possible direct neurotoxic effect of parathyroid hormone in uremia.

Bacterial flora may play an important role in production of uremic toxins, and consequently in the pathogenesis of the encephalopathy.[175,183]

Gut bacteria transform substrates reaching the gut; they can be reabsorbed and accumulate, since renal function is impaired, with toxic implications for the brain.[184] Some of these bacterial products such as phenols, aliphatic and aromatic amines, polyamines, and indoles are increased in uremic patients.[175] Bacterial overgrowth is related to an increased generation of aliphatic amines.[185] Their cerebral toxicity is demonstrated by the correlation of abnormal neurobehaviroial parameters with serum amine levels[183]; clinical improvement can be obtained following intestinal sterilization with nonadsorbable antibiotics. Phenols and phenolic acids (2,4- and 3,4-dihydroxybenzoic acids) are indicated to be toxic for the brain.[185] In experimental paradigms they produce EEG slow-wave activity, generalized convulsions, myoclonic jerks, alterations in seizure thresholds.[167] Phenols and phenolic acids may affect brain function because they inhibit several enzymes such as nucleotidase, aminoxidase, and lactic dehydrogenases.[167] Furthermore, their structural similarity to the catecholamines suggests a possible interference with the noradrenergic and/or dopaminergic system.[186]

Abnormalities of plasma and brain concentrations of metals such as calcium, tin, rubidium, cadmium, strontium, zinc, arsenic, aluminum, and molybdenum have been found in uremic patients, especially those undergoing dialysis.[187–190] At present, the significance of the changes in brain levels of metals and their relation to the ME is incompletely understood.

Beside uremic encephalopathy, patients with renal insufficiency may develop other neurological disturbances following dialysis or renal transplantation. These abnormalities connote three different syndromes known as disequilibrium syndrome, dialysis encephalopathy or dementia, and rejection encephalopathy.

3.2.1. Disequilibrium Syndrome

The disequilibrium syndrome may occur during the hemodialysis or may be delayed 8–24 hr after the procedure. The neurological symptoms include headache, nausea, hypertension, fasciculation, asterixis, convulsions, and coma. Reversion to the normal status occurs in 1–2 days. The most often alleged cause of the brain involvement is intracellular hyperosmolality. It has been suggested that a too-rapid dialysis may remove urea from the plasma but less easily from the brain because of the blood–brain barrier.[191] The higher cerebral urea concentration may lead to a shift of water into the cells, with consequent edema, rise in intracranial pressure, and cerebral function derangement. An osmotic gradient between brain and plasma may be produced also by an acute correction of serum pH.[192] However, other factors still unknown may be causative.

3.2.2. Dialysis Dementia

Dialysis dementia is a progressive neurological disorder complicating long-term hemodialysis.[188,193,194] This complication has no satisfactory treatment. The mean duration of hemodialysis before the onset of the encephalopathy is about 2.5 years, and the mean duration of the clinical course until death is 6.3 months. The clinical features include dyspraxia, dysphasia, gait difficulty, myoclonic seizures, dementia.

Nonfluent speech is the most characteristic and usually the earliest sign. The EEG shows diffuse waves and paroxysm of monomorphic slow waves, bilaterally synchronous mostly in the frontal areas. The factor most often indicated is aluminum, which can accumulate in the brain and may act as an offending neurotoxin: aluminum was found in high concentration in plasma and brain of dialyzed patients with encephalopathy.[188,189]

Cerebral accumulation may depend upon an overload of aluminum following prolonged therapy with aluminum gel or dialysis with water containing this metal.[188,195]

3.2.3. Rejection Encephalopathy

Renal transplantation is the most effective treatment in patients with renal insufficiency. However, transplantation may produce neurological disturbances including convulsions, headache, confusion, disorientation, and papilloedema. These alterations may depend upon hypertension, altered drug metabolism, superimposed infections, or effect of steroid therapy.

If there is a severe necrosis of the transplanted kidney, a toxic encephalopathy may develop. Accordingly, in a recent report Gross *et al.*[196] have emphasized that rejection encephalopathy is related to the severity of the rejection crisis and not to other features such as blood pressure, fever, steroid therapy, or plasma electrolyte derangement.

3.3. Diabetic Ketoacidosis

Diabetic ketoacidosis (DKA) is a clinical emergency that may occur as a complication of diabetes mellitus. The severity of such a condition is demonstrated by the fact that it can lead to mortality in 5–15% of the patients affected. Usually, triggering factors may antecede and predispose the onset of DKA. They include omission or reduction of insulin therapy, viral or bacterial infections, trauma, pregnancy, acute pancreatitis, myocardial infarction; however, in some subjects the etiology remains enigmatic.[197] The ketotic state probably results from lack of insulin and an excess of insulin counterregulatory hormones (GH, cortisol, catecholamines, and glucagon in particular) secreted in response to various stressful conditions.[198,199]

The unbalanced effects of insulin and glucagon on lipid metabolism may trigger the formation of ketone bodies, which can reach plasma concentrations exceeding 3 mM per liter. Patients with DKA exhibit polyuria, polydipsia, nausea, vomiting, weight loss, abdominal pain, tiredness, dehydration, Kussmaul's respiration with smell of acetone in the breath, impairment of consciousness progressing to coma. Common laboratory findings are hyperglycemia and ketonemia with related glycosuria and ketonuria, acidosis, azotemia, and hyperuriacidemia.[200] Brain metabolic derangement is demonstrated by several lines of evidence: for example, Kety *et al.*[8] observed a remarkable decrease of cerebral oxygen consumption in patients with DKS and coma. This phenomenon may be related to the impaired glucose utilization found in DKA. Altered glucose metabolism is suggested by the increase in cerebral levels of fructose-6-phosphate, citrate, lactate, and glycogen found in diabetic ketotic animals.[201–203] The mechanism producing these metabolic changes and the alterations in mental status is not known; however, it is likely that several factors may act in concert; impairment of cerebral circulation, acidosis, hyperosmolality, and ketone body excess are the most often alleged causes. Plasmatic hyperviscosity or intravascular deposition of fibrin have been observed to occur during DKA.[204–206] Both conditions may lead to an impaired brain circulation and tissue damage. Therefore, an ischemic mechanism has been implicated in the causation of coma. Nevertheless, no reduction of CBF has been reported in patients with DKA.[200] Metabolic acidosis is the most common disorder complicating diabetes. It derives from several causes[207]; in DKA it may depend upon ketone body excess. It is well known that acidosis can affect cerebral metabolism and consequently brain function, since reduction of cerebral glucose utilization has been observed during respiratory and metabolic acidosis.[208,209] However, some authors consider improbable this cause–effect relation between acidosis and DKA coma.[210,211] Based on clinical evidences, Fulop *et al.*[210] have emphasized that consciousness depression is related to hyperglycemia and consequent hyperosmolality and not to the degree of ketosis or acidemia. Although hyperosmolality seems to be important in producing depression of the sensorium,[212] in patients with ketoacidosis it is not usually as high as those levels associated with coma.[213] Probably hyperosmolality is necessary but not sufficient *per se* to induce coma; additional factors, such as how acutely hyperosmolality develops, may be determinant.[211]

As stated, DKA is associated with considerable plasma levels of ketone bodies. These compounds can enter the brain according to their arterial concentration, and they can be rapidly utilized by the CNS as an alternative source of energy.[202] Ketone body excess may induce metabolic changes (directly or indirectly through acidosis) with toxic implications for the brain. Ketone body utilization may lower cerebral glycolysis; it enhances brain glycogen content and inhibits the oxidation of pyruvate.[202,214] An increased brain ATP/ADP ratio is the central feature of chronic ketosis, and it may account for the increased neuronal stability associated with this state.[214] The "depressive" effect on the CNS produced by ketone bodies is further supported by the fact that a ketogenic diet augments, in rats, the convulsive threshold to electroshock.[215] In addition, infusion of acetoacetic acid (but not of hydroxybutyrate) may induce coma in animals.[216] It can be hypothesized that ketone bodies may alter brain function by lowering cerebral pH. Although acute ketosis may decrease plasma and brain pH, elevated plasma levels of ketone bodies maintained throughout the day, did not provoke significant alterations of plasma and brain pH.[217] Compensatory mechanisms may intervene, preserving brain pH in spite of elevated concentrations of circulating ketone bodies.

At present it is not clear whether ketone bodies are responsible for the cerebral derangement during DKA, and the mechanisms whereby they could produce such alterations is not known.

At times after therapy and improvement of DKA coma, some patients may exhibit seizures followed by fatal coma.[218,219] Cerebral edema is the common autopsy finding. This phenomenon could be explained by the fact that a too-rapid correction of body fluids can lead to an extracellular hypotonicity, which in turn elicits water movement into the brain cell, and cellular edema.[219] Furthermore, during diabetes brain cells may accumulate sorbitol and fructose (and/or other substances osmotically active) in response to plasma hypertonicity.[218] The resulting intracellular hyperosmolality may produce cellular swelling when plasma osmolality is corrected. Hence, cerebral functional derangement ensues.

3.4. Respiratory Encephalopathy

Several chronic pulmonary diseases such as emphysema or fibrositic lung diseases can cause an impaired gas transport to and from the environment leading to oxygen deficiency and accumulation of CO_2. Failure of respiratory function may produce neurological signs and symptoms that depend upon respiratory insufficiency *per se* and not upon the nature of the lung disease.

Respiratory encephalopathy is characterized by a variety of clinical features[220–223]; they include headache, often intense and persistent; loss of recent memory; disorientation in time and place; hallucinations; anxiety; agitation; lethargy; stupor; and coma. Motor abnormaltities such as tremors, asterixis; restless leg syndrome; myoclonic jerks; less frequently, vertigo; tinnitus; and seizures have been reported. Papilloedema may occur, raising an erroneous suspicion of intracranial expanding process. The EEG shows diffuse and progressive slowing of the rhythms in the θ and δ range, and it may even become isoelectric.[224]

Hypoxemia, hypercapnia, and acidosis are the most often implicated causes of the respiratory encephalopathy. However, hypoxemia and hypercapnia may or may not coexist; hypercapnia may or may not be associated with acidosis, depending on the severity of the lung disease and on the efficacy of the homeostatic mechanisms.

Hypercapnia, directly or indirectly through acidosis, might play an important role in producing cerebral metabolic abnormalities and related neurological features, which, in turn, depend upon the "type" of hypercapnia, i.e., acute, chronic, or severe.

CO_2 has a complex effect on the CNS: it powerfully and rapidly enhances CBF,[46,225] increases the rate of CSF production,[226] and may alter blood–brain barrier permeability.[227] Functionally, CO_2 augments seizure threshold in rats,[228] causes clouding in mentation processes,[229] and induces retrograde amnesia.[230]

Depending on the duration and severity, hypercapnia alters cerebral metabolites; for instance, moderate reductions of pyruvate, lactate, citrate, α-ketoglutarate, and malate have all been reported.[231–235] It is likely that such alterations could be ascribed to the inhibition of pH-dependent cerebral enzymes.

The impairment of cerebral metabolism is related to the EEG changes seen during severe hypercapnia. Hypercapnia does not affect the cerebral metabolic rate for O_2,[234,236,237] whereas glucose utilization is depressed[238,239] owing to the inhibition of phosphofructokinase, probably mediated by CO_2-induced acidosis.[240] However, other mechanisms should be invoked since glucose-6-phosphate returns to its normal level after 60 min of hypercapnia, when glucose utilization is still inhibited.[240]

Since the brain energy balance, i.e., ATP and ADP concentrations, is unchanged,[238] it appears that the brain utilizes other metabolites than glucose for its energy requirements. The reduced brain concentration of glutamic and aspartic acids, along with an enhance production of ammonia and glutamine, may indicate that oxidation of those amino acids could represent the source of energy for the brain.[238,240,241]

In some cases ammonia levels were also found to be increased in the plasma of patients with chronic cor pulmonale and neurological symptoms, suggesting a role of ammonia in producing respiratory encephalopathy,[242,243] although up to now no significant correlation has been clearly established. All the above alterations may contribute to the neurological derangement observed during respiratory failure. Individuals with chronic respiratory diseases may adapt to chronic hypercapnia, and its effects may be different from those produced by the acute form; for instance, in pneumopathic patients, CBF does not increase in proportion to hypercapnia, whereas it does occur in subjects without pneumopathies.[225] In addition, during prolonged experimental hypercapnia, which more closely mimics hypercapnia induced by pulmonary insufficiency, alterations of brain oxygen consumption, CBF, pH, and blood–brain barrier do not occur.[240,244,245] Chronic hypercapnia enhances glutamine and GABA cerebral content, decreases glutamate and aspartate, and induces moderate brain edema.[240,245,246] These changes may modify brain excitability and may be causative of the neurological disturbances associated with chronic pulmonary failure.

Usually, this symptomatology is reversible and it abates on return to normocapnia. Accordingly, hypercapnia does not elicit cerebral damage. Experimentally, even extreme hypercapnia (at a Pa_{CO_2} of 300 mm Hg), with associated grave tissue acidosis, produces minor brain alteration (such as slight chromatin clumping and mitochondrial swelling) so that irreversible cell damage is unlikely.[247]

3.5. Pancreatic Encephalopathy

The term *pancreatic encephalopathy* was first coined by Rottermich and von Haam who, in 1941, reported eight cases of this complication.[248] Encephalopathy associated with acute pancreatitis is relatively common, brain involvement usually becoming evident a few days after the onset of pancreatitis. The clinical features are various, embracing psychomotor agitation, confusion, hallucinations (mainly auditory), coma, and multifocal neurological signs such as aphasia, hemiparesis, and focal convulsions.[249] In terms of neuropathology, cerebral edema, perivascular demyelination, hemorrhagic foci, and gliodystrophic reactions have been described.[250,251] Several factors related to the pancreatitis have been considered to explain the pathogenesis of the neurological symptoms such as calcium, mangesium, or potassium imbalance, dehydration and hyperosmolality, ischemia, avitaminosis, fat embolism, excess of CSF lipase, and mumps virus infection.[249]

The liver and its metabolism seem to be functionally affected during acute pancreatitis, in spite of a normal liver function test.[252] Related alterations of amino acid and urea metabolism probably account for the brain disturbances.[252] Estrada *et al.*[253] in a recent study on 17 cases of acute pancreatitis, not alcohol induced, concluded that a constant parallelism exists between increase of CSF lipase, EEG aspecific changes, and encephalopathy. Conversely, no relation was found between neurological symptoms and severity of the disease.

From the data of Estrada *et al.*,[253] it appears that pancreatic encephalopathy can be ascribed to the effects of the enzyme lipase on the cerebral tissue; brain damage may regress consonant with amelioration of the pancreatic condition.

3.6. Gastrointestinal Encephalopathy

This very aspecific term is used in this chapter to indicate a few MEs somehow related to the digestive system. Obviously this is only a piecemeal examination of the pathogenetic link between gut and brain, as it would require a more extensive and detailed description. It is well known that a great variety of gastrointestinal illnesses such as Whipple's, Chaga's, and celiac disease are associated with sometimes predominant neurological symptoms. However, the etiology, causes, and mechanisms producing CNS involvement are poorly understood; nutritional, toxic, metabolic, or infectious factors may be involved.

Here, jejunoileostomy and parenteral nutrition will be briefly discussed as an example of altered intestinal function producing cerebral metabolic derangement.

Patients who have undergone jejunoileostomy, in order to achieve weight loss through controlled malabsorption, may develop a recurrent episodic syndrome characterized by memory loss, slurred speech, poor concentration, visuo-spatial disorientation, and even coma.[254,255] No brain lesions have been revealed by computed tomography. The frequency of neurological complications following jejunoileostomy is about 5% or more: the symptoms may subside with appropriate food intake or reconstitution of the normal intestinal anatomy. Thiamin or vitamin B_{12} deficiencies seem unlikely to be the cause of this syndrome.[254] Bypass procedures may produce hepatic alterations probably related to malnutrition. Liver changes were seen during the first year after bypass, when rapid weight loss and amino acid malabsorption occur.[256,257] In addition, hepatic failure and coma have been reported following jejunoileostomy.[258] Consequently, the encephalopathy after jejunoileal bypass could result from the related hepatic derangement. In this case, the same pathogenetic mechanisms producing the neurological disturbances in hepatopathic patients might be involved.

Herling *et al.*[259] reported changes in amino acid concentration in rats with jejunoileal bypass similar to those resulting from portal-systemic shunting. The observed high concentrations in the brain of phenylalanine and tyrosine may affect catecholamine synthesis and may lead to the formation of false neurotransmitters such as phenylethanolamine and octopamine. These alterations may account for the cerebral metabolic derangement. During total parenteral nutrition or realimentation after starvation, some patients may develop a ME characterized by confusion, convulsions, and coma.[260–262] Neuromuscular disturbances such as tremors and weakness may be present.

Magnesium deficiency, related to an imbalanced dietary intake, has been indicated as a cause of these neurological disturbances.[263] However, other electrolytes or metals have been implicated; phosphate depletion may be causative in some cases.[264]

In a patient suffering from confusion after total parenteral nutrition following bowel resection, chromium was found at the lowest normal serum level. Ion replenishment resulted in the disappearance of the encephalopathy.[265] Some patients on long-term total parenteral nutrition had enhanced aluminum levels in bone, urine, and plasma.[266] This observation deserves interest since aluminum seems to be responsible for the mental abnormalities encountered in the course of renal insufficiency and long-term dialysis treatment.

3.7. Endocrine Encephalopathies

Several endocrinopathies may be accompanied by pronounced CNS disturbances that may dominate the clinical picture. In this chapter the term *endocrine encephalopathies* refers to mental states and neurological abnormalities that occur in the course of hyper-hypoparathyroidism, hypo-hyperthyroidism, adrenocortical failure, Cushing's syndrome. Two pathogenetic mechanisms have been indicated to explain cerebral involvement. Brain function may be affected (1) directly by the lack or excess of hormone influence upon the brain and (2) indirectly by metabolic alterations (i.e., electrolyte imbalance). Obviously the two conditions may coexist, both being contributory to the resulting MEs. As an example, hypercalcemia may account for mental state changes during hyperparathyroidism, whereas a mismatched glucocorticoid brain receptor interaction may be the main cause of the ME encountered in adrenal failure. Since brain alterations observed during the clinical setting of hyper-hypoparathyroidism mainly depend upon lack or excess of calcium, they were described earlier in this chapter.

3.7.1. Adult Hyper- and Hypothyroidism

Lack or excess of thyroid hormone exerts several effects upon the brain; these changes underlie the neurological signs and symptoms associated with hyper- and hypothyroidism. The mechanism by which these neurological dysfunctions are caused is poorly understood. Metabolically, cerebral oxygen and glucose consumption have been reported unchanged in hyper- and hypothyroid patients, whereas a reduction of CBF and an increase in cerebrovascular resistance have been observed in patients with myxedema.[267–269]

Evidence that thyroid hormones have neurophysiological effects on the brain is demonstrated by the fact that modifications in EEG evoked potentials have been described in normal subjects after exposure to exogenous T_3, and EEG alterations are associated with a hyperthyroid state.[270,271]

Absence of hypoglycemia, osmolality, or electrolyte changes, which may be caused by hyperthyroidism, suggests a primary effect of thyroid hormone on the brain.[272] In addition, in rats, thyroxine decreases the threshold for cardiazol- or electroshock-induced seizures, whereas thyroidectomy has the opposite effect.[273]

Brain functional derangement may result from alterations of cerebral neurotransmitters. Several lines of evidence indicate that thyroid hormone can influence metabolism and effects of catecholamines. Prange *et al.*[274] reported decreased catecholamine turnover in the brain and heart of hyperthyroid rats; Lipton *et al.*[275] found hypothyroid rats to be hypoadrenergic and to have increased synthesis of norepinephrine in the heart and brain. Brain stem cell response to iontophoretically applied noradrenaline varies, depending on the thyroid state of the animal.[276] The inhibition of the Purkinje cells, induced by stimulation of the locus coeruleus–cerebellum tract (noradrenergic pathway), is increased in hyperthyroid rats.[277]

Dopamine neurotransmission is also modified during hyper- and hypothyroidism. In hyperthyroid patients, CSF levels of homovanillic acid were decreased,[278] probably as a result of an exaggerated dopamine turnover, reflecting dopamine receptor hypersensitivity.[279,280] Accordingly, hyperthyroid guinea pigs are more sensitive to apomorphine.[279] This hypersensitivity may account for the choreic phenomena observed in the course of hyperthyroidism. Thyroid hormone can modulate brain concentrations of other substances, such as peptides, which may have physiological roles. Dynorphin levels were reported increased in the cortex of hypothyroid animals and lowered in the hypothalamus of hyperthyroid rats.[281] In addition, hypoparathyroidism is associated with augmented concentrations of β-endorphin in the hypothalamus and decreased levels in the pituitary gland and striatum; whereas hyperthyroidism lowers peptide levels in the hypothalamus.[282] At present the pathophysiological implications of these alterations are not known. Nevertheless, they deserve interest since, for example, dynorphin seems to be involved in the production of spontaneous feeding, and in raising body temperature.[282,283] Undoubtedly, weight and temperature regulation are both deranged in hyper- and hypothyroid states.

Lack of thyroid hormone, as a result of thyroidectomy, can determine anatomical brain alterations in adult rats. Modifications in the morphology of cortical neurons, a decreased number of dendritic spines and their abnormal distribution along the shaft, have been reported.[284]

These abnormalities, reversible once the euthyroid state is achieved, may correlate with the alterations of learning capacity and EEG activity found in hypothyroid rats.[285,286] Hyperthyroidism is accompanied by a great variety of signs and symptoms. Besides the signatures of thyrotoxicosis (ocular, cutaneous, cardiovascular abnormalities) patients may present neurological disturbances such as tremor, hyperreflexia, myopathy, seizures, and chorea.[287] Focal or generalized convulsions are associated with EEG evidence of cerebral hyperexcitability.

Chorea has been reported by several authors and may be continuous or, rarely, paroxymal.[288–291] Usually, this motor disorder subsides when the euthyroid state is reached, but it can outlast the endocrine disease.[292] Psychological abnormalities, such as anxiety, irritability, depression, decreased friendliness, cognitive impairment, and sleep abnormalities, also contribute to the clinical picture.[287,293]

These mental disturbances may be seen in normal individuals after treatment with triiodothyronine[294]; they can be ascribed to a primary effect of thyroid hormone on the brain.[293] There is a so-called "apathetic hyperthyroidism" that occurs in elderly patients, and it is characterized by the lack of thyrotoxicosis stigmata, various neurological disturbances, and dementia.[295–297] The underlying pathophysiological mechanism is not understood; however, the treatment of hyperthyroidism leads to a resolution of the neurological and mental deficits.

Adult hypothyroidism may result from an idiopathic atrophy of the thyroid gland or may be secondary to treatment with radioactive iodine, thyroidectomy, or ingestion of goitrogens. The general clinical features are quite suggestive[287]; they include hair loss, constipation, hoarse voice, and intolerance to cold. Furthermore, hypothyroid patients may present neurological abnormalities such as ataxia, intention tremor, nystagmus, peripheral neuropathy, delayed reflex relaxation, occasionally deafness, and depression of taste and smell. Stupor and coma may also occur. Mental symptoms are prominent, ranging from apathy, depressed mood, confusion, defective memory, to psychotic behavior and frank madness. Mental deterioration may be slowly progressive, and thus might go unnoticed. Its recognition is crucial since adequate therapy (with thyroid hormone) may reverse the psychiatric abnormalities.

3.7.2. ACTH and Corticosteroids: Their Relationship to Brain Function

Lack of excess of glucocorticoids and/or ACTH are directly involved in the pathogenesis of the cerebral dysfunctions observed in the course of adrenocortical failure and Cushing's syndrome.

The aim of this section is to provide a view of the effects produced by these hormones upon the brain. However, the extreme complexity of the relation between brain, glucocorticoids, and ACTH precludes exhaustive explanations for the CNS functional derangement. Indeed, the exact mechanism whereby such derangement is produced is still a matter of speculation. ACTH and ACTH-derived peptides are widely distributed throughout the brain. Immunoreactive ACTH has been found in the cerebral cortex, cerebellum, hippocampal formation, hypothalamus, thalamus, amygdala, periaqueductal gray, and some reticular nuclei.[298,299] ACTH and its fragments influence several biochemical processes of the brain, such as protein and polyamine synthesis,[300,303] protein kinase activity, and protein phosphorylation.[302,303] It is appropriate to mention that the hormone affects serotonin, dopamine, norepinephrine, and ACh metabolism, since derangement of these neurotransmitters could be associated with affective disorders.[304–307]

Electrophysiologically, $ACTH_{4\text{-}10}$ alters EEG frequencies and $ACTH_{4\text{-}9}$ changes cortical evoked potentials.[308,309]

Experimental evidence indicates that ACTH and ACTH-derived peptides may modify brain function and behavior such as sexual behavior, pain perception, learning, and retention of learned behavior.[310–312]

The latter effect may result from the activation of regions such as the cingulate cortex, anterior thalamic nuclei, and subcular cortex, which are components to the limbic system.[313,314] Interestingly, other drugs such as oxotremorine, scopolamine, and phencyclidine, which affect memory or behavior, act on the same areas of the limbic system.[315–317]

Glucocorticoid hormone target sites have been localized in the brain: radio labeled neurons have been found in the septum, hippocampus, amygdala, cortex and hypothalamus.[318] The metabolic effects of glucocorticoid upon the CNS are protean. They modify the activity of enzymes such as RNA polymerase, fatty acid synthetase, acetyl-CoA carboxylase, arylsulfatase, and β-galactosidase.[319–321] ACh and serotonin metabolism is influenced by glucocorticosteroids, since increased plasma levels of corticosterone enhance tryptophan hydroxylase and cholinesterase activity.[304,322,323] Furthermore, the levels of serotonin are augmented in specific cerebral regions: amygdala, mesencephalon, and hypothalamus.[324] Cortisol enhances brain concentrations of free amino acids with associated decreases of GABA, whereas adrenalectomy produces opposite effects.[325]

Thus, lowered concentration of GABA may account for the increased brain excitability following acute and chronic administration of cortisol. Corticosteroid binding to neurons induces immediate modifications of the cerebral electrical activity, as suggested by the fact that cortisone and cortisol increase electroshock seizure threshold in rats.[326] In addition, hydrocortisone changes the amplitude of brain stem evoked potentials[327]; cortisol affects spontaneous firing of the cells of the hypothalamus, hippocampus, and reticular formation.[328] Corticosteroids may also affect brain functional activity, since, for example, they can facilitate the retention of passive and active avoidance in rats.[329] On the other hand, stressful conditions may affect the brain by altering the binding of glucocorticoids: in this regard, socially deprived and crowding-stressed mice showed a decreased uptake of corticosterone in the septum and hippocampus.[318]

Thus, it is likely that, as has been suggested, the binding and the effects of corticosteroids to some components of the limbic system may constitute a key to explain the behavioral modifications associated with alterations of the pituitary–adrenal axis.[318]

3.7.3. Adrenocortical Insufficiency (Addison's Disease) and Cushing's Syndrome

Adrenocortical failure is a rare disease characterized by an intolerance to any shift in the environment.

In the past, tuberculosis was the most frequently alleged cause. Actually, autoimmune processes, and less frequently fungal infections, or trophic hormone deficiency, account for almost all of the reported cases.[330]

The clinical picture includes hyperpigmentation of the skin, hypotension, muscular weakness, nausea, vomiting, diarrhea, and weight loss.

Neuropsychiatric disturbances are frequent and various, ranging from nervousness and depression to severe personality changes, delirium, psychosis. Neurological symptoms such as hiccough, dizziness, convulsions, choreiform movements, stupor, coma, and, rarely, papilloedema may occur.[330,331]

Cerebral dysfunction is attributable to adrenal hormone deficiency; however, water–electrolyte changes, i.e., hyponatremia, hyperkalemia, hypercalcemia, hypoglycemia, or dehydration, may contribute.

EEG abnormalities, such as slowing of the α rhythms and diffuse slow-wave activity, parallel the intensity of the encephalopathy and can be reversible following glucocorticosteroid therapy.

The term *Cushing's syndrome* embraces any form of hypercortisolism. Most frequently it ensues on prolonged therapy with ACTH or glucocorticoids, or it may be secondary to pituitary microadenoma, ACTH-secreting tumors, adenoma, or carcinoma of the adrenal gland.[330,332] The clinical signs and symptoms are typical, including moon face, truncal obesity, muscular atrophy and weakness, striae rubrae, osteoporosis, hypertension, hyperglycemia. Abnormalities in the EEG may be present, consisting of diffuse slow-wave activity or fast waves in the occipital regions.

The incidence of mental disturbances, sometimes the main presenting symptoms, is very high. They may be relatively mild, such as nervousness, irritability, and apathy, or severe, such as maniac behavior, schizophrenic reactions, or frank psychosis.[330–332] However, pessimism, self-reproach, and grave depression, which may lead to suicide attempts, are predominant.

4. CONCLUSIONS

Some observations must be reemphasized as concluding remarks of this chapter.

There is an extreme polymorphism of the clinical picture of MEs, with no simple symptomatology related to a classic scheme of semiology, but clinical pictures that are variable among the various MEs and also within a single ME. Such heterogeneity is related to the great variety of physiopathogenetic factors as well as to the anatomophysiological and metabolic complexity of the nervous system. Common to all these forms is the global nature of the clinical picture, with the concomitant presence of psychiatric and neurological symptoms and signs, including affective disorders, dementia, disorders of consciousness, and others, and such neurological symptoms and signs of pyramidal, extrapyramidal, and cerebellar disorder, with the extrapyramidal ones somewhat more frequent.

Another point of interest is the fact that in almost all MEs we find mixed together the two modalities of expression of the CNS, i.e., the excitatory and the inhibitory; in various MEs we can see the presence of seizures, focal or generalized, and of paresis of various areas.

Finally, of perhaps the most importance to both clinicians and neurobiologists is the fact that most MEs are initially potentially reversible; therefore, the point is to know when the metabolic damage, which is mainly functional, becomes structural and, therefore, no longer reversible.

As stated, MEs can be complications of very common human diseases; however, from this review it appears that, in spite of the great amount of research done, little is known about them. Because of the frequency of the MEs, the therapeutic goals may alone justify the efforts already made and in progress to clarify the mechanisms underlying brain functional and/or anatomical involvement.

Beside the therapeutic purposes, the study of the pathogenetic mechanisms of the MEs may also provide further information about brain function. A great variety of neurological and psychiatric disorders characterize the clinical picture of the MEs; therefore, the experimental models of such diseases may be useful for investigating the relation between specific metabolic alterations and the resulting clinical signs and symptoms.

A functional "mapping" of the brain is the most important goal in neurological sciences, and the study of the MEs may provide some keys to its realization.

REFERENCES

1. Gibson, G. E., and Blass, J. P., 1983, *Handbook of Neurochemistry*, Volume 3 (A. Lajtha, ed.), Plenum Press, New York, pp. 633–651.
2. Plum, F., and Posner, J. B., 1980, *The Diagnosis of Stupor and Soma,* 3rd ed., Davis, Philadelphia.
3. Siesjo, B. K., and Agardh, C. H., 1983, *Handbook of Neurochemistry,* Volume 3 (A. Lajtha, ed.), Plenum Press, New York, pp. 353–379.
4. Sokoloff, L., Fitzgerald, G. G., and Kaufman, E. E., 1977, *Nutrition and the Brain,* Volume 1 (R. J. Wurtman and J. J. Wurtman, eds.), Raven Press, New York, pp. 87–139.
5. Marks, V., 1972, *Br. Med. J.* **1**:430–432.
6. Marks, V., and Rose, F. C., 1965, *Hypoglycemia,* Blackwell, Oxford.
7. Wilkinson, D. S., and Prockop, L. D., 1976, *Handbook of Clinical Neurology,* Volume 27 (P. J. Vinken and G. H. Bruyn, eds.), North-Holland, Amsterdam, pp. 53–78.
8. Kety, S. S., Polis, B. D., Nadler, C. S., and Schmidt, C. F., 1948, *J. Clin. Invest.* **27**:500–510.

9. Creutzfeldt, O. D., and Meisch, J. J., 1963, *Electroencephalogr. Clin. Neurophysiol. Suppl.* **24**:158–171.
10. Agardh, C. D., and Rosen, I., 1979, *Neurosci. Lett. Suppl.* **3**:45.
11. Finley, K. H., and Brenner, C., 1941, *Arch. Neurol. Psychiatry* **45**:403–438.
12. Hicks, S. P., 1950, *Arch. Pathol.* **49**:111–137.
13. Brierley, J. B., Brown, A. W., and Meldrum, B. S., 1971, *Brain Res.* **25**:483–499.
14. Lewis, L. D., Ljunggren, B., Ratcheson, R. A., and Siesjo, B. K., 1974, *J. Neurochem.* **23**:673–679.
15. Ghajar, J. B. G., Plum, F., and Duffy, T. E., 1982, *J. Neurochem.* **38**:397–409.
16. Lewis, L. D., Ljunggren, B., Norberg, K., and Siesjo, B. K., 1974, *J. Neurochem.* **23**:659–671.
17. Norberg, K., and Siesjo, B. K., 1976, *J. Neurochem.* **26**:345–352.
18. Agardh, C. D., Chapman, A. G., Nilsson, B., and Siesjo, B. K., 1981, *J. Neurochem.* **36**:490–500.
19. Ratcherson, R. A., Blank, A. C., and Ferrendelli, J. A., 1981, *J. Neurochem.* **36**:1952–1958.
20. McCandless, D. W., 1981, *Brain Res.* **215**:225–233.
21. Himwich, H. E., 1951, *Brain Metabolism and Cerebral Disorders,* Williams and Wilkins, Baltimore, pp. 257–307.
22. Ferrendelli, J. A., and Chang, M. M., 1973, *Arch. Neurol.* **28**:173–177.
23. Hinzen, D. H., Becker, P., and Müller, U., 1970, *Pflügers Arch.* **321**:1–14.
24. Agardh, C. D., Kalimo, H., Olsson, Y., and Siesjo, B. K., 1980, *Acta Neuropathol.* **50**:31–41.
25. Kalimo, H., Agardh, C. D., Olsson, Y., and Siesjo, B. K., 1980, *Acta Neuropathol.* **50**:43–52.
26. Agardh, C. D., Westerberg, E., and Siesjo, B. K., 1980, *Acta Physiol. Scand.* **109**:115–116.
27. Pressman, B. G., and Lardy, H. A., 1956, *Biochim. Biophys. Acta* **21**:548–566.
28. Lochner, A., Kotze, J. C. N., Benade, A. J. S., and Gevers, W., 1978, *J. Mol. Cell. Cardiol.* **10**:857–875.
29. Chan, P. H., and Fishman, R. A., 1978, *Science* **201**:358–360.
30. Gaudet, R. J., and Levine, L., 1979, *Biochem. Biophys. Res. Commun.* **86**:285–291.
31. Agardh, C. D., and Siesjo, B. K., 1981, *J. Cereb. Blood Flow Metab.* **1**:267–275.
32. Fahan, S., Davis, J. N., and Rowland, L. P., 1979, *Cerebral Hypoxia and Its Consequences,* Raven Press, New York.
33. Cohen, P., Alexander, S., Smith, R., Reivich, M., and Wollman, H., 1967 *J. Appl. Physiol.* **23**:183–189.
34. Duffy, T. E., Nelson, S. R., and Lowry, O. H., 1972, *J. Neurochem.* **19**:959977.
35. Salford, L. G., Plum, F., and Siesjo, B. K., 1973, *Arch. Neurol.* **29**:227–238.
36. Siesjo, B. K., Johannsson, H., Ljunggren, B., and Norberg, K., 1974, *Brain Dysfunction in Metabolic Disorders* (F. Plum, ed.), Raven Press, New York, pp. 75–112.
37. Gibson, G. E., and Blass, J. P., 1976, *J. Neurochem.* **27**:37–42.
38. Gibson, G. E., and Duffy, T. E., 1981, *J. Neurochem.* **36**:28–33.
39. Drachman, D. A., 1978, *Psychopharmacology: A Generation of Progress* (M. A. Lipton, A. DiMascio, and K. F. Killam, eds.), Raven Press, New York, pp. 651–662.
40. Davis, J. N., and Carlsson, A., 1973, *J. Neurochem.* **20**:913–915.
41. Davis, J. N., and Carlsson, A., 1973, *J. Neurochem* **21**:783–790.
42. Davis, J. N., Giron, L. T., Stanton, E., and Maury, W., 1979, *Adv. Neurol.* **26**:219–223.
43. Light, K. E., Schetlar, D., Strahlendorf, J. C., Huges, M. J., and Barnes, C. D., 1981, *J. Neurochem.* **36**:773–775.
44. Yoshino, Y., and Elliott, K. A. C., 1970, *Can. J. Biochem.* **48**:228–235.
45. Gibson, G. E., Peterson, C., and Sansone, J., 1981, *J. Neurochem.* **37**:192–201.
46. Kety, S. S., and Schmidt, C. F., 1948, *J. Clin. Invest.* **27**:484–492.
47. Artru, A. A., and Michenfelder, J. D., 1981, *J. Cereb. Blood Flow Metab.* **1**:277–283.
48. Siesjo, B. K., and Nilsson, L., 1971, *Scand. J. Clin. Lab. Invest.* **27**:83–96.
49. Ginsberg, M. D., Myers, R. E., and McDonagh, B. F., 1974, *Arch. Neurol.* **30**:209–216.
50. Brierley, J. B., Prior, P. F., Calverley, J., and Brown, A. W., 1977, *J. Neurol. Sci.* **31**:133–157.

51. van Harreveld, A., 1972, *The Structure and Function of Nervous Tissue,* Volume IV (G. H. Bourne, ed.), Academic Press, New York, pp. 447–511.
52. Hossmann, K. A., 1982, *J. Cereb. Blood Flow Metab.* **2**:275–297.
53. Siesjo, B. K., 1978, *Brain Energy Metabolism*, Wiley, New York.
54. Rehncrona, S., Rosen, I., and Siesjo, B. K., 1981, *J. Cereb. Blood Flow Metab.* **1**:297–311.
55. Miller, J. R., and Myers, R. E., 1970, *Neurology* **20**:715–724.
56. Nordstrom, C. H., Rehncrona, S., and Siesjo, B. K., 1978, *J. Neurochem.* **30**:479–486.
57. Rehncrona, S., Abdul-rahman, A., and Siesjo, B. K., 1979, *Acta Neurol. Scand* **60**(Suppl. 72):294–295.
58. Nordstrom, C. H., Rehncrona, S., and Siesjo, B. K., 1978, *Stroke* **9**:335–343.
59. Rehncrona, S., Mela, L., and Siesjo, B. K. 1979, *Stroke* **10**:437–446.
60. Kalimo, H., Rehncrona, S., Soderfeldt, B., Olsson, Y., and Siesjo, B. K., 1981, *J. Cereb. Blood Flow Metab.* **1**:313–327.
61. Rehncrona, S., Westerberger, E., Akesson, B., and Siesjo, B. K., 1982, *J. Neurochem.* **38**:84–93.
62. Levy, D. E., van Uitert, R. L., and Pike, C. L., 1979, *Neurology* **29**:663–665.
63. Hallenbeck, J. M., and Furlow, T. W., Jr., 1979, *Stroke* **10**:629–637.
64. Jorgensen, M. B., and Diemer, N. H., 1982, *Acta Neurol. Scand.* **66**:536–546.
65. Yoschida, S., Inoh, S., Asano, T., Sano, K., Kubota, M., Schimazaki, H., and Ueta, N., 1980, *J. Neurosurg.* **53**:323–331.
66. Sotos, J. F., Dodge, P. R., Meara, P., and Talbot, N. B., 1960, *Pediatrics* **26**:525–538.
67. Arieff, A. I., and Kleeman, C. R., 1973, *J. Clin. Invest.* **52**:571–583.
68. Fulop, M., Rosenblatt, A., Kreitzer, S. M., and Gerstenhaber, B., 1975, *Diabetes* **24**:594–599.
69. Pappius, H. M., and Dayes, L. A., 1965, *Arch. Neurol.* **13**:355–402.
70. Fishman, R. A., 1974, *Brain Dysfunction in Metabolic Disorders* (F. Plum, ed.), Raven Press, New York, pp. 159–172.
71. Chan, P. H., and Fishman, R. A., 1977, *J. Neurochem.* **29**:179–181.
72. Fishman, R. A., Reiner, M., and Chan, P. H., 1977, *J. Neurochem* **28**:1061–1067.
73. Shank, R. P., and Baxter, C. V., 1973, *J. Neurochem.* **21**:301–313.
74. Arieff, A. I., and Kleeman, C. R., 1974, *J. Clin. Endocrinol. Metab.* **38**:1057–1067.
75. Chan, P. H., and Fishman, R. A., 1979, *Brain Res.* **161**:293–301.
76. McDowell, M. E., Wolf, A. V., and Steer, A., 1955, *Am. J. Physiol.* **180**:545–558.
77. Arieff, A. I., and Guisardo, R., 1976, *Kidney Int.* **10**:104–116.
78. Prockop, L. D., 1976, *Handbook of Clinical Neurology,* Volume 27 (P. J. Vinken and G. W. Bruyn, eds.), North-Holland, Amsterdam, pp. 79–98.
79. Arieff, A. I., and Carroll, H. J., 1972, *Medicine* **51**:73–94.
80. Schwartz, T. B., and Apfelbaum, R. I., 1965-66, *Year Book of Endocrinology* (T. B. Schwartz, ed.), Year Book, Chicago, pp. 165–181.
81. Dibenedetto, R. J., Crocco, J. A., and Soscia, J. L., 1965, *Arch. Intern. Med.* **116**:74–82.
82. Maccario, M., 1968, *Arch. Neurol* **19**:525–534.
83. Espinas, O. E., and Poser, C. M., 1969, *Arch. Neurol.* **20**:182–186.
84. Singh, B. M., Gupta, D. R., and Strobos, R. J., 1973, *Arch. Neurol.* **29**:187–190.
85. Singh, B. M., and Strobos, R. J., 1980, *Ann. Neurol.* **8**:155–160.
86. Venna, N., and Sabin, T. D., 1981, *Arch. Neurol.* **38**:512–514.
87. Vastola, E. F., Maccario, M., and Homan, R., 1967, *Neurology* **17**:520–526.
88. Juul-Jensen, P., and Denny-Brown, D., 1966, *Arch. Neurol.* **15**:563–578.
89. Chatrian, G. E., Shaw, C. M., and Leffman, H., 1964, *Electroencephalogr. Clin. Neurophysiol.* **17**:177–193.
90. Maccario, M., Messis, C. P., and Vastola, E. F., 1965, *Neurology* **15**:195–206.
91. Fishman, R., 1976, *Handbook of Clinical Neurology,* Volume 28 (P. J. Vinken and G. W. Bruyn, eds.), North-Holland, Amsterdam, pp. 495–505.
92. Dodge, P. R., Crawford, J. D., and Probst, J. H., 1960, *Arch. Neurol.* **3**:513–529.
93. Rymer, M., and Fishman, R. A., 1973, *Arch. Neurol.* **38**:49–54.
94. Swinyard, E. A., 1949, *Am. J. Physiol.* **156**:163–169.
95. Wakim, K. G., 1969, *Mayo Clin. Proc.* **44**:433–460.

96. Lajtha, A., and Sershen, H., 1975, *J. Neurochem.* **24**:667–672.
97. Katzman, R., and Pappius, H. M., 1973, *Brain Electrolytes and Fluid Metabolism,* William and Wilkins, Baltimore.
98. Dila, C. J., and Pappius, H. M., 1972, *Arch. Neurol.* **26**:85–90.
99. Carvalho, A. P., 1982, *Handbook of Neurochemistry,* Volume 1, 2nd ed. (A. Lajtha, ed.), Plenum Press, New York, pp. 69–116.
100. Knapp, S., Mandel, A. J., and Ballard, W. P., 1976, *Life Sci.* **16**:1583–1593.
101. Whittam, R., 1968, *Nature* **219**:610.
102. Bozler, E., and Lavine, D., 1958, *Am. J. Physiol.* **195**:45–49.
103. Holloszy, J. D., and Karakara, T., 1967, *J. Gen. Physiol.* **50**:551–562.
104. Seeman, P., 1972, *Pharmacol. Rev.* **24**:583–655.
105. Winterstein, H., 1961, *Pharmacol. Rev.* **13**:71–107.
106. Geschwind, I., 1969, *Frontiers in Neuroendocrinology* (W. F. Ganong and L. Martini, eds.), Raven Press, New York, pp. 389–431.
107. Douglas, W. W., and Rubin, R. P., 1963, *J. Physiol.* (*London*) **167**:288–310.
108. Randic, M. and Padjen, A., 1967, *Nature* **215**:990.
109. Frame, B., 1976, *Handbook of Clinical Neurology,* Volume 27 (P. J. Vinken and G. W. Bruyn, eds.), North-Holland, Amsterdam, pp. 283–320.
110. Wills, M. R., 1974, *Br. J. Hosp. Med.* **11**:270–288.
111. Dwards, G. A., and Daum, S. M., 1959, *Arch. Intern. Med.* **10**:29–36.
112. Breslau, N. A., and Pak, C. Y. C., 1979, *Metabolism* **28**:1261–1276.
113. Mateo, D., and Roldan, G. G., 1982, *Arch. Neurol.* **39**:424–425.
114. Plum, F., and Hindfelt, B., 1976, *Handbook of Clinical Neurology,* Volume 27 (P. J. Vinken and G. W. Bruyn, eds.), North-Holland, Amsterdam, pp. 349–377.
115. Victor, M., 1973, *Brain Dysfunction in Metabolic Disorders* (F. Plum, ed.), Raven Press, New York, p. 1–12.
116. MacGilliray, B. B., 1976, *Handbook of Electroencephalography and Clinical Neurophysiology,* Volume 15 (A. Remond, ed.), Elseiver, Amsterdam, pp. 26–44.
117. Cavanagh, J. B., 1974, *Brain Dysfunction in Metabolic Disorders* (F. Plum, ed.), Raven Press, New York, pp. 13–38.
118. Fisher, J. E., 1974, *Brain Dysfunction in Metabolic Disorders* (F. Plum, ed.), Raven Press, New York, pp. 53–73.
119. Dodsworth, J. M., Cummings, M. G., James, J. H., and Fisher, J. E., 1973, *Gastroenterology* **64**:881.
120. Phear, E. A., Scherlock, S., and Summerskill, W. H. J., 1955, *Lancet* **2**:867–870.
121. Fisher, J. E., and Faloon, W. W., 1957, *N. Engl. J. Med.* **256**:1030–1034.
122. Duffy, T. E., Hindfelt, B., and Plum, F., 1975, *Clin. Res.* **23**:393A.
123. Cole, M., Rutherford, R. B., and Smith, F. O., 1972, *Arch. Neurol.* **26**:130–136.
124. Breen, K. J., and Schenker, S., 1972, *Progress in Liver Diseases,* Volume 4 (H. Popper and F. Schaffner, eds.), Grune and Stratton, New York, pp. 301–332.
125. Benjamin, A. M., Okamoto, K., and Quastel, J. H., 1978, *J. Neurochem.* **30**:131–143.
126. Hawkins, R. A., Miller, A. L., Nielsen, R. C., and Veech, R. L., 1973, *Biochem. J.* **134**:1001–1008.
127. Benjamin, A. M., 1982, *Handbook of Neurochemistry,* Volume 1, 2nd ed. (A. Lajtha, ed.), Plenum Press, New York, pp. 117–137.
128. Kline, D. G., Doberneck, R. C., Chun, B. K., and Rutherford, R. B., 1966, *Ann. Surg.* **164**:1003–1012.
129. Victor, M., Adams, R. D., and Cole, M., 1965, *Medicine* **44**:345–396.
130. Cavanagh, J. B., and Kyu, M. H., 1971, *J. Neurol. Sci.* **12**:63–75.
131. Zamora, A. J., Cavenegh, J. B., and Kyu, M. H., 1978, *J. Neurol. Sci.* **18**:25–41.
132. Norenberg, M. D., and Lapham, L. W., 1974, *J. Neuropathol. Exp. Neurol.* **33**:422–435.
133. Vergara, F., Plum, F., and Duffy, T. E., 1974, *Science* **183**:81–83.
134. Duffy, T. E., Vergara, F., and Plum, F., 1974, *Brain Dysfunction in Metabolic Disorders* (F. Plum, ed.), Raven Press, New York, pp. 39–52.
135. Caesar, J., 1962, *Clin. Sci.* **22**:33–41.
136. Davidson, C. S., and Gabuzda, G. J., 1969, *Disease of the Liver* (L. Schiff, ed.), Lippincott, Philadelphia, pp. 378–409.

137. James, J. H., Hodgman, J. M., Funovics, J. N., Yoschimura, N., and Fischer, J. E., 1976, *Metabolism* **25**:471–476.
138. Bloxam, D. L., and Curzon, G., 1978, *J. Neurochem.* **31**:1255–1263.
139. Fischer, J. E., Yoschimura, N., Aguirre, A., James, J. H., Cummings, M. G., Abel, R. M., and Deindorfer, F., 1974, *Am. J. Surg.* **127**:40–47.
140. Battistin, L., and Zanchin, G., 1980, *Neurochemistry and Clinical Neurology* (L. Battistin, G. Hashim, and A. Lajtha, eds.), Liss, New York, pp. 315–326.
141. Zanchin, G., Rigotti, P., Dussini, N., Vassanelli, P., and Battistin, L., 1979, *J. Neurosci. Res.* **4**:301–310.
142. James. J. H., Jeppson, B., Zipara, V., and Fisher, J. E., 1979, *Lancet* **2**:772–775.
143. Mans, A. M., Biebuyck, J. F., Saunders, S. J., Kirsch, R. E., and Hawkins, R. A., 1979, *J. Neurochem.* **33**:409–418.
144. Mans, A. M., Biebuyck, J. F., Shelly, K., and Hawkins, R. A., 1982, *J. Neurochem.* **38**:705–717.
145. Curzon, B., Kantameneni, B. D., Fernando, J. C., Woods, M. S., and Cavanagh, J. B., 1975, *J. Neurochem.* **24**:1065–1070.
146. Cummings, M. G., Soeters, P. B., James, J. H., Keane, J. M., and Fischer, J. E., 1976, J. Neurochem. **27**:501–509.
147. Cummings, M. G., James, J. H., Soeters, P. B., Keane, J. M., Foster, J., and Fischer, J. E., 1976, *J. Neurochem.* **27**:741–746.
148. Smith, A., Rossi Fanelli, F., Ziparo, V., James, J. G., Perelle, B., and Fischer, J. E., 1978, *Ann. Surg.* **187**:343–350.
149. James, J. H., Hodgman, J. M., Funovics, J. M., and Fischer, J. E., 1976, *J. Neurochem.* **27**:223–227.
150. Baldessarini, R. J., and Fischer, J. E., 1977, *Arch. Gen. Psychiatry* **34**:958–964.
151. Boulton, A. A., and Juorio, A. V., 1982, *Handbook of Neurochemistry,* Volume 1, 2nd ed (A. Lajtha, ed.), Plenum Press, New York, pp. 189–222.
152. Faraj, B. A., Bowen, P. A., Isaacs, J. W., and Rudman, D., 1976, *N. Engl. J. Med.* **294**:1360–1364.
153. Faraj, B. A., Bether, A. R., Ali, F. M., Rudman, D., and Galambos, J., 1980, *Noncatecholic Phenylethylamines* (A. D. Mosnaim and M. E. Wolf, eds.), Dekker, New York, Part 2, pp. 81–95.
154. Faraj, B. A., Ali, F. M., Ansley, J. D., and Malveaux, E. J., 1978, *Gastroenterology* **75**:1041–1044.
155. Faraj, B. A., Camp, V. M., Ansley, J. D., Scott, J., Ali, F. M., and Malveaux, E. J., 1981, *J. Clin. Invest.* **67**:395–402.
156. Walker, C. O., McCandless, D. W., McGarry, J. D., and Schenker, S., 1970, *J. Lab. Clin. Med.* **76**:569–573.
157. Samson, F. E., Dahl, N., and Dahl, D. R., 1956, *J. Clin. Invest.* **35**:1291–1299.
158. Zieve, L., Doizaki, W. M., and Zieve, F. J., 1974, *J. Lab. Clin. Med.* **83**:16–28.
159. Phear, E. A., Ruebner, B., Sherlock, S., and Summerskill, W. H. J., 1956, *Clin. Sci.* **15**:97–117.
160. Bloch, P., Delorme, M. L., Rapin, J. R., Granger, A., Boschat, M., and Opolon, P., 1978, *Surg. Gynecol. Obstet.* **146**:551–558.
161. Denis, J., Opolon, P., Nusinovici, V., Granger, A., and Darnis, F., 1978, *Gut* **19**:787–793.
162. Denis, J., Delorme, M. L., Boschat, M., Nordlinger, B., and Opolon, P., 1983, *J. Neurochem.* **40**:10–19.
163. Tyler, H. R., 1971, *Diseases of the Kidney* (M. B. Strauss, and L. G. Welt, eds.), Little, Brown, Boston, pp. 335–342.
164. Tyler, H. R., 1976, *Handbook of Clinical Neurology,* Volume 27 (P. J. Vinken and G. W. Bruyn, eds.), North-Holland, Amsterdam, pp. 321–348.
165. Noriega-Sanchez, A., Martinez-Maldonado, M., and Haiffe, R. M., 1978, *Neurology* **28**:667–669.
166. Olsen, S., 1961, *Acta Psychiatr. Scand.* **36**(Suppl. 156):1–128.
167. Glaser, G. H., 1974, *Brain Dysfunction in Metabolic Disorders* (F. Plum, ed.), Raven Press, New York, pp. 173–198.

168. Pye, I. F., McGale, E. H. F., Stonier, C., Hutchinson, E. C., and Aber, G. M., 1979, *Clin. Chim. Acta* **92**:65–72.
169. Sullivan, P. A., Murnaghan, D., Callaghan, N., Kantamaneni, B. D., and Curzon, G., 1978, *J. Neurol. Neurosurg. Psychiatry* **41**:581–588.
170. Jellinger, E., Irsigler, K., Kothbauer, P., and Riederer, P., 1977, *Excerpta Med.* **27**:169.
171. Deferrari, G., Garibotto, G., Robaudo, C., Ghiggeri, G. M., and Tizianello, A., 1981, *Kidney Int.* **20**:505–510.
172. Scheinberg, P., 1954, *Neurology* **4**:101–105.
173. Heyman, A., Patterson, J. L., and Jones, R. W., Jr., 1951, *Circulation* **3**:558–563.
174. Taylor, W. H., 1982, *Lancet* **1**:630.
175. Bergström, J., and Fürst, P., 1978, *Kidney Int.* **13**(Suppl. 3):59–512.
176. Lascelles, P. T., and Taylor, W. H., 1966, *Clin. Sci.* **31**:403–413.
177. Merrill, J. P., Legrain, M., and Hoigne, R., 1953, *Am. J. Med.* **14**:519–520.
178. Katz, A. D., and Kaplan, L., 1973, *Arch. Surg.* **107**:51–55.
179. Arnaud, C. D., 1973, *Kidney Int.* **4**:89–95.
180. Massry, S. G., and Goldstein, D. A., 1978, *Kidney Int.* **13**(Suppl. 8):539–542.
181. Nathan, E., and Pedersen, S. E., 1980, *Acta Paediatr. Scand.* **69**:793–796.
182. Goldstein, D. A., Feinstein, E. I., Chui, L. A., Pattabhiraman, R., and Massry, S. G., 1980, *J. Clin. Endocrinol. Metab.* **51**:130–134.
183. Simenhoff, M. L., Saukkonent, J. J., Burke, J. F., Schaendler, R. W., Vogel, W. H., Bovee, K., and Lasker, N., 1978, *Kidney Int.* **13**(Suppl. 8):516–519.
184. Simenhoff, M. L., 1975, *Kidney Int.* **7**:5314–5317.
185. Hicks, J. M., Young, D. S., and Wootton, I. D. P., 1964, *Clin. Chim. Acta* **9**:228–234.
186. Wardle, E. N., 1978, *Kidney Int.* **13**(Suppl. 8):513–515.
187. Alfrey, A. C., Rudolph, H., and Smythe, W. R., 1974, Proceedings of the Adequacy of Dialysis Symposium, Monterey, California.
188. Alfrey, A. C., Le Gendre, G. R., and Kaehny, N. D., 1976, *N. Engl. J. Med.* **294**:184–188.
189. Alfrey, A. C., Hegg, A., and Crashwell, P., 1980, *Am. J. Clin. Nutr.* **33**:1509–1516.
190. Larsen, N. A., Pakkenberg, H., Damsgaard, E., Heydorn, K., and Wold, S., 1981, *J. Neurol. Sci.* **51**:437–446.
191. Raskin, N. H., and Fishman, R. A., 1976, *N. Engl. J. Med.* **294**:204–210.
192. Arieff, A. J., Guisardo, R., Massry, S. G., and Lazarowitz, V. L., 1976, *J. Clin. Invest.* **58**:306–311.
193. Niessenson, A. R., Lavin, M. L., and Klawans, H. L., 1977, *J. Chronic. Dis.* **30**:705–733.
194. Lederman, R. J., and Henry, C. E., 1978, *Ann. Neurol.* **4**:199–204.
195. Elliot, H. L., Dryburg, F., Fell, G. S., Sabet, S., and MacDougall, A. I., 1978, *Br. Med. J.* **1**:1101–1103.
196. Gross, M. L. P., Sweny, P., Pearson, R. M., Kennedy, J., Fernando, O. N., and Moorhead, J. F., 1982, *J. Neurol. Sci.* **56**:23–34.
197. Genuth, S., 1980, *Clinical Diabetes: Modern Management* (S. Podolski, ed.), Appleton-Century-Crofts, New York, pp. 173–207.
198. Schade, D. S., and Eaton, R. P., 1977, *Diabetes* **26**:596–599.
199. McGarry, J. D., and Foster, D. W., 1977, *Arch. Intern. Med.* **137**:495–501.
200. Baruh, S., Sherman, L., and Markowitz, S., 1981, *Med.Clin. North Am.* **65**:117–132.
201. Blachshear, P. J., and Alberti, K. G., 1974, *Biochem. J.* **138**:107–117.
202. Ruderman, N. B., Ross, P. S., Berger, M., and Goodman, M. N., 1974, *Biochem. J.* **138**:1–10.
203. Sosula, L., Beaumosnt, P., Hollows, F. C., Jonson, K. M., and Regtop, H. I. M., 1974, *Diabetes* **23**:221–231.
204. Reubi, F. C., 1953, *Circ. Res.* **1**:410–413.
205. Anderson, J. M., Machin, G. A., McKinley, I., and Thistlethwaite, D., 1974, *Lancet* **1**:1341–1342.
206. Timplerey, W. R., Preston, F. E., and Ward, J. D., 1974, *Lancet* **1**:952–956.
207. Halperin, M. L., Bear, R. A., Hannaford, M. C., and Goldstein, M. B., 1981, *Diabetes* **30**:781–787.
208. Kuschinsky, W., Suda, S., and Sokoloff, L., 1981, *Am. J. Physiol.* **241**:H772–H777.

209. Miller, A. L., Hawkins, R. A., and Veech, L., 1975, *J. Neurochem.* **25**:553–558.
210. Fulop, M., Tannenbaum, H., and Dreyer, N., 1973, *Lancet* **2**:635–639.
211. Fulop, M., Rosenblatt, A., Kreitzer, S. M., and Gerstenhaber, B., 1975, *Diabetes* **24**:594–599.
212. Arieff, A. I., and Carrol, H. J., 1974, *Diabetes* **23**:525–531.
213. Guisardo, R., and Arieff, A. I., 1975, *Metabolism* **24**:665–675.
214. De Vivo, D. C., Leckie, M. P., Ferendelli, J. S., and McDougal, D. B., Jr., 1978, *Ann. Neurol.* **3**:331–337.
215. Appleton, D. B., and De Vivo, D. C., 1974, *Epilepsia* **15**:211–227.
216. Schneider, R. and Droller, H. Q., 1938, *Q. J. Exp. Physiol.* **28**:323–333.
217. Dividian, N. M., Butler, T. C., and Poole, D. T., 1978, *Epilepsia* **19**:369–378.
218. Winegard, A. I., and Clements, R. S., Jr., 1971, *Med. Clin. North Am.* **55**:899–911.
219. Young, E., and Bradley, R. F., 1967, *N. Engl. J. Med.* **276**:665–669.
220. Ortiz-Vazquez, J., 1979, *Handbook of Clinical Neurology,* Volume 38 (P. J. Vinken and G. W. Bruyn, eds.), North-Holland, Amsterdam, pp. 285–307.
221. Lamisse, F., Gautier, J., Kiffer, A., and Rouzand, M., 1970, *Presse Med.* **78**:1925–1928.
222. Austen, F. K., Carmichael, M. W., and Adams, R. D., 1957, *N. Engl. J. Med.* **257**:579–590.
223. Spillane, J. D., 1970, *Brit. Med. J.* **4**:796–798.
224. Labram, C., Bursaux, M., Baudouin, R., and Gaillot, M. J., 1966, *Presse Med.* **74**:2465.
225. Patterson, J. L., Heyman, A., and Ferguson, R. W., 1955, *J. Clin. Invest.* **34**:1857–1864.
226. Westlake, E. K., and Kaye, M., 1954, *Br. Med. J.* **1**:302–304.
227. Cameron, I. R., Davson, H., and Segal, M. B., 1969, *Yale J. Biol. Med.* **42**:241–247.
228. Woodbury, D. M., Rollins, L. T., Gardner, M. D., Hirschi, W. L., Hogan, G. R., Rallison, M. L., Tanner, G. S., and Brodie, D. A., 1958, *Am. J. Physiol.* **192**:79–90.
229. Schaffer, K. E., 1958, *J. Appl. Physiol.* **13**:1–13.
230. Taber, R. I., and Banauazizi, A., 1966, *Psychoparmacologia* **9**:382–391.
231. Granholin, L., and Siesjo, B. K., 1969, *Acta Physiol. Scand.* **75**:257–266.
232. Leusen, I., and Demeester, G., 1965, *Arch. Int. Physiol. Biochim.* **74**:25–34.
233. Messeter, K., and Siesjo, B. K., 1971, *Acta Physiol. Scand.* **83**:21–30.
234. Folbergrova, J., MacMillan, V., and Siesjo, B. K., 1972, *J. Neurochem.* **19**:2497–2505.
235. Folbergrova, J., MacMillan, V., and Siesjo, B. K., 1972, *J. Neurochem.* **19**:2507–2517.
236. Xanalatos, C., and James, I. M., 1972, *Clin. Sci.* **42**:63–68.
237. Nilsson, B., and Siesjo, B. K., 1976, *Acta Physiol. Scand.* **96**:72–82.
238. Miller, A. L., Hawkins, R. A., and Veech, R. L., 1975, *J. Neurochem.* **25**:553–558.
239. Borgstrom, L., Norberg, K., and Siesjo, B. K., 1976, *Acta Physiol. Scand.* **96**:569–574.
240. Miller, A. L., 1981, *International Review of Neurobiology,* Volume 22 (J. R. Smythies and R. J. Bradley, eds.), Academic Press, New York, pp. 47–82.
241. Folbergrova, J., Norberg, K., Quistorff, B., and Siesjo, B. K., 1975, *J. Neurochem.* **25**:457–467.
242. Dutton, E., Nicholas, W., Fisher, C. S., and Renzetti, R., 1959, *N. Engl. J. Med.* **261**:1369–1373.
243. Szam, I., Vass, A., and Wein, I., 1969, *Cardiologia* **54**:321–328.
244. Betz, E., 1965, *Acta Neurol. Scand.* **41**(Suppl. 14):121–128.
245. Matakas, F., Birkle, J., and Cervos-Navarro, J., 1978, *Acta Neuropathol.* **41**:207–210.
246. Weyne, J., Van Leuven, F., Kazemi, H., and Leusen, I., 1978, *Appl. Physiol. Resp. Environ. Exercise Physiol.* **44**:333–339.
247. Paljarvi, L., Soderfeldt, B., Kalimo, H., Olsson, Y., and Siesjo, B. K., 1982, *Acta Neuropathol.* **58**:87–94.
248. Rottermich, N. O., and von Haam, E., 1941, *J. Clin. Endocrinol.* **1**:873.
249. Sharf, B., and Levy, N., 1976, *Handbook of Clinical Neurology,* Volume 27 (P. J. Vinken and G. W. Bruyn, eds.), North-Holland, Amsterdam, pp. 444–458.
250. Vogel, F. S., 1951, *Arch. Pathol.* **52**:355–362.
251. Vogel, F. S., 1951, *J. Exp. Med.* **93**:297–304.
252. Sjaastad, O., Giessing, L., and Ritland, S., Blichfeldt, P., and Sandnes, K., 1979, *Nor. J. Neurol.* **220**:83–94.
253. Estrada, R. V., Moreno, J., Martinez, E., Hernandez, M. C., Gilsanz, G., and Gilsanz, V., 1979, *Acta Neurol. Scand.* **59**:135–139.

254. Reynolds, A. F., Villar, H. V., and Kasniak, A. W., 1981, *Neurosurgery* **9:**153–155.
255. Ayub, A., Faloon, W. W., and Heining, R. E., 1981, *J. Am. Med. Assoc.* **246:**970–973.
256. Juhl, E., Chrostoffersen, P., Baden, H., and Quaade, F., 1971, *N. Engl. J. Med.* **285:**543–547.
257. Moxley, R. I., III, Pozefsky, T., and Lockwood, D. H., 1974, *N. Engl. J. Med.* **290:**921–926.
258. Weisemann, R.E., and Johnson, R. E., 1977, *Am. J. Surg.* **134:**253–258.
259. Herling, P. M., James, J. H., Joffe, S. N., Kulneff-Herling, A. E. A., and Fischer, J. E., 1982, *J. Neurochem.* **38:**1170–1173.
260. Manelli, J. C., 1977, *Ann. Anesthesiol. Fr.* **18:**992–996.
261. Knochel, J. P., 1979, *Nouv. Presse Med.* **8:**121–124.
262. Chudley, A. E., Ninan, A., and Young, G. B., 1981, *Can. Med. Assoc. J.* **125:**604–607.
263. Paymaster, N. J., 1975, *Br. J. Anaesth.* **47:**85–87.
264. Knochel, J. P., 1977, *Clin. Nephrol.* **7:**131–137.
265. Freund, H., Atamian, S., and Fischer, J. E., 1979, *J. Am. Med. Assoc.* **5:**496–498.
266. Klein, G. L., Alfrey, A. C., Miller, N. L., Sherrad, D. J., Hazlet, T. K., Ament, M. E., and Coburn, J. W., 1982, *Am. J. Clin. Nutr.* **35:**1425–1429.
267. Senseback, W., Madison, L., Eisenberg, S., and Ochs, L., 1954, *J. Clin. Invest.* **33:**1434–1440.
268. Fazekas, J. F., Graves, F. B., and Alman, R. W., 1951, *Endocrinology* **48:**169.
269. Sokoloff, L., Wechler, R. L., Magnold, R., Balls, K., and Ketty, S., 1953, *J. Clin. Invest.* **32:**202–208.
270. Wilson, W. P., Johnson, J. E., and Feist, F. W., 1964, *Electroencephalogr. Clin. Neurophysiol.* **16:**329–331.
271. Skanse, B., and Nyman, E., 1956, *Acta Neuroendocrinol. (Copenhagen)* **22:**246–263.
272. Jabbari, B., and Huott, A. D., 1980, *Epilepsia* **21:**91–96.
273. Woodbury, D. M., Hurley, R. E., Lewis, N. G., McArthur, M. W., Copeland, N. W., Kirzchvink, J. F., and Goodman, L. S., 1952, *J. Pharmacol. Exp. Ther.* **106:**331–340.
274. Prange, A. J., Meek, J. L., and Lipton, M. A., 1970, *Life Sci.* **9:**901–907.
275. Lipton, M. A., Prange, A. J., Diarman, W., and Undenfriend, S., 1968, *Fed. Proc.* **27:**399.
276. Gonzales-Vegas, J. A., and Fuenmayor, D., 1978, *Experientia* **34:**1527–1528.
277. Fuenmayor, D., and Gonzales-Vegas, J. A., 1980, *Experientia* **36:**841–842.
278. Klawans, H. L., and Shenker, D. M., 1972, *J. Neural. Transm.* **33:**73–81.
279. Klawans, H. L., Goetz, C., and Weiner, W. J., 1973, *J. Neural Transm.* **34:**187–193.
280. Klawans, H. L., Shenker, D. M., and Weiner, W. J., 1973, *Adv. Neurol.* **1:**543–549.
281. Morley, J. E., Elson, M. K., Levine, A. S., and Shafer, R. B., 1982, *Eur. J. Pharmacol.* **78:**125–127.
282. Gambert, S. R., Garthwaite, T. L., Pontzer, C. M., and Hagen, T. C., 1980, *Horm. Metab. Res.* **12:**345–346.
283. Morley, J. E., and Levine, A. S., 1981, *Life Sci.* **18:**1901–1903.
284. Ruiz Marcos, A., Sanchez-Toscano, F., Del Rey, F. E., and Morreale de Escobar, G., 1980 *Brain Res.* **185:**91–102.
285. Eayrs, J. T., and Levine, S., 1963, *J. Endocrinol.* **25:**505–513.
286. Bradley, P. B., Eayrs, J. T., and Schalmbach, K., 1980, *Electroencephalogr. Clin. Neurophysiol.* **12:**467–477.
287. Greene, R., 1976, *Handbook of Clinical Neurology,* Volume 27 (P. J. Vinken and G. W. Bruyn, eds.), North-Holland, Amsterdam, pp. 255–278.
288. Dhar, S. K., and Nair, C. P. V., 1974, *Ann. Intern. Med.* **80:**426.
289. Heffron, W., and Eaton, R. P., 1970, Ann. Intern. Med. **76:**425–428.
290. Fidler, S. M., O'Rourke, R. A., and Buchsbaum, H. W., 1971, *Neurology* **21:**55–57.
291. Fischbeck, K. H., and Layzer, R. B., 1979, *Ann. Neurol.* **6:**453–454.
292. Delwaide, P. J., and Schoenen, J., 1978, *Acta Neurol. Scand.* **58:**309–312.
293. MacCrimmon, D. J., Wallace, J. E., Goldberg, W. M., and Streiner, D. L., 1979, *Psychosom. Med.* **41:**331–340.
294. Wilson, W. P., Johnson, J. E., and Smith, R. B., 1962, *Recent Adv. Biol. Psychiatry* **4:**234–243.

295. McGee, R. R., 1959, *Ann. Intern. Med.* **50**:1418–1432.
296. Thomas, F. B., Mazzaferri, E. L., and Skillmann, T. G., 1970, *Ann. Intern. Med.* **72**:679–685.
297. Bulens, C., 1981, *Arch. Neurol.* **38**:669–670.
298. Krieger, D. T., Liotta, A., and Brownstein, M. J., 1977, *Endocrinology* **100**:227–237.
299. Watson, S. J., Richard, C. W., and Barchas, J. D., 1978, *Science* **200**:1180–1182.
300. Tintner, R., Dunn, A. J., Iuvone, P. M., Shukla, J. B., and Rennert, O. M., 1979, *J. Neurochem.* **33**:1067–1073.
301. Schotman, P., von Heuven Nolsen, D., and Gispen, W. M., 1980, *J. Neurochem.* **34**:1661–1670.
302. Zwiers, H., Tonnaer, J., Wiegant, V. M., Schotman, P., and Gispen, W. M., 1979, *J. Neurochem.* **32**:247–256.
303. Zwiers, H., Wiegant, V. M., Schotman, P., and Gispen, W. M., 1978, *Neurochem. Res.* **3**:455–463.
304. Azmitia, E. C., and McEwen, B., 1969, *Science* **166**:1274–1276.
305. Davis, J., 1970, *Int. Rev. Neurobiol.* **12**:145–175.
306. Botticelli, L. J., and Wurtman, R. J., 1981, *Nature* **289**:75–76.
307. Delanoy, R. L.; Kramarcy, N. R., and Dunn, A. J., 1982, *Brain Res.* **231**:117–129.
308. Miller, C., Nakamura, J., and Leung, C. Y., 1975, *Neuropharmacology* **14**:385–396.
309. Rockstroh, B., Elbert, T., Lutzenberger, W., Birbaumer, N., Fehm, H. L., and Voigt, K. M., 1981, *Psychoneuroendocrinology* **6**:301–310.
310. Thody, A. J., Wilson, C. A., and Everard, D., 1981, *Psychopharmacology* **74**:153–156.
311. Katz, R. J., 1979, *Eur. J. Pharmacol.* **53**:383–385.
312. De Wield, D. A., Witter, A., and Greven, H. M., 1975, *Biochem. Pharmacol.* **24**:1463–1468.
313. Brush, F. R., and Fraley, S. M., 1979, *Acta Neurobiol. Exp.* **39**:453–457.
314. McCulloch, J., Kelly, P. A. T., and Van Delft, A. M. L., 1982, *Eur. J. Pharmacol.* **78**:151–158.
315. Meibach, R. C., Glicks, D., Cox, R., and Maayani, S., 1979, *Nature* **282**:625–626.
316. Weinberger, J., Greenberg, J. M. Waldman, M. T. G., Sylvestro, A., and Reivich, M., 1979, *Brain Res.* **177**:337–345.
317. Dam, M., and London, E. D., 1984, *Brain Res.* **295**:137–144.
318. Reinis, S., and Goldman, J. M., 1982, *The Chemistry of Behavior,* Plenum Press, New York, pp. 449–463.
319. Stith, R. D., and Weingarten, D., 1978, *Neuroendocrinology* **26**:129–140.
320. Farooqui, A. A., Elkoudy, A., and Mandel, P., 1977, *J. Neurochem.* **29**:365–369.
321. Volpe, J. J., and Marasa, J. C., 1976, *J. Neurochem.* **27**:841–845.
322. Griauzde, M., and Radulovacki, M., 1976, *J. Neurochem.* **26**:1301–1302.
323. Naik, R., and Kugelberg, E., 1975, *Indian J. Med. Res.* **63**:695–700.
324. Telegdy, G., and Vermes, I., 1975, *Neuroendocrinology* **13**:16–26.
325. Woodbury, D. M., 1972, *Handbook of Neurochemistry,* Volume 7 (A. Lajtha, ed.), Plenum Press, New York, pp. 255–287.
326. Woodbury, D. M., 1954, *Recent Prog. Horm. Res.* **10**:65–107.
327. Feldman, S., Todt, J. C., and Porter, R. W., 1961, *Neurology* **11**:109–115.
328. Dafny, N., Phillips, M. I., Newman-Taylor, A., and Gilman, S., 1973, *Brain Res.* **59**:257–263.
329. Flood, J. F., Vidal, D., Bennett, E. L., Orme, A. E., and Rosenzweig, M. R., 1978, *Pharmacol. Biochem. Behav.* **8**:81–87.
330. Van Der Sande, J. J., and Van Seters, A. P., 1980, *Handbook of Clinical Neurology,* Volume 39 (P. J. Vinken and G. W. Bruyn, eds.), North-Holland, Amsterdam, pp. 465–515.
331. Sachar, E. J., 1974, *Brain Dysfunction in Metabolic Disorders* (F. Plum, ed.), Raven Press, New York, pp. 239–251.
332. Starkman, M. N., and Schteingart, D. E., 1981, *Arch. Intern. Med* **141**:215–219.

27

Brain Iron Metabolism

Biochemical and Behavioral Aspects in Relation to Dopaminergic Neurotransmission

M. B. H. Youdim

1. INTRODUCTION

Three hundred years ago, Syndenham recognized the use of iron as a specific remedy for chlorosis, now recognized as iron-deficiency (ID) anemia. Thus, iron was the first trace element known to be essential, and is the most abundant trace element in the body.[1] The relatively large amount of iron found in the brain is unevenly distributed and shows a close similarity to the distribution of dopamine (DA) in various regions. In the brain, as in the periphery, iron is associated with hemin, heme-containing enzymes, is found as nonheme iron in many enzymes and structural proteins, and is also important for protein synthesis.[2–5]

The purpose of this chapter is not to review the literature on iron metabolism in the brain. However, special consideration will be given to the effect of alteration of brain iron on neurotransmitter metabolism function and behavioral abnormalities resulting from ID in human subjects and experimental animals. This special emphasis is warranted in light of the fact that ID is the most prevalent nutritional disorder known[6] and is associated with severe behavioral and certain neurological disorders in infants and children.[7–9] Furthermore, recent studies with the animal models of nutritional ID have indicated a correlation between behavioral changes and modified neurotransmission.[10–12]

Iron is present in all tissues in two forms—heme and nonheme compounds—and can exist in two stable oxidation states (Fe^{2+} and Fe^{3+}). It can form many complexes and its special properties make it a vital constituent of every mammalian cell and an essential element in biological processes. In biological systems, almost all iron is involved in processes related to oxygen metabolism.[2]

M. B. H. Youdim • Rappaport Family Institute for Research in the Medical Sciences, Department of Pharmacology, Technion, Haifa, Israel.

As in other tissues of the body, the brain iron compounds can be classified into several categories. The *heme iron* compounds in the brain include the mitochondiral cytochromes, heme-containing enzymes, and hemoglobin. Under normal conditions, hemoglobin is present only within the red blood cells in the vascular bed. *Nonheme iron* compounds in the brain, as elsewhere in the body, fall into two categories: compounds that are known to have metabolic or enzymatic function such as the flavoproteins, sulfur-containing proteins, and those that are believed to serve only a storage function, such as ferritin and hemosiderin. The third category of compounds do not contain iron but require iron or heme as a cofactor. This group includes many enzymes such as aconitase, tryptophan hydroxylase, tryosine hydroxylase, and succinic deyhdrogenase.[2,3] The iron in these compounds together with that of the ferritin comprise only a fraction of the measurable iron in the brain; the greater portion of the brain iron occurs in yet unidentified or functional form.[4,5]

2. DETERMINATION OF IRON IN THE CNS

It is possible to measure quantitatively the amount of total, heme, and nonheme iron in the brain.

2.1. Total Iron

For measuring total iron, the tissue is combusted or dissolved. Combustion of the tissue can be carried out either by wet ashing using concentrated acids or by dry ashing in a furnace or under vacuum.[13–15] The mineralized iron is then dissolved in less-concentrated hydrochloric acid and can be analyzed by colorimetric methods using iron chelators that form colored complexes with iron, or by more sensitive methods including atomic absorption, neutron activation, emission spectrometry. With the wide spread use of the atomic absorption method to determine trace elements, including iron, it has been found that Soluen-100, a quaternary ammonium hydroxide tissue solubilizer, gives homogeneous and aspirable tissue samples in which one can assay iron and other metals. This procedure is simple and quick and more reliable than dry or wet ashing treatment, and is likely to be less contaminated by iron from other sources.[16]

2.2. Heme Iron

After homogenization of the tissue, the sample is extracted by a mixture of ethyl acetate and glacial acetic acid. This procedure extracts exclusively heme-bound iron as hemin.[17] After washing the hemin with 1 N hydrochloric acid followed by centrifugation, two distinctly separate phases appear in addition to a sediment: the upper ethyl acetate layer is colored and contains hemin, and the lower phase is a colorless one. The ethyl acetate phase is used for the determination of iron at an absorbance wavelength of 330–366 nm.[18]

2.3. Nonheme Iron

Direct determination of nonheme iron is based on its extraction from the tissue using hydrochloric acid, either in the presence of sodium pyrophosphate, or sulfur solution containing a reducing agent (for review see 19). The iron extracted by one of these methods then can be analyzed by one of the principal methods described above for total iron.

Iron in the CSF can be determined by the same method for serum iron by simple colorimetry after splitting the protein-bound iron by acid and chelating it with compounds (e.g., O-phenanthroline or 3-(2-pyridyl)-5,6-bis(4-phenyl-sulfonic acid)-1,2-4 triazine) that form colored complexes. Other methods of iron determination in the CSF are based on wet ashing and reaction with O-phenanthroline staining or by activation analysis.[20] Considering the high iron concentration in the blood, great care should be taken to avoid contamination with blood while taking a CSF sample.

2.4. Histochemical Determination of Iron

Since the description of the Prussian blue reaction by Vichow (1847) and its use as a histochemical procedure by Perl,[21] this method has provided consistent histochemical localization of ferric iron in a variety of tissues including the brain. Negative staining techniques[22] and bismuth subnitrate staining[23] have facilitated high-resolution demonstration of ferritin. Intracellular iron has been localized with X-ray microprobe[24] and ultrastructural autoradiographic techniques.[25,26] Iron in tissues can also be stained with a variety of compounds that form colored iron complexes.[27]

Although there are many procedures available for the histological localization of iron,[28,29] Perl's reaction (Prussian blue) has been found to be most effective for iron from brain tissue. Recently, Nguyen-Legros *et al.*[30] have shown that the latter reaction can be highly intensified with diaminobenzidine (DAB).

Perl's reaction localizes ferric iron (Fe^{3+}) only by forming blue-colored ferric ferrocyanide when the tissue is acidified in the presence of potassium ferrocyanide. The acid (hydrochloric) serves to extract some of the Fe^{3+} from the proteins to which it is bound. In contrast, Turnbull blue reaction[29] measures ferrous iron (Fe^{2+}) and interestingly enough produces no visible reaction in rat or human brain tissue, thus suggesting that most of the available iron in the brain is in the ferric form.

3. DISTRIBUTION OF IRON IN THE BRAIN

3.1. Human

Iron in the brain was first histochemically detected by Zaleski[31] in 1886. Guizzetti[32] used an ammonium sulfite solution on thick human brain slices and found a characteristic distribution of iron in the brains of human and various

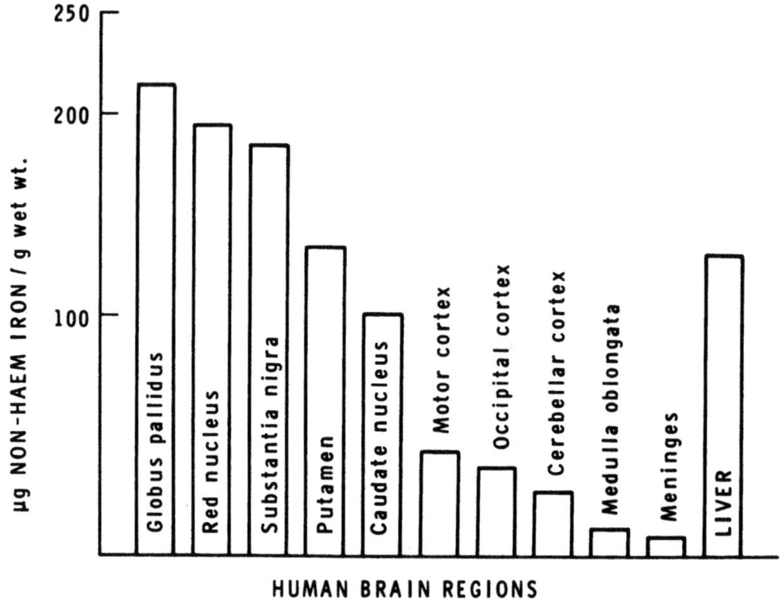

Fig. 1. The distribution of nonheme iron in various regions of the human brain and liver (from Hallgren and Sourander[37]). Note the high concentrations of iron in the extrapyramidal regions. Similar values have been reported by other investigators.[35–41]

mammals. The globus pallidus, substantia nigra, red nucleus, and dentate nucleus were most intensely stained. Using a similar technique, Spatz[33,34] was the first to study the distribution of iron in the brain systematically. The uneven and characteristic distribution of iron made it feasible for him to divide the brain structures into four groups according to the concentration of iron. The most intense iron reactions were found in the globus pallidus and substantia nigra, followed by the red nucleus, caudate, putamen, thalamus, subthalamus, and dentate nucleus. The third group included the cerebral cortex, cerebellum, anterior thalamus, mammillary body, midbrain tectum, cerebellar cortex, and central gray. The fourth group, consisting of the medulla oblongata, gray matter, spinal cord, spinal and sympathetic ganglia and the white matter of the CNS, showed no staining for iron. Light microscopic studies show that iron is deposited as fine granular structures in neutrophils, oligodendrocytes, and nerve cells of the globus pallidus and substantia nigra.

All subsequent studies using a variety of other techniques including quantitative chemical procedures to measure iron in the brain[35–41] have confirmed the results of Spatz[33,34] but the actual amounts reported vary considerably (see Fig. 1). Thus, Hallgren and Sourander[37] reported that the iron content of the globus pallidus, red nucleus, putamen, and substantia nigra is higher than that of the liver, the latter organ being the main site of iron metabolism (Fig. 1). However, the concentration of iron in the brain is about a fifth of that in the liver.

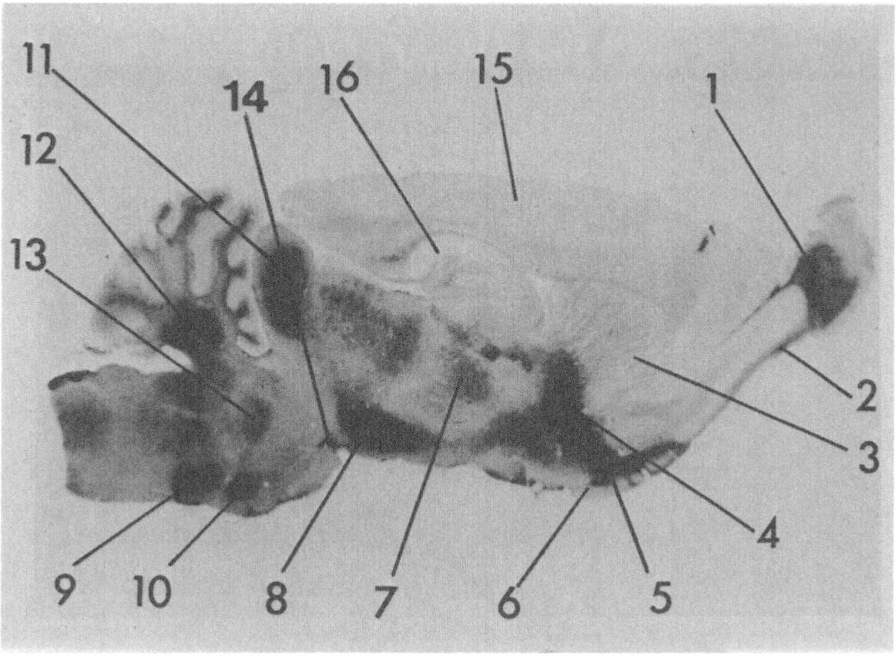

Fig. 2. Histochemical demonstration of distribution of iron in parasagittal view of female rat brain. The dark-stained areas represent accumulation of iron as identified by Perl's diaminobenzidine reaction.[49,137] 1, olfactory bulb; 2, olfactory tract; 3, caudate putamen nucleus; 4, globus pallidus; 5, ventral pallidum; 6, olfactory tubercle; 7, ventral thalamus; 8, substantia nigra; 9, facial nucleus; 10, superior olivary nucleus; 11, inferior colliculus; 12, cerebellar nucleus; 13, spinal trigeminal nucleus; 14, interpeduncular nucleus; 15, cortex; 16, hippocampus.

3.2. *Animal*

There are a few reports of the systematic determination of brain iron by histochemical or quantitative procedures. According to Guizetti[32] and Spatz,[33,34] animal brain tissue contains less iron than human brain. Mouse brain reportedly gave no histochemical iron reaction, but strong positive reactions were found in the globus pallidus and substantia nigra of the monkey, rabbit, cat, and dog. In the monkey, brain iron was visible in the dentate nucleus but not in the red nucleus.[32]

In the rat, the distribution of nonheme iron measured chemically showed large differences between the various regions. The striatum, caudate and hypothalamus had the highest concentrations.[10,42–44] Measurement of total iron in the different brain areas of the rabbit[45] and guinea pig[46] has not revealed such differences. Because of the different methods employed and brain areas studied, a quantitative or qualitative comparison with human brain could not be made.[47,48]

On the face of it, larger differences are seen in human brain than in lower mammals. However, the recent elegant and systematic histochemical study of Hill and Switzer[49] has revealed a parallelism between the distribution of iron

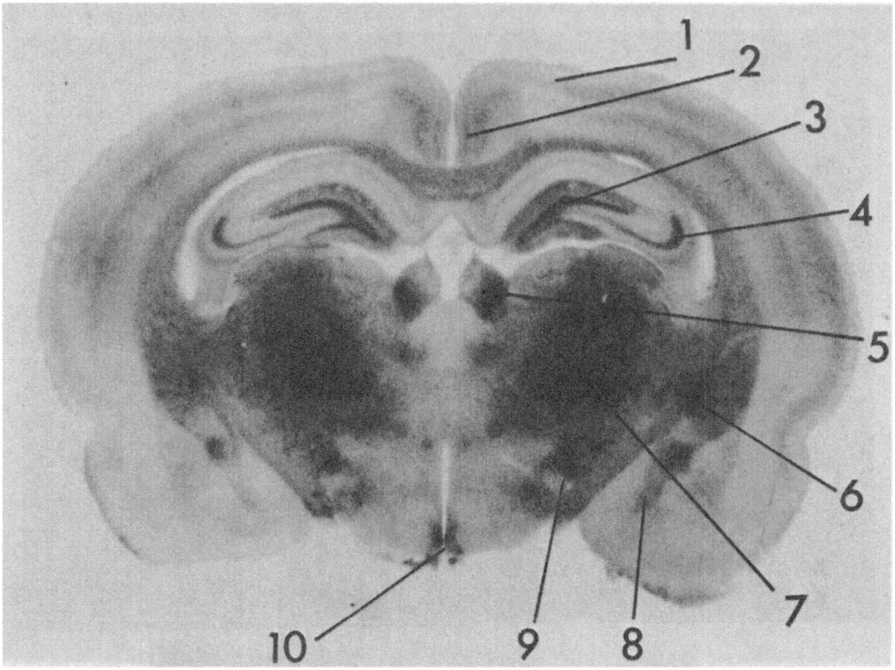

Fig. 3. Distribution of iron in the rat forebrain at the level of the thalamus. Stained areas represent accumulation of iron as identified by Perl's DAB histochemical method.[49] 1, cortex; 2, cingulate cortex; 3, dentate gyrus; 4, hippocampus; 5, lateral habenular nucleus; 6, ventral thalamus; 7, caudate putamen nucleus; 8, central amygdalar nucleus; 9, entopeduncular nucleus; 10, ventromedial hypothalamus.

in the rat and human brain confirming the biochemical determination of Youdim and Green.[10] Perl's DAB histochemical procedure used by Hill and Switzer[49] indicated that in the rat brain, iron is also unevenly distributed similar to the pattern found in the human brain, and is present in different types of cells and structures in these areas. The dark-stained (Perl's DAB reaction) areas in Fig. 2–4 represent accumulation of iron. In the high-iron area (globus pallidus, substantia nigra, ventral pallidum, dentate nucleus, and caudate putamen nucleus), iron occurs in glial cells and nerve fibers; in the circumventricular organs, iron occurs in granules, tanycytes, fiberlike processes, and amorphous accumulations. In areas low in iron, namely the lateral septum, red nucleus, stria terminales, and cortex, iron occurs in boutonlike structures or in the perikarya of neurons. In the supraoptic, paraventricular, and suprachiasmatic nuclei, iron occurs as a fine dusting of grains within the perikarya of cells.

4. SUBCELLULAR DISTRIBUTION OF IRON

Hallgren and Sourander[37] have measured nonheme iron in subcellular fractions of the dog and human brain. The mitochondria and microsomes together

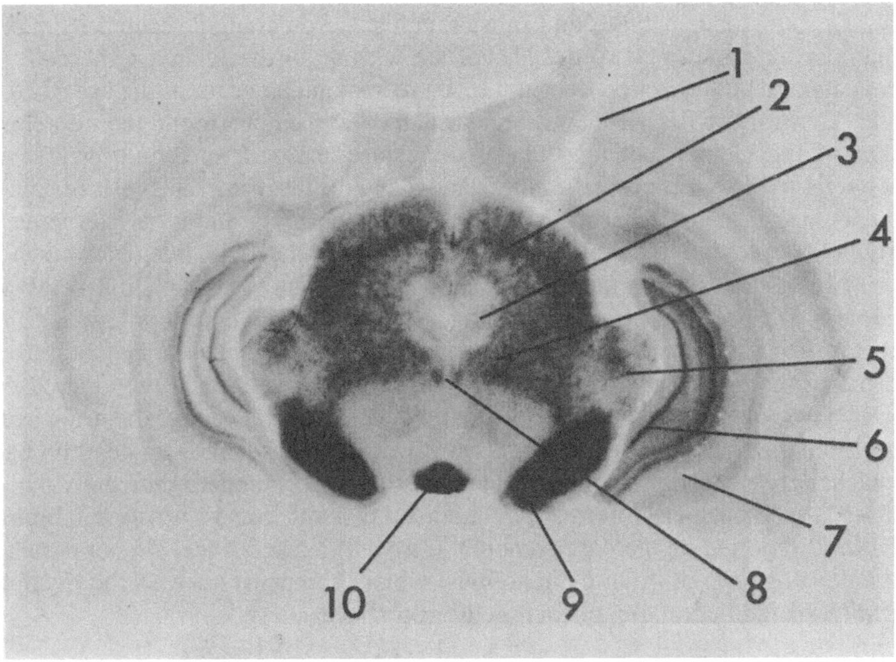

Fig. 4. Distribution of iron in the rat midbrain at the level of the substantia nigra. The dark-stained areas represent iron deposits as identified by Perl's DAB histochemical method.[30,49] 1, cortex; 2, superior colliculus; 3, central gray; 4, reticular formation; 5, medial geniculate nucleus; 6, dentate gyrus; 7, hippocampus; 8, oculomotor nucleus; 9, substantia nigra; 10, interpeduncular nucleus.

contained almost half of the total nonheme iron; about 14% was found in the nuclear fraction, and the remainder in the soluble (myelin) fraction. A similar distribution of honheme iron was found in the rat brain.[10,42–44,50] Using a slightly different separation procedure for the rat brain, Rajan *et al.*[44] measured fractions of most brain areas and found the synaptosomal and the myelin fractions to contain the highest levels of iron. In rat whole brain, about 80% of the iron was in the myelin fraction with the rest divided between mitochondrial and synaptosomal fractions.[44] This pattern of distribution, however, could not be confirmed in brain areas such as the brain stem, striatum, cerebral cortex, hypothalamus, and cerebellum. The presence of large amounts of iron in the myelin fraction may suggest a physiological role for this trace element in the neuronal pathways. It is interesting to note that the postnatal development of uptake of radioactive iron by rat brain coincides with the beginning of myelination.[16] In both cases their accumulation in the brain continues late into adulthood.[37,51,52] Both iron and myelin are very stable and have a very slow rate of turnover.[53,54]

5. *EFFECT OF AGE ON IRON CONTENT*

A number of studies have shown that in human brain, iron increases in the first three decades of life,[32,33,35–38] and may remain unchanged thereafter.

The rate of iron accumulation differs in the various brain regions. In areas[17] such as the globus pallidus and substantia nigra, iron content reaches its maximum level in the first two decades of life with no further increase thereafter; caudate and putamen iron does not reach its maximal level until the age of 50–60.[37] Contrary to these reports, Sundermann and Kempf[38] found the iron content of the globus pallidus, caudate nucleus, and red nucleus to increase throughout life, whereas in the cortex and thalamus the maximal level is reached by the age of 30–40. Except in ID anemia, when brain and liver iron values are both depressed, the brain iron levels do not correlate with liver iron values.

Most studies using rat[47,54,55] or mouse[56] brain have shown similar development of iron concentrations with age. At birth, iron content is low and increases to its maximal level by the age of 8 weeks. In mouse brain, total iron drops in the first 5 days and remains constant for 3 weeks and then increases until 8 weeks.[56] As a whole, the postnatal developmental changes of brain iron are similar in male and female rats. However, in all cases the female brain has a higher concentration than the male[57] and it can accumulate significantly more brain iron than male littermates.[58] Measuring total iron in whole rat brain, Kofod[55] reported an increased amount of iron with age but this did not parallel the development of brain cytochromes, which is concurrent with the myelinization or the fluctuation of total body iron stores.[59]

6. *THE BEHAVIORAL AND BRAIN BIOCHEMICAL CHANGES IN IRON DEFICIENCY*

6.1. *Human Studies*

ID in man is the most prevalent nutritional disorder and anemia in the world, a condition unrelated to the socioeconomic status of the subject.[6,60–64] In addition to the well-known extrahematological manifestations, the suggestion has been made that ID causes behavioral changes among infants, children, and adults[65–70] that are reversible by iron therapy.[7,9,65–67] The connection between ID and the behavioral disturbances had often been implied but until recently the evidence was largely circumstantial. Even so, the previous clinical studies had provided evidence that the behavioral changes include unusual lethargy, irritability, apathy, listlessness, fatigue, lack of ability to concentrate, pagophagia and pica (pathological craving for ice and pica), inattention, hypoactivity, and decreased IQ.[65–73] More recently, there have been a number of well-controlled studies[65–67] that largely confirm the behavioral deficit of ID subjects indicated by the earlier studies of Oski and others.[67,68,71–74] Thus, Pollitt *et al.*[66] reported that children (3–6 years old) with a hemoglobin level greater than 10.5 g/100 ml take more trials to reach a learning criterion in three discrimination-learning tasks than those children with a hemoglobin level greater than 11.5 g/100 ml. In oddity learning tasks, the number of correct responses among ID children was significantly lower than the control group of children. The children with ID showed relative performance deficits in simple tasks. All differences between the two groups were absent after iron repletion therapy.[66] In an extremely socioeconomically homogeneous population of children in which ID anemia was the sole factor distinguishing the anemic group from the control, it was found that anemic infants tend to be less active, persistent, responsive, and reactive, but were more tense and fearful than the nonanemic infants.[65] The initial mean Mental Development Index (MDI) and Psychomotor Development Index (PDI) were significantly lower in the anemic group as compared to the control group. These deficits were most marked in infants 19 to 24 months old.[65] Although the results of previous studies have been challenged for serious methodological shortcomings, these findings suggest that the motivation to persist in intellectually challenging tasks may be lowered, attention span shortened and overall intellectual performance diminished in ID children.[65] Thus, ID has an adverse effect on cognition, and this is reversible following iron therapy in some cases.

It is most likely that behavioral abnormalities as identified in subjects with ID are related to some underlying biochemical defect in CNS function. However, very few studies have been carried out to elucidate the biochemical changes associated with this CNS function. The importance of iron in protein synthesis, as a cofactor for many heme and nonheme enzymes, and its structural role in many membrane- and non-membrane-bound proteins are well documented (for reviews see 2,3,74).

Iron has been implicated as a cofactor for tryptophan hydroxylase and tyrosine hydroxylase, the respective rate-limiting enzymes in the biosynthesis of serotonin (5-HT) and catecholamines [dopamine (DA) and norepinephrine (NE)].[75–77] The enzyme monoamine oxidase (MAO), which is responsible for the inactivation of the above monoamine neurotransmitters, has been shown to be functionally and qualitatively lowered in platelets and probably in other tissues of subjects with ID anemia.[10,42,78–80] This may explain the increased urinary excretion of NE in ID.[81] Treatment with iron restores the activity of MAO in platelets and decreases urinary NE when serum iron concentrations have returned to normal.[80,81]

Given the putative function of various biogenic monoamines in neurotransmission in the brain and their involvement in various behaviors, psychiatric disorders, and neurological diseases, the study of the metabolism and functional activity of 5-HT, and catecholamines is essential for the delineation of any behavioral abnormalities that might occur in ID[82] or iron overload.

There are few studies on the functional activity of monoamine neurotransmitters in human subjects with ID. Those that are available have been limited to the study of platelets since this organelle has been used as a limited model of the 5-HT neuron.[82] Its 5-HT uptake system has a close similarity to that found in the brain, and the receptors from ID children and adults show abnormal platelet 5-HT-induced aggregation as well as reduced 5-HT uptake.[84,85] The alteration in platelet aggregation responses in ID patients is due to a platelet defect and not the plasma in which it is suspended.[84] Successful treatment with iron(II) sulfate restores the 5-HT-induced aggregation responses.[84] In this respect it is interesting to note that Kaladhar and Narasingsa Rao[86] reported recently a diminution of 5-HT uptake by brain synaptosomal preparations from nutritionally ID rats. Whether the same phenomena may occur in human subjects is not known since no one has examined the metabolism or urinary excretion of 5-HT in ID patients.

6.2. Animal Studies

6.2.1. Brain Biochemical Changes

Animal behavioral models have been widely used to investigate changes in CNS monoamine neurotransmitter function and one might expect that any behavioral changes induced by ID would prove amenable to investigation with such models. However, until recently there were few studies on the effect of ID on the behavior of experimental animals that dissociated the peripheral effects (e.g., muscle working capacity and availability of oxygen) from those initiated by the CNS. Those that did were confined to studies of maze-learning, treadmill activity, and behavioral attitudes toward food; moreover, none of these studies investigated the role of either a possible change in iron storage in the brain or changes in the concentrations and functions of central neurotransmitter substances (see 7,9,87–89).

The interest in the role of iron in theCNS has been growing in the past few years. Dallman *et al.*[57] first reported that a brief period of severe ID in young rats (1–21 days old) resulted in a 30–40% deficit of brain iron and that after 45 days of iron therapy, the brain nonheme iron remained 15% below that of control rats. These data have been confirmed in recent studies by Youdim and Ben-Shachar.[154] Dallman *et al.*[57] did not examine the behavior of these rats prior to or after iron supplementation. These latter results differ from those of Youdim and Green[10,42] who reported that the diminished iron in brains of adult ID rats could be restored within 1 week of iron repletion when ferrous sulfate was introduced in the diet. These two sets of results may not be contradictory since the reversal of brain iron content may be an age-dependent phenomenon. Thus, it is possible that if ID occurs at an earlier age, there may be a greater, irreversible damage in the development of those neuronal and nonneuronal cells of the brain with which iron is associated. Nevertheless, further systematic studies are required. However, it is worth remembering that ID and its induction of behavioral changes are more prominent in children than in adults.[65,66]

The first systematic studies of the effect of ID on brain biochemical changes at the level of neurotransmitter metabolism and monoamine neurotransmitter-induced behavior were performed by Youdim and Green.[10,42] These authors' premise was that, since the monoamine neurotransmitter enzymes are iron dependent (as discussed earlier), ID could alter their activities via the lowered brain iron, thus affecting the metabolism of the neurotransmitter and the induction of behavior. The nutritional lack of iron, even though brain nonheme iron content was significantly lowered by 40%, was without effect on brain MAO and tyrosine or tryptophan hydroxylases, succinate dehydrogenase, and aldehyde dehydrogenase activities *in vitro* or *in vivo*.[10,42] Neither was a change in brain content or turnover of NE, DA, and 5-HT seen. These results are in marked contrast to those reported recently by Mackler *et al.*,[90,91] who reported a 35% reduction in brain aldehyde oxidase activity, a 10% increase in brain 5-HT, and a 5% decrease in brain 5-hydroxyindoleacetic acid in ID rats. The results reported by Mackler *et al.*[90] are very difficult to explain since aldehyde oxidase is not considered to be an iron-dependent enzyme and, furthermore, this enzyme, unlike MAO, does not regulate brain cytoplasmic 5-HT content.[92] In contrast, Youdim and Green[42] reported a slight but significant reduction of endogenous 5-HT in ID rats, without a change in 5-HT turnover rate or in brain tryptophan. The latter findings are supported by the studies of Tamir *et al.*,[93] which show that in comparison to other metals, Fe^{2+} selectively affects the binding of 5-HT to a partially purified serotonin-binding protein (SBP) isolated from rat brain synaptosomes and human platelets, thus suggesting a role for iron 5-HT storage in the brain.[93–95] Further evidence for the role of iron in 5-HT storage in rat brain comes from studies on uptake of 5-HT by synaptosomal preparations from control and ID rats. 5-HT uptake by brain synaptic vesicles prepared from brain of the latter group of animals shows a significant reduction as compared to control values.[86] Parallel changes in hemoglobin and brain nonheme iron levels during ID were also observed, although changes in brain nonheme iron concentration occur more slowly and to a lesser degree. In agreement with Dallman *et al.*,[57] the brain nonheme iron concentrations do not return to control values even after 4 weeks of iron repletion, although 5-HT uptake does.[86]

The reports that brain NE is not changed in ID rats[10,47] are not matched by the increased plasma and urinary NE.[96,97] This is understandable since brain NE makes a very little contribution to the total excretion of this neurotransmitter in the urine. Thus, the increased urinary and plasma NE[96,97] may originate from peripheral adrenergic innervation in which MAO activity is lowered due to ID. Evidence to support this hypothesis is the observation that in contrast to brain, the peripheral tissue (liver, heart, adrenal glands, and spleen) MAO activity is significantly reduced.[10,42] To some extent these results confirm what has been observed in infants and adults with ID anemia.[80,81,98]

6.2.2. Brain Iron and DA Sub- and Supersensitivity

6.2.2a. Behavioral Consequences of ID and DA Receptor Subsensitivity. The behavioral manifestations of ID anemia are becoming increasingly more apparent. However, many of the animal experimental and human studies have flaws in design. Frequently, there is no information on statistical analyses, methods for testing, or definition of tasks, and groups are often not matched. Furthermore, it is often unclear whether the results obtained are solely due to ID and not also to malnutrition.[7,99] Strictly speaking, when one considers behavioral changes, it is envisaged that they have their origin in the CNS. Even so, most behavioral studies in ID have neglected this aspect and have concentrated on maximum work capacity tests on a treadmill in humans[100–104] and rats.[105–109] ID anemia has also been shown to result in elevated heart rate and blood lactate at a given work load and reduced maximum oxygen uptake (for review see 104) and work capacity.[106] The relationship between ID anemia and physical activity thus appears to be firmly established. However, the results

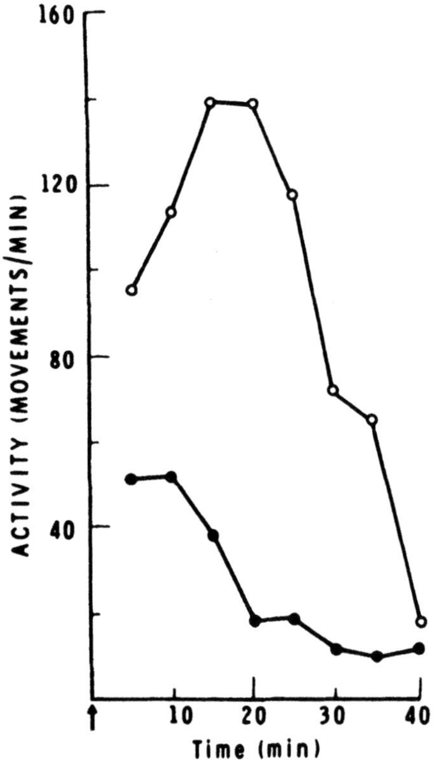

Fig. 5. The behavioral response of iron-deficient (●—●) and control (○—○) rats to apomorphine. Rats were injected with 2 mg/kg of apomorphine and hyperactivity was measured as movements/min.[11,42,112]

from these experiments cannot be considered as behavioral changes since it has been shown that they result from peripheral muscle dysfunction due to reduced muscle α-glycerophosphate dehydrogenase activity, an iron-containing enzyme.[99,104,106]

Therefore, plausible animal behavioral models that indicate a neural dysfunction would be valuable for validating neurotransmitter function in ID. Youdim and Green[10,42] were the first to examine the functional activity of DA, NE, and 5-HT in ID rats in relation to behavioral syndromes for the possible presynaptic and postsynaptic changes. Although ID did not alter the activities of the brain monoamine-metabolizing enzymes *in vitro* or *in vivo* or modify the brain levels of the above neurotransmitters, the behavioral syndromes produced by increasing brain synthesis of 5-HT (via treatment of rats with either L-tryptophan or 5-hydroxytryptophan plus the irreversible MAO inhibitor tranylcypromine) and DA (L-dihydroxphenylalanine plus tranylcypromine) were significantly diminished. The behavioral responses produced by the suggested 5-HT agonist 5-methoxy-*N*′,*N*-dimethyltryptamine, the DA agonist apomorphine, and the DA-releasing drug amphetamine were also diminished (Figs. 5 and 6). The normal behavioral responses to the above drugs were restored when rats were fed an iron-supplemented diet for 8 days. It was suggested that the most likely explanation for the diminution of behavioral responses was that it is the result of altered postsynaptic response to the released neurotransmitters

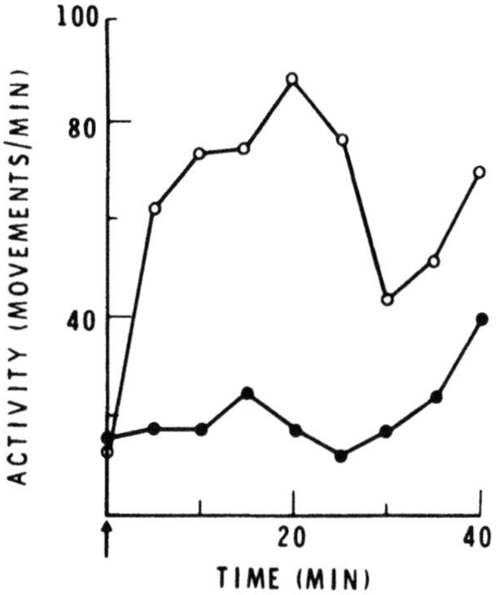

Fig. 6. The behavioral response of iron-deficient (●—●) and control (○—○) rats to 5-methoxy-*N*,*N*-dimethyltryptamine (5-MeODMT), a serotonin agonist. Rats were injected with 5-MeODMT (2 mg/kg) 30 min after administration of the irreversible monoamine oxidase inhibitor tranylcypromine (20 mg/kg) and hyperactivity was recorded as movements/min.[11,42,110,111]

and not due to the inhibiton of release of 5-HT or DA into the synaptic cleft. Because the behavioral changes that were observed after the administration of various postsynaptic-acting drugs (5-methoxy-*N*,*N*-dimethyltrptamine or apomorphine) were diminished in ID animals, Youdim and Green[10] could not rule out either that the DA receptors are modified by some conformational changes or that the lack of iron diminishes the rate of receptor synthesis. It is well known that in the production of the 5-HT-induced behaviors, a dopaminergic system is involved, since α-methyl-*p*-tyrosine-treated rats do not exhibit 5-HT-induced behavior.[110,111] The examination of DA-sensitive adenylate cyclase activity, part of the presumed DA D_1 receptor in the caudate nucleus of both ID and control rats, showed no difference between the two groups. This finding suggested that either the abnormality is distal to the supposed postsynaptic receptor or that DA-sensitive adenylate cyclase was not that receptor.[10,42]

In contrast, but complementing the above results, it has recently been shown that ID in rats results in reduced brain iron content, DA D_2 receptors (the binding site using the radioligand [^{3}H] spiroperidol) (Table I), and DA-dependent behaviors and functions, which can be restored to normal if the animals are placed on an iron-supplemented diet for 8–14 days.[10,42,112–114] This would suggest either that iron is directly involved in the biochemical and functional aspect of D_2 receptors or that iron acts via other chemical processes (e.g., hormones) involved in maintaining a proper functioning or synthesis of these receptors. This contention is supported by results obtained from animals made hemolytically anemic with phenylhydrazine.[11] These animals have extremely low hemoglobin (<7.0 g/dl) but have normal serum iron, [^{3}H] spiroperidol-binding sites in the caudate nucleus, and apomorphine-induced behaviors.[11] Therefore, the suggestion by Leibel *et al.*[99] that the diminished drug-induced behavioral responses in ID rats are due to a peripheral effect of ID

Table I
The Effect of Iron Deficiency (ID) in Rats on Neurotransmitter-Binding Sites of Brain[a]

Neurotransmitter	Control	ID
Dopamine (D_2), [^{3}H]spiroperidol		
K_D	0.75 ± 0.11	6.75 ± 0.4
B_{max}	46.2 ± 7.5	18.6 ± 2.0
α-Adrenoreceptor, [^{3}H]-WB4101		
K_D	0.34 ± 0.02	0.37 ± 0.02
B_{max}	8.89 ± 0.46	9.02 ± 0.31
β-Adrenoreceptor, [^{3}H]dihydroaloprenolol		
K_D	0.89 ± 0.07	0.99 ± 0.04
B_{max}	6.81 ± 0.58	7.89 ± 0.16
Acetylcholine (muscarinic), [^{3}H]quinuclidinyl benzylate		
K_D	0.24 ± 0.04	0.21 ± 0.05
B_{max}	82.1 ± 13.8	74.8 ± 20
Serotonin, [^{3}H]serotonin		
K_D	7.1 ± 0.06	6.54 ± 1.61
B_{max}	55.39 ± 13.06	50.86 ± 15.75
Cortex	420	42

[a] K_D and B_{max} of receptors (binding sites) were calculated from Scatchard plot analysis. K_D and B_{max} are expressed as nM and pmol of the labeled ligand bound per g wet weight of brain regions. DA receptor (caudate nucleus); α- and β-adrenoreceptors (cortex); muscarinic receptor (hippocampus); and serotonin receptor (cortex). Data from Youdim *et al.*[11] and Ashkenazi *et al.*[121]

anemia can be discounted. Although this theory might be partially true for the diminished forced behavior seen in locomotor activity studies, it could not explain the inhibition of *d*-amphetamine-induced hypothermia (Fig.7), or the increase in the barbiturate sleeping time in ID rats.[12,114–116] The latter phenomena are dependent on intact dopaminergic systems.[12,117] DA receptor-blocking drugs, neuroleptics, not only inhibit *d*-amphetamine-induced hypothermia but also increase the barbiturate sleeping time in rats.[117,118] As far as the *d*-amphetamine-induced hypothermia is concerned, Yehuda[118] using lesioned animals has shown it to be the feature of dopaminergic neurons in the nucleus accumbens. Thus, it can be assumed that the reduction of the behavioral responses and *d*-amphetamine-induced hypothermia of ID rats to centrally acting drugs (apomorphine and *d*-amphetamine) is due to a central defect at the level of diminished (down-regulation) D_2 receptor number.[11,12,112–114,117,118]

Further evidence that lowered brain iron due to ID causes the down-regulation of dopaminergic activity (D_2 receptor), also seen in the pituitary, comes from recent studies on the regulation of prolactin receptors in the liver and sex organs of male and female rats. It has been demonstrated that chronic treatment of rats with ovine prolactin can up-regulate prolactin-binding sites in the liver and lung.[119,120] Chronic treatment with neuroleptics (e.g., haloperidol and fluphenazine), which block the D_2 receptor and cause a significant rise in serum prolactin, markedly induces the liver prolactin-binding sites as measured by the binding of ^{125}I-labeled ovine prolactin.[113,139] Similar increases (500%) in

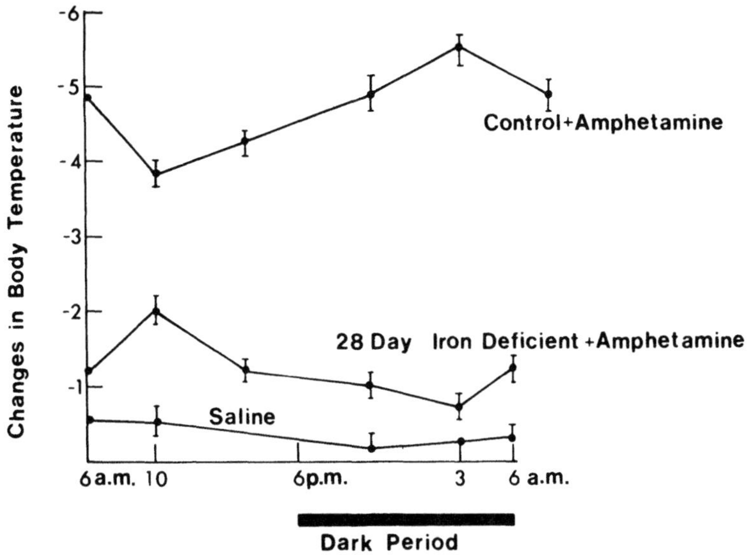

Fig. 7. *d*-Amphetamine-induced hypothermia in iron-deficient and control rats. Colonic temperatures (°C) were recorded using a telethermometer after the rats had received *d*-amphetamine (15 mg/kg) or saline and placed in a chamber at 4°C with humidity of 40%. Note that the colonic temperatures show a circadian rhythm in both control and iron-deficient rats but that of iron-deficient rats is reversed. Furthermore, the hypothermic effect of *d*-amphetamine is significantly diminished in iron-deficient rats.[114] The latter results parallel what is observed in control rats pretreated with neuroleptics (e.g., haloperidol).[117]

the ^{125}I-labeled ovine prolactin-binding site of livers and serum prolactin of ID rats were observed.[113]

So far, in all aspects examined, locomotor and stereotype behavioral responses involving dopaminergic activity, thermoregulatory response to *d*-amphetamine, barbiturate sleeping time, the induction of prolactin receptors in the liver, and increased serum prolactin, the ID rat resembles a neuroleptic-treated animal, with diminished dopaminergic activity.

The mechanism by which nutritional iron maintains D_2 receptor function is not well understood. ID may result in some conformational changes in the receptor that would prevent the binding of other DA or [^{3}H]spiroperidol. This suggestion seems unlikely since the affinity of [^{3}H]spiroperidol for its binding site is unchanged in the caudate nucleus of ID rats. It is possible that iron may be part of the receptor to which the agonists or antagonists bind or that iron is important for the processes that attach the receptor to the postsynaptic neuron. Finally, iron could be concerned with the synthesis of the receptor and this may be one reason for the slow rate of reappearance of D_2 receptors after the ID rats were placed on an iron-supplemented diet for 8–14 days. There is very good evidence that iron plays an important role in the synthesis of certain proteins,[2,87] and more recently Weisenberg *et al.*[87] reported that ID as well as copper deficiency in rats results in the impairment of protein synthesis as measured by incorporation of various ^{14}C-labeled amino acids into the protein of brain and liver. They further demonstrated that the diminution of brain protein synthesis due to copper deficiency was the result of ID in these animals and that it could be corrected by iron supplementation. Copper-deficient rats have a marked decrease of iron in the serum and liver and there is a depletion of hepatic ferritin.[87] Considering that hemolytic anemia, induced in rats by chronic phenylhydrazine treatment, is without any effect on DA receptor or DA-dependent behaviors, it is interesting to note that Weisenberg *et al.*[87] reported brain and liver protein synthesis to be normal in this condition. Thus, it cannot be ruled out that the synthesis of D_2 receptors may require an adequate supply of iron.

The selectivity of the down-regulation of D_2 receptor B_{max} in ID rats has been examined by measuring the B_{max} and K_D of α- and β-adrenoreceptors, cholinergic muscarinic, benzodiazepine, γ-aminobutyric acid (GABA), and 5-HT-binding sites. The B_{max} and K_D of the latter receptors were unchanged in ID rats.[11,121] Thus, in this respect the effect of ID is specific for D_2 receptors (Table I).

Tucker and Sandstead[122,123] have reported profound EEG changes in human subjects with different iron status. Considering that iron is unevenly distributed in the brain and high concentrations are found in the globus pallidus and DA-rich brain areas, it can be speculated that iron may be particularly important to these dopaminergic processes and functions. Thus, given the reduction in brain D_2 receptor-binding sites in ID[11,121] and the essential role of adequate DA systems to attentional processes,[124–126] it is more than possible that the DA system may be a route mediating the effect of ID on altered cognition and attentional modulatory systems in human subjects. Tucker[127] has interpreted evidence of asymmetrical neurophysiological function in schizophrenia and amphetamine psychoses as an indication that DA processes may be particularly important to the activation and cognition operation of the left hemisphere in humans. In the studies of Tucker and Sandstead,[122,123] both ERP (even-related potentials) and spectral EEG measures were in the direction of greater left hemisphere activation and cognitive processes associated with higher iron status, a finding that would be expected if poor iron status impaired the dopaminergic system that was, in turn, especially critical to left hemisphere arousal.[123]

Because of the limitations of behavioral studies in human subjects, Weinberg *et al.*[58,128,129] have begun to develop animal models in which environmental and nutritional variables can be more carefully examined during early development of ID in the rat, reflecting the behavioral and functional consequences of early ID reported in children. A period of severe early ID (birth to 28 days of age) produced a persistent reduction of brain nonheme iron in adulthood when they were fed an iron-rich diet. ID rats were less responsive than controls in a mildly aversive novel situation (openfield) and ambulated less in an exploratory task (the holeboard). ID rats also exhibited longer reentry latencies.[58,128] Although the plasma levels of corticosterone were higher in rats as compared to controls, the ID rats exhibited a smaller stress increment when exposed to combined stress of ether and cardiac puncture. Thus, it would appear that ID may reduce the animals' general responsiveness to environmental stimuli.[128–130] These results may well complement the findings of Youdim and co-workers,[11,42,112–114,121] in that there is a significant diminution of behavioral response in ID rats to drugs that act pre- and postsynaptically, due to down-regulation of dopaminergic activity. Youdim *et al.*[131] have suggested that in many respects the ID rat resembles a neuroleptic (e.g., haloperidol)-treated animal. Thus, in both above conditions, (1) the behavioral response to presynaptic (amphetamine) and postsynaptic (e.g., apomorphine or 5-methoxy-*N'*,*N*-dimethyltryptamine) acting drugs are significantly diminished; (2) the amphetamine-induced hypothermia is blocked; (3) sleeping time to phenobarbitone is significantly increased; (4) plasma prolactin is elevated; and (5) the prolactin receptor-binding sites (B_{max}) in the liver are increased. However, unlike the neuroleptic-treated animals, ID rats do not exhibit catatonia.

In addition, the data clearly indicate that the long-term consequences of early mild and severe ID do persist after the iron status of young (1–28 days old) rats has been restored to normal (rehabilitation) by feeding the animals an iron-supplemented diet.[58,128,129] In contrast, older animals (28–80 days old) made ID and then rehabilitated show all the normal behavioral responses as compared to their matched controls.[11,42,112–114,121] Thus, from animal studies it would appear that early ID may have a permanent effect on the neuronal pathways and neurotransmitter functions that govern behavior. Support for this contention has come from studies[154] where young (10 days old) and adult (48 days old) rats were made ID and then rehabilitated with iron supplementation. In both groups, nonheme brain iron was diminished by 30–40%, as previously described by Dallman *et al.*[57] and Weinberg *et al.*[58] Furthermore, there was a significant diminution of D_2 receptor number ([^{3}H]spiroperidol-binding sites) in the caudate nucleua as well as diminished behavioral response to apomorphine. Adult ID rats fed an iron-supplemented diet for 8 days had all three parameters restored to values found in control animals. However, the young rats never recovered from the effect of early ID even after 5–7 weeks of an iron-supplemented diet.[154] To date, follow-up studies have not been carried out in children with early ID anemia to determine the long-term developmental effects of early ID. Findings in animals indicating long-term irreversible changes in arousal or responsiveness due to early ID in young (1–28 days old) rats[58] may be explained by the changes observed in DA neurotransmission.

6.2.2b. Effect of ID on Circadian Rhythms. One of the most interesting and exciting aspects of nutritional ID in rats is its effect on circadian rhythms. Glover and Jacobs[132] showed that rats made ID and kept in a 12-hr-light–12-hr-dark cycle have lower 24-hr spontaneous activity than control rats fed the same diet with added iron. They further reported that, in contrast to control rats, the ID anmials were more active in the light period as compared to the dark period, i.e., there was a reversal of the 24-hr circadian rhythm (Fig. 8). Youdim *et al.*[114,133] have confirmed these results in ID rats and extended their studies to the examination of other behaviors.[133] Similar changes were also noted in stereotypic behaviors (head movement, circling, and biting) and motor activity produced by apomorphine (2.5 mg/kg) and *d*-amphetamine (2.5 mg/kg), respectively. As mentioned earlier, *d*-amphetamine (15 mg/kg) causes hypoth-

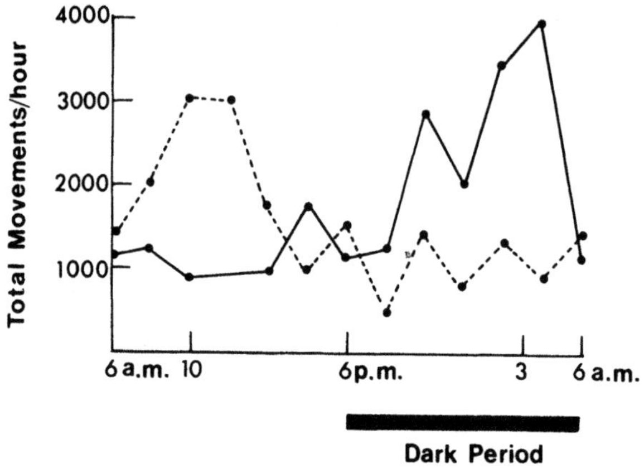

Fig. 8. Twenty-four-hour spontaneous motor activity of control (●—●) and ID rats (●----●) recorded on an activity meter.[114]

ermia in control rats kept at 4°C,[11] and the largest and smallest effects occur during the dark and light period, respectively. This action of *d*-amphetamine is thought to be mediated by mesolimbic DA pathways.[118] ID rats exhibit a markedly diminished *d*-amphetamine response, i.e., a greater hypothermic effect is observed in the light as compared to the dark period[114] (Fig. 7). It is important to note that, in this respect, the effect of ID is similar to the action of DA-blocking neuroleptics (e.g., pimozide and haloperiodol). Thus, it appears that ID is the most effective method available for reversing the circadian rhythm of the rat. The biochemical lesion responsible for it is not known but it may be related to the observed down-regulation of the D_2 receptor.[12,114,121] The normal circadian rhythms can be reestablished if ID adult animals are placed on the ID diet supplemented with ferrous sulfate for at least 8 days.[114]

6.2.2c. Increased Brain Iron and DA Receptor Supersensitivity. It is well documented that DA D_2 receptors increase in density, as measured by the binding of [^{3}H]sprioperidol in the striatum, after long-term neuroleptic administration.[155] The consequence of this phenomenon is an increased behavioral response to the dopamine agonist, apomorphine. These animal behavioral and biochemical changes have been cited as an explanation for neuroleptic-induced DA supersensitivity and tardive dyskinesia.[155] No adequate biochemical evidence has been provided for these phenomena. The possibility that neuroleptics may bring about these changes by affecting (increasing) brain iron has been recently investigated. ID in rats failed to abolish the increase in [^{3}H]spiroperidol binding in the caudate nucleus or the increase in behavioral response to apomorphine produced by chronic haloperidol treatment.[113] In normal circumstances, it is extremely difficult to increase brain levels of iron by nutritional or other means. It is more than possible that haloperidol and other neuroleptics (namely phenothiazines) may affect iron turnover in the brain, since chronic administration of chlorpromazine, which also causes DA supersensitivity, increases iron concentrations in the caudate nucleus of the guinea pig.[46] Thus, an excess of brain iron may prevent the decrease in D_2 receptor-binding sites caused by ID.[113] Being highly lipophilic, the neuroleptics may be able to mobilize iron stores from the periphery to the CNS, since liver nonheme iron is significantly diminished in ID as well as normal (control) animals treated chronically with haloperidol, or prevent iron efflux from the brain due to their strong iron-chelating property.[156] Phenothiazines chelate iron more avidly than any other metal.[156]

Further evidence that brain iron metabolism may be the one underlying cause of DA supersensitivity (increased [^{3}H]spiroperidol-binding sites) has come from the studies of Csernansky *et al.*,[157] who injected $FeCl_3$ into the left amygdaloid nucleus of the rat. Subsequent repeated apomorphine challenges in these animals produced a steady development of behavioral DA supersensitivity, well above the saline-injected controls, reaching a maximal effect on day 23. The behavioral supersensitivity observed in $FeCl_3$-injected animals was accompanied by increases in [^{3}H]spiroperidol binding in the contralateral amygdala, striatum, and nucleus accumbens. In contrast, saline-injected controls did not show dopaminergic supersensitivity.[157] When the anterior commissive was severed in groups of $FeCl_3$-injected animals, the apomorphine-induced behavior as well as [^{3}H]spiroperidol binding were similar to saline-injected animals, suggesting that the transport of iron from the amygdala and its possible accumulation in other brain areas may be the underlying cause of DA supersensitivity. If a change in dopaminergic function underlies schizophrenic systems and affective disorders, the models of ID anemia in rats as well as the above model may have relevance in understanding these diseases. These data may also explain the unique distribution of iron in the brain, being high in DA-rich regions, and why it is important to maintain the brain level of iron constant.

7. *NEUROPATHOLOGICAL CONDITIONS AND IRON STATUS OF BRAIN*

In normal circumstances it is extremely difficult to increase brain levels of iron. However, a number of reports describe increased brain iron associated with certain neurological disorders (see 47,134,135). Haller–Vorden–Spatz disease,[33,134,136,137] which is a progressive degenerative disorder involving the basal ganglia and is manifested by various hyperkinetic states of rigidity and progressive dementia, has been associated with increased brain level of iron. Although general iron metabolism is normal, there is a slow turnover of brain iron. A three-fold increase of iron was found in the globus pallidus.[138] Szanto and Gallyas[140] administered ^{59}Fe to patients with this syndrome and reported normal iron metabolism. However, Vakili *et al.*,[138] using the same techniques, showed increased uptake of ^{59}Fe in the region of the basal ganglia. Hematochromatosis is a rare disorder of iron metabolism characterized by excessive deposits of iron in the body including the brain.[141,142] Accumulation of ferrogenous material was observed in the choroid plexus, olfactory bulb, pineal, posterior pituitary, and infundibulum.[143] In Huntington's chorea, increased positive staining for iron, particularly in the caudate and globus pallidus, was reported.[24,144–146] Serum iron and its turnover in these patients were not elevated, so that the origin of the observed increase is unknown. In the case of Parkinson's disease, there is disagreement as to whether brain iron changes. Larger than normal amounts of iron were found by Earle.[145] X-ray fluorescence spectroscopy of formalin-fixed brain tissue from Parkinsonian patients revealed that iron was consistently increased by a factor of two or more above control values. In surgical specimens removed at the time of ventrolateral thalomotomy in Parkinsonian patients, abnormally high deposits of iron were found in the glial cells as well as in nerve fibers.[135] These results are in sharp contrast to those reported by Riederer *et al.*[147] The latter investigators, using atomic absorption, measured six trace elements, including iron, in 23 brain areas of Parkinsonian and control subjects and could not show any differences between them. However, they did confirm the original data of Spatz[33,34] with regard to the distribution of iron in the brain. The procedures used for the detection of iron may account for the contradiction. Increased iron in the basal ganglia, globus pallidus, and caudate nucleus has been reported for Kaschin–Bech's disease, a disease caused by increased dietary iron,[148] dementia paralytica,[34,35] and psychotic (schizophrenic) patients.[34,47,149] The epileptic discharges occurring after head injury and hemorrhagic infarction have been traced to the deposition of iron in the neurophil.[150,151] Neuropathological studies of human posttraumatic epileptic foci[15,152] have shown hemosiderosis as well as leptomeningeal fibrosis, neuronal changes, and astrocytic gliosis. Hemorrhagic cotricl infarction and intracerebral hemorrhage cause an increased incidence of early and late seizures. These observations suggest that the presence of metal compounds in blood may be important in epileptogenesis. Iontophoresis of ferrous and ferric chloride results in epileptiform discharges. The damaging effects of iron are thought to be due to its catalytic role in lipid peroxidation. The peroxidation of lipids is a chain reaction caused by the hydroxyl radical rearranging lipid bonds and destroying membranes. The superoxide radical (O_2^-) formed from the reaction of ferrous iron (Fe^{2+}) and molecular oxygen (O_2) reacts with H_2O_2 to produce the very reactive hydroxyl radical (see 153 for review). Although there are many *in vivo* enzymatic processes to prevent lipid peroxidation, neurological disorders could result from interaction of iron and oxygen leading to the disturbance of membrane structure and function (see 135). Thus, excess brain iron could be one process by which certain neurological disorders can be elicited.

8. CONCLUSION

The uneven distribution of iron in brain and particularly its high concentrations in extrapyramidal areas would suggest a function for this trace metal in the CNS. The increasing evidence that ID in human subjects and animals produces behavioral abnormalities supports this suggestion. Until very recently, few behavioral studies were performed on ID animals that dissociated the peripheral (muscle function) effects from those of the CNS. The studies of Youdim *et al.*[10,11,114,133,154] and Weinberg *et al.*[58,128,129] clearly indicate that, besides peripheral muscle dysfunction as reported by Edgerton *et al.*,[104–106] there are CNS-mediated behavioral changes. These include diminution of apomorphine-induced behavior, reversal and reduction of DA-dependent circadian rhythms, increased sleeping time to phenobarbitone (exclusive of the increased drug-metabolizing features),[116] and decreased performance in exploratory and learning tasks. Furthermore, if ID is induced in the first 3 weeks of postnatal development, many of these behavioral features remain once the animals have been rehabilitated (their tissue iron states returned to normal).[128,129] However, brain iron remains significantly low[57,154] as does D_2 receptor number.[154] One important biochemical feature of ID is the selective down-regulation of the D_2 receptor. In the light of the latter findings and involvemen of DA in many of the above behaviors and learning processes, abnormalities as reported in ID children and adults may be a manifestation of the receptor function changes. However, further studies are required to establish at what level brain iron maintains the normal functions of DA receptors[12] since injection of $FeCl_3$ into the brain induces DA receptor supersensitivity.[157]

ACKNOWLEDGMENT. The author is grateful to Dr. Joana Hill for her kind permission to reproduce Figures 1–3.

REFERENCES

1. Widdowson, E. M., 1969, *Mineral Metabolism in Paediatrics* (D. Barltrop and W. L. Burland, eds.), Adland, Dorking, p. 87.
2. Beutter, E., and Fairbanks, V. F., 1981, *Iron in Biochemistry and Medicine II* (A. Jacobs and M. Worwood, eds.), Academic Press, New York, pp. 394–425.
3. Jacobs, A., and Worwood, M. (ed.), 1974, *Iron in Biochemistry and Medicine,* Academic Press, New York.
4. Frieden, E., 1974, *Protein–Metal Interactions* (M. Friedman, ed.), Plenum Press, New York, pp. 1–32.
5. Yasunobu, K. T., Mower, H. F., and Hayashi, O., 1975, *Iron and Copper Proteins,* Plenum Press, New York.
6. Garby, L., 1973, *Clinics in Haematology* (S. T. Callender, ed.), Saunders, Philadelphia, pp. 245–257.
7. Pollitt, E., and Leibel, R. L., 1976, *J. Pediatr.* **88**:372–381.
8. Youdim, M. B. H., and Iancu, T. C., 1977, *Br. J. Haematol.* **36**:298–299.
9. Pollitt, E., and Leibel, R. L. (eds.), 1982, *Iron Deficiency, Brain Biochemistry and Behavior,* Raven Press, New York.
10. Youdim, M. B. H., and Green, A. R., 1977, *Iron Metabolism* (R. Porter and W. Fitzsimons, eds.), Elsevier, Amsterdam, pp. 201–226.
11. Youdim, M. B. H., Yehuda, S., Ben-Shachar, D., and Ashkenazi, R., 1982, *Iron Deficiency, Brain Biochemistry and Behavior* (E. Pollitt and R. L. Leibel, eds.), Raven Press, New York, pp. 39–57.
12. Ashkenazi, R., Ben-Shachar, D., and Youdim, M. B. H., 1982, *Pharmacol. Biochem. Behav.* **17**:43–48.
13. Bogniard, R. P., and Whipple, G. H., 1932, *J. Exp. Med.* **55**:653–658.
14. Austoni, M. E., Rabinovitch, A., and Greenberg, D. M., 1940, *J. Biol. Chem.* **134**:17–26.
15. Caraway, W. T., 1963, *Clin. Chem.* **9**:188–196.
16. Jackson, A. J., Michael, L. M., and Schumacher, H. J., 1972, *Anal. Chem.* **44**:1064–1065.
17. Thunell, S., 1965, *Clin. Chim. Acta* **11**321–333.
18. Vacha, J., Dungel, J., and Kleinwachter, V., 1978, *Exp. Haematol.* **6**:718–724.

19. Weinfeld, A., 1964, *Acta Med. Scand.* **177(Suppl. 427)**:13–20.
20. Kejellin, K. G., 1966, *J. Neurochem.* **13**:413–421.
21. Perl, M., 1867, *Virchows Arch. A* **39**:42–48.
22. Cameron, B. F., 1970, *Anal. Biochem.* **35**:515–517.
23. Highman, B., 1942, *Arch. Pathol.* **33**:937–943.
24. Jackson, A. J., Michael, L. M., and Schunacher, H. J., 1972, *Anal. Chem.* **44**:1064–1065.
25. Parmley, R. T., Spicer, S. S., and Alvarez, C. J., 1978, *J. Histochem. Cytochem.* **26**:729–741.
26. Thunell, S., 1965, *Clin. Chim. Acta* **11**:321–333.
27. Hukill, P. B., and Putt, F. A., 1962, *J. Histochem. Cytochem.* **10**:490–494.
28. Pearse, A. G. E., 1961, *Histochemistry,* 2nd ed., Little, Brown, Boston.
29. Humason, G. L., 1979, *Animal Tissue Techniques,* 4th ed., Freeman, San Francisco.
30. Nguyen-Legros, J., Bizot, J., Bolesse, J., and Pulicani, P., 1980, *Histochemistry* **66**:239–244.
31. Zaleski, S., 1887, *Arch. Exp. Pathol. Pharmakol.* **23**:77–99.
32. Guizzetti, P., 1915, *Riv. Pat. Nerv. Ment.* **20**:103–117.
33. Spatz, H., 1922, *Z. Gesamte Neurol. Psychiatr.* **77**:261–290.
34. Spatz, H., and Metz, A., 1926, *Z. Gesamte Neurol. Psychiatr.* **100**:428–449.
35. Tingey, A. H., 1937, *J. Ment. Sci.* **83**:451–460.
36. Cumings, N. J., 1948, *Brain* **71**:410–415.
37. Hallgren, B., and Sourander, P., 1958, *J. Neurochem.* **3**:41–51.
38. Sundermann, A., and Kempf, G., 1961, *Z. Alternsforsch.* **15**:97–105.
39. Musil A., Bertha, H., Haas, W., and Waurschinek, O., 1962, *Monatsh.* **93L**:536–540.
40. Courville, C. B., Nusbaum, R. E., and Butt, E. M., 1963, *Arch. Neurol.* **8**:481–489.
41. Harrison, W. W., Netsky, M. G., and Brown, M. D., 1968, *Clin. Chim. Acta* **21**:55–61.
42. Youdim, M. B. H., Green, A. R., Bloofield, M. R., Mitchell, B., Heal, D. J., and Grahame-Smith, D. G., 1980, *Neuropharmacology* **19**:259–267.
43. Rajan, K. S., Colburn, R. W., and Davis, J. M., 1971, *J. Neurochem.* **18**:345–364.
44. Rajan, K. S., Colburn, R. W., and Davis, J. M., 1976, *Life Sci.* **18**:423–432.
45. Hanig, R. C., and Aprison, M. H., 1967, *Anal. Biochem.* **21**:169–177.
46. Weiner, W. J., Nausiela, P. A., and Klawans, H., 1977, *Life Sci.* **20**:1181–1185.
47. Rafaelson, O. J., and Kofod, B., 1969, *Handbook of Neurochemistry,* Volume 6 (A. Lajtha, ed.), Plenum Press, New York, pp. 261–271.
48. Williamson, A. M., and Nj, K. T., 1980, *Physiol. Behav.* **24**:561–567.
49. Hill, J. M., and Switzer, R. C., III, 1984, *Neuroscience* **11**:595–603.
50. Colburn, R. W., and Mass, J. W., 1965, *Nature* **208**:37–41.
51. Columbo, J. A., and Saporta, A., 1980, *Exp. Neurol.* **70**:417–437.
52. Norton, W. J., and Poduslo, S. E., 1973, *J. Neurochem.* **21**:759–762.
53. Davison, A. N., and Peters, A., 1970, *Myelination,* Thomas, Springfield, Illinois.
54. Dallman, P. R., and Spirito, R. A., 1977, *J. Nutr.* **107**:1075–1081.
55. Kofod, B., 1970, *Eur. J. Pharmacol.* **13**:40–45.
56. Keen, O. L., and Hurley, L. S., 1980, *Mech. Ageing Dev.* **13**:161–176.
57. Dallman, P. R., Simes, M. A., and Manies, E. C., 1975, *Br. J. Haematol.* **31**:209–215.
58. Weinberg, J., Dallman, P. R., and Levine, S., 1979, Pharmacol. Biochem. Behav. **12**:493–502.
59. Dallman, P. R., 1974, *Iron in Biochemistry and Medicine* (A. Jacob and M. Worwood, eds.), Academic Press, New York, pp. 437–475.
60. W.H.O. Scientific Group on Nutritional Anaemias, 1968, *W.H.O. Tech. Res. Ser.* **405**.
61. Ten-State Nutritional Survey 1968–1970, Volume IV, U.S. Department of Health, Education and Welfare, Publication (HSM) 72-8132.
62. Nutrition Canada, 1973, Nutritional Canada National Survey, Canadian Department of National Health and Welfare, Publication H 58-36-1973.
63. Kessner, J., and Kalk, A., 1973, Strategy for evaluating health services, Institute of Medicine, National Academy of Sciences, Washington, D.C.
64. HANES, 1974, 1971–1972 Dietary Intake and Biochemical Findings, U.S. Department of Health, Education and Welfare, Publication (HRA) 74-1219-1.

65. Lazoff, B., Britteenham, G., Viteri, F. E., and Urrutia, J. J., 1982, *Iron Deficiency, Brain Biochemistry and Behavior* (E. Pollitt and R. L. Leibel, eds.), Raven Press, New York, pp. 183–194.
66. Pollitt, E., Viteri, F., Saco-Pollitt, C., and Leibel, R. L., 1982, *Iron Deficiency, Brain Biochemistry and Behavior* (E. Pollitt and R. L. Leibel, eds.), Raven Press, New York, pp. 195–208.
67. Oski, F. A., and Honig, A.M., 1978, *J. Pediatr.* **92**:21–25.
68. Leibel, R., 1977, *J. Am. Diet. Assoc.* **71**:398–404.
69. Leibel, R. L., Greenfield, D. B., and Pollitt, E., 1979, *Nutrition Pre- and Postnatal Development* (M. Winick, ed.), Plenum Press, New York.
70. Pollitt, E., Lewis, N., Leibel, R. L., and Greenfield, D. B., 1981, *Human Nutrition: Clinical and Biochemical Aspects* (P. J. Garry, ed.), American Association for Clinical Chemistry, Washington, D. C., pp. 290–301.
71. Oski, F. A., 1979, *Am. J. Dis. Child.* **133**:315–322.
72. Webb, T., and Oski, F. A., 1973, *J. Pediatr.* **82**:827–830.
73. Cantwell, R. J., 1974, *Pediatr. Res.* **8**:342.
74. Saltman, P., and Hegenauer, J. (eds.), 1982, *The Biochemistry and Physiology of Iron,* Elsevier, Amsterdam.
75. Hamon, M., Bourgoin, S., and Youdim, M. B. H., 1979, *Aromatic Amino Acid Hydroxylases and Mental Disease* (M. B. H. Youdim, ed.), Wiley, New York, pp. 191–233.
76. Weiner, N., 1979, *Aromatic Amino Acid Hydroxylases and Mental Disease* (M. B. H. Youdim, ed.), Wiley, New York, pp. 141–190.
77. Lovenberg, W., 1983, *Handbook of Neurochemistry,* Volume 4 (A. Lathja, ed.), Plenum Press, New York, pp. 133–150.
78. Symes, A. L., Sourkes, T. L., Youdim, M. B. H., Gregoriades, G., and Birnbaum, H., 1969, *Can. J. Biochem* **47**:999–1003.
79. Symes, A. L., Missala, K., and Sourkes, T. L., 1971, *Science* **174**:153–155.
80. Youdim, M. B. H., Woods, H. F., Mitchel, D. B., Grahame-Smith, D. G., and Callender, D. S., 1975, *Clin. Sci. Mol. Med.* **48**:289–295.
81. Voorhess, M. L., Stuart, M. J., Stockman, J. A., and Oski, F. A., 1975, *J. Pediatr.* **86**:542–547.
82. Grahame-Smith, D. G. (ed.), 1982, *Psychopharmacology,* Volumes 1 and 2, Excerpta-Medica, Amsterdam.
83. Pletscher, A., 1978, *Essays in Neurochemistry and Neuropharmacology,* Volume 4 (M. B. H. Youdim, W. Lovenberg, D. F. Sharman, and J. R. Lagnado, eds.), Wiley, New York, pp. 49–103.
84. Graf, M., and Pletscher, A., 1979, *Br. J. Pharmacol.* **65**:601–608.
85. Woods, H. F., Youdim, M. B. H., Boullin, D., and Callender, D. S., 1977, *Ciba Found. Symp.* **51**:227–248.
86. Youdim, M. B. H., Ishalom, N., and Iancu, T., 1984, *Br. J. Haematol.* (in press).
87. Kaladhar, M., and Narasinga Rao, B. S., 1982, *J. Neurochem.* **38**:1576–1581.
88. Weisenberg, E., Halbreich, A., and Mager, J., 1980, *Biochem. J.* **188**:633–641.
89. Massaro, T. F., 1982, *Iron Deficiency, Brain Biochemistry and Behavior* (E. Pollitt and R. L. Leibel, eds.), Raven Press, New York, pp. 125–140.
90. Edgerton, V. R., Ohira, Y., Gardner, G. W., and Senewiratne, B., 1982, *Iron Deficiency, Brain Biochemistry and Behavior* (E. Pollitt and R. L. Leibel, eds.), Raven Press, New York, pp. 141–160.
91. Mackler, B., Person, R., Miller, L. R., Inamdar, A. R., and Finch, C. A., 1978, *Pediatr. Res.* **12**:217–220.
92. Mackler, B., and Finch, C., 1982, *Iron Deficiency, Brain Biochemistry and Behavior* (E. Pollitt and R. L. Leibel, eds.), Raven Press, New York, pp. 31–38.
93. Youdim, M. B. H., and Ashkenazi, R., 1982, *Serotonin in Biological Psychiatry* (B. Ho and E. Usdin, eds.), Raven Press, New York, pp. 35–51.
94. Tamir, H., Klein, A., and Rapport, M., 1976, *J. Neurochem.* **26**:871–878.
95. Tamir, H., and Gershen, M. D., 1979, *J. Neurochem.* **33**:35–44.
96. Tamir, H., Brunner, W., Casper, D., and Rapport, M., 1980, *J. Neurochem.* **34**:1719–1724.

97. Dillman, E., Gale, C., Green, W., Johnson, D. G., Mackler, B., and Finch, C. A., 1980, *Am. J. Physiol.* **238**:377–381.
98. Dillman, E., Johnson, D. G., Martin, J., Mackler, B., and Finch, C. A., 1979, *Am. J. Physiol.* **237**:R297–300.
99. Dillman, E., Mackler, B., Johnson, D. G., Brengelman, G., Green, W., Gale, C., Martin, J., Layrisse, M., Martinez-Torres, C., and Finch, C., 1982, *Iron Deficiency Brain Biochemistry and Behavior* (E. Pollitt and R. L. Leibel, eds.), Raven Press, New York, pp. 57–62.
100. Leibel, R., Greenfeld, D., and Pollitt, E., 1979, *Br. J. Haematol.* **41**:145–150.
101. Anderson, H. T., and Barkve, H., 1970, *Scand. J. Clin. Lab. Invest.* **25**:1–62.
102. Gardner, G. W., Edgerton, V. R., Senewiratne, B., Barnard, R. J., and Ohira, Y., 1977, *Am. J. Clin. Nutr.* **30**:910–917.
103. Ohira, Y., Edgerton, V. R., Gardner, G. W., Senewiratne, B., Barnard, R. J., and Simpson, D. R., 1979, *Br. J. Haematol.* **41**:365–372.
104. Viteri, F. E., and Torun, B., 1974, *Clin. Haematol.* :609–626.
105. Edgerton, V. R., Ohira, Y., Gardner, G. W., and Senewiratne, B., 1982, *Iron Deficiency, Brain Biochemistry and Behavior* (E. Pollitt and R. L. Leibel, eds.), Raven Press, New York, pp. 141–160.
106. Edgerton, V. R., Diamon, L. B., and Olson, J., 1977, *J. Nutr.* **102**:381–399.
107. Edgerton, V. R., Diamon, L. B., and Olson, J., 1977, *J. Nutr.* **107**:595–601.
108. Finch, C. A., Gollnick, P. D., Hlastada, M. P., Miller, L. R., Dillman, E., and Mackler, B., 1979, *J. Clin. Invest.* **64**:129–137.
109. Finch, C. A., Miller, L. R., Inamdar, A. R., Person, R., Seiler, K., and Mackler, B., 1976, *J. Clin. Invest.* **58**:447–453.
110. Ohira, Y., Edgerton, V. R., Gardner, G. W., Gunawardena, K. A., Senewiratne, B., and Ikawa, S., 1981, *J. Nutr. Sci. Vitaminol.* **27**:87–96.
111. Green, A. R., and Grahame-Smith, D. G., 1974, *Neuropharmacology* **13**:949–959.
112. Youdim, M. B. H., Green, A. R., and Grahame-Smith, D. G., 1977, *Vth International Symposium on Parkinson's Disease* (W. Birkmayer and O. Hornykiewiz, eds.), Roche, Basel, pp. 127–135.
113. Ashkenazi, R., Ben-Shachar, D., and Youdim, M. B. H., 1982, *Br. J. Pharmacol.* **74**:762–763.
113a. Ashkenazi, R., Ben-Shachar, D., and Youdim, M. B. H., 1982, *Pharmacol. Biochem. Behav.* **17**:43–47.
114. Youdim, M. B. H., Yehuda, S., and Ben-Uriah, Y., 1981, *Eur. J. Pharmacol.* **74**:295–301.
115. Becking, G. C., 1972, *Biochem. Pharmacol.* **21**:1585–1594.
116. Youdim, M. B. H., Green, A. R., and Aronson, J. K., 1977, *Anaemia and Haematinics* (R. G. Richardson, ed.), Abbott Laboratories Press, London, pp. 37–42.
117. Yehuda, S., and Wurtman, R. J., 1975, *Eur. J. Pharmacol.* **30**:154–158.
118. Yehuda, S., 1979, *Commun. Psychopharmacol.* **3**:115–120.
119. Amit, T., Ben-Harari, R. R., and Youdim, M. B. H., 1981, *Br. J. Pharmacol.* **74**:955–956P.
120. Barkey, R. J., Shani, J., Amit, T., and Youdim, M. B. H., 1981, *Mol. Cell. Endocrinol.* **21**:129–138.
121. Ashkenazi, R., Ben-Shachar, D., and Youdim, M. B. H., 1982, *The Biochemistry and Physiology of Iron* (P. Saltman and J. Hegenauer, eds.), Elsevier, Amsterdam, pp. 575–583.
122. Tucker, D. M., and Sandstead, H. H., 1981, *Physiol. Behav.* **26**:439–449.
123. Tucker, D. M., and Sandstead, H. H., 1982, *Iron Deficiency, Brain Biochemistry and Behavior* (E. Pollitt and R. L. Leibel, eds.), Raven Press, New York, pp. 161–181.
124. Iversen, S. D., 1977, *Handbook of Psychopharmacology,* Volume 8 (L. L. Iversen, S. D. Iversen, and S. H. Snyder, eds.), Plenum Press, New York, pp. 333–384.
125. Margolin, D. I., 1978, *J. Clin. Psychiatry* **39**:120–130.
126. Shaywitz, B. A., Yager, R. D., and Klopper, J. H., 1976, *Science* **194**:305–308.
127. Tucker, D. M., 1981, *Psychol. Bull.* **89**:19–46.
128. Weinberg, J., Dallman, P. R., and Levine, S., 1980, *Pharmacol. Biochem. Behav.* **12**:493–502.
129. Weinberg, J., Brett, L. P., Levine, S., and Dallman, P. R., 1981, *Pharmacol. Biochem. Behav.* **14**:447–453.

130. Massaro, T. F., 1982, *Iron Deficiency, Brain Biochemistry and Behavior* (E. Pollitt and R. L. Leibel, eds.), Raven Press, New York, pp. 125–140.
131. Youdim, M. B. H., Ben-Shachar, D., Ashkenazi, R., and Yehuda, S., 1983, *CNS Receptors from Behavior to Pharmacology* (P. Mandel and F. De Fredius, eds.), Raven Press, New York, pp. 309–331.
132. Glover, J., and Jacobs, A., 1972, *Br. Med. J.* **2**:627–628.
133. Yehuda, S., and Youdim, M. B. H., 1984, *Eur. J. Pharmacol.* (in press).
134. Youdim, M. B. H., Ben-Shachar, D., Ashkenazi, A., and Yehuda, S., 1984, *7th Parkinson's Disease Symposium*, Raven Press, New York (in press).
135. Switzer, R. C., III, 1982, *The Biochemistry and Physiology of Iron* (P. Saltman and J. Hegenauer, eds.), Elsevier, Amsterdam, pp. 569–575.
136. Meyer, A., 1958, *Neuropathology* (W. Blackwood, W. H. McMenemey, A. Meyer, and R. M. Norman, eds.), Arnold, London, pp. 525–537.
137. Dobling, E. C., Schoene, W. C., and Richardson, E. P., 1974, *Arch. Neurol.* **30**:70–83.
138. Vakili, S., Drew, A. L., Von Schuching, S., Becker, D., and Zeman, W., 1977, *Arch. Neurol.* **34**:729–738.
139. Borkey, R., Ben-Schachar, D., Amit, T., and Youdim, M. B. H., 1984, *Eur. J. Pharmacol.* (in press).
140. Szanto, J., and Gallyas, F., 1966, *Arch. Neurol.* **14**:438–442.
141. Sheldon, J. H., 1927, *Q. J. Med.* **21**:123–127.
142. Cammermeyer, J., 1947, *J. Neuropathol. Exp. Neurol.* **6**:111–127.
143. Klinworth, G. k., 1969, *Progress in Neurogenetics* (A. Barbeau and J. R. Brunette, eds.), Excerpta Medica, Amsterdam, pp. 589–596.
144. Klinworth, G. K., 1972, *Adv. Neurol.* **1**:353–368.
145. Earle, K. M., 1968, *J. Neuropathol. Exp. Neurol.* **27**:1–14.
146. Barbeau, A., and Brunette, J. R., 1967, *Progress in Neurogenetics* (A. Barbeau and J. R. Brunette, eds.), Excerpta Medica, Amsterdam, pp. 509–682.
147. Riederer, P., Birkmayer, W., Jellinger, K., and Youdim, M. B. H., 1984, *J. Neural. Transm.* (in press).
148. Hsiang, N. S., 1941, *J. Oriental Med.* **33**:119–162.
149. Strassman, G., 1945, *J. Neuropathol. Exp. Neurol.* **4**:393–401.
150. Rubin, J. J., and Willmore, L. J., 1980, *Exp. Neurol.* **67**:472–480.
151. Willmore, L. J., Hurd, R. W., and Sypert, G. W., 1978, *Brain Res.* **152**:406–410.
152. Willmore, L. J., Sypert, G. W., Munson, J. V., and Hurd, R. W., 1978, *Science* **200**:1501–1503.
153. Porter, R., and Elliot, K. C., (eds.), 1978, *Oxygen, Free Radicals and Tissue Damage*, Elsevier, Amsterdam.
154. Glover, J., and Jacobs, A., 1972, *Br. Med. J.* **2**:627–628.
155. Seeman, P., 1980, *Pharmacol. Rev.* **32**:229–313.
156. Rajan, K. S., Manian, A. A., Davis, J. M., and Skriplus, A., 1974, *Phenothiazines and Structurally Related Drugs* (I. S. Forrest, C. J. Carr, and E. Usdin, eds.), Raven Press, New York, pp. 344–358.
157. Csernansky, J. G., Holman, C. A., Bonnet, K. A., Grabowski, K., King, R., and Hollister, L. E., 1982, *Life Sci.* **32**:385–390.

28

The Neurochemistry of Vitamin Deficiencies

Pierre M. Dreyfus

1. INTRODUCTION

While enormous technological strides have been made in recent decades, nutritional depletion and vitamin deficiencies continue to plague large segments of the world's population. In spite of a century of intensive research by prominent scientists, huge gaps exist in our knowledge of how the function and the structure of the nervous system are affected by alterations in vitamin metabolism. In this chapter an attempt will be made to review pertinent information concerning the role of vitamins in the metabolic activity of the nervous system and to give an overview of current thinking about the pathogenesis of neurologic disorders engendered by a lack of vitamins.

A few general introductory remarks about vitamins are appropriate. The term "vitamin" was coined by Funk in 1914,[1] following extensive research on substances that were discovered to be vital to the well-being of experimental animals. In view of the fact that some of these substances were found to be amines, the word "vitamin" seemed quite appropriate. Subsequent research led to the conclusion that these essential nutrients are not synthesized by mammalian tissue and must therefore be ingested as part of the diet. However, while vitamins are indeed indispensable, or "vital," only a few are in fact amines.

During the past half-century much has been learned about the structure, synthesis, metabolism, and fundamental biochemical functions of vitamins. A great deal of information concerning the neurochemical consequences of vitamin deficiency has been gathered over the years, yet it is virtually impossible to specify which of several well-documented, specific "biochemical lesions" is responsible for the initial dysfunction and the ultimate destruction of parts of the nervous system.

Almost all vitamins are ingested as part of the diet. They are absorbed by the gastrointestinal tract, then transported by the plasma to vital organs, where

---

Pierre M. Dreyfus • Department of Neurology, University of California School of Medicine, Davis, California 95616.

they gain entry into cells. In the nervous system, vitamins have to deal with the blood–brain barrier and are probably first absorbed by glial cells (astrocytes) rather than by neurons. Subsequently, vitamins are converted into their active cofactor or coenzyme form, at which point they interact with specific apoenzyme proteins to form the holoenzymes that are involved in vital biochemical reactions. Certain toxic substances, such as ethanol, when ingested to excess, are known to effectively interfere with one or another of these steps of vitamin absorption and transformation. Genetically determined metabolic disorders characterized by a faulty step, or steps, of vitamin utilization are continually being identified.[2] Many of these disorders affect the nervous system, resulting in seizures, ataxia, and mental retardation. In some of the genetic errors there appears to be an alteration of the structure of the apoenzyme or a reduced affinity of the apoenzyme for its specific coenzyme—the vitamin.[3] Some degree of improvement of these disorders can be achieved with pharmacologic rather than nutritional doses of the corresponding vitamin. Many of these genetically determined, vitamin-dependent states need to be more clearly elucidated. In this chapter an attempt will be made to clarify the neurochemistry of vitamin deficiencies by discussing a few relevant vitamin-dependent states.

Whereas the role of vitamins as active coenzymes in key biochemical reactions has been well defined, experimental evidence gathered suggests that vitamins may have pharmacologic properties that are distinctly different from their function as coenzyme.

Most of what is known about the neurochemistry of vitamin deficiencies has been gleaned from experimental animals in which a state of deficiency has been produced by artificial means, i.e., by the feeding of synthetic diets that contain little or none of the vitamin under study, or by the administration of specific vitamin analogs. Neither situation bears much resemblance to the naturally occurring disease states in man; thus, extrapolation of animal data to normal or diseased humans must, of necessity, be interpreted with great caution.

2. *VITAMIN TRANSPORT INTO THE NERVOUS SYSTEM*

Until recently it was assumed that water-soluble B vitamins reached the nervous system by the process of simple diffusion. It now seems clear that under normal circumstances the CNS is protected from undue fluctuations in vitamin concentrations by homeostatic mechanisms that regulate the concentrations of these substances.[4] Vitamins such as thiamin, ascorbic acid, pyridoxine, niacin, folate, riboflavin, and, probably, biotin traverse the blood–brain barrier, reaching the extracellular space of the brain either directly from the bloodstream by way of the cerebral capillaries or from CSF by way of the choroid plexus.[5–8] To date, three distinct transport systems for vitamins have been identified: (1) an active, carrier-mediated, energy-dependent system (in the choroid plexus) that transports the vitamin against a concentration gradient from blood into CSF, for ascorbic acid, riboflavin, and methyltetrahydrofolate[7]; (2) a facilitated diffusion transport system between blood and CSF, for niacin and nicotinamide[8]; and (3) a complex system that probably involves phosphorylation–dephosphorylation following facilitated diffusion of a nonphosphorylated vitamin through various membranes, for thiamin and pyridoxine.[5,6] It is tempting to postulate that the improper function of one or another of these transport systems could account for obscure neuropsychiatric disorders in man. This idea is heavily favored by proponents of the megavitamin, or orthomolecular, treatment for psychiatric disease. To date, however, insufficient objective data are available either to prove or to disprove this contention.

The method of transport of vitamins into the PNS is virtually unknown.

3. *THIAMIN*

3.1. *Thiamin and Its Esters*

In mammalian tissue, including the brain, thiamin (vitamin B_1) is found in four forms: free thiamin; thiamin monophosphate (TMP); thiamin diphosphate (TDP), the active coenzyme form of the vitamin (formally known as thiamin pyrophosphate or cocarboxylase); and thiamin triphosphate (TTP), a non-coenzyme form of the vitamin. TDP achieves the highest concentrations in both the CNS and the PNS. In the brain, total thiamin concentrations tend to be quite low in comparison to those in other tissues.[9,10]

Normal rat brain contains 10–12 μg/g dry wt of total thiamin (all forms), the spinal cord (white matter) 8–10 μg/g, and the sciatic nerve 4–5 μg/g. Most of the thiamin (80%) is in the diphosphate or pyrophosphate form, 10% consists of TTP, and the remainder consists of either free thiamin or TMP. In the nervous system, thiamin concentrations are considerably lower than in other organs, such as the liver, heart, and kidney, each of which contains almost three times as much.[9]

A magnesium-dependent enzyme, thiamin pyrophosphokinase, catalyzes a conversion of free thiamin to its mono- and diphosphate forms. TDP is further converted to TTP by enzymatic transphosphorylation, an important biochemical step that may be essential to the normal function of excitable membranes. The phosphorylated forms of thiamin are dephosphorylated to monophosphothiamin and free thiamin by ubiquitous thiamin phosphatases. The thiamin molecule can be destroyed enzymatically by thiaminases, a group of enzymes known to cleave the thiamin molecule between the pyrimidine and the thiazole moieties. Thiaminases have been identified in the tissues of freshwater fish, particularly carps, mollusks, and crayfish, and have also been found in ferns. Certain strains of *E. coli* have been shown to elaborate the enzyme.[11]

3.2. *Thiamin-Dependent Enzyme Systems*

TDP acts as coenzyme for two major enzyme systems identified in all mammalian tissue, including the nervous system. The first system is involved in the oxidative decarboxylation of pyruvate and α-ketoglutaric acid, two essential steps of glycolysis. These enzymes are found predominantly in mitochondria. The second enzyme system involves two transketolation steps of the phosphogluconate pathway (hexose monophosphate shunt), an alternate pathway of glucose degradation essential in the generation of NADPH, a function especially prominent in tissues that actively carry out reductive synthesis of fatty acids and steroids. The phosphogluconate pathway also generates pentoses essential in the synthesis of nucleic acid. This enzyme system is considered to be cytoplasmic.

Studies of thiamin-dependent enzyme systems in normal adult rat brain have shown that the activity of pyruvate dehydrogenase—the enzyme system responsible for initiating the decarboxylation of pyruvate—tends to be highest in parts of the brain that are most heavily populated with neurons, such as the cerebral and cerebellar cortex. The activity of the enzyme system is considerably lower in white matter.[12] By contrast, brain transketolase activity in normal adult rat brain is highest in white matter in the brain stem; areas more richly populated with neurons, such as the cerebral cortex and caudate nucleus, have the lowest enzymatic activity.[13]

A limited study performed on normal human brain has revealed very high transketolase activity in the mammillary body. This is the anatomic structure that is invariably involved in patients who are afflicted with Wernicke's disease, the human counterpart of thiamin deficiency in animals. However, other anatomic areas that are usually not affected by Wernicke's disease, such as the brain stem and diencephalon, reveal almost equally high levels of enzyme activity.[13]

Thiamin deficiency in animals and man results in neurologic symptoms and signs usually referred to as encephalopathy. Characteristic lesions are found in vulnerable parts of the brain; the location of these lesions varies from species to species. To date, it has not been possible to clearly elucidate the neurochemical events that lead to symptoms and the eventual destruction of tissue. In spite of the many, well-documented, neurochemical abnormalities caused by thiamin deficiency, it remains essentially impossible to determine which of these is responsible for failing function and tissue destruction.

During progressive depletion, brain thiamin begins to decline gradually within 2 weeks following the inception of the state of deficiency. Symptoms and signs characteristic of vitamin depletion and striking histologic changes in selective areas of the brain occur after an 80% drop in total brain thiamin. Reversal of neurologic symptoms occurs following a 10% replenishment of brain thiamin, usually within a few hours. During progressive depletion, TDP primarily is depleted, while free thiamin and TMP remain relatively constant; TTP tends to increase. Most anatomic areas of the brain yield their thiamin content in an even manner except for the pons, the site of major histologic changes, where the loss of TDP is greatest when compared to other parts of the brain.[14]

Although the content of total thiamin in various areas of the normal rat brain may not vary considerably and the loss of the vitamin during progressive depletion appears to be quite even, turnover rates of thiamin seem to differ from one area of the brain to another. According to recent studies the turnover rates may in fact reflect the sensitivity of certain parts of the brain to the state of deficiency. Thus, the pons and the medulla, which have relatively high thiamin contents and turnover rates comparable to the cerebellum, are particularly vulnerable to thiamin depletion.[15]

When the two major thiamin-dependent enzyme systems are investigated during progressive depletion, it has been observed that pyruvate decarboxylase activity decreases minimally, even at the most advanced stage of deficiency, when the animals show striking clinical manifestations and pathologic lesions. By contrast, pyruvate decarboxylase is sharply decreased in other organs, such as the heart, liver, and kidney.[12,16] The enzymatic defect correlates with tissue pyruvate and lactate levels.[16] Recent observations on thiamin-deficient rat brain have shown that the symptoms of deficiency correlate with decreased glucose utilization in the lateral vestibular nucleus of the pons, the most common site of histopathologic changes. Similar changes have been noted in structures that are not visibly affected by the state of deficiency, i.e., the fornix and the pyramidal tract. The administration of thiamin results in increased glucose utilization in the lateral vestibular nuclei, apparently mediated by glial cells. In these experiments the 2-deoxyglucose autoradiographic method was utilized.[17] It might be expected that a failure in pyruvate and α-ketoglutaric acid decarboxylation as a consequence of thiamin deficiency would result in impaired acetylcholine metabolism. Some investigations have revealed normal regional acetylcholine levels and reduced acetylcholine utilization[18]; however, others have found acetylcholine and acetyl-CoA levels to be significantly reduced.[19] Although the actual turnover rates of acetylcholine in the deficient CNS have not been established with great accuracy, it appears that they may be markedly reduced at the synapse following nerve stimulation and that the incorporation of labeled choline and glucose into acetylcholine is decreased as a result of vitamin depletion.[20] It has in fact been suggested that the deficiency state induced an early functionally significant central muscarinic cholinergic lesion.[21]

The pronounced decrease in transketolase activity appears to have very little effect on the activity of the pentose phosphate shunt when that activity is measured directly.[22]

Recent studies carried out in cultured glial (C_6) and neuroblastoma cells grown in thiamin-deficient medium have revealed that the synthesis of fatty acid is significantly impaired.[23] The effect appears to be most pronounced in glial cells. This is most likely caused by a marked decrease in the activity of acetyl-CoA carboxylase and fatty acid synthetase, two key lipogenic enzymes. Impaired fatty acid synthesis can be promptly reversed by the addition of the vitamin to the culture medium. Furthermore, thiamin deficiency produced *in vitro*, affects cholesterol synthesis. This is presumably due to a reduction in the activity of the key regulatory enzyme 3-hydroxy-3-methylglutaryl-CoA (HMG-CoA) reductase. In these studies the disturbances of fatty acid and cholesterol biosyntheses were accompanied by reduced cell growth. The relationship of lipid synthesis and cell growth seems to indicate that the glial cell may be the site of the early lesion in thiamin deficiency and that thiamin nutrition within the nervous system may be mediated by glial cells.[23,24]

While the *in vivo* synthesis of RNA appears to be normal, a decrease in DNA synthesis, most pronounced in the brain stem, has been measured in symptomatic, thiamin-deficient rat brain. Studies of the effects of thiamin deficiency on rat cerebral protein synthesis have shown some degree of inhibition. In mouse brain, acute vitamin depletion leads to an increased rate of degradation of total protein whereas a more chronic state of deficiency has the opposite effect. Alterations of protein metabolism may be due in part to the severity of hypothermia and decreased nutrient assimilation and utilization frequently engendered by the lack of thiamin.[25]

Since the tissue changes in the CNS and PNS produced by thiamin deficiency reveal demyelination as well as other changes, it has been assumed that the vitamin is essential to the maintenance of myelin. Studies performed on thiamin-deficient rats during a period of development when they are most susceptible to nutritional insults and at a time when myelination is most active have revealed no significant changes in either myelin or other lipids.[26,27] No correlation between the content of thiamin and the state of myelination of the nervous system of developing animals can be established. Thiamin deficiency in the mother results in offspring with reduced body and brain size; however, no major changes in nucleic acid concentrations can be demonstrated. Brain lipids, such as gangliosides, cerebrosides, and other sphingolipids, reflect a state of undernutrition. Total phospholipids and the distribution of individual phospholipids tend to be normal, as are cholesterol levels. None of the changes can be ascribed specifically to thiamin deficiency. Such studies support the notion that thiamin is not related to the development of myelin or its maintenance in nervous tissue. Thiamin-dependent metabolism appears to be different in the nervous system of the developing rat than it is in that of mature rodents. During the first 10 days of life there appears to be a sharp rise in the activity of brain transketolase. This correlates with the rise of enzymes involved in oxidative phosphorylation and glycolysis, which suggests that transketolase activity may reflect an increasing glial cell duplication, proliferation, and migration.[27,28]

3.3. Thiamin and Membrane Function

Recently, it has been shown that thiamin deficiency affects synaptic transmission, i.e., a significant decrease in the uptake of serotonin by a synaptosomal preparation of the cerebellum in thiamin-deficient rats occurs. This finding suggests that some of the neurologic manifestations of the deficiency state may in fact be due to a failure of serotoninergic neurons, specifically in the cerebellum, which would account for the symptoms of ataxia invariably observed in depleted animals.[29]

Evidence of marked impairment of tissue transketolase activity has been obtained by enzyme measurements of blood, reflecting in a highly sensitive and specific manner the state of thiamin nutrition in man. These observations provide both clinical and biochemical evidence that at least some of the signs and symptoms of Wernicke's disease (thiamin deficiency in man) result from a specific lack of thiamin and that their prompt reversibility is due to the presence of a biochemical lesion that antedates tissue changes.[30] A most intriguing observation was recently made on cultured fibroblasts of patients afflicted with

Wernicke–Korsakoff's syndrome. These fibroblasts revealed an aberration of transketolase activity that consists of reduced avidity for TDP binding. This abnormality of coenzyme binding persists through several generations of fibroblasts, suggesting that the defect is genetically determined rather than induced by either alcohol or general malnutrition. If verified, this observation would point to a genetic predisposition to the development of neurologic signs of thiamin deficiency at times when the dietary supply of the vitamin is marginal.[31]

In the nervous system thiamin most likely plays another physiologic role, one that is totally different and separate from its function as a coenzyme in biochemical reactions. Since early observations, made 40 years ago,[32] a number of important contributions have been made that may explain the neurologic symptoms engendered by a deficiency of the vitamin. It has been shown that electrical stimulation of peripheral nerves and spinal cord results in the release of free thiamin. The addition of physiologic concentrations of acetylcholine, tetrodotoxin, serotonin, and LSD to the preparation has a similar effect. These agents cause the release of free thiamin and monophosphate either from perfused nerve fibers or from membrane preparations obtained from either brain or peripheral nerves. Within the nervous system, most of the existing thiamin is in the diphosphate form within mitochondria, and only 10% is associated with the membrane fraction. Thiamin released by neuroactive drugs originates in cell membranes and does not seem to come from either the axoplasm or the mitochondria.[33,34] It is generally assumed that the neuroactive form of thiamin is its triphosphate form, which is rapidly and readily interconvertible with TDP. The enzyme thiamin pyrophosphatase most likely regulates the concentrations of TDP and TTP, both of which form stable complexes with calcium.[35] TTP most probably occupies a site on the nerve membrane and it appears to be involved in the sodium-gating mechanisms of excitable membranes. Thiamin analogs, such as pyrithiamin, an inhibitor of thiamin pyrophosphokinase that freely crosses the blood–brain barrier and readily induces neurologic symptoms in experimental animals, have been found to interfere with membrane sites that catalyze the voltage-dependent changes in membrane permeability to sodium.[36] The effects of thiamin deficiency on the function of excitable membrane may explain the rapid and dramatic evolution of the clinical events observed in severe thiamin deficiency in both man and animals. It has been postulated that the thiamin-deficient membrane is unable to maintain osmotic gradients and that, as a result, a failure of glial electrolyte and water transport ensues. The precise role of TTP, a substance that is not a coenzyme, in the function of excitable membranes remains essentially unknown.[34]

3.4. Thiamin-Dependent Disorders

Three genetically determined disorders affecting the nervous system have been shown to involve thiamin-mediated metabolism. In Leigh's disease (subacute necrotizing encephalomyelopathy), a disease that affects infants and children and that resembles Wernicke's encephalopathy, a total absence of triphosphothiamin in the brain has been demonstrated. Patients and normal carriers of the disease have been shown to elaborate a factor that can be detected in blood, urine, and CSF. The factor inhibits thiamin pyrophosphate-ATP phosphoryltransferase, the enzyme responsible for the synthesis of TTP in the nervous system.[37,38] In rare cases of Leigh's disease, the inborn error of metabolism consists of a deficiency of pyruvate carboxylase (not dependent on thiamin). Variants of the disease appear to exist.[39]

In an inherited disorder characterized by intermittent ataxia and in some cases by spinocerebellar degeneration of the Friedreich's ataxia type, a defect in pyruvate oxidation has been detected. The abnormality involves the thiamin-dependent first enzyme of the pyruvate dehydrogenase complex.[40]

It has also been demonstrated that TDP increases the activity of the branched-chain α-ketoacid dehydrogenase complex of liver and fibroblasts cultured from the skin of individuals affected with maple syrup urine disease, an inherited disorder characterized by mental retardation and neurologic deterioration. In some cases the administration of thiamin and the dietary restriction of branched-chain amino acids have been shown to be beneficial.[41] The branched-chain ketoacid, α-ketoisocaproic acid, a derivative of leucine, one of the three branched-chain amino acids that pile up in the plasma of individuals affected with maple syrup urine disease, tends to inhibit the decarboxylation of pyruvate.[42]

4. *PYRIDOXINE (VITAMIN B_6)*

4.1. *Pyridoxine-Dependent Enzyme Systems*

Pyridoxine refers to a group of pyridine derivatives, such as pyridoxol, pyridoxal, and pyridoxamine, each of which has a similar physiologic action within mammalian tissue, including the brain. These various forms of vitamin B_6 are referred to as vitamers. Their phosphorylated forms, specifically pyridoxal and pyridoxamine phosphates, are active as coenzyme. The enzyme pyridoxal 5-phosphotransferase catalyzes the phosphorylation of the vitamin in brain and other organs. Such drugs as isonicotinic acid hydrazide and hydralazine (used in the treatment of tuberculosis and hypertension, respectively) interfere with the phosphorylating enzyme in such a way as to cause decreased levels of the pyridoxal phosphate in tissues. The phosphorylated forms of vitamin B_6 catalyze a number of basic enzymatic reactions related to the synthesis, catabolism, and transport of amino acids and the metabolism of glycogen and unsaturated fatty acids.[43] In the nervous system the vitamin is directly involved in the synthesis of the putative neurotransmitters 5-hydroxytryptamine, dopamine, and noradrenaline from tryptophan and tyrosine.[44]

Pyridoxine-dependent enzymes in the nervous system fall into two major categories: amino acid decarboxylases and transaminases. Some of these enzymes are intimately involved in the GABA shunt, an oxidative pathway restricted to the nervous system in which α-ketoglutaric acid is metabolized to succinate by way of glutamate and GABA. It appears that the affinity of the coenzyme for its apoenzyme varies from enzyme to enzyme. Thus, decarboxylases have a lower affinity for the cofactor than do transaminases. Decarboxylases may be partially saturated with cofactor and are therefore more susceptible to vitamin deficiency than are transaminases. Severe vitamin deprivation also results in a decrease in enzyme protein by virtue of impaired amino acid metabolism. The addition of excessive amounts of coenzyme to a vitamin-deficient enzyme preparation *in vitro* fails to restore complete activity; apoenzyme production stimulated by the addition of excess vitamin B_6 to a normal tissue extract can be inhibited by puromycin. Pyridoxal phosphate appears to regulate intracellular enzyme synthesis. Organs or cells with a high rate of protein turnover tend to be most sensitive to vitamin depletion. Cellular proliferation and the synthesis of specific proteins involved in immunologic reactions have been shown to require the presence of vitamin B_6. Two pyridoxine-dependent decarboxylases play an important role in neuronal activity. The first enzyme, glutamic decarboxylase, which appears to be restricted to neurons, controls the production of the neuroinhibitor GABA from glutamic acid. In the rat nervous system, enzymatic activity is highest in the hypothalamus and midbrain and lowest in the spinal cord. While significant enzymatic activity can be demonstrated in human cerebral cortex, none can be detected in white matter. Recent investigations suggest that there exist regional variations, ranging from 10% to 40%, in the rate of cofactor saturation of the enzyme, whereas the regional distribution of the coenzyme pyridoxal phosphate varies only slightly from one area of the brain to another.[45]

The second enzyme, 5-hydroxytryptophan decarboxylase, appears to be localized to nerve terminals and is involved mainly in the synthesis of serotonin [5-hydroxytryptamine (5-HT)]. In rat brain the activity of this enzyme is highest in the hypothalamus, midbrain, and caudate nucleus, areas of the brain rich in serotoninergic terminals. The enzymatic activity is considerably lower in cerebral and cerebellar cortex. The same decarboxylase may be involved in the decarboxylation of L-dopa to dopamine. Exogenous pyridoxine reduces the effectiveness of L-dopa in the treatment of patients afflicted with Parkinson's disease, presumably because of increased peripheral decarboxylation of L-dopa.

The dietary deficiency of vitamin B_6 decreases the seizure threshold in both experimental animals and man; the young nervous system appears to be more susceptible to convulsions than does the older and more mature system. Induction of vitamin B_6 deficiency in immature, newborn rats produces convulsive seizures, whereas these tend to be rare in mature weanling animals, which display only unusual irritability and occasional motor weakness as a result of vitamin B_6 deficiency. When the depletion of pyridoxal phosphate, reflected by the B_6-dependent SGOT activity, is reduced by 50% of normal, glutamic decarboxylase and 5-HTP decarboxylase activities are sharply reduced in the adult brain. The loss of glutamic decarboxylase activity is 36–65%, and that of 5-HTP decarboxylase is 46–84%, depending on the area of the nervous system. Decreased enzymatic activity is even greater in deficient neonatal animals that convulse; this suggests that there is a correlation between the seizure threshold and the level of activity of these two brain enzymes that are dependent on vitamin B_6.[46] The DNA, RNA, and protein content of the immature, pyridoxine-deficient rat brain is reduced significantly, although it tends to remain normal in deficient adult animals.[47]

A lack of pyridoxine in the maternal rat results in cytoarchitectonic alterations and decreased myelination in the brains of the offspring.[47,48] The latter is reflected by a reduction in the activity of 2′,3′-cyclic nucleotide-3′-phosphohydrolase, an enzyme frequently used as a marker for active myelination.[49]

Vitamin B_6 deficiency induced in immature animals during a period of rapid myelination appears to have a significant effect on lipid metabolism. Deficient brain reveals decreased incorporation of labeled acetate into total lipids and into the cholesterol, glycolipid, and phospholipid fractions. Cerebroside and sulfatide levels are markedly decreased.[50] Furthermore, the deficiency causes a decrease in the polyunsaturated fatty acid content of myelin phosphatides. It appears that the vitamin is essential to the activity of the enzyme serine palmitoyltransferase (3-ketohydrosphingosine synthetase), which is reduced in the depleted brain. Vitamin B_6 deficiency also reduces the biosynthesis of long-chain polyunsaturated fatty acids. Thus, the state of deficiency delays or reduces the rate of myelin formation in the CNS.

Evidence of vitamin B_6 deficiency in man can be obtained from a number of biochemical determinations on blood and urine. None of these reflects involvement of the nervous system.[51] Vitamin estimations in blood and CSF are of limited value. Under normal circumstances, pyridoxine and its derivatives can be measured in the urine; in the case of severe deficiency, none can be demonstrated. Vitamin B_6 depletion causes increased urinary excretion of the tryptophan metabolites and the xanthurenic and kynurenic acids, after the administration of tryptophan. Decreased activity of 5-HTP decarboxylase results in decreased excretion of 5-HIAA in the urine, and faulty activity of cysteine sulfinic acid, or cysteic acid decarboxylase, causes decreased taurine levels in the urine. The deficiency state also affects cystathionine cleavage, with resultant cystathioninuria.[52] Reduced activity of two pyridoxine-dependent enzymes, SGOT and glutamic-pyruvate transaminase, can be demonstrated as can the presence of oxalic acid in the urine.

4.2. Pyridoxine Dependency

The effects of vitamin B_6 deficiency on myelination of the brain are reflected in the observations made in the case of a child who died as a consequence of vitamin B_6-dependent convulsive seizures. Examination of the brain revealed an abnormally sparse quantity of myelinated fibers in the cerebral hemispheres. Glutamic acid concentrations were elevated and GABA levels reduced in the frontal and occipital cortices, but not in the spinal cord. With the exception of increased levels of cystathionine in the occipital cortex, the concentration of amino acids measured was normal. Pyridoxal-5-phosphate was reduced in the frontal cortex. Glutamic acid decarboxylase activity was normal when the tissue preparation contained more than 0.05 mM pyridoxal-5-phosphate. It appears that pyridoxine-dependent seizures in infants are associated with decreased GABA levels caused by an abnormality of glutamic acid decarboxylase and hypoplasia of central myelin.[53]

5. VITAMIN B_{12}

5.1. Vitamin B_{12} Coenzymes and Vitamin B_{12}-Dependent Enzymes

The neurochemistry of vitamin B_{12}, of its coenzyme forms, and of the various cobalamins has not been studied extensively. Mammalian tissues, including the brain and spinal cord, do not synthesize vitamin B_{12}, and most animals, including man, depend on a dietary source of the vitamin for normal metabolism. In order to be metabolically useful, the natural vitamin cyanocobalamin must first be transformed into one of its active coenzyme forms, 5-deoxyadenosylcobalamin or methyl-B_{12}. In the nervous system, two major enzymatic reactions have been shown to require one of these two coenzyme forms. The first of these reactions is responsible for the isomerization of methylmalonyl-CoA to succinyl-CoA; the second involves the transmethylation of homocysteine to methionine. The enzymes catalyzing these reactions are methylmalonyl-CoA mutase and *N*-methyltetrahydrofolate homocysteine methyltransferase, respectively.[54] Both of these enzyme sytsems have been estimated in the nervous system. It is not known whether these systems are glial or neuronal, just as nothing is known concerning their specific localization and distribution in the nervous system.

5.2. Vitamin B_{12} Deficiency and Its Consequences

The insufficient absorption of or a severe dietary lack of the vitamin results in striking lesions that result in spongy demyelination of the cerebral white matter, optic nerves, spinal cord, and peripheral nerves. The pathologic process is frequently referred to as subacute degeneration. The characteristics of the lesions in the nervous system consist of large, irregular, spongy, honeycomblike zones of demyelination. The fibers with the largest diameter appear to be affected first and foremost, while axons tend to be spared. It is presumed that demyelination is a primary lesion and that axonal degeneration, when it occurs, is a secondary manifestation. Although the initial insult appears to involve glial cells, a disturbance of axonal metabolism has not been excluded. Biochemical studies on the nervous system should assume greater significance in view of the fact that it is now possible to produce in rhesus monkeys neurologic manifestations of vitamin B_{12} deficiency and demyelinating lesions in the CNS and PNS indistinguishable from those observed in man.[55] The lesions become evident when the animals are maintained on a deficient diet for at least 4 years. In man, vitamin B_{12} depletion may occur for more than a decade before clinical manifestations are demonstrated.

It appears to be established that vitamin B_{12} deficiency in experimental animals and humans is responsible for decreased conversion of methylmalonyl-CoA to succinyl-CoA in the tissues, resulting in the increased urinary excretion of methylmalonic acid. The presence of this organic acid in the urine of animals and humans is highly specific and is in fact an extremely sensitive index of vitamin B_{12} deficiency. Methylmalonic acid levels in CSF have been shown to exceed those found in plasma. These observations suggest that the acid may be elaborated by cerebral tissue, spinal cord, and nerve roots as a consequence of defective propionate metabolism.[56]

Vitamin B_{12}-deficient liver, kidney, and brain have all revealed reduced activity of methylmalonyl-CoA mutase; methylmalonyl-CoA is hydrolyzed to methylmalonic acid rather than to succinyl-CoA and the acid appears in the urine. Since the rate of disappearance of methylmalonyl-CoA is essentially the same in normal and deficient tissue, there must exist adaptive mechanisms that rid the organism of the high levels of methylmalonyl-CoA that may in fact be toxic to various tissues.[57]

A number of interesting investigations have been carried out on vitamin B_{12}-deficient peripheral nerves and tissue cultures of neural origin that may shed light on the pathogenesis of the characteristic demyelination in the CNS and PNS caused by a state of deficiency.

Investigations of fatty acid metabolism in sural nerve samples obtained from patients afflicted with neurologic manifestations have revealed the presence of branched-chain (C:15) and odd-chain (C:17) fatty acids. In addition, the total lipid and fatty acid content and the rate of synthesis of fatty acids have been found to be markedly reduced in nerve samples obtained from deficient patients when compared to those from normal controls. It has been postulated that the abnormally accumulated methylmalonyl-CoA might replace malonyl-CoA in the early steps of fatty acid synthesis.[58] Experiments on fatty acid patterns in cultured glial cell lines grown in serum-free medium lacking vitamin B_{12} or in medium containing serum obtained from vitamin B_{12}-deficient pigs have in fact demonstrated the presence of higher proportions of C:15:0, C:17:0, and C:17:1 fatty acids, with no evidence of branched-chain fatty acids. When the deficient cells were grown in the presence of [^{14}C]propionic acid in the medium, the label was found only in the 15:0 and 17:0 fatty acid peaks. The levels of propionate were elevated in the deficient cells and the abnormal fatty acids persisted through six generations of cultured cells. The relative amount of 15:0 fatty acid increased the longer the cells were grown in the deficient medium without added vitamin.[59] Similar observations have been made on samples of brain, spinal cord, and sciatic nerve gleaned from an infant who died of methylmalonic aciduria caused by a genetic metabolic disorder in which vitamin B_{12} cannot be converted to 5-deoxyadenosylcobalamin. In these experiments, the accumulation of branched-chain and odd-number fatty acids was measured. The branched-chain fatty acids were identified as methyl-hexadecanoic acid. The phosphatides separated from the spinal cord contained methyl-substituted palmitic acid, with the highest concentration occurring in the β position of the phosphatidylcholine.[60]

In has also been shown that methylmalonyl-CoA inhibits both acetyl-CoA carboxylase and fatty acid synthetase activity. These biochemical findings may further explain the structural changes observed in both central and peripheral myelin and the consequent neurologic symptoms seen in long-standing vitamin B_{12} deficiency.[61]

5.3. Vitamin B_{12} and Nitrous Oxides

Other explanations for the clinical and pathologic changes have been sought. The prolonged inhalation of the anesthetic gas nitrous oxide may lead to bone marrow changes and peripheral neuropathy indistinguishable from those encountered in patients afflicted with a deficiency of vitamin B_{12}, suggesting that the gas interferes with some aspect of vitamin B_{12} metabolism.[62–64] The exposure of rats to the gas results in an early and marked decrease in methionine synthetase activity, which recovers very slowly once the offending agent has been removed. It appears that the gas displaces the vitamin from the enzyme when radioactive cobalt-labeled vitamin B_{12} is used. This displacement parallels the changes in methionine synthetase activity. A decrease in tissue methylcobalamin content and the presence of vitamin B_{12} coenzyme analogs have also been detected. Changes in methylmalonyl-CoA mutase activity, on the other hand, seem to occur after very prolonged exposures to the gas, suggesting that it does not directly affect the latter enzyme as it does methionine synthetase. It has been suggested that the rapid inhibition of methionine synthetase may be responsible for the hematologic abnormalities and that the development of neurologic symptoms and signs may be caused by the appearance of cobalamin analogs.[65] In another study, nitrous oxide has been shown to induce subacute demyelination of the spinal cord of monkeys. This is thought to be caused by the trapping of folate as 5-methyltetrahydrofolate and by the inhibition of methionine synthesis, resulting in a deficiency of methyl groups. Some of the animals receiving methionine supplements appear to escape damage to the spinal cord. Thus, it has been postulated that reduced methionine synthetase activity is the primary cause of subacute demyelination observed in the nervous system.[66,67]

There remain huge gaps in our understanding of the role of vitamin B_{12} and related substances in the metabolic activity of myelinated structures.

6. FOLIC ACID

That there is a direct link between the biochemistry of folic acid and vitamin B_{12} is now established.[68] A deficiency of either vitamin leads to the increased excretion of formiminoglutamic acid. The excretion of formate and aminoimidazolecarboxamide is also increased in either folate or vitamin B_{12} deficiency. The three compounds are substrates for folate-dependent enzymatic reactions. Their urinary excretion is promptly reduced by the administration of methionine to vitamin B_{12}-deficient animals. Both vitamin B_{12} and folate deficiency lead to decreased incorporation of [^{14}C]formate into serine and to reduced oxidation of the β carbon of serine to respiratory CO_2.[68]

The most plausible theory thus far advanced to explain the effect of vitamin B_{12} on folate metabolism consists of the methyl trap theory, according to which methylfolates are converted to nonmethyl forms by reacting with homocysteine to produce methionine in the cobalamin-dependent methyltransferase reaction. Whether these mechanisms exist in nervous tissue has not been established.[68]

The distribution of folates in the nervous system varies from one anatomic area to another, being highest in the raphe nuclei of the midbrain and the corpus striatum and lowest in the olfactory bulb, substantia nigra, locus coeruleus, and spinal cord. It has been postulated that the concentration of 5-methyltetrahydrofolate correlates with the presence of indoleamine neurons.[69]

The active coenzyme form of folate contains four additional hydrogen atoms (tetrahydrofolate); folic acid reductase permits the reduction of folate to tetrahydrofolate. Various derivatives of tetrahydrofolate have been identified, including methyl-, methenyl-, methylene-, hydroxymethyl-, formyl-, and formiminotetrahydrofolate. In general, the various forms of folate are instrumental in the transfer of single carbon units.

The specific role of folic acid in cerebral metabolism has not been completely elucidated as yet, although it is interesting to note that folate levels in CSF are 2–3 times those in serum, even in states of folate deficiency.[70]

The developing and immature nervous system may be more susceptible to the deficiency of folate than is its adult counterpart. There is ample experimental evidence that folate plays an important role in the early development of the CNS. A number of folate-dependent enzymes have been shown to be increased during early brain development, pointing to a role of the vitamin in the synthesis of purines and pyrimidines and hence DNA, RNA, and protein synthesis. Since folate is also involved in the synthesis of methionine, the vitamin controls methylating processes. Folate appears to be essential to some aspects of catecholamine and indoleamine metabolism.[69,71]

It appears that there exists a relationship between certain anticonvulsant agents, such as phenytoin, phenobarbital, primidone, and folic acid. The interaction seems to be independent of the primary epileptogenic properties of folic acid. Phenobarbital interferes with the uptake of folic acid by the brain, whereas phenytoin does not. By contrast, phenytoin enhances the entry of folic acid into the brain. The administration of folic acid, on the other hand, reduces the levels of phenytoin in the brain. A depletion of brain folic acid has been noted following the experimental induction of seizures, suggesting that the vitamin or its metabolism may be involved in the epileptogenic process. The literature suggests that folic acid and its derivatives have significant epileptogenic properties that may reverse the anticonvulsant effects of phenytoin, particularly when the blood–brain barrier is damaged. Experimental evidence suggests the possible blockade of GABA receptors by folic acid, reducing the animals' seizure threshold.[69,72,73]

7. NIACIN

Niacin, or nicotinic acid, is a constituent of two coenzymes that transfer hydrogen, or electrons. These coenzymes are nicotinamide adenine dinucleotide (NAD) and nicotinamide adenine dinucleotide phosphate (NADP). Both nucleotides are essential to a number of enzymatic reactions of carbohydrate and fatty acid metabolism. Thus, lactic dehydrogenase and glycerophosphate dehydrogenase—enzymes of glycolysis—require the reduced form of NAD, while in the citric acid cycle, two isocitrate dehydrogenases require some form of nucleotide as cofactor. Reduced NADP assumes an important role in the early steps of the pentose phosphate shunt and the subsequent reductive synthesis of lipids. While they are unable to synthesize other B vitamins, most mammalian cells, probably including those of the nervous system, are able to synthesize nicotinic acid from the amino acid tryptophan. A deficiency of niacin is known to produce severe neurologic and psychologic disease in humans. Pellagra, an encephalopathy associated with mental symptoms and signs of spinal cord and peripheral nerve involvement, is usually caused by a dietary deficiency of the niacin precursor tryptophan. The main neuropathologic change observed in this disease is degeneration of the large pyramidal cells (Betz cells) of the motor cortex, an alteration known as central chromatolysis, or central neuritis. The spinal cord and peripheral nerve lesions consist principally of demyelination. Niacin deficiency brings about a reduction in cerebral NAD and NADP levels and decreased activity of all of the enzymes dependent upon these nucleotides. How these biochemical changes cause disease and tissue destruction remains unknown.

In patients afflicted with pellagra, NAD levels in red blood cells are reduced and the urinary excretion product of niacin (*N*-methylnicotinamide) is diminished. The administration of niacin to pellagrins and to niacin-deficient animals brings about a marked elevation, to well above normal, of blood and tissue nucleotide levels. Indirectly, by virtue of increased nucleotide tissue levels, the synthesis of norepinephrine and dopamine is significantly augmented by the administration of nicotinic acid.[74]

The analog of nicotinamide, an amide of nicotinic acid, 6-aminonicotinamide, has been shown to cause an experimental affliction of the spinal cord in rats, cats, and rabbits. The animals develop weakness of their extremities and unsteadiness of gait. Pathologic changes consist of loss of neurons in the gray matter of the spinal cord, medulla, midbrain, and roof nuclei of the cerebellum. Occasionally other sites, such as the geniculate bodies, thalamus, and corpus striatum, are involved. The white matter of the spinal cord and other areas is spared. Structural changes in glial cells have also been noted. The lesions do not resemble those reported with niacin deficiency.[75] Presumably the analog produces 6-amino-NADP that inhibits the activity of 6-phosphogluconate levels. Thus, the vitamin analog interferes with the pentose phosphate pathway, resulting in significant abnormalities of nucleic acid, cerebral, and cerebellar RNA synthesis, as well as changes of lipid and protein metabolism within glial cells and neurons.[76]

8. PANTOTHENIC ACID

Pantothenic acid, or one of its higher forms, is a dietary essential for man. At least two metabolically active forms have been recognized. Pantothenic acid is a constituent of CoA and acyl carrier protein. Whether the latter occurs in mammalian tissue, and the nervous system in particular, has not been established. In addition to pantothenic acid, CoA contains ribose, adenine, phosphoric acid, and β-mercaptoethylamine. Its sulfhydryl group is the site that links acid and acetyl groups.[77]

Pantothenic acid, as a constituent of CoA, participates in a variety of biochemical reactions involved in fatty acid synthesis and the metabolism of steroids and acetylcholine. The enzymatic reactions that utilize CoA are either acetokinases or transacetylases.[78] Pantothenic acid appears to be essential to amide and peptide linkages.

As is the case with many other water-soluble vitamins, very little is known about the neurochemistry of pantothenic acid. Relatively high concentrations of the vitamin exist in brain, which is not easily depleted of this particular vitamin. Pantothenic acid deficiency in animals has been shown to be responsible for demyelinating lesions of peripheral nerves. Similar lesions may be encountered in humans deprived of the vitamin, in whom numbness and tingling of hands and feet and occasionally a "burning-foot" syndrome may develop. Patients afflicted with pantothenic acid deficiency have an impairment of *p*-aminobenzoic acid acetylation and a significant decline in blood levels of cholesterol and cholesterol esters.[79] Evidence of adrenocortical hypofunction has also been described. In experimental animals, pantothenic acid deficiency results in deranged acetylcholine, cholesterol, glucosamine, and galactosamine synthesis, in faulty fatty acid oxidation, and in reduced energy production.

9. *VITAMIN E*

α-Tocopherol, one of the major components of vitamin E, is generously distributed in mammalian tissue. In the nervous system, gray matter appears to have the highest levels of α-tocopherol and the spinal cord the lowest levels. In general, blood levels of α-tocopherol do not reflect tissue stores. Although the precise metabolic role of tocopherols has not been clearly elucidated, recent observations have shown that vitamin E is involved in a number of important biologic processes. It seems to act as a nonspecific antioxidant, preventing the peroxidation of polyunsaturated fatty acids, a nonenzymatic reaction that occurs normally in the course of intracellular metabolism. The resultant products of peroxidation are highly damaging to cell membranes.[80] Immature human and animal brain, which tends to be relatively abundant in highly unsaturated fatty acids, contains less tocopherol than do other tissues, including adipose tissue.[81] Thus, the requirements for vitamin E may be higher in infants than in adults and may render infants more susceptible to a deficiency state. A lack of antioxidants, including vitamin E, leads to changes in the fatty acid composition of phospholipids, in particular phosphatidylethanolamine.[82] Furthermore, a deficiency state may affect glutathione stability by reducing it, since the vitamin normally maintains sulfhydryl bonds. Vitamin E, in addition to preventing lipid peroxidation, has been shown to interact with reactive free radicals, acting as a "scavenger" substance; the vitamin is also involved in prostaglandin metabolism.

Experimentally induced vitamin E deficiency in animals leads to alterations in the nervous system consisting of encephalomalacia, axonal dystrophy (particularly pronounced in the dorsal funiculi of the spinal cord), and ceroid (a form of lipochrome pigment) deposits in both neurons and muscles. The persistent inability to absorb fats is the most likely mechanism by which human subjects become deficient in this fat-soluble vitamin. In children, the most common underlying causes of deficiency are disorders in which fats are poorly absorbed, such as celiac disease, cystic fibrosis, and biliary atresia. In the adult, sprue and chronic pancreatitis appear to be the most common causes. In both adults and children, a progressive myelopathy, axonal dystrophy, ceroid deposition in smooth muscles, and skeletal muscle lesions that resemble muscular dystrophy have been described as a consequence of vitamin E deficiency.

10. BIOTIN

Biotin is a water-soluble vitamin that is involved in a number of important anaplerotic enzymatic reactions in mammalian tissue. Recent investigations suggest that biotin may be significant to the integrity of the nervous system of animals and man. A state of deficiency appears to be relatively rare, probably because of the fact that enteric organisms are able to synthesize the vitamin, which is rapidly absorbed from the gastrointestinal tract.[83] A deficiency of biotin has been shown to result from copious ingestion of raw eggs or egg whites, from prolonged administration of tube feedings, and in patients in whom gastrointestinal absorption is impaired and who are being treated by means of total parenteral nutrition and broad-spectrum antibiotics. Neurologic manifestations have been described in animals and man. In the former, spasticity and ataxia have been described; in man, irritability, lethargy, hypotonia, ataxia, and paresthesia have been reported.[84,85]

In order to be biologically active, biotin must be transformed to biocytin, in which the vitamin is covalently linked to the ϵ-amino group of a terminal lysine residue in the apoprotein (ϵ-*N*-biotinyllysine). Biotin is involved as a cofactor in a number of important CO_2 fixation reactions (carboxylations): carboxylation of acetyl-CoA, an important step in fatty acid synthesis; propionyl-CoA carboxylase, important in propionate oxidation; 3-methylcrotonyl-CoA carboxylase, involved in leucine oxidation; and pyruvate carboxylase, of importance in gluconeogenesis.[83,86,87] Since these biotin-dependent carboxylases occupy key positions in major metabolic pathways, a state of biotin deficiency can be expected to affect the metabolism of the nervous system.

The effect of biotin deficiency on mammalian tissues has been the subject of extensive investigation. However, specific observations on the effects of biotin deficiency on CNS metabolism remain very scanty. As is the case with many water-soluble vitamins, the brain appears to preserve its total biotin content during progressive depletion to a greater degree than do other organs. Biotin deficiency induced in pregnant rats causes ataxia and underdeveloped fur in the offspring and a 70% decrease in pyruvate carboxylase activity that under normal circumstances appears to have its highest activity in the brain stem.[88]

Increased lactate and hypoglycemia in blood and brain can be measured in the biotin-deficient offspring. Biotin deficiency induced in postweanling rats results in reduction in the incorporation of [^{14}C]leucine *in vivo* in both brain and liver proteins.[89]

11. CONCLUSION

In this chapter an attempt has been made to review what is currently known about the neurochemistry of vitamin deficiencies. It is evident that, in spite of the variety and quantity of biochemical data gathered over half a century of research, huge gaps exist in our understanding of the effects of specific vitamin deficiencies of the nervous system and the mechanisms underlying the resultant symptoms and signs of disease and tissue destruction. A multitude of biochemical and physiologic functions are known to be dependent upon vitamins and their cofactors, yet only a few vitamins have been shown to play a primary role in the metabolic activity of the nervous system. Data concerning the neurochemistry of vitamin deficiencies have tended to be fragmentary; in most states of deficiency the correlation between known biochemical lesions and clinical manifestations is tenuous. Only a few experimentally induced vitamin deficiencies resemble both clinically and pathologically the naturally occurring diseases in man. It should be reiterated that deficiency syndromes in man are the result of multifactorial nutritional depletion or disease states; they are frequently accompanied by multiple metabolic abnormalities not necessarily related to vitamin deficiency. Therefore, results obtained from experiments on animals in which deficiency has been produced artificially may have little or no application to the nutritional problems of humans.

More complete exploration of the role of presently known water- and fat-soluble vitamins in the metabolic activity of the nervous system could reveal the existence of vitaminlike substances whose presence is essential to the maintenance of the nervous system. There may also be other forms of active cofactor, or cofactors, of vitamins than are now known. That there are interrelationships between vitamins has been demonstrated in the metabolism of homocysteine and methionine (vitamin B_{12}, folic acid, and pyridoxine) or the metabolism of tryptophan (thiamin, pyridoxine, and niacin). Further examples of vitamin interdependencies remain to be elucidated.

Very little is known about the coenzyme binding sites on apoenzyme proteins. It has been postulated that in certain vitamin-dependent states, e.g., pyridoxine dependency, there exist abnormal coenzyme binding sites on the glutamic acid decarboxylase apoenzyme that require excessive amounts of pyridoxine coenzyme for normal enzyme activity. Other similar, genetically determined, vitamin-responsive disorders need to be explored.

Very little is known about apoenzymes and the binding sites of their respective coenzymes. Genetically determined diseases in which coenzyme binding sites are abnormal and that affect the nervous system primarily or exclusively may not be uncommon. Finally, vitamins or a vitamin derivative may possess pharmacologic properties in the nervous system quite different from their function as cofactor in enzymatic reactions. TTP (a noncoenzyme form of thiamin) may have such properties.

There remain a host of areas in the field of vitaminology that may, when they are elucidated, shed further light on the biochemistry of diseases affecting the nervous system.

REFERENCES

1. Funk, C., 1914, *Die Vitamine*, Bergmann, Wiesbaden.
2. Rosenberg, L. E., 1974, *Brain Dysfunction in Metabolic Disorders* (F. Plumb, ed.), Raven Press, New York, pp. 263–271.
3. Arakawa, T., 1970, *Am. J. Med.* **48**:594–598.
4. Spector, R., 1977, *N. Engl. J. Med.* **296**:1393–1398.
5. Spector, R., 1978, *J. Neurochem.* **30**:881–887.
6. Spector, R., and Greenwald, L. L., 1978, *J. Biol. Chem.* **253**:2373–2379.
7. Spector, R., 1979, *Neurology, Psychiatry and Internal Medicine* (I. M. Botez and E. H. Reynolds, eds.) Raven Press, New York, pp. 187–194.
8. Spector, R., 1979, *J. Neurochem.* **33**:895–904.
9. Dreyfus, P. M., 1959, *J. Neurochem.* **4**:183–190.
10. Pincus, J. H., and Grove, I., 1970, *Exp. Neurol.* **28**:477–483.
11. Dreyfus, P. M., and Geel, S. E., 1981, *Basic Neurochemistry* (G. J. Siegel, R. W. Albers, B. W. Agranoff, and R. Katzman, eds.), 3rd ed., Little, Brown, Boston, pp. 661–679.
12. Dreyfus, P. M., and Hauser, G., 1965, *Biochim. Biophys. Acta* **164**:78–84.
13. Dreyfus, P. M., 1965, *J. Neuropathol. Exp. Neurol.* **24**:119–129.
14. Dreyfus, P. M., 1961, *J. Neurochem.* **8**:139–145.
15. Rindi, G., Patrini, C., Comincioli, V., and Reggiani, C., 1980, *Brain Res.* **181**:369–380.
16. McCandless, D. W., and Schenker, S., 1968, *J. Clin. Invest.* **47**:2268–2280.
17. Sharp, F. R., Evans, K., and Bolger, E., 1982, *Neurology* **32**:808–814.
18. Voorhees, C. V., Schmidt, D. E., Barrett, R. J., and Schenker, S., 1977, *J. Nutr.* **107**:1902–1908.
19. Heinrich, C. P., Stadler, H., and Weiser, H., 1973, *J. Neurochem.* **21**:1273–1281.
20. Sacchi, O., Ladinsky, H., Prigioni, I., Consolo, S., Peri, G., and Perri, V., 1978, *Brain Res.* **151**:609–614.
21. Barclay, L. L., Gibson, G. E., and Blass, J. P., 1981, *J. Pharmacol. Exp. Ther.* **217**:537–543.
22. McCandless, D. W., Curley, A. D., and Cassidy, C. E., 1976, *J. Nutr.* **106**:1144–1151.
23. Volpe, J. J., and Marasa, J. C., 1978, *J. Neurochem.* **30**:975–981.
24. Schwartz, J. P., and McCandless, D. W., 1976, *Mol. Cell. Biochem.* **13**:49–53.
25. Henderson, G. I., Hoyumpa, A. M., Jr., and Schenker, S., 1978, *Biochem. Pharmacol.* **27**:1677–1683.
26. Barchi, R. L., and Braun, P. E., 1971, *Brain Res.* **35**:622–624.
27. Geel, S. E., and Dreyfus, P. M., 1975, *J. Neurochem.* **24**:353–360.
28. Trostler, N., and Sklar, D., 1977, *J. Nutr. Sci. Vitaminol.* **24**:105–111.
29. Plaitakis, A., Nicklas, W. J., and Berl, S., 1978, *Neurology* **28**:691–698.
30. Dreyfus, P. M., 1962, *N. Engl. J. Med.* **267**:596–598.
31. Blass, J. P., and Gibson, G. E., 1977, *N. Engl. J. Med.* **297**:1367–1370.
32. von Muralt, A., 1962, *Ann. N.Y. Acad. Sci.* **98**:499–507.
33. Itokawa, Y., Schulz, R. A., and Cooper, J. R., 1972, *Biochim. Biophys. Acta* **266**:293–299.
34. Cooper, J. R., and Pincus, J. H., 1979, *Neurochem. Res.* **4**:223–239.
35. Itokawa, Y., and Cooper, J. R., 1968, *Biochim. Biophys. Acta* **158**:180–182.
36. Barchi, R. L., 1976, *Thiamine* (C. J. Gubler, M. Fujiwara, and P. M. Dreyfus, eds.), Wiley, New York, pp. 283–305.
37. Pincus, J. H., 1972, *Dev. Med. Child Neurol.* **14**:87–101.
38. Pincus, J. H., Cooper, J. R., Piros, K., and Turner, V., 1974, *Neurology* **24**:885–890.
39. Tang, T. T., Good, T. A., Dyken, P. R., Johnson, S. D., McCreadie, S. R., Sy, S. T., Lardy, H. A., and Rudolph, F. B., 1972, *J. Pediatr.* **81**:189–190.
40. Blass, J. P., Avigan, J., and Uhlendorf, B. W., 1970, *J. Clin. Invest.* **49**:423–432.
41. Elsas, L. J., Danner, D. J., and Rogers, B. L., 1976, *Thiamine* (C. J. Gubler, M. Fujiwara, and P. M. Dreyfus, eds.), Wiley, New York, pp. 335–349.
42. Bowden, J. A., McArthur, C. L., and Fried, M., 1971, *Biochem. Med.* **5**:101–108.
43. Williams, M. A., 1964, *Vitam. Horm. (N.Y.)* **22**:561–579.
44. Lajtha, A. L., Maker, H. S., and Clarke, D. D., 1981, *Basic Neurochemistry* (G. J. Siegel, R. W. Albers, B. W. Agranoff, and R. Katzman, eds.), 3rd ed., Little, Brown, Boston, pp. 329–353.

45. Nitsch, C., 1980, *J. Neurochem.* **34**:822–830.
46. Wiss, O., and Weber, F., 1964, *Vitam. Horm. (N.Y.)* **22**:495–501.
47. Bhagavan, H. N., and Coursin, D. B., 1971, *Int. J. Vitam. Nutr. Res.* **41**:419–423.
48. Morre, D. M., Kirksey, A., and Das, G. D., 1978, *J. Nutr.* **108**:1250–1259.
49. Morre, D. M., and Kirksey, A., 1978, *Brain Res.* **146**:200–204.
50. Stephens, M. C., and Dakshinamurti, K., 1975, *Neurobiology* **5**:262–269.
51. Kurtz, D. J., and Kanfer, J. N., 1973, *J. Neurochem.* **20**:963–968.
52. Hsia, E. Y., and Wolf, B., 1981, *Basic Neurochemistry* (G. J. Siegel, R. W. Albers, B. W. Agranoff, and R. Katzman, eds.), Little, Brown, Boston, pp. 563–600.
53. Lott, I. T., Coulombe, T., Di Paolo, R. V., Richardson, E. P., and Levy, H. L., 1978, *Neurology* **28**:47–54.
54. Stadtman, T. C., 1971, *Science* **171**:859–867.
55. Agamanolis, D. P., Victor, M., Chester, E. M., Kark, J. A., Hines, J. D., and Harris, J. W., 1976, *Neurology* **26**:905–914.
56. Girwood, R. H., 1968, *Proc. Nutr. Soc.* **27**:101–107.
57. Cardinale, G. J., Dreyfus, P. M., Auld, P., and Abeles, R. H., 1969, *Arch. Biochem. Biophys.* **131**:92–99.
58. Frenkel, E. P., 1973, *J. Clin. Invest.* **52**:1237–1245.
59. Barley, F. W., Sato, G. H., and Abeles, R. H., 1972, *J. Biol. Chem.* **247**:4270–4276.
60. Kishimoto, Y., Williams, M., Moser, H. W., Hignite, C., and Biemann, K., 1973, *J. Lipid Res.* **14**:69–77.
61. Frenkel, E. P., Kitchens, R. L., and Johnston, J. M., 1973, *J. Biol. Chem.* **248**:7540–7546.
62. Amess, J. A. L., Burman, J. F., Rees, G. M., Nancekievill, D. G., and Mollin, D. L., 1978, *Lancet* **2**:339–342.
63. Lassen, H. C. A., Henricksen, E., Neukirch, F., and Kirsten, H. S., 1956, *Lancet* **1**:527–530.
64. Layzer, R., Fishman, R. A., and Schafer, J. A., 1978, *Neurology* **28**:504–506.
65. Kondo, H., Osborne, M. L., Kolhouse, J. F., Binder, M. J., Podell, E. R., Utley, C. S., Abrams, R. S., and Allen, R. H., 1981, *J. Clin. Invest.* **67**:1270–1283.
66. Scott, J. M., Wilson, P., Dinn, J. J., and Weir, D. G., 1981, *Lancet* **2**:334–337.
67. Scott, J. M., and Weir, D. G., 1981, *Lancet* **2**:337–340.
68. Stokstad, E. L. R., 1976, *Present Knowledge in Nutrition*, 4th ed., The Nutrition Foundation, New York, pp. 204–216.
69. Ordonez, L. A., 1979, *Folic Acid in Neurology, Psychiatry, and Internal Medicine* (M. I. Botez and E. H. Reynolds, eds.), Raven Press, New York, pp. 129–145.
70. Girwood, R. H., 1968, *Proc. Nutr. Soc.* **27**:101–107.
71. Meller, E., and Friedhoff, A. J., 1979, *Folic Acid in Neurology, Psychiatry, and Internal Medicine*, (M. I. Botez and E. H. Reynolds, eds.), Raven Press, New York, pp. 157–177.
72. Hommes, O. R., Hollinger, M. J. T., Jansen, M., Schoofs, M., Vanderwiel, T., and Kok, J. C. N., 1979, *Folic Acid in Neurology, Psychiatry, and Internal Medicine* (M. I. Botez and E. H. Reynolds, eds.) Raven Press, New York, pp. 285–316.
73. Smith, D. B., and Obbens, E. A. M. T., 1979, *Folic Acid in Neurology, Psychiatry, and Internal Medicine* (M. I. Botez and E. H. Reynolds, eds.), Raven Press, New York, pp. 267–283.
74. Darby, W. J., McNutt, K. W., and Todhunter, E. N., 1976, *Present Knowledge in Nutrition*, 4th ed., The Nutrition Foundation, New York, pp. 162–174.
75. Wolf, A., Cowen, D., and Geller, L. M., 1959, *Trans. Am. Neurol. Assoc.* **84**:140–145.
76. Herken, H., Lange, K., Kolbe, H., and Keller, K., 1974, *Central Nervous System—Studies on Metabolic Regulation and Function* (E. Genazzani and H. Herken, eds.), Springer, Berlin, pp. 41–54.
77. Wright, L. D., 1976, *Present Knowledge in Nutriton*, 4th ed., The Nutrition Foundation, New York, pp. 226–231.
78. Novelli, G. D., 1953, *Physiol. Rev.* **33**:525–543.
79. Bean, W. B., and Hodges, R. E., 1954, *Proc. Soc. Exp. Biol. Med.* **86**:693–698.
80. Sun, A. Y., and Sun, G. Y., 1978, *Adv. Exp. Med. Biol.* **97**:295–289.
81. Vatassery, G. T., and Younoszai, R., 1978, *Lipids* **13**:828–831.
82. Witting, L. A., 1969, *J. Neurochem.* **16**:1253–1256.
83. McCormick, D. B., 1979, *Present Knowledge in Nutrition*, 4th ed., The Nutrition Foundation, New York, pp. 217–225.

84. Mock, D. M., DeLorimer, A. A., Liebman, W. M., Sweetman, L., and Baker, H., 1981, *N. Engl. J. Med.* **304:**820–823.
85. Tanaska, K., 1981, *N. Engl. J. Med.* **304:**839–840.
86. Murthy, P. N., and Mistry, S. P., 1977, *Prog. Food Nutr. Sci.* **2:**405–455.
87. Bonjour, J. P., 1977, *Int. J. Vitam. Nutr. Res.* **47:**107–118.
88. Schrijver, J., Dias, T., and Hommes, O. R., 1979, *Nutr. Metab.* **23:**179–191.
89. Bhagavan, H. N., and Coursin, D. B., 1974, *Int. J. Vitam. Nutr. Res.* **44:**363–369.

Epilogue

The publication of these ten volumes stirs my wonder and imagination about general aspects of the development of the sciences in the changing world. My knowledge of history is superficial but it seems that experimental chemistry started with alchemy, the search for the magic of gold, though the idea of ultimate particles was in the minds of classical Greek thinkers. Chemical study of biological materials started with von Liebig and others but got largely diverted into particular interest in carbon compounds though that subject was and is still called organic chemistry. Biological chemistry and related subjects started later in medical and agricultural faculties and developed rapidly as research interest and support came to medical scientists, particularly physiologists. Nevertheless, biochemistry, including neurochemistry, has its foundation in organic chemistry; some of the early work on brain chemistry, by Thudichum, Irvine Page, and others, was confined to the structure of prominent substances in brain. I even feel a little relevance of my own work in 1927 in organic chemistry on steric hindrance. Knowledge of space relations of atoms is now basic to descriptions of nucleic acids, receptor groups, etc.

The more recent impetus to neurochemistry came particularly from neurophysiologists and pharmacologists. I regard neurochemistry as a branch of physiology and my interest in it, or in any aspect of biochemistry, has significance to me to the extent that it illuminates biology. (A tendency exists to treat biochemical substances and reactions as of "pure interest" in themselves. One otherwise excellent textbook of biochemistry omits whole fields that should interest a biologically oriented student.)

As F. O. Schmitt, in his Preface to Volume 7, has emphasized, we should remember that neurochemistry is one aspect of the multidisciplinary field of neuroscience. In our thinking we should be aware of our fellow disciplines and be ready to move from one to another. I like to note that, of the two successors in my old job as a neurochemist at the Montreal Neurological Institute, one was a biochemist who passed through neurochemistry to become a distinguished experimental neuropathologist and the other, a medical scientist previously involved in insect behavior, is contributing neurochemical studies of clinical interest and is also a major successor to Garrod in the field of inborn errors of metabolism.

Neurochemistry, our branch of neuroscience, has developed enormously during my lifetime. Encouragement by the provision of funds for research from a variety of sources has been essential. Of critical importance has been the development of techniques and the commercial provision of materials and highly sophisticated instruments. (In my early days I had to prepare chemicals myself and build some of my own instruments. I took some pride in a little chemical ability and in some glass blowing. I don't really envy those now who can buy pure obscure chemicals, have to rely on black boxes and computers, and are deprived of the excitement of calculating, with slide rule, the results of experiments!)

Of course, it is the interested, dedicated work by many bright, even brilliant people that has produced the great achievements of neurochemical knowledge and understanding disclosed in these volumes. The increase in productivity of such people has been phenomenal. In 1940, in Philadelphia, I organized a symposium on Biochemistry and Physiology in Relation to Mental Diseases. I think this was the first-ever neurochemical symposium. Among the speakers were Detlev Bronk, Ralph Gerard, and Harold Himwich. We covered the subject in a half-day. In July 1983 there was a week-long annual meeting of the International Society for Neurochemistry in Vancouver where about 1400 presentations were made. In 1944 I started a small research laboratory in the Montreal Neurological Institute. The name of every department of that institute had the prefix "neuro-", so my domain was neurochemistry and I was listed, in the annual report, as a neurochemist. So the term was born. Of course, there were people ranging from Thudichum to Quastel who were already well known in the field. Now the 1983–84 directory of the International Society for Neurochemistry lists about 1100 neurochemists.

The first nearly modern description of brain chemistry was *Chemistry of the Brain* edited by I. H. Page in 1937. In 1955 Quastel, Page, and I put together a more or less comprehensive set of aricles in a single volume called *Neurochemistry* and we produced a second edition in 1962. The latter edition contained 41 articles. For the present set of 10 volumes Abel Lajtha has assembled about 190 articles involving 290 authors.

An enormous number of very diverse facets of neurochemistry have been covered in these volumes and no doubt there are other facets that never reached the editor or have been developed since the publication. All this outburst of detailed knowledge is wonderful and exicting but it is also alarming. No one mind can absorb it all. To do valid work each investigator has to specialize quite narrowly in knowledge and the techniques appropriate to his/her field. This tends to limit the overview, consideration of the relation of the findings to the general view of neuroscience. Perhaps the time has come to follow the example of physicists and to recognize somewhat different approaches by experimental and theoretical neuroscientists. The job of the latter groups would be to integrate the different kinds of information and perhaps indicate fields that particularly need experimental studies.

I maintain, further, that neuroscience should recognize itself as the integrator or central science of existence! Existence consists of the three domains—the physical, life, and the mental. All our chemical substances and reactions are expressions of physical particles, energy, and the physical forces that govern them. We study the role of these physical conditions and events in the various forms of life. In living creatures, we observe the immense complexity and the utterly astonishing perfection by which these physical substances and reactions result in life processes and activities. Personally, I cannot accept evolution as simply the result of individual chance mutations that empowered species to arise and reproduce faster than they are destroyed in the world of claw and fang competition. I think there must be, or have been, some principle whereby massive integrated changes occur as if as a result of some Plan or Purpose. This perhaps mystical idea need not, however, inhibit or study of tangible events. Neuroscience is also basic to our approach to the mental world. Though we may not be able to conceive how chemical reactions can give rise to consciousness, we certainly know that by chemicals we can affect thought, memory, behavior, pain, and sleep. We are learning much about transmitters, modifiers, electrical potentials, etc. that affect various aspects of the mind and behavior and about the particular regions of the brain that are involved. We can purposefully (or accidentally) affect these aspects by artificially applied chemicals and physical conditions.

So what are our goals? In the first place there is the beauty, excitement, and delight in finding things out—research for its own sake. But we are glad, and socially justified, in finding out useful knowledge. In his preface to Volume 10, Donald Tower has powerfully outlined our involvement in the prevention, control, or amelioration of neurological diseases. I feel awe for our role in mental diseases. To me it is astonishing what beneficial effects a few little pills can have on a disturbed mind, though I feel rather strongly that personal help and compassion must not be omitted.

Neurochemistry is a powerful force but powerful forces are not always benign. We must remember that neurotoxins have been, and can still be, used as weapons of war or of personal violence. Drug addictions and certain chemical treatments can cause degradation of personality. Neurochemists and all other scientists are also persons and citizens. We should share the concerns of our fellows and be ready to share our knowledge of dangerous as well as valuable applications of neurochemistry. Members of our central science discipline should be as aware of the danger, to our existence and of all of the great works of mankind, of that product of science and engineering, the atom bomb. Let us join with all our fellows in all parts of the world in seeking to control the madness among those who govern us. When I consider the possible extinction of mankind, my concerns are particularly for my grandchildren, for continued performances by Les Grands Ballets Canadiens and other such delights, and for the hope that Abel Lajtha, or his successor, will be able to summarize for us the advances in neurochemistry into the 21st century. New technology of communication may by then have been developed, if we are still around!

Allan (K. A. C.) Elliott

Index

GPSR Compliance

The European Union's (EU) General Product Safety Regulation (GPSR) is a set of rules that requires consumer products to be safe and our obligations to ensure this.

If you have any concerns about our products, you can contact us on ProductSafety@springernature.com

In case Publisher is established outside the EU, the EU authorized representative is:

Springer Nature Customer Service Center GmbH
Europaplatz 3
69115 Heidelberg, Germany

Batch number: 10838053

Printed by Printforce, the Netherlands